KIRK-OTHMER

ENCYCLOPEDIA OF CHEMICAL TECHNOLOGY

Third Edition

VOLUME 20

Refractories
to
Silk

KIRK-OTHMER

ENCYCLOPEDIA OF CHEMICAL TECHNOLOGY

THIRD EDITION

VOLUME 20

REFRACTORIES
TO
SILK

1807 1982

A WILEY-INTERSCIENCE PUBLICATION

John Wiley & Sons

NEW YORK · CHICHESTER · BRISBANE · TORONTO · SINGAPORE

Library of Congress Cataloging in Publication Data:

Main entry under title:
Encyclopedia of chemical technology.

At head of title: Kirk-Othmer.
"A Wiley-Interscience publication."
Includes bibliographies.
1. Chemistry, Technical—Dictionaries. I. Kirk, Raymond
Eller, 1890–1957. II. Othmer, Donald Frederick, 1904—
III. Grayson, Martin. IV. Eckroth, David. V. Title:
Kirk-Othmer encyclopedia of chemical technology.

TP9.E685   1978   660'.03   77-15820
ISBN 0-471-02073-7

CONTENTS

EDITORIAL STAFF
FOR VOLUME 20

Executive Editor: **Martin Grayson**
Associate Editor: **David Eckroth**
Production Supervisor: **Michalina Bickford**
Editors: **Caroline I. Eastman**   **Carolyn Golojuch**   **Anna Klingsberg**
Goldie Wachsman   **Mimi Wainwright**

CONTRIBUTORS
TO VOLUME 20

David Adler, *Massachusetts Institute of Technology, Cambridge, Massachusetts,* Semiconductors, amorphous

Barry Arkles, *Petrarch Systems, Inc., Bristol, Pennsylvania,* Silanes; Silicon ethers and esters; both under Silicon compounds

Robert R. Barnhart, *Uniroyal Chemical, Naugatuck, Connecticut,* Rubber compounding

A. M. Bueche†, *General Electric Co., Fairfield, Connecticut,* Research management

Rointan F. Bunshah, *University of California at Los Angeles, California,* Refractory coatings

Chuan C. Chang, *Bell Laboratories, Murray Hill, New Jersey,* Semiconductors, fabrication and characterization

Ward Collins, *Dow Corning Corporation, Midland, Michigan,* Silicon halides under Silicon compounds

K. W. Cooper, *York Division, Borg-Warner Corporation, York, Pennsylvania,* Refrigeration

T. D. Coyle, *National Bureau of Standards, Washington, D.C.,* Introduction under Silica

Paul Danielson, *Corning Glass Works, Corning, New York,* Vitreous silica under Silica

Reg Davies, *E. I. du Pont de Nemours & Co., Inc., Wilmington, Delaware,* Sampling

† Deceased.

vii

E. M. Elkin, *Noranda Mines Limited, Montreal East, Quebec, Canada,* Selenium and selenium compounds

Stephen H. Erickson, *The Dow Chemical Company, Midland, Michigan,* Salicylic acid and related compounds

James S. Falcone, Jr., *The PQ Corporation, Lafayette Hill, Pennsylvania,* Synthetic inorganic silicates under Silicon compounds

Robert Haddon, *Bell Laboratories, Murray Hill, New Jersey,* Semiconductors, organic

Bruce B. Hardman, *General Electric Company, Waterford, New York,* Silicones under Silicon compounds

K. E. Hickman, *York Division, Borg-Warner Corporation, York, Pennsylvania,* Refrigeration

Arnold J. Hoiberg†, *Manville Corporation, Denver, Colorado,* Roofing materials

James S. Johnson, Jr., *Oak Ridge National Laboratory, Oak Ridge, Tennessee,* Reverse osmosis

Martin L. Kaplan, *Bell Laboratories, Murray Hill, New Jersey,* Semiconductors, organic

E. D. Kolb, *Bell Laboratories, Murray Hill, New Jersey,* Synthetic quartz crystals under Silica

Thomas J. Kucera, *Consultant, Evanston, Illinois,* Reprography

R. A. Laudise, *Bell Laboratories, Murray Hill, New Jersey,* Synthetic quartz crystals under Silica

H. D. Leigh, *C-E Basic, Inc., Beltsville, Ohio,* Refractories

W. G. Lidman, *Kawecki Berylco Industries, Inc., Revere, Pennsylvania,* Rubidium and rubidium compounds

Charles D. Livengood, *North Carolina State University, Raleigh, North Carolina,* Silk

Ernest G. Long, *Manville Corporation, Denver, Colorado,* Roofing materials

James Martin, *William Zinsser & Co., Somerset, New Jersey,* Shellac

W. C. Miller, *Manville Corporation, Denver, Colorado,* Refractory fibers

Nancy R. Passow, *C-E Lummus, Bloomfield, New Jersey,* Regulatory agencies

William R. Peterson, Jr., *Petrarch Systems, Inc., Bristol, Pennsylvania,* Silanes under Silicon compounds

Percy E. Pierce, *PPG Industries, Inc., Allison Park, Pennsylvania,* Rheological measurements

Edwin P. Plueddemann, *Dow Corning Corporation, Midland, Michigan,* Silylating agents under Silicon compounds

Walter Runyan, *Texas Instruments, Inc., Dallas, Texas,* Pure silicon under Silicon and silicon alloys

L. McDonald Schetky, *International Copper Research Association, New York, New York,* Shape-memory alloys

Clifford K. Schoff, *PPG Industries, Inc., Allison Park, Pennsylvania,* Rheological measurements

S. A. Schwarz, *Bell Laboratories, Inc., Murray Hill, New Jersey,* Semiconductors, theory and application

J. D. Seader, *University of Utah, Salt Lake City, Utah,* Separation systems synthesis

Raymond B. Seymour, *The University of Southern Mississippi, Hattiesburg, Mississippi,* Sealants

Ronald F. Silver, *Elkem Metals Company, Niagara Falls, New York,* Metallurgical under Silicon and silicon alloys

P. N. Son, *BF Goodrich Company, Brecksville, Ohio,* Rubber chemicals

David R. St. Cyr, *Goodyear Tire & Rubber Co., Akron, Ohio,* Rubber, natural

L. W. Steele, *General Electric Co., Research and Development, Fairfield, Connecticut,* Research management

Ladislav Svarovsky, *University of Bradford, Bradford, West Yorkshire, UK,* Sedimentation

Ray Taylor, *BF Goodrich Company, Brecksville, Ohio,* Rubber chemicals

A. S. Teot, *Dow Chemical U.S.A., Midland, Michigan,* Resins, water-soluble

Arnold Torkelson, *General Electric Company, Waterford, New York,* Silicones under Silicon compounds

Paul M. Treichel, *University of Wisconsin, Madison, Wisconsin,* Rhenium and rhenium compounds

John C. Weaver, *Case Western Reserve University, Cleveland, Ohio,* Resins, natural

Frederick B. White, Jr., *Kawecki Berylco Industries, Inc., Revere, Pennsylvania,* Rubidium and rubidium compounds

Joan D. Willey, *University of North Carolina at Wilmington, Wilmington, North Carolina,* Amorphous silica under Silica

Fred Wudl, *Bell Laboratories, Murray Hill, New Jersey,* Semiconductors, organic

NOTE ON CHEMICAL ABSTRACTS SERVICE REGISTRY NUMBERS AND NOMENCLATURE

Chemical Abstracts Service (CAS) Registry Numbers are unique numerical identifiers assigned to substances recorded in the CAS Registry System. They appear in brackets in the *Chemical Abstracts* (CA) substance and formula indexes following the names of compounds. A single compound may have many synonyms in the chemical literature. A simple compound like phenethylamine can be named β-phenylethylamine or, as in *Chemical Abstracts*, benzeneethanamine. The usefulness of the *Encyclopedia* depends on accessibility through the most common correct name of a substance. Because of this diversity in nomenclature careful attention has been given the problem in order to assist the reader as much as possible, especially in locating the systematic CA index name by means of the Registry Number. For this purpose, the reader may refer to the CAS Registry Handbook-Number Section which lists in numerical order the Registry Number with the *Chemical Abstracts* index name and the molecular formula; eg, **458-88-8,** Piperidine, 2-propyl-, (S)-, $C_8H_{17}N$; in the *Encyclopedia* this compound would be found under its common name, coniine [*458-88-8*]. The Registry Number is a valuable link for the reader in retrieving additional published information on substances and also as a point of access for such on-line data bases as Chemline, Medline, and Toxline.

In all cases, the CAS Registry Numbers have been given for title compounds in articles and for all compounds in the index. All specific substances indexed in *Chemical Abstracts* since 1965 are included in the CAS Registry System as are a large number of substances derived from a variety of reference works. The CAS Registry System identifies a substance on the basis of an unambiguous computer-language description of its molecular structure including stereochemical detail. The Registry Number is a machine-checkable number (like a Social Security number) assigned in sequential order to each substance as it enters the registry system. The value of the number lies in the fact that it is a concise and unique means of substance identification, which is independent of, and therefore bridges, many systems of chemical nomenclature. For polymers, one Registry Number is used for the entire family; eg, polyoxyethylene (20) sorbitan monolaurate has the same number as all of its polyoxyethylene homologues.

Registry numbers for each substance will be provided in the third edition cumulative index and appear as well in the annual indexes (eg, Alkaloids shows the Registry Number of all alkaloids (title compounds) in a table in the article as well, but the intermediates have their Registry Numbers shown only in the index). Articles such as Analytical methods, Batteries and electric cells, Chemurgy, Distillation, Economic evaluation, and Fluid mechanics have no Registry Numbers in the text.

Cross-references are inserted in the index for many common names and for some systematic names. Trademark names appear in the index. Names that are incorrect, misleading or ambiguous are avoided. Formulas are given very frequently in the text to help in identifying compounds. The spelling and form used, even for industrial names, follow American chemical usage, but not always the usage of *Chemical Abstracts* (eg, *coniine* is used instead of *(S)-2-propylpiperidine*, *aniline* instead of *benzenamine*, and *acrylic acid* instead of *2-propenoic acid*).

There are variations in representation of rings in different disciplines. The dye industry does not designate aromaticity or double bonds in rings. All double bonds and aromaticity are shown in the *Encyclopedia* as a matter of course. For example, tetralin has an aromatic ring and a saturated ring and its structure appears in the

Encyclopedia with its common name, Registry Number enclosed in brackets, and parenthetical CA index name, ie, tetralin, [*119-64-2*] (1,2,3,4-tetrahydronaphthalene). With names and structural formulas, and especially with CAS Registry Numbers the aim is to help the reader have a concise means of substance identification.

CONVERSION FACTORS, ABBREVIATIONS, AND UNIT SYMBOLS

SI Units (Adopted 1960)

A new system of measurement, the International System of Units (abbreviated SI), is being implemented throughout the world. This system is a modernized version of the MKSA (meter, kilogram, second, ampere) system, and its details are published and controlled by an international treaty organization (The International Bureau of Weights and Measures) (1).

SI units are divided into three classes:

BASE UNITS

length	meter[†] (m)
mass[‡]	kilogram (kg)
time	second (s)
electric current	ampere (A)
thermodynamic temperature[§]	kelvin (K)
amount of substance	mole (mol)
luminous intensity	candela (cd)

[†] The spellings "metre" and "litre" are preferred by ASTM; however "-er" are used in the Encyclopedia.

[‡] "Weight" is the commonly used term for "mass."

[§] Wide use is made of "Celsius temperature" (t) defined by

$$t = T - T_0$$

where T is the thermodynamic temperature, expressed in kelvins, and $T_0 = 273.15$ K by definition. A temperature interval may be expressed in degrees Celsius as well as in kelvins.

SUPPLEMENTARY UNITS

| plane angle | radian (rad) |
| solid angle | steradian (sr) |

DERIVED UNITS AND OTHER ACCEPTABLE UNITS

These units are formed by combining base units, supplementary units, and other derived units (2–4). Those derived units having special names and symbols are marked with an asterisk in the list below:

Quantity	Unit	Symbol	Acceptable equivalent
*absorbed dose	gray	Gy	J/kg
acceleration	meter per second squared	m/s^2	
*activity (of ionizing radiation source)	becquerel	Bq	1/s
area	square kilometer	km^2	
	square hectometer	hm^2	ha (hectare)
	square meter	m^2	
*capacitance	farad	F	C/V
concentration (of amount of substance)	mole per cubic meter	mol/m^3	
*conductance	siemens	S	A/V
current density	ampere per square meter	A/m^2	
density, mass density	kilogram per cubic meter	kg/m^3	g/L; mg/cm^3
dipole moment (quantity)	coulomb meter	C·m	
*electric charge, quantity of electricity	coulomb	C	A·s
electric charge density	coulomb per cubic meter	C/m^3	
electric field strength	volt per meter	V/m	
electric flux density	coulomb per square meter	C/m^2	
*electric potential, potential difference, electromotive force	volt	V	W/A
*electric resistance	ohm	Ω	V/A
*energy, work, quantity of heat	megajoule	MJ	
	kilojoule	kJ	
	joule	J	N·m
	electron volt[†]	eV[†]	
	kilowatt-hour[†]	kW·h[†]	

[†] This non-SI unit is recognized by the CIPM as having to be retained because of practical importance or use in specialized fields (1).

Quantity	Unit	Symbol	Acceptable equivalent
energy density	joule per cubic meter	J/m^3	
*force	kilonewton	kN	
	newton	N	$kg \cdot m/s^2$
*frequency	megahertz	MHz	
	hertz	Hz	1/s
heat capacity, entropy	joule per kelvin	J/K	
heat capacity (specific), specific entropy	joule per kilogram kelvin	$J/(kg \cdot K)$	
heat transfer coefficient	watt per square meter kelvin	$W/(m^2 \cdot K)$	
*illuminance	lux	lx	lm/m^2
*inductance	henry	H	Wb/A
linear density	kilogram per meter	kg/m	
luminance	candela per square meter	cd/m^2	
*luminous flux	lumen	lm	$cd \cdot sr$
magnetic field strength	ampere per meter	A/m	
*magnetic flux	weber	Wb	$V \cdot s$
*magnetic flux density	tesla	T	Wb/m^2
molar energy	joule per mole	J/mol	
molar entropy, molar heat capacity	joule per mole kelvin	$J/(mol \cdot K)$	
moment of force, torque	newton meter	$N \cdot m$	
momentum	kilogram meter per second	$kg \cdot m/s$	
permeability	henry per meter	H/m	
permittivity	farad per meter	F/m	
*power, heat flow rate, radiant flux	kilowatt	kW	
	watt	W	J/s
power density, heat flux density, irradiance	watt per square meter	W/m^2	
*pressure, stress	megapascal	MPa	
	kilopascal	kPa	
	pascal	Pa	N/m^2
sound level	decibel	dB	
specific energy	joule per kilogram	J/kg	
specific volume	cubic meter per kilogram	m^3/kg	
surface tension	newton per meter	N/m	
thermal conductivity	watt per meter kelvin	$W/(m \cdot K)$	
velocity	meter per second	m/s	
	kilometer per hour	km/h	
viscosity, dynamic	pascal second	$Pa \cdot s$	
	millipascal second	$mPa \cdot s$	
viscosity, kinematic	square meter per second	m^2/s	

Quantity	Unit	Symbol	Acceptable equivalent
	square millimeter per second	mm^2/s	
volume	cubic meter	m^3	
	cubic decimeter	dm^3	L(liter) (5)
	cubic centimeter	cm^3	mL
wave number	1 per meter	m^{-1}	
	1 per centimeter	cm^{-1}	

In addition, there are 16 prefixes used to indicate order of magnitude, as follows:

Multiplication factor	Prefix	Symbol	Note
10^{18}	exa	E	
10^{15}	peta	P	
10^{12}	tera	T	
10^9	giga	G	
10^6	mega	M	
10^3	kilo	k	
10^2	hecto	h[a]	[a] Although hecto, deka, deci, and centi
10	deka	da[a]	are SI prefixes, their use should be
10^{-1}	deci	d[a]	avoided except for SI unit-mul-
10^{-2}	centi	c[a]	tiples for area and volume and
10^{-3}	milli	m	nontechnical use of centimeter,
10^{-6}	micro	μ	as for body and clothing
10^{-9}	nano	n	measurement.
10^{-12}	pico	p	
10^{-15}	femto	f	
10^{-18}	atto	a	

For a complete description of SI and its use the reader is referred to ASTM E 380 (4) and the article Units and Conversion Factors which will appear in a later volume of the *Encyclopedia*.

A representative list of conversion factors from non-SI to SI units is presented herewith. Factors are given to four significant figures. Exact relationships are followed by a dagger. A more complete list is given in ASTM E 380-79(4) and ANSI Z210.1-1976 (6).

Conversion Factors to SI Units

To convert from	To	Multiply by
acre	square meter (m^2)	4.047×10^3
angstrom	meter (m)	1.0×10^{-10}†
are	square meter (m^2)	1.0×10^2†
astronomical unit	meter (m)	1.496×10^{11}
atmosphere	pascal (Pa)	1.013×10^5
bar	pascal (Pa)	1.0×10^5†
barn	square meter (m^2)	1.0×10^{-28}†

† Exact.

To convert from	*To*	*Multiply by*
barrel (42 U.S. liquid gallons)	cubic meter (m^3)	0.1590
Bohr magneton (μ_β)	J/T	9.274×10^{-24}
Btu (International Table)	joule (J)	1.055×10^3
Btu (mean)	joule (J)	1.056×10^3
Btu (thermochemical)	joule (J)	1.054×10^3
bushel	cubic meter (m^3)	3.524×10^{-2}
calorie (International Table)	joule (J)	4.187
calorie (mean)	joule (J)	4.190
calorie (thermochemical)	joule (J)	4.184†
centipoise	pascal second (Pa·s)	1.0×10^{-3}†
centistoke	square millimeter per second (mm^2/s)	1.0†
cfm (cubic foot per minute)	cubic meter per second (m^3/s)	4.72×10^{-4}
cubic inch	cubic meter (m^3)	1.639×10^{-5}
cubic foot	cubic meter (m^3)	2.832×10^{-2}
cubic yard	cubic meter (m^3)	0.7646
curie	becquerel (Bq)	3.70×10^{10}†
debye	coulomb·meter (C·m)	3.336×10^{-30}
degree (angle)	radian (rad)	1.745×10^{-2}
denier (international)	kilogram per meter (kg/m)	1.111×10^{-7}
	tex‡	0.1111
dram (apothecaries')	kilogram (kg)	3.888×10^{-3}
dram (avoirdupois)	kilogram (kg)	1.772×10^{-3}
dram (U.S. fluid)	cubic meter (m^3)	3.697×10^{-6}
dyne	newton (N)	1.0×10^{-5}†
dyne/cm	newton per meter (N/m)	1.0×10^{-3}†
electron volt	joule (J)	1.602×10^{-19}
erg	joule (J)	1.0×10^{-7}†
fathom	meter (m)	1.829
fluid ounce (U.S.)	cubic meter (m^3)	2.957×10^{-5}
foot	meter (m)	0.3048†
footcandle	lux (lx)	10.76
furlong	meter (m)	2.012×10^{-2}
gal	meter per second squared (m/s^2)	1.0×10^{-2}†
gallon (U.S. dry)	cubic meter (m^3)	4.405×10^{-3}
gallon (U.S. liquid)	cubic meter (m^3)	3.785×10^{-3}
gallon per minute (gpm)	cubic meter per second (m^3/s)	6.308×10^{-5}
	cubic meter per hour (m^3/h)	0.2271
gauss	tesla (T)	1.0×10^{-4}
gilbert	ampere (A)	0.7958
gill (U.S.)	cubic meter (m^3)	1.183×10^{-4}
grad	radian	1.571×10^{-2}
grain	kilogram (kg)	6.480×10^{-5}
gram force per denier	newton per tex (N/tex)	8.826×10^{-2}

† Exact.
‡ See footnote on p. xiv.

To convert from	To	Multiply by
hectare	square meter (m^2)	$1.0 \times 10^{4\dagger}$
horsepower (550 ft·lbf/s)	watt (W)	7.457×10^2
horsepower (boiler)	watt (W)	9.810×10^3
horsepower (electric)	watt (W)	$7.46 \times 10^{2\dagger}$
hundredweight (long)	kilogram (kg)	50.80
hundredweight (short)	kilogram (kg)	45.36
inch	meter (m)	$2.54 \times 10^{-2\dagger}$
inch of mercury (32°F)	pascal (Pa)	3.386×10^3
inch of water (39.2°F)	pascal (Pa)	2.491×10^2
kilogram force	newton (N)	9.807
kilowatt hour	megajoule (MJ)	$3.6^\dagger$
kip	newton (N)	4.48×10^3
knot (international)	meter per second (m/s)	0.5144
lambert	candela per square meter (cd/m^2)	3.183×10^3
league (British nautical)	meter (m)	5.559×10^3
league (statute)	meter (m)	4.828×10^3
light year	meter (m)	9.461×10^{15}
liter (for fluids only)	cubic meter (m^3)	$1.0 \times 10^{-3\dagger}$
maxwell	weber (Wb)	$1.0 \times 10^{-8\dagger}$
micron	meter (m)	$1.0 \times 10^{-6\dagger}$
mil	meter (m)	$2.54 \times 10^{-5\dagger}$
mile (statute)	meter (m)	1.609×10^3
mile (U.S. nautical)	meter (m)	$1.852 \times 10^{3\dagger}$
mile per hour	meter per second (m/s)	0.4470
millibar	pascal (Pa)	1.0×10^2
millimeter of mercury (0°C)	pascal (Pa)	$1.333 \times 10^{2\dagger}$
minute (angular)	radian	2.909×10^{-4}
myriagram	kilogram (kg)	10
myriameter	kilometer (km)	10
oersted	ampere per meter (A/m)	79.58
ounce (avoirdupois)	kilogram (kg)	2.835×10^{-2}
ounce (troy)	kilogram (kg)	3.110×10^{-2}
ounce (U.S. fluid)	cubic meter (m^3)	2.957×10^{-5}
ounce-force	newton (N)	0.2780
peck (U.S.)	cubic meter (m^3)	8.810×10^{-3}
pennyweight	kilogram (kg)	1.555×10^{-3}
pint (U.S. dry)	cubic meter (m^3)	5.506×10^{-4}
pint (U.S. liquid)	cubic meter (m^3)	4.732×10^{-4}
poise (absolute viscosity)	pascal second (Pa·s)	$0.10^\dagger$
pound (avoirdupois)	kilogram (kg)	0.4536
pound (troy)	kilogram (kg)	0.3732
poundal	newton (N)	0.1383
pound-force	newton (N)	4.448
pound per square inch (psi)	pascal (Pa)	6.895×10^3
quart (U.S. dry)	cubic meter (m^3)	1.101×10^{-3}

† Exact.

To convert from	To	Multiply by
quart (U.S. liquid)	cubic meter (m^3)	9.464×10^{-4}
quintal	kilogram (kg)	$1.0 \times 10^{2\dagger}$
rad	gray (Gy)	$1.0 \times 10^{-2\dagger}$
rod	meter (m)	5.029
roentgen	coulomb per kilogram (C/kg)	2.58×10^{-4}
second (angle)	radian (rad)	4.848×10^{-6}
section	square meter (m^2)	2.590×10^{6}
slug	kilogram (kg)	14.59
spherical candle power	lumen (lm)	12.57
square inch	square meter (m^2)	6.452×10^{-4}
square foot	square meter (m^2)	9.290×10^{-2}
square mile	square meter (m^2)	2.590×10^{6}
square yard	square meter (m^2)	0.8361
stere	cubic meter (m^3)	$1.0^{\dagger}$
stokes (kinematic viscosity)	square meter per second (m^2/s)	$1.0 \times 10^{-4\dagger}$
tex	kilogram per meter (kg/m)	$1.0 \times 10^{-6\dagger}$
ton (long, 2240 pounds)	kilogram (kg)	1.016×10^{3}
ton (metric)	kilogram (kg)	$1.0 \times 10^{3\dagger}$
ton (short, 2000 pounds)	kilogram (kg)	9.072×10^{2}
torr	pascal (Pa)	1.333×10^{2}
unit pole	weber (Wb)	1.257×10^{-7}
yard	meter (m)	$0.9144^{\dagger}$

Abbreviations and Unit Symbols

Following is a list of commonly used abbreviations and unit symbols appropriate for use in the *Encyclopedia*. In general they agree with those listed in *American National Standard Abbreviations for Use on Drawings and in Text (ANSI Y1.1)* (6) and *American National Standard Letter Symbols for Units in Science and Technology (ANSI Y10)* (6). Also included is a list of acronyms for a number of private and government organizations as well as common industrial solvents, polymers, and other chemicals.

Rules for Writing Unit Symbols (4):

1. Unit symbols should be printed in upright letters (roman) regardless of the type style used in the surrounding text.

2. Unit symbols are unaltered in the plural.

3. Unit symbols are not followed by a period except when used as the end of a sentence.

4. Letter unit symbols are generally written in lower-case (eg, cd for candela) unless the unit name has been derived from a proper name, in which case the first letter of the symbol is capitalized (W,Pa). Prefix and unit symbols retain their prescribed form regardless of the surrounding typography.

5. In the complete expression for a quantity, a space should be left between the numerical value and the unit symbol. For example, write 2.37 lm, *not* 2.37lm, and 35 mm, *not* 35mm. When the quantity is used in an adjectival sense, a hyphen is often used, for example, 35-mm film. *Exception:* No space is left between the numerical value and the symbols for degree, minute, and second of plane angle, and degree Celsius.

6. No space is used between the prefix and unit symbols (eg, kg).

7. Symbols, not abbreviations, should be used for units. For example, use "A," not "amp," for ampere.

8. When multiplying unit symbols, use a raised dot:

$$N \cdot m \text{ for newton meter}$$

In the case of W·h, the dot may be omitted, thus:

$$Wh$$

An exception to this practice is made for computer printouts, automatic typewriter work, etc, where the raised dot is not possible, and a dot on the line may be used.

9. When dividing unit symbols use one of the following forms:

$$m/s \ or \ m \cdot s^{-1} \ or \ \frac{m}{s}$$

In no case should more than one slash be used in the same expression unless parentheses are inserted to avoid ambiguity. For example, write:

$$J/(mol \cdot K) \ or \ J \cdot mol^{-1} \cdot K^{-1} \ or \ (J/mol)/K$$

but *not*

$$J/mol/K$$

10. Do not mix symbols and unit names in the same expression. Write:

$$joules \ per \ kilogram \ or \ J/kg \ or \ J \cdot kg^{-1}$$

but *not*

$$joules/kilogram \ nor \ joules/kg \ nor \ joules \cdot kg^{-1}$$

ABBREVIATIONS AND UNITS

A	ampere	AIME	American Institute of Mining, Metallurgical, and Petroleum Engineers
A	anion (eg, H*A*); mass number		
a	atto (prefix for 10^{-18})		
AATCC	American Association of Textile Chemists and Colorists	AIP	American Institute of Physics
		AISI	American Iron and Steel Institute
ABS	acrylonitrile–butadiene–styrene	alc	alcohol(ic)
abs	absolute	Alk	alkyl
ac	alternating current, *n.*	alk	alkaline (not alkali)
a-c	alternating current, *adj.*	amt	amount
ac-	alicyclic	amu	atomic mass unit
acac	acetylacetonate	ANSI	American National Standards Institute
ACGIH	American Conference of Governmental Industrial Hygienists	AO	atomic orbital
		AOAC	Association of Official Analytical Chemists
ACS	American Chemical Society		
AGA	American Gas Association	AOCS	American Oil Chemist's Society
Ah	ampere hour		
AIChE	American Institute of Chemical Engineers	APHA	American Public Health Association

API	American Petroleum Institute	cm	centimeter
aq	aqueous	cmil	circular mil
Ar	aryl	cmpd	compound
ar-	aromatic	CNS	central nervous system
as-	asymmetric(al)	CoA	coenzyme A
ASH-		COD	chemical oxygen demand
RAE	American Society of Heating, Refrigerating, and Air Conditioning Engineers	coml	commercial(ly)
		cp	chemically pure
		cph	close-packed hexagonal
ASM	American Society for Metals	CPSC	Consumer Product Safety Commission
ASME	American Society of Mechanical Engineers	cryst	crystalline
		cub	cubic
ASTM	American Society for Testing and Materials	D	Debye
		D-	denoting configurational relationship
at no.	atomic number		
at wt	atomic weight	**d**	differential operator
av(g)	average	*d-*	dextro-, dextrorotatory
AWS	American Welding Society	da	deka (prefix for 10^1)
b	bonding orbital	dB	decibel
bbl	barrel	dc	direct current, *n.*
bcc	body-centered cubic	d-c	direct current, *adj.*
BCT	body-centered tetragonal	dec	decompose
Bé	Baumé	detd	determined
BET	Brunauer-Emmett-Teller (adsorption equation)	detn	determination
		Di	didymium, a mixture of all lanthanons
bid	twice daily		
Boc	*t*-butyloxycarbonyl	dia	diameter
BOD	biochemical (biological) oxygen demand	dil	dilute
		DIN	Deutsche Industrie Normen
bp	boiling point	*dl-*; DL-	racemic
Bq	becquerel	DMA	dimethylacetamide
C	coulomb	DMF	dimethylformamide
°C	degree Celsius	DMG	dimethyl glyoxime
C-	denoting attachment to carbon	DMSO	dimethyl sulfoxide
		DOD	Department of Defense
c	centi (prefix for 10^{-2})	DOE	Department of Energy
c	critical	DOT	Department of Transportation
ca	circa (approximately)		
cd	candela; current density; circular dichroism	DP	degree of polymerization
		dp	dew point
CFR	Code of Federal Regulations	DPH	diamond pyramid hardness
cgs	centimeter–gram–second	dstl(d)	distill(ed)
CI	Color Index	dta	differential thermal analysis
cis-	isomer in which substituted groups are on same side of double bond between C atoms		
		(*E*)-	entgegen; opposed
		ϵ	dielectric constant (unitless number)
cl	carload	*e*	electron

ECU	electrochemical unit	GRAS	Generally Recognized as Safe
ed.	edited, edition, editor	grd	ground
ED	effective dose	Gy	gray
EDTA	ethylenediaminetetraacetic acid	H	henry
		h	hour; hecto (prefix for 10^2)
emf	electromotive force	ha	hectare
emu	electromagnetic unit	HB	Brinell hardness number
en	ethylene diamine	Hb	hemoglobin
eng	engineering	hcp	hexagonal close-packed
EPA	Environmental Protection Agency	hex	hexagonal
		HK	Knoop hardness number
epr	electron paramagnetic resonance	HRC	Rockwell hardness (C scale)
		HV	Vickers hardness number
eq.	equation	hyd	hydrated, hydrous
esp	especially	hyg	hygroscopic
esr	electron-spin resonance	Hz	hertz
est(d)	estimate(d)	^{i}(eg, Pr^i)	iso (eg, isopropyl)
estn	estimation	i-	inactive (eg, i-methionine)
esu	electrostatic unit	IACS	International Annealed Copper Standard
exp	experiment, experimental		
ext(d)	extract(ed)	ibp	initial boiling point
F	farad (capacitance)	IC	inhibitory concentration
F	faraday (96,487 C)	ICC	Interstate Commerce Commission
f	femto (prefix for 10^{-15})		
FAO	Food and Agriculture Organization (United Nations)	ICT	International Critical Table
		ID	inside diameter; infective dose
		ip	intraperitoneal
fcc	face-centered cubic	IPS	iron pipe size
FDA	Food and Drug Administration	IPTS	International Practical Temperature Scale (NBS)
FEA	Federal Energy Administration		
		ir	infrared
fob	free on board	IRLG	Interagency Regulatory Liaison Group
fp	freezing point		
FPC	Federal Power Commission	ISO	International Organization for Standardization
FRB	Federal Reserve Board		
frz	freezing	IU	International Unit
G	giga (prefix for 10^9)	IUPAC	International Union of Pure and Applied Chemistry
G	gravitational constant = 6.67×10^{11} N·m^2/kg^2		
		IV	iodine value
g	gram	iv	intravenous
(g)	gas, only as in H$_2$O(g)	J	joule
g	gravitational acceleration	K	kelvin
gem-	geminal	k	kilo (prefix for 10^3)
glc	gas-liquid chromatography	kg	kilogram
g-mol wt; gmw	gram-molecular weight	L	denoting configurational relationship
GNP	gross national product	L	liter (for fluids only)(5)
gpc	gel-permeation chromatography	l-	levo-, levorotatory
		(l)	liquid, only as in NH$_3$(l)

LC_{50}	conc lethal to 50% of the animals tested	mxt	mixture
LCAO	linear combination of atomic orbitals	μ	micro (prefix for 10^{-6})
		N	newton (force)
LCD	liquid crystal display	N	normal (concentration); neutron number
lcl	less than carload lots		
LD_{50}	dose lethal to 50% of the animals tested	N-	denoting attachment to nitrogen
LED	light-emitting diode	n (as n_D^{20})	index of refraction (for 20°C and sodium light)
liq	liquid		
lm	lumen	^{n} (as Bu^n), n-	normal (straight-chain structure)
ln	logarithm (natural)		
LNG	liquefied natural gas	n	neutron
log	logarithm (common)	n	nano (prefix for 10^9)
LPG	liquefied petroleum gas	na	not available
ltl	less than truckload lots	NAS	National Academy of Sciences
lx	lux		
M	mega (prefix for 10^6); metal (as in MA)	NASA	National Aeronautics and Space Administration
M	molar; actual mass	nat	natural
$\overline{M}_w$	weight-average mol wt	NBS	National Bureau of Standards
$\overline{M}_n$	number-average mol wt		
m	meter; milli (prefix for 10^{-3})	neg	negative
m	molal	NF	*National Formulary*
m-	meta	NIH	National Institutes of Health
max	maximum		
MCA	Chemical Manufacturers' Association (was Manufacturing Chemists Association)	NIOSH	National Institute of Occupational Safety and Health
		nmr	nuclear magnetic resonance
MEK	methyl ethyl ketone	NND	New and Nonofficial Drugs (AMA)
meq	milliequivalent		
mfd	manufactured	no.	number
mfg	manufacturing	NOI-(BN)	not otherwise indexed (by name)
mfr	manufacturer		
MIBC	methyl isobutyl carbinol	NOS	not otherwise specified
MIBK	methyl isobutyl ketone	nqr	nuclear quadruple resonance
MIC	minimum inhibiting concentration	NRC	Nuclear Regulatory Commission; National Research Council
min	minute; minimum		
mL	milliliter	NRI	New Ring Index
MLD	minimum lethal dose	NSF	National Science Foundation
MO	molecular orbital	NTA	nitrilotriacetic acid
mo	month	NTP	normal temperature and pressure (25°C and 101.3 kPa or 1 atm)
mol	mole		
mol wt	molecular weight		
mp	melting point	NTSB	National Transportation Safety Board
MR	molar refraction		
ms	mass spectrum	O-	denoting attachment to oxygen

o-	ortho	ref.	reference
OD	outside diameter	rf	radio frequency, *n.*
OPEC	Organization of Petroleum Exporting Countries	r-f	radio frequency, *adj.*
		rh	relative humidity
o-phen	*o*-phenanthridine	RI	Ring Index
OSHA	Occupational Safety and Health Administration	rms	root-mean square
		rpm	rotations per minute
		rps	revolutions per second
owf	on weight of fiber	RT	room temperature
Ω	ohm	^{s} (eg, Bu^{s}); *sec*-	secondary (eg, secondary butyl)
P	peta (prefix for 10^{15})		
p	pico (prefix for 10^{-12})	S	siemens
p-	para	(*S*)-	sinister (counterclockwise configuration)
p	proton		
p.	page	*S*-	denoting attachment to sulfur
Pa	pascal (pressure)		
pd	potential difference	*s*-	symmetric(al)
pH	negative logarithm of the effective hydrogen ion concentration	s	second
		(s)	solid, only as in H_2O(s)
phr	parts per hundred of resin (rubber)	SAE	Society of Automotive Engineers
p-i-n	positive-intrinsic-negative	SAN	styrene–acrylonitrile
pmr	proton magnetic resonance	sat(d)	saturate(d)
p-n	positive-negative	satn	saturation
po	per os (oral)	SBS	styrene–butadiene–styrene
POP	polyoxypropylene	sc	subcutaneous
pos	positive	SCF	self-consistent field; standard cubic feet
pp.	pages		
ppb	parts per billion (10^9)	Sch	Schultz number
ppm	parts per million (10^6)	SFs	Saybolt Furol seconds
ppmv	parts per million by volume	SI	Le Système International d'Unités (International System of Units)
ppmwt	parts per million by weight		
PPO	poly(phenyl oxide)		
ppt(d)	precipitate(d)	sl sol	slightly soluble
pptn	precipitation	sol	soluble
Pr (no.)	foreign prototype (number)	soln	solution
pt	point; part	soly	solubility
PVC	poly(vinyl chloride)	sp	specific; species
pwd	powder	sp gr	specific gravity
py	pyridine	sr	steradian
qv	quod vide (which see)	std	standard
R	univalent hydrocarbon radical	STP	standard temperature and pressure (0°C and 101.3 kPa)
(*R*)-	rectus (clockwise configuration)	sub	sublime(s)
r	precision of data	SUs	Saybolt Universal seconds
rad	radian; radius		
rds	rate determining step	syn	synthetic

^{t} (eg, Bu^t), t-, $tert$-	tertiary (eg, tertiary butyl)	Twad	Twaddell
		UL	Underwriters' Laboratory
		USDA	United States Department of Agriculture
T	tera (prefix for 10^{12}); tesla (magnetic flux density)	USP	*United States Pharmacopeia*
		uv	ultraviolet
t	metric ton (tonne); temperature	V	volt (emf)
		var	variable
TAPPI	Technical Association of the Pulp and Paper Industry	*vic*-	vicinal
		vol	volume (not volatile)
tex	tex (linear density)	vs	versus
T_g	glass-transition temperature	v sol	very soluble
tga	thermogravimetric analysis	W	watt
THF	tetrahydrofuran	Wb	Weber
tlc	thin layer chromatography	Wh	watt hour
TLV	threshold limit value	WHO	World Health Organization (United Nations)
trans-	isomer in which substituted groups are on opposite sides of double bond between C atoms		
		wk	week
		yr	year
TSCA	Toxic Substance Control Act	(*Z*)-	zusammen; together; atomic number
TWA	time-weighted average		

Non-SI (Unacceptable and Obsolete) Units		*Use*
Å	angstrom	nm
at	atmosphere, technical	Pa
atm	atmosphere, standard	Pa
b	barn	cm^2
bar†	bar	Pa
bbl	barrel	m^3
bhp	brake horsepower	W
Btu	British thermal unit	J
bu	bushel	m^3; L
cal	calorie	J
cfm	cubic foot per minute	m^3/s
Ci	curie	Bq
cSt	centistokes	mm^2/s
c/s	cycle per second	Hz
cu	cubic	exponential form
D	debye	C·m
den	denier	tex
dr	dram	kg
dyn	dyne	N
dyn/cm	dyne per centimeter	mN/m
erg	erg	J
eu	entropy unit	J/K
°F	degree Fahrenheit	°C; K
fc	footcandle	lx
fl	footlambert	lx
fl oz	fluid ounce	m^3; L
ft	foot	m
ft·lbf	foot pound-force	J

† Do not use bar (10^5Pa) or millibar (10^2Pa) because they are not SI units, and are accepted internationally only for a limited time in special fields because of existing usage.

Non-SI (Unacceptable and Obsolete) Units		*Use*
gf den	gram-force per denier	N/tex
G	gauss	T
Gal	gal	m/s^2
gal	gallon	m^3; L
Gb	gilbert	A
gpm	gallon per minute	(m^3/s); (m^3/h)
gr	grain	kg
hp	horsepower	W
ihp	indicated horsepower	W
in.	inch	m
in. Hg	inch of mercury	Pa
in. H_2O	inch of water	Pa
in.-lbf	inch pound-force	J
kcal	kilogram-calorie	J
kgf	kilogram-force	N
kilo	for kilogram	kg
L	lambert	lx
lb	pound	kg
lbf	pound-force	N
mho	mho	S
mi	mile	m
MM	million	M
mm Hg	millimeter of mercury	Pa
$m\mu$	millimicron	nm
mph	miles per hour	km/h
μ	micron	μm
Oe	oersted	A/m
oz	ounce	kg
ozf	ounce-force	N
η	poise	Pa·s
P	poise	Pa·s
ph	phot	lx
psi	pounds-force per square inch	Pa
psia	pounds-force per square inch absolute	Pa
psig	pounds-force per square inch gauge	Pa
qt	quart	m^3; L
°R	degree Rankine	K
rd	rad	Gy
sb	stilb	lx
SCF	standard cubic foot	m^3
sq	square	exponential form
thm	therm	J
yd	yard	m

BIBLIOGRAPHY

1. The International Bureau of Weights and Measures, BIPM (Parc de Saint-Cloud, France) is described on page 22 of Ref. 4. This bureau operates under the exclusive supervision of the International Committee of Weights and Measures (CIPM).

2. *Metric Editorial Guide* (*ANMC-78-1*) 3rd ed., American National Metric Council, 1625 Massachusetts Ave. N.W., Washington, D.C. 20036, 1978.

3. *SI Units and Recommendations for the Use of Their Multiples and of Certain Other Units* (*ISO 1000-1981*), American National Standards Institute, 1430 Broadway, New York, N. Y. 10018, 1981.

4. Based on *ASTM E 380-82* (*Standard for Metric Practice*), American Society for Testing and Materials, 1916 Race Street, Philadelphia, Pa. 19103, 1982.

5. *Fed. Regist.*, Dec. 10, 1976 (41 FR 36414).

6. For ANSI address, see Ref. 3.

R. P. LUKENS
American Society for Testing and Materials

R continued

REFRACTORIES

Refractories are materials that resist the action of hot environments by containing heat energy and hot or molten materials (1). The type of refractories that are used in any particular application depends upon the critical requirements of the process. For example, processes that demand resistance to gaseous or liquid corrosion require low porosity, high physical strength, and abrasion resistance. Conditions that demand low thermal conductivity may require entirely different refractories. Indeed, combinations of several refractories are generally employed. There is no well established line of demarcation between those materials that are and those that are not refractory although the ability to withstand temperatures above 1100°C without softening has been cited as a practical requirement of industrial refractory materials (see also Ceramics).

Physical Forms

Refractories may be preformed (shaped) or formed and installed on site.

Brick. The standard dimensions of a refractory brick are 23 cm long by 11.4 cm wide and 6.4 cm thick (straight brick). Quantities of bricks are given in brick equivalents, that is, the number of standard 23-cm (9-in.) bricks with a volume equal to that of the particular installation. The actual shape and size of bricks depends upon the design of the vessel or structure in question and may vary considerably from the standard 23-cm straight brick. For example, bricks for basic oxygen furnaces (BOF vessels) may be in the shape of a key 65.6 cm long, 7.6 cm thick, and tapering in width from 15.2–10.2 cm. Numerous other shapes are available from manufacturers as standard items as well as custom made or special ordered shapes.

Bricks may be extruded or dry-pressed on mechanical or hydraulic presses.

Formed shapes may be burned before use or, in the case of pitch, resin or chemically bonded brick, cured.

Setter Tile and Kiln Furniture. These products are formed in a similar manner to bricks and are used to support ware during firing operations. The wide variety of available shapes and sizes include flat slabs, posts, saggers, and car-top blocks.

Fusion-Cast Shapes. Refractory compositions are arc-melted and cast into shapes, eg, glass-tank flux blocks as large as $0.33 \times 0.66 \times 1.33$ m. After casting and annealing, the blocks are accurately ground to ensure a precise fit.

Cast and Hand-Molded Refractories. Large shapes such as burner blocks, flux blocks, and intricate shapes such as glass feeder parts, saggers, and the like are produced by either slip or hydraulic cement casting or hand-molding techniques. Because these techniques are labor intensive, they are reserved for articles that cannot be satisfactorily formed in other ways.

Insulating Refractories. Insulating refractories in the form of brick are much lighter than conventional brick of the same composition by virtue of the brick porosity. The porosity may be introduced by means of lightweight grog or additives that create porosity by a foaming action or by evolution of combustion products upon burn-out. Refractory fiber made from molten oxides may be formed into bulk fiber, blankets, boards, or blocks. Such fibers are used as back-up thermal insulation and low heat-capacity linings in kilns and reheat furnaces (see Insulation, thermal; Refractory fibers).

Castables and Gunning Mixes. Castables consist of refractory grains to which a hydraulic binder is added. Upon mixing with water, the hydraulic agent reacts and binds the mass together. Gunning mixes are designed to be sprayed through a nozzle under water and air pressure. The mixture may be slurried before being shot through the gun or mixed with water at the nozzle. In addition to refractory grains, the mix may contain clay and nonclay additives to promote adherence to the furnace wall. Unlike casting mixes, gunning mixes may be used to make maintenance repairs without removing the furnace from service.

Plastic Refractories and Ramming Mixes. Plastic refractories are mixtures of refractory grains and plastic clays or plasticizers with water. Ramming mixes may or may not contain clay and are generally used with forms. The amount of water used with these products varies but is held to a minimum.

Mortars. These consist of finely ground refractory grain and plasticizers that can be thinly spread on brick during construction (see Cement). For air-setting mortars, sodium silicates or phosphates provide strength at room temperature. Heat-setting mortars contain no additives and develop strength only when a ceramic bond is formed at high temperatures.

Composite Refractories. Although many refractories may be considered composites, examples of composites analogous to metallurgical and organic composites are not common (see Composite materials). Recently, however, shaped and unshaped refractories containing metallic reinforcements have appeared on the market, as well as setter tile made with a sandwich construction, ie, one internal layer of deformation-resistant material such as silicon carbide bounded by two layers of oxidation-resistant material.

Refractory Coatings. Refractory coatings are applied either by painting or by spraying to a fine-grained refractory mix at room temperature. By heating, a dense sintered coating is formed. Other techniques include flame or plasma spraying. In the former, the powdered coating material is fed into a burner and sprayed at elevated temperatures. The pyroplastic grains form a dense monolithic coating when they impinge on the substrate. Plasma spraying is carried out in essentially the same manner, except that an electrically ionized gas plasma heats the coating powder to temperatures up to 16,700°C (see Refractory coatings).

Raw Materials

In the past, refractory raw materials were selected from a variety of available deposits and used essentially as mined minerals (2–7). Selective mining yielded materials of the desired properties and only in cases of expensive raw materials, such as magnesite, was a beneficiation process required. Today, however, high purity natural raw materials are increasingly in demand as well as synthetically prepared refractory grain made from combinations of high purity and beneficiated raw materials (see Tables 1 and 2). The material produced upon firing raw as-mined minerals or synthetic blends is called grain, clinker, co-clinker, or grog.

Silica. The most common refractory raw materials are ganister, which is a dense quartzite, and silica gravels (see Silica; Silicon compounds). The latter are generally purer than the former and are often further beneficiated by washing to give the raw material for superduty silica brick with no more than 0.5% impurities (alkalies, Al_2O_3, and TiO_2). Quartzite and gravel deposits are widespread throughout the world. The most important U.S. deposits are found in Pennsylvania, Ohio, Wisconsin, Alabama, Colorado, and Illinois. Synthetically produced electrofused silica is a thermally stable, shock-resistant, high purity raw material.

Fireclay. Fireclays consist mainly of the mineral kaolinite [1318-74-7], $Al_2O_3.2SiO_2.2H_2O$, with small amounts of other clay minerals, quartzite, iron oxide, titania, and alkali impurities. Clays can be used in the raw state or after being calcined. Raw clays may be coarsely sized or finely ground for incorporation in a refractory mix. Some high purity kaolins like those that occur in Georgia are slurried, classified, dried, and air-floated to achieve a consistent, high quality. The classified clays also may be blended and extruded or pelletized, and then calcined to produce burned synthetic kaolinitic grog, or coarsely crushed raw kaolinite may be burned to produce grog. Upon calcination or burning, kaolinite decomposes to mullite and a siliceous glass incorporating mineral impurities associated with the clay deposit (eg, quartzite, iron oxide, titania, and alkalies), and is consolidated into dense hard granular grog at high temperatures. Fireclay deposits are widely distributed; in the United States they occur in Pennsylvania, Missouri, and Kentucky.

High Alumina. The naturally occurring raw materials are bauxites, sillimanite [12141-45-6] group minerals, and diaspore clays (see Aluminum compounds). Other high alumina raw materials are made by beneficiation, blending, and other processing techniques.

Bauxites. Bauxites [1318-16-7] consist mainly of gibbsite [14762-49-3], $Al(OH)_3$, with varying amounts of kaolinite, and iron and titania impurities. Because the loss on ignition is high, bauxite must be calcined to high temperatures before use. During calcination, it is converted to a dense grain consisting mainly of corundum [12252-63-0], Al_2O_3, and mullite. Refractory-grade bauxite is relatively rare since a high iron content makes most bauxites unsuitable for refractory use. Commercially mined deposits are in South America, especially Guyana and Surinam, and the People's Republic of China. Other deposits occur in India and Central Africa but are not mined for refractory grades at present.

Table 1. Composition of Refractory Raw Materials, %[a,b]

Name	Location	SiO$_2$	Al$_2$O$_3$	Fe$_2$O$_3$	TiO$_2$	CaO	MgO	Cr$_2$O$_3$	Alkalies	ZrO$_2$
silica raw materials										
ganister	Bwlehgwyn (N. Wales, UK)	97.4	0.73	0.78	0.1					
gravel	Sharon Conglomerate, Ohio	98.0	0.3	0.5	0.1					
clays										
flint clays	Pennsylvania	50.40	34.58	1.42	2.06	0.45	1.00		1.58	
	Missouri	43.80	38.29	0.60	2.33	0.07	0.20		0.39	
	Kentucky	44.94	35.17	1.56	3.12	0.11	0.18		1.45	
	People's Republic of China	51.20	46.62	0.87	0.87				0.15	
plastic clays	Pennsylvania	54.00	27.94	1.39	2.45	0.07	1.22		3.33	
	Missouri	55.12	28.65	1.66	1.55	0.11	0.10		2.54	
	Kentucky	56.42	26.92	2.05	1.95	0.18	0.51		1.75	
kaolin	Georgia	43.00	37.55	0.85	2.10	0.09	0.18		0.15	
	Florida	46.5	37.62	0.51	0.36	0.25	0.16		0.42	
fireclay	Stourbridge (Scotland)	68.1	27.2	1.95	1.1	0.72	0.35		1.28	
	Pfalz (FRG)	45.1	36.3	2.21	1.12	0.08	0.11		2.25	
plastic	Chasov-Yar (USSR)	51.6	33.3	0.9	1.37	0.53	0.57		3.28	
semikaolin	Suvorov (USSR)	46.1	33.9	2.14	1.52	0.41	0.23		0.44	
kaolin	Vladimirovka (USSR)	48.3	36.7	0.83	0.78	0.3	0.33		0.77	
high-alumina										
natural										
siliceous bauxite[c]										
ca 70% Al$_2$O$_3$	Eufaula, Alabama	25.9	70.1	1.13	2.9	0.05	0.03		0.13	
ca 60% Al$_2$O$_3$	Eufaula, Alabama	34.9	60.6	1.26	2.5	0.07	0.12		0.11	
ca 60% Al$_2$O$_3$	People's Republic of China	32.40	63.50	1.50	2.20				0.20	
ca 50% Al$_2$O$_3$	People's Republic of China	43.30	52.80	1.42	1.90				0.19	
South American bauxite[c]	Guyana	7.0	87.5	2.00	3.25	trace	trace		trace	
Chinese bauxite[c]	People's Republic of China	6.0	87.5	1.50	3.75				0.50	
kyanite[d]	Virginia		59–61	0.2–0.9	0.67	0.03	0.01		0.50	
sillimanite[d]	India	38.6	59–61						0.4	

4

synthetic								
fused alumina								
black		0.48	97.3	0.15	2.45	0.07	0.11	0.05
gray		0.06	99.5+	0.15	0.06	trace	trace	0.07
sintered alumina		0.06	99.5	0.06	trace	trace	trace	0.05
sintered mullite	Georgia	27.90	68.00	1.33	2.61	0.06	0.04	0.06
sintered magnesium aluminate	Japan	0.2	67.90	0.2		0.3	31.9	
fused mullite		22.0	77.7	0.12	0.05	trace	trace	0.35
calcium aluminate cement								
low purity		8.4	42.0	10.7		37.0	1.2	
high purity		0.1	79.0	0.3	trace	18.0	0.1	0.5
zirconium								
zircon		32.5	0.08	0.05	0.06	trace	trace	trace
baddeleyite	Brazil			3–5				trace
basic raw materials								
calcined magnesias								
natural magnesite	Austria	0.3	0.3	5.4		2.7	91.3	
natural magnesite	Greece	1.2	0.3	0.3		2.8	95.4	
natural magnesite	U.S.	1.00	0.36	0.65		3.02	94.97	
natural magnesite	People's Republic of China	0.40	0.13	1.73		1.31	96.36	
natural magnesite	Turkey	1.42	<0.02	<0.02		1.58	96.95	
seawater	Japan	0.37	0.04	0.01		1.23	98.32	
seawater	UK	0.70	0.20	0.10		2.30	96.66	
seawater	U.S.	0.80	0.12	0.14		2.22	96.70	
seawater	U.S.	0.80	0.15	0.22		2.65	96.12	
seawater	Ireland	0.83	0.18	0.17		1.07	97.75	
brine	Israel	0.14	0.02	0.04		0.77	99.45	
dolomite	UK	1.8	1.1	5.9		54.6	36.4	
dolomite, low flux	U.S.	0.7	0.3	0.9		57.7	40.4	
chrome ore	Africa	3.8	14.9	27.1		0.3	9.9	44.0
chrome ore	Philippines	5.5	31.0	15.5		0.5	16.0	31.5
							66.7	
							70.80	

[a] Refs. 2 and 5.
[b] Difference between total analysis and 100 = loss on ignition, %.
[c] Calcined.
[d] Raw.

5

Table 2. Physical Properties of Refractory Raw Materials [a]

Material	Pyrometer cone equivalent	Main crystalline phases	Specific gravity, g/cm^3 Bulk	True	Apparent porosity, %
silica					
ganister		quartz		2.66	1.6
gravel		quartz		2.61	0.3
clays					
flint clays	32–33				
	34	kaolinite, illite, quartz			
	33	kaolinite, quartz, illite			
			2.55		
plastic clays	31	kaolinite, quartz, illite			
	30	kaolinite, quartz, illite			
	30	kaolinite, quartz, illite			
kaolin	33–34	kaolinite			
fireclay	33–34	kaolinite			
plastic	1720°C	kaolinite, illite			
semikaolin	1740°C				
kaolin	1750°C				
high-alumina					
natural					
siliceous bauxite [b]					
ca 70% Al$_2$O$_3$	38–39	mullite	2.85–2.95	3.1–3.2	4–8
ca 60% Al$_2$O$_3$		mullite	2.75–2.85	2.95–3.05	3–7
ca 60% Al$_2$O$_3$			2.70		
ca 50% Al$_2$O$_3$			2.65		
South American bauxite [b]		corundum, mullite	3.1	3.6–3.7	15–20
Chinese bauxite [b]			3.20		
kyanite [c]	36–37	kyanite		3.5–3.7	
sillimanite [c]		sillimanite		3.23	
synthetic					
fused alumina					
black	42	α-alumina	3.87	4.01	3.49
gray	42+	α-alumina	3.95	3.98	0.5–1.0
sintered alumina	42+	α-alumina	3.45–3.6	3.65–3.80	5.0
sintered mullite	39		2.85		
sintered magnesium aluminate		spinel, periclase	3.33		
fused mullite	39	mullite	3.1	3.45	0.1
calcium aluminate cement					
low purity		calcium monoaluminate			
high purity	34	α-alumina, calcium monoaluminate			
zirconium					
zircon		zircon		4.2–4.6	
baddeleyite (ZrO$_2$)		baddeleyite		5.5–6.5	
basic raw materials					
calcined magnesias					
natural magnesite		periclase	3.2		
natural magnesite		periclase	3.4		
natural magnesite			3.39		
natural magnesite		magnesite, dolomite, calcite	3.40		
natural magnesite			3.39		
seawater		periclase	3.44		
seawater		periclase	3.35		
seawater		periclase	3.40		

Table 2 (*continued*)

Material	Pyrometer cone equivalent	Main crystalline phases	Specific gravity, g/cm^3		Apparent porosity, %
			Bulk	True	
seawater		periclase	3.44		
seawater		periclase	3.22		
brine		periclase	3.41		
dolomite		periclase + CaO			
dolomite, low flux		periclase + CaO			
chrome ore		chromite spinel	4.2		
chrome ore		chromite spinel	3.9		

[a] Refs. 2 and 5.
[b] Calcined.
[c] Raw.

Sillimanite Minerals. This group includes sillimanite, andalusite [12183-80-1], and kyanite [1302-76-7], all of the formula $Al_2O_3.SiO_2$. Upon heating, a mixture of mullite, silica, and a siliceous glass is obtained. The specific gravity of sillimanite and andalusite is ca 3.2, and that of kyanite 3.5. Thus, kyanite expands upon conversion to mullite by ca 16–18 vol %, but sillimanite and andalusite expand only slightly. Kyanite is found in India and South Africa, and, in the United States, in Virginia and South Carolina. Large-grained kyanite is rare, and the largest size commercially available from U.S. sources is ca 500 μm (35 mesh). Refractory-grade sillimanite occurs in India and South Africa.

Diaspore Clays. These consist of a mixture of kaolin and the mineral diaspore [14457-84-2], $Al_2O_3.H_2O$. They were actively mined in Missouri, but these deposits are now largely depleted (see Clays).

Mullite. Although mullite is found in nature, for example, as inclusions in lava deposits on the island of Mull, Scotland, no commercial natural deposits are known. It is made by burning pure sillimanite minerals or sillimanite–alumina mixtures. Fused mullite of high purity is obtained by arc melting silica sand and calcined alumina. High purity sintered mullite is made from alumina and silica, but requires mineralizing agents and very high temperatures.

Alumina. A pure grade of alumina is obtained from bauxite (not necessarily refractory grade) by the Bayer process. In this process, the gibbsite from the bauxite is dissolved in a caustic soda solution and thus separated from the impurities. Alumina, calcined, sintered, or fused, is a stable and extremely versatile material used for a variety of heavy industrial, electronic, and technical applications.

Calcined alumina is a reactive powder which is used to make synthetic grain. It also may be used as a bonding or fine component in batched refractory mixes, as a raw material for molten cast refractories, or for refractory casting slips.

Sintered alumina, also known as tabular alumina, is formed by burning aggregates made from reactive calcined alumina to high temperatures to obtain a stable high purity corundum grain.

Fused alumina is obtained by fusing either calcined alumina or bauxite. Bauxite is beneficial during fusion as iron and silica are removed as ferrosilicon. A special grade of fused alumina is obtained by blowing air through a stream of molten alumina. Thus, small bubbles of alumina are formed, an excellent lightweight aggregate of high purity and refractoriness.

Calcium Aluminate Cements. Low purity calcium aluminate [12042-78-3] cements are obtained by sintering or fusing bauxite and lime in a rotary or shaft kiln. A high purity calcium aluminate cement, $2CaO.5Al_2O_3$, capable of withstanding service temperatures of 1750°C can be prepared by the reaction of high purity lime with calcined or hydrated alumina.

Zirconia. Zircon, the most widely occurring zirconium-bearing mineral, is dispersed in various igneous rocks and in zircon sands. The main deposits are in New South Wales, Australia; Travancore, India, and Florida in the United States. Zircon can be used as such in zircon refractories or as a raw material to produce zirconia. The zircon structure becomes unstable after about 1650°C, depending upon its purity, and decomposes into ZrO_2 and SiO_2 rather than melting (see Zirconium and zirconium compounds).

Zirconia occurs as the mineral baddeleyite [12036-23-6], for instance, in a deposit around Sao Paulo, Brazil. However, these baddeleyite deposits generally contain large amounts of impurities and only about 80% ZrO_2. High purity zirconia can be obtained from baddeleyite by leaching with concentrated sulfuric acid or chlorination at high temperatures. The zirconium sulfates and chlorides thus formed are readily separated from the impurities. Zirconia also is made from zircon by electric-arc melting under reducing conditions. Here, the silica is separated from the zirconia by adding iron to form ferrosilicon. The most remarkable property of zirconia is its volume instability. Upon heating, unstabilized ZrO_2 transforms from the low temperature monoclinic form to the tetragonal form at about 1000°C, resulting in a 9% volume contraction. By adding impurities with a cubic structure, such as magnesia, calcia, or yttria, zirconia can be transformed into a cubic crystal phase stable at all temperatures, although often only enough impurities are added for partial stabilization, ie, both cubic and either the tetragonal or monoclinic form exist.

Basic Raw Materials. *Magnesite.* Calcined or dead-burned magnesite [13717-00-5] is obtained by firing naturally occurring magnesium carbonate to 1540–2000°C. This treatment produces a dense product composed primarily of periclase [1309-48-4], MgO. Large sedimentary deposits of magnesite occur in Austria, Manchuria, Greece, the Ural Mountains, and in Washington, Nevada, and California. Calcia, silica, alumina, and iron-bearing phases occur as accessory minerals (see also Magnesium compounds).

Seawater contains approximately 1294 ppm Mg (seawater magnesite); higher concentrations occur in magnesium-rich brines (see Chemicals from brine). Treatment with hydrated lime, $Ca(OH)_2$ precipitates $Mg(OH)_2$:

$$CaO + H_2O + MgCl_2 \rightarrow Mg(OH)_2 + CaCl_2$$

The $Mg(OH)_2$ precipitate is filtered, dried, and calcined. For high quality refractory grain (98% MgO), an initial calcination is followed by mechanical compaction and final high temperature calcination to form a dense product. Large seawater or brine plants are located in the United States (Gulf of Mexico, California, Michigan, and New Jersey), Mexico, the United Kingdom, Ireland, Israel, Italy, Japan, and the USSR.

Magnesium hydroxide also occurs in sedimentary deposits as the mineral brucite [1317-43-7], eg, in Quebec and Nevada.

Dolomite. Dolomite [17069-72-6], $CaMg(CO_3)_2$, occurs in widespread deposits in many areas including southern Austria, the UK, the USSR, and the United States. Raw dolomite may be used for certain refractories, but in most instances it is calcined to form a grain consisting primarily of MgO (periclase) and CaO [1305-78-8]. Calcined dolomite absorbs H_2O and CO_2 from the atmosphere and eventually disintegrates. Fluxes such as SiO_2, Fe_2O_3, and Al_2O_3 increase hydration resistance but also sharply reduce the fusion point of the dolomite. High-purity, low-flux dolomites with less than 2% impurities are produced by high-temperature calcination of natural dolomites in Ohio and Pennsylvania (see Lime and limestone).

Forsterite. Pure forsterite is rare in nature; most natural magnesium orthosilicates form solid solutions of fayalite (Fe_2SiO_4) and forsterite. Forsterite refractories are usually made by calcining magnesium silicate rock such as dunite, serpentine, or olivine with sufficient magnesia added to convert all excess silica to fosterite and all sesquioxides to magnesia spinels.

Chrome Ore. Chromia-bearing spinel materials, although considered neutral, are generally used in combination with basic magnesite. Chrome ores consist essentially of a complex solid-solution series of spinels including hercynite, $FeO.Al_2O_3$; ferrous chromite, $FeO.Cr_2O$; magnesioferrite, $MgO.Fe_2O_3$; picrochromite, $MgO.Cr_2O_3$; spinel, $MgO.Al_2O_3$; and magnetite $FeO.Fe_2O_3$. Silicate phases such as serpentine, talc, and enstatite are commonly associated with the spinel grains. The principal deposits of chrome ore occur in Africa (Transvaal, South Africa, Zimbabwe), the USSR, Turkey, Greece, Cuba, and the Philippines; African and USSR chrome ores are high in iron content, and the Cuban and Philippine ores are higher in alumina.

Silicon Carbide. Silicon carbide is made by the electrofusion of silica sand and carbon. Silicon carbide is hard, abrasion resistant, and has a high thermal conductivity. It is relatively stable but has a tendency to oxidize above 1400°C. The silica thus formed affords some protection against further oxidation.

Beryllia and Thoria. These are specialty oxides for highly specialized applications that require electrical resistance and high thermal conductivity. Beryllia is highly toxic and must be used with care. Both are very expensive and are used only in small quantities.

Carbon and Graphite. Carbon and graphite [7782-42- 5] have been used alone to make refractory products although they are commonly used in conjunction with other refractory raw materials. Carbon blacks are commercially manufactured, whereas graphites have to be mined.

General Properties

Oxides. A number of simple and mixed refractory oxide materials are described in Table 3. At present, Al_2O_3 is the most widely used simple oxide; it has moderate thermal-shock resistance, good stability over a wide variety of atmospheres, and is a good electric insulator at high temperatures. The strength of ceramics is influenced by minor impurities and microstructural features. In general, polycrystalline alumina has reasonably good and nearly constant strength up to about 1000–1100°C; at higher temperatures, the strength drops to much less than one half of the room temperature value over a 400°C temperature increment. Single-crystal alumina is stronger than polycrystalline Al_2O_3 and actually increases in strength between 1000 and 1100°C. Fused silica glass has excellent thermal-shock properties but devitrifies on long heating above 1100°C and loses much of its shock resistance.

Table 3. Properties of Pure Refractory Materials[a]

Material	CAS Registry Number	Formula	Mp, °C	True specific gravity, g/cm³	Mean specific heat J/(kg·K)[b]	Temp range, °C	Thermal conductivity, W/(m·K) at 500°C	at 1000°C	Linear thermal expansion coefficient per °C × 10^6, from 20–1000°C
aluminum oxide	[1344-28-1]	Al_2O_3	2015	3.97	795.5	25–1800	10.9	6.2	8.6
beryllium oxide	[1304-56-9]	BeO	2550	3.01	1004.8	25–1200	65.4	20.3	9.1
calcium oxide	[1305-78-8]	CaO	2600	3.32	753.6	25–1800	8.0	7.8	13.0
magnesium oxide	[1309-48-4]	MgO	2800	3.58	921.1	25–2100	13.9	7.0	14.2
silicon dioxide[c]	[7631-86-9]	SiO_2		2.20	753.6	25–2000	1.6	2.1	0.5
thorium oxide	[1314-20-1]	ThO_2	3300	10.01	251.2	25–1800	5.1	3.0	9.4
titanium oxide	[13463-67-7]	TiO_2	1840	4.24	711.8	25–1800	3.8	3.3	8.0
uranium oxide	[1344-58-7]	UO_2	2878	10.90	251.2	25–1500	5.1	3.4	
zirconium oxide[d]	[1314-23-4]	ZrO_2	2677	5.90	460.6	25–1100	2.1	2.3	
mullite	[55964-99-3]	$3Al_2O_3.2SiO_2$	1850[e]	3.16	628.0	25–1500	4.4	4.0	4.5
spinel	[1302-67-6]	$MgO.Al_2O_3$	2135	3.58	795.5		9.1	5.8	0.2
forsterite	[15118-03-3]	$2MgO.SiO_2$	1885	3.22	837.4		3.1	2.4	9.5
zircon	[10101-52-7]	$ZrO_2.SiO_2$	2340–2550[f]	4.60	544.3		4.3	4.1	4.0
carbon	[7440-44-0]	C		2.10	1046.7	25–1300	13.4	9.9	4.0
silicon carbide	[409-21-2]	SiC	3990[g]	3.21	795.5	25–1300	22.5	23.7	5.2

[a] Refs. 8–10.
[b] To convert J to cal, divide by 4.184.
[c] Silica glass.
[d] Cubic, stabilized with CaO.
[e] Congruent.
[f] Incongruent.
[g] Dissociates above 2450°C in reducing atmosphere and is readily oxidized above 1650°C.

Beryllium and magnesium oxides are stable to very high temperatures in oxidizing environments. Above 1700°C MgO is highly volatile under reducing conditions and in vacuum, whereas BeO exhibits better resistance to volatilization but is readily volatized by water vapor above 1650°C. Beryllia has good electrical insulating properties and high thermal conductivities; however, its high toxicity restricts its use. Calcium oxide, and to a lesser extent uranium oxide, hydrate readily; UO_2 can be oxidized to lower melting U_3O_8. Zirconia in pure form is rarely used in ceramic bodies; however, stabilized or partially stabilized cubic ZrO_2 is currently the most useful simple oxide for operations above 1900°C. Thorium oxide exhibits good properties for high temperature operation but is very expensive. Since it is a fertile nuclear material, it is under the control of the U.S. NRC. Titanium oxide is readily reduced to lower oxides and cannot be used in neutral or reducing atmospheres.

Carbon, Carbides, and Nitrides. Carbon (graphite) is a good thermal and electrical conductor. It is not easily wetted by chemical action, which is an important consideration for corrosion resistance. As an important structural material at high temperature, pyrolytic graphite has shown a strength of 280 MPa (40,600 psi). It tends to oxidize at high temperatures, but can be used up to 2760°C for short periods in neutral or reducing conditions. The use of new composite materials made with carbon fibers is expected, especially in the field of aerospace structure.

When heated under oxidizing conditions, silicon carbide and silicon nitride, Si_3N_4 [12033-89-5], form protective layers of SiO_2 and can be used up to ca 1700°C. In reducing or neutral atmospheres, their useful range is much higher. Silicon carbide has very high thermal conductivity and can withstand thermal shock cycling without damage. It also is an electrical conductor and is used for electrical heating elements. Other carbides have relatively poor oxidation resistance, but under neutral or reducing conditions, they have potential usefulness as technical ceramics in aerospace application; eg, the carbides of B, Nb, Hf, Ta, Zr, Ti, V, Mo, and Cr. Ba, Be, Ca, and Sr carbides are hydrolyzed by water vapor.

Silicon nitride has good strength retention at high temperature and is the most oxidation resistant nitride. Boron nitride [10043-11- 5] has excellent thermal-shock resistance and is in many ways similar to graphite, except that it is not an electrical conductor.

Borides and Silicides. These materials do not show good resistance to oxidation; however, some silicides form SiO_2 coatings upon heating which retards further oxidation. Molybdenum disilicide [1317-33-5], $MoSi_2$, is used widely, primarily as an electrical heating element.

Metals. The highest-melting refractory metals are tungsten (3400°C), tantalum (2995°C), and molybdenum (2620°C); all show poor resistance to oxidation at high temperatures. Hafnium–tantalum alloys form a tightly adhering oxide layer that gives partial protection up to 2200°C; the layer continues to grow with extended use.

Phase Equilibria. Phase diagrams represent the chemical equilibria that exist among one, two, or three components of a system under the influence of temperature and pressure (11–12). Reference to a phase diagram permits the determination of the amount and composition of solid and liquid phases that coexist under certain specified conditions of temperature and pressure for a particular system. With such information, the occurrence of physical and chemical changes within a system or between systems at high temperatures can be predicted. Systems containing more than three components are difficult or impossible to present graphically; however, mathematical methods have been suggested (13–14).

Phase diagrams can be used to predict the reactions between refractories and various solid, liquid, and gaseous reactants. However, phase diagrams derived from phase equilibria of relatively simple pure compounds and real systems are highly complex and may contain a large number of minor impurities that significantly affect equilibria. Furthermore, equilibrium between the reacting phases may not be reached under actual service conditions, and the physical environment of a product may be more influential to its life than the chemical environment (15–19).

Physical Properties. The important physical properties of some refractories are listed in Tables 4–7. Brick bulk density depends mainly on the specific gravity of the constituents and the porosity. The latter is controlled by the porosity of the raw materials and the brick texture. Coarse, medium, and fine-sized material contribute to the degree of packing; usually the highest density possible is desired. Upon firing, the grains and matrix form glassy, direct, or solid-state ceramic bonds. Sintering is generally accompanied by shrinkage, unless new components are formed that may cause expansion. Differential volume change between the coarse and fine fraction caused by differential sintering rates and formation of addition phases may create stresses. Particle size distribution, forming method, and firing process contribute to texture, whereas permeability is related to porosity, which in turn is dependent upon texture.

Mechanical Properties. The physical properties of a particular refractory product depend upon its constituents and manner in which they were assembled. The physical properties may be varied to suit specific applications; for example, for thermal insulations highly porous products are employed, whereas dense products are used for slagging or abrasive conditions.

The strength (modulus of rupture) at room temperature is determined by the degree of bonding. Fine-grained refractories generally are stronger than coarse-grained types and those with a low porosity are stronger than those of high porosity. However, the room temperature strength of a refractory is not necessarily indicative of the strength at high temperature because the bond strength may be due to a glassy phase that softens upon heating.

Generally, high temperature strength is lower than room temperature strength. The former is a measure of the degree of solid-state bonding between refractory grains, whereas high temperature creep indicates the amount of associated liquid or glassy phases and their viscosity. The development of solid-state or direct-bonded basic brick requires high firing temperatures, and is impeded by glassy phases. By referring to phase diagrams, refractory compositions may be designed that avoid the development of such phases.

The modulus of elasticity is related to the strength and can be used as a nondestructive quality-control test on high cost special refractory shapes such as slide gate valves employed in the pouring of steel. Generally speaking, the strength is related directly to slide-gate performance.

Thermal Properties. Refractories, like most other solids, expand upon heating, but much less than most metals. The degree of expansion depends upon the chemical composition. A diagram of the thermal expansion of the most common refractories is shown in Figure 1.

Irreversible Thermal Expansion. Most refractory bricks are not chemically in equilibrium before use. During the prolonged heating in service, additional reactions occur that may cause the brick to shrink or expand. For instance, 70% Al_2O_3 bricks

Table 4. Composition of Alumina, Silica, and Zirconia Refractories, %[a]

Type	SiO_2	Al_2O_3	Fe_2O_3	TiO_2	CaO	MgO	Alkalies	ZrO_2
silica	95–96	0.2–1.5	0.8–1.0	0.3–0.1	3.3–2.7	<0.1	<0.2	
fireclay								
semisilica	70–79	17–27	0.6–2.0	0.8–1.6	0.1–0.4	0.1–0.4	0.2–0.4	
medium duty	56–70	25–36	1.8–3.4	1.3–2.7	0.2–0.4	0.4–0.6	1.0–2.7	
high duty	51–59	35–41	1.5–2.5	1.1–3.0	0.2–0.6	0.1–0.6	1.3–2.6	
super duty	50–54	40–46	0.8–2.3	1.1–2.5	0.1–0.5	0.4–0.6	0.1–1.4	
high alumina								
50% Al_2O_3	40–47	47–57	0.9–1.6	2.2–2.4	0.5–0.6	0.3–0.6	0.8–1.3	
60%	27–37	58–67	0.9–2.7	1.7–3.0	0.1–0.3	0.2–0.7	0.2–1.2	
70%	19–28	68–77	0.9–2.2	2.0–3.3	0.1–0.3	0.1–0.7	0.2–1.2	
80%	8–17	78–87	0.7–1.7	2.5–3.2	0.1–0.2	0.1–0.4	0.1–1.0	
85%[b]	6–10	82–86	1.0–2.5	2.0–3.0	trace–0.2	trace–0.4	0.1–0.3	
90%	3–10	88–96	0.1–1.4	0.1–2.6	0.1–0.2	0.1–0.3	trace–0.9	
100%	0.2–1.1	97–99	0.1–0.3	trace–0.3	0.1–0.3	0.1–0.3	0.1–0.3	
zircon	32	1	0.1	1.8	trace	0.1	0.02	64
molten cast								
Al_2O_3–ZrO_2–SiO_2	11–13	50–51	trace–1.0	trace–1.0	trace	trace	1.0–1.5	33–35
Al_2O_3–SiO_2	18–22	57–70	1–4	3.0–4.5	0–1	0–1	0.0–1.5	
Al_2O_3 high soda	0.04	93.8	0.1	0.05	0.1	0.1	5.6	
Al_2O_3 low soda	0.10	99.2	0.1	0.05	0.1	0.1	0.3	

[a] Refs. 2, 5, 20–23.
[b] Phosphate bonded.

13

Table 5. Typical Ranges of Physical Properties of Alumina, Silica, and Zirconia Refractory Brick[a]

Type	PCE[b]	Modulus of rupture MPa[c]	Deformation under 172 kPa[c] load, % linear change after 1½ hours		Linear reheat change[d]		Thermal spalling loss[e]		Bulk density, g/cm³	Porosity, %
			Temperature, °C	Change, %	at °C	%	at °C	%		
silica		2.8–11.2	1650	0					1.60–1.80	20–30
fireclay										
semisilica	27–31	2.1–4.2	1450	0.1–2			1480	0–10	1.80–2.10	20–30
medium duty	29–31	7.0–11.2	1450	1–6			1600	2–6	2.11–2.20	17–21
high duty	31–33	2.8–21.0	1450	0.5–15	1600	0–1 S	1600	3–7	2.13–2.30	4–30
super duty	33–34	2.8–24.0	1450	0–9	1600	0–1 S	1650	2–8	2.28–2.48	5–22
high alumina										
50% Al₂O₃	34–35	7.0–11.2	1450	2–6	1600	0–1 E	1650	8–12	2.27–2.43	14–18
60%	36–37	7.0–11.2	1450	1–4	1600	0–5 E	1650	1–5	2.10–2.49	13–28
70%	37–38	7.0–11.2	1450	1–3	1600	1–7 E	1650	3–7	2.20–2.66	14–28
80%	38–39	7.0–12.6							2.50–2.90	14–29
85%[f]	38–39	21.0–35.0	1450	0.3–3	1600	0–2 E	1650	0–5	2.70–2.92	12–17
90%	40–41	14.0–35.0	1700	0–1	1700	0–1 E	1650	0–1	2.67–3.10	11–27
100%	41–42	12.6–21.0	1650	1–2	1700	0–0.5 S	1650	0–5	2.84–3.10	19–29
zircon			1600	2–5	1540	0			3.77	20
molten cast										
Al₂O₃–ZrO₂–SiO₂									3.70–3.74	0.8–3
Al₂O₃–SiO₂									3.00–3.20	1–3
Al₂O₃ high soda									2.89	
Al₂O₃ low soda									3.50	

[a] Refs. 2, 5, 20–23.
[b] Pyrometer cone equivalent, as determined by ASTM C 24.
[c] To convert MPa to psi, multiply by 145; for kPa, multiply by 0.145.
[d] S = Shrinkage; E = expansion.
[e] As determined by ASTM C 38.
[f] Phosphate bonded.

14

Table 6. Composition of Basic Refractory Bricks, %[a]

Type	SiO$_2$	Al$_2$O$_3$	Fe$_2$O$_3$	CaO	MgO	Cr$_2$O$_3$	Residual carbon
magnesite							
burned	0.4–4.5	0.1–1.0	0.1–2.3	0.6–3.8	91.0–98.0	0.0–0.9	
burned, tar impregnated	0.4–4.5	0.1–1.0	0.1–2.3	0.6–3.8	91.0–98.0	0.0–0.9	2.0–2.5
tar bonded, tempered	0.5–3.8	0.2–1.0	0.1–3.0	0.9–6.2	88.0–98.0	0.0–0.4	4.2–4.7
resin bonded	0.5–3.8	0.2–1.0	0.1–0.5	1.5–2.7	96.0–97.0	0.0–0.2	3.8–4.7
high carbon	0.7–2.3	0.1–0.7	0.2–0.9	0.9–3.0	94.0–98.0		7.0–20.0
magnesite–chrome							
burned	1.8–7.0	6.0–13.0	2.0–12.0	0.6–1.5	50.0–82.0	6.0–15.0	
direct bonded	1.0–2.6	3.0–16.0	3.0–10.0	0.6–1.1	50.0–80.0	7.0–20.0	
chemical bonded	2.0–8.0	5.0–14.0	2.0–8.0	1.0–2.0	50.0–80.0	6.0–15.0	
chrome							
burned	5.0–8.0	27.0–29.0	12.0–20.0	0.4–1.0	15.0–23.0	29.0–35.0	
chrome–magnesite							
burned	3.0–5.0	8.0–20.0	8.0–12.0	0.7–1.1	40.0–50.0	18.0–24.0	
dolomite							
burned	0.8–1.5	0.3–0.8	0.6–1.5	38.0–58.0	38.0–58.0		0.0–1.4
tar bonded	0.3–1.5	0.1–0.6	0.2–2.0	50.0–58.0	38.0–43.0	0.0–0.3	4.0–5.0

[a] Refs. 2, 5, 7, 20–24.

15

Table 7. Typical Ranges of Physical Properties of Basic Brick[a]

Type	Bulk density, g/cm³	Apparent porosity, %	Modulus of rupture, MPa[b]			Refractoriness under load, shear temp, °C	Linear reheat change[d] at 1650°C, %
			at 20°C	at 1260°C	at 1480°C		
magnesite							
burned	2.8–3.0	15–19	7.0–24.5	3.5–18.5		1590–1760	0–0.4 S
burned, tar impregnated	3.0–3.2	(13–17)[c]	20.0–35.0	10.5–21.0	7.0–16.0	1700+	
tar bonded, tempered	3.0–3.1	3–7	7.0–11.0				
resin bonded	2.9–3.1	4–7	8.0–27.0				
high carbon	2.7–3.0	1–6	7.0–9.0				
magnesite–chrome							
burned	2.9–3.0	17–20	3.0–4.9	0.7–2.0		1500–1700	0–0.3 S
direct bonded	2.9–3.2	14–19	5.6–13.9	8.4–17.5	2.0–4.2	1700+	0.7 S–1.0 E
chemical bonded	3.0–3.2	18–20	7.0–14.0	0.7–2.8		1650–1760	1 S–5.0 S
chrome							
burned	3.1–3.3	18–20	6.0–14.0	0.4–1.1		1260–1370	0–0.6 S
chrome–magnesite							
burned	3.0–3.2	19–21	5.6–8.4	2.8–11.2		1650–1700	0–0.1 E
dolomite							
burned	2.7–3.1	6–19	7.0–32.8			1700+	0–0.2 S
tar bonded	2.8–3.0	(8–12)[c]	7.0–11.0				

[a] Refs. 2, 5, 7, 20–23, and 25.
[b] To convert MPa to psi, multiply by 145.
[c] Obtained on ignited sample.
[d] S = shrinkage; E = expansion.

16

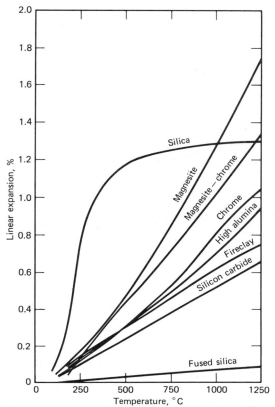

Figure 1. Thermal expansion values of some materials. Courtesy of *Chemical Engineering*.

consist of a mixture of corundum grains, mullite, and siliceous glass. In service, the siliceous glass reacts with corundum to form additional mullite, causing expansion. Mullite bricks that contain only mullite are volume-stable since equilibrium conditions have been reached during initial firing. Considerable expansion also may be caused by gas formation during heating, for instance, by the decomposition of sulfates. In the presence of a viscous siliceous glass, these gases cannot escape and expansion occurs. This mechanism explains the bloating behavior of common bloating ladle brick. The linear reheat changes of various refractories during an ASTM reheat test are shown in Tables 5 and 7.

In general, basic brick exhibits good volume stability at high temperatures. Some slight shrinkage may occur in chemically bonded or low fired brick; this shrinkage is most pronounced in high silicate compositions. In high fired, high purity brick containing chrome ore, some expansion on reheat is generated by periclase–spinel reactions, usually with very complex interdiffusion mechanisms, where the spinel constituents are dissolved in the periclase lattice at high temperature and the spinel phases precipitate from the periclase solid solution on cooling.

Thermal Conductivity.   The refractory thermal conductivity depends on the chemical composition of the material and increases with decreasing porosity. The thermal conductivities of some common refractories are shown in Figure 2.

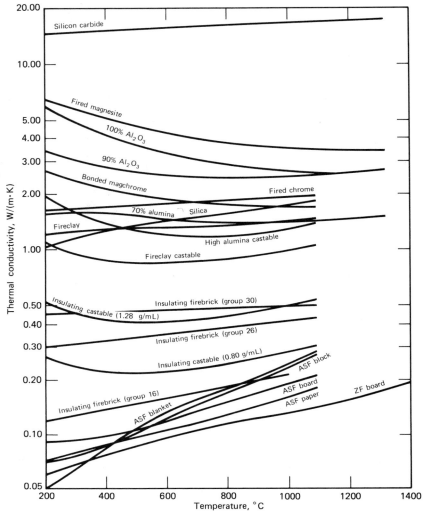

Figure 2. Thermal conductivity of typical refractories (5,25). ASF = aluminosilicate fiber; ZF = zirconia fiber. See Table 13 for group classification.

Specific Heat. In some applications, refractories are used for heat-exchange purposes on the regenerative principle, for instance, in blast-furnace stoves. High heat capacity is required in such applications (see Table 8).

Thermal Spalling. Refractories are brittle and stresses caused by sudden variations in temperatures can cause cracking and destruction. The susceptibility to thermal cracking and spalling depends upon certain characteristics of the raw material and the macrostructure of the particular refractory. Fireclay and high alumina refractories usually have a higher resistance to thermal shock than periclase refractories. Dense strong bodies withstand high stress and transmit it over large volumes; when failure occurs it is serious. Weak porous bodies, however, tend to crack before catastrophically large stresses are generated and thus are much less seriously damaged and generally remain intact.

Table 8. Mean Specific Heats of Refractory Brick and Minerals, Between 0°C and the Indicated Temperature, J/(kg·K) [a,b]

Temperature, °C	Brick				Forsterite	Mullite	Cristobalite	Periclase	Corundum
	Fireclay	Silica	Magnesite	Chrome					
0	249.5	218.5	268.9	219.8	232.9	237.9	213.3	268.9	221.1
93	257.3	243.0	283.1	227.5	258.5	248.2	236.6	293.5	253.4
204	266.4	272.8	300.0	235.3	279.3	276.7	263.8	310.3	276.7
316	274.1	296.1	312.9	243.1	297.4	288.3	309.0	324.5	292.2
427	284.5	307.7	324.5	250.8	310.3	296.1	324.5	332.3	303.9
538	293.5	318.1	333.6	257.3	318.1	301.3	333.6	338.8	312.9
649	302.6	325.8	340.1	259.9	323.3	306.4	341.4	345.2	320.7
760	311.6	333.6	346.5	268.9	328.4	310.3	346.5	349.1	325.8
871	320.7	336.2	353.0	274.1	333.6	312.9	350.4	354.3	332.3
982	327.1	341.4	359.5	279.3	338.8	316.8	353.0	358.2	336.2
1093	333.6	346.5	365.9	284.5	343.9	319.4	355.6	362.0	341.5
1204	338.8	351.7	372.4	287.0	349.1	322.0	358.2	364.6	345.2
1316	343.9	356.9	378.8	289.6	354.3	324.5	359.5	368.5	349.1
1427	347.8	360.7	384.0	292.2	359.5	327.1	360.7	372.4	353.0
1538						329.7	362.0	375.0	356.9
1642						331.0	363.3	377.6	360.7
1760						333.6		380.1	364.6

[a] Ref. 26.

[b] To convert J/(kg·K) to Btu/(lb·°F), multiply by 2.39×10^{-4}.

The resistance against thermal spalling of fireclay and high alumina brick is indicated in Table 5. No standard test has been adopted for basic brick. Refractories composed of 100% magnesia exhibit poor thermal-shock resistance, which is improved by addition of chrome ore.

Refractoriness. Most refractories are mixtures of different oxides, sometimes with significant quantities of impurities. Thus, they do not have sharp melting points but a softening range. Refractoriness is the resistance to physical deformation under the influence of temperature. It is determined by the pyrometric cone equivalent (PCE) test (see under Analytical and Test Methods).

Manufacture

Processing. Initial processing may include an extensive survey of the deposit, selective mining, stockpiling by grade, and beneficiation techniques such as weathering, grinding, washing, heavy-media separation, froth flotation, etc. Some materials can be used without further processing although many must be subjected to heat treatment. In the case of synthetic grain, the selected and beneficiated raw materials are blended in the desired proportions and formed into suitable shapes for calcination by briquetting, pelletizing, or extrusion. Slurries may be calcined; however, this practice is avoided in the interest of fuel economy. Originally, calcination referred specifically to the treatment of calcareous minerals to remove CO_2. Today, the term is often used to indicate heat treatment to sinter or burn (dead burn) the refractory grain to a stable dense material as well as to decompose minerals. Calcination may be carried out in rotary kilns, shaft kilns, multiple-hearth furnaces, or fluidized-bed reactors; the last two devices are reserved for relatively light calcining. The particular feed is dictated by the kiln type and the precalcination processes to which the raw materials or mixes may have been subjected. These hard, burned materials are called grain, clinker, or grog (term that is also used for ground firebrick). Low density or expanded aggregates can be made by burning clay or clay mixtures that evolve gas during burning and thereby expand or produce porosity. The burn-out material may be naturally present in the clay or may be mixed with the clay before burning.

Both raw and processed materials can be fused or melted in electric-arc furnaces. These materials can be melted and then cast into shapes, formed in the furnace itself as an ingot or formed into fibers. Partially fused ingots are crushed into grains. Since these ingots are not homogeneous, the grains have to be graded; the middle portion is purer than the outside material which is less well fused. Melts used for casting or fiber forming are homogeneous.

Crushing and Grinding. Some raw materials, such as hard clay and quartzite, must first be crushed to grains small enough for the grinding equipment. In general, a jaw, gyratory, or roll crusher is employed (see Size reduction).

Almost all raw materials require grinding after primary crushing. For coarse grinding, a dry pan or occasionally a wet pan is used. The dry pan is similar to a grist mill but has a perforated bottom through which the crushed material is continuously removed. The wet pan is similar, but has a solid bottom. For very fine grinding, a ring-roll, ball, or impact mill is employed.

Screening. To obtain a high density product, the mix is grain-sized. In a continuous screening operation, the ground raw materials are generally fed to vibrating high capacity screens that may be heated. Material that does not pass the screen is returned to the grinding system for further size reduction. A single screen with uniform openings produces a single product referred to as a straight grind. In gap-grain sizing, a series of two, three, or more vibrating screens are set in stacks yielding a coarse band, an intermediate band, and a fine band. These fractions are blended into the mix in the proper proportions to provide optimum properties. Gap-grain-sized material was originally used for the manufacture of basic brick; however, today it has been extended to many types of high alumina, fireclay, and silica brick. Coarse and medium-fine grain sizing is accomplished by the aforementioned methods, whereas fine-sized materials generated in rod mills, ball mills, ring-roll mills, and the like are classified by air separators (see Size separation).

Mixing. As in other ceramic processes, more than one type of raw material is often required for a refractory product. The purpose of mixing is to homogenize the various ingredients (see Mixing and blending). Although the specific steps and equipment involved in the mixing of batches for fireclay, high alumina, and basic refractories are somewhat different, the general principles are similar. Mixes that are to be dry-pressed contain 2–6% binding liquid, depending upon the plasticity of the raw material–bond system and the fineness of the mix. The ingredients may be blended in a pug mill, dry pan, or other type mixer and tempered with the bonding ingredients. Tempering, in the sense used here, denotes the kneading action produced on the mix, usually in a muller mixer. Mixes to be extruded or hand formed contain 10–20% liquid. These mixes may be prepared in a pug mill or wet pan.

The mixing of tar-, resin-, and chemically bonded brick material present special problems. Tar- and resin-bonded mixes are usually basic compositions. Because coal tar and petroleum pitches have ring-and-ball values of ca 100°C, ie, the softening point as determined by ASTM D 30, provisions for mixing the ingredients at high temperatures must be made. Typically, the grain is heated to ca 150°C and maintained at that temperature while the hot pitch is added. After a preset mixing time, the batch is transferred to the pressing equipment. For resin-bonded mixes, the grain does not have to be heated, although liquid resins may be heated to enhance their flow characteristics. Phenolic (phenol–formaldehyde), alkyd-oil–urethane, urea–formaldehyde, furan resins, and the like are used (see Phenolic resins; Alkyd resins; Amino resins). Resin-bonding systems are more expensive than pitches, but recent concerns regarding potential health risks of pitch fumes have renewed interest in resins.

Chemically bonded basic bricks are blended much the same as burned brick mixes except that a bonding agent, eg, magnesium sulfate or magnesium chloride is added to the mix as well as tempering water to form oxysulfates or oxychlorides.

Forming. Most refractory shapes are formed by mechanical equipment, but some very large or intricate shapes require hand molding in wooden, steel-lined molds with loose liners to permit easy removal.

A few fireclay refractories are produced by the stiff-mud process with an auger machine that pugs, de-airs, and continuously extrudes a clay column. A wire cutter cuts the clay into blanks which are then sized, shaped, and branded by a repress machine.

Refractory shapes are generally produced on a mechanical toggle press, screw press, or hydraulic press. In this operation, a mold cavity is filled with the damp mix. During the pressing cycle, some products such as fireclay brick are de-aired by applying a vacuum to the top and bottom press heads which contain a series of small openings;

de-airing promotes a denser product and reduces laminations. In some cases bricks are encased in steel or have internal steel plates which are placed in the mold box before charging; the plates become an integral part of the brick upon pressing. Plates may also be bonded to the brick after pressing. From one to four bricks may be pressed at a time, depending on the size. Pressures range from ca 17 (2500) to ca 98 MPa (14,000 psi) for plastic firebrick or nonplastic basic mixes, respectively. Pitch-bonded brick must be formed hot. Initial heating of the press may be required, but the residual heat from the mix is usually sufficient to maintain the necessary temperature. Another type is the impact- or jolt-mold press that consists of a mold box and press heads activated by pneumatic cylinders (eg, jack hammers). Some special shapes are produced by air-ramming which is similar to hand molding, except that reinforced steel molds are required and the damp mix is slowly fed to the mold while molders manually compact the material with pneumatic rammers. Special shapes can also be formed by slip casting and hot pressing.

Fusion-cast refractories are formed by first melting the material in an electric arc. The liquid is poured into a mold and allowed to cool. Various annealing processes and refining techniques ensure a uniform and structurally sound shape. The cast form is then cut or ground to size. Fused-cast refractory is very dense but may contain a system of closed pores and large, highly oriented grains may exist in a particular casting. The size and distribution of the pore and grain phases must be controlled.

Isostatic pressing gives a highly uniform product, although the production rate is somewhat low; it typically contains very small grains and little or no porosity. In this process, a rubber sock or bag of the desired shape is filled with the refractory mix. The sock is then subjected to extremely high pressure in a hydraulic pressure chamber.

Large and small shapes may be slip cast from both plastic and nonplastic mixes by the usual techniques. Precise shapes, such as glass feeder parts, are made in this way as well as large flux blocks. The process requires the formulation of a slip of suitably stable character to be poured into a plaster mold to be dewatered. After it solidifies, the mold is removed and dried further before firing.

Drying. The drying step for large shapes is critical; extremely large fireclay and silica shapes are sometimes allowed to dry on a temperature-controlled floor heated by steam or air ducts embedded in the concrete. Smaller shapes are generally dried in a tunnel dryer. The ware is placed on cars that enter the cold end and exit at the hot end. These dryers may be humidity-controlled and are heated by their own source or with waste heat from the burning operation. Microwave and infrared drying are being investigated.

Pitch-bonded and resin-bonded bricks are treated or cured in special ovens at temperatures higher than that used for drying other types of bricks. Pitch-bonded brick is cured at 230–320°C; this process is called tempering which is not to be confused with tempering during mixing. Tempering removes some of the volatiles from the pitch and eliminates or reduces thermoplastic or slumping behavior in service (see Drying).

Burning. Bricks are fired or burned in kilns to develop a ceramic bond within the refractory and attain certain desired properties. This step does not apply to chemically or organically bonded products. Preferred for this purpose is the continuous or tunnel kiln, a structure of narrow cross section and 61–183 m in length. The bricks are set on cars that move slowly through the kiln, which is divided into the preheating, burning, and cooling zones. The length of the zones controls the rates of heating and cooling, and the time at temperature (soak time). The temperature in the firing zone ranges from 1000 to 1700°C. Gas, oil, or coal may be used as a fuel.

Silica brick and large fireclay shapes are fired in circular down-draft kilns. These kilns vary in diameter and can accommodate up to 150,000 23-cm brick or their equivalent in other sizes. The complete burning cycle for a typical periodic kiln ranges from 21 to 27 days as compared with four to seven days for a tunnel kiln.

The shuttle kiln consists of a firing chamber with two or more kiln cars on which the bricks to be fired are set. While one load of brick is being fired, a second is being set. Somewhat similar is the bell top or top hat kiln which is raised and lowered above and over the kiln cars to be fired. These kilns are more expensive to operate than tunnel kilns but provide flexibility in burning conditions and production schedules.

Burned brick may be impregnated with tar or pitch to improve corrosion resistance. The heated bricks are placed in the impregnation unit which is sealed and evacuated to remove air from the pores of the refractory. The evacuated chamber is then filled with hot pitch and the vacuum is released. The treated product is allowed to drain free of excess pitch and is ready for shipment. Although this treatment is primarily used on basic refractories, it can be extended to other classes. The benefits derived from impregnation are lost if the pitch is burned off at high temperature; therefore, a reducing atmosphere is required such as is encountered in a basic oxygen steelmaking furnace.

Specialty Refractories. Bulk refractory products include gunning, ramming, or plastic mixes, granular materials, and hydraulic-setting castables, and mortars. These products are generally made from the same raw materials as their brick counterparts.

Granular materials are shipped raw or calcined and usually have been ground to a specified screen size or size distribution. The additives depend upon the application and service conditions. These materials are used in construction, repair, or maintenance of furnaces and vessels. Refractory mortars are used to lay brick of the same composition. They are manufactured wet premixed or dry.

Fireclay, ramming mixes, and high alumina and chrome ore plastics are manufactured similarly to brick; that is, the coarse refractory aggregate is added to a wet pan and combined with a small amount of raw clay for plasticity. Water is added to the batch; refractory plastics generally contain more water and raw clay than ramming mixes. After discharge from the wet pan, plastic mixes are formed by extrusion into 45-kg blocks and sliced into 9–12-kg slabs. The material is packaged damp and is ready for installation with pneumatic rammers. Ramming mixes are generally shredded into small lumps and placed damp in airtight steel drums ready for installation with pneumatic rammers. When installed, ramming mixes generally require forms to contain the material as it is rammed. Plastics, however, are soft and cohesive enough to be installed without forms.

Basic raw materials are susceptible to hydration and therefore specialty products are shipped dry and are mixed with water on site for gunning or ramming. Certain basic specialties are offered with organic vehicles such as oils and can be used without on-site mixing.

Information on manufacturing can be found in references 26–30.

Table 9.   Distribution of U.S. Refractory Sales [a]

Industry	Percentage of total U.S. sales		
	1964	1977	1979
iron and steel	61.2	47.2	51.6
nonferrous metals	5.4	6.1	7.5
cement	2.0	3.6	4.9
glass	5.3	4.6	5.1
ceramics	5.0	8.7	9.7
chemical and petroleum	3.0	2.5	2.1
public utilities	1.1	0.7	0.9
export	5.3	6.5	7.4
all other and unspecified	11.7	20.1	10.9
Total, 10^8 $	*4.97*	*12.99*	*16.95*

[a] Ref. 31.

Economic Aspects

As can be seen from Table 9, the iron and steel industry is the principal consumer of refractories. The distribution, however, has changed since 1964; the decrease in refractories consumption coincides with the change from open-hearth steelmaking to BOF (Bessemer oxygen furnace) practice (see Steel).

The values of refractories shipped from 1967–1980 are given in Table 10. These data were obtained from the Bureau of Census current industry reports on refractories which have been issued on a quarterly basis since 1957. These reports provide reliable data on industry trends and give a fairly accurate economic picture.

The number of 23-cm brick equivalents of shaped refractories and the dollar sales values are given in Table 11. The dollar value for one brick equivalent of each type of refractory has increased by 76% for fireclay, 45% for high alumina, 305% for silica brick, and 173% for basic bricks in the period shown.

Table 10.   Refractories Shipped, 1967–1980, 10^6 $ [a,b]

Year	Constant dollar, total	Current dollar		
		Total	Clay refractories	Nonclay refractories
1967	524	524	225	299
1970	495	599	256	342
1973	572	780	327	453
1976	589	1084	464	621
1979	738	1727	712	1015
1980 [c]	475	1211	512	698

[a] Ref. 31.
[b] The data were adjusted for price changes using the Producer Price Index for refractories as published by the Bureau of Labor Statistics. The base year is 1967.
[c] For nine months only.

Table 11.   Shipments of Brick[a]

Year	Fireclay brick[b]		High alumina brick[c]		Silica brick		Basic brick[d]	
	1000[e]	10[6] $	1000[e]	10[6] $	1000[e]	10[6] $	1000[e]	10[6] $
1965	614	102	56	41	110	21	140	124
1975	507	142	129	137	41	32	126	279
1977	484	191	114	153	29	28	131	320

[a] Ref. 31.

[b] Includes regular fireclay, semisilica superduty fireclay, ladle brick, and IFB below 23.

[c] Includes high alumina, mullite, and extra-high alumina brick.

[d] Includes magnesite, magnesite–chrome, chrome, chrome–magnesite brick, and dolomitic brick.

[e] Brick equivalents.

ASTM Classifications

In addition to testing methods, ASTM publishes a list of classifications covering a wide variety of refractory types (32). The various brands from numerous producers and producing districts are grouped into classes with a nomenclature indicative of their chemical composition, heat resistance, and service properties.

Fireclay and High-Alumina Brick.   ASTM designation C 27 covers fireclay and high alumina brick (see Table 12). High-alumina brick is classified according to alumina content, starting at 50% and continuing up to 99% Al_2O_3. Manufacturers are allowed ±2.5% of the nominal alumina content, except for 85 and 90% (±2.0%) and 99% (min 97%). An additional requirement for alumina bricks with Al_2O_3 content of 50, 60, 70, and 80%, are PCEs of 34, 35, 36, and 37, respectively.

Basic Bricks.   Chrome brick, chrome–magnesite brick, magnesite–chrome brick, and magnesite brick are classified under ASTM C 455. The six classes of chrome–magnesite and magnesite–chrome brick start with 30% MgO and increase in 10% increments to 80% MgO; the minimum requirement for MgO is 5% less than the nominal percentage MgO specified for each class. For the three magnesite classes of 90, 95, and 98% MgO the minimum requirement is 86, 91, and 96% MgO, respectively. A chrome brick is manufactured entirely of chrome ore.

Insulating Brick.   ASTM classifies insulating firebrick under C 155 by group; the group number indicates the service temperature multiplied by 100 degrees Fahrenheit (37.8°C) (see Table 13); eg, group 16 corresponds to a test temperature of ca 1600°F (871°C).

Mullite Refractories.   Mullite refractories are classified under ASTM C 467. This brick must have an Al_2O_3 content between 56 and 79% and less than 5% impurities. Impurities are considered metal oxides other than those of aluminum and silicon. The hot-load subsidence is 5% max at 1593°C.

Silica Brick.   Under ASTM C 416, types A and B silica bricks are classified according to chemical composition and strength. Silica brick must have an average modulus of rupture of 3.5 MPa (500 psi); they must have <1.50% Al_2O_3, no more than 0.20% TiO_2, <2.50% Fe_2O_3, and <4.00% CaO. Type A brick must have a flux factor equal to or <0.5. The flux factor is equal to percent alumina plus twice the percent of alkalies. Type B are all other silica brick covered by the standard chemical and strength specifications.

Table 12. ASTM C 27 Fireclay Brick Classification[a]

Classification	Type	PCE[b]	Panel spalling loss, max, %	Hot-load subsidence, max, %	Reheat shrinkage, at 1600°C, max, %	Cold modulus of rupture, MPa[c]	Other test requirements
superduty	regular	33	8 at 1650°C		1.0	4.14	
	spall-resistant	33	4 at 1650°C		1.0	4.14	
	slag-resistant	33				6.89	bulk density, min, 2.243 g/cm³
high duty	regular	31½	10 at 1600°C				
	spall-resistant	31½				3.45	bulk density, min, 2.195 g/cm³, or max porosity 15%
	slag-resistant	31½				8.27	silica content, min, 72%
semisilica				1.5 at 1350°C		2.07	
medium duty		29				3.45	
low duty		15				4.14	

[a] Ref. 33.
[b] Pyrometer cone equivalent.
[c] To convert MPa to psi, multiply by 145.

Table 13. ASTM C 155 Insulating Brick Classification[a]

Group identification	Reheat change of no more than 2% at test temperature, °C (°F)	Bulk density, g/cm³
group 16	815 (1550)	<0.545
group 20	1065 (1950)	<0.641
group 23	1230 (2250)	<0.768
group 26	1400 (2550)	<0.865
group 28	1510 (2750)	<0.961
group 30	1620 (2950)	<1.089
group 32	1730 (3150)	<1.522
group 33	1790 (3250)	<1.522

[a] Ref. 33.

Zircon Refractories. ASTM C 545 classifies zircon refractories in two types. Types A and B have the same chemical requirements of not less than 60% ZrO_2 and not less than 30% SiO_2. Type A (regular) must have a density of less than 3.85 g/cm³ and type B (dense) more than 3.85 g/cm³.

Castable Refractories. Hydraulic-setting refractory castables are classified under ASTM C 401 into three groups: normal strength with a modulus of rupture of 2.07 MPa (300 psi); high strength with a modulus of rupture of 4.14 MPa (600 psi); and insulating classified on the basis of bulk density. In addition, the normal and high strength groups are divided into classes A through G (see Table 14). The classification of insulating castables is given in Table 15.

Refractories used in steel-pouring pits are classified under ASTM C 435 (see Table 16).

Specifications

Among the many specifications covering refractory products, the best known are those published by ASTM. In addition, specifications are issued by the Federal Government and the armed forces. The former are generally preceded by the prefix HH and the latter by the prefix MIL. The ASTM refractory specifications always suggest a use, whereas Federal and military specifications are inconsistent in this respect.

Table 14. Classification of Normal and High Strength Refractories[a]

Class	No more than 5% shrinkage permitted after firing for 5 h at °C
A	1095
B	1260
C	1370
D	1480
E	1595
F	1705
G	1760

[a] Ref. 33.

Table 15. Classification of Insulating Castables[a]

Class	No more than 5% shrinkage permitted after firing for 5 h at °C	Density, g/cm³
N	925	0.88
O	1040	1.04
P	1150	1.20
Q	1260	1.44
R	1370	1.52
S	1480	1.52
T	1595	1.60
U	1650	1.68
V	1760	1.68

[a] Ref. 33.

Table 16. ASTM C 435 Classification For Steel-Pouring-Pit Refractories[a]

Class	Type	PCE[b]	Porosity, %	Reheat change above 1350°C, %[c]
nozzle	A	15–20	8 min	1.0, min
	B	20–29	8 min	1.0, min
	C	29 min	10 min	1.0, max
sleeve	A	15–20	10 min	1.0, min
	B	20–29	10 min	1.0, min
	C	29 min	10 min	1.0, max
laddle brick	A[d]	15 min	18 max	5.0, above 1290°C min
	B[d]	15 min	18 max	2.5, above 1350°C min
	C[d]	26 min	18 max	0.5, above 1500°C max

[a] Ref. 33.
[b] Pyrometer cone equivalent.
[c] Except ladle brick which is heated as stated; min refers to expansion, and max to shrinkage of diameter.
[d] Modulus of rupture of 4.83 MPa (700 psi) required.

No less important are specifications issued by industrial consumers. With the increasing emphasis on quality, this type of specification is becoming more and more prevalent in recent years. Where export is concerned, suppliers are faced with foreign specifications. Standard refractory samples are available from the Office of Standard Reference Materials, National Bureau of Standards, Washington, D.C., for chemical analysis and standardization, and for pyrometer cone equivalent (PCE) standardization from The Refractories Research Center, Ohio State University, Columbus, Ohio.

Analytical and Test Methods

The test methods applicable to refractories are found in the Annual Book of ASTM Standards, Part 17 (see Table 17). The chemical composition generally is reported on an oxide basis. ASTM designations C 571, C 572, C 573, C 574, C 575, and C 576 apply to the various types of refractories.

Table 17.   ASTM Test Methods for Refractories[a]

Material	Test identification	Properties
burned brick	C 20, C 830	apparent porosity, water adsorption, bulk density
brick, various shapes	C 133, C 607, C 93	crushing strength, modulus of rupture
basic brick	C 456	hydration resistance
brick and tile	C 154	warpage
granules	C 357, C 493	bulk density
periclase grains	C 544	hydration
mortar	C 198	cold-bonding strength
air-setting plastics	C 491	modulus of rupture
castables	C 298	modulus of rupture
granular dead-burned dolomite	C 492	hydration
fireclay plastics	C 181	workability index
castables	C 417	thermal conductivity
plastics	C 438	thermal conductivity
general refractories	C 288	disintegration in CO atmosphere
	C 135	true specific gravity
	C 201	thermal conductivity
	C 92	sieve analysis and water content

[a] Ref. 33.

Refractoriness.   Refractoriness is determined by several methods. The PCE test (ASTM C 24) measures the softening temperature of refractory materials. Inclined trigonal pyramids (cones) are formed from finely ground material, set on a base, and heated at a specific rate. The time and temperature (heat treatment) required to cause the cone to bend over and touch the base is compared to that for standard cones.

The standard ASTM PCE test is relative and used extensively only for alumina-silica refractories (see Table 5). However, the upper service limit is generally several hundred degrees below the nominal PCE temperature since some load is generally applied to the refractory during service. In addition, chemical reactions may occur that alter the composition of the hot face and, therefore, the softening point. The relationship between PCE numbers and temperature is described in ASTM C 24.

Another measure of refractoriness is the hot-compressive strength or hot-load test for refractory bricks or formed specialties. The specimen carries a static load from 69 kPa (10 psi) to 172 kPa (25 psi); it is heated at a specific rate to a specific temperature which is then held for $1\frac{1}{2}$ h, or it is heated at a specific rate until it fails. The percent deformation or the temperature of failure is measured. The procedure is described in ASTM C 16. In a variation (ASTM C 546), the specimens are held at a specified temperature under a static load for 50 h. The results are reported as percent deformation or the time taken to achieve 13-mm deformation from the original length. A load of 345 kPa (50 psi) is applied to silicon carbide specimens and of 172 kPa (25 psi) to all other refractories.

Thermal Strength and Stability.   Dimensional changes that occur upon reheating can be determined by ASTM C 605, C 436, C 210, C 179, or C 113. Specimens are selected and cut or formed to an appropriate size and measured before and after being heated at an appropriate temperature schedule to a specified temperature for 5 to 24 hours. The linear, diametral, or volume percentage change is noted. High temperature strength of refractory materials is determined on rectangular prisms $25 \times 25 \times 150$ mm cut from the product being tested. The specimens are placed in a furnace, heated to a desired temperature, and the modulus of rupture is determined. A detailed description is given in ASTM C 583. Thermal spalling resistance is determined by ASTM C 439, C 180, C 122, C 107, and C 38. The last four methods apply to fireclay, high alumina bricks, and plastics using the apparatus described in C 38. The specimens are weighed and built into a panel (wall section) that was preheated for 24 h at a specific temperature. The panels are then subjected to twelve cycles of heating to 1400°C followed by cooling with a water-and-air spray. The panel is dismantled, the loose spalls are removed, and the weight loss is recorded as percent spalling loss. Thermal spalling of silica brick is determined on six specimens placed on a guarded hot plate. The specimens are heated and cooled at a specified rate. The heating rate is reported and any cracking that may have occurred is described.

Special Tests. Even though the American Society for Testing and Materials offers a wide range of test methods, there are other special tests that are imposed upon the manufacturer by consumers, the military, the Federal Government, and in some cases local or municipal governments. These tests are generally very specific and are oriented toward particular service conditions. In many instances, the producers develop special tests within their laboratories to solve customer problems. Many of these tests are adopted subsequently by ASTM.

Health and Safety Factors

Because industrial refractories are by their very nature stable materials, they usually do not constitute a physiological hazard. This statement does not apply however to unusual refractories that might contain heavy metals or radioactive oxides such as thoria, urania, and plutonia, or to binders or additives that may be toxic.

Inhalation of certain fine dusts may constitute a health hazard. For example, exposure to silica, asbestos, and beryllium oxide dusts over a period of time results in the potential risk of lung disease. OSHA regulations specify the allowable levels of exposure to ingestible and airborne particulate matter. Material Safety Data Sheets, OSHA form 20, available from manufacturers, provide information about hazards, precautions, and storage pertinent to specific refractory products.

Selection and Uses

Any manufacturing process requiring refractories depends upon proper selection and installation. When selecting refractories, the environmental conditions are evaluated first, then the functions to be served, and finally the expected length of service. All factors pertaining to the operation, service, design, and construction of equipment must be related to the physical and chemical properties of the various classes of refractories (34).

Service conditions that impare the effectiveness of refractories include chemical attack (ie, slags, fumes, gases, etc); operating conditions (ie, temperatures and cycling); and mechanical forces (ie, abrasion, erosion, and impact).

Design factors that influence the selection include type of equipment and its construction (ie, brick or monolithic material); refractory function (material containment, flow deflection, heat storage or release); heat environment (ie, exposure to constant or variable temperatures); refractory strength (ie, exposure to varying stress conditions); and thermal function (ie, insulation, dissipation, or transmission of heat).

The effects of processing conditions on refractories are given in Table 18.

Various standardized applications of different refractory types are listed in an excellent series of "Industrial Surveys of Refractory Service Conditions," compiled by Committee C-8 of the ASTM. These surveys cover the principal industrial applications of refractories and furnish a description of furnace operations and destructive influences, such as slagging, erosion, abrasion, spalling, and load deformation.

By far the most common industrial refractories are those composed of single or mixed oxides of single or mixed oxides of Al, Ca, Cr, Mg, Si, and Zr. These oxides exhibit relatively high degrees or stability under both reducing and oxidizing conditions. Carbon, graphite, and silicon carbide have been used both alone and in combination with the oxides. Refractories made from the above materials are used in ton-lot quantities in industrial applications. Other refractory oxides, nitrides, borides, and silicides are used in relatively small quantities for specialty applications in the nuclear, electronic, and aerospace industries.

The common industrial refractories are classified into acid, SiO_2 and ZrO_2, basic, CaO and MgO, and neutral, Al_2O_3 and Cr_2O_3. Oxides within each group are generally

Table 18.　Effect of Processing Conditions on Refractories [a]

Service condition	Chemical resistance
oxidizing atmosphere	oxides and combinations of oxides (ie, silicates, fireclays) are unaffected; carbon and graphite oxidize; silicon carbide is fairly stable to 1650°C
steam or water vapor	can cause hydration of magnesite refractories at low temperatures and will oxidize carbon and graphite above 705°C
hydrogen	silica and silica-containing refractories are attacked above 1100°C; high-alumina, ZrO_2, MgO, and calcium-aluminate refractories show good resistance
sulfur and sulfates	above 870°C sulfur reacts with refractories containing silica; carbon and high purity oxides show good resistance; sulfates react to some degree; calcium-aluminate cement is more resistant than Portland cement
fuel ash	alkali and vanadium attack from ash can be severe on fireclay; high alumina resists
reducing atmosphere	most refractories are stable; however, iron oxide impurities, when reduced, can cause destruction, particularly if cycled
carbon monoxide	iron impurities can act as a catalyst to cause deposition of carbon in fireclay refractories; CO can oxidize graphite and SiC, and cause destructive changes in basic refractories
chlorine and fluorine	chlorine attacks silicates above 650°C; F attacks all refractory materials except graphite; basic refractories have poor resistance to both
acids	basic refractories have fair to poor resistance, fireclay and high alumina good resistance, except for HF; zircon, zirconia, and silicon carbide have good resistance; carbon and graphite do not react
alkalies	fireclay and high alumina perform well at low temperatures, magnesite refractories fair to good, chrome refractories poor, and graphite excellent

[a] Ref. 34.

compatible with each other whereas mixtures of acid and basic oxides often give low melting products; neutral oxides are generally compatible with both acidic and basic oxides.

Reactions between Refractories and Liquids.   Molten metals are generally much less reactive than slags; therefore, the response of a refractory to chemical environments generally depends upon its slag resistance which, in turn, depends upon the compositions and properties of slag and refractory. Other factors include temperature, severity of thermal cycling or shock of the process, velocity and agitation of the slag in contact with the refractory, and the abrasion to which the refractory is subjected. Considering these factors, it is not surprising that similar refractories placed in similar furnaces can wear at vastly different rates if the operation practices are different.

As a general rule, acid slags (CaO/SiO_2 <1) require acid refractories, whereas basic slags (CaO/SiO_2 >1) require basic refractories. Fireclay and aluminosilicate refractories perform best for slags with a C/S ratio <1; however, acid slags containing considerable amounts of iron or manganese oxides, high alumina refractories are required. High-alumina refractories are also superior to fireclay refractories when the basicity-to-acidity ratio = 1. When the basicity of the slag >1, basic refractories should be employed, such as MgO, MgO–CaO, and $MgO.Cr_2O_3$. Magnesium oxide resists slags of a wide range of compositions; however, in actual practice all basic refractories contain some silica which usually occurs in various silicate phases at the grain boundaries. As slag attack proceeds, liquid phases migrate from the hot face to the cooler regions and attack of the grain-boundary silicate phase precedes solution of the periclase grain. For this reason, both the character of the bonding silicate phase and the porosity of the refractory are very important. Penetration of the slag may be impeded by materials such as carbon, pitch, or resins by lowering the porosity and changing the wetting characteristics.

The thermal conductivity of refractories in high wear areas of steelmaking vessels is increased with internal metal plates and by incorporating flake graphite with magnesite refractories. Slags penetrate refractories until their viscosity becomes too high for further migration. Refractories with high thermal conductivity can cause the slag to chill and reduce the penetrated volume.

Slag penetration alters the structure of the refractory by changing its porosity, density, and strength. If the altered refractories are subject to thermal cycling or if volume changes occur upon crystallization of the slag, stress concentrations build up immediately behind the densified zone and spalling or cracking may result. An example of structural alteration is the iron oxide bursting phenomenon in magnesite–chromite brick. Iron oxide contained in the chromite ore, or that which is allowed to penetrate the brick, can cause excessive expansion of the lattice because of the unequal diffusion of iron and chromium ions in magnesia chrome spinel, which leads to the production of pores in the iron-rich phase.

Reactions between Refractories and Gases.   Reactions with gases can be quite destructive as the gases generally penetrate the pores of the refractory and destroy its structure. The refractory may either expand and crack because of the formation of new, low density compounds, or its refractoriness may be drastically reduced because of the formation of low melting compounds. An example is the disintegration of alumina silicates in blast furnaces caused by carbon monoxide. The deposition of carbon is catalyzed by iron in the bricks. The growth of the carbon deposit causes the brick to rupture and more surface is exposed. Therefore, a brick of low iron and alkali content with a dense, low permeability is preferred.

Reactions between Refractories. In Table 19, the compatibilities between various refractories are given for a range of temperatures. Generally, dissimilar refractories react vigorously with each other at high temperature. Phase diagrams are an excellent source of information concerning the reactivity between refractories.

Silica Refractories. This type consists mainly of silica in three crystalline forms: cristobalite [14464-46-1], tridymite [1546-32-3], and quartz [14808-60-7]. Quartzite sands and silica gravels are the main raw material ingredients, although lime and iron oxide are added to increase the mineral content. Uses include open-hearth roof linings, refractories for coke ovens, coreless induction foundry furnaces, and fused-silica technical ceramic products. Consumption of silica refractories has declined dramatically over the last 20 years, mainly because of the increasing use of oxygen in steelmaking.

Semisilica Refractories. Semisilica refractories consist essentially of silica, bonded by a glassy matrix. They contain 75–93% SiO_2 and may be made from siliceous clay or sand and fireclay. They tend to resist slag attack owing to a self-glazing tendency. Uses include shapes for open hearth stoves and checkers.

Fireclay Refractories. These products are made from clay minerals containing ca 17–45% Al_2O_3; pure kaolin has the highest alumina content. Fireclay refractories

Table 19. Approximate Initial Temperature, °C, at which Refractories React[a,b]

Refractory	Magnesite, 92% MgO	Magnesite–chrome, CB[b]	Chrome–magnesite, CB[c]	Chrome–magnesite, fired	Forsterite, stabilized	Chrome, fired	90% alumina	70% alumina
magnesite, 92% MgO		>1700	>1700	>1700	(1700)	1700	>1700	1600
magnesite–chrome, CB[c]	>1700		>1700	>1700	1650	>1700	1450	1450
chrome–magnesite								
CB[c]	>1700	>1700		>1700	1650	>1700	1650	1600
fired	>1700	>1700	>1700		1650	>1700	1650	1650
forsterite, stabilized	(1700)	1650	1650	1650		1650	1650	1600
chrome refractories, fired	1700	>1700	>1700	>1700	1650		1650	1600
alumina								
90%	>1700	1600	1600	1600	1650	1650		>1700
70%	1650	1600	1600	1500	1650	1600	>1700	
zircon	>1700	1600	1650	1600	1600	>1700	>1700	>1700
fireclay								
superduty	1400	(1700)	1650	1650	1600	1500	(1700)	(1700)
high-duty	1400	1650	1600	1600	1650	1500	(1700)	(1700)
semisilica	1500	1400	1400	1500	1500	1500	(1500)	(1500)
silicon carbide, clay-bonded	1500	1400	1400	1400	1600	1500	1650	1650
silica								
type B[d]	1500	1600	1600	1500	1650	1650	1650	1500
superduty	1500	1600	1600	1600	1650	1650	1650	1500

[a] Ref. 26.
[b] Max temperature tested = 1700°C.
[c] Chemically bonded (not fired); one or both materials not sufficiently refractory for test.
[d] Ref. 33.

are used in kilns, ladles, and heat-regenerators, acid-slag-resistant applications, boilers, blast-furnaces, and rotary kilns. They are generally inexpensive.

High Alumina Refractories. These refractories have alumina contents from 100% to just above 45%. The desired alumina content is obtained by adding bauxites, synthetic aluminosilicates, and synthetic aluminas to clay and other bonding agents. These refractories are used in kilns, ladles, and furnaces that operate at temperatures or under conditions for which fireclay refractories are not suited. Phosphate-bonded alumina bricks have exceptionally high strength and are employed in aluminum furnaces. High alumina and mullite are used in furnace roofs and petrochemical applications.

Chrome Refractories. Naturally occurring chrome ore composed mainly of chromite [53293-42-8], may be made into a brick or may be blended with fine calcined magnesite to obtain the desired chrome-to-magnesia ratio. When blended with magnesite, the products containing more chrome than magnesia are denoted chrome magnesite. These refractories are used in nonferrous metallurgical furnaces, rotary kiln linings, secondary refining vessels, such as argon–oxygen decarborizers (AODs) and glass-tank regenerators.

Magnesite Refractories. These refractories do not contain magnesite as their name implies, but rather periclase. The term magnesite refers to the ore from which the periclase was made, although magnesite brick can be made from synthetic periclase, ie, seawater MgO. Shaped magnesite refractories may be impregnated or bonded with pitch or resin to improve their resistance to slag attack. Chrome ore can be added to magnesite to produce magnesite–chrome refractories which are used in lining and maintenance of steelmaking and refining vessels and checkers.

Dolomite Refractories. These refractories contain dead-burned dolomite and possibly fluxes such as millscale, serpentine, or clay. Shaped refractories may be bonded or impregnated with pitch to improve slag resistance and inhibit hydration. Addition of magnesite gives magnesite–dolomite or magdol refractories. Dolomite refractories are primarily used in linings of BOF vessels and refining vessels, and in ladels and cement kilns.

Spinel Refractories. These refractories contain synthetic spinel, $MgAl_2O_4$. They exhibit good strength at high temperatures and thermal shock resistance. Chromium and magnesium oxides crystallize in the same structure and are referred to as chrome–magnesite spinels. Spinel refractories are used in cement kilns and steel-ladle linings.

Forsterite Refractories. These refractories are made from forsterite, Mg_2SiO_4. They resist alkali attack and have good volume stability, high temperature strength, and fair resistance to basic slags. Uses include nonferrous metal furnace roofs and glass-tank refractories not in contact with the melt, ie, checkers, ports, and uptakes.

Silicon Carbide Refractories. Silicon carbide has a wide range of refractory uses including chemical tanks and drains, kiln furniture, abrasion-resistant linings, blast-furnace linings, and nonferrous metallurgical crucibles and furnace linings.

Zirconia Refractories. The most common zirconia-containing refractories are made from zircon sand and are used mostly for glass-tank paver brick. Refractory blocks made from a composition of zircon and alumina, used to contain glass melts, are generally electromelted and then cast. They exhibit excellent corrosion resistance but are subject to thermal shock. Refractories made from pure ZrO_2 are extremely expensive and are reserved for extra high temperature service above 1900°C. Additives such as yttria, or CaO and MgO prevent deterioration during heating and cooling.

BIBLIOGRAPHY

"Refractories" in *ECT* 1st ed., Vol. 11, pp. 597–633, by L. J. Trostel and R. P. Heuer, General Refractories Co.; "Refractories" in *ECT* 2nd ed., Vol. 17, pp. 227–267, by W. T. Bakker, G. D. Mackenzie, G. A. Russell, Jr., and W. S. Treffner, General Refractories Co.

1. A. F. Greaves-Walker, *Bull. Am. Ceram. Soc.* **6**, 20, 213 (1941).
2. F. Singer and S. S. Singer, *Industrial Ceramics*, Chemical Publishing Company, Inc., New York, 1964.
3. R. D. Pehlke and co-eds., "Refractories" in *Basic Oxygen Furnace Steelmaking*, Vol. 4, The Iron and Steel Society of AIME, New York, 1977, Chapt. 11, pp. 1–58.
4. F. H. Norton, *Refractories*, 3rd ed., McGraw-Hill Company, Inc., New York, 1949.
5. J. J. Suec and co-eds., *Ceramic Data Book 1981 Suppliers' Catalog and Buyers' Directory*, Cahners Publishing Company, Denver, Col., 1981.
6. W. D. Kingery, *Introduction to Ceramics*, John Wiley & Sons, Inc., New York, 1960.
7. A. Alper, ed., *High Temperature Oxides*, (*1–4*) Vol. 5 of *Refractory Materials*, Academic Press, New York, 1970.
8. E. Ryshkewitch, *Oxide Ceramics*, Academic Press, Inc., New York, 1960.
9. H. Salmang, *Ceramics, Physical and Chemical Fundamentals*, Butterworth and Company, Ltd., London, 1961.
10. J. R. Hague, J. F. Lynch, A. Rudnick, F. C. Holden, and W. H. Duckworth, eds., *Refractory Ceramics for Aerospace: A Materials Selection Handbook*, The American Ceramic Society, Inc., Columbus, Ohio, 1964.
11. B. Phillips, *Res. Dev.* **18**, 22 (1967).
12. E. M. Levin, C. R. Robbins, and H. F. McMurdie, *Phase Diagrams for Ceramists*, The American Ceramic Society, Inc., Columbus, Ohio, 1964.
13. L. A. Dahl, "Rock Products, Sept. to Dec., 1938" in *PCA Research Bulletin*, Vol. 1, 1939.
14. L. A. Dahl, *J. Phys. Chem.* **52**, 698 (1948).
15. C. N. Fenner, *Am. J. Sci.* **36**(4), 383 (1913).
16. J. W. Greig, *Am. J. Sci.* **13**(5), 1 (1927).
17. A. Muan and E. F. Osborn, *Phase Equilibria Among Oxides in Steelmaking*, Addison-Wesley Publishing Company, Inc., Reading, Mass., 1965.
18. N. L. Bowen and J. F. Schairer, *Am. J. Sci.* **29**(5), 153 (1935).
19. P. Duwez, F. Odell, and F. H. Brown, Jr., *J. Am. Ceram. Soc.* **35**(5), 109 (1952).
20. H. E. McGonnon, ed., *The Marking, Shaping and Treating of Steel*, 9th ed., U.S. Steel Corporation, Pittsburgh, Pa., 1970.
21. *Refractories for Industry*, C-E Refractories, Valley Forge, Pa., 1976.
22. *Product Literature*, Kaiser Refractories, Oakland, Calif., 1977.
23. *Criterion I and Other Glass Furnace Refractories*, C-E Refractories, Valley Forge, Pa., 1977.
24. C. O. Fairchild and M. F. Peters, *J. Am. Ceram. Soc.* **9**, 700 (1926).
25. *Fibrous Ceramic Thermal Insulation for Ultra-High Temperature Use*, Zircar Products, Inc., New York, 1976.
26. *Modern Refractories Practice*, Harbison-Walker Refractories Company, Pittsburgh, Pa., 1961.
27. P. P. Budnikov, *The Technology of Ceramics and Refractories*, The MIT Press, Cambridge, Mass., 1964.
28. J. E. Neal and R. S. Clark, *Chem. Eng.*, 56 (May 4, 1981).
29. *Refractories*, General Refractories Company, Philadelphia, Pa., 1949.
30. A. A. Litvakovski, *Fused Cast Refractories*, Israel Program for Scientific Translations, Jerusalem, 1961.
31. *Refractories*, Current Industrial Reports, U.S. Bureau of Census, Industry Division, Annual Reports, Washington, D.C., 1965–1979.
32. *Manual of ASTM Standards on Refractory Materials*, 8th ed., American Society for Testing and Materials, Philadelphia, Pa., 1957.
33. "Refractories, Glass, Ceramic Materials; Carbon and Graphite Products" in *ASTM Annual Book of ASTM Standards Part 17*, 1981.
34. J. F. Burst and J. A. Spieckerman, *Chem. Eng.*, 85 (July 31, 1967).

General References

J. R. Rait, *Basic Refractories, Their Chemistry and Their Performance*, Ilife & Sons, Ltd., London, England, 1950.

J. H. Chesters, *Steel Plant Refractories, Testing Research and Development*, The United Steel Companies, Ltd., Sheffield, England, 1963.

J. R. Coxey, *Refractories*, The Pennsylvania State College, State College, Pa., 1950.

L. R. McCreight, H. W. Rauch, and W. H. Sulton, eds., *Ceramic and Graphite Fibers and Whiskers*, Vol. 1 of *Refractory Materials*, Academic Press, New York, 1970.

E. K. Storms, ed., *The Refractory Carbides*, Vol. 2 of *Refractory Materials*, Academic Press, New York, 1970.

A. M. Alper, ed., *Phase Diagrams, Materials Science and Technology (1–3)*, Vol. 6 of *Refractory Materials*, Academic Press, New York, 1970.

A. K. Kulkarni and V. K. Moorthy, ed., *Proceedings of the 1st Symposium on Material Science Research*, Series 2, Chemical Metallurgy Commission, Department of Atomic Energy, Bombay, India, 1970, pp. 208–218.

G. R. Belton, ed., *Proceedings of the International Conference of Metal Material Science, 1969*, Plenum Press, New York, 1970.

H. Bibring, G. Seibel, and M. Rabinouitch, eds., *2nd International Conference of Strength Metals Alloys*, Series 3, Conference Proceedings, ASM, Metals Park, Ohio, 1970, pp. 1178–1182.

G. H. Criss and A. R. Olsen, ed., *Proceedings of the 3rd National Incinerator Conference*, American Society of Mechanical Engineering, New York, 1968, pp. 53–68.

C. Brosset, ed., *Trans. Int. Ceram. Congr.* **10**, (1967).

R. M. Fulrath and J. A. Pask, eds., *Ceramic Microstructures, Proceedings of the 3rd International Material Symposium*, John Wiley & Sons, Inc., New York, 1968.

G. C. Kuczynski, ed., *Sintering Related Phenomena, Proceedings of the 2nd International Conference*, Gordon and Breach Science Publishers, New York, 1967.

H. H. Hausner, ed., *Fundamental Refractory Compounds*, Plenum Press, New York, 1968.

R. C. Bradt, D. P. H. Hasselman, and F. F. Lange, eds., *Mech. Ceramics, Proceedings of the Symposium*, Plenum Press, New York, 1974.

S. J. Lefond, *Industrial Mineral Rocks*, 4th ed., American Institute of Mechanical Engineers, New York, 1975.

C. S. Tedmon, Jr., *Corrosion Problems in Energy Conversion Generators*, Electrochemical Society, Princeton, N.J., 1974.

J. J. Burke, A. E. Gorum, and N. R. Katz, eds., *Ceramic High-Performance Applications, Proceedings of the 2nd Army Materials Technology Conference*, Brook Hill Publishing Company, Chestnut Hill, Mass., 1974.

R. C. Bradt and R. E. Tressler, eds., *Deformation of Ceramic Materials, Proceedings of the 1974 Symposium*, Plenum Press, New York, 1975.

R. F. S. Fleming, ed., *Proceedings of the Industrial Mineral International Congress*, Met. Bull. Ltd. London (1975).

Z. A. Foroulis and W. W. Smeltzer, eds., *Met. Slag-Gas React. Processes*, Electrochemical Society, Inc., Princeton, N.J., 1975.

S. Modry and M. Svata, eds., *Pore Structures Prop. Materials, Proceedings of the International Symposium*, Scademia Prague, Czechoslovakia, 1974.

F. V. Tooley, *Handbook of Glass Manufacturing*, Books Ind., Inc., New York, 1974.

10th International Congress of Glass, Ceramic Society of Japan, Tokyo, 1974.

N. Standish, ed., *Alkalis Blast Furnaces, Proceedings of the 1973 Symposium*, Department of Metallurgical Material Science, McMaster University, Hamilton, Ontario, Canada, 1973.

R. C. Marshall, ed., *Silicon Carbide, Proceedings of the 3rd International Conference, 1973*, University of South Carolina Press, Columbia, S.C., 1974.

B. Cockayne, ed., *Mod. Oxide Materials, Prep., Prop. Device Applications*, Academic Press, London, 1972.

R. R. M. Johnston, ed., *Corrosion Technology in the Seventies*, Technical Paper, 12th Annual Conference of the Australiasian Corrosion Association, Parkville, Victoria, Australia, 1972.

L. D. Pye, ed., *Introduction to Glass Science, Proceedings of a Tutorial Symposium*, Plenum Press, New York, 1972. Review of glass melting refractory corrosion.

High Temperature Material, Proceedings of the 3rd Symposium on Material Science Research, Department of Atomic Energy, Bombay, India, 1972. Papers cover a variety of oxide and nonoxide high temperature materials like spinel, alumina and silicon nitride.

J. J. Burke, ed., *Powder Metal High-Performance Applications, Proceedings of the 18th Sagamore Army Material Research Conference*, Syracuse University Press, Syracuse, N.Y., 1972. Review on silicon carbide silicon nitride ceramics.

Mechanical Behavior of Materials, Proceedings of the 1st International Conference, Society of Material Science, Kyoto, Japan, 1972.

J. D. Buckley, ed., *Advanced Materials, Composite Carbon. Pap. Symposium*, American Ceramics Society, Inc., Columbus, Ohio, 1972. Papers on composite materials including carbon, graphite of nitride composites.

J. I. Duffy, *Refractory Materials, Developments since 1977*, Chemical Technology Review, No. 178, Noyes Data Corporation, Park Ridge, N.J., 1980.

I. Ahmad and B. R. Noton, eds., *Advanced Fibers Compositions at Elevated Temperatures, Proceedings of the Symposium of the Metallurgical Society*, American Institute of Mechanical Engineers, Warrendale, Pa., 1980.

G. V. Samsonov and I. M. Vinitskii, *Handbook of Refractory Compounds*, Plenum Press, New York, 1980.

A. V. Levy, ed., *Proceedings of the Corrosion/Erosion Coal Conversion Systems Materials Conference*, National Association of Corrosion Engineers, Houston, Texas, 1979.

Basic Oxygen Steelmaking—New Technology Emerges?, Proceedings of the Conference, Metallurgical Society of London, 1979. Contains papers on refractories for conventional and new bottom blown vessels.

M. R. Louthan, Jr. and R. P. McNitt, eds., *Environmental Degradation of Engineering Materials, Proceedings of the Conference*, Virginia Polytechnical Institute, Blacksburg, Va., 1978.

19th International Refractories Colloquium, Institute Gesteinshuttenkunde RWTH Aachen, Aachen, Federal Republic of Germany, 1976. Contains a number of good papers covering a wide range of refractory materials, properties and problems.

Contin. Cast. Steel, Proceedings of the International Conference, Metallurgical Society of London, 1977.

3rd U.S.—U.S.S.R. Colloquium or Magnetohydrodynamic Electric Power Generation, National Technical Information Service, Springfield, Va., 1976.

P. Vincenzini, ed., *Advanced Ceramic Processes, Proceedings of the 3rd International Meeting of Modern Ceramics Technology*, National Research Council, Research Laboratory of Ceramics Technology, Faenza, Italy, 1978.

G. Y. Onoda and L. L. Hench, eds., *Ceramic Processing Before Firing*, John Wiley & Sons, Inc., New York, 1978.

V. I. Matkovich, ed., *Boron Refractory Borides*, Springer-Verlag, Berlin, Germany, 1977.

R. M. Fulrath and J. A. Pask, *Ceramic Microstructures, Proceedings of the 6th International Materials Symposium*, Westview Press, Boulder, Col., 1977.

H. D. LEIGH
C-E Basic, Inc.

REFRACTORY COATINGS

Refractory coatings denote those metallic, refractory-compound (ie, oxides, carbides, nitrides) and metal-ceramic coatings associated with high temperature service as contrasted to coatings used for decorative or corrosion-resistant applications. They also denote coatings of high melting materials that are used in other than high temperature applications. A coating may be defined as a near-surface region with properties that differ significantly from the bulk of the substrate (see Ceramics; Metallic coatings; Metal surface treatments).

The highest melting refractory metals are tungsten, tantalum, molybdenum, and niobium, although titanium, hafnium, zirconium, chromium, vanadium, platinum, rhodium, ruthenium, iridium, osmium, and rhenium may be included. Many of these metals do not resist air oxidation; hence, very few, if any, are used in their elemental form for high temperature protection. However, bulk alloys based on nickel, iron, and cobalt with alloying elements such as chromium, titanium, aluminum, vanadium, tantalum, molybdenum, silicon, and tungsten are used extensively in high temperature service. Some modern high temperature oxidation- and corrosion-resistant coatings have compositions similar to the high temperature bulk alloys and are applied by thermal spraying, evaporation, or sputtering. The protection mechanism for these high temperature alloy coatings is based on adherent impervious surface films of the Al_2O_3, SiO_2, CrO_2, or a spinel-type that grow on high temperature exposure to air.

Refractory coatings also include materials with high melting points, eg, silicides, borides, carbides, nitrides, or oxides, and combinations such as oxy-carbides, etc. In addition, mixtures of metals and refractory compounds (sometimes called metallides) of various microstructural configurations (ie, laminates, dispersed phases, etc) can also be classified as refractory coatings. Other terms used in the past to describe such multiphase coatings are multilayer coating, multicoating, composite coating, reinforced coating, and the like.

In some cases, a coating is a new material that is deposited onto the substrate by a variety of methods. It is then called a deposited or overlay coating. In other cases, the coating may be produced by altering the surface material to produce a surface layer composed of both the added and substrate materials. This is called a conversion coating, cementation coating, diffusion coating, or chemical-conversion coating when chemical changes in the surface are involved. Coatings may also be formed by altering the properties of the surface by melting and quenching, mechanical deformation, or other processes that change the properties without changing the composition.

Coating technology has developed extensively in the past 20 years. In this article, emphasis is given to new processes or processes that have undergone extensive development in the past decade.

All coating methods consist of three basic steps: synthesis or generation of the coating species or precursor at the source; transport from the source to the substrate; and nucleation and growth of the coating on the substrate. These steps can be completely independent of each other or may be superimposed on each other, depending on the coating process. A process in which the steps can be varied independently and controlled offers great flexibility, and a larger variety of materials can be deposited.

Numerous schemes can be devised to classify deposition processes. The scheme used here is based on the dimensions of the depositing species, ie, atoms and molecules, liquid droplets, bulk quantities, or the use of a surface-modification process (1–2) (see Table 1).

The coating has to adhere to the substrate. The bonding may be mechanical as a result of the interlocking between the asperities on the surface and the coating. Thus, the surface roughness, ie, the average distance between asperities, must be equal to or larger than the dimensions of the depositing particles. Consequently, plasma-sprayed coatings, where the material is deposited as droplets, do not adhere to a polished metal surface because the surface roughness is on a much smaller scale than the dimensions of the liquid droplets. The substrate surface has to be coarsened by techniques such as grit blasting that not only provide the required concave asperities, but also clean the surface (removal of oxide layers and scales) and provide high energy sites for a denser nucleation of coating crystallites and, to some extent, increase the real interfacial area.

In diffusion or chemical bonding, the substrate and the coating material inter-diffuse at the interface. The latter depends on the equilbrium between the two materials at the effective temperature of deposition. If the two materials have no solid solubility, a sharp or abrupt interface forms. The bond strength depends on the affinity between the materials and increases with the degree of wetting of the substrate with the coated material. If the two materials exhibit extensive solid solubility, the interface is a solid solution. If intermetallic phases are indicated on the equilibrium diagram, they may also form at the interface as may gas-metal compounds, depending on the degree of contamination present on the substrate or arriving at the substrate from the environment.

Residual stresses, which are always present in coatings, arise from thermal-expansion mismatch between coating and substrate, or growth stresses caused by imperfections in the coating that are built-in during the process of film growth. Growth stresses are usually significant for coating processes carried out near room temperature. Failure or debonding of the coating may be caused by the residual stresses in the coating and substrate. The location of the failure may be in the substrate, at the interface, or in the coating, depending on plastic deformation and resistance to nucleation and growth of cracks, ie, fracture toughness. For example, a coating may fail at the interface because of crack propagation caused by a brittle intermetallic phase. A postfailure surface analysis of the substrate and the coating is recommended; the use of surface chemical-analysis techniques, eg, scanning electron microscopy (SEM) with energy-dispersive x-ray analysis (edax) capabilities, Auger spectroscopy, or esca analysis can determine the chemical composition at the failed interfaces and deduce its cause.

Coatings may be permeable either to the atmosphere or the substrate material. Diffusion of oxygen through a coating can result in gaseous products that may rupture the coating. Even if the gas can escape through the substrate, as with graphite, the substrate can be consumed and the bond weakened. Coatings permeable to the substrate material by outward diffusion can suffer similarly, as oxidation of the diffused species may occur at the atmosphere–coating interface.

The design of a coating that might equilibrate with its substrate during use is based on the phase diagram of the system. Extensive regions of solid solubility may indicate that rapid interdiffusion can be expected. For multicomponent coatings on

Table 1. Coating Methods[a]

Atomistic deposition	Particulate deposition	Bulk coatings	Surface modification
electrolytic environment	thermal spraying	wetting processes	chemical conversion
electroplating	plasma-spraying	painting	electrolytic
electroless plating	D-gun	dip coating	anodization (oxides)
fused-salt electrolysis	flame-spraying	electrostatic spraying	fused salts
chemical displacement	fusion coatings	printing	chemical–liquid
vacuum environment	thick-film ink	spin coating	chemical–vapor
vacuum evaporation	enameling	cladding	thermal
ion-beam deposition	electrophoretic	explosive	plasma
molecular-beam epitaxy	impact plating	roll bonding	leaching
plasma environment		overlaying	mechanical
sputter deposition		weld-coating	shot peening
activated reactive evaporation		liquid-phase epitaxy	thermal
plasma polymerization			surface enrichment
ion plating			diffusion from bulk
chemical-vapor environment			sputtering
chemical-vapor deposition			ion implantation
reduction			
decomposition			
plasma enhanced			
spray pyrolysis			

[a] Ref. 2.

40

multicomponent substrates, a large number of possibilities exists for formation of intermetallic compounds. Limited compound formation at the coating-substrate interface may be preferred if it decreases the permeability of the coating. Wetting or, more strictly speaking, high affinity between coating and substrate is conducive to high temperature service.

A reasonably close match of the thermal expansion of the coating and substrate over a wide temperature range to limit failure caused by residual stresses is desired for coatings. Since temperature gradients cause stress even in a well-matched system, the mechanical properties, strength, and ductility of the coating as well as the interfacial strength must be considered.

The simple case of protecting tungsten or molybdenum with platinum illustrates the above point. A wide range of solid solubility exists that contributes to interdiffusion between coating and base metal to form brittle intermetallic compounds. Both solutions and compounds oxidize preferentially, losing tungsten or molybdenum. Platinum is also somewhat permeable to oxygen, which permits subcoating oxidation. A thermal-expansion mismatch exists between platinum and these intermetallic compounds and adds another complication where thermal cycling is required. Silicides and beryllides exhibit sufficient conductivity to be electroplated and could well be used as diffusion barriers. However, the bond strengths and mechanical compatibility of two layers instead of one must be considered.

Atomic Deposition Processes

Electrodeposition. *Aqueous Electrodeposition.* The theory of electrodeposition is well known (see Electroplating). Of the numerous metals used in electrodeposition, only ten have been reduced to large-scale commercial practice. The most commonly plated metals are chromium, nickel, copper, zinc, rhodium, silver, cadmium, tin, and gold followed by the less frequently plated metals iron, cesium, platinum, and palladium and the infrequently plated metals iridium, ruthenium, and rhenium. Of these, only platinum, rhodium, iridium, and rhenium are refractory.

The electrodeposition of tungsten alloys of iron, nickel, and cobalt is commercially feasible but has remained largely experimental, although their properties should be of sufficient interest for engineering applications.

Cermets, ie, materials containing both ceramic and metal, eg, TiC–Ni and Al_2O_3–Cr, can be deposited from plating baths if the particulate matter is suspended by air agitation or stirring (see Ceramics; High temperature composites). Particle sizes from 1 to 50 μm may be deposited to concentrations over 20%. Chromium-based cermets with zirconium and tungsten borides, zirconium nitride, and molybdenum carbide can be plated on refractory metals and graphite. However, at 1650°C, protection is afforded only for a few minutes. Cermets also suffer from porosity; therefore, such coatings find application where exposure times are short and erosion conditions severe, and where they can be used to bridge the expansion mismatch between a metallic substrate and a ceramic coating.

Fused-Salt Electrodeposition. Molten-salt electrolysis has been used since Davy's and Faraday's pioneering experiments, but application as a method for coating metals has remained of academic interest. Fused-salt baths may be used to plate ruthenium, platinum, and iridium with improved coating soundness. Recent developments in this technology may lead to commercial plating of other refractory metals, such as zirconium, hafnium, vanadium, niobium, tantalum, molybdenum, and tungsten. Molten-salt electrolysis is based on the same principles as aqueous electrolysis processes, but the vehicle, ie, the bath, is a molten salt. The coating material deposits as a layer on the substrate.

Electroless Deposition. Electroless plating (qv) is defined as a controlled, autocatalytic chemical-reduction process for depositing metals. It resembles electroplating because it can be run continuously to build up a thick coating. Unlike displacement processes, it does not involve a chemical reaction with the substrate metal; and unlike the well-known silver-reduction process (Tollens reaction) for the silvering of optical glass, it is selective, ie, deposits form only on a catalytic surface. The metals, nickel, cobalt, platinum, palladium, and gold, and nickel–cobalt alloys can be deposited. Chromium, iron, and vanadium are often claimed, but electroless plating of active metals must be viewed with suspicion, since most such processes are of the displacement type.

Physical Vapor-Deposition Processes (PVD). The three physical vapor-deposition processes are evaporation, ion plating, and sputtering (see Film deposition techniques).

The materials that are deposited by PVD techniques include metals, semiconductors, alloys, intermetallic compounds, refractory compounds, ie, oxides, carbides, nitrides, borides, etc, and mixtures thereof. The source material must be pure and free of gases and inclusions; otherwise spitting may occur.

Metals and Elemental Semiconductors. Evaporation of single elements can be carried out from various evaporation sources subject to the restrictions with regard to melting point, container reactions, deposition rate, etc. A typical arrangement is shown in Figure 1 for electron-beam heating (3). This type of source is ideal for refractory coatings, since the material to be deposited is contained in a noncontaminating water-cooled copper crucible and the surface of the material can easily be heated to >3000°C.

Alloys. Alloys consist of two or more elements with different vapor pressures and hence different evaporation rates. As a result, the vapor phase and, therefore, the deposit constantly vary in compositions. This problem can be solved by multiple sources or a single rod-fed or wire-fed electron-beam source fed with the alloy. These solutions apply equally to evaporation or ion-plating processes.

Multiple Sources. Multiple sources offer a more versatile system. The number of sources evaporating simultaneously is equal to or less than the number of constituents in the alloy. The material evaporated from each source can be a metal, alloy, or compound. Thus, it is possible to synthesize a dispersion-strengthened alloy, eg, $Ni-ThO_2$. However, the evaporation rate from each source has to be monitored and controlled separately. The source-to-substrate distance would have to be sufficiently large (38 cm for 5-cm-dia sources) to blend the vapor streams prior to deposition, which decreases the deposition rate (see Fig. 2). Moreover, if the density of two vapors differs greatly, it may be difficult to obtain a uniform composition across the width of the substrate because of scattering of the lighter vapor atoms.

Evaporating each component sequentially produces a multilayered deposit that is homogenized by annealing. High deposition rates are difficult to obtain.

Single Rod-Fed Electron-Beam Source. The disadvantages of multiple sources for alloy deposition can be avoided by using a single wire-fed or rod-fed source (see Fig. 3) (3). A molten pool of limited depth is above the solid rod. If the equilibrium

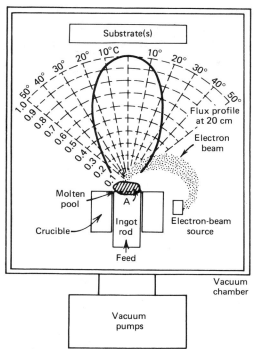

Figure 1.   Vacuum-evaporation process with use of electron-beam heating.

vapor pressures of the components of an alloy A_1B_1 are in the ratio of 10:1 and the composition of the molten pool is A_1B_{10}, under steady-state conditions, the composition of the vapor is the same as that of the solid being fed into the molten pool. One can start the procedure with a pellet of appropriate composition A_1B_{10} on top of a rod A_1B_1 to form the molten pool initially, or one can start with a rod of alloy A_1B_1 and evaporate the molten pool until it reaches composition A_1B_{10}. The temperature and volume of the molten pool must be constant to obtain a constant vapor composition. A theoretical model has been developed and confirmed by experiment, and deposits of Ni–20 wt % Cr, Ti–6 wt % Al–4 wt % V, Ag–5 wt % Cu, Ag–10 wt % Cu, Ag–20 wt % Cu, Ag–30 wt % Cu, and Ni–Cr–yAl–zY alloy have been successfully prepared. This method can be used with a 5000-fold vapor-pressure difference between components. It cannot be used when one of the alloy constituents is a compound, eg, Ni–ThO_2.

Sputtering.   Sputtering deposits alloys by means of an alloy target. The surface composition of the target changes in the inverse ratio of the sputtering yields of the individual elements, as in the alloy evaporation from a single rod-fed electron-beam source. Alternatively, the sputtering target can be made of strips of the components of the alloy with the respective surface areas inversely proportional to their sputtering yield (see also Film deposition techniques).

Refractory Compounds.   Refractory compounds resemble oxides, carbides, nitrides, borides, and sulfides in that they have a very high melting point. In some cases, they form extensive defect structures, ie, they exist over a wide stoichiometric range. For example, in TiC, the C:Ti ratio can vary from 0.5 to 1.0, which demonstrates a wide range of vacant carbon-lattice sites.

In direction evaporation, the evaporant is the refractory compound itself, whereas

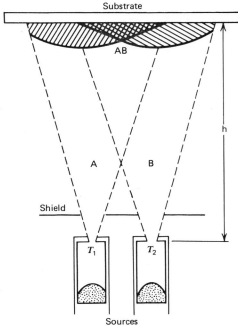

Figure 2.   Two-source evaporation.

in reactive or activated reactive evaporation (ARE), a metal or a low-valency metal compound is evaporated in the presence of a partial pressure of a reactive gas to form a compound deposit, eg, Ti is evaporated in the presence of N_2 to form TiN, or Si or SiO is evaporated in the presence of O_2 to form SiO_2.

Direct Evaporation.   Evaporation can occur with or without dissociation of the compound into fragments. The observed vapor species show that very few compounds evaporate without dissociation. Examples are MgF_2, B_2O_3, CaF_2, SiO, and other Group-IV divalent oxides (SiO homologues such as GeO and SnO).

In general, when a compound is evaporated or sputtered, the material is not transformed to the vapor state as compound molecules but as fragments thereof. Subsequently, the fragments recombine, most probably on the substrate, to reconstitute the compound. Therefore, the stoichiometry (anion:cation ratio) of the deposit depends on the deposition rate, the ratios of the various molecular fragments, the impingement of other gases present in the environment, the surface mobility of the fragments which, in turn, depends on their kinetic energy and substrate temperature, the mean residence time of the fragments of the substrate, the reaction rate of the fragments on the substrate to reconstitute the compound, and the impurities present on the substrate. For example, direct evaporation of Al_2O_3 results in a deposit deficient in oxygen, ie, that had the composition Al_2O_{3-x}. This O_2 deficiency could be compensated for by introducing O_2 at a low partial pressure into the environment.

Reactive Evaporation.   In reactive evaporation (RE), metal or alloy vapors are produced in the presence of a partial pressure of reactive gas to form a compound either in the gas phase or on the substrate as a result of a reaction between the metal vapor and the gas atoms, eg,

$$2\,Ti + C_2H_2 \rightarrow 2\,TiC + H_2$$

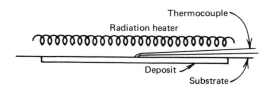

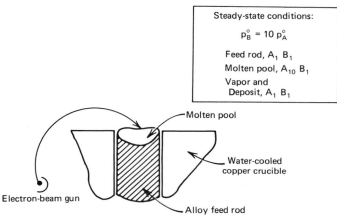

Figure 3. Alloy evaporation from a single rod-fed source. P_B° = the equilibrium vapor pressure of component B, P_A° = the equilibrium vapor pressure of component A.

If the metal and gas atoms are activated or ionized in the vapor phase, which activates the reaction, the process is called the activated reactive process (ARE), as illustrated in Figure 4 (4). The metal is heated and melted by a high acceleration-voltage electron beam. The melt has a thin plasma sheath on top from which low energy secondary electrons are pulled upward into the reaction zone by an electrode placed above the

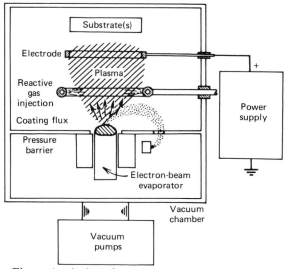

Figure 4. Activated reactive evaporation process (4).

pool; the electrode is biased to a low positive d-c potential (20–100 V). These low energy electrons have a high ionization cross-section, thus ionizing or activating the metal and gas atoms and increasing the reaction probability on collision. Titanium carbide was synthesized with this process by reaction of Ti metal vapor and C_2H_2 gas with a carbon:metal ratio approaching unity. Moreover, by varying the partial pressure of either reactant, the carbon:metal ratio of carbides could be varied at will. This process has been applied recently to the synthesis of the five different Ti oxides (5). With the ARE process (ie, with a plasma), as compared to the RE process (ie, without a plasma), a higher oxide formed for the same partial pressure of O_2, which demonstrates a better gas utilization in the presence of plasma (see Plasma technology, Supplement volume).

Reactive-Ion Plating.   In reactive-ion plating (RIP), as in the reactive evaporation process, the metal atoms and reactive gases form a compound aided by the presence of a plasma. Since the partial pressures on the gases are much higher (>1.3 Pa or 10^{-2} mm Hg) than in the ARE process (13 mPa or 10^{-4} mm Hg), the deposits may be porous or sooty. In the simple diode ion-plating process, the plasma cannot be supported at a lower pressure. Therefore, an auxiliary electrode adjusted to a positive low voltage, as originally conceived for the ARE process, is used to initiate and sustain the plasma at a low pressure (ca 0.13 Pa or ca 10^{-3} mm Hg), as shown in Figure 5 (6).

Reactive Sputtering.   Reactive sputtering is very similar to reactive evaporation and reactive-ion plating in that at least one coating species enters the system in the gas phase. Examples include sputtering Al in O_2 to form Al_2O_3, Ti in O_2 to form TiO_2, In–Sn in O_2 to form tin-doped In_2O_3, Nb in N_2 to form NbN, Cd in H_2S to form CdS, In in PH_3 to form InP, and Pb–Nb–Zr–Fe–Bi–La in O_2 to form a ferroelectric oxide.

By reactive sputtering, many complex compounds can be formed from relatively easy-to-fabricate metal targets; insulating compounds can be deposited with a d-c

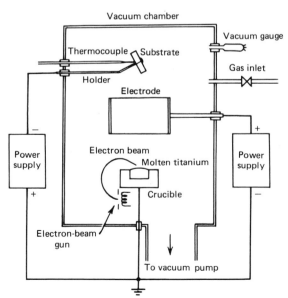

Figure 5.   Reactive-ion plating with auxiliary electrode for low pressure operation in deposition of compounds (6). Courtesy of Kobayashi and Doi.

power supply; and graded compositions can be formed, as described in the preceding section. However, the process is complicated.

Chemical Vapor Deposition (CVD) and Plasma-Assisted Chemical Vapor Deposition (PACVD). In CVD, thin films or bulk coatings up to 2.5 cm in thickness are deposited by means of a chemical reaction between gaseous reactants passing over a substrate. Temperatures can be anywhere between 200 and 2200°C but are usually between 500 and 1100°C. The optimum for a given reaction often lies within a very narrow range, and the process needs to be tailored to the substrate and the intended application. The substrate's melting point and susceptibility to chemical attack by the reacting gases or by their side-products has to be considered. Coatings have application in a wide array of corrosion- and wear-resistant uses, but also in decorative layers, semiconductors, and magnetic and optical films (see Film deposition techniques).

In most cases, CVD reactions are activated thermally, but in some cases, notably in exothermic chemical-transport reactions, the substrate temperature is held below that of the feed material to obtain deposition. Other means of activation are available (7), eg, deposition at lower substrate temperatures is obtained by electric-discharge plasma activation. In some cases, unique materials are produced by plasma-assisted CVD (PACVD), such as amorphous silicon from silane where 10–35 mol % hydrogen remains bonded in the solid deposit. Except for the problem of large amounts of energy consumption in its formation, this material is of interest for thin-film solar cells. Passivating films of SiO_2 or SiO_2–Si_3N_4 deposited by PACVD are of interest in the semiconductor industry (see Semiconductors).

Plasma-assisted CVD processes use deposition temperatures lower than CVD processes; the desired deposition reaction is aided by the energy present in the plasma. An example of a PACVD process apparatus is shown in Figure 6 (8). The plasma greatly extends the utility of CVD processes, eg, the ability to deposit films on substrates that cannot withstand the temperature needed for the CVD process in reactions such as polymerization, anodization, nitriding, deposition of amorphous silicon, amorphous carbon (diamondlike carbon), etc. The films deposited by PACVD are generally more

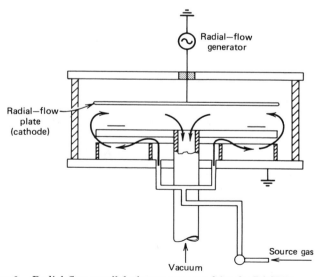

Figure 6.   Radial-flow parallel-plate reactor used for the PACVD process (8).

complex than their analogues deposited by normal CVD methods. For example, silicon nitride films have a composition $Si_xN_yH_z$ and may contain as much as 15 to 30 at. % hydrogen as contrasted to the Si_3N_4 composition of the film deposited by the normal CVD process. Other advantages of PACVD are substrate surface etching and activation to produce good bonding at low deposition temperatures.

Ion Implantation. High energy ion implantation is a highly successful process for doping semiconductors to very precisely controlled concentrations (see Ion implantation). The part's surface is bombarded with high energy ions (about 50 keV), which results in a layer of implanted ions at an approximate depth of 8 nm. The potential for producing refractory coatings by implantation of specific metal ions or by reactions within the substrate to form refractory compounds exists but has not been exploited to date.

Particulate Deposition Processes

Melt Spraying. In melt spraying, a combustion flame, electric arc, or arc plasma or explosive-wave front heats particles of the refractory-coating material to a temperature sufficient to achieve sintering or cohesive solidification when the particles impinge on a substrate. Penetration of the substrate often accompanies the coalescence. The techniques include flame plating (D-gun), torch spraying, arc-plasma spraying, and induction-plasma spraying. These processes generate considerable noise, and acoustical protection is needed.

Flame plating employs oxyacetylene fuel. In this method, developed by the Linde Division of the Union Carbide Corporation, the oxyacetylene gas mixture is detonated by an electric spark at four detonations per second. The powders are fed under control into a chamber from which they are ejected when detonation occurs. The molten, 14–16-μm particles are sprayed at a velocity of 732 m/s at distances of 5.1–10.2 cm from the surface. The substrate is moved past the stationary gun.

Torch spraying is the most widely used melt-spraying process. In the power-feed method, powders of relatively uniform size ($<44\ \mu$m or 325 mesh) are fed at a controlled rate into the flame. The torch, which can be held by hand, is aimed a few cm from the surface. The particles remain in the flame envelope until impingement. Particle velocity is typically 46 m/s, and the particles become at least partially molten. Upon impingement, the particles cool rapidly and solidify to form a relatively porous, but coherent, polycrystalline layer. In the rod-feed system, the flame impinges on the tip of a rod made of the material to be sprayed. As the rod becomes molten, droplets of material leave the rod with the flame. The rod is fed into the flame at a rate commensurate with melt removal. The torch is held at a distance of ca 8 cm from the object to be coated; particle velocities are ca 185 m/s.

Arc-plasma-spraying equipment has been commercially available since 1958. This process does not utilize a fully ionized gas (plasma). Uniform particles ($<74\ \mu$m or <200 mesh) are fed into the jet that emerges through the nozzle of a high pressure d-c arc. The particles are melted and ejected at controllable velocities of 30–210 m/s. The composition of the atmosphere can be varied over a wide range, ie, wet, dry, oxidizing, reducing ($\leq80\%$), or inert. A 40–50 kW, 600-A d-c source supplies the average power requirements. The uv radiation generated during operations presents a safety hazard.

Induction-plasma spraying is still an experimental process. A graphite susceptor is inductively heated until the gas temperature creates a plasma. This plasma subsequently interrupts the inductive field and acts as the susceptor. Powders are fed into the plasma through a duct near the inlet of the plasma where they vaporize. Particle residence time can be varied over a greater range than in the arc-plasma process since gas velocity can be varied by adjusting the nozzle size. The flame temperature decreases rapidly with increasing gas velocity, and this affects the extent to which the particle melts. This apparatus is powered by a 50-kW, high frequency a-c source and cannot be held by hand. As in flame plating, the object is moved in the path of the gun.

Any refractory material that does not decompose or vaporize can be used for melt spraying. Particles do not coalesce within the spray. The temperature of the particles and the extent to which they melt depend on the flame temperature, which can be controlled by the fuel:oxidizer ratio or electrical input, gas flow rate, residence time of the particle in the heat zone, the particle-size distribution of the powders, and the melting point and thermal conductivity of the particle. Quenching rates are very high, and the time required for the molten particle to solidify after impingement is typically 10^{-4}–10^{-2} s.

A broad range of materials can be handled by plasma spraying. However, each material requires some optimization of conditions, such as modification of carrier gas, power, particle-size distribution, and substrate. In some melt-spray processes, a preferential loss of material may occur that leads to a change in stoichiometry. It can be compensated for by adding an appropriate additional gas to the gas stream. Thus, additions of oxygen can keep the reduction of TiO_2 and HfO_2 within acceptable limits, whereas methane reduces the loss of carbon from carbides. A wide selection of coating materials is available. Except for particle-velocity limitations, spray conditions are excellent because of the inert, clean carrier gases used. Inert atmospheres reduce the degradation or oxidation of the particles, preclude oxidation of the substrate, and diminish contamination of the sprayed material resulting from gas adsorption.

Torch spraying and arc-plasma spraying are the least expensive. Flame plating requires little substrate preparation, but is otherwise expensive. The adherence of melt-sprayed ceramic coatings onto metallic substrates varies widely from process to process and depends on procedure and substrate preparation (see Table 2). The high strength of flame-plated coatings is a consequence of the higher particle velocity that results in penetration of the steel surface. For other melt-spray processes, the substrate must be roughened to provide a suitable surface. Chemical bonding contributes to adhesion but the mechanical bond is preferred. Advantages and disadvantages of melt-spray processes are tabulated in Table 3.

Table 2.   Tensile-Bond Test for Melt-Sprayed Al_2O_3

Method	Particle velocity, m/s	Bond strength, MPa[a]
torch spraying	46–83	3.4–13.8
flame plating	732	69
arc-plasma process	30–213	6.9–69
induction-plasma process	6–>300	na[b]

[a] To convert MPa to psi, multiply by 145.

[b] Na = not applicable.

Table 3. Melt-Spray Processes[a]

Advantages	Disadvantages
produces coatings on any substrate	undercuts, deep blind holes, and small internal
versatility in configuration size, shape, and	diameters are difficult to coat
structure	properties are generally inferior
good dimensional control, no shrinkage	nonstoichiometry may result
mixed compositions	residual stresses can limit coating thickness
fine grain size	
permits patching and repairing	
alternating layers	
graded coatings	
by masking, localized areas can be coated	

[a] Porosities are 0.5–20.0%.

Electrophoretic Processes. Electrophoretic coatings are obtained through the migration of charged particles when a potential is applied to electrodes immersed in a suitable suspension of the particles (see Electromigration). This process is particularly suited for applying uniform layers on complex bodies. Projecting edges become insulated and the current shifts to bare surfaces as the deposit builds up on them. Particles, not ions, are deposited. The early stages of coating are powdery, and densification is required to produce adherence. Electrophoresis can be used to apply metals and alloys, ceramics, and cermets. The three steps are preparation of the dispersion, deposition and conversion, and bonding. In the preparation phase, particle size is critical and diameters may be from 1 to 50 μm; however, too small a size leads to cracking after thick coatings are dried and consolidated.

Several suspension media with suitable viscosities, densities, conductivities, dielectric constants, and chemical stabilities are available. Binders improve the green strength of the coating. Deposition is effected rapidly with potentials of 50–1000 V dc. Simple d-c sources with no special filtering are sufficient. All inorganic materials are amenable to deposition and most are amenable to codeposition, even though they have different densities; for example, NiO (7.5 g/cm^3) and WC (15.7 g/cm^3) have been codeposited in the same ratio as they were dispersed. The practical limit of thickness is about 0.5 mm as set by shrinkage cracking. Conversion and bonding techniques vary with the type of coating. Hydrogen reduction, hydrostatic pressing, and sintering are used for most metal-base coatings. The oxides of iron, nickel, cobalt, molybdenum, and tungsten can be reduced in hydrogen at less than 600°C. For example, NiO–Cr green deposits are reduced in H$_2$ at 320°C to produce metallic nickel, with no particle sintering. After pressing at 69 MPa (10,000 psi), further densification is achieved by sintering to 90% of theoretical density at 1090°C. Vacuum sintering is preferred for titanium, zirconium, niobium, and tantalum. Glass and ceramic coatings can be either vitrified or fused.

Bulk Coating

In bulk-coating processes, bulk materials are joined to the substrate either by a surface-melt process or by attachment of the solid material. A recent example of the latter is the application of heat-resistant tiles of silica-type material to the aluminum-alloy skin of a space-shuttle vehicle; this enables the vehicle to withstand the reentry heat.

In cladding, one metal is coated with another by rolling or extruding the two metals in close contact to each other. Coherence of the two metals is induced by soldering, welding, or casting one in contact with the other prior to the rolling operation. By this mechanial process, steel can be coated with copper, nickel, or aluminum. The coextrusion process, which is used for lightweight rifle barrels made of a steel core and titanium alloy wrap, is a good example of the cladding process.

Another familiar commercial method is the immersion or hot-dipping process. The article to be coated is immersed in a molten metal bath. Usually little else is done to change the properties of the coating, which adheres to the surface upon removal of the article from the bath. For a successful coating, an alloying action must take place between the components to some extent. Zinc and tin coatings are applied to sheet steel by hot-dipping.

Surface coatings are also applied by welding processes, such as manual arc welding (oxy–acetylene, gas–tungsten), gas–metal arc welding, submerged-arc welding, spray welding, plasma-arc welding, and electro-slag welding (see Welding).

Thin coatings, 0.25–0.75 mm, are applied by spray welding. For heavier coatings in the range of 6 to 65 mm or more, other welding processes are used.

These welding processes are not suitable for the application of refractory coatings of reactive metals such as tungsten, tantalum, niobium, and chromium, although electro-slag welding is being studied for this purpose. On the other hand, such refractory-metal, alloy, and cermet coatings can be applied by electron-beam melting to form thick coatings, since this process is carried out in vacuum.

A new development in this area is the use of high power laser beams to surface-melt refractory-metal or cermet coatings onto substrates in a controlled atmosphere to prevent contamination of the reactive metals. This process is sometimes referred to as laser glazing (see Lasers).

Enameling meets decorative as well as protective requirements. Ceramic enamels are mainly based on alkali borosilicate glasses. The part to be enameled is dipped into or sprayed with a slip, ie, a water suspension of glass fragments called frit. The slip coating is dried and fused in a enameling furnace under careful heat control (see Enamels, porcelain or vitreous).

Troweling and painting methods are used to apply thick protective coatings of a refractory paste or cermet onto a variety of substrates for high temperature service. Fiberfrax (The Carborundum Co.) coating cements, composed of aluminosilicate fibers bonded with air-setting temperature-resistant inert binders, is commonly used as a coating for reducing the oxidation rate of graphite.

In the microelectronics industry, powdered metals and insulating materials that consist of nonnoble metals and oxides are deposited by screen printing in order to form coatings with high resistivities and low temperature coefficients of resistance. This technique may be useful in depositing oxide–metal refractory coatings.

Surface-Modification Processes

Cementation (Diffusion) Coatings. Cementation is defined as the introduction of one or more elements into the outer portion of a metal object by means of diffusion at elevated temperatures. Cementation was first used to convert iron to steel, and copper to brass, but today it is considered only as a method of surface treatment. The coating produced by cementation is formed by an alloying or chemical combination of the diffusing elements with the substrate material. Cementation coatings enjoy wide metallurgical application. A prime example is the case hardening of steel whereby a soft, ductile, low-carbon steel is heated to 810–900°C in a packing of carbonaceous material to produce a high carbon steel surface (see Steel). The coating formed in this manner is limited to use at temperatures of about 600°C or lower. Truly refractory coatings are alloyed onto molybdenum, tantalum, niobium, and tungsten by cementation. Such coatings provide short-term protection at 1650°C or higher, and long-term protection at 1370°C.

Cementation coatings are produced at temperatures well below the melting points of either the coating or substrate material by means of a vapor-transport mechanism. The coating material and the substrate form definite chemical compounds, such as silicides, aluminides, beryllides, or chromides. On the other hand, some cementation coatings can be solid solutions of indefinite composition. Oxidation-resistant coatings act as diffusion barriers for both the inward diffusion of oxygen and the outward diffusion of the substrate. The diffusion of the coating components into the substrate is generally the rate-controlling step. Cementation coatings are applied by pack cementation, activated or nonactivated slurry processes, and the fluidized-bed technique.

In pack cementation, the part to be coated is placed in a retort and surrounded with a powdered pack consisting of the coating component and an activator; the latter reacts with the coating component to form the carrier vapor, usually a halide or an inert diluent, to prevent the pack from sintering together and to permit vapor transport of the alloying component through the pack.

The slurry process requires less coating component. The latter is suspended in a vehicle, eg, lacquer or water, and is painted onto the substrate. The coated part is heated in an alumina retort containing a layer of activator at the bottom. The coating component forms a halide and is deposited onto and diffused into the substrate. Slurry processes can be either activated or nonactivated. In the latter case, development of the coating relies purely upon diffusion without the possible benefits of vapor deposition.

The fluidized-bed technique combines aspects of pack cementation and vapor deposition. A fluidized bed consists of a mass of finely divided solids contained in a column. The solids are brought into a fluidized state by the lifting action of a gas as it rises through the column. A vaporized halide may be carried by a gas into the bed, where it reacts with the fluidized coating powder to form the coating component halide, which then thermally decomposes and deposits on the substrate contained within the retort. Alternatively, the metal coating particles may be fluidized before entering the bed, whereby the halide gas permeates the fluidized particles to form the coating halide vapor. This vapor is carried into the bed of inert particles, where it thermally decomposes to deposit the coating (see Fluidization).

The chromizing of iron is described by three mechanisms:

Displacement	$Fe(alloy) + CrCl_2(g) \rightarrow FeCl_2(g) + Cr(alloy)$
Reduction	$H_2(g) + CrCl_2(g) \rightarrow 2\ HCl(g) + Cr(alloy)$
Thermal Decomposition	$CrCl_2(g) \rightarrow Cl_2(g) + Cr(alloy)$

The displacement mechanism involves placing the iron alloy packed in chromium powder, NH_4Cl, and Al_2O_3 in a sealed retort, which is heated to promote vapor-deposition and diffusion processes. The exact chemistry is not known, but the following steps probably occur:

$$NH_4Cl(g) \rightarrow NH_3(g) + HCl(g)$$
$$2\,NH_3(g) \rightarrow N_2(g) + 3\,H_2(g)$$
$$Cr(powder) + 2\,HCl(g) \rightarrow CrCl_2(g) + H_2(g)$$
$$CrCl_2(g) + Fe(alloy) \rightarrow FeCl_2(g) + Fe-Cr(alloy)$$

Several thermodynamic and kinetic requirements must be fulfilled for the reactions to proceed as shown. First, the vapor pressure of the activators or carriers must be sufficiently high at the coating temperature for the reaction to proceed. The activator, ie, the chloride, should have a boiling point slightly above the temperature of the coating to provide a reservoir for reaction without volatilizing too rapidly. The coating-component halide should have a boiling point below the coating temperature to saturate the pack. The volatility of the by-product of the coating reaction must be high; fast removal prevents the formation of a barrier to continued deposition of the coating element and also prevents contamination of the coating.

Another example is the silicidizing of tantalum, basically an oxidation–reduction reaction. The packing is sodium fluoride and silicon. After deposition, the coating diffuses continuously into the substrate, according to the following reactions:

$$6\,NaF(g) + 2\,Si(l) \rightarrow Si_2F_6(g) + 6\,Na(g)$$
$$Ta + 2\,Si_2F_6(g) \rightarrow 3\,SiF_4(g) + Si(TaSi_2)(alloy)$$
$$3\,SiF_4(g) + Si(l) \rightarrow 2\,Si_2F_6(g)$$

If the rules for volatilities and thermodynamics of the halides are followed, the reaction can be used for aluminizing, silicidizing, chromizing, and similar processing.

Silicide coatings of refractory metals may contain as much as three to five coating components other than silicon. A mixture of halide carriers is selected containing the best carrier for each component.

The outstanding characteristics of a fluidized bed are its high heat-transfer coefficient and its turbulence, which yield optimum temperature uniformity throughout the bed. These factors contribute to the successful treatment of large, complex objects, which might not be possible by other means.

Analogous to the carburization of steel is the nitrogen case-hardening of steel in which surfaces are hardened by heating the steel in the presence of nascent nitrogen. The nitrogen reacts with impurities or alloy constituents, eg, aluminum, chromium, vanadium, tungsten, etc, to form dispersed nitrides. Other examples of cementation coatings include the surface-hardening of steel by immersion in a molten sodium cyanide bath at 850–870°C to promote codiffusion of both carbon and nitrogen into the surface. Iron and steel are cemented with silicon at 800°C, and iron, nickel, and copper with tungsten at 800–1350°C. Similarly, iron and nickel are cemented with molybdenum by heating in ferromolybdenum. In the same temperature range, titanium coatings are obtained on ferrous metals by heating them for 1.5 h in a powdered mixture of sponge titanium containing 0.5–6.0% Fe at 800–1200°C. Boride coatings are produced by heating iron, cobalt, or nickel substrates in boron powder at 950°C in a vacuum of 67 mPa (5×10^{-4} mm Hg).

The most advanced cementation coatings are the intermetallic coatings, specifically silicides and aluminides, that protect refractory metals. The earliest and simplest of these is the molybdenum silicide coating developed for molybdenum substrates. It consists largely of $MoSi_2$ but is modified by additions of boron, manganese, titanium, chromium, beryllium, etc, singly or in combinations. Coating systems for niobium are more complex, but oxidation-resistant coatings result by diffusing silicon, chromium, aluminum, boron, titanium, and beryllium singly or in combinations into the substrate. Diffusion coatings for tantalum are based on the formation of $TaAl_3$, an oxidation-resistant material. Silicide coatings on tantalum also provide significant oxidation protection. Aluminum, boron, or manganese are often added as modifiers. Tungsten presents a more difficult problem, but various silicide coatings provide a measure of protection at temperatures as high as 1815°C. After 10 h at this temperature, WSi_3 is converted to the glassy $W_5Si_4O_2$ composition.

An activated-slurry process can be used to place impervious silicon carbide coatings on graphite. A slurry of silicon carbide, carbon, and appropriate organic binders is applied to the surface of the graphite by spraying, dipping, or painting. After a low temperature treatment to drive off the binder, the coated substance is heated in the presence of silicon vapor, which diffuses into the surface to form a SiC layer on the graphite. The coating composition can vary from self-bonded silicon carbide containing only a few percent of uncombined silicon to SiC crystals bonded with a continuous silicon matrix. Successful SiC coatings on complex graphite shapes have been obtained with isotropic graphite substrates possessing compatible thermal-expansion coefficients. Resistance to oxidation at 1400°C for 100 h or more has been accompanied by successful service in nuclear-reactor environments. Coatings of SiC on graphite fail under compressive loads or impact at levels that do not damage un-coated graphite. This occurs because the graphite is less brittle and deforms under load, whereas the thin, more rigid SiC coating does not.

Aluminide and silicide cementation coatings such as $TaAl_3$ on tantalum and $MoSi_2$ on molybdenum oxidize at slow rates and possess some inherent self-repair characteristics. Fine cracks that appear and are common to these coatings can be tolerated because stable, protective oxides form within the cracks and seal them. Thermal cycling, however, accelerates failure because of thermal-expansion mismatch that ultimately disrupts the protective oxide coating.

An important application is the aluminizing of air foils of gas-turbine engines made of high temperature Ni- or Co-base alloys. The aluminizing can be carried out either in a pack process or in an out-of-the-pack process.

Cementation coatings rely on diffusion to develop the desired surface alloy layer. Not only does the coating continue to diffuse into the substrate during service, thereby depleting the surface coating, but often the substrate material diffuses into the surface where it can be oxidized. Since the diffusion rate is temperature dependent, this may occur slowly at lower service temperatures.

The substrate has to be prepared for cementation. The surface must be clean and free of oxide. Corners and edges are particularly important in diffusion-type coatings; sharp edges are usually detrimental. Barrel finishing, ie, tumbling in a barrel with abrasive media, may result in the desired shape.

The quality of cementation coatings does not necessarily equal that produced by other techniques. There are cases, however, where a moderate degree of corrosion resistance is useful and where other requirements are best met by the application of cementation. With other processes, it may be difficult to coat porous surfaces or pre-serve the contour of machined surfaces. Cementation may then be the method of choice. Coatings for molybdenum, niobium, tantalum, and tungsten not easily obtained by other means can be achieved by this method. Coated-refractory-metal systems have been applied to leading edges, skins, and structural members of space vehicles, rocket combustion chambers, nozzle inserts, extension skirts, and to vanes or blades for advanced gas-turbine engines.

Metalliding. Metalliding, a General Electric Company process (9), is a high temperature electrolytic technique in which an anode and a cathode are suspended in a molten fluoride salt bath. As a direct current is passed from the anode to the cathode, the anode material diffuses into the surface of the cathode, which produces a uniform, pore-free alloy rather than the typical plate usually associated with electrolytic processes. The process is called metalliding since it encompasses the interaction, mostly in the solid state, of many metals and metalloids ranging from beryllium to uranium. It is operated at 500–1200°C in an inert atmosphere and a metal vessel; the coulombic yields are usually quantitative, and processing times are short; controlled uniform coatings from a few μm to many μm are obtained, many of which are unavailable by other techniques. Diffusion rates are high and the process can be run continuously. Boron and silicon anodes can be diffused into most metals of groups VB, VIB, VIIB, VIII, and IB of the periodic table.

The borides are extremely hard (9.8–29 GPa or 1000–3000 kgf/mm^2, Knoop) and, in the case of molybdenum, >39 GPa (4000 kgf/mm^2) (see Hardness). However, oxidation resistance is usually poor unless a subsequent coating is formed, such as silicidizing or chromizing, which imparts oxidation resistance. Silicides are generally very oxidation resistant, but not as hard as borides. Silicide coatings formed on molybdenum (51 μm in 3 h) at 675°C have superior oxidation resistance. At these low temperatures, the molybdenum substrate does not embrittle and the coatings are quite flexible.

The metals that can be aluminized act similarly with boron, as do the metals in group IVB; resistance to oxidation is the principal benefit. Titanizing and zirconizing are extremely sensitive to oxygen impurities, and when the salts are competely free of oxides and blanketed by high purity argon, excellent diffusion coatings can be formed in LiF at 900–1100°C. The most promising area for applications appears to be with nickel- and iron- based alloys where intermetallic-compound formation gives rise to many unique coatings that are tough and oxidation resistant. Beryllide coatings can be formed on approximately forty metals ranging from titanium to uranium and with compositions such as $TiBe_{12}$, Ni_5Be_{21}, and UBe_{13}. They are very hard, usually oxidation resistant, and easily formed up to several mils in thickness.

Moving further to the right in the periodic table, the scope of the metalliding processes becomes much more limited. Iron, cobalt, and nickel are restricted to approximately twelve metals into which they can be diffused. Iron can be diffused into cobalt and nickel with very good results, but the reverse is unsuccessful. The diffusion of nickel into molybdenum, tungsten, and copper has been very successful. Germanium occupies a slightly higher position in the electromotive series than nickel and therefore diffuses into nickel; it is below cobalt and iron, however, as determined by voltage measurements in the salts, and cannot be diffused into either of these metals, even at high voltages and current densities. It is readily diffused into molybdenum, palladium, copper, platinum, gold, and Monel.

Microstructure of Coatings

The microstructure of bulk coatings resembles the normal microstructure of metals and alloys produced by melt solidification. The microstructure of particulate-deposited materials resembles a cross between rapidly solidified bulk materials with severe deformation and powder compacts produced by pressing and sintering. A special feature of particulate coatings is a significant degree of porosity (ca 2–20 vol %) that strongly affects the properties of the deposit.

The microstructure and imperfection content of coatings produced by atomistic deposition processes can be varied over a very wide range to produce structures and properties similar to or totally different from bulk-processed materials. In the latter case, the deposited materials may have high intrinsic stress, high point-defect concentration, extremely fine grain size, oriented microstructure, metastable phases, incorporated impurities, and macro- and microporosity, all of which may effect the physical, chemical, and mechanical properties of the coating.

Microstructure

PVD Condensates. Physical-vapor-deposition condensates can deposit as single-crystal films on certain crystal planes of single-crystal substrates, ie, by epitaxial growth or, in the more general case, the deposits are polycrystalline. In the case of films deposited by evaporation techniques, the main variables are the nature of the substrates; the temperature of the substrate during deposition; the rate of deposition; and the deposit thickness. Contrary to what might be expected, the deposit does not initially form a continuous film of one monolayer and grow. Instead, three-dimensional nuclei are formed on favored sites on the substrates, such as cleavage steps on a single-crystal substrate. These nuclei grow laterally and in thickness (growth state), ultimately impinging on each other to form a continuous film. The average thickness at which a continuous film forms depends on the nucleation density and the deposition temperature and rate; both influence the surface mobility of the adatom. This thickness varies from 1 nm for Ni condensed at 15 K to 100 nm for Au condensed at 600 K.

The microstructure and morphology of thick single-phase films have been extensively studied for a wide variety of metals, alloys, and refractory compounds. The structure model first proposed is shown in Figure 7; it was subsequently modified as shown in Figure 8 (10–11).

At low temperatures, the surface mobility of the adatoms is limited and the

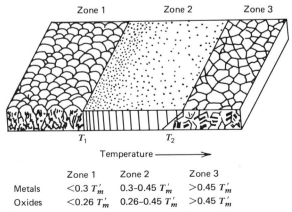

	Zone 1	Zone 2	Zone 3
Metals	$<0.3\,T_m'$	$0.3\text{-}0.45\,T_m'$	$>0.45\,T_m'$
Oxides	$<0.26\,T_m'$	$0.26\text{-}0.45\,T_m'$	$>0.45\,T_m'$

Figure 7. Structural zones in condensates at various substrate temperatures (10).

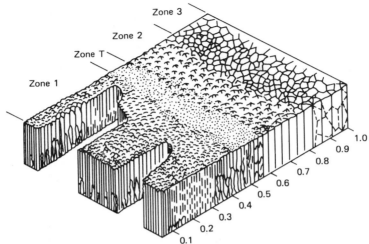

Figure 8. Structural zones in condensates showing the effect of gas pressure (11).

structure grows as tapered crystallites from a limited number of nuclei. It is not a full-density structure but contains longitudinal porosity on the order of a few tens of nm width between the tapered crystallites. It also contains numerous dislocations with a high level of residual stress. Such a structure has also been called botryoidal and corresponds to Zone 1 in Figures 7 and 8.

As the substrate temperature increases, the surface mobility increases and the structural morphology first transforms to that of Zone T, ie, tightly packed fibrous grains with weak grain boundaries, and then to a full-density columnar morphology corresponding to Zone 2 (see Fig. 8).

The size of the columnar grains increases as the condensation temperature increases. Finally, at still higher temperatures, the structure shows an equiaxed grain morphology, Zone 3. For pure metals and single-phase alloys, T_1 is the transition temperature between Zone 1 and Zone 2 and T_2 is the transition temperature between Zone 2 and Zone 3. According to the original model (10), T_1 is 0.3 T_m for metals and 0.22–0.26 T_m for oxides, whereas T_2 is 0.4–0.45 T_m for both (T_m is the melting point in K).

The modification shows that the transition temperature may vary significantly from those stated above and in general shift to higher temperatures as the gas pressure in the synthesis process increases. The transition from one zone to the next is not abrupt, but smooth. Hence, the transition temperatures should not be considered as absolute but as guidelines. Furthermore, not all zones are found in all types of deposit. For example, Zone T (Fig. 8) is not prominent in pure metals, but becomes more pronounced in complex alloys, compounds, or in deposits produced at higher gas pressures. Zone 3 is not seen very often in materials with high melting points.

CVD Coatings. As in PVD, the structure of the deposited material depends on the temperature and supersaturation, roughly as pictured in Figure 9 (12); however, in the case of CVD, the effective supersaturation (ie, the local effective concentration in the gas phase of the materials to be deposited, relative to its equilibrium concentration) depends not only on concentration, but on temperature, since the reaction is thermally activated. Since the effective supersaturation for thermally activated

Figure 9. Morphological effects of supersaturation and temperature on vapor-deposited materials (12).

reactions increases with temperature, the opposing tendencies can lead in some cases to a reversal of the sequence of crystalline forms listed in Figure 9, as temperature is increased (12).

Growth of columnar grains is characteristic of many materials in certain ranges of conditions. This structure results from uninterrupted growth toward the source of supply. Where growth in one crystallographic direction is preferred over others, grains having that orientation engulf those of other orientations.

Electrodeposits. *Columnar* structures are characteristic of deposits from solutions (especially acid solutions) containing no additives, high metal-ion concentration solutions with high deposition rates, or from low metal-ion-concentration solutions at low deposition rates. They usually exhibit lower tensile strength, elongation, and hardness than other structures, but are generally more ductile. Such deposits are usually of highest purity (high density) and low electrical resistivity.

Fibrous structures represent a grain refinement of columnar structure. Stress-relieving additives, eg, saccharin or coumarin, promote such refinement, as do high deposition rates. These may be considered intermediate in properties between columnar and fine-grained structures.

Fine-grained deposits are usually obtained from complex-ion solutions, eg, cyanide, or with certain addition agents. These deposits are less pure, less dense, and exhibit higher electrical resistivities because of the presence of foreign material.

Banded structures are characteristic of some alloy deposits and of bright deposits resulting from brightening addition agents. Plating-current modifications (P.R., periodic reverse, IC, interrupted current, pulse) favor the conversion of normal structure from a solution to a banded structure. These deposits generally possess higher tensile strength, hardness, and internal stress and lower ductility than the other structures.

Grain size varies widely, from 10 to 5000 nm; the grain size of fine-grained or banded deposits is usually 10–100 nm. Some metals, notably copper, nickel, cobalt and gold, can be deposited in all four types of grain structure, depending on the solution composition and plating conditions.

Characterization and Testing

Evaluation for high temperature service has been less reliable and less standardized than evaluations for conventional service. This is partly the result of the difficulty of reproducing in the laboratory the severe service conditions capable of yielding acceptable correlations. Test conditions do not necessarily simulate the geometrical and environmental conditions of the service. Nevertheless, screening tests yield primary behavioral parameters that usually define the limits of operation (see Analytical methods). Various tests are summarized in Tables 4 and 5.

Selection Criteria

The selection of a particular deposition process depends on the material to be deposited and its availability; rate of deposition; limitations imposed by the substrate, eg, maximum deposition temperature; adhesion of deposit to substrate; throwing power; apparatus required; cost; and ecological considerations. Criteria for CVD, electrodeposition, and thermal spraying are given in Table 6 (13).

Applications

Coatings can be classified into six categories: chemically functional, mechanically functional, optically functional, electrically functional, biomedical, and decorative. In addition, there are some unique applications in the aerospace program, such as the

Table 4.   Characterization of Refractory Coatings

Test	Factors evaluated
optical and electron microscopy metallographic and microscopic observation	substrate and coating structures; coating thickness; bond characteristics; detection of inclusions
physical electron microscope, SEM and TEM[a]	detection of injurious inclusions in substrate or coating
x-ray diffraction electron diffraction	effects of processing on coating composition; composition of coating at various depths within coating can be determined by controlled polishing followed by x ray
chemical microscopy edax aes[b] esca[c] sims[d]	chemical analysis of surface and subsurface layers; resolution can be as low as 5-nm segregation at imperfections
mechanical bend tests tensile tests	adhesion; ductility effect of coating or processing on base-material strength; ductility; elongation
microhardness transverse	cross-sectional hardness of coating and substrate; effect of processing on substrate; ductility
fatigue tests	effect of coating or coating process on substrate; fatigue properties; fatigue strength of coated-metal system
thermal chemical oxidation thermal cycling with sustained load in air	oxidation and thermal-shock resistance; effect of coating on ductility of base and of specimen creep on coating; integrity; thermal-shock and oxidation resistance
plasma-arc or oxyacetylene-torch test	coating emissivity; thermal shock and oxidation resistance; melting point

[a] Scanning electron microscopy; transmission electron microscopy.
[b] Auger electron spectroscopy.
[c] Electron spectroscopy for chemical analysis.
[d] Secondary-ion mass spectroscopy.

Table 5.   Nondestructive Tests for Refractory Coatings

Method	Factors evaluated
evaporgraph[a]	discontinuities on flat panel surfaces[c]
electromagnetic inspection	pores, cracks, pits[c]
ultrasonic inspection	surface and subsurface flaws; unsuitable for thin skins[c]
fluorescent particle inspection[b]	surface cracks, pits, and similar coating defects
red-dye penetrant	surface flaws
radiographic inspection	small coating flaws in assembled structures; sensitivity is difficult to control
microscopy	flaws by observation of oxidation and weakness in bond strength resulting from thermal shock[d]

[a] Detection of residual moisture left behind in cracks, pores, or flaws.

[b] Finely divided, fluorescent-coated magnetic particles are attracted to and outline the pattern of any magnetic-leakage fields created by discontinuities.

[c] Unsuitable for corner or edge defects.

[d] Very reliable after an exposure test.

ablative coatings of pyrolytic carbon and graphite- and silica-based materials for protection of nose cones and the space shuttle during reentry (see Ablative materials). Another unique energy related application is the coating of low-Z elements such as TiC for the first wall of thermonuclear reactors to minimize contamination of the plasma.

Chemically Functional.   Refractory coatings are used for corrosion-resistant high temperature service in gas turbine and diesel engines, components such as crucibles, thermocouple protection tubing, valve parts, etc.

Blades and vanes used in the hot-end of a gas turbine are subject to high stresses in a highly corrosive environment of oxygen, sulfur, and chlorine-containing gases. A single or monolithic material such as a high temperature alloy cannot provide protection against both. A bulk alloy designed for its mechanical properties provides the corrosion resistance by means of an overlay coating of an M–Cr–Al–Y alloy where M stands for Ni, Co, Fe or Ni + Co. In production, the coating is deposited by electron-beam evaporation; in the laboratory, by sputtering or plasma-spraying. These overlay coatings have several advantages over diffusion aluminide coatings. The latter lose their effectiveness at higher temperatures because of interactions with the substrate. The composition and properties of overlay coatings can be more easily tailored to the needs of specific applications. In addition, coatings of stabilized zirconia are used as thermal barriers in diesel engines and gas turbines, experimentally at this time, to raise operating temperature of the engine and protect it from corrosive fuels.

Boron nitride and titanium diboride coatings are used on graphite for the evaporation of liquid aluminum.

Mechanically Functional.   Refractory coatings are used in engine parts, landing gears, soft-film lubricants, and cutting and forming tools (see Tool materials).

A large and rapidly growing application is the coating of cutting and forming tools and industrial knives with carbides, nitrides, oxides, or multiple layers of these materials. These coatings are deposited by CVD and PVD methods and increase the tool life by factors ranging from 2 to 10, depending on the operating conditions. Cermet coatings such as TiB_2–Ni are also deposited electrolytically to provide wear resistance.

Table 6. Characteristics of Deposition Processes

Characteristic	Evaporation	Ion plating	Sputtering	Chemical vapor deposition	Electro-deposition	Thermal spraying
mechanism of production of depositing species	thermal energy	thermal energy	momentum transfer	chemical reaction	deposition from solution	deposition from flames or plasmas
deposition rate	can be up to 75,000 nm/min	can be up to 25,000 nm/min	low except for pure metals[a]	moderate, 20–2500 nm/min	low to high	very high
depositing specie	atoms and ions	atoms and ions	atoms and ions	atoms	ions	droplets
throwing power for complex shaped object	poor line-of-sight coverage except by gas scattering	good, but thickness distributions nonuniform	good, but thickness distribution nonuniform	good	good	none
into small blind holes	poor	poor	poor	limited	limited	very limited
deposition of metal	positive	positive	positive	positive	positive but limited	positive
alloy	positive	positive	positive	limited	limited	positive
refractory compound	positive	positive	positive	positive	limited	positive
energy of depositing species	ca 0.1–0.5 eV	1–100 eV	1–100 eV	high with PACVD	can be high	can be high
bombardment of substrate and deposit by inert gas ions	normally not	yes	possible, depending on geometry	possible	none	positive
growth interface perturbation	normally not	yes	yes	yes, by rubbing	none	none
substrate heating by external means	normally yes	yes	generally not	no	none	normally not

[a] For copper, 1000 nm/min.

61

These coatings are also employed as solid lubricants in engine components. Refractory materials such as MoS_2 and WSe_2 are lamellar compounds and provide very effective solid-state lubrication in spacecraft bearings and components used in radiation environments where conventional organic liquid lubricants are not stable (see Lubrication and lubricants).

Optically Functional. Laser optics, layer architectional glass panels (up to 3×4.3 m), lenses, TV-camera optical elements, and similar applications require optically functional coatings.

A large and expanding operation is coatings on architectural glass panels used in buildings to alter the transmission and reflection properties of glass in various wave-length ranges and thus conserve energy usage. Furthermore, attractive visual effects, such as a bronze appearance, can be achieved. These coatings consist of multiple layers of oxides and metals (the precise compositions are proprietary) and are deposited in large in-line sputtering and evaporation systems. Future applications will utilize the selective transmission of coatings of nitrides, carbides, and borides in solar-thermal applications.

Another growing application that overlaps the electrically functional area is the use of transparent conductive coatings or tin oxide, indium–tin oxide, and similar materials in photovoltaic solar cell and various optic electronic applications (see Photovoltaic cells). These coatings are deposited by PVD techniques as well as by spray-pyrolysis, which is a CVD process.

Electrically Functional. Refractory coatings are used in semiconductor devices, capacitors, resistors, magnetic tape, disk memories, superconductors, solar cells, and diffusion-barriers to impurity contamination from the substrate to the active layer.

Thin-film capacitors and resistors contain such dielectric materials as silicon monoxide and dioxide, tantalum oxide, silicon nitride, and the like. These coatings are deposited by a variety of atomic deposition techniques (PVD, CVD) and thick-film methods. In some cases coatings, eg, SiO, are formed by thermal oxidation. An important application is the deposition of passivating layers of SiO_2 or SiO_2–Si_3N_4 by plasma-assisted CVD techniques. A new area is insulation coatings for GaAs devices formed by plasma-assisted oxidation methods and the deposition of amorphous silicon by CVD techniques. The fabrication of tungsten-emitter elements for thermionic convertors by CVD is another application.

Biomedical. Heart-valve parts are fabricated from pyrolytic carbon, which is compatible with living tissue. Such parts are produced by high temperature pyrolysis of gases such as methane. Other potential biomedical applications are dental implants and other prostheses where a seal between the implant and the living biological surface is essential (see Prosthetic and biomedical devices).

Decorative. Titanium nitride has a golden color and is used extensively to coat steel and cemented carbide substrates for watch cases, watch bands, eyeglass frames, etc. It provides excellent scratch resistance as well as the desired esthetic appearance, and it replaces gold coatings used previously.

Economic Aspects

Diffusion aluminide and silicide coatings on external and internal surfaces for high temperature corrosion protection in parts such as gas-turbine blades is estimated at 40×10^6/yr in North America and about 50×10^6 worldwide.

Overlay coatings onto gas-turbine blades and vanes of M–Cr–Al–Y type alloys by electron-beam evaporation is estimated at 10×10^6 to coat 200,000 parts at an average cost of $50 per part.

Hard facing of various components in the aircraft gas-turbine engine and in industrial applications for textile machinery parts, oil and gas-machinery parts, paper-slitting knives, etc, is estimated at 400×10^6 in 1978 with an estimated growth rate of 7–10% annually. The mix is approximately 60% aerospace applications, 40% industrial applications. Additionally, repair coatings for gas-turbine blades and vanes is estimated at 300×10^6. These coatings are primarily deposited by plasma-spray and detonation-gun techniques.

Refractory compound coatings of carbides, nitrides, and oxides on cemented-carbide cutting tools, mainly by the CVD process, is estimated at 200×10^6 annually worldwide.

A current application is the coating of complex shaped high speed steel cutting tools such as drills, holes, gear-cutters, etc, with a titanium nitride coating deposited by the methods of plasma-aided reactive evaporation and ion plating. A very rough estimate of the add-on value of the coating is $(5–10) \times 10^6$ in 1981 but with a very rapid growth rate of 100–300% expected over the next few years. The same processes are also used to deposit gold-colored wear resistant titanium nitride coatings on watch bezels, watchbands and other decorative jewelry items. A very rough estimate is 5×10^6 annually. The last two applications are practiced principally in Japan, Europe, and the United States.

Finally, refractory coatings are extensively used in the semiconductor and optical-materials industry. The annual add-on value of such coatings exceeds 10^9.

BIBLIOGRAPHY

"Refractory Coatings" in *ECT* 2nd ed., Vol. 17, pp. 217–284, by Bruno R. Miccioli, The Carborundum Company.

1. H. H. Hausner, ed., *Coatings of High Temperature Materials*, Plenum Publishing Corp., New York, 1966.
2. R. F. Bunshah and D. M. Mattox, *Phys. Today* **33**, 50 (May 1980).
3. R. F. Bunshah in *New Trends in Materials Processing*, American Society of Metals, Metals Park, Ohio, 1974, pp. 200–269.
4. R. F. Bunshah and A. C. Raghuram, *J. Vac. Sci. Technol.* **9**, 1385 (1972).
5. W. Grossklaus and R. F. Bunshah, *Int. Vac. Sci. Technol.* **13**, 532 (1975).
6. M. Kobayashi and Y. Doi, *Thin Solid Films* **54**, 57 (1978).
7. K. K. Yee, *Int. Metal. Rev.* 1(226), 19 (1978).
8. T. Bonifield, "Plasma Assisted Chemical Vapor Deposition" in R. F. Bunshah, ed., *Films and Coatings for Technology*, Noyes Data Corp., Park Ridge, N.J., 1982, Chapt. 9.
9. U.S. Pats. 3,024,175–3,024,177, 3,232,853 (Mar. 6, 1962), N. C. Cook (to General Electric Co.).
10. B. A. Movchan and A. V. Demchishin, *Fizika Metall.* **28**, 653 (1969).
11. J. A. Thornton, *J. Vac. Sci. Technol.* **12**, 830 (1975).
12. J. A. Blocher, "Chemical Vapor Deposition" in ref. 8, Chapt. 8.
13. R. F. Bunshah and co-workers in ref. 8.

General References

Ref. 8 is a general reference.
H. H. Hausner, ed., *Coatings of High Temperature Materials*, Plenum Publishing Corp., New York, 1966.

J. Huminik, Jr., ed., *High-Temperature Inorganic Coatings*, Reinhold Publishing Corp., New York, 1963.

D. H. Leeds, "Coatings on Refractory Metals" in J. E. Hove and W. C. Riley, eds., *Ceramics for Advanced Technologies*, John Wiley & Sons, Inc., New York, 1965, Chapt. 7.

F. A. Lowenheim, ed., *Modern Electroplating*, 2nd ed., John Wiley & Sons, Inc., New York, 1963.

C. F. Powell, J. H. Oxley, and J. M. Blocher, Jr., *Vapor Deposition*, John Wiley & Sons, Inc., New York, 1966.

R. F. Bunshah, ed., *Techniques of Metals Research*, Vol. I, Parts 1–3; Vol. VII, Part 1, John Wiley & Sons, Inc., New York, 1968.

L. Holland, *Vacuum Deposition of Thin Films*, Chapman & Hall, 1968.

L. I. Maissel and R. Glang, *Handbook of Thin Film Technology*, McGraw-Hill, Inc., New York, 1970.

B. Chapman and J. C. Anderson, eds., *Science and Technology of Surface Coatings*, Academic Press, Inc., New York, 1974.

J. L. Vossen and W. Kern, eds., *Thin Film Processes*, Academic Press, Inc., New York, 1978.

R. F. Bunshah, *Materials Coating Techniques*, Agard Lecture Series, No. 106, NATO, 1980.

A. R. Reinberg, "Plasma Deposition of Their Films," *Annual Rev. Mat. Sci. Technol.* **9,** 341 (1979).

R. W. Haskell and J. G. Byrne in H. Herman, ed., "Studies in Chemical Vapor Deposition," in *Treatise on Materials Science and Technology*, *VI*, Academic Press, Inc., New York.

R. Bakish, ed., "Electron and Ion Beam Science and Technology," a series of International Conferences.

R. Sard, H. Leidheiser, Jr., and F. Ogburn, eds., *Properties of Electrodeposits—Their Measurement and Significance*, Electrochemical Society, Princeton, N.J.

D. L. Hildenbrand and D. D. Cubicciotti, eds., *High Temperature Metal Halide Chemistry*, a 1977 symposium.

R. G. Frieser and C. J. Mogab, eds., *Plasma Processing*, a 1980 symposium.

K. K. Yee, "Protective Coatings for Metals by Chemical Vapor Deposition," *Int. Met. Rev.* **1,** 19 (1978).

K. C. Mittal, ed., *Adhesion Measurements of Thin Films, Thick Films and Bulk Coatings*, ASTM STP **640,** (1978).

J. I. Duffy, ed., *Electroless and Other Non-electrolytic Plating Techniques*, Noyes Data Corp., Park Ridge, N.J., 1980.

W. Kern and G. L. Schnable, "Low Pressure Chemical Vapor Deposition for Very Large Scale Integration Processing—A Review," *IEEE Trans.* **ED-26,** 647 (1979).

G. Wahl and R. Hoffman, "Chemical Engineering with CVD," *Rev. Int. Temp. Refract. Fr.* **17,** 7 (1980).

R. S. Holmes and R. G. Loasby, *Handbook of Thick Film Technology*, Electrochemical Publications, 1976.

"Heat Treating, Cleaning and Finishing, *ASTM Metals Handbook*, Vol. 2, American Society for Metals, 1964.

K. L. Mittal, ed., *Surface Contamination: Genesis, Detection and Control*, Vols. 1 and 2, Plenum Press, Inc., New York, 1979.

Journals

Thin Solid Films, Elsevier Sequoia S.A., Lausanne, Switzerland.
Journal Vac. Sci. Technol., American Vacuum Society.
Proceedings of International Conference on Chemical Vapor Deposition, Electrochemical Society.
J. Electrochem. Soc.

Rointan F. Bunshah
University of California, Los Angeles

REFRACTORY FIBERS

The term refractory fiber defines a wide range of amorphous and polycrystalline synthetic fibers used at temperatures generally above 1093°C (see also Fibers, chemical; Refractories). Chemically, these fibers can be separated into oxide and nonoxide fibers. The former include alumina–silica fibers and chemical modifications of the alumina–silica system, high silica fibers (>99% SiO_2), and polycrystalline zirconia, and alumina fibers. The diameters of these fibers are 0.5–10 μm (av ca 3 μm). Their length, as manufactured, ranges from 1 cm to continuous filaments, depending upon the chemical composition and manufacturing technique. Such fibers may contain up to ca 50 wt % unfiberized particles. Commonly referred to as shot, these particles are the result of melt fiberization usually associated with the manufacture of alumina–silica fibers. The presence of shot reduces the thermal efficiency of fibrous systems. Shot particles are not generated by the manufacturing techniques used for high silica and polycrystalline fibers, and consequently, these fibers usually contain <5 wt % unfiberized material. Refractory fibers are manufactured in the form of loose wool. From this state, they can be needled into flexible blanket form, combined with organic binders and pressed into flexible or rigid felts, fabricated into rope, textile, and paper forms, and vacuum-formed into a variety of intricate, rigid shapes.

The nonoxide fibers, silicon carbide, silicon nitride, boron nitride, carbon, or graphite [7782-42-5] have diameters of ca 0.5–50 μm. Generally, nonoxide fibers are much shorter than oxide fibers except for carbon, graphite, and boron fibers which are manufactured as continuous filaments. Carbon, graphite, and boron fibers are used for reinforcement in plastics in discontinuous form and for filament winding in continuous form. In addition to temperature resistance, these fibers have extremely high elastic modulus and tensile strength. Carbon and graphite fibers cannot be accurately classified as refractory fibers because they are oxidized above ca 400°C. This is also true of boron fibers which form liquid boron oxide at approximately 560°C. Most silicon carbide, silicon nitride, and boron nitride fibers are relatively short, ranging in length from single-crystal whiskers less than 1 mm to fibers as long as 5 cm. Diameters are ca 0.1–10 μm. These fibers are used mostly to reinforce composites of plastics, glass, metals, and ceramics. They also have limited application as insulation in and around rocket nozzles and in nuclear fusion technology requiring resistance to short-term temperatures above 2000°C. For inert atmosphere or vacuum applications, carbon-bonded carbon fiber composites provide effective thermal insulation up to 2500°C.

Until the early 1940s, natural asbestos was the principal source of insulating fibers (see Asbestos). In 1942, Owens-Corning Fiberglass Corporation developed a process for leaching and refiring E glass fibers to give a >99% pure silica fiber (1). This fiber was used widely in jet engines during World War II. Shortly after the war, the H.I. Thompson Company promoted insulating blankets for jet engines under the trade name Refrasil (2). The same material was manufactured in the UK by the Chemical and Insulating Company, Ltd. (3), and a similar material called Micro-Quartz was developed by Glass Fibers, Inc. (now Manville Corporation) from a glass composition specifically designed for the leaching process (4). Leached-glass fibers can be produced with extremely fine diameters using flame attenuation (see Glass).

Several techniques were developed by Engelhard Industries, Inc. in the early 1960s for manufacturing fused silica fibers by rod-drawing techniques (5). High thermal-efficiency insulating felts with very low densities are produced by rod-drawing. They are essentially free of shot, and were used as blankets for jet engines and space vehicles (see Ablative materials). High purity silica fiber was chosen for the tiles which form the first reusable thermal-protection system on the space shuttle Orbiter. Many of the leading edge and underside areas on the Orbiter are insulated with replaceable carbon-bonded carbon–fiber tile capable of protecting the metal skin beneath from the 1200–1800°C temperatures developed in those areas. In the early 1940s techniques for blowing molten kaolin [1318-74-7] clay ($Al_2O_3.2SO_2$) into fibers were also developed by Babcock & Wilcox Company (6) (see Clays). At about the same time, it was discovered that by blowing molten mullite [1302-98-8] ($3Al_2O_3.2SiO_2$) with air and steam jets, a portion of the melt is converted into glassy fibers (7).

The Johns-Manville Corporation became interested in alumina–silica fibers in 1946, establishing a cooperative development with Carborundum. Chemical modifications of the basic alumina–silica composition have resulted in three materials of interest. In the Carborundum Company process addition of ca 5 wt % zirconia to the basic 50 wt % alumina–50 wt % silica system produces a longer fiber for use at 1260°C. In 1965, Johns-Manville developed a chromia-modified, alumina–silica fiber with a temperature limit of 1427°C (8). More recently, the 3M Company has introduced a boria-modified alumina–silica fiber for continuous use at 1427°C. This fiber, trade named Nextel, is particularly suitable for fabrication into flexible high temperature textiles.

The next technological advance was the development of the precursor process by Union Carbide Corporation. An organic polymeric fiber is used as a precursor to absorb dissolved metal oxides. A subsequent heat treatment burns out the organic fiber, leaving a polycrystalline refractory metal oxide in the shape of the host polymeric material (9). With this technique Union Carbide developed the first commercially available 1650°C zirconia fiber, trade-named Zircar. At about the same time, Babcock & Wilcox, along with others, discovered that inorganic fibers could be made by spinning, blowing, or drawing a viscous, aqueous solution of metal salts into fibers followed by a heat treatment to convert the salts to the oxide form. The resulting fibers were polycrystalline and usually contained 5–10% porosity. This process allowed the fiberization of metallic oxides and combinations of metallic oxides, whose viscosity in the molten state would not allow fiberization by ordinary methods.

From this new technology, Babcock & Wilcox developed the first true mullite fiber, which was used experimentally in the development of tile insulation for the space shuttle Orbiter. In 1972, ICI introduced the first commercial fibers made by the sol process. Under the trade names Saffil alumina and Saffil zirconia, fibers were offered for continuous use at 1400 and 1600°C, respectively. The sol process eliminated the need for the relatively costly organic precursor fiber and thereby significantly reduced the cost of manufacturing relatively pure oxide fibers. The zirconia fiber, however, was dropped from production.

During the mid-1960s, ultralightweight high strength composites replaced traditional metallic structural members and other parts of the early space vehicles, and cast metals were reinforced with short fibers consisting of alumina, silicon carbide, silicon nitride, or boron nitride (see High temperature composites). Their strengths are 1.4–20 GPa (($2–29$) $\times 10^5$ psi) with elastic moduli of 70–700 GPa (ca ($1–10$) $\times 10^7$

psi). Silicon carbide whiskers (single-crystal fibers) are of special interest because they offer not only high strength and stiffness but are useful up to 1800°C. Numerous companies worldwide became involved in carbide–fiber research. In many manufacturing processes these fibers are grown as reaction products of gaseous silicon monoxide and carbon monoxide at high temperatures. In some cases, carbon fibers were allowed to react with silicon monoxide. By 1968, both Thermokinetic Fibers, Inc. and Carborundum sold short silicon carbide fibers at ca $500/kg. A recent innovation is the production of high quality short silicon carbide fibers by the pyrolysis of rice hulls in an inert or ammonia atmosphere.

Boron nitride exhibits exceptional resistance to thermal shock and can be used in inert or nonoxidizing atmospheres to 1650°C (11). In a Carborundum process, boron oxide fibers are converted to polycrystalline boron nitride at elevated temperatures in an ammonia atmosphere. A 1971 Lockheed Aircraft patent describes a similar process in which a boron filament is heated first in air and then converted to the nitride in a nitrogen atmosphere (12).

Properties

Refractory-fiber insulating materials are generally used in applications above 1063°C. Table 1 gives the maximum long-term use temperatures in both oxidizing and nonoxidizing atmospheres. For short exposures, some of these fibers can be used at temperatures much closer to their melting temperature without degradation.

The most important properties of these fibers are thermal conductivity, resistance to thermal and physical degradation at elevated temperatures, and tensile strength and elastic modulus. The thermal conductivity, in W/(cm·K), is actually the inverse of the R factor that expresses the insulating value of building insulations (see Insulation, thermal). Fibrous insulations with increasingly lower thermal conductivities require reliable test methods, and currently ASTM C 201 and ASTM C 177 are the standard methods. In addition, ASTM C 892-78 defines maximum thermal conductivities and unfiberized contents. Thermal conductivity is affected by the bulk density of the material, the fiber diameter, the amount of unfiberized material (shot), and the mean temperature of application. Typical silica fibers have a mean fiber diameter of

Table 1. Maximum Use Temperatures of Refractory Fibers

Fiber type	CAS Registry Number	Mp, °C	Maximum use temperature, °C Oxidizing atmosphere	Nonoxidizing atmosphere
Al_2O_3	[1344-28-1]	2040	1540	1600
ZrO_2	[1314-23-4]	2650	1650	1650
SiO_2	[7631-86-9]	1660	1060	1060
Al_2O_3–SiO_2	[37287-16-4]	1760	1300	1300
Al_2O_3–SiO_2–Cr_2O_3	[65997-17-3]	1760	1427	1427
Al_2O_3–SiO_2–B_2O_3		1740	1427	1427
C	[7440-44-0]	3650	400	2500
B	[7440-42-8]	1260	560	1200
BN	[10043-11-5]	2980	700	1650
SiC	[409-21-2]	2690	1800	1800
Si_3N_4	[12033-89-5]	1900	1300	1800

1–2 μm and less than 5% unfiberized material. Alumina–silica fibers range in mean diameter from 2.5 to 5.5 μm and have 20–50 wt % shot, which significantly reduces their thermal efficiency (increases thermal conductivity). Alumina and zirconia fibers are manufactured with a uniform diameter of ca 3 μm and contain less than 5% unfiberized particles. With such a wide range of shot contents and fiber diameters, the thermal conductivity of equivalent bulk-density blankets or felts made from different fiber can differ significantly. A thermal conductivity-versus-mean-temperature plot is given for several fibers in Figure 1. The effect of density on the thermal conductivity of silica and alumina–silica fibers is shown in Figure 2. For reinforcing purposes, tensile strength measured at room temperature and stress-to-strain ratio measured as Youngs modulus are of primary importance (see Table 2). Alumina-silica fibers are not used as reinforcements, and consequently, strength and modulus measurements are not important.

At the relatively low densities of these materials, solid conduction of heat is negligible when considering total heat transfer (see Heat-exchange technology). Heat transfer is affected mainly by radiation, which at 1260°C, is reponsible for approximately 80% of all heat transfer. Heat transfer by convection is small, and gas conduction within the insulation is responsible for the balance of heat transfer through the material. Consequently, most efficient insulations are those that effectively block radiation. The temperature ratings given these fibers are generally the temperatures at which the linear shrinkage stabilizes within 24 h and does not exceed 5%. The 5% linear shrinkage value has become a design factor for furnace insulation and other high temperature applications.

Alumina–silica and leached-glass fibers have been converted from amorphous glass to their crystalline forms. The rate of crystallization, sometimes called devitrification, depends on time and temperature (see Crystallization). For some time, the degradation of mechanical strength of refractory fiber products was attributed to this crystallization. In alumina–silica fibers, crystallization is completed relatively quickly,

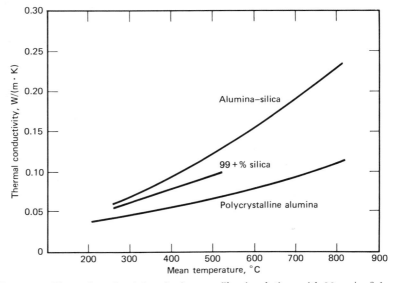

Figure 1.   Thermal conductivity of refractory fiber insulations with 96 mg/cm³ density.

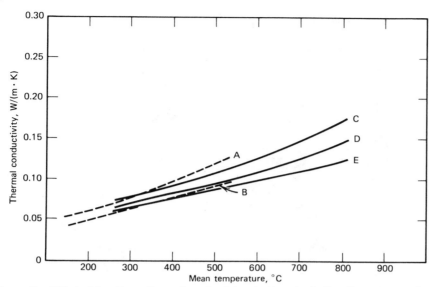

Figure 2. Effect of density on thermal conductivity. A, 48 mg/cm³ silica fiber; B, 96 mg/cm³ silica fiber; C, 128 mg/cm³ alumina–silica fiber; D, 192 mg/cm³ alumina–silica fiber; E, 384 mg/cm³ alumina–silica fiber.

within 24 h, at rated use temperatures; further degradation of physical properties with time is due to grain growth in the fiber (13). Sintering, the bonding together of these fibers at temperatures well below their softening point, is also responsible for linear shrinkage. At contact points, solid-state diffusion of molecules has a bridging effect whereby previously flexible material is transformed into a rigid structure. A scanning electron micrograph of alumina–silica–chromia fiber after 120 h exposure to 1427°C is shown in Figure 3. Both the formation of mullite and cristobalite crystals in the originally amorphous fiber and the sintering at fiber intersections can be clearly seen. The diffusion of oxides in the fibers toward the contact point has a slight shortening effect upon the fiber that ultimately contributes to the overall shrinkage of the product. In general, with increased sintering the resistance to abrasion or mechanical vibration decreases.

Table 2. Mechanical Properties of Oxide and Nonoxide Fibers

Fiber type	Density, g/cm³	Tensile strength, GPa[a]	Young's modulus, GPa[a]
SiO_2	2.19	5.9	72
Al_2O_3	3.15	2.1	170
ZrO_2	4.84	2.1	345
carbon	1.50	1.4	210
graphite[b]	1.66	1.8	700
BN	1.90	1.5	90
SiC[b]	3.21	2.0	480
Si_3N_4[b]	3.18	1.4	380

[a] To convert GPa to psi, multiply by 145,000.
[b] Single-crystal whiskers.

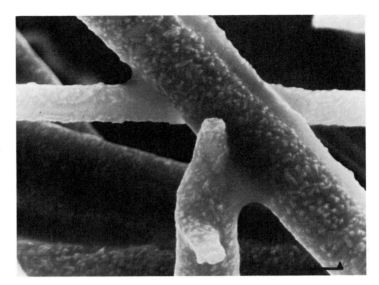

Figure 3.   Alumina–silica–chromia fiber after 120 h at 1426°C showing crystallization and sintering at contact points. ×5000.

Alumina–Silica Fibers

Current refractory-fiber production consists mainly of melt-fiberized alumina–silica, and modified alumina–silica fibers. The 1260°C-alumina–silica fibers are produced by melting high purity alumina and silica or calcined kaolin in electric resistance furnaces. In either case, the resulting fiber contains ca 50 ± 2 wt % SiO_2 and Al_2O_3. The higher temperature grade, for use to 1427°C, is made by adding ca 3 wt % Cr_2O_3 to the basic composition or increasing the Al_2O_3 content to 55–60 wt %.

The raw materials are melted in a three-phase electric furnace that operates on the electrical resistance of the pool of molten material in which the electrodes are immersed. The initial melt is established by passing a current through graphite or coke granules that ultimately burn off, leaving the molten alumina–silica to serve as the conductive medium. Graphite is the common electrode material, although refractory metals, such as molybdenum and tungsten, are also used. The latter must be kept submerged in the molten pool to prevent oxidation. Some electrode loss occurs, but the refractory metal electrodes are generally not attacked by long-term use (14). The molten material is discharged through a temperature-controlled orifice at the bottom of the vessel (14–15), or the furnace is tilted to deliver the melt via a refractory trough to the fiberizing unit.

In the steam-blowing process, the liquid material is dropped in the path of a high velocity blast from a steam jet. Air jets are also used for fiberizing. Normally, the fiberizing pressures in the blowing processes are ca 700 kPa (ca 100 psi), but they can be much higher. The steam fiberizing of molten materials is an art long associated with the manufacture of mineral–wool insulations from low temperature melts. Ultra-high-speed photography has revealed that the blast initially shreds the molten stream into tiny droplets. As each droplet picks up momentum from the velocity of the steam blast, it elongates into a teardrop shape and attenuates into a fiber with a spherical globule of molten material as its head, called shot. Several examples of fiber with the spherical shot particles still attached are shown in Figure 4. Each fiber is therefore the tail from a droplet of nonfibrous material. However, the amount of fiber in the crude product on a volume basis is very large compared with that of nonfibrous material.

In the melt-fiberization or spinning process, the molten material is dropped on the periphery of a vertically oriented, rotating disk. The molten material bonds to the high-speed rotating disk surface from which the melt droplets are ejected. The fiber is attenuated both by centrifugal action on the ejected droplet and the fact that the fiber tail frequently is still attached to the disk's surface. Generally, spinning produces a longer fiber than steam blowing, and converts more of melt to fiber, depending on the melt-delivery rate.

High Purity Silica Fibers

Refractory fibers of essentially >99 wt % silica are not made by conventional melt-fiberizing techniques. Thus, to achieve the high purity required for operation at 1093–1260°C, a leaching process is employed (16). A glass of ca 75 wt % SiO_2 and 25 wt % Na_2O is melted in a typical glass furnace at ca 1100°C. Filaments with a diameter of ca 0.3 mm are drawn from orifices in the furnace bottom. These filaments are then passed in front of a gas flame and attenuated to a diameter of approximately 1.5 μm. The loose fibers are subjected twice to an acid-leaching process to remove the Na_2O. The fiber is thoroughly rinsed, and dried at ca 315°C. The resulting fiber has a maximum of 0.01 wt % Na_2O and K_2O, 0.20 wt % Al_2O_3, and 0.04 wt % MgO and CaO.

The direct manufacture of pure silica fibers requires either a fused or extruded silica particulate rod of 5–25-mm dia. The fused silica rods with diameters of 6–7 mm are formed from molten high purity silica and mounted in groups of 20 on a motor-driven carriage. The rods are reduced to ca 2 mm in dia by being forced through a graphite guide and over a vertically oriented oxyhydrogen burner operating at 1800°C.

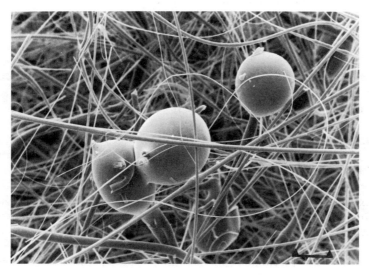

Figure 4.   Refractory fiber (1260°C) and unfiberized shot particles. ×1000.

These relatively large fibers are passed through another graphite block containing 20 holes. As the fibers exit, they are hit by an axial oxyhydrogen jet attenuating the fibers toward a rotating drum which collects them. Fibers with a diameter of 4–10 μm and a silica content of 99.95 wt % are produced (17).

Chemically Produced Oxide Fibers

Ceramic oxide fibers are difficult to produce by common melt technology because oxides such as zirconia and alumina have very high melting points and low viscosities (see also Ceramics). In 1969, Union Carbide marketed the first 1650°C zirconia refractory fiber manufactured by a chemical technique termed the precursor process. The precursor can be any organic fiber containing extremely small crystallites of polymer chains held together in a matrix of amorphous polymer. When immersed in a solvent such as water, the fiber swells, thus expanding the spaces between the crystallites. Although a number of organic fibers have this swelling characteristic, eg, wool, cotton, and cellulose acetate, rayon is preferred because of its structural uniformity and high purity. After an initial swelling and dewatering by centrifuge, the rayon fiber is immersed in a 2 M aqueous solution of zirconyl chloride containing a small amount of yttrium salt. The excess solution is centrifuged, and the fibers are dried and heated to ca 400°C in an atmosphere containing <10 vol % oxygen. This treatment pyrolyzes the rayon to a carbonaceous residue; the zirconyl chloride is converted to microcrystalline zirconia fiber containing yttrium oxide for phase stabilization and to prevent embrittlement.

Imperial Chemical Industries produces 1600–1650°C fibers by the sol process (18). Both silica-stabilized alumina and calcia-stabilized zirconia fibers are marketed as Saffil. First a metal salt, such as aluminum oxychloride in the case of alumina fiber, is mixed with a medium molecular weight polymer such as 2 wt % poly(vinyl alcohol). This solution is then slowly evaporated in a rotary evaporator to a viscosity of ca 80 Pa·s (800 P). This solution is then extruded through a 100-μm spinneret; the fibers are collected on a drum where they are fired to a temperature of 800°C. This action burns the organic component away and a fine-grained aluminum oxide fiber is formed with a porosity of 5–10 vol % and a fiber diameter of 3–5 μm. These fibers are used as filter media because of their inherent porosity. For refractory application, they are heated to 1400–1500°C, just long enough to eliminate the porosity, which would result in 3–4% linear shrinkage in application. The same process is followed for zirconium oxide fibers, starting with zirconium oxychloride, zirconium acetate, and calcium oxide. The only truly continuous alumina–silica-type fiber is made by the 3M Company employing a similar solution process (19). In this case, a 10-μm dia alumina–silica–boria fiber is manufactured for use to 1427°C. Basic aluminum acetate is dissolved in water and the solution is mixed with an aqueous dispersion of colloidal silica and dimethylformamide. The resulting solution is concentrated in a Rotavapor flask and centrifuged. The solution is then extruded through a 75-μm spinneret at 100 kPa (1 atm). The resulting fibers are collected on a conveyor chain and passed through a furnace at 870°C converting the filaments to metallic oxides. Heating in another furnace at 1000°C produces a glassy aluminum borosilicate [12794-54-5] with the calculated composition $3Al_2O_3 \cdot B_2O_3 \cdot 3SiO_2$.

Although fiber production by this technique is more expensive than melt fiberization, the sol process offers many other advantages. In melt fiberization, the effect upon viscosity and surface tension must be taken into account when adding even small amounts of other desirable oxides. In the sol process, however, the viscosity is controlled independently of the metals added and thus any number of metal salts can be easily added without adverse effect. The controlled addition of other metals can also serve as grain-growth inhibitors, sintering aids, phase stabilizers, or catalysts.

Nonoxide Refractory Fibers

The most important nonoxide refractory fibers are silicon carbide fibers and single-crystal whiskers. In the early work on silicon carbide, 1 part by volume SiO was mixed with 3 parts CO at 1300–1500°C in an inert or reducing carrier gas. The fibers formed on the colder parts of the reactor tube had 4–6-μm dia and were 50-mm long (20). In a similar but less complicated technique developed by Corning Glass Works, the SiO and CO gases are obtained by heating a mixture of carbon and silica in a molar ratio of 2:1 to 1300–1550°C in a hydrogen fluoride or hydrogen chloride atmosphere (21). The fibers grow more quickly than in an inert carrier gas and the process can be run continuously. These fibers are made up of beta silicon carbide crystals with a surface sheath of silica which prevents oxidation.

Since 1960, a number of processes for high quality silicon nitride fibers have been developed. These fibers are a by-product in the production of silicon nitride powder from silicon metal and nitrogen at high temperatures. Additions of a reducing agent greatly increased the fiber yield by converting silica first to silicon monoxide and then to silicon metal for the reaction with nitrogen (22).

$$SiO_2 + R \rightarrow SiO + RO$$

$$3\,SiO + R + 2\,N_2 \rightarrow 3\,RO + Si_3N_4$$

R is the reducing agent. Any silicate that forms thermally and chemically stable residual compounds as its SiO_2 content is reduced and provides a suitable source of silicon for the reaction. Alternate aluminum–silica and graphite plates are stacked in a graphite-lined alumina tube where the plates are separated by 2–4-cm graphite spacers. This tube is heated to 1400°C for 12 h in a nitrogen atmosphere in a silicon carbide-resistance furnace. After approximately 6 h the tube is cooled and the fibers are removed.

Boron nitride fibers are produced by nitriding boron filaments obtained by chemical vapor deposition of boron on heated tungsten wire. The tungsten wire is passed through a reactor containing boron trichloride in a carrier of hydrogen at 1100–1300°C. The boron trichloride is reduced and boron is deposited on the tungsten wire. The continuous boron filament exits the furnace and is wound upon a spool. In nonoxidizing atmospheres, boron filaments can be considered refractory fibers. When heated to 560°C, they develop a liquid boron oxide surface coating. For pure boron nitride fiber, the boron oxide filaments are further heated to 1000–1400°C in an ammonia atmosphere for ca 6 h. The following reactions are assumed to take place:

$$4\,B(s) + 3\,O_2(g) \rightarrow 2\,B_2O_3(l)$$

$$B_2O_3(l) + 2\,NH_3(g) \rightarrow 2\,BN(s) + 3\,H_2O(g)$$

Economic Aspects

Refractory fibers are 50–70% more efficient insulators than conventional brick linings at equivalent thicknesses. Although the initial cost of installing a fiber lining in a high temperature furnace is higher than that of installing brick lining, the difference can be quickly recovered in energy savings. As the cost of energy continues to increase, refractory-fiber furnace lining is becoming more and more attractive. In 1980, the average cost of 1260°C alumina–silica refractory fiber was ca $3.75/kg, whereas the 1970 price was ca $5/kg. The decrease in cost reflects the advancement in technology.

The 1981 selling price for the >99% silica fibers was approximately $55/kg in both bulk and felted forms. This price has remained relatively constant over the past ten years. The more refractory Saffil alumina fiber sells for approximately $33/kg. When first introduced in 1974, both the Saffil zirconia and alumina fibers sold for approximately $24/kg. Production quantities are proprietary.

A substantial reinforced-composites market for continuous filaments has developed in the past few years. Carbon, graphite, and boron fibers are produced commercially at prices of $40–400/kg, depending upon their properties. Silicon carbide whiskers have been produced commercially over the last 20 yr by several processes. Prices for these fibers have been based upon pilot plant quantities and have ranged from $2000/kg to as low as $150/kg in 1980. The price of silicon nitride fibers has been $200–500/kg. Boron nitride fibers have not been produced commercially until about 1975. Their price has remained at ca $50/kg since that time.

Health and Safety Factors

Synthetic or man-made refractory fibers do not appear to pose the health hazards of naturally occurring mineral fibers like asbestos. Because the diameter of most refractory fibers is <3.5 μm, they are considered respirable. Above 3.5 μm, fibers are not able to penetrate the functional components of the lung (23). Even though a portion of these fibers are respirable, they are considered a nuisance dust, because studies have shown they are not biologically active in living tissue. The Thermal Insulation Manufacturers Association is currently conducting animal inhalation studies on a number of man-made fibers including refractory fibers. Regardless of the results of these studies, proper respirators should be worn in environments of excessive exposure to refractory fibers.

Silica and alumina–silica refractory fibers, that have been in service above 1100°C, undergo partial conversion to cristobalite, a form of crystalline silica that can cause silicosis, a form of pneumoconiosis. The amount of cristobalite formed, the size of the individual crystallites, and the nature of the matrix in which they are embedded are time and temperature dependent. Under normal use conditions, refractory fibers are generally exposed to a temperature gradient. Consequently, it is most probable that only the fiber nearest the hot surface has an appreciable content of cristobalite. It is also possible that fiber containing devitrified cristobalite is more friable and therefore may generate a larger fraction of dust when it is removed from a high temperature furnace. Hence, removal of old furnace lining offers the greatest risk of exposure, and adequate protection against respiration should be provided. Fibers with diameters of ≥5 μm cause irritation to skin and mucous membranes. This is usually not a serious problem and can be avoided by wearing proper clothing and respirators. When heated in an oxidizing atmosphere, silicon carbide fibers form a surface coating of silica that converts to cristobalite. In addition, silicon carbide whiskers have been found to be biologically active in animal lung tissue and may cause pneumoconiosis. Adequate respiratory protection is advised when dealing with these fibers. The diameter of carbon and boron filaments places them above the respirable range, but they can cause skin irritation from penetration during handling. Silicon nitride and boron nitride fibers are not considered hazardous but may irritate the skin.

Uses

Blankets. Today, ca 60% of the refractory fiber production is used for 48–128-mg/cm^3 flexible needled blankets. The fiber is collected on a moving conveyor, run through compression rolls, and penetrated by barked needle boards. This needling has the effect of tying the fibers together; subsequent compression and heating increase the tensile strength of the product. Flexible needled blankets are commercially available in widths of 1.3 and 0.65 m, lengths ≤33 m, and thicknesses of 6 to 50 mm. They are primarily used as furnace-wall and roof insulations either as the exposed hot face or as back-up insulation behind refractory brick. Because of their lightweight flexible nature, blankets offer no structural support to the furnace wall and have to be anchored in place. Typically, the blankets are applied to furnace walls and roofs in overlapping layers by impaling them on metallic or ceramic studs fixed to the supporting metal framework. They can also be applied over existing brick walls using high temperature cement or mechanical anchoring systems providing a new, more insulating furnace interior (see Insulation, thermal; Furnaces). Because of their relatively low heat storage and thermal conductivity, they have replaced brick linings in most industrial kilns in order to reduce energy costs. Other applications for these blankets include insulation for automotive catalytic convertors and aircraft and space vehicle engines and a wide variety of uses in the steel-making industry.

Felts. Felts (qv) contain an organic binder, generally a phenolic resin or in some cases a latex material. As the fibers are collected after fiberization, they are mixed with a dry phenolic resin. The fiber is then passed through an oven and a constant pressure is applied by flight conveyors. Alternatively, the uncured fiber and resin are compressed between heated press plates to form felts with densities as high as 380 mg/cm^3. These high density felts are now used extensively in ingot-mold operations in steel foundries. Felts provide excellent expansion joints in high temperature applications because the fibers tend to expand after the organic binder has been burned out (see also Heat-resistant polymers).

Bulk Fibers. Bulk refractory fiber is used as a general-purpose high temperature filler for expansion joints, as stuffing wool, for furnace and oven construction, in steel mills and aluminum and brass foundries, in glass manufacturing operations, as a loose-fill insulation, and as a raw material for vacuum-formed shapes (see Fillers).

Vacuum-Formed Shapes. Approximately 20% of fiber production is converted to rigid shapes by a vacuum-forming process. The bulk fibers are mixed in aqueous suspension with clays, colloidal metal oxide particles, and organic binders. Molds with fine-mesh screen surfaces are used to accrete the solids into special shapes as the water is drawn through the screen by vacuum. These products are then dried at 100–200°C to obtain rigid shapes having densities generally between 200 and 300 mg/cm^3. In recent years, the more expensive 1650°C alumina and zirconia fibers have been mixed with small amounts of less expensive and lower temperature alumina–silica fibers in the vacuum-forming process to increase the use temperature. The uses of vacuum-formed refractory fiber products are extremely varied. The metal industry makes extensive use of vacuum-formed shapes from tap hole cones in furnaces and ladles to insulating risers in metal castings. The tiles of the space shuttle thermal protection system are made of 99.7 wt % silica fiber, trade named Q Fiber, made by a highly specialized vacuum-forming process. The largest application for vacuum-formed products is as lightweight board insulation for furnace linings. In the rigid vacuum-formed state, the fibers are less susceptible to abrasion from direct gas impingement emanating from burners or flue fans.

Modules. Lightweight refractory fiber modules are a recent innovation in furnace lining (24). The refractory fiber blanket is folded in an accordion fashion and then compressed and banded to form modules 0.3 by 0.3 m, and 0.2–0.3-m thick. Metallic hardware, attached to the back of each module as it is folded and compressed, is welded to the metal shells of furnace roofs and walls. The modules are snapped into place in a parquet fashion forming the complete insulation system for the furnace. This attachment technique offers distinct advantages over layered blanket construction by reducing installation time and labor and eliminating the need for metal or ceramic anchoring pins. The shrinkage effect is minimized because the resilient fibers expand somewhat when their compression banding is released during construction. This concept of furnace lining is gaining widespread acceptance in the industry because of the ease of installation and other technical advantages.

Fibers and Yarns. The continuous refractory oxide fibers, ie, silica and alumina–silica–boria can be twisted into yarns from which fabrics are woven. These fabrics are used extensively in heat-resistant clothing, flame curtains for furnace openings, thermocouple insulation, and electrical insulation. Coated with Teflon, these fibers are used for sewing threads for manufacturing speciality high temperature insulation shapes for air craft and space vehicles. The cloth is often used to encase other insulating fibers in flexible sheets of insulation. The spaces between the rigid tile on the space shuttle Orbiter are packed with this type of fiber formed into tape.

Nonoxide fibers in continuous form, such as carbon, graphite, and boron are primarily used in filament winding and the manufacture of high strength, high modulus fabrics. Lightweight, high strength pressure vessels are made by running these fibers through an epoxy impregnant and then winding them around plaster mandrels. After curing the resin, the plaster is dissolved leaving a thin-walled vessel. The drill stems used for the collection of lunar core samples were a combination of boron cloth with an epoxy–graphite filament. The carbon fibers in loose fill and cloth form are currently used in nonoxidizing processes as insulation up to 1800°C and higher temperatures for short duration. The shorter nonoxide fibers are primarily used as strength-enhancing reinforcements in resins, some ceramics, and metals. Silicon carbide and silicon nitride short fibers are dispersed in a number of resins that are then cast into various shapes. Applications for these cast parts are generally found in high technology areas such as the fabrication of specialty electrical parts, aircraft parts, and radomes (microwave windows). These fibers are also used as reinforcing inclusions in metals. Boron nitride is of particular value as a reinforcing fiber for cast aluminum parts because it is one of the few materials totally wet by molten aluminum. Silicon carbide and boron nitride fibers as well as short single-crystal alumina fibers are used to reinforce gold and silver castings.

The growth and commercialization of the nonoxide fiber industry parallels the growth of the high strength-composite industry. When these fibers can be produced in lengths of 2–10-cm in the $10–20 per kilogram price range, a large >1600°C insulation market will open up.

BIBLIOGRAPHY

"Refractory Fibers" in *ECT* 2nd ed., Vol. 17, pp. 285–295, by T. R. Gould, Johns-Manville Corporation.

1. U.S. Pat. 2,461,841 (Feb. 15, 1949), M. E. Nordberg (to Corning Glass Works).
2. *Chem. Eng. News* **25**(18), 1290 (1947).
3. C. Z. Carroll-Porczynski, *Advanced Materials*, 1st ed., Astex Publishing Company, Ltd., Guildford, England, 1959.
4. U.S. Pat. 2,823,117 (Feb. 11, 1958), D. Labino (to Glass Fibers, Inc.).
5. U.S. Pat. 3,177,057 (Aug. 2, 1961), C. Potter and J. W. Lindenthal (to Engelhard Industries, Inc.).
6. U.S. Pat. 2,467,889 (Apr. 19, 1949), I. Harter, C. L. Horton, Jr., and L. D. Christie, Jr. (to Babcock & Wilcox Company).
7. *Ceram. Age* **78**(2), 37 (1962).
8. U.S. Pat. 3,449,137 (June 10, 1969), W. Ekdahl (to Johns-Manville Corporation).
9. U.S. Pat. 3,385,915 (May 28, 1968), B. Hamling (to Union Carbide Corporation).
10. H. W. Rauch, Sr., W. H. Sutton, and L. R. McCreight, *Ceramic Fibers and Fibrous Composite Materials*, Academic Press, New York, 1968, p. 24.
11. U.S. Pat. 3,429,722 (July 12, 1965), J. Economy and R. V. Anderson (to Carborundum Company).
12. U.S. Pat. 3,573,969 (Apr. 10, 1971), J. C. Millbrae and M. P. Gomez (to Lockheed Aircraft Corporation).
13. L. Olds, W. Miiller, and J. Pallo, *Am. Ceram. Soc. Bull.* **59**(7), 739 (1980).
14. U.S. Pat. 2,714,622 (Aug. 2, 1955), J. C. McMullen (to The Carborundum Company).
15. U.S. Pat. 3,066,504 (Dec. 4, 1962), F. J. Hartwig and F. H. Norton (to Babcock & Wilcox Company).
16. U.S. Pat. 4,200,485 (Apr. 29, 1980), G. Price and W. Kielmeyer (to Johns-Manville Corporation).
17. Brit. Pat. 507,951 (June 23, 1939).
18. Brit. Pat. 1,360,197 (July 17, 1974), M. Morton, J. Birchall, and J. Cassidy (to Imperial Chemical Industries, Ltd.).
19. U.S. Pat. 3,760,049 (Sept. 18, 1973), A. Borer and G. P. Krogseng (to Minnesota Mining and Manufacturing Company).
20. U.S. Pat. 3,246,950 (Apr. 19, 1966), B. A. Gruber (to Corning Glass Works).
21. U.S. Pat. 3,371,995 (March 5, 1968), W. W. Pultz (to Corning Glass Works).
22. U.S. Pat. 3,244,480 (Apr. 5, 1966), R. C. Johnson, J. K. Alley, and W. H. Warwick (to The United States of America).
23. J. Leineweber, *ASHRAE J.* **3**, 51 (1980).
24. U.S. Pat. 3,952,470 (Apr. 27, 1976), C. Byrd, Jr. (to J. T. Thorpe Company).

W. C. MIILLER
Manville Corporation

REFRIGERATION

Refrigeration is the use of mechanical or heat-activated machinery for cooling purposes. The production of extremely low temperatures, $< -150°C$ (1), is usually thought of as cryogenics (qv), and the use of refrigeration equipment to provide human comfort is known as air conditioning (qv). This article deals only with refrigeration which, in its broadest sense, covers such diverse uses as food processing and storage, supermarket display cases, skating rinks, ice manufacture, and biomedical applications such as blood and tissue storage or hypothermia used in surgery. The first commercial machine used for refrigeration was built in 1844 (2). In this article, chemical industry applications are emphasized, although the basic principles are valid for other uses of refrigeration.

The chemical industry uses refrigeration in installations covering a broad range of cooling capacities and temperature levels. The variety of applications results in a diversity of mechanical specifications and equipment requirements. Nevertheless, the methods for producing refrigeration are well-standardized.

Basic Principles

Thermodynamic principles govern all refrigeration processes. For a review of the basic laws of thermodynamics and the nature of the reverse Carnot cycle, ie, the refrigeration cycle, see the article on Thermodynamics (see also Heat-exchange technology, heat transfer). Refrigeration is accomplished in this cycle by using a fluid that evaporates and condenses at suitable pressures for practical equipment designs. The vapor cycle is illustrated by a pressure–enthalpy diagram, Figure 1.

The compressor raises the pressure of the refrigerant vapor so that its saturation temperature is slightly above the temperature of an available cooling medium, eg, air or water. This difference in temperature allows transfer of heat from the vapor to the cooling medium so that the vapor can condense. The liquid next expands to a pressure such that its saturation temperature is slightly below the temperature of the product to be cooled. This difference in temperature allows transfer of heat from the product to the refrigerant, causing the refrigerant to evaporate. The vapor formed must be removed by the compressor at a sufficient rate to maintain the low pressure in the evaporator and keep the cycle operating.

Many of today's chemical plants use pumped recirculation of refrigerant rather than direct evaporation of refrigerant to service remotely located or specially designed heat exchangers. This technique provides the users with wide flexibility in applying refrigeration to complex processes and greatly simplifies operation. Secondary refrigerants or brines are also commonly used today for simple control and operation. Direct application of ice and brine storage tanks may be used to level-off batch cooling loads and reduce equipment size. This approach provides stored refrigeration where temperature control is vital as a safety consideration to prevent runaway reactions or pressure buildup.

All mechanical cooling results in the simultaneous production, somewhere else, of a greater amount of heat. It is not always realized that this heat can be put to good use by applying the heat-pump principle. This requires provision to recover the heat

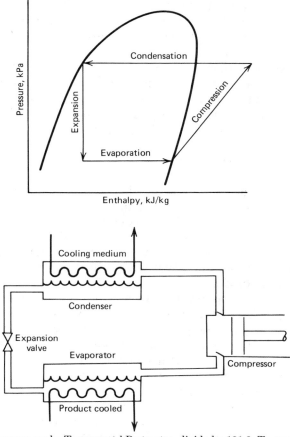

Figure 1. Basic vapor cycle. To convert kPa to atm, divide by 101.3. To convert kJ/kg to Btu/lb, multiply by 0.4302.

normally rejected to cooling water or air in the refrigeration condenser. Recovery of this waste heat at temperatures up to 65°C is frequently employed in modern plants to achieve improved heat balance and operating economy (see Energy management).

If a steady supply of waste heat is available, an absorption machine can be chosen for continuous-cooling duty. Heat at elevated temperatures, ie, steam or hot water, is required in the generator to drive off the high pressure refrigerant. The heat rejected to the cooling-tower water from both the absorber and the condenser is the sum of the waste heat supplied plus the cooling produced. This requires a larger cooling tower and greater cooling-water flow than for the vapor-compression cycle (see Air conditioning).

For water-chilling service, absorption (qv) systems generally use water as the refrigerant and lithium bromide as the absorbent solution. For process applications that require chilled fluid <7°C, the ammonia–water pair is used, with ammonia serving as the refrigerant. Single-stage systems are most common with generator heat input <95°C. The coefficient of performance (COP) of a system is the cooling achieved in the evaporator divided by the heat input to the generator. The COP of a lithium bromide machine is generally 0.65–0.70 for water-chilling duty.

Absorption machines can be built with a two-stage generator for heat input temperatures ≥150°C. Such machines are called dual-effect machines. Coefficients of performance above 1.0 can be obtained, but at the expense of increased design and operating complexity.

Historically, capacities of mechanical refrigeration systems have been stated in tons of refrigeration, a unit of measure related to the ability of an ice plant to freeze one short ton (907 kg) of ice in 24 h. Its value is 3.51 kW_t (12,000 Btu/h). Often a kilowatt of refrigeration capacity is identified as kW_t to distinguish it from the amount of electricity (kW_e) required to produce the refrigeration.

Refrigeration Cycles and System Overview

Refrigeration can be accomplished in either closed-cycle or open-cycle systems. In a closed cycle, the refrigerant fluid is confined within the system and recirculates through the processes in the cycle. The system shown at the bottom of Figure 1 is a closed cycle. In an open cycle, the fluid used as the refrigerant passes through the system once on its way to be used as a product or feedstock outside the refrigeration process. An example is the cooling of natural gas to separate and condense heavier components (see Gas, natural).

In addition to the distinction between open- and closed-cycle systems, refrigeration processes are also described as simple cycles, compound cycles, or cascade cycles. Simple cycles employ one set of components and a single refrigeration cycle as in Figure 1. Compound and cascade cycles employ multiple sets of components and two or more refrigeration cycles. The cycles interact to accomplish cooling at several temperatures or to allow a greater span between the lowest and highest temperatures in the system than can be achieved with the simple cycle.

Closed-Cycle Operation. For a simple cycle, the lowest evaporator temperature that is practical in a closed-cycle system (Fig. 1) is set by the pressure-ratio capability of the compressor and by the properties of the refrigerant. Most high speed reciprocating compressors are limited to a pressure ratio of 9:1, so that the simple cycle is used for evaporator temperatures of 2 to −50°C. Below these temperatures, the application limits of a single reciprocating compressor are reached. Beyond that there is a risk of excessive heat which may break down lubricants, high bearing loads, excessive oil foaming at startup, and inefficient operation because of reduced volumetric efficiency.

Centrifugal compressors with multiple stages can generate a pressure ratio up to 18:1, but their high discharge temperatures limit the efficiency of the simple cycle at these high pressure ratios. As a result, they have practical evaporator temperatures in the same range as reciprocating compressors.

The compound cycle (Fig. 2) achieves temperatures of ca −100°C by using two or three compressors in series and a common refrigerant. This keeps the individual machines within their application limits. A refrigerant gas cooler is normally used between compressors to keep the final discharge temperature at a satisfactory level.

Below −100°C, most refrigerants with suitable evaporator pressures have excessively high condensing pressures. For some refrigerants, the specific volume of refrigerant at low temperature may be so great as to require compressors and other equipment of uneconomical size. In other refrigerants, even though the specific volume

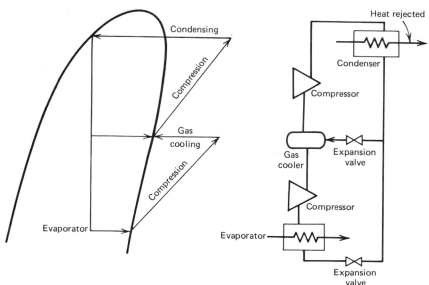

Figure 2. Compound cycle.

of refrigerant is satisfactory at low temperature, the specific volume may become too small at the condensing condition. In some circumstances, although none of the above limitations is encountered and a single refrigerant is practical, the compound cycle is not used because of oil-return problems or difficulties of operation.

To satisfy these conditions, the cascade cycle is utilized (Fig. 3). This consists of two or more separate refrigerant cycles. The cascade condenser–evaporator rejects heat to the evaporator of the high temperature cycle, which condenses the refrigerant of the low temperature cycle. This makes possible the use of a refrigerant such as R-13 (see Table 1) in the low stage, with pressure–temperature–volume characteristics well-suited to the low temperature. Refrigerants with pressure–temperature–volume characteristics more favorable at higher temperatures, eg, R-12 or R-22, are used in one or more higher stages. For extremely low temperatures, more than two refrigerants may be cascaded as, for example, methane–ethylene–propane to produce −160°C evaporator temperatures. Expansion tanks, sized to handle the low temperature refrigerant as a gas at ambient temperatures, are used during standby to hold pressure at levels suitable for economical equipment design.

Compound cycles using reciprocating compressors or any cycle using a multistage centrifugal compressor allow the use of economizers or intercoolers between compression stages. Economizers reduce the discharge gas temperature from the preceding stage by mixing relatively cool gas with discharge gas before entering the subsequent stage. Either flash-type economizers, which cool refrigerant by reducing its pressure to the intermediate level, or surface-type economizers, which subcool refrigerant at condensing pressure, may be used to provide the cooler gas for mixing. This keeps the final discharge gas temperature low enough to avoid overheating of the compressor and improves compression efficiency.

Compound compression with economizers also affords the opportunity to provide refrigeration at an intermediate temperature. This provides a further thermodynamic efficiency gain because some of the refrigeration is accomplished at a higher temper-

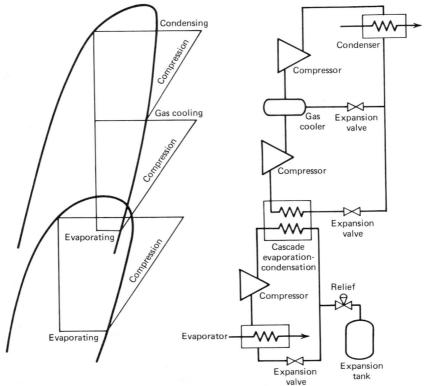

Figure 3. Cascade cycle.

ature, and less refrigerant must be handled by the lower temperature stages. This reduces the power consumption and the size of the lower stages of compression.

Figure 4 shows a typical system schematic with flash-type economizers. Process loads at several different temperature levels can be handled by taking suction to an intermediate compression stage as shown. The corresponding pressure–enthalpy (P–H) diagram illustrates the thermodynamic cycle.

Flooded refrigeration systems are a version of the closed cycle which may reduce design problems in some applications. In flooded systems, the refrigerant is circulated to heat exchangers or evaporators by a pump. Figure 5 shows the flooded cycle, which can employ any of the simple or compound closed-refrigeration cycles.

The refrigerant-recirculating pump pressurizes the refrigerant liquid and moves it to one or more evaporators or heat exchangers which may be remote from the receiver. The low pressure refrigerant may be used as a single-phase heat-transfer fluid as in (**a**) of Figure 5, which eliminates the extra heat-exchange step and increased temperature difference encountered in a conventional system that uses a secondary refrigerant or brine. This approach may simplify the design of process heat exchangers where the large specific volumes of evaporating refrigerant vapor would be troublesome. Alternatively, the pumped refrigerant in the flooded system may be routed through conventional evaporators as in (**b**) and (**c**), or special heat exchangers as in (**d**) (see Evaporation).

The flooded refrigeration system is helpful when special heat exchangers are

Table 1. Refrigerant Numbering System (ASHRAE 34-78) [a]

Refrigerant number designation	Chemical name	Chemical formula	Molecular weight	Normal boiling point, °C	Group classification [b]
10	carbon tetrachloride	CCl_4	153.8	77	2
11	trichlorofluoromethane	CCl_3F	137.4	24	1
12	dichlorodifluoromethane	CCl_2F_2	120.9	−30	1
13	chlorotrifluoromethane	$CClF_3$	104.5	−81	1
22	chlorodifluoromethane	$CHClF_2$	86.5	−41	1
23	trifluoromethane	CHF_3	70.0	−82	
113	1,1,2-trichlorotrifluoro-ethane	CCl_2FCClF_2	187.4	48	1
114	1,2-dichlorotetrafluoro-ethane	$CClF_2CClF_2$	170.9	4	1
Hydrocarbons					
50	methane	CH_4	16.0	−161	3
170	ethane	CH_3CH_3	30	−89	3
290	propane	$CH_3CH_2CH_3$	44	−42	3
600	butane	$CH_3CH_2CH_2CH_3$	58.1	0	3
600a	isobutane	$CH(CH_3)_3$	58.1	−12	
Inorganic compounds					
717	ammonia	NH_3	17.0	−33	2
732	oxygen	O_2	32.0	−183	
744	carbon dioxide	CO_2	44.0	−78	1
764	sulfur dioxide	SO_2	64.1	−10	2
Unsaturated organic compounds					
1140	vinyl chloride	$CH_2{=}CHCl$	62.5	−14	c
1141	vinyl fluoride	$CH_2{=}CHF$	46	−72	
1150	ethylene	$CH_2{=}CH_2$	28.1	−104	3
1270	propylene	$CH_3CH{=}CH_2$	42.1	−48	3
Azeotropes [d]	*Composition*	*Azeotropic temp, °C*			
500	R-12/152a (73.8/26.2 wt %)	0	99.31	−33	1
502	R-22/115 (48.8/51.2 wt %)	19	112	−45	1

[a] Ref. 3. Courtesy of ASHRAE, Inc., Atlanta, Ga.

[b] Classification is discussed under Refrigerants.

[c] Normally only used where already present in the process.

[d] All azeotropic refrigerants, by their nature, exhibit some segregation of components at conditions of temperature and pressure other than those at which they were formulated. The exact extent of this segregation depends on the particular azeotrope and hardware-system configuration.

necessary for process reasons, or where multiple or remote exchangers are required.

Refrigerant Selection for the Closed Cycle. In any closed cycle, the choice of the operating fluid is unrestricted and is based on the one whose properties are best suited to the operating conditions. The choice depends on a variety of factors, some of which may not be directly related to the refrigerant's ability to remove heat. For example, flammability, toxicity, density, viscosity, availability, and similar characteristics are often deciding factors. The suitability of a refrigerant also depends on factors such as the kind of compressor to be used (ie, centrifugal, rotary or reciprocating), safety in application, heat-exchanger design, application of codes, size of the job, and temperature ranges. The factors below should be taken into account when selecting a refrigerant.

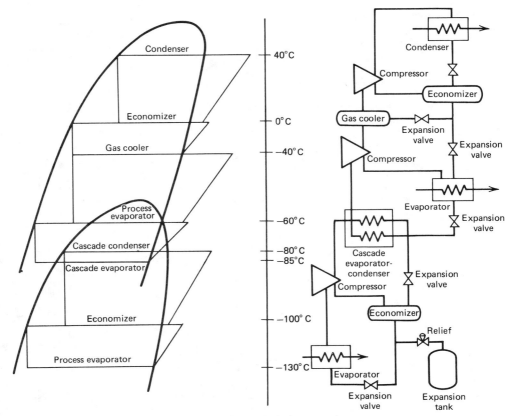

Figure 4.   System with flash economizers.

Discharge (condensing) pressure should be low enough to suit the design pressure of commercially available pressure vessels, compressor casings, etc. However, discharge pressure, ie, condenser liquid pressure, should be high enough to feed liquid refrigerant to all the parts of the system that require it.

Suction (evaporating) pressure should be above ca 3.45 kPa (0.5 psi) for a practical compressor selection. When possible, it is preferable to have the suction pressure above atmospheric to prevent leakage of air and moisture into the system. Positive pressure normally is considered a necessity when dealing with hydrocarbons because of the explosion hazard presented by any air leakage into the system.

Standby pressure (saturation at ambient temperature) should be low enough to suit equipment design pressure unless there are other provisions in the system for handling the refrigerant during shutdown, eg, inclusion of expansion tanks.

Critical temperature and pressure should be well above the operating level. As the critical pressure is approached, less heat is rejected as latent heat compared to the sensible heat from desuperheating the compressor discharge gas, and cycle efficiency is reduced. Methane (R-50) and chlorotrifluoromethane (R-13) usually are cascaded with other refrigerants because of their low critical points.

Suction volume sets the size of the compressor. High suction volumes require centrifugal compressors and low suction volumes dictate the use of reciprocating compressors. Suction volumes also may influence evaporator design, particularly at

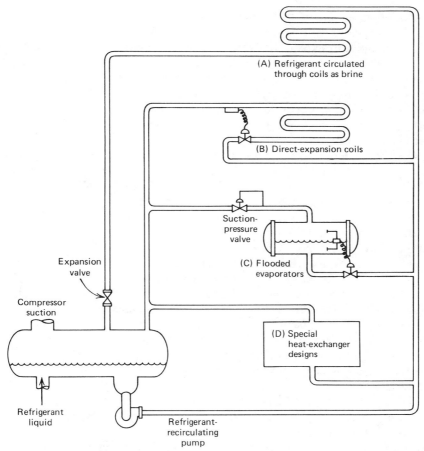

Figure 5.   Liquid recirculation.

low temperatures, since they must include adequate space for gas–liquid separation.

Freezing point should be lower than minimum operating temperature. This generally is no problem unless the refrigerant is used as a brine.

Theoretical power required for adiabatic compression of the gas is slightly less with some refrigerants than others. However, this is usually a secondary consideration offset by the effects of particular equipment selections, eg, line-pressure drops, etc, on system power consumption.

Vapor density (*or molecular weight*) is an important characteristic when the compressor is centrifugal because the lighter gases require more impellers for a given pressure rise, ie, head, or temperature lift. On the other hand, centrifugal compressors have a limitation connected with the acoustic velocity in the gas, and this velocity decreases with the increasing molecular weight.

Low vapor densities are desirable to minimize pressure drop in long suction and discharge lines.

Liquid density should be taken into account. Liquid velocities are comparatively low so that pressure drop is usually no problem. However, static head may affect evaporator temperatures, and should be considered when liquid must be fed to elevated parts of the system.

Latent heat should be high because it reduces the quantity of refrigerant that needs to be circulated. However, large flow quantities are more easily controlled because they allow use of larger, less sensitive throttling devices and apertures.

Refrigerant cost depends on the size of the installation and must be considered both from the standpoint of initial charge and of composition owing to losses during service. Although a domestic refrigerator contains only a few dollars worth of refrigerant, the charge for a typical chemical plant may cost thousands of dollars.

Other desirable properties. Refrigerants should be stable and noncorrosive. For heat-transfer considerations, a refrigerant should have low viscosity, high thermal conductivity, and high specific heat. For safety to life or property, a refrigerant should be nontoxic and nonflammable, should not contaminate products in case of a leak, and should have a low leakage tendency through normal materials of construction.

Refrigerants. No one refrigerant meets all these requirements for the wide range of temperature and the multitude of applications required by modern chemical processing. However, the tendency in the chemical industry is towards the use of, even insistence upon, low cost fluids such as propane and butane whenever they are available in the process. These rediscovered hydrocarbon refrigerants, often thought of as too hazardous because of flammability, are entirely suitable for use in modern compressors and frequently add no more hazard than already exists in the oil refinery or petrochemical field. These low cost refrigerants are used in simple, compound, and cascade systems, depending on operating temperatures.

With a flammable refrigerant, extra precautions may have to be taken in the engineering design if it is required to meet the explosion-proof classification. It may be more economical to use a higher cost, but nonflammable, refrigerant.

Most of the nontoxic, nonflammable refrigerants are halogenated hydrocarbons containing one or more of the halogens: fluorine, chlorine, and occasionally bromine (see Chlorocarbons and chlorohydrocarbons; Fluorine compounds, organic).

A standard numbering system, shown in Table 1, has been devised to identify refrigerants without the use of the cumbersome chemical name (see also Aerosols). Numbers assigned to the hydrocarbons and halohydrocarbons of the methane, ethane, propane and cyclobutane series are such that the number uniquely specifies the refrigerant compound. ASHRAE Standard 34-78 describes the method of coding (3).

The American National Standards Institute (ANSI) (4) groups refrigerants in three classes, depending on toxicity and flammability. Group 1 refrigerants are nontoxic and nonflammable; Group 2 are slightly toxic or flammable; Group 3 are highly toxic or flammable. The most commonly used refrigerants are shown in Table 2.

The Group 1 refrigerants generally fulfill the basic requirements for an ideal refrigerant with considerable flexibility as to refrigeration capacity. They are ideal for comfort air conditioning since they are nontoxic and nonflammable. These refrigerants are used selectively in terms of compressor displacement and operating temperature levels desired.

Refrigerant 12, dichlorodifluoromethane, is at present the most widely known and used refrigerant. It is ideal for close-coupled or remote systems ranging from small reciprocating to large centrifugal units. It has been used for temperatures as low as

−90°C although −85°C is a more practical lower limit because of the high gas volumes necessary for attaining these temperatures. It is suited for single-stage or compound cycles using reciprocating and centrifugal compressors.

Refrigerant 22, chlorodifluoromethane, is used in many of the same applications as R-12, but its lower boiling point and higher latent heat permit the use of smaller compressors and refrigerant lines than R-12. The higher pressure characteristics also extend its use to lower temperatures in the range of −100°C.

Refrigerant 11, trichlorofluoromethane, has a low pressure–high volume characteristic suitable for use in close-coupled centrifugal compressor systems for water or brine cooling. Its temperature range extends no lower than ca −7°C.

Refrigerant 114, dichlorotetrafluoroethane, is similar to R-11 but its slightly higher pressure and lower volume characteristic than R-11 extend its use to ca −17°C and higher capacities.

Refrigerant 13, chlorotrifluoromethane, is used in low temperature applications to ca −126°C. Because of its low volume, high condensing pressure, or both, and because of its low critical pressure and temperature, R-13 is usually cascaded with other refrigerants at a discharge pressure corresponding to a condensing temperature in the range of −56°C to −23°C.

Of the Group 2 refrigerants, only ammonia is still used to any extent. Although toxic and flammable within specific limits, ammonia is one of the best and most widely used refrigerants. Ammonia liquid has a high specific heat, an acceptable density and viscosity, and high conductivity. This makes it an ideal heat-transfer fluid with reasonable pumping costs, pressure drop, and flow rates. As a refrigerant, ammonia provides high heat transfer except when affected by oil at temperatures below ca −29°C where oil films become viscous. To limit the ammonia-discharge-gas temperature to safe values, its normal maximum condensing temperature is 38°C. Generally, ammonia is used with reciprocating compressors, although relatively large centrifugal compressors ($\geq$3.5 MW$_t$ or 1.2 × 10^6 Btu/h) with 8–12 impeller stages required by its low molecular weight are in use today. Materials that are to contain ammonia should contain no copper (with the exception of Monel metal).

Refrigerants in Group 3 are generally applicable where a flammability or explosion hazard is already present and their use does not add to the hazard. These refrigerants have the advantage of low cost and, although they have fairly low molecular weight, they are suitable for centrifugal compressors of larger sizes. Because of the high acoustic velocity in Group 3 refrigerants, centrifugal compressors may be operated at high impeller tip speeds, which partly compensates for the higher head requirements than Group 1 refrigerants.

These refrigerants should be used at pressures greater than atmospheric to avoid increasing the explosion hazard by the admission of air in case of leaks. In designing the system, it also must be recognized that these refrigerants are likely to be impure in refrigerant applications. For example, commercial propane liquid may contain about 2 wt % ethane, which in the vapor phase, might represent as much as 16–20 vol %. Thus, ethane may appear as a noncondensable. Either this gas must be purged or the compressor displacement must be increased ca 20% if it is recycled from the condenser; otherwise, the condensing pressure will be higher than required for pure propane and the power requirement will be increased.

Propane is the most commonly used Group 3 refrigerant. It is well suited for use with reciprocating and centrifugal compressors in close-coupled or remote systems. Its temperature range extends to ca −40°C (see Table 2).

Table 2. Properties of Common Refrigerants

Refrigerant number[a]	Chemical formula	Normal temp range, °C	System capacity	Cycle type	Compressor type	Relative cost	Recommended equipment design pressure, kPa[b]	
							High side	Low side
Group 1, Nontoxic and nonflammable								
R-11	CCl_3F	−5 to 60	low	simple	centrifugal	medium	210	210
R-114	$CClF_2CClF_2$ CCl_2FCF_3	−20 to 65	low	simple	centrifugal	medium	450	450
R-12	CCl_2F_2	−90 to 60	medium	simple and compound	centrifugal and reciprocating	medium	1650	1140
R-22	$CHClF_2$	−100 to 60	high	simple and compound	centrifugal and reciprocating	medium	2170	1140
R-13	$CClF_3$	−125 to −20	high	cascade	centrifugal and reciprocating	high	2170[c]	2170[c]
Group 2, Slightly toxic and flammable								
R-717	NH_3	−30 to 40	high		reciprocating	medium	2170	1140
Group 3, Highly toxic or flammable								
R-600	$CH_3CH_2CH_2CH_3$	2 to 60	low	simple	centrifugal	low	760	450
R-290	$CH_3CH_2CH_3$	−40 to 60	medium	simple	centrifugal and reciprocating	low	2170	1140
R-1270	$CH_3CH{=}CH_2$	−45 to 60	high	simple	centrifugal and reciprocating	low	2170	1140
R-170	CH_3CH_3	−90 to −5	medium	cascade	centrifugal and reciprocating	low	2170[c]	2170[c]
R-1150	$CH_2{=}CH_2$	−105 to −30	high	cascade	centrifugal and reciprocating	low	2170[c]	2170[c]
R-50	CH_4	−160 to −110	high	cascade	centrifugal and reciprocating	low	2170[c]	2170[c]

[a] For corresponding chemical names see Table 1.
[b] To convert kPa to psi, multiply by 0.145.
[c] Expansion tank to allow all liquid to vaporize is required when design pressure as shown is used.

88

Propylene is similar to propane but has slightly higher pressure characteristics.

Butane occasionally is used for close-coupled systems in the medium temperature range of 2°C. It has a low pressure and high volume characteristic suitable for centrifugal compressors where the capacity is too small for propane and the temperature is within range.

Ethane normally is used for close-coupled or remote systems at −87 to −7°C. It must be used in a cascade cycle because of the high pressure characteristics.

Ethylene is similar to ethane but has a slightly higher pressure, lower volume characteristic which extends its use to −104 to −29°C. Like ethane it must be used in the cascade cycle.

Methane is used in an ultralow range of −160 to −110°C. It is limited to cascade cycles. Methane condensed by ethylene, which is in turn condensed by propane, is a cascade cycle commonly employed to liquefy natural gas.

Refrigerant Mixtures. Because the bubble point and dew point temperatures are not the same for a given pressure, nonazeotropic mixtures may be used to help control the temperature differences in low temperature evaporators. This may be seen in the lowest stage of some LNG plants (5); although undesirable mixtures or impurities such as are commonly found in hydrocarbon processing can have the opposite effect and raise the condensing pressure or lower the evaporating pressure if the impurities are not properly accounted for (see Liquefied petroleum gas).

Open-Cycle Operation. In many chemical processes, the product to be cooled can itself be used as the refrigerating liquid. An important example of this is in the gathering plants for natural gas (see Gas, natural). Gas from the wells is cooled, usually after compression and after some of the heavier components are removed as liquid. This liquid may be expanded in a refrigeration cycle to further cool the compressed gas, which causes more of the heavier components to condense. Excess liquid not used for refrigeration is drawn off as product. Another example is in ammonia synthesis. The product is cooled, condensed, and held in spheres or other tanks at low temperature (−33°C) and atmospheric pressure. In the transportation of liquefied petroleum gas and of ammonia (qv) in ships and barges, the open refrigeration system can be used. The LPG or ammonia is compressed, cooled, and expanded. The liquid portion after expansion is passed on as product until the ship is loaded (see also Transportation).

Open-cycle is similar to closed-cycle operation, except that one or more parts of the closed cycle may be omitted. For example, the compressor suction may be taken directly from gas wells, rather than from an evaporator. A condenser may be used, and the liquefied gas may be drained to storage tanks.

Compressors may be installed in series or parallel for operating flexibility or for partial standby protection. With multiple reciprocating compressors, or with a centrifugal compressor, gas streams may be picked up or discharged at several pressures (see Fig. 6) if there is refrigerating duty to be performed at intermediate temperatures. It always is more economical to refrigerate at the highest temperature possible.

Principal concerns in the open cycle involve dirt and contaminants, wet gas, compatibility of materials and lubrication circuits, and piping to and from the compressor. The possibility of gas condensing under various ambient temperatures either during operation or during standby must be considered. Beyond these considerations, the open-cycle design and its operation are governed primarily by the process re-

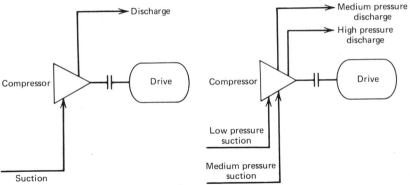

Figure 6.   Open-cycle compressors.

quirements. The open system can use standard refrigeration hardware. It is not necessary to go to the expensive specialty field for the equipment.

Open-Cycle Refrigerant Selection.   Process gases used in the open cycle include chlorine, ammonia, and mixed hydrocarbons. These create a wide variety of operating conditions and corrosion problems. Gas characteristics affect both heat exchangers and compressors, but their impact is far more critical on compressor operation. All gas properties and conditions should be clearly specified to obtain the most economical and reliable compressor design. Many installations are greatly overspecified, resulting in design features that not only add significant cost but also complicate the operation of the system and are difficult to maintain. Specifications should consider the following:

Composition.   Molecular weight, enthalpy–entropy relationship, compressibility factor, and operating pressures and temperatures influence the selection and performance of compressors. If process streams are subject to periodic or gradual changes in composition, the range of variations must be indicated.

Corrosion.   Special materials of construction and types of shaft seals may be necessary for some gases. Gases that are not compatible with lubricating oils or that must remain oil-free may necessitate reciprocating compressors designed with carbon rings or otherwise made oilless, or the use of centrifugal compressors designed with isolation seals. However, these features are unnecessary on most installations. Standard designs usually can be used to provide savings in first cost, simpler operation, and reduced maintenance.

Dirt and liquid carryover.   Generally, the carryover of dirt and liquids can be controlled more effectively by suction scrubbers than by costly compressor design features. Where this is not possible, all anticipated operating conditions should be stated clearly so that suitable materials and shaft seals can be provided.

Polymerization.   Gases that tend to polymerize may require cooling to keep the gas temperature low throughout compression. This can be handled by liquid injection or by providing external cooling between stages of compression. Provision may be necessary for internal cleaning with steam.

These factors are typical of those encountered in open-cycle gas compression. Each job should be thoroughly reviewed to avoid unnecessary cost and obtain the simplest possible compressor design for ease of operation and maintenance. Direct coordination between the design engineer and manufacturer during final stages of system design is strongly recommended.

Indirect Refrigeration (Brine). The process fluid is cooled by an intermediate liquid, water or brine, that is itself cooled by evaporating the refrigerant as shown in Figure 7. In the chemical industry, process heat exchangers must frequently be designed for corrosive products, high pressures, or high viscosities, and are not well suited for refrigerant evaporators. Other problems preventing direct use of refrigerant are remote location, lack of sufficient pressure for the refrigerant liquid feed, difficulties with oil return, or inability to provide traps in the suction line to hold liquid refrigerant. Use of indirect refrigeration simplifies the piping system; it becomes a conventional hydraulic system.

The brine is cooled in the refrigeration evaporator and then is pumped to the process load. The brine system may include a tank, either open or closed but maintained at atmospheric pressure through a small vent pipe at the top, or may be a closed system pressurized by an inert, dry gas.

The brines commonly used are

1. Brines with a salt base. These are water solutions of various concentrations and include the most common brines, ie, calcium chloride and sodium chloride.

2. Brines with a glycol base. These are water solutions of various concentrations, most commonly ethylene glycol or propylene glycol.

3. Brines with an alcohol base. Where low temperatures are not required, the alcohols are occasionally used in alcohol-water solutions.

4. Brines for low temperature heat transfer. These usually are pure substances such as methylene chloride, trichloroethylene, trichlorofluoromethane, acetone, methanol, and ethanol.

Table 3 shows the general areas of application for the commonly used brines. Criteria for selection are discussed in the following paragraphs. The order of importance depends on the specific application.

Costs of brines vary widely. Even considering the labor cost of mixing and the cost of water treating and inhibitors at temperatures where calcium or sodium chloride solutions can be used, these solutions are much less expensive than brines such as R-11.

Corrosion problems with sodium chloride and calcium chloride brines limit their use. Nevertheless, when properly maintained in a neutral condition and protected

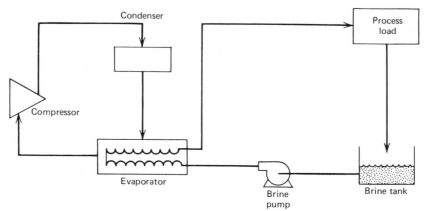

Figure 7. Secondary brine system.

Table 3.   Common Brines

Brines	Minimum practical temperature, °C	Toxic	Explosive	Corrosive	Cost
Salts					
calcium chloride	−40	no	no	yes	low
sodium chloride	−10	no	no	yes	low
Glycols					
propylene	−10	no	no	some	medium
ethylene	−15	yes	no	some	medium
Alcohols					
methanol	−35	yes	yes	some	high
ethanol	−30	yes	yes	some	high
Low temperature brines					
methylene chloride	−85	no	no	no	high
trichloroethylene	−75	no	no	no	high
trichlorofluoromethane	−100	no	no	no	high
acetone	−100	no	yes	no	high

with inhibitors, they will give 20–30 years of service without corrosive destruction of a closed system. Glycol solutions and alcohol–water solutions are generally less corrosive than salt brines, but they require inhibitors to suit the specific application for maximum corrosion protection. Methylene chloride, trichloroethylene, and trichlorofluoromethane do not show general corrosive tendencies unless they become contaminated with impurities such as moisture (see Corrosion and corrosion inhibitors).

However, methylene chloride and trichloroethylene must not be used with aluminum or zinc; they also attack most rubber compounds and plastics. Alcohol in high concentrations will attack aluminum. Reaction with aluminum is of concern because, in the event of leakage into the refrigeration compressor system, aluminum compressor parts will be attacked.

Toxicity is an important consideration in connection with exposure to some products and to operating personnel. Where brine liquid, droplets, or vapor may contact food products, as in an open spray-type system, sodium chloride and propylene glycol solutions are acceptable because of low toxicity. All other brines are toxic to some extent or produce odors which require that they be used only inside of pipe coils or a similar pressure-tight barrier.

Flash-point and explosive-mixture properties of some brines require precautions against fire or explosion. Acetone, methanol, and ethanol are in this category but are less dangerous when used in closed systems.

Specific heat of a brine determines the mass rate of flow that must be pumped to handle the cooling load for a given temperature rise. The low temperature brines, such as trichloroethylene, methylene chloride, and trichlorofluoromethane have specific heats approximately one third to one fourth those of the water-soluble brines. Consequently, a significantly greater mass of the low temperature brines must be pumped to achieve the same temperature change.

Stability at high temperatures is important where a brine may be heated as well as cooled. Above 60°C, methylene chloride may break down to form acid products. Trichloroethylene can reach 120°C before breakdown begins.

Viscosities of brines vary greatly. The viscosity of propylene glycol solutions, for example, makes them impractical for use below −7°C because of the high pumping costs and the low heat-transfer coefficient at the concentration required to prevent freezing. Mixtures of ethanol and water can become highly viscous at temperatures near their freezing points, but 190-proof ethyl alcohol has a low viscosity at all temperatures down to near the freezing point. Similarly, methylene chloride and trichlorofluoromethane have low viscosities down to −73°C. In this region, the viscosity of acetone is even more favorable.

Since a brine cannot be used below its freezing point, certain brines are not applicable at the lower temperatures. Sodium chloride's eutectic freezing point of −20°C limits its use to ca −12°C, even though the viscosity is not unreasonable. The eutectic freezing point of calcium chloride brine is −53°C, but achieving this limit requires such an accuracy of mixture that −40°C is a practical low limit of usage.

Water solubility in any open or semi-open system can be important. The dilution of a salt or glycol brine, or of alcohol by entering moisture, merely necessitates strengthening of the brine. But for a brine that is not water-soluble, such as trichloroethylene or methylene chloride, precautions must be taken to prevent free water from freezing on the surfaces of the heat exchanger. This may require provision for dehydration or periodic mechanical removal of ice, perhaps accompanied by replacement with fresh brine.

Vapor pressure is an important consideration for brines that will be used in open systems, especially where the brine may be allowed to warm to room temperature between periods of operation. It may be necessary to pressurize such systems during periods of moderate temperature operation. For example, at 0°C the vapor pressure of R-11 is 39.9 kPa (299 mm Hg); that of a 22% solution of calcium chloride is only 0.49 kPa (3.7 mm Hg). The cost of vapor losses, the toxicity of the escaping vapors, and their flammability should be carefully considered in the design of the semi-closed or open system.

Energy requirements of brine systems may be greater because of the power required to circulate the brine and because of the extra heat-transfer process which necessitates the maintenance of a lower evaporator temperature. Figure 8 shows the approximate efficiency of the system using various brines.

Use of Ice. Where water is not harmful to a product or process, ice may be used to provide refrigeration. Direct application of ice or of ice and water is a rapid way to control a chemical reaction or remove heat from a process. The rapid melting of ice furnishes large amounts of refrigeration in a short time and allows leveling-out of the refrigeration capacity required for batch processes. This stored refrigeration also is desirable in some processes where cooling is critical from the standpoint of safety or serious product spoilage.

Large ice plants such as the block-ice plants built during the 1930s are not being built today. However, ice still is used extensively, and equipment to make flake or cube ice at the point of use is commonly employed. This method avoids the loss of crushing and minimizes transportation costs.

System Components, Construction, and Operation

Compressors. Reciprocating or centrifugal compressors can be used singly or in parallel and series combinations for process compression and refrigeration applications (see High pressure technology).

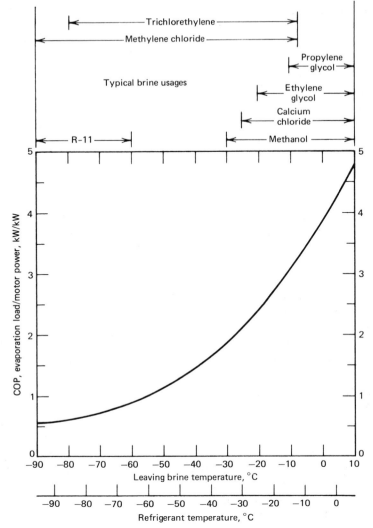

Figure 8. Power requirements at various temperature levels. Centrifugal brine chilling systems based on 29.5°C condenser water. Includes gear losses.

Modern high speed reciprocating compressors with displacements up to 0.283–0.472 m³/s (600–1000 cfm) generally are limited to a pressure ratio of about 9. The reciprocating compressor is basically a constant-volume variable-head machine. It handles various discharge pressures with relatively small changes in inlet-volume flow rate as shown by the heavy line in Figure 9.

Open systems and many processes require nearly fixed compressor suction and discharge pressure levels. This load characteristic is represented by the horizontal typical open-system line on Figure 9. In contrast, condenser operation in many closed systems is related to ambient conditions, eg, through cooling towers, so that on cooler days the condenser pressure can be reduced. When the refrigeration load is lower, less

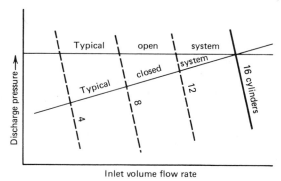

Figure 9. Volume–pressure relationships in a reciprocating compressor.

refrigerant circulation is required. The resulting load characteristic is represented by the typical closed-system line on Figure 9.

The compressor must be capable of matching the pressure and flow requirements imposed upon it by the system in which it operates. The reciprocating compressor matches the imposed discharge pressure at any level up to its limiting pressure ratio. Varying flow requirements can be met by providing devices that unload individual or multiple cylinders. This unloading is accomplished by blocking the suction or discharge valves that open either manually or automatically. Speed control also can be used to effect changes in capacity.

Most reciprocating compressors have a lubricated design. Oil is pumped into the process gas stream or refrigeration system during operation. Thus, systems must be designed carefully to return oil from various parts of the system to the compressor crankcase to provide for continuous lubrication and also to avoid contaminating heat-exchanger surfaces. At very low temperatures (ca −50°C or lower, depending on refrigerants used) oil becomes too viscous to return, and provision must be made for periodic plant shutdown and warmup to allow manual transfer of the oil.

Compressors usually are arranged to start unloaded so that normal torque motors are adequate for starting. When gas engines are used for reciprocating compressor drives, careful torsional analysis is essential.

The centrifugal compressor is preferred whenever the gas volume is high enough to allow its use, because it offers better control, simpler hookup, minimal lubrication problems and lower maintenance. Single-impeller designs are directly connected to high speed drives or driven through an internal speed increaser. These machines are ideally suited for clean, noncorrosive gases in moderate-pressure process or refrigeration cycles in the range of 0.236–1.89 m³/s (500–4000 cfm). Multistage centrifugal compressors are built for direct connection to high speed drives or for use with an external speed increaser. Designs available from suppliers generally provide for two to eight impellers per casing covering the range of 0.236–11.8 m³/s (500–25,000 cfm) depending on the operating speed. A wide choice of materials and shaft seals to suit any gas composition, including dirty or corrosive process streams, is available.

The centrifugal compressor has a more complex head–volume characteristic than reciprocating machines. Changing discharge pressure may cause relatively large changes in inlet volume, as shown by the heavy line in Figure 10. Adjustment of variable inlet vanes or of a diffuser ring allows the compressor to operate anywhere below the heavy line to match conditions imposed by the system. A variable-speed driver offers

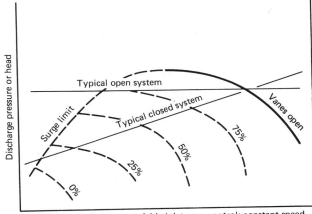

Inlet-volume flow rate, variable inlet-vane control; constant speed

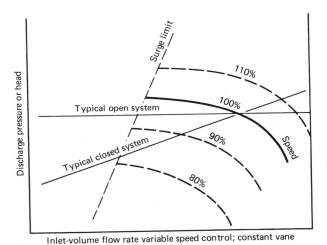

Inlet-volume flow rate variable speed control; constant vane

Figure 10.   Volume–pressure relationships in a centrifugal compressor.

an alternative way to match the compressor's characteristics to the system load, as shown in the lower half of Figure 10. The maximum head capability is fixed by the operating speed of the compressor. Both methods have advantages: generally, variable inlet vanes or diffuser rings provide a wider range of capacity reduction; variable speed usually is more efficient. Maximum efficiency and control can be obtained by considering both methods of control.

The centrifugal compressor has a surge point, ie, a minimum-volume flow below which stable operation cannot be maintained. The percentage of load at which the surge point occurs depends on the number of impellers, design-pressure ratio, operating speed, and variable inlet-vane setting. The system design and controls must keep the inlet volume above this point by artificial loading if necessary. This is accomplished with a bypass valve and gas recirculation. Combined with a variable inlet-vane setting, variable diffuser ring, or variable speed control, the gas bypass allows stable operation down to zero load.

Provision for no-load operation is strongly recommended for all chemical-plant installations, regardless of expected fluctuations in plant load. This permits the refrigeration system to be started up and thoroughly checked out independently of the chemical process.

The contrast between the operating characteristics of the reciprocating compressor and the centrifugal compressor are important considerations in plant design to achieve satisfactory performance. Unlike a reciprocating compressor, the centrifugal compressor will not rebalance abnormally high system heads. The drive arrangement for the centrifugal compressor must be selected with sufficient speed to meet the maximum head anticipated. The relatively flat head characteristic of the centrifugal compressor necessitates different control approaches than for reciprocating machines, particularly when parallel compressors are utilized. These differences, which account for most of the troubles experienced in centrifugal-compressor systems, cannot be overlooked in the design of a plant.

A system that uses centrifugal compressors designed for high pressure ratios and that requires the compressors to start with high suction density existing during standby poses a special problem of starting torque. This situation can be handled by the driver if it has sufficient starting torque. If not, the system must have provisions to reduce the suction pressure at startup. This problem is particularly important when using single-shaft gas-turbine engines, or reduced-voltage starters on electric drives. Split-shaft gas turbines are preferred for this reason.

Drive ratings that are affected by ambient temperatures, altitudes, etc, must be evaluated at the actual operating conditions. Refrigeration installations normally require maximum output at high ambient temperatures, a factor that must be considered when using drives such as gas turbines and gas engines.

Condensers. The refrigerant condenser is used to reject the heat of compression and the process heat load picked up in the evaporator. This heat can be rejected to cooling water or air, both of which are commonly used.

The heat of compression depends upon the compressor horsepower and becomes a significant part of the load on low temperature systems affecting the size of condensers. The condenser heat-rejection rate can be expressed as follows:

$$\text{rejection rate} = \text{evaporator load} + \text{motor power}$$

where both rejection rate and evaporator load are in kilowatts (kJ/s). Approximate condenser heat-rejection rates are summarized in Table 4 for various evaporator temperatures.

Table 4. Approximate Condenser Heat-Rejection Rates

Evaporator temperature, °C	Approximate kJ rejection per kJ evaporator load
10	1.2
−20	1.5
−50	1.8
−80	2.2
−110	2.5
−140	2.8
−160	3.1

Water-cooled shell-and-tube condensers designed with finned tubes and fixed tube sheets generally provide the most economical exchanger design for refrigerant use. Figure 11 shows a typical refrigerant condenser. Commercially available condensers conforming to ASME Boiler and Pressure Vessel Code (6) construction adequately meet both construction and safety requirements for this duty. Design practices essential to many special process exchangers used in refineries and chemical industries add unnecessary expense and, in most cases, result in exchangers less suited for this application than standard condensers specifically developed for refrigerant condensing.

It is common practice in chemical plants to tie in refrigeration equipment with a water-cooling system serving the entire plant needs. The temperature and water flow rate variation must be carefully surveyed. Separate controls on the cooling water for the refrigeration condenser may be needed. For example, water temperatures, which control condensing pressure, may have to be maintained above a minimum value to ensure proper refrigerant liquid feeding to all parts of the system. Separate valving may be necessary to shut off the condenser water supply when the refrigeration unit is shut down.

Cooling towers and spray ponds are frequently used for water-cooling systems. These generally are sized to provide 29°C supply water at design load conditions. Circulation rates typically are specified so that design cooling loads are handled with a 5.6°C cooling-water temperature rise. Pump power, tower fans, makeup water (ca 3% of the flow rate) and water treatment should be taken into account in operating cost studies (see Water, industrial water treatment).

River or well water, when available, provides an economical cooling medium. Quantities circulated will depend on initial supply temperatures and pumping cost but are generally selected to handle the cooling load with 8.3–16.6°C water temperature range. Water treatment and special exchanger materials frequently are necessary because of the corrosive and scale-forming characteristics of the water. Well water, in particular, must be analyzed for corrosive properties, with special attention given to the presence of dissolved gases, eg, H_2S and CO_2. These are extremely corrosive to many exchanger materials, yet difficult to detect in sampling. Pump power, water treatment, and special condenser material should be evaluated when considering costs.

Allowances must be made in heat-transfer calculations for fouling or scaling of exchanger surfaces during operation. This ensures sufficient surface to maintain rated performance over a reasonable interval of time between cleanings. Scale-factor allowances are expressed in $(m^2 \cdot K)/kW$ as additional thermal resistance.

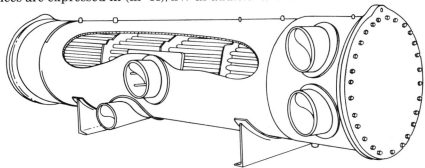

Figure 11.   Typical refrigerant condenser.

Commercial practice normally includes a scale-factor allowance of 0.088. However, the long hours of operation usually associated with chemical-plant service and the type of cooling water frequently encountered generally justify a greater allowance to minimize the frequency of downtime for cleaning. Depending on these conditions, an allowance of 0.18 or 0.35 is recommended for chemical-plant service. Scale allowance can be reflected in system designs in two ways—as more heat-exchanger surface or as higher design condensing temperatures with attendant increase in compressor power. This is illustrated in Figure 12. Generally, a compromise between these two approaches is most economical. For extremely bad water, parallel condensers, each with 60–100% capacity, may provide a more economical selection and permit cleaning one exchanger while the system is operating.

Use of air-cooled condensing equipment for all chemical-plant service is on the increase. With tighter restrictions on the use of water, air-cooled equipment is used even on larger centrifugal-type refrigeration plants, although it requires more physical space than cooling towers. The service required of an air-cooled condenser in a chemical-plant atmosphere frequently dictates the use of the more expensive alloys in tube construction or of conventional materials with higher wall thickness to give acceptable service life. A battery of air-cooled condensers, with propeller fans located at the top, pull air over the condensing coil. Circulating fans and exchanger surface are usually selected to provide design condensing temperatures of 49–60°C with 35–38°C ambient dry-bulb temperature.

The design dry-bulb temperature should be carefully considered since most weather data reflect an average or mean maximum temperature. If full load operation must be maintained at all times, care should be taken to provide sufficient condenser capacity for the maximum recorded temperature. This is particularly important when the compressor is centrifugal because of its flat-head characteristics and the need for adequate speed. Multiple-circuit or parallel air-cooled condensers must be provided with traps to prevent liquid backup into the idle circuit at light load. Pressure drop through the condenser coil must also be considered in establishing the compressor discharge pressure.

In comparing water-cooled and air-cooled condensers, the compression horsepower at design conditions is invariably higher with air-cooled condensing. However, ambient air temperatures are considerably below the design temperature most of the time, and operating costs frequently compare favorably over a full year. In addition,

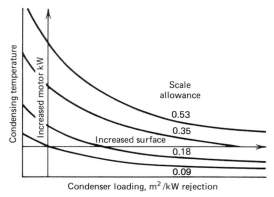

Figure 12.   Effect of different scale factors.

air-cooled condensers usually require less maintenance, although dirty or dusty atmospheres may affect this.

Evaporators. The broad range of chemical-plant cooling requirements has led to an extremely wide variety of refrigerant evaporator designs. The chemical-plant design engineer is concerned with heat-exchanger design for nearly every process. Familiarity with heat-exchanger requirements may tempt the chemical engineer to treat the design of the evaporator for a refrigeration system in a fashion identical to evaporators for nonrefrigeration service. Although the general laws of heat transfer apply in both cases, there are special requirements for evaporators in refrigeration service that are not always present in other types of heat-exchanger design. These include problems of oil return, flash-gas distribution, gas–liquid separation, and submergence effects (see Evaporation).

Oil Return. When the evaporator is used with reciprocating-compression equipment, it is necessary to assure adequate oil return from the evaporator. If oil will not return in the refrigerant flow, it is necessary to provide an oil reservoir for the compression equipment and to remove oil mechanically from the low side of the system on a regular basis. Evaporators used with centrifugal compressors do not normally require oil return from the evaporator since centrifugal compressors pump very little oil into the system. However, even with centrifugal equipment, low temperature evaporators eventually may become contaminated with oil which must be reclaimed.

Flash-Gas Distribution. As a general rule, refrigerants are introduced into the evaporator by expanding liquid from a higher pressure. In the expansion process, a significant amount of refrigerant flashes off into gas. This flash gas must be introduced properly into the evaporator for satisfactory performance. Improper distribution of this gas can result in liquid carryover to the compressor and in damage to the exchanger tubes from erosion or from vibrations.

Gas–Liquid Separation. The suction gas leaving the evaporator must be dry to avoid compressor damage. The design should provide adequate separation space or include mist eliminators. Liquid carryover is one of the most common sources of trouble with refrigeration systems.

Submergence Effect. In flooded evaporators the evaporating pressure and temperature at the bottom of the exchanger surface is higher than at the top of the exchanger surface owing to the liquid head. This static head or submergence effect significantly affects the performance of refrigeration evaporators operating at extremely low temperatures and low suction pressures.

Beyond these basic refrigeration-design requirements, the chemical industry imposes many special conditions. Exchangers frequently are applied to cool highly corrosive process streams; consequently, special materials for evaporator tubes and channels of particularly heavy wall thicknesses are dictated. Corrosion allowances, ie, added material thicknesses, in evaporator design may be necessary in chemical service.

High pressure and high temperature design, particularly on the process side of refrigerant evaporators, is frequently encountered in chemical-plant service. Process-side construction may have to be suitable for pressures seldom encountered in commercial service, and differences between process inlet and leaving temperatures $\geq 55°C$ are not uncommon. In such cases, special consideration must be given to thermal stresses within the refrigerant evaporator. U-tube construction or floating-tube-sheet construction may be necessary. Minor process-side modifications may permit use of less expensive standard commercial fixed-tube-sheet designs. However, coordination between the equipment supplier and chemical-plant designer is necessary to tailor the evaporator to the intended duty. Relief devices and safety precautions common to the refrigeration field normally meet chemical-plant needs but should be reviewed against individual plant standards. Here again, it must be the mutual responsibility of the refrigeration equipment supplier and the chemical-plant designer to evaluate what special features, if any, must be applied to modify commercial equipment for chemical-plant service.

Process requirements in the chemical plant may call for very sudden or unexpected load changes on the refrigeration evaporator. Possible thermal shocks, with attendant stresses, must be evaluated in the design phase and the evaporator suitably designed to meet all conditions of its particular service.

Evaporators in chemical-plant duty normally require inspection and cleaning on an annual basis. For this reason, it is sometimes necessary to modify commercial designs for increased accessibility and ease of tube replacement.

Possible contamination of the process stream by leakage from the refrigerant side, or contamination of the refrigerant side by process fluid leakage, should be evaluated in the design of chemical-plant evaporators. Special means of leak detection may be dictated.

Low temperature refrigeration in the chemical-plant industry frequently results in extremely high viscosities on the process side of the equipment. When this is the case, special evaporator designs may be necessary to minimize process-side pressure drops and maintain optimum heat transfer. In general, the process stream is on the tube side of the evaporator to simplify cleaning. Although commercial practice in evaporator design calls for tubes of small inside diameter to achieve optimum heat-transfer performance, the small tube diameters may be incompatible with the process stream. Extra large tube sizes may be required in some chemical-plant evaporators. In extremely low temperature and extremely viscous chemical-plant cooling duties, evaporators may be provided with rotating internal scrapers within the tubes to provide for continual scraping of the heat-transfer wall and to assure flow of high viscosity fluid through the evaporators.

Refrigeration evaporators are usually designed to meet the ASME Boiler and Pressure Vessel Code (6), which provides for a safe reliable exchanger at economical cost. In refrigeration systems, these exchangers generally operate with relatively small temperature differentials for which fixed-tube-sheet construction is preferred. Refrigerant evaporators also operate with simultaneous reduction in pressure as temperatures are reduced. This relationship results in extremely high factors of safety on pressure stresses, eliminating the need for expensive nickel steels from -29 to $-59°C$. Most designs are readily modified to provide suitable materials for corrosion problems on the process side.

The basic shell-and-tube exchanger with fixed tube sheets (Fig. 13) is most widely used for refrigeration evaporators. Most designs are suitable for process fluids up to ca 2170 kPa (300 psig) and for operation with up to 38°C temperature differences. Above these limits, specialized heat exchangers generally are used to suit individual requirements.

With the process fluid on the tube side, the shell side is flooded with refrigerant for efficient wetting of the tubes (see Fig. 14). Designs must provide for distribution

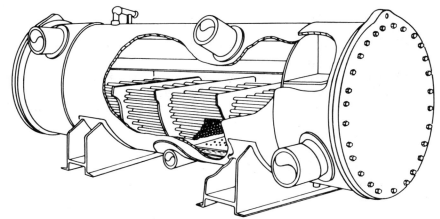

Figure 13. Typical fixed-tube-sheet evaporator.

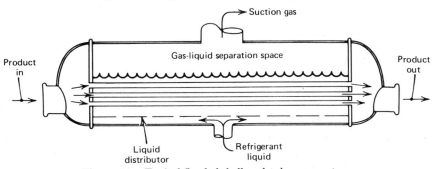

Figure 14. Typical flooded shell-and-tube evaporator.

of flash gas and liquid refrigerant entering the shell and for separation of liquid from the gas leaving the shell before it reaches the compressor.

In low temperature applications and large evaporators, the exchanger surface may be sprayed rather than flooded. This eliminates the submergence effect or static-head penalty, which can be significant in large exchangers, particularly at low temperatures. The spray cooler (Fig. 15) is recommended for some large coolers to

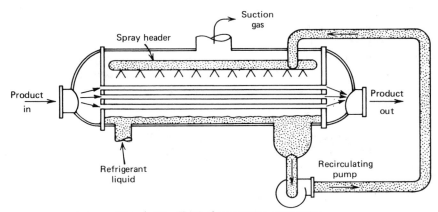

Figure 15. Typical spray-type evaporator.

offset the cost of refrigerant inventory or charge which would be necessary for flooding.

Where the Reynolds number in the process fluid is low, as for a viscous or heavy brine, it may be desirable to handle the fluid on the shell side to obtain better heat transfer. In these cases, the refrigerant must be evaporated in the tubes. On small exchangers, commonly referred to as direct-expansion coolers, refrigerant feeding is generally handled with a thermal-expansion valve.

On large exchangers, this can best be handled by a small circulating pump to ensure adequate wetting of all tubes (see Fig. 16). An oversize channel box on one end provides space for a liquid reservoir and for effective liquid–gas separation.

System Design Considerations

Chemical processes may be continuous or batch. Continuous processing is characterized by temperatures, pressures, flow levels, compositions, and other parameters that do not change with time. Associated with continuous operation are refrigeration start-up and shutdown conditions that invariably differ, sometimes widely, from those of the process itself. These conditions, although they occupy very little time in the life of the installation, must be properly accommodated in the design of the refrigeration system. Consideration must be given to the amount of time required to achieve design operating conditions, the need for standby equipment, etc.

In batch processing, operating conditions are expected to change with time, usually in a repetitive pattern. The refrigeration system must be designed for all extremes. Use of brine storage or ice banks can reduce equipment sizes for batch processes.

Cooling loads can generally be classified in three types in refrigeration systems: sensible cooling, condensing, and partial condensing.

Many chemical cooling loads extend over a wide temperature range. With sensible cooling, the load is almost directly proportional to temperature reduction. Particularly in larger plants, applying refrigeration in several stages improves operating economy. If all of the cooling is done at the final lowest temperature, the compression power required is a maximum. Because compression power increases substantially as temperatures are lowered, the power can be reduced by breaking the load into several stages and cooling part of it at higher temperatures. The number of stages must be

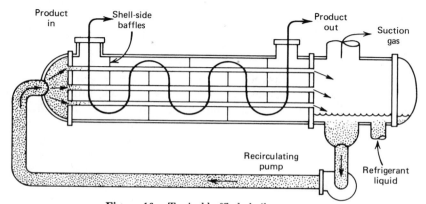

Figure 16. Typical baffled-shell evaporator.

balanced against the first cost of equipment, which increases as the number of stages is increased.

Condensing loads are essentially isothermal. Refrigeration must be applied at the required condensing temperature.

Partial condensing, which is frequently encountered in chemical processes because of impurities such as air or high pressure constituents, justifies staged cooling. In most situations, the process stream requires a small amount of sensible cooling. Condensing then generates a significant cooling load as the temperature is reduced through the next few degrees. From this point, the load diminishes. Significant temperature reduction is required to accomplish the final stages of condensing, however, because of the increasing percentage of noncondensables which rapidly reduce the partial pressure and required condensing temperature of the condensing constituent.

Closed-cycle operation involves both liquid and gas phases. System designs must take into account liquid-flow problems in addition to gas-flow requirements and must provide for effective separation of the liquid and gas phases in different parts of the system. These factors require careful design of all components and influence the arrangement or elevation of certain components in the cycle.

Liquid pressures must be high enough to feed liquid to the evaporators at all times, especially when evaporators are elevated or remotely located. In some cases, a pump must be used to suit the process requirements. The possibility of operation with reduced pressures caused by colder condensing temperatures than the specified design conditions must also be considered. Depending on the types of liquid valves and relative elevation of various parts of the system, it may be necessary to maintain condensing pressures above some minimum level, even if doing so increases the compression power.

Provision must be made to handle any refrigerant liquid that can drain to low spots in the system upon loss of operating pressure during shutdown. It must not be allowed to return as liquid to the compressor upon startup.

The operating charge in various system components fluctuates depending on the load. For example, the operating charge in an air-cooled condenser is quite high at full load but is low, ie, essentially dry, at light load. A storage volume such as a liquid receiver must be provided at some point in the system to accommodate this variation. If the liquid controls permit the evaporator to act as the variable storage, the level may become too high, resulting in liquid carryover to the compressor.

Abnormally high process temperatures may occur either during startup or process upsets. Provision must be made for this possibility, for it can cause damaging thermal stresses on refrigeration components and excessive boiling rates in evaporators, forcing liquid to carry over and damage the compressor.

Factory-designed and built packages, which provide cooling as a service or utility, can require several thousand kilowatts of power to operate, but in most cases, they require no more installation than connection of power, utilities, and process lines. As a result, there is a single source of responsibility for all aspects of the refrigeration cycle involving the transfer and handling of both saturated liquids and saturated vapors throughout the cycle, oil return, and other design requirements. These packages are custom engineered, including selection of components, piping, controls, base designs, torsional and critical speed analysis, and individual chemical process requirements. Large packages are designed in sections for shipment but are readily interconnected in the field.

As a general rule, field-erected refrigeration systems should be close-coupled to minimize problems of oil return and refrigerant condensation in suction lines. Where process loads are remotely located, pumped recirculation or brine systems are recommended. Piping and controls should be reviewed with suppliers to assure satisfactory operation under all conditions.

Refrigeration System Specifications

To minimize costly and time-consuming alterations owing to unexpected requirements, the refrigeration specialist who is to do the final design must have as much information as possible before the design is started. Usually, it is best to provide more information than thought necessary, and it is always wise to note where information may be sketchy, missing, or uncertain. Carefully spelling out the allowable margins in the most critical process variables and pointing out portions of the refrigeration cycle that are of least concern is always helpful to the designer.

A checklist of minimum information (Table 5) needed by a refrigeration specialist to design a cooling system for a particular application may be helpful.

Process flow sheets. Seeing the process flow sheets is the best overall means for the refrigeration engineer to become familiar with the chemical process for which the refrigeration equipment is to be designed. In addition to providing all of the information shown in Table 5, they give the engineer a feeling for how the chemical plant will operate as a system and how the refrigeration equipment fits into the process. This will allow the engineer to be on the lookout for any areas where savings or more reliable operation in the refrigeration equipment can be achieved or where problem areas may occur. Inability to see the complete process flow sheets, eg, for reasons of process secrecy, will limit the refrigeration engineer's effectiveness in providing the proper refrigeration system for the job. In this case, it is extremely important that the refrigeration engineer be given all of the information specified in the checklist.

Basic specifications. This portion of Table 5 fills in the detailed mechanical information that is not found on the flow sheets. The information requested tells the refrigeration engineer how the equipment should be built, where it will be located, and specific safety requirements. This determines which standard equipment can be used and what special modifications need to be made.

Instrumentation and control requirements. These tell the refrigeration engineer how the system will be controlled by the plant operators and how it will connect to the rest of the chemical-process controls. Particular controller types as well as control sequencing and operation must be spelled out to avoid misunderstandings and costly redesign. The refrigeration engineer needs to be aware of the degree of control required for the refrigeration system; for example, the process may require remote starting and stopping of the refrigeration system from the central control room. This could influence the way in which the refrigeration safeties and interlocks are designed (see also Instrumentation and control).

Off-design operation. It is likely that the most severe operation of the refrigeration system will occur during startup or shutdown of the chemical process. The rapidly changing pressures, temperatures, and loads experienced by the refrigeration equipment can cause motor overloads, compressor surging, or loss of control if they are not anticipated during design.

Table 5.   Necessary Information for Design of a Cooling System

Complete process flow sheets
  type of process
    batch
    continuous
  normal heat balances
  normal material balances
  normal material composition
  normal operating pressures and temperatures
  normal refrigeration loads
  energy recovery possibilities
  manner of supplying refrigeration; ie, primary or
    secondary refrigerant
Basic specifications
  mechanical system details
    construction standards
      industry
      company
      local plant
    insulation requirements
    special corrosion prevention requirements
    special sealing requirements
      process streams to the environment
      process stream to refrigerant
  operating environment
    indoor or outdoor location
    extremes
    special requirements
  special safety considerations
    known hazards of process
    toxicity and flammability constraints
    maintenance limitations
  reliability requirements
    effect of loss of cooling on process safety
    maintenance intervals and types that may be

performed
    redundancy requirement
    acceptance test requirements
Instrumentation and control requirements
  safety interlocks
  process interlocks
  special control requirements
    at equipment
    central control room
  special or plant standardized instruments
  degree of automation: interface requirements
  industry and company control standards
Off-design operation
  process startup sequence
    degree of automation
    refrigeration loads vs time
    time needed to bring process onstream
    frequency of startup
    process pressure, temperature, and composition
      changes seen by refrigeration equipment
      during startup
    special safety requirements
  minimum load
  need for standby capability
  peak-load pressures and temperatures
  composition extremes
  process shutdown sequence
    degree of automation
    refrigeration load vs time
    shutdown time span
    process pressure, temperature, and composition
      changes
    special safety requirements

BIBLIOGRAPHY

"Refrigeration" in *ECT* 2nd ed., Vol. 17, pp. 295–327, by J. R. Chamberlain and R. A. Dorwart, York Division, Borg-Warner Corporation.

1. *ASHRAE Handbook*, American Society of Heating, Refrigerating, and Air Conditioning Engineers, Inc., Publication Department, Atlanta, Ga., 1978, Chapt. 50.
2. W. R. Woolrich, *The Men Who Created Cold*, Exposition Press, Inc., New York, 1967, p. 23.
3. *ASHRAE Standard 34-78, Number Designations of Refrigerants*, ASHRAE, Inc., Publications Department, Atlanta, Ga., 1978.
4. *ANSI Standard B9.1, Safety Code for Mechanical Refrigeration*, New York.
5. G. G. Haselden, *Mech. Eng.*, 44 (Mar. 1981).
6. *ASME Boiler and Pressure Vessel Code*, Sect. VIII, Div. 1, The American Society of Mechanical Engineers, New York, 1980.

General References

F. C. McQuiston and J. D. Parker, *Heating, Ventilating and Air Conditioning, Analysis and Design*, John Wiley & Sons, Inc., New York, 1977.
W. F. Stoecker, *Design of Thermal Systems*, Industrial Press, Inc., New York, 1968.

ASHRAE Handbooks, American Society of Heating, Refrigerating and Ventilating Engineers, Inc., Publications Department, Atlanta, Ga., 4 Volumes: Fundamentals, Equipment, Systems, Applications, one volume revised each year.

ASHRAE Journal, ASHRAE, Inc., Atlanta, Ga., monthly.

ASHRAE Transactions, ASHRAE, Inc., Atlanta, Ga., biannually.

Compressor Handbook for the Hydrocarbon Processing Industries, Gulf Publishing Company, Book Division, Houston, Texas, 1979.

Klima & Kalte Ingenieur, Verlag C. F. Muller, Karlsruhe, Germany, monthly.

Chemical Engineering, McGraw-Hill, Inc., New York, biweekly.

Heating, Piping, and Air Conditioning, Reinhold Publishing Division, Chicago, Ill., monthly.

IIR Bulletin, International Institute of Refrigeration, Paris, France.

V. M. Solomonov, A. Yu. Goikhamn, and B. A. Ustinnikov, *Kholod. Tekh.* (1), 18 (1981).

W. Duscha, *Chem. Tech. (Heidelberg)* **7**(8), 323 (1978).

B. Fricke, *Chem. Tech. (Heidelberg)* **7**(4), 137 (1978).

K. Ackroyd, *Chem. Eng. (London)* **332**, 366 (1978).

G. Holldorf, *Kaelte Klimatech.* **30**(12), 497 (1977).

M. Sekine, *Nippon Kikai Gakkaishi* **80**(709), 1268 (1977).

H. Noack, *Lehrb. Kaeltetch.* **1**, 395 (1975).

H. J. Bauder, *Lehrb. Kaeltetech.* **1**, 227 (1975).

Yu. R. Mehra, *Chem. Eng.* **85**(28), 97 (Dec. 1978).

D. P. Shea and J. E. Shelton, *Industrial Application of a Direct Refrigeration System for Energy Conservation*, ASME 80-Pet-1, ASME, New York, Feb. 3–7, 1980.

G. C. Briley, *Hydrocarbon Process.* **55**(5), 173 (May 1976).

K. W. Cooper
K. E. Hickman
York Division
Borg-Warner Corporation

REGULATORY AGENCIES

The U.S. government must provide for the welfare of its citizens; thus, the maintenance of a clean and healthy environment, especially as related to chemical engineering, is regulated by government agencies. In the attempt to meet changing needs, the regulations change rapidly, and the specifications of these regulations quickly become outdated. Therefore, this article presents an overview of the Federal laws that affect human health and environment, especially as they apply to industrial facilities, with a discussion of the general intent of the more important regulations. Relevant nomenclature is listed at the end of the article.

The two main Federal agencies involved in the protection of human health and environment are the Environmental Protection Agency (EPA) and the Occupational Safety and Health Administration (OSHA). The EPA's principal concern is the protection of the environment, especially as it affects the quality of life. There are ten regional offices that carry out the regulatory functions of the agency (see Table 1).

The principal function of OSHA is the protection of people in the workplace. Regional offices are in the same areas as those of the EPA (see Table 1). There are a number of other Federal agencies involved in related work. Pertinent agencies and their areas of concern are listed in Table 2.

The General Services Administration of the Federal government publishes a daily document, the *Federal Register*, which contains proposed regulations, final regulations, and notices for all of the Federal agencies. Subscriptions to the *Federal Register* can be made through the Superintendent of Documents, U.S. Government Printing Office, Washington, D.C. 20402. At the end of each month, the Office of the Federal Register publishes separately a List of Code of Federal Regulations (CFR) Sections Affected (LSA), which lists parts and sections affected by documents published since the revision date of each title. The first issue of each month of the *Federal Register* gives all CFRs issued within the past year, including price. All documents can be ordered from the Superintendent of Documents, U.S. Government Printing Office, Washington, D.C. 20402. Some large libraries maintain a CFR file. Regulations are listed by agency in the Code of Federal Regulations. The listing first gives the number for

Table 1. Federal Regions for Various Regulatory Agencies Including the EPA and OSHA

Region	States, etc	Location of regional office
1	Conn., Maine, Mass., N.H., R.I., Vt.	Boston, Mass.
2	N.J., N.Y., Puerto Rico, Virgin Islands	New York, N.Y.
3	Del., Wash. D.C., Md., Pa., Va., W. Va.	Philadelphia, Pa.
4	Ala., Fla., Ga., Ky., Miss., N.C., S.C., Tenn.	Atlanta, Ga.
5	Ill., Ind., Mich., Minn., Ohio, Wisc.	Chicago, Ill.
6	Ark., La., N.M., Okla., Texas	Dallas, Texas
7	Iowa, Kan., Mo., Neb.	Kansas City, Mo.
8	Colo., Mont., N.D., S.D., Utah, Wyo.	Denver, Colo.
9	Ariz., Calif., Nev., Hawaii, Trust Territory of the Pacific Islands (Carolina Is., Mariana Is., Marshall Is.)	San Francisco, Calif.
10	Alaska, Idaho, Ore., Wash.	Seattle, Wash.

Table 2.　**Federal Agencies and Their Functions**

Agency (acronym)	Function
Agriculture Department (USDA)	agriculture, animal- and plant-health inspection, forest service, food safety, Rural Electrification Administration, soil conservation service
Commerce Department (Commerce)	Census Bureau, economics, trade, National Oceanic and Atmospheric Administration, Patent and Trademark Office
Department of Defense (DOD)	Engineers Corps: dredge and fill operations, waterways, etc
Energy Department (DOE)	all aspects of energy use, conservation, costs, etc
Health and Human Services Department (HHS)	Food and Drug Administration (FDA), National Institute for Occupational Safety and Health (NIOSH)
Interior Department (Interior)	land management, fish and wildlife, Geological Survey, mines, surface mining and reclamation
Justice Department (Justice)	Antitrust Division
Labor Department (Labor)	employment standards and statistics, OSHA, Mine Safety and Health Administration
Transportation Department (DOT)	Coast Guard; Federal Aviation Administration (FAA); highway, railroad, and maritime administration; hazardous material shipping
Treasury Department (Treasury)	Alcohol, Tobacco, and Firearms Bureau (ATF); Customs Service, Internal Revenue Service
Council on Environmental Quality (CEQ)[a]	provides policy advice on environmental matters; implements NEPA; coordinates environmental concerns among other agencies
Consumer Product Safety Commission (CPSC)[a]	
Equal Employment Opportunity Commission (EEOC)[a]	
Interstate Commerce Commission (ICC)[a]	
Nuclear Regulatory Commission (NRC)[a]	ensures that radioactive materials are used and nuclear facilities are conducted with regard for environment and public health, safety, and security
Office of Management and Budget (OMB)[a]	reviews regulations to ensure that they are cost-effective; approves paperwork or other information-gathering requirements; Regulatory Flexibility Act
Synthetic Fuels Corporation (SFC)[a]	
Tennessee Valley Authority (TVA)[a]	
Water Resources Council (WRC)[a]	encourages conservation, development, and utilization of water and related land resources

[a] Independent agencies.

the agency, then CFR, and then the chapter number. An example is the Environmental Protection Agency (EPA) regulations on National Primary and Secondary Ambient Air Quality Standards: 40 CFR 50. Regulations of the EPA are listed under 40, then each set of regulations has a different chapter number.

In addition to the Federal agencies, there are many state agencies that regulate the environmental and health areas. When state regulations are not as stringent as Federal regulations, the Federal rulings usually take precedence.

The difference between laws and regulations is important. For the former, Congress first passes a bill, which is then signed by the President, and thereby made a law. The act describes what Congress wants regulated, the general method to be used, and the ultimate results expected. It is then the responsibility of the designated agency to write and administer regulations to meet these requirements. First, the agency issues draft regulations that are mainly for internal review. It also tries to obtain the response of the affected portions of the public and industry. Next, proposed regulations are issued in the *Federal Register*. A specific comment period is allowed, during which public hearings may be held. The Office of Management and Budget also reviews the regulation. Once the comments are received and revisions are made, the final regulations are issued in the *Federal Register*. Every few years, Congress reviews a particular law and its regulations to see how well it is working. At that time, it can be amended to make any improvements.

Figure 1 shows the main Federal and state environmental permits and regulatory approvals.

Water

For a long time in the United States, the approach to water-pollution control was through the establishment of water-quality standards, with most limits established on a state-by-state basis (see Water, water pollution). There was no effective, national, legal authority to limit the discharge of pollutants. In the late 1960s, the U.S. government revived an old law, the Rivers and Harbor Act of 1899 (the Refuse Act) (1). The law prohibited the discharge of anything into navigable waters unless a permit was obtained from the Corps of Engineers. This provided a first step toward control of industrial discharges. In addition, some preliminary laws were passed: the Federal Water Pollution Control Act of 1956, the Pollution Control Act Amendments of 1961, the Water Quality Act of 1965, the Clean Water Restoration Act of 1966, and the Water Quality Improvement Act of 1970.

Definite limits on water-pollutant discharges were enacted by the passage of the Federal Water Pollution Control Act Amendments (FWPCA) of 1972 (2) and the Clean Water Act (CWA) of 1977 (3); the latter is also known as the Federal Water Pollution Control Act Amendments of 1977.

The stated objective of the FWPCA of 1972 was to "restore and maintain the chemical, physical, and biological integrity of the nation's waters." In order to achieve this, specific deadlines were set and certain programs were required. Water quality was to be sufficiently improved by July 1, 1983, to protect aquatic life and to allow for recreational use of the waters. In addition, by July 1, 1985, the discharge of pollutants was to be eliminated. This is known as zero discharge. The Environmental Protection Agency was instructed to develop a number of programs to implement these goals. One general goal was to have a nationwide basis of controls rather than individual state controls.

Water-Quality Standards. The FWPCA expanded the water-quality-standards program, which was first established in the Water Quality Act of 1965. The first step in water-quality standards is "stream-use classification" (4). The individual states must decide what the uses of their water will be. The four categories, as defined by the EPA, are Class A, primary water contact recreation; Class B, propagation of desirable aquatic life; Class C, public water supplies prior to treatment; and Class D, agricultural and industrial uses. States may vary the definition of these classes to meet their own needs.

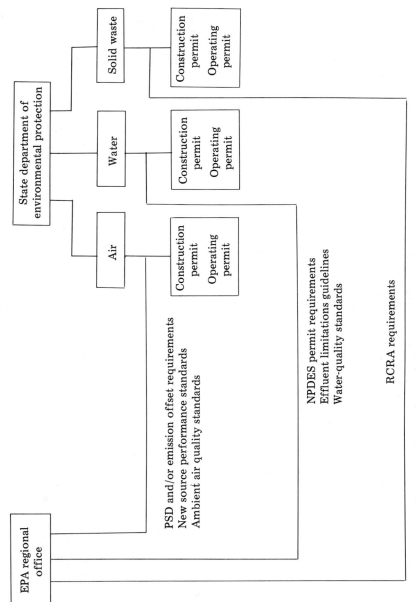

Figure 1. Main Federal and state environmental permits and regulatory approvals.

The second step is to develop "water-quality criteria" (4). This is the specific concentration of a pollutant that is allowable for the designated use. The EPA has been developing most of these criteria for use by the states.

Effluent Guidelines and Standards.   Of special concern to industry are the Effluent Guidelines and Standards for 42 industrial categories (see Table 3). These standards limit the discharge of pollutants, usually in terms of a unit weight of pollutant per unit of either product or feed. Mathematical modeling must be used to determine the water quality resulting from a discharge. As most state standards are water-quality based, mathematical modeling may be necessary.

Table 3.   Sources Subject to Effluent Guidelines and Standards

Source	CFR
dairy products	40 CFR 405
grain mills	40 CFR 406
canned and preserved fruits and vegetables	40 CFR 407
canned and preserved seafood	40 CFR 408
sugar processing	40 CFR 409
textile mills	40 CFR 410
cement manufacturing	40 CFR 411
feedlots	40 CFR 412
electroplating	40 CFR 413
organic chemicals manufacturing	40 CFR 414
inorganic chemical manufacturing	40 CFR 415
plastics and synthetic plants	40 CFR 416
soaps and detergents plants	40 CFR 417
fertilizer manufacturing	40 CFR 418
petroleum refining	40 CFR 419
iron and steel manufacturing	40 CFR 420
nonferrous-metals manufacturing	40 CFR 421
phosphate manufacturing	40 CFR 422
steam electric power generation	40 CFR 423
ferroalloy manufacturing	40 CFR 424
leather tanning and finishing plants	40 CFR 425
glass manufacturing	40 CFR 426
asbestos manufacturing	40 CFR 427
rubber-processing plants	40 CFR 428
timber products	40 CFR 429
pulp, paper, and paperboard manufacturing	40 CFR 430
builders' paper and board mills	40 CFR 431
meat products	40 CFR 432
coal mining	40 CFR 434
offshore oil and gas extraction	40 CFR 435
mineral mining and processing	40 CFR 436
pharmaceutical manufacturing	40 CFR 439
ore mining and dressing	40 CFR 440
paving- and roofing-materials manufacturing	40 CFR 443
paint formulating	40 CFR 446
ink formulating	40 CFR 447
gum and wood-chemicals manufacturing	40 CFR 454
pesticides-chemicals manufacturing	40 CFR 455
explosives manufacturing	40 CFR 457
carbon-black manufacturing	40 CFR 458
photographic processing	40 CFR 459
hospitals	40 CFR 460

The effluent standards originally required two levels of treatment: best practicable control technology currently available (BPCTCA), often referred to as BPT, which had to be met by all existing plants by July 1977; and best available technology economically achievable (BATEA), referred to as BAT, which was to be met by all sources by July 1, 1983. By July 1, 1977, ca 90% of all industrial sources and ca 50% of all municipal sources had achieved BPT. Some extensions were granted under the CWA.

The CWA also revised the BAT standards. A new category, best conventional pollutant control technology (BCT), was set up as BAT for conventional pollutants. Conventional pollutants are "oxygen-demanding substances, solids, or nutrients" (5). The two criteria for listing as a conventional pollutant are a substance occurring in nature or one that is similar to it, that has an impact on water quality; and a substance that has "traditionally been a primary focus of waste-water control" (5). As of the beginning of 1982, those pollutants listed by the EPA as conventional are biochemical oxygen demand (BOD), total suspended solids (TSS), fecal coliform bacteria, pH, and oil and grease. Other substances may be added, including chemical oxygen demand (COD) and phosphorus.

Standards for BCT are supposed to be more stringent than those for BPT, but not as stringent as those for BAT. Whereas BAT had to be economically achievable, the cost for BCT must be reasonable in relation to the resultant pollutant reductions and comparable to an equivalent reduction at a publicly owned treatment work (POTW). The BCT standards must be met by July 1, 1984.

The discharge of pollutants that are considered nonconventional, such as nontoxic organics and chemical or thermal pollution, will still be limited under BAT standards. These standards are to be issued by July 1981 and must be met by July 1987. In addition, a consent decree issued June 8, 1976, requires the EPA to promulgate effluent and pretreatment guidelines for 65 classes of 129 specific toxic or hazardous compounds from 21 industries (see Table 4). The standards for these priority pollutants must be met by July 1, 1984.

The EPA has also been developing pretreatment standards for industrial facilities that discharge directly to publicly owned treatment works (POTWs) (40 CFR 403). The three types of pollutants of principal concern are pollutants that interfere with the operation of the POTW, pollutants that contaminate the sludges produced in the POTW, and pollutants that pass through the POTW or are otherwise incompatible.

National Pollutant Discharge Elimination System Permit Program. To ensure adherence to standards, the EPA developed the National Pollutant Discharge Elimination System (NPDES) permit program (40 CFR 125). Any source discharging or planning to discharge to any U.S. water must obtain an NPDES permit. The NPDES permit application must list all pollutants to be discharged, including any priority pollutants, and it must indicate the proposed treatment methods and resultant effluent. The NPDES permit indicates the allowable discharge. As a minimum, the effluent standards must be met, but more stringent limits may be required. Where no standards for a pollutant or industry exist, limits are decided on an individual basis.

Each state is expected to develop its own NPDES permit program; the Federal EPA retains the right of review. The EPA applies the program if the state program has not yet been approved.

Table 4.   Alphabetical Listing of Priority Pollutants [a]

acenaphthene	endosulfan and metabolites
acrolein	endrin and metabolites
acrylonitrile	ethylbenzene
aldrin	fluoranthene
antimony and compounds	haloethers
arsenic and compounds	halomethanes
asbestos	heptachlor and metabolites
benzene	hexachlorobutadiene
benzidine	hexachlorocyclohexane
beryllium and compounds	hexachlorocyclopentadiene
cadmium and compounds	isophorone
carbon tetrachloride	lead and compounds
chlordane (technical mixture and metabolites)	mercury and compounds
chlorinated benzenes (other than	naphthalene
dichlorobenzenes)	nickel and compounds
chlorinated ethanes	nitrobenzene
chloroalkyl ethers	nitrophenols
chlorinated naphthalene	nitrosamines
chlorinated phenols	pentachlorophenol
chloroform	phenol
2-chlorophenol	phthalate esters
chromium and compounds	polychlorinated biphenyls (PCBs)
copper and compounds	polynuclear aromatic hydrocarbons
cyanides	selenium and compounds
DDT and metabolites	silver and compounds
dichlorobenzenes	TCDD (2,3,7,8-tetrachlorodibenzo-p-dioxin)
dichlorobenzidine	tetrachloroethylene
dichloroethylenes	thallium and compounds
2,4-dichlorophenol	toluene
dichloropropane and -propene	toxaphene
dieldrin	trichloroethylene
2,4-dimethylphenol	vinyl chloride
dinitrotoluene	zinc and compounds
diphenylhydrazine	

[a] 40 CFR 401.15.

Drinking Water.   Additional protection of the public water supply is provided by the Safe Drinking Water Act, passed by Congress in 1974 (6). The EPA must develop primary standards to protect public health and secondary standards to protect public welfare.

Underground injection controls (UIC) are also required under this law to protect underground drinking sources.

At this time, interim primary drinking water regulations (40 CFR 142) and secondary drinking water regulations (40 CFR 143) have been issued. Also, criteria and standards for the underground injection program (40 CFR 146) and regulations on the review of projects affecting sole-source aquifers (40 CFR 149) have been promulgated.

Other Laws.   Many other laws have been passed to protect wetlands, coastal areas, and the oceans. These include the Water Research and Development Act of 1978 (7); the Water Resources Planning Act (8); the Port and Tanker Safety Act of 1979 (9); the Marine Protection, Research, and Sanctuaries Act of 1972, which includes control of ocean dumping (10); the Deepwater Port Act of 1974 (11); the National Ocean Planning Act of 1978 (12); and the Outer Continental Shelf Lands Act (13).

Another Federal agency besides the EPA that regulates water pollution is the Army Corps of Engineers. The Corps of Engineers' main concern is navigation. They require permits for water-intake facilities, dredge-and-fill operations, construction of docks, and any other activities that may in any way inhibit navigation (40 CFR 320 and 322–329).

Air

As long ago as the thirteenth century, air pollution (qv) was linked to the burning of coal (qv). The main concern in these early times was the smell from the sulfur in the coal and the effects of the soot. It was not until many years later that the effects of air pollution on people's health were discovered. Various laws have been passed in the United States to control air pollution, including the Air Pollution Control Act of 1955 (14), the Clean Air Act of 1963 (15), and the Air Quality Act of 1967 (16). The first law that had any real effect was the Clean Air Act of 1970 (CAA) (16) followed by the Clean Air Act Amendments of 1977 (16).

National Ambient Air Quality Standards (NAAQS). Under the Clean Air Act of 1970 and the Clean Air Act Amendments of 1977, seven criterion pollutants, ie, pollutants of special concern, were established by the EPA: sulfur oxides (SO_x), particulates, carbon monoxide (CO), hydrocarbons (HC), nitrogen oxides (NO_x), ozone (O_3) (photochemical oxidants), and lead. Criteria documents for each pollutant, which describe the basis for establishing limits, were issued by the EPA. In addition, National Ambient Air Quality Standards (NAAQS) were developed by the EPA. Primary and secondary NAAQS were originally issued April 30, 1972, except for lead, which was issued in October 1978. Primary NAAQS are to protect public health, without regard for "cost, technical feasibility, or any other factors." Secondary NAAQSs are to protect public welfare, ie, the protection of "soil, vegetation, man-made materials, visibility, and personal comfort and well-being." The NAAQS are listed in 40 CFR 50.

The NAAQS are expressed in the form of ground level concentrations (GLC), which are the concentrations of pollutant in the ambient air as measured at ground level, in units of either micrograms per cubic meter or ppm. In order to convert a source's emission in kilograms per hour to a GLC, dispersion modeling must be used.

The Clean Air Act required the establishment of air quality control regions (AQCR) (40 CFR 51 and 40 CFR 52). These are either inter- or intrastate areas that, because of the effects of meteorological, industrial, or socioeconomic factors, would tend to have a uniform air-pollution pattern. Each AQCR was then monitored to determine whether it was in compliance with the NAAQS. Originally, compliance with primary NAAQS was set for 1975, with secondary NAAQS compliance to be achieved within a reasonable time. The 1977 CAAA revised this schedule to attainment by December 31, 1982; a possible exemption was made for ozone and CO to December 31, 1987.

The Clean Air Act requires a state implementation plan (SIP) in each state; the SIP indicates how that state intends to meet the NAAQS. According to Congress, this role was given to the states because the "prevention and control of air pollution at its source is the primary responsibility of state and local government." State implementation plans can include such ideas as emission limitations, economic incentives or disincentives, plant closings or relocations, and changes in either operating schedules or methods. The requirements for SIPs and the specific state SIPs are listed at 40 CFR 51 and 40 CFR 52.

New Source Performance Standards.   In order to have a nationwide basis for air-pollution-emission controls and to set a minimum emission limit, the EPA developed New Source Performance Standards (NSPS). The NSPS set specific pollutant emission limits or describe the best available control technology (BACT) that should be applied at that source. The EPA has issued NSPS for 30 different sources (see Table 5). New or revised NSPS are being prepared for such sources as synthetic organic-chemicals manufacturing, the petroleum industry, industrial boilers, mineral products, polymers and resins manufacturing, industrial incinerators, and fuel-conversion facilities. New Source Performance Standards apply to new construction as well as to large modifications. Clean-air areas, ie, where the air quality is better than the NAAQS, and areas where the NAAQS are not met, ie, nonattainment areas, are of special concern to the EPA.

Table 5.   Sources Subject to New-Source Performance Standards [a]

Source
fossil-fuel fired steam generators, 73 MW (250 × 10^6 Btu/h) heat input
incinerators, 45 metric ton per day
portland cement plants
nitric acid plants
sulfuric acid plants
asphalt concrete plants
petroleum refineries
storage vessels for petroleum liquids, 151,400 L (40,000 gal)
secondary lead smelters
secondary brass and bronze ingot production plants
iron and steel plants
sewage treatment plants (sludge incinerator)
primary copper smelters
primary zinc smelters
primary lead smelters
primary aluminum reduction plants
phosphate fertilizer industry
wet-process phosphoric acid plants
superphosphoric acid plants
diammonium phosphate plants
triple superphosphate plants
granular triple superphosphate storage facility
coal-preparation plants
ferroalloy production facilities
steel plants: electric-arc furnaces
kraft pulp mills
glass-manufacturing plants
grain elevators
stationary gas turbines
lime-manufacturing plants
ammonium sulfate manufacture

[a] 40 CFR 60.

Prevention of Significant Deterioration. The EPA originally issued regulations for Prevention of Significant Deterioration (PSD) in December 1974 to protect clean air areas (40 CFR 51.24 and 40 CFR 52.21). These regulations were revised and expanded by the CAAA of 1977 and again following a Federal court case in August 1980 (17). Three air-quality classes were designated: Class I to protect pristine areas, Class II to allow moderate development, and Class III to permit more intensive development. For each class, allowable increments over the particulate and SO_2 baseline concentrations were set. Controls for the other criterion pollutants and for the protection of visibility are being developed. Most areas in the United States have been initially designated as Class II. Many large national parks and wildlife areas have been classified as Class I.

Regulations for the Prevention of Significant Deterioration apply to the construction or modification of 28 specific sources (see Table 6) that emit at least 90 t/yr of any pollutant regulated under CAA (see Table 7) and to any nonlisted source that emits at least 227 t/yr of any air pollution regulated under the CAA. If the cutoff is exceeded for any air pollutant, the PSD review must be done for all the air pollutants emitted. The only exception is any pollutant for which the area is nonattainment (see Emission Offset Policy). The new source or modification must cause a significant increase in net emissions or have an impact on ambient air quality. For all of the pollu-

Table 6.   Sources Subject to Prevention of Significant Deterioration Review [a]

Source
fossil-fuel steam electric plants of more than 73 MW (250×10^6 Btu/h)
coal-cleaning plants with thermal dryers
kraft pulp mills
portland cement plants
primary zinc smelters
iron and steel mills
primary aluminum ore reduction plants
primary copper smelters
municipal incinerator capable of charging more than 225 t/24 h
sulfuric acid plants
petroleum refineries
lime plants
phosphate-rock processing plants
coke oven batteries
sulfur-recovery plants
carbon-black plants (furnace process)
primary lead smelters
fuel-conversion plants
secondary-metal production facilities
fossil-fuel boilers of more than 73 MW (250×10^6 Btu/h) heat input
hydrofluoric acid plants
nitric acid plants
sintering plants
chemical-process plants
petroleum storage and transfer facilities with capacities of 48,000 m^3 (3×10^5 bbl)
taconite-ore processing facilities
glass-fiber processing plants
charcoal production facilities

[a] 40 CFR 51.18, 40 CFR 51.24, and 40 CFR 52.21.

Table 7.   Air Pollutants Regulated Under the Clean Air Act [a]

carbon monoxide (CO)	beryllium
nitrogen oxides (NO_x)	mercury
sulfur dioxide (SO_2)	vinyl chloride
total suspended particulates (TSP)	fluorides
ozone (VOC)	sulfuric acid mist
lead	total reduced sulfur (including H_2S)
asbestos	hydrogen sulfide (H_2S)

[a] 40 CFR 51.24 and 40 CFR 52.21.

tants shown in Table 7, the EPA has listed *de minimis* values in tons per year and micrograms per cubic meter (40 CFR 51.24 and 40 CFR 52.21).

As part of the PSD review, the applicant must show that BACT has been applied to all sources. Items to be evaluated include energy, environmental, economic, and other costs associated with each alternative technology as well as the associated benefits of reduced emissions. Another requirement is an ambient air quality analysis to show that the new emissions will not exceed either the NAAQS or PSD increments. Existing air quality data must be available in order to conduct this analysis. The EPA requires at least one year of ambient air quality monitoring data.

Emission Offset Policy. In January 1979, the EPA issued final rules for the Emission Offset Policy governing development in nonattainment areas (40 CFR 51, Appendix S and 40 CFR 52.24). These regulations were also revised in August 1980, based on the Alabama Power vs EPA court case (17). The Emission Offset Policy rules apply to any source emitting at least 90 t/yr of any air pollutant regulated under the CAA. The new source must apply the lowest achievable emission rate (LAER) for the problem pollutant and must obtain a more than equivalent offsetting emission reduction from existing sources. The existing sources can either be owned by the same company or the reduction can be bought from other companies. In this way, new growth is allowed while air-quality improvement is achieved.

National Emission Standards for Hazardous Air Pollutants. The EPA has also issued National Emission Standards for Hazardous Air Pollutants (NESHAP) located at 40 CFR 61. These apply to new and existing sources that handle materials designated as hazardous air pollutants. The following substances have been listed: asbestos, benzene, beryllium, mercury, and vinyl chloride. The regulations are a combination of emission limits and specified control technologies.

The EPA has also proposed a general cancer policy for the control of airborne carcinogens. The proposed policy is designed to identify, assess, and regulate airborne substances that increase the risk of human cancer. The EPA will indicate potential industrial sources of such substances. Draft generic standards have been proposed by the EPA, but the program is still very much in the formative stage.

Solid Waste

Regulation of pollution resulting from solid-waste disposal has been formulated at a much slower pace than regulation of air or water pollution (see Wastes). The Solid Waste Disposal Act was passed in 1965 (18), but it was not until the Resource Conservation and Recovery Act (RCRA) of 1976 was passed (19), which supersedes the Solid Waste Disposal Act of 1965, that substantial controls were authorized.

The main objectives of the RCRA are to "protect public health and the environment" and to "conserve natural resources." The act requires the EPA to develop and administer the following programs: solid-waste-disposal practices providing acceptable protection levels for public health and the environment; transportation, storage, treatment, and disposal of hazardous wastes practices that eliminate or minimize hazards to human health and the environment; whenever technically and economically feasible, the use of resource conservation and recovery; and Federal, state, and local programs to achieve these objectives.

At this time, the section of the RCRA of most concern to industry is Subtitle C, the hazardous waste management regulations (20). The purpose of this section is to regulate hazardous wastes from their generation to their disposal. Facilities that generate, treat, store, or dispose of hazardous wastes are covered by these regulations. Eventually, each state will have an authorized hazardous waste program; until then, the EPA has full jurisdiction.

The RCRA definition of solid waste covers a wide range of materials, including "solid, liquid, semisolid, or contained gaseous material resulting from industrial, commercial, mining, and agricultural operations" (40 CFR 261). A hazardous waste is a substance that must either be listed by the EPA or have a hazardous characteristic. A mixture containing a hazardous waste is considered hazardous. Several types of solid wastes are specifically excluded from hazardous waste regulation because of their great volume or for other reasons. These include household wastes and some agricultural, mining, and fossil-fuel combustion and exploration wastes (40 CFR 261).

The EPA has issued three categories of lists (40 CFR 261): specific types of chemical compounds; specific sources of wastes, ie, types of equipment; and certain discarded commercial products. There are four hazardous waste characteristics: ignitability, corrosivity, reactivity, and extraction procedure (EP) toxicity. The latter refers to the leachability of a waste and the resultant toxicity in the groundwater.

Generators of small quantities of waste, ie, <1000 kilograms per calendar month, are exempt from Federal regulations. However, many of the state programs have omitted this exemption. There are specific requirements for the different categories of facilities. Many of these regulations were issued during 1980 and 1981 and will be amended and supplemented in the next few years.

Regulations for generators are given in 40 CFR 262. It is the generator's responsibility to determine if a substance is hazardous or not. Requirements include recordkeeping, proper labeling, use of proper containers, providing information to transporters, following the manifest system, and periodic reports to the EPA. The manifest system is a set of papers that is passed from the generator to the transporters of the waste, to those responsible for final disposal, and back to the generator to signify that the waste has been disposed of properly.

The transporter regulations are mainly concerned with the manifest system (40 CFR 263). In addition, the EPA has coordinated its efforts with the Department of Transportation (DOT) to avoid conflicting rules. The DOT regulations are described in 49 CFR 171–179.

Regulations for owners and operators of hazardous waste treatment and storage and disposal facilities (TSDFs) cover both interim and general status (40 CFR 264 and 40 CFR 265) Any facility operating prior to November 30, 1980, was considered existing and was granted interim status if notification was sent to the EPA. The requirements cover operating methods and location, design, and construction of TSDFs.

These include tanks; surface impoundments; waste piles; land treatment; landfills; incinerators; thermal treatment; chemical, biological, and physical treatment; and underground injection. Groundwater and air-quality monitoring are required for all facilities that have the potential to generate emissions. There are also requirements for contingency plans in the case of accidents, closure and postclosure plans, and financial requirements to ensure that closure plans can be followed. The regulations in 40 CFR 122 and 40 CFR 124 describe the permit program required for TSDFs. The permit application must include an estimate of the composition, quantity, concentration, and time frequency or rate of disposal, treatment, transport, or storage.

One other law of interest is the Comprehensive Environmental Response, Compensation, and Liability Act of 1980, known as the "superfund" legislation (21). This act provides a means for the Federal government to collect money from industry for use in cleaning existing and abandoned hazardous waste disposal sites and spills. 65% of industry's share of the fees come from taxes on primary petrochemicals, 20% from taxes on organic raw materials, and 15% from taxes on crude oil.

Chemicals

The control of the manufacture, use, and exposure to hazardous or toxic chemicals is mainly divided between the EPA and OSHA. In addition, the Food and Drug Administration (FDA) has control over chemicals in food, drugs, and cosmetics (qv). The Consumer Product Safety Commission (CPSC) is concerned with the safety of all consumer products, including child-resistant packaging regulations. The U.S. Department of Agriculture (USDA) maintains strict controls over chemicals in food.

Toxic Substance Control Act. In 1976, Congress passed the Toxic Substance Control Act (TSCA), which is administered by the EPA (22). The two main goals of TSCA are acquisition of sufficient information to identify and evaluate potential hazards from chemical substances and regulation of the production, use, distribution, and disposal of these substances.

One important aspect of TSCA is the premarket notification program (40 CFR 720). Before a manufacturer produces a new chemical substance, the EPA must be given 90 days' notice. The notice includes all testing done by the manufacturer to determine the health effects of the chemical. The EPA then has 45 days from the end of the review period to determine if the production or distribution of the chemical should be restricted or prohibited because the chemical presents an unreasonable risk. The manufacturer has an additional 30 days to object to the EPA's decision. The EPA established an inventory of chemicals manufactured or processed in the United States (23). Any substance not on the list by August 30, 1980, is considered a new chemical and must be described in the premarket notification.

As part of TSCA, the EPA can require the testing of any chemical if there is the possibility of an unreasonable risk to health or environment or if there is significant human or environmental exposure. If the substance poses an unreasonable risk, the EPA can prohibit the manufacture, processing, or distribution of the substance; limit the amount of the substance that can be manufactured, processed, or distributed; prohibit a particular use for the substance; limit the concentration of the substance during manufacture, processing, or distribution; regulate disposal methods for the substance; and require manufacturers to maintain records of process and to conduct tests to assure compliance with EPA rules.

Another section of TSCA requires the manufacturer to notify the EPA if there is any indication of substantial risk from any chemical. Failure to do so by the manufacturer within a specified time period may result in civil penalties or, possibly, criminal prosecution.

Also, TSCA addresses the problem of polychlorinated biphenyls (PCBs) and chlorinated fluorocarbons (CFCs) (40 CFR 761 and 40 CFR 762). The EPA has developed regulations banning the future use of PCBs and is developing controls for the handling of existing PCBs. The manufacture and use of CFCs has been banned for all but essential uses.

Industry has a number of concerns about TSCA, particularly that of confidentiality. Manufacturers are worried about the amount of information that must be included in a premanufacture notification. In theory, this information is passed only to the EPA, but there may be some cases where the information could be released to other parties.

Occupational Safety and Health Act. OSHA has very broad responsibilities for protecting the workplace. The Occupational Safety and Health Act is administered by the Occupational Safety and Health Administration under the Department of Labor (24). The act covers all health and safety aspects of a worker's environment. Subpart Z of the Act, Toxic and Hazardous Substances, lists allowable employee exposure to many different chemical substances (29 CFR 1910.1000). These are given as ambient air concentrations over a certain time period, which usually is an 8-h TWA. Sometimes a ceiling concentration is given as well. Certain substances, eg, vinyl chloride, benzene, etc, are discussed in terms of necessary controls and limits.

The National Institute of Occupational Safety and Health (NIOSH) (under the Department of Health and Human Services, previously the Department of Health, Education, and Welfare) is also involved with OSHA. It is NIOSH's responsibility to determine safe exposure limits for chemical substances and to recommend to OSHA that these limits be adopted as standards.

The Occupational Safety and Health Administration issued its cancer policy in January 1980 (29 CFR 1900). The policy establishes a classification system for candidate and potential carcinogens, defines how OSHA will determine potential carcinogenicity of workplace substances, and sets forth rule-making procedures. In this way, OSHA hopes to avoid a lengthy procedure each time a substance is considered a carcinogen; instead, the substance is simply added to a list. This policy was originally developed by the Interagency Regulatory Liaison Group (IRLG), made up of OSHA, the EPA, the FDA, and the CPSC. It was decided that OSHA would take the lead. However, the other three agencies continue to coordinate with OSHA. Two priority categories were established under the policy. A substance may be placed in Category I if it is a potential carcinogen in humans or a single mammalian species; the determination is based on a long-term study, with other supporting evidence. Category II has the same criteria with the modification that the evidence is only suggestive, and the single mammalian species case has no supporting evidence.

Controls for a Category I substance include lowering worker exposure to a level that is technically feasible and cost-effective. If possible, a less hazardous substitute should be used. Reduced exposure should be achieved through engineering controls and work-practice controls, eg, protective clothing, remote handling, change in work schedules, etc. Category II substances require worker exposure to be reduced on a case-by-case basis. Again, this should be through engineering and work-practice controls. OSHA can issue emergency temporary standards for Category I substances. These would then become permanent standards for confirmed carcinogens.

A "list of substances that may be candidates for further scientific review and possible identification, classification, and regulation as potential occupational carcinogens" has been issued by OSHA (40 CFR 1990). This list is not a "preclassification warning." It contains 257 substances.

Pesticides. Exposure to pesticides during their manufacture is regulated under the EPA effluent standards and OSHA standards (40 CFR 455 and 29 CFR 1910.1000). In addition, there are the Federal Insecticide, Fungicide, and Rodenticide Act (FIFRA) (25) and the Federal Environmental Pesticide Control Act (FEPCA) of 1972 (26). A number of regulations have been issued under these Acts, which are administered by EPA (40 CFR 162–167, 169–172, 173, 180). The composition of the pesticides are regulated, and all pesticides must be registered with the EPA. Methods of pesticide application and the types of plants, etc, to which they can be applied are regulated. Proper labeling to explain use limitations are required. There also is a regulation protecting agricultural workers. The disposal and storage of pesticides and pesticide containers is covered by one set of regulations. One very extensive regulation limits the residue of specific pesticides allowed on agricultural products.

Noise

Noise has long been considered a nuisance if not a definite health hazard (see Noise pollution). Nonetheless, until recently, its regulation has not been as vigorous as for other environmental concerns. The 1969 Amendment to the Walsh-Healy Public Contracts Act established a maximum sound-pressure level of 90 dBA (decibels measured on the A scale) for noise exposure in the workplace as an eight-hour TWA. This standard was incorporated under OSHA in 1970 (29 CFR 1910.95).

The Noise Control Act of 1972 is under the jurisdiction of the EPA (27). Its purpose is to "promote an environment for all Americans free from noise that jeopardizes their health or welfare." It was further amended by the Quiet Communities Act of 1978 (28).

The jurisdiction of OSHA and that of the EPA differs in that OSHA protects workers' hearing in the plant, whereas the EPA is concerned with the effects of noise on the "quality of life," ie, the effect of noise emitted from the plant to the surrounding communities. The EPA has been setting standards for specific equipment; however, many of these regulations are still being developed.

OSHA has been trying to amend its occupational noise exposure standards for several years. Amendments to the standards were issued in January 1981 (29 CFR 1910, proposed rules 46 FR 4078, *et seq.*, and 46 FR 42622 *et seq.*). The new rules establish a hearing-conservation program for all employees who are exposed to occupational noise equal to or greater than an 8-h TWA of 85 dBA. This includes exposure monitoring, annual audiometric testing, proper selection of hearing protectors, education and training of employees, warning signs, and the keeping of records pertaining to monitoring and testing.

The permissible 8-h TWA exposure limit is still 90 dBA. Where this is exceeded, engineering controls or administrative controls must be used to reduce employee exposure. If this is not feasible, the controls must be supplemented by personal protection equipment. Engineering controls involve reducing the amount of noise produced by the equipment; administrative controls include limiting the amount of time an employee spends working in an area.

Conservation, Land Use, and Other Areas Subject to Regulation

A new facility should be located in an area that is economically suitable. At the same time, the social, environmental, and esthetic effects must be considered. These decisions are part of land-use policies. At this time, there is no national land-use program. However, there are some laws that cover portions of this decision-making process and some states have such programs.

National Environmental Policy Act. The principal goal of the National Environmental Policy Act (NEPA) is to establish a national policy that ensures continued growth and technological advancement while maintaining the quality of the environment (29). Any Federal agency sponsoring action that significantly affects the environment, eg, granting a construction permit or introducing new legislation, must issue a detailed report describing the environmental impact of the proposed action and alternatives. This report is called an environmental impact statement (EIS).

The NEPA process is administered by the Council on Environmental Quality (CEQ), which originally issued regulations in 1970 and in 1973 on the preparation of EISs. In November 1978, CEQ revised these regulations to try to improve the NEPA process (40 CFR 1500–40 CFR 1508).

An EIS is written and issued by the Federal agency involved with a particular project. The agency, however, relies on the owners of the proposed facility to provide the information contained in the EIS. The section on alternatives is considered the most important part of the EIS. The proposed project and its alternatives are usually described in detail. The environmental impacts of the proposal and the alternatives are presented in tabular form.

The affected environment section provides a brief description of the existing environment that will be affected by the proposed project. Topics covered are air quality, climatology, meteorology, and noise; water quality and hydrology; geology, topography, soils, and seismology; biology, ie, terrestrial and aquatic life; socioeconomic factors, ie, demography and land use; and historical and archeological features.

The environmental consequences section provides the scientific background for the alternatives section. Items covered include environmental impact of the alternatives; any unavoidable adverse environmental effects; relation between short-term uses of the environment and the maintenance and enhancement of long-term productivity; and any irreversible and irretrievable commitment of resources.

Based on all of this information, the Federal agency must determine if the proposed project is environmentally acceptable.

Wetlands and Coastal-Zone Management. The 1972 Coastal Zone Management Act was passed by Congress to protect the coastal areas of the United States (30). This Act is administered by the National Oceanic and Atmospheric Administration (NOAA). States with coastlines, either along oceans or large lakes, must have approved state programs. Such a program must include identification of the boundaries of the coastal zone; definition of what shall constitute permissible land and water uses; inventory of areas of particular concern; identification of the means by which the state proposes to exert control over the land and water uses; broad guidelines on priority of uses in particular areas; and a description of the organizational structure proposed to implement the management program. Some states have coastal-zone management programs that are not yet part of the Federal program. Many of the state programs require additional permits.

The Corps of Engineers (CE) has defined navigable waters broadly to include all wetlands. The Fish and Wildlife Coordination Act requires that the Fish and Wildlife Service (under the Department of the Interior) be consulted whenever any project is expected to affect wetlands.

Others. The Marine Mammal Protection Act (31) and the Endangered Species Act of 1973 (32) are designed to protect plants and animals. As part of the environmental impact assessment of a project, it must be ascertained that no endangered or threatened species of the flora and fauna will be harmed.

The Department of the Interior is involved in environmental reviews. Some of the agencies within this department are the Fish and Wildlife Service; the Land Management Bureau; the Geological Survey; the Heritage, Conservation, and Recreation Service, which is involved with historical and archeological impacts; the Bureau of Land and Water Resources; the Bureau of Mines; and the Office of Surface Mining Reclamation and Enforcement. All of these agencies can become involved in the environmental impact assessment of a project.

The Nuclear Regulatory Commission regulates the environmental and health aspects of nuclear use. They license nuclear power plants, spent-fuel storage facilities, fuel-manufacturing facilities, etc (see Nuclear reactors).

If the above mentioned regulations are or are expected to be used on a daily basis, it is recommended that one of the subscription services available be used (see Table 8). This article should not be used in lieu of legal services, but only as an introduction and summary of a complex subject.

Table 8. Subscription Services

Service/address	Publications and comments
Bureau of National Affairs, Inc. (BNA) 1231 25th St., N.W. Washington, D.C. 20037	*Environment Reporter* (Federal and state regulations for air, water, solids, and others) *Chemical Regulation Reporter* (Federal: TSCA, RCRA, and OSHA) *Hazardous Materials Transportation* *Noise Regulation Reporter* *Occupational Safety and Health Reporter*
P.O. Box 7167 Ben Franklin Station Washington, D.C. 20044	*Inside EPA Weekly Report* (weekly newsletter covering Federal EPA)
The Washington Monitor, Inc. 499 National Press Building Washington, D.C. 20045	*Federal Yellow Book* (loose-leaf directory of Federal departments and agencies)

Nomenclature

AQCR	= air quality control region
BACT	= best available control technology (air)
BAT	= best available technology (wastewater)
BATEA	= best available technology economically available
BCT	= best conventional pollutant-control technology

BOD	= biochemical oxygen demand
BPT	= best practicable technology (wastewater)
BPCTCA	= best practicable control technology currently available
CAAA	= Clean Air Act Amendments
CE	= Corps of Engineers
CEQ	= Council of Environmental Quality
CFC	= chlorinated fluorocarbons
CFR	= Code of Federal Regulations
CO	= carbon monoxide
COD	= chemical oxygen demand
CPSC	= Consumer Product Safety Commission
CWA	= Clean Water Act of 1977
DOT	= Department of Transportation
EIS	= environmental impact statement
EP	= extraction procedure (toxicity)
EPA	= Environmental Protection Agency
FEPCA	= Federal Environmental Pesticide Control Act
FIFRA	= Federal Insecticide, Fungicide, and Rodenticide Act
FWPCA	= Federal Water Pollution Control Act
GEP	= good engineering practice
GLC	= ground-level concentration
HC	= hydrocarbon
HEW	= Department of Health, Education, and Welfare (now known as Dept. of Health and Human Services)
IRLG	= Interagency Regulatory Liaison Group
LAER	= lowest achievable emission rate
NAAQS	= National Ambient Air Quality Standard
NEPA	= National Environmental Policy Act
NESHAP	= National Emission Standards for Hazardous Air Pollutants
NIOSH	= National Institute for Occupational Safety and Health
NO_x	= nitrogen oxides
NPDES	= national pollutant-discharge-elimination system
NS/EQ	= New Source Environmental Questionnaire
NSPS	= New Source Performance Standard
OSHA	= Occupational Safety and Health Act (and Administration)
PCBs	= polychlorinated biphenyls
POTW	= publicly owned treatment works
PSD	= Prevention of Significant Deterioration
RCRA	= Resource Conservation and Recovery Act
SIP	= state implementation plan
SO_x	= sulfur oxides
TSCA	= Toxic Substance Control Act
TSDF	= treatment, storage, and disposal facilities
TSP	= total suspended particulates (air)
TSS	= total suspended solids (wastewater)
TWA	= time-weighted average
UIC	= underground-injection controls
USDA	= U.S. Department of Agriculture
VOC	= volatile organic compounds

BIBLIOGRAPHY

1. The Rivers and Harbor Act of 1899, 33 USC §§401–413, and Executive Order 11574.
2. FWPCA of 1972 superseded all previous laws. 33 USC §1251 *et seq.* PL92-500 and further amendments.
3. Clean Water Act of 1977, amendment to (4), PL95-217; most recent amendment, PL96-510.
4. 33 USC §1251, sect. 101.

5. 33 USC §1251, sect. 304(a)4 and 40 CFR 401.16.
6. Safe Drinking Water Act, 42 USC §300F *et seq.*, PL93-523; most recent amendment, PL96-502.
7. 42 USC §7801 *et seq.*, PL95-467.
8. 42 USC 1962, PL89-80; most recent amendment, PL95-404.
9. PL95-474.
10. 33 USC §1401 *et seq.*, PL92-532; most recent amendment, PL97-16; 40 CFR 220–225, 227–229.
11. PL93-627; most recent amendment, PL95-36.
12. PL95-273; most recent amendment, PL96-255.
13. 43 USC §§1331–1351 and 1811–1866, PL95-372.
14. 79 Stat. 903, PL89-234.
15. 77 Stat. 392, PL88-206.
16. Clean Air Act 42 USC §1857 *et seq.;* Air Quality Act of 1967, PL90-148; CAA of 1970 42 USC §7401 *et seq.*, PL91-604; CAA of 1977 PL95-95; most recent amendment, PL97-23.
17. *Alabama Power Company v. Costle*, 13 ERC 1225.
18. Solid Waste Disposal Act of 1965, 42 USC §3251 *et seq.*
19. RCRA of 1976, 42 USC §6901, *et seq.*, PL94-580; most recent amendment, PL96-510.
20. 42 USC §6901, Subtitle C.
21. Superfund, 42 USC §9601, PL96-510.
22. TSCA, 15 USC §2601 *et seq.*, PL94-69.
23. *TSCA Chemical Substances Inventory.* Initial Inventory, May 1979 (43,641 chemicals); Supplement I, Oct. 1979 (3000 chemicals); Cumulative Supplement, July 30, 1980 (55,103 chemicals), Government Printing Office No. 055-000-00189-8. Further information can be obtained from Industry Assistance Office, Pesticides and Toxic Substances, EPA, 401 M St., S.W., Washington, D.C., 20460.
24. OSHA of 1970, 29 USC §655 and 29 USC §657.
25. Federal Insecticide, Fungicide, and Rodenticide Act, 7 USC §135, *et seq.*
26. Federal Environmental Pesticide Control Act of 1972 amended #25, 7 USC §7401 *et seq.*, PL92-516; most recent amendment, PL96-539.
27. 42 USC §4901 *et seq.*
28. 42 USC §4912a.
29. National Environmental Policy Act 42 USC §4341, PL91-190; most recent amendment, PL94-83.
30. 16 USC §1451 *et seq.*, PL92-583; most recent amendment, PL96-464.
31. 16 USC §1361, PL92-522; most recent amendment, PL96-470.
32. PL93-205; most recent amendment, PL96-159.

General References

N. R. Passow, "U.S. Environmental Regulations Affecting the Chemical Process Industries," *Chem. Eng.*, 173 (Nov. 20, 1978).
N. R. Passow, "U.S. EPA Regulations and the CPI," *paper presented at AIChE 86th National Meeting*, Houston, Tex., Apr. 14, 1979.
R. A. Young, ed., *Environmental Law Handbook*, Government Institutes, Inc., 6th ed., Bethesda, Md., 1979.
J. E. Granger, "Plantsite Selection," *Chem. Eng.*, 88 (June 15, 1981).
J. Quarles, "Federal Regulation of New Industrial Plants: A Survey of Environmental Regulations Affecting the Siting and Construction of New Industrial Plants and Plant Expansion," *Environment Reporter*, Monograph No. 28, Bureau of National Affairs, Inc., Washington, D.C., 1979. (Also available for $5.00 from New Plant Report, P.O. Box 998, Ben Franklin Station, Washington, D.C. 20044.)

Water

J. H. Robertson, W. F. Cown and J. Y. Longfield, "Environmental Outlook for the '80's—Water Pollution Control," *Chem. Eng.*, 102 (June 30, 1980).
"A Citizen's Guide to Clean Water," prepared by Izaak Walton League of America for U.S. EPA, June, 1973.
"A Decade of Water Pollution Control," *Chem. Eng.*, 95, 125 (Aug. 15, 1977).

Air

A. J. Buonicore, "Environmental Outlook for the '80's—Air Pollution Control," *Chem. Eng.*, 81 (June 30, 1980).
P. Brimblecombe, "Attitudes and Responses Towards Air Pollution in Medieval England," *Air Pollut. Control Assoc.* (Oct. 1976).

Solid waste

J. W. Lynch, "Environmental Outlook for the '80's—The New Hazardous Waste Regulations," Part I, *Chem. Eng.*, 55 (July 28, 1980).
A. Grodet and W. L. Short, "Environmental Outlook for the '80's—Managing Hazardous Waste Under RCRA," Part II, *Chem. Eng.*, 60 (July 28, 1980).

Noise

"Downing the Plant's Din," *Chem. Eng.*, 30 (Dec. 24, 1973).
A. F. Meyer, Jr., "EPA Implementation of the Noise Control Act," *Pollut. Eng.*, 42 (Nov. 1974).
"Toward a National Strategy for Noise Control," U.S. EPA, Apr. 1977.
Industrial Noise Control Manual, rev. ed., NIOSH Report No. 79-117, Dec. 1978.

Conservation, land use and general

N. R. Passow, "Preparing the Environmental Impact Statement," *Chem. Eng.*, 69 (Dec. 15, 1980).

NANCY R. PASSOW
C-E Lummus

RENEWABLE RESOURCES.   See Chemurgy.

REPELLENTS.   See Supplement Volume.

REPROGRAPHY

Reprography is the art and science of reproducing documents. It includes copying, which is the reproduction of one or more copies by processes that involve the complete formation of an image for every copy, and duplicating, which is the multiple reproduction from an intermediate-imaged master. Copying by some methods is usually restricted to one or a few copies but, with the advent of high speed copiers, hundreds of copies are routinely produced on copier–duplicators (CDs). These machines are either high speed electrophotographic copiers or high speed integrated master makers and duplicators that produce copies with the same automatic operation as copiers. Duplicating overlaps the CD range: some processes are used for as few as a dozen copies and, in others, masters are used to print 10,000 copies or more (see Printing processes).

Copying, duplicating, and printing have been profoundly affected by electronically generated, stored, and transmitted information. The production of documents through word and data processing has not substituted display for copies, but rather has expanded the use of most of the reprographic methods used to reproduce copies plus many new techniques to produce or compose copies and masters where no nonelectronic original exists.

Reprography and printing make up the graphic arts industry. Reprography is growing faster than printing; copying (including copying/duplicating) is growing faster than conventional duplicating. The fastest-growing segment is the reprographic production of copy on paper of digital information (ie, represented by a code or a sequence of discrete elements) from computer-base files, facsimile transmission and word-processing sources, and to a lesser extent copies from microform produced directly from documents or from computers, ie, computer-output microfilm (COM). Facsimile is copy that has been scanned for conversion to electrical signals that are transmitted by telephone or by radio or microwave communication links and recreated by reprographic means. Reprography is the means by which information is conveyed. According to the U.S. Department of Commerce, 75% of the U.S. work force is in the service sector with two thirds concerned with information. With about half of the U.S. work force and a significant fraction of the work force of most countries involved in the information industries, reprography is now a multibillion ($n \times 10^9$) dollar industry, and reprographic equipment and processes have become as much a part of the office as the typewriter was before World War II. As the office incorporates more data and word processing, reprography continues to provide the seemingly necessary tangible paper output.

A myriad of new reprographic processes and materials have been developed and applied since World War II. In addition to the considerations of quality and convenience, ie, greater rapidity and less effort on the part of the operator, the costs of copying and of high volume duplicating account for the change from chemically based materials to processes that depend on physical phenomena. For the most part, the transfer of costs from supplies to the more complicated mechanical–electrostatic machines has resulted in lower cost per copy, not only because of the lower cost papers but the high volume realized as a result of the higher speed of the new equipment.

In Figure 1, the relative cost characteristics of some copying, duplicating, and computer printers are shown. The diffusion-transfer-reversal (DTR) system does not include office copying because of the costly silver halide paper and liquid chemical developer (see Diffusion Transfer Reversal). Diazo plan copying requires the least

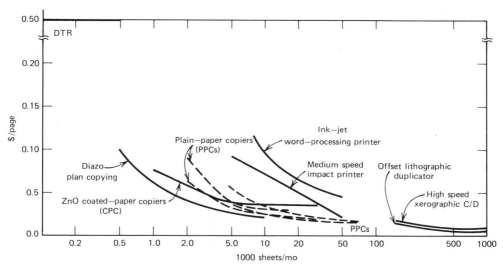

Figure 1. Comparison of total costs, ie, paper, supplies, machine, of various copying, duplicating, and computer printout prints (21.6 × 28 cm or [8.5 × 11]) over monthly volume from 150 to 1,000,000 copies per month.

expensive chemical papers and so maintains a place in the engineering reproduction market despite its significant shortcomings in copying letters and documents for office applications (see Diazo Processes). The more expensive ZnO paper is used in a more complicated machine processing sequence, but it does have a place in the office copy market. However, it is not used for high volume work because of the coating on the paper which increases the cost of the paper over plain paper and which is unappealing in terms of appearance.

The four plain-paper electrophotographic copier costs are examples of scores of machines available for use in the middle volume range. The place on the monthly volume scale is mostly a function of the basic speed of the equipment. The costs of the more complex machines vary within the range for individual models because of the great variety of ways of paying for the equipment, ie, renting, leasing, meter charge, or writing off the purchase price. In addition, the maintenance of this type of equipment is a significant cost factor. There are also minor variations in quality because of the nature of photoconductors and the methods of development and of fixing the image onto the plain paper.

The high speed duplicators, whether the xerographic (ie, dry copying by transfer electrophotography) or integrated plate maker–offset duplicator, are expensive and complicated equipment; but the plain paper they use and the very high volume market they serve bring the copy cost almost down to the basic paper cost (see Electrophotography). The print quality, which is always high in good offset lithography, is being approached or equalled by the xerographic CDs. Costs of high performance nonimpact printer operation in the same volume range also are ca $0.01 per print (1) (see Nonimpact Printing).

The medium-range printers, ie, the impact and the nonimpact ink-jet printers used for word processing, illustrate the sharp dependency of cost on volume (see Ink-Jet Printing). The print cost is for equipment and maintenance with the paper and ink adding relatively little to the overall cost.

The cost variability of low- and middle-range duplicating methods, eg, stencil, spirit, and offset, are too great to be shown easily on the graph (see Duplicating). In addition to duplicating speed and cost, the cost of the master and especially the length of run can give a very wide range of print cost.

In many reprographic duplicating processes, an intermediate stage is wrong-reading, that is to say, the mirror images of the letters are printed. In such processes, there must be a final product that is right-reading. In photography (qv), negative refers to a reproduction in which black becomes white and white becomes black; in a positive, white remains white and black remains black. Since the overwhelming majority of writing is done with letters that are dark upon a lighter ground, positive often connotes dark letters on a light ground. Handwriting with ink is therefore positive in this sense, although it is not a reproduction. In color photography (qv), a positive gives a true color rendition, whereas in a negative, the colors are complementary to the original.

The object of reprography is to produce right-reading, preferably positive, reproductions of originals or originals that are usually positive and universally right-reading. Originals and copies that have a type size common for ordinary reading are generically referred to as hard copy.

Copying

Copying primarily provides copies of documents for office use. It produces legible copy, ie, hard copy on paper and other substrates, from documents and sources such as electronic output from computers, word processors, facsimile, and microforms. It can produce many copies conveniently and cheaply, or it can be used to produce image carriers or masters for duplicators and printing presses.

Some of the first methods of copying were based on silver halide photography and did not always give black-on-white copies, but rather white-on-black photostats (2). In nonoffice applications, eg, for plans or graphics, some unconventional photographic techniques, despite their slow speed, were in general use, particularly in engineering and construction. After 1950, silver halide photocopying progressed to direct copies provided by diffusion transfer reversal (DTR), and gelatin-dye transfer (Kodak Verifax). Other nonsilver photocopy methods, mainly thermographic (3M Company, Thermofax) and electrophotographic, the latter of which includes both direct (RCA Electrofax) and indirect xerography (Haloid-Xerox), started to make unconventional methods conventional for photocopying.

SILVER HALIDE PHOTOCOPYING

Until the 1960s, photocopying by methods based on silver halide photography were economically significant. Despite the development of methods which increased the convenience of using basically wet chemistry, first dry thermography and then the dry or at least nonaqueous electrophotography that yielded stable copies in increasingly convenient equipment replaced silver halide photocopying for office work. Before the rapid escalation in the cost of silver-based materials, silver halide was used primarily for some offset applications (see Silver Halide Camera Masters) and micrographics.

The disadvantage of ordinary silver halide processing, which requires exposure, development, short stop, rinsing, fixing, washing, drying, exposure, development, short stop, rinsing, fixing, washout, and drying, arises from the necessity of using a negative and essentially repeating the steps to obtain a positive copy. Half of the steps were eliminated by producing a negative, ie, white letters on a black background, in the Itek Photostat, Xerox Rectigraph, and Anken Photoclerk machines, which automated the processing sequence. Although the materials are still available, the production of Photostat equipment stopped in 1974. The Kodak Q-System based on autopositive paper with a two-bath stabilization processor was also removed from the market in the early 1970s.

Gelatin-Dye-Transfer Process. The gelatin-dye-transfer process was an important silver halide photocopy method. The Kodak-produced material and equipment, which was sold by many companies as Verifax, was no longer marketed in 1975. The process involved tanning gelatin in a silver halide photocoating and the transfer of unhardened gelatin containing dye to a plain copy sheet to give positive copies of positive originals.

Diffusion Transfer Reversal (DTR). Diffusion-transfer-reversal processes (also known as chemical-transfer processes) were very successful in office photocopying during the 1950s and 1960s. It was originally introduced in Europe by Agfa and by Gevaert, Inc. which, as Agfa-Gevaert, is the principal manufacturer of DTR photomaterials along with Eastman Kodak. Its principles inspired the development of the Polaroid-Land film processes. Two sensitized sheets are used in DTR processes: a photosensitive silver halide sheet and a nonphotosensitive sodium thiosulfate sheet. The photosensitive sheet is exposed in contact with the original to give a negative image. Both the exposed negative or doner sheet and a receiver sheet are passed through a single solution developer. The imaged sheet is blackened in the nonimage areas and the image areas contain undeveloped silver salts. When the imaged sheet is next contacted to the transfer or receiver sheet while both sheets are moist, the salts in the image areas are transferred and diffused by thiosulfate solution to the receiver sheet; the salts are reduced to silver metal with the aid of developing nuclei in the colloidal layer of the receiver sheet or film to give a fixed positive image (3–6). The image is not durably fixed, as there is no washout step; after ca 30 s of contact, the two sheets are peeled apart to give one wrong-reading negative-image sheet and one right-reading positive-image copy.

Diffusion transfer reversal is no longer an important photocopy method but is used to prepare copy quickly for screened prints, paste-up, overhead-projection cells, stencils, diazo intermediates, and proofs. In Agfa-Gevaert copyproof, both high contrast and continuous negatives transfer to receiver sheets and films designed with adhesive backs for paste-up (7). A significant use is transferring from contact and camera-speed negatives to paper or metal lithograph plates as in the Kodak PMT system, the Agfa-Gavaert Transferlith or Rapilith, and National Graphics Transfer Reversal.

Usually the negative sheet is discarded but it can be fixed, washed, dried, and used as a contact master for reflex copying (5). Some positive copies can be used to image added transfer sheets. The developer need not be in the solution; it can be incorporated in the sheets, and an activator solution can be used for the production copy.

There once was a score of manufacturers of equipment for DTR copying. The relatively simple processors are employed for specialty applications, but the processing chemicals and materials are only produced by Eastman Kodak and Agfa-Gevaert. The cost for the materials used for transfer to paper has increased to $0.50 and, for material used for transparencies, to $0.80 per 21.6 × 28 cm (8.5 × 11) sheet.

Polaroid Process. The Polaroid process is a DTR in which camera-contained elements of a two-sheet system are activated for development and transfer after exposure by advancement of the film to open a pod between pressure rollers. The pod provides the processing solution (3,6,8). Many process-camera manufacturers, eg, Itek and Kenro, have film-head adapters for Polaroid film to provide developing within the camera. These cameras are used with high contrast film (Polacolor 2, Type 801) and an extended-range, low contrast film (Polacolor ER, Type 809). Polaroid film materials include black-and-white panchromatic film, high speed recording material, lantern-slide and projection materials, color film, and Type 55P/N photomaterial; the latter provides a positive print and reusable negatives. Color film for overhead transparencies (Colorgraph, Type 891) is available in a 20.3 × 28 cm (8 × 11) format.

The portability of the Polaroid-Land cameras, which can cost as little as $30.00, the permanence of films treated with Polaroid print coater, the excellent reproduction of fine details and continuous tones, and the controllability of contrast often offset the disadvantages of copy costs (over $1.25 for each 10.2 × 12.7-cm (4 × 5) sheet) and the manual effort involved. Polaroid makes the Polaroid Copymaker Model 208 for 28 × 35.6-cm (11 × 14) copying and the Polaroid MP-4 Industrial View Camera for various in-plant uses (see Color photography, instant).

Stabilization Process. In the stabilization process, the classical four-bath method, ie, developing, rinsing, fixing, and washing, is reduced to two steps, ie, developing and stabilization, in two or even one (monobath) baths (4). Developer can be included in the film, as in the Agfa-Gevaert Rapidoprint system, and autopositive emulsions can be used if positives are desired and recopying is not desired. However, the stabilization process became outmoded with the switch to dry, nonsilver systems. Materials that cost only $0.05 in the 1960s cost ca $0.40 in the early 1980s and are produced by Agfa-Gevaert, Eastman Kodak, and Itek Business Systems.

THERMAL SILVER COPYING

Two thermographic processes involving silver and heat development were marketed by 3M in the 1960s. The direct dual-spectrum process was used in low volume photocopying but lost its market share to electrophotography. The dry-silver, negative-working process is used in computer-output microfilm and for phototypesetting.

Dual-Spectrum Process. In the original Workman process, transparent film base coated with 4-methoxy-1-naphthol and erythrosine or another photoreducible dye is placed in contact with the original, which may be colored, and is irradiated from the film side with a high intensity tungsten lamp (9). When faced to paper sensitized with silver behenate ($C_{21}H_{43}COOAg$) in a thermographic machine, the coating images the copy paper blue-black wherever it has not been inhibited as a developer by photoexposure. Thus the process makes a direct positive.

Dry Silver Process. The dry silver process also involves a silver soap or alkanoate ($C_nH_{2n+1}COOAg$), eg, silver behenate, as a physical-development imaging material, but the photosensitive material is also a silver material, ie, a silver halide, which functions as a light-sensitive catalyst (10–11). After photoexposure, photolytic silver from the halide photolysis catalyzes an image-forming redox reaction between the silver soap and developer for 100°C development. The process is negative-working. Resolution is 1500 line pairs per millimeter at ASA 0.001; a speed of ASA 0.1 has been attained (see Photography). These photomaterials are used for a variety of rapid-access, heat-developable photographic products, including line and continuous-tone films, papers for contact and projection printing, cathode-ray-tube and electron-beam recording, and for production of hard copy from microtransparencies at high speed. Type 770 dry-silver paper is used in Microcord Corporation Model El-4 microfiche enlarger printers for producing one print every five seconds.

Copy materials cost only a few cents per microimage but machines cost $1,800–18,000. The poor permanence of copies, their fair quality for tone reproduction, and their curling are more than offset by the dryness of the process, the quick access time compared to that attained with silver halide, excellent reproduction of detail, high contrast and resolution, and low waste. Also, up to 5000 prints can be produced per eight-hour shift per machine and Federal Microfiche Standards are met (12).

THERMOCOPYING

In thermocopying, an image is produced by heat. Radiation thermocopying with infrared-sensitive photomaterials, mainly the 3M Thermofax, was very popular in the 1950s and 1960s. It offered a dry, less expensive, and quick alternative to DTR copying. Any original image with little or no infrared absorption, eg, unpigmented dye lines, light colors, or stamp-pad ink, did not reproduce on the thermosensitive paper nor were the copies fixed or in any way permanent, as any subsequent inadvertent heating caused development of the paper and obliterated the image. Thermosensitive copy paper is supplied to many thermal copiers by 3M, Speed-O-Print, Heyer, Deltex, and others, but electrophotographic copiers supplanted most thermal copiers for office communication. Other than producing inexpensive disposable copies, the thermal units are used to image infrared transparency films for overhead projection. Direct-working films are available in black, red, yellow, blue, and green and in negative or reverse films which give images with a black, red, or blue background. The same thermal-imaging equipment is used for rapid preparation of thermal stencils and thermal spirit masters (see Thermal Stencils; Facsimile Spirit Masters).

Thermorecording by heat conduction preceded the radiation process by several decades (13). Thermo images formed by contact with hot styli or thermal printheads are very important in electronically driven thermal printers (see Thermal Printers). Other thermographic processes, ie, Eichner, Kodak, Extafax, Games, and Adherography or thermoadherography, are based on radiative techniques but are no longer marketed. A form of raised-letter printing or duplicating used for wedding invitations and the like is also called thermography, but it bears no relation to either of the above heat-radiative or heat-conductive imaging methods.

Thermofax. An imaging process of differential thermal absorption was conceived in 1940 (2), and thermodifferentially imageable paper and a machine for its use were patented (14–15). Both physical, ie, blush-coating, and chemical papers were made for thermocopying (10,12–13). Imaging was provided with the original and the copy sheet was irradiated but thermally insulated from the environment; subsequently, darkening of background areas was reduced by providing contact of the original sheet-copy-sheet sandwich with a heat-conducting surface during scanning exposure (16).

In the blush-coated or frosted paper, an opaque and white coating covered a dark underlayer and differential heating fused the top coating, which in the fused amorphous state reveals the dark undercoating. The piezosensitivity of the top coating tended to reduce image quality as the background was mechanically developed during handling and passage through the copier. The chemical papers were favored for this reason. The 3M patents describe the use of ferric phenol, heavy metal sulfide, and anion–cation acceptor–doner combinations (17).

Other Thermal Processes. In addition to Thermofax, two processes, ie, the Eichner Dry Copy and Kodak Ektafax processes, that also used contact printing with an ir source were introduced. Both involved the transfer to a paper-receiving sheet for stable copies but suffered from some of the limitations of Thermofax, eg, insensitivity to noninfrared-absorbing colored originals and poor copy quality and poor tone rendition. Both processes were discontinued in the 1970s.

The Games Process (Thermal Vaporization). The image process is a thermal process that relies on the selective evaporation of a light oil from an image (17–18). Oil is applied to the original and evaporates from the image on the original (consequent to ir radiation) to a receiving sheet to form a positive, right-reading oil image. This latent oil image is developed by a toner adhering to the sticky oil and then is fixed. The process was limited to images that had good ir absorption, and it does modify the original in a manner reminiscent of the early letter-book copying process in which a portion of the ink was stripped from a printed page (19–20). This process was of great interest in the early 1960s because of the high density and contrast of copies from originals with high carbon-black ink, eg, newsprint; however, it was never commercialized (21).

PLAN-COPY PROCESSES

Prior to the reprography revolution of the 1960s (2), several unconventional, ie, nonsilver halide, systems were well established for copying of plans, engineering drawings, and charts (22–23). Although limited by their low quantum yield, which was as much as 10^{-6} that of silver halide systems, to contact copying of translucent or transparent originals, the relatively low cost and convenience of copying large originals has permitted these chemical processes to survive along with the extremely popular electrophotographic techniques.

The photochemical methods of blueprinting and brownprinting, which date from the first half of the 19th century, are obsolete and are only of interest historically to show the rapid product utilization of chemical discoveries. Diazotype expansion into the office copy market in the 1950s was tempered by the need for translucent originals and the inconvenience of aqueous solutions, ammonia, or almost-stable thermal systems, and was severely inhibited by the silver systems, Thermofax, and the emergent electrophotography. By the end of the 1970s, the production of diazo-coated products virtually leveled at almost 8.4×10^5 m^2 (10^7 yd^2) in the United States in 1979. However, binary-coated diazo sheet was growing 8% annually because of the pressure-development technique and other applications with film, which were increasing at 23%/yr.

Blueprinting. The photosensitivity of iron salts and the blueprinting process were discovered in 1842. The mechanisms of blueprinting reactions are incompletely understood in terms of modern photochemical principles (24–31). The basic photoreaction of blueprinting is a photoreduction requiring the presence of an electron doner:

$$Fe^{3+} + e + h\nu \rightarrow Fe^{2+} \tag{1}$$

Reaction 1 is the basic photoreaction of many iron processes which include blueprint processes and brownprint processes, the latter including sepia, Van dyke (13), and other similar physical development processes, eg, kallitype, platinotype, palladiotype, aurotype, mercurotype, etc.

In the blueprint process, the photoproduct Fe^{2+} is developed and fixed by reaction with ferricyanide ion, which is in the coating.

$$3\,Fe^{2+} + 2\,Fe(CN)_6^{3-} \rightarrow Fe_3[Fe(CN)_6]_2\downarrow \tag{2}$$

The initial precipitate is weakly paramagnetic Prussian blue; this undergoes charge-transfer rearrangement to isomorphous, strongly paramagnetic Turnbull's blue. Since blue develops in the areas exposed to light (unchanged Fe^{3+} does not yield a precipitate with ferricyanide $Fe(CN)_6^{3-}$; these ions are washed away by water development to leave a white line image on the blue background), the process is negative-working. If ferrocyanide $Fe(CN)_6^{4-}$ is used instead of ferricyanide, the process is positive-working.

In field engineering, particularly in construction, blueprints have several advantages over diazo prints. Blueprints resist fading in direct sunlight, withstand extremely hard usage as they are coated on 50–100% rag-base paper and, under dirty conditions, the blue background does not show dirt as does the white background of diazo prints. Burnout processes are slow photographically, and most require uv arcs or equivalently energetic light sources and thus they are unsuited for convenience copying.

Coatings. Advantages of aqueous coatings and aqueous development are that they are inexpensive, inflammable or toxic solvents are not used, and coating and printing installation wastes can be compatibly drained into existing sewage systems. The iron must be ionic and, thus, the only way that an electron doner can be intimately proximate to Fe^{3+} atoms is if it is in an anion with which the ferric ion can be complexed by oxalate, tartarate, or citrate.

The blueprinting reactions 1 and 2 are illustrated below in terms of molecular oxalato-system equations (13):

$$Fe_2(C_2O_4)_3 + h\nu \rightarrow 2\,FeC_2O_4 + 2\,CO_2\uparrow \tag{3}$$

$$3\,FeC_2O_4 + 2\,K_3Fe(CN)_6 \rightarrow Fe_3(Fe(CN)_6)_2\downarrow + 3\,K_2C_2O_4 \tag{4}$$

The correct formulation is most probably

$$2\,Fe(C_2O_4)_3^{3-} + h\nu \rightarrow 2\,Fe(C_2O_4)_2^{2-} + 2\,CO_2 \tag{5}$$

$$3\,Fe(C_2O_4)_2^{2-} + 2\,Fe(CN)_6^{3-} \rightarrow Fe_3(Fe(CN)_6)_2\downarrow + 6\,C_2O_4^{2-} \tag{6}$$

Blueprinting and related processes have been stagnated technologically by the rise of other plan-copying processes, eg, xerography and diazotypy.

Sensitizers. Ferric oxalate, usually formulated as $Fe_2(C_2O_4)_3.6H_2O$, but most probably an inner complex $[Fe(H_2O)_6][Fe(C_2O_4)_3]$, prints at fair speed in $K_3Fe(CN)_6$ coatings and has excellent latitude and fair washing and keeping properties. Ferric ammonium oxalate $(NH_4)_3[Fe(C_2O_4)_3].3H_2O$ has poor washing and keeping properties. Washing and keeping properties are much improved with the use of ferric sodium oxalate $Na_3[Fe(C_2O_4)_3].4.25H_2O$, but printing speed is considerably reduced and latitude is greatly reduced, perhaps because of the lessened solubility of the sensitizer. However, relatively insoluble ferric potassium oxalate $K_3[Fe(C_2O_4)_3] \cdot 3H_2O$ provides good printing speed.

Changing the anion changes the blueprinting properties. None of the citrates, eg, ferric, ferric potassium, green ferric ammonium, and brown ferric ammonium, have overall properties as good as the oxalates; although ferric chloride $FeCl_3 \cdot 6H_2O$ can be used to improve the speed of other slower sensitizers.

Brownprint Processes.   The brownprint processes are part of a large group of physical development processes in which the Fe^{2+} photoproduct of reaction 1 is involved in a subsequent redox reaction with metal cation M^{n+} to precipitate a metal M to form the image. The redox reaction is

$$n\,Fe^{2+} + M^{n+} \rightarrow n\,Fe^{3+} + M\downarrow \tag{7}$$

where M can be Pt, Pd, Au, Hg, Cu, etc. Although M^{n+} salts are often photosensitive, eg, $AgNO_3$ and Cu_3Cl_3, their photosensitivity is not involved here (only the photosensitivity of Fe^{3+} is); their only involvement is to act as supplies of image former M to a reductant. All the Fe^{3+} processes except argentotype, ie, silver systems, are obsolete, and the best known argentotypic systems, ie, Van Dyke, sepia print, and brownprint, are obsolescent, though still are used for making negative intermediates for subsequent washoff or blueprint reproduction.

Diazo Processes.   The photosensitivity of diazo compounds $RN{=}NX$, where R is a hydrocarbon radical and X is any electronegative substituent, has given rise to many processes used in photography, printing, and reprography (5,32–40). In light-struck areas of diazo coatings, the potential for coupling to form diazo dyes is lost (33,39–40). Immersing the exposed coating in a coupler solution develops in color the nonilluminated or image area giving a positive print. In 1920, a German monk put both diazo and the coupling component on the same paper, thereby conceiving the dry process (2). Diazotypy, diazocopy, and diazoreprography are often classified as autopositive photocopying processes (5).

The anhydrous photoreaction is ideally formulated as a photolysis of the diazo group (41).

$$ArN_2X + h\nu \rightarrow ArX + N_2 \tag{8}$$

In the presence of water, reaction 8 is often complicated by hydrolysis:

$$ArN_2X + H_2O + h\nu \rightarrow ArOH + N_2 + HCl \tag{9}$$

The presence of HCl can interfere with the coupling of ArN_2X to form an azo dye (see Azo dyes). More important, in some cases ArOH can couple with ArN_2X to give a color different from that desired for the particular coupler used. The nitrogen evolved on photolysis is used in image formation in the vesicular process (see Vesicular Process).

Couplers.   Use of different couplers produces different colors. By choice of couplers, red, orange, yellow, yellow-brown, maroon, red-brown, violet-brown, sepia, brown-black, blue, blue-violet, violet, and red-violet images can be obtained. Coupling components can be variously classified. One classification with the associated image colors is: derivatives of phenols; derivatives of catechols (see Hydroquinone, resorcinol, and catechol); derivatives of resorcinols; condensates of resorcinol; derivatives of hydroquinone; trihydroxybenzenes; diphenyl derivatives; naphthols; naphthoic acid derivatives; naphthalenediol and naphthalenesulfonic acid derivatives, mostly blue, blue-violet, or violet; diketones, mostly red; acetonitriles, mostly yellow; cyanacetylamides, $NCCH_2CONR_2$, mostly sepia; sulfonamides; acetoacetic acid derivatives; alkylmalonamates, mostly yellow; pyridinones, mostly blue or violet; hydroxypyridinones, mostly purple; 4-oxyquinolinones, blue, violet, or maroon; pyrazolinones, mostly red; thiophenes, red, blue, or violet; miscellaneous (33).

The preceding compounds are water-soluble. The diketones, eg, cyclohexanediones, are usually soluble in acetone and cellulose acetate as well; an exception is 1,1-bis[2-(5,5-dimethyl-1,3-dioxocyclohexanyl)]butylene, which must dissolve in methyl alcohol to impregnate cellulose acetate.

Acetonitriles are soluble in isopropyl alcohol as well as in water. Some couplers, eg, α-resorcylamide, N-phenyl-α-resorcylamide, and 6-ethyl-4-hydroxy-1-methyl-2(1)-pyridinone, should be dissolved in dimethylformamide. These couplers are about the same as those used in color photography and can be used with many diazo compounds (26,28,38–39,42–43).

Diazo Compounds. Many diazo compounds have been used in diazotypy (39,44). They can be classified into the following groups: mono-, di-, and trisubstituted diazobenzenes; substituted diazodiphenyls; diazo-p-aminobenzenes; diazo-o-aminobenzenes; diazo-p-amino esters; diazomercaptobenzenes (33). Diazo compounds like these should always be treated as if they were explosives. Diazoium halides can be stabilized as halozincates, halocadmates, halostannates, or fluoroborates by formation of double salts with halides, eg, zinc chloride, cadmium chloride, stannic chloride, or boron trifluoride. Typical stabilized salts are of the $[R\overset{+}{N}{\equiv}N][BF_4^-]$ or $[R\overset{+}{N}{\equiv}N]_2[ZnCl_4^{2-}]$ types.

The photosensitivity of diazo compounds is usually near 375–397 nm and uv exposure is usually required in diazotypy. Usually m-diazobenzenes are more photosensitive in the optical range (ir plus visible plus near uv) than p-diazobenzenes, and o-diazobenzenes are even more photosensitive. Some can be exposed with incandescent lamps, eg, 2-methyl-6-(dimethylamino)benzenediazonium chloride, 2-(dimethylamino)-5-ethoxy-4-benzoylaminobenzenediazonium chloride, and 3-chloro-2-(dimethylamino)benzenediazonium chloride. The replacement of the chlorine atom by bromine or iodine atoms further shifts absorption to longer wavelengths and increases the coupling ability. For high stability to accompany high imaging and light sensitivity, the most electropositive substituents are needed, eg, dialkylamino, phenylamino, benzoylamino, morpholino, or piperidino groups. At best, quantum yields are ca 0.50, and diazotypy basically remains a contact speed process that is essentially color-blind.

Commercialization. Although diazotypy was discovered in the 1890s, it was not until after the close of World War I and after derivatives of diazo oxides (diazophenols or quinone diazides), eg, naphthalene-2,1-diazo-oxide-4-sulfonic acid and naphthalene-1,2-diazo-oxide-4-sulfonic acid, were used that commercialization by Kalle, A.G. began in earnest (33). Faster developing dialkylaminobenzenediazonium salts soon replaced the diazo oxides introduced by Kalle in 1923 in the original Ozalid process, which involved the use of resorcinol (1,3-benzenediol), phloroglucinol (1,3,5-benzenetriol), or methylphenylpyrazolone couplers in stoichiometric proportions. The original Ozalid paper was stable because diazo oxides couple only after alkaline opening of the chelate ring. However, development requires 10–15 min in an ammonia atmosphere or the use of strong, discoloring, alkaline solutions.

Ammonia-Developable Diazos.   In view of the limitations of the original Ozalid process, Kalle searched for papers that could be developed in a few seconds, even if shelf-life thereby became poor. In 1925, the p-dimethylaminobenzenediazonium chloride and the p-diethylaminobenzenediazonium chloride sensitizers were discovered and these are still used today as are 2,5-dialkoxy-4-morpholinobenzenediazonium compounds, usually as the hemizincated salts, in ammonia-developable papers (45).

The ammonia process has been called dry development to distinguish it from wet development which has been called the semiwet process. In the ammonia process, after photoexposure to burn out the background, the image develops by vapor from diluted, warm, aqua ammonia. Thus the coupler must be present in the paper with a means to prevent precoupling or blue spoilage.

Two-Component Dry Diazo.   In two-component dry diazo paper, the diazo compound and the coupler are different materials and their precoupling is prevented by the presence of organic acids, eg, oxalic, citric, or tartaric acids, or by weak inorganic acids, eg, boric acid or an acid salt of phosphoric acid (46). Too much acid can yield background discoloration, called pink spoilage for blueline diazos. Some acids, ie, stabilizing acids, eg, 2,7-, 2,6-, or 1,5-naphthalenedisulfonic acids, which often are buffered with their monosodium salts, tend to give less pink spoilage and less background yellowing in ammoniacal development. However, the use of acids slows development, and a balance must be struck between shelf-life impaired by pink spoilage and slow processing on the one hand and fast processing capability and shelf-life impaired by blue spoilage on the other hand. The shelf-life can be lengthened considerably by conditioned-diazomaterial packaging and refrigerated storage and better formulation including the use of humectant coating, parting agents for keeping reactants separated until exposure, thiourea and other reductants to prevent phenolic oxidation of the diazo or coupler, and sulfonated phenolic ion-exchange resin to insolubilize the diazo and reduce migration of the coating components, thereby improving quality and life with minimal slowing of development.

One-Component Dry Diazo.   In the original Ozalid coatings, diazo-oxides that were noncoupling until chemically changed by developer alkalinity were used. By selection of diazo oxides whose hydrolysates were couplers, autocoupling development not requiring a separate coupler component was possible. In ca 1930, compounds that were stable in acid but autocoupling on addition of alkali were developed. Burnout of these one-component systems followed by conventional development gives positive copies. Polynuclear compounds with the diazo group on one nucleus and a hydroxyl or amino group on the other nucleus may be used, eg, diazotates of II acid (1-amino-8-hydroxynaphthalene-3,6-sulfonic acid), J acid (2-amino-5-hydroxynaphthalene-7-sulfonic acid), or 4-(2-piperidyl-4-aminobenzoyl)amino-5-hydroxy-2,7-naphthalenedisulfonic acid (33).

One-Component Semiwet Diazo.   If aqueous alkali is used for development instead of ammonia vapor, the coupler can be used in the developer and one-component papers can be used where there is no danger of precoupling before exposure. If the products or by-products of photoreaction 8 or 9 are not couplers, color fidelity can be maintained by proper development control. For high coupling capacity and improved photosensitivity, compounds such as 2,5-dibutoxy-4-morpholino-1-benzenediazonium chloride were introduced and stability was improved by use of dicarboxylic acids, eg, succinic and glutaric acids. However, the high coupling speeds demanded for semiwet papers have made these one-component papers no more stable than two-component dry-process papers because the diazos desired in the trade are unstable.

Developers are made alkaline with basic salts or volatile organic amines; the latter can be objectionably odorous, but they do not deteriorate paper after development since they evaporate. High energy couplers, eg, resorcinol, phloroglucinol, and 2-naphthol, can be used in the semi-wet process; for blacker prints, caffeine, and other xanthines can be used in the developer as color reinforcers for phloroglucinol. Development is usually from a metered roller rotating in the developer, and prints that are dry to the touch can be assured by the use of a wringing roller and heat after development.

Thermal-Development Diazo. Thermal development of diazo papers was investigated in the 1950s and 1960s to avoid the inconvenience of odorous gas or fluid and to afford a market niche in the office photocopy market. Thermally sensitive ammonium salts (47), encapsulated systems (48), and separate layer coatings (49) were tried but were not stable or convenient enough for office copying.

Pressure-Development (PD) Diazocopy. In the 1970s, a roller-type developing apparatus with a wiping control for two-component diazo sheets was introduced by AM-Bruning and did not involve the inconvenience of ammonia or a wet sheet (50). Accurate control is achieved in a developing zone between pressure and applicator rolls. The applicator roll has random microrecesses, which deliver a controlled amount of developing fluid. The developing fluid or activator contains a fixed base, ethanolamines, alkali-metal or alkaline-earth metal salts of amino acids in water, or glycol and is applied at 0.5–3.0 grams of liquid per square meter of paper to provide dry-to-the-touch copies without heat (51). Although this great improvement in convenience did not really reintroduce the diazo process to the office, it did offer an advantage in a compact machine for copying large drawings as there was no need to exhaust or absorb odors or dissipate heat. Its rapidly growing use in drafting and engineering is in part responsible for the increase in two-component diazo paper production.

Diazocopy Characteristics. Copies are not commonly accepted as archival, but British Admiralty ship plans have survived several decades of cabinet storage. Excellent resolution and rendition of fine detail is inherent in diazotypy and, recently, contrast control has been vastly improved. Ability to reproduce colored lines is limited by the color-blindness of diazo materials, but otherwise, especially with anticurl stocks, diazocopying has become an improved quality process.

Diazocoating. Opaque or transparent paper, cloth, transparent foils or films, and metal foils or sheets are coated with diazo sensitizers for a variety of uses in the graphic arts. For uv transparency, special fabrications are used and uv-absorbing base components are undesirable. Reduced catalysis of denitrogenation of diazos and oxidation of phenolics or amines is achieved by keeping iron, copper, lead, and other metals or their salts out of bases or coating solutions. Precoating is used to minimize diffusion into the paper and base discoloration by impurities and for more efficient display of the developed images; this may be done with alumina, silica polymer, aluminum silicate, baryta, rice starch, and many other compounds.

Coating machines usually consist of a feed-roll stand, an applicator, drying train, and a rewind arbor. The applicator, often a rubber-coated metal roll, rolls in the sensitizer and feeds to the web underside. It may be followed by an air-knife doctor, which controls the thickness of the applied coating. After excess liquid has been removed by doctoring, moisture is reduced to 30 wt % of web in a drying chamber, and subsequent drying to 3.5 wt % is conducted with close monitoring of temperature and moisture. Chill-roll cooling is followed by rewinding unless the sheeting and packaging are to be done immediately (see Coating processes).

Photoexposure. The high energy requirements for diazo photoprinting can be met by using high pressure mercury-vapor arc lamps. The 366- and 405-nm Hg lines do most of the burnout. The 436-nm Hg peak is important because of its intensity but is better suited for superfast diazos, eg, morpholinos. Whatever the wavelength, burnout under steady illumination is constant in velocity, and once the paper speed has been set by formulation and coating, a single speed number classifying the paper suffices for properly setting an exposing machine (33).

Diazo Reflex Copying. Use of diazo material to copy two-sided originals was made possible with the development of a copying method involving the use of light reflected from the pass of the original on a reflex foil (33). Electrophotographic equipment, which normally exposes by reflection, solved this problem for office copying and eliminated any need to have a diazo system that would copy opaque originals.

Whiteprinters. Machines for exposing diazos were called whiteprinters in the engineering drawing reproduction shops in analogy to blueprinters of like function. The manual work required is limited to the introduction of a transparent or a translucent original, and a diazo sheet if the machine is not web-fed, into the machine feeder. An endless-belt conveyor carries the original copy sandwich around a glass cylinder with a central rod light source monitored by a photocell. Bleaching of a yellow diazo, eg, 4-(4-methylbenzyl)amino-2-ethoxybenzenediazonium chloride, can also be monitored visibly to check the setting. After exposure, the original is separated by a vacuum and is discharged from the machine. The exposed paper is passed into the developing chamber (detail depends on whether ammonia vapor or aqueous solution is used) and is converted and stacked for delivery. The contact copying process requires a transparent or translucent intermediate, which can be one-sided and hence is not suited for book or common office copying. Special typing on uv-transparent paper helped the marketability of diazo for office copying, especially in Europe and Japan, but it was eclipsed by the success of electrophotography.

Diazocopier Characteristics. Depending on the required output volume and input capabilities, diazocopiers in 1981 cost \$895–30,000. Materials costs start at \$.01 per copy and apparently cannot be made competitive with high volume, plain-paper reprography. Ordinary copying is positive-working and, with fast, labile papers, copies can be obtained in 0.3 s (12,000 copies/h). However, edition copying is not economical; more dependable papers have slower output (up to 7200 copies/h). Large machines can process wide webs at 38 m/min, and even higher speeds are possible. The equipment is made by AM Bruning Company, Blue Ray Diazo, Dietzgen, GAF, Teledyne Post, and Teledyne Rotolite.

Translucent and Transparent Masters. Transparent masters can be prepared by various means (5). Besides writing, typing, and printing with uv-opaque images on uv-transparent bases, diazo intermediates can be prepared by silver halide, stabilization autopositive, DTR, xerography, and washoff processes including diazo (4).

Diazo color films in red, green, blue-green, and double-coated black on clear acetate base are direct-working as are color thermal films. They are also used for overhead projection, but they are not uv-opaque and thus are unsuitable as diazo intermediates.

The introduction of flexible, dimensional, stable transparent materials with sensitized coatings revolutionized the template and lofting arts. Glass cloth and especially polyester bases provided superior stability and low hygroscopic and thermal expansion. Cellulose acetate base is particularly sensitive to moisture and vinyl-base materials have high coefficients of thermal expansion.

Vesicular Process. The submicroscopic bubbles of nitrogen gas given off by diazonium compounds during photo exposure and photolysis are expanded by heat to give light-scattering images by vesicular trapping in a suitable plastic matrix (52–58). The diazo compound, eg, p-(dialkylamino)benzenediazonium chlorozincate, p-(diphenylamino)benzenediazonium sulfate, sodium 4-diazo-3,4-dihydro-3-oxo-1-naphthalenesulfonate, or 4-(cyclohexylamino)-3-methoxybenzenediazonium p-chlorobenzenesulfonate, is coated on film base in a hydrophobic thermosetting or thermoplastic resin matrix, which protects the diazo compound from ambient moisture and extends shelf-life (59). After cold exposure with near uv light, a reversal image is developed in a thermal process that expands the latent image vesicles. Subsequent overall exposure stabilizes the image by decomposing the diazo in the background. The nitrogen produced by blanket exposure is lost from the matrix if heat is not applied quickly. This sequence yields a very stable, light-scattering bubble image with a clear background and is widely used to produce positive duplicates of the primary negative-silver microforms. A direct duplicate can be obtained by cold exposure with time allowed for the nitrogen to diffuse out of the film, and blanket exposure followed by heat development (60).

The cloudy white interference image projects black which makes it particularly useful for duplicates of silver halide microfilm. Exposure and thermal processors are distributed by Canon U.S.A., Novamedia Corporation, and Xidex Corporation. The process was introduced by the Kalvar Corporation in the 1950s and was once known as the Kalvar process (55–56); however, the Kalvar Corporation no longer manufactures nor sells vesicular photomedia. Vesicular microfilm and processing equipment costs (1981) were $670–1000. Material in larger format for photomechanical reproduction is available from some suppliers.

Washoff Process. The washoff process is used for reproducing engineering drawings. Polyester film and waterproofed tracing cloth or paper may be coated with a layer of hydrophilic gelatin to improve adhesion of the photosensitive coating that is to be applied. This intermediate coating is called subbing. The photosensitive coating consists of a colloid of glue, gelatin, or poly(vinyl alcohol); a dichromate, ie, the photosensitizer and photohardener; and silver halide, which provides nonphotoacting image material, in this case in water. After the coating is dry, it is exposed under a photographic negative, which may be conventional or a brownprint. After exposure, the coating is washed with warm water leaving a positive, photohardened image. The image is visible because some of the silver prints out. The printout image is developed to yield a black image on a clear background.

Washoff reproduction films are available with a drafting surface on one or both sides. They are stable with a long shelf-life, as many have emulsion-bleaching and antifogging agents (13). The image is moist-erasable, which permits major drawing revisions. Matte finish for pencil and ink acceptances have been balanced and their translucency permits fast printing speed with blueprint, diazo, and other photocopying processes. Washoff, moist-erasable film is supplied by DuPont (Crovex) and Keufel and Esser (Photact). Erasable diazo film and sepia-paper intermediates can be corrected with any abrasive rubber eraser or a liquid eradicator.

ELECTROPHOTOGRAPHY

The extremely rapid expansion of office copying in the 1960s and 1970s has been called a revolution, albeit a revolution in dry-copying methods (2). The sharp increase of the service and white-collar segment of the work force meant an even greater need for and growth of copying and duplicating other than in the commercial print shop, which led to the popularity of electrophotography. By the 1980s, reprographic prints were produced at over 10^{12} copies, and well over 50% of the copies were being made directly on electrophotographic copiers or copier–duplicators or indirectly on duplicators with masters prepared by electrophotographic means.

The main electrophotographic imaging processes have been described (61–68). Chemical technology contributed significantly to the development of electrophotographic materials, mainly photoreceptors and developers, but the overwhelming success of electrophotography came from the combination of equipment and materials that virtually eliminate any chemistry or any technical input by the user. Copying preceded this revolution, but of primary importance is convenience in copying and duplicating. The convenience of masterless duplicating blurred the traditional distinction between copying and duplicating, as the cost of making copies in an edition directly from an original became competitive with the cost of short-run and longer run duplicating.

Office Copy Machines. In the 1960s, the first automated xerographic machine, the Xerox 914, was the forerunner of a long line of increasingly automated, faster, and more versatile photocopying machines (69). The first machines were marketed by Xerox but, after a Federal Trade Commission order in July 1975 making the thousands of Xerox patents available for license, numerous manufacturers in the United States, Europe, and particularly Japan, introduced machines that were called PPCs (plain-paper copiers). Plain-paper copiers better describe this type of equipment, for although they are often called xerographic because of the development and transfer steps, liquid development belies dry writing.

During the early electrophotographic era, only Xerox and the manufacturers of materials and equipment using direct electrophotography, ie, the RCA Electrofax process, were available (70) There were numerous manufacturers of these coated-paper copiers (CPCs) (71).

Later, the use of an indirect process by transferal of the electrostatic image from a photoreceptor drum to a coated, dielectric sheet on which the image was subsequently developed and fixed, ie, transfer of electrostatic images (TESI), by Minolta combined indirect processing and a coated paper in a compact copier. This process, also known as electrographic printing, is used for computer printout and tone reproduction (72). During this period, the main reprographic application of electrophotography was office copying. Indirect electrophotography or xerography started as a manual process to produce offset masters, but once it was automated, the materials and process steps were extensively modified to produce better copies faster and to produce less expensive machinery with less maintenance. By the 1980s, PPCs were manufactured and sold by Xerox, IBM, Nashua, Savin, Saxon, Van Dyk, 3M, Yorktown, and Apeco among others in the United States. The Japanese manufacturers, Canon, Copyer, Konishiroku, Minolta, Mita, Panasonic, Ricoh, Sharp, and Toshiba, introduced as many as 24 new PPC models in 1981. In Europe, Agfa-Gevaert, Develop, Océ, Eskofot, Geha, Kalle/Infotec, and Lumoprint also produced machines for the office-copy market.

Practically all of the Japanese and European PPC manufacturers also had CPC models, although most did not produce the zinc-oxide-coated paper or toner, ie, developing material, for use in them. In the United States, only Xerox, IBM, and Van Dyk did not enter the market with CPCs. Apeco introduced the first Electrofax office copier, which was a table-top unit with dry development, in 1961. SCM followed in 1962 with a liquid-immersion-development (LID) unit, which was commonly used in direct electrophotography with ZnO-coated paper. Bruning also had a dry-developing console in the same year. Many LID machines, which were mainly desk-top models, by Savin Business Machines, Dennison Manufacturing, Bohn Business Machines, Formfoto Manufacturing Company, Speed-O-Print, Old Town Corporation, and the above mentioned Japanese and European manufacturers followed.

An extension of electrophotographic office copiers as a substitute for diazo in engineering-plan reproduction was easily achieved for small drawings (A-Size and 21.6×28 cm [8.5×11]). The relatively high cost of PPCs, especially those that could accommodate the larger size drawings, made them more costly than most diazo prints. However, the high photo speed of selenium photoconductors permits projection exposure and the ability to reduce or to enlarge. Also the exposure by transmitted light in diazo restricts copying to transparent or translucent originals, whereas exposure with reflected light permits copying from opaque originals or, perhaps more importantly, from composite or corrected originals. A PPC can transfer the image not only to plain bond paper but to translucent vellum or transparent film or to offset master stock. Once transferred, the plans can be further copied on diazo whiteprinters, or they can be duplicated on offset presses.

The Xerox 2080 printer can copy originals up to 91 cm wide and 7.62 m long and can produce prints up to 61 cm wide at a rate of seven D-size (55.9×86.4 cm [22×34]) prints per minute. It can produce copies reduced to 45% and enlarged to 141% of the original and it costs (1981) $100,000. The Océ 7200 at $50,000 can copy originals up to 61 cm wide and 3.05 m long, and it prints at 1:1 size or 1:0.667 reduction at a rate of five D-size prints per minute. Both use selenium photoconductor drums and dry toner development with thermal fusing.

Although the CPCs, especially the LIDs, are significantly less costly than the PPCs, the lowering of prices of PPCs through technological improvements and the combination of competition and high volume production has shifted the market heavily in favor of the PPCs. The introduction of copiers and duplicators with auxiliary equipment continues, and monthly reports on the characteristics and performance of these systems are published (73).

Materials and Process Steps. The evaluation of electrophotographic materials and techniques as they relate to photocopying and later to the nonimpact printing and duplicating applications is given in refs. 74 and 75.

Photoreceptors. The first commercial photoreceptor used by Xerox for indirect or transfer electrophotography was selenium. Selenium that was modified with tellurium to enhance red response, as well as arsenic alloys (ie, $As_2Se_{2.7}Te_{0.3}$) and arsenic triselenide, succeeded vitreous selenium. In 1980, more PPC models used selenium photoreceptors than any other type including those manufactured in Japan, only a third of which in 1976 had selenium-based photoreceptors. As the processing speeds of the machines were increased, the higher speed and recovery characteristics of the selenium alloys outweighed any mechanical problems of these coatings encountered in handling and cleaning. Multilayered photoreceptors with improved characteristics that have been used by Matsushita and others are Se-based (76). Improvement in dark decay and surface potential of selenium coatings has occurred through the use of better raw materials, including superfinished substrates (77), and control of the manufacturing process. As the need for greater speed, service life, tonal rendition, and spectral response grows, the development of structured photoreceptors, many of which are based on selenium and selenium alloys, continues.

Organic photoconductors (OPC) were introduced in higher speed copier–duplicators first by IBM and later by Kodak. The IBM copiers had a roll with a coating of a 1:1 molar complex of polyvinylcarbazole and trinitrofluorenone. Sections of this web advance over the surface to a drum support. Kodak, however, used pyrylium dye salts (78). Kodak, Xerox, AM International, and 3M among others, and the Japanese photocopier manufacturers have been active in investigating OPCs. Many classes of saturated aromatic compounds, eg, quinoid benzene resins bridged by cycloalkanes, polyphenyls, triarylalkanes, polycyclic aromatics, triaminobenzenes, heterocyclics, eg, 2,3,4,5-tetraarylpyrrole, and hydrazones, have been patented for use in electrophotography (79). Combinations of organic layers to form multilayered systems, such as a conductive epoxy-cross-linked intermediate layer containing pyridinium salts, draw upon the many OPCs to form new families of photoreceptors.

Dye-sensitized ZnO coatings on paper were introduced commercially in the 1961 Apeco Electrostat, the first direct electrophotographic office copier. The blending of the dyes to get good color sensitivity with a new white appearance was the prime requisite for this coated photocopy paper (70). Selection of the insulating polymeric binder to give thinner, brighter coatings with ZnO binder systems to give a copy with better feel and lower cost helped to establish Electrofax for low volume copying. Zinc oxide for electrophotography was studied intensively in the 1950s and 1960s (80–81). The base paper was treated with electroconductive polymers to provide reliable conduction over a wide humidity range (82). In addition, the paper base for liquid-immersion development (LID) is treated to improve hydrocarbon holdout characteristics.

In the 1970s, ZnO-coated paper masters for use in the photoreceptors in PPCs were developed by Japanese manufacturers. In 1976, almost a third of the Japanese PPCs had ZnO photoconductors, although by 1980 none of the new models used this approach. Changing from the single use of ZnO coating in the direct process to the multiple use in the PPCs changed the characteristics for both coating and the base and was analogous to the changes necessary for a good ZnO offset master. The color of the coating was no longer restricted to off-white, but the coating's ability to recover from the previous exposure cycle became important as was surface resistance to abrasion and wear during development and cleaning. The life of these paper-based masters is a few hundred to a thousand copies in low speed copiers, which either have master changing mechanisms or require the operator to change the master manually. Océ produced a high speed machine in the 1970s that had a fan-folded ZnO-coated master belt, which permitted time for recovery before automatic recycling (83). The dependence of ZnO-binder coating on an oxygen absorption–desorption mechanism in addition to the mechanical properties of the binder are limitations for ZnO coatings in high speed equipment, which requires photoreceptor life of >100,000 images (80,84).

Cadmium sulfide dispersed in an organic polymeric binder with a polyester overlayer was used by Canon in the NP Process and later by Saxon. The extended spectral response as well as the surface, which made it suitable for both dry and liquid development, made it a more desirable substitute for selenium than the ZnO system. Minolta also used a CdS–binder system for PPCs and machines based on transfer of electrostatic images (TESIs) and investigated the use of CdS modified with manganese (85).

Cadmium sulfide in lead-sealing glass binders has been developed to provide a smooth glass-enamel surface that would be very durable and resist damage from development, cleaning, and handling (86). Coulter Information Systems utilized a sputtered CdS thin layer to achieve a high speed, high resolution process (87). Canon claims to have developed a sputtered, amorphous doped silicon (Si:H) photoconductive layer that is superior to ZnO binder in having low fatigue and high speed (88). These hydrogenated silicon films have excellent photosensitivity over the spectral range and are deposited by glow discharge of SiH_4 in multilayers (89) and monolayers (90).

Photoreceptors are incorporated into a machine as part of a system; thus, manufacturers tend to design series of machines without changing from a familiar type. However, substitutions of photoreceptors in essentially the same models to CdS from an OPC drum and to OPC master from a ZnO master have occurred. Selenium-based systems are regaining their dominance, especially as their durability is increased; CdS, because of its spectral range, is also growing; and the OPCs appear to be firmly established with at least two manufacturers for high speed CDs (75).

Development. The first methods of development in both indirect and direct electrophotography-automated equipment were in dry two-component toning systems: cascade in Xerox 914 and magnetic brush in the Apeco Electrostat. By the late 1960s, CPCs principally involved liquid immersion development (LID) as it offered much simpler and smaller development and fixing equipment more suitable for low cost desk-top machines than the magnetic brush with a radiant fuser. Coated-paper copiers in the 1970s were designed for single-component, magnetic-toner development with either cold-pressure roll or heat-roll fusing. Some LID machines were converted with a field-kit to single-component dry systems. The completely dry, high contrast copies (albeit on coated paper) afforded by this change in technology slowed the introduction of PPCs in the slower speed desk-top copier market. The liquid system was not particularly suitable to medium speed CPC machines (15–30 copies/min) because of the difficulty of drying at these rates and the amount of hydrocarbon vapor generated. The rapid rise in oil prices exacerbated the tendency to avoid the use of liquid hydrocarbon-based developers. Recently, however, a liquid-development apparatus for high speed microcrystalline CdS was patented (91).

The two-component dry-toning system in PPCs has remained the most important type of development, especially in high speed CDs. The disadvantages of the cascade system for fill-in and solid-image areas and size of the development section were overcome with development electrodes and later by use of magnetic-brush development. Beginning in the 1970s, Ricoh/Savin low and medium speed PPCs employed liquid-toning development with great success. The carryout of the dielectric isoparaffin carrier and the need for a hard-finished plain paper for best density and resolution did not hinder its acceptance. Canon achieved very low hydrocarbon carryout by using a conductive sponge-rubber roll that developed and cleaned (75).

Single-component magnetic-brush development in PPCs followed its use in ZnO-coated paper machines. Eliminating the need for toner concentration monitoring and control, which is important in two-component systems, and the need to change the carrier periodically plus a sharp reduction in the size and complexity of the toner-cleaning station by using the developer magnet or another magnet as the cleaners, offered a great advantage, especially in compact desk-top copiers. Among others, Sharp, Mita, and Apeco introduced machines with the single-component, magnetic-toning systems. The main drawback was the transfer of the inductively charged toner image to plain paper at medium-to-high relative humidity, but this can be overcome by the use of modified paper that is electrically resistive at high humidity. However, this paper was often considered a special rather than a plain paper. In an effort to reduce the need for special paper, Canon embodied a single-component toner and development mechanism in the NP-200 copier, which charges the toner triboelectrically and develops by propelling the charged toner with 1000-V, 400-Hz a-c bias to the CdS photoreceptor and then by transferring the charged, toned image to plain paper with a regular transfer corona (92). Océ makes use of single-component magnetic toner by transferring to a preheated copy sheet from a heated silicone transfer belt (83).

Fixing. The single-component toners in use are also formulated for cold-pressure roll fusing, thus reducing the energy required as compared to heat fusing and permitting an instant-on feature, which is important in convenience copiers. A disadvantage of this method is calendering of the copy sheet by high pressure of the fusing roller, which imparts a glossy appearance to the sheet and the image. One approach to eliminate this problem is to encapsulate a liquid toner or ink in the toner particles or to encapsulate an aggregate of pressure-sensitive adhesive-toner granules with magnetic pigment (93). The toned image can thus be fixed at the low pressure necessary to break the granules without calendering the sheet.

Sublimation Transfer. Based on the sublimation transfer of dyes to textiles (see Printing and Presswork), a copy process utilizing direct electrophotography with single-component toners containing one or more sublimable dyestuffs was developed (94–95). The sublimable dyes are monoazo, anthraquinone, quinophthalone, or styryl dyestuffs (96) (see Azo dyes; Dyes, anthraquinone).

Changing the optics on a dry, single-component-toner copy machine so that an inverted or wrong-reading image is produced on the zinc-oxide-coated sheet effects a right-reading image by transfer to a receiving surface with application of heat to the back of the toned sheet (97). The heat transfer is usually in a heat-transfer press and takes 15–>60 s at 200°C, depending on the receiving surface. The dyes in the toner sublime and deposit on the receiving surface to form a right-reading image on polyester sheet; polyester-coated metal decorative foils; textiles, eg, T-shirts and wall hangings; and lacquered metal, eg, anodized aluminum or brass. The dyes are not substantive to natural fibers, which must be sized with suitable thermoplastic resin to permit the formation of a dense image that withstands washing. The single-component toners are made with a full spectrum of dyes, including black, and the developer units are easily substituted in the photocopy machine in order to change colors. Multicolor transfer can be made by combining images made with different toners and transferring to a single receiving surface. The system including both equipment and materials is supplied by Spectra Corporation.

Screen Ion Projection. Electrostatic stencil printing was developed at Stanford Research Institute in the late 1950s (98). As first conceived, the process consisted of propelling dry toner through a screen and onto a printing surface by setting up an electrostatic field between the two. Dainippon Ink and Chemicals developed, through a joint venture, a device that prints on china, steel ingots, and electronic components. Later, the process was developed for a multilayered screen at which ions were directed. The screen apertures were controlled by a segmented layer of conductive material modulated by applied potentials to block or to allow complete or partial passage of the ions, depending on the fringing fields at the apertures (99). The screen was further modified with a photoconductor, eg, selenium, so that the aperture fields were modulated by the light pattern on the selenium (100). A wire-mesh screen with an organic photoconductor and overcoated with an insulative material, eg, poly(diphenyl siloxane), retained charge and permitted multiple-ion image projections and, thus, many copies of the original image (101).

Electroprint Process. In the Electroprint process, the usual procedure was to form a latent electrophotographic image on the screen by charging and exposure and then to activate a projection corona. The Xerox and Kodak methods require exposure of the photoconductor-coated screen and ion projection to occur simultaneously, whereas the AM arrangement involves blanket exposure, charging image exposure, charging, a second blanket exposure, and then multiple-ion projections (102). The projected ions are deposited on a dielectric sheet, which is developed with either dry or liquid toner and then is fixed. The pattern ions can also be deposited on a dielectric-surfaced drum; the image develops and transfers to plain paper (103). As the toner does not come in contact with the photoreceptor screen, color prints are produced by exposing to give color-separated images and then by developing the successive charge patterns with corresponding-colored toners.

PLASTIC–POLYMER IMAGING

Plastic-Deformation Imaging. Thermal deformation of coatings for thermoplastic and photoplastic recording as well as the frost or thermoplastic process is a significant reprographic technology, since the deformation images are really not legible (see Deformation recording media). The frost process, however, is used in one system of updatable microimaging (see Updatable Micrographics).

Photopolymerization. The main use of photopolymerization in imaging is the production of printing plates for letterpress and lithography beyond the needs of duplication runs. The DuPont Crolux Photopolymer Reproduction film is no longer on the market. Cross-linking printing-ink polymers make up another special case of uv-curing a surface coating (104).

MICROGRAPHICS

The information explosion not only required better and faster methods of producing documents and of copying them, but also promoted the use of most of the same reprographic technologies to reduce these documents to microform and to retrieving them. Microform is a document in reduced format, such that the text cannot be read with the naked eye. Microfilm is silver halide film on which documents are recorded sequentially on a strip. One of the most widely known examples of this is newspaper microfilm. Aperture cards with frames of microfilm and legible descriptions or identification of the frames could be punched, located, and handled by card sorters. Microcards are contact prints from a master negative and contain 60–80 pages in reduced form with identifying catalogue entries at the top of each page. Microfiche is similar to the microcard in format, except that the former is on a transparent film. Microfiche was introduced in Europe before it became popular in the United States; it is a form that greatly facilitates access by X–Y–Z addressing and affords enlarged viewing and printing by projection. Records in microform occupy as little as 2% of the space occupied by documents on paper. Distribution cost savings with microforms are also significant with increasing postal costs.

Micrographic Recording. Equipment for projection viewing and retrieval as aperture cards, microcards, and microfiche was developed for storage and distribution (105). Silver halide, both conventional and dry-thermal, is used to record computer output on microfilm and microfiche by cathode ray tube (CRT) and laser beam (106). Kodak, 3M, Datagraphics, and Bell & Howell supply equipment and materials for computer output microfilm. The microfiche with 30 or more 21.6 × 28-cm (8.5 × 11) pages on a 10 × 15.2-cm (4 × 6) sheet lends itself to ready access by X–Y location and easy viewing in a reader or reader–printer (107).

Ultramicrofiche with images reduced 75–200 times have been used for microform catalogues for mail-order houses and parts suppliers. Ultrastrip recording at 200–300 times reduction in two steps with silver halide microfilm for a positive image is the basis for ICOT Corporation's Carms/11 (computer assisted records management system). This is a high capacity (ca 1,000,000 21.6 × 28-cm pages), high speed (0.5–4-s retrieval time) data station that requires 10 characters to retrieve any page.

NCR's PCMI system has image reduction to $1/_{50,000}$ and really is a micropublishing system. Documents are reduced and copied onto 35-mm film and the image is further reduced and recorded as a photochromic layer on glass. Multiple microfiche copies can be made from a photographic microimage master (PMI).

Microform Duplication. Diazo and vesicular microform on polyester base are used extensively for duplicating silver-based microimages (108). These diazonium-based media are ca one-third the cost of silver-based systems, are easily processed, and are relatively insensitive to radiation prior to use. Two-component diazo film produces direct grainless microform duplication on high speed duplicators with ammonia development. Use of a base with a built-in yellow filter produces a nonreproducible image; unauthorized duplication on diazo or vesicular equipment by autogeneration is thus prevented. Diazo microfilm material is manufactured by AM Bruning, Citiplate, Kodak, James River Graphics, Kimoto U.S.A., Inc., Kleer-vu, and Xidex Corporation.

Vesicular films have two added advantages over diazo: greater image permanence than diazo type or silver print and the ability to produce either direct print or reversal prints (1). The same exposure device can be used for vesicular and diazo duplicates; however, thermal development of the exposed vesicular film requires 3 s at 135°C in a dry-heat processor.

Updatable Micrographics. The usefulness of microfiche is extended by technology that allows entry posting for an active or open file. Micrographic recording makes it possible to maintain multiple access, record integrity, efficient retrieval, savings in space, and updated records.

Technologies are available to add documents conveniently to a file on microfiche through special step-and-repeat camera–processors. The A. B. Dick System 200 (earlier known as Scott System 200) is an electrophotographic process in which a frame of organic photoconductor-coated film is corona-charged, exposed, and then developed with a liquid toner. The system does not allow erasure of a frame but a document can be removed by overprinting "VOID." In 1981, the System 200 Record Processor cost $32,975, the file film costs $0.60, and toner costs ca $0.01 per frame. A similar system with dry toner has been developed by Imtex, Inc.

The Microx System of Bell & Howell utilizes thermoplastic recording. A conductive layer on a polyester base is overcoated with a photoplastic. A frame of the film is corona-charged and exposed, and the charge is conducted to ground in the lightstruck area. The film is then heated to the softening point of the thermoplastic coating and a frost image forms as the charge in unexposed areas forces grooves or surface distortion into the plastic surface. The resultant positive, light-scattering image of microfiche has good resolution and low contrast and can be produced in seven seconds. A frame can be deleted or erased by heating the film, thereby allowing the surface to flow and become smooth relative to its original configuration. The delete cycle takes seven seconds as does the replace or reimage step. The camera–processor–monitor sells (in 1981) for $13,000 and the photothermoplastic film costs $0.60/sheet.

Both 3M and A. B. Dick have worked with Energy Conversion Devices, Inc. on the MicrOvonics dry process for production of archival and updatable microforms (109). It is a two-step process involving a dry silver image that is used in the flash exposure of a thin opaque coating. The coating clears from the exposure to yield a positive image. The process has not yet been commercialized.

A very small portable microfilm camera–processor–viewer with instant, updatable, Xerox dry microfilm has been described (110). The microfilm is a matrix that can soften and that is embedded with a monolayer of photosensitive Se particles. The Se particles migrate selectively after the film is charged, exposed by a flash, and softened by heat. All these operations are carried out in an $18 \times 15 \times 8.5$-cm unit with internal batteries.

Another approach to updatable storage and retrieval is the laser file (111). An optical video disk records and stores, and is accessible for plain-paper printout with a laser scanner–printer (see Laser Beam Recording). The capacity of one laser-file disk is 4000 documents and posting entry and printout times are three seconds each.

Micrographic Retrieval. The need for enlarged hard copy from microforms has led to the use of most of the reprographic technologies available for one-to-one copying. Kodak has transparent electrophotographic films that can be used for both microfilm and full-size copies. Plain-paper enlarger–printers have been developed by Xerox and others. Bimodal, electrophotographic reader–printers by Kodak and Bell & Howell can produce either positive or negative images without requiring a change of either the zinc oxide paper or the toner. A zinc-oxide-coated sheet with a thin metal interlayer was introduced by 3M in the 1960s for the Filmac reader–printers, which are no longer marketed (112). This process produced negative images by electrolytic development with a reducible metal salt (Ag^+) in those areas made photoconductive by light. Other metals, eg, nickel, can be used in this photoconductographic method (113). Dry-silver DTR and stabilization processes have been used to produce enlarged copies from microform.

ELECTRONIC PRINTOUT

The printout of information generated and transmitted electronically with digital signals increased as computers expanded in the post-World War II period. In the

1970s to the present, the numbers and types of printers expanded with the users needs include prints from distributed to data processing and word processing, graphics, and electronic communications (eg, facsimile). By 1980, there were 1250 different commercially available printers which used 10–15 different printing technologies, including most of the optical phototechniques used for copying as well as direct nonoptical methods, ie, thermal, magnetic, thermal-magnetic, ink-jet, and electrostatic. Revenues for computer printers in 1980 and 1981 increased 25–30% annually with sales of \$6 $\times 10^9$ (114).

The digital information can be recorded by impact printing with full type characters or by the formation of characters or graphics with a matrix of closely packed dots. The use of coded information for dates and text alphanumeric is efficient and is common for storage, processing, and printing of full characters. The combination of noncoded with matrix technologies made possible the handling of ideographic languages, eg, Chinese or Japanese; hand-written material, eg, signatures; and graphics in general. This use of digital, noncoded information requires much more intelligence than coded information, and there is an increase of one-hundred- to one-thousand-fold in storage capacity and a corresponding increase in time for processing and transmission (114).

The use of nonimpact methods to produce hard copy involves dot matrix systems (115). The print quality and resolution depend on the matrix and the number of picture elements or pels describing the character. A nonimpact dot matrix yielding the quality of this article would have a density of 120–200 dots or pels/cm; this high resolution level requires 0.75–1.22 megabytes for a 21.6 × 28-cm (8.5 × 11) page. (One byte is the smallest addressable unit in a computer or word processor and it consists of eight sequential binary digits or bits.) In any given dot-matrix printing system, the lower the concentration of dots delivered, the greater the output per unit of time, because much less central-processing-unit (CPU) time is used. The print quality also depends on the qualities of the dots, eg, shape, density, density distribution used, and contrast to background, which are functions of the printing method and equipment (116).

Print quality is measured against the well-defined solid-character print of the fully formal character printers, ie, typewriters. Dot-matrix methods can easily produce draft quality for rough drafts, internal office documents, etc, whereas closely packed dot matrix is sometimes acceptable for correspondence quality for common business use and is characterized by good-quality, legible print. For letter-quality word processing and formal business applications, the basic dot matrix system needs enhancements, such as multipass printing and overlapping printhead elements.

Impact Printing.   Impact printers at first were an extension of the typewriter and became more advanced in step with the computer input sources. The imaging of whole characters serially by a hammering member striking paper through inked ribbon involved typical typewriter print elements, ie, keys and ball impacters. The high speed, full-character daisy wheel has a circular series of flexible spokes at the ends of which are raised characters. The hub is rotated into position and the desired character is hammered forward onto the ribbon and paper. As many as 30–55 characters can be produced per second. A variation of the daisy wheel is the thimble, which has two characters per spoke instead of one; the spokes rotate and move up and down for selection of a single character from the upper or lower row. The thimble is faster than the daisy wheel and operates at 55 characters per second.

Much faster at 60–90 characters per second are the dot-matrix printers. The printhead contains banks of wires arranged in matrix formats [12.7 × 17.8 cm, 17.8 × 17.8 cm, 22.9 × 17.8 cm or 5 × 7, 7 × 7, 9 × 7, respectively]. A dot grouping or matrix in the desired character shape is formed by firing the wires at high speeds against an inked ribbon and paper. The matrix wires can be fired to form characters serially or to print a whole line on a line printer.

The mechanical bases for impact line printers are varied, ie, band, drum, chain, and train, to position a whole line of embossed characters that are either struck by hammers into inked ribbon and carbon or have the ribbon and back of the paper struck towards them. Their print speed varies from 150 lines per minute to almost 4000 lines per minute for a band line printer.

The disadvantage common to all impact printers is noise. A typical daisy-wheel printer produces over 80 dB, and the operating noise of the line printer often is well above the OSHA target of 55–60 dBs for offices (see Noise pollution). The dot-matrix printer can produce numerous character sets. Changing type fonts is relatively easy in a daisy wheel and even easier with a thimble, but it is much more of a limitation in the impact line printers. On the other hand, all impact printers can make multiple copies with carbon sets, whereas the primary output of nonimpact printers is a single copy or master. The 1981 prices for the serial impact printer vary from $600 for dot matrix to $6000 for some daisy wheel printers to $3000–$112,000 for line printers.

Nonimpact Printing. Other than the use of digital electronic information to form dot-matrix light images with lasers, light-emitting diodes (LEDs) (see Light-emitting diodes and semiconductor lasers), and CRTs and to print these images by electro-photographic techniques, electronic, magnetic, and thermal signals are used to form images directly. Compared to impact printers, all of these methods are virtually noiseless. The cost of nonimpact printers range from the relatively low speed, low cost thermal printers to high speed, electrographic and electrophotographic page printers. The cost per print including equipment costs (rental plus use and maintenance), supplies or consumables, and labor is lowest for very high volume copying (500,000–1,200,000 prints per month). The prints from high speed impact printers are of very low cost, but middle and low performance nonimpact prints cost as much or more than those from photocopiers (1). The nonimpact printers do offer mixable character sets and fonts, graphic printing including nonalphabetic language, and noiseless, high speed, high quality printing.

Thermal Printers. Thermal printing is the most popular nonimpact printing technique. Electrical signals are converted to thermal energy and are deposited through a microdot-matrix printhead to thermal paper to form text or graphics in one step (117). The three types of thermal printheads in use (118) are thick film (119), thin film (120), and semiconductor (121) types. The solid-state semiconductor head is a monolithic silicon construction in which each mesa contains a transistor–resistor-pair element. Each element is selectively energized so that the resistor causes the top of the selected mesa to become hot. Such a matrix can print up to 30 characters per second. The equipment is small and relatively inexpensive, ie, $1000–5000 and less than $200 for narrow-format calculators. The image dots are somewhat diffuse because of the nature of the heat transfer from heated elements but is satisfactory for low volume, medium-low quality applications, eg, printing calculators and master-format printer plotters.

Thermal printing is used in compact desk-top printer units, eg, the Panafax Transverter, which receives in a facsimile mode and a data mode. The machine can be connected to telex, TWX, communicating word processors, and computers. Electronically, the machine can determine the kind of transmitting equipment and whether the message is data or facsimile, digital or analogue, and it can print on thermal paper at a selected speed and resolution of either 80 or 40 lines per centimeter.

Thermal image formation. Color image formation in thermosensitive paper occurs at temperatures much lower than the melting points of the dye precursor and coreactant and usually at 70–80°C. A combination of 2-(2-ethoxyethyl)amino-3-chloro-6-(diethylamino)fluorene as the dye precursor and 2,2-bis(4-hydroxyphenyl)-propane (bisphenol A) in poly(vinyl alcohol) binder is commonly used as a coating (122). Benzoheterocycloammonium salts (both indolinic and benzothiazolenic condensation products) with salicylic aldehydes have been investigated as color precursors with bisphenol A to provide increased color range (123).

Shelf-life of these systems and image fading has led to the study of photosensitive thermal-imaging material. A diazothermographic system consisting of diazosulfonate and a coupler in a binder that can be fixed is used in commutation-ticket-issuing machines. A diazosulfonate on the paper is activated by light exposure, a thermal head causes a coupling imaging reaction, and finally a pulsed xenon flash decomposes the remaining diazo compound and thereby fixes the images (124).

Thermal ink transfer. To avoid the instability, appearance, and feel associated with thermal paper and the activated chemical reaction, a thermal printing method by transfer of solid ink to plain paper has been designed. A traveling thermal element is used to transfer the solid ink from a thermographic ribbon to plain paper by simultaneous application of heat and pressure (123). Thermal ink-transfer imaging techniques involve a thermal imaging sheet with an ink coating, which produces both direct thermal print and direct transfer copy on plain paper as the combination of plain paper and thermal-ink paper is passed over a thermal head. Another variant of this technique is the ink-film method, which transfers ink to plain paper from an ink-coated film and produces a clear reverse image on the film as the ink is transferred (125).

Electrosensitive Printing. Electrosensitive printing forms an image in a single step by localized electrical discharge which selectively removes an opaque metallized (aluminum) or pigmented (ZnO or TiO_2) top coating and thereby exposes an underlying black layer. The electric discharge matrix elements can record much faster than corresponding thermal elements, and characters can be formed at up to 6600/s, albeit with the generation of odor, smoke, and dust. The 1981 price of the compact equipment was $500–3000. Electric-discharge prints provide mediocre copy quality. However, in the ANAC digital graphics printer, a 16-level gray scale is achieved by changing the intensity of the picture element rather than by increasing the number of dots per centimeter (126). The TiO_2 coating affords a whiter and more paperlike appearance and has supplanted metallized coating in most applications.

Magnetic Printing. Magnetography has long been studied as a means of high, nonimpact, multicopy printing on plain paper (127–128). The initial image is formed by a magnetized recording head on a magnetic tape or drum. The magnetic image is developed by a dry magnetic toner. The toned image is transferred and fixed to plain paper by heated rollers (129) or by applying a d-c bias for transfer with subsequent pressure fusing (130).

The Iwatsu magnetographic copier–duplicator was the only commercial magnetic graphic printer available in 1980–1981 and then only in Japan (130), despite an earlier introduction by Data Interface in 1972. The d-c transfer is from a magnetic drum, and a metallic blade cleans the drum and prevents fouling of the recording head. The Iwatsu copier produces up to 9999 copies at 100 copies per minute from a single latent image. The quality of the image is like that produced by electrophotography for regular text, including Chinese ideographs, and is sold in Japan for $18,000.

The AM International PXL-6 is a slower magnetographic printer and has a 95 × 190 dot per square centimeter print-matrix head capable of producing letter quality at 6 copies per minute. This magnetic recording head is driven from a replaceable joint cartridge and transmits the magnetic signal onto a continuous-loop magnetic tape, which is developed with magnetic toner. The toned image is corona-charged and then transferred full-format to paper by electrostatic charge and pressure. The tape is then cleaned for residual toner and is magnetically erased for the next cycle. The printer for word-processing applications is compact and cost (in 1981) ca $10,000–15,000 (131).

Research involving magnetographic process and materials is concerned with toners and the toner transfer and cleaning steps as well as with the magnetic media. The Iwatsu drum is plated with Co–Ni–P and is said to have a life of over 10^6 copies. Commercially available CrO_2 tape was investigated as the recording medium (132) and special alloys $Fe_{35}Ni_{60}Co_5$ and $Fe_{40}Co_{60}$, were used in piezoelectric transducer recording heads (133).

Thermomagnetic printing. The thermomagnetic process as described for the DuPont Cirtak screenless image-transfer system is magnetic printing that forms the latent image by thermal means. It resembles indirect electrophotography in that the latent magnetic image forms by demagnetization of a premagnetized CrO_2 coating with heat as light forms the latent-charge image by discharge of the precharged photoreceptor. A coating of chromium oxide in polyurethane is heated above the Curie temperature of CrO_2 (120°C) by a laser or by flash exposure. Wet or dry magnetic toners are used to develop the image, which is transferred to a heated substrate. The Cirtak system is used in printed circuit-board manufacture, although it can be used for document printing (134).

Ink-Jet Printing. In the 1970s, commercial ink-jet devices that form images by controlled projection of ink provided a nonimpact and noncontact means to produce high quality images at low to very high speeds (135). The use of ink-jet printing has grown rapidly as it is a reliable printing method for business forms, product coding, computer hard-copy printout, and chart recorders (136).

There are three basic systems: a high frequency, pressurized, continuous synchronous jet that is electrostatically deflected; a low frequency, pressurized electrostatically generated–gated, hence a synchronous or intermittent-jet; and a low-to-medium frequency, nonpressurized impulse jet, which is also known as ink-on-demand or drop-on-demand.

Droplet generation from a single pressurized jet depends upon the operating parameters of the jet diameter and the velocity and stimulation frequency, which is ultrasonic (137). Maintenance of uniform and synchronous distribution of the ultrasonic stimulation energy requires special arrangements if large jet arrays are used (138).

Continuous jet. Electrostatic scanning of continuous jets is described in ref. 139. The continuous streams of droplets are modulated by applying d-c potential and then deflecting the charged drops electrostatically and allowing the uncharged drops to deposit and print (140–141). This binary system is depicted in Figure 2. In a scanning

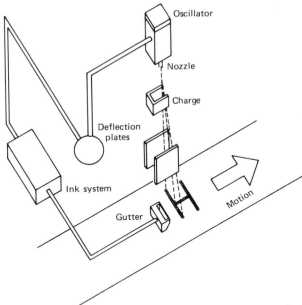

Figure 2.   A continuous deflected-drop ink-jet printer. Courtesy of the Society of Photographic Scientists and Engineers (135).

or analogue jet system, the charging voltage is stepped in synchronization with drop formation. The deflection field has a scanning effect as these various charged drops pass through to form a row of characters.

Scanning by means of a mechanically oscillated ink jet is suitable for alphanumeric character printing in single line scan at ca 30 characters per second (142). A scanning system that moves a jet, which is integrated with its control and deflection field electrodes on a carriage across the face of a revolving drum, was also developed (142). Three-color ink-jet heads on a single scanning carriage are used to plot color maps and charts (143). This type of drum plotter is used in most graphic applications, eg, in the Applicon color plotter (144).

Full-width web scanning is achieved with an array of ink jets. With two rows of closely spaced arrays, the Mead Dijit continuous jet-ink system can print the entire width of a continuously moving web at a potential speed of 600 m/min.

Intermittent jet.   The intermittent-jet systems are also employed in single-jet scanning with the valving electrode used to produce drops of ink. The ink in the intermittent jet is pulled out of the nozzle by electrostatic force generated at the drop-emission electrode. A second electrode acts as a gate providing the additional voltage gradient to produce the stream of drops. Scanning, as effected by the charge on the drops and by sets of deflection electrodes which direct the drops to form alphanumeric characters, was used in the Teletype Inktronic terminal printer in the early 1970s. Later, Toshiba used the intermittent jet for facsimile units, and Minolta choose this electrostatically generated jet for a X–Y plotter (145–146). Both the continuous and the intermittent jets require a device to catch the charged ink drops deflected from the stream of uncharged drops directed to the image-receiving surface. This catcher or gutter returns the ink to the jet with little or no misting or accumulation (147).

Impulse jet (drop-on-demand). The impulse ink jet produces drops individually by a transient pressure pulse, so that the drops are generated digitally on demand and are directed at the receiving surface. There is no ink circulation in the impulse ink-jet system. The drop ejection by pressure is caused by volume changes in an ink cavity behind the ink-ejection orifice. The volume change is created by a piezoelectric crystal transducer that, upon application of a voltage pulse (ca 50 V), radially compresses or flexes a wall diaphragm. The Zoltan impulse ink jet has a cylindrical transducer surrounding a tubular glass nozzle (see Fig.3) (148–149). In the Silonics structure, a monomorph piezoelectric driver is on one side of a thin rectangular chamber (150). Another rectangular device makes use of a monomorph and diaphragm with ink between two orifice plates (151). Matsushita developeed a version similar to the preceding device, except that an air jet is formed downstream of the ink orifice (152). The combination of amplitude modulation and air aspiration is said to afford the highest drop rate (20 kHz) for an impulse jet.

Impulse ink-jet printers are used in small arrays, eg, in the Silonics Quietype printhead which contains a vertical array of seven ink orifices. Under computer control, ink droplets are ejected from a selected orifice to the paper as the array is scanned horizontally. Small arrays of impulse jets are capable of speeds up to 200 characters per second compared to 2000 characters per second with small arrays of continuous-scanning jets. However, the impulse-jet system offers the simplest and potentially the lowest cost since it requires no ink circulation, no deflection electrodes or ink catchers, and no significant pressurization. Impulse-jet printers for low and medium speed printer terminals and facsimile as well as color and half-tone printing have been developed by United States, Japanese, and European firms.

Inks. Most ink-jet inks are water-based and pH-controlled to keep the color of the dyes from shifting and to prevent any corrosion of the mechanical component of the printer by the ink. The ink for continuous and intermittent systems that rely on electrostatic deflection must have a conductivity of ca 400 ohm·cm. Polyols, ie, ethylene glycol, glycerol, and polyglycol, are added up to 20 wt % to control viscosity for good aerodynamically stable drops (152–153) and to prevent this ink from drying in nozzles and orifices. Fast-drying ir-absorptive inks contain polyethyleneimine which controls viscosity to 1–10 mPa·s (= cP) at 25°C (154). Control of the fluid properties of ink for an impulse jet is more critical because of the necessary balance between ink refill and drop ejection. Special alcohol-based inks are produced for code printers, eg, the A. B. Dick Videojet and Dennison printers. The ink must be able to adhere to nonabsorbant surfaces, eg, beer cans, and they must be fast drying. Magnetic ink developed by IBM for magnetic ink-jet printing contains surfactants and additives to prevent precipitation of Fe_3O_4 in an aqueous system (153).

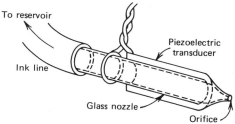

Figure 3. The Gould impulse ink jet. Courtesy of the Society of Photographic Scientists and Engineers (135).

The inks must be filtered free of any particles that might block the jet orifices (155). Pigmented inks cannot normally be used because the pigment particles clog the filters in the recycle systems and eventually stop ink flow. A compound-jet method increases the image density produced by soluble dye inks (143). The compound jet has a nozzle 0.5 mm below the surface of the ink so that the high speed jet essentially entrains a cylindrical layer of ink as it emerges from the liquid and, thus, is able to deliver significantly more ink and colorant to the paper without much loss in resolution (see Inks).

Electrostatic Printing.   The deposition of charge from an array or matrix of corona-discharge styli is used to form electrostatic latent images, which are subsequently developed with electroscopic toners. The image is formed directly on dielectric-coated paper and is developed or the image is formed indirectly by placement of the charge pattern on an electroreceptor surface, development of the latent image with toner, and transfer of the toner image to plain paper (156).

Direct electrostatic printing.   Printer–plotter printheads contain 2–4 offset rows of styli to give 40–80 and >80 lines per centimeter, respectively, on dielectric paper in the Benson-Varian equipment. High speed, medium quality dot-matrix printers, such as the Honeywell Page Printer, generate up to 18,000 lines per minute on dielectric paper with liquid-toner development (157). Better dot coverage is obtained from the two offset rows of styli by the dragged-image technique, by which the styli remain energized as the copy paper moves a short distance. At high speed and high volume (ca 2,000,000 copies per month), the liquid-toner copies require an efficient heated platen, forced air blowing for rapid drying, and a complete recovery system for the isoparaffinic toner carrier which evaporates from the copy sheets.

Electrostatic transfer printing.   A modified Minolta 301 Copier, called the EC301 Intelligent Copier, incorporates a dot matrix that deposits charge directly to the electrographic paper at 800 lines per minute. The basic electrophotographic, latent-image transfer to the dielectric paper can be operated at 10 copies per minute. One command to the machine can integrate photogenerated latent-charge image by transfer and the data-dot matrix by direct charge to the same sheet of paper before development and fixing.

Electrostatic plain-paper transfer printing.   A high speed electrostatic recording technique that prints on plain paper and involves toned image transfer from a dielectric-coated drum was introduced in 1977 (158). The printer (Fig. 4) can be used for data and facsimile at 3000 lines per minute and for image resolution of 95 dots per centimeter. Application of negative voltage to the pin electrodes forms a latent electrostatic image by selectively removing the blanket positive charge on the dielectric drum. The positive toner adheres to points discharged by the pin electrode to form a toner image, which is transferred to plain paper with a transfer corona. The toner image is fixed in a radiant fuser. The drum has a 25-nm thick TiO_2-acrylic–melamine dielectric coating, which is cleaned by another blanket as corona discharge and a fur-brush cleaner. A similar multistylus electrographic machine has been described by Fuji-Xerox (159). Both printers can produce prints at high speed with resolution good enough for both alphanumeric and Japanese kana characters.

In another electrostatic drum-transfer printing technique called ion projection, the matrix printhead does not deposit the charge directly on a dielectric-coated drum (160). Rather, ions generated by gated, high voltage oscillators are selectively extracted by print voltage pulses on a screen electrode. The latent electrostatic image so formed

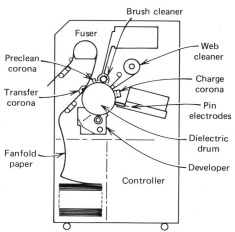

Figure 4. Electrostatic pin–electrode transfer printer, Fujitsu FY-3 Graphics Printer (schematic). Courtesy of the Society of Photographic Scientists and Engineers (158).

is developed with single-component magnetic toner and the toner image is simultaneously transferred and fixed to plain paper in the nip formed between the dielectric-coated cylinder and a pressure roller. Printers have been operated at 120 dots per centimeter and 60 pages per minute. The simultaneous transfer and the fusing technique has also been applied to an electrophotographic system, in which the latent electrostatic image is transferred to the dielectric cylinder from a photoreceptor drum and develops before the transfer–fuser step (161).

An electrostatic recording technique, which directly connects the electric signals to visible image without forming a latent electrostatic image, has been developed (162). Electrophoretic continuous-tone recording can produce pictorial facsimile prints. A 50–100-nm gap between an array of recording styli and a conductive ground drum is filled with flowing liquid toner, which is an electrophoretic suspension of charged toner particles in an isoparaffinic carrier. The toner image forms on the drum surface as the control voltage is modulated, and the image is then transferred to smooth plain paper or, for improved resolution, to a coated paper stock.

Magnetic stylus recording. Magnetic stylus recording or the conductive-toner bridge method is an electrographic printing process that images electronic signals directly on plain paper without transfer (163). An electronically conductive, magnetic-toner brush that forms between the array of magnetically permeable styli and a dielectric layer deposits a dot of toner in response to electronic signals. The magnetic styli serve the same function as the magnets in electrophotographic developers, ie, providing a magnetic counterforce to a selective electronic developing force as well as a means of transporting the toner to the printing zone (Fig. 5). The styli in this process are also the source of an electronic signal that forms a visible toner image. The dot image deposited on the drum is then transferred to plain paper and is fixed.

Each stylus is driven by its own low voltage (30 V) electronic-signal source with common integrated circuits, and high speed printing equivalent to over 500 characters per minute has been achieved. The resolution depends on the stylus density in the linear array. Excellent micrographic images are obtained at 113 styli per centimeter and half-tone images at 160 styli per centimeter. At a dot-stylus density of 50/cm, which is adequate for electronic printing (1), the 1981 cost for a 21.6-cm wide letter array

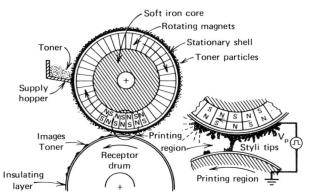

Figure 5. Rotating magnet toner–transport and magnetic styli for magnetic stylus recording. Courtesy of the Society of Photographic Scientists and Engineers (163).

could be $20,000–40,000. Besides very high speed electronic printing, the process could be used for plain-paper facsimile and copying, and with transfer to special stock, for offset plates and labels.

Electrophotographic Printing. The adaption of systems of electrophotographic copiers to computer printout involves a variety of digitally controlled light inputs. Although a direct electrophotographic process could be used, the transfer plain-paper techniques are extensively adopted for computerized printing equipment.

Laser-beam recording. A laser-beam scanning raster is projected directly onto the charged photoreceptor to photodischarge it, followed by the usual process steps (see Lasers). The scanning is accomplished by deflecting the modulated laser beam with a rotating polyhedral mirror (Xerox 9700, IBM 3800, Canon LBP-10), an oscillating galvanometer-mounted mirror (164) (Fig. 6), or a solid-state deflector (165). The gas or solid-state laser (He–Ne, He–Cd, GaAlAs) is matched to the photoreceptors, which in some cases are specially adapted to the output of the laser. The scanner optics and drum are critical to accurate placement of the raster spots and the image quality at 120 dots per centimeter. High speed printers, eg, the Siemens ND2, also have optical form-overlay features in addition to a printing speed of 146 pages per minute. The Xerox 9700 images forms, logos, charts, etc, electronically from data stored within the printer. Laser-beam scanning has also been proposed for recording and retrieving text on an optical video disk.

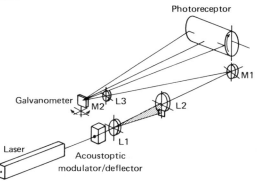

Figure 6. Optical schematic of laser scan printer with horizontal galvanometer deflection and vertical deflection with an acoustoptic cell (wobble scanning). L = lens, M = mirror. Courtesy of the Society of Photographic Scientists and Engineers (164).

Cathode-ray-tube recording. Electrophotographic printing with exposure of the photosensitive element with light from a cathode-ray tube (CRT) has an advantage over laser-beam exposure because the former is an all electronic method with no moving parts, thereby providing an inertialess movement of a spot of light (1). The lower levels of light energy from the phosphor make the spectral matching of photoreceptor sensitivity and the phosphor output even more important than in the case of the laser. The use of a fiber-optic CRT compensates for the relative inefficiency of the phosphor, and several firms use this combination for copier printers (166).

Light-emitting-diode recording. Other inertialess light-scanning systems for use with a xerographic device have been proposed for printer and high speed facsimile. An array of scanning light-emitting diodes (LEDs) ($GaAs_{0.35}P_{0.65}$) that emit at 630 nm was combined with a three-layer coating with Te added to increase the photosensitivity at the LED wavelength in a monocomponent, magnetic developing machine (167). Bismuth iron garnet thin film has been used to provide high resolution dot matrix of high speed (<1 μs) magnetic-optic switches in an electrophotographic recorder. The light can also be modulated on a time-length basis to provide graph–scale recording (168).

Intelligent Copier. The intelligent copier is a concept that arose in the 1970s as the electronic impact printers roughly provided the quality and speed of electrographic and electrophotographic copiers. The rapid advances in microelectronics, laser diodes, charge-coupled devices, word-processing information storage, digital codes, fiber optics, and data compression and transmission increased the feasibility of an electronic office environment, the need to provide high quality copier-type hard copy from data and word processing, and basic image processing. Other than the high speed printers with xerographic devices introduced by Xerox, IBM, and Siemens during this period, some of the first lower speed equipment shown, such as the Konishiroku copier nonimpact printer had regular 12 copy-per-minute copy modes and 21 copy-per-minute print modes from a host computer, and the Ricoh combination copier–facsimile apparatus (169) had the basic electrographic module with two scanning inputs.

The question of compatible signals from diverse sources, as word processors, facsimile machines (170), and hardware computer input is not completely answered with standard codes such as the American Standard Code for Information Interchange (ASCII) and the United Nation's Consultative Committee on International Telephone and Telegraph (CCITT) Group II Transmission Code for facsimile (= AM at three minutes per page). However, the ever-increasing cost-effective microcircuits provide conversion between one encoding system and another that should bridge any compatibility gap.

Color Prints

The vast majority of the images that are copied or duplicated are in black and white. However, the work to extend most of the reprographic technologies to produce copies with color, if not to provide color reproduction, started in many cases before black monochrome was commercialized (171).

ELECTROPHOTOGRAPHIC COLOR

The 3M Color-in-Color, which is a combined electrophotographic, thermal-field color copier was introduced in 1969 but was withdrawn in 1975. Numerous color copiers based on direct electrophotography with zinc oxide were demonstrated during this period and one, the Hitachi, was marketed for a few years in Japan. The Xerox 6500 was marketed in 1973 and this machine with both document and 35-mm-color-slide input has been principally marketed as a color-copy service through Xerox Reproduction Centers. In 1980, Canon demonstrated a full-color large-format [28 × 43 cm (11 × 17)], indirect electrophotographic machine, ie, the Canon NP Color, which produces 15 copies per minute (172). Konishiroku also announced the development of a full-color PPC.

The market results in the 1970s did raise a doubt as to an extensive need for full color in office applications. Both Ricoh and Konishiroku introduced two-color PPCs that copy in black or add a distinguishing red to the copy to denote emphasis or urgency. The Ricoh DC520 unit forms a bipolar electrostatic image on a photoconductive drum. The two colors then develop sequentially in two developers with colored toners of opposite polarity (173). Further work on a Ricoh system involves two-layer photoconductive layers and multiple exposure with filtered light to achieve color separation. The charge polarities of this two-layer receptor is arranged so that development with up to six toners yields a three-color print (174).

The multicolor print possibilities with the Electroprint process and the sublimation transfer technique utilizing direct electrophotographic ZnO paper to prepare the transfer medium have been developed. Addition of optical filters to the single-component tone machine to achieve color separation has made possible full-color, reverse prints by combining the color-toned prints for thermal sublimation transfer with a right-reading image (175–176). This approach is similar to the integrated combination of electrophotocopying and thermal transfer that was used in the 3M color copier.

SILVER HALIDE COLOR COPY

Using silver halide color technology, Ilford introduced the Ilford Cibachrome color copying system in 1981 (177). This combined camera–processor can produce, per hour, 50 of either color prints on coated paper or color transparencies and does not require a darkroom or darkroom operator. This silver dye bleach system gives high fidelity prints at high cost relative to xerographic color, but lower than the traditional silver chromogenic color prints. The (1981) cost of Cibachrome copy is $4.10 including labor, utilities, and $1.10 for materials. A Xerox 6500 copy costs ca $2.00 and high quality Polaroid 8410 color prints are $7.00–8.00 per print. A regular slow process (four-hour) Kodak lithographic C print costs $15.00–25.00. The Cibachrome Mark I that can copy only from opaque originals costs $20,000 and the Mark II with an enlargement–reduction feature costs $60,000.

NONPHOTO COLOR PRINTS

The use of computer intelligence rather than an optical imaging technique has been used to form color prints with color ink jets in the Applicon color plotter (144). Impulse ink-jet technology in the Print A Color Corporation IS 8001 and GP 1024 color ink-jet printer terminals with intelligence supplied by a computer gives the hard-copy form to the display on a color CRT terminal. The yellow, cyan, and magenta jet has a resolution of 35 dots per centimeter and produces prints in 2 minutes. The continuous ink jet with computer control of color shading and intensity reportedly produces very high quality color prints (178). On the other end of the imaging-method range is the Watanabe Intelligent Digi-Plot on an X–Y plotter that drives a fiber-tip pen plotter. Manual change of the pen gives color graphics and drawings.

Duplicating

Duplicating has grown rapidly in large part because of the use of high speed copiers as xerographic copier–duplicators producing hundreds and thousands of copies in a run. Copier–duplicators, eg, Xerox 9200, 9400, 9500; Van Dyk 8000; and Eastman Kodak 150s, do not duplicate from a master, they reimage at every cycle and for every copy. They are customarily used with a panoply of feeders, sorters, staplers, etc.

OFFSET DUPLICATING

Offset lithography is by far the fastest growing sector of duplicating. It represents an advance in plate or master preparation and, with improved presses, it is the most prolific and least expensive high volume reprographic method. Stencil printing and spirit duplicating remain significant reprographic methods with continued modification maintaining their niche in low volume, low cost installations and in special applications.

Offset lithography has the same relation to direct lithography that letterset has to letterpress printing. It has outmoded direct lithography because it was more easily controlled to give better quality prints (179–182). Small offset duplicating blossomed in the latter years of World War II as inexpensive and easy-to-make direct-image plates became available. The advent of improved photolithographic plates after the war, followed by the development of electrostatic masters, maintained the growth of offset duplicating during the electrophotographic boom.

DIRECT-IMAGE OFFSET PLATES

In direct-image plates, a hydrophilic surface is imaged with oleophilic material, eg, the oily ink of a typewriter ribbon. On a lithographic press, the oleophilic image accepts the oily printing ink, and the unimaged hydrophilic surface, when wet with aqueous fountain solution, repels the hydrophobic ink. A direct-image plate is right-reading and hence is printed by offset. The direct plate has been supplanted by higher quality prints from copiers either by simply copying or making a master on a copier from typescript or another original.

Photolithographic Plates. Photolithographic plate coatings have had a long genesis (13). Diazo-contact photohardening systems were developed and did not have the dark decay reaction of the dichromates but required contact exposure with a high energy light source. Dark, stable camera-speed systems based on silver halide chemistry and, later, direct electrophotography, produced camera-direct masters. Scanning or flash-projection exposure and processing is provided in direct and indirect electrophotographic copying machines.

Dichromate Plates. Photolithographic plates were sensitized with bichromated colloid coatings for use in offset printing (183). These plates had grained metal surfaces and were neither inexpensive nor easy to use. Cellulose-acetate-coated paper plates were the first inexpensive photoplates for small offset duplication. On photoexposure, the bichromate photosensitizer made exposed areas nonsaponifiable and, hence, nonhydrolyzable; such areas remained durably hydrophobic to treatments that hydrolyzed unexposed cellulose acetate areas to hydrophilic cellulose. On the press, the image area was relatively oleophilic; it attracted ink by repelling fountain solution. Instead of dichromates, ferric or silver salts were also useful for sensitizing cellulose acetate plates.

Dichromate Photolytes. Photochemical studies have yielded no tenable explanation of the photolysis the dichromate ion undergoes in solutions containing colloids (26,28,30–31,183). Nor are the mechanisms of hardening or insolubilization of colloids by the photolytic products understood. Studies indicate that the photogelatin systems important in gravure, screen, and collotype printing become cross-linked by chromium(III) bonding with acid groups of two or more gelatin molecules. According to this view, gelatin in a zwitterionic poly(amino acid) polyelectrolyte of high polybasicity, and the formation of chromic polysalts is the insolubilization mechanism and provides structural rigidity.

In the late 19th century, it was assumed that photoalkalinolysis of dichromate to chromate ion was simultaneous with photoreduction of dichromate to a mixed-valence oxide, $m\mathrm{CrO_3}.n\mathrm{Cr_2O_3}$, from which it was presumed that chromic oxide was hydroextracted in a form that hardened the colloid. Later, it was presumed that $m = 3$, and $n = 4$ and that uncharged oxygen atoms were the photolytes that insolubilized gelatin. It was suggested that the gelatin was hardened by cross-linking and the photolyte $(\mathrm{CrO})_2\mathrm{CrO_4}$ (= $3\mathrm{CrO_2}$) was proposed as the hardening agent. None of these explanations were compatible with the known dark decay of bichromate–colloid mixtures. Later this dark-decay hardening was inhibited by adding Cr(III) chelants. This inhibition is in accord with the dark decay stipulated in 1951 (183). The net dark reaction is described as

$$\mathrm{Cr_2O_7^{2-}} + 6\,\mathrm{H^+} \rightarrow \mathrm{Cr^{3+}} + \mathrm{CrO_4^{2-}} + 3\,\mathrm{H_2O} - 3\,e \tag{10}$$

Its redox component is taken as

$$\mathrm{Cr_2O_7^{2-}} + 14\,\mathrm{H^+} \rightarrow 2\,\mathrm{Cr^{3+}} + 7\,\mathrm{H_2O} - 6\,e \tag{11}$$

The alkalinolysis reaction then would be one of the following:

$$\mathrm{Cr_2O_7^{2-}} + \mathrm{H_2O} \rightarrow 2\,\mathrm{CrO_4^{2-}} + 2\,\mathrm{H^+} \tag{12}$$

$$\mathrm{Cr_2O_7^{2-}} + 2\,\mathrm{OH^-} \rightarrow 2\,\mathrm{CrO_4^{2-}} + \mathrm{H_2O} \tag{13}$$

Presensitized Photoplates (PSPs). Factory-manufactured photoplates are diazo plates, which are dark-stable enough to be used in offices and in-plant reprography departments without platemaking equipment. Although many photosensitive materials are dark-stable, they are not all stable with specific colloids nor do their photolytes always harden colloids (28). Diazoresin salts are poor photohardeners, but Kalle A.G. succeeded in photohardening using the diazonovolak resin, Diazoresin No. 4. The photolysis of the diazogroup N=N of the dark-stable material yields a phenol by the reaction

$$\mathrm{ArN_2OH} \rightarrow \mathrm{ArOH} \tag{14}$$

with a consequent photohardened coating.

Diazoresin No. 4 is at most a dimer (183). Gradually, a positive-working PSP was developed. Since Diazoresin No. 4 is hardened by light, its use requires photographic negatives so that light can pass through clear areas that correspond to the original image; a photoplate must be exposed to a negative image so it can print a positive image called a negative-working plate. The positive-working PSPs are photosolubilized where light strikes, thus the unexposed areas print black where the photographic transparency is black; photoplates exposed under a positive are called positive-working plates (184–185).

Presensitized photoplates are sold by AM Multigraphics, Azoplate Corporation, Citiplate, Inc., A. B. Dick Company, ATF Davidson, Lithoplate, Inc., 3M Company, Polychrome Corporation, Richardson Graphics Company, S. D. Warren Company, and Western Litho Plate, among others. Plate sizes range from 25×38 cm (10×15) for small offset duplicators used in offices to 147×197 cm for full-size offset presses in printing plants. Negative-working plates tend to be much more common. Aluminum plates may be grainless, grained and sensitized on both surfaces, fine-grained, fine-grained and sensitized on one or both surfaces, ball-grained, brush-grained, mechanochemically grained, mechanically grained, chemically grained, electrochemically grained, dry-blast-grained, dry-blast- and brush-grained, or anodized. The different surface treatments provide for different ink-dampener balances. Instead of sheet aluminum, aluminum foil that is variously grained may be laminated to paper or plastic to make a low cost plate. Paper plates may be plastic-coated, grainless, fine-grained, plasticized, or kraft with a fine-grained plastic surface; cellulose acetate plates are also used.

Wipe-On Plates. Just as it seemed that all shop-made plates based on bichromate sensitization were going to be replaced by factory-manufactured photoplates, web-fed offset printing became very successful and longer-run litho plates were needed. Presensitized photoplates do not give long runs; ca 50,000 copies is the upper limit for most plates, although some PSPs can produce 10^6 copies. On the other hand, certain kinds of shop-made plates give runs of hundreds of thousands but they are expensive and require large whirlers to spread and dry the viscous colloid coatings. Some printers thought that PSP coatings applied in their shops might be less expensive than factory-made PSPs. Diazo resin solutions are thin because diazo resins are not resins and make for easy wipe-on platemaking.

The success of this *ad hoc* approach is attested to by the fact that wipe-on solutions are sold by many graphic-arts dealers and suppliers, including suppliers of PSPs, and by the availability of plates up to 147×197 cm for shop-coating. All of the grains and bases sold for PSP manufacture are available, including additional varieties of laminates and deep-grained, foil-grained, marble-grained, extra-fine marble-brush-grained, etch-grained, as well as aluminum types.

Offset Duplicators. Offset duplicators are manufactured by A. B. Dick Company, AM Multigraphics, ATF Davidson, Didde Graphics Systems Corporation, Gestetner Corporation, Hamada of America, Inc., Heidelberg USA, Itek Graphic Products, Rotoprint, Inc., Royal Zenith Corporation, Solna Corporation, and Ryobi. Web-fed offset can produce 3000 sheets per hour to 50,000 web copies per hour. The weight of the stock can be $7.5–208$ g/m^2 ($1.5–43$ lb/ft^2), and card stocks and envelopes can be printed on duplicators.

Sheet sizes are 7.6×12.7 cm (3×5) to 44.5×57 cm with a maximum image area for duplicators of 44.5×56.5 cm. For a given plate size, sheet and print sizes may vary widely; thus a 45.7 × 38.1 cm plate press may have a print area of 44.5 × 33 cm and may print a paper size of 7.6 × 12.7 cm. The cost of duplicators usually is $1200–10,000; only one other company, Didde Graphics with a 25,000-copy-per-hour web model at $38,000, offers a higher speed above the usual range of less than 10,000 copies per hour. This press was once sold by IBM, but its speed and price removed it from the usual list of offset duplicators.

Unusual Offset Processes. Several offset processes print quite differently from those based on direct-image or common photoplates. Direct-image plates produce line copies (186). Although photography can record continuous variations of tones, printing can only simulate continuous tones. In printing, an area is either printed, ie, black or monochrome, or it is not, ie, white. Continuous-tone grays are simulated by printing small dots of certain sizes and at a certain number per area; the more there are per unit area and the larger each dot is, the darker the gray that the eye perceives. To space and size the dots appropriately, photographic transparencies that are used to make photoplates are halftone-screened. The unusual offset processes attain printing effects, eg, continuous tones, that are different from those of conventional offset.

Collotype. Collotype is the oldest offset process and involves bichromated gelatin photosensitization. Gelatin is hydrophilic and photohardened gelatin is relatively oleophilic. Once a collotype plate is made, it is soaked in glycerol or a glycol and then is inked. As long as the humidity is high enough (otherwise rewetting is needed), the applied polyol acts as a humectant and the nonimage areas remain wet and oleophobic enough to repel ink. The image areas accept ink and print. The amount of ink accepted is directly proportional to the amount of photohardening which is directly proportional to the amount of light passed through the transparency that is used to make the collotype plate. Thus collotype is a continuous-tone negative-working process. It is believed that gelatin shrinks on the photohardening and that the fissures produced thereby determine the amount of ink retained.

Continuous-Tone Lithography. Wipe-on coatings are very thin; their average coverage may be only 23-nm thick on a metal plate. Thus, wiping on a metal plate tends to leave a random pattern of thinly coated and uncoated spots on the plate surface, as well as sloping areas where the coating thickness normal to the plane of the plate can be many times that of the average coating thickness. Printers tend to overdevelop plates they overexpose and underdevelop plates they underexpose to obtain a standard gray scale in imaging. Thus, a further randomization of random effects of photohardening occurs.

These random effects are on a submicroscopic scale; therefore, they are not significant when line copy or halftone copy is printed. However, if a lithographic plate, and not necessarily a wipe-on plate, is exposed under a continuous-tone transparency without a halftone screen, the plate prints as if it were collotype. Unobstructed by the imposition of strong line and halftone dot images, the random effects of the plate appear as an integration of the light intensity by the exposed coating. The exact mechanics are not understood, but printers who have learned how to use this very critical randomization are quite happy with the continuous-tone high color effects.

Camera-Direct Masters. Although the contact plates have the highest quality and longest run length on aluminum, camera-direct masters can be prepared very rapidly, do not require a dark room, yield high quality line reproduction, and can be enlarged or reduced.

Silver Halide Camera Masters. There are three basic orthochromatic projection plates based on silver halide and available for camera preparation of masters. The Kodak Verilith plate is a positive-working paper master made by direct exposure of the master. The master consists of a paper base coated with developer which is overcoated with an unexposed silver halide emulsion which in turn is overcoated with a preexposed silver halide emulsion. Light from the nonimage area of an original passes through the top layer, solarizing it, and through the unexposed layer, exposing it. Thus, on placement of the exposed master in an activating solution, developer diffuses into the lower layer. Where the lower layer has been exposed, its silver halide is reduced, thereby exhausting the developer and preventing it from altering the uppermost layer. In the image areas, where there has been no illumination of the master, the developer is not consumed on diffusing through the unexposed layer and develops a positive image in the uppermost layer. In the uppermost layer, the gelatin is tanned or hardened by the diffusing developer and is altered so that it is ink-receptive after a stop-bath treatment. Where developer has not reached the gelatin, the master remains water-receptive. Approximately two minutes is required to make an offset master by the photo-direct process (4). Cameras made by A. B. Dick Company, AM International, Itek Business Systems, and Madax Graphic Products have image-reversing devices so that a right-reading positive master is obtained. Both Itek and A. B. Dick use plate material that is cut to size from a roll and the others use precut masters that are loaded into and processed with the camera at the rate of three masters per minute. Verilith masters are capable of 10,000 impressions.

The 3M MR412 camera plate is made of a silver halide coating on a polyester base and uses activator–developer four-bath processing instead of the Verilith activator–stabilizer system. The MR412 camera–processor can also make positive paper stock as well as positive transparencies with a change in roll material.

Diffusion-transfer plates are manufactured by Agfa-Gevaert for the DTR Rapilith process (187). Paper plates are made by direct exposure, processing, and transfer in Visual Graphics Corporation's VGC Platemaker 200 at a rate of two per minute. The paper plate can make up to 5000 impressions. The direct-exposure negative can also be transferred to a master in the Agfa-Gevaert Translith T42 processing unit. Contact exposure of a negative can be made with subsequent transfer to paper or to metal litho masters.

Electrostatic Camera Masters. Copies on any zinc-oxide-coated paper used in direct electrophotography can be used to make at least a few lithographic prints. The toned image is resinous, as is the xerographic image, and is ink receptive. The background can be converted and made oleophobic and nonreceptive to ink by treatment with slightly acidic potassium ferrocyanide solution (188). Most masters made by electrophotography are made on zinc-oxide-coated paper, albeit on special paper with water-resistant bases to give longer runs with modified coatings to enhance conversion; these are electrostatic masters.

Most zinc oxide, lithographic masters are made in equipment specially designed to utilize the same type of master stock that is used in photocopiers, but with large copy boards up to 61×91 cm for duplicate sizes, variable focus applications, and reduction and enlargement. Production rates are up to 500 plates per hour. In some equipment, the masters are automatically converted and emerge press-ready. The masters are made on equipment manufactured by A. B. Dick Company, AM Multigraphics, ATF Davidson, Itek Graphic Systems, Madax, 3M, and Rotoprint. These camera–platemakers cost \$7250–13,500, depending on copyboard size and other features (73).

Information International, Inc. manufactures a machine that makes zinc oxide masters directly from computer-output microfilm (COM). The negative-film input is reversed to a positive offset plate by a reversal liquid development. Direct imaging of a photoconductive master by CRT has also been proposed (189). Zinc oxide masters are made by A. B. Dick Company, AM Multigraphics, Gestetner Company, Itek Graphics Systems, Madax, Mita, Polychrome, Ricoh, Rotoprint, and Tomoegawa Paper Company.

An improved electrostatic master of binder-free, fully crystalline cadmium sulfide photoconductor, which after charging, exposing, and developing with liquid toner can be converted to a long run (>100,000) lithographic master, has been patented (190). The conversion is accomplished by treating the photoconductive surface with acidified potassium permanganate or chromic acid solution (191–192).

Integrated Electrostatic Platemaker–Duplicator.   In the 1960s, integrated plate-maker–duplicators with direct zinc oxide imaging and direct lithographic duplicating in the same housing were produced by AM International and Saxon Industries. Later the direct lithographic systems were replaced by a combination of an electrostatic master imager and a high speed offset duplicator. The AM Multigraph Total Copy Systems 4-45 and 5-45 and the A. B. Dick 1600 processes involve exposure of the original copy, development, fixing, and automatic insertion of the master onto the cylinder of the duplicator, delivery of the copy to a tray, and ejection of the used master.

Copier-Made Masters.   Masters are made on almost all the office copiers that use zinc-oxide-coated paper. Some of these machines that are roll-paper fed have a single sheet bypass to introduce special coated sheets, eg, offset masters. The offset master stock in roll form can also be substituted for the normal copy paper. For machines that make masters, liquid development as well as single- and two-component dry development are used. Normal drying or thermal fusing, both radiant and roll, are satisfactory for producing masters. Pressure-roll fusing tends to modify adversely the surface of the zinc oxide coating, which makes the conversion of the plate difficult.

The first equipment made by Haloid-Xerox and based on xerography was for the production of offset plates or masters if they were paper-based. As the process was mechanized from a camera with auxiliary processing equipment to a series of fully integrated machines, the transfer of the image to paper-offset masters continued. All of the plain-paper copiers can transfer to offset master stock instead of to plain paper. The relatively low speed copiers can thus supply the duplicator press with masters to run at high speed at usually a lower unit cost. Even high speed models, eg, the Xerox 9200 at 9200 copies per hour, which approaches the speed of an offset duplicator, are used in some cases to image masters for a battery of duplicators.

STENCIL PRINTING

Stencil printing is an ancient form of duplication in which ink flows through openings on a porous sheet to form an image on a receiving surface; no ink can pass through unopened blocked areas. Screen-process printing, or silk screen, rotary screen, or serigraphy, is based on woven screens. In the other type of stencil printing, ie, mimeography, the stencil is coated on a special nonwoven medium.

Screen-Process Printing. Screen-process printing is for specialty printing on almost any surface and for fine-art printing (19,193). The screens are mainly monofilament nylon, mono- and multifilament polyester, metalized polyester, and stainless-steel mesh fabrics (194). Monofilament screens are available in sizes of ca 2.35–0.03 mm (9–500 mesh) so that fine detail can be produced. The thickness of the ink deposit depends on ink viscosity, printing speed, the construction of the screen, and the shape and flexibility of the squeegee blade. Screen printing is commonly used for art prints, decals, nameplates, greeting cards, and program covers, and is particularly well adapted for printing on fabrics, felt, leather, glass, ceramic, and plastic materials (195).

Early in the 20th century, photomechanical methods, initially with dichromated colloids and later with diazo systems, were applied to stencil making. By 1980, there were 8000 screen printers in the United States and Canada with a total market size of ca 8×10^9.

Screen Photoresists. Presensitized or hand-sensitized photographic emulsion screens are the most commonly used for screen-process printing (196). The emulsion is coated on a screen, sensitized by dichromate, diazo, poly(vinyl cinnamate)s, and other such sensitizers; dried; exposed with high intensity uv light; and washed to give an imaged screen. A positive transparency of the original is used, because where the coating is light-struck, it becomes insolubilized and blocks the ink. Ready-made, presensitized screens are commonly used, but a pour-on emulsion that is spread evenly on the screen, exposed, and washed is also used.

In 1977, the wet-direct system was introduced. This method utilizes acrylate-terminated prepolymer moieties. The film positive is coated, while in contact with the screen mesh, with a photopolymer fluid which holds the film in place during exposure. Exposure of the liquid film can be immediate, as no drying is required, which reduces the time to make the screen. The use of one such fluid carrier-free radiation curable compositions results in an imaged screen with only the background cured and solid and the uncured compound in the image washed away with ethyl acetate (197). A variant of the presensitized indirect film is the capillary direct film and was developed in 1980 (198). The method allows adhering of the presensitized film prior to exposure and development with water instead of 2 wt % hydrogen peroxide solution. With the ease of preparation of sensitized screens, the need for dark-stable sensitized film, whether indirect or transfer, that can be transferred to a screen after exposure and washoff, diminished, so that such systems, eg, Kodak Ektagraph and Rotofilm, are no longer available commercially (see also Photoreactive polymers).

Carbon Tissues. Carbon tissue consisted of tissue paper coated with gelatin and loaded with carbon black and, later, pigment such as red Fe_2O_3. The pigment scatters the light during exposure after the coating is sensitized with 2 wt % potassium dichromate.

In the wet method, the wet, freshly sensitized tissue is exposed with the positive and is transferred to the screen. The double-transfer method involves drying the sensitized tissue, exposure to the positive, transferring to a temporary support, and developing in hot water. After transfer and drying to anchor to the screen, the support and backing are stripped. This method is rarely used commercially.

Unsensitized Film. Unsensitized films are supported on polyester fiber. Sensitizing is achieved by brushing with dichromate solution. Double transfer can be avoided by exposure through the backing sheet.

Electrostatic Screen Printing. In electrostatic propulsion on charged, dry toner particles through a screen with an opposite-charge-attracting plate behind the substrate, liquid ink is not used and the image must be fused to the substrate. In a combination of screen and electrostatic techniques called ion projection, the ink does not go through the screen, but rather the screen is used to block or allow passage of ionized air to a dielectric surface. The latent electrostatic image is developed with either dry or liquid toner.

Screens that block or permit a lesser or greater stream of ions to flow through the mesh in a pattern controlled by a photoconductor coated on a wire mesh are fundamental to the ion-modulator Electroprint process. Development is by a toner on a dielectric sheet or on a dielectric drum which is transferred to a receiving sheet, as in indirect electrophotography, and then fixed.

Printing and Presswork. Screen-printing presses are available in many forms to adapt to any type or conformation of surface. Hand presses with stationary screens with pressing of ink through the screens by a hand squeegee each sell for under $30 to a few hundred dollars. A second type of press uses a fixed squeegee and moving rotary screen. In a rotary press, a magnetic rod rolls within a rotating cylindrical screen over a stationary electromagnet. By far the most common types of commercial screen-printing presses are the flatbed press and the flat-screen cylinder press. In both cases, sheets are held by vacuum during the printing impression, which is by automated flood and squeegee stroke. The screen is raised and lowered into position automatically (194).

A variety of presses are commercially available for printing large sheets (up to 127×203 cm) in one or more colors. Flatbed presses with 51×76-cm format can operate up to 5000 prints per hour with a 43×56-cm sheet. One type of multicolor screen press uses a vacuum instead of a squeegee to pull ink through the screen. Instead of separate screens per color; this method can print any number of spot colors.

For printing on molded, irregular surfaces, an ink spray is forced through the stencil. As important as the photoscreen to the growth of screen printing is the use of uv-curing inks (104,199). The almost instantaneous drying of the ink reduces the space required for the press and eliminates the health and fire hazard of the ink solvents. Water-based inks also reduce the solvent vapor problem but require water-resistant stencil systems and drying on a receiving cloth surface of 150°C for several minutes (200). Inks containing dispersions of sublimable dyes can be printed or screened onto a release paper, ie, an uncoated, smooth-finish stock such as offset, typing, copy paper, or newsprint, and then transferred by sublimation by heating the intermediate sheet in contact with the surface (usually a textile) to be imaged. The sublimation inks cost more than twice water-based inks and solvent inks with non-sublimable pigments (200).

Mimeograph Duplicating. Mimeography first became popular in ca 1885 when the office typewriter provided a convenient means of punching through a wax covering on a fibrous-baselike Yoshino paper. The mimeograph was an A. B. Dick Company machine; the term has since become popular and generic. Electronic facsimile stencil cutters and thermal stencils have extended the range beyond the alphanumerics afforded by typing. Imaged stencils consist essentially of ink-impervious coatings, which are cut to expose a permeable sheet through which ink can pass. The stencils are wrapped around the rotating drum of a duplicator and the drum is perforated to permit brushed-on ink to pass through it to the stencil and through the images on the stencil to the image paper feeding against the stencil on the drum.

Setoff of wet images and spreading of images have been minimized by improvement in mimeograph inks, paper, stencils, and machines (201). The machines have been improved mechanically, especially in terms of inking, and an increasingly larger proportion are automatic and high in speed.

Automatic Machines. A. B. Dick Company, Deltek Business Machines, Gestetner Corporation, Standard Rex-Rotary, Roneo Alcatel, and Speed-O-Print Corporation make electrically operated mimeograph machines that give 4000–12,000 copies per hour. However, since stencil sizes vary from 7.6 × 12.7 cm (3 × 5) to 21.6 × 35.6 cm (8.5 × 14), the automatic machines provide less printing output than is given by larger size offset duplicators. However, 1981 machine costs are $300–3000, so the least expensive offset press is more costly than the most expensive automatic mimeograph. Thus the small office or school tends to use mimeography as its workhorse duplicator.

Hand-Operated Machines. A. B. Dick Company, Gestetner Corporation, Heyer, Inc., Roneo Alcatel, Standard Rex-Rotary America, and Speed-O-Print Corp. make hand-operated mimeograph machines, some with manual and some with automatic feeds, for a variety of card and small-sheet reproduction. Weber Marking Systems manufactures small-format mimeograph machines for printing on roll label stock. 1981 costs of machines were $250–900, and automatic feed machines can give up to 6600 copies per hour. Hand-operated machines are most useful for offices that make short runs of many originals.

Stencil Preparation. Stencils may be prepared by stylus drawing, typing, electronic facsimile scanning, or thermal imaging. Typing preparation is most common in offices, but with stencils prepared by electronic scanners and thermal imagers, artwork and composition can be combined (195).

Stylus Drawing. In trypography, a metal stylus was used to cut into the stencil placed wax side up on a rough surface, ie, a file plate. This method was patented in 1880s and 1890s and is still used in China to produce ideographs for mimeograph duplication.

Typed Stencils. A typed stencil is a manifold of three sheets: a stencil sheet, consisting of dyed wax on a strong ink-impervious fiber and overprinted with layout and space guide marks; an interleaf sheet, which colors from the key-expressed wax to give a proof copy of the typing; and a backing sheet, which facilitates handling of the stencil and transferring it to the machine drum. With care, the ordinary stencil can run ≥3000 copies. High quality stencils can give considerably longer runs on good and well-managed machines. Automatic inking increases the quality and uniformity of output and the life of the stencil.

Typing on a manual machine should be done with firm, even strokes; however, this technique is obviated with electric typewriters. Top film, ie, ribbon or sheet cellophane, can be used to sharpen the copy, to avoid cutout of interiors of letters, and to keep typewriter keys from fouling with stencil coating. Paint-over correcting fluids are available for stencils, and these can be retyped with no indication of prior error. Electric typewriters with narrow type faces give the best stencils. Tabulators and teleprinters can also cut stencils, and stencils can be cut by computer chain printers. A stencil that can be cut by the high speed, low impact daisy-wheel word-processor printer is available from A. B. Dick Company, Gestetner Corporation, Frankel Manufacturing Corporation, and Repeat-O-Type Stencil Manufacturing Company.

Electronic Stencils. The facsimile-scanning principle can be applied to the cutting of stencils. The original to be copied is mounted on the same cylinder that holds the stencil to be cut, thereby assuring synchrony. The cylinder is rotated and the original is scanned by light beam. The photoreceptor generates electrical pulses that actuate a thermoelectric cutter, which is a mechanized version of Thomas Edison's electric pen invented for stencil cutting (202). Although a photograph can be reproduced by this Stenafax process, a halftone print of a photograph produces a much better likeness. Scanners for electronic stencil cutting are manufactured by A. B. Dick Company, Gestetner Corporation, Heyer, Inc., Rex-Rotary, and Roneo. A stencil can be imaged in 3–15 min, depending on the original used.

Thermal Stencils. By far the most rapid method to prepare imaged stencil masters is in an infrared copying machine. The original is typed or printed with line graphics and logos and is simply passed through the machine with the stencil master and its tabbed carrier sheet in less than ten seconds. The thermal-sensitive double coating of cellulose acetate butyrate, which is heavily plasticized to reduce the melting point to 100–160°C, separates into two phases during imaging and cooling (203). The melted coating is absorbed into an overlay tissue and any unabsorbed coating in the image areas is left as a noncontinuous film that is pervious to ink. The thermal stencil is normally capable of reproducing 2.5 line pairs per millimeter but the original image must have good infrared absorption, as in all Thermofax materials. Thermal mimeograph stencil masters are made for all the mimeograph duplicators mentioned previously and cost about double that of nonthermal stencils. On these machines, the stencil masters are capable of producing a minimum of 1000 acceptable duplicate copies.

Color and Specialty Mimeography. Colored inks and paper can be used for various effects in mimeography but not as conveniently as in spirit duplicating. Full-color reprography, especially that involving halftone work, is better done on a color offset press than on a mimeograph machine where rerun and register problems are difficult. A special mimeograph ink places a concealed starch–oxidizing agent image on duplicated copies for self-instruction tests and sales-promotion material. The application of marker pen with a fluid containing a soluble iodide over an area with latent imaging produces an intense visible image, which forms as the iodide oxidizes to iodine, which reacts with the starch (204). This system is available from A. B. Dick Company.

SPIRIT DUPLICATING

In spirit duplicating, a dye image on a master is contacted to a surface that absorbs the dye. The surface gives up the dye where contacted with a moist copy sheet.

Hectography. In its earliest and simplest form, spirit duplicating was called hectography and was practiced using flat trays into which hot gelatin solution had been poured and allowed to set. Paper bearing methyl violet writing or typing ink was then contacted to the gelatin, flattened thereon, and allowed to rest until the gelatin had a substantial image pickup. Then successive moist copy sheets were pressed flat against the tray image to secure 10–20 copies. Tray hectography became popular with the invention of hectograph ribbon for typewriters, and when flatbed and rotary-type duplicators were introduced.

Flatbed Gelatin Hectography. The old flatbed hectographic machines permitted replication of many masters at one time. Gelatin was used in sheet or in roll form, and the gelatin was advanced each time a new master was replicated. Up to 15 copies per minute could be made with the flatbed duplicator. There were no interruptions in terms of changing masters as was the case for the single-tray method, where the image had to be washed out or the gelatin recast before a new original could be copied. If the dye image was not washed out, it vanished by diffusion in about a day and the gelatin could then be used anew.

Rotary Gelatin Duplicators. Thin sheet gelatin was fastened to the cylinder of the rotary hectographic duplicator. The rotary duplicator was more than three times as fast as the flatbed machines and gave up to 50 copies per minute. This type of equipment is outmoded and has been suceeded by spirit duplicators.

Spirit Duplication. Spirit duplicators are rotary machines and are hand-operated or electrically driven and use a wraparound master. The master, prepared by typing or thermal means, consists of a manifold composed of a glazed paper sheet and a backing sheet coated with pigment on the side adjacent to the glazed sheet. Thus, typing on the glazed sheet produces a right-reading typed proof image on the front of the sheet and a wrong-reading transfer image on the back side of the glazed sheet. The take-off image, which is thus generated in the manner of a direct-image lithoplate image, is called a carbon image, and the backing sheet is called a carbon sheet or transfer sheet. As is usual for carbon paper, the dyed and/or pigmented wax or polymer breakaway coating is called carbon ink.

The carbon ink is compounded so that only a portion of it is transferred to a copy sheet on moistening with an alcohol solvent or spirit, methanol is usually used because of its low cost. However, denatured ethanol formulations are increasing in use because of their lower toxological properties. The master is wrapped around the drum of the machine and is run against moist copy paper. Each copy removes part of the ink, ultimately exhausting the master. Conventional masters give 100–200 copies; improved masters give up to 300 copies.

The more evenly the spirit duplicator can be run and inked, the better are the copies and the greater is the acceptable output. Thus, as in mimeography, the manual machine has been progressively displaced by machines in which the feeding of paper and of solvent and the operation are automatic. Spirit machines have replaced the gelatin duplicators because the former can use inks that contain colorants that need not be transferable to and from gelatin. Brighter colors, a variety of colors, and more even reproduction of longer runs are possible. However, correction of masters in typing is awkward, since two-sided correction is involved.

Spirit Duplicators. Speeds of electrically operated spirit duplicators are 4000–6600 copies per hour and cost $700–1200, whereas hand-operated equipment at lower speeds cost $250–700. Both types of machines are manufactured by A. B. Dick Company, Deltek Business Machines, Heyer, Inc., and Standard Rex-Rotary.

Facsimile Spirit Masters. Spirit masters are conveniently made from originals in an infrared copier. The original can be typed, printed, drawn, or handwritten. The original and a thermal spirit master construction of a master sheet and an ink-coated, 0.013-mm polyester film with a backup carrier sheet are passed through a Thermofax machine in a matter of seconds. The spirit ink transfers thermally in the image areas to the master sheet to form the wrong-reading master with a resolution of 20 lines per centimeter. Thermal spirit masters are produced to yield a minimum of 100 acceptable copies. The familiar spirit purple is the most popular because it gives the most copies, ie, up to 300.

Color and Specialty Spirit Duplicating. Spirit duplicating has a unique capability for multicolor replication. The master can be part-imaged using one color of carbon. Then it can be imaged further with a second color-transfer sheet. As many different carbons as desired may be used to effect desired artwork. Although a multicolor original can be reproduced by a number of other processes, only spirit duplicating can do so with a single master on a single impression.

The concealed-image process used in mimeo duplication is also adaptable for spirit duplication (204). An oxidizing agent and ethyl cellulose are incorporated in the transfer sheet and are transferred to the imaged master by typing or heat and are leached onto the copy sheet with spirit fluid. The image is rendered visible by the marking material containing soluble iodide. A mixture of colors and latent invisible images can be produced with a single master.

Outlook

The shift in office reprography from duplicating to copying to production of copy from electronic sources has been one of changing emphasis. Although some processes have become obsolete, many are being used or adapted in new applications. The shift to electronic records and information have led to more selective reprographic retrieval and to a greater perceived need for hard copy. The shift has not resulted in less paper, but to an overall increase in paper output. The advance of electronics has increased the scope of reprography.

The uses of reprographic methods as copier–electronic recorders are the focus of extensive work. Lithographic techniques for the production of low cost, integrated microcircuits have resulted not only in computers with vastly increased computing power but minicomputers with capabilities that rival the giants of earlier years. These small computers in the office, shop, and home are creating a further need for reprographic prints.

Fiber optics (qv) are part of CRT recording, but arrays of focused fiber optics are also used in compact photocopy machines in place of the familiar lens–mirror arrangement. Even more important is the use of fiber optics for high speed, high capacity transmission of data information (205). Additional advantages of this nonelectrical transmission are security from nondetectable tapping and safety in transmission of record signals in hazardous, ie, flammable, chemical areas.

Many new technologies as well as combinations of old and new techniques will be tried to serve the increasing need of business communications. There are now, however, reprographic methods that satisfy the output requirements of business. Factors other than the primary one of producing legible hard copy at a reasonable cost are important. How much the equipment affects the office environment became important in 1970s and may be the deciding factor in the choices of the future. Energy consumption, heat and noise generation, and air pollution from solvents, ozone, and deposition of used materials, eg, photoreceptors, are considered in selection of techniques and the design of machines. With the application of relatively mature and proven methods, engineering to produce equipment that is reliable with low maintenance and downtime is probably the most significant factor in developing new reprographic equipment.

The single most important reprographic development of the 20th century is regularly used for secret and illegal copying of documents. As computer-assisted makeup and imaging systems (CAMIS) further extend the conversion of electronic information through an ever increasing number of processes to hard copy, the conflict between access and security and between copyright and private retrieval takes on increasing legal and public-policy implications that require answers to assure fair and widespread use of the expanding technology (see Trademarks and copyrights). The improvement in color copying also poses a resistant counterfeiting problem for the Treasury Department and others, and this may only be obviated if a checkless–moneyless era is reached (206).

The more efficient storage or filing of information requires microforms and even more dense methods capable of media recording, eg, magnetic and optical video disks with laser scanning. The new networks and increased ability of electronic information sources to interface may favor display instead of hard copy, but there is no sign of elimination of the need for reprography for copy and retrieval of information.

BIBLIOGRAPHY

"Reprography" in *ECT* 2nd ed., Vol. 17, pp. 328–378, by Jack J. Bulloff, State University of New York at Albany.

1. F. J. Neema, *1st International Congress on Advances in Non-Impact Printing Technologies*, June 22–26, 1981, Venice, Italy, Society of Photographic Scientists and Engineers.
2. *Chem. Eng. News* **42,** 115, 85 (July 13 and 20, 1964).
3. L. E. Varden in C. B. Neblette, ed., *Photography: Its Materials and Processes*, D. Van Nostrand Company, New York, 1962, Chapt. 28.
4. W. R. Hawken, *Copying Methods Manual*, American Library Association, Chicago, Ill., 1966.
5. H. R. Verry, *Document Copying and Reproduction Processes*, Fountain Press, London, 1958.
6. A. Tyrell, *Basics of Reprography*, Focal Press, London, 1972, Chapt. 7, p. 113.
7. D. F. Schultze, *J. Appl. Photogr. Eng.* **5,** 163 (1979); J. Van Den Houte and L. Vermeulen, *J. Appl. Photogr. Eng.* **5,** 167 (1979); M. R. V. Sanyun, *J. Appl. Photogr. Eng.* **5,** 32 (1979).
8. A. A. Newman, *Brit. J. Phot.* **109,** 212 (1962).
9. U.S. Pat. 3,094,417 (June 18, 1963), W. R. Workman (to 3M Company).
10. B. R. Harriman, *Proceedings of the 2nd Symposium on Unconventional Photographic Systems*, Oct. 29–31, 1964, Washington, D.C., Society of Photographic Scientists and Engineers.
11. U.S. Pat. 4,123,282 (Oct. 31, 1978), J. Winslow (to 3M Company).
12. *Federal Microfiche Standards*, *P B Rept. 167,630*, National Technical Information Service, Springfield, Va., 1968.
13. J. Kosar, *Light-Sensitive Systems: Chemistry And Application of Nonsilver Halide Photographic Processes*, John Wiley & Sons, Inc., New York, 1965.
14. U.S. Pat. 2,740,896 (Apr. 3, 1956), C. S. Miller (to 3M Company).
15. U.S. Pat. 2,740,895 (Apr. 3, 1956), C. S. Miller (to 3M Company).
16. U.S. Pat. 2,891,165 (June 16, 1959), C. A. Kuhrmeyer, R. Owen, and J. R. Favorite (to 3M Company).
17. U.S. Pats. 2,663,654–2,663,657 (Dec. 22, 1953), C. S. Miller and B. L. Clark (to 3M Company).
18. A. Games, *J. Phot. Sci.* **10,** 100 (1962); Brit. Pats. 943,401–943,403 (Dec. 4, 1963) (to Imagic Processes, Ltd.).
19. W. B. Proudfoot, *The Origin of Stencil Duplicating*, Hutchinson & Co., London, 1972.
20. Ref. 3, p. 355.
21. Neth. Pat. Appl. 6,401,297 (Aug. 17, 1964), N. V. Unilever; Belg. Pat. 666,125.
22. *Reprographic Guide*, International Reprographic Association, Franklin Park, Ill.
23. Ref. 13, Chapt. 1.3, pp. 27–50.
24. K. Bonhoeffer and P. Harteck, *Grundlagen der Photochemie*, Steinkopf, Dresden, Germany, 1933.
25. J. G. Calvert and J. N. Pitts, Jr., *Photochemistry*, John Wiley & Sons, Inc., New York, 1966.

26. P. Glafkides, *Photographic Chemistry*, Fountain Press, London, 1960.
27. J. N. Pitts, Jr., F. Wilkinson, and G. S. Hammond, *Adv. Photochem.* **1,** 1 (1963).
28. H. O. Dickinson, O. J. Fry, and co-workers, *Rept. Progr. Appl. Chem.* **44,** 95 (1959); **48,** 116 (1963); **50,** 99 (1965).
29. J. P. Simons, *The Principles of Photochemistry*, John Wiley & Sons, Inc., New York, 1954.
30. H. Staude, *Photochemie*, Bibliographisches Institut, Mannheim, Germany, 1962.
31. E. Teller, E. W. R. Steachie, W. West, and H. S. Taylor, eds., *Ann. N.Y. Acad. Sci.* **41,** 169 (1941).
32. A. Harder and A. Matheson, *Perspective* **6**(1), 5 (1964).
33. Ref. 13, Chapt. 6, pp. 201–248.
34. J. Kosar in ref. 10.
35. M. P. Doss, ed., *Information Processing Equipment*, Reinhold Publishing Corporation, New York, 1955.
36. J. Kosar, *Perspective* **7**(1), 5 (1965).
37. W. H. Cliffe, *Chem. Ind.*, 1248 (1958).
38. R. Holzbach, *Die Aromatischen Diazoverbindungen*, F. Enke, Stuttgart, Germany, 1947.
39. K. H. Saunders, *The Aromatic Diazo Compounds and Their Technical Applications*, 2nd ed., Edward Arnold & Co., London, 1949.
40. H. Zollinger, *Azo and Diazo Chemistry*, Interscience Publishers, Inc., New York, 1961.
41. M. Tsuda and S. Oikawa, *Photogr. Sci. Eng.* **23,** 177 (1979).
42. T. H. James, ed., *Mees' Theory of the Photographic Process*, 3rd ed., Macmillan, Inc., New York, 1966.
43. *Focal Encyclopedia of Photography*, 2nd ed., Focal Press, Ltd., London, 1965.
44. M. S. Dinsburg, *Photoactive Diazo Compounds and Their Use*, Khimiya, Moscow, 1964.
45. U.S. Pat. 1,628,279 (May 10, 1927), M. P. Schmidt and W. Krieger (to Kalle & Co.).
46. U.S. Pat. 4,273,850 (June 16, 1981), H. D. Frommeld (to Hoesch Actiengesellshaft).
47. D. P. Habib and G. R. Hodgkins in ref. 10, pp. 113–136.
48. S. Shinozaki and K. Itano in ref. 10, p. 142.
49. J. Weiner and L. Roth, *Diazotype Papers*, Institute of Paper Chemistry, Appleton, Wisc., 1965 (rev. ed. 1967).
50. U.S. Pat. 3,446,620 (May 27, 1969), K. Parker (to Addressograph-Multigraph Corporation); U.S. Pat. 3,578,452 (May 11, 1971), K. Parker (to Addressograph-Multigraph Corporation); U.S. Pat. 3,809,599 (May 7, 1974), H. J. Neuman (to Addressograph-Multigraph Corporation); U.S. Pat. 4,155,762 (May 22, 1979), T. Matsuda, Y. Hagiwara, Y. Arai, and T. Hirabaya (to Ricoh Company, Ltd.).
51. U.S. Pat. 3,640,203 (Feb. 8, 1972), L. R. Raab and D. E. Toby (to Addressograph-Multigraph Corp.).
52. Ref. 13, pp. 276–282.
53. R. T. Nieset, W. A. Seifert, and W. F. Elbrecht, *Phot. Sci. Eng.* **5,** 239 (1961).
54. Caps Consultants, Ltd., *Brit. J. Phot.* **11,** 976 (1963).
55. U.S. Pat. 2,911,299 (Nov. 3, 1959), A. Baril, I. H. De Barberis, and R. T. Nieset (to Kalvar Corporation).
56. U.S. Pat. 3,032,414 (May 1, 1962), R. W. James and R. B. Parker (to Kalvar Corporation).
57. R. T. Nieset, *Proc. Tech. Assoc. Graphic Arts* **16,** 203 (1964).
58. R. T. Nieset, *J. Phot. Sci.* **10,** 188 (1962).
59. U.S. Pat. 4,272,603 (June 9, 1981), D. J. Chenevert, J. R. Grawe, and J. C. McDaniel; U.S. Pat. 4,302,524 (Nov. 24, 1981), W. L. Mandella and J. R. Kuszewski (to GAF Corporation).
60. U.S. Pat. 3,120,437 (Feb. 4, 1964), R. M. Lindquist (to IBM Corporation).
61. J. H. Dessauer and H. E. Clark, *Xerography and Related Processes*, Focal Press, London, 1965.
62. R. M. Schaffert, *Electrophotography*, 2nd ed., Focal Press, London, 1975.
63. V. M. Fridkin, *Physical Principles of Electrophotogrpahy*, Focal Press, London, 1973.
64. V. M. Fridkin and I. S. Zheludev, *Photoelectrets and the Electrophotographic Process*, Consultants Bureau, New York, 1961.
65. J. W. Weigl, *J. Chem. Phys.* **24,** 370 (1956); J. H. Dessauer, G. R. Mott, and H. Bogdonoff, *Photogr. Eng.* **6,** 250 (1955).
66. R. B. Commizzoli, G. S. Moser, and D. A. Ross, *Proc. IEEE* **60,** 348 (1972).
67. J. W. Weigl, *Angew. Chem. Int. Ed.* **16,** (1977).
68. W. F. Berg and K. Hauffe, eds., *Current Problems in Electrophotography*, de Gruyter, Berlin, Germany, 1972.
69. J. Brooks, *The New Yorker* (Apr. 1, 1967); *Chemtech* **11,** 466 (Aug. 1981).

70. C. J. Young and H. G. Greig, *RCA Rev.* **15,** 469 (1954); U.S. Pat. 3,403,023 (Sept. 24, 1968), H. R. Carrington and F. Gaesser (to GAF Corporation).

71. *The Wall Street Journal*, 1, (Oct. 13, 1964).

72. P. G. Roetling and T. M. Holladay, *J. Appl. Photogr. Eng.* **5,** 179 (1979).

73. *Datapro Reports On Office Systems, Copiers & Duplicators*, Datapro Research Corporation, Delran, N.J.

74. J. K. Ghosh and W. E. Bixby, *J. Appl. Photogr. Eng.* **6,** 109 (1980).

75. E. S. Baltazzi, *J. Appl. Photogr. Eng.* **6,** 147 (1980).

76. U.S. Pat. 3,573,906 (Apr. 6, 1971), W. L. Goffe (to Xerox Corporation); U.S. Pat. 3,725,058 (Apr. 3, 1973), Y. Hayashi and M. Hasegawa (to Matsushita Electric Industrial Company); U.S. Pat. 3,816,117 (June 11, 1974), J. Y. Kaukeinen (to Eastman Kodak Company); U.S. Pat. 3,837,851 (Sept. 24, 1974), M. D. Shattuck and W. J. Weiche (to IBM Corporation); U.S. Pat. 3,839,034 (Oct. 1, 1974), W. Wiedemann (to Kalle Aktiengesellshaft); U.S. Pat. 3,850,630 (Nov. 26, 1974), P. J. Regensburger and J. J. Jakubowski (to Xerox Corporation); U.S. Pat. 4,140,529 (Feb. 20, 1979), D. M. Pai, J. F. Yanus, and M. Stolka (to Xerox Corporation); U.S. Pat. 4,278,746 (July 14, 1981), S. Goto, Y. Takei, I. Imaho, and N. Nomori (to Konishiroku Photo Industry, Ltd.); U.S. Pat. 4,287,279 (Sept. 1, 1981), G. A. Brown, L. A. Relyea, M. E. Scharfe, and H. W. Prinsler (to Xerox Corporation); U.S. Pat. 4,296,191 (Oct. 20, 1981), R. L. Jacobsen and T. T. Lin (to 3M Company).

77. U.S. Pat. 4,134,763 (Jan. 16, 1979), I. Fujimura and K. Endo (to Ricoh Company, Ltd.); U.S. Pat. 4,277,551 (July 7, 1981), T. J. Sonnonstine and K. G. Kneipp (to Minnesota Mining and Manufacturing Company).

78. U.S. Pat. 3,250,615 (May 10, 1966), J. A. Van Allen, C. C. Natale, and F. J. Rauner (Eastman Kodak Company); U.S. Pat. 3,586,500 (June 22, 1971), L. E. Contois and D. P. Specht (to Eastman Kodak Company); U.S. Pat. 3,938,994 (Feb. 17, 1976), G. A. Reynolds, J. A. Van Allan, and L. E. Contois (to Eastman Kodak Company); U.S. Pat. 4,167,412 (Sept. 11, 1979), M. T. Regan, G. A. Reynolds, D. P. Specht, and J. A. Van Allan (to Eastman Kodak Company).

79. U.S. Pat. 3,485,625 (Dec. 23, 1969), C. J. Fox (to Eastman Kodak Company); U.S. Pat. 3,542,544 (Nov. 24, 1970), E. J. Seus and M. Goldman (to Eastman Kodak Company); U.S. Pat. 3,754,986 (Aug. 28, 1973), E. A. Perez-Albuerne (to Eastman Kodak Company); U.S. Pat. 3,820,989 (June 28, 1974), N. G. Rule and R. C. Riordan (to Eastman Kodak Company); U.S. Pat. 3,966,468 (June 29, 1976), L. E. Contois and N. G. Rule (to Eastman Kodak Company); U.S. Pat. 4,018,607 (Apr. 19, 1977), L. E. Contois (to Eastman Kodak Company); U.S. Pat. 4,150,987 (Apr. 24, 1979), H. W. Anderson and M. T. Moore (to IBM Corporation); U.S. Pat. 4,152,152 (May 1, 1979), L. E. Contois and J. F. Jones (to Eastman Kodak Company); U.S. Pat. 4,278,747 (July 14, 1981), T. Murayama, S. Otsuka, and T. Tajima (to Mitsubishi Chemical Industries, Ltd.); U.S. Pat. 4,279,981 (July 21, 1981), M. Ohta, K. Sakai, M. Hashimoto, A. Kosima, and M. Sasaki (to Ricoh Company).

80. *Zinc Oxide Rediscovered*, New Jersey Zinc Company, New York, 1957; G. Heiland, E. Mollwo, and F. Stockman, *Solid State Phys.* **8,** 191 (1959); K. Hauffe, *Z. Anorg. Allgem. Chem.* **316,** 190 (1962); T. Elder, *J. Appl. Phys.* **33,** 2804 (1962).

81. V. I. Gaidyalis, N. N. Markevich, and E. A. Montrimas, *Physical Processes In Electrophotographic ZnO Layers*, Vilnius, Lithuania, 1968 (in Russian).

82. U.S. Pat. 3,011,918 (Dec. 5, 1961), L. H. Silvernail and M. W. Zembal (to The Dow Chemical Company); U.S. Pat. 3,830,655 (Aug. 20, 1974), E. Rothwell and G. Smalley (to Allied Colloids Manufacturing Company); U.S. Pat. 4,148,638 (Apr. 10, 1979), G. D. Sinkovitz and K. W. Dixon (to Calgon Corporation); U.S. Pat. 4,222,901 (Sept. 16, 1980), G. D. Sinkovitz (to Calgon Corporation).

83. Model 1900 Copier by Oce-Van Der Grinten, N.V.

84. W. Maenhout-Van Der Vorst and F. Craenest, *Phys. Status Solidi* **5,** 357 (1964); K. Hauffe and R. Stechemesser, *Photogr. Sci. Eng.* **11,** 145 (1967).

85. U.S. Pat. 4,275,135 (June 23, 1981), T. Tomonash (to Minolta Camera Kabushiki Kaisha).

86. U.S. Pat. 3,151,982 (Oct. 6, 1964), L. Corrsin (to Xerox Corporation); U.S. Pat. 3,288,603 (Nov. 29, 1966), L. Corrsin (to Xerox Corporation); U.S. Pat. 3,754,965 (Aug. 28, 1973), J. B. Mooney (to Varian Associates); U.S. Pat. 3,830,648 (Aug. 20, 1974), S. L. Rutherford and M. Feinleig (to Varian Associates); U.S. Pat. 4,053,309 (Oct. 11, 1977), G. A. Marlor (to Varian Associates, Inc.); U.S. Pat. 4,061,599 (Dec. 6, 1977), G. A. Marlor; U.S. Pat. 4,221,855 (Sept. 9, 1980), D. Manabe, S. Asai, and M. Suga (to Nippon Electric Company, Ltd.).

87. M. Kuehnle, *J. Appl. Photogr. Eng.* **4,** 155 (1978); U.S. Pat. 4,025,339 (May 24, 1977), M. R. Kuehnle (to Coulter Information Systems); U.S. Pat. 4,155,640 (May 22, 1979), M. R. Kuehnle and A. K. Ha-

genlocher (to Coulter Information Systems); U.S. Pat. 4,170,475 (Oct. 9, 1979), M. R. Kuehnle and A. K. Hagenlocher (to Coulter Information Systems); U.S. Pat. 4,241,160 (Dec. 23, 1980), M. R. Kuehnle (to Coulter Information Systems); U.S. Pat. 4,242,433 (Dec. 30, 1980), M. R. Kuehnle and A. K. Hagenlocher (to Coulter Information Systems); U.S. Pat. 4,269,919 (May 26, 1981), M. R. Kuehnle (to Coulter Information Systems).

88. U.S. Pat. 4,265,991 (May 5, 1981), Y. Hirai, T. Komatsu, K. Nakagawa, T. Misumi, and T. Fukuda (to Canon Kabushiki Kaisha).

89. I. Shimizu, T. Komatsu, K. Saito, and E. Inoue, *J. Non-Cry. Solids* **35/36,** 773 (1980); J. Mort, S. Gramatica, J. C. Knights, and R. Lujan, *Photogr. Sci. Eng.* **24,** 241 (1980).

90. N. Yamamoto, K. Wakita, Y. Nakayama, and T. Kawamura, *Jpn. J. Appl. Phys. Suppl.* **20**(1), 305 (1981); U.S. Pat. 4,226,897 (Oct. 7, 1980), J. H. Coleman (to Plasma Physics Corporation).

91. U.S. Pat. 4,259,005 (Mar. 31, 1979), M. R. Kuehnle (to Coulter Systems Corporation); U.S. Pat. 4,271,785 (June 9, 1981), S. R. DiNallo, Sr. and L. K. Najarian (to Coulter Systems Corporation).

92. U.S. Pat. 4,292,387 (Sept. 29, 1981), J. Kanbe, T. Toyono, N. Hosono, and T. Takahashi (to Canon Kabushiki Kaisha).

93. U.S. Pat. 4,254,201 (Mar. 3, 1981), Y. Sawai, H. Ushiyama, H. Tsuiki, T. Fujii, E. Akutsu, and I. Ikeda (to Ricoh Company, Ltd.).

94. U.S. Pat. 4,145,300 (Mar. 20, 1979), D. Hendriks (to Sublistatic Holding S.A.).

95. U.S. Pat. 4,251,616 (Feb. 17, 1981), D. Hendriks (to Sublistatic Holding S.A.).

96. U.S. Pat. 4,246,331 (Jan. 20, 1981), W. Mehl and D. Hendriks (to Sublistatic Holding S.A.).

97. U.S. Pat. 4,119,374 (Oct. 10, 1978), W. Mehl and R. Monti (to Sublistatic Holding S.A.); U.S. Pat. 4,129,376 (Dec. 12, 1978), T. Yotsukura (to Hitachi, Ltd.).

98. U.S. Pat. 3,339,469 (Sept. 5, 1967), S. B. McFarlane (to Sun Chemical Corporation); U.S. Pat. 3,689,935 (Sept. 5, 1972), G. L. Pressman and J. V. Casanova (to Electroprint, Inc.); U.S. Pat. 3,779,166 (Dec. 18, 1973), G. L. Pressman, H. Frohbach, and D. E. Blake (to Electroprint, Inc.); U.S. Pat. 3,850,628 (Nov. 26, 1974), G. L. Pressman (to Electroprint, Inc.).

99. U.S. Pat. 3,796,409 (Mar. 12, 1974), G. L. Pressman (to Electroprint, Inc.); U.S. Pat. 3,977,323 (Aug. 31, 1976), G. L. Pressman, D. E. Blake, and H. Frohbach (to Electroprint, Inc.).

100. U.S. Pat. 4,014,694 (Mar. 29, 1977), H. D. Crane, G. L. Pressman, and G. J. Eilers (to Electroprint, Inc.); U.S. Pat. 4,016,813 (Apr. 12, 1977), G. L. Pressman and J. V. Casanova (to Electroprint, Inc.); U.S. Pat. 4,277,550 (July 7, 1981), M. Nishikawa and N. Ameemiya (to Olympus Optical Company, Ltd.).

101. U.S. Pat. 3,966,871 (Oct. 19, 1976), J. D. Blades and J. E. Jackson (to Addressograph-Multigraph Corporation).

102. U.S. Pat. 3,220,324 (Nov. 30, 1965), C. Snelling (to Xerox Corporation); J. D. Blades and J. E. Jackson, *J. Appl. Photogr. Eng.* **4,** 97 (1978).

103. U.S. Pat. 3,811,765 (May 21, 1974), D. E. Blake (to Electroprint, Inc.).

104. G. E. Green and B. P. Stark, *Chem. Br.* **17,** 228 (1981).

105. D. Costinan, ed., *Guide to Microproduction Equipment*, National Micrographics Association, Silver Springs, Md., 1979.

106. S. H. Boyd, *TAPPI 2nd Reprography Conference*, 1972, St. Charles, Ill., pp. 123–131; I. Brodie, *TAPPI 1st Reprography Conference*, 1971, Minneapolis, Minn., pp. 239–248.

107. J. B. Freedman, *J. Appl. Photo. Eng.* **5,** 45 (1979).

108. L. Lessin, *Photomethods* **24**(5), 23 (May 1981).

109. U.S. Pat. 3,966,317 (June 29, 1976), H. H. Wachs, P. H. Klose, S. R. Ovshinsky, and R. W. Hallman (to Energy Conversion Devices, Inc.).

110. P. S. Vincett, A. L. Pundsack, G. C. Hartmann, R. K. Hunter, Jr., and W. D. McCrary, *J. Appl. Photogr. Eng.* **6,** 97 (1980).

111. P. J. Rice, H. F. Frohbach, N. A. Peppers, J. R. Young, L. F. Schaefer, and G. A. Pierce, *J. Appl. Photogr. Eng.* **6,** 62 (1980).

112. W. R. Hawken, *Enlarged Prints From Library Microforms*, American Library Association, Chicago, Ill., 1963.

113. R. L. Carlson, *Phot. Sci. Eng.* **8**(3), 167 (1964); S. Takumoto and co-workers, *Phot. Sci. Eng.* **7,** 218 (1963); R. D. Weiss, *Phot. Sci. Eng.* **11**(4), 287 (1967).

114. A. H. Sporer, *J. Appl. Photogr. Eng.* **6,** 89 (1980).

115. S. Shaw, *Electronic Business* **7,** 34 (June, 1981).

116. J. Savit, *TAPPI* **64**(5), 83 (1981).

117. W. H. Puterbaugh and S. P. Emmons, *Proc. Spring Joint Comp. Conf.* 121 (1967).

118. W. Peng in ref. 1; U.S. Pat. 4,268,838 (May 19, 1981), K. Nakano, T. Iwabushi, and I. Yamamoto (to Oki Electric Industry Company, Ltd.).

119. U.S. Pat. 3,161,457 (Dec. 15, 1964), H. Schroeder, W. H. Puterbaugh, and R. C. Meckftroth (to NCR Corporation); R. Brescia in ref. 1.

120. *Hewlett-Packard J.* 22 (1978); H. H. Busta and J. W. White in ref. 1.

121. T. R. Payne and H. R. Plumlee, *IEEE J. Solid-State Circuits SC* **8,** 71 (1973).

122. U.S. Pat. 3,746,675 (July 7, 1973), J. H. Blose and S. G. Talvalkar (to NCR Corporation).

123. R. Gugielmetti, C. Trebaul, and J. Brelivet in ref. 1.

124. F. Knirsch in ref. 1.

125. Y. Tokunaga and K. Sugiyama, *IEEE Trans. Electron Devices* **ED-27,** 218 (1980); Y. Tokunaga and R. Takano, *J. Appl. Photogr. Eng.* **7,** 10 (1981).

126. ANAC Model 911/961, Auckland, New Zealand.

127. E. L. Nemirovshi, *New Printing Processes*, Moscow, 1956, pp. 100–104 (in Russian).

128. U.S. Pat. 2,932,278 (Apr. 12, 1960), K. C. Sims (to Sperry-Rand Corporation); W. H. Meiklejohn, *AID Conf. Proc.* (10), 1102 (1972); U.S. Pat. 4,122,456 (Oct. 24, 1978), A. E. Berkowitz and J. A. Lahut (to General Electric Corporation).

129. A. K. Berkowitz, J. A. Lahut, W. H. Meikejohn, R. E. Skoda and J. M. Wang in ref. 1.

130. N. Kokaji, K. Kinoshita, T. Urano, and K. Saito in ref. 1; U.S. Pat. 4,268,872 (May 19, 1981), N. Kokaji, K. Kinoshita, T. Urano, and K. Saito (to Iwatsu Electric Company, Ltd.).

131. *Mod. Office Proc.* **26**(9), 55 (1981).

132. D. F. Blossey in ref. 1.

133. P. Bernstein and J. P. Lazzari in ref. 1.

134. E. F. Haugh in ref. 1.

135. L. Kuhn and R. A. Myers, *Sci. Am.* (Apr. 1979); I. G. Doane, *J. Appl. Photogr. Eng.* **7,** 121 (1981).

136. F. J. Kamphoefner, *IEEE Trans. Electron Devices* **ED-19,** 584 (1972); R. D. Carnahan and S. L. Hou, *IEEE Trans. Ind. Appl.* **A-13,** 95 (1977).

137. J. W. S. Rayleigh, *Proc. London Math. Soc.* **10,** 4 (1878).

138. U.S. Pat. 3,882,508 (May 6, 1975), L. G. Stoneburner (to Mead Corporation); U.S. Pat. 3,891,121 (June 24, 1975), L. G. Stoneburner (to Mead Corporation); U.S. Pat. 4,095,232 (June 13, 1978), C. L. Cha (to Mead Corporation); U.S. Pat. 4,138,687 (Feb. 6, 1979), C. L. Cha (to Mead Corporation).

139. R. G. Sweet, Technical Report, Stanford Electrical Laboratory, Stanford University, Stanford, Ca., 1964; U.S. Pat. 3,373,437 (Mar. 12, 1968), R. G. Sweet and R. C. Cumming.

140. U.S. Pat. 3,936,135 (Feb. 3, 1976), P. L. Duffield (to Mead Corporation); U.S. Pat. 3,836,914 (Sept. 17, 1964), P. L. Duffield (to Mead Corporation).

141. U.S. Pat. 4,274,100 (June 16, 1981), S. F. Pond (to Xerox Corporation); U.S. Pat. 4,302,761 (Nov. 24, 1981), Y. Yamomoto (to Sharp Kabushiki Kaisha).

142. C. H. Hertz and A. Mansson, *Rev. Sci. Instr.* **43,** 413 (1972); R. Eriksson, *Measurements*, Report 1, Lund Institute of Technology, Dept. E1, 1975.

143. B. Hermanrud and C. H. Hertz, *J. Appl. Photogr. Eng.* **5,** 220 (1979).

144. P. Duffield, *J. Appl. Photogr. Eng.* **5,** 248 (1979).

145. M. Mutoh, S. Kaieda, and K. Kamimura, *J. Appl. Photogr. Eng.* **6,** 78 (1980).

146. U.S. Pat. 4,085,408 (Apr. 18, 1978), M. Muto, N. Tashima, S. Kaieda, and K. Kamimura (to Minolta Camera Kabushiki Kaisha); U.S. Pat. 4,183,030 (Jan. 8, 1980), S. Kaieda, M. Mutoh, and K. Kamimura (to Minolta Camera Kabushiki Kaisha); U.S. Pat. 4,215,353 (July 29, 1980), S. Kaieda, M. Mutoh, and K. Kamimura (to Minolta Camera Kabushiki Kaisha).

147. U.S. Pat. 4,268,836 (May 19, 1981), D. H. Huliba and L. P. Robinson (to Mead Corporation).

148. U.S. Pat. 3,683,212 (Aug. 8, 1972), S. J. Zoltan (to Clevite Corporation).

149. T. E. Johnson and K. W. Bower, *J. Appl. Photogr. Eng.* **5,** 174 (1979).

150. U.S. Pat. 3,940,398 (Feb. 24, 1976), E. L. Kyser and S. B. Sears (to Silonics).

151. U.S. Pat. 3,747,120 (July 17, 1973), N. G. E. Stemme; E. Stemme and S. Larsson, *IEEE Trans. Electron Devices* **ED-20,** 14 (Jan. 1973).

152. U.S. Pat. 3,940,773 (Feb. 24, 1976), A. Mizoguchi, K. Wamamori, and Y. Hiromori (to Matushita Electrical Industrial Company); U.S. Pat. 4,072,958 (Feb. 7, 1978), H. Hayami, H. Tsuchiya, K. Yoshioa, Y. Tsuda, and Y. Kanno (to Matushita Electrical Industrial Company).

153. Z. Kovac and C. Sambucetti, *Division of Colloid and Surface Chemistry 182nd ACS National Meeting*, Aug. 23–28, 1981, New York.

154. U.S. Pat. 4,299,630 (Nov. 10, 1981), K. S. Hwang (to The Mead Corporation).

155. C. T. Ashley, E. Edds, and D. L. Elbert, *IBM J. Res. Dev.* **21,** 69 (1977).

156. R. H. Windhager, *TAPPI Conf. Papers, Printing and Reprography—Testing*, 105 (1977); I. Brodie, J. A. Dahlquist, and A. Sher, *J. Appl. Phys.* **39**, 1618 (1968); D. Lanheer, *J. Appl. Phys.* **51**, 1809 (1980).
157. R. F. Borelli, R. B. Bayless, and E. R. Truax, *Honeywell Computer J.* **8**, 67 (1974).
158. M. Amaya, K. Aikawa, and M. Horie, *J. Appl. Photogr. Eng.* **6**, 58 (1980).
159. S. Wakoh, T. Toyoshima, H. Todo, T. Kimoto, K. Nakano, and J. Tomiyama in ref. 1.
160. U.S. Pat. 4,267,556 (May 12, 1981), R. A. Fotland and J. J. Carrish (to Dennison Manufacturing Company); R. A. Fotland and J. J. Carrish in ref. 1.
161. U.S. Pat. 4,195,527 (Apr. 1, 1980), R. A. Fotland and J. J. Carrish (to Dennison Manufacturing Company).
162. H. D. Hinz, H. Löbl, and U. Rothgorth, *J. Appl. Photogr. Eng.* **6**, 69 (1980).
163. A. R. Kotz, *J. Appl. Photogr. Eng.* **7**, 44 (1981); "3M Magnestylus Printing; Graphic Arts Application" in ref. 1.
164. R. N. Blazey and B. E. Cates, *J. Appl. Photogr. Eng.* **6**, 144 (1980).
165. U.S. Pat. 4,274,101 (June 16, 1981), K. Kataoka, K. Tatsumi, and S. Saito (to Hitachi, Ltd.).
166. J. E. Wurtz, *J. Appl. Photogr. Eng.* **6**, 73 (1980); W. E. Haas, F. G. Genovese, and J. W. Lannom in ref. 1; U. Schieber and K. Carl in ref. 1.
167. Y. Hoshino, K. Tateishi, Y. Ikeda, Y. Tokunaga, and H. Miyata in ref. 1.
168. B. Hill and K. P. Schmidt, *Phillips J. Res.* **33**, 211 (1978); "High-resolution Magneto-optic Pattern Generator for Fast Optical Line-printing" in ref. 1.
169. U.S. Pat. 4,277,805 (July 7, 1981), Y. Sato (to Ricoh Company, Ltd.).
170. K. Nakano, S. Nakaya, M. Katsuta, and K. Saito, *J. Appl. Photogr. Eng.* **7**, 21 (1981).
171. T. J. Kucera, *Perspective* **4**, 133 (1962).
172. U.S. Pat. 4,275,134 (June 23, 1981), W. Knechtel (to Canon Kabushiki Kaisha).
173. U.S. Pat. 4,264,185 (Apr. 28, 1981), W. Ohta (to Ricoh Company, Ltd.).
174. U.S. Pat. 4,281,051 (July 28, 1981), K. Sakai (to Ricoh Company, Ltd.).
175. U.S. Pat. 4,017,176 (Apr. 12, 1977), R. Beguin and R. Monti (to Sublistatic Holding S.A.).
176. U.S. Pat. 4,251,611 (Feb. 17, 1981), W. Mehl, D. Hendriks, and R. Decombe (to Sublistatic Holding S.A.).
177. *Business Week* (May 11, 1981).
178. W. Crooks, A. B. Jaffe, and T. K. Niweigha, *2nd International Conference, Business Graphics, Society of Photographic Scientists and Engineers*, Nov. 4–7, 1979, Washington, D.C.; W. Crooks, E. W. Luttman, and A. B. Jaffe in ref. 1.
179. J. Klasnic, *In-Plant Printing Handbook*, GAMA Communications, Salem, N.H., 1981.
180. *Graphic Arts Manual*, Arno Press, New York, 1981.
181. W. P. Spence and D. G. Vequist, *Graphic Reproduction*, Chas. A. Bennett Company, Peoria, Ill., 1980.
182. *Pocket Pal A Graphic Arts Production Handbook*, 12th ed., International Paper Company, New York, 1979.
183. Ref. 13, Chapt. 2.
184. U.S. Pat. 4,263,392 (Apr. 21, 1981), T. H. Jones (to Richardson Graphics Company).
185. U.S. Pat. 4,272,604 (June 9, 1981), J. D. Meador and E. H. Parker (to Western Litho Plate & Supply Company).
186. V. Strauss, *The Printing Industry*, Printing Industries of America, Washington, D.C., 1967.
187. A. Poot and A. DeJaeger, *J. Appl. Photogr. Eng.* **5**, 169 (1979).
188. U.S. Pat. 2,988,988 (June 30, 1961), P. F. Kurz (to Haloid Xerox); U.S. Pat. 3,001,872 (Sept. 26, 1961), P. F. Kurz (to Xerox Corporation).
189. U.S. Pat. 4,270,859 (June 2, 1981), W. E. Galbraith, A. L. Kaufman, and H. Klepper (to Eltra Corporation).
190. U.S. Pat. 4,266,869 (May 12, 1981), M. R. Kuehnle, R. E. Cox, G. B. Harris, J. Forrest, and C. D. Haroy (to Coulter Systems Corporation).
191. U.S. Pat. 4,263,387 (Apr. 21, 1981), F. Martinez (to Coulter Systems Corporation).
192. U.S. Pat. 4,265,987 (May 5, 1981), T. M. Lawson (to Coulter Systems Corporation).
193. F. Eichenberg, *Lithogrpahy and Silkscreen*, Harry N. Abrams, New York, 1978.
194. W. M. Gilgore, *Am. Printer Lithographer*, 35 (Oct. 1979).
195. *Screen Printing Bibliography*, Screen Printing Association International, Fairfax, Va., 1981 (includes audio-visual aids as well as books).
196. D. McCoy, *Ink, Inc.* **1**(2), 10 (June 1981); D. Boyce, *Ink, Inc.* **1**(6), 10 (Oct. 1981).

197. U.S. Pat. 4,262,084 (Apr. 14, 1981), L. C. Kinney (to Imaging Sciences, Chicago).
198. Ref. 195, p. 85.
199. U.S. Pat. 4,270,985 (June 2, 1981), M. A. Lipson and D. W. Knoth (to Dyna Chemical Corporation).
200. P. Fresener, *The Press* **3**(6), 6 (Oct. 1981).
201. U.S. Pat. 4,268,576 (May 19, 1981), F. H. Montmarquet, Jr. (to Repeat-O-Type Stencil Manufacturing Company).
202. U.S. Pat. 180,857 (Oct. 26, 1875), T. A. Edison.
203. U.S. Pat. 4,065,595 (Dec. 27, 1977), M. L. Schick and B. E. Anderson (to Weber Marking Systems, Inc.); U.S. Pat. 4,074,003 (Feb. 14, 1978), M. L. Schick and B. E. Anderson (to Weber Marking Systems, Inc.).
204. U.S. Pat. 3,632,364 (Jan. 4, 1972), R. E. Thomas, R. H. Dalal, and R. I. Scheuer (to A. B. Dick Company); U.S. Pat. 3,788,863 (Jan. 29, 1974), R. I. Scheuer (to A. B. Dick Company); U.S. Pat. 4,051,283 (Sept. 27, 1977), R. E. Thomas, R. T. Florence, R. H. Dalal, and R. I. Scheuer (to A. B. Dick Company).
205. R. Haavind, *High Technol.* **1**(2), 35 (1981).
206. *The Wall Street Journal*, 13 (Feb. 10, 1977).

THOMAS J. KUCERA
Consultant, Evanston, Ill.

RESEARCH MANAGEMENT

The term research and development (R&D) has largely replaced the more limited label of Research in industrial usage. The term is being broadened to include the full range of activities required to introduce new technology into commercial application. The brief treatment possible in this article necessitates focusing on the salient features of greatest concern to industrial management now and in the near-term future. Engineering management is described quite fully in ref. 1.

Research and development, as it is practiced in industry, is not an end in itself; it is part of a larger process whose objective is to contribute to growth and profitability by providing opportunities for innovation through new and improved technology. Although this larger process of technical innovation typically begins with the initiating discovery in research and development, it includes not only the succeeding activities of preliminary development and design of prototypes, but also the development of applicable processes for manufacture, market R&D, and eventually investment in new plants and facilities and human resources to produce and sell the new product. At each stage, the resource requirements escalate as does the financial risk associated with continued investment.

The focus of attention in R&D management depends on the stage of maturation of the practice of research and development in any given business or industry. It is also influenced by the rate-limiting parts of the process that seem to be most in need of attention at any particular time. Industrial R&D is not a new activity. The DuPont Company created a formal laboratory in 1902, but an explosives lab existed in the

1880s. Corporate R&D at General Electric began with the research laboratory in 1900. The Bell Telephone Laboratories were established in 1925. The Exxon Research and Engineering Company had its origins in 1882. These companies have been among the elite industrial enterprises for decades. The focus of attention in creating these early industrial laboratories was to bring science to bear on the problems of industry. Nevertheless, the rate of market penetration prior to World War II was quite small, the aggregate expenditures were limited, and the number of companies involved was limited. For example, the Industrial Research Institute (IRI) was formed under the sponsorship of the National Research Council in 1938 with 14 charter members.

The events of World War II constituted a watershed for industrial R&D. Not only were the resources of the existing laboratories focused on the problems of developing technology for the war effort; more importantly, the heightened awareness of the general public and of businessmen of the benefits that could be obtained from new technology set the stage for a period of rapid growth after the war. This growth involved not only significantly increased resources and manpower devoted to industrial R&D, but a greatly expanded base of participation. The number of member companies in IRI increased from 14 in 1938 to 260 in 1980. This greatly expanded effort on research and development is reflected in the percentage of the gross national product devoted to R&D activities. As shown in Table 1, research and development as a percentage of GNP increased dramatically from 1.4% in 1953 to 3.0% in 1966. Actual expenditures increased more than ten-fold from 5.1×10^9 in 1953 to ca 54.3×10^9 in 1980.

This doubling of the percentage of effort devoted to R&D represented a massive shift of resources to the task of developing new technology. The relative level of expenditure has decreased since the peak in 1964. This deterioration has been cause for widespread concern associated with the apparent decline in the U.S. technical position (3).

Changing Concerns

During this postwar period and extending probably up until the mid-1960s, the focus of R&D management attention was on the creation of the R&D function itself. Creating a new function requires a redistribution of resources and a restructuring of relationships, which take considerable time and are likely to generate some conflict. The role of the R&D function had to be defined and accepted. The process involved a reshuffling of activities previously distributed among other functions as well as the creation of new activities and new objectives not previously sought. The nature of the interface between research and development and the other functions had to be worked out and to be generally accepted by the other functions: for example, the relationship

Table 1.   Total R&D Expenditures as a Percent of GNP, 1953–1980[a]

1953–1957	1958–1962	1963–1967	1968–1972	1973–1977	1978–1980
1.4	2.4	2.9	2.8	2.3	2.2
1.5	2.5	3.0	2.7	2.3	2.2
1.5	2.7	2.9	2.6	2.3	2.3
2.0	2.7	2.9	2.5	2.3	
2.2	2.7	2.9	2.4	2.2	

[a] Ref. 2.

of R&D to marketing with respect to product planning, to manufacturing with respect to the introduction of new materials or new processes, and to finance regarding estimates of costs and levels of investment.

This insertion of the research and development activity has necessitated a reshuffling of the power structure. During this period, the R&D managers focused their attention on defining the direction of the activity and seeking to ensure that the direction was understood and accepted by other members of the business. They also focused a great deal of attention on learning how to manage this suddenly increased scale of activity in an effective fashion. The management of research and development, with its inherently larger element of risk and longer time scale, differs significantly from much industrial management activity. Furthermore, R&D managers had to evolve the precepts and modes of management behavior needed to deal with a process whose success depends intimately on motivating the work of highly creative people.

Perhaps because of this intense concentration on the internal functioning of the R&D process and possibly because of the generally favorable attitude of society toward technology and the high status accorded scientists and engineers, R&D managers tended to devote less time and attention to, and to operate on simplistic assumptions regarding the difficulties involved in ensuring that the fruits of their effort, ie, new scientific discoveries and new technological capabilities, were adopted into commercial practice. Some underestimated the difficulty of creating technological innovation. They failed to appreciate the significant amount of additional effort that is required on the part of the R&D organization to ensure that the technology is ready for commercial application, and they failed to appreciate the amount of resistance and resentment that innovation almost inevitably creates.

Consequently, in about the middle of the 1960s, the attention of R&D management tended to shift so as to provide greatly increased effort and attention on solving the problems of achieving commercial application of the results of R&D. More recently this effort has been broadened to include increased attention to the process of defining the goals, strategies, and management characteristics of the enterprise. Managers of R&D have come increasingly to recognize that success or failure frequently depends not only on the attractiveness of the idea and the effectiveness of the R&D effort, but also on the way in which the enterprise defines its purposes and the extent and firmness of its commitment to achieving growth through technological innovation.

During this period in which managers of research and development have come to view their work in a new light and assigned different priorities to the tasks that require their attention, the attitudes of society toward technology have changed. Fundamentally, R&D exists and prospers in a democratic society to the extent that it is responsive to the requirements and desires of a majority of the people. U.S. technology has been remarkably sensitive to the moods of the public, and as these moods and expressions of wants and needs have changed, the direction of technology has changed accordingly. This pattern of responsiveness is well established and surely will continue. A ready example is environmental technology. In recent years technical progress has been blamed, for example, for polluting air and water. However, the ecological situation cannot be separated from the past desires of people, ie, from decisions made democratically over the years with respect to what the majority thought they wanted and needed.

In the absence of public concern about the environment during the earlier decades of this century, it is not surprising that the technologists produced for people what they showed that they wanted most, ie, powerful cars, more things to make jobs easier and life more comfortable, new activities to occupy leisure time, and more energy to keep those things running.

With the realization that people do want an improved environment, that they are concerned with the cost and availability of energy, that they want more reassurance with respect to some of the long-term, unanticipated consequences of technological innovation, scientists and engineers are responding to the tenor of the times. Many of them have taken the lead in recent years in trying to communicate improved understanding of the problems, but scientists and engineers in a democratic society find it difficult to accomplish that which the mood of the people does not support. This response to changing public attitudes and priorities has affected the directions of R&D effort, and has magnified the amount of information that must be obtained with respect to product safety, environmental effects of technology, and long-term societal effects resulting from the changes that new technology creates.

The attention of R&D management is much more focused on the interface between research and development and the rest of society and between the research and development function and the rest of the corporation. Higher priority is being given to ensure that the mission and charter of the R&D organization are appropriate, that the objectives and strategies of the corporation make allowance for the contributions that can be obtained from new technology, and that R&D managers as well as the programs of the laboratory contribute to the formulation of corporate objectives and to their realization.

Consequently, those aspects of management that focus on the internal operation of a laboratory are not discussed in this article. The management work associated with recruiting competent people, motivating highly creative scientists and engineers, managing projects successfully, achieving the necessary effective internal communication, providing administrative support, etc, has been thoroughly covered in other literature. The following discussion focuses on managing R&D in the context of the total innovation process.

Innovation

Innovation is the introduction of change. Successful R&D management requires understanding and fostering the circumstances that lead to change, both on the part of the R&D managers and the managers of the enterprise or organization supporting the R&D effort. Any study of the history of attempts at innovation, both successes and failures, underscores the critical importance of a supportive environment if change is to be successful. Without this supportive environment, complex techniques of planning and managing are likely to be unproductive no matter how skillfully they are applied.

Innovation is intrinsically disruptive: it alters traditional economic and social relationships; it changes people's fortunes for better or for worse; it creates anxiety on the part of those caught up in the process. While the process is underway, it frequently is impossible to determine whether the outcome will be successful or unsuccessful; whether the results will be constructive or destructive. The process also tends to generate controversy. Under these circumstances and unless a culture provides customs, values, and statuses that attribute great worth and high rewards to the creators of change, it is exceedingly unlikely to nurture the few people who are able to create change.

Change can be born out of desperation. If an enterprise is in serious trouble, it is more receptive to radical prescriptions for salvation. However, these painful circumstances are not appealing as the most suitable way to create innovation. The other approach is to adopt innovation as one of the indispensable ingredients for the future prosperity and survival of the enterprise and, therefore, deliberately seek to establish the circumstances that ensure successful innovation.

Unfortunately, many of the techniques of conventional management are antithetical to the circumstances necessary to nurture innovation. Unless these basic nurturing cultural requirements are recognized and established, attempts to improve the more visible features of successful management, involving program planning, project management, control of expenditures, review of performance, etc, are unlikely to lead to successful R&D management.

Successful R&D management rests on two key cultural ingredients. The first is understanding the meaning of risk, which implies a willingness to accept the necessity for occasional failure as an inherent feature of the process of innovation. The second is providing sufficient incentive to induce individuals to undertake the risk that is inevitable if innovation is to occur. People, not organizations, create innovation.

The first and the largest category of managerial work is the requirement to provide stewardship for the resources of the enterprise. This stewardship requires that the successful manager produce profitable results and a growth in assets from the use of the resources of the enterprise, but also it places a high premium on the preservation of those resources. The second category of managerial work is to innovate so that the enterprise can generate profits and growth above the level that stewardship alone would produce. Although both these features of management may be present in the responsibilities of any given manager, the overwhelming majority of managers concentrate their attention to the stewardship role. Furthermore, much of the effort devoted to improving management methods and techniques is devoted to improving the stewardship function, because this area of responsibility lends itself more readily to the use of rigorous methodology.

One important goal of managers emphasizing the stewardship role is essentially to eliminate risk, ie, to create certainty. If one collects appropriate information, analyzes it properly, makes sound decisions, and installs the mechanisms for follow-up and review, the result is the establishment and attainment of goals that fulfill the stewardship objective. The requirement to achieve established goals tends to constrain the business, because it leads the manager to avoid situations in which adequate information is not available and in which, therefore, unattainable goals might be established.

Viewed from this perspective, failure to achieve goals is a failure in management, because sound planning and execution ensure that goals are achievable and that they are achieved. A manager attuned to performing the stewardship role regards risk as something that is manipulatable. Reducing risk to trivial considerations is regarded as one of the manager's functions. A prudent manager tends to look upon situations in which high risk is an inescapable feature as undesirable, as territory into which the enterprise should not venture. An orderly world in which organizational relationships are carefully defined is sought, responsibilities are unambiguous and do not overlap, work flows smoothly, and conflict is minimized.

A manager attempting innovation or providing support for an innovative effort must recognize the intrinsically probabilistic nature of risk with respect to innovation.

The manager must recognize that it is impossible to plan soundly enough or execute shrewdly enough to reduce risk to trivial proportions. The manager knows or discovers that statistically an appreciable percentage of attempts at successful innovation must fail. Viewed from this perspective, the task of a manager committed to achieving innovation is twofold. First, an environment with incentives such that the potential rewards for success are adequate to compensate for the risks and costs of failure must be created. Second, insoluble barriers must be identified as early as possible so as to abort the effort with minimum expenditure or to discover new goals that appear to be attainable.

Many managers are unprepared for the shock of discovering that an innovation must be aborted. Managers in this frame of mind tend to look for a culprit, to assert that the failure results from incompetent management and that somebody must be blamed for the fiasco. Such a point of view is wholly counterproductive to the environment necessary to nurture innovation. Failure must be recognized as not necessarily a failure in management. If ambitious, capable people discover that association with an aborted innovation labels one as incompetent, they rapidly lose interest in pursuing career advancement through innovation.

Under the most supportive circumstances, innovation exacts its price. No matter what compensatory or reassuring steps a sympathetic manager takes to recompense people associated with an aborted innovation, there is little that can be done to make up for the time lost from their careers. Consequently, potential rewards must appear to be worth the intrinsic risk. There are two general courses of action that a manager can take to stimulate attempts to create innovations. The first, as noted, is to provide sufficient reward to induce people to accept the risk. The other is to understand the nature of the innovation process sufficiently to minimize the avoidable barriers that increase the probability of failure. The pace of innovation is such that increases in salary are difficult to use effectively as the sole means of rewarding outstanding performance. The problem is one of timing. If one waits until there is convincing demonstration of a significant success before granting large increases in salary, it is unlikely to provide timely incentive for the ambitious, action-oriented people necessary to accomplish innovation. Conversely, the magnitude of success in early years may not be sufficient to justify large salary increases. Consequently, salary increases are probably not the most important form of incentive.

The use of increased status as a reward is more subtle. Increased status and recognition from peers is an important source of satisfaction for scientists and engineers as well as for R&D managers. It is important for an R&D manager to ensure that the technical people have the opportunity to establish their reputations for publishing and presenting papers. However, this incentive tends to take a different form when applied to people seeking to create an innovation. These people tend to associate status with position in an organization. Furthermore, the intrinsic characteristics of successful managers of innovation are such that they place a premium on having control of an operation. Consequently, the opportunity for accelerated promotion becomes perhaps the most attractive incentive to induce people with the requisite skills and personal attributes to attempt innovative activities. Of course, an increase in income accompanies a promotion to higher responsibility in the organization.

In addition to providing incentives to accept risk, management must also reduce needless barriers to success. Innovation typically generates a competitive battle between the old and the new. From society's point of view, this competitive battle is advantageous because it represents the only valid means for demonstrating which technical approach is superior. Competitive superiority cannot be demonstrated by analysis and simulation; it must be proved in actual use. Consequently, the new technology succeeds in some applications and not others. It also stimulates increased R&D for conventional technology and accelerates its advancement. Consequently, there is great need for aggressive, competent, persistent protagonists for both the old and the new. Society benefits by the competition.

The manager seeking to encourage an environment for successful research and development must recognize the inevitability of a considerable degree of conflict which cannot be managed away by careful planning and organizing. It is important to try to ensure that this kind of conflict does not become so disruptive that it unduly impedes effective progress. However, the manager should be equally and possibly more concerned with ensuring that those who are striving to introduce innovation are at least strong enough to be worthy opponents for those defending the *status quo* of conventional technology. Where innovation is ineffective, the problem may result from inadequate strength on the part of those seeking to innovate rather than too much conflict between the old and the new. The operating principle that should be followed in seeking to foster or nurture innovation is that one should not expect the advocates of a present technology simultaneously to nurture a new competing technology. The protagonists of conventional technology are likely to be stewardship-oriented managers. They try to arrange the introduction of the innovation so that it represents a smooth, unbroken, risk-free, untroubled transition from a dominant conventional technology to a dominant new technology, with maximum advantage being extracted from the conventional technology as it decreases in importance and is outmoded.

Unfortunately, innovation rarely occurs in this smooth fashion. Innovation succeeds because it is pushed by aggressive, capable, persistent people who are impatient with the *status quo.* They are very unlikely partners in a smooth transition from old technology to new. Moreover, a critical reversal in status has to occur, with the protagonists of the old technology dropping from a position of dominance to a position of equality or subordination. It is unrealistic to expect such a change in status to occur without conflict. When an attempt is made to introduce an innovation in such fashion, one is creating almost insuperable barriers to success.

Technology

It is important to recognize the various ways in which technology can contribute to technological innovation, because the various dimensions of technology should be taken into account in defining the direction of the R&D organization. Technology can contribute to technological innovation by providing the basis for new and improved products that can lead to increased sales or improved profit margins. This product-focused technology is perhaps the most widely recognized dimension of technological innovation. In addition to creating new products, there is a dimension which is becoming increasingly important: contributing to the internal efficiency and effectiveness of the enterprise. This can lead to improved productivity, greater flexibility in operations, and shorter response time in reacting to a competitive threat or in capitalizing on a new market opportunity. This aspect of technology has two dimensions: the first is process technology associated with the manufacture of products; the second is the information processing and communications technologies that provide the basis for improved effectiveness and efficiency in all aspects of the management of the enterprise in all functions. It also provides the technological foundation for the improved control of processes and automation associated with improved productivity in manufacture.

The chemical industry has long given sustained attention to the process technology that must be mastered to produce chemical products with the desired properties and the requisite yields. This awareness of the importance of process technology is not so well-developed in many other industries. All segments of industry face the challenge of learning how to capitalize on the rapid advances in information processing and data communications.

These three dimensions of technology: product, process, and information communication, provide one scale for defining the nature of R&D work. The determination of the most suitable functional home for information processing and communications is much less well understood and exhibits much greater variability within industry than is true for product and process technology.

The other main dimension required to specify the charter of the R&D organization is to determine the extent to which its efforts will be focused on providing support for present businesses of a company, ie, to improving present technologies or to introducing new technologies into present businesses as opposed to providing opportunities for growth from technological discovery. Some R&D organizations are created solely to provide support for present businesses.

The fundamental strategic decision facing R&D management is to determine the balance between these two activities: supporting present businesses and providing opportunities for growth. Establishing this balance calls for continuous fine-tuning by two relatively independent forces. The first and probably most important is the changing needs and priorities of the business. Achieving a sound balance in this dimension calls for a fine political sense and a thorough knowledge of day-by-day business activities, because to some extent the R&D organization should act as a counterbalance to other forces and priorities in the enterprise. For example, if the enterprise is deeply committed to achieving improved results from present businesses, the R&D organization may be the only place in the enterprise in which some effort can be devoted to providing the basis for future growth. As management in the company swings its attention from preserving present businesses to concern over growth opportunities, the R&D organization needs to have available new growth opportunities in which to invest. It will have them only if it has been pursuing a different set of priorities from the rest of the company. Conversely, if company management is heavily focused on achieving success with heavy current commitments to innovation, the R&D organization may need to pay increased attention to the technological health of present businesses, because they are otherwise liable to be neglected.

Another dimension involved in the balance of supporting present businesses versus providing opportunities for growth results from the internal dynamics of the laboratory. As technical work advances from the early phases of exploratory or search activity through the various stages of development to application of the activities of business, it requires an increasingly larger commitment of financial and human resources in order to maintain the requisite momentum. Thus, variations in the stage of maturity of the R&D programs can affect the need for changes in the balance of technical effort.

Figure 1 indicates one way of viewing this spectrum of technical effort. Technical

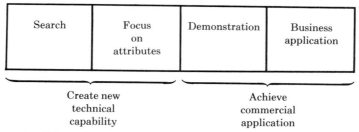

Figure 1. Schematic representation of efforts involved with technical innovation.

work typically begins at the left, with the search for new opportunity. As interesting new opportunities are discovered, effort is focused on these promising opportunities to try to determine more of their nature and to begin to perceive possible areas of application. If the discoveries continue to be promising, they then move to the demonstration phase where their economic utility in specific commercial applications is demonstrated. Finally they move into the business application phase and become either a new business operation or part of a present one. The search and discovery activities are associated with creating a new technical capability, whereas their demonstration and application are associated with achieving practical application. It has been the experience of certain corporate R&D departments that approximately equal efforts are required in creating a new opportunity and in achieving practical application. The total effort associated with the research and discovery phase is very much larger than the demonstration and application phase, because as new technology advances to demonstration and business application, the business operations of the company become involved and the resources they require to commercialize the technical advance are very great.

Roughly 80% of the technical effort in the R&D component is devoted to focused work and demonstration. The distribution of effort among programs must reflect their positions among these areas. As effort is directed to demonstration and application, increasingly greater resources are required.

Centralized vs Decentralized Organization

Another basic decision that affects the management of research and development is the determination of appropriate organization. An excellent case can be made for the theory that all advanced technical work should be done as close to the user, ie, the production, engineering, and manufacturing operations, as possible. This approach encourages close coordination in planning, reduces communication barriers inherent in any technology transfer situation, and fosters a team spirit in those involved.

On the other hand, organizational and geographic decentralization sacrifices the advantages of interdisciplinary synergism and the ability to assemble a critical mass of technical expertise at a single location. Centralized research has a better chance of being protected from the kind of perpetual problem solving that dilutes its real objectives. Larger centralized laboratories can be made more attractive to outstanding young talent and thus can be an excellent means of hiring good scientists and engineers who subsequently may realize that their careers can best be furthered by moving into the decentralized business operations. Centralized laboratories can serve as a catalyst for the interchange of ideas and skills between various diverse operations of the supporting company. Centralized laboratories also have a better chance of being protected from the vicissitudes of business that cause undesirable swings in the levels of effort in operating components. Perhaps most important of all, centralized laboratories are less committed to the applications of conventional technology and thus are better adapted to creating and introducing the revolutionary new technology which can replace the technology presently being used. The choice of organization depends on the particular company, its size, the diversity of its business, the extent of its geographic dispersal, and the mode of its overall organization, ranging from strictly functional components to multiproduct divisions.

Ford established a corporate laboratory to perform research and to develop the basis for a new business but with limited responsibility to protect or to extend present businesses that are more or less self-sufficient technically. General Motors, DuPont, and Monsanto have created central technical centers where the corporate laboratory is the main technical component, but additional laboratories managed by operating groups share the common site and some facilities. Bell Telephone Laboratories have centralized the administration of all technical work for AT&T. One organization performs all technical programs associated with new components and systems, ensures coordination of communication, and provides the new technology needed by the company. Many of the large oil companies follow similar patterns. Many conglomerates or holding companies perform virtually all technical work on a decentralized basis, with each group being self-sufficient in meeting technical needs with little or no corporate R&D support. General Electric, Westinghouse, and RCA are examples of companies that have established corporate laboratories to perform a mixture of research, development, and new business development. The interface with operations varies depending on history, the leverage technology exerts on business success, geography, technical capability of operating components, etc.

An important responsibility of management is to identify the inherent advantages and disadvantages in a chosen organizational arrangement and to take actions through staffing and procedures to compensate for the disadvantages. The criteria that should be kept in mind in evaluating the effectiveness of any given organizational arrangement are

1. Caliber and availability of human and equipment resources; centralization has advantages here.

2. Relative effectiveness in planning and establishing the technical program; decentralization has advantages, but management effort can counteract difficulties in centralized laboratories.

3. Effectiveness in performing work in terms of costs, timeliness, quality, effectiveness of transition, etc; there is some advantage for decentralization, but management action can improve performance in centralized laboratories.

4. Kinds of output sought from R&D, ie, new businesses, basic understanding underlying key technologies, introduction of new technologies, or evolutionary improvements; the choice for centralization or decentralization depends on company strategy.

During the period after World War II when management attention was focused on creating the R&D function, there was a tendency to emphasize the need for physical and organizational separation in the R&D organization. This created a bias in favor of isolated locations and reporting at the top levels of the company. In the more recent past, with the increasing attention to commercializing research results, management priorities have shifted to more emphasis on coupling and sometimes with closer physical association with operating activities but certainly with closer integration in program planning and execution.

Program Planning and Evaluation

Irrespective of the mode of organization chosen, one of the critical tests of R&D management is in strategic program planning, which begins with the identification of the main areas of technology that are important to achieving the objectives of the enterprise. This work requires analysis of the various businesses of the company for their technical requirements and consideration of the external trends in science and technology that may provide opportunities for the company. In defining the strategic program areas, attention should be focused on the output required and not on the disciplines needed to generate the output. The primary audience for this undertaking is the managers of the enterprise. They are frequently not much interested in what it takes to generate the needed technical output; they regard that as a responsibility of the R&D manager. Rather, they want to be assured that the future technical needs of the company are being provided for.

The chief problem in program evaluation typically is not to determine the technical soundness and relevance of particular candidate programs. With the high level of technical competence and technical training available in most industrial R&D organizations, one does not frequently encounter program proposals that are inherently unsound technically. Given an array of technically sound and relevant programs, each of which has promise for contributing to company growth and prosperity, and given limited resources of money and people, the question remains as to which programs should be supported and which discarded.

Strategic program evaluation requires addressing three questions: First, if the program is successful, what impact will it have on the company? Given a choice, one would prefer to work on those programs that have the largest potential impact. Second, what is the likelihood of achieving the desired and needed technical success on this program? Third, what level of effort is required to provide a rate of progress that has a reasonable chance of competitive R&D success?

In making the program evaluation it is important to recognize the practical constraints that are inherent in the process. A typical program evaluation exercise may well involve the evaluation of several dozen or maybe a few hundred programs. By their very nature, some of these programs address market opportunities for which detailed relevant market information is not available nor, in the early phases of the R&D process, would it be cost-effective to make the investment required to generate extensive market data. An additional constraint is the inherent uncertainties associated with the evaluation of both possible impact and probability of success that no amount of study or appraisal will eliminate. The exact level of technical capability that will be achieved, how long it will take, and what resources will be required are largely unknowable. Consequently, in practical terms, the R&D manager needs a screening process or a series of screening processes that supplement and extend his or her managerial judgment without requiring an inordinate commitment of time and resources. Obviously, these processes should take advantage of existing market data.

The nature of this program evaluation process will of necessity change and become more elaborate and complicated as programs proceed from the search-focus phases to the demonstration and business application phases. As the level of committed resources increases, the amount of market research, cost projection, and competitive analysis that are required increase markedly. Concurrently, the number of programs that survive to this stage decrease markedly and the task becomes more manageable.

The R&D manager faces the problem of evaluating potential impact among a variety of programs in some consistent manner where adequate data for doing so are frequently lacking and where an extensive data-gathering process is not warranted. One way that has proved useful is to draw analogies to present businesses in the company. In some cases, it is not necessary to work by analogy because the technology is focused on existing businesses. In other cases, the R&D manager may be able to project that the project, if successful, could lead to a market approximately equal in size to that of an already existing business in the company and thus by analogy approximate likely market size. The R&D manager also needs to be able to project the likely rate of growth of the market because more rapidly growing markets are inherently more attractive than the slowly growing or declining ones. Frequently, in the early phases, the R&D manager is limited to evaluations of the rate of growth of a particular segment of industry rather than the precise market for the new product. It is also desirable to take into account the likely ability of the company to penetrate the particular market, which is influenced by competition, ie, the number of competitors and their degree of commitment. It is also influenced by the skills and resources of the company that can be brought to bear on this particular market opportunity, including managerial skills and attitudes about change. In none of these dimensions, ie, market size, rate of growth, or extent of penetration, is technology a critical factor. The final consideration is the sensitivity of the given market to technological advance. Some markets are much more technology-driven than others. Improvements in product attributes, eg, efficiency, size, life, maintainability, cost, etc, largely determine success for some classes of products. For others, style, price, advertising, availability, etc, are key factors.

Various techniques can be used to establish an approximate order of value for programs based on these criteria. The limited accuracy of the data precludes elaborate schemes. One easily applied system is to establish an approximate dollar value for the size of the potential market using relatively large increments as the step function and then to apply the other criteria, ie, rate of market growth, likely level of penetration, and sensitivity to technology, as a series of discounts to reduce the value of the market opportunity.

In a similar fashion, it is necessary to evaluate the likelihood of success for a particular program. Here the factors that need to be evaluated include:

1. The degree of difficulty of the technical problem being attacked: is it an exceedingly difficult technical challenge or one that can almost certainly be met if resources are applied?

2. The competitive status of the particular approach being proposed: is it a situation in which there is a likely winner among other technologies being pursued, or is it a race in which the proposed approach has at least as good a chance as any other, or is the proposed approach regarded as the favorite?

3. The fit of the resources of the organization with the requirements of the program: are the necessary skills and facilities already in place or at least readily obtainable, or are they quite foreign to the traditional work of the laboratory and could one expect some considerable difficulty in obtaining such resources and employing them productively?

4. The likelihood of being able to transfer the result, if successful, into commercial application. Does an operating component already exist that would be able to take the new technology in hand? Does the component have the requisite technical competence to undertake the transition? Is its financial position such that it would be able to divert resources to undertaking such a development? Perhaps most important, how willing is the manager to undertake change? Does the technology face the problem of having no likely place in the company?

For each element of probability a scale can be devised with definitions for step functions from most difficult or improbable to least difficult. The precise method of quasi-quantification is less important than the thought process that the exercise imposes on the manager or planner.

It also is necessary to consider programs in terms of their contribution to the basic charter and mission of the organization and the fundamental priorities that have been established for its work. Clearly, certain types of work must be undertaken and completed before resources can be diverted to work of lower priority. Typically, the highest priority is for activities to protect present businesses of the company to ensure that they stay healthy or that businesses that were in trouble are returned to a state of success. Only after these needs have been satisfied can resources be devoted to pursuit of additional opportunities or to probing for breakthroughs that could lead to new programs in the future.

The program evaluation procedure may be performed in two different ways. One is to establish a specialized program-planning-and-evaluation group to carry out all such analyses. The other is to insist that the analyses be prepared by the R&D managers. The former approach is likely to lead to more consistent and probably more rigorously applied criteria in the evaluation of programs. On the other hand, it constitutes a remote process that frequently invites the criticism and even hostility of the R&D managers. A preferable technique is to insist that the managers do it. Even though the answers may not be so consistent or so rigorously developed, the process of developing the evaluation data provides a learning experience for the managers and it encourages them to be mindful while planning technical work of the criteria critical to the success of the organization. The requirements to perform this screening process also encourages R&D managers to solicit help from marketing people in operations.

Level of Effort

The third critical step in evaluation is the determination of the minimum size of effort necessary to provide a reasonable likelihood of competitive success. This evaluation is judgmental, but it is frequently made unrealistically. R&D managers confront many competing demands for their resources. They often try to take on too many programs in an attempt to satisfy many clients. In this situation, an independent appraisal from the planning component can be helpful in calling attention to the range of effort being applied in other organizations and in questioning the effectiveness of the proposed effort. Strenuous efforts are required by higher levels of R&D management to focus adequate resources on key programs and to stop programs of more limited potential.

As programs move from the search to the focus phase and particularly as they become candidates for demonstration, a more rigorous and extensive type of evaluation is necessary. The manager now considers specific, practical demonstrations of economic and technical feasibility. Thus it is necessary to develop more extensive and quantified market data, to establish more realistic and detailed cost projections, and to specify more completely the product attributes that are needed.

Again, two modes of operation are possible: one is for the R&D organization to undertake such work itself; the other is to insist that the work be done wherever possible by the operations managers who are the likely recipients of the new technology. The latter approach is much preferred. Market studies and cost projections are typically not activities in which R&D organizations are expert. They rarely have the requisite skills and resources. More important, no matter how competently the work may be performed, its credibility is subject to criticism by operations. The answers are much more likely to be accepted as realistic if they are developed by those in operations. Furthermore, their involvement in the process at this early stage is one way of beginning to enlist their participation. If operations personnel become enthusiastic about the prospects for a project, its likelihood of success is markedly improved.

The early contact with market opportunities can provide valuable guidance for continued R&D effort. Frequently the attributes of the technology that are regarded as most attractive in the early phases of development are not necessarily the most attractive ones to the prospective customer. Consequently, continuing active interaction with potential customers as the project develops is an important element of eventual commercial success. The single most important cause of failure in innovation is lack of adequate timely market input.

The program-planning-and-evaluation techniques are particularly useful in the earlier phases of the evaluation-and-program-planning process. They help the individual R&D manager establish an array of programs from those with the greatest attraction and most warranting support to those that border on the marginal. The final array of programs for the operating plan for the entire organization must of course be fitted to budget realities as perceived by the senior officer for research and development. This final array must reflect informed judgment by technical managers, but the previously described techniques or analogous ones developed locally can help focus management attention on the questionable programs that warrant management scrutiny.

Communication Between Operations and Corporate Management

Effective communication in research and development almost always requires a two-way interchange of information. Research and development generates new information and creates opportunities for change. Consequently, the level of uncertainty in language and substance inherent in the communication is large compared with most management communication. The possibility of error or misinterpretation is large, therefore communication must occupy a significant fraction of the time of the R&D manager if he or she is to ensure understanding of the program and support for the objectives.

Communication with operations involves three elements. The first is a flow, primarily from operations to the laboratory, of inputs regarding problems and opportunities as perceived by operations management. These inputs are requisite to program planning, because they help to identify areas of potential importance to operations and they provide a valuable stimulus to technical people to generate new ideas. Although not all problems or opportunities can or should be pursued, R&D managers must give evidence to operations management of their intense interest in the problems it faces.

The second element of communication involves the collaborative planning of joint programs. As noted above, this process begins by soliciting the leadership of operations management in developing initial market and cost evaluations on programs being considered for transition. The joint program should capitalize on the specialized resources in operations and in the laboratory. The eventual success of the project hinges on the continuing refinement of marketing and manufacturing objectives as work progresses. Communication of this type requires extensive face-to-face contact. Documentation serves to maintain a record but it is not an adequate basis for achieving the required level of understanding and agreement. A typical communication pattern involves monthly or perhaps quarterly meetings of all the key people involved on a project, supplemented by daily communication by telephone.

The third element of communication with operations involves selling the entire R&D program and function. This activity begins with continued communication regarding the goal of the R&D organization and the nature of its interface with operations. Effective understanding of the appropriate role for the R&D component can best be achieved by case-by-case consideration of specific examples. A second dimension of selling the program involves formulation and discussion of the priorities used in choosing programs and allocating resources. In making these judgments, R&D management must balance responsiveness to the problems of operations management against the need to perform the longer range work that can provide the basis for significant new opportunities.

Communication with corporate management also must begin with an attempt to clarify the basic mission of research and development. Initiating the proposed mission for corporate R&D may be the responsibility of R&D management, but it is a subject of sufficient complexity and importance that it warrants extensive dialogue with corporate management to ensure agreement on purposes and to educate those at the corporate level to the various possible roles for R&D.

Such discussion of the direction or mission of R&D provides the foundation for determining the basic priorities that are to be used by R&D management in allocating resources. The R&D organization rarely, if ever, has the resources to do all the things it could consider doing. The critical strategic balance between support and opportunities for growth must reflect the mission of R&D; the basic strategy of the enterprise, whether it should be directed toward internal growth, diversification, improved productivity, resolving severe business difficulties, etc; and current topical concerns of corporate management. It must also reflect the promise inherent in current R&D programs and the need to maintain reasonable stability in R&D efforts.

The task of selling the program at the corporate level is somewhat different from that associated with individual operations components. Corporate management does not have the time and frequently lacks sufficient technical background to review all of the technical programs being carried out. Consequently, it is necessary to find ways of aggregating programs into larger groupings. The particular mode of classifying programs into larger aggregates typically varies with circumstances and with the nature of the communication task as perceived by the R&D managers. The programs may be grouped by the principal operations components, or they may be classed according to key strategic efforts identified by the corporation, or they may be classified into key technology output categories, for example, energy efficiency and conservation, productivity improvement, environmental compliance, materials availability, communication and control, etc. It is unlikely that any chosen classification scheme is useful for all purposes.

The guiding principle that should be followed is to identify a relatively small number of categories that provide a unifying theme to characterize the work of the organization. The objective is to elicit the enthusiastic support of corporate management. In so doing, the R&D manager must walk a fine line between two somewhat incompatible objectives. On the one hand, the manager wants to evoke the image of an R&D manager who is acting as a responsible member of the corporate management team, ie, aiding the corporation to achieve its chosen objectives as successfully and expeditiously as possible. On the other hand, it is necessary also to attempt to act as the conscience of the organization, constantly supporting the importance of the long-term objectives and the value of effort aimed at new opportunities.

Transition to Commercial Application

Participation by operations in evaluating commercial potential is one way of effecting early participation of operations in research management. The transition process cannot be considered to have begun, however, until operations management commits its own resources. This commitment may involve the allocation of funds to the R&D organization for the project and the commitment of human resources to work jointly on the development as it proceeds. Obtaining this commitment of resources is the most important single step to commercial application. It requires creativity and persistence on the part of R&D managment as well as flexibility in reshaping program objectives to reflect the wishes and perceptions of operations management. Eventual success, however, requires the creation of a truly joint program.

Technology transfer is an exceedingly complex and demanding task. It rarely proceeds satisfactorily without direct participation of the recipient. Without such participation it is virtually impossible for individuals to achieve the level of personal involvement and enthusiasm that is necessary to accomplish innovation. The one universal rule in transition to commercial application is that it occurs through people and it requires competence, dedication, and enthusiasm to succeed.

Transition is a complex communication process that requires an extraordinarily high level of communication among a group of diverse people. Advances in technology are providing additional tools for this task. Interactive graphics, rapid and easy worldwide voice communication, telecopying, and video help to bring increased precision to the task, reduce the chances of error, and provide the kind of redundancy and rapid feedback that is necessary for this type of communication to be effective.

Probably the most effective single mode of communication is transferal of people. People from operations may be sent to the laboratory on a temporary assignment to participate in a joint program; simultaneously or sequentially people from the laboratory may be sent to operations; and in many cases people from the laboratory transfer permanently with a project. This arrangement helps to ensure that the enthusiasm that was associated with the original program is carried over to operations and that information that is difficult to verbalize and to document is made available by people who have been personally involved in the work.

One of the striking features of R&D management in the last ten years is the greatly increased attention being devoted to the transition of results into commercial application. It is not unusual for an R&D organization to devote half or more of its total human resources and funds to the attainment of commercial application. R&D managers recognize that the development of a new technical capability that does not achieve application is a waste of valuable resources. The goal of research and development is the introduction of technological innovation, not simply the invention of new technical capability.

Outlook

The success or failure of the R&D component depends heavily on the concept of corporate management with respect to the basic nature of the enterprise itself, ie, how it seeks to ensure long-term success and growth, and to the strategies it devises to fulfill its mission. Consequently, R&D managers will increasingly expand their horizons to become active participants in the formulation of the fundamental philosophy of corporate management and the development of the strategies to implement that basic philosophy.

BIBLIOGRAPHY

"Research Management" in *ECT* 2nd ed., Suppl. Vol., pp. 855–875, C. A. Stokes, Consultant.

1. J. M. Amos and B. R. Sarchet, *Management for Engineers*, Prentice-Hall, Inc., Englewood Cliffs, N.J., 1981.
2. Personal communication, S. Kramer, Industrial Research Institute, 1981.
3. S. Ramo, *America's Technology Slip*, John Wiley & Sons, Inc., New York, 1980.

General References

W. J. Abernathy, *The Productivity Dilemma: Roadblock to Innovation in the Automobile Industry*, Johns Hopkins, Baltimore, Md., 1978.
W. J. Abernathy and J. M. Utterback, *Technol. Rev.*, 40 (June 1978).
D. Allison, ed., *The R&D Game: Technical Men, Technical Managers, and Research Productivity*, MIT Press, Cambridge, Mass., 1969.
T. J. Allen, *Managing the Flow of Technology: Technology Transfer and the Dissemination of Technological Information Within the R&D Organization*, MIT Press, Cambridge, Mass., 1977.
J. R. Bright, *Technological Forecasting for Industry and Government*, Prentice-Hall, Englewood Cliffs, N.J., 1968.
Bus. Week, 56 (Feb. 16, 1976).
M. J. Cetron and B. Bartocha, *The Methodology of Technology Assessment*, Gordon & Breach, New York, 1972.
M. J. Cetron and co-workers, *Technical Resource Management: Quantitative Methods*, MIT Press, Cambridge, Mass., 1969.
D. Crane, "Science Policy Studies" in P. T. Durbin, ed., *A Guide to the Culture of Science, Technology and Medicine*, Free Press, New York, 1980, Chapt. 9.
P. F. Drucker, *Harvard Business Review* **41**(1), 103 (Jan.–Feb. 1963).
P. F. Drucker, *Science* **204**, 806 (May 25, 1979).
D. Fishlock, *The Business of Science: The Risks and Rewards of Research and Development*, John Wiley & Sons, Inc., New York, 1975.
W. R. Fusfeld, *Technol. Rev.*, 51 (May 1978).
F. W. Gluck and R. N. Foster, *Harvard Business Review* **53**, 139 (Sept. 1975).
W. A. Gruber and D. Marquis, *Factors in the Transfer of Technology*, MIT Press, Cambridge, Mass., 1969.

Interactions of Science and Technology in the Innovative Process: Some Case Studies, Battelle Columbus Laboratories, Columbus, Ohio, March 19, 1973.

J. Langrish, *Innovation in Industry, Some Results of the Queen's Award Study*, Research Report No. *15*, University of Manchester, Manchester, England, Sept. 1969.

T. Levitt, *Harvard Business Review* 41(3), 45 (May–June 1963).

J. W. Lorsch and P. R. Lawrence, *Harvard Business Review* 43(2), 109 (Jan.–Feb. 1965).

"Managing Advancing Technology," *Innovation*, American Management Association, New York, 1972.

E. Mansfield, *The Economics of Technological Change*, Norton, New York, 1968.

J. A. Morton, *Organizing for Innovation: A Systems Approach to Technical Management*, McGraw-Hill, New York, 1971.

C. D. Orth, J. C. Bailey, and F. W. Wolek, *Administering Research and Development: The Behavior of Scientists and Engineers in Organization*, Richard D. Irwin & Co., Homewood, Ill., 1969.

J. B. Quinn, *Harvard Business Review* 38(2), 67 (Mar.–Apr. 1969).

J. B. Quinn, *Harvard Business Review* 39(9), 88 (July–Aug. 1961).

J. B. Quinn, *Strategies for Business Growth and Change*, Richard D. Irwin & Co., Homewood, Ill., 1980.

J. B. Quinn, *Yardsticks for Industrial Research: The Evaluation of Research and Development Output*, Ronald Press, New York, 1959.

J. B. Quinn and J. A. Mueller, *Harvard Business Review* 41(1), 49 (Jan.–Feb. 1963).

E. Roberts, *The Dynamics of Research and Development*, Harper & Row, New York, 1964.

E. Roberts and A. L. Frohman, *Technol. Rev.*, 32 (Jan. 1978).

D. A. Schon, *Technology and Change; The New Hiraclitus*, Delacorte Press, New York, 1967.

Seminar on R&D Management (Istanbul, 1970; Paris, Organization for Economic Cooperation and Development, 1972).

I. Spiegel-Rosing and D. des. Price, *Science, Technology and Society*, Sage Press, London, and Beverly Hills, Calif., 1970.

Success and Failure in Industrial Innovation, Report on Project Sappho, Science Policy Research Unit, Center for the Study of Industrial Innovation, London, 1972.

TRACES: Technology in Retrospect and Critical Events in Science, Vols. 1 and 2, Illinois Institute of Technology Research Institute, Chicago, Ill., 1968, 1969.

E. A. Von Hippel, *Technol. Rev.*, 30 (Jan. 1978).

M. Y. Yoshino, *Japan's Managerial System: Tradition and Innovation*, MIT Press, Cambridge, Mass., 1968.

Journals

Research Policy, North Holland Publishing Company, Amsterdam and New York.

A. M. BEUCHE
L. W. STEELE
General Electric Co.

RESINS, NATURAL

Resins are organic solids that break with a conchoidal fracture. This classic definition lies between crystalline planarity and the dubious coherence of formless gums (qv) and waxes (qv). It antedates the modern concept of the glassy state and the glass-transition temperature T_g, which occurs below the melting point. Resins are organic glasses above T_g.

Natural resins are derived from many sources, and they have diverse properties and uses (see Table 1) (1). Natural resins are mainly oleoresins from tree saps and related fluids, and they occur secondarily from other sources, eg, shellac (qv), insect exudations, and mineral hydrocarbons, eg, gilsonite (qv) and Utah coal resin. Oleoresins from tree saps are mixtures of single- and fused-ring compounds with various oxygenated pendant groups. A wide range of molecular weights and chemical structures results in very different solubilities and chemical and physical properties among the principal commercial sources, eg, Congos, damars, East Indias, elemi, kauri, Manilas, pontianak, sandarac, rosin (see Terpenoids), and others.

The classification and properties of natural resins depend on their uses, which can be reviewed in terms of three periods of history: ancient and medieval; modern, ie, ca 1500–1940; and recent, ie, from ca 1920, with a twenty-year overlap with the modern period. The recent period is marked by a rapid rise of polymer science and technology and of synthetic chemistry in medicine, perfumery, etc.

It was estimated that the annual U.S. consumption of natural resins prior to 1932 peaked at 18,000–23,000 metric tons. The worldwide annual consumption of rosin, however, is ca 9×10^5 t.

Recent and immense advances in medicine, science, and technology in general, and polymer synthesis in particular, have altered greatly the roles of natural resins. Some will dwindle in supply and use, but others could experience a resurgence because of their unique values that would be analogous to the resurgence in natural rubbers (see Rubber, natural). It is premature to predict the demise of most natural resins despite severe competition from rosin and synthetic polymers.

Amber [8002-67-3] is a fossil resin from an extinct conifer and is still gathered or dug along the Baltic shore. It is used as a gemstone, eg, by the ancient Greeks in the sixth century AD and by contemporary artisans in necklaces and other ornaments. It may be the hardest of the historic natural resins. In ancient Egypt, where there were no local trees to serve as resin sources, natural oleoresin was used (perhaps mastic from Lebanon or Asia Minor) for embalming because of its aroma and hydrophobic protective properties in and around abdominal cavities. Artists from early times melted natural oleoresins and resins with pigments for liquid application. Artists also used linseed oil as a paint medium, but it was relatively soft, and they attempted to incorporate natural resins for harder paint. They sought optimum hardness in a coating through modification of the hard resin component by adding either the oil in an oleoresin, linseed oil, or both.

The terms resin from the Latin *resina* and lac from the Hindustani *lakh* imply overlapping physical properties and are used interchangeably in the commercial parlance of Western languages. Shellac is exuded by the *Kerria lacca* insect as a protective cover onto host twigs in India. Its polyhydroxy fatty acid structure confers

Table 1. Sources and Properties of Natural Resins

Resin	CAS Registry No.	Class	Country of origin	Properties			
				Mp, °C	Acid no.	Saponification no.	Iodine no.
accroides (yacca)	[9000-20-8]	amber	Australia	130		65	200
Congo	[9000-14-0]	dark	Zaire	170	100	125	125
Congo	[9000-14-0]	white, pale	Zaire	150	110	125	120
damar	[9000-16-2]	Batavia	Indonesia	105	28	34	104
damar	[9000-16-2]	Singapore	Indonesia	115	30	38	113
East India	[9000-16-2]	batu	Indonesia	174	19	33	81
		black	Indonesia	162	20	32	83
		pale, Macassar	Indonesia	140	18	40	103
		pale, Singapore	Indonesia	152	24	35	84
elemi, gum	[9000-74-2]		Philippines	plastic	30	30	118
gilsonite	[12002-43-6]		Utah, U.S.	170			
kauri		brown	New Zealand	160	70	90	120
		pale	New Zealand	130	70	90	140
Manila	[9000-14-0]	Boea, Loba	Indonesia	130	123	160	130
		Macassar	Indonesia	121	136	180	121
		Philippine	Philippines	129	115	145	120
mastic, gum	[61789-92-2]		Greece	76	65	75	100
pontianak			Indonesia	135	118	150	130
rosin		gum	U.S.	70	165	174	220
		tall oil	U.S.	80	170	175	
		wood	U.S.	60	156	166	215
sandarac	[9000-57-1]		Morocco	140	135	150	130
shellac	[9000-59-3]		India			150	130
Utah coal resin			Utah, U.S.	170	7		145

solubility in ethanol and isopropanol for spirit lacquers, whereas competitive lacquers from Manila resin must be used with mixtures of alcohols and hydrocarbon solvents. Chinese and Japanese lacquer artware is formed from the juice of the *Rhus verniciflua* lac tree, and its higher alkyl phenolic structure, ie, urushiol, resinifies on exposure to air. Cashew-nut-shell oil [*8007-24-7*] from *Anacardium occidentale* forms industrial resins from its higher alkyl phenolic structures. Both of these are structurally related to poison-ivy sap, and all three cause high allergic reactions in people. Utah coal resin is in the form of flakes and veins of yellow hydrocarbons, which dissolve readily in aliphatic and aromatic volatile solvents and in linseed and other drying oils for use as a paint medium (see Paint). Gilsonite (qv) is a black hydrocarbon resin from a vertical east–west mineral vein extending across the Utah–Colorado border. Like the paler Utah coal resin, gilsonite is soluble in hydrocarbons and drying oils and is used in protective coatings.

Disparity of technical advances and shifts from hand-gathered to industrially produced natural resins is shown among ca 200 technical-journal articles in the 1970s. The importance of natural resins, in terms of the percentage of reports on them, are 55% on rosin and associated naval-stores processes and products, 20% on shellac, 12% on Japan lac, 3% on cashew-nut-shell oil, and 10% on all others with 0.5% on Manila copal.

Commercial, political, and technological changes from World War II onward have greatly altered the competitive and technical relationships among these classes of natural resins, as has the use of synthetic polymers. Hand gathering of natural resins is relatively more costly, eg, hand-gathered gum rosin versus tall-oil rosin in the United States, kauri in New Zealand, and Congo resin in Zaire. Lumber and paper-pulp operations in both the Western Hemisphere and the various islands southeast of Asia have infringed on harvesting of natural resins there. Political transitions from colonial to autonomous governments have altered the routes and customs of gathering natural resins and affected their commerce, particularly in Indonesia.

Synthetic polymers are made in great diversity from petrochemical monomers, eg, vinyl chloride at $0.49 per kilogram, styrene at $0.88, and methyl methacrylate at $1.28, for a host of specialized and uniform qualities which compete favorably with traditional uses of natural resins. Shifts from traditional oilcloth, linoleum, and slow air-drying paints to factory-made films produced from synthetic polymers have narrowed the range of natural-resin uses. However, price escalations of petroleum and natural gas could shift commercial interests back toward natural resins. Typical 1981 prices of natural resins and of monomers in the synthesis of polymers that are competitive with natural resins are listed in Table 2.

Gum, eg, kauri as a varnish gum, is an alternative commercial term for various natural resins, but it should not be confused with nonresinous natural gums, which are less hard or less hydrophobic, eg, chewing gum or hydratable polysaccharides. Oleoresin is a classic term for the viscous-to-semisolid precursors of a natural resin and describes varnishes made by combining drying oils, eg, linseed oil, with resins. Oleoresin exudes naturally as drops, flakes, or lumps or it can flow from taps, hacks, or natural injuries to trunks and branches of trees and bushes. The chemical compositions and physical properties of oleoresins are diverse and, in the early 1900s, they were investigated in detail (1) (see also Vegetable oils; Driers).

In the late 1900s, rosin and synthetic polymers were competitive with natural resins, which led to a reduction of scientific study of compositions of natural resins

Table 2.   Typical 1981 Prices of Natural Resins and of Competitive Monomers

Type	Price, $/kg
Resins	
damar	1.43
East India, pale	1.43
East India, black	0.99
pontianak	1.54
Manila	1.54
elemi	1.43
yacca (accroides)	0.99
shellac	4.85
gilsonite	0.26
rosin, gum	1.21
rosin ester, gum	1.65
rosin, wood	1.10
rosin ester, wood	1.43
rosin, tall oil	0.88
Monomers	
styrene	0.88
vinyl acetate	0.88
methyl methacrylate	1.32

except to complement the expanding uses of wood and tall-oil rosins (see Tall oil). The great diversity of botanic and chemical mechanisms of formation of natural resins is beyond the scope of this article, except to use rosin as a norm or illustration among natural resins.

The natural separation and gradual conversion of some of the hydrophilic components of sap and related plant fluids from the cambium layer of a tree into increasingly hydrophobic solids is the generic process of forming diverse gums, resins, and waxes. The oleoresin intermediate in this process is typified in pine gum, which flows from hacks on the trunks of southern yellow pine in southeastern United States, in France, and in other countries. Pine gum contains ca 80% rosin and 20% turpentine.

Rosin is mainly a mixture of C_{20}, fused-ring, monocarboxylic acids typified by levopimaric and abietic acids; both are susceptible to numerous chemical transformations, as indicated in Figure 1 (2). Other natural resins, eg, Congo resins, contain dicarboxylic acids, tannins, and benzoic, cinnamic, *para*-coumaric, and other acids that have widely varying solubilities and reactivities. They are used in varnishes, flavors, perfumes (qv), and medicines. Turpentine includes mainly C_{10} terpenes, especially α-pinene and β-pinene, and is typical of the volatile oil portions of various oleoresins.

α-Pinene is a starting material in the synthesis of dozens of compounds used in perfumery and flavors (see Flavors and spices). The biochemical origins of natural resins have been postulated primarily as oligomerizations of isoprene through terpenes to sesquiterpenes and diterpenes and numerous hydroxyl and carboxyl derivatives thereof.

Resinification from oleoresin can result from either natural evaporation of oil from an exudate or slow collection in ducts in sapwood and heartwood; it is visible as resin streaks in planed lumber. *Pinus* stumps are valuable enough to be harvested,

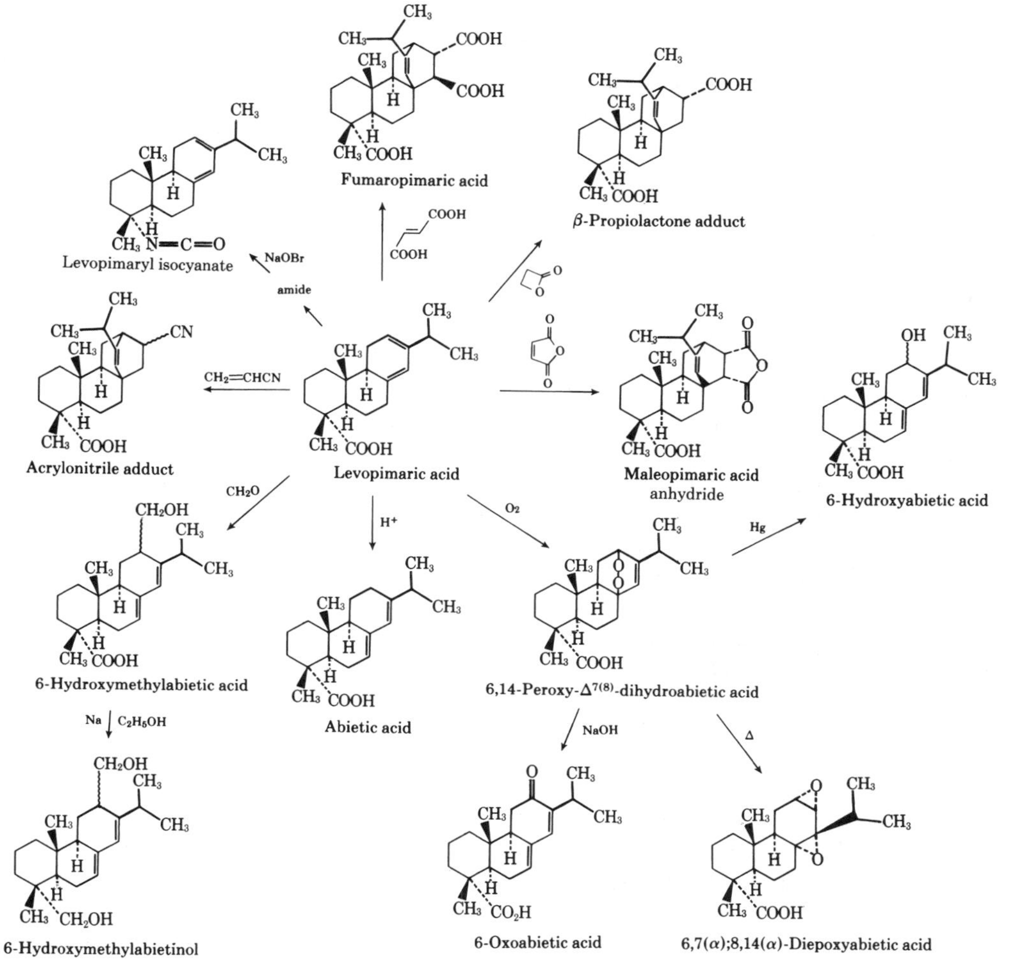

Figure 1.   Levopimaric acid reaction products.

chipped, and extracted with hexane or higher-boiling paraffins to yield wood rosin, wood turpentine, and other terpene-related compounds by fractional distillation. In the kraft, ie, sulfate, paper process, pinewood is digested with alkali, and crude tall oil and crude sulfate turpentine are by-products (see Pulp). Fractionation of crude tall oil yields tall-oil rosin, and fatty acids and fractionation of crude sulfate turpentine gives terpenes. Condensation in natural resins occurs over months, years, or millenia in slow hardening to semifossil and fossil grades, which are sometimes designated as copals. For example, amber and kauri resins are extremely aged and hard natural resins.

Copal also now describes hard oleoresins, particularly those of fossil origin; such resins are useful in varnishmaking. Copal became a designation for quality in an attempt to distinguish between botanic and commercial points of origin, and it applies mainly to Congo, kauri, pontianak, and Manila resins. Amber and Utah coal resins may be the oldest of the copals.

Properties and Uses in Varnishes and Other Coatings

The importance of one property versus another, eg, hardness, refractive index, acid value, etc, is highly judgmental according to the use. Elaborate tabulations of chemical and physical properties have been published (1,3–5). Approximately 38 variants among a dozen general classes of natural resins have ir spectra at 2.5–50 μm (6).

Varnish. Varnish and its modern language variants vernis, berniz, and others, may be a term derived from Berenice, Queen of Cyrene, whose golden hair resembled amber which may have been the first natural resin used as a coating medium by the early Greeks. Varnish is a clear liquid that solidifies as a thin film. When it was first processed from natural resins, it was melted on a surface, which allowed volatile oils to evaporate from the oleoresin; later the resin was combined with linseed oil, which oxidized to the solid state and conferred flexibility to the hard resin component. Adding turpentine as a volatile solvent to reduce viscosity was practiced in recent centuries. Thus, varnish became a combination of various proportions of resin, drying oil, volatile solvent, and minor additives, eg, metal-containing drier catalysts, antioxidants, etc. Uses of natural resins flourished in the 1800s and early 1900s through combinations of resins, oils, and volatile solvents.

Natural resins of various ages and botanic species have various limits of solubility in linseed oil and other drying oils and in volatile solvents (see Resins, water-soluble). The early varnishmaker's art required a skillful pyrolysis or running of the resin at 250–350°C to reduce molecular weights and to simplify the molecular structure for improved solubility and to preserve enough of the hardness of the resin to obtain sufficient film hardness for a particular use. After a 10–20% resin loss from pyrolysis, the resin-to-oil ratios of typical varnishes might be 50 wt % resin–50 wt % oil for furniture varnish for factory use, 33 wt % resin–67 wt % oil for floor varnish, or 25 wt % resin–75 wt % oil for wooden ship spars or for building exteriors. Similar varnishes are pigmented to make paints and enamels.

Dissolving natural resins into linseed or other drying oils and into turpentine or other volatile solvents is known as fusing, running, sweating, melting, or cracking. In addition to a physical solution, partial pyrolysis of the resin yields water, carbon dioxide, and low molecular weight compounds, which ranges from easily volatile solvents such as turpentine to nonvolatile plasticizers and softeners of the resin. A kauri–linseed varnish formulation, which was published in 1928, is given below (7):

Durable-wearing body varnish, No. 1922
(density, 0.925 g/cm^3)
45 kg pale kauri no. 1
132 L varnish-making linseed oil
2 kg litharge
0.5 kg manganese borate
<u>170 L</u> turpentine
341 L
<u>−15 L</u> loss
326 L product

This hard-gum varnish is fairly pale, works easily under the brush, has a good-luster, dries in ca 12 h, and is durable. It is designed as a carriage-body finishing varnish.

From ca 1930 to ca 1950, stainless-steel-lined reactors with capacities of up to 19,000 L (5000 gal) and with mechanical stirring, sparging, closed vents, fume condensers, and sludge collectors were developed. They were designed for the manufacture of alkyds (see Alkyd resins) and other synthetic resins and varnishes; they did not accommodate the special heat-transfer needs for large lumps of natural resins and the partial reflux retention of their volatile products upon pyrolysis. From ca 1960, increasing environmental concern and control of atmospheric contaminants made the earlier open, portable varnish kettles obsolete. Natural resin that is used now in making varnishes is mainly cut in unheated, closed reactors with high speed cutting blades to break and dissolve resin lumps in volatile solvents and other varnish components.

Each of the principal natural resins of past or present commercial importance as used in varnishes is described briefly below as well as in refs. 1 and 2.

Accroides resin, also called yacca gum, accumulates at the base of dead, rushlike leaves on 1-m stems of 2.5-m *Xanthorrhoea* trees throughout Australia and especially in Tasmania. A yellow variety (mp 97°C) occurs in *Xanthorrhoea hostalis* and a red resin (mp 110°C) occurs in ca 16 other varieties, but mainly in *Xanthorrhoea tateana* in South Australia. Accroides' high content of tannic and other aromatic acids makes it soluble in higher alcohols and volatile esters for use as lacquers on paper and wallboard and in mahogany stains for wood.

Amber is the hardest fossil resin and is still collected along the Baltic shores as lumps from exudations of the extinct *Pinus succinifera*. Amber's small, scattered supply and superior hardness and clarity limits its current use to ornamental beads for necklaces and other ornamental pieces. Ancient and medieval varnishes were sometimes made by heat-fluxing amber with linseed oil.

Canada balsam [8007-47-4] is an oleoresin containing much volatile terpene. It exudes from the balsam fir *Abies balsamea* and is gathered by hand in the northern United States and Canada in small quantities. Its clarity, though slightly greenish, and refractive index of 1.5180–1.5210 render it highly suitable for mounting microscopy specimens and for cementing optical glass elements. It may be used neat or diluted with xylene or other volatile solvents.

Congo is one of the hardest natural resins and was named for its mainly fossil origin in the swamps of the Congo River basin in Zaire, mainly between Kinshasa and Kisangani. Harvesting of congo resin by handdigging in the late spring after floods receded was a large industry while Belgium governed this area, but was stopped during political turmoil in Zaire in the 1960s and 1970s. A classification system of lump sizes and melting ranges, based on a great diversity of colors from nearly colorless to deep yellow and brown, is described in ref. 1. Chemical analyses of these resins from Zaire and from other tropical areas of Africa and South America are also described in ref. 1. All of the Congo resins have fused-ring structures and large and variable proportions of mono- and dicarboxylic acids. Their very limited solubility in drying oils and volatile organic solvents require either thermal treatment, strong mastication in roll mills heated to ca 100°C, esterification with glycerol, or combinations of these treatments to incorporate Congo resin into useful varnishes.

Damar resin is not a fossil resin but is gathered or tapped from cuts that are made in the sapwood of many species, eg, *Hopea* and *Shorea* of the *Dipterocarpaceae* family of trees on the Borneo and Sumatra islands and the Malay peninsula. Damar resins are graded in terms of pale to yellow color, size, cleanliness, and commercial gathering stations, notably Djakarta, Thailand, and Singapore. The chemical composition is any of diverse mixtures of several mono- and dicarboxylic acids and other oxyorganic compounds of up to 56 carbons, most of which are probably terpenic. The acid value range of 20–35 is low relative to those of rosin and other natural resins. It is easily soluble in aromatic, aliphatic, and terpene hydrocarbons and has long been used with linseed oil in varnishes, in which it does not need thermal degradation, and in lacquers, where its initial paleness, and subsequent bleaching by sunlight make it useful in clear top coatings. Its limited solubility in alcohols and esters, from which a waxy β-resene separates, requires that it be fractionated before use in cellulosic lacquers. Damar resin was a long-favored resin modifier of nitrocellulose.

Pale batu and black East India resins are recent fossil resins from the same family of trees that yields nonfossil damars. The resins are gathered from near the trees of origin and from alluvial deposits downstream. The lumps need more scraping and sorting than do damars so that they can be separated into about eight size and color grades. With processing, solubility and durability of East Indies are similar but harder and generally superior to damars, except for being darker. They have been used in recent decades in traffic paints and general-purpose industrial enamels. The differences between damar and East India resins in terms of moisture content, solubility in alcohols versus hydrocarbons, and utility in various practical coatings are delineated in ref. 1.

Elemi is an oleoresin and, like pine gum, contains much volatile oil, ie, 20–25 wt %. It is gathered from successive scrapes or hacks through the bark of the *Canarium luzonium* trees of the *Burseraceae* family throughout the Philippines. Unlike those of pine gum, its low acid values (20–25) indicate that it contains more neutral resinous compounds related to terpenes and phenanthrene (1). Elemi is a solvent and a pale plasticizer in lacquers, for which volatile aromatic and ester solvents are typically used.

Kauri fossil resin was dug by Maori workers from soil containing lumps of kauri that exuded from generations of *Agatha australis* which is a conifer. Economic shifts have sharply reduced this labor-intensive source of the resin. Later harvesting of kauri resin was by hydraulic mining in Tasmania. Its reputation is preserved in two reagents in contemporary testing: the kauri-butanol test for practical volatile hydrocarbon solvents, which are complex mixtures of aromatic cycloparaffins (naphthenes), olefins, paraffins, etc and the kauri-reduction test for elasticity of spar varnishes and similarly clear and pigmented coatings. Kauri contains fewer terpene volatiles and more oil-soluble monocarboxylic resinous acids than congo resin, and it has greater chemical stability than the abietic acid family of monocarboxylic acids in rosin. Thermal processing of kauri resin with linseed oil yields varnishes and enamels, which were of the highest quality in the first quarter of the twentieth century.

The kauri-butanol value of hydrocarbon solvents is a test method of worldwide use and is standardized in ASTM D 1133 (8). The reagent contains ca 17 wt % clean, pale kauri resin and 83 wt % n-butanol; the two are very precisely combined and adjusted for butanol content so that titration to cloud point with pure toluene gives an arbitrary kauri butanol (k-b) value standard of 105. The lower k-b value scale limit is ca 25 for n-octane, and typical mineral spirits (called white spirits in Europe) have k-b values of 25–40.

The kauri-reduction value for toughness of varnishes evolved in the early 1900s as a reference point of superior quality and is standardized in ASTM D 1642-70 (8). The reagent contains ca 33 wt % thermally processed pale kauri resin in 67 wt % turpentine (α- and β-pinenes). The nonvolatile content of a good quality spar varnish can be diluted, ie, reduced, with 100% of this reagent. After a test film preparation on a 0.227 mm thick bright tin panel, the film can be bent rapidly over a 3 mm rod without appearance of cracks in the deformed test film.

Manila resins are exuded from *Agathis alba* and related giant conifers in the *Araucariaceae* family, which grow widely in the Philippines and other islands of the Indonesian archipelago. Oleoresin flows from cuts in the cambium layer and is harvested at various ages: weeks for soft melengket Manila, months for half-hard loba Manila, years for hard Manila or pontianak, or many years for buried hard fossil Manila or boea. The broad range of color, acid value, melting point, and hardness is summarized in ref. 1. Because of their alcohol solubility, Manilas are used in lacquer formulations, and their high acid values enable them to be neutralized by amines in waterborne coatings and polishes.

Mastic resin is harvested mainly on the Greek island of Chios as drops or tears exuded from taps or wounds through the bark of *Pistacia lentiscus*, which is a small tree of the *Anacardiaceae* family. Mastic is soluble in aromatics and alcohols, and its solutions have been used as clear coatings over artistic paintings.

Oriental lacquers, both Chinese and Japanese, are formed *in situ* as decorative and protective surfaces when handcrafted from the gray-brown viscous sap tapped from the *Rhus vernififera* tree in China. Research in Japan reveals the compositional detail of the urushiol and related constituent alkyl phenolic compounds and of the enzymatic and oxidative transformations of the sap into durable lacquers (9).

Sandarac resin is harvested in northwest Africa as globules exuded from *Tetraclinis articulata* trees of the *Coniferae* family. Like mastic, it dissolves in aromatic and alcoholic volatile solvents for use in various artistic crafts.

Utah coal resin occurs as flakes and veins in vast coal beds in the Salina and Huntington regions of central Utah. It was separated and sold in the 1940s and 1950s as RBH Resin 510 by the Interchemical Corporation, now the Inmont Division of United Technologies. The resin was dissolved from coal by and fractionally precipitated from, aliphatic volatile solvents in yellow and brown grades. Its relatively high melting point, low acid value, and intermediate iodine number indicate that it is a hydrocarbon resin with easy solubility in linseed and related drying oils. Varnishes and enamels made from Utah coal resin competes in properties with oleoresinous varnishes from the natural resins typified by Congo and the hydrocarbon resins (qv) synthesized from indene, coumarone, styrene, terpenes, etc. Utah coal resin use continues to depend primarily on the economies of coal mining and processing.

Natural Resins in Medicines, Perfumes, and Flavors. From the dawn of history, natural resins have affected the senses and sensibilities in medicinal, religious, and social applications, eg, as used in frankincense and myrrh, Egyptian embalming, perfumery, and wines. The efficacy of ancient and more recent medical uses of resins is challenged by great advances in pharmacology; by legal regulations, eg, the Toxic Substances Control Act; and by approval procedures of the FDA. Allowances and uses of natural resins in medicine are described in current medical and regulatory literature and in editions of *Pharmacopeia* (10). The names, origins, and Merck Index Numbers (11) of historically important natural resins used in medicine are given below:

Balm of Gilead [8022-26-2] or Mecca balsam is derived from the *Commiphora opobalsamum* tree in Arabia and Abyssinia (Merck No. 951).

Balsam of Peru [8007-00-9] is derived from the bark of the *Myroxylon pereirae* trees in Central America (Merck No. 952).

Balsam of Tolu [*9000-64-0*] exudes from incisions into the *Myroxylon punctatum* tree in equatorial America (Merck No. 953).

Benzoin [*9000-05-9*] occurs in various species of the *Styrax* tree in Thailand, and Sumatra, and Java (Merck No. 1102). Benzoin contains up to 38 wt % benzoic acid and up to 30 wt % cinnamic acid.

Copaiba [*8001-61-4*] is obtained from the trunk of the *Copaifera landsdorfi* tree in the West Indies and Amazon region (Brazil) (Merck No. 2492).

Frankincense [*8050-07-5*] or olibanum is from the *Boswellia* tree of the *Burseraciae* family in Arabia and neighboring areas. Early Greeks, Romans, and Arabs used it in medicines and religious ceremonies. It is still used as incense (Merck No. 6679).

Guaiac [*9000-29-7*] exudes from the wood of the *Guajacum officinale* tree in the West Indies and on the northern coast of South America (Merck No. 4396).

Labdanum [*8016-26-0*] is from the *Cistus ladaniferus* and related shrubs. Its odor is suggestive of ambergris, and the resin is used in perfumes.

Myrrh [*9000-45-7*] is from the *Commiphora* tree of the *Burseraceae* family in East Africa and Arabia and has been used in medicine and perfumes (Merck No. 6164).

Opopanax [*9000-78-6*] is from one of the family of *Burseraceae* and has long been used in perfumes and in medicine.

Storax [*8046-19-3*] is obtained from the inner bark of the *Liquidambar orientalis* tree in Asia Minor and is purified by alcohol extraction. It is used in medicine and in perfumes (Merck No. 8063).

BIBLIOGRAPHY

"Resins, Natural" in *ECT* 1st ed., Vol. 11, pp. 666–687, by E. L. Kropa, Battelle Memorial Institute; "Resins, Natural" in *ECT* 2nd ed., Vol. 17, pp. 379–391, by Frank A. Wagner, B. F. Goodrich Chemical Company.

1. C. L. Mantell and co-workers, *The Technology of Natural Resins*, John Wiley & Sons, Inc., New York, 1942.
2. *U.S. Dept. Agric. Agric. Res. Serv.*, ARS **72-35** (Jan. 1965).
3. C. L. Mantell, ed., *Natural Resins Handbook*, American Gum Importers Association, New York, 1939. (Now inquire of the O. G. Innes Corp., 10 East 40th Street, New York, N.Y. 10016.)
4. F. J. Martinek in C. R. Martens, eds., *Technology of Paints, Varnishes and Lacquers*, Reinhold Book Corp., New York, 1968, Chapt. 15, pp. 258–272; reprinted by R. E. Krieger Publishing Co., Huntington, N.Y., 1974.
5. G. G. Sward, ed., *Paint Testing Manual*, 13th ed., American Society for Testing and Materials, Philadelphia, Pa., 1972, pp. 76–91.
6. *An Infrared Spectroscopy Atlas for the Coatings Industry*, 2nd ed., Federation of Societies for Coatings Technology, Philadelphia, Pa., 1980.
7. Winfield G. Scott, *Formulas and Processes for Manufacturing Paints, Oils and Varnishes, a Laboratory Manual*, Trade Review Co., Chicago, Ill., 1928.
8. *Book of Standards*, ASTM D 1133 in Part 29, D 1642 in Part 27, reissued annually and separately, Philadelphia, Pa.
9. *Macromol. Chem.* **179**(1), 47 (1978).
10. *The United States Pharmacopeia XX (USP XX–NF XV)*, The United States Pharmacopeial Convention, Inc., Rockville, Md., 1980.
11. M. Windholz, ed., *Merck Index*, IX ed., 1976, Merck and Company, Rahway, N.J.

JOHN C. WEAVER
Case Western Reserve University

RESINS, WATER-SOLUBLE

Water-soluble resins are polymeric materials that are completely soluble or swell substantially in water. They include natural, modified natural, and completely synthetic materials. Water-soluble resins are of commercial interest because they alter the properties of aqueous solutions or form films. Functions include dispersion, rheology control, binding, coating, flocculation, emulsification, foam stabilization, and protective colloid action (see Flocculating agents; Emulsions).

Natural

Natural gums (qv) are produced principally in Africa or Asia and are obtained from seeds, seaweed, or exudates from trees (see also Resins, natural). All hydrocolloids from plants or microbes are polysaccharides. The hydrophilic groups can be either nonionic, anionic, or cationic. Chemically, the polysaccharides are composed of pentoses or, more commonly, hexoses. The stereochemistry of the hydroxyl groups on C-1 and C-2 of the glycoside ring is either cis (α) or trans (β), as shown in the following structures (1):

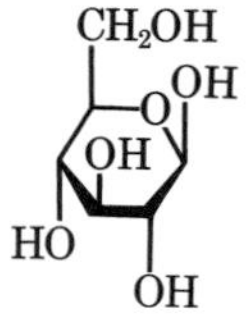

β-D-glucose 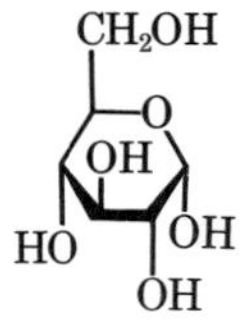 α-D-glucose

Different polysaccharides are made by enzymatic reaction of the OH group of C-1 with the hydroxyl group of a second ring at any position except C-1. Thus, in amylose (linear) starch [9005-82-7], the ether linkage is between C-1 and C-4 on adjacent pyranose rings, which form a uniform α-D (1→4) structure:

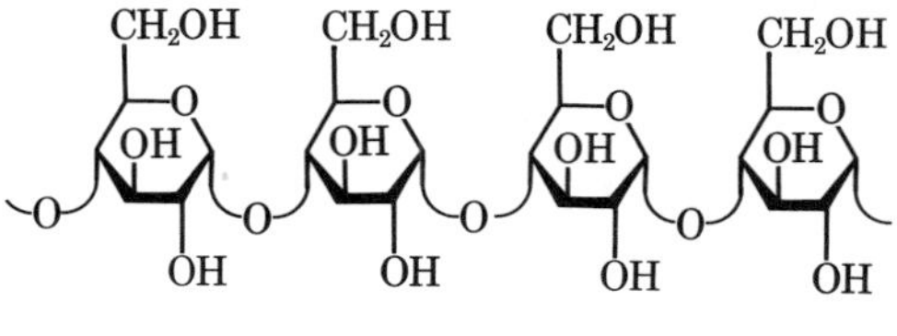

Branched polysaccharides occur when two sugar units are glycosidically condensed with two OH groups other than C-1 on a third sugar unit. In addition, functional groups, eg, carboxyl, sulfate, amine, etc, are present in some polysaccharides (1–2). Table 1 lists the source, chemical composition, principal uses, and production of the important natural water-soluble resins (3–4). Collection, isolation, and purification methods vary with the material. In the United States, corn starch is by far the largest volume, natural, water-soluble polymer produced (see Starch). Some chemically modified starches and celluloses are of commercial importance (see Table 2) (see Cellulose; Cellulose derivatives; Carbohydrates).

Table 1. Principal Natural Water-Soluble Resins [a]

Resin	CAS Registry No.	Chemical composition	Source	1979 Production, 10^3 metric tons	Uses
agar	[9002-18-0]	polygalactose sulfate ester	seaweed extract	0.70	food, dentistry, medicine, microbiology
carragenan	[9000-07-1]	anhydrogalactose–galactose polymer sulfate ester	seaweed extract	0.48	food, pharmaceuticals, cosmetics
corn starch	[9005-25-8]	poly(1,4-D-glucopyranose) (amylose) + amylopectin (branched)	grain extract	2489 [b]	food, sizing
guar gum	[9000-30-0]	mannose polymer with galactose branches on every 2nd unit	seed extract	47.08	food, papermaking, mining, petroleum production
gum arabic	[9000-01-5]	highly branched polymer of galactose, rhamnose, arabinose, and glucuronic acid	plant exudate	9.88	food, pharmaceuticals, printing inks
gum karaya	[9000-36-6]	polymer of galactose, rhamnose, and partially acetylated glucuronic acid	plant exudate	2.74	food, pharmaceuticals, paper binder, printing dyes
gum tragacanth	[9000-65-1]	polymer of fucose, xylose, arabinose, and glucuronic acid	plant exudate	0.17	food, cosmetics, printing pastes, pharmaceuticals
locust bean gum	[9000-40-2]	mannose polymer with galactose branches on every 4th unit	seed extract	3.28	food, papermaking, textile sizing, cosmetics
potato, wheat, and rice starches	[9005-25-8]	poly(1,4-D-glucopyranose) (amylose) + amylopectin (branched)	grain and root extract	77.1 [b]	food
tapioca	[9005-25-8]	poly(1,4-D-glucopyranose) (amylose) + amylopectin	root extract	32.9	food

[a] Refs. 3–4.
[b] 1977.

Water-soluble polymers from animal sources are proteins, ie, amphoteric polyelectrolytes, which have the following general polypeptide structure:

$$\left[\begin{array}{ccc} H & R & O \\ | & | & \| \\ N & CHC \end{array}\right]_n$$

Industrially, the most important products are casein [9000-71-9], gelatin (qv) [9000-70-8], and animal glue (see Glue). Casein is the chief constituent in the milk

Table 2.   Some Modified, Natural, Water-Soluble Polymers[a]

Resin	CAS Registry No.	Chemical composition, if not given as resin name	Preparation	1979 Production, 10^3 t	Uses
cationic starch	[9043-45-2]	aminoalkyl starch	starch + chlorethylamine or epoxy alkylamine		textiles, paper retention and dry strength, flocculant
dextran	[9004-54-0]	poly(1,6-α-D-glucopyranose)	enzyme fermentation of sucrose	62.6	blood-plasma extender
hydroxyalkyl starches	[9005-27-0]		starch + ethylene oxide or propylene oxide		paper coating, textile size adhesives, laundry starch
hydroxyethyl and hydroxypropyl cellulose	[9004-62-0] [9004-64-2]		sodium cellulose + ethylene oxide or propylene oxide	24	coatings, petroleum production, cement, paper, textiles, adhesives, asphalt emulsions
methyl cellulose	[9004-67-5]		sodium cellulose + methyl chloride	12	pharmaceuticals, cosmetics, adhesives, concrete, gypsum, polymerizations
sodium carboxymethyl cellulose	[9004-32-4]		sodium cellulose + sodium chloroacetate	28.7[b]	detergents, textiles, foods, drilling muds
xanthan gum	[11138-66-2]	glucose, mannose, glucuronic acid polymer with acetyl pyruvic acid side chains	microbial fermentation	5–6	petroleum production, food, pharmaceuticals

[a] Refs. 1 and 3.
[b] 1975.

of mammals, which is its only source (see Milk and milk products) (5). It is a complex mixture of proteins (qv) containing ca 18 amino acids with different R groups and is isolated from skim milk by precipitation with acid or the enzyme rennin. Principal uses are in wood glues, paper adhesives (qv), and sizing and as a binder in water-based paints (see Paint). Gelatin is the purified form of a partial hydrolysate of collagen, and animal glue is the technical grade (6). Collagen occurs in bone, connective tissue, skin, etc, of animals. Gelatin contains ca 18 amino acids, although half the molecule is made up of glycine, proline, and hydroxyproline. The principal uses are in food, photographic film, and pharmaceuticals (see Photography).

The price of free-market products has always been related inversely to their volume of use and production. Table 3 shows the prices in the United States for natural and modified natural products.

Table 3.   1979 Prices of Natural and Modified, Natural, Water-Soluble Resins in the United States[a]

Resin	CAS Registry No.	$/kg
Natural		
agar		12.67–14.87
casein	[9000-71-9]	1.60–1.75
corn starch		0.17–0.18
guar gum		0.99–1.76
gum arabic		1.58–2.53
gum tragacanth		18.73–74.90
karaya gum		3.30–3.75
locust bean gum		2.31–2.42
pectin	[9000-69-5]	16.96
sodium alginate	[9005-38-3]	4.08–7.71
xanthan gum		6.37–8.82
Modified natural		
cationic starches		
gelatin	[9000-70-8]	2.09–2.42
hydroxyethyl cellulose		3.37–6.17
methyl cellulolose		3.37–4.63
sodium carboxymethyl cellulose		1.68–2.54
sodium carboxymethyl starch	[9063-38-1]	

[a] Ref. 3.

Synthetic

Completely synthetic, water-soluble polymers are prepared by polymerizations, in which chain growth results from breaking of ring structures, opening of double bonds, or condensation reactions involving elimination of a small molecule. Since many monomeric materials are available, a broad range of compositions may be synthesized, including homopolymers, copolymers, and copolymers with monomers whose homopolymers are not water-soluble.

Poly(Vinyl Alcohol).   Poly(vinyl alcohol) [9002-89-5] (PVA) is the largest volume, completely synthetic, water-soluble resin produced in the world (see Vinyl polymers, poly(vinyl alcohol)). The polymer can only be made by hydrolysis of other vinyl polymers, since the theoretical monomer is the enol form of acetaldehyde. Commercially, the term poly(vinyl alcohol) includes all water-soluble resins made from poly-(vinyl acetate).

$$\left[\text{CH}_2\text{CH} \atop \text{OH} \right]_n$$

Properties.   Poly(vinyl alcohol) is a group of products that vary in molecular weight, polymer chain branching, the number of residual acetate groups, etc. The degree of polymerization of commercial poly(vinyl alcohol)s manufactured worldwide is 350–2500 (7). Some properties of the commercial poly(vinyl alcohol) grades offered by one manufacturer are listed in Table 4 (8–9).

Table 4. Properties of Various Grades of Poly(Vinyl Alcohol) [a]

Trade name	Viscosity, mPa·s (= cP) [b]	Hydrolysis, mol % [c]	Solution pH	Volatilities, % max
Elvanol 90-50	13–15	99.0–99.8	5–7	5
Elvanol 71-30	28–32	99.0–99.8	5–7	5
Elvanol HV	55–65	99.0–99.8	5–7	4
Elvanol T66	13–15	98.0–99.8	5–7	5
Elvanol T25	25–31	99.0–99.8	5–7	5
Elvanol 85-50	1100–1500 [d]		3.6–5.1	
Elvanol 85-60	25–57 [e]		5.9–6.6	
Elvanol 85-80	3200–4200 [f]		3.1–4.1	

[a] Refs. 8 and 9.
[b] Viscosity of a 4% aqueous solution at 20°C determined by Hoeppler falling-ball method.
[c] Mole % hydrolysis of acetate groups, dry basis.
[d] Brookfield viscosity of 10% aqueous solution at 25°C; model LVF, 60 rpm, No. 3 spindle.
[e] Brookfield viscosity of 4% aqueous solution at 25°C; model LVF, 60 rpm, No. 1 spindle.
[f] Brookfield viscosity of 10% aqueous solution at 25°C; model LVF, 60 rpm, No. 4 spindle.

Maximum water solubility of PVA occurs at ca 88% hydrolysis; at high hydroxyl content, crystallization takes place and hot water must be used to dissolve the polymer. Water solutions are quite temperature-stable and show no permanent viscosity loss after boiling for several days. The polymer may be salted from solution by addition of small amounts of Na_2SO_4 (8). Viscous gels may be formed by addition of borax and other materials (8).

The dry polymer is quite hygroscopic and absorbs 25–30 wt % water at 100% rh (8). Films of PVA show gradual discoloration, embrittlement, and reduced water solubility upon heating in air above 100°C. However, films show excellent stability to sunlight. Water, glycerol, and other polyhydroxy compounds are good plasticizers (qv) for poly(vinyl alcohol) films.

Manufacture. Vinyl alcohol polymers are prepared by a two-step process (7). The first step is polymerization of vinyl acetate to poly(vinyl acetate), which is converted to PVA by alcoholysis in the presence of an acidic or basic catalyst. The PVA precipitates from solution and is recovered as a powder by filtration and drying.

The properties of the poly(vinyl alcohol) depend to a large extent on the parent poly(vinyl acetate) (7). Azobisisobutyronitrile (AIBN) and a relatively small group of peroxides have been used commercially as initiators for polymerization (10) (see Initiators).

Alkali-catalyzed alcoholysis in methanol is usually used industrially to hydrolyze poly(vinyl acetate). Both batch and the more frequently practiced continuous process are employed. As the alcoholysis proceeds, the reaction mixture becomes a very viscous gel (11).

Economic Aspects. Japan traditionally leads the world in the production of poly(vinyl alcohol) (7). In addition to applications as a water-soluble resin, PVA is extensively used as a fiber (Vinylon) and in packaging film. The fiber is rendered water-insoluble either by heat treatment, ie, crystallization, or by reaction with formaldehyde.

In the United States, Air Products and Chemicals, Inc., E. I. du Pont de Nemours & Co., and Monsanto Company market poly(vinyl alcohol) under the respective trade

names of Vinol, Elvanol, and Gelvatol. 1981 prices of the various PVA grades were
$2.05–$4.14/kg. Total use, excluding the PVA made into poly(vinyl butyral), was
70–75,000 metric tons (3).

Health and Safety Factors. Poly(vinyl alcohol) is a safe and nonhazardous material
when properly handled. Feeding studies with laboratory animals indicate a very low
order of oral toxicity. Technical-grade poly(vinyl alcohol)s are not recommended for
inclusion in any food or other preparation that might be taken internally (8). The resins
are not primary skin irritants and do not produce skin sensitization. Additional information on poly(vinyl alcohol) toxicity and handling is available from the manufacturers.

Uses. Poly(vinyl alcohol) offers a combination of excellent film-forming and
binder characteristics along with insolubility in cold water and organic solvents. In
adhesives, poly(vinyl alcohol) is used in paperboard; corrugated cartons; general-
purpose household adhesives; industrial products for bonding paper, textiles, and
wood; and for laminating solid fiberboard, linerboard, and spiral-wound tubes and
cores (8) (see Packaging, industrial; Textiles).

Poly(vinyl alcohol) is used extensively as a warp size for spun yarns and as a paper
surface size or release coating in paper specialties (8,12). Most poly(vinyl acetate)
emulsions contain PVA as the stabilizer (13). Poly(vinyl alcohol) is used in emulsion
and suspension polymerizations as a colloidal stabilizer (14–15).

Polyacrylamides. Polyacrylamides are a versatile family of synthetic polymers
used worldwide (see Acrylamide polymers). The main constituent in the nonionic,
anionic, and cationic polymers is acrylamide (qv).

Properties. Polyacrylamides are marketed in the form of a dry solid, an aqueous
solution, and a dispersed phase, ie, hydrated polymer particles dispersed in hydrocarbon. The properties of some anionic [9003-06-9], nonionic [9003-05-8], and cationic
[41222-47-3] polyacrylamides are summarized in Table 5 (16). The nonionic homopolymer has an absolute density of 1.302 g/cm³ at 23°C and a glass-transition temperature of 153°C (17–18). It has good thermal stability when compared with other
water-soluble polymers, ie, degradation in the absence of oxygen begins at greater than
210°C (16).

Polyacrylamide (PAM) is infinitely soluble in water although it is limited in
practical values by its viscosity. In general, the rheological behavior of aqueous polyacrylamide solutions is pseudoplastic, ie, the apparent viscosity decreases with increasing shear rate at relatively low shear rates. Also, solutions show a permanent
decrease in viscosity, ie, shear degradation, at higher shear rates (19). The average
molecular weight decreases and its distribution narrows during shear degradation.

Anionic and cationic polyacrylamides are polyelectrolytes (qv) and exhibit the
expected sensitivity to salts and pH. Excess NaCl causes an exponential decrease in
viscosity with increasing salt concentration (20). For anionic polyacrylamide, the

Table 5. Properties of Some Anionic, Nonionic, and Cationic Polyacrylamides[a]

Ionic character	Separan[b] designation	Mol wt	Form	pH	Freezing point, °C	Density, g/cm³
slightly anionic	NP10	medium	white solid	7.0[c]		0.51
anionic	AP273 premium	very high	white solid	10.1[d]		0.76
nonionic	MGL	high	white solid	7.5[c]		0.51
slightly anionic	MG200	very high	white solid	8.2[c]		0.51
anionic	MG500	medium	white solid	10.5[c]		0.62
slightly anionic	PG6	high	white solid	7.0[c]		0.51
cationic	CP-7HS	medium	7% soln	10–11	−1	1.003
anionic	87D	low	20% soln		−1	1.066

[a] Ref. 16.
[b] Trademark of The Dow Chemical Company.
[c] 10% aqueous solution.
[d] 0.5% aqueous solution.

viscosity increases at intermediate and alkaline pH. However, at pH <3.8, the viscosity decreases as the number of anionic groups increases until an insoluble precipitate forms above ca 28 wt % acrylic acid groups in the copolymer (21).

Manufacture. Aqueous solution polymerization with free-radical initiators is the most common commercial method of preparing PAM (22). Acrylamide readily forms high molecular weight polymers with common initiators, eg, peroxides, redox pairs, and azo compounds. The principal method used to prepare dry polymer is thermal sheet drying.

In the mixed-solvent polymerization process, the monomers dissolve in a mixture of water and an organic liquid that is a nonsolvent for the polymer (23). The dispersed-phase polymerization process is an alternative method for producing high molecular weight polymer (24–25). Droplets of an aqueous acrylamide–catalyst mixture suspend in an immiscible organic solvent and polymerize by heating.

Anionic polyacrylamide is made by either *in situ* hydrolysis of acrylamide homopolymer with alkali or copolymerization with acrylic acid (26–27). The copolymerization conditions are similar to those for preparing the homopolymer. Two general processes are employed for preparing cationic polyacrylamides: postreaction of PAM with formaldehyde and dimethylamine (Mannich reaction) and copolymerization (28–29). A variety of comonomers have been employed to impart cationic functionality to polyacrylamide. Some of the more important products are diallyldimethylammonium chloride–acrylamide copolymer [9022-17-7], aminomethylmethacrylate–acrylamide copolymer [25568-01-4], and dimethylaminoethyl methacrylate–acrylamide copolymer [25568-39-2] (30–32).

Economic Aspects. Currently, 15–20 main suppliers in the world produce in excess of 45,000 t/yr of polyacrylamides. Some anionic and cationic acrylamide polymers that are commercially available in the United States are listed in Table 6. 1981 prices in the United States were $3.86–$8.18/kg on a polymer basis (16). Generally, cationics command the highest price ($4.48–$8.18/kg), particularly if they are to be used as paper additives. High molecular weight anionics cost $3.68–$6.48/kg and nonionics (<4% hydrolyzed) are priced similarly to the anionics. The U.S. International Trade Commission reported U.S. production to be 27,000 t (3).

Table 6.   Some Commercial Anionic and Cationic Polyacrylamides Available in the United States

Trade name	Manufacturer	Ion type
Aerofloc 550	American Cyanamid Co.	anionic
Gafloc	GAF Corp.	cationic
Hercofloc 816-823	Hercules, Inc.	anionic
Hydroaid 776	Merck & Co.	cationic
Magnifloc 521C, 523C	American Cyanamid Co.	cationic*
Nalco 633	Nalco Chemical Co.	anionic
Polyfloc 1260	Betz Laboratories, Inc.	cationic
Polyhall 295	Stein Hall and Co.	anionic
Reten 205, 210	Hercules, Inc.	cationic
Separan CP7	The Dow Chemical Co.	cationic
Separan NP10	The Dow Chemical Co.	anionic

Health and Safety Factors.   Polyacrylamides appear to represent little hazard to public health or the environment. The polymers are nonirritating to skin under normal use conditions, have low single-dose oral toxicity, and should present no problem from ingestion under normal conditions of industrial use (33).

Acrylamide monomer is a severe neurotoxin and a cumulative toxicological hazard. Residual monomer in the polymer appears to be the main toxicity concern and is generally controlled to a level considered to be of minimal hazard (34).

Uses.   The largest application for anionic polyacrylamides is as a flocculant in wastewater clarification for municipal sewage, industrial plants, and mining (35,16) (see Water, municipal water treatment). Anionic polymers are also used as mobility-control agents in enhanced oil recovery and as paper dry-strength resins and retention aids (36–38) (see Petroleum, chemicals for enhanced recovery; Papermaking additives). Important uses for cationic polyacrylamides are as paper wet strength and drainage aids and in sewage-sludge dewatering (16,37).

Cationic Resins.   Cationic resins are the commercially important resins of high cationic character. They include two general types of products: polymeric amines and quaternary ammonium polymers (see Quaternary ammonium compounds). The first group is comprised of polyethylenimine [9002-98-6]; high Mannich-substituted polyacrylamides [25765-48-4]; polymers of cationic monomers, especially poly(dimethylaminoethyl methacrylate) [25119-82-8]; and polyalkylene polyamines [26913-06-4]. The principal quaternary ammonium polymers are poly(vinylbenzyltrimethylammonium chloride) [9017-80-5], poly(diallyldimethylammonium chloride) [26062-79-3], poly(glycidyltrimethylammonium chloride) [51838-31-4], and poly(2-hydroxypropyl-1,1-N-dimethylammonium chloride) [39660-17-8].

Properties.   Cationic resins are supplied in the form of aqueous liquids or dry solids (39). There are so many different chemical types of cationic polymers that it is not practical to describe each in detail; the manufacturer's brochures should be consulted for these data. In addition to polymer structure, the most important parameters are molecular weight, percent cationic groups, and type of cationic groups, ie, quaternary, primary, secondary, or tertiary amine.

Molecular weights of various commercial cationic polymers are ca (2×10^4)–10^6 and are usually reported as Brookfield viscosities of dilute aqueous solutions (39). For polyamines, the viscosity varies significantly with pH; the highest viscosities occur under acidic conditions.

Manufacture. Polyethylenimine is prepared by ring opening of aziridine monomer (see Imines, cyclic) (40).

$$n \quad \underset{\underset{H}{\overset{|}{N}}}{\overset{H_2C\text{———}CH_2}{\diagdown \diagup}} \quad \xrightarrow{\text{acid}} \quad +CH_2CH_2NH+_n$$

The polymer is highly branched and contains ca 30% primary, 40% secondary, and 30% tertiary amine groups. Aziridine monomer is extremely toxic, unlike the polymer (41).

Polyalkalene polyamines and quaternized polyamines are important products and are made by a number of different processes (see Table 7) (42–48). Aminoethyl methacrylate and dimethylaminoethyl methacrylate homopolymers and copolymers are made in acidic aqueous solution with conventional free-radical initiators (49–51). Another commercially useful type of amine polymer is the polyacrylamide Mannich products that are made in high conversion (>75% cationic groups).

Poly(diallyldimethylammonium chloride) is prepared by a process that gives an unusual cyclic structure (52–53).

$$n \quad \underset{\underset{CH_3 \quad CH_3}{}}{\overset{CH_2 \diagdown \qquad \diagup CH_2}{\underset{|}{\overset{+}{N}}}} \quad Cl^- \quad \xrightarrow{t\text{-}CH_3(CH_2)_3OOH} \quad \left[\underset{\underset{CH_3 \quad CH_3}{}}{\overset{+}{N}} Cl^- \right]_n$$

The polymerization is carried out in dilute aqueous solution at 50–75°C and yields a water-soluble, high molecular weight resin.

Poly(vinylbenzyltrimethylammonium chloride) can be made either by free-radical polymerization of the monomer or by postreaction of polystyrene with CH_2O and HCl followed by quaternization with trimethylamine (54). Molecular weight can be readily varied. Also poly(2-hydroxypropyl-1,1-*N*-dimethylammonium chloride) is an important product prepared from dimethylamine and 1-chloro-2,3-epoxypropane.

Poly(glycidyltrimethylammonium chloride) can be prepared by postquaterni-

Table 7. Selected Polyalkalene Polyamine Processes

Reactants	Manufacturer	Ref.
NH_3 + $ClCH_2CH_2Cl$ + CH_3Cl; tetraethylenepentamine + epichlorohydrin + CH_3Cl	Nalco Chemical Co.	42
NH_3 + epichlorohydrin	Nalco Chemical Co.	43
polyepichlorohydrin + a trialkylamine	The Dow Chemical Co.	44
polyethylene polyamine + epichlorohydrin + aziridine	The Dow Chemical Co.	45
epichlorohydrin + CH_3NH_2	American Cyanamid Co.	46
alkyl dihalides + polyalkylene polyamines	Nalco Chemical Co.	47
polyalkylene polyamines + $ClCH_2CH_2Cl$ + epichlorohydrin	Calgon Corp. (owned by Merck and Co., Inc.)	48

zation of polyepichlorohydrin with trimethylamine (55). Polycondensation quaternaries are relatively low molecular weight (usually $<5 \times 10^4$) and do not readily give reproducible charge density (56).

Economic Aspects. The trade names and manufacturers of representative polyamines, quaternary ammonium polymers, and polyethylenimine produced in the United States are listed in Table 8. 1981 U.S. prices of cationic water-soluble resins as obtained from the manufacturers were \$4.85–\$7.61/kg for polyethylenimine and \$3.30–\$11.50/kg for the polyamine and poly(quaternary ammonium) resins, depending on type and volume. It has been estimated that the total consumption of polyamine and quaternary ammonium resins in the United States was 9,000–11,000 metric tons (3).

Health and Safety Factors. Cationic polymers are eye irritants, skin irritants, and have low acute oral toxicity. As with any polymer, the toxicology often depends on residual low molecular weight impurities, eg, monomer. Manufacturers' product bulletins should be consulted for specific toxicological data.

Uses. The largest use of polyamines is probably in wastewater and potable-water treatment. Various polymeric amines are employed as flocculants to thicken or concentrate the dilute sludge and to improve dewatering of sludge during vacuum filtrations (16,57). In the paper industry, the polyamines are competitive with cationic polyacrylamides as retention and drainage, ie, dewatering, aids (58). They also are used as wet-strength additives and as flocculants. There are a number of smaller but significant markets in petroleum recovery, mineral processing, and miscellaneous applications, eg, film tiecoats (59–60). Electroconductive coating for paper used in office copying is the largest use for the quaternary polymers (61) (see Reprography). There are some sales of quaternary polymers in water treatment.

Poly(Acrylic Acid) and Derivatives.   *Properties.* Poly(acrylic acid) [9003-01-4], poly(sodium acrylate) [9003-04-7], and poly(ammonium acrylate) [9003-03-6] are supplied in a range of molecular weights by various companies. The trade names and properties of some polyacrylic acids and salts are listed in Table 9 (62–63). Dilute aqueous solutions of poly(acrylic acid)s and poly(methacrylic acid)s have low reduced viscosities. More concentrated solutions do show a marked difference in behavior, eg, a 5% solution of a ca 800,000-mol-wt polyacrylic acid is thixotropic, whereas the poly(methacrylic acid) solution gels with shearing (64).

Another class of products is the acrylic or methacrylic acid emulsion copolymers, which are supplied as liquids or dry solids, eg, Carboset from B. F. Goodrich and the Acrysol ASE series from Rohm and Haas. Some of these products are alkali-soluble and can be used as temporary coatings; other materials are thermosetting (62–63). These polymers have very good adhesion to most surfaces and, in general, give high gloss, water-resistant nonblocking films.

Table 8.   Trade Names and Manufacturers of Representative Polyamine, Polyethylenimine, and Quaternary Ammonium Polymers

Trade name	Chemical type	Manufacturer
Cat-Floc T	quaternary	Merck and Co.
Corcat P-12, P-18	polyethylenimine	Cordova Chemical Co.
Magnifloc 521C	polyamine	American Cyanamid Co.
Nalco 603	polyamine	Nalco Chemical Co.
Purifloc C31	polyamine	The Dow Chemical Co.

Table 9. Properties of Some Commercial Polyacrylic Acid Resins

Trade name[a]	Composition	Solids, wt %	Moisture, wt %	Mol wt	pH	Viscosity at 25°C, mPa·s (= cP)	Sp gr	Density, g/cm³
Acrysol A-1	polyacrylic acid	25		5×10^4	2	320		
Acrysol A-3	polyacrylic acid	25		1.5×10^5	2	300		
Acrysol GS	sodium poly-acrylate	12.5			8.5–9.8	$(1–2) \times 10^4$	1.07	
Acrysol A-5	ammonium polyacrylate	12.5		3×10^5	2	250		
Goodrite K-702	polyacrylic acid	25		24.3×10^4	2–3	500–1500	1.09	
Goodrite K-722	polyacrylic acid	45		10.4×10^4	1.5–2	$(0.5–1.5) \times 10^4$	1.15	
Goodrite K-732	polyacrylic acid	50		5100	2–2.5	140–340	1.18	
Goodrite K-752	polyacrylic acid	65		2100	2–2.5	300–1300	1.23	
Goodrite K-739	sodium poly-acrylate	powder	10–15	6000	5.5–6.5			0.5
Goodrite K-759	sodium poly-acrylate	powder	10–15	2100	5.5–6.5			0.5

[a] Acrysol produced by Rohm and Haas Co.; Goodrite produced by B. F. Goodrich Co.

A third type of commercial product is the thickeners, which are lightly cross-linked copolymers that swell many times their original volume but do not truly dissolve. This class of products is typified by the well-known Carbopol resins; some of their properties are listed in Table 10 (65). Maximum viscosity is achieved at pH 5–11. Water solutions are pseudoplastic, exhibit high rheological yield values useful in stabilizing suspensions, and are shear-sensitive.

Manufacture. Poly(acrylic acid) and poly(methacrylic acid) can be made by polymerization and copolymerization of the monomers or by hydrolysis of poly(acrylate ester)s or of polyacrylonitrile.

$$\left[CH_2CH \atop \big| \atop CN \right]_n \xrightarrow{\text{NaOH}} \left[CH_2CH \atop \big| \atop \underset{O}{CONa} \right]_n$$

Products made by the latter two processes usually contain unhydrolyzed ester or nitrile groups. Hydrolysis of methacrylic esters is more difficult than for acrylates (64).

Polymerization of acrylic and methacrylic acid monomers is often carried out in ≤25% aqueous solution with conventional free-radical initiators, ie, peroxydisulfate, at 90–100°C. The rates of polymerization are fastest at pH 2–5 (66).

Copolymers can be prepared by emulsion-polymerization techniques; one important family of products made by this technique is the alkali-soluble acrylic emulsions. These products are copolymers of methacrylic acid and the lower acrylate and methacrylate esters, eg, methacrylic acid–butyl acrylate copolymer [25035-82-9] (67–68).

Table 10.   Some Properties of Carbopol Thickeners[a]

Type	Mol wt (approximate)	Special properties	Viscosity[b] mPa·s (= cP)	Viscosity at 25°C[c], mPa·s (= cP)	
				10^2 s^{-1}	10^4 s^{-1}
907	4.5×10^5	exceptional water solubility; lubricity without viscosity		72	18
910	7.5×10^5	good ion tolerance, long rheology; effective at low (ppm) concentrations	2600	910	61
941	12.5×10^5	highly stable emulsions and suspensions at relatively low viscosities even with ionic systems; more efficient than Carbopol 934 and 940 at low-to-moderate concentrations	6200		
934	3×10^6	excellent stability at high viscosity; produces thick formulations, eg, heavy gels, emulsions, and suspensions	3.5×10^4	4200	150
940	4×10^6	excellent thickening efficiency at high viscosities; shortest rheology	5.2×10^4	4600	180

[a] Trademark of B. F. Goodrich Co.
[b] 0.5% Aqueous Carbopol; Brookfield viscosity, 20 rpm, NaOH-neutralized at 25°C.
[c] 1% Aqueous Carbopol; NaOH-neutralized to pH 7, Ferranti Viscometer.

The particulate, swellable thickening agents are prepared by copolymerization of an ethylenically unsaturated acid, an optional second monomer, and the polyallyl ether of sucrose in an organic solvent with an oil-soluble peroxide initiator (69). Other monomers and cross-linking agents can be used (70). Thickening agents that perform in the presence of inorganic salts can be made by cross-linking a copolymer of a carboxylic acid monomer with a C_{10-30} alkyl acrylate ester (71).

Economic Aspects.   The principal suppliers of poly(acrylic acid) derivatives are B. F. Goodrich Co., American Cyanamid Co., Rohm and Haas Co., W. R. Grace, and Borden Chemical Company. Prices in 1981 for poly(acrylic acid) and its salts were $2.96–$3.62/kg, and the 1981 prices for the acrylic emulsion copolymers were $2.65–3.53/kg. Thickener grades are substantially more expensive at $8.16–$10.00/kg. Annual U.S. consumption of poly(acrylic acid)s is ca 15,000–18,000 t (3). The polyacrylamide volume could be 5,000–10,000 t higher and the poly(acrylic acid) sales correspondingly lower if a different polymer classification is used. Commercially, the acrylic acid products are more significant than the methacrylic acid products.

Health and Safety Factors.   All rabbits survived dermal application of Carbopol 910 at doses of 3 g/kg with no unusual behavior or systemic reactions (65). Transient eye irritation in rabbits was produced by instillation of Carbopol resin; thus, the resin should be considered an eye irritant. The rat inhalation LC_{50} for a 4-h exposure to Carbopol 910 dust was 1.7 mg/L. No unusual reactions during exposure or for the 14-d observation period were noted.

The rat acute oral LD_{50} is 10.25 g/kg. At low doses, hypoactivity and diarrhea were observed and, at high doses, additional symptoms were observed. Gross pathology of the rats that died revealed pale livers, kidneys, and spleens.

Uses. Polyacrylic acid derivatives are used in the textile, petroleum, paper, cosmetic, water-treatment, paint, and foundry industries. Sodium polyacrylate is used as a thickener in latexes, as a textile size, and to retard deposition of scale in boilers (72–74). Soluble acrylic emulsions are employed as binders and coatings in paper (75). High molecular weight sodium polyacrylate in combination with starch is useful as a flocculant for bauxite-ore clarification (76). There are many applications for cross-linked polyacrylic acid, eg, cosmetics (qv) and carpet-back coatings (65,62).

Poly(Ethylene Oxide). Poly(ethylene oxide) [25322-68-3] (PEO) refers to the high molecular weight [(1–50) $\times 10^5$] polyethers as opposed to the low molecular weight (200–15,000) products, which are customarily called polyglycols. Poly(ethylene oxide) has been commercially available since 1959 from Union Carbide Corporation, the sole producer in the United States, under the trade name Polyox (see Polyethers).

Properties. Some of the properties of commercially available Polyox resins are summarized in Table 11 (77). The polymers designated WSR-N have a narrower molecular weight distribution than normal. The solid polymer is highly crystalline (mp 66°C) and is a tough, ductile thermoplastic with a glass-transition temperature of −50°C. It is soluble in many polar organic solvents and in water. The apparent viscosity is a function of shear rate. Poly(ethylene oxide), like other high molecular weight linear polymers, is shear-degradable.

Polyox resins exhibit an inverse solubility relationship with temperature. The cloud point depends on the concentration, molecular weight, and the presence of other materials in the water phase but is typically 97–99°C. Inorganic salts have a marked effect on cloud point, eg, the cloud point of 0.5 wt % PEO is 98°C, 88°C, and 48°C, respectively, in water, 0.2 M KF, and 0.2 M Na_3PO_4.

Manufacture. Poly(ethylene oxide) is made by polymerization of ethylene oxide (qv) monomer with a heterogeneous catalyst at 70–110°C.

$$H_2C\!-\!\!-\!\!CH_2 \xrightarrow[\text{solvent}]{\text{catalyst}} -\!(CH_2CH_2O)_n\!-$$

The polymerization process includes the use of a hydrocarbon solvent, in which the monomer is soluble but the polymer is insoluble. Effective catalysts include Fe_2O_3,

Table 11. Some Properties of Commercial Polyox Resins [a]

Grade	Mol wt (approximate)	Viscosity (5% soln at 25°C), mPa·s (= cP)	Viscosity (5% soln at 80°C), mPa·s (= cP)
WSR N-10	10×10^5	12–38	5.4–11
WSR N-80	2×10^5	81–105	16–26
WSR N-750	3×10^5	600–1000	170–270
WSR N-3000	4×10^5	2250–3350	650–900
WSR-205	6×10^5	4500–8800	
WSR-1105	9×10^5	8800–17600	
WSR-301	4×10^6	1500–4500[b]	330–750[b]
coagulant	5×10^6	4500–6500[b]	1500–1850[b]

[a] A trademark of Union Carbide Corp.
[b] 1% Aqueous solution.

$Zn(C_2H_5)_2$, aluminum alkyls, $BaCO_3$, $SrCO_3$, etc (78). Careful control of purity, water content, particle-size distribution, etc, is required to control polymer molecular weight (79).

Economic Aspects. In 1981, the prices of high molecular weight poly(ethylene oxide) were $7.28–$9.15/kg, depending on type and volume. It has been estimated that 1981 sales were 2,000–3,000 t (3).

Health and Safety Factors. High molecular weight poly(ethylene oxide) products are of apparent low oral toxicity (80). A concentration of 5 wt % in the diet of rats for 2 yr caused no detectable harm. Poly(ethylene oxide) is believed to be nonirritating to human skin and to have a low potential for skin sensitization; eye injury is also believed to be slight. A study in humans has indicated that Polyox resin in 3% aqueous solution exerts no primary cutaneous irritant or allergenic action.

Uses. Poly(ethylene oxide) is commercially used in adhesives, especially in applications where a high degree of wet tack and nontacky dry conditions are required (80). It is also used as a thickener in cosmetic formulations and industrial cleaning solutions. Very high molecular weight product is employed in the mining industry as a flocculant for clarification of wastewater.

Planting of seeds uniformly mounted on strips of PEO film accelerates germination and, most importantly, eliminates the need for handthinning crops. The well-known drag-reduction property of PEO in turbulent flow is utilized in fire fighting (81).

Poly(N-Vinyl-2-Pyrrolidinone). Poly(N-vinyl-2-pyrrolidinone) [9003-39-8] (PVP) is a nonionic polymer that is probably best known for its unusual complexing ability and physiological inertness. PVP was first developed in Germany in the 1930s and was widely employed by the Germans as a blood-plasma extender during World War II (see Vinyl polymers, N-vinyl polymers).

$$\left[\!\!\begin{array}{c} CH_2CH \\ | \\ N \diagdown_{\!O} \end{array}\!\!\right]_n$$

Properties. The homopolymer of N-vinyl-2-pyrrolidinone (see Acetylene-derived chemicals) is readily soluble in water and many organic solvents (82). The solid polymer is hygroscopic and quite stable when stored under ordinary conditions. The molecular weights of commercial products are $2,500-7 \times 10^5$. The properties of the homopolymers supplied by GAF Corporation and BASF are listed in Table 12. The viscosity of aqueous solutions is low and shows little change at pH 1–10. Some viscosity increase occurs in high concentrations of HCl, but the polymer precipitates in strong alkali.

Vinyl acetate–N-vinyl-2-pyrrolidinone copolymers [25086-89-9] (VA–VP) are the most important products commercially. The VA:VP ratio varies from 70:30 to 30:70. The GAF and BASF trademarks for these compositions are listed in Table 13 (83–84).

Manufacture. N-Vinyl-2-pyrrolidinone polymerizes in aqueous solution with ammonia–H_2O_2 or azobisisobutyronitrile catalysts (85–86). Polymerizations are best carried out under neutral or basic conditions at 50–60°C. The copolymers are prepared in alcohol solution and usually with an azo catalyst (87).

Table 12.   Commercial Grades of Poly(N-2-Vinyl Pyrrolidinone)

Trade name GAF	BASF	Form	K value[a]	Water, wt %	Residual monomer, wt %	Mol wt (av)
	Kollidon 12PF	powder	11–14	<5	<0.2	2500
PVP K-15		powder	12–18	<5	<1.0	8000
Plasdone C-15	Kollidon 17	powder	16–18	<5	<0.2	$(1–1.1) \times 10^4$
Plasdone K29/32	Kollidon 30	powder	26–35	<5	<0.2	$(3.5–4) \times 10^4$
Plasdone K25	Kollidon 25	powder	24–27	<5	<0.2	$(2.5–3.2) \times 10^4$
PVP K-60		solution	50–62	55	<1.0	2.16×10^5
PVP K-90		powder	80–100	<5	<1.0	6.3×10^5
	Kollidon 90	powder	85–95	<5	<0.2	6.3×10^5
Polyclar AT		powder	cross-linked	<5		
	Divergan	powder	cross-linked	<5		

[a] The Fikentscher K value corresponds to the inherent viscosity and is closely related to the average molecular weight (88).

Table 13.   Commercial N-Vinyl-2-Pyrrolidinone–Vinyl Acetate Copolymers

Trade name GAF	BASF	Composition Vinylpyrrolidinone, wt %	Vinyl acetate, wt %	Solvent
PVP/VA E335	Luviskol VA37E	30	70	ethanol
PVP/VA E535	Luviskol VA55E	50	50	ethanol
PVP/VA E635		60	40	ethanol
PVP/VA E735	Luviskol VA73I	70	30	ethanol
PVP/VA I335	Luviskol VA37I	30	70	2-propanol
PVP/VA I535	Luviskol VA55I	50	50	2-propanol
PVP/VA I735	Luviskol VA73I	70	30	2-propanol
PVP/VA S630	Luviskol VA64 Powder	60	40	

Economic Aspects.   GAF Corporation is the only producer of poly(N-vinyl-2-pyrrolidinone) in the United States, although BASF sells resin imported from the FRG. The volume of PVP produced in the United States has been estimated to be 2,000–4,000 t (3). 1981 prices were \$3.20–\$3.50/kg for vinyl pyrrolidinone–vinyl acetate copolymers and \$10.00–\$16.35/kg for beverage and pharmaceutical grades.

Health and Safety Factors.   Poly(N-vinyl-2-pyrrolidinone) is essentially an inert material (82). The acute toxicity is extremely low with rats and guinea pigs, which tolerate 100 g/kg administered orally. If PVP is injected intravenously, the LD_{50} is 12–15 g/kg. The polymer is also nonhazardous when inhaled and is not a primary irritant, skin-fatiguing material, or sensitizer.

Long-term chronic and radiotracer studies conducted by independent laboratories have demonstrated that over 99% of the PVP ingested is excreted unchanged (89). Carcinogenic properties are not inherent in high polymers, and PVP may even have a hindering effect on the development of spontaneous tumors (90–91).

Uses.   Important uses for PVP are in cosmetics and toiletries, especially hair sprays (92). The preferred polymer for the latter application is a VA-VP copolymer, which has good film-forming ability, excellent adhesion to hair, and is less hygroscopic than PVP homopolymer.

Another use for PVP is in pharmaceuticals. The purified resin is employed as a tablet binder and a coating as well as in injectible preparations of antibiotics, hormones, analgesics, etc (93–94).

In disinfectants (qv), the ability of PVP to form a stable complex with iodine has led to the marketing of PVP–iodine. This product retains the germicidal properties of I_2 and drastically reduces iodine's toxicity to mammals (95). The complexing ability of PVP with tannins has been used to clarify beers and wines (96) (see Beer; Wine).

Vinyl Ether Polymers.   The homopolymers of alkyl vinyl ethers exhibit limited water solubility. The important products are alkyl vinyl ether–maleic anhydride copolymers, especially the methyl vinyl ether resin [9011-16-9]:

$$\left[\begin{array}{c} CH_2CHOCH_3 \\ \\ O{\diagdown}{}_O{\diagup}O \end{array}\right]_n$$

Properties.   Methyl vinyl ether–maleic anhydride copolymer (PVM–MA) is sold under the trade name Gantrez AN (97). The copolymer is a white fluffy powder and has a softening point of 200–225°C and a bulk density of 0.32 g/cm^3. The PVM–MA resin dissolves slowly in water at RT with opening of the anhydride bond to give the maleic acid form. Dissolution is much more rapid in the presence of bases, and the resultant solution viscosity is considerably higher. The aqueous solution viscosities of three molecular weight grades of PVM–MA copolymer as the free acid and as the monosodium salt are listed in Table 14. The viscosities of aqueous solutions decrease on storage, but the degradation can be retarded with a water-soluble antioxidant (see Antioxidants and antiozonants).

Alcoholysis of PVM–MA copolymer with lower molecular weight alcohols gives the half esters, which are soluble in aqueous NaOH and alkanolamines. These materials are important commercial products used in hair sprays. Gantrez ES-335 and ES-225 are the monoisopropyl [31307-95-6] and monoethyl [25087-06-3] esters of poly(methyl vinyl ether-co-maleic acid), respectively.

Manufacture.   Polymerization of methyl vinyl ether and maleic anhydride is carried out in benzene by free-radical polymerization, and an alternating copolymer forms (98). The polymer precipitates from solution during the polymerization and is recovered by filtration. The half esters of PVM–MA are obtained by heating the copolymer with an excess of alcohol (99). Usually, it is convenient to carry out the esterification at the reflux temperature of the alcohol at an amount calculated to give a 10–20 wt % solution of the half ester.

Table 14.   Viscosity of Acid and Monosodium Salts of Gantrez AN at 25°C and 5 wt %

Product	Brookfield viscosity, (mPa·s = cP)	
	Acid form	Monosodium salt
Gantrez AN-119	ca 15	ca 50
Gantrez AN-139	ca 50	ca 600
Gantrez AN-169	ca 250	ca 15,000

Economic Aspects. GAF Corporation is the only producer of vinyl ether polymers in the United States. The products are produced in Europe by BASF and Imperial Chemical Industries. Sales of methyl vinyl ether–maleic anhydride copolymers are ca 2000 t/yr with 1981 prices being $5.49–$9.37/kg (3).

Health and Safety Factors. The results of acute oral toxicity tests indicate that Gantrez AN is relatively nonhazardous (LD_{50} = 8–9 g/kg) (97). Neither the dry powder nor its aqueous solutions appears to be a primary irritant or sensitizer. The use of a dust mask is recommended, however, as is common with all powdered materials. Humans given patch tests showed no skin reactions after two applications of Gantrez AN 139.

Uses. The dominant market for PVM–MA resins is as a film former in hair sprays (100). The preferred products for this use are the monoethyl and monobutyl esters, which impart high gloss, good adhesion, and resistance to high humidity. Also, PVM–MA copolymers are used in cosmetic lotions and creams, as a protective colloid in suspension polymerization, and as a binder in adhesives and textile finishes (101–104).

Styrene–Maleic Anhydride Copolymers. Styrene–maleic anhydride resins [9011-13-6] (SMA) are made in two molecular weight ranges by two different companies. The molecular weights of SMA (trademark of Arco Chemical Co.) products are 1600–2500, and those of Scripset resins made by Monsanto Company are 10,000–50,000 (105–106) (see Styrene plastics).

$$m = 1\text{--}3$$

Properties. Some of the properties of commercially available styrene–maleic anhydride copolymers are summarized in Table 15. The SMA resins are usually used as the aqueous ammonium salt [29299-70-5], which is formed by heating the base anhydride resin in NH_4OH solution. SMA 1440, 17352, and 2625 are partial esters of the lower alcohols, eg, butyl ester. The viscosities of the ester forms are higher than that of the base polymer.

In general, styrene–maleic anhydride resins have a number of properties that are useful in a variety of applications. Some of these are pale color, high gloss, excellent leveling and low solution viscosity. Although readily soluble in water as the NH_4^+ salts, dried films are quite water-insensitive.

Manufacture. The low molecular weight styrene–maleic anhydride copolymers are made by a precipitation process at high temperatures (250–276°C) in high boiling, highly methylated aromatic hydrocarbons, eg, 2-propylbenzene or 4-isopropyl-1-methylbenzene, which chain-terminate the polymer (107). After polymerization, the solvent is stripped under vacuum, and the molten polymer is removed from the reactor and is cooled and ground.

Half-esters of the low molecular weight styrene–maleic anhydride copolymer are prepared by heating the copolymer with the appropriate stoichiometric quantities

Table 15. Some Properties of Commercial Styrene–Maleic Anhydride Copolymers

Trade name	Styrene:maleic anhydride	Mol wt, number average	Melting range, °C	Chemical composition
SMA				
1000	1:1	1600	150–170	anhydride form
2000	2:1	1700	140–160	anhydride form
3000	3:1	1900	115–130	anhydride form
1440	1:1	2500	55–75	partial ester of SMA 1000
17352	1:1	1700	160–170	partial ester of SMA 1000
2625	2:1	1900	135–150	partial ester of SMA 2000
Scripset				
520	1:1	5×10^4		anhydride form
540	>1:1	2×10^4		partial ester
550	>1:1	10^4		partial ester
808	1:1	5×10^4		half-amide partial ammonium salt

of alcohol (108). The partial ester is insoluble in water but is readily soluble in aqueous NH_4OH. The higher molecular weight copolymers may be manufactured by mass polymerization, solvent and nonsolvent polymerization, or both.

Economic Aspects. Manufacturers' 1981 prices for styrene–maleic anhydride resins were $2.18–$3.53/kg.

Health and Safety Factors. In general, the toxicity of SMA resins is very low (105–106). The resins have been tested for acute oral and inhalation toxicity. Breathing of the powdered resin dust should be avoided. The resins are acceptable as adhesives, components of paper and paperboard in contact with dry food, and components of paper and paperboard in contact with aqueous and fatty foods (109–111).

Uses. A large application for the low molecular weight partial esters (SMA 17352 and 2625) is as a leveling agent in floor-polish emulsions of the acrylic or styrene–acrylic type. Scripset 550 also is used in this application (see Polishes). Another important use for the low molecular weight resins is as an embrittlement additive in rug shampoos; the unesterified polymer, eg, SMA 100, is preferred for this application (105). Other applications include dispersants and coatings.

The largest market for the higher molecular weight SMA copolymers is probably as a pigment dispersant, especially in vinyl acetate latex paints (106). Other uses include textile and paper sizes, a protective colloid in polymerizations, and an adhesion promotor.

Ethylene–Maleic Anhydride Copolymer. Properties. Ethylene–maleic anhydride copolymers [9006-26-2] are available in a range of molecular weights in both linear and cross-linked varieties (112). Typical physical properties are listed in Table 16. EMA (trademark of Monsanto Company) linear resins are soluble in hot and cold water when hydrolyzed. Cross-linked grades form colloidal gels on hydrolysis. The viscosities of these solutions vary over a wide range and they exhibit pseudoplasticity.

Manufacture. EMA polymers are prepared by free-radical polymerization in toluene or xylene under pressures of 6.2–8.6 MPa (900–1250 psi) at 80–150°C (113). The product precipitates from solution and is recovered as a white solid after filtration

Table 16.   Typical Physical Properties of Ethylene–Maleic Anhydride Resins

Property	Value
softening point, °C	170
mp, °C	235
decomposition temperature, °C	247
true density, g/cm^3	1.54
bulk density, g/cm^3	0.37
pH (1.0 wt % soln)	2.3
appearance[a]	fine, white, free-flowing powder
dichloroethane as Cl, wt %	2.0 max
screen size[a] (USSS), %	
1.4 mm (14 mesh)	0.1 max[b]
0.84 mm (20 mesh)	5.0 max

	Linear grades[a]			Cross-linked grades[a]	
	EMA 1103	EMA 21	EMA 31	EMA 61	EMA 91
mol wt (estd av)	8×10^3	25×10^3	1×10^5	unmeasurable	
specific viscosity	0.28–0.38	0.5–0.7	0.9–1.2		
(1 wt % in DMF)					
gel viscosity, mPa·s (= cP)[c]				$(1.0–2.2) \times 10^4$	$(1.6–2.1) \times 10^5$
moisture, wt %	1.5 max	1.5 max	1.5 max		

[a] Monsanto specification.

[b] Except for Monsanto's EMA 31, 0.5% max.

[c] Gel viscosity is measured with a Brookfield Model RVT Viscometer at 25°C with a No. 6 spindle at 5 rpm in a 1 wt % soln adjusted to pH 9.0 with ammonia.

and drying. Low molecular weight products are prepared with higher concentrations of benzoyl peroxide and an active chain-transfer agent (114). The cross-linked polymers are terpolymers with dienes, eg, 1,5-hexadiene; the extremely high molecular weight polymers are useful as thickeners (115).

Economic Aspects.   Monsanto is the sole producer of EMA resins in the United States. The 1981 prices for the linear grades were $7.27–$8.49/kg and $5.18–$7.27/kg for the cross-linked grades.

Health and Safety Factors.   The EMA resins are considered to be practically nontoxic to slightly toxic, depending on the specific EMA, by ingestion in single doses, and they are practically nontoxic by single dermal applications (112). EMA resins placed into the conjunctival sac of the rabbit eye are nonirritating to mildly irritating, depending on the EMA. No irritation resulted after the resins were held in continuous 24-h contact with abraded rabbit skin. For eye protection, workers should wear chemical safety goggles where eye contact is likely. Protective gloves should be worn to protect hands. NIOSH-approved equipment should be used when airborne exposure is a problem or when airborne exposure limits are exceeded. Sufficient ventilation helps to minimize exposure; local exhaust ventilation is preferred.

Uses.   The dominant application of EMA copolymers is as a thickener in textile-printing pastes. The high molecular weight, cross-linked products primarily are employed. Other applications are in liquid-shampoo thickeners, ceramic-clay binders, and polymerization aids (116).

Hydrophilic Gels

Hydrophilic gels are lightly cross-linked hydrophilic polymers that undergo substantial volume increase upon the absorption of water. These do not include soft contact lenses, which are heavily cross-linked to provide dimensional stability (see Contact lenses).

Particulate thickeners, which absorb water and swell to many times their dry volume, have been industrially available for a number of years. There is an area of emerging commercialization with large economic potential for absorbent polymers in disposable diapers, sanitary napkins, surgical pads, bath mats, etc (117). The required properties are demanding for some of these applications. For example, disposable diapers must absorb 30–60 g of urine per gram of polymer without feeling wet. The desired rate of absorption is very fast, ie, 5–7 s. The products may be particulate but are often fabricated into fibers or film laminated between cellulose, foams, etc. Many cross-linked polymers have been proposed for this use, eg, carboxymethyl cellulose, hydrolyzed polyacrylonitrile-grafted starch (see Cyanoethylation), polyvinylpyrrolidinone, copolymers of maleic and acrylic acids, and partially hydrolyzed polyacrylamide (118–122).

Various products are offered on a commercial and semicommercial scale. One of these materials is CLD, a cross-linked carboxymethyl cellulose from the Buckeye Cellulose Corp. Acrylic-based hydrophilic gels are offered by National Starch and Chemical Company, The Dow Chemical Company, and B. F. Goodrich. Hercules and the Swedish firm Svenska market a fiber (Aqualon) and a sheet (Aquasorb) form of hydrophilic polymer. The Super Slurpers of the U.S. Department of Agriculture also have received considerable publicity (117). 1981 prices for these products were $5.50–$11.00/kg.

Polymeric Surface-Active Agents

Polymeric surface-active agents make up a special class of water-soluble polymers that is beginning to achieve commercial status in advanced emulsion polymerization and emulsification technology. The oldest members of the class are polysoaps, in which the repeat units have the characteristic structure of a soap, ie, hydrophobic tail–hydrophilic head (123). The original polysoaps were made by alkylating poly(2-vinylpyridine) with fatty alkyl bromides. Poly(2-vinylpyridine) and poly(4-vinylpyridine) are commercially available from Reilly Tar and Chemical Corp. Even when partially alkylated, the polysoaps appear to exist in solution as permanent micelles. Weak-acid anionic polysoaps have also been made (124). Another type of polysoap is the polyionenes, in which both charged and alkyl groups are in the backbone of the polymer chain (125).

Alkyl-sulfide-terminated oligomer surfactants are commercially available under the trade name Polywet (trademark of Uniroyal, Inc.) (126). These products have the following structure (127–128):

$$RS(CH_2CR'Y)_a(CH_2CR''Z)_bH$$

where R = C_6–C_{20} alkyl; R' = H, CH_3, C_2H_5; Y = CN, C_1 – C_8 ester, etc; R'' = H, CH_3, CH_2COOH; Z = COOH, $CONH_2$, CH_2OH etc; a = moles of more hydrophobic units; b = moles of more hydrophilic units; $a + b$ = 6–50. It has been reported that usually

R is dodecyl and the oligomer groups are random copolymers of acrylonitrile and acrylic acid. The molecular weights are 1200–1500.

Polymeric surfactants are surface-active water-soluble polyelectrolytes in which the monomer units are not surface-active (129). The polymeric surfactants are random copolymers of water-insoluble and water-soluble ethylenically unsaturated monomers. The water-soluble monomer can be anionic or cationic (130–131). The molecular weight must be low to achieve surface activity. Sales figures for these products are unavailable. In some cases, the materials are used in-plant and in others the monomers instead of the polymeric surfactant are sold to the user (see Surfactants and detersive systems).

BIBLIOGRAPHY

"Resins, Water-Soluble" in *ECT* 2nd ed., Vol. 17, pp. 391–410, by Frank A. Wagner, B. F. Goodrich Chemical Company.

1. W. Jarowenko in N. M. Bikales, ed., *Encyclopedia of Polymer Science and Technology*, Vol. 12, Interscience Publishers, a division of John Wiley & Sons, Inc., New York, 1970, p. 788.
2. R. L. Whistler in ref. 1, Vol. 11, 1969, p. 396.
3. *Water-Soluble Resins*, *Chemical Economics Handbook*, SRI International, Menlo Park, Calif., 1981.
4. *Gum Brochure*, Meer Corp., North Bergen, N.J., 1980.
5. H. K. Salzberg in ref. 1, Vol. 2, 1965, p. 859.
6. H. H. Young in ref. 1, Vol. 7, 1967, p. 446.
7. M. K. Lindemann in ref. 1, Vol. 14, 1971, p. 149.
8. *Elvanol® 71-30 Bulletin, E29691*, E. I. du Pont de Nemours & Co., Inc., Wilmington, Del., 1979.
9. *Elvanol® Bulletin, E-37651*, E. I. du Pont de Nemours & Co., Inc., Wilmington, Del., 1980.
10. K. Noro in C. A. Finch, ed., *Polyvinyl Alcohol*, John Wiley & Sons, Inc., New York, 1973, p. 67.
11. *Ibid.*, p. 124.
12. C. R. Williams and D. D. Donermeyer, *Am. Dyes. Rep.* **57,** 440 (1968).
13. J. N. Coker, *Ind. Eng. Chem.* **49,** 382 (1957).
14. M. K. Lindemann in G. E. Ham, ed., *Vinyl Polymerizations*, Vol. I, Part I, Marcel Dekker, Inc., New York, 1967, p. 288.
15. F. H. Winslow and W. Matregek, *Ind. Eng. Chem.* **43,** 1108 (1954).
16. *Separan Polymers Settle Process Problems*, and specific product data sheets, 1975–1981, The Dow Chemical Company, Midland, Mich.
17. *Chemistry of Acrylamide*, American Cyanamid Company, Stamford, Conn., 1966.
18. K. Illers, *Kolloid Z.* **190**(1),.16 (1963).
19. A. H. Abdel-Alim and A. E. Hamielec, *J. Appl. Poly. Sci.* **17,** 3769 (1963).
20. G. D. Jones and S. J. Goetz, *J. Poly. Sci.* **25,** 201 (1957).
21. R. E. Friedrich and co-workers, *36th Midland Section Amer. Chem. Soc. Fall Scientific Meeting*, Midland, Mich., 1980.
22. W. M. Thomas in ref. 1, Vol. 1, 1964, p. 177.
23. Neth. Pat. 6,504,429 (Apr. 7, 1965), D. Monagle and W. Shuluk (to Hercules Powder Co.).
24. U.S. Pat. 3,284,393 (Nov. 18, 1966), J. W. Vanderhoff and R. M. Wiley (to The Dow Chemical Co.).
25. U.S. Pat. 3,923,756 (Dec. 2, 1975), J. Svarz (to Nalco Chemical Co.).
26. U.S. Pat. 2,820,777 (Jan. 12, 1958), T. J. Suen and A. M. Schiller (to American Cyanamid Co.).
27. U.S. Pat. 3,022,279 (Feb. 20, 1962), A. C. Proffitt (to The Dow Chemical Co.).
28. U.S. Pat. 3,539,535 (Nov. 10, 1970), R. L. Wisner (to The Dow Chemical Co.).
29. U.S. Pat. 3,979,348 (Sept. 7, 1976), E. G. Ballweber and K. G. Phillips (to Nalco Chemical Co.).
30. W. H. Schuller and co-workers, *J. Chem. Eng. Data* **4**(3), 273 (1959).
31. Brit. Pat. 1,111,105 (Apr. 15, 1967), C. G. Humiston and D. P. Sheetz (to The Dow Chemical Co.).
32. U.S. Pat. 3,336,269 (Aug. 15, 1967), D. J. Monagle and W. P. Shyluk (to Hercules Powder Co.).
33. D. D. McCollister, C. L. Hake, S. E. Sadek, and V. K. Rowe, *Toxicol. Appl. Pharmacol.* **7**(5), 639 (1965).

34. D. D. McCollister, F. Oyen, and V. K. Rowe, *Toxicol. Appl. Pharmacol.* **6**(5), 172 (1964).
35. Brit. Pat. 805,526 (Apr. 1, 1959), D. J. Pye and G. F. Schurz (to The Dow Chemical Co.).
36. U.S. Pat. 3,039,529 (June 19, 1962), K. McKennon (to The Dow Chemical Co.).
37. W. F. Reynolds, *Tappi* **44**(2), 177A (1961).
38. U.S. Pat. 2,972,560 (Feb. 21, 1961), E. K. Stilbert, M. W. Zembal, and L. H. Silvernail (to The Dow Chemical Co.).
39. Manufacturer's Technical Bulletins from American Cyanamid, Calgon Corp., Cordova Chemical Co., The Dow Chemical Co., and Nalco Chemical Co., 1976–1980.
40. U.S. Pat. 2,182,306, (Dec. 5, 1939), H. Ulrich and W. Harz (to I. G. Farben Industrie).
41. Technical Bulletin, Cordova Chemical Co., North Muskegon, Mich., 1980.
42. U.S. Pat. 3,409,547 (Nov. 5, 1968), M. T. Kajani (to Nalco Chemical Co.).
43. U.S. Pat. 3,147,928 (Mar. 23, 1965), G. T. Kekish (to Nalco Chemical Co.).
44. U.S. Pat. 3,320,317 (May 16, 1967), W. A. Rogers and J. E. Woehst (to The Dow Chemical Co.).
45. U.S. Pat. 3,275,588 (Sept. 27, 1966), D. C. Garms (to The Dow Chemical Co.).
46. U.S. Pat. 3,493,502 (Feb. 3, 1970), A. T. Coscia (to American Cyanamid Co.).
47. U.S. Pat. 3,219,578 (Nov. 23, 1965), G. A. Cruickshank and C. E. Johnson (to Nalco Chemical Co.).
48. U.S. Pat. 3,523,892 (Aug. 11, 1970), D. L. Schiegg (to Calgon Corp.).
49. U.S. Pat. 2,677,679 (May 4, 1954), A. L. Barney (to E. I. du Pont de Nemours & Co., Inc.).
50. U.S. Pat. 2,838,397 (June 10, 1958), I. J. Bruntfest and D. B. Fordyce (to Rohm and Haas Co.).
51. U.S. Pat. 3,385,839 (May 28, 1968), H. L. Honig and co-workers (to Farbenfabriken Bayer).
52. U.S. Pat. 3,288,770 (Nov. 29, 1966), G. B. Butler (to Penisular ChemResearch, Inc.).
53. J. E. Lancaster, L. Baccei, and H. P. Panzer, *J. Poly. Sci. Polym. Lett. Ed.* **14**, 549 (1976).
54. U.S. Pat. 2,694,702 (Nov. 16, 1954), G. D. Jones (to The Dow Chemical Co.).
55. U.S. Pat. 3,320,317 (May 16, 1967), W. A. Rogers and J. E. Woehst (to The Dow Chemical Co.).
56. M. F. Hoover, *J. Macromol. Sci. Chem.* **A4**(6), 1327 (1970).
57. U.S. Pat. 3,171,805 (Mar. 2, 1965), T. J. Suen and A. M. Schiller (to American Cyanamid Co.).
58. W. A. Foster in N. M. Bikales, ed., *Water Soluble Polymers*, Plenum Publishing Corp., New York, 1973, p. 3.
59. U.S. Pat. 3,023,162 (Feb. 27, 1962), D. B. Fordyce, F. J. Glavis, and S. Melamed (to Rohm and Haas Co.).
60. U.S. Pat. 2,828,237 (Mar. 25, 1958), C. M. Rosser (to American Viscose).
61. U.S. Pat. 3,011,918 (Dec. 5, 1961), L. H. Silvernail and M. W. Zembal (to The Dow Chemical Co.).
62. *Additive and Specialty Polymer Data*, Bulletin Sp-1, B. F. Goodrich Co., Cleveland, Ohio, 1980.
63. Technical Bulletins on Individual Acrysols, Rohm and Haas Co., Philadelphia, Pa., 1973–1975.
64. M. L. Miller in ref. 1, Vol. 1, 1964, p. 217.
65. *Carbopol Water Soluble Resins*, Bulletin GC-67, B. F. Goodrich Co., Cleveland, Ohio, 1980.
66. G. Blauer, *Trans. Faraday Soc.* **56**, 606 (1960).
67. U.S. Pat. 3,035,004 (May 15, 1962), F. J. Glavis (to Rohm and Haas Co.).
68. D. B. Fordyce, J. Dupré, and W. Toy, *Ind. Eng. Chem.* **51**(2), 115 (1959).
69. U.S. Pat. 2,798,053 (July 2, 1957), H. P. Brown (to B. F. Goodrich).
70. U.S. Pat. 2,923,692 (Feb. 2, 1960), J. F. Ackerman and J. F. Jones (to B. F. Goodrich Co.).
71. U.S. Pat. 3,915,921 (Oct. 28, 1975), R. K. Schlatzer (to B. F. Goodrich Co.).
72. G. L. Brown and B. S. Garrett, *J. Appl. Poly. Sci.* **1**, 287 (1959).
73. U.S. Pat. 2,807,865 (Oct. 1, 1957), F. B. Shippee and M. D. Hurwitz (to Rohm and Haas Co.).
74. *Good-rite® K700 Polyacrylates for Deposit Control in Water Treatment*, Bulletin GC-62, B. F. Goodrich Co., 1980.
75. J. J. Latimer, *Tappi* **51**(11), 142A (1968).
76. U.S. Pat. 3,575,868 (Apr. 20, 1971), T. J. Galvin and F. A. Hughes (to Atlas Chemical Industries Inc.).
77. *Polyox® Water Soluble Resins*, technical bulletin, Union Carbide Corp., New York, 1978.
78. R. A. Miller and C. C. Price, *J. Poly. Sci.* **34**, 161 (1959).
79. F. N. Hill, F. E. Bailey, and J. T. Fitzpatrick, *Ind. Eng. Chem.* **50**(1), 5 (1958).
80. D. B. Braun in R. L. Davidson, ed., *Handbook of Water Soluble Gums and Resins*, McGraw-Hill Book Co., New York, 1980, Chapt. 19.
81. *Ucar Rapid Water Additive*, bulletin, Union Carbide Corp., New York, 1979.
82. GAF Corp., *Technical Bulletin 9642-070*, New York, 1970.
83. *Polyvinylpyrrolidone (PVP)*, bulletin, BASF Aktiengesellschaft, Ludwigshafen, FRG, 1979.

84. A. S. Wood, "Polyvinylpyrrolidone" in A. Standen, ed., *Encyclopedia of Chemical Technology*, 2nd ed., Vol. 21, Interscience Publishers, a division of John Wiley & Sons, Inc., New York, 1970, p. 427.
85. C. E. Schildknecht, *Vinyl and Related Polymers*, John Wiley & Sons, Inc., New York, 1952, p. 662.
86. O. F. Soloman and co-workers, *J. Appl. Poly. Sci.* **12**, 1835 (1968).
87. U.S. Pat. 3,296,231 (Jan. 3, 1967), R. Resz and H. Bartl (to Farbenfabriken Bayer).
88. G. M. Kline, *Mod. Plast.*, 157 (Nov. 1945).
89. L. W. Burnett, *Proc. Sci. Sect. Toilet Goods Assoc.* **38**(1), (1962).
90. J. Lindner, *Trans. Ger. Soc. Path. 44th Meeting*, Munich, FRG, 1960, p. 272.
91. K. Stearn and co-workers, *Proc. Am. Assoc. Cancer Res.* **2**, 150 (1956).
92. P. Lenthen, *Am. Perfume Cos.* **81**, 53 (1963).
93. M. A. Lesser, *Drug Cosmet. Ind.* **75**, 1 (1954).
94. U.S. Pat. 2,820,741 (Jan. 21, 1958), C. J. Endicott, T. A. Prickett, and A. A. Dallavis (to Abbott Laboratories).
95. U.S. Pat. 2,739,922 (Mar. 27, 1956), H. A. Shelanski (to GAF Corp.).
96. *Polyclar® AT Stabilizer in Brewing*, GAF Corp., New York, 1974.
97. GAF Corporation, *Technical Bulletin 9653-023*, New York, 1965.
98. U.S. Pat. 2,782,182 (Feb. 19, 1957), R. M. Verburg (to GAF Corp.).
99. U.S. Pat. 2,047,398 (July 14, 1936), A. Voss and E. Dickhauser (to GAF Corp.).
100. Brit. Pat. 944,439 (Dec. 11, 1963), (to Gillette Co.).
101. U.S. Pat. 2,702,277 (Feb. 15, 1955), P. W. Kinney (to GAF Corp.).
102. U.S. Pat. 2,998,400 (Aug. 29, 1961), D. M. French (to Wyandotte Chemical Co.).
103. U.S. Pat. 2,895,866 (July 21, 1959), W. F. Amon and F. Claudi-Magnussen (to GAF Corp.).
104. U.S. Pat. 2,609,350 (Sept. 2, 1952), C. Spatt (to GAF Corp.).
105. *SMA® Resins*, general bulletin, Arco Chemical Co., New York, 1979.
106. *Scripset® Resins Technical Bulletins SC-1–SC-13*, Monsanto Company, St. Louis, Mo., 1979–1980.
107. U.S. Pat. 3,451,979 (June 24, 1969), I. E. Muskat (to Sinclair Research Inc.).
108. U.S. Pat. 3,342,787 (Sept. 19, 1967), I. E. Muskat (to Sinclair Research Inc.).
109. CFR, Title 21, par. 175.105.
110. CFR, Title 21, par. 176.180.
111. CFR, Title 21, par. 176.170.
112. *Technical Bulletin FP-7*, Monsanto Company, St. Louis, Mo., 1978.
113. U.S. Pat. 2,378,629 (June 19, 1945), W. E. Hanford (to E. I. du Pont de Nemours & Co., Inc.).
114. U.S. Pat. 2,857,365 (Oct. 21, 1958), J. H. Johnson (to Monsanto Co.).
115. U.S. Pat. 3,060,155 (Oct. 23, 1962), R. H. Reinhard (to Monsanto Co.).
116. U.S. Pat. 2,824,862 (Feb. 25, 1958), R. L. Longley and R. H. Martin (to Monsanto Co.).
117. *Chem. Week*, 21 (July 24, 1974).
118. U.S. Pat. 3,589,364 (June 29, 1971), W. L. Dean and G. N. Ferguson (to The Buckeye Cellulose Corp.).
119. U.S. Pat. 3,661,815 (May 9, 1972), T. Smith (to Grain Processing Corp.).
120. U.S. Pat. 3,810,468 (May 14, 1974), B. G. Harper, R. N. Bashaw, and B. L. Atkins (to The Dow Chemical Co.).
121. U.S. Pat. 3,966,679 (June 29, 1976), J. R. Gross (to The Dow Chemical Co.).
122. U.S. Pat. 3,686,024 (Aug. 23, 1972), R. J. Nankee, J. C. Lamphere, and R. A. Nelson (to The Dow Chemical Co.).
123. U. P. Strauss in K. L. Mittal, ed., *Micellization, Solubilization and Microemulsions*, Vol. 2, Plenum Press, Inc., New York, 1977, p. 896.
124. K. Ito, H. Omo, and Y. Yamashita, *J. Coll. Sci.* **19**, 28 (1964).
125. A. J. Sonnessa, W. Cullen, and P. Ander, *Macromolecules* **13**, 196 (1980).
126. Uniroyal Chemical Division of Uniroyal Inc., *Bulletin ASP-4139*, Naugatuck, Conn., 1979.
127. U.S. Pat. 3,498,943 (Mar. 3, 1970), L. E. Dannals (to Uniroyal Inc.).
128. U.S. Pat. 3,646,099 (Feb. 29, 1972), L. E. Dannals (to Uniroyal Inc.).
129. R. A. Wessling and D. M. Pickelman, *paper presented at 54th Colloid and Surface Science Symposium*, June 15–18, 1980, Lehigh University, Bethlehem, Pa.; *J. Dispersion Sci. Technol.* **2**, 281 (1981).
130. U.S. Pat. 3,917,574 (Nov. 4, 1975), D. S. Gibbs and co-workers (to The Dow Chemical Co.).

131. U.S. Pat. 3,965,032 (June 22, 1976), D. S. Gibbs, R. D. Vandell, and R. A. Wessling (to The Dow Chemical Co.).

General References

Refs. 1, 2, 5, 6, 7, 10, 56, 64, 80, 85, and 123 are general references.
M. Glicksman, "Gums" in *Handbook of Food Additives*, 2nd ed., CRC Press, Inc., Cleveland, Ohio, 1972.
R. L. Whistler and J. N. BeMiller, *Industrial Gums*, Academic Press, Inc., New York, 1973.
R. H. Yocum in E. B. Nyquist, ed., *Functional Monomers—Their Preparation, Polymerization and Application*, Vols. 1 and 2, Marcel Dekker, Inc., New York, 1974.

A. S. Teot
Dow Chemical U.S.A.

RESORCINOL.   See Hydroquinone, resorcinol, and catechol.

REVERSE OSMOSIS

In reverse osmosis, solutions of salt or other low molecular weight solutes are contacted with a membrane and subjected to pressure. A solution lower in solute concentration emerges from the other side of the membrane. The term reverse osmosis was coined in connection with seawater processes in which any difference in concentrations of feed and filtrate corresponding to practical desalting resulted in a substantial osmotic pressure difference across the membrane. To reverse the normal osmotic flow from the low to the high concentration side, a pressure difference that is greater than the difference in osmotic pressures of the solutions adjacent to the interfaces of the membrane (or of the salt-rejecting layer of it) is needed. The synonym hyperfiltration stresses continuity with ultrafiltration with the distinction that ultrafiltration refers to filtering coarser particles in the colloidal range rather than to filtering dissolved solutes of low enough molecular weight to affect appreciably the osmotic pressure (see Ultrafiltration).

Membranes that can filter salt have been known for some time. For example, salt removal was observed in the course of ultrafiltration in the 1920s and 1930s and was attributed to ion-exclusion properties of ion exchangers (1–2). Salt rejection by cellophane was also reported (3). These early observations are reviewed in ref. 4.

Reverse osmosis gained the status of a potential industrial separation process in the late 1950s, when a variety of organic films were tested at the University of Florida, Gainesville, with desalination as the motivation (5). This study showed that cellulose acetate had sufficient salt rejection for potential production of potable water from seawater in a single pass. Potable water is by convention 500 ppm max salts. A group at the University of California in Los Angeles also focused on cellulose acetate and developed a procedure to cast membranes not only of high salt rejection but also with fluxes in a range of interest for economical desalination (6–7). Salt rejection was obtained by ion-exchange membranes, but the fluxes through the thick electrodialysis membranes tested were low (4) (see Ion exchange; Water, supply and desalination).

These developments triggered extensive research and development, which has in less than a generation culminated in a substantial industry (see also Hollow-fiber membranes; Membrane technology; Film and sheeting materials; Film deposition techniques; Chemicals from brine).

For orientation, some basic aspects of the reverse osmosis process are shown schematically in Figure 1. The membrane is in contact with the solution to be treated which is circulated under pressure at a solute concentration c_α. The solution emerging from the membrane is of a lower concentration c_ω, and passes unchanged through the porous layer or layers underneath. The initial porous layer may be an integral part of a membrane cast by an appropriate method, eg, asymmetric cellulose acetate cast by the Loeb-Sourirajan procedure (6–7), or a thin salt-rejecting layer on a separate porous material. In either case, other materials that pass product water are needed to support the pressure on the membrane array, unless the asymmetric or homogeneous membrane is in the form of a hollow fiber of small enough diameter to withstand the force.

The rejected salt causes the solution concentration near the α interface to increase above that of the bulk solution, in the case of an unstirred solution, rapidly to the point where c_ω is essentially equal to the concentration of the bulk on the high-pressure side, c_t (8). Stirring is required to reduce this concentration polarization. The case illustrated in Figure 1 is controlled by circulating the pressurized solution at a velocity v_t high enough to produce turbulence.

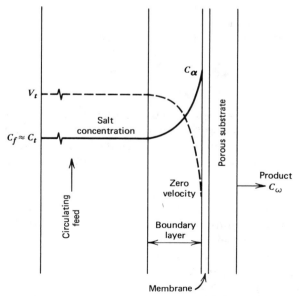

Figure 1. Schematic salt and velocity distribution near membrane in reverse osmosis–turbulent flow.

Transport through Membranes

Removal of large particles from permeating fluid can be attributed to their size relative to the dimensions of passages through films. With NaCl and other solutes in the same size range, such an explanation is less plausible because of the similarity of sizes of solvent and solute molecules, although pore models based on exclusion of solute from surface layers penetrating the membrane have been advanced (9–12). More commonly, the membrane is treated as a homogeneous layer of immobilized solvent extractant. Separations are dependent on differing extraction of components, differing diffusion rates in the membrane, and coupling between flows. Frequently, discussions of membrane transport, reverse osmosis, and similar subjects start with irreversible thermodynamics (13–14) (see, as examples of a large population, refs. 15–17). For present purposes, an approximate and intuitive discussion that can be shown to be consistent with the Onsager equation suffices; a more complete treatment is given in ref. 18 (see Thermodynamics).

The two limiting cases are an uncharged membrane, ie, with no ionizable groups fixed to the matrix, and an idealized ion-exchange membrane in which the distribution of ionic species between solution and membrane phases can be attributed completely to ion-exchange capacity. For a neutral membrane, the flux of a solute component J_s [mol/(m^2·s)] through the membrane is the sum of a diffusion term depending on the concentration gradient of the solute and a term describing the coupling of water (more generally, solvent) flow J_w [kg/(m^2·s)] with solute.

$$J_s = -\mathcal{D}^*(dm^*/dx) + \beta J_w m^* \tag{1}$$

where the solute concentration m^* is in mol/kg solvent, the asterisk denotes the membrane phase and its absence the solution phase, and x is the distance (m) in the direction of flow (the feed-membrane interface α is taken as $x = 0$ and the direction through the membrane as positive). $\mathcal{D}^*$ is the diffusion coefficient consistent with molality (the usual diffusion coefficient $\mathcal{D}_v^*$ (m^2/s) for volumetric concentration units is $(1/\rho_m f_w)\mathcal{D}^*$, where ρ_m is the density of wet membrane (kg/m^3) and f_w the weight fraction of solvent in the membrane), and β is the coupling coefficient for solvent and solute flow,

$$(J_s/J_w)_{dm^*/dx\,=\,0} = \beta m^* \tag{2}$$

The flux of solvent can be approximated as proportional to the difference between the pressures at the two faces of the membrane minus the difference in the osmotic pressure of the solutions in contact. By continuity, when a steady state has been established, J_s/J_w in the membrane is constant and equal to m_ω, the concentration of the effluent solution.

Several restrictions on the system have been implied: constancy of activity coefficients as a function of concentration within the membrane and within the solution phases (although values are different in the two phases, if referred to the same reference state, because distribution coefficients $D_\alpha^* = m_\alpha^*/m_\alpha$ and $D_\omega^* = m_\omega^*/m_\omega$ are usually other than unity); homogeneity of the membrane; incompressibility of the membrane (constant ρ_m); and $\mathcal{D}^*$ independent of x. The assumption of constancy of activity coefficients requires that $D_\alpha^* = D_\omega^*$.

For neutral membranes which have been studied in detail these restrictions seem to be obeyed to a good approximation. Ion-exchange membranes obey other rules. In part, differences arise from differences in salt chemical potential μ_s. In a neutral membrane with NaCl as an example of a single solute, the concentration of cations and anions is the same, whereas in an ion-exchange membrane they differ by the ion-exchange capacity C^*. The concentration of the solute component m^* in this case is that of the co-ion, the ion having the same charge sign as the fixed ion-exchange groups.

Neutral

$$(1/RT)d\mu_s/dx = 2\, d \ln m^* \gamma_\pm^* /dx \tag{3}$$

Ion-exchange

$$(1/RT)d\mu_s/dx = (d \ln m^*/dx) + d \ln (C^* + m^*)/dx + 2\, d \ln \gamma_\pm^* /dx \tag{4}$$

Equation 1 for salt flow is consequently more complicated for an ion-exchange membrane, even if activity coefficients are constant.

Equation 1 and its analogues for ion-exchange membranes may be integrated to give a profile of m^* across the membrane as a function of J_w, $\mathcal{D}^*$, and β. From values of m^* and distribution coefficients D_α^*, the salt rejection

$$\mathbf{R} \equiv (m_\alpha - m_\omega)m_\alpha = 1 - (m_\omega/m_\alpha) \tag{5}$$

as a function of the specified membrane properties may be calculated (17–18). For purposes of illustration, the simple example of a neutral membrane for which $\beta = 0$ is discussed (17). In practice, this is an important case because flow through the neutral membranes, which currently dominate applications, is to a good approximation un-coupled (19–20), and the "solution–diffusion" model (19) is therefore applicable. In-tegration gives a concentration profile in the membrane

$$\frac{m^*}{m_\omega} = D + [J_w l/\mathcal{D}^*][(1 - (x/l))] \tag{6}$$

and from this at $x = 0$ and $x = l$,

$$\mathbf{R} = \frac{1}{1 + (D^*\mathcal{D}^*/J_w l)} \tag{7}$$

From equation 7, as pressure and consequently solvent flux through the mem-brane increase, rejection approaches unity, ie, no salt is in the effluent. More generally, if coupling between solvent and solute flux takes place, the limiting rejection at high flux is given by

$$\mathbf{R}_\infty = 1 - \beta D_\alpha^* \tag{8}$$

The degree to which the limiting rejection is approached depends on the dimensionless group $J_w l/\mathcal{D}^*$. Under the restrictions here, the solvent flux is inversely proportional to membrane thickness. As a result, for a given membrane material, the thinner the membrane or its active layer, the higher is the flux corresponding to a given fraction of the attainable rejection, $\mathbf{R}_\infty$.

Membrane Properties

The preceding simplified discussion implies that product concentrations vary with flux. At equilibrium, for zero pressure and flux, concentrations are the same as both sides of the membrane ($R = 0$). As pressure is exerted and flux begins, concentration on the downstream side is smaller than on the high-pressure side for a salt-rejecting membrane. An asymptotic rejection R_∞ is approached at high flux. The value of R_∞ depends on the degreee of coupling between solvent and solute fluxes and the distribution coefficient between solution and membrane at the high-pressure interface.

Distribution coefficients follow different rules for neutral and for ion-exchange membranes. For neutral films they usually vary relatively little with solution concentration, but depend strongly on the water content of the membrane. In the formalism used here, with concentrations in terms of mol/kg water in both phases and the same reference states, $D^* = m^*/m = \gamma_\pm/\gamma_\pm^* \equiv 1/\Gamma$; values of the activity coefficient ratios for membranes of a given material, eg, cellulose acetate or polyamide, correlate fairly well with values measured of salt in solutions modeling the membrane, eg, in water solutions of glycol or glycerol acetates (21) or of amides (22).

With ion-exchange membranes, distribution coefficients are primarily determined by the exclusion of co-ions. Equality of activities of a 1,1 electrolyte such as NaCl between an aqueous phase and a cation-exchange membrane (symmetrical for an anion-exchanger) is expressed by

$$m_{Na}m_{Cl} = m^2 = m_{Na}^* m_{Cl}^* \Gamma^2 = (C^* + m_{Cl}^*)m_{Cl}^* \Gamma^2 \tag{9}$$

where C^* is the ion-exchange capacity in equivalents per kilogram water in the membrane. The distribution coefficient $D^* = m_{Cl}^*/m$ for this example ($m^* = m_{co\text{-}ion}^*$) is given by the equation

$$D^{*2} + (C^*/m)D^* - (1/\Gamma^2) = 0 \tag{10}$$

and is seen to vary with the concentration of solution in contact with the membrane, even if $1/\Gamma^2$ is constant or unity.

Measurements of salt distribution and water uptakes on ion-exchange beads have shown that, if concentrations are expressed as mol/kg water in both phases, Γ is frequently constant to a good approximation, even when there is an order of magnitude or more variation in the values of $\gamma_\pm$ in the aqueous phase over the concentration range in question. Furthermore, with assignment of the same reference state to the two phases, values of Γ frequently fall within a factor of two of unity (23).

Diffusion coefficients $\mathcal{D}_v^*$ of NaCl in cellulose acetate membranes range from 3 $\times 10^{-6}$ m^2/s for 33.6 % acetyl content to 4 $\times 10^{-15}$ for 43.2% (19). Water content of the membranes also declines with acetyl content. In general, diffusion coefficients in neutral materials are strongly dependent on water content and log $\mathcal{D}_v^*$ is approximately linear with the logarithm of the volume fraction of the membrane material, ie, one minus the volume fraction of water (24–25) with a slope of ca −5.

Diffusion coefficients in ion exchangers decrease with increasing cross-linking, from which a dependence on water content may also be inferred (26). Effective degrees of cross-linking are low in most ion-exchange membranes considered for reverse osmosis, and the salt diffusion coefficients are therefore usually higher than those in neutral membranes.

Water and salt flux through neutral membranes of cellulose acetate, and presumably also polyamide, is essentially uncoupled. In the original results (19) obtained with asymmetric membranes, there appeared to be a small contribution of coupling to salt flux (β = ca 0.05). However, later measurements with ultrathin membranes cast under rigorously controlled conditions (20) attained the theoretical rejections up to 99.8% predicted for the experimental conditions by the solution–diffusion model (β = 0). Leakage of brine into effluent through membrane imperfections results in a rejection-flux pattern similar to that of a homogeneous layer in which the actual flow is coupled.

In discussions of electrokinetic processes and of transport in electrodialysis, it has frequently seemed a plausible approximation that the coupling coefficients for individual ions in ion exchangers are unity (27). If this assignment is made, and if the diffusion coefficients of the ions are equal, the coupling coefficient of the salt component in reverse osmosis should approach two as the solution concentration and co-ion concentration decrease (18,28). However, extensive studies of electrokinetic properties of an organic ion-exchange membrane, Zeo-Karb 315 (29), indicate that for NaBr, the value of β_{NaBr} is considerably closer to one than to two (30).

Transport parameter values typical of the two classes are given in Table 1, whereas the rejection dependence on flux is illustrated in Figure 2. Highest rejection $\mathbf{R}_\infty$ is realized at lower flux with the neutral than with the ion-exchange class primarily because the salt diffusion coefficient is much lower in the neutral class, and the parameter $J_\text{w}/l\mathcal{D}^*$ is consequently larger for a given flux. The curves are computed from the equations appropriate to the membrane in question, with transport parameters selected to fit the results (18,31–32). The curve for cellulose acetate includes a small contribution of coupling to compensate for presumed leakage of the membranes. For both classes, the values of parameter groups appear reasonable.

Rejections as a function of concentration are compared in Figure 3. Those for cellulose acetate are not exactly comparable with one another nor with those for the ion-exchange membrane, because the effect of concentration polarization was not eliminated and the membranes were prepared by different methods. They serve, however, to illustrate that rejection is relatively independent of salt concentration with this type of membrane, which is typical of the neutral class, and are in general higher for moderate-to-high salt concentrations than those attained by ion exchangers. The curves ($1 - \mathbf{R}_\infty = D_\alpha^*$) for the hydrous oxide, an anion exchanger at this pH (33),

Table 1. Equilibrium and Transport Properties: Comparison of Typical Neutral Membranes with Typical Ion Exchangers

Property	Neutral (cellulose acetate)	Ion exchange
distribution coefficient, $D^* = m^*/m$	ca 0.2^a, $\Gamma > 1$	^{b} Γ ca 1
diffusion coefficient, $\mathcal{D}_v^*$ solute in membrane (m²/s)	ca 10^{-7}	ca 5×10^{-4}
coupling, β	ca 0	ca 1
water permeability m/(s·100 kPa) [ca m/(s·atm)] for 1 μm thickness	ca 2×10^{-8}	ca $(1–50) \times 10^{-6}$

^{a} Independent of solution concentration.

^{b} Dependent on solution concentration and ion-exchange capacity.

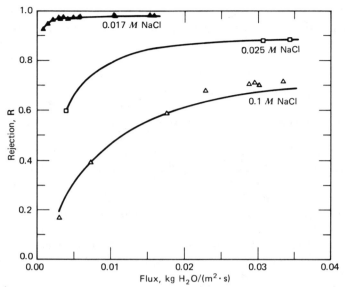

Figure 2. Salt rejection as a function of flux. ▲, Neutral membrane (cellulose acetate), pressure 0.5–10.6 MPa (5–105 atm) (19); curve computed for $\beta = 0.046$, $D^* = 0.22$ $(l/\mathcal{D}^*) = 425$ (m²·s)/kg (18). □△, Ion-exchange membrane (hydrous zirconium(IV) oxide), pressure 0.7–6.2 MPa (7–61 atm) (31); curve computed for $C^* = 0.45$ equiv/kg H_2O, $l/\mathcal{D}^* = 29.2$ (m²·s)/kg, $\beta = 1$ (32).

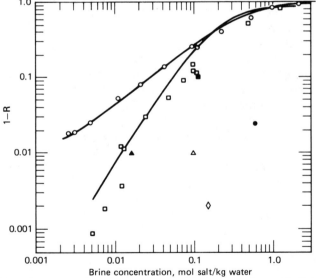

Figure 3. Effect of concentration on salt rejection. Neutral membranes (cellulose acetate), NaCl solution: ▲, asymmetric (19); △, homogeneous (5); ◇, thin-layer (20); ●, asymmetric (7). Ion-exchange membrane (hydrous Zr(IV) oxide) (31); □, $MgCl_2$; ○, NaCl; ■, $MgCl_2$ after NaCl; curve, computed from ion-exclusion model, $\beta = 1$, $C^* = 0.45$ mol/kg H_2O.

were computed by equation 10 (or its analogue for $MgCl_2$), with activity coefficient ratios estimated from their dependence on ionic strength in aqueous solutions (31).

Salt rejection is reasonably well described by current models, once the transport properties of the membrane material and its configuration are known. The equilibrium distribution properties are analogous to those of solutions, and outstanding questions are targets of the generations-old efforts to understand the chemistry of water–organic solute–electrolyte systems. The values of kinetic parameters, particularly coupling, are less predictable as yet.

Concentration Polarization and Fouling

In Figure 1, it is shown that concentration at the brine–membrane interface is higher than the average concentrations on the high pressure side because of build-up of rejected salt at the interface. A velocity profile in a boundary layer frequently assumed in analysis of systems with turbulent flow was indicated. Concentration polarization penalizes performance by decreasing product quality because only a fraction of salt in contact with the membrane at the interface is rejected, and by lowering the concentration of slightly soluble components that can be tolerated without fouling by precipitation on the membrane.

For the permeabilities of present conventional membranes in systems of simple flow geometry, theory describes salt concentration polarization adequately. Fouling, especially by high-molecular-weight species, is more complex and unpredictable and frequently involves slow or irreversible reactions and perhaps specific interactions with the membrane surface.

Equations for salt concentration polarization have been presented in ref. 34 both for laminar and turbulent flow in tubular channels, and for laminar flow in a two-dimensional channel for membranes completely rejecting salt. These analyses have been extended to membranes of less-than-complete rejection (35–36) and to various configurations (18). To illustrate the magnitudes of concentration polarization effects that may be encountered even with turbulent flow, some results obtained with a dynamically formed hydrous zirconium(IV) oxide membrane are summarized in Figure 4 (37). Plotted are values of a function of the observed rejection

$$\mathbf{R}_{\text{obs}} \equiv 1 - (c_\omega/c_t) \tag{11}$$

vs the permeation rate v^*, in cm/s, divided by the circulation velocity v past the membrane in $(\text{cm/s})^{0.75}$. Concentrations are in volume units, similar to molalities here, and c_t, the concentration in the turbulent core of the solution, is essentially the same as feed concentration in these experiments. It can be seen that values of $\mathbf{R}_{\text{obs}}$ different by 20% can occur by changing only the circulation velocity from ca 1–7 m/s. The function graphed is derived (37) from equations in refs. 34–35,

$$\ln\left(\frac{1 - \mathbf{R}_{\text{obs}}}{\mathbf{R}_{\text{obs}}}\right) = k\left[\left(\frac{d}{\eta_k}\right)^{1/4}\left(\frac{\eta_k}{\mathcal{D}_v}\right)^{2/3}\right]\frac{v^*}{v^{0.75}} + \ln\left(\frac{1 - \mathbf{R}}{\mathbf{R}}\right) \tag{12}$$

where d is the inside diameter of the carbon tube on which the membrane was formed, η_k is the kinematic viscosity (cm^2/s), $\mathcal{D}_v$ is the diffusion coefficient of $MgCl_2$ in water (cm^2/s), and k is a semiempirical constant arising from expressions in ref. 38; by analogy with heat-transfer results, the expected value of k is about 25 (see Heat-exchange technology). The function plotted is approximately linear and allows extrapolation to infinite circulation velocity to obtain $\mathbf{R}$. When allowance is made for the fact that rejection by this membrane is a function of concentration (Fig. 3), the observed slope agrees adequately with predictions of equation 12 (37).

Concentration polarization in laminar flow is more complicated because it in-

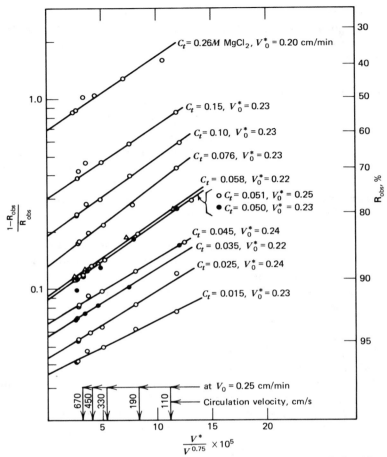

Figure 4. Rejection of MgCl$_2$ at various circulation velocities (hydrous Zr(IV) oxide membrane on porous carbon tube) (37). R_{obs} = 1 − product concentration/feed concentration; v^*, permeation flow through membrane, cm/s; v, circulation through tube, cm/s.

creases with distance down the channel (parallel to the membrane surface) from the feed entrance until an asymptotic concentration profile is established. The asymptotic value of the ratio of c_α to the mixing cup (ie, overall average) concentration at that point depends on flux times thickness of the channel divided by solute diffusion coefficient; the units are chosen to give the dimensionless Péclét number. For usual fluxes and circulation velocities, the limit is not reached even when the flow path parallel to the membrane is very long, unless the membrane channel is very thin.

In investigating the transition region between laminar and turbulent flow, it was found that plots of j-factors, or Stanton numbers divided by Schmidt numbers, as a function of Reynolds number allowed delineation of the regimes, in a manner reminiscent of similar plots of heat-transfer results (39).

Flow patterns in configurations preferred for applications, such as hollow fine fibers surrounded by brine and spiral-wound membranes, are complicated and difficult to analyze quantitatively. Because of the irregularities in geometry caused by the dense packing of fibers and the screen spacers in spiral-wound systems, which serve as turbulence promoters, the flow probably is effectively turbulent in most cases. In analysis of the performance of hollow-fiber systems, concentration polarization has been assumed to be negligible (40).

To avoid confusion, it may be worthwhile to emphasize that concentration polarization discussed here refers to the increase in concentration at the high-pressure membrane–solution interface over that of the brine in the vicinity (cup-mixing average for laminar flow or that in the turbulent region). There is also, of course, an increase in brine concentration though a module or a plant as dilute filtrate is removed. A third definition of rejection, based on average filtrate concentration compared to inlet feed concentration, is sometimes encountered. It evaluates the overall salt-removal capability of the plant or of individual modules in it.

Fouling penalizes flux by adding resistance to permeation and decreases rejection by increasing concentration polarization from a thickening of the unstirred layer at the interface. Amelioration is possible in some cases by circulation velocity or use of turbulence promoters to increase shear at the interface. Periodic washes or antiscaling chemicals are also used (see Water, industrial water treatment). Pretreatment of feeds is perhaps the most effective control. Selection of washes and pretreatments depends heavily on the nature of the feed. Various pretreatment methods, eg, softening, are employed. The control of pH prevents formation of carbonate or hydroxide precipitates and is necessary for some membranes with a limited range of pH stability. Chlorine or other agents prevent biological attack on the membranes as well as fouling. Activated carbon removes, among other materials, chlorine which attacks many membranes. Various types of filtration, such as with sand or multimedia, remove colloidal matter. The plant investment for pretreatment can approach that of the membrane system itself, and the expense can contribute a large fraction of product cost. So far, the methods used appear to be those standard in water treatment. Modification of old techniques or development of new ones specifically for reverse osmosis is one possibility for reducing costs.

Membranes

The discussion of transport through membranes indicated that a thin homogeneous membrane layer gives better production rates than thick membranes without sacrifice in rejection. Whatever mechanism of salt removal is assumed, the distances over which forces between molecular species are appreciable is so small that, if a membrane material is to discriminate between solution components, the passages through which they are forced must be small. Even for coulombic interactions, which vary with the inverse square of distance, 10 nm would be a long distance (41–42). Even 1 nm is rather large for interactions with uncharged membrane materials for which the dependence may be more like the inverse sixth power of distance. Consequently, membrane materials of high salt-rejection properties are likely to be of low intrinsic permeability (flux through a layer of given thickness for a unit pressure drop). The development of thin-layer configurations that can stand the pressure and other operating conditions has been a concern as much as, if not more than, the selection of materials with favorably low values of D^*, $\mathcal{D}^*$, and β, along with high permeabilities for water. Although there has been considerable progress in correlating fundamental chemistry with desirable properties, *a priori* selection of neutral materials, usually polymeric, for good transport and casting characteristics is not yet feasible in a quantitative sense. In addition to obvious requirements like insolubility in the medium of use, certain characteristics, such as a degree of hydrophilicity, are generally recognized as desirable.

Thin membrane films can be attained by several approaches.

Casting. *Asymmetric Membranes.* The development of the Loeb-Sourirajan casting procedure for high-flux membranes brought economical desalination within reach. The original formulation involved a cellulose acetate–acetone–water–magnesium perchlorate mixture and a procedure involving evaporation, coagulation by immersion in cold water, and annealing at ca 75°C, all for controlled times. Salt rejection could be traded for higher water permeability by variation in time and temperature of annealing. An all-organic casting mix of cellulose acetate, acetone, and formamide was later developed in the same laboratory.

Intensive research and optimization has been stimulated by these developments, and there are many models proposed to explain the fundamental processes, still in dispute in important aspects. Examples are given in refs. 43–46. Many improvements have resulted, such as membranes that can be stored dry. The product is an asymmetric film, with a relatively thin homogeneous layer on one side (the air side in the Loeb-Sourirajan case), and on the other a thick porous layer with channels of the order of 100 nm. The presence of the rejecting layer was inferred from the fact that good rejection was obtained only with one of the sides toward the feed (otherwise, the slowness of diffusion of rejected salt through the porous layer caused serious concentration polarization) and the fact that scratching this side damaged performance. The existence of the rejecting layer was confirmed by electron microscopy (47), from which the thickness of the active layer was estimated to be in the range of 200–300 nm, or <1% of a typical overall thickness.

Thin Homogeneous Layers and Composite Membranes. Various techniques have been developed to coat porous substrates either directly or by transfer with cellulose acetate films as thin as 20 nm. An example is the formation of a cellulose acetate membrane on polysulfone-lined fiberglass tubes (48) by floating a casting solution on top of a water column, which is allowed to run out of the tube. In another example, porous cellulose nitrate–cellulose acetate sheets are coated with poly(acrylic acid) to prevent intrusion of membrane material into pores before dipping the sheets into casting solution. The poly(acrylic acid) is later leached out. Variations of this procedure of forming composite membranes have been adapted for continuous casting of membranes to be incorporated in spiral-wound modules (49).

Composite membranes are also prepared on porous supports by monomer polymerization or polymer cross-linking. An important example is PA-300, prepared on polysulfone from a thin layer of an aqueous solution of an epichlorohydrin–polyethylenediamine condensate by contact with a water-immiscible solution of isophthaloyl dichloride (50). This membrane, in spiral-wound configuration, is used in the first large reverse-osmosis seawater desalination plant, an installation of ca 12,000 t/d (3×10^6 gal/d) at Jidda, Saudi Arabia (51).

An interesting variation, plasma polymerization, is the formation of a thin polymer film on a porous substrate by passing monomer over the substrate under conditions giving a glow discharge (52). Membranes may be formed of saturated, as well as unsaturated, monomers. Commercialization of this approach has not yet been attempted.

In general, the composite-membrane approach allows a selection of a porous substrate of better properties and strength than, for example, the porous sublayer of asymmetric membranes.

Ion-Exchange Membranes. Much less attention has been given to development of this class than to membranes of neutral materials. Salt rejections by ion-exchange membranes from feeds of high concentration are usually less, and performance is sometimes affected by the presence of polyvalent counterions in feeds. Membranes have been prepared by grafting polyelectrolytes onto films of cellulose or synthetic organic polymers. However, their thickness prevents the attainment of fluxes that the high intrinsic permeabilities should allow. Several more promising developments are no longer pursued. For example, dip-coating porous films with polyacrylate produced membranes of reasonable rejection and high flux (53). Sulfonated poly(phenylene oxide) has been cast in thin films of interesting properties (54). Ion-exchange membranes are also formed by plasma polymerization.

Dynamically Formed Membranes. Another approach to preparation of thin ion-exchange layers is dynamic formation. A solution or slurry of a membrane-forming substance is circulated under pressure past a porous support. A thin filtercake forms that removes a fraction of solute, sometimes with fluxes of the order of a cm/min [$cm^3/(cm^2 \cdot min)$ or ca 350 gal/($ft^2 \cdot d$)]. Dynamic membranes are formed by many substances, eg, hydrolyzable ions or hydrous oxides (55), synthetic organic polyelectrolytes (56), natural polyelectrolytes, or clays (57–58). Materials contained in waste streams such as sewage and wood-pulping effluents also form salt-rejecting membranes, although prior formation with selected materials frequently gives better performance. Membranes of nonelectrolyte polymers can be formed if they are soluble enough in water, but films of the high rejection of conventional cellulose acetate or polyamide neutral membranes have not been attained. Neutral polymers such as polyvinylpyrrolidinone do, however, form promising dynamic membranes in ultrafiltration applications. *In-situ* formation of cellulose acetate membranes from organic solutions have been reported (59–60), although these are not strictly in the dynamic class.

The chemical nature of the support seems of secondary importance. Porous metals, carbon, ceramics (61), woven fabrics such as fire-hose jackets (62), and plastic films, such as those marketed as Millipore and Acropor, can be employed if the pore sizes are in the proper range; the optimum is usually several hundred nanometers. If the pore sizes are too large, the support may be adapted by precoating with particulates such as filter aids (63) or by circulating very fine particulates in the membrane-forming solution (64).

Where substantial salt rejection is required, the best option so far is a membrane with a sublayer of hydrous zirconium(IV) oxide and a polyacrylate top layer (65). Even these are suitable, at least in single-stage operation, only for brackish waters. The dynamic class appears to have most potential in waste-treatment and recovery applications where both moderate solute removal and ultrafiltration of colloidal matter is needed. Such applications are attractive because of the ability to operate at elevated process temperatures and in difficult chemical environments, as well as the possibility of removal and reformation without disassembly of the system when performance becomes unsatisfactory owing to fouling.

Modules

The design of equipment for applications of reverse osmosis membranes must include several features. The thin fragile layers have to be supported against process pressures, which range up to ca 10 MPa (100 atm). The ratios of pressurized volumes to membrane surface area should be kept low in order to minimize the cost of heavy-walled equipment. There should be a continuous flow path for brine past the membrane, open enough to allow flows sufficient to keep concentration polarization and fouling to an acceptable level. The last two features are contradictory to an extent and trade-offs are necessary for specific applications and different membranes.

At present, four configurations are offered commercially: plate-and-frame, tubular, spiral-wound, and hollow-fine-fiber types.

In the plate-and-frame configuration, the effluent sides of the membranes are separated by flat rigid plastic or metal porous support; these plates are stacked in a filter-press type of arrangement, and brine circulates between them. The spiral-wound system (66) is also comprised of membrane sandwiches; the separator in this case is a flexible screen. The effluent side is connected to an axial tube, and the membrane pairs are wound in a spiral about it, with spacers between the pairs for brine flow (see Membrane technology). The assembly is enclosed in a tube that contains the feed pressure. A high packing of sheet-membrane surface area per unit pressurized volume is realized.

Tubes may be perforated with a porous collector between membrane and support (67) or may be porous, such as fiber glass (68). Ordinarily, the membrane is on the inside of the tube. With materials that are stronger under compression than under tension, eg, ceramics, pressure may be exerted on the outside at the expense of less favorable hydrodynamic flow patterns if the tubes are arranged in bundles rather than in individual jackets. Hollow fibers (69–70) are small enough (typically <0.1 mm OD, with ca 50% wall thickness in reverse osmosis applications) to support pressure without extraneous backing, with pressurized brine usually on the outside. For a schematic of their arrangement in a module, see Hollow-fiber membranes (Vol. 12, p. 492). Compared to sheet membranes, hollow-fiber membranes compensate for their approximate order of magnitude lower flux {0.04–0.2 m/d [m^3/(m^2·d)] or 1–5 gal/(ft^2·d)} by offering an improvement of about the same factor in lower cost per unit area. Because of large surface area, fibers may be homogeneous, although techniques for spinning asymmetric fibers have been developed.

Production rate per unit pressurized volume varies with the flux of the membranes used. Hollow-fiber units are usually highest; a value for a typical 0.04 m/d membrane flux is ca 600 times the pressurized volume [(600 m^3/(m^3·d) or 4500 gal/(ft^3·d)]. Spiral-wound units follow; the flux density with a membrane of 0.6 m/d flux is 400 times the volume. A tubular system equipped with a membrane of 0.8 m/d flux would be less and plate-and-frame units would be lower.

Primarily because of lower total capital cost per unit membrane area and per unit of product per day, spiral-wound and hollow-fiber configurations now dominate the market. However, applications have so far been mostly to relatively clean feeds to produce industrial process water or potable water from brackish or sea sources. The close spacing in brine-flow channels, particularly with hollow fibers, often increases the need for pretreatment and may entail prohibitively expensive difficulties with dirty waste streams. Implementations of the latter applications may raise the competitive standing of the more open tubular configurations, particularly with high-flux film and dynamic membranes. The intricate sealing and complex flow controls needed in plate-and-frame arrangements restrict the use of this configuration.

Economic and Energy Aspects

Since the first publications on reverse osmosis ca 20 yr ago (5–7), an industry estimated at >10^8/yr has been established (71). Although less than 4000 metric tons (ca 10^6 gal) capacity per day of product water had been sold up to 1970, that number had increased to 1.4×10^6 t/d by the end of 1979 (72). A parallel growth picture is given in successive inventories of plants. The plants operating or under construction on Jan. 1, 1977 and June 30, 1980 are compared in Table 2.

The market has been dominated by the United States and Middle Eastern

Table 2. Desalination Plants Operating or Under Construction January 1, 1977[a] and June 30, 1980[b], 1000 t/d[c]

Location	Municipal[d]	Industrial	Power	Discharge	Other[e]	Total
U.S. and possessions						
1977	60.2	99.9	32.6		6.1	198.8
1980	123.0	168.4	93.1	362.2	8.7	755.4
other Western Hemisphere						
1977	6.8	11.4	4.9			23.1
1980	11.4	23.1	17.0			51.5
Europe, except USSR						
1977	1.9	10.2	6.1		1.5	19.7
1980	4.9	52.6	8.3		14.8	80.6
USSR						
1977						
1980		12.9				12.9
Africa						
1977	2.3	5.7	0.8		1.9	10.7
1980	7.6	82.9	0.4		2.6	93.5
India and Far East[f]						
1977	0.4	12.5	48.1		0.8	61.8
1980	2.3	28.0	48.4	0.4	1.9	81.0
Saudi Arabia						
1977	254.4	14.0	2.6		0.4	271.4
1980	279.3	54.9	1.5			335.7
other Asian Mid East						
1977	9.1	26.1	2.6			37.8
1980	18.2	45.0	10.2		1.9	75.3
Total[g]						
1977	335.1	179.8	97.7		10.7	623.3
1980	446.7	467.8	178.9	362.6	29.9	1485.9

[a] Desalination Inventory #6 of 518 plants (73).

[b] Desalination Inventory #7 of 929 plants (74).

[c] To convert metric tons to gallons, multiply by 264.2.

[d] Includes tourist.

[e] Includes military, demonstration, waste, and irrigation.

[f] Over 80% in Japanese installations.

[g] Although grand totals vary from those in the references (632 in 1977 and 1478 in 1980) because of roundoff and other arithmetical errors here, the numbers are adequate for the present discussion.

countries; if the Yuma, Arizona (360,000 t/d) plant for irrigation discharge waters, still under construction, is disregarded, the total capacities of these two areas are similar. In the Middle East, desalinated water is used mostly for municipal supply, whereas in the United States, clean water for industrial and power (boiler feed) applications are also important. An important early market, for example, was the preparation of pure water for the semiconductor industry, for which the feed might be ordinary tap water. The original objective was to lower costs of regeneration chemicals for ion-exchange demineralizers, but the concurrent removal of other substances also improved production reliability. In municipal supply applications, most U.S. installations are located along the coast, particularly in Florida resort communities, and operate on brackish feeds. In such locations, concentrated brine can be disposed of in the ocean, and a severe constraint on inland desalination plants is avoided.

The feeds are mostly of low salinity. Seawater was the source for 2% of capacity in 1977 (3 installations) and of 5% in 1980 (34 installations). The first seawater plant of considerable size was the 12,000 t/d facility at Jidda (51). The reverse osmosis capacity has increased, relative to other processes, as shown below.

	Capacity × *reverse osmosis capacity*	
Year	*Distillation*	*Electrodialysis*
1977	4.5	0.24
1980	3.7	0.19

In general, membranes dominate processing of brackish feeds, and electrodialysis is competitive with reverse osmosis only for salinities up to a few thousand ppm. Reverse osmosis appears to be approaching competitiveness with distillation for seawater desalination. Energy requirements are less for reverse osmosis: seawater distillation requires 214 kJ/L or 59.6 W·h/L (770 Btu/gal), whereas reverse osmosis requires 94.7 kJ/L or 26.3 W·h/L (340 Btu/gal) at assumed generating efficiency (75). The potential for lowering energy requirements appears better for membrane processes than for evaporation as improvements are made in membranes and pumps, devices to recover energy in discharge of pressurized waste brines are adopted, and pretreatments to lower chemical cost are developed. Brackish waters usually require only about one fifth to one third of the kW·h per unit product needed for seawater. Scale and corrosion problems are lessened in reverse osmosis because it operates at ambient temperatures.

Distillation can be carried out with relatively low quality steam, and many desalination plants utilize sources from electric-power generation. For water-only plants, using high temperature steam for shaft power for reverse osmosis and low temperature steam for distillation may be optimal.

Growth of desalination is expected to continue. For the year 2000, a U.S. market of over 100×10^6 t/d and an overseas market of about 20×10^6 t/d is projected (75), with over half of this production estimated to be by membrane processes. Included in feeds allocated to membranes for the year 2000 are about 18×10^6 t/d of difficult brackish waters, eg, industrial return flows and recycle. Table 2 indicates that in spite of long-standing needs for water management in this category, reverse osmosis has as yet made only a small contribution. Ultrafiltration has proved useful in specific applications, particularly when a valuable constituent is recovered, for example in electrocoating operations, in effluents from textile sizing, and in cheese whey. The typical feeds are severely fouling, and the more-open tubular configurations are generally used in spite of the much lower cost per unit of membrane area of hollow fibers or spiral-wound modules.

However high market-size projections are evaluated, there are obvious needs in many areas, such as the pulp-and-paper and textile industries, for water treatments requiring a combination of removal of colloidal substances and moderate rejection of salts or low molecular weight organic solutes. Upgrading of municipal sewage for at least some categories of reuse is another potential application (see Water, sewage). The largest installation so far for this purpose is a 19,000 t/d spiral-wound plant in Orange County, Calif., where the effluent from tertiary treatment of sewage is desalted for ground-water recharge (76). It is the second largest after the Yuma plant installation whose feed is classified as waste in ref. 74, although it is included in the municipal category. Before being fed to the membrane plant, the effluent from secondary treatment has been chemically clarified, stripped of ammonia, recarbonated, and passed through multimedia filters and through activated carbon, plus some further steps specifically for the membrane unit. Although some of these steps are believed to be unnecessary (76), they illustrate the rather stringent pretreatment needs for present systems in processing feeds likely to foul membranes.

The small impact reverse osmosis has had so far in these areas suggests that the present membrane systems, which were mostly developed for the production of fresh water by the desalination of relatively clean feeds, are not adequate for the "difficult brackish" category. Advances in pretreatment may overcome these difficulties. Another possibility, previously suggested, is that tubular systems, with dynamic or other membranes of higher flux and greater stability to aggressive measures for removal of fouling layers, may fill this need.

Nomenclature

c	= concentration
C^*	= ion-exchange capacity, equivalents/kg water in membrane phase
d	= inside diameter of tube, cm
D	= distribution coefficient
D^*	= distribution coefficient solute between membranes and solution, m^*/m
$\mathcal{D}_v$	= diffusion coefficient of solute in solution phase, m^2/s or cm^2/s
$\mathcal{D}_v^*$	= diffusion coefficient of solute in membrane phase, m^2/s
$\mathcal{D}^*$	= diffusion coefficient of solute in membrane phase consistent with molality concentration units ($= \rho_m f_w \mathcal{D}_v^*$)
f_w	= wt fraction of water in membrane
j-factor	= Stanton numbers divided by Schmidt numbers
J_s	= flux, solute through membrane, mol/(m²·s)
J_w	= flux, water through membrane, kg/(m²·s)
k	= semiempirical constant in concentration polarization equations (see eq. 12)
l	= thickness of membrane or its active layer
m	= molality, mol/kg water (or solvent), solution phase
m^*	= molality, mol/kg water, membrane phase
m_ω	= concentration of effluent solution
R	= gas constant
R	= solute rejection $(m_\alpha - m_\omega)/m_\alpha$
R$_{obs}$	= observed salt rejection $(m_t - m_\omega)/m_t$
R$_\infty$	= limiting value of R approached at high flux
T	= temperature

v	= circulation velocity of brine parallel to membrane, cm/s
v^*	= permeation through membrane, cm/s
v_t	= velocity high enough to produce turbulence
x	= distance perpendicular to membrane $x = 0$ at high pressure membrane-feed
β	= coefficient for coupling of solute and solvent flux through membranes (eq. 2)
γ_k	= mean ionic activity coefficient of solute, solution phase, concentration in molality; reference state, infinite dilution
$\gamma_\pm^*$	= mean ionic coefficient of solute, membrane phase, same reference state as solution
Γ	= $\gamma_\pm^*/\gamma_\pm$ (for 1,1 electrolyte)
η_k	= kinematic viscosity (cm²/s)
μ_s	= chemical potential of solute
ρ_m	= wet membrane density, kg/m³

Superscript

*	= membrane phase

Subscripts

f	= feed
t	= turbulent core
α	= high pressure solution–membrane interface
ω	= membrane–effluent interface

Acknowledgment

Preparation of this review was supported by the Division of Chemical Sciences, Office of Basic Energy Sciences, U.S. Department of Energy, under contract W-7405-eng-26 with Union Carbide Corporation.

BIBLIOGRAPHY

"Osmosis and Osmotic Pressure" in *ECT* 1st ed., Vol. 9, pp. 643–660, by E. H. Immergut and K. G. Stern, Institute of Polymer Research, Polytechnic Institute of Brooklyn; "Osmosis, Osmotic Pressure, and Reverse Osmosis" in *ECT* 2nd ed., Vol. 14, pp. 345–356, by Betram Keilin, Amicon Corporation.

1. B. Erschler, *Kolloid-Z* **68**, 289 (1934).
2. J. W. McBain and R. F. Stuewer, *J. Phys. Chem.* **40**, 1157 (1936).
3. S. Trautman and L. Ambard, *J. Chim. Phys.* **49**, 220 (1952).
4. J. G. McKelvey, K. S. Spiegler, and M. R. J. Wyllie, *Chem. Eng. Progr. Symp. Ser.* **55**(24), 199 (1959).
5. C. E. Reid and E. J. Breton, *J. Appl. Polym. Sci.* **1**, 133 (1959); *Chem. Eng. Progr. Symp. Ser.* **55**(24), 171 (1959).
6. S. Loeb and S. Sourirajan, *University of California at Los Angeles Department of Engineering Report*, 60-60, Los Angeles, Calif., 1960.
7. S. Loeb and S. Sourirajan, *Adv. Chem. Ser.* **38**, 117 (1963).
8. R. J. Raridon, L. Dresner, and K. A. Kraus, *Desalination* **1**, 210 (1966).
9. T. R. Yuster, S. Sourirajan, and K. Bernstein, *University of California at Los Angeles Department of Engineering Report 7*, 58-26, Los Angeles, Calif., 1958.
10. S. Sourirajan, *Ind. Eng. Chem. Fundam.* **2**, 51 (1963).
11. E. Glueckauf, *Proc. 1st Int. Symp. on Water Desalination, Washington* **1**, 143 (1965).
12. C. P. Bean in G. Eisenman, ed., *Membranes: Macroscopic Systems and Models*, Marcel Dekker, Inc., New York, 1972, Chapt. 1.
13. A. J. Staverman, *Rev. Trav. Chim.* **70**, 83 (1951).
14. P. Mazur and J. T. G. Overbeek, *Rec. Trav. Chim.* **70**, 83 (1951).
15. J. G. Kirkwood in H. T. Clarke, ed., *Ion Transport Across Membranes*, Academic Press, New York, 1954, p. 119.
16. O. Kedem and A. Katchalsky, *Biochem. Biophys. Acta* **27**, 229 (1958).
17. J. S. Johnson, Jr., L. Dresner, and K. A. Kraus in K. S. Spiegler, ed., *Principles of Desalination*, 1st ed., Academic Press, New York, 1966, Chapt. 8.
18. L. Dresner and J. S. Johnson in K. S. Spiegler and A. D. K. Laird, eds., *Principles of Desalination*, 2nd ed., Academic Press, New York, 1980, Chapt. 8.
19. H. K. Lonsdale, U. Merten, and R. L. Riley, *J. Appl. Polym. Sci.* **9**, 1341 (1965).

20. R. L. Riley, H. K. Lonsdale, C. R. Lyons, and U. Merten, *J. Appl. Polym. Sci.* **11**, 2143 (1967).
21. K. A. Kraus, R. J. Raridon, and W. H. Baldwin, *J. Am. Chem. Soc.* **86**, 2571 (1964); R. D. Lanier, *J. Phys. Chem.* **69**, 2697 (1965).
22. C. F. Coleman, *J. Phys. Chem.* **69**, 1377 (1965).
23. Ref. 18, Table 8.5.
24. Ref. 18, Fig. 8.19.
25. H. Yasuda and C. E. Lamaze, *Office of Saline Water Report 473*, U.S. Department of the Interior, Washington, D.C., 1969.
26. F. Helfferich, *Ion Exchange*, McGraw-Hill, Inc., New York, 1962.
27. R. Schlögl, *Fortschr. Physik. Chem.* **9**, (1964).
28. E. Hoffer and O. Kedem, *Desalination* **2**, 25 (1967).
29. P. Meares and T. Foley, *Office of Saline Water Report 584*, U.S. Department of the Interior, Washington, D.C., 1970; P. Meares and J. F. Thain, *J. Phys. Chem.* **72**, 2789 (1968); P. Meares, D. G. Dawson, A. H. Sutton, and J. F. Thain, *Ber. Bunsengesellschaft Phys. Chem.* **71**, 765 (1967); W. J. McHardy, P. Meares, A. H. Sutton, and J. F. Thain, *J. Coll. Interf. Science* **29**, 116 (1969); W. J. McHardy, P. Meares, and J. F. Thain, *J. Electrochem. Soc.* **116**, 920 (1969); D. G. Dawson, W. Dorst, and P. Meares, *J. Poly. Sci. Part C* **22**, 901 (1967); D. G. Dawson and P. Meares, *J. Coll. Interf. Sci.* **33**, 117 (1970).
30. Ref. 18, Table 8.7.
31. A. J. Shor, K. A. Kraus, W. T. Smith, Jr., and J. S. Johnson, Jr., *J. Phys. Chem.* **72**, 2200 (1968).
32. *Annual Report, Oak Ridge National Laboratory, Chemistry Division, for period ending 20 May 1969,* ORNL-4437, Oak Ridge, Tenn., p. 78.
33. K. A. Kraus, H. O. Phillips, T. A. Carlson, and J. S. Johnson, Jr., *Proc. 2nd Int. Conf. on Peaceful Uses of Atomic Energy, Geneva* **28**, 3 (1958).
34. T. K. Sherwood, P. L. T. Brian, R. E. Fisher, and L. Dresner, *Ind. Eng. Chem. Fundam.* **4**, 113 (1965).
35. P. L. T. Brian, *Proc. 1st Int. Symp. on Water Desalination, Washington* **1**, 349 (1965).
36. P. L. T. Brian, *Ind. Eng. Chem. Fundam.* **4**, 439 (1965); *MIT Desalination Research Laboratory Report 295-7*, Cambridge, Mass., 1965.
37. A. J. Shor, K. A. Kraus, J. S. Johnson, Jr., and W. T. Smith, Jr., *Ind. Eng. Chem. Fundam.* **7**, 44 (1968).
38. R. B. Bird, W. E. Stewart, and E. N. Lightfoot, *Transport Phenomena*, John Wiley & Sons, Inc., New York, 1962.
39. D. G. Thomas, *Ind. Eng. Chem. Fundam.* **12**, 189 (1973).
40. M. S. Dandavati, M. R. Doshi, and W. N. Gill, *Chem. Eng. Sci.* **30**, 877 (1975).
41. L. Dresner and K. A. Kraus, *J. Phys. Chem.* **67**, 990 (1963).
42. L. Dresner, *J. Phys. Chem.* **69**, 2230 (1965).
43. H. Strathman, P. Scheible, and R. W. Baker, *J. Appl. Polym. Sci.* **15**, 811 (1971); H. Strathman, K. Koch, P. Amar, and R. W. Baker, *Desalination* **16**, 79 (1975).
44. M. A. Frommer and R. M. Messalem, *Ind. Eng. Chem. Prod. Res. Dev.* **12**, 328 (1973).
45. G. B. Tanny, *J. Appl. Polym. Sci.* **18**, 2149 (1974).
46. E. Klein and J. K. Smith in H. K. Lonsdale and H. E. Podall, eds., *Reverse Osmosis Membrane Research*, Plenum Publishing Corporation, New York, 1972, p. 263.
47. R. L. Riley, J. O. Gardner, and U. Merten, *Science* **143**, 801 (1964).
48. L. T. Rozelle, J. E. Cadotte, A. J. Senechal, W. L. King, and B. R. Nelson in ref. 46, p. 419.
49. R. L. Riley, G. R. Hightower, C. R. Lyons, and M. Tagmi, *Proc. 4th Int. Symp. on Fresh Water from the Sea* **4**, 333 (1973).
50. R. L. Riley, R. L. Fox, C. R. Lyons, C. E. Milstead, M. W. Seroy, and M. Tagami, *Desalination* **19**, 113 (1976).
51. A. Al-Gholaikah, N. El-Ramly, I. Jamjoom, and R. Seaton, *Desalination* **27**, 215 (1978).
52. H. Yasuda, *Office of Saline Water Report 811*, U.S. Department of the Interior, Washington, D.C., 1972; H. Yasuda and C. Lamaze, *J. Appl. Polym. Sci.* **17**, 201 (1973); H. Yasuda, H. C. Marsh, and J. Tsai, *J. Appl. Polym. Sci.* **19**, 2157 (1975); H. Yasuda and H. C. Marsh, *J. Appl. Polym. Sci.* **19**, 2981 (1975).
53. S. B. Sachs and H. K. Lonsdale, *J. Appl. Polym. Sci.* **15**, 797 (1971).
54. A. B. LaConti, P. J. Chludzinski, and A. P. Fickett in ref. 46, p. 263.
55. A. E. Marcinkowsky, K. A. Kraus, H. O. Phillips, J. S. Johnson, Jr., and A. J. Shor, *J. Am. Chem. Soc.* **88**, 5744 (1966).

56. K. A. Kraus, H. O. Phillips, A. E. Marcinkowsky, J. S. Johnson, Jr., and A. J. Shor, *Desalination* **1**, 225 (1966).
57. K. A. Kraus, A. J. Shor, and J. S. Johnson, Jr., *Desalination* **2**, 243 (1967).
58. U.S. Patent 3,331,772 (July 18, 1967), E. R. Brownscombe and L. R. Kern (to Atlantic Refining Company).
59. A. Gollan and M. P. Tulin in ref. 46, p. 341.
60. F. E. Littman, H. K. Bishop, and G. Belfort, *Desalination* **11**, 17 (1972).
61. D. G. Thomas and W. R. Mixon, *Desalination* **15**, 287 (1974).
62. J. A. Dahlheimer, D. G. Thomas, and K. A. Kraus, *Ind. Eng. Chem. Process Des. Dev.* **9**, 565 (1970).
63. J. S. Johnson, Jr., K. A. Kraus, S. M. Fleming, H. D. Cochran, Jr., and J. J. Perona, *Desalination* **5**, 359 (1968).
64. *Annual Report, Oak Ridge National Laboratory, Chemistry Division, for period ending 31 August 1978*, ORNL-5485, Oak Ridge, Tenn., p. 74.
65. J. S. Johnson, Jr., R. E. Minturn, and P. A. Wadia, *J. Electroanal. Chem.* **37**, 267 (1972).
66. U. Merten, *Office of Saline Water Report 165*, U.S. Department of the Interior, Washington, D.C., 1966.
67. S. Loeb, *Desalination* **1**, 35 (1966).
68. G. G. Havens and D. B. Guy, *Chem. Eng. Progr. Symp. Ser.* **64**(90), 299 (1968).
69. H. I. Mahon, *National Academy of Sciences-National Research Council Publication 942*, Washington, D.C., 1963, p. 345.
70. W. P. Cooke, *Desalination* **7**, 31 (1969).
71. H. K. Lonsdale, *J. Membr. Sci.* **5**, 263 (1979).
72. *Water Desalination Report*, Tracys Landing, Md., Jan. 3, 1980.
73. N. A. El-Ramly and C. F. Congdon, *Desalting Plants Inventory No. 6*, Office of Water Research and Technology, U.S. Department of the Interior, Washington, D.C., 1977.
74. N. A. El-Ramly and C. F. Congdon, *Desalting Plants Inventory No. 7*, National Water Supply and Improvement Association, Ipswich, Mass., 1981.
75. Fluor Engineers and Constructors, Inc., *Desalting Plans and Progress. An Evaluation of the State-of-the-Art and Future Research and Development Requirements. Final report to OWRT on Contract 14-34-001-7707*, Irvine, Calif., Jan. 1978; P. J. Shroeder, *Desalination* **30**, 5 (1979).
76. P. K. Allen and G. L. Elser, *Desalination* **30**, 23 (1979).

General References

Ref. 18 is also a general reference.
Ref. 27 is also a general reference. It is a particularly useful treatment of membrane transport.
U. Merten, *Desalination by Reverse Osmosis*, MIT Press, Cambridge, Mass., 1966.
S. Sourirajan, *Reverse Osmosis*, Academic Press, New York, 1970.
Desalination (journal).
S. Sourirajan, ed., *Reverse Osmosis and Synthetic Membranes*, NRCC No. 15627, National Research Council of Canada, Ottawa, 1977.
A. F. Turbak, ed., *Synthetic Membranes*, 2 vols., American Chemical Society Symposium Series, Vol. 154, Washington, D.C., 1981.
J. McDermott, *Desalination by Reverse Osmosis*, Noyes Data Corporation, Park Ridge, N.J., 1970. Patents are reviewed.
P. R. Keller, *Membrane Technology and Industrial Separation Techniques*, Noyes Data Corporation, Park Ridge, N.J., 1976. Patents are reviewed.

JAMES S. JOHNSON, JR.
Oak Ridge National Laboratory

REYNOLDS' NUMBER.   See Fluidization; Fluid mechanics; Rheological measurements; Sedimentation.

RHENIUM AND RHENIUM COMPOUNDS

Rhenium

Rhenium [*7440-15-5*], the seventy-fifth element in the periodic table, is the heaviest element in Group VIIB. Its congeners in this periodic group are manganese and technetium, elements 25 and 43, respectively. Rhenium has an atomic weight of 186.2, and occurs in nature as two nuclides: ^{185}Re [*14391-28-7*], mass 184.9530, abundance 37.500%; and ^{187}Re [*14391-29-8*], mass 186.9560, abundance 62.500%. The latter isotope is radioactive and emits very low energy radiation (^{187}Re $\rightarrow$ ^{187}Os + 0.002 MeV β), with a half-life of ca $(4.3 \pm 0.5) \times 10^{10}$ yr. The radioactivity of ^{187}Re is at so low a level that detection is possible only with complicated laboratory equipment. This radioactive decay is of some current interest in dating the time of the earth's formation. In addition to ^{185}Re and ^{187}Re, sixteen other radioactive isotopes are known; none has an appreciable half-life (see Radioisotopes).

Although its existence had been predicted much earlier by Mendeleev, rhenium was not discovered until 1925. The element was identified, based on x-ray spectral data, in samples of columbite, gadolinite, molybdenite, and in platinum ores (1). The element was named after the German province Rhineland. Several other reports claiming identification of this element were made in subsequent months from other laboratories (2). The toxicities of rhenium compounds have not been established, but users are advised to handle them with care (3).

Occurrence. Rhenium is one of the least abundant of the naturally occurring elements. Various estimates of its abundance in the earth's crust have been made; the most widely quoted figure is 0.027 atoms per 10^6 atoms of silicon (0.05 ppm by weight) (4). However, this number has a high uncertainty; it is based on analyses for the commonest rocks, ie, granites and basalts. The abundance of rhenium in stony meteorites is approximately the same value; the average abundance in siderites is 0.5 ppm. The abundance of rhenium in lunar materials is discussed in ref. 5. Rhenium-187 appears to be enriched by 1.4–29% relative to terrestial abundance. It is suggested that this is a consequence of a nuclear reaction sequence involving neutron capture by ^{186}W (^{186}W$(n,\gamma)^{187}$W) followed by β decay of ^{187}W with a half-life of 24 h (^{187}W $\rightarrow$ ^{187}Re + β) (5).

Rhenium is apparently present in seawater (probably as the perrhenate ion) in very low concentrations.

Rhenium occurs in one mineral, dzhezkazganite, which occurs with copper deposits in the USSR. This substance is believed to be a solid solution of MoS_2 and ReS_2, with copper not actually being a component of the mineral. The current source of rhenium is a molybdenite concentration obtained from porphyry copper deposits. Rhenium also occurs in trace amounts in a variety of other minerals including gadolinites; alvites; columbites; certain manganese, platinum, and uranium ores; and certain uraniferous ores and vanadiferous shales; none of which contains rhenium in high enough concentration to make commercial recovery feasible. Rhenium occurs as Re_2O_7 [*1314-68-7*] in uranium and ilsemannite ores from the Colorado plateau; otherwise, it occurs as a sulfide.

The common natural occurrence of rhenium and molybdenum together is a consequence of the similarities of these elements. Both elements have a high affinity for sulfide ion and occur in sulfide ores. Moreover, the radii of Re^{4+} and Mo^{4+} are almost identical, 0.74 nm and 0.70 nm, respectively, so that ReS_2 [12038-63-0] and MoS_2 have similar crystal structures with almost identical dimensions.

Isolation. Rhenium is obtained from molybdenite concentrates from porphyry copper ores. Rhenium, although present in small amounts in the original ore, is concentrated in the molybdenite by-products to as much as 2000 ppm. During 1980, porphyry copper mines in Canada (Island Copper Mine of Utah International, Inc.), Chile (Chuquicamata, El Teniente, and El Salvador mines, owned by the Chilean government and managed by a state agency, the Corporacion Nacional de Cobre de Chile (CODELCO)), the USSR (mines in the Tadzhikistan and Uzbekistan regions of central Asia), Peru, and the United States (in Utah, Arizona, Nevada, and New Mexico) supplied rhenium. Additional sources of rhenium are being developed in Mexico and Yugoslavia. A copper-smelting operation involving one of the world's largest copper deposits in The People's Republic of China is scheduled to begin operation in 1984; this deposit also contains recoverable rhenium.

The Chilean deposits have an average grade of rhenium in concentrations of 230–570 ppm. The El Teniente mine is said to contain the world's largest reserves of rhenium, ca 680 metric tons. This ore averages 440 ppm Re with 1.5 wt % Cu and 0.04 wt % Mo.

Most of the processing of the molybdenite by-product to obtain rhenium is carried out in the United States and the FRG. Some Canadian concentrate is processed in the United States and is then returned to the owners in Canada for resale.

The traditional route by which rhenium is obtained involves roasting of the molybdenite concentrate thereby converting most of the rhenium to Re_2O_7. This oxide is volatile and exits the system with some MoO_3 and sulfur oxides. These water-soluble compounds are concentrated in a wet-scrubbing system, to which sodium hydroxide is added to react with the sulfur oxides. Processing involving anion exchange or solvent extraction permit separation of ReO_4^- from other materials. Treatment of solutions of perrhenate ion with HCl and hydrogen sulfide gas produces ReS_2 as a black mud, which is separated. Addition of H_2O_2 and NH_4OH produces a solution of NH_4ReO_4 [13598-65- 7]; this salt crystallizes upon evaporation of the solvent and is either sold as is or converted to the metal by reduction with hydrogen gas.

In earlier procedures, the ReO_4^- anion was precipitated from water as the relatively insoluble potassium salt $KReO_4$. Reduction of this substance with hydrogen gas gives rhenium metal contaminated with ca 0.4 wt % potassium, which cannot be easily separated. Although suitable for some purposes, rhenium formed in this way is unsatisfactory in specific applications, eg, in filaments in mass spectrometer systems. The processing route via NH_4ReO_4 avoids this problem.

Before scrubbing procedures were established, most of the rhenium was lost as the volatile Re_2O_7. A small portion (ca 10%) was retained in the flue dust which was processed to give the metal.

A commercial flotation process for the recovery of the molybdenite by-product permits high recovery of molybdenum and rhenium. This process will be used in the Mexican operation at the new Caridad mine.

Physical Properties. Selected physical properties of rhenium are summarized in Table 1. Rhenium is silvery-white with a metallic luster. Only platinum, iridium, and osmium have higher densities than rhenium. Its melting point is higher than that of any other element except tungsten (3410°C) and carbon (3550°C).

Rhenium metal exhibits a very low paramagnetism, which is field-independent and almost temperature-independent. Data for two independent samples at 79–471 K revealed magnetic susceptibilities corresponding to the equations $\chi_g \times 10^6 = 0.280 + (7.5 \times 10^{-6})$ T and $\chi_g \times 10^6 = 0.310 + (17 \times 10^{-6})$ T, where T (temperature) is in kelvins (6).

A significant property of rhenium is its ability to alloy with molybdenum and tungsten. Phase diagrams for Mo–Re and W–Re are given in Figures 1 and 2, respectively (7). Molybdenum alloys containing up to 50 wt % are bcc (body-centered-cubic) solid solutions. An abrupt increase in ductility occurs just below the solubility limit, ca 47 wt % Re. An alloy with 50 wt % Re can be fabricated by either warm or cold working. Unlike molybdenum, this alloy is ductile, even at temperatures <−196°C. The alloy can be welded. The addition of rhenium to tungsten gives alloys with improved properties. The alloy of composition 25 wt % Re is a bcc solution. This alloy has improved ductility and a lower ductile-to-brittle transition temperature than pure tungsten.

The addition of thoria to a Re–W alloy produces 74 wt % W–24 wt % Re–2 wt % ThO_2, which is useful in heated cathodes in electron tubes. This material has good ductility and high resistance to breaking by mechanical shock.

Nickel–rhenium alloys containing thoria or other additives have been developed for use as cathodes on electrovacuum devices. Rhenium improves the strength prop-

Table 1. Selected Physical Properties of Rhenium

Property	Value
at no.	75
at wt	186.2
mp, °C	3180
bp, °C	5926 est
density, g/cm³	21.02
crystal structure	hexagonal close-packed
lattice constants, nm	a, 0.2760
	c, 0.4458
metal radius (12-coordinate), nm	0.1372
ΔH°_{subl}, kJ/mol[a]	791
S° (crystal), J/(mol·K)[a]	37.2 ± 1.2
specific heat (at 20°C), J/(mol·K)[a]	25.1
ionization potentials, kJ[a]	757
	1597
	2502
electrical resistivity (at 20°C), $\mu\Omega\cdot$cm	19.3
thermal coefficient of electrical resistivity, °C⁻¹	3.95×10^3
Richardson constant, kJ (eV)	462 (4.8)
thermal conductivity (at 0–100°C), W/(cm·K)	0.400
module of elasticity, Pa[b]	0.46

[a] To convert J to cal, divide by 4.184.
[b] To convert Pa to psi, divide by 6895.

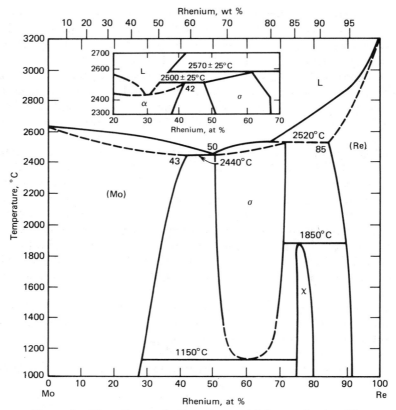

Figure 1. Phase diagram for the system molybdenum–rhenium (7).

erties substantially: at 1000°C, the strength of 90 wt % Ni–10 wt % Re alloys exceeds the strength of a Ni–V cathode material by 90% and exceeds its rigidity by 150–200%.

Chemical Properties. Rhenium does not react with atmospheric oxygen at ambient temperatures but, at higher temperatures, it burns giving Re_2O_7. This volatile yellow crystalline compound dissolves in water to give perrhenic acid [13768-11-1], $HReO_4$. Reactions of the metal occur with chlorine or bromine to produce Re_2Cl_{10} or Re_2Br_{10} [30937-53-2] and with fluorine to yield ReF_6 and ReF_7. Rhenium metal is not affected by water and hydrohalic acids, but it reacts quickly with HNO_3 to produce $HReO_4$. Fusing the metal with NaOH and oxidizing agents, eg, KNO_3 or Na_2O_2, produces perrhenate salts. The metal also oxidizes to $HReO_4$ in a reaction with 30 wt % H_2O_2; the rate of this reaction depends on the nature of the sample.

The oxidation potentials for rhenium in aqueous acidic solution are summarized below (8).

$$\text{Re} \xrightarrow{0.260\ V} \text{ReO}_2 \xrightarrow{0.386\ V} \text{ReO}_3 \xrightarrow{0.768\ V} \text{ReO}_4^-$$

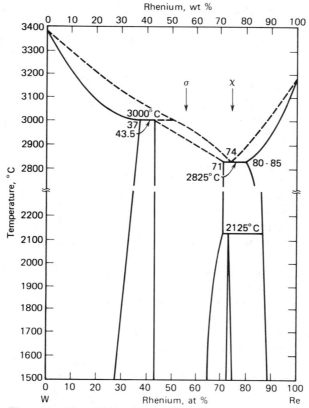

Figure 2. Phase diagram for the system tungsten–rhenium (7).

The potential for the oxidation of rhenium to the perrhenate ion is low. Perrhenate is thus a much weaker oxidizing agent than permanganate, which accounts for its formation in most oxidizing systems and its inability to serve as a strong oxidizing agent. The absence of a divalent ion in this sequence, Re^{2+}, is significant.

Volumetric, gravimetric, and spectrophotometric methods for the analysis of rhenium, which are mainly of historical interest at this time, have been described in some detail (9).

Rhenium may be electroplated from solutions of ReO_4^- in sulfuric acid. To obtain a durable plating of rhenium, it is desirable to deposit the rhenium in thin layers with annealing at high temperature after each layer is applied.

Processing. Rhenium and alloys with molybdenum and tungsten can be consolidated by powder metallurgy (qv). The process can be carried out with blended powders, but better results occur with prealloyed material. After screening, sizing, and blending of the powder, bars are pressed either mechanically or isostatically at pressures of ca 210–340 MPa (30,000–50,000 psi). The material then must be sintered at 1200°C in hydrogen or under vacuum. With W–Re alloys, for which diffusion rates are slower, the sintering temperature should be as high as possible to avoid formation of a rhenium-rich σ phase which is brittle. Use of prealloyed particles permits sintering temperatures as low as 2200°C.

The Mo–Re and W–Re alloys can also be produced by electron-beam or arc melting. However, the resulting grain size is much larger and fabrication by forging or extrusion is desirable. The main advantage of melting is the resultant homogeneity.

Fabrication of pure rhenium is done cold with recrystallization annealing after each 10–40% reduction, depending on the operation and size. Annealing of Mo–Re and W–Re products is also necessary.

Economic Aspects. The world production of rhenium in 1978 and 1979 was ca 7.1 metric tons and 7.3 t, respectively. The price of rhenium metal in 1978 was ca ($660–1050)/kg. In 1979 there was a substantial increase to $4400/kg, which reflected the increased use of rhenium as a catalyst material. The price in 1967 was ca $1300/kg.

Uses. Certain properties of rhenium permit its use at high temperatures as filaments in electron tubes, light bulbs, and photoflash bulbs. It is used in filaments in mass spectrometers, since it is inert to many materials in these systems. Alloys with molybdenum or tungsten are used as heating elements. These uses require either evacuated systems or inert atmospheres since rhenium oxidizes in atmospheric oxygen at high temperatures. Approximately 8% of the rhenium produced is used in these applications.

The high melting point and low vapor pressure of rhenium are prerequisites to its use at high temperatures. Although the electrical resistance of rhenium is ca 3.5 times higher than tungsten at 20°C, this difference is reduced at higher temperatures because of rhenium's lower temperature coefficient, so that at 2500°C its resistivity is only ca 20% greater. The higher resistance at low temperatures and its lower temperature coefficient lead to rapid heating of a filament.

Rhenium exhibits a greater resistance than tungsten to the water-cycle effect, a phenomenon that is marked by the blackening of lamps and electron tubes by deposition of metal. This phenomenon involves catalysis by small quantities of water which react with the metal in a hot filament to produce a volatile metal oxide and hydrogen. The oxide condenses on the surface of the bulb and is reduced back to the metal by hydrogen.

The thermionic work function for rhenium 4.8 eV is slightly higher than tungsten but is low enough so that it can be used in thermionic devices. Alloys with thoria and tungsten or nickel are often used; these have higher emission, improved ductility, and greater shock resistance (see Thermoelectric energy conversion).

Two alloys with tungsten are commercially available. One contains ca 3 wt % rhenium and is used for heating filaments; the rhenium contributes resistance to thermal and mechanical shock. The other alloy contains ca 25 wt % rhenium; it is sold as sheet, rod, and heavy wire and may be fabricated for various uses. An important use of these alloys is in the construction of thermocouples. Various combinations (3 wt % Re–97 wt % W or 5 wt % Re–95 wt % W vs 25 wt % Re–75 wt % W) are useful for measurement of temperatures to 2500°C (see Temperature measurement).

Rhenium coatings are used on some electrical contacts because of the element's resistance to arc-erosion wear. Rhenium is also used in contacts for engine magnetos. Interruption of the current effects the formation of a thin oxide coating that prevents the contacts from sticking or welding together. The oxide coating also conducts and does not impair further efficient operation (see also Electrical connectors).

By far the largest use of rhenium is in bimetallic petroleum-reforming catalysts in the production of unleaded and low-lead gasoline. These catalysts contain ca 0.3 wt % platinum and 0.3 wt % rhenium; in newer catalyst systems a higher proportion of rhenium may be used. Such systems have an advantage over monometallic systems, which usually contain platinum alone, in that the former can be regenerated almost indefinitely. Platinum–rhenium catalysts are also used in the production of benzene, toluene, and xylenes by reforming (see BTX processing). Recently, the use of rhenium catalysts for the liquid-phase reduction of nitrobenzenes has been studied. Such catalysts appear comparable to palladium systems and much superior to nickel. Rhenium catalysts have also been studied in olefin metathesis reactions.

A rhenium–molybdenum alloy is reported to be superconducting at 10 K (see Superconducting materials).

Rhenium Compounds

As a general rule, the elements in the second and third transition series have similar chemical properties which contrast substantially with the properties of the first member of the series. This pattern of behavior is seen in Group VIIB: rhenium and technetium differ considerably from manganese.

Compounds of rhenium exist with metal oxidation states between -1 and $+7$. Lower oxidation states, ie, -1, 0, $+1$, occur only with carbonyl groups and other π-acceptor ligands. The higher oxidation states are much more frequently encountered with rhenium than with manganese; indeed many rhenium complexes with oxidation states of $+4$, $+5$, $+6$, and $+7$ have been identified. In contrast to manganese, there is little chemistry associated with a $+2$ ion. For the intermediate oxidation states, there is an abundance of polynuclear compounds of rhenium in which metal–metal bonding occurs. A survey of known types of rhenium compounds is presented in Table 2.

Carbonyls and Related Compounds. The parent compound of the rhenium carbonyls is $Re_2(CO)_{10}$. This white crystalline compound (mp 177°C) is volatile and is soluble in most organic solvents. Its preparation in a high pressure reaction between Re_2O_7, H_2, and CO was reported in 1941. Its molecular structure has two square pyramidal $Re(CO)_5$ groups linked by a metal–metal bond. It is available commercially as a specialty chemical (1981 price, ca $20/g). It is the precursor to other low valent rhenium carbonyl compounds, including the halides, $ReX(CO)_5$ (X = Cl, Br, I); alkyl, aryl, and acyl compounds, $Re(R)(CO)_5$; the hydride complexes $ReH(CO)_5$ [16457-30-0], $Re_2H_2(CO)_8$ [38887-05-7], and $Re_3H_3(CO)_{12}$ [12146-47-3]; and hydrocarbon complexes including $Re(CO)_3(\eta\text{-}C_5H_5)$ and $[Re(NO)(CO)_2(\eta\text{-}C_5H_5)]PF_6$ [12306-73-9]. Recent research interests in this area have focused on the photochemical activation of H_2 by $Re_2(CO)_{10}$ and on the reduction of a coordinated carbon monoxide ligand to methane or methanol. The latter reaction provides mechanistic information about the Fischer-Tropsch reaction (see Fuels, synthetic).

Pyrolysis of $Re_2(CO)_{10}$ at 400°C *in vacuo* or in an inert atmosphere has been used to obtain pure rhenium metal.

Oxides and Sulfides. Rhenium reacts with O_2 at moderate temperatures to give yellow solid Re_2O_7 (mp 220°C). The solid-state structure of this material contains tetrahedral and octahedral rhenium atoms in a lattice structure, which is so arranged that the formation of molecular Re_2O_7 upon vaporization can occur with ease. Perrhenic acid is a strong acid and forms when Re_2O_7 dissolves in water; the acid may be isolated upon evaporation of the solvent as crystalline $Re_2O_7.2H_2O$ [41017-57-6]. Lower oxides of rhenium, ie, ReO_3 [1314-28-9] and ReO_2 [12036-09-8], form upon heating of mixtures of Re_2O_7 and Re in the proper mole ratio. The hydrated compound

Table 2. Examples of Rhenium Compounds

Oxidation state	Electronic configuration	Coordination number	Geometry	Examples[a]	CAS Registry Number
Re(−I)	d^8	5		$Na[Re(CO)_5]$	[33634-75-2]
Re(0)	d^7	6	octahedral	$Re_2(CO)_{10}$	[14285-68-8]
Re(I)	d^6	6	octahedral	$ReCl(CO)_5$	[14099-01-5]
				$ReBr(CO)_5$	[14220-21-4]
				$ReI(CO)_5$	[13821-00-6]
				$K_5Re(CN)_6$	[39700-07-7]
				$[Re(CNC_6H_5)_6]I$	[81195-37-1]
				$Re(CO)_3(\eta\text{-}C_5H_5)$	[12079-73-1]
				$ReCl(dppe)_2(N_2)$	[25349-29-5]
Re(II)	d^5	6	octahedral	$ReCl_2(diars)_2$	[14325-13-4]
				$[ReCl(N_2)(dppe)_2]PF_6$	[40740-27-0]
Re(III)	d^4	6	octahedral	$ReCl_3(PR_3)_3$	
				$[Re_3Cl_9]_x$	[13569-63-6]
				$Cs_3[Re_3Cl_{12}]$	[19568-81-1]
		5	square pyramid	$K_2Re_2Cl_8$	[13841-78-6]
		6	trigonal prism	$Re[S_2C_2(C_6H_5)_2]_3$	[14264-08-5]
		7		$K_4[Re(CN)_7]\cdot2H_2O$	[59370-53-5]
				$ReH_3(dppe)_2$	[17300-50-4]
Re(IV)	d^3	6	octahedral	K_2ReCl_6	[16940-97-9]
				$ReCl_4$	[13569-71-6]
				$K_4Re_2OCl_{10}$	[19979-02-3]
				$ReCl_4(PR_3)_2$	
Re(V)	d^2	5		Re_2Cl_{10}	[39368-69-9]
				ReF_5	[30937-52-1]
		6	octahedral	$ReOCl_3[P(C_6H_5)_3]_2$	[17442-18-1]
				K_2ReOCl_5	[17443-52-6]
		8		$ReH_5[P(C_6H_5)_2C_2H_5]_3$	[12104-30-2]
Re(VI)	d^1	5	square pyramid	$ReOCl_4$	[13814-76-1]
				$ReO(CH_3)_4$	[53022-70-1]
		6	octahedral	ReF_6	[10049-17-9]
				$Re(CH_3)_6$	[56090-02-9]
		8	square antiprism	K_2ReF_8	[57300-90-0]
Re(VII)	d^0	4	tetrahedral	$KReO_4$	[10466-65-6]
				ReO_3Cl	[7791-09-5]
		4,6	tetrahedral and octahedral	$(Re_2O_7)_x$	[1314-68-7]
		7	pentagonal bipyramid	ReF_7	[17029-21-9]
		9	tricapped trigonal prism	$[(C_2H_5)_4N]_2(ReH_9)$	[25396-44-5]

[a] dppe = 1,2-bis(diphenylphosphino)ethane; diars = o-bis(dimethylarsino)benzene; PR_3 = tertiary phosphine.

$ReO_2\cdot2H_2O$ [54706-20-6] forms from the hydrolysis of $ReCl_6^{2-}$. A black heptasulfide Re_2S_7 [12038-67-4] is prepared when a ReO_4^- solution in HCl is saturated with H_2S. This may be reduced to ReS_3 [12166-08-4] by H_2 or to ReS_2 upon heating at 750°C under nitrogen; ReS_2 is also the product from the reaction of rhenium and sulfur.

Salts of perrhenic acid may be obtained in acid–base reactions and may include the tetrahedral ReO_4^- anion or the octahedral anion ReO_6^{5-}, eg, in $Ba_5(ReO_6)_2$ [13598-09-9]. Ammonium perrhenate and perrhenic acid, as well as rhenium metal, are the forms of rhenium that are sold by the primary suppliers of this element.

Rhenium oxides have been studied as catalyst materials in oxidation reactions, eg, sulfur dioxide to trioxide, sulfite to sulfate, nitrite to nitrate, but there has been no commercial development in this area. These compounds have also been used as catalysts for reductions but do not have exceptional properties. Rhenium sulfide catalysts have been used for hydrogenations of organic compounds, including benzene and styrene, and for dehydrogenations of alcohols to give aldehydes and ketones. The significant property of these catalyst systems is that they are not poisoned by sulfur compounds.

Halides and Halide Complexes. Rhenium reacts with chlorine at ca 600°C to produce rhenium pentachloride, Re_2Cl_{10}, a volatile species which is dimeric via bridging halide groups. Bromine reacts with rhenium in a similar fashion, but iodine is unreactive toward the metal. The compounds $ReCl_4$, $ReBr_4$ [36753-03-4], and ReI_4 [59301-47-2] can be prepared by careful evaporation of a solution of $HReO_4$ and HX. Lower oxidation state halides are also prepared from the pentavalent or tetravalent compounds by thermal decomposition or chemical reduction.

Various coordination complexes of these halides are known; the more common examples have the formulas ReX_4L_2, $Re_2X_8^{2-}$, $Re_3X_9L_3$, and $Re_3X_{12}^{3-}$, where L indicates a ligand. The rhenium(IV) complexes form in reductive reactions between the ligand and Re_2Cl_{10}; eg, reduction occurs spontaneously when Re_2Cl_{10} dissolves in acetonitrile, giving a product $ReCl_4(CH_3CN)_2$ [16853-53-5]. The complexes of rhenium(III) are obtained from $(Re_3Cl_9)_x$, $(Re_3Br_9)_x$ [13569-49-8], and $(Re_3I_9)_x$ [15622-42-1] and a ligand. The rhenium(III) complexes are generally regarded as classic examples of compounds containing metal–metal bonds. The structure of $Re_2Cl_8^{2-}$ is two square-planar $ReCl_4$ units linked by a short (0.2237 nm) rhenium–rhenium quadruple bond. The basic unit in $(Re_3X_9)_x$ and in salts of $Re_3X_{12}^{3-}$ is a triangle of rhenium atoms linked by metal–metal single bonds (ca 0.25 nm) and halide bridging atoms.

Hydrides and Alkyls. Reduction of $KReO_4$ by potassium in ethanol gives K_2ReH_9 [25396-46-7]. The identity of this compound was unknown until its structure was defined by neutron diffraction study. The anion contains rhenium with nine hydride ligands in a tricapped, trigonal prismatic arrangement. Reactions of this species with various phosphines give $[ReH_8L]^-$, ReH_7L_2, and ReH_5L_3 (L = tertiary phosphines). The neutral complexes are also formed from phosphine–rhenium halide complexes with $LiAlH_4$.

The compounds $Re(CH_3)_6$, $ReO(CH_3)_4$, $Li_2Re_2(CH_3)_8$ [60975-25-9], $ReO_2(CH_3)_3$ [56090-01-8], and ReO_3CH_3 [70197-13-6] have been prepared. The first two compounds were obtained from reaction of rhenium halides or oxyhalides and methyllithium; the last three were formed from the species by oxidation or reduction.

The use of these hydride and alkyl complexes as catalysts is being studied.

BIBLIOGRAPHY

"Rhenium and Rhenium Compounds" in *ECT* 1st ed., Vol. 11, pp. 721–730, by C. F. Hiskey, Polytechnic Institute of Brooklyn; "Rhenium" in *ECT* 2nd ed., Vol. 17, pp. 411–423, by John H. Port, Cleveland Refractory Metals Division, Chase Brass & Copper Co., Inc.

1. W. Noddack, I. Tacke, and O. Berg, *Naturwissenshaften* **13**, 567 (1925).
2. M. E. Weeks and H. M. Leicester, *The Discovery of the Elements*, 7th ed., Chemical Education Press,

Easton, Pa., 1968, pp. 823–827; J. G. F. Druce, *Rhenium*, Cambridge University Press, Cambridge, UK, 1948.

3. N. I. Sax, *Dangerous Properties of Industrial Materials*, 4th ed., Van Nostrand-Reinhold Company, New York, 1975, p. 1079.

4. L. H. Aller, *The Abundance of the Elements*, Vol. 7, Interscience Publishers, Inc., New York, 1961, Table 2-2, p. 32–32; L. H. Ahrens and S. R. Taylor, *Spectrochemical Analysis*, Addison Wesley Press, Reading, Mass., 1961, p. 93; K. Rankama, *Isotope Geology*, Pergamon Press, London, 1954, p. 135.

5. R. Michel, U. Herpers, H. Kulus, and W. Herr, *J. Radioanal. Chem.* **17,** 177 (1973).

6. R. W. Asmussen and H. Soling, *Acta Chem. Scand.* **8,** 563 (1954).

7. R. P. Elliot, *Constitution of Binary Alloys*, 1st Suppl., McGraw-Hill, Inc., New York, 1965, pp. 629, 776.

8. J. P. King and J. W. Cobble, *J. Am. Chem. Soc.* **79,** 1550 (1957).

9. R. Colton, *The Chemistry of Rhenium and Technetium*, Interscience Publishers, a division of John Wiley & Sons, Inc., New York, 1965, pp. 20–26; R. D. Peacock, *The Chemistry of Technetium and Rhenium*, Elsevier Publishers, Amsterdam, 1966, pp. 122–125.

General References

R. D. Peacock in A. F. Trotman-Dickinson, ed., *Comprehensive Inorganic Chemistry*, Vol. 3, Pergamon Press, London, 1973, p. 905.

F. A. Cotton and G. Wilkinson, *Comprehensive Inorganic Chemistry*, 4th ed., John Wiley & Sons, Inc., New York, 1980, p. 883.

L. J. Alverson, in *Mineral Yearbook*, Vol. 1, Bureau of Mines, U.S. Government Printing Office, Washington, D.C., 1980, p. 743.

D. F. C. Morris and E. L. Short, *Handbook of Geochemistry*, Springer-Verlag Publishers, Heidelberg, Federal Republic of Germany, 1965, p. 75-1.

R. D. Peacock, *The Chemistry of Technetium and Rhenium*, Elsevier Publishing Company, Amsterdam, The Netherlands, 1966.

R. Colton, *The Chemistry of Rhenium and Technetium*, Interscience Publishers, a division of John Wiley & Sons, Inc., New York, 1965.

PAUL M. TREICHEL
University of Wisconsin

RHEOLOGICAL MEASUREMENTS

Rheology is the science of the deformation and flow of matter. It is concerned with the response of materials to mechanical force. That response may be irreversible flow, reversible elastic deformation, or a combination of the two (see Fluid mechanics). An understanding of rheology and the ability to measure rheological properties is necessary before rheology can be controlled, and control is essential for the manufacture and handling of a great many materials, eg, foods, cosmetics (qv), plastics, paints, drilling muds, etc (see Paint; Petroleum, drilling fluids). For example, in the coatings industry, rheology in the form of flow is crucial from the time that raw materials are pumped into the paint plant to the time that the finished paint is applied to a house or to manufactured goods.

The study of flow and elasticity dates to antiquity. Practical rheology existed for centuries before Hooke and Newton proposed the basic laws of elastic response and simple viscous flow, respectively, in the 17th Century. Further advances in understanding came in the mid-19th Century with the proposed models for viscous flow in round tubes. The introduction of the first practical rotational viscometer in 1890 was another milestone. In recent years the science of rheology has grown rapidly.

This article is concerned with rheological principles as well as measurements. The emphasis is on flow, ie, permanent deformation, of liquids and solids. The flow properties of a liquid are defined by its resistance to flow, ie, viscosity, and may be measured by determining the rate of flow through a capillary, the resistance to flow when the fluid is sheared between two surfaces, or the rate of motion of a bubble or ball moving through the fluid. The mechanical properties of an elastic solid may be studied by applying a stress and measuring the deformation or strain. These two parameters are sufficient to define the behavior of an ideal elastic solid, such as a metal, but many solids, such as polymers, undergo flow in addition to recoverable elastic deformation. Also, a number of liquids show elastic as well as flow behavior. These materials are viscoelastic, and additional techniques beyond those indicated for solids and liquids are needed for complete characterization. Examples of such methods are the measurement of response to sinusoidal oscillatory motion; measurement of flow with time after application of stress, ie, creep; and the measurement of the rate and degree of recovery after removal of stress, ie, stress relaxation.

Microscopes are just as important as viscometers for defining and solving flow problems (1). This is particularly true for formulated materials, eg, paints, inks, many food products, cosmetics, and pharmaceuticals, where use of a microscope allows the investigator to see the physical structure of the material and its effect on flow.

Deformation is the relative displacement of points of a body and can be divided into two general types: flow and elasticity. Flow is irreversible deformation; when the stress is removed, the material does not revert to its original configuration. This irreversibility means that work is converted to heat. Elasticity is reversible deformation; the deformed body recovers its original shape and the applied work is largely recoverable. Viscoelastic materials show both flow and elasticity. A good example is Silly Putty which bounces like a rubber ball when dropped but slowly flows to form a puddle if allowed to stand. Viscoelastic materials provide special challenges in terms of modeling behavior and devising techniques for measurement.

The usual way of defining the rheological properties of a material is to determine the resistance to deformation. This may be done in qualitative terms, ie, a thin or thick liquid or a soft or hard solid, which indicates something about the ease of deformation but is not very useful. What is needed is an index that provides an actual measure of the resistance to deformation, preferably in terms of the stress required to produce a unit deformation or rate of deformation. The index for flow is viscosity, which is the resistance to flow under mechanical stress. The index for elastic deformation is elastic modulus.

Viscosity

A liquid is a material that continues to deform as long as it is subjected to a tensile or shear stress. For a liquid under shear, the rate of deformation or shear rate is proportional to the shearing stress. The original exposition of this relationship is Newton's law, which essentially states that the ratio of the stress to the shear rate is a constant. That constant is viscosity. Under Newton's law, viscosity is independent of shear rate. This is true for ideal or Newtonian liquids, but the viscosities of many liquids, particularly a number of those of interest to industry, are not independent of shear rate. These non-Newtonian liquids may be classified according to their viscosity behavior as a function of shear rate. Some exhibit shear thinning, whereas others give shear thickening. Some liquids at rest behave like solids until the shear stress exceeds a certain value, called the yield stress, after which point they flow.

Some commonly observed types of flow behavior are shown in Figure 1, in which the shear stress is plotted vs shear rate. These plots are called flow curves and are frequently used to express the rheological behavior of liquids.

The viscosity of a fluid is equal to the slope of the shear stress–shear rate curve ($\eta = d\tau/d\dot\gamma$). The quantity $\tau/\dot\gamma$ is the absolute viscosity η for a Newtonian liquid and the apparent viscosity η_a for a non-Newtonian liquid. The kinematic viscosity is the viscosity coefficient divided by the density, $\nu = \eta/\rho$. The fluidity is the reciprocal of the viscosity, $\phi = 1/\eta$. The most common units for viscosity η are (dyn·s)/cm^2 or g/(cm·s), which are called poise and often are expressed as centipoise cP. These units

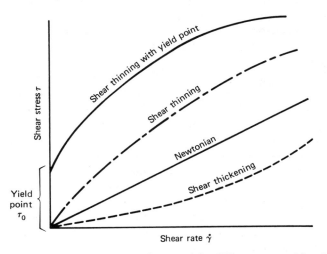

Figure 1.   Flow curves (shear stress vs shear rate) for different types of flow behavior.

are being superseded by the SI units of pascal seconds (Pa·s) and mPa·s (1 mPa·s = 1 cP). In the same manner, the shear stress units of dyn/cm^2 are being replaced by pascals (10 dyn/cm^2 = 1 Pa) and newtons per square meter (N/m^2 = Pa). Units of shear rate are s^{-1} in both systems. The common units for kinematic viscosity ν are stokes (St) and centistokes (cSt), and the exactly equivalent SI units are cm^2/s and mm^2/s, respectively.

Many flow models have been proposed and are useful for treatment of experimental data or for describing flow behavior (see Table 1). It is probable that no given model fits the rheological behavior of a material over an extended shear rate range. Nevertheless, these models are useful for summarizing rheological data and are frequently encountered in articles relating to the rheology of liquids.

A number of the models in Table 1 can be applied to shear thinning in situations where time-dependent effects are absent. This situation encompasses many technically important materials from polymer solutions to latexes, pigment slurries, and blood. At high shear rates, most of these materials tend to a Newtonian viscosity limit. At intermediate shear rates, the power law or the Casson equation are useful approximations. At lower shear rates, these materials tend either to a yield point or to a low shear Newtonian limiting viscosity. The power law with yield point and the Casson equation are good approximations for materials with yield points. The Williamson model and Cross equation are useful models for systems with limiting low and high shear Newtonian viscosity behavior. Most models are only approximate and are less accurate if extended over a large range of shear rates. Serious errors may occur if models are used to extrapolate data outside the range of experimental determination.

Non-Newtonian fluids are characterized by measuring viscosity or shear stress at a number of different shear rates, usually with a rotational viscometer. The quantity measured may then be plotted against shear rate to identify behavior, eg, shear thinning or thickening. If a sufficiently wide range of shear rates is measured, it may be possible to identify high and low shear Newtonian regions. However, this usually takes several viscosity sensors or even several different viscometers. The various flow models may be tested by suitable plots: log shear stress vs log shear rate for the power law model, $\tau^{1/2}$ vs $\dot{\gamma}^{-1/2}$ for the Casson model, etc. Yield stress τ_0 and plastic viscosity $\eta_{\mathrm{p}} = (\tau - \tau_0)/\dot{\gamma}$ may be determined from the intercept and slope beyond the intercept, respectively, of a shear-stress vs shear-rate plot.

Table 1. Flow Equations for Various Flow Models

Model	Flow equation
Newtonian	$\tau = \eta\dot{\gamma}$
plastic body or Bingham body	$\tau - \tau_0 = \eta\dot{\gamma}$
power law	$\tau = k\lvert\dot{\gamma}\rvert^{n}$
power law with yield value	$\tau - \tau_0 = k\lvert\dot{\gamma}\rvert^{n}$
Casson fluid	$\tau^{1/2} - \tau_0^{1/2} = \eta_\infty^{1/2}\,\dot{\gamma}^{1/2}$
Williamson	$\eta = \eta_\infty + \dfrac{(\eta_0 - \eta_\infty)}{1 + \dfrac{\lvert\tau\rvert}{\tau_{\mathrm{rel}}}}$
Cross and extended Williamson	$\eta = \eta_\infty + \dfrac{(\eta_0 - \eta_\infty)}{1 + \alpha\dot{\gamma}^{n}}$

Thixotropy and Other Time Effects. In addition to the nonideal behavior described, many fluids exhibit time-dependent effects. Some fluids increase in viscosity (rheopexy) or decrease in viscosity (thixotropy) with time when sheared at a constant shear rate. These effects can occur in fluids with or without yield values. Rheopexy is a comparatively rare phenomenon, but thixotropic fluids are fairly common. Examples of thixotropic materials are starch pastes, gelatin (qv), mayonnaise, drilling muds, and latex paints. The thixotropic effect is shown in Figure 2, where the curves are for a sample exposed first to increasing and then to decreasing shear rates. Because of the decrease in viscosity with time as well as with shear rate, the up-and-down flow curves do not superimpose. Instead, they form a hysteresis loop, the so-called thixotropic loop. Since flow curves for thixotropic or rheopectic liquids depend on the shear history of the sample, different curves for the same material can be obtained depending on the experimental procedure.

Experimentally, it is sometimes difficult to detect differences between a shear-thinning liquid, in which the viscosity coefficient decreases with increasing shear, and a thixotropic material, in which the viscosity decreases with time, because of the combined shear and time effects that occur during a series of measurements. This is especially true if only a few data points are collected. In addition, a number of materials, eg, paints and printing inks, are both thixotropic and shear-thinning.

Viscosity–time measurements during shear or after shearing can be used to show time-dependent effects. The plots in Figure 3 are representative of the results of such measurements on a thixotropic material. The viscosity drops sharply at first, but the rate of change continually decreases until a constant or nearly constant level is reached. This is sometimes referred to as the sheared state. When the shearing action stops, the viscosity increases quickly at first, then more slowly and approaches the original level asymptotically. A good example of such behavior is a latex house paint. The shearing plot represents the brushing action, whereas the recovery plot shows what happens when the brushing stops. The thixotropic behavior allows the paint to be easily brushed to a thin film and gives a short period of time for the brushmarks to level; then the viscosity increase prevents running and sagging.

Causes of time-dependent behavior include irreversible changes, eg, cross-linking, coagulation, degradation, and mechanical instability, and reversible changes involving the breaking and reforming of colloidal aggregations and networks. Models of time-dependent behavior are less satisfactory and are more controversial than for shear-

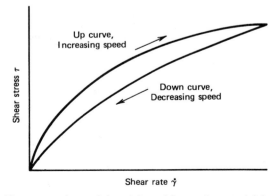

Figure 2.   Flow curves (up and down) for a thixotropic material: hysteresis loop.

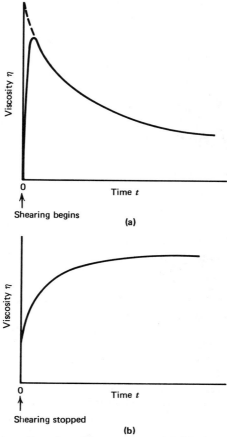

Figure 3.   Viscosity–time effects for a thixotropic material: (**a**), shearing; (**b**), recovery. A nonthixotropic material would give horizontal lines in both cases.

dependent behavior. Also, few comprehensive investigations of the viscosity–shear–time profiles of thixotropic and rheopectic materials have been published. However, there are a few good discussions of thixotropy in the literature, particularly in references 2 and 3. The complexities and history of investigation of rheopexy are described in references 4 and 5.

One method for measuring time-dependent effects is to determine the decay of shear stress as a function of time at one or more constant shear rates (6). The results of such an experiment are shown in Figure 4. A rotational viscometer, which is connected to a recorder, is used. After the sample is loaded and allowed to come to mechanical and thermal equilibrium, the viscometer is turned on and the rotational speed is increased in steps starting from the lowest speed. The resultant shear stress is recorded with time. On each speed change, the shear stress reaches a maximum value, then decreases exponentially toward an equilibrium level. The peak shear stress, which is obtained by extrapolating the curve to zero time, and the equilibrium shear stress are indicative of the viscosity-shear behavior of unsheared and sheared material, respectively. The stress-decay curves are indicative of the time-dependent behavior. A rate constant for the relaxation process can be determined at each shear rate. In

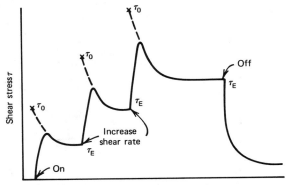

Figure 4. Decay of shear stress during steady shear at various shear rates. Determination of zero-time shear stresses or yield stresses and equilibrium shear stresses.

addition, zero time and equilibrium shear-stress values can be used to construct a hysteresis loop which is similar to that shown in Figure 2 but, unlike that plot, is independent of acceleration and time of shear.

Another method for estimating thixotropy involves the hysteresis of the thixotropic loop. There are two different techniques. One method involves calculating or measuring the area of the thixotropic loop and has been shown to work well with printing inks (1). A variation is to determine the up curve on an undisturbed sample, shear the sample at high shear (>2000 s^{-1}) for 30–60 s, then determine the down curve (7). The data are then plotted as Casson-Asbeck plots, $\eta^{1/2}$ vs $\dot{\gamma}^{-1/2}$ (8), as shown in Figure 5. Such plots are best used for comparison and ranking, but a measure of the degree of thixotropy can be gained by measuring the angle formed by the two lines or by determining the area of the triangle formed by the lines and a vertical line through a given value for $\dot{\gamma}^{-1/2}$. Additional methods for determining the degree of thixotropy are described in ref. 9.

Also useful is the time necessary for the thixotropic structure of the sample to rebuild on standing. This time period can range from minutes to several hours or even days. However, in general, it is difficult to relate the time-decay characteristics of the shear stress under constant shear to the rate of rebuilding the structure on standing at zero shear rate.

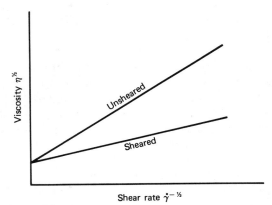

Figure 5. Casson plots of sheared and unsheared paints. The degree of divergence of the lines is used to estimate thixotropy.

Results from measurements of time-dependent effects depend on sample history and experimental conditions and should be considered approximate. For example, the state of an unsheared or undisturbed sample is a function of its previous shear history and the length of time since it underwent shear. The area of a thixotropic loop depends on the shear range covered, the rate of acceleration of shear, and the length of time at the highest shear rate. However, measurements of time-dependent behavior can be very useful in evaluating and comparing a number of industrial products and in solving flow problems.

Dilute Polymer Solutions. The measurement of dilute solution viscosities of polymers is widely used for polymer characterization. Very low concentrations reduce intermolecular interactions and allows measurement of polymer–solvent interactions. These measurements usually are made in capillary viscometers, some of which have provisions for direct dilution of the polymer solution in the viscometer. The key viscosity parameter for polymer characterization is the limiting viscosity number or intrinsic viscosity $[\eta]$. It is calculated by extrapolation of the viscosity number, ie, reduced viscosity, or the logarithmic viscosity number, ie, inherent viscosity, to zero concentration.

The viscosity ratio or relative viscosity η_{rel} is the ratio of the viscosity of the polymer solution to the viscosity of the pure solvent. With capillary viscometer measurements, the relative viscosity (dimensionless) is the ratio of the flow time for the solution t to the flow time for the solvent t_0

$$\eta_{\mathrm{rel}} = t/t_0 = \eta/\eta_0$$

The specific viscosity (dimensionless) is defined as

$$\eta_{\mathrm{sp}} = (\eta - \eta_0)/\eta_0 = \eta_{\mathrm{rel}} - 1$$

The viscosity number or reduced viscosity is defined in units of m^3/kg or dL/g as

$$\eta_{\mathrm{red}} = \eta_{\mathrm{sp}}/C = (\eta_{\mathrm{rel}} - 1)/C$$

and the logarithmic viscosity number or inherent viscosity, in units of m^3/kg or dL/g is

$$\eta_{\mathrm{inh}} = (\ln \eta_{\mathrm{rel}})/C$$

where C is the concentration of polymer in convenient units (traditionally $g/100\ cm^3$ but kg/m^3 in SI units). The viscosity number and logarithmic viscosity number vary with concentration, but each can be extrapolated (see Fig. 6) to zero concentration to give the limiting viscosity number (intrinsic viscosity)

$$[\eta] = \lim_{C \to 0} \frac{\eta_{\mathrm{sp}}}{C} = \lim_{C \to 0} \frac{\ln \eta_{\mathrm{rel}}}{C}$$

Extrapolation to infinite dilution requires viscosity measurements at usually four or five concentrations. For relative measurements or rapid determinations, a single-point equation often may be used. A very useful expression is the following (10):

$$[\eta] = 2(\eta_{\mathrm{sp}} - \ln \eta_{\mathrm{rel}})^{1/2}C$$

An even simpler but equally useful method is to approximate the limiting viscosity number by the logarithmic viscosity number of a single sufficiently dilute solution ($C = 1$ or $2\ kg/m^3$ (0.1 or $0.2\ g/100\ cm^3$)), ie,

$$[\eta] \cong \eta_{\mathrm{inh}} = (\ln \eta_{\mathrm{rel}})/C \text{ at low values of } C$$

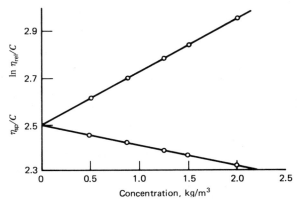

Figure 6. Plots of viscosity number ($\eta_{\text{red}} = \eta_{\text{sp}}/C$) and the logarithmic viscosity number ($\eta_{\text{inh}} = \ln \eta_{\text{rel}}/C$) vs concentration. Extrapolations to zero concentration give the limiting viscosity number $[\eta]$.

The limiting viscosity number or intrinsic viscosity is an important measure of the characteristics of the polymer and of its interactions with the solvent in which it is dissolved. The limiting viscosity number depends on the polymer, solvent, and temperature, but under a given set of conditions it is related to the molecular weight by the relation $[\eta] = KM^a$, where K and a are constants and M is the molecular weight of the polymer. Tables of K and a for a large number of polymers and solvents are available in the literature (11–12).

For an excellent summary of equations, techniques, and references relating to the viscosity of dilute polymer solutions, see reference 13.

Concentrated Polymer Solutions. Knowledge of the viscosity behavior of concentrated solutions is important to the manufacture and application of a number of commercial materials, eg, caulks, adhesives, inks, paints, and varnishes. This knowledge may be gained by a variety of methods including the use of simple capillary viscometers, extrusion rheometers, and rotational viscometers. Unlike dilute solutions, concentrated polymer solutions show a great deal of interaction between the macromolecules. The degree of interaction is governed by the concentration, the characteristics of the chains, and the nature of the solvent. A convenient measure of concentration is the dimensionless reduced concentration $\tilde{C}$, which consists of the product of the concentration and the limiting viscosity number (intrinsic viscosity) $C[\eta]$ (14). The transition from dilute to concentrated solutions occurs at a critical concentration C_c and corresponds to $\tilde{C}$ values of several units. In addition, there is a critical molecular weight such that at $M > M_c$ and $C > C_c$ a fluctuating entanglement network forms. For a concentrated solution, properties above and below M_c may be quite different. For example, the dependence of viscosity on molecular weight, which is much greater in concentrated solutions than in dilute solutions, changes from a value on the order of unity below M_c to one of 3.4–3.5 above M_c (15). That is

$$\eta = KM \text{ below } M_c$$

$$\eta = KM^{3.4–3.5} \text{ above } M_c$$

Viscosity in these expressions should really be the zero shear viscosity η_0, but since the relationships hold for low shear measurements in many cases, the notation remains in the more general form η.

Dilute solutions have low viscosities and are Newtonian. Depending on the concentration, solvent used, and the shear rate of measurement, concentrated polymer solutions may give wide ranges of viscosity and appear to be Newtonian or non-Newtonian. This is illustrated in Figure 7 where solutions of a styrene–butadiene–styrene block copolymer are Newtonian and quite viscous at low shear rates but become shear-thinning at high shear rates and shear to relatively low viscosities beyond 10^5 s^{-1} (16). The shear rate at which the break in behavior occurs depends on the concentration and on the solvent. The viscosity of concentrated polymer solutions is discussed in detail in ref. 14 as well as in refs. 15 and 17.

Melt Viscosity. The study of the viscosity of polymer melts is important for the manufacturer who must supply suitable materials and for the fabrication engineer who must select polymers and fabrication methods. Thus, melt viscosity as a function of temperature, pressure, rate of flow, and polymer molecular weight and structure is of considerable practical importance. It should be noted that polymer melts exhibit elastic as well as viscous properties. This is evident in the swell of the polymer melt upon emergence from an extrusion die, behavior which results from the recovery of stored elastic energy.

A number of experimental methods have been applied to measure the melt viscosity of polymers, but capillary extrusion techniques are very popular. Rotational methods are also used, and some permit the measurement of normal stress effects resulting from elasticity as well as of viscosity. Oscillatory shear measurements also are useful for measuring elasticity. A special rotational rheometer has been designed to measure both viscosity and elasticity of polymer melts (18).

Polymer melts show a low shear-rate Newtonian limit and a region of diminishing viscosity with increasing shear rate. Although it is likely that a high shear-rate Newtonian region exists, this has generally not been observed experimentally because of the effects of heat generation and polymer degradation at high shear rates.

The limiting low shear or zero shear viscosity η_0 of the molten polymer can be related to its weight-average molecular weight $\overline{M}_w$ by the same relations noted for concentrated solutions; ie, $\eta_0 = K_1 \overline{M}_w$ for a low molecular weight, and $\eta_0 = K_2 \overline{M}_w^{3.5}$ for a high molecular weight. The transition between two forms of behavior occurs at

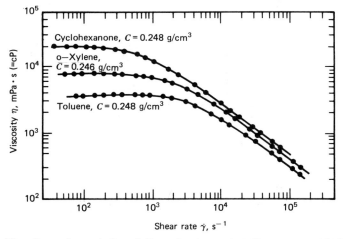

Figure 7.   Viscosity vs shear rate for solutions of a styrene–butadiene–styrene block copolymer (16). Courtesy of the Society of Plastics Engineers, Inc.

a critical molecular weight M_c, which corresponds to a critical chain length Z_c. The transition is clearly shown in Figure 8, which is a plot of Newtonian viscosity vs chain length in terms of carbon atoms for a series of molten polyethylenes (17). The above relationships are true for narrow molecular-weight distribution polymers. For polymers with broad molecular-weight distributions, viscosity should depend on a molecular weight average between $\overline{M}_w$ and the next higher average $\overline{M}_z$ (Z average), approaching $\overline{M}_z$ as the distribution broadens. If branching is presented, $g\overline{M}_w$ should be substituted for $\overline{M}_w$, where g is the branching index (19).

The temperature dependence of melt viscosity at temperatures considerably above the glass-transition temperature T_g approximates an exponential function of the Arrhenius type. However, near the glass transition the viscosity–temperature relationship for many polymers is in better agreement with the Williams–Landel–Ferry (WLF) treatment (20). These authors showed that, with a proper choice of reference temperature T_s, the ratio of the viscosity to the viscosity at the reference temperature could be expressed by a single universal equation:

$$\log (\eta/\eta_s) = \frac{-8.86\,(T - T_s)}{101.6 + (T - T_s)}$$

where T_s often is defined as $T_g + 50°C$. In general the WLF equation holds over the temperature range T_g–($T_g + 100°C$).

Pressure also affects melt viscosity. The compression of a melt reduces the free volume and, therefore, raises the viscosity. Low density polyethylene gives a viscosity increase by a factor of roughly ten over a static pressure range of 34–170 MPa (5,000–25,000 psi). More highly compressible polystyrene is affected even more by pressure, showing an increase in viscosity by a factor of ca 500 over the same pressure range (21).

The rheology of molten polymers is reviewed in references 22–24.

Dispersed Systems. Many fluids of commercial and biological importance are dispersed systems, eg, solids suspended in liquids, ie, dispersions, and liquid–liquid suspensions, ie, emulsions. Examples of dispersions include inks, paints, pigment

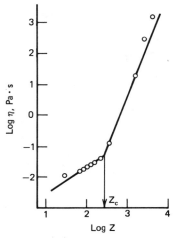

Figure 8. Newtonian viscosity vs chain length in terms of the number of carbon atoms for a series of molten polyethylenes (17). To convert Pa·s to P, multiply by 10. Courtesy of Springer Verlag.

slurries, and concrete; examples of emulsions include oil-and-vinegar salad dressings, margarine, butter, mayonnaise, and milk. Blood seems to fall in between as it is a suspension of deformable but not liquid particles, and it does not behave exactly like either a dispersion or an emulsion (25) (see Emulsions).

When a solid or liquid is dispersed in a liquid, the viscosity rises. In many cases, the addition transforms a Newtonian fluid into one showing non-Newtonian flow behavior. Shear thinning is very common, but shear thickening, which is often called dilatancy, also is possible. Shear thinning results from the ability of the solid particles or liquid droplets to come together to form network structures when at rest or under low shear. With increasing shear, the interlinked structure gradually breaks down and the resistance to flow decreases. The viscosity of a dispersed system depends on hydrodynamic interactions between particles or droplets and the liquid, particle–particle interactions, eg, bumping, and interparticle attractions that promote the formation of aggregates, flocs, and networks.

Dispersions have been studied in much greater detail than emulsions, probably because of the greater complexity of the latter. Emulsions tend to be unstable. In many cases the droplets begin to coalesce soon after the emulsion forms. Thus the emulsion is continually changing. In addition, there is the possibility of the droplets deforming under shear or when they are packed tightly together. This makes it difficult if not impossible to apply theories developed for solid spheres or other well-defined geometries of shape. A detailed treatment of the viscosity of colloidal suspensions, including dispersions and emulsions, is given in ref. 26. A recent review of the viscous and elastic behavior of dispersions is given in ref. 27.

Viscosity–Concentration Relationship for Dilute Dispersions. The viscosities of dilute dispersions have received considerable theoretical and experimental treatment, partly because of the similarity between polymer solutions and small particle dispersions at low concentration. Nondeformable spherical particles usually are assumed in the cases of molecules and particles. The key viscosity quantity for dispersions is the relative viscosity or viscosity ratio η_{rel}

$$\eta_{rel} = \frac{\eta}{\eta_0} = \frac{\text{dispersion viscosity}}{\text{viscosity of liquid}}$$

since the effect of the dispersed solid rather than the dispersing medium is usually more significant. However, the latter should not be ignored. Many industrial problems involving unacceptably high viscosities in dispersed systems are solved by substituting lower viscosity solvents.

The relative viscosity of a dilute dispersion of rigid spherical particles is given by

$$\eta_{rel} = 1 + a\phi$$

where a is equal to $[\eta]$ the limiting viscosity number (intrinsic viscosity) in terms of volume concentration, and ϕ is the volume fraction. Einstein showed that, provided the particle concentration is low enough and certain other conditions are met, $[\eta] = 2.5$ and the viscosity equation is

$$\eta_{rel} = 1 + 2.5\,\phi$$

This expression is usually called the Einstein equation.

For higher concentrations ($\phi > 0.05$) where particle–particle interactions are noticeable, the viscosity is higher than predicted by the Einstein equation. The viscosity–concentration equation becomes

$$\eta_{rel} = 1 + 2.5\,\phi + b\phi + c\phi + \ldots$$

where b and c are additional constants. For a more detailed discussion and experimental values for b and c, see reference 28. The deviation from the Einstein equation at higher concentrations is represented in Figure 9, which is typical of many systems; see refs. 29–30 for examples of such curves. The relative viscosity tends to infinity as the concentration approaches the limiting volume fraction of close packing $\phi_m\,(\phi \sim 0.7)$. The above equation has been modified (31–32) to take this into account such that the expression for η_{rel} becomes

$$\eta_{rel} = \frac{1 + 2.5\,\phi + b\phi + c\phi + \ldots}{1 - (\phi_m/\phi)}$$

which is a more universally applicable equation.

Other Factors Affecting the Viscosity of Dispersions. There are factors other than concentration that affect the viscosity of dispersions. For example, nonspherical particles tend to give viscosities greater than predicted, if Brownian motion is great enough to maintain a random orientation of the particles. However at low temperatures or high solvent viscosities, the Brownian motion is small and the particles tend to line up in flow giving lower viscosities than predicted. This is a form of shear thinning.

If the dispersion particles are attracted to each other, they tend to aggregate and form a structure. In many cases the structure is so pronounced that the mixture behaves like a solid when it is at rest. Shearing breaks up the structure and the viscosity

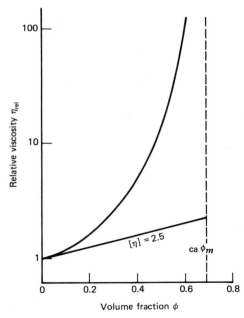

Figure 9. Relative viscosity vs volume fraction for a typical dispersion (curved line). The solid straight line is for the Einstein relationship $[\eta] = 2.5$. The dotted line is an approximate value for the limiting volume fraction ϕ_m.

decreases. If the structure rebuilds immediately after shearing stops, then the material is considered to be shear-thinning only. If the structure rebuilds slowly, then the material is thixotropic; however, it also may be shear-thinning as in the case of many latex paints.

If the volume concentration of the solid in the dispersion is high enough, shearing may produce an increase in viscosity rather than a decrease. Such behavior is called shear thickening or dilatancy and is common in dispersions of certain pigments and other powders. Such dispersions are closely packed but usually are shear-thinning up to moderate shear rates, ie, a few hundred to a few thousand s^{-1}. Higher shear introduces irregularities in the packing with bridging effects occurring between the particles. The overall packing loosens, which implies that the total space between particles increases. The liquid is not able to fill this space, so it can no longer wet all the particles. Much of the lubricating effect of the liquid therefore is lost, and internal friction rises to a high level, producing a high viscosity. This is true dilatancy as the volume actually increases.

Shear thickening can also occur in dilute suspensions. In such cases there is no volume expansion, but instead there is buildup of particle aggregates that ultimately produces a network that behaves like a gel. This sort of behavior is not dilatancy as it does not involve an increase in volume. In general, shear thickening is a better term because it is more descriptive of the phenomenon, and in most cases it is not known whether there has been volume expansion or not. Examples of shear thickening dispersions include clay, titanium dioxide, and zinc oxide slurries; concrete; and certain paints and printing inks. For an excellent and detailed discussion of shear thickening, particularly dilatancy, see reference 2.

Emulsions. Since emulsions are quite different from dispersions, different viscosity–concentration relationships must be used. For example, the Einstein equation must be modified. In an emulsion, the droplets are not rigid and their viscosity can vary over a wide range. Several equations have been proposed to account for this. One of the more useful is an extension of the Einstein equation and includes a factor that allows for the effect of variations in fluid circulation within the droplets and subsequent distortion of flow patterns around them (33). The magnitude of this effect depends on the ratio η_i/η_0, where η_i is the viscosity of the fluid in the droplet and η_0 is the viscosity of the continuous medium. The equation is

$$\eta_{\text{rel}} = 1 + a \left(\frac{1 + \dfrac{2}{5}\dfrac{\eta_0}{\eta_i}}{1 + \dfrac{\eta_0}{\eta_i}} \right)$$

When $\eta_i \gg \eta_0$, this expression reduces to the Einstein equation, but under all other conditions η_{rel} is lower than for a dispersion of solid particles at the same volume fraction. Detailed discussions on the viscosity of emulsions are given in references 26 and 28.

Extensional Viscosity. In addition to the shear viscosity η, two other rheological constants can be defined for fluids. They are the bulk viscosity κ and the extensional or elongational viscosity η_e. The bulk viscosity relates the hydrostatic pressure to the rate of deformation of volume, whereas the extensional viscosity relates the tensile stress to the rate of extensional deformation of the fluid. Because of the difficulty in measuring these viscosities, relatively little has appeared in the literature concerning them. However, interest in extensional viscosity has been growing recently as a result of the discovery that it is important to a number of industrial processes and problems (14,34–36) and the realization that shear properties alone are not sufficient for characterization of many fluids, particularly polymer melts (37).

Extensional flows occur when fluid deformation is the result of a stretching motion. The extensional viscosity is related to the stress required for the stretching. This is the stress necessary to increase the normalized distance between two material entities in the same plane when the separation is s and the relative viscosity is ds/dt. The deformation rate is the extensional strain rate, which is given by (34):

$$\dot{\epsilon} = \frac{1}{s}\frac{ds}{dt}$$

Unlike the shear viscosity, extensional viscosity has no meaning unless the type of deformation is specified. Three types of extensional viscosity have been identified and measured: uniaxial or simple, biaxial, and pure shear. Uniaxial viscosity is the only one used to characterize fluids. It has been employed mainly in the study of polymer melts but also for other fluids. For a Newtonian fluid, the uniaxial extensional viscosity is three times the shear viscosity

$$(\eta_e)_{\text{uni}} = 3\,\eta$$

This is confirmed at very low shear rates in Figure 10, which provides a typical example of the extensional viscosity behavior of a polymer (37). The other two extensional viscosities are used to study elastomers in the form of films or sheets. Uniaxial and biaxial extensions are important in industry, the former for the spinning of textile fibers and roller spatter of paints and the latter for blow molding, vacuum forming, film blowing, and foam processes (34–36,38) (see Foamed plastics; Plastics processing). The theoretical aspects of extensional viscosity are described in refs. 14, 35, 39 (see also Viscosity Measurements).

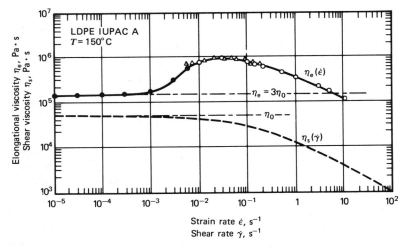

Figure 10.   Shear viscosity η_s and extensional viscosity η_e as a function of deformation rate for IUPAC A low density polyethylene (LDPE) at 150°C (37). To convert Pa·s to P, multiply by 10. Courtesy of *Rheologica Acta*.

Elasticity and Viscoelasticity

Elastic deformation is a function of stress and is expressed in terms of relative displacement or strain. Strain may be in terms of relative change in volume, length, or other measurement depending on the nature of the stress. An ideal elastic body is a material that deforms reversibly and for which the strain is proportional to the stress (Hooke's law) with immediate recovery to the original volume and shape when the stress is released. The relationship between stress σ and strain ϵ may be written as

$$\sigma = k\epsilon$$

where k is a proportionality constant called the modulus of elasticity. For a homogeneous, isotropic, Hookean solid, three moduli may be defined. Young's modulus E relates tensile stress to tensile strain. The shear modulus G relates shear stress to shear strain $(G = \tau/\gamma)$. The bulk modulus K relates hydrostatic pressure to the change in volume. Another elastic constant needed for complete specification of behavior in tension is Poisson's ratio μ, which relates change in volume to change in shape. For an incompressible solid, $\mu = 0.5$; but for real materials, $\mu < 0.5$. Young's modulus is related to shear modulus by the equation $E = 2G(1 + \mu)$. If μ is 0.5, then Young's modulus is three times the shear modulus. Values for the various moduli and Poisson's ratio for some representative materials are given in Table 2. Information on purely elastic bodies and techniques for characterization is given in refs. 40–41.

Although many common materials, eg, metals, are strictly elastic under use conditions, many more materials are both viscous and elastic. Most thick fluids, ie, >1 Pa·s (>10 P), and soft solids are viscoelastic. Analysis of the behavior of such materials is important to their manufacture and use.

Mechanical Models.   Because the complex rheological behavior of viscoelastic bodies is difficult to visualize, mechanical models often are used to represent it. In these models the viscous response to applied stress is assumed to be that of a Newtonian fluid and is represented by a dashpot, ie, a piston operating in a cylinder of Newtonian fluid. The elastic response is idealized as a Hookean solid and is represented by a spring. The dashpot represents the dissipation of energy in the form of heat, whereas the spring represents a system storing energy. Stress–strain diagrams for the two elements are given in Figure 11. With the dashpot, the stress is relieved by viscous flow and is independent of strain. The result is a plot with a constant value for stress. With the spring there is a direct dependence of stress on strain and the ratio of the two is the modulus E (or G).

Behavior of materials may be approximated by combinations of springs and dashpots. These may be in series so that the strains or deformations are additive or they may be combined in parallel, in which case the stresses are additive. Usually, the inertial effect of the mass is neglected. It is also the practice to fit experimental data with models that are the sum of many simple viscoelastic elements each with ideal coefficient, which are independent of stress or strain. The notations below are given in terms of shear measurements, but the principles apply equally well to tensile deformation.

A few mechanical models are shown in Figure 12. The Kelvin or Voigt model consists of a spring and dashpot in parallel. This model is of a solidlike material, since the equilibrium deformation produced by a given force depends only on the spring. The Maxwell model is a spring and dashpot in series. A Maxwell body is liquidlike,

Table 2.   Measured Values of Some Elastic Constants at Small Extensions and at 25°C[a]

Material	Young's modulus Y, GPa[b]	Approximate proportionality limit, % extension	Shear modulus G, GPa[b]	Poisson's ratio, μ	Bulk modulus K, GPa[b]
vitreous silica	70.0	3	30.5	0.14	37.4
barium borosilicate glass, Jena No. 1299	78.1	3	30.8		57.0
barium aluminophosphate glass, Jena No. 270	62.0	3	24.8		41.8
lead silicate glass (high density), Jena No. 500	53.8	3	21.8		34.5
mild steel	200–220	2.5	76–83	0.29	160.0
brass	80–100	2	26–38	0.25–0.4	61
constantan	163	2	61	0.33	160.0
nickel	200–220	2	78–80	0.30	170.0
tin	39–55		17–20	0.33	52
silver	60–80	2	24–28	0.38	100
granite	ca 30	0.5	ca 10	ca 0.3	ca 30
gelatin gel containing 80% H_2O	2×10^{-4}	10		0.50	
dry wood, axial piece	4–18	1			
dry wood, radial piece	ca 1	1			
silk thread	6.4	1			
natural rubber	8.6×10^{-4}	400–600	2.9×10^{-4}	0.49	0.019
hard rubber	0.36	3			
mineral-filled phenolic resin	2.4	2			
nylon	0.31	3			
polyethylene	0.010	2			

[a] Ref. 6.
[b] To convert GPa to dyn/cm², multiply by 10^{10}; to psi, multiply by 145,000.

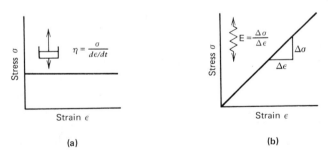

Figure 11.   Stress–strain diagrams. (**a**), Dashpot of viscosity η; the intercept indicates the force resisting the motion and is proportional to the speed of testing. (**b**), Spring of modulus E; the slope is the modulus, which is independent of the speed of testing.

since the application of a stress results in a permanent deformation. The Burgers model is a combination of the Kelvin and Maxwell models in series and is a reasonably good model of a linear viscoelastic material. Table 3 contains relevant quantities for these three models under a variety of operating conditions from extension under constant stress to forced sinusoidal oscillations.

With the parallel elements of the Kelvin model, the applied stress is shared between the elements and each is subjected to the same deformation. The stress–strain

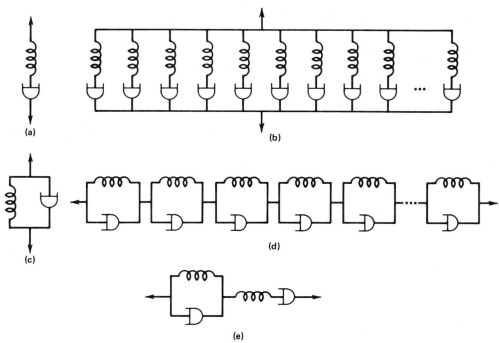

Figure 12. Mechanical models for viscoelastic behavior (6): (**a**), Maxwell; (**b**), Weichert (generalized Maxwell); (**c**), Kelvin; (**d**), Becker (generalized Kelvin); (**e**), Burgers.

diagram in Figure 13**a** reflects this by the straight-line dependence of stress on strain, but it also reflects the viscous resistance to stress by the intercept. The behavior may be more obvious in the strain–time diagram in Figure 13**b**, which shows what happens when a stress is suddenly applied to a Kelvin model. The model creeps, ie, extends, with time. The dashpot supplies the greatest resistance initially, but gradually the spring provides increasing resistance until the spring is fully extended. The strain as a function of time is given by the expression

$$\gamma = \frac{\tau}{G}\left[1 - e^{-(G/\eta)t}\right]$$

where G is the modulus of the spring and η is the viscosity of the dashpot. The quantity $e^{-(G/\eta)t}$ is a strain-retardation factor and the key parameter is η/G which is the retardation time which is a measure of the time delay in the strain after imposition of the stress. The retardation time is long when the viscosity is high. If the stress is suddenly relaxed, the strain of the Kelvin element exponentially decays to zero as the spring returns the system to its original configuration as fast as the dashpot allows. The retardation time also is an indication of the rate of this recovery.

With the Maxwell model combination of a spring and dashpot in series, the stress is the same on both elements and the total strain is the sum of the strains on each element. The stress–strain diagram is a curve, as shown in Figure 14**a**, which results from initial instantaneous elastic deformation followed by viscous flow.

If a Maxwell model is quickly stretched to some preselected strain γ_0 and held there, it exhibits stress relaxation (see Fig. 14**b**). At first the spring extends to the required strain, which puts a force on the dashpot. Then the dashpot slowly extends,

Table 3. Equations for Selected Mechanical Models[a]

Operation	Maxwell series	Kelvin parallel	Burgers series (2) and parallel (1)
extension	$\gamma = \tau/G + \tau t/\eta$	$\gamma = (\tau/G)$ $\times (1 - e^{-t/(\eta/G)})$	$\gamma = \tau/G_2 + \tau t/\eta_2$ $+ \tau/G_1(1 - e^{-tG_1/\eta_1})$
held at constant strain, γ_0	$\tau = \gamma_0 G e^{-t/(\eta/G)}$	$\tau = \gamma_0 G$ = constant	$\tau = \gamma_0 G_2 e^{-tG_2/\eta_2 + \gamma_0 G_1}$
release of stress at time θ	$\tau_0\gamma = \tau_0 t/\eta$ = constant	$\gamma = (\tau_0/G)e^{-t/(\eta/G)}$	$\gamma = \tau_0 t/\eta_2 + (\tau_0/G_1)e^{-tG_1/\eta_1}$
sinusoidal oscillation			
elastic component	$\eta''(\omega) = G\omega(\eta/G)^2/(1 + \omega^2(\eta/G^2)$	$\eta''(\omega) = G/\omega$	$\eta''(\omega) = \dfrac{\eta_2^2/G_2}{1 + \omega^2\eta_2^2/G_2^2} + G_1/\omega$
viscous component	$\eta'(\omega) = \eta/(1 + \omega^2(\eta/G)^2)$	$\eta'(\omega) = \eta$	$\eta'(\omega) = \dfrac{\eta_2}{1 + \omega^2\eta_2^2/G_2^2} + \eta_1$
phase angle	$\delta = \arctan (1/\omega(\eta/G))$	$\delta = \arctan \omega(\eta/G)$	$\delta = \arctan \eta'/\eta''$

[a] Ref. 6.

thereby allowing the spring to contract and the stress, ie, the load required to sustain deformation, to decay to zero. The stress relaxes according to the relationship

$$\tau = \gamma_0 \, G e^{-(G/\eta)t}$$

The ratio η/G is the relaxation time, which is the time taken for the stress to fall to $1/e$ of its initial value. The phenomenon of stress relaxation is typical of a viscoelastic fluid.

The behavior of real materials is not accurately modeled by either the Kelvin or the Maxwell model. A number of more complex models have been devised to approximate more closely real behavior; the Burgers model is one example. It is the simplest model that exhibits the type of creep response observed for real materials, eg, plastics and flour dough. It has a relaxation time and a retardation time; the ratio of the two is called the springiness number (5). Springiness, as subjectively assessed, depends on the extent and rate of elastic recovery when the material is squeezed and then released. The extent of recovery is measured by the relaxation time and the rate is a function of the reciprocal of the retardation time.

The stress–strain diagram of the Burgers model depends on the strain rate, as shown in Figure 15. A slowly applied strain gives a plot that resembles that of the Maxwell model, whereas the rapidly applied strain gives the straight line of the Hookean spring. A creep curve for the Burgers model is shown in Figure 16. If a constant stress is applied to this system, there is a certain amount of instantaneous deformation caused by the Maxwell spring D followed by slow deformation from the Maxwell dashpot C and the Kelvin element A/B. After the Kelvin element is fully

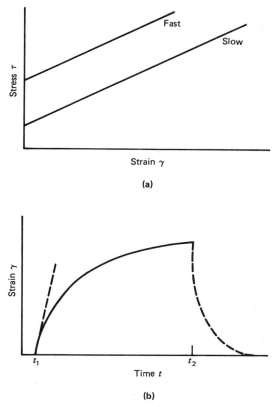

Figure 13.   The Kelvin model. (**a**), Stress–strain diagram; the slope is the modulus of the spring, and the height of the intercept is proportional to the speed of testing. (**b**), Strain–time diagram (creep curve); the stress is applied at time t_1 and removed at t_2 and the initial slope (strain rate) is proportional to η.

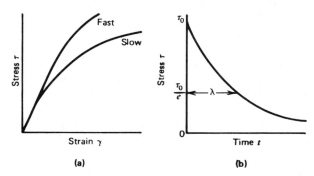

Figure 14.   The Maxwell model. (**a**), Stress–strain diagram; the initial slope gives the modulus, and the height of the curve depends on the speed of testing. (**b**), Stress relaxation; τ_0 = initial stress, and λ = relaxation time or the time to reach stress of τ_0/e.

extended, the overall extension increases in a linear manner as a result of the viscous element C. If the stress is removed, the system contracts instantaneously by the amount originally contributed by the Maxwell spring D. This is followed by slow recovery of the Kelvin element. The viscous flow of element C is not recoverable and represents a permanent deformation of the material.

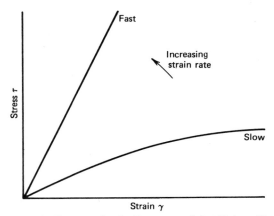

Figure 15.   Stress–strain diagrams for the Burgers model at high and low strain rates.

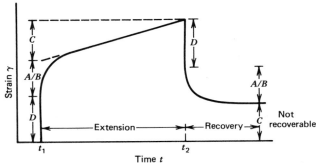

Figure 16.   Creep curve for the Burgers model. Similar to curves for many real viscoelastic materials. Stress applied at time t_1 and removed at t_2.

Whether a viscoelastic material behaves as a viscous liquid or an elastic solid depends on the relation between the time scale of the experiment and the time required for the system to respond to either stress or deformation. Although the concept of a single relaxation time λ is not generally applicable to real materials, a mean characteristic time can be defined as the time required for a stress to decay to $1/e$ of its elastic response to a step change in strain. The ratio of this characteristic time to the time scale of the experiment t_e is called the Deborah number. A material at high Deborah numbers responds elastically and at low Deborah numbers exhibits viscous behavior, ie, at $t_e \ll \lambda$, the material behaves like an elastic solid and at $t_e \gg \lambda$ it behaves like a viscous liquid. These effects can be seen in geological strata, where rock has flowed to relieve the stresses imposed by geological events. The time scale is very long, so the material appears to be viscous.

Dynamic Behavior.   Knowledge of how the mechanical models would behave if they underwent stress–strain and time-dependent measurements is important for the characterization of real materials, particularly to determine their response under conditions of processing or use. For this reason, stress–strain, creep, and stress-relaxation measurements are commonly made to define material properties. The dynamic response of viscoelastic materials to cyclic stresses or strains also is important, partly because cyclic motion occurs in many processing operations and applications and partly because so much rheological information can be gained from dynamic measurements.

Dynamic methods depend on measuring the response of a viscoelastic material to periodic stresses or strains. By subjecting a specimen to a sinusoidal stress at a given frequency and determining the response, both the elastic and viscous or damping characteristics can be obtained. Elastic materials store energy, ie, convert mechanical work into potential energy, which is recoverable. Liquids do not store energy when stressed but dissipate it as heat as they flow. This dissipation gives highly damped motion. Viscoelastic materials give both elastic and damping behavior. The latter causes the deformation to be out of phase with the sinusoidal stress applied in the dynamic measurement.

When a linear viscoelastic material is stressed sinusoidally at a frequency f, the amplitude of the strain is proportional to that of the stress. However, as noted above, the strain response is not in phase with the stress but lags behind by some angle δ. The stress and strain can be represented by complex variables. The complex strain is given by

$$\gamma^* = \gamma_0 e^{i\omega t}$$

and the complex stress by

$$\tau^* = \tau_0 e^{i(\omega t + \delta)}$$

where i is the operator $\sqrt{-1}$, ω is the angular frequency ($\omega = 2\pi f$), and δ is the phase angle. Complex strain and complex stress are vectors in complex planes. They can be resolved into real, ie, in-phase, and imaginary, ie, 90° out-of-phase, components such that

$$\gamma^* = \gamma' + i\gamma''$$
$$\tau^* = \tau' + i\tau''$$

The shear modulus also can be represented by a complex variable, ie, the complex dynamic modulus G^*, which is the ratio of the complex stress and complex strain

$$G^* = \tau^*/\gamma^*$$

The dynamic modulus also can be resolved into two components

$$G^* = G' + iG''$$

where

$$|G^*| = \sqrt{(G')^2 + (G'')^2}$$

and

$$G' = |G^*| \cos \delta$$
$$G'' = |G^*| \sin \delta$$

G' is the storage modulus. It is in phase with the real components of γ^* and τ^* and is associated with the energy stored in elastic deformation. G' is approximately equal to the elastic modulus determined in creep and stress-relaxation experiments. G'' is the loss modulus. It arises from the out-of-phase components of γ^* and τ^* and is associated with viscous energy dissipation, ie, damping. The ratio of G'' and G' gives another measure of damping called the dissipation factor or loss tangent:

$$\tan \delta = G''/G'$$

A viscoelastic material also possesses a complex dynamic viscosity

$$\eta^* = \eta' + i\eta''$$

It can be shown that

$$\eta^* = G^*/i\omega$$

and

$$\eta' = G''/\omega$$

$$\eta'' = G'/\omega$$

η' is the dynamic viscosity and is related to G'' the loss modulus. The plot of shear viscosity η of a viscoelastic fluid as a function of shear rate is numerically identical to the plot of the absolute value of the complex dynamic viscosity $|\eta^*|$ vs angular frequency (42).

Table 3 also gives results derived for oscillatory measurements in selected mechanical models. In the case of real materials, the oscillatory behavior, like the creep or stress-relaxation behavior, requires the superposition of elements for modeling of the observed mechanical behavior. However the real significance of the various parameters G'', G', $\tan \delta$, η', and η'' is that they can be determined experimentally and used to characterize real materials. They all depend on frequency and temperature and this dependence can be used to define behavior. For example, viscoelastic fluids often are characterized by log–log plots of one or more of these quantities vs angular frequency ω, as shown in Figure 17, which illustrates the behavior of a polymer melt (43).

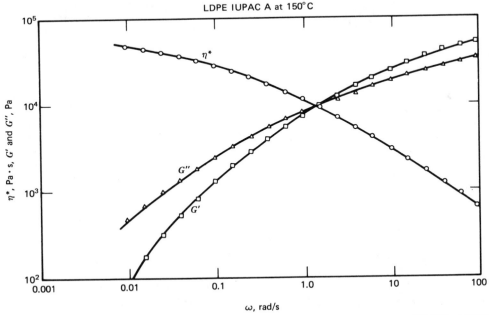

Figure 17.   Dynamic viscoelastic properties of IUPAC A low density polyethylene (LDPE) at 150°C: complex dynamic viscosity η^*, storage modulus G', and loss modulus G'', vs angular velocity (43). To convert Pa·s to P, multiply by 10; to convert Pa to dyn/cm^2, multiply by 10. Courtesy of Rheometrics, Inc.

Normal Stress (Weissenberg Effect). Many viscoelastic fluids flow in a direction normal to the direction of shear stress in steady-state shear. This normal stress effect was first analyzed and linked to elasticity by Weissenberg (44). Manifestations of the effect include flour dough climbing up a beater, polymer solutions climbing up the inner cylinder in a concentric-cylinder viscometer, and paints forcing apart the cone and plate of a cone–plate viscometer. The normal stress effect has been put to practical use in certain screwless extruders designed in a cone–plate or plate–plate configuration, such that the polymer enters at the periphery and exits at the axis.

There are two normal stress functions $N_1(\dot\gamma)$ and $N_2(\dot\gamma)$, which are referred to as the first and second normal stress differences, respectively. $N_1(\dot\gamma)$ is positive and increases with increasing shear rate, as shown in Figure 18 (43), which describes steady-shear behavior of the same polymer melt as in Figure 17. $N_2(\dot\gamma)$ is smaller in absolute value than $N_1(\dot\gamma)$ and is at least sometimes negative. The first normal stress difference is a useful quantity as it often gives a good quantitative measure of visco-elasticity. It can be determined from the normal force, which is measurable with several of the commercially available rotational viscometers. In highly elastic liquids, it is not unusual for $N_1(\dot\gamma)$ to be considerably larger than the shear stress.

Viscosity Measurement

Whether a rheologist encounters a flow problem or just wishes to characterize a given fluid better, the decision of which instrument to use must be faced. There are so many commercial viscometers to choose from with such a variety of geometries and such wide ranges of viscosity and shear rates that it rarely is necessary to construct an instrument. However, in choosing a commercial viscometer, a number of criteria

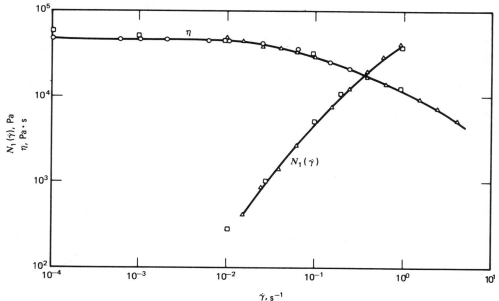

Figure 18. Shear viscosity η and first normal stress difference $N_1(\dot\gamma)$ vs shear rate for IUPAC A low density polyethylene at 150°C (43). O, Parallel plate; △, cone and plate; □, J. Meissner, 1972, 1975. To convert Pa to dyn/cm², multiply by 10; to convert Pa·s to P, multiply by 10. Courtesy of Rheometrics, Inc.

must be considered. One of the most important is the nature of the material to be tested: whether it is of high or low viscosity, whether it is elastic or not, the temperature dependence of its viscosity, etc. Other important considerations are the accuracy and precision required and whether the measurements are for quality control or research. It is important that the investigator match the viscometer to the materials and processes of interest; otherwise the results may be misleading. For example, there is not much point in using a low shear rate viscometer to determine the sprayability of a shear-thinning paint, since spraying is a high shear rate process.

Most early viscometers, many of which have been incorporated into standard industrial tests, give single-point measurements, ie, instead of describing viscosity or shear stress over a range of shear rates, only a single point on the flow curve is produced; the rest of the curve is unknown. This is not a problem with Newtonian liquids because viscosity is independent of shear rate, but it can be very misleading in the case of non-Newtonian materials. The latter situation is shown in Figure 19 where two fluids give viscosity profiles that cross each other. Measurements carried out at the shear rate corresponding to the intersection would indicate that the two materials were identical even though a simple examination, eg, shaking or pouring, would show that they were not. A non-Newtonian fluid cannot be adequately characterized by a single-point measurement. The use of multipoint measurement techniques with such materials is strongly recommended.

Another important point is temperature control. Viscosity is highly dependent on temperature, and accurate, precise measurements can only be made if temperature is carefully controlled. More errors are made and more disagreements over viscosity results arise because of incorrect or drifting temperature than for any other reason. Good temperature control can be achieved with thermostatted baths or circulators, many of which are available commercially. Unfortunately, some viscometers cannot be adequately thermostatted and others lose temperature control due to heat generation at very high shear rates and with high viscosity materials. Such a temperature increase makes a material appear to lose viscosity under shear. A natural but erroneous conclusion would be that the material is thixotropic or shear-thinning.

There are three basic viscometer types: capillary, rotational, and moving body. Most of the commonly used instruments or techniques fit in one of these classifications, although there are other types usually for special applications. In most cases there are several different instruments available based on any given type. The choice depends

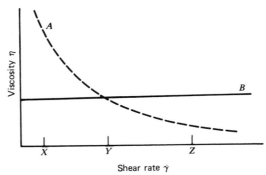

Figure 19.   Viscosity vs shear-rate curves for two fluids showing the fallacy of a single-point measurement. Fluid A would appear to be more viscous than fluid B if measured only at point X, the same viscosity at point Y, and less viscous if measured only at point Z.

on the particular requirements of the investigator and what can be afforded. For detailed information on any given viscometer, the reader is referred to ref. 6. Many commercially available viscometers are listed and described in ref. 45.

Capillary Viscometers. Capillary flow measurement is one of the most popular methods for measuring viscosity; it is also the oldest. In this method, a liquid drains or is forced through a fine-bore tube and the viscosity is determined from the measured flow, applied pressure, and tube dimensions. The basic equation is the Hagen-Poiseuille expression

$$\eta = \frac{\pi r^4 \Delta p t}{8\,VL}$$

where η is the viscosity, r is the radius of the capillary, Δp is the pressure drop through the capillary, V is the volume of liquid that flows in time t, and L is the length of the capillary. Steady-state, laminar, isothermal flow is assumed. For a given viscometer with similar fluids and a constant pressure drop, the equation reduces to $\eta = Kt$ or, more commonly, $\nu = \eta/\rho = Ct$, where ν is the kinematic viscosity. Therefore, viscosity can be determined by multiplying the efflux time by a suitable constant.

Capillary viscometers are particularly useful for measuring precise viscosities of Newtonian liquids at 1–20,000 mPa·s (= cP). Shear rates range very widely and depend on the instrument and the liquid being studied. The shear rate at the capillary wall for a Newtonian fluid may be calculated from the equation

$$\dot{\gamma}_w = \frac{4\,Q\pi}{r^3}$$

where Q is the volumetric flow rate and r is the radius of the capillary. The shear stress at the wall is $\tau_w = r\Delta p/2L$. Absolute viscosities are difficult to measure with capillary viscometers, but viscosities relative to some standard fluid of known viscosity, eg, water, are readily determined. The viscometer is calibrated with the reference fluid, then viscosities of other fluids relative to the reference are determined from their flow times.

For highly accurate work, corrections must be made for kinetic energy losses, incomplete drainage, any turbulence that occurs, and possible surface tension and heat effects. These corrections are discussed in considerable detail elsewhere (6). The largest correction is that resulting from loss of effective pressure because of the appreciable kinetic energy of the issuing stream. The next most important correction, but much smaller, is that for energy loss resulting from end effects, ie, viscous resistance caused by velocity gradients as the liquid enters and leaves the capillary. When terms for these two correction factors are incorporated into the viscosity equation, it is changed to

$$\nu = \eta/\rho = Ct - B/t$$

where t is the time of flow and B and C are instrument constants generated from measurements of fluids of known viscosity. For much industrial capillary viscometer work, these corrections are not used since the Poiseuille equation adequately expresses the flow. However, corrections are absolutely necessary for determining even approximate viscosities with short capillary devices, eg, orifice viscometers. Use of a long capillary (>10 × diameter) and long efflux times (>300 s) helps minimize the need for corrections.

In addition to the orifice type, there are two other main classifications of commercially available capillary viscometers: glass and cylinder–piston types.

Glass-Capillary. The glass-capillary viscometer has been widely used to measure the viscosity of Newtonian fluids. The driving force usually is the hydrostatic head of the test liquid. Kinematic viscosity is measured directly, and most of the viscometers are limited to low viscosity fluids, ca 0.4–16,000 mm^2/s (= cSt). However, external pressure can be applied to many glass viscometers to increase the range of measurement and allow non-Newtonian behavior to be studied. Glass-capillary viscometers are low shear stress instruments: 1–15 Pa (10–150 dyn/cm^2) if operated by gravity only. The rate of shear is ca 1–20,000 s^{-1} based on 200–800-s efflux time.

The basic design is that of the Ostwald viscometer: a U-tube with two reservoir bulbs separated by a capillary, as shown in Figure 20**a**. The liquid is added to the viscometer, pulled into the upper reservoir by suction, then allowed to drain by gravity back into the lower reservoir. The time that it takes for the liquid to pass between two etched marks, one above and one below the upper reservoir, is a measure of the viscosity. In U-tube viscometers, the effective pressure head and, therefore, the flow time depend upon the volume of liquid in the instrument. Therefore, the conditions must be the same for each measurement.

The original Ostwald viscometer has been modified in many ways, and a number of different versions are sold commercially. Most are available with a wide choice of capillary diameters and, therefore, a number of viscosity ranges. A suitable collection of viscometers would allow coverage of a very wide range of viscosities. Some commercial glass-capillary viscometers are listed in Table 4 with brief comments on viscosity ranges, capillary diameters, etc (see also ref. 6). A number of these viscometers are described in ASTM D 445, which also lists detailed recommendations on dimensions and methods of use.

The Cannon-Fenske viscometer (Figure 20**b**) is excellent for general use. A long capillary and small upper reservoir result in a small kinetic energy correction, and the large diameter of the lower reservoir minimizes head errors. Because the upper and

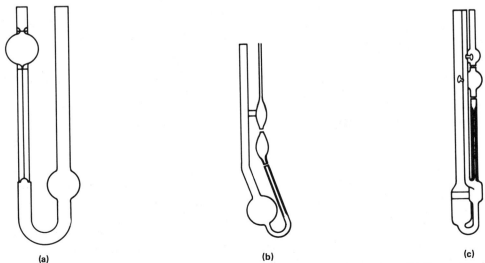

      (a)   (b)   (c)

Figure 20. Various types of viscometers: (**a**), Ostwald glass capillary viscometers; (**b**) Cannon-Fenske viscometer; (**c**), Ubbelohde viscometer.

Table 4. Glass-Capillary Viscometers[a]

Approx. constant k	Viscosity range, mm²/s	Size number[b] (= cSt)	Series number[c]	Respective capillary diameters, mm					
				Cannon-Fenske	Ubbelohde	FitzSimons	SIL	Atlantic	Zeitfuchs
0.003	0.6–3.0	25, 50	1	0.31 ± 0.02			0.43 ± 0.01	0.41 ± 0.02	0.41 – 0.42	0.28
0.005	1.0–5.0			0.42					0.47 ± 0.01	
0.01	2.0–10	100	2	0.63	0.58 ± 0.02	0.61	0.61	0.56	0.38
0.03	6–30	150	3	0.78	0.77	0.81 ± 0.03	0.73	0.74 ± 0.02	0.50
0.05	10–50				0.87		0.91	0.84	
0.1	20–100	200	4	1.02	1.10 ± 0.03	1.05	1.14 ± 0.03	1.00	0.67
0.3	60–300	300	5	1.26	1.43	1.32 ± 0.04	1.50	1.31 ± 0.02	0.88
0.5	100–500	350		1.48	1.64		1.71	1.48	
1	200–1,000	400	6	1.88	1.95	1.96	2.03	1.77 ± 0.03	1.20
3	600–3,000	450	7	2.20 ± 0.05	2.67 ± 0.04		2.80	2.34	1.42
5	1,000–5,000	500		3.10	3.06		3.06		
10	2,000–10,000	600	8	4.0	3.62		3.79 ± 0.04	2.65 ± 0.04	1.93
30	6,000–30,000		9						2.52
100	$(2-10) \times 10^4$		10						3.06

	Respective sizes					
	Cannon-Fenske	Ubbelohde	FitzSimons	SIL	Atlantic	Zeitfuchs
Volume of bulb, mL	3.15 ± 0.15[d]	4.6[e]	3.75 ± 0.15	4–6[e]	3.2 ± 0.2	0.35
Length of capillary, mm	73 ± 3	90 ± 5	120 ± 5	{145 ± 1, 127 ± 3[f]	100	{210, 165[g]

[a] Ref. 6.
[b] Applies only to the Cannon-Fenske viscometers.
[c] Applies only to the Zeitfuchs viscometers.
[d] All except #600 for regular flow (vol = 4.3 ± 0.1 mL) and #25 for reverse flow (vol = 1.6 ± 0.1 mL).
[e] Bulb volume for the three smallest viscometers (first in above listing) is 4 mL, for the medium ones, 5 mL, and for the three larger ones, 6 mL.
[f] The capillary length of 127 mm applies only to the three larger viscometers.
[g] The capillary length of 165 mm applies only to the four larger viscometers.

lower bulbs lie on the same vertical axis, there are minimum variations in the head when the viscometer is used in positions that are not perfectly vertical. There also is a reverse-flow Cannon-Fenske viscometer for opaque liquids. In this type of viscometer, the liquid flows upwards past the timing marks rather than downward as in the normal direct-flow instrument. Thus, the position of the meniscus is not obscured by the film of liquid on the glass wall.

The Ubbelohde viscometer is shown in Figure 20c. This type of viscometer is particularly useful for measurements at several different concentrations as flow times are not a function of volume and, therefore, dilutions can be made in the viscometer. There are several modifications of this design, particularly the Cannon-Ubbelohde, semimicro, and dilution viscometers, but the essential features are shown in the figure. The Ubbelohde viscometer is also called a suspended-level viscometer. This term arises from the fact that the liquid emerging from the lower end of the capillary flows only down the walls of the reservoir directly below it. Therefore, the lower liquid level always coincides with the lower end of the capillary, and the volume initially added to the instrument need not be precisely measured. This also eliminates the temperature correction for glass expansion necessary for Cannon-Fenske viscometers.

The SIL viscometer has the advantage of an overflow gallery, which is on the open arm. This permits very precise establishment of volume. The Zeitfuchs cross-arm viscometer is a reverse-flow type suitable for both transparent and opaque liquids. The FitzSimons instrument also is a suspended-level viscometer. As with the Ubbelohde, filling, flow measurement, and cleaning can be done without removing the viscometer from the temperature bath. Single- and double-capillary models are available. The latter has two different-sized capillaries giving two separate efflux times.

Great care must be taken when using a glass-capillary viscometer to maintain the capillary absolutely clean. This is critical for accurate, precise measurements. The viscometer must be cleaned thoroughly after each series of operations. Samples being tested and cleaning solvents should be filtered to remove particles that could clog the capillary.

All glass-capillary viscometers should be calibrated carefully; for details of techniques see ref. 6. The standard method of calibration is to determine the efflux time of distilled water at 20°C. Unfortunately, because of its low viscosity, water can only be used to standardize the smaller capillary instruments. However, a calibrated viscometer can be used to determine the viscosity of a higher viscosity liquid, eg, a mineral oil. This oil can then be used to calibrate a viscometer with a larger capillary. Another method is to calibrate directly with two or more certified standard oils differing in viscosity by a factor of approximately five. Such oils are useful for calibrating virtually all types of viscometers. Since viscosity is very temperature-dependent, particularly in the case of standard oils, good temperature control is absolutely necessary for accurate calibration.

In recent years several commercial instruments that allow some degree of automation in the operation of capillary viscometers have become available. For example, there are devices that measure the time of flow through the interruption of a light beam focused on a photoelectric cell. Such a device can be attached to an existing viscometer to make measurements more easily and with more precision. There also are completely automatic instruments that control the temperature, fill the viscometer, take several readings (sometimes with dilutions), and clean, rinse, and dry the viscometer before the next filling. The instrument then prints efflux times and, in some cases, can be programmed to print viscosities. Manufacturers of automated viscosity systems include Schott and Lauda; Wescan supplies timer and photocell combinations.

Glass-capillary viscometers can be operated by applying external pressure. The viscometer is connected to an air reservoir or other source of constant pressure and the liquid is forced through the capillary under pressure. The principles are basically the same as for gravity-driven viscometers. Pressure-driven viscometers have some advantages, including a wider viscosity range for a given size capillary and minimization of surface-tension effects and other head corrections. Disadvantages include the fact that pressure-driven viscometers are inconvenient to use compared to gravity instruments, the difficulty in maintaining constant pressure, the need for drainage corrections to allow for variations in the rate of emptying of reservoirs, and the errors caused by dissolved gases. Other pressure-driven capillary viscometers are cylinder-piston types.

Orifice. Orifice viscometers, which are often called efflux or cup viscometers, are commonly used to measure and control flow properties in the manufacture, processing, and use of many fluids, eg, inks, paints, adhesives, and lubricating oils. Their design grew from needs for simple, easy-to-operate viscometers in areas where precision and accuracy were not particularly important. In these situations, knowledge of a true viscosity is not necessary, so that efflux time of a fixed volume of liquid is a sufficient indication of the fluidity of the material. Examples of orifice viscometers include the Ford, Zahn, and Shell cups used for paints and inks and the Saybolt Universal and Furol instruments used for oils. These and others are listed in Table 5 with some of their characteristics and means of application.

Orifice viscometers usually have extremely short capillaries. The typical orifice viscometer is a cup with a hole in the bottom. The cup is filled and the time required for the liquid to flow out is measured. The hydrostatic head decreases as the liquid flows and there is a large kinetic energy effect. Analysis of flow in orifice viscometers shows that it does not follow the Hagen-Poiseuille law and that efflux times are not related to viscosities in any simple manner. Therefore, it is better not to convert efflux times to viscosities except during calibration with standard oils. Efflux times should be accepted as arbitrary measures of viscosity and noted in terms of the viscometer being used, ie, Saybolt seconds, Ford seconds, etc. In addition to this restriction, the precision of orifice viscometers is poor because of the lack of temperature control, wear during use, and variations in manufacture. However, these viscometers are widely used in industry. They are popular because they are inexpensive, robust, and easy to use. Some of these instruments can be dipped into or left in the material to be tested, which is very convenient. Dip cups are used to determine approximate or relative viscosities in resin reactors, ink reservoirs, paint dip tanks, adhesive mixing tanks, etc. These are applications where the limitations of orifice cups are not important.

Orifice viscometers should not be used for setting product specifications, as this demands better precision than is possible with these instruments. Since they are designed for Newtonian and near-Newtonian fluids, they should not be used with thixotropic or highly shear-thinning materials; such fluids should be characterized with a multispeed rotational viscometer.

Orifice viscometers with relatively long capillaries are available. Examples are the Shell dip cup and the European ISO cup, which is like a Ford cup with a capillary. These cups need smaller kinetic-energy corrections and definitely give better precision

Table 5. Some Orifice Viscometers Used in Industry[a]

Viscometer	Orifice length, mm	Diameter, cm	Viscosity units	Viscosity ranges, mm²/s (= cSt)	Thermostat	Approximate constants[a] k	K	Main applications
Barbey (France)			mL/h	1–4000	no			petroleum products
Demmler			s	4–20,000	no			wire enamel
Engler (FRG)	20.0 ± 0.10	2.90 (top) 2.80 (bottom)	s/200 mL or Engler degree	1–1500	yes	0.073	0.0631	tar, petroleum products
Ford (Nos. 2–4)		0.25–0.41	s	25–500	no	[b]	[b]	paint, varnish
ISO	20 ± 0.05	0.3–0.6	s	10–700	no	[c]	[c]	paints, inks
Marsh funnel			s/946 mL					drilling mud
Redwood (UK) No. 1	10 ± 0.05	1.62 min	s/60 mL	1–500	yes	0.00264 / 0.00247	1.9 for 40 < t < 85 s / 0.65 for 85 < t < 200 s	petroleum products
Redwood No. 2	50 ± 0.2	3.8	s/60 mL	500–5000	yes	0.026	0.40	petroleum products
Saybolt Universal	12.2	1.76	s/60 mL	1–400	yes	0.00226	1.95 for t < 100 s	petroleum products
Saybolt Furol	12.2	3.15	s/60 mL	400–4000	yes	0.00220	1.30 for t > 100 s	petroleum products
Shell (Nos. 1–5)	25	0.18–0.58	s	4–1600	no	[d]	[d]	paints, inks
Zahn (Nos. 1–5)		0.2–0.53	s	ca 20–1500	no	[b]	[b]	varnish, paints, inks

[a] Constants k and K are based on $\nu = kt - K/t$, ν in mm²/s (= cSt).

[b] Zahn and Ford cup constants are given in ref. 46.

[c] ISO cup constants are given in ISO International Standard 2431.

[d] Shell cup constants are given in proposed ASTM method for viscosity by dip cups being developed by Subcommittee 24 of ASTM Committee D-1 on Paint and Related Products.

than the corresponding short-capillary viscometers. They still are not precision instruments, however, and should only be used for control purposes.

If it is necessary to calculate kinematic viscosities from efflux times, such as in a calibration procedure, the following equation should be used:

$$\nu = kt - K/t$$

Approximate values for some constants are listed in Table 5, others are listed in ref. 46. In most cases it is sufficient to be able to convert from one viscometer seconds value to another or to approximate kinematic viscosities. For this purpose, charts such as that shown in Figure 21 or tables of viscosities as given in many textbooks and literature from viscometer manufacturers are useful.

Piston–Cylinder (Extrusion). Pressure-driven piston–cylinder capillary viscometers, ie, extrusion rheometers, are primarily used to measure the melt viscosity of polymers and other viscous materials. In this type of viscometer, a reservoir is connected to a capillary tube. Molten polymer or other material is extruded through the capillary by means of a piston on which a constant force is applied. The viscosity can be determined from the volumetric flow rate and the pressure drop along the capillary. Polymer melts frequently are non-Newtonian. In such a case, the earlier expression given for the shear rate at the capillary wall $\dot{\gamma}_w$ does not hold. A correction factor $(3\,n + 1)/4\,n$, called the Rabinowitsch correction, must be applied such that

$$\dot{\gamma}_{tw} = \frac{(3\,n + 1)\dot{\gamma}_w}{4\,n}$$

where $\dot{\gamma}_{tw}$ is the true shear rate at the wall and n is a power-law factor

$$n = \frac{d \log \tau_{tw}}{d \log \dot{\gamma}_w}$$

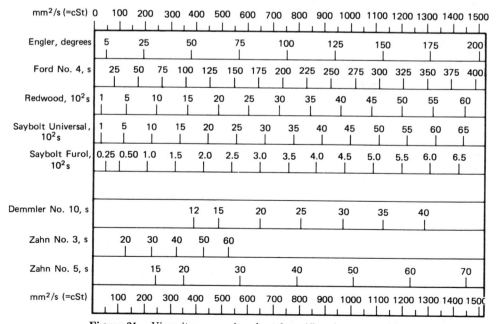

Figure 21.   Viscosity conversion chart for orifice viscometers (6).

determined from the slope of a log–log plot of the true shear stress at the wall τ_{tw} vs $\dot{\gamma}_w$. For a Newtonian liquid, $n = 1$. A true apparent viscosity η_t can be calculated from

$$\eta_t = \frac{\tau_{tw}}{\dot{\gamma}_{tw}}$$

Several instruments based on the extrusion principle are available commercially. Examples are the Instron Merz-Colwell rheometer (MCR), Instron Model 3211, Monsanto Automatic, Sieglaff-McKelvey, and Burrell-Severs capillary rheometers; Standard Oil high pressure capillary viscometer; and the melt indexer or extrusion plastometer (ASTM D 1238) (6,22). Many of these instruments are large and expensive. They are very useful when large numbers of quality-control or other test measurements for melt viscosity are needed for batches of a single material or very similar materials. When melt viscosities of a wide range of materials must be measured, rotational viscometers are a better choice.

Rotational Viscometers. Rotational viscometers consist of two basic parts separated by the fluid being tested. The parts may be concentric cylinders (cup-and-bob), plates, a low angle cone and a plate, or a disk, paddle, or rotor in a cylinder. One of the parts rotates relative to the other and produces a shearing action on the fluid. The torque required to produce a given angular velocity or the angular velocity resulting from a given torque is a measure of the viscosity.

Rotational viscometers are more versatile than capillary viscometers, thus they are more useful for many industrial purposes. They can be used with a wide range of materials since opacity, settling, and non-Newtonian behavior do not cause difficulties as they do with most capillary viscometers. Most rotational viscometers can measure viscosities over a range of shear rates and as a function of time. Therefore, they are useful for characterizing shear thinning and time-dependent behavior. They are considerably more complicated mechanically than most capillary viscometers. In price and complexity they range from relatively simple, inexpensive (<\$1000), low shear devices, eg, the Brookfield Synchro-Lectric and the Stormer, to very complicated and expensive ($\geq$\$100,000) instruments capable of both steady-state and oscillatory motion, eg, the Weissenberg Rheogoniometer and Rheometrics Fluids Rheometer. A number of commercial instruments are briefly described in Table 6.

Rotational viscometers often are not considered for absolute or highly accurate measurements because of problems with end effects. However, corrections can be made, and highly accurate measurements are possible. Because rotational viscometers operate under steady-state conditions, they can closely approximate industrial process conditions, eg, stirring, dispersing, pumping, and metering. In addition to research regarding these processes, rotational viscometers are widely used for routine evaluations and quality-control measurements. A number of instruments are commercially available and they are effective over a wide range of viscosities and shear rates.

The equations and methods for determining viscosity vary greatly with the type of instrument, but in many cases calculations may be greatly simplified by determining the instrument constant or using one supplied by the manufacturer. The constant is obtained by calibration of the viscometer with a standard fluid, the viscosity of which is known for the conditions involved. The constant is then used with stress and shear-rate terms to determine viscosity:

$$\eta = K \text{ (stress term/shear-rate term)}$$

Table 6. Some Commercial Rotational Viscometers

Viscometer	Manufacturer	Price range[a]	Approximate viscosity range, mPa·s (= cP)	Shear rate capabilities	Temperature control
Rotovisco	Gebruder Haake Berlin, Federal Republic of Germany[b]	D	$2-10^9$	wide range, low to high shear	good
Rheomat	Contraves A.G. Zurich, Switzerland[c]	D	$2-10^9$	wide range, low to high shear	good
Rheotron	Brabender OHG Duisberg, Federal Republic of Germany[d]	D	$2-10^9$	wide range, low to high shear	good
MacMichael	Fisher Scientific Pittsburgh, Pa.	A	$\leq 10^6$	low shear	fair
Fann V-G	Fann Instrument Co. Houston, Texas	A–B	$0.5-(5 \times 10^6)$	low to moderate shear	none to fair
Ferranti-Shirley cone–plate	Ferranti Ltd. Moston, Manchester, UK[e]	D–E	$20-(3 \times 10^7)$	wide range, low to high shear	good
ICI cone–plate	Research Equipment (London) Ltd. Hampton, Middlesex, UK[f]	B	$\leq 10^4$	high shear	good
Rheogoniometer	Sangamo Schlumberger Bognor Regis, Sussex, UK[g]	G	$10^{-1}-(5 \times 10^9)$	wide range, low to high shear	excellent
Mechanical Spectrometer	Rheometrics, Inc. Union, N.J.	F	10^3-10^9	wide range, low to high shear	excellent
Fluids Rheometer	Rheometrics, Inc. Union, N.J.	F	$10-10^5$	wide range, low to high shear	excellent
Brookfield Synchro-Lectric	Brookfield Engraving Co. Stoughton, Mass.	A	$1-10^8$	low shear	none to good, depending on system used
Stormer	A. H. Thomas Co. Philadelphia, Pa.	A	$10-10^6$	low shear (high shear attachment available)	none to fair
Mooney Disk	Monsanto Co. Akron, Ohio	C	$\leq 10^8$	low shear	fair

[a] Price ranges: A, <$2000; B, $2000–5000; C, $5000–10,000; D, $10,000–25,000; E, $25,000–50,000; F, $50,000–100,000; G, >$100,000.

[b] U.S. representative is Haake, Inc., Saddle Brook, N.J.

[c] U.S. representative is Tekmar Company, Cincinnati, Ohio.

[d] U.S. representative is C. W. Brabender, South Hackensack, N.J.

[e] U.S. representative is Ferranti Electric, Commack, N.Y.

[f] U.S. representative is Gardner/Neotech Instrument Division, Pacific Scientific Co., Silver Spring, Md.

[g] U.S. representative is Sangamo Transducers, Grand Island, N.Y.

where the stress term may be torque, load, deflection, etc and the shear rate may be in rpm, rps (revolutions per second), s^{-1}, etc. Where factors are supplied by the manufacturer, separate constants may be given for converting the stress and shear-rate terms to the correct quantities and units.

Often a constant is determined from measurements with a Newtonian oil; this is particularly true of calibrations supplied by manufacturers. This constant is valid only for Newtonian specimens and, if used with non-Newtonian fluids, gives an apparent viscosity based on an inaccurate shear rate. Thus the viscosity is really an apparent viscosity. However, for relative measurements, this value can be useful. Employment of an instrument constant can save a great deal of time and effort and can increase accuracy in many cases because end and edge effects, slippage, turbulent interferences, etc are included.

Principles, analyses, and descriptions of various rotational viscometers are given in refs. 6, 45–49.

Coaxial (Concentric-Cylinder).   The earliest and most common type of rotational viscometer is the coaxial or concentric-cylinder instrument. It consists of two cylinders, one within the other (cup-and-bob), with the sample between them, as shown in Figure 22. The first practical rotational viscometer consisted of a rotating cup with an inner cylinder supported by a torsion wire. There have been many variations on this design, in a number of which the inner cylinder rotates rather than the outer one. Instruments of both types are available commercially and are useful for a variety of applications.

The relationship between viscosity and angular velocity and torque for a Newtonian fluid in a concentric-cylinder viscometer is given by the Margules equation (6):

$$\eta = \left(\frac{M}{\Omega\, 4\, \pi h}\right)\left(\frac{1}{R_i^2} - \frac{1}{R_o^2}\right) = \frac{kM}{\Omega}$$

where M is the torque on the inner cylinder, h is the length of the inner cylinder, Ω is the relative angular velocity of the cylinder in rad/s, R_i is the radius of the inner cylinder wall, R_o is the radius of the outer cylinder wall, and k is an instrument constant. Therefore viscosity can be determined from the torque and angular velocity.

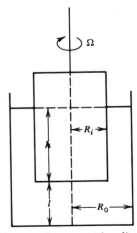

Figure 22.   Diagram of a concentric cylinder viscometer.

However one is more apt to use shear rate and shear stress to calculate viscosity. These two parameters can be obtained from the Margules equation. The shear rate is given by

$$\dot{\gamma} = \frac{(2\,\Omega/r^2)(R_i^2 R_o^2)}{(R_o^2 - R_i^2)}$$

where r is any given radius. The shear stress is given by

$$\tau = \frac{M}{2\,\pi r^2 h}$$

The shear rate and shear stress can be calculated for any radius r from these equations. In most cases the radius used is R_i, since the shear stress and shear rate of interest are at the inner, torque-sensing cylinder. Thus

$$\dot{\gamma} = \frac{2\,\Omega R_o^2}{(R_o^2 - R_i^2)} \quad \text{and} \quad \tau = \frac{M}{2\,\pi R_i^2 h}$$

The viscosity of a Newtonian fluid may be determined from the Margules equation or from the slope of a shear-stress–shear rate plot. Non-Newtonian fluids give intercepts and curves with such plots. Apparent viscosities can be calculated, but accurate values depend on including correction factors for yield points and shear thinning, ie, shear-rate corrections, in the above equations (6). The shear-rate correction may be minimized by using a very small gap size, ie, the ratio of the inner-to-outer radius should be as close to unity as possible. In practical terms, this means maintaining a ratio of 0.95, which is not possible with many sensors used with commercial rotational viscometers. With highly shear-thinning materials, even this is not sufficient and corrections must be made regardless of the gap size.

Besides non-Newtonian flow, the main correction necessary for concentric-cylinder measurements is that caused by end effects. Since the inner cylinder is not infinitely long, there is drag on the ends as well as on the face of the cylinder. The correction appears as an addition h_0 to the length h. The best method to determine the correction is to measure the angular velocity and torque at several values of h, that is, at various depths of immersion. The data are plotted as M/Ω vs h and extrapolation is made to a value of h_0 at $M/\Omega = 0$. The quality $(h + h_0)$ is substituted for h in the various equations.

Cone-and-Plate. In a cone-and-plate viscometer (see Fig. 23), a low angle ($\leq 3°$) cone rotates against a flat plate with the fluid sample between them. The cone–plate instrument is a simple, straightforward device that is easy to use and extremely easy

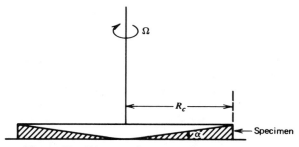

Figure 23. Diagram of a cone-and-plate viscometer.

to clean. It lends itself well to routine work as measurements are very rapid and no tedious calculations are necessary. However with careful calibration and good temperature control, a cone-and-plate viscometer can be very useful for research as well. Heated instruments can be used for melt-viscosity measurements.

In most rotational viscometers, the rate of shear varies with the distance from a wall or the axis of rotation. However, in a cone-and-plate viscometer, the rate of shear across the conical gap is essentially constant, because the linear velocity and the gap between the cone and plate both increase with increasing distance from the axis. No tedious correction calculations are required for non-Newtonian fluids. The relevant equations, respectively, for viscosity, shear stress, and shear rate at small angles α are:

$$\eta = \frac{3 \, \alpha M}{2 \, R_c^3}$$

$$\tau = \frac{3 \, M}{2 \, \pi R_c^3}$$

$$\dot{\gamma} = \frac{dv}{dr} = \frac{\Omega}{\alpha}$$

for Newtonian fluids, where M is the torque, R_c is the radius of the cone, v is the linear viscosity, and r is the distance from the axis. Cone-and-plate geometry has several advantages over concentric-cylinder geometry, including smaller sample size, homogeneous shear rate, and easy conversion of data to apparent viscosities. Possible drawbacks include sample drying or evaporation, slinging of material from the gap, and the distinct possibility of viscous heating, particularly at high shear rates. The latter problem is compounded by the fact that temperature control with commercial instruments is not always as good as it should be.

Dynamic. A dynamic viscometer is a special type of rotational viscometer used for characterizing viscoelastic fluids. It measures elastic as well as viscous behavior by determining the response to both steady-state and oscillatory shear. The geometry may be cone-and-plate, plate-and-plate, or concentric-cylinder (see Viscoelasticity).

Other. Some viscometers of industrial importance involve neither the concentric-cylinder nor cone-and-plate geometry. Some employ a disk as the inner member or bob, eg, in the Brookfield Synchro-Lectric and Mooney viscometers. Others use paddles (one geometry of the Stormer). These nonstandard geometries are very difficult to analyze, particularly for an infinite bath, as is normal with the Brookfield and Stormer. The Brookfield disk has been analyzed for Newtonian and non-Newtonian fluids and shear-rate corrections have been developed (7). Other nonstandard geometries are best handled by determining instrument constants by calibration with standard fluids.

Specific Commercial Viscometers. The Haake Rotovisco system includes a number of rotational viscometers that share the same wide range of sensors, both concentric-cylinder and cone-and-plate. Some of the instruments have only fixed speeds and are designed mainly for quality-control work. The RV-3 (Fig. 24) is a research instrument with a wide range of speeds, both fixed and variable. The specified viscosity and shear rates are 2–10^9 mPa·s (= cP) and 4×10^{-4} to 4×10^4 s^{-1}, respectively. Haake recently introduced several new instruments: a control unit–programmer–recorder (RV100),

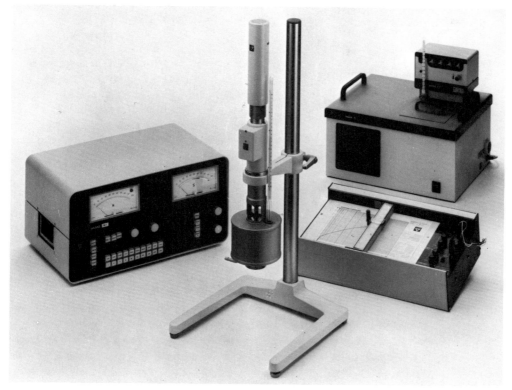

Figure 24.   Haake Rotovisco RV-3 rotational viscometer. Courtesy of Haake, Inc.

a low shear Couette-type concentric-cylinder device (CV100), and an improved cone-and-plate device (PK100). The CV100 is of particular interest as it can operate in an oscillatory mode to measure the elastic properties of liquids. Another improvement in the Rotovisco system is that hardware and software are available for interfacing to a calculator or microcomputer. The Contraves Rheomat line also is made up of a number of viscometers with many different sensors. Some of the models are designed for process and quality control, but the Rheomat-30 (Fig. 25) is a very versatile research instrument. Its viscosity and shear-rate ranges are comparable to the Rotovisco RV-3. A calculator interface is available for automation of experiments. The Brabender Rheotron is a true Couette instrument, ie, the outside cylinder is the rotating element. A number of concentric-cylinder sensors and a cone-and-plate system allow measurement of a viscosity range similar to that of the Rotovisco RV-3 and Rheomat-30 over a slightly narrower shear-rate range (5×10^{-2} to 2×10^4 s^{-1}). An advantage of the Rheotron is that normal force and elastic modulus resulting from fluid elasticity can be measured as well as shear viscosity.

The MacMichael viscometer is the most straightforward of the rotational viscometers. The outer cup rotates and the inner cylinder is suspended from a torsion wire. The drag on the inner cylinder is measured as degree of twist on the wire. The viscosity range increases to ca 10^6 mPa·s (= cP). The shear rate range is limited, ca 2–12 s^{-1}, but with modification higher shear rates could be attained. The instrument is best suited for slow time-dependent effects and low shear viscosities. Another

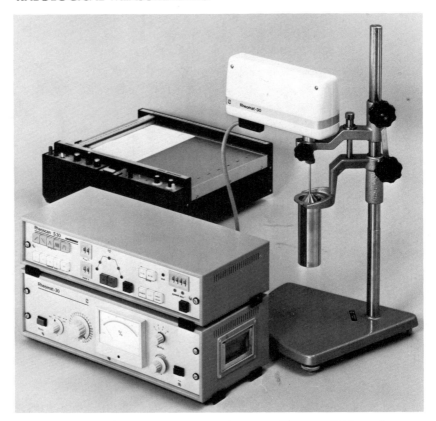

Figure 25. Contraves Rheomat-30 viscometer. Courtesy of Tekmar Co.

Couette viscometer is the Fann V-G, one model of which is shown in Figure 26, which was originally designed for measuring apparent viscosities and yield values for drilling muds. It should be applicable to a variety of other materials however. Models range from two-speed hand-cranked and six-speed motorized versions suitable for process and quality control to instruments with continuously variable speeds (0–600 rpm) and shear stress–shear rate plotting capability. A number of torsion springs and cylinder combinations are available. The overall possible viscosity range is $0.5–(5 \times 10^6)$ mPa·s (= cP) and the shear rate range is $0.24–3254$ s^{-1}.

The Ferranti-Shirley (Fig. 27) was the first commercially available, general-purpose cone-and-plate viscometer and the design still is one of the best available. Viscosities of $20–(3 \times 10^7)$ mPa·s (= cP) can be measured over a shear-rate range of $1.8–18,000$ s^{-1} and at up to 200°C with special ceramic cones. Features include accurate temperature measurement and good temperature control (thermocouples are embedded in the water-jacketed plate), electrical sensing of cone-plate contact, and a means of adjusting and locking the position of the cone and plate so that they just touch. The instrument can be interfaced with a computer or microprocessor. The ICI cone-and-plate device (Fig. 28) is a relatively simple, inexpensive, one-speed instrument designed for routine determination of paint viscosities at a shear rate comparable to that of brushing or spraying ($10,000$ s^{-1}) (50–51). The viscosity range at that shear rate is limited to $5–1000$ mPa·s (= cP). Use of a wider angle cone allows measurement

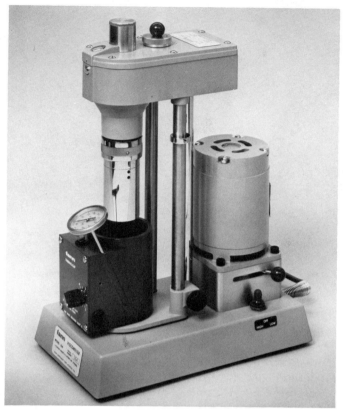

Figure 26.   Fann V-G viscometer. Courtesy of Fann Instruments Operations, Dressler Industries, Inc.

to 10,000 mPa·s (= cP) but at a lower shear rate of 3,000 s^{-1}. The instrument can be obtained with a temperature capability of 25–200°C in 25° steps, which makes it suitable for melt viscosities.

The Weissenberg Rheogoniometer (Fig. 29) is a very complex dynamic viscometer that can measure elastic behavior as well as viscosity. It can be used for complete characterization of viscoelastic materials. Capabilities include steady-state rotational shear measurements within a viscosity range of 10^{-1} to 5×10^{9} mPa·s (= cP) at shear rates of 8×10^{-3} to 10^{4} s^{-1}, measurement of normal forces (elastic effect) exhibited by the material being sheared, and oscillatory shear from which elastic modulus and dynamic viscosity can be determined. These measurements can be made over a wide range of temperature.

Rheometrics, Inc. produces a family of complex dynamic viscometers incorporating microprocessors which provide test control and data analysis. The viscometer of choice for characterization of low-to-medium viscosity fluids is the Fluids Rheometer (RFR) shown in Figure 30. This instrument is capable of measuring viscosity and elasticity (normal force, dynamic modulus) over a wide range of steady-state and dynamic shear. Cone-and-plate, parallel-plates, and coaxial cylinder geometries can be used. Shear rates at steady shear are 10^{-2}–10^{4} s^{-1} and oscillatory shear frequencies are 10^{-2}–500 radian/s. Viscosities of 10–10^{5} mPa·s (= cP) can be measured. Use of this instrument enables the rheologist to characterize the viscous and elastic behavior of fluids, eg, paints, oils, food products, and cosmetics.

Figure 27.   Ferranti-Shirley cone-and-plate viscometer. Courtesy of Ferranti Electric, Inc.

The Brookfield Synchro-Lectric viscometer (Fig. 31) is a widely used rotational viscometer for moderately low shear rates. It is rugged, easy to use, inexpensive, and can be used to characterize a wide range of materials. It is specified or is mentioned in 16 different ASTM methods. The original and most common sensor is a disk rotating in an infinite sea of fluid. This can be replaced with a concentric-cylinder system, eg, Small Sample, UL Adapter, or Thermosel System, which provides better defined geometry and, in the case of the Small Sample adapter, good temperature control. A cone-and-plate version also is available. This and the concentric-cylinder systems are preferred for accurate measurements. The new digital model with an output for a recorder is even easier to use than the older dial model and extends the application of the viscometer for the characterization of time-dependent behavior. With the several springs, which function as torque gauges, and a number of spindles, the overall range of measurable viscosities is very wide, ca $1-10^8$ mPa·s (= cP). The shear rates depend on the viscometer model and the sensor system but are ca $0.1-100$ s^{-1} for the disk spindles, ≤ 132 s^{-1} for the concentric cylinders, and ≤ 1500 s^{-1} for the cone-and-plate for a low viscosity sample. Viscosities at very low shear rates (ca $10^{-3}-1$ s^{-1}) can be measured with either the concentric-cylinder or cone-and-plate system by a relaxation technique (49,52–53).

The Stormer viscometer (Fig. 32) is an unusual but useful low shear rotational viscometer that can be converted to high shear with a suitable attachment. It is one of the very few instruments in which shearing stress rather than rate of shear is kept constant. With this instrument, a constant torque is applied to an inner cylinder or paddle and the rate of rotation is measured. The shear stress is applied by attaching weights to a string and permitting free fall through a vertical distance. The instrument accepts many different types of rotors and cups, including some from other viscometers such as the MacMichael. Therefore, many different geometries are possible, including a narrow-gap ($R_i/R_o = 0.99$) concentric-cylinder for high shear: the CRGI/Glidden

Figure 28. ICI cone-and-plate viscometer. Courtesy of Gardner/Neotech Instruments Div., Pacific Scientific Co.

Miniviscometer (54–55). Varying weights and concentric-cylinder gaps makes it possible to measure non-Newtonian behavior over a range of shear rates. Another Stormer-type viscometer is the Cannon Mini-Rotary viscometer, which is used to measure viscosity and yield stress of motor oils at −40–0°C (56).

In the Mooney Shearing Disk viscometer, a serrated disk is rotated in a sample fixed in a pressurized cavity. The instrument was developed for application to rubber and other elastomeric materials and is a standard quality-control instrument in the rubber industry (ASTM D 1646). The Mooney viscometer is used to measure high viscosities (given in arbitrary Mooney units but usually ca 7.5×10^7 mPa·s or cP) at low shear rates (ca 1.5 s^{-1}).

Moving-Body Viscometers. Moving body viscometers are instruments or techniques by which the motion of a ball, bubble, plate, or rod through a material is monitored to determine viscosity. Falling-ball viscometers are based on Stokes' law, which relates the viscosity of a Newtonian fluid to the velocity of the falling sphere. If a sphere is allowed to fall freely through a fluid, it accelerates until the viscous force exactly balances the gravitational force. Stokes' equation relating viscosity to the fall of a solid body through a liquid may be written as

$$\eta = \frac{2\, r^2 g (d_s - d_l)}{9\, v}$$

Figure 29.   Weissenberg Rheogoniometer. Courtesy of Sangamo Transducers.

where r is the radius of the sphere, d_s and d_l are the density of the sphere and the liquid, respectively, g is the gravitational force, and v is the velocity of the sphere.

Viscometers of the falling-ball type can be used over an extremely wide viscosity range but generally are employed for fairly viscous materials, because small balls and small differences in density are needed to obtain a suitably slow rate of fall in a low viscosity fluid. The devices are limited to measurements of Newtonian fluids since no practical formula has been developed for non-Newtonian materials. Although it is theoretically possible to use falling-ball viscometers to make highly accurate measurements, rarely is the uniformity of the sphere and of the tube sufficient for such measurements. However, falling-ball viscometers are excellent instruments for routine work.

The speed at which a sphere rolls down a cylindrical tube filled with a fluid or down an angled plate covered with a film of the fluid also gives a measure of the viscosity of the fluid. For the cylindrical tube geometry, the following equation for any given instrument is a generalized form of the Stokes equation

$$\eta = \frac{k(d_s - d_l)}{v}$$

where v is the translational velocity of the rolling sphere and k is the instrument constant, which is determined by calibration with standard fluids.

The Hoeppler viscometer probably is the most widely used falling/rolling-sphere viscometer. It is composed of a water-jacketed, precision-bore glass tube mounted at an angle 10° from the vertical. A series of balls of different diameters allow measurement of viscosities at $0.5–10^6$ mPa·s (= cP). A simple falling-ball viscometer may be constructed from a test tube or graduated cylinder with a rubber stopper and a short guide tube for insertion of the balls, as shown in Figure 33. Miniature ball bearings

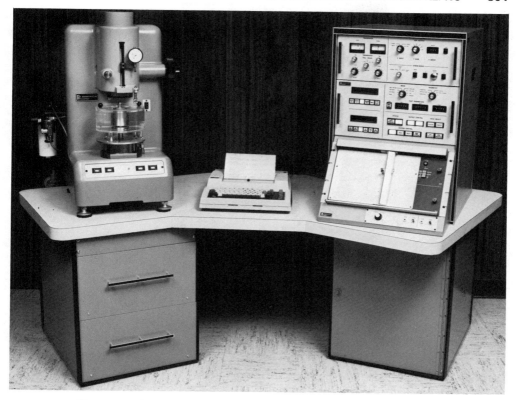

Figure 30. Rheometrics Fluids Rheometer. Courtesy of Rheometrics, Inc.

generally are suitable spheres. They may be obtained in different sizes and in materials of various densities.

The technique of determining viscosity from the velocity of a rolling ball on a plate or panel has been used largely for following the flow, drying, and curing of paints and other coatings, but it could be used for other materials as well (57–58). The system is calibrated with standard oils or other fluids of known viscosity. The geometry is ill-defined as the ball often slides as well as rolls, but the technique has proven useful.

Viscosity also can be determined from the rate of rise of an air bubble through a liquid. This simple technique is widely used for routine viscosity measurements of Newtonian fluids and especially polymer solutions. A bubble-tube viscometer consists of a glass tube of specific size to which liquid is added until a small air space remains at the top. The tube is then capped. When the tube is inverted, the air bubble rises through the liquid. The rise time in bubble seconds may be taken as a measure of viscosity or an approximate viscosity in mm^2/s (= cSt) may be calculated from it. An older method that is commonly used is to match the rate of rise to that of a member of a series of standards, eg, with Gardner-Holdt bubble tubes. Unfortunately, this technique gives viscosities in terms of a nonlinear scale of letter designations and does nothing to improve communications between users and rheologists.

If a broad definition of a moving body is adopted, then certain other viscometers can be considered to be of this type, including the band and falling-rod viscometers

Figure 31. Brookfield Synchro-Lectric viscometer with Small Sample Adapter and water circulator. Courtesy of Brookfield Engineering Laboratories, Inc.

which are used to test printing inks. The band viscometer includes an arrangement of parallel plates, the basic geometry used to define viscosity. The band, which is a strip of Mylar film, is sandwiched between two fixed plates. The fluid is placed between the band and the plates and the band is pulled through the fluid. For a given gap and force on the band, the speed of the band is a measure of viscosity. The falling-rod viscometer is quite similar but is based on movement of a rod rather than a plate through the fluid (59–60). This design is a form of a falling coaxial cylinder viscometer. Most of these devices have been used for semiempirical studies of materials, such as bitumens and rosins, but more recent versions are capable of precise measurements of polymer melts and solutions (61).

Other Viscometers. A number of other viscometers exist, many of which are custom-built for specific research or product applications. One design involves the use of ultrasonic vibrational techniques to measure viscosity. At least three viscometers, ie, Bendix Ultra-Viscoson, Automation Products Dynatrol, and Nametre, are based on this principle. Since the rate of shear is not easily determined or changed, these instruments are best used for controlling or studying processes in which viscosity changes with time or temperature.

Extensional Viscosity. Methods exist for measuring all three types of extensional viscosity: uniaxial, biaxial, and pure shear. There are very few commercial instruments available, however, and most measurements are made with improvised equipment.

Most uniaxial measurement techniques involve extending a strand or cylindrical rod of the material and measuring the force necessary to do so. A good method for fluids consists of extruding the fluid from a spinneret nozzle and extending it by rolling it

Figure 32.   Stormer viscometer. Courtesy of Arthur H. Thomas Co.

up on a rotating drum (36,62). The generated force is measured by the deflection of
the nozzle or tube as it is pulled by the filament of fluid. This method is the basis for
a commercial instrument, the Sangamo Schlumberger elongation viscometer, which
is shown in Figure 34. The fluid is in a reservoir which is pressurized by a clean gas.
The pressure forces fluid along a calibrated, thin-walled tube and out through a
spinneret nozzle. The fluid falls vertically, then is picked up tangentically on a 50 mm
diameter rotating drum, and is elongated. The fluid is cut away from the drum into
a container below. The length of the fluid filament is variable. The force is measured
by the deflection of the thin-walled tube as determined by a transducer. The whole
assembly is housed in an environmental chamber and the sample can be heated to
100°C.

   A method used for measuring the uniaxial extensional viscosity of polymer solids
and melts consists of a tensile tester in an oil bath and involves cylindrical rods as
specimens (63–64). The extruder for the rod may be part of the apparatus and be
combined with a device for grabbing the extruded material (65). However, most of
the recent versions use prepared rods, which are placed in the apparatus and then
heated to soften or melt the polymer (37,66–67). A constant stress is applied and the

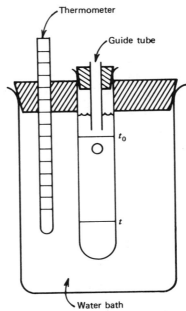

Figure 33.   A simple falling-ball viscometer. Lines t_0 and t are the timing lines for velocity determinations.

Figure 34.   Sangamo Schlumberger Elongational Viscometer. Courtesy of Sangamo Transducers.

resultant extensional strain rate is measured or a constant strain rate is applied and the resultant stress is measured. This method also has been applied to two commercial instruments, both of which are based on a test-frame design by BASF (37,66–67). The Rheometrics RER 9000 extensional rheometer combines this test frame with a Rheometrics analyzer–controller. The cylindrical specimen is suspended vertically in a hot-oil bath. A load cell is fixed to the lower end of the specimen and a moveable rod to the upper end. The rod is attached to a servo drive with an integral displacement transducer, which indicates sample length and acts as a feedback element for the servo mechanism. The transducer signal as a function of time gives a direct measure of elongation. This instrument can also be used for creep and stress-relaxation measurements. It provides a wide range of strain rates (10^{-3}–5 s^{-1}) and stresses (10^3–10^6 Pa (10^4–10^7 dyn/cm^2)) and has temperature capabilities of 30–250°C. There is another instrument based on the BASF test frame, but in which the sample is mounted horizontally rather than vertically. It is produced in the Federal Republic of Germany by the Göttfert Company.

There are several methods for measuring biaxial extensional viscosity, the most common being bubble inflation in which a thin polymer sheet is inflated with an inert gas or liquid (68–70). A more recent method involves stretching a sheet of polymer with eight rotating clamps placed in an octagonal pattern (71). A complex servo control system is required to coordinate the motion of the clamps. An even newer technique is lubricated squeezing flow, by which displacement is measured as the material is squeezed between two disks with lubricated surfaces (38). Extensional viscosity that results purely from shear deformation seems to be of less interest to investigators but it has been measured (34,72).

Viscoelasticity Measurement

There are a number of methods for measuring the various quantities that describe viscoelastic behavior. Some of them require expensive commercial instruments, others depend on customized research instruments, and still others require only very simple devices. Even qualitative observations can be quite useful in the case of liquids, eg, polymer melts, paints, and resins, where elasticity may mean a bad batch or unusable formulation. Examples include observing die swell of the material from a syringe with a microscope and noting the Weissenberg effect as seen in the forcing of a cone and plate apart during viscosity measurements or the climbing of a resin up the stirrer shaft during polymerization or mixing.

Creep. One of the simplest experiments to describe viscoelastic behavior is creep. Creep experiments involve measurement of deformation as a function of time after a given load has been applied. Such measurements may be made on samples in tension, compression, or shear. A simple apparatus for tensile creep measurements is shown in Figure 35 (73). The specimen is held between two film clamps; the upper one is attached to a support and the lower one bears a weight. The apparatus can easily be placed inside a test tube and immersed in a water bath for temperature control. Experiments involve measuring the extension of the sample as a function of time. A graduated scale can be used to measure large extensions, but a cathetometer is necessary for measuring smaller ones. The data may be presented in the form of a strain–time curve or the modulus may be calculated and a modulus–time curve constructed (see Fig. 16). The material undergoes initial elastic deformation followed by a slow increase in strain with time.

Compliance J is the reciprocal of modulus and often is used to characterize creep. The creep compliance at any time t is the ratio of the strain at that time to the constant shear stress. Because of the large ranges of compliance and time normally covered, creep compliance curves are logarithmic, as shown in Figure 36 which is a typical curve

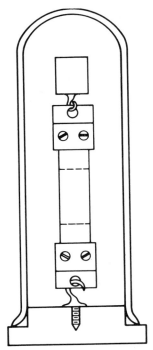

Figure 35.   Simple apparatus for creep measurements (73). Courtesy of Marcel Dekker, Inc.

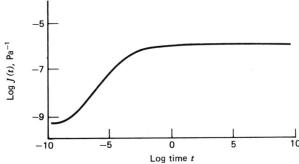

Figure 36.   Creep compliance $J(t)$ vs time for a lightly vulcanized rubber (74). To convert Pa^{-1} to cm^2/dyn, divide by 10.

for a lightly cross-linked rubber (74). Creep measurements are practical tests. They are useful for evaluating the behavior of structural materials, fastenings, gaskets, pipe, and other items that are subject to loading while in use.

Stress-Relaxation.   In a stress-relaxation experiment, deformation is held constant and the resulting stress in the sample is measured as a function of time. The deformation of the sample produces an initial stress, which decays with time in the case of viscoelastic materials. A diagram of a stress-relaxation apparatus is shown in Figure 37. The sample is held by a set of clamps. The upper clamp is attached to a strain gauge or balance arm. The lower clamp is attached to a device for rapidly elongating the sample by a fixed amount. In an experiment, the sample is rapidly deformed and the resulting forced measured as a function of time. The data may be plotted as stress,

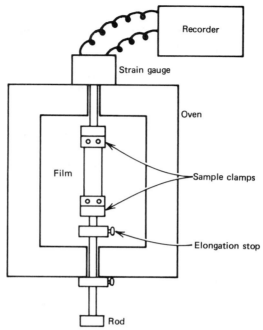

Figure 37.   Schematic diagram of a stress–relaxation apparatus (73) Courtesy of Marcel Dekker, Inc.

stress/initial stress, or modulus. An example of a stress-relaxation curve is shown in Figure 38. The material is the same rubber shown in the creep-compliance curve in Figure 36. Stress-relaxation experiments are generally preferred to creep experiments by investigators interested in structure–property relationships and testing the theory of viscoelastic materials.

Penetration–Indentation.   Various penetration and indentation tests have long been used to characterize viscoelastic materials, eg, asphalt, rubber, plastics, and coatings. There are many variations but the basic test consists of pressing an indentor of prescribed geometry against the test surface. Most instruments have an indenting tip, ie, a cone, needle, hemisphere, etc, attached to a short rod that is held vertically. The load is controlled at some constant value and the time of indentation is specified. The size or depth of the indentation is then measured.

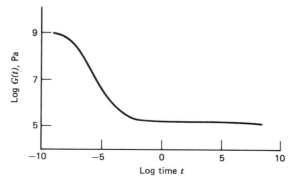

Figure 38.   Stress–relaxation curve for a lightly vulcanized rubber (74). To convert Pa to dyn/cm², multiply by 10.

The main disadvantages of this technique are that the area of contact between the indentor and the specimen does not remain constant throughout the test and many of the indentors have geometries, eg, diamonds, pyramids, cones, and cylinders, such that basic viscoelastic quantities cannot be calculated from the results. Penetrometers of this type include a number of instruments, eg, the Pfund, Rockwell, Tukon, and Buchholz testers which are used to measure indentation hardness, which is a function of modulus, of various materials (see Hardness).

Other penetrometer–indentometers include transducers to sense the position of the load, recorders and temperature controllers–programmers, and in some cases, microprocessors for data reduction. A diagram of such an instrument is shown in Figure 39. These instruments are used to measure softening points, which usually is ca T_g, as well as measure indentation hardness, creep, creep recovery, and modulus. Examples of such instruments include the DuPont, Mettler, and Perkin-Elmer thermomechanical analyzers (TMAs) and the ICI Microindentometer. They can be used to generate modulus and modulus–temperature data from indentation–time plots through use of the Hertz equation (41,75).

$$\frac{E}{1 - \mu^2} = \frac{3}{4} \frac{P}{r^{1/2} h^{3/2}} = H_k$$

where E is the elastic or Young's modulus, μ is Poisson's ratio, r is the radius of the hemispherical indentor, P is the force on the indentor (mass load $\times$ g), h is the indentation, and H_k is the indentation hardness. Since the indentation varies with time, the modulus must be specified for a certain indentation time, eg, a 10-s modulus. The Hertz equation holds only for purely elastic materials. However, considerable application has been made to viscoelastic materials, including polymers and coatings, with excellent results (75–81). An example of an indentation–hardness-versus-temperature curve generated using this technique is shown in Figure 40 (75,77). Indentation is not

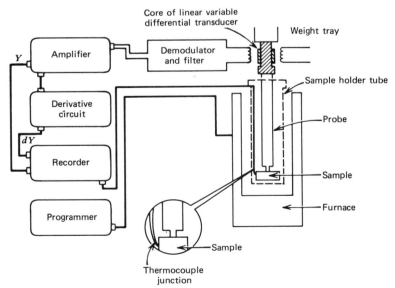

Figure 39.   Schematic of the DuPont Thermal Mechanical Analyzer. Courtesy of E. I. du Pont de Nemours & Co., Inc.

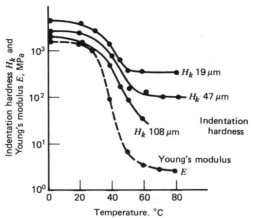

Figure 40. Indentation hardness at different thicknesses and Young's modulus of a free film (as functions of temperature) of an acrylic coating. These results show the dependence on thickness that occurs with thin films. To convert MPa to dyn/cm^2, multiply by 10^7; to psi, multiply by 145. Courtesy of *Farbe + Lack.*

usually thought of as a method for measuring viscoelastic properties, but the technique can be very useful in this area. It may become more popular with the introduction of Mettler's new TMA, which allows setting and changing of force by a key board rather than by loading and unloading weights. It also has a mode by which periodic and very short (12 s) indentation creep measurements can be made while the specimen is being heated. The latter speeds up modulus–temperature measurements considerably.

Tensile Testing. The commonest and one of the best methods for measuring the viscoelastic properties of solids is the tensile tester or stress–strain instrument which extends a sample at constant rate and records the stress. Creep and stress–strain measurements also can be made. A large number of commercial instruments of various sizes and capacities are available. They vary a great deal in terms of complication and degree of automation, ie, from hand-operated to completely microprocessor-controlled. Some have temperature chambers so that measurements can be made over a range of temperatures. Well-known manufacturers include Instron, MTS, Tinius Olsen, Applied Test Systems, Thwing-Albert, Shimadzu, and Monsanto.

A typical stress–strain curve as might be generated by a tensile tester is shown in Figure 41. Creep and stress-relaxation results are essentially the same as those described in the section on simple versions of these measurements. Regarding stress–strain diagrams and from the standpoint of measuring viscoelastic properties, the investigator is most interested in the early part of the curve, the region of small deformations. At small strain levels, stress is proportional to strain and the modulus is independent of strain. At higher stresses and corresponding strains, a point is reached where modulus depends on strain. Eventually, irreversible changes take place in the sample resulting in physical breaking. These forms of behavior are shown in Figure 41. The linear relationship between stress and strain is shown in the region between the start of the experiment and point *A*. The latter is the yield point common to many materials, in which the specimen yields with no increase in or with a decrease in stress. At higher levels, the sample breaks at break point *B*. The tensile strength *b* and elongation at break *d* are important mechanical properties but are outside the scope of this article.

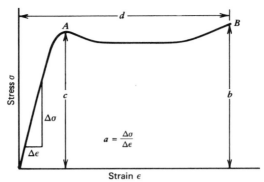

Figure 41.   Typical stress–strain curve. Point A is the yield point of the material. The sample breaks at point B. Mechanical properties are identified as follows: a, modulus; b, tensile strength; c, yield strength; d, elongation at break. The toughness or work to break is the area under the curve.

Dynamic Measurements.   Dynamic methods are required for investigating the response of a material to rapid processes, for studying fluids, or for examining a solid as it passes through a transition region (T_g, melting, etc). The techniques impart cyclic motion to a specimen and measure the resultant response. Dynamic techniques are used to determine storage and loss moduli G' and G'' and the loss tangent tan δ. Some instruments are sensitive enough for the study of liquids and can be used to measure the dynamic viscosity η'. Measurements are made as a function of temperature, time, or frequency and results can be used to determine transitions and chemical reactions as well as the properties noted above. Dynamic mechanical techniques for solids can be grouped into three main areas: free vibration, resonance-forced vibrations, and nonresonance-forced vibrations. Dynamic techniques are described in detail in refs. 74, 77, 82–90.

Free-Vibration Methods.   Free-vibration instruments subject a specimen to a displacement and allow it to vibrate freely. The oscillations are monitored for frequency and damping characteristics as they disappear. The displacement is repeated again and again as the specimen is heated or cooled. The results are used to calculate storage and loss modulus data. The torsional pendulum and torsional-braid analyzer (TBA) are examples of free-vibration instruments.

The torsional pendulum (Fig. 42) is a simple apparatus for making dynamic tests in the frequency range 10^{-1}–10 Hz. The specimen is rigidly clamped at its lower end and the upper end is clamped to an inertia bar or disk that is free to rotate. The suspension wire supporting the system is passed over a pulley and the weight is counterbalanced to prevent tensile stresses on the specimen. The experiment is begun by setting the inertia member into free oscillation. The natural frequency and decay of these oscillations are measured. The storage modulus G' is calculated from the square of the frequency. The loss tangent tan δ is calculated from the logarithmic decrement of the sample. The logarithmic decrement is the natural logarithim of the ratio of two successive amplitudes.

A torsional-braid analyzer is a torsional pendulum with which a composite specimen instead of a free film is used (85). The specimen is prepared by soaking a multifilamented glass braid in a polymer solution and, after mounting the specimen in the apparatus, removing the solvent by heating. Measurements are similar to those with the torsional pendulum. Rigidity (proportional to G') and damping (proportional to tan δ) values are calculated.

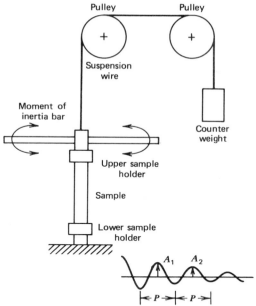

Figure 42.   A schematic diagram of a torsion pendulum and a typical damped sine wave output. P is the period of the motion; A_1 and A_2 are successive amplitudes (73). Courtesy of Marcel Dekker, Inc.

With both instruments, data can be plotted as a function of time or temperature to give dynamic mechanical spectra. An example is shown in Figure 43, where rigidity and damping from TBA experiments are shown as functions of temperature for a styrene–butadiene–styrene radial block copolymer (91). Such spectra are particularly useful for the study of polymers, including the identification of physical transitions and the onset and kinetics of cross-linking or embrittlement, as well as for measuring various viscoelastic parameters. The curves in Figure 43 show that the material has two glass-transition temperatures, each indicated by a downturn in the rigidity curve and a damping peak. The material is glassy (has a high modulus) below the lower T_g, is rubbery or leathery (moderate modulus) between the two transitions, and is almost fluid (low modulus) above the upper T_g. Knowledge of such characteristics is very important for prediction of behavior under use conditions.

Resonance Forced Vibration.   An example of the resonance-forced-vibration technique is the vibrating reed (88). A specimen in the form of a thin strip, ie, the reed, is clamped firmly at one end with the other end free. The clamped end is then forced to vibrate and the amplitude of the oscillating free end is measured. As the driver frequency is varied, the natural resonance of the reed is reached and the amplitude passes through a maximum. A schematic of the device and a typical curve is shown in Figure 44 (73). Young's modulus can be calculated from the square of the resonant frequency and the geometry of the sample. The loss tangent can be calculated from the width of the resonance peak.

Another resonant-frequency instrument is the DuPont Dynamic-Mechanical Analyzer (DMA). A barlike specimen is clamped between two pivoted arms and sinusoidally oscillated at its resonant frequency with an amplitude selected by the operator. An amount of energy equal to that dissipated by the specimen is added on each cycle to maintain a constant amplitude. Young's modulus is calculated from the

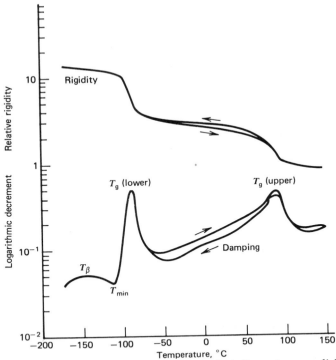

Figure 43.   Thermal mechanical behavior of a styrene–butadiene–styrene radial block copolymer in nitrogen at −180–150°C (91).

resonant frequency and the makeup energy represents a damping function which can be related to loss modulus G''.

Nonresonance Forced Vibration.   A simplified schematic of the Rheovibron (Imass, Inc.) is shown in Figure 45. The specimen is attached to strain gauges at each end. One of the gauges G_1 is a stress transducer, which measures the force applied by the driver; the other G_2 records the deformation of the sample. A sinusoidal tensile strain is applied at a given frequency to one end of the sample by the driver. The stress response is measured at the other end. The instrument gives a direct reading of the absolute dynamic modulus $|E^*|$, ie, the ratio of maximum stress amplitude to maximum strain amplitude, and the phase angle δ between stress and strain. From these quantities, the tensile storage E' and loss E'' moduli and the loss tangent $\tan \delta$ can be calculated since $E' = E^* \cos \delta$, $E'' = E^* \sin \delta$, and $\tan \delta = E''/E'$. The Autovibron is an automatic version of the Rheovibron.

Fluids.   Rheometrics, Inc. supplies several devices designed especially for characterizing viscoelastic fluids (see also Viscosity Measurement). All of the instruments measure the response of a liquid to sinusoidal oscillatory motion to determine dynamic viscosity and storage and loss moduli. The Rheometrics line includes a Viscoelastic Tester (RVE-M), a Dynamic Spectrometer (RDS), the Mechanical Spectrometer (RMS), and the new System Four which is the top of the line. The Viscoelastic Tester has limited capabilities and is designed primarily for quality and process control. The Dynamic Spectrometer is more of a research instrument as it gives a wider frequency range and can test solids and liquids. However, it is a stripped-down version of the

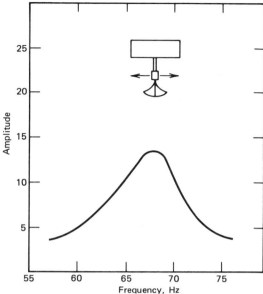

Figure 44.   Schematic of vibrating reed and amplitude–frequency curve for an epoxy ester varnish film (73). Courtesy of Marcel Dekker, Inc.

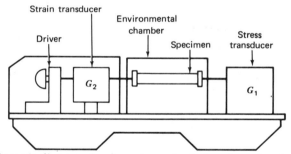

Figure 45.   Schematic diagram of a dynamic mechanical analyzer based on the nonresonance-forced-vibration principle (Rheovibron type).

extremely versatile Mechanical Spectrometer, which is a dynamic viscometer and a dynamic mechanical testing device. The RMS can carry out measurements under rotational shear, oscillatory shear, torsional motion, and tension–compression. It can be used to characterize a wide range of materials, including adhesives, various pastes, rubber, processed food, and plastics. The System Four is a laboratory in a single test frame. Four independent test heads are mounted on a turret carousel linked to a common microcomputer system for control and data analysis. One of the heads provides transient and steady rotation for the study of low viscosity fluids in the same manner as the Fluids Rheometer. The other three heads provide steady rotation, transient rotation, and transient linear motion, which can be used for viscosity, viscoelastic, and tensile–compression measurements, respectively.

Time–Temperature Superposition, Master Curves.   Since the modulus of a viscoelastic material varies with time and temperature, it is necessary to make measurements over wide ranges of these variables for full characterization. Often this is not possible, particularly for very long and very short times. Even if it were possible, the amount of experimental work involved would be prohibitive. Therefore, techniques have been developed to determine modulus values and curves at times or temperatures not attainable experimentally. A series of stress–relaxation curves measured at different temperatures can be shifted on the log time axis to give a single modulus-time master curve that covers a very wide time range (Fig. 46) (92). This shifting is possible because time and temperature have equivalent effects on modulus (20,74,89,93).

Practical Rheology

The technologist is interested in rheology for quality and process control, to improve field performance of existing products, to develop new products to meet specific needs, etc. Rarely is the technologist trained in rheology and rarely is the rheologist trained in the technology of the area to be improved. In the practical application of rheology, one of the first hurdles that must be overcome is for the technologist to learn enough rheology to understand the rheologist and for the rheologist to learn enough technology to make a meaningful application of talent to the requirements of the technical problem.

Because of the imprecise nature of language, the technologist may use words such as flow to describe effects that are not connected with rheology in the way that a trained rheologist would interpret them. For example, a paint technologist may talk about a paint having poor flow, meaning that the surface has imperfections or defects. An inexperienced rheologist may make elaborate measurements before finding that the sample flows in a rheological sense very easily during application and cure. The solution of the technologist's problem may not involve alteration of the viscosity of the sample

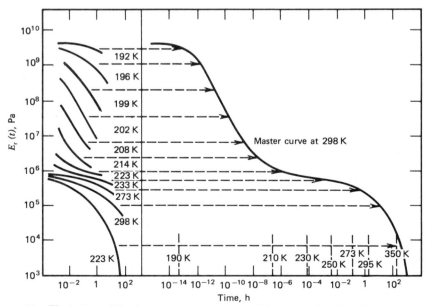

Figure 46.   Illustration of the time–temperature superposition principle as based on stress–relaxation data for polyisobutylene. To convert Pa to dyn/cm^2, multiply by 10. Courtesy of International Textbook Co., Ltd. (92). Data from ref. 93.

at all. Addition of a fraction of a percent of silicone surfactant may improve the surface appearance of the paint film.

A good working rule in the correlation of viscosity measurements with processing conditions is to select and compare viscosities of materials at the shear rate that corresponds to the process in question. Often a simple ordering of good and bad materials yields an insight into the desired range of viscosity required for satisfactory processing. This strategy assumes that the viscosity of the material is the significant controlling parameter in the process. This assumption should always be tested. Often other factors have an equal or even greater influence on the system. For example, in the case of thin films and surfaces, the surface tension and physics of wetting can play a role equal to that of viscosity.

Few processes involve a single shear rate or set of mechanical conditions. Typical processes involve low, intermediate, and high shear regions or sections or stages. For example, a paint or coating may be pumped (intermediate shear rate), sprayed (high shear rate), coalesce and flow to form a uniform film (intermediate to low shear rate), and sag or run under gravity (low shear rate) during application. Thus a complete rheological evaluation of a material to determine its processing characteristics requires consideration of the viscosity of the material over an extended shear-rate range. A given material may process well at low or intermediate shear rates but fail completely at higher shear rates.

Additional complications can also occur if the mode of deformation of the material in the process differs from that of the measurement method. Most rheological measurements are made under certain defined conditions of shear. If the material is extended or drawn into filaments, the extensional viscosity may be a more appropriate quantity for correlation with performance. This is the case in the parting nip of a roller in which the filamenting paint can cause roller spatter if the extensional viscosity exceeds certain limits (36).

The preceding examples indicate that the correlation of rheological measurements with product and process performance require as much study, knowledge, and hard work as carrying out the rheological measurements themselves. Useful surveys of some existing correlations are available (1,28,47,49,94). Extensive application of rheological measurements have been made to food products, biological fluids, lubricants, paints, printing inks, rubber, plastics, soaps, pharmaceuticals, slurries and suspensions, etc.

Rheological measurements are very important for the production and application of paints and inks. Viscosity measurements are also employed extensively in the development of new or improved products. A variety of measuring instrumentation are used, including bubble-tube viscometers; Ford or Zahn viscosity cups; rotational viscometers, eg, the Stormer and Brookfield instruments; cone-and-plate high shear-rate viscometers, eg, the ICI instrument; and research rotational viscometers, eg, the Rotovisco and Ferranti-Shirley. Single-point measurements are used for most development and quality-control work. However, the non-Newtonian nature of the products and the wide range of shear rates to which they are exposed in handling and application demand multipoint techniques.

The consistency of fats, margarine, and butter has been thoroughly measured by a variety of techniques. The creep compliance behavior of these materials has been studied at low shearing stresses by torsional measurements. Margarine exhibits linear viscoelastic behavior at low shearing stresses, but at higher stresses the creep behavior becomes nonlinear, the compliance increasing as the shear stress increases. The initial elastic modulus is ca 10^5 Pa (10^6 dyn/cm^2) and increases rapidly over 10 d approaching a value of ca 10^6 Pa (10^7 dyn/cm^2). The retardation spectrum also changes with time, becoming much flatter and with diminishing contributions from the lower end of the time scale (see Vegetable oils).

The cone penetrometer can be used to provide information on the consistency of butter and margarine. The penetrometer data for worked and nonworked butter can be combined to provide a spreadability index and to measure the initial hardness and change in hardness on working. The hardness of working samples stored at constant temperature increases slowly with time. This is attributed to the slow renewal of the fat crystal lattice and the formation of secondary bonds. The rheological properties of fats, margarine, and butter have also been studied by extrusion through capillary tubes. In these tests, the fat is subjected to much higher shearing forces than when tested by the cone penetrometer and undergoes a greater degree of structural change. Since the consistency of fatty materials depends greatly on mechanical and thermal treatment, a standardized preliminary treatment is necessary for meaningful comparisons (see Fats and fatty oils; Milk and milk products).

Lubricating oils of either the mineral or synthetic type are characterized by their viscosity and viscosity–temperature characteristics. The viscosity of lubricating oils is measured by glass-capillary viscometers according to standardized procedures. Ostwald, suspended-level, and reverse-flow glass-capillary instruments are used. Measurements at 40 and 100°C are carried out to establish the variation of viscosity with temperature, which is expressed as a viscosity index V.I. Addition of small amounts of certain polymers, eg, polyisobutylene, improve the viscosity index. The modified oils are non-Newtonian and exhibit shear thinning behavior. The low shear capillary measurements do not reflect the viscosity of the oil under cranking speeds (shear rates ca 10^4 s^{-1}) or under normal engine operation (10^6 s^{-1}). Therefore, a high shear concentric-cylinder viscometer (cold-cranking simulator) has been developed to simulate cranking under severe starting conditions in an automobile engine.

Various capillary and concentric-cylinder viscometers have been developed to measure the viscosity of oils under high pressure and high shear rate conditions that prevail in bearings. High frequency oscillatory measurements under high ambient pressure conditions have been proposed to provide information on the shear modulus of lubricants under temperature and pressure conditions characteristic of contact.

A grease is a lubricating oil thickened with a gelling agent, typically a metallic soap. Greases are graded or classified by a penetration test in which the degree of penetration of a standard penetrometer cone is measured after 5 s at 25°C. In addition, an apparent viscosity is measured at various shear rates (10–15,000 s^{-1}) in a standardized, motor-driven, piston capillary instrument to ensure that the grease can be applied and distributed. Greases are non-Newtonian shear-thinning materials. The high shear viscosity approximates that of the ungelled oil from which it is prepared. The viscosity of a grease depends on its history so that the condition of the sample tested, whether undisturbed, unworked, worked, etc, must be specified.

The rubber and plastics industries use a wide variety of rheological measurements for research, quality control, and testing. Two examples of standard instruments are the Mooney viscometer and the melt indexer developed by DuPont. The latter is the standard quality-control instrument for the measurement of melt viscosity in the plastics industry. Creep and stress–relaxation measurements provide useful engineering information and are used to develop structure–property relationships.

BIBLIOGRAPHY

"Rheology" in *ECT* 1st ed., Vol. 11, pp. 730–748, by John R. Van Wazer, Monsanto Chemical Company; "Viscometry" in *ECT* 1st ed., Vol. 14, pp. 756–776, by F. H. Stross and P. E. Porter, Shell Development Corporation; "Rheology" in *ECT* 2nd ed., Vol. 17, pp. 423–444, by John W. Lyons, Monsanto Company; "Viscometry" in *ECT* 2nd ed., Vol. 21, pp. 460–484, by Richard S. Stearns, Sun Oil Company.

1. H. Green, *Industrial Rheology and Rheological Structures*, John Wiley & Sons, Inc., New York, 1949.
2. W. H. Bauer and E. A. Collins in F. R. Eirich, ed., *Rheology*, Vol. 4, Academic Press, Inc., New York, 1967, p. 423.
3. O. C. C. Lin, *J. Appl. Polym. Sci.* **19**, 199 (1975).
4. R. Roscoe in J. J. Hermans, ed., *Flow Properties of Disperse Systems*, North Holland Publishing Company, Amsterdam, The Netherlands, 1953, p. 1.
5. G. W. Scott Blair, *Elementary Rheology*, Academic Press, Inc., New York, 1969.
6. J. R. Van Wazer, J. W. Lyons, K. Y. Kim, and R. E. Colwell, *Viscosity and Flow Measurement, A Laboratory Handbook of Rheology*, Interscience Publishers, a division of John Wiley & Sons, Inc., New York, 1963.
7. P. E. Pierce, *J. Paint Technol.* **43**(557), 35 (1971).
8. W. K. Asbeck, *Off. Dig. Federation Soc. Paint Technol.* **33**(432), 65 (1961).
9. M. R. Rosen, *Polym. Plast. Technol. Eng.* **12**(1), 1 (1979).
10. O. F. Solomon and I. Ciuta, *J. Appl. Polym. Sci.* **6**, 683 (1962).
11. M. Kurata and co-workers in J. Brandrup and E. H. Immergut, eds., *Polymer Handbook*, 2nd ed., Wiley-Interscience, New York, 1975, p. IV-1.
12. S. H. Aharoni, *J. Appl. Polym. Sci.* **21**, 1323 (1977).
13. J. F. Rabek, *Experimental Methods in Polymer Chemistry*, Wiley-Interscience, New York, 1980.
14. G. V. Vinogradov and A. Ya. Malkin, *Rheology of Polymers*, Springer-Verlag Inc., New York, 1980.
15. T. G. Fox, S. Gratch, and S. Loshaek in F. R. Eirich, ed., *Rheology*, Vol. 1, Academic Press, Inc., New York, 1956, p. 431.
16. D. R. Paul, J. E. St. Lawrence, and J. H. Troell, *Polym. Eng. Sci.* **10**, 70 (1970).
17. G. Berry and T. G. Fox, *Adv. Polym. Sci.* **5**, 261 (1968).
18. B. Maxwell and M. Nguyen, *Polym. Eng. Sci.* **19**, 1140 (1979).
19. F. Bueche, *J. Chem. Phys.* **40**, 484 (1964).
20. M. L. Williams, R. F. Landel, and J. D. Ferry, *J. Am. Chem. Soc.* **77**, 3701 (1955).
21. J. E. Carley, *Mod. Plast.* **39**, 123 (Dec. 1961).
22. G. Pezzin, *Materie Plastiche ed Elastomeri*, Aug. 1962, translated and reprinted as *Application Series*, *PC-12*, Instron Corporation, Canton, Mass., 1972.
23. R. A. Mendelson in N. M. Bikales, ed., *Encyclopedia of Polymer Science and Technology*, Vol. 8, Interscience Publishers, a division of John Wiley & Sons, Inc., New York, 1968, pp. 587–620.
24. J. L. White in K. Walters, ed., *Rheometry: Industrial Applications*, Research Studies Press, a division of John Wiley & Sons, Inc., New York, 1980, pp. 209–280.
25. H. L. Goldsmith and S. G. Mason in F. R. Eirich, ed., *Rheology*, Vol. 4, Academic Press, Inc., New York, 1967, p. 232.
26. H. L. Frisch and R. Simha in ref. 15, p. 525.
27. J. Mewis in G. Astarita and co-eds., *Principles*, Vol. 1 of *Rheology*, Plenum Press, New York, 1980, p. 149.
28. P. Sherman, *Industrial Rheology*, Academic Press, Inc., New York, 1970.
29. D. Quemeda, *Rheol. Acta* **16**, 82 (1977).
30. N. L. Ackermann and H. T. Shen, *AIChE J.* **25**, 327 (1979).
31. M. Mooney, *J. Colloid Sci.* **6**, 162 (1951).
32. D. C.-H. Cheng, *Chem. Ind.* (*London*) **1980**, 403 (May 17, 1980).
33. G. I. Taylor, *Proc. R. Soc. London Ser. A* **138**, 41 (1932); P. Sherman, *Industrial Rheology*, Academic Press, Inc., New York, 1970, p. 137.
34. C. D. Denson, *Polym. Eng. Sci.* **13**, 125 (1973).
35. J. W. Hill and J. A. Cuculo, *J. Macromol. Sci. Rev. Macromol. Chem.* **C14**, 107 (1976).
36. J. E. Glass, *J. Coat. Technol.* **50**(641), 56 (1978).
37. H. M. Laun and H. Münstedt, *Rheol. Acta* **17**, 415 (1978).
38. Sh. Chatrei, C. W. Macosko, and H. H. Winter, *J. Rheol.* **25**, 433 (1981).

39. J. M. Dealy, *Polym. Eng. Sci.* **11**, 433 (1971).
40. C. Zener, *Elasticity and Anelasticity of Metals*, University of Chicago Press, Chicago, Ill., 1948.
41. S. P. Timoshenko and J. N. Goodier, *Theory of Elasticity*, 3rd ed., McGraw-Hill, Inc., New York, 1969.
42. M. H. Wohl, *Chem. Eng.* **74**(4), 130 (Feb. 12, 1968).
43. J. Starita in *Fluids*, Vol. 2 of G. Astarita and co-eds., *Rheology*, Plenum Press, New York, 1980, p. 229; Rheometrics, Inc., *Rheometrics System Four*, Union, N.J., 1980.
44. K. Weissenberg, *Nature* **159**, 310 (1974).
45. R. D. Athey, Jr., *Chem. Technol.* 207 (May 1981).
46. P. E. Pierce, *J. Paint Technol.* **41**(533), 383 (1969).
47. K. Walters, ed., *Rheometry: Industrial Applications*, Research Studies Press, a division of John Wiley & Sons, Inc., New York, 1980.
48. R. W. Whorlow, *Rheological Techniques*, Halsted Press, a division of John Wiley & Sons, Inc., New York, 1980.
49. T. C. Patton, *Paint Flow and Pigment Dispersion*, 2nd ed., John Wiley & Sons, New York, 1979.
50. C. J. H. Monk, *J. Oil Colour Chem. Assoc.* **49**, 543 (1966).
51. P. S. Pond and C. J. H. Monk, *J. Oil Colour Chem. Assoc.* **53**, 876 (1970).
52. T. C. Patton, *J. Paint Technol.* **38**(502), 656 (1966); *Cereal Sci. Today* **14**, 178 (1969).
53. P. A. Sandford, J. E. Pittsley, P. R. Watson, K. A. Burton, M. C. Cadinus, and A. Jeanes, *J. Appl. Polym. Sci.* **22**, 701 (1978).
54. D. M. Gans, *J. Paint Technol.* **44**(571), 68 (1972).
55. A. Quach and P. E. Pierce, *J. Paint Technol.* **45**(586), 69 (1973).
56. J. F. Hutton in ref. 47, p. 149.
57. A. Quach and C. M. Hansen, *J. Paint Technol.* **46**(592), 40 (1974).
58. W. Goring, N. Dingerdissen, and C. Hartmann, *Farbe + Lack* **83**, 270 (1977).
59. R. Bassemir, *Am. Inkmaker* **39**(4), 24 (1961); ASTM Method D 4040 has been developed for this instrument by Subcommittee 56 of ASTM Committee D-1 on Paint and Related Products.
60. J. Mewis in ref. 47, p. 324.
61. K. K. Chee, K. Sato, and A. Rudin, *J. Appl. Polym. Sci.* **20**, 1467 (1976).
62. J. Ferguson in G. Astarita and co-eds., *Proceedings of the VIII International Congress of Rheology*, Naples, Italy, Plenum Press, New York, 1980.
63. R. L. Ballman, *Rheol. Acta* **4**, 137 (1974).
64. F. N. Cogswell, *Rheol. Acta* **8**, 187 (1969).
65. J. Meissner, *Rheol. Acta* **8**, 78 (1969).
66. H. Münstedt and H. M. Laun, *Rheol. Acta* **20**, 211 (1981).
67. H. Münstedt, *J. Rheol.* **23**, 421 (1979).
68. C. D. Denson and J. R. Gallo, *Polym. Eng. Sci.* **11**, 174 (1971).
69. D. D. Joye, G. W. Poehlein, and C. D. Denson, *Trans. Soc. Rheol.* **16**, 421 (1972).
70. J. M. Maerker and W. R. Schowalter, *Rheol. Acta* **13**, 627 (1974).
71. S. E. Stephenson and J. Meissner in ref. 62, Vol. 2, p. 431.
72. C. D. Denson and D. L. Crady, *paper presented at the Annual Meeting of the Society of Rheology*, Knoxville, Tenn., 1971.
73. P. E. Pierce in *Characterization of Coatings*, Vol. 2 (Pt. I) of R. R. Myers and J. S. Long, eds., *Treatise on Coatings*, Marcel Dekker, Inc., New York, 1969, p. 99.
74. J. D. Ferry, *Viscoelastic Properties of Polymers*, 2nd ed., John Wiley & Sons, Inc., New York, 1970.
75. A. Zosel, *Farbe + Lack* **82**, 15 (1976).
76. A. Zosel, *Farbe + Lack* **80**, 1130 (1974); **83**, 804 (1977).
77. A. Zosel, *Progr. Org. Coat.* **8**, 47 (1980).
78. A. G. Epprecht, *Farbe + Lack* **80**, 505 (1974); **82**, 685 (1976).
79. R. L. J. Morris, *J. Oil Colour Chem. Assoc.* **53**, 761 (1970).
80. A. T. Riga, *Polym. Eng. Sci.* **14**, 764 (1974).
81. K. T. Gillen, *J. Appl. Polym. Sci.* **22**, 1291 (1978).
82. T. Alfrey, Jr., *Mechanical Behavior of High Polymers*, Interscience Publishers, Inc., New York, 1948.
83. J. J. Aklonis, W. J. MacKnight, and M. Shen, *Introduction to Polymer Viscoelasticity*, Wiley-Interscience, New York, 1972.
84. R. F. Boyer, *Preprints, Org. Coat. Plast. Div. Am. Chem. Soc.* **44**, 492 (1981).
85. J. K. Gillham, *AIChE J.* **20**, 1066 (1974).

86. S. Ikeda, *Progr. Org. Coat.* **1**, 205 (1972/73).
87. N. G. McCrum, B. E. Read, and G. Williams, *Anelastic and Dielectric Effects in Polymer Solids*, John Wiley & Sons, Inc., New York, 1967.
88. L. E. Nielsen, *Mechanical Properties of Polymers and Composites*, Vols. 1 and 2, Marcel Dekker, New York, 1974.
89. A. V. Tobolsky, *Properties and Structures of Polymers*, John Wiley & Sons, Inc., New York, 1960.
90. I. M. Ward, *Mechanical Properties of Solid Polymers*, John Wiley & Sons, Inc., New York, 1971.
91. C. K. Schoff, *J. Coat. Technol.* **49**(633), 62 (1977).
92. J. M. G. Cowie, *Polymers: Chemistry and Physics of Modern Materials*, International Textbook Company, Ltd., Aylesbury, England, 1973.
93. E. Catsiff and A. V. Tobolsky, *J. Colloid Sci.* **10**, 375 (1955).
94. J. L. White, *Rubber Chem. Technol.* **42**, 257 (1969).

General References

J. R. Van Wazer, J. W. Lyons, K. Y. Kim, and R. E. Colwell, *Viscosity and Flow Measurement, A Laboratory Handbook of Rheology*, Interscience Publishers, a division of John Wiley & Sons, Inc., New York, 1963. Although out of date, this is the single most useful reference in this area.

Principles, Vol. 1, *Fluids*, Vol. 2, and *Applications*, Vol. 3, of G. Astarita and co-eds., *Rheology*, Plenum Press, New York, 1980.

R. B. Bird, W. E. Stewart, and E. N. Lightfoot, *Transport Phenomena*, John Wiley & Sons, Inc., New York, 1960.

J. M. Burgers, J. J. Hermans, and G. W. Scott Blair, eds., *Deformation and Flow*, Interscience Publishers, Inc., New York, Vols. I, II, 1952; Vols. III–V, 1953; Vol. VI, 1955; Vol. VII, 1956.

B. D. Coleman, H. Markovitz, and W. Noll, *Viscometric Flow of Non-Newtonian Fluids*, Springer-Verlag Inc., New York, 1966.

J. M. G. Cowie, *Polymers: Chemistry and Physics of Modern Materials*, International Textbook Company, Ltd., Aylesbury, England, 1973.

A. Dinsdale and R. Moore, *Viscometry and Its Measurement*, The Institute of Physics and the Physical Society, Reinhold Publishing Corp., New York, 1963.

F. R. Eirich, ed., *Rheology: Theory and Applications*, Academic Press, Inc., New York, Vol. 1, 1956; Vol. 2, 1958; Vol. 3, 1960; Vol. 4, 1967; Vol. 5, 1970.

J. D. Ferry, *Viscoelastic Properties of Polymers*, 2nd ed., John Wiley & Sons, Inc., New York, 1970.

A. G. Fredrickson, *Principles and Applications of Rheology*, Prentice-Hall, Inc., Englewood Cliffs, N.J., 1964.

V. A. Kargin and G. L. Slonimsky in N. M. Bikales, ed., *Encyclopedia of Polymer Science and Technology*, Vol. 8, Interscience Publishers, a division of John Wiley & Sons, Inc., New York, 1968, pp. 445–516.

S. Middleman, *The Flow of High Polymers*, Interscience Publishers, Inc., a division of John Wiley & Sons, Inc., New York, 1968.

C. C. Mill, ed., *Rheology of Disperse Systems*, Pergamon Press, Inc., New York, 1959.

L. E. Nielsen, *Mechanical Properties of Polymers and Composites*, Vols. 1 and 2, Marcel Dekker, New York, 1974.

P. F. Onyon, "Viscometry," in P. W. Allen, ed., *Techniques of Polymer Characterization*, Butterworths Scientific Publications, London, 1959.

M. Reiner, *Deformation, Strain and Flow*, 2nd ed., Interscience Publishers, Inc., New York, 1960.

M. Rosen, *Polym. Plast. Technol. Eng.* **12**(1), 1 (1979).

W. R. Schowalter, *Mechanics of Non-Newtonian Fluids*, Pergamon Press, Inc., New York, 1977.

G. W. Scott Blair, *Elementary Rheology*, Academic Press, Inc., New York, 1969.

G. V. Vinogradov and A. Ya. Malkin, *Rheology of Polymers*, Springer-Verlag, New York, Inc., New York, 1980.

K. Walters, *Rheometry*, Halsted Press, a division of John Wiley & Sons, Inc., New York, 1975.

R. W. Whorlow, *Rheological Techniques*, John Wiley & Sons, Inc., New York, 1980.

PERCY E. PIERCE
CLIFFORD K. SCHOFF
PPG Industries, Inc.

RHEOLOGY. See Rheological measurements.

RIBOFLAVIN. See Vitamins.

ROOFING MATERIALS

Roofs are a basic element of shelter from inclement weather. Natural or hewn caves, including those of snow or ice, are early evidence of human endeavors for protection from cold, wind, rain, and sun. Nomadic man, before the benefits of agriculture had been discovered and housing schemes were developed, depended upon the availability of natural materials to construct shelters. Portable shelters, eg, tents, probably appeared early in history. Later, more permanent structures were developed from stone and brick, with courses stepped inward to form the roof. Salient features depended strongly upon the availability of natural materials. The Babylonians used mud to form bricks and tiles that could be bonded with mortars or natural bitumen. Ancient buildings in Egypt were characterized by massive walls of stone with closely spaced columns to carry stone lintels to support a flat roof, often made of stone slabs.

As larger and larger structures were desired, building design became an important element in construction. Roofs evolved from a simple covering to roofing systems designed to perform a number of functions to separate indoor and outdoor environments. Selection of roof design depended not only upon factors of economy and comfort, but also on the availability of materials and structural and aesthetic factors (1). Thus, a modern design normally includes a structure to carry loads, insulation to control heat flow, a barrier to control air and vapor flow, and a roofing element to prevent water penetration (1–5).

The bituminous types are by far the most common roof coverings in the United States. The built-up roofing membrane of bituminous materials is estimated to cover over 90% of the commercial and industrial roof surfaces in North America. Asphalt-based materials predominate over coal-tar materials. Various types of roofing materials are employed for residential structures, both for new construction and reroofing, especially in the western area of the United States. Wood and miscellaneous materials account for 20% of residential roofing in the western region of the United States, but only for 5% in the northeast and central areas (see Wood); asphalt roofing accounts for the rest (6–7) (see Asphalt).

Continuous or Jointed Systems

Built-Up Roofings. Built-up roofing (BUR) is a continuous-membrane covering manufactured on site from alternate layers of bitumen, bitumen-saturated felts or asphalt-impregnated glass mats, saturated and coated felts, and surfacings. These membranes are usually applied with hot bitumens or by a cold process utilizing bituminous-solvent or water-emulsion cements (see Adhesives; Cements; Felts).

The deck may be nailable, eg, wood or light concrete, or not, eg, steel or dense concrete. The felts or mats may be organic (cellulose), asbestos, or fiber glass (see Cellulose; Asbestos; Glass). The ply adhesives may be hot-melt or cold-process bitumens (emulsion or cutback). The roof slope ranges from dead level (0–2.1 cm/m or 0–1.2°) to flat (2.1–12.5 cm/m or 1.2–7.1°) to steep (12.5–25 cm/m or 7.1–14°).

Specifications and Application Rates. Specifications as published in catalogs of roofing manufacturers or contractor associations appear complicated because of the many variations possible. Application methods depend on the type and slope of the deck, the types of insulation and roofing membrane, and the fastening method (see Table 1) (8–15). Built-up roofs are generally not applied to slopes steeper than 14° (25 cm/m) (14%) since application of hot asphalt to such roofs is difficult, whereas other forms of roofing, such as shingles, are easily applied.

Common BUR systems are installed in different ways: membrane adhered to deck without insulation; insulation adhered to deck with membrane applied to insulation; base sheet adhered to deck, insulation to base sheet, and top membrane to insulation; and membrane adhered to deck and insulation applied over the membrane, the so-called protected-membrane roof.

The components must be anchored as protection against wind uplift, slippage, and membrane movement.

Application rates for BUR membranes are given in Table 2 (16–19). Membrane strength is related to felt or glass-mat strengths and the number of plys. Continuous bonding is inadvisable if crack development or joint movement can be expected in the membrane substrate, since the elastic limit of the membrane will be exceeded and result in rupture failure. Nailing or spot adhesion provides relief from concentrated strains.

Properties of Ply Felts and Asphalt-Saturated and Coated Felts. The properties of asphalt-saturated and coated felts and impregnated fiber-glass mats are given in Table 3 (9,11,20–21). In addition, ASTM D 3672 (22) describes an asbestos-saturated or glass-asphalt-saturated and coated venting-base sheet. It has an embossed-granule

Table 1. Simplified Hot-Applied Membrane Specifications [a]

Type	ASTM specification
organic felt membrane [b]	D 226, Types I–IV
base sheet	D 2626 or D 3158
asbestos ply felts [b]	D 250, Types I–IV
organic base sheet	D 2626
asbestos base sheet	D 3378
fiber-glass mats [b]	D 2178, Types I–IV
base sheet	D 2178, Type V
surfacings [c]	
organic base sheet	D 371
fiber-glass-mat base sheet	D 3909
bitumens [d]	D 1683
emulsified asphalts	D 1227

[a] From refs. 8–15.

[b] Smooth surface or top pour of bitumen plus gravel or slag.

[c] Cap sheet covered with mineral granules.

[d] Usually with gravel or slag.

Table 2.   Application Rates for BUR Membranes

Material	ASTM specification	L/m^2 [a]	kg/m^2 [b]
primer	D 41[c]	0.4	
bitumen ply	D 312		1.0
adhesive	D 450		
bitumen surfacing	D 312, D 450		2.4–2.9
gravel covering	D 1863		19.5
slag covering	D 1863		14.6
asphalt-liquid surfacings[d]		0.6–2.0	

[a] To convert L/m^2 to $gal/100\ ft^2$, multiply by 2.45.
[b] To convert kg/m^2 to $lb/100\ ft^2$, multiply by 20.5.
[c] Ref. 16.
[d] See refs. 16–19 for ASTM specifications.

design on the bottom surface to permit lateral relief of pressure that might be caused by moisture movement.

Table 4 describes properties of ASTM ply felts used for BUR. For additional details, the appropriate ASTM specifications should be consulted. Periodic revisions are made by ASTM committees to keep up with current practices.

Cold-Process Coatings. Cold-process BUR applications do not require heating of the bitumen on the job. Simple application and economical maintenance are primary considerations. Bitumens are liquefied either by dissolving in a solvent or by emulsifying in water. With both types, the base cutbacks or emulsions can be modified by addition of other components, commonly mineral fillers, to form specialized coatings. At ordinary temperatures, the solvent or water evaporates and adhesive bonds or weather-resistant surface coatings are formed.

The compositions of solvent-type asphalt coatings are illustrated in Figure 1. These coatings range from fluid cutbacks used for priming deck surfaces to viscous mineral-filled compositions used as flashing cements. The ASTM specifications in refs. 16–19 should be consulted for details (see Coatings; Coating processes). Lap cements used with asphalt-roll roofings are specified by ASTM D 3019 (23). This specification covers cements consisting of asphalt and petroleum solvent with or without asbestos fiber.

Emulsified asphalt used as a protective coating is specified by ASTM D 1227 (24). These emulsions are applied above freezing by brush, mop, or spray and bond to either damp or dry surfaces. Such application is not recommended for inclines <4° (4%) to avoid the accumulation of water. However, curing by water evaporation can be slow, and these emulsions may remain water susceptible.

Protected-Membrane Roofs. Primitive roofs covered with earth and sod over sloping wood decks shingled with bark were of the protected-membrane type and gave excellent service. Grass and earth provided insulation and protected the shingled deck from inclement weather (25).

In modern construction, insulation is needed that is unaffected by water or that can be kept dry in some manner and that stays in place over the roof membrane. In the United States, extruded polystyrene-foam insulation boards are commonly employed (see Insulation, thermal). They are placed over the roof membrane, which is attached to the decks. The insulation may or may not be bonded to the membrane.

Table 3. Properties of Asphalt-Saturated and Coated Felts for BUR

Property	Organic base sheet[a] ASTM D 2626		Asbestos base sheet[a] ASTM D 3378		Asphalt-impregnated fiber-glass mat[b], ASTM D 2178	Organic felt, asphalt[c], ASTM D 3158
	Type I	Type II	Type I	Type II	Type V	
mass, min, kg/m²[d]	1.8	1.9	1.8	1.9	0.73	1.41
mass dry felt, min, g/m²[d]	254	341	415	537	49	254
mass saturant, min, g/m²[d]	356	478	166	215	454	356
mass coating and surfacing, min, g/m²[d]	878	878	878	878	461	585
breaking strength, kN/m[e]						
with grain	6.1	7.9	5.3	7.0	5.3	6.1
across grain	3.5	3.5	2.6	3.5	5.3	3.5
permeance, max, pg/(Pa·s·m²)[f]	1.7	1.7	1.7	1.7		
pliability radius, 5 pass (25°C)	12.7	12.7	12.7	12.7	1.27	12.7

[a] First ply or as vapor barrier under insulation; coated on both sides with asphalt and surfaced on top side with fine mineral granules; unrolls at or above 10°C (9,20).

[b] Combination base sheet; faced on one side with Kraft paper; unrolls at or above 10°C (11).

[c] Saturated and coated; perforated or not; perforations provide vents for gases liberated during applications; surfaced on top side with fine mineral granules; unrolls at or above 10°C (21).

[d] To convert g/m² to lb/100 ft², multiply by 0.0205.

[e] To convert kN/m to lbf/in., divide by 0.175.

[f] To convert pg/(Pa·s·m²) to grain/(h·in. Hg·ft²), divide by 5.72.

323

Table 4. Properties of Roofing and Waterproofing Ply Felts[a] for BUR

Property	Organic felt Asphalt-saturated[b,c], ASTM D 226, type			Coal-tar saturated[c], ASTM D 227	Asbestos felt, asphalt-saturated[b,c], ASTM D 250, type		Fiber-glass mat, asphalt impregnated[c,d], ASTM D 2178, type		
	I	II	III		I	II	I	III	IV
nominal weight, g/m²[e]	732	1460	976	732	732	1460	356	474	342
felt or fiber-glass-mat mass, min, g/m²[e]	254	488	332	254	415	854	49	73	83
saturant mass, min, g/m²[e]	356	732	469	356	166	342	225	308	146
breaking strength, min, kN/m[f]									
with grain	5.25	7.00	6.13	5.25	3.50	7.00	2.63	3.85	7.71
across grain	2.63	5.25	2.98	2.63	1.75	3.50	2.63	3.85	7.71
pliability radius, pass at 25°C, mm	12.7	19.1	12.7	12.7	12.7	19.1	12.7	12.7	12.7
ash, %	10 max	10 max	10 max		70.0 min	70.0 min	75–88	75–88	75–88
loss on heating at 105°C, 5 h max, %	4	4	4	4.0	5	5	1.0	1.0	1.0

[a] All are nonsticking.
[b] Perforated or nonperforated.
[c] Unrolls above 10°C.
[d] Faced on one side with Kraft paper; combination base sheet.
[e] To convert g/m² to lb/100 ft², multiply by 0.0205.
[f] To convert kN/m to lbf/in., divide by 0.175.

324

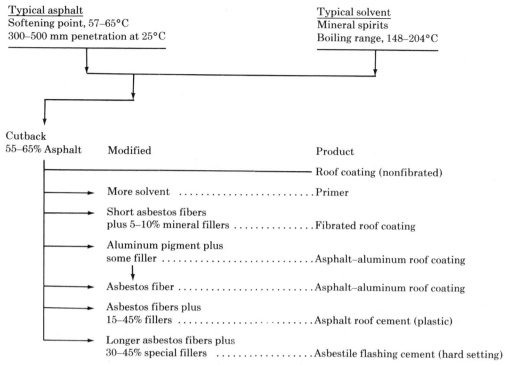

Figure 1. Production of asphaltic coatings, primer, and cements.

Gravel or slag at the rate of 48.8 kg/m² (1000 lb/100 ft²) holds the insulation in place and offers protection from the sun. The insulation joints are open and drainage must be provided. Various other materials, eg, patio blocks, mortar, and concrete slabs, are also used as surfacings and ballast. The extra weight imposes more exacting requirements on construction. Although of relatively recent design, the protected-membrane roof is finding a place in roof systems (26–28).

Elastomeric Products Applied as Single- or Limited-Ply Roofing Membranes. New materials have proliferated during the last decade; their claimed advantages include light weight, high chemical and weather resistance, high elasticity, and ease of application and repair. These new roofing systems depend primarily on the application of single-ply-sheet membranes giving continuous coverage (see Film and sheeting materials). Materials applied as liquids prepared from a variety of base materials are often used in combination with glass-mat or plastic reinforcements (29–31) (see Plastic building products).

Although ASTM test methods are given in ref. 32, standards are currently not available to assess the merits of these materials. Elastic sheets are likely to have high extensibility. However, if the sheet is solidly attached and the substrate develops a crack, the latter is transmitted through the ply. The lapping of watertight and secure joints has presented a problem; excellent technology and careful application are necessary. Since usually only one ply is applied, a roof disaster can occur if the material or application is faulty. Until adequate field experience has been gained, particular material, back-up, and support from the supplier are essential.

Although estimates vary, the single-ply and liquid systems have probably captured

5–10% of the BUR market in terms of dollar volume; an increase of 50% is projected over the next 10 years (33).

In 1980, 36 manufacturers offered 42 sheet-membrane materials, 33% of which were poly(vinyl chloride), followed by modified asphalts and ethylene propylene diene monomer at 20% each; neoprene sheeting, chlorinated polyethylene, and polyisobutylene accounted for the rest. New materials and combinations of materials are continuously entering the market (34).

Single-ply loose-laid membranes are usually ballasted with smooth stones [22.0–48.8 kg/m^2 (450–1000 lb/100 ft^2)].

Various liquid-applied single or two-component systems provide roofing membranes. Some are used with polymeric fabrics or glass reinforcements to obtain single- or multilayered membranes. Compositions are based on asphalt (solvent blends or emulsions), acrylic emulsions, urethane two-component systems, or chlorinated rubber (neoprene) with a top coat of chlorosulfonated polyethylene (Hypalon). Specification ASTM D 3468 describes solutions of neoprene and Hypalon for roofing and waterproofing applications (35) (see Elastomers, synthetic).

Other Types.   Asbestos-cement corrugated sheets show excellent strength and durability properties in industrial and commercial roofing applications. Described by ASTM C 221, the sheets are lapped and usually applied over steel purlins (36). The usual span is 1.37 m; special products are suitable for spans up to ca 3 m.

Metal roofs are used on industrial and farm buildings and are usually made of corrugated aluminum or galvanized steel; a variety of other metals and coated metals is also applied, eg, ASTM A 755 specifies a zinc-coated steel sheet (37) and B 101 a lead-coated copper sheet (38). In the 1800s, standing-seam metal roofs of flat sheets, crimped at each edge, were popular in New England and the Midwest. Steel roofs with 92-cm wide panels, designed with trapezoidal ribs and a factory-applied finish, are used in some residential areas (39). Metal roofing systems also are promoted for plant reroofing (40).

Concrete-slab roofs that do not require a membrane save both on cost and maintenance (41–42). Cracks are prevented by posttensioning the roof slabs during installation.

Fabric roofs are either air-supported or tensioned. In the 1940s, air-supported structures were first used for radar-system enclosures. Later, a variety of military and recreational uses were developed. Both vinyl- and Teflon-coated fiber-glass fabrics have been used (43–44). A resin-covered glass-fiber fabric cover for a domed stadium is much less expensive than a steel-and-concrete roof (45).

Plastic sheets have been installed in roof-support systems. Because of their excellent strength properties, polycarbonate sheets have been reported to be highly successful in severe environments (46).

Shingle and Unit Construction

Shingling is highly effective for shedding water. Natural materials on hand have been used since ancient times, and even today straw-thatch constructions are employed in some countries.

Asphalt Shingles.   For many years, organic-felt-base asphalt shingles have been the standard for residential roofing; specification ASTM D 225 was originally issued in 1925 (47). In general, the individual three-tab strip shingle is 91.4 by 30.5-cm wide.

Maximum exposure of 12.7 cm to the weather in application provides a head lap of not less than 18 cm to give double coverage. Other shapes and sizes are available, such as hexagonal tab strips. Seal-down shingles are treated with a factory-applied bonding adhesive after installation to prevent wind damage. Type II of ASTM D 225 has a thick butt, which is the portion exposed to the weather. The typical organic-felt shingle has a felt base saturated with an asphalt of a 60.0–68.3°C softening point and a penetration of 3–5 mm. The asphalt coating usually contains 60 wt % of a finely divided mineral stabilizer, eg, limestone or silica, which is applied to both sides of the saturated felt, with the heavier coating on the weather side. The coating asphalt usually has a softening point of 88–113°C and a penetration of 1.5 mm, min at 25°C. These coating properties also are specified for fiber-glass shingles by ASTM D 3462 (48). The mass of dry felt in organic shingles is specified at a minimum of 53.6 kg/m^2 (11 lb/100 ft^2); a minimum of saturant, coating, and mineral matter is also required. The marketing unit is the sales square, ie, the area of shingles to cover 9.3 m^2 (100 ft^2) of roof. Sales squares are usually packaged in three bundles with a total weight of 107 kg (see Felts).

ASTM D 3462 specifies asphalt shingles with a glass-mat base (48). The glass mat is composed of fine fibers deposited in a nonwoven pattern that may be reinforced with glass yarns. A water-insoluble binding agent holds the fibers together. In shingle-manufacturing, one or more thicknesses of mat are impregnated with an asphalt that is usually compounded with a mineral stabilizer. Asbestos or glass fibers are sometimes added or used in place of the mineral stabilizer. The weather side of the shingle is surfaced with mineral granules and the reverse side with a pulverized mineral to prevent sticking. A self-sealing adhesive is specified, and requirements for wind resistance and tear strength are also given. The mass of the glass mat is specified at 65.9 g/m^2 (1.35 lb/100 ft^2) min.

Other Shingles and Units. The preference for wood, slate, asbestos-cement, or tile roofing shingles and units varies in different parts of the United States. Nationwide, the average is ca 87% asphalt, 7% wood, and 6% miscellaneous (tile, concrete, and metal). The average in the western states is 65% asphalt, 20% wood, and 15% miscellaneous residential applications (6).

Specification ASTM C 222 describes and gives minimum requirements for asbestos-cement shingles (49); three grades of slate roofing units are specified by ASTM C 406, with expected service of 20–40, 40–75, and 75–100 yr (50). Tiles of fired clay with glazed surfaces are attractive and durable. Wood shingles and hand-split shakes have considerable esthetic appeal in certain areas (51). However, fire resistance has been a problem (52), and these materials have been banned in some localities unless a UL Class-C fire requirement is met (53).

Asphalt Roofing Components

The production of asphalt coatings, primer, and cements is illustrated in Figure 1.

Bitumens. Although native bitumens have been used since ancient times for waterproofing, in North America the use of saturated felts dates back to the 1850s. Coal tar and, to some extent, pine tar are the saturants. Coal tar, a by-product from gasworks, is also used as an adhesive in BURs (see Tar and Pitch). The petroleum industry, rapidly developed in the late 1800s, supplied residue asphalts for pavements.

Table 5. Properties of Roofing Asphalt, ASTM D 312[a]

Property	Type I[b] Min	Type I[b] Max	Type II Min	Type II Max	Type III Min	Type III Max	Type IV Min	Type IV Max
softening point, °C	57	66	70	80	85	96	99	107
penetration[c], mm								
at 0°C, 200 g, 60 s	0.3		0.6		0.6		0.6	
at 25°C, 100 g, 5 s	1.8	6.0	1.8	4.0	1.5	3.5	1.2	2.5
at 46°C, 50 g, 5 s	9.0	18.0		10.0		9.0		7.5
flash point, COC[d], °C	225		225		225		225	
ductility at 25°C, cm	10.0		3		2.5		1.5	
solubility in trichloro-ethylene, %	99		99		99		99	
slope								
%	up to 4.2		4.2–12.5		8.3–25		16.7–50	
cm per horizontal line	up to 3.8		3.8–11.4		7.6–22.8		15.2–45.6	

[a] Ref. 56.
[b] Self-healing, generally with slag or gravel surface.
[c] Of a certain weight, in g, for a period of time, in s.
[d] Cleveland open cup.

After the discovery of the air-blowing process, these residue asphalts became another source of bitumen for roofings. Roofing asphalts were developed that are highly resistant to flow after application and that retain sufficient fluidity in application (54–55).

Coal tars when used as ply adhesives are generally limited to slopes of ≤4° (4%) without nailing and ≤15° (15%) with back-nailing. The permissible slopes for asphalts without nailing range up to 50° (50%), depending on the asphalt grade specified. Properties of hot-applied bitumens for use in BUR as specified by ASTM are given in Tables 5 and 6 (56–58).

Table 6. Coal-Tar Bitumen Used in Roofing, Dampproofing, and Waterproofing, ASTM D 450[a]

Property	Type I[b] Min	Type I[b] Max	Type II[c] Min	Type II[c] Max	Type III[d] Min	Type III[d] Max
softening point, °C	52	60	41	52	56	64
water, %		0		0		0
ash, %		0.5		0.5		0.5
flash point, COC, °C	120		120		120	
specific gravity, 25°/25°C	1.22	1.34	1.22	1.34	1.22	1.34
solubility in CS$_2$, %	72	85	72	85	72	85
distillate, up to 300°C, %		10		10		5[e]
specific gravity	1.03		1.03			
softening point of residue, °C		80		80		

[a] Ref. 57.
[b] For BUR with ASTM D 227 felts (58).
[c] For membrane waterproofing systems.
[d] For BUR, but less volatile than Type I.
[e] From 315–360°C.

Felts. Roofing felts consist of fiber mats impregnated or partially saturated and sometimes coated with bitumen. Usually, the felt is manufactured in sheets from wood-fiber pulp and some scrap-paper pulp. Earlier felts contained rag fibers thought to be necessary to impart durability. Extensive weathering tests and field experience demonstrated that Asplund wood-fiber felts were satisfactory for BUR and prepared roofings (59). Rag fibers are added only rarely. Organic felts are formed from a water slurry of pulp; the fibers are picked up on a rotating screened drum or by a traveling screen. The sheet is dried by vacuum suction and steam-heated drums and then calendered and wound into large rolls for subsequent trimming and slitting to the required width. The felt is saturated by dipping it into hot bitumen and passing it under and over a series of rolls. The saturated felts are cooled, marked with guide lines for application, and wound into rolls of convenient size. Small perforations let air and moisture escape during application. The saturant improves the bond to the bitumen used in the BUR process and to asphalts used in coated felts. Furthermore, the saturant reduces the rate of water absorption. Vapor permeability and water absorption are further reduced by asphalt coatings stabilized with mineral fillers (see Fillers). A heavier coating is placed on the weather side. A parting agent, eg, sand, mica, or talc, is spread on the coating to prevent sticking (see Abherents).

Asbestos felts are produced essentially in the same manner, except that the mix contains an 85 wt % min of asbestos fibers; the rest is glass or organic fibers. This increases the bulk of the asbestos and permits better saturation with bitumen. Asbestos felts are highly resistant to fungal attack and absorb water to a lesser extent than wood-fiber felts. On direct exposure, the asbestos felts show superior weather properties and excellent dimensional stability because of resistance to moisture and to other weather influences.

Glass felts are manufactured from fibrous glass prepared from a molten mixture of sand and other ingredients; the latter are drawn or blown into filaments. The filaments may be short or may be used in combination with continuous-strand filaments. They may be deposited dry or from a dilute water slurry on a screen conveyor. In either process, a binder is applied. Heat is applied to dry the mat and set the binder. Rolls of glass felt, as received at the roofing plant, are run directly to the coater to be impregnated with asphalt, spread with a mineral parting agent, cooled, marked with guide lines, and rolled to desired lengths. High dimensional stability and low moisture absorption are obtained with fiber-glass roofing felts.

For the preparation of mineral-surfaced roofings and shingles, mineral granules are applied to the heavier-coated side of the felt (see Fig. 2). Variations in manufacture also produce cap sheets, venting base sheets, and flashings.

Roofing Performance

Performance (effectiveness) of a roof covering is determined not only by environmental conditions, but also by the covering of the system. The design should be such that no component plays a critical part in the overall performance. Thus, each component should be lasting or durable in service to contribute to the overall roof performance (60).

Procedures for accelerated system testing are still in the development stage. They are largely based on durability tests of components, determination of system design, and the proper combination of the roof elements.

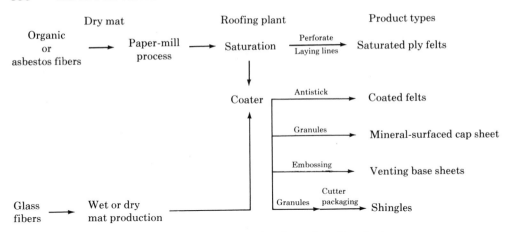

Figure 2. Manufacture of roof coverings, flow sheet.

Roof coverings must be protected from external abuse, eg, foot or other traffic. Penetration of BUR membranes by air-conditioning installations or other roof devices usually voids roof guarantees by the manufacturer, unless specific permission has been received.

Weathering. Bitumen films harden progressively owing to oxidation reactions and, to some extent, loss of plasticizing oils, especially with coal tars. Temperature changes may result in expansion and contraction of the roof covering. Other weather factors include moisture, ice, and hail. Wind affects shingles as well as BUR membrane performance.

In an accelerated weather study of fifteen coating-grade asphalts, a sixfold variation in durability was reported (61). Performance is correlated to properties and composition. Low asphaltene and high resin contents are desirable (62). As weathering progresses, the brittle-point temperature, taken by the Fraass procedure, increases (63). Blending of fractions or base stocks from different crudes improves performance more than antioxidants (64).

In a comprehensive review of theoretical and practical aspects of asphalt durability, six weathering factors were considered (65). A test apparatus for the accelerated weathering of organic materials and procedures was developed at the National Bureau of Standards; a carbon arc was used as the energy source (66). However, outdoor weather exposures are standard procedures for prepared roofings. The opaque granules and mineral stabilizers in the coating asphalts offer excellent protection against the destructive effects of radiant energy and, even in the accelerated weatherometer, an excessive number of daily cycles is required for failure.

Finely divided mineral stabilizers affect the weatherability of coating-grade roofing asphalts. Additions up to 60 wt % of mineral additives increase durability substantially (61). Platy minerals used as ground slates are the most effective. The nature of the asphalt, however, is the most important consideration (67–68). The results of weatherometer and outdoor exposure on combinations of four stabilizers and two asphalts are given in ref. 69.

A number of laboratory procedures that are more rapid than use of the accelerated weatherometer have been described for the determination of expected weatherability of coating asphalts (70–71). Recent work sponsored by the Asphalt Roofing Manufacturers Association describes a stepwise procedure to determine changes in the crude-asphalt source (72) (see Asphalt).

Thermal Effects. Temperature influences the oxidation rate, whereas temperature changes impart a mechanical stress to roof coverings. Daily temperature changes, sometimes quite abrupt, can result in a fatigue mechanism causing tensile failure. Repeated cycles of straining, especially at a high level, diminish the strength of roofing felts to the rupture point. Strain reduction by partial bonding of the membrane to the deck offers the biggest improvement in performance (73).

Data for thermal movement of various bitumens and felts and for composite membranes are given in refs. 74–75, which describe the development of a thermal-shock factor based on strength factors and the linear thermal-expansion coefficient.

Tensile and flexural fatigue tests on roof membranes were taken at 21 and −18°C, and performance criteria were recommended (76).

A study of four types of fluid-applied roofing membranes under cyclic conditions showed that they could not withstand movements of ≤1.0 mm over joints (77).

The limitations of present test methods for new roofing materials, such as prefabricated polymeric and elastomeric sheets and liquid-applied membranes, are described in ref. 78. For evaluation, both laboratory and field work are needed.

Water Effects. Water in its different forms (liquid, vapor, hail, and ice) profoundly influences the performance of roof coverings. Moisture migrations in roof insulation, eg, vapor that can accumulate and later liquefy, or water leakage through the roof covering, reduces insulating efficiency and leads to physical deterioration of the roofing material. Moisture can be detected by direct cut tests, electrical-resistance moisture meters, ir-scanning techniques, or nuclear moisture meters.

Flashing flaws in some roofs can be directly responsible for water leakage. The National Roof Contractors Association has developed criteria for curb and edge detail. The recommended flashings include those for stacks, pipes, and expansion joints (79). Factory Mutual Systems (FM) prescribe perimeter-flashing details dealing primarily with wind-uplift forces that can be encountered on built-up roofs (80). Roof-edge damage in windstorms can be extensive and would permit rain leakage and interior damage.

In roof coverings, organic felts are highly susceptible to expansion–contraction movements with moisture change. Shingles, with a low degree of asphalt saturation, especially if the felt has not been uniformly saturated, show warping, described as fish mouthing or clawing. Erratic movements in built-up roofing can be produced by moisture–thermal effects and may cause ridging and failure of the membrane by cracking (81).

The absorbed equilibrium moisture content of built-up roofing membranes varies with changing environmental conditions. In faulty designs, moisture accumulates during late fall and winter in amounts exceeding those that the system can accommodate in the summer. The weakening effects of moisture are substantially less on asbestos or glass with low equilibrium moisture contents (82).

Organic-felt laminates show more movement owing to humidity changes than to temperature changes. Asbestos felts respond moderately to both temperature and humidity and, as with the organic felts, the changes were larger in the longitudinal than in the transverse direction. Glass felts showed small and moderate dimensional changes in response to humidity and temperature changes, respectively; these changes are essentially nondirectional (83).

Hail damage occurs at a minimum impact energy (84). Icing is a function of the roof slope, roof-covering composition, and amount of sun exposure (85–86).

Fire and Wind Hazards. Weather resistance of roof coverings is not necessarily correlated to fire and wind resistance. Underwriter's Laboratories (UL) and the Factory Mutual Systems (FM) are nonprofit organizations that test and rate fire- and wind-hazard resistance.

Organic felt or fiber-glass-mat base shingles are commonly manufactured to meet minimum UL requirements which, in addition to minimum mass, require wind- and fire-resistance properties (87–89).

Fire Ratings. Above-deck fire hazards are rated by UL based on propagation of the flame along the roof covering and on the penetration of the fire into the deck or structure (90). Surface-burning characteristics are measured by the spread-of-flame test. The test deck, either of shingle or BUR construction, is set at a specific incline or slope and exposed to a standard gas flame or burning brand. Resistance Class A is considered effective against severe test exposure and affords a high degree of fire protection to roof decks; Class B affords a moderate degree of protection and Class C a light degree of protection. A list of systems and their classifications is published by UL. Metal deck assemblies are tested by UL with the tunnel test. The assembly, exposed to a gas flame, must protect the roofing membrane on top.

The FM uses a calorimeter fire-test chamber to evaluate the hazard of an under-deck fire. The deck is exposed to a gas flame and the rate of heat release is measured to relate to the rate of flame propagation. Another FM test assesses the damage to roof insulations exposed to radiant heat (91–92).

Wind Testing. Underwriters Laboratory employs two wind-resistance tests, one for shingles and another for BUR assemblies (93); ASTM D 3161 is a standard procedure for measuring the wind-resistance of asphalt shingles (89). In the latter and the UL shingle test, the conditioned test deck is placed at a specified slope and exposed to a wind velocity of 97 km/h for 2 h or until failure by tab or shingle lifting is shown.

Factory Mutual explains how to protect buildings from wind damage (94). Pressure coefficients are given for various structures and roofs. A laboratory uplift-pressure test checks roof assemblies and a field uplift test checks Class-I insulated steel-roof deck construction (93–94). An uplift pressure of 2.9 kPa (0.42 psi) must be withstood in either test under FM conditions to meet the Class-I requirements. The FM guide is revised periodically.

Economic Aspects

In the continental United States, 31 companies have 118 roofing manufacturing plants. As of 1977, 110 plants were in operation. The number of operating plants has remained relatively constant from 1954 through 1977, between 100 and 116 (95). They vary in size, with production in 53 plants ranging from 7,260–408,200 metric tons per year (96). In 1977, 40% of the plants were located in the south and produced 34% of the total production, exclusive of ply felts.

The prices of asphalt roofing products tripled between 1969 and 1978. These increases are attributed primarily to rising material costs, especially asphalt. The cost of felt components has also increased. Granules, parting agents, and stabilizers were estimated to constitute ca 16% of the total cost of materials in 1979. The cost of these materials does not have an appreciable effect on roofing products (97).

Table 7.   Roll Roofing and Saturated Ply Felt, U.S. Shipments[a]

| Year | Smooth roll, 10^6 sales squares | | Mineral or grit, 10^6 sales squares | | Saturated felt[b], 10^3 t |
	Organic and inorganic	Inorganic	Organic and inorganic	Inorganic	
1970	21.2	na	13.2	na	834.3
1975	17.8	na	13.5	na	671.5
1977	17.4	na	12.3	na	826.7
1978	23.7	3.2	13.1	2.4	807.4
1979	22.1	6.7	15.0	3.3	857.8
1980	24.9	9.6	15.6	3.1	605.7

[a] From ref. 100.
[b] Asphalt or tarred.

A new trend is the replacement of organic felt by fiber-glass mats in shingles and roll roofings. By the mid-1980s, fiber-glass shingles are expected to account for half of the market (98).

The asphalt roofing industry is not seasonal like the housing industry. Supply and demand are well balanced, primarily since there is a large reroofing segment that comprises 50–70% of the market, depending on new construction activity. Although a roof is an indispensible part of a building, it represents less than 1% of its cost. Wood shingles are considerably more expensive than asphalt shingles. Asphalt roofing production has shown a 2% annual growth rate from 1969 to 1977 and is expected to continue at this rate into the mid-1980s (99).

Sales trends for strip shingles, roll roofings, and saturated felts are shown in Tables 7 and 8 (100).

Health and Safety Factors

OSHA enforces standards established by NIOSH to protect workers against safety and health hazards (101). In addition to Federal requirements, state and local agencies may have more stringent health and safety regulations. Health-hazard evaluation reports for various plants are issued after inspection by NIOSH (102–103). According to the EPA, the impact of OSHA regulations on the roofing industry seems to be minimal (104).

Asphalt derived from petroleum differs from coal tar, which is a condensation

Table 8.   Strip Shingles, U.S. Shipments, 10^6 Sales Squares[a]

| Year | Organic base | | Fiber-glass base | Total strip shingles |
	Self-sealing	Standard		
1970	26.2	20.0	na	46.2
1975	55.6	1.5	na	57.1
1977	60.2	2.4	na	62.6
1978	67.0	6.3	6.2	79.5
1979	64.3	9.3	9.3	82.9
1980	47.5	2.2	11.3	61.0

[a] From ref. 100.

by-product obtained from the carbonization of coal. The predominant emissions from asphalts are paraffinic or cycloparaffinic in character, whereas coal-tar emissions are predominantly aromatic in character. Low temperatures for handling are suggested as a precaution (105). The Asphalt Institute has issued reports describing emissions from asphalt-paving plants (106) and asphalt built-up roofing operations (107). The emissions from paving asphalts correspond to those expected from saturants used in the asphalt roofing industry. No significant air pollution or health problems were reported in paving operations. In order to reduce worker exposure to emissions from roofing kettles, it was suggested that the kettles operate at the lowest possible temperature and that the lid remain closed.

BIBLIOGRAPHY

"Roofing Materials" in *ECT* 2nd ed., Vol. 17, pp. 459–475, by G. W. Berry, Johns-Manville Corporation.

1. T. Eastwood, *Proc. Geol. Assoc. London* **62,** 6 (1951).
2. "Roof" in *Encyclopedia Britannica*, Vol. 19, Chicago, Ill., 1962, pp. 527–533.
3. M. C. Baker, *Roofs*, Multiscience Publications Ltd., Montreal, Can., 1980, pp. 1–6.
4. S. D. Probert and T. J. Thirst, *Appl. Energy* **6,** 79 (1980).
5. H. W. Busching, *Int. J. Hous. Sci. Appl.* **3,** 21 (1979).
6. Unpublished reports, Asphalt Roofing Manufacturers Association, Washington, D.C.
7. Ref. 3, p. 34.
8. *Asphalt-Saturated Organic Felt Used in Roofing and Waterproofing*, ASTM D 226.
9. *Asphalt-Saturated and Coated Asbestos Felt Base Sheet Used in Roofing*, ASTM D 3378.
10. *Asphalt-Saturated Asbestos Felt Used In Roofing and Waterproofing*, ASTM D 250.
11. *Asphalt-Impregnated Glass (Felt) Mat Used in Roofing and Waterproofing*, ASTM D 2178.
12. *Wide-Selvage Asphalt Roll Roofing (Organic Felts) Surfaced with Mineral Granules*, ASTM D 371.
13. *Asphalt Roll Roofing (Glass Mat) Surfaced with Mineral Granules*, ASTM D 3909.
14. *Emulsified Asphalt Used as a Protective Coating for Roofing*, ASTM D 1227.
15. *Mineral Aggregate Used in Built-Up Roofs*, ASTM D 1863.
16. *Asphalt Primer Used in Roofing, Dampproofing, and Waterproofing*, ASTM D 41.
17. *Asphalt Roof Coatings*, ASTM D 2823.
18. *Aluminum-Pigmented Asphalt Roof Coatings*, ASTM D 2824.
19. *Asphalt Roof Cement*, ASTM D 2822.
20. *Asphalt-Saturated and Coated Organic Felt Base Sheet Used in Roofing*, ASTM D 2626.
21. *Asphalt-Saturated and Coated Organic Felt Used in Roofing*, ASTM D 3158.
22. *Venting Asphalt-Saturated and Coated Inorganic Felt Base Sheet Used in Roofing*, ASTM D 3672.
23. *Lap Cement Used with Asphalt Roll Roofing*, ASTM D 3019.
24. *Emulsified Asphalt Used as a Protective Coating for Roofing*, ASTM D 1227.
25. Ref. 3, p. 194.
26. C. P. Hedlin in D. E. Brotherson, ed., *Moisture Content in Protected Membrane Roof Insulations*, ASTM Special Technical Publication 603, ASTM, Philadelphia, Pa., 1976.
27. K. A. Epstein and L. E. Putnam, *J. Thermal Insul.* **1,** 149 (Oct. 1977).
28. Ref. 3, p. 197.
29. W. J. Rossiter, Jr., and R. G. Mathey, *Nat. Bur. Stand. U.S. Tech. Note 972*, (July, 1978).
30. D. A. Davis and M. P. Krenick in ref. 26, p. 30.
31. Ref. 3, p. 55.
32. *Elastomeric and Plastomeric Roofing and Waterproofing Materials*, ASTM D 3105.
33. "Single Ply Penetrates Roofing Market" in *Roofing, Siding, Insulation*, **55,** 57, 92, 108 (Aug. 1978).
34. "Elasto/Plastic Products," unpublished memorandum, National Roofing Contractors Association, Dec. 1980.

35. *Liquid-Applied Neoprene and Chlorosulfonated Polyethylene Used in Roofing and Waterproofing*, ASTM D 3468.
36. *Corrugated Asbestos-Cement Sheets*, ASTM C 221.
37. *General Requirements for Steel Sheet Zinc-Coated (Galvanized) by the Hot-Dip Process and Coil Coated for Roofing and Siding*, ASTM A 755.
38. *Lead-Coated Copper Sheets*, ASTM B 101.
39. "Steel Roofs Gain Foothold," *Denver Post*, (Dec. 7, 1980).
40. R. C. Baldwin, *Plant Eng.* **33**, 118 (1979).
41. B. Starkovich, *Concr. Int.* **2**, 65 (May 1980).
42. I. Martin, *Concr. Int.* **1**, 59 (Jan. 1979).
43. A. Morrison, *Civ. Eng. ASCE* **50**, 60 (Aug. 1980).
44. B. Mead, *Architectural Fabric Roofs*, ACS Preprints, 179th Meeting, Houston, Tex., Mar. 1980.
45. *Chem. Week*, 60 (Sept. 2, 1981).
46. *Plast. World* **39**, 30 (Feb. 1981).
47. *Asphalt Shingles (Organic Felts) Surfaced with Mineral Granules*, ASTM D 225.
48. *Asphalt Shingles Made from Glass Mat and Surfaced with Mineral Granules*, ASTM D 3462.
49. *Asbestos-Cement Roofing Shingles*, ASTM C 222.
50. *Roofing Slate*, ASTM C 406.
51. Sweet's Division, McGraw-Hill Information Services, New York (issued annually).
52. C. A. Holmes, *Evaluation of Fire-Retardant Treatment for Wood Shingles*, Forest Service Research Paper, FSRP-FPL-158 (May 1971).
53. *Houston Chronicle and Houston Post* (July 31, 1979); *Los Angeles Times* (Nov. 25, 1980).
54. L. W. Corbett, "Manufacture of Petroleum Asphalt," in A. J. Hoiberg, ed., *Bituminous Materials*, Vol. 2, 1965, reprint ed., Krieger Publishing Co., Huntington, N.Y., 1979, p. 81.
55. E. O. Rhodes in ref. 54, Vol. 3, p. 1.
56. *Asphalt Used in Roofings*, ASTM D 312.
57. *Coal-Tar Bitumen Used in Roofing, Dampproofing, and Waterproofing*, ASTM D 450.
58. *Coal-Tar-Saturated Organic Felt Used in Roofing and Waterproofing*, ASTM D 227.
59. S. H. Greenfield, *Nat. Bur. Stand. U.S. Tech. Note 477*, (Mar. 1969).
60. A. J. Hoiberg, *Nat. Bur. Stand. U.S. Spec. Publ. 361*, **1**, Vol. 1, 1972, pp. 777–787; *NBS Special Publication 361*, **1**; *Proc. Joint RELEM-ASTM-CIB Symposium*, 777 (May 1972).
61. S. H. Greenfeld, *Nat. Bur. Stand. J. Res.* **64C**, 299 (Oct.–Dec. 1960).
62. *Ibid.*, p. 297.
63. P. M. Jones, *I.E.C. Prod. Res. Dev.* **4**, 57 (Mar. 1965).
64. B. D. Beitchman, *Nat. Bur. Stand. J. Res.* **64C**, 13 (Jan.–Mar. 1960).
65. J. R. Wright in ref. 54, p. 249.
66. *Accelerated Weathering Test of Bituminous Materials*, ASTM D 529.
67. S. H. Greenfield, "Effect of Mineral Additives on the Durability of Coating-Grade Roofing Asphalts, *Nat. Bur. Stand. Bldg. Mat. Struct. Rep. 147* (Sept. 1956).
68. S. H. Greenfeld, *Natural Weathering of Mineral Stabilized Asphalt Coatings*, NBS Bldg. Science Ser. No. 24, Oct. 1969, 15 pp.
69. D. A. Davis and E. J. Bastian, Jr. in Sereda and Litvan, eds., *Durability of Building Materials and Components*, ASTM Special Technical Publication 691, ASTM, Philadelphia, Pa., 1980, p. 767.
70. C. D. Smith, C. C. Schuetz, and R. S. Hodgson, *I.E.C. Prod. Res. Dev.* **5**, 153 (1966).
71. S. H. Greenfeld and J. R. Wright, *Mat. Res. Stand.*, 738 (Sept. 1962).
72. A. J. Hoiberg, *I.E.C. Prod. Res. Dev.* **19**, 450 (1980).
73. K. G. Martin, "Rupture of Built-Up Roofing Components," in *Engineering Properties of Roofing Systems*, ASTM Special Technical Publication 409, ASTM, Philadelphia, Pa., 1967.
74. W. C. Cullen, *Nat. Bur. Stand. U.S. Monogr. 89* (Mar. 1965).
75. W. C. Cullen and T. H. Boone, *Thermal-Shock Resistance for Built-Up Membranes*, NBS Bldg. Science Ser. 9, Aug. 1967.
76. G. F. Sushinsky and R. G. Mathey, *Fatigue Tests of Bituminous Membrane Roofing Specimens*, Report No. NBS-TN-863, Apr. 1975.
77. M. Koike in ref. 73, p. 26.
78. J. O. Laaly and P. J. Sereda in ref. 69, p. 757.
79. *Criteria for NRCA Roof Curb Approval*, National Roofing Contractors Assoc., Oak Park, Ill.
80. *Perimeter Flashing*, Factory Mutual Engineering Corp., Norwood, Mass.
81. E. C. Shuman in ref. 73, p. 41.

82. K. Tator and S. J. Alexander, in ref. 73, p. 187.
83. E. G. Long, in ref. 73, p. 71.
84. S. H. Greenfeld, *Hail Resistance of Roofing Products*, NBS Bldg. Science Ser., BSS-23, Aug. 1969.
85. J. W. Lane, S. J. Marshall, and R. H. Mumis, *Cold Req. Res. Eng. Lab.*, *Rep. 79-17*, (July, 1979).
86. B. J. Dempsey in ref. 69, p. 779.
87. *Building Materials Directory*, Underwriters Laboratories, Oak Park, Ill. (issued annually).
88. *Fire Tests of Roof Coverings*, ASTM E 108.
89. *Wind Resistance of Asphalt Shingles*, ASTM D 3161.
90. *Tests for Fire Resistance of Roof Covering Materials*, UL 790, Underwriters Laboratories, Oak Park, Ill., 1978.
91. *Approval Guide*, Factory Mutual System, Norwood, Mass.
92. *Approval Standard for Class I Insulated Steel Deck Roofs*, Factory Mutual Research Corp., Norwood, Mass.
93. *Wind Resistance of Prepared Roof-Covering Materials*, UL 997, Underwriters Laboratories, Oak Park, Ill., 1973.
94. *Wind Forces on Buildings and Other Structures*, FM System, Norwood, Mass.
95. "Asphalt Roofing Manufacturing Industry—Background Information for Proposed Standards," U.S. EPA, Research Triangle Park, N.C., 1980, pp. 8–21.
96. *Ibid.*, pp. 3–6.
97. *Ibid.*, pp. 8–38.
98. *Ibid.*, pp. 8–129.
99. *Ibid.*, pp. 8–126.
100. From asphalt and tar roofing and side products. U.S. Department of Commerce, Bureau of Census, M4-29A Series.
101. "General Industry Standards," U.S. Dept. of Labor, "OSHA Safety and Health Standards" (29 CFR 1910), OSHA 2206, Washington, D.C. (Nov. 7, 1978).
102. A. G. Apol and M. Okawa, *Health Hazard Evaluation Determination Report No. 76-55-443*, NIOSH, National Technical Information Service, Springfield, Va. (Nov. 1977).
103. M. T. Okawa and A. G. Apol, *Health Hazard Evaluation Determination Report No. 77-56-467*, NIOSH, National Technical Information Service, Springfield, Va. (Feb. 1978).
104. Ref. 95, pp. 8–122.
105. V. P. Puzinauskas and L. W. Corbett, *Differences between Petroleum Asphalt, Coal-Tar Pitch and Road Tar*, RR-78-1, Asphalt Institute, College Park, Md., 1978.
106. V. P. Puzinauskas and L. W. Corbett, *Report on Emissions from Asphalt Hot Mixes*, RR-75-1A, Asphalt Institute, College Park, Md., 1975.
107. V. P. Puzinauskas, *Emissions from Asphalt Roofing Kettles*, RR-79-2, Asphalt Institute, College Park, Md., 1979.

General References

Refs. 1–3 are general references.

ARNOLD J. HOIBERG
ERNEST G. LONG
Manville Corporation

ROSANILINE. See Triphenylmethane and related dyes.

ROSIN AND ROSIN DERIVATIVES. See Resins, natural; Terpenoids.

RUBBER CHEMICALS

Accelerators of Vulcanization

The main applications of elastomers require that the polymer chains be cross-linked after being formed into a desired shape (see Elastomers, synthetic). After cross-linking of the polymer chains, which is called curing or vulcanization, the article is elastic. It deforms under stress but returns to the shape it had when vulcanization occurred if the stress is removed. The most common method of cross-linking elastomeric polymers is through the use of sulfur. Vulcanization was first discovered in 1839 with the observation of physical change in a mixture of natural rubber (NR), sulfur, and basic lead carbonate [1319-46-8] after heating (1) (see Rubber, natural). Prior to that discovery, natural rubber had limited utility. The rate of the uncatalyzed reaction of sulfur with unsaturated polymers is slow and the physical properties of the vulcanizates are inferior to those attained with modern vulcanization accelerators.

The accelerating effect of metal oxides, eg, lead oxide [1317-36-8], calcium oxide [1305-78-8], and magnesium oxide [1309-48-4], was recognized very early. Organic accelerators, which are indispensable to today's technology, were developed with the discovery that aniline [62-53-3] accelerated vulcanization (2). Although aniline is no longer used because of toxicity and technical reasons, derivatives of aniline are the principal materials in use.

Natural rubber was the first elastomer to be vulcanized, but a wide variety of new synthetic elastomers have been commercialized in the last 50 yr. Their different vulcanization characteristics and physical-property requirements have produced a great variety of vulcanization accelerators and curing agents in the marketplace. The principal elastomer applications require a variety of operations that produce and require heat, eg, mixing, extruding, calendering, molding, etc, during which cross-linking of the polymer to an appreciable extent cannot be tolerated. When cross-linking occurs during processing, the stock is said to have scorched. The need to prevent premature cross-linking has been met by the development of delayed-action accelerators. These materials are not accelerators initially but undergo chemical reactions during processing to produce the active accelerator species in a delayed manner. The main accelerators in use exhibit some degree of delayed action. When more delay than the accelerator furnishes is required, a vulcanization retarder can be incorporated.

Approximately 43,000 metric tons of organic vulcanization accelerators was produced in the United States in 1980 (3). A large portion of this volume was comprised of derivatives of 2-mercaptobenzothiazole (MBT). There is a large variety of cross-linked polymers. They range from polymers with high levels of unsaturation to those with very low levels or no unsaturation. Other polymers are cured through other functional groups, eg, halogens. This variety in polymers and the wide variety of vulcanizate physical-property and processing requirements has resulted in the industrial use of a large number of different accelerators and curing agents. A list of commercially available materials is given in Table 1.

Table 1. Selected Commercial Accelerators and Curing Agents

Structure no.	Compound	CAS Registry No.	Structure	Acute oral LD$_{50}$, mg/kg^a	Uses	Trade name	Supplier
Aldehyde-amine reaction products							
(1)	butyraldehyde–aniline reaction product (3,5-diethyl-1,2-dihydro-1-phenyl-2-propylpyridine)	[34562-31-7]		2,250	semi ultra-accelerator for NR, SBR, CR, IIR	Vanax 808 Beutene	Vanderbilt Uniroyal
(2)	*N,N′*-dicinnamylidene-1,6-hexane-diamine	[140-73-8]			safe curing agent for Viton fluoroelastomer	Diak No. 3	DuPont
(3)	heptaldehyde–aniline reaction product				high temperature accelerator for NR; gives fast cure and high modulus	Hepteen Base	Uniroyal
(4)	hexamethylenetetra-mine	[100-97-0]			slow accelerator for NR and SBR; for activating thiazoles and thiurams	Aceto HMT Rhenogran Hexa Sanceler H	Aceto Chemical Mobay Sanshin
Benzothiazoles							
(5)	2-mercaptobenzo-thiazole (MBT)	[149-30-4]		3,000	primary accelerator for NR and synthetic rubbers	Akrochem MBT Nocceler M Pennac MBT Captax Sanceler M Soxinol M-G Thiotax Vulcafor MBT Vulkacit Mercapto Naugex MBT	Akron Chemical Ouchi Shinko Pennwalt Vanderbilt Sanshin Sumitomo Monsanto Vulnax Mobay Uniroyal

338

No.	Name	CAS No.	LD50	Uses	Trade names	Manufacturers
(6)	bis(2,2'-benzothiazolyl) disulfide (MBTS)	[120-78-5]	>5,000	primary and scorch-modifying secondary accelerator for NR and SBR	Akrochem MBTS; Altax; Nocceler DM; Pennac MBTS; Sanceler DM; Soxinol DM-G; Thiofide; Vulcafor MBTS; Vulkacit DM; Naugex MBT	Akron Chemical; Vanderbilt; Ouchi Shinko; Pennwalt; Shanshin; Sumitomo; Monsanto; Vulnax; Mobay; Uniroyal
(7)	zinc salt of 2-mercaptobenzothiazole (2(3H)-benzothiazolethione, zinc salt)	[155-04-4]	540	primary accelerator for NR and synthetic rubbers; secondary accelerator for latex foam	Akrochem ZMBT; Bantex; Nocceler MZ; Pennac ZT; Soxinol MZ; Vulcafor ZMBT; Vulkacit ZM; Zetax; Oxaf	Akron Chemical; Monsanto; Ouchi Shinko; Pennwalt; Sumitomo; Vulnax; Mobay; Vanderbilt; Uniroyal; Pennwalt; Ouchi Shinko
(8)	2-benzothiazolyl-N,N-diethylthiocarbamyl sulfide	[95-30-7]	2,700	delayed-action primary accelerator for NR and synthetic rubbers	Ethylac; Nocceler 64	Pennwalt; Ouchi Shinko

Benzothiazolesulfenamides

No.	Name	CAS No.	LD50	Uses	Trade names	Manufacturers
(9)	N-tert-butyl-2-benzothiazolesulfenamide	[95-31-8]		delayed-action accelerator for NR and synthetic rubbers	Akrochem BBTS; Pennac TBBS; Santocure NS; Vanax NS; Vulkacit NZ; Delac NS	Akron Chemical; Pennwalt; Monsanto; Vanderbilt; Mobay; Uniroyal
(10)	N-cyclohexyl-2-benzothiazolesulfenamide	[95-33-0]	7,000	delayed-action accelerator for NR and synthetic rubbers	Akrochem CBTS; Durax; Nocceler CZ; Pennac CBS; Sanceller CM; Santocure; Soxinol CZ-G; Vulcafor CBS; Vulkacit CZ; Delac S	Akron Chemical; Vanderbilt; Ouchi Shinko; Pennwalt; Sanshin; Monsanto; Sumitomo; Vulnax; Mobay; Uniroyal

Table 1 (*continued*)

Structure no.	Compound	CAS Registry No.	Structure	Acute oral LD$_{50}$, mg/kg^a	Uses	Trade name	Supplier
(11)	N,N-dicyclohexyl-2-benzothiazolesulfenamide	[4979-32-2]		10,000	delayed-action accelerator for NR and synthetic rubbers	Noceler DZ Vulcafor DCBS Bulkacit DZ/C	Ouchi Shinko Vulnax Mobay
(12)	N,N-di-isopropyl-2-benzothiazolesulfenamide	[92-29-4]			delayed-action accelerator for NR and SBR	DIBS Dipac	American Cyanamid Pennwalt
(13)	2-(4-morpholinylthio)benzothiazole (MTB)	[102-77-2]		>10,000	delayed-action accelerator for NR, SBR, IR, and SBR	Akrochem OBTS Amax Nocceler MSA Sanceller NOB Santocure MOR Soxinol NBS-G Vulcafor MBS Vulkacit MOZ Delac MOR	Akron Chemical Vanderbilt Ouchi Shinko Sanshin Monsanto Sumitomo Vulnax Mobay Uniroyal
(14)	2-(4-morpholinyldithio)benzothiazole	[95-32-9]		>16,000	accelerator and sulfur donor from NR and synthetic rubbers	Morfax Nocceler MDB	Vanderbilt Ouchi Shinko
Dithiocarbamates							
(15)	activated dithiocarbamate			4,200	accelerator for room-temperature vulcanization	Butyl Eight	Vanderbilt
(16)	bismuth dimethyldithiocarbamate	[21260-46-8]		>3,000	accelerator for NR, IR, BR, and SBR at high temperature and speed	Accelerator Bismet Bismate Robac BiDD	Akron Chemical Vanderbilt Robinson Brothers
(17)	cadmium diamyldithiocarbamate	[19010-65-2]			for improved dynamic properties in NR and IR	Amyl Cadmate	Vanderbilt

No.	Name	CAS	Structure	Toxicity (LD₅₀, mg/kg)	Application	Trade name	Manufacturer
(18)	cadmium diethyldithiocarbamate	[14239-68-0]	$\left[\begin{array}{c} C_2H_5 \\ C_2H_5 \end{array}\!N\!-\!\overset{S}{\overset{\|}{C}}\!-\!S\right]_2 Cd$	7,200 (mice), may cause pulmonary diseases	primary accelerator for NR, EPDM, and SBR	Ethyl Cadmate	Vanderbilt
(19)	copper dimethyldithiocarbamate	[137-29-1]	$\left[\begin{array}{c} CH_3 \\ CH_3 \end{array}\!N\!-\!\overset{S}{\overset{\|}{C}}\!-\!S\right]_3 Cu$		ultra-accelerator for SBR and IR	Methyl Cumate, Robac CuDD	Vanderbilt, Robinson Brothers
(20)	dimethylcyclohexylammonium dibutyldithiocarbamate	[149-82-6]	$\begin{array}{c} C_4H_9 \\ C_4H_9 \end{array}\!N\!-\!\overset{S}{\overset{\|}{C}}\!-\!S^-\ {}^+NH(CH_3)_2$ (cyclohexyl)	11,383 (mice)	for room-temperature cure of NR and SBR cements and latexes	RZ-100	Monsanto
(21)	lead diamyldithiocarbamate	[36501-84-5]	$\left[\begin{array}{c} C_5H_{11} \\ C_5H_{11} \end{array}\!N\!-\!\overset{S}{\overset{\|}{C}}\!-\!S\right]_2 Pb$	>10 mL/kg	for improved dynamic properties in NR and IR	Amyl Ledate	Vanderbilt
(22)	lead dimethyldithiocarbamate	[19010-66-3]	$\left[\begin{array}{c} CH_3 \\ CH_3 \end{array}\!N\!-\!\overset{S}{\overset{\|}{C}}\!-\!S\right]_2 Pb$	>10 mL/kg	for high speed and high temperature cure	Methyl Ledate, Robac LMD	Vanderbilt, Robinson Brothers
(23)	lead pentamethylenedithiocarbamate	[81195-36-0]	$\left[N\!-\!\overset{S}{\overset{\|}{C}}\!-\!S\right]_2 Pb$ (piperidyl)		for the continuous vulcanization of cable compounds	Robac LPD	Rosinson Brothers
(24)	3-methyl-2-thiazolidinethione	[1908-87-8]	thiazolidinethione ring, N–CH₃		good scorch safety and compression set for neoprene	Vulkacit CRV	Mobay
(25)	piperidinium pentamethylenedithiocarbamate	[98-77-1]	$N\!-\!\overset{S}{\overset{\|}{C}}\!-\!S^-\ {}^+H_2N$ (piperidyl / piperidinium)	>500	for low temperature cures of latex carpet backing, etc	Nocceler PPD, Robac PPD, Vanax 552, Vulkacit P	Ouchi Shinko, Robinson Brothers, Vanderbilt, Mobay
(26)	selenium dimethyldithiocarbamate	[19632-73-6]	$\left[\begin{array}{c} CH_3 \\ CH_3 \end{array}\!N\!-\!\overset{S}{\overset{\|}{C}}\!-\!S\right]_4 Se$	104 (mice)	accelerator and vulcanizing agent for NR, SBR, and IR	Methyl Selenac	Vanderbilt

Table 1 (*continued*)

Structure no.	Compound	CAS Registry No.	Structure	Acute oral LD$_{50}$, mg/kg^a	Uses	Trade name	Supplier
(27)	sodium dibutyldithiocarbamate	[136-30-1]	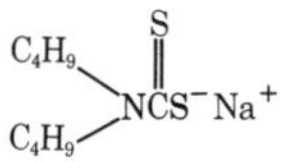 (C$_4$H$_9$)$_2$N–C(=S)S$^-$Na$^+$		for fast precure of latex	Butyl Namate	Vanderbilt
(28)	tellurium diethyldithiocarbamate	[20941-65-5]	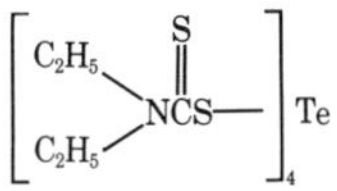 [(C$_2$H$_5$)$_2$N–C(=S)S]$_4$Te	17,000 (mice)	provides high modulus vulcanizates of NR, SBR, NBR, and EPDM	Ethyl Tellurac Soxinol TE-G	Vanderbilt Sumitomo
(29)	zinc diamyldithiocarbamate	[15337-18-5]	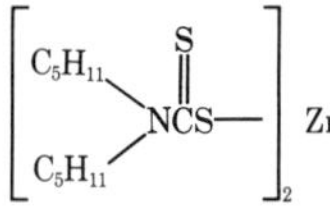 [(C$_5$H$_{11}$)$_2$N–C(=S)S]$_2$Zn	14,900	for improved dynamic properties in NR and synthetic rubbers	Amyl Zimate	Vanderbilt
(30)	zinc di-*n*-butyldithiocarbamate	[136-23-2]	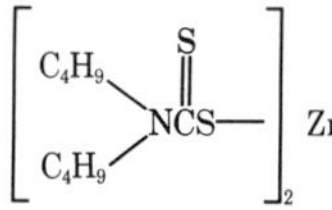 [(C$_4$H$_9$)$_2$N–C(=S)S]$_2$Zn	>10,000	for EPDM and natural and synthetic latexes	Butazin Butyl Zimate Nocceler BZ Robac ZBUD Sanceller BZ Soxinol BZ Vulcafor ZDBC Vulkacit LBD Butazate	Pennwalt Vanderbilt Ouchi Shinko Robinson Brothers Sanshin Sumitomo Vulnax Mobay Uniroyal
(31)	zinc dibenzyldithiocarbamate	[14726-36-4]	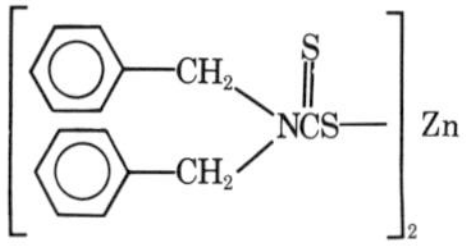 [(C$_6$H$_5$CH$_2$)$_2$N–C(=S)S]$_2$Zn		for the manufacture of butyl rubber cable insulation	Accelerator ZBED Robac ZBED Arazate	Akron Chemical Robinson Brothers Uniroyal

342

No.	Name	CAS No.	Structure	LD50	Application	Trade names	Manufacturers
(32)	zinc diethyldithiocarbamate	[14324-55-1]	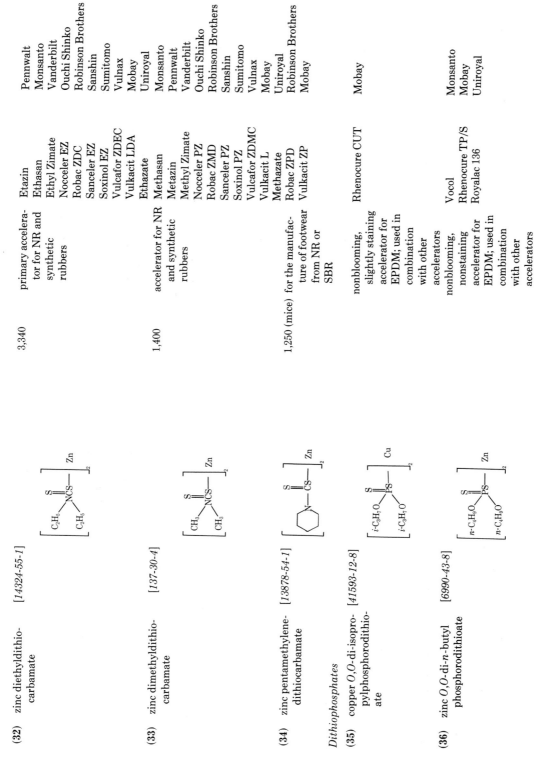	3,340	primary accelerator for NR and synthetic rubbers	Etazin Ethasan Ethyl Zimate Nocceler EZ Robac ZDC Sanceler EZ Soxinol EZ Vulcafor ZDEC Vulkacit LDA	Pennwalt Monsanto Vanderbilt Ouchi Shinko Robinson Brothers Sanshin Sumitomo Vulnax Mobay
(33)	zinc dimethyldithiocarbamate	[137-30-4]		1,400	accelerator for NR and synthetic rubbers	Ethazate Methasan Metazin Methyl Zimate Nocceler PZ Robac ZMD Sanceler PZ Soxinol PZ Vulcafor ZDMC Vulkacit L	Uniroyal Monsanto Pennwalt Vanderbilt Ouchi Shinko Robinson Brothers Sanshin Sumitomo Vulnax Mobay
(34)	zinc pentamethylene-dithiocarbamate	[13878-54-1]		1,250 (mice)	for the manufacture of footwear from NR or SBR	Methazate Robac ZPD Vulkacit ZP	Uniroyal Robinson Brothers Mobay

Dithiophosphates

No.	Name	CAS No.	Structure	LD50	Application	Trade names	Manufacturers
(35)	copper O,O-di-isopropylphosphorodithioate	[41593-12-8]			nonblooming, slightly staining accelerator for EPDM; used in combination with other accelerators	Rhenocure CUT	Mobay
(36)	zinc O,O-di-n-butyl phosphorodithioate	[6990-43-8]			nonblooming, nonstaining accelerator for EPDM; used in combination with other accelerators	Vocol Rhenocure TP/S Royalac 136	Monsanto Mobay Uniroyal

Table 1 (*continued*)

Structure no.	Compound	CAS Registry No.	Structure	Acute oral LD$_{50}$, mg/kg[a]	Uses	Trade name	Supplier
(37)	zinc *O,O*-di-isopropyl phosphorodithioate, biscyclohexylamine complex	[52585-16-7]	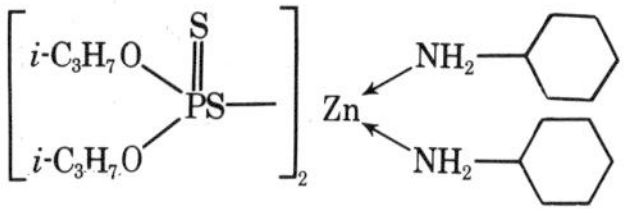		nonstaining accelerator for EPDM; provides good heat resistance and compression set	Rhenocure ZAT	Mobay
Guanidines							
(38)	1,3-diphenylguanidine	[102-06-7]	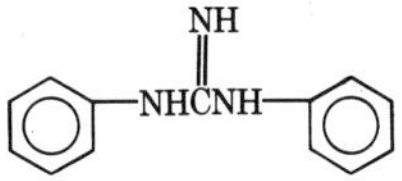	375–850	secondary accelerator for thiazoles, sulfenamides, and thiurams	DPG Nocceler D Sanceler D Soxinol D Vanax DPG Vulcafor DPG Vulkacit D/C	Akron Chemical, American Cyanamid, Monsanto Ouchi Shinko Sanshin Sumitomo Vanderbilt Vulnax Mobay
(39)	di-*o*-tolylguanidine	[97-39-2]	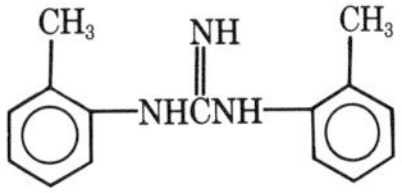	500	slow-curing accelerator for NR, SBR, and NBR; activates thiazole accelerators	DOTG Soxinol DT Vanax DOTG Vulcafor DOTG Vulkacit DOTG/C	American Chemical, American Cyanamid, DuPont Sumitomo Vanderbilt Vulnax Mobay
(40)	*N,N*-bis(2-methyl-phenyl)guanidinium biscatecholborate	[16971-82-7]	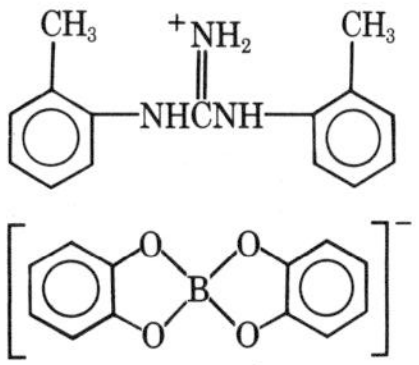		fast-curing accelerator for neoprene; activator for NR and SBR	Nocceler PR Permalux	Ouchi Shinko DuPont

344

Thioureas

No.	Compound	CAS	LD50 / toxicity	Use	Trade names	Manufacturers
(41)	N,N'-dibutylthiourea	[109-46-6]	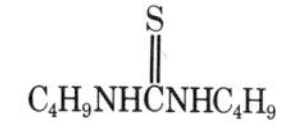 350 (rodent)	accelerator for neoprene, EPDM, and chlorobutyl	Robac DBTU Thiate U	Robinson Brothers Vanderbilt
(42)	N,N'-diethylthiourea	[105-55-5]	 316, carcinogenic in rats	for CR, EPDM, and chlorobutyl	Pennzone E Robac DETU Sanceller EUR Thiate H	Pennwalt Robinson Brothers Sanshin Vanderbilt
(43)	ethylenethiourea (2-imidazolidinethione)	[96-45-7]	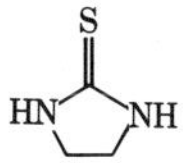 1,832 (rodent), carcinogenic in rats and mice	nonstaining accelerator for CR and epichlorohydrin	Akrochem ETU-22 Robac 22 Sanceller 22 Vulkacit NPV/C	Akron Chemical Robinson Brothers Sanshin Mobay
(44)	1,1'-(1,4-piperazinediyldimethylene)di(2-imidazolinethione)	[14764-02-4]	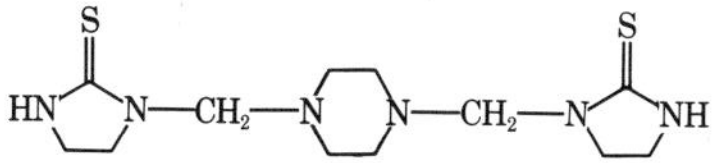	for halogenated polymers; provides scorch safety	Vulcafor 322	Vulnax
(45)	thiocarbanilide (N,N'-diphenylthiourea)	[102-08-9]	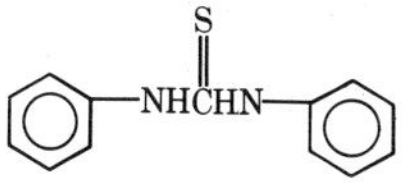	nonstaining secondary accelerator for CR and EPDM	A-1 Thiocarbanilide Stabilizer C	Monsanto Mobay
(46)	trimethylthiourea	[2489-77-2]	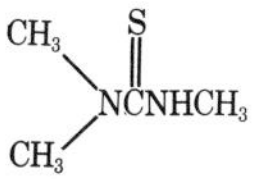 670 ± 110	fast-curing accelerator for neoprene; gives low compression set	Thiate EF-2	Vanderbilt

Thiurams

No.	Compound	CAS	LD50 / toxicity	Use	Trade names	Manufacturers
(47)	dipentamethylenethiuram disulfide	[94-37-1]	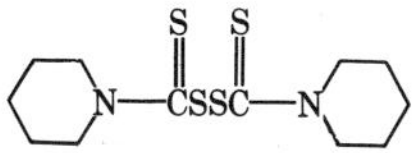 5,000–6,000 (mice)	for the manufacture of latex-dipped gloves	Robac PTD	Robinson Brothers

345

Table 1 (*continued*)

Structure no.	Compound	CAS Registry No.	Structure	Acute oral LD$_{50}$, mg/kg^a	Uses	Trade name	Supplier
(48)	dipentamethylene-thiuram hexasulfide	[971-15-3]			ultra-accelerator and sulfur donor for NR and synthetic rubbers; primary accelerator for Hypalon synthetic rubber and butyl	Akrochem DPTT Sulfads Tetrone A	Akron Chemical Vanderbilt DuPont
(49)	dipentamethylene-thiuram tetrasulfide	[120-54-7]	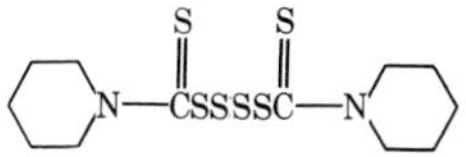	>500	for hypalon cable insulation; also used in nitrile and butyl rubbers	Nocceler TRA Robac P25 Sanceller TRA Soxinol TRA	Ouchi Shinko Robinson Brothers Sanshin Sumitomo
(50)	dipentamethylene-thiuram monosulfide	[725-32-6]	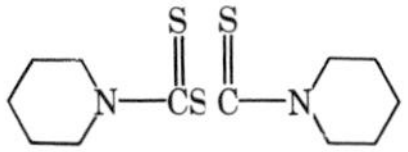		for chemically blown cellular rubber products from NR and SBR	Robac PTM	Robinson Brothers
(51)	tetrabutylthiuram disulfide	[1634-02-2]	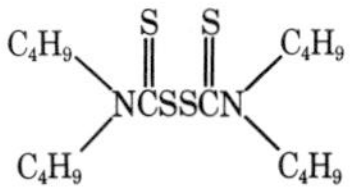		for the production of sulfurless-cured articles, which must be free from bloom	Butyl Tuads Nocceler TBT Robac TBUT Soxinol TBT	Vanderbilt Ouchi Shinko Robinson Brothers Sumitomo
(52)	tetraethylthiuram disulfide	[97-77-8]	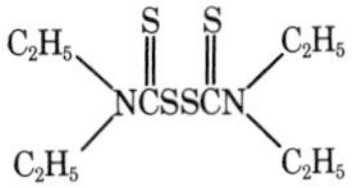	1,300; LD$_{Lo}$ (humans), 160	excellent for fast press cures; less scorching than TMTD (53)	Ethyl Tuads Etiurac Nocceler TET Robac TET Sanceler TET Soxinol TET Ethyl Tuex	Vanderbilt Pennwalt Ouchi Shinko Robinson Brothers Sanshin Sumitomo Uniroyal

346

No.	Compound	CAS	Structure	LD / toxicity	Description	Trade names	Manufacturers
(53)	tetramethylthiuram disulfide (TMTD)	[137-26-8]	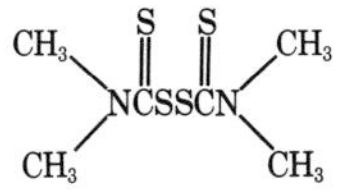	780–1,300; LD_{Lo} (humans), 50	excellent for fast press cures; especially good for IIR and CR	Akrochem TMTD, Methyl Tuads, Metiurac 0, Nocceler TT, Robac TMT, Sanceler TT, Soxinol TTG, Thiurad, Vulcafor TMTD, Vulkacit Thiuram/C, Tuex	Akron Chemical, Vanderbilt, Pennwalt, Ouchi Shinko, Robinson Brothers, Sanshin, Sumitomo, Monsanto, Vulnax, Mobay, Uniroyal
(54)	tetramethylthiuram monosulfide	[97-74-5]	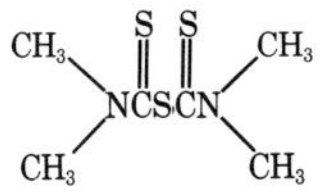	1,250–1,390	booster for thiazoles, especially in nitrile rubbers	Akrochem TMTM, Mono Thiurad, Nocceler TS, Pennac MS, Robac TMS, Sanceller TS, Soxinol TS-G, Unads, Vulcafor TMTM, Vulkacit Thiuram MS, Monex	Akron Chemical, Monsanto, Ouchi Shinko, Pennwalt, Robinson Brothers, Sanshin, Sumitomo, Vanderbilt, Vulnax, Mobay, Uniroyal

Thiocarbamyl sulfenamide

No.	Compound	CAS	Structure	Description	Trade names	Manufacturers
(55)	N-oxydiethylenethio-carbamyl-N'-oxy-diethylenesulfen-amide	[13752-51-7]	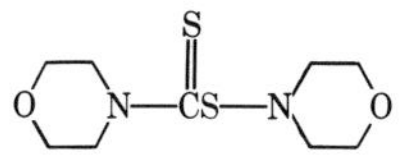	primary accelerator for NR and synthetic rubbers; provides fast cure and scorch safety	Cure-rite 18	Akron Chemical, BF Goodrich

Curing (vulcanizing) agents

No.	Compound	Structure	Description	Trade names	Manufacturers
(56)	alkylphenol-formaldehyde resin		very effective as resin curing agent for IIR; provides good heat resistance	Tackirol 201	Sumitomo
(57)	alkylphenol disulfides	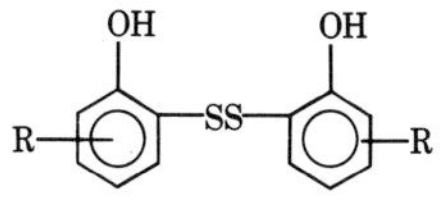	vulcanizing agent	Vultac 2,3,4,5	Pennwalt

Table 1 (*continued*)

Structure no.	Compound	CAS Registry No.	Structure	Acute oral LD$_{50}$, mg/kg[a]	Uses	Trade name	Supplier
(58)	N,N'-caprolactam disulfide	[23847-08-7]	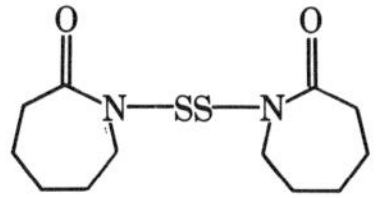		nonblooming and nonstaining sulfur donor in the vulcanization of NR and synthetic rubbers	Rhenocure S	Mobay
(59)	p-quinone bis(benzoyloxime)	[120-52-5]	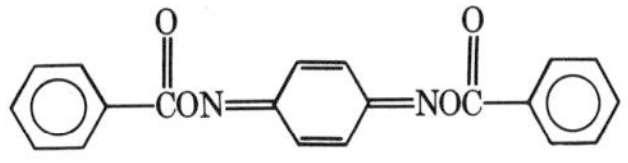		less scorching than p-quinonedioxime; an effective coagent of peroxide cure	Dibenzo GMF Vulnox DGM	Uniroyal Ouchi Shinko
(60)	4,4'-dithiobismorpholine	[103-34-4]	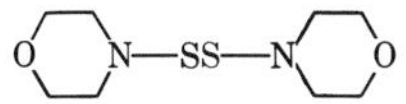	3,690	vulcanizing agent for NR and synthetic rubbers; provides excellent heat aging	Naugex SD-1 Sulfasan R Vanax A Vulnoc R	Uniroyal Monsanto Vanderbilt Ouchi Shinko
(61)	hexamethylenediamine carbamate	[143-06-0]	H$_3$N(CH$_2$)$_6$NHCOO$^-$ (with $^+$ on H$_3$N)		curing agent for Viton fluoroelastomers	Diak No. 1	DuPont
(62)	p-quinone dioxime	[105-11-3]	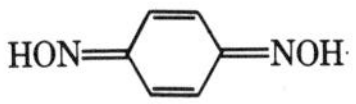 HON=⬡=NOH		used for vulcanizing IIR	GMF Vulnoc GM	Uniroyal Ouchi Shinko
(63)	sulfur	[10544-50-0]	S$_8$		for vulcanizing NR and synthetic rubbers	sulfur	Akron Chemical, FMC, C. P. Hall, Harwick, Stauffer
(64)	insoluble sulfur				minimizes sulfur bloom	Crystex insoluble sulfur	Stauffer

[a] Toxicity data are for rats unless noted otherwise.
[b] Abbreviations: IIR, butyl rubber; CR, chloroprene rubber; EPDM, ethylene–propylene–diene terpolymer; IR, isoprene rubber; NR, natural rubber; SBR, styrene–butadiene rubber; BR, butadiene rubber.

348

Mechanism of Accelerated Sulfur Vulcanization. Although a good deal of effort has been directed at understanding the chemical mechanism of sulfur cross-linking, the exact steps are in contention. The main uncertainty is whether the steps are ionic or free radical in nature (4–8). The evidence indicates that the following general steps occur: the accelerator, which generally is an organic sulfur anion or radical, attacks S_8 to form a polysulfide adduct; this adduct becomes attached to the polymer chain through sulfur as a pendant polysulfide and is capped by the accelerator moiety; the cross-link forms through these pendant polysulfide groups by one of several possible mechanisms; and the average number of sulfur atoms in the cross-link decreases as the vulcanization process continues (this decline is probably a continuation of attack on the polysulfide groups by the accelerator).

All of the sulfur does not result in cross-links. A portion results in internal cyclic structures and pendant groups (see Fig. 1). These alternative structures are more prevalent in unaccelerated or weakly accelerated vulcanization and can result in inferior properties. The vulcanizate properties are also affected by the cross-link density and the average number of sulfur atoms per cross-link. As the cross-link density increases, the modulus increases and elongation to break decreases.

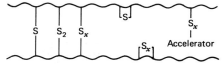

Figure 1.   Structures formed during accelerated vulcanization of elastomers.

Since less energy is required to cleave S–S bonds than C–C or C–S bonds, those vulcanizates with high percentages of polysulfidic cross-links generally have lower thermal stability and poorer aging. Higher accelerator levels combined with lower sulfur levels and systems based on organic sulfur donors and organic curing agents, which donate small amounts of sulfur, produce primarily mono- and disulfide cross-links and are called efficient vulcanization (EV) systems.

Figure 2 depicts a typical cure curve attained with a delayed-action accelerator and sulfur vulcanization. Cure curves of this type can be obtained from any of several curometers, which are widely used to monitor factory operations. A curometer operates in the following manner. A sample of rubber compound containing the cure ingredients is placed in a preheated, temperature-controlled, stationary chamber and the polymer

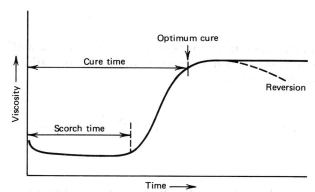

Figure 2.   Cure curves for delayed-action vulcanization.

is compressed against a rotor. The rotor is then oscillated at a predetermined amplitude and frequency. The peak stress required to achieve each strain cycle is measured and plotted against time. The induction period before the viscosity rises appreciably is the scorch time. In actual factory operations, the processing heat history uses up most of this scorch time. The total cure time is the scorch time plus the cross-linking time during which the maximum viscosity or shear modulus is attained.

In some systems, the viscosity or modulus of the stock decreases after passing through a maximum; this is called reversion. This phenomenon is especially troublesome in natural rubber at elevated cure temperatures, especially when a vulcanization system producing predominately polysulfidic cross-links is employed.

2-Mercaptobenzothiazole and Derivatives. Derivatives of 2-mercaptobenzothiazole (MBT) (5) are the most important group of accelerators. The discovery of this compound in ca 1920 led to the delayed-action accelerators that are in wide use (9). 2-Mercaptobenzothiazole is prepared by heating aniline, carbon disulfide, and sulfur at 250°C and 3.1 MPa (450 psi) (10). The product is weakly acidic and is generally purified by dissolving in aqueous caustic, which removes undissolved organics, and then by reprecipitating MBT by the addition of acid.

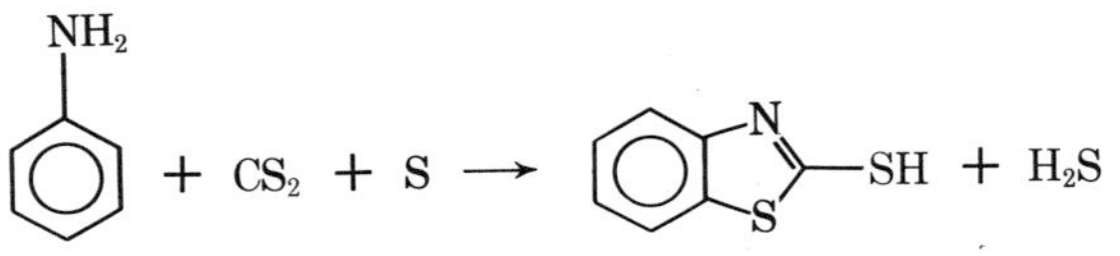

$$\text{(aniline)} \quad + \quad CS_2 \ + \ S \ \longrightarrow \quad \text{(5)} \ + \ H_2S \tag{1}$$

2-Mercaptobenzothiazole is a fairly weak accelerator with little if any delay. Metal salts of MBT are also used as accelerators. It can be easily oxidized to the disulfide (MBTS) (6) by oxidizing agents, eg, NaOCl, HNO$_2$, H$_2$O$_2$, etc. This derivative has moderate scorch delay and a respectable cure rate. It is used widely and especially in natural-rubber compounds. The most important derivatives are the sulfenamides, which have very long scorch delays coupled with good cure rates.

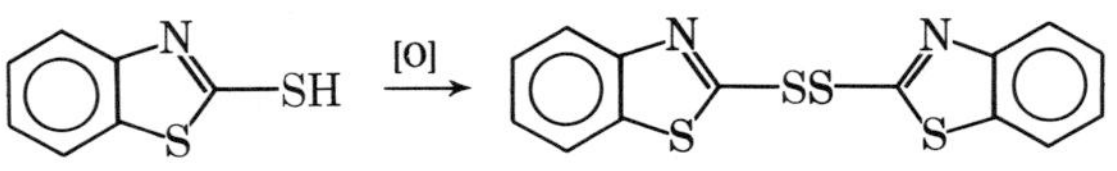

$$\text{—SH} \quad \xrightarrow{\text{[O]}} \quad \text{—SS—} \tag{2}$$

(6)

The sulfenamides are formed by oxidation of a mixture of MBT and the amine (11–16). For example, the following equation shows the formation of *N*-cyclohexyl-2-benzothiazolsulfenamide (10).

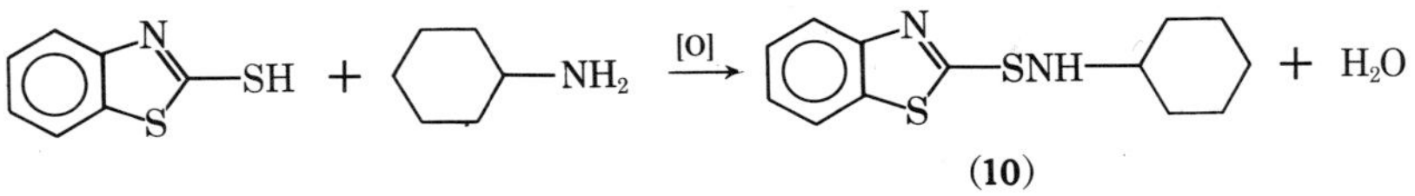

$$\text{—SH} \ + \ \text{—NH}_2 \quad \xrightarrow{\text{[O]}} \quad \text{—SNH—} \ + \ H_2O \tag{3}$$

(10)

An alternative route is the reaction of the *N*-chloroamine with the sodium salt of MBT.

The sulfenamides in commercial use are generally derived from secondary amines or from primary amines that are somewhat hindered; ie, the nitrogen attachment is to a secondary or tertiary carbon.

The scorch safety and cure rates of the various sulfenamides vary over a broad range depending on the amine used. The basicity of the amine and the steric hindrance around the amine affect these characteristics (7). Scorch safety is increased by hindrance or by using less basic amines, eg, morpholine. Cure rates are generally decreased by increased hindrance and lower basicity of the amines. Commercially important sulfenamides are derived from morpholine, t-butylamine, cyclohexylamine, diisopropylamine, dicyclohexylamine, and diethylamine.

Dithiocarbamates and Derivatives.   Dithiocarbamates and their derivatives are sometimes called ultra-accelerators because of their high cure rates. The dithiocarbamates are prepared by the reaction of a secondary amine with carbon disulfide in the presence of a base or excess amine.

$$
2 \; \begin{matrix} R \\ \\ R \end{matrix} \!\! NH \; + \; CS_2 \; \longrightarrow \; \begin{matrix} R \\ \\ R \end{matrix} \!\! N\overset{\overset{\textstyle S}{\|}}{C}S^- \; {}^+H_2N \!\! \begin{matrix} R \\ \\ R \end{matrix}
\tag{4}
$$

If the reaction is carried out with one mole of amine in the presence of a base, such as sodium hydroxide, the sodium salt of the dialkyl dithiocarbamate is produced. The main derivatives are generally produced from sodium salts. The salts are very active accelerators but give little or no delay. A wide variety of heavy metal salts is available for use alone or in combination with other accelerators.

The sodium dithiocarbamate salts can be oxidized to yield another important class of accelerators called tetraalkylthiuram disulfides. The oxidizing agents are usually sodium hypochlorite or hydrogen peroxide. This type of accelerator gives a short delay followed by a high cure rate. They also function as curing agents; ie, they cure without addition of sulfur since some of the sulfur in the accelerator is available for cross-linking. When they are used without sulfur, very short cross-links are produced.

$$
2 \; \begin{matrix} R \\ \\ R \end{matrix} \!\! N\overset{\overset{\textstyle S}{\|}}{C}S^-Na^+ \; \overset{[O]}{\longrightarrow} \; \begin{matrix} R \\ \\ R \end{matrix} \!\! N\overset{\overset{\textstyle S}{\|}}{C}SS\overset{\overset{\textstyle S}{\|}}{C}N \!\! \begin{matrix} R \\ \\ R \end{matrix}
\tag{5}
$$

Another class of derivatives of the dithiocarbamates is thiuram monosulfides. These compounds can be produced in several ways. One route is the reaction of the dithiocarbamate salt with phosgene (17).

$$
2 \; R_2N\overset{\overset{\textstyle S}{\|}}{C}S^-Na^+ \; + \; COCl_2 \; \longrightarrow \; R_2N\overset{\overset{\textstyle S}{\|}}{C}S\overset{\overset{\textstyle S}{\|}}{C}NR_2 \; + \; COS \; + \; 2 \; NaCl
\tag{6}
$$

This type of compound has more scorch safety than the disulfides but appreciably less than most benzothiazolesulfenamides. The thiuram monosulfides have high cure rates like the other dithiocarbamate derivatives and can be used as primary accelerators or in conjunction with other less active accelerators.

The dithiocarbamate derivatives with the greatest scorch delay are thiocarbamylsulfenamides. They can be produced in a variety of ways (18–22), but the simplest route involves reaction of carbon disulfide with a mixture of amine and N-chloroamine in the presence of a base (23). For instance, the commercially available morpholine derivative can be produced as follows:

$$\text{O}\diagdown\text{NH} + \text{O}\diagdown\text{N—Cl} + \text{CS}_2 + \text{NaOH} \longrightarrow \text{O}\diagdown\text{N—}\overset{\overset{\displaystyle S}{\|}}{\text{CS}}\text{—N}\diagdown\text{O} + \text{NaCl} + \text{H}_2\text{O} \qquad (7)$$

$$(55)$$

In addition to scorch safety similar to that of the benzothiazole sulfenamides, these compounds produce high cure rates (24).

Guanidines and Aldehyde–Amine Reaction Products.   Guanidines and aldehyde–amine reaction products are rarely used as primary accelerators, but they are frequently used as activators for thiazole accelerators. 1,3-Diphenylguanidine (**38**) can be prepared by the reaction of aniline with cyanogen chloride as follows:

$$2\,\text{C}_6\text{H}_5\text{—NH}_2 + \text{CNCl} \longrightarrow \text{C}_6\text{H}_5\text{—NH}\overset{\overset{\displaystyle NH}{\|}}{\text{C}}\text{NH—C}_6\text{H}_5 + \text{HCl} \qquad (8)$$

$$(38)$$

The aldehyde–amine condensation products are formed by reaction of aniline, p-toluidine, or n-butylamine with butyraldehyde, acetaldehyde, formaldehyde, or 2-ethylhexenal in the presence of a weak acid as the catalyst. A Schiff base forms and polymerizes through the carbon–nitrogen double bond to condensation products of indefinite structure (25).

Dialkyl Phosphorodithioates.   Dialkyl phosphorodithioates have the following general structure:

$$\left[(RO)_2\overset{\overset{\displaystyle S}{\|}}{P}S \right]_2 M$$

where R is butyl or isopropyl and M is copper or zinc. They are used primarily in rubber made from ethylene–propylene–diene monomer (EPDM). They generally are used in combination with other accelerators to achieve high cure rates and to reduce blooming, which is the formation of crystals of accelerator residues on the surface of the article. Blooming is especially troublesome in EPDM where high accelerator levels are used to compensate for the low unsaturation of the polymer.

Cure Systems for Halogenated Polymers. Certain special-purpose elastomers, eg, polychloroprene (CR) and polyepichlorohydrin, require different cure systems than the diene polymers. Ethylenethiourea (2-imidazolidinethione) (43) has been the most widely used accelerator for these polymers. However, there is much concern about the potential health hazard of this material. It is generally used in the form of a pre-mixed rubber–accelerator blend, ie, as the masterbatch, to minimize exposure to personnel in factory operations. Concern about this material has brought about market introduction of several new accelerators as alternatives (see Table 1), as well as the development of new unsaturated polymers, which allow the use of general-purpose vulcanization systems as alternatives.

Vulcanizing Agents. As was previously mentioned, efficient vulcanization (EV) giving predominately short cross-links can be achieved by using high accelerator-to-sulfur ratios. However, the most efficient systems include organic sulfur donors, which frequently generate an accelerator group. Thiuram disulfides are very active and have very little scorch delay. The most common is tetramethylthiuram disulfide (53). Dipentamethylenethiuram tetra- and hexasulfides, (49) and (48), respectively, are also used; they contain higher levels of available sulfur.

Another compound in common use is 2-(4-morpholinodithio)benzothiazole (14). This compound can be prepared in a number of ways. The following equation illustrates a simple amine-catalyzed sulfur insertion into the sulfenamide.

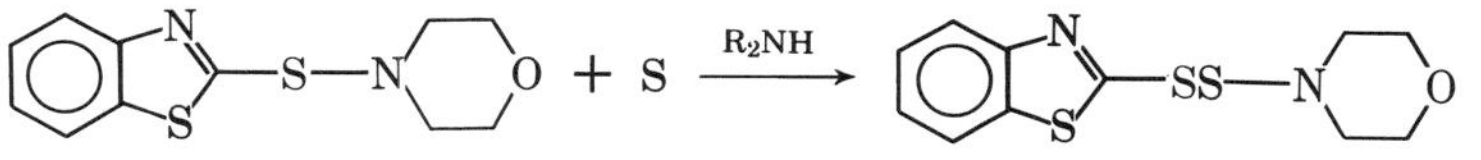

(9)

(13)   (14)

This vulcanizing agent has much more delay than the thiuram disulfides. It can also serve as a vulcanization accelerator with sulfur. In this use it has somewhat less scorch delay than the corresponding benzothiazole sulfenamide of morpholine when typical sulfur concentrations are used.

Another common vulcanizing agent is 4,4′-dithiobismorpholine (60). It can be prepared by the following general reaction (26):

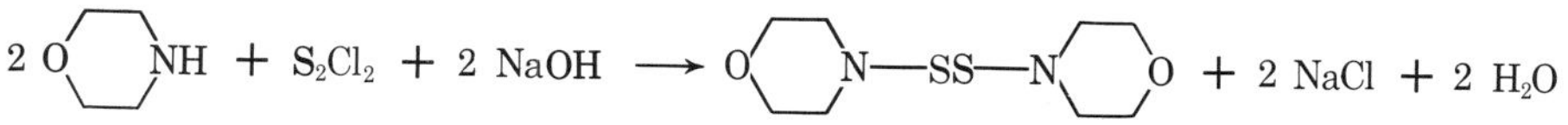

(10)

(60)

This vulcanizing agent is generally used in combination with delayed-action accelerators to give good scorch delay.

Peroxide Cures. In 1914, it was discovered that dibenzoyl peroxide [94-36-0] cross-links rubber. The use of the more effective dialkyl peroxides to cross-link polymers started shortly after 1950. The cross-link is attained through free radicals formed by homolytic decomposition of the peroxide. This produces alkoxy radicals, which can abstract hydrogen from the polymer chain, producing alkyl radicals on the polymer chain. Coupling of these radicals leads to cross-links, as shown in the following simplified scheme, where R represents an alkyl group and P represents a polymer chain.

$$ROOR \xrightarrow{\text{heat or } h\nu} 2\,RO\cdot$$

$$RO\cdot + P—H \rightarrow ROH + P\cdot$$

$$2\,P\cdot \rightarrow P—P \text{ (cross-link)}$$

The carbon–carbon cross-links produced by this process have appreciably higher bond energies (343 kJ/mol or 82 kcal/mol) than sulfur–sulfur or carbon–sulfur bonds. Consequently, the cross-links are thermally more stable than typical sulfur vulcanizates and even more stable than an EV sulfur system. The physical properties of the resultant vulcanizates have unique features that are not attained in the EV sulfur system.

The peroxide system has marked advantages over sulfur systems in several situations, ie, when the polymer to be cross-linked has little or no unsaturation or when two polymers that have markedly different degrees of unsaturation or rates of cure with sulfur are blended. In sulfur systems, one phase is much more highly cross-linked than the other. Since peroxides are more reactive and therefore less selective, they give more uniform cures and better properties. If it is necessary to maximize the physical properties that are associated with very stable cross-links, eg, maximum thermal stability, minimum compression set, etc, the peroxide system should be used.

A great variety of organic peroxides have been made and used to cross-link polymers. They exhibit a wide range of efficiencies, ie, cross-links per mole of peroxide, and cure rates. The cure rate is related to the decomposition rate of the peroxide and is generally given in terms of a half-life. Half-life is the time required to decompose one half of the peroxide at a given temperature in a given medium. The cure of a polymer is essentially complete after 6–7 half-lives, at which time over 99% of the original peroxide has decomposed. Dialkyl peroxides are the class most frequently used for cross-linking elastomers, and dicumyl peroxide (**68**) is the most widely used peroxide in this class. However, a variety of other peroxides are used (see Table 2).

All polymers do not undergo peroxide cure with the same efficiency. One of the principal reasons is that radicals on polymer chains can cause chain cleavage or beta cleavage as well as cross-linking:

$$R\overset{\cdot}{C}HCH_2CH_2R' \longrightarrow RCH{=}CH_2 + \cdot CH_2R'$$

This process is especially prevalent if the polymers contain appreciable amounts of hydrogen on tertiary carbon atoms. Polymers that tend to undergo undesirable cleavage are polypropylene, butyl rubber, and polyepichlorohydrin.

A number of other factors can have negative effects on peroxide cures, eg, the presence of oxygen which reacts with carbon radicals to form hydroperoxides. This can lead to chain cleavage and to autoxidation during the life cycle of the products. Thus, cures should not be attempted in air or when an article is removed from the mold prior to completed peroxide decomposition and subsequently heated in the presence of oxygen.

Peroxides are unstable in the presence of acids and undergo ionic decomposition, which does not produce radicals for cross-linking. Radicals can also be trapped in noncross-linking reactions by other ingredients in the elastomer, especially antioxidants and antiozonants (see Antioxidants and antiozonants). Among antioxidants, the quinoline type has the least effect on peroxide cures. Aromatic amine antioxidants are next, followed by hindered phenols. The p-phenylenediamine antiozonants are particularly effective radical traps and greatly inhibit peroxide cures.

Large amounts of nonpolymeric diluents also decrease cross-linking efficiency by consuming radicals in noncross-linking side reactions. Among extending oils, the least problems are encountered with the paraffinic types followed by cycloparaffinics and aromatics in that order.

Peroxide cross-linking efficiency can also be improved and the vulcanizate properties modified by certain additives. A variety of polyfunctional active olefins are effective for this purpose. They appear to function through rapid polymerization of the active double bonds to produce a second polymer network, which can be initiated by an alkyl radical from the peroxide or from a radical on the base polymer. In the latter case, the new polymer is grafted on the parent polymer chain. This process also reduces beta cleavage, since the effective concentration of radicals on the original polymer chain is reduced by addition to the active olefins. Examples of compounds that function this way are ethylene glycol dimethacrylate [97-90-5], trimethylolpropane trimethacrylate [3290-92-4], triallyl cyanurate [101-37-1], zinc acrylate [14643-87-9], N,N'-m-phenylenedimaleimide [3006-93-7], and polybutadiene [9003-17-2].

Other Cure Systems. A large variety of other types of cure systems has been described (27); of these, the most important are resin cures. Phenol–formaldehyde resins are used to cure elastomers; however, their cure mechanism has not been resolved (28–30). They give thermally stable cross-links with properties similar to peroxide cures, and they have been used extensively to cure latex tire-cord dips and butyl-curing bladders.

Toxicity of Vulcanization Systems. The long-term biological effects of most of the compounds listed in Table 1 are not known. Use of premixed masterbatches, oiling, pelletizing of powders, etc, is becoming commonplace to minimize exposure. Two areas of concern are the possible carcinogenicity of several of the thioureas previously mentioned and the presence of secondary alkyl nitrosamines in the air of certain rubber-processing plants (31). Many nitrosamines produce cancers in test animals.

Volatile secondary amines are produced as decomposition products from many accelerators. These amines can be converted to N-nitrosamines by nitrosating agents in the polymer or in the atmosphere. Of particular concern is the use of N-nitrosodiphenylamine (**72**) as a retarder, which can transnitrosate secondary aliphatic amines which may be in the same rubber mix. Also, the presence of oxides of nitrogen in the atmosphere can nitrosate amines.

Retarders

Retarders are chemicals that prevent the premature vulcanization of rubber compounds during mixing, calendering, and other processing steps. In the absence of the processing safety provided by retarders, scorched stocks, and consequently,

Table 2. Peroxides

Structure no.	Compound	CAS Registry No.	Structure	Acute oral LD$_{50}$ (rat), mg/kg	Uses[a]	Trade name	Supplier
(65)	benzoyl peroxide	[94-36-0]	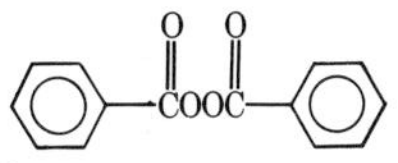		cross-linking silicone and fluorosilicone elastomers	Luperco AST Peradox BS	Pennwalt Noury
(66)	1,1-bis(*t*-butylperoxy)-3,3,5-trimethylcyclohexane	[6731-36-8]	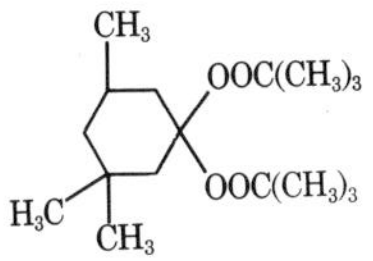		cross-linking SBR, NBR, EPR, EPDM, and silicones at low operating temperatures	Luperox 231[b] Percadox 29	Pennwalt Noury
(67)	bis(2,4-dichlorobenzoyl) peroxide	[133-14-2]	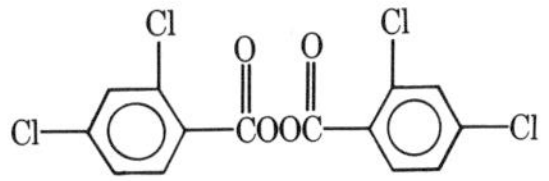		cross-linking silicone or fluorosilicone elastomers	Luperco CST	Pennwalt
(68)	dicumyl peroxide	[80-43-3]	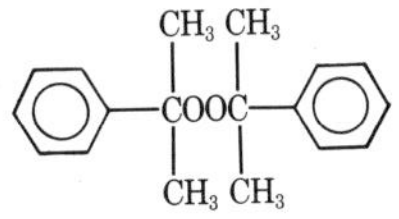	4,100	nonsulfur curing agent for NR, IR, SBR, BR, NBR, EPDM, and CR	Di-cup T[c] Di-cup R[d] Di-cup 40C[e] Di-cup 40KE[f] Luperox 500T[c] Luperox 500R[d] Luperox 50040C[e] Luperox 50040KE[f] Varox DCP-T[c] Varox DCP-R[d] Varox DCP40C[e] Varox DCP40KE[f]	Hercules Pennwalt Vanderbilt
(69)	2,5-dimethyl-2,5-bis-(*t*-butylperoxy)hexane	[78-63-7]	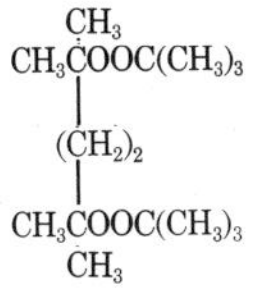	32,000	vulcanization or cross-linking elastomers and polyolefins, eg, EPDM, EPM, PE, and NBR	Luperco 101[b] Varox DBPH	Pennwalt Vanderbilt

[a] Abbreviations: NBR, nitrile–butadiene rubber; EPR, ethylene–propylene rubber; PE, polyethylene (see also footnote [a] in Table 1).
[b] Also available on a mineral carrier for both suppliers.
[c] Technical grade.
[d] Recrystallized material.
[e] 40 wt % dicumyl peroxide on 60 wt % calcium carbonate.
[f] 40 wt % dicumyl peroxide on 60 wt % burgess clay.

356

waste results either during the processing steps or during storage of the fully compounded green stocks. Retarders are often called antiscorching agents, scorch inhibitors, cure retarders, or prevulcanization inhibitors. Retarders have been classified
(32–33). Retarders with a sulfenamide group are called prevulcanization inhibitors,
whereas such conventional retarders as salicylic acid (**74**), phthalic anhydride (**73**),
and N-nitrosodiphenylamine (NDPA) (**72**) are simply called retarders. This classification is by no means a rigid one, and retarders and prevulcanization inhibitors are
used synonymously throughout this article. Some commercially available retarders
are listed in Table 3.

Conventional retarders include benzoic acid (**70**), phthalic anhydride, and N-
nitrosodiphenylamine (NDPA). More recent ones include a sulfonamide derivative
Vulkalent E (Mobay) and N-(cyclohexylthio)phthalimide (CTP) (**71**), Santogard PVI
(Monsanto). The total annual consumption of these retarders in the U.S. is ca
1800–2300 t. Of this, ca 80% of the retarders is used in the manufacture of tires, and
the rest is used in miscellaneous mechanical goods, footwear, and sheet and calendered
goods.

N-Nitrosodiphenylamine (NDPA) has been widely used for many decades. Its
production in the United States peaked in 1974, reaching ca 1600 metric tons, and then
decreased gradually to 184 t in 1980. The decline in the use of NDPA and the increasing
demand for the new type of retarders resulted because N-nitrosodiphenylamine is
a slightly staining retarder with a peptizing effect which works well with sulfenamide
accelerators, but it is not efficient in the presence of alkyl–aryl or dialkyl-substituted
p-phenylenediamine antidegradants (32). The gaseous decomposition products of
NDPA during vulcanization have been identified as a cause of porosity in thick
cross-section extrusions. In addition, NDPA is suspected as a nitrosating agent of
secondary amines which are suspected to be animal carcinogens (34–35) (see N-Nitrosamines). In contrast, CTP and other prevulcanization inhibitors are nonstaining,
effective with the full range of sulfenamide accelerators and antidegradants, and
relatively nontoxic. Thus, consumption of the new types of retarders is expected to
surpass 90% of the total usage of retarders in the near future.

Conventional. The exact mechanism by which acid retarders, eg, salicylic acid,
extend scorch time is uncertain. According to one explanation, the retardation by acidic
retarders is caused by ionic scission of the rubber molecule at the initial stage of vulcanization (36). Another explanation is the delay of interaction of sulfur with rubber,
which is brought about by the formation of neutral products between the acidic retarder and basic material in the rubber, eg, NaOH and Na_2CO_3 in polybutadiene (37).
The third mechanism describes the acidic retarders as interacting with the degradation
products of accelerators to form compounds that decrease the rate of vulcanization
(38).

The acidic retarders can be used in conjunction with 2-mercaptobenzothiazole
(MBT) and bis(2,2'-benzothiazolyl) disulfide (MBTS) (**6**), but they are generally ineffective with sulfenamide accelerators. They also tend to decrease the vulcanization
rate, and their use is limited to footwear, mechanical goods, and other miscellaneous
categories.

The chemistry of phthalic anhydride (PA) in rubber compounds during vulcanization is also controversial, as shown in the following observations: phthalic anhydride
hinders the formation of a MBT–ZnO–stearic acid complex (39); PA chemically adds
to rubber (39); PA forms a PA–ZnO–MBT complex, which causes less scorching than

Table 3. Retarders

Structure no.	Compound	CAS Registry No.	Structure	Acute oral LD$_{50}$ (rat), mg/kg	Uses[a]	Trade name	Supplier
(70)	benzoic acid	[65-85-0]	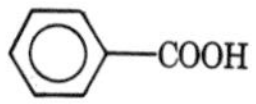	2530	retarder for NR and synthetic rubbers	Retarder BA Vulkalent GK	Akron Chemical Mobay
(71)	N-(cyclohexylthio)phthal-imide (CTP)	[17796-82-6]	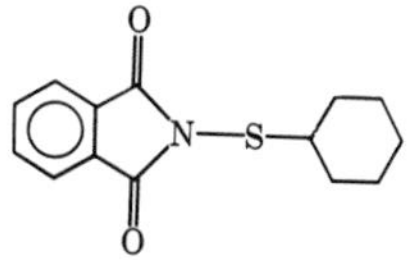		retarder for NR and synthetic rubbers; it works best with sulfenamide type accelerators	Santogard PVI	Monsanto
(72)	N-nitrosodiphenylamine (NDPA)	[86-30-6]	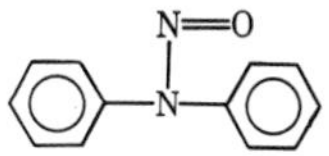	3000 (rats developed tumor)	staining retarder for NR and synthetic rubbers	Goodrite Vultrol Redax Sconoc Vulcatard A Vulkalent A	BF Goodrich Vanderbilt Ouchi Shinko Vulnox Mobay
(73)	phthalic anhydride (PA)	[85-44-9]	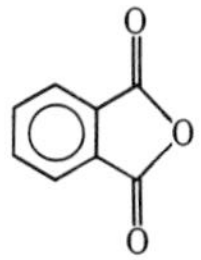	4020	nonstraining retarder for NR, SBR, and NBR	Retarder B-C Retarder PX Sconoc 7 Sumitard BC Vulkalent B/C Retarder ESEN	Sanshin Akron Chemical Ouchi Shinko Sumitomo Mobay Uniroyal
(74)	salicylic acid	[69-72-7]	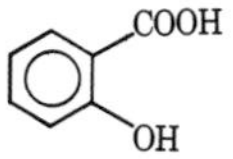	890	retarder for NR and SBR; accelerator for W types of Neoprene	Retarder SAX Vulcatard SA	Akron Chemical Vulnax
(75)	sulfonamide derivative				nonstaining retarder for NR and synthetic rubbers	Vulkalent E	Mobay

[a] Abbreviations: NR, natural rubber; SBR, styrene–butadiene rubber; NBR, nitrile–butadiene rubber.

358

MBT alone (40); and PA converts sulfenamide accelerators into bis(sulfenamide)s (41). The main use of PA is in footwear.

The retardation mechanism by NDPA is based on *in situ* formation of nitric oxide (NO). Nitric oxide is believed to react with sulfur-containing radicals during vulcanization, thereby disrupting the chain reaction (37). It is also claimed that the scorch delay by NDPA results from the slower rate of disappearance of the sulfenamide accelerator and sulfur (42). The harmful side effects from the use of NDPA have been explained in terms of the formation of a stable nitroxyl radical between rubber and NO_2 (43). The use of NDPA is limited to mechanical goods and other special applications.

New Retarders. A prototype of a prevulcanization inhibitor was first reported in the USSR in 1964 (44). According to the inventors' claims, *N*-(trichloromethylthio)phthalimide *[133-07-3]* **(76)** is a good retarder for natural and synthetic rubbers and 0.2 part of it can be as effective as 0.5 part of phthalic anhydride.

(76)
N-(trichloromethylthio)phthalimide

Five years later, a dramatically efficient new retarder was disclosed (45). Subsequently, a number of patents and papers have been published, including an extensive review article on prevulcanization inhibitors (33). At present, *N*-(cyclohexylthio)phthalimide (CTP) (Santogard PVI) is by far the most widely used retarder in the United States. Approximately 1400 t of CTP is estimated for domestic usage. It can be synthesized in high yield by the reaction of phthalimide with cyclohexylsulfenyl chloride in the presence of a base and suitable solvent.

Cyclohexylsulfenyl chloride is in turn prepared by reaction of cyclohexyl mercaptan with chlorine (33). An alternative method is electrosynthesis of CPT from phthalimide and cyclohexyl disulfide in acetonitrile. After 18 h of reaction time, a 99% yield of CTP can be isolated (46). The recrystallized CTP melts at 93–94°C.

S,*S*-Diisopropyl-*N*-(*p*-toluenesulfonyl)sulfilimine **(77)** has been patented by Goodyear (47).

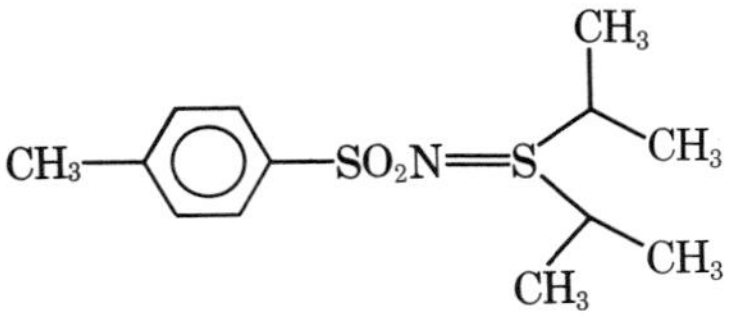

(77)

S,S-diisopropyl-*N*-(*p*-toluenesulfo-
nyl)sulfilimine [*18922-54-8*]

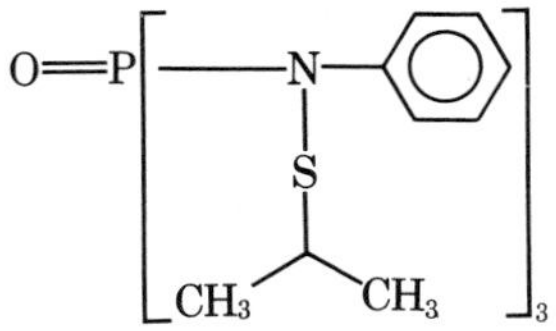

(78)

N,N',N''-phosphinylidynetris[*N*-
phenyl(isopropyl)sulfenamide] [*51877-44-2*]

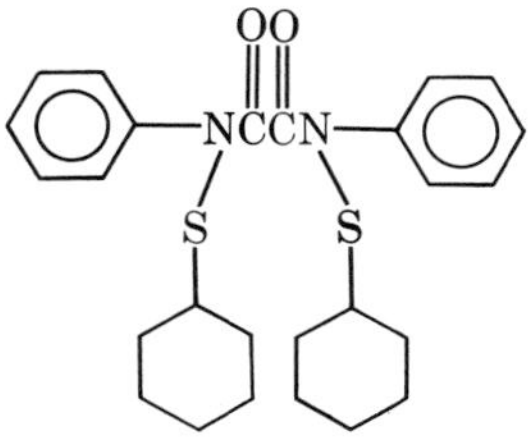

(79)

N,N'-bis(cyclohexylthio)oxalanilide
[*50863-05-3*]

The prevulcanization inhibitor is particularly effective with 2-(4-morpholinyldi-thio)benzothiazole (14) (Morfax) in a natural rubber (NR) stock. In the presence of *N*-cyclohexyl- (10) and 2-(4-morpholinylthio)benzothiozolesulfenamide (13) twice the weight of the Goodyear product is required to produce scorch delay comparable to that achieved with CTP. However, the sulfilimine can be used at higher levels, up to 1.5 phr (parts per hundred rubber) in NR, without producing bloom on the vulcanizates and can be produced continuously from sodium *N*-chloro-*p*-toluenesulfo-namide (chloramine-T) and diisopropyl sulfide without requiring anhydrous conditions (48). *N,N',N''*-Phosphinylidynetris[*N*-phenyl(isopropyl)sulfenamide] [*56616-05-8*] (78) is available as Vulcatard PRS (Vulnax) (49). *N,N*-Bis(cyclohexylthio)oxalanilide (BCTO) [*50863-05-3*] (79) patents were issued to the B. F. Goodrich Company (50). In certain rubber stocks, BCTO is more efficient and bloom-resistant than CTP.

Efficiency and Chemistry.   Unlike the conventional retarders, new retarders have little if any effect on the rate of cure, physical properties, or aging performance. In general, CTP increases the processing safety of sulfur-curable elastomers in direct proportion to the amount used and is effective with a wide range of curing systems. This broad-based application gives the compounders flexibility in solving scorch-delay problems. Use of CTP and other prevulcanization inhibitors can bring about the following benefits (51–53): improved green-stock storage stability; recovery of stocks with marginal processing safety; increased productivity of factory process with the use of either kickers (secondary accelerators) or higher processing temperatures; and, in certain stocks, the prevulcanization inhibitors can enhance the adhesion between brass cord and rubber.

Development and commercialization of the unusually efficient and versatile new prevulcanization inhibitors has been matched with very interesting results concerning the chemistry involved in the scorch-delay mechanism. In 1970, it was proposed that the function of *N*-(cyclohexylthio)phthalimide (CTP) (71) is to remove 2-mercapto-benzothiazole (MBT) (5) from the autocatalytic sequence of reaction; namely, sul-fenamide converts to bis(2,2-benzothiazolyl) disulfide (MBTS) (6) followed by the formation of cross-link precursors, ie, bisbenzothiazole polysulfide (80) and ben-zothiazole-S_x-rubber (49,54). The manner by which CTP removes MBT is shown in reaction 13.

$$(6) + S_8 \longrightarrow \text{(80)} \longrightarrow \longrightarrow \text{vulcanization} \qquad (12)$$

The rate of reaction 13 has to be much faster than that of competing reaction 11, which leads eventually to the vulcanization of rubbers as shown by reaction scheme 12. Since the N—S bond of CTP is much more labile than that of MTB (13), MBT preferentially attacks CTP rather than MTB or other sulfenamide accelerators to form 2-cyclo-hexyldithiobenzothiazole [28084-58-4] (CDB) (81).

The fast reaction between MBT and a compound with a labile N—S bond does not necessarily mean that the compound is an excellent prevulcanization inhibitor (55). For example, whereas CTP takes more than two hours to convert 30% of MBT

(82)

N-(cyclohexylthio)-*o*-benzoic sulfimide
[*39922-89-9*]

to CDB at room temperature, *N*-(cyclohexylthio)-*o*-benzoic sulfimide [*39922-89-9*] (CTBS) (82) reacts instantaneously and completely with MBT at the same temper-ature. Based on these observations, CTBS would be expected to be a better prevul-canization inhibitor than CTP. Unexpectedly, CTP is six times more efficient than CTBS in a natural-rubber compound. This contradiction is clarified by the finding that CTBS is thermally unstable under the vulcanization conditions.

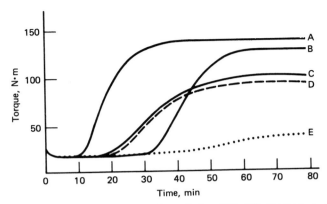

Figure 3. Curometer curves at 140°C (55). Formulation: NR, 100; HAF black, 50; ZnO, 5; stearic acid, 3; S, 2.5. A, MTB = 1.0; B, MTB = 1.0, CTP = 1.0; C, CTP = 1.0, MBT = 0.64; D, CDB = 1.1; E, CTP = 1.0. To convert N·m to dyn·cm, multiply by 1.0×10^7.

In addition to two main retarder criteria, ie, fast reaction with MBT and thermal stability of a retarder, an excellent retarder is capable of converting the fast-curing MBT into a relatively nonscorching accelerator, eg, CDB (56–57). As shown in the curometer curves in Figure 3, CDB is much less scorching than MTB (18.2 vs 7.5 min). A similar cure curve C can be obtained by substituting CDB (81) in stock D with equimolar amounts of MBT (5) and CTP (71). The similarity between cure curves C and D suggests that the reaction between MBT and CTP is very facile even in natural rubber.

By analogy, CTP (71) could also react with cross-link precursors to form less reactive species, as shown below:

The validity of the above speculation is supported by the observation that CTP can stop further cross-linking if CTP is added to a partially cross-linked rubber (56). Since CTP does not devulcanize the cured stocks, the immediate cessation of cure upon the addition of CTP can be explained in terms of trapping the cross-linking precursors and converting them to less scorching precursors. Cyclohexyldithio derivatives, including CDB, extend scorch time of benzothiazolesulfenamide accelerators. Thus, one way to interpret the scorch delay mechanism by prevulcanization inhibitors is the distribution of available sulfur in the form of cyclohexyldithio derivatives, thereby reducing the number of labile sulfur atoms of the reactive polysulfur intermediate (57). More recently, the possibility of the acceleration of the cross-link degradation by CTP was reported and is being studied further to substantiate that the long scorch delay obtained by the use of CTP is indeed the result of cross-link degradation (58).

In summary, several plausible mechanisms can explain how the new class of retarders or prevulcanization inhibitors provides excellent scorch delay in the vulcanization of various rubber compounds:

Prevulcanization inhibitors remove MBT during the cure and delay the autocatalytic conversion of benzothiazolesulfenamide to cross-link precursors.

The fast reaction between prevulcanization inhibitor and MBT forms a disulfide, which is a nonscorchy accelerator. This disulfide in turn acts as a retarder for the benzothiazolesulfenamide accelerator.

Prevulcanization inhibitors can remove cross-link precursors and stop further cross-linking at the early stage of vulcanization.

Prevulcanization inhibitors can distribute available sulfur into many forms, thereby reducing the number of labile sulfur atoms of the polythiobenzothiazoles.

Prevulcanization inhibitors and amine from sulfenamide accelerators can produce another form of sulfenamide, which acts as a retarder.

BIBLIOGRAPHY

"Rubber Chemicals" in *ECT* 1st ed., Vol. 11, pp. 870–892, by R. A. Mathes and A. E. Juve, The B. F. Goodrich Research Center; "Rubber Chemicals" in *ECT* 2nd ed., Vol. 17, pp. 509–542, by Forrest W. Shaver, The B. F. Goodrich Company.

1. U.S. Pat. 3,633 (June 15, 1844), C. Goodyear.
2. G. Oenslager, *Chem. Ind. (London)*, 91 (Feb. 1933).
3. *Synthetic Organic Chemicals, United States Trade Commission Publications 766 and 1099*, U.S. Government Printing Office, Washington, D.C., 1974 and 1980.
4. L. Bateman, C. G. Moore, M. Porter, and B. Saville, *Chemistry and Physics of Rubber-like Processes*, John Wiley & Sons Inc., New York, 1963.
5. C. R. Parks, D. K. Parker, and D. A. Chapman, *Rubber Chem. Technol.* **45**, 467 (1972).
6. W. Scheele, *Rubber Chem. Technol.* **34**, 1306 (1961).
7. R. S. Kapur, J. L. Koenig, and V. R. Shelton, *Rubber Chem. Technol.* **47**, 911 (1974).
8. E. Morita, *Rubber Chem. Technol.* **53**, 393 (1980).
9. L. B, Sebrell and C. W. Bedford, *Ind. Eng. Chem.* **13**, 1034 (1921); **15**, 1009 (1923).
10. U.S. Pat. 1,631,871 (June 7, 1927), W. J. Kelly.
11. E. L. Carr, G. E. P. Smith, Jr., and G. Alliger, *J. Org. Chem.* **14**, 921 (1949).
12. J. J. D'Amico, M. W. Harmon, and R. H. Cooper, *J. Am. Chem. Soc.* **79**, 5270 (1957).
13. J. J. D'Amico, *J. Org. Chem.* **26**, 3426 (1961).
14. J. J. D'Amico, E. Morita, and E. J. Young, *Rubber Chem. Technol.* **41**, 704 (1968).
15. U.S. Pat. 2,822,367 (Feb. 4, 1958), G. Alliger (to Firestone Tire and Rubber Company).
16. S. Torii, H. Tonaka, and M. Ukida, *J. Org. Chem.* **43**, 3223 (1978).
17. U.S. Pat. 3,133,966 (May 19, 1964), W. R. Trutna (to E. I. du Pont de Nemours & Co., Inc.).
18. G. E. P. Smith, G. Alliger, E. L. Carr, and K. C. Young, *J. Org. Chem.* **14**, 935 (1949).
19. U.S. Pat. 2,285,813 (June 9, 1942), R. S. Hanslick (to United States Rubber Company).
20. U.S. Pat. 3,417,086 (Dec. 17, 1968), G. E. P. Smith and W. S. Cook (to Firestone Tire and Rubber Company).
21. J. J. D'Amico and E. Morita, *Rubber Chem. Technol.* **44**, 881 (1971).
22. U.S. Pat. 3,892,741 (July 1, 1975), R. D. Taylor (to B. F. Goodrich Company).
23. U.S. Pat. 3,985,743 (Oct. 12, 1976), R. D. Taylor (to B. F. Goodrich Company).
24. R. D. Taylor, *Rubber Chem. Technol.* **47**, 906 (1974).
25. M. M. Sprung, *Chem. Rev.* **26**, 297 (1940).
26. M. C. Throdahl and M. W. Harmon, *Ind. Eng. Chem.* **43**, 421 (1951).
27. M. C. Kirkham, *Prog. Rubber Technol.* **41**, 61 (1978).
28. J. I. Cunneen, E. H. Former, and H. P. Koch, *J. Chem. Soc.* 472 (1943).
29. A. I. Vasilev, Y. K. Dzhagarova, and R. G. Dobreyeva, *Polym. Sci. USSR* **17**(10), 2557 (1975).
30. S. Van der Meer, *Rubber Chem. Technol.* **18**, 853 (1945).
31. *Tripartate Meeting on Nitrosamines in the Rubber Industry*, National Institute of Occupational Safety and Health, May 14, 1980, Washington, D.C.
32. J. J. Luecken and H. L. Paris, *Rubber World*, 47 (Nov. 1977).
33. C. D. Trivette, Jr., E. Morita, and O. W. Maender, *Rubber Chem. Technol.* **50**, 570 (1977).
34. L. Fishbein, *Sci. Total Environ.* **13**, 157 (1979).
35. J. M. Fajen, G. A. Carson, D. P. Rounbehler, T. Y. Fan, R. Vita, U. E. Goff, M. H. Wolf, G. S. Edwards, D. H. Fine, V. Reinhold, and K. Biemann, *Science* **205**, 1262 (1979).

36. D. Craig, *Rubber Chem. Technol.* **30,** 1291 (1957).
37. B. A. Dogadkin, A. V. Dobromyslova, and O. N. Belyatskaya, *Rubber Chem. Technol.* **35,** 501 (1962).
38. A. E. Grinberg, V. F. Chertkova, V. Z. Smolyanitski, A. R. Makeeva, and N. P. Runyantseva, *Sov. Rubber Technol.* **18**(1), 21 (1959).
39. A. A. Kashina, B. A. Dogadkin, and A. V. Dobromyslova, *Polym. Sci. USSR* **14**(2), 309 (1972).
40. P. N. Son, *Rubber Chem. Technol.* **49,** 118 (1976).
41. Tyre Research Institute in Moscow, *Eur. Rubber J.* **157,** 41 (1975).
42. A. Dibbo, D. G. Lloyd, and J. Payne, *Rubber Chem. Technol.* **36,** 911 (1963).
43. D. G. H. Ballard, J. Myatt, and J. F. P. Richter, *J. App. Polym. Sci.* **16,** 2647 (1972).
44. USSR Pat. 164,670 (Aug. 19, 1964), I. I. Eitingon and co-workers.
45. U.S. Pat. 3,427,319 (Feb. 11, 1969), U.S. Pat. 3,546,185 (Dec. 8, 1970), U.S. Pat. 3,752,824 (Aug. 14, 1973), and U.S. Pat. 3,855,262 (Dec. 17, 1974), A. Y. Coran and E. Kerwood (to Monsanto Company).
46. S. Torii, H. Tanaka, and M. Ukida, *J. Org. Chem.* **44,** 1554 (1979).
47. U.S. Pat. 4,085,093 (Apr. 18, 1978), R. J. Hopper (to Goodyear Tire & Rubber Company).
48. R. J. Hopper, *Rubber Chem. Technol.* **53,** 1106 (1980).
49. U.S. Pat. 3,932,403 (Jan. 13, 1976), S. Ashton, V. J. Sharma, and J. A. Taylor (to ICI Ltd.).
50. U.S. Pat. 3,780,001 (Dec. 18, 1973) and U.S. Pat. 3,855,261 (Dec. 17, 1974), P. N. Son (to B. F. Goodrich Company).
51. S. J. Stewart, R. I. Lieb, and J. E. Kerwood, *Rubber Age*, 56 (Oct. 1970).
52. H. Roebuck, *Eur. Rubber J.* **155,** 10 (August 1973).
53. E. R. Rodger and H. Roebuck, *J. Elast. Plast.* **8,** 381 (1976).
54. R. I. Leib, A. B. Sullivan, and C. D. Trivette, Jr., *Rubber Chem. Technol.* **43,** 1188 (1970).
55. P. N. Son, K. E. Andrews, and A. T. Schooley, *Rubber Chem. Technol.* **45,** 1513 (1972).
56. P. N. Son, *Rubber Chem. Technol.* **46,** 999 (1973).
57. J. J. D'Amico and co-workers, *Rubber Chem. Technol.* **46,** 1299 (1973).
58. R. Anand, D. C. Blackley, and K. S. Lee, *Rubbercon 77*, The International Rubber Conference (May 16–20, 1977), Brighton, England, Vol. II, The Plastics and Rubber Institute, 1977, p. 22/1.

RAY TAYLOR
P. N. SON
BF Goodrich Company

RUBBER COMPOUNDING

Rubber is defined in ASTM D 1566 as a material that is capable of recovering from large deformations quickly and forcibly, and that can be or already is modified to a state in which it is essentially insoluble but can swell in boiling solvent, eg, benzene, methyl ethyl ketone, and ethanol–toluene azeotrope. A rubber in its modified state and free of diluents retracts within one minute to less than 1.5 times its original length after being stretched at 18–29°C to twice its length and held for one minute before release.

With a few exceptions, raw rubber in the dry state has few commercial applications. For the great majority of uses, the rubber must be modified, usually by the addition of vulcanizing agents and other materials followed by vulcanization (see Rubber chemicals). The exceptions include such uses as crepe-rubber shoe soles; cements, eg, in rubber adhesives; and adhesives and masking tape (see Elastomers, synthetic; Rubber, natural). The various grades of carbon black and the abbreviations used in the rubber industry to designate them are shown in Table 1 (see Carbon, carbon black).

Synthetic Rubbers

Polybutadiene rubbers (BRs) are solution-polymerized polymers. There are three types of polymers in use: very high cis-1,4-polybutadiene (97% cis-1,4) which is made with cobalt catalyst, high cis-1,4-polybutadiene (92–93% cis-1,4) which is made with titanium halide-based catalyst, and a polybutadiene prepared with an alkyllithium catalyst with a 40% cis-1,4 content. These polymers are generally used in blends with styrene–butadiene (SBR) and natural rubbers for their resistance to abrasion and flex cracking.

Polyisoprene rubbers (IRs) are solution-polymerized polymers with ca 92% cis-1,4-polyisoprene. They are used as replacements for natural rubber except where tack-imparting properties are important.

The 1000 series of styrene–butadiene rubber (SBR) or hot SBR is an emulsion

Table 1. Carbon Blacks

Grade	ASTM D 1765 classification	Older industry classification
High reinforcing furnace blacks		
super abrasion furnace	N 110	SAF
super abrasion furnace, high structure	N 121	SAF-HS
intermediate super abrasion furnace, low structure	N 219	ISAF-LS
intermediate super abrasion furnace	N 220	ISAF
intermediate super abrasion furnace, low modulus	N 231	ISAF-LM
intermediate super abrasion furnace	N 234	
intermediate super abrasion furnace, high structure	N 242	ISAF-HS
high abrasion furnace, low structure (channel replacement furnace)	N 326	CRF
high abrasion furnace, low structure	N 327	HAF-LS
high abrasion furnace	N 330	HAF
high abrasion furnace	N 332	
high abrasion furnace, high structure	N 347	HAF-HS
high abrasion furnace	N 339	
super processing, furnace	N 358	SPF
high abrasion furnace	N 351	
high abrasion furnace	N 375	
Medium reinforcing furnace blacks		
fine extrusion furnace, low structure	N 539	FEF-LS
fine extrusion furnace	N 550	FEF
fine extrusion furnace, high structure	N 568	FEF-HS
high modulus furnace	N 601	HMF
general purpose furnace, high structure	N 650	GPF-HS
general-purpose furnace	N 660	GPF
all-purpose furnace	N 683	APF
semireinforcing furnace, low structure	N 754	SRF-LS
semireinforcing furnace	N 762	SRF-LM
semireinforcing furnace, high structure	N 765	SRF-HS
semireinforcing furnace	N 770	SRF
semireinforcing furnace	N 774	SRF-NS
multipurpose furnace	N 785	MPF
semireinforcing furnace	N 787	SRF-HM
Thermal blacks		
fine thermal	N 880	FT
medium thermal	N 990	MT
Conductive blacks		
conductive furnace	N 293	CF
superconductive furnace	N 294	SCF
extra-conductive furnace	N 472	ECF
Channel blacks		
easy-processing channel	S 300	EPC
medium processing channel	S 301	MPC

copolymer of butadiene and styrene. The usual limits are 71–77 parts butadiene and 29–23 parts styrene. The combined styrene in the final polymer is less than that charged and varies with the charging ratio from 23.5 to 20%. The polymerization is conducted at 50°C with sodium tallow soap or rosin soap as the emulsifier. This rubber and its variations are general-purpose rubbers. Cold SBR is an emulsion copolymer of butadiene and styrene prepared at a low temperature, usually 5°C. The nonpigmented polymers are designated as the SBR 1500 series. The numbering system for other SBR polymers and latices are as follows:

1600 series	cold black masterbatch with 14 or less parts of oil per 100 parts SBR
1700 series	cold oil masterbatch
1800 series	cold oil, black masterbatch with more than 14 parts of oil per 100 parts SBR
1900 series	miscellaneous dry-polymer masterbatches, generally emulsion, high styrene, resin-rubber masterbatches
2000 series	hot latices
2100 series	cold latices

Nitrile rubbers (NBRs) are emulsion copolymers of butadiene and acrylonitrile with the proportions varying from 55 wt % butadiene–45 wt % acrylonitrile to 82 wt % butadiene–18 wt % acrylonitrile. These are used primarily for their resistance to swelling in oils and solvents.

Butyl rubber (IIR) is a copolymer of isobutylene (97–98 wt %) and isoprene (2–3 wt %) prepared at ca −100°C with a Friedel-Crafts catalyst. The principal use of this rubber is for inner tubes, sealants (qv), and extrusions.

Ethylene–propylene terpolymer (EPDM) is a low gravity rubber polymer based on ethylene–propylene and a controlled amount of a nonconjugated diene. The unsaturation in the polymer is pendant to the saturated polymer chain and permits sulfur vulcanization. The EPDM vulcanizates have exceptional low temperature properties and are resistant to oxygen, ozone, and heat degradation.

Silicone rubbers, eg, types MQ, VMQ, PMQ, and PVMQ, are siloxane polymers composed of a central chain of alternating silicon and oxygen atoms with alkyl groups attached to the silicon atoms. They are useful in an extremely wide temperature range.

Types GN, GNA, GW, FB, and GRT neoprenes (CRs) are emulsion copolymers of 2-chloro-1,3-butadiene and sulfur (1). These polychloroprenes from DuPont are peptizable. The GRT type also contains 2,3-dichloro-1,3-butadiene as a monomer. Types W-M1, W, WHV-100, WB, TW, and TW-100 are emulsion polymers of 2-chloro-1,3-butadiene and are not peptizable. Types WRT, WD, WK, and TRT are emulsion copolymers of 2,3-dichloro-1,3-butadiene and 2-chloro-1,3-butadiene and are not peptizable. These types have very slow crystallization rates. All these types are general-purpose neoprenes. The adhesive neoprenes are AC, AD, CG, and AF. Similar polychloroprene types are available as Baypren (Bayer), Butaclor (Distugil), Denka (Denki-Kagaku, Denka Chemical USA), and Skyprene (Toyo Soda). These rubbers are used primarily for oil, solvent flexing and weathering resistance.

Polysulfides, eg, types OT and EOT, are prepared by treating sodium polysulfide with dihalogenated materials. Thiokol FA is based on ethylene dichloride and dichloroethyl formal and Thiokol ST is based on trichloropropane and dichloroethyl formal. Thiokol LPs are organic polysulfide liquid polymers. These rubbers are used for their oil and solvent resistance.

Polyacrylate rubbers, eg, type ACM, are specialty elastomers prepared by emulsion polymerization. They are copolymers of alkyl acrylic acid esters and other monomers which effect cross-linking. The commonly used esters are ethyl and butyl acrylate and the cross-linking comonomer may be acrylonitrile or a chlorinated vinyl derivative. They are used in products requiring high resistance to elevated temperatures and to the solvent or swelling effects of oil.

The amorphous polyepichlorohydrin homopolymer (CO) and the 1:1 copolymer of epichlorohydrin and ethylene oxides (ECO) are the most promising epichlorohydrin rubbers. These polymers provide good high temperature aging resistance, low temperature flexibility, and ozone resistance.

Fluoroelastomers (FDM) are rubbers containing fluorine, hydrogen, and carbon. They are extremely resistant to chemicals and other deteriorating fluids and to extremes of temperature.

Hypalon, a chlorosulfonated polyethylene (CSM) manufactured by DuPont, is prepared by chlorosulfonation of polyethylene. Types of Hypalon contain 27–45 wt % chlorine and 1.0–1.4 wt % sulfur. These rubbers have extremely good color stability and resistance to weathering and ozone.

Halogenated butyl, eg, chlorobutyl (CIIR) and bromobutyl (BIIR), rubbers are prepared by the reaction of small amounts of chlorine or bromine with butyl rubber. The addition of these halogen materials greatly improves cure versatility and compatibility with high unsaturated rubbers. These rubbers retain the low gas permeability and good ozone resistance of regular butyl and have improved cure adhesion to itself and to other rubbers. These rubbers are used in tire inner liners, sidewalls, and cover strip compounds.

Chlorinated polyethylene rubbers (CPEs) are produced by the random chlorination of high density polyethylene. These rubbers have excellent ozone, weather, and heat resistance.

Polyurethanes, eg, types AU and EU, are combinations of low molecular weight polyester or polyether polymers that are terminated with hydroxy groups that have reacted with di- or polyfunctional isocyanates and usually with low molecular weight polyfunctional alcohols or amines. Variations in the characteristics of the liquid starting polymer and the concentration, type, and arrangement of the isocyanate and other small molecules used for chain extension produce a wide range of different polyurethane rubbers. Some of the commercial polyurethanes are Arcon (Polycom, Inc., formerly Allied Resin Corporation), Adiprene (DuPont), Castethane (Upjohn Company), Catepol (Arnco), Cyanaprene (American Cyanamid), Fastcast Arpro (Arnco), Q-Thane (K. J. Quinn & Co.), Vibrathane (Uniroyal), and Rucothane (Hooker). These rubbers produce excellent abrasion and solvent resistance.

Thermoplastic rubbers, eg, YSBR, YSIR, YEPM, YAU, YEU, YACM, and YEAM, are intermediate between rubbers and plastics (2). At moderate ambient temperatures they have the properties of vulcanized rubber, whereas at elevated temperatures they are melt-processable like thermoplastics. There are several classes of these materials. The styrene–butadiene block copolymers (YSBR) are represented by Kraton (Shell) and Solprene (Phillips). The styrene–isoprene rubber (YSIR) is Kraton 1107 (Shell). The polyesters are represented by Hytrel (DuPont). The vinyl acetate-ethylene copolymers (YEAM) are represented by Vynathane (U.S. Industrial

Chemicals). The polyolefins (YEPM) are represented by TPR (Uniroyal). These rubbers can be easily processed with standard thermoplastic equipment including injection molding and extrusion; no postcure is needed, which eliminates the need for steam, ovens, or continuous-vulcanization (CV) lines; the scrap can be reprocessed directly. However, they soften when reheated to elevated temperatures, which limits their usefulness.

Compounding

The object of compounding is to select the most suitable combination of materials in their correct proportion and to determine the treatment that the chosen combination shall undergo in the processes of mixing, forming, and vulcanization so that the finished rubber product is of the required quality and is produced at the lowest possible cost. The operations of mixing, forming, and vulcanizing are the essential fabrication steps.

Basic Recipes. The terms pure gum or gum compound are used for a rubber admixture containing only those ingredients necessary for vulcanization, processing, coloring, and resistance to aging. The simplest of these is natural rubber, 100 parts by wt; sulfur, 8–10 parts by wt.

The rate of vulcanization of rubber and sulfur with heat alone is extremely slow and is affected by the type and quality of the rubber used as well as the temperature at which the curing process or vulcanization takes place. Thus, this simple gum compound, when heated at 141°C for one hour, develops a tensile strength of only 2.8 MPa (410 psi). It may take 3–4 h to toughen the compound to the point where it is a very elastic vulcanizate. After 4 h at 141°C, this compound produces a tensile strength of ca 20.7 MPa (ca 3000 psi) and an ultimate elongation of 900–1000%. Compounds such as this have been used for many years and are still used for products such as golf-ball thread rubber. The first accelerators were inorganic chemicals. The use of basic lead carbonate was largely replaced by oxides of lead, ie, litharge, hydrated lime, and magnesium oxide. These inorganic materials increase the vulcanizing rate and produce a slight improvement in tensile strength and a reduction in breaking elongation.

When it is desired to produce mixtures with lower sulfur ratios and shorter curing times, organic accelerators combined with metal oxides are used. The lowered sulfur ratio is desirable because: it permits the production of nonblooming compounds, ie, the excess, uncombined sulfur left following the vulcanization is insufficient to cause crystallization or bloom to occur at the surface of the article; lower sulfur ratios confer better aging properties; and better physical properties in general are obtained.

A typical composition, A, might be as follows: natural rubber, 100 parts by wt; sulfur, 3 parts by wt; zinc oxide, 5 parts by wt; accelerator (an aldehyde–amine), 0.3 parts by wt. This mixture, when vulcanized for various intervals at 138°C, exhibits the properties listed in Table 2. Note the tendency to reversion at the 90-min cure.

Most organic accelerators require the presence of fatty acids for their most efficient functioning. In the above instance, the naturally occurring fatty acids in the natural rubber are sufficient for this purpose. As an illustration of the essential part played by the fatty acids, mixture A is made in the same way except that the rubber is extracted with acetone to remove the fatty acids. This resultant mixture B on vulcanization gives the properties listed in Table 2.

The thiazoles, which are the most widely used class of accelerators, require more

Table 2. Properties of Mixtures A and B

Time of vulcanization, min	500% modulus, MPa[a]	Tensile strength, MPa[a]	Elongation, %
Mixture A			
10	1.3	19.7	915
20	2.9	29.5	805
30	3.5	31.5	765
40	3.6	30.4	775
60	2.8	31.3	835
90	2.4	19.5	855
Mixture B			
30	0.7	6.6	925
45	0.8	8.4	940

[a] To convert MPa to psi, multiply by 145.

fatty acid for their maximum effectiveness than is normally present in crude natural rubber. Table 3 illustrates the role of fatty acid by comparing stocks C, D, and E based on, respectively, acetone-extracted crude rubber which is essentially free of fatty acids, normal crude containing the naturally occurring fatty acids, and normal crude with two parts of additional fatty acid as stearic acid. The recipe for mixture C, in parts by wt, is extracted pale crepe, 100.0; zinc oxide, 5.0; sulfur, 3.0; mercaptobenzothiazole (MBT), 0.6; stearic acid, 0. The recipe for mixture D is pale crepe, 100.0; zinc oxide, 5.0; sulfur, 3.0; MBT, 0.6; stearic acid, 0. The recipe for mixture E is pale crepe, 100.0; zinc oxide, 5.0; sulfur, 3.0; MBT, 0.6; stearic acid, 2.0.

Because the proportion of naturally occurring fatty acids varies from one grade of rubber to another, it is common to add additional fatty acid to the composition when thiazole accelerators are used. Oleic and lauric acids can be used in place of stearic acid. It is believed that the fatty acid reacts with the metallic oxide during vulcanization to form a rubber-soluble salt or soap, which reacts with the accelerator enabling it to exert its full effect.

The other essential ingredient in this base recipe is the metallic oxide. Although the oxides of calcium, magnesium, and lead can be and occasionally are used, zinc oxide is by far the most widely used material. The effect of varying the zinc oxide concentration in the particular composition of mixture F is illustrated in Table 4. The recipe

Table 3. Properties of Mixtures C, D, and E

Cure time (at 138°C), min	600% modulus, MPa[a]	Tensile strength, MPa[a]	Elongation, %
Mixture C			
40	2.1	4.2	695
80	2.7	3.3	640
Mixture D			
40	2.5	11.5	840
80	2.4	13.4	885
Mixture E			
40	4.1	19.8	790
80	5.2	21.0	810

[a] To convert MPa to psi, multiply by 145.

Table 4. Properties of Mixture F[a] with Varying Zinc Oxide Concentrations

Amount of zinc oxide, parts by wt	600% modulus, MPa[b]	Tensile strength, MPa[b]
1.0	2.1	16.1
2.0	5.3	20.1
2.5	5.7	22.1
3.0	6.2	24.1
5.0	6.1	24.2
10.0	6.3	24.0

[a] Cured 40 min at 138°C.
[b] To convert MPa to psi, multiply by 145.

for mixture F, in parts by wt, is pale crepe, 100.0; stearic acid, 2.0; sulfur, 3.0; MBT, 0.6; zinc oxide, varied.

In the case of mixture F, a concentration of 3 pph rubber provides maximum effect. The role of zinc oxide is not the same for all sulfur vulcanizing systems, and the above data apply in general only to the thiazole acceleration system. Some organic accelerators function quite well in the absence of a metal oxide, eg, the aldehyde amines and the guanidines. The omission of zinc oxide from stocks containing these two types of accelerators, however, causes a decrease in their heat-aging resistance.

Natural-rubber vulcanizates deteriorate during aging. In addition to reducing the sulfur ratio, the use of certain organic accelerators greatly improves the aging properties of the vulcanizates in which they are used. In Table 5, a comparison is made of the aging properties of compositions G and H, which are identical except that G is accelerated with MBT, a good aging accelerator, whereas H is accelerated with an aldehyde–amine, a poor aging accelerator.

In addition to the advantages obtained by the use of low sulfur ratios and good aging accelerators, antioxidants further improve resistance to deterioration during aging. The effect produced by the addition of one part of dioctylated diphenylamine (Octamine, Stalite-S) to compositions G and H is shown in Table 6.

Table 5. Aging Properties of Compositions G[a] and H[b]

Cure time (at 138°C), min	400% modulus, MPa[c]		Tensile strength, MPa[c]		Elongation, %	
	G	H	G	H	G	H
Original properties						
40	6.0	8.5	21.7	26.6	667	633
80	4.0	8.1	16.2	26.2	707	620
After aging 24 h at 100°C in test tubes						
40	6.2	4.7	15.0	7.6	613	473
80	3.8	6.0	10.7	6.7	627	440
After aging 24 h in oxygen bomb at 70°C and 2 MPa[c]						
40	7.6	8.3	20.9	13.8	613	487
80	5.2	[d]	12.4	[d]	593	[d]

[a] MBT acceleration.
[b] Aldehyde–amine acceleration.
[c] To convert MPa to psi, multiply by 145.
[d] Too brittle to test.

Table 6.   Aging Properties of G[a] and H[b] Compounded with Dioctylated Diphenylamine

Cure time (at 138°C), min	400% modulus, MPa[c]		Tensile strength, MPa[c]		Elongation, %	
	G	H	G	H	G	H
Original properties						
40	4.5	9.1	19.3	29.3	720	640
80	4.1	9.7	20.9	21.0	707	647
After aging 24 h at 100°C in test tubes						
40	5.7	6.0	17.6	13.8	633	507
80	6.0	5.7	14.1	8.3	580	480
After aging 24 h in oxygen bomb at 70°C and 2 MPa[c]						
40	6.0	8.6	14.3[d]	24.5	573	613
80	5.0	6.9	13.5	6.9	613	400

[a] MBT acceleration.
[b] Aldehyde–amine acceleration.
[c] To convert MPa to psi, multiply by 145.
[d] Short, premature break.

A typical accelerated gum compound in which the various features discussed above are incorporated and which is suitable for use in certain products is composed, in parts by wt, as follows: natural rubber, 100.0; sulfur, 3.0; zinc oxide, 5.0; bis(2,2′-benzothiazolyl) disulfide (MBTS), 1.0; stearic acid, 1.0; BLE-25 (a high temperature reaction product of diphenylamine and acetone produced by Uniroyal, Inc.), 1.0. This compound develops the following stress–strain properties when vulcanized for 45 min at 141°C:

modulus (at 500%), MPa (psi)	2.5 (360)
modulus (at 700%), MPa (psi)	10.8 (1570)
tensile strength, MPa (psi)	24.7 (3580)
ultimate elongation, %	840

This basic compound can be the source of innumerable products simply by modifying the amount and kind of the ingredients or by adding reinforcing agents, fillers, colors, softeners, and extenders or other materials. The following modifications can be made without changing the general characteristics of a pure gum compound:

Use of different grades of natural rubber, eg, clean pale crepe when a light amber color is required; a less expensive grade where color is not important or some dirt can be tolerated; or a well-masticated grade, that is, one having a high plasticity, to permit easier processing.

Variation of the sulfur ratio. Higher ratios for higher modulus and stiffer vulcanizates and lower ratios for better aging or lower modulus. Such modifications are usually accompanied by suitable adjustments in the accelerator ratio.

Accelerator variation. The kind and amount of accelerator might be varied to attain a faster or slower rate of vulcanization; to obtain a stock with less odor, taste, or lighter color; or to obtain high or low modulus or other effects.

The fatty acid might be omitted if the accelerator used does not require more than what is normally present in natural rubber.

The zinc oxide might be omitted if an accelerator were used which did not require it, or its concentration might be reduced to about 1 part if transparency is desired, or another metal oxide might be substituted for it.

The kind and amount of antioxidant might be varied depending on the color, staining, or aging requirements of the service.

The elemental sulfur might be omitted and a material, eg, tetramethylthiuram disulfide (TMTD), substituted which during vulcanization would supply sufficient sulfur to effect vulcanization.

Auxiliary materials might be added, eg, retarders for increasing processing safety, lubricants to facilitate extrusion or calendering, or tackifying resins to improve tack.

Other modifications of this basic recipe that involve the addition of reinforcing agents, fillers, softeners, or other modifying agents to produce specific properties need not involve any changes in the proportions of the various ingredients unless the added materials have an effect on the curing rate, in which case the proportion of curing agents may be adjusted.

Similar gum compounds of SBR have very poor physical properties. Whereas natural-rubber gum compounds have tensile strengths of 20.7–34.5 MPa (3000–5000 psi), SBR gum compounds have tensile strengths of ca 2.1 MPa (300 psi); therefore, they are of no commercial value. The butadiene–acrylonitrile copolymers, *cis*-polybutadiene, and EPDM polymers likewise exhibit very poor physical properties in gum formulations. Polyisoprene and neoprene, on the other hand, develop excellent gum properties. Butyl rubber also develops fairly good properties in gum compounds, although with slight overcures the good properties disappear. Typical gum recipes for several of these polymers and the stress–strain properties at the best cures are listed in Table 7.

Table 7. Stress–Strain Properties of Selected Polymers

Property	Value
Neoprene[a], cure for 40 min at 139°C	
300% modulus, MPa[b]	2.5
700% modulus, MPa[b]	12.4
tensile strength, MPa[b]	28.3
ultimate elongation, %	800
Butyl rubber[c], cure for 30 min at 150°C	
300% modulus, MPa[b]	0.3
700% modulus, MPa[b]	3.2
tensile strength, MPa[b]	17.7
ultimate elongation, %	900
Polyisoprene[d], cure for 10 min at 145°C	
300% modulus, MPa[b]	1.4
tensile strength, MPa[b]	25.3
ultimate elongation, %	930

[a] Neoprene recipe, in parts by wt: neoprene GN 100.0; magnesium oxide, 4.0; zinc oxide, 5.0; stearic acid, 0.5.

[b] To convert MPa to psi, multiply by 145.

[c] Butyl rubber recipe, in parts by wt: butyl, 100.0; zinc oxide, 5.0; stearic acid, 3.0; TMTD, 1.0; sulfur, 2.0.

[d] Polyisoprene recipe, in parts by wt: polyisoprene, 100.0; zinc oxide, 3.0; stearic acid, 3.0; nondiscoloring antioxidant, 1.0; Trimene Base (reaction product of ethyl chloride, formaldehyde, and ammonia, Uniroyal), 0.3; zinc salt of 2-mercaptobenzothiazole (ZMBT) (OXAF, Uniroyal, Inc.; Zetax, R. T. Vanderbilt), 0.7; sulfur, 1.5.

Reinforcing Agents and Fillers

Dry pigments other than those added to rubber as vulcanizing agents or vulcanizing aids are loosely classified either as reinforcing agents or fillers (qv). The former improve the properties of the vulcanizates and the latter serve primarily as diluents. There is no general agreement among rubber technologists as to what reinforcement is, how it can be measured, or precisely where the dividing line is which distinguishes the reinforcing materials from the fillers. There is general agreement that all grades of carbon black with the exception of N 880 (FT, fine thermal) and N 990 (MT, medium thermal) are reinforcing agents. The degree of reinforcement increases with a decrease in particle size. There is more disagreement with respect to the nonblack fillers, but it is generally conceded that some reinforcement results from the use of fine-particle zinc oxides; fine, precipitated calcium carbonates; calcium silicates; amorphous, hydrated silicon dioxide; pure silicon dioxides; fine clays; and fine-particle magnesium carbonate. There also is general agreement that none of these nonblack materials are as good for reinforcement as the finer furnace or channel carbon blacks (see Calcium compounds, calcium carbonate; Carbon, carbon black; Clays).

Reinforcement has been defined as the incorporation into rubber of small-particle substances which give to the vulcanizate high abrasion resistance, high tear and tensile strength, and some increase in stiffness (3). In general, when a reinforcing agent is added to a base pure gum recipe, such as those previously described for natural or synthetic rubbers, that agent imparts greater stiffness and higher ultimate tensile strength than would be obtained by using an equal volume of a recognized filler, eg, coarse particle-size whiting. The usual quantities of these materials are ca 10–50 parts per 100 parts by volume of rubber.

The most important characteristic required of a reinforcement agent is small particle size. Data on the particle size and surface area of some of the more common reinforcing and nonreinforcing agents are listed in Table 8. The reason for the use of such a wide variety of materials is that properties of the unvulcanized mix and those of the vulcanizate can be varied over a wide range to fit the processing requirements, the service, and the economic demands of the product. In general, the finer reinforcing agents require more energy for their dispersion into the rubber and the plasticity of the resultant mix is lower. Therefore, it is more difficult to process in operations following mixing. This effect of reinforcing agents on the properties of the rubber-filler mixture is of great practical importance and, for the manufacture of some products, it may be a more important factor in the selection of reinforcing agents and fillers than their effect on the vulcanizate properties.

The effect of varying surface area of carbon black on the Mooney viscosity of several rubber compounds is illustrated in Table 9. In this study, particle size had a lesser effect than carbon black structure. Mooney viscosity is mainly dependent on carbon black structure and loading in most polymers. The Mooney viscosity rises rapidly with increasing loadings for all compounds except those containing thermal blacks, for which loading does not seem to have much effect.

Associated with this toughening is the decrease in solubility of the compound in solvents. The time interval following the incorporation of the filler influences both effects. On standing, the compound becomes less plastic and less soluble. Also associated with these phenomena is the tendency towards drying of the uncured stock by the finer particle-size fillers and rendering the stock less tacky so that fabrication operations following calendering or tubing are more difficult.

Table 8.   Particle Size of Fillers

Grade	Surface area, m^2/g	Average particle diameter, nm
Carbon black		
CC	250–125	10–20
S-301 MPC	100–90	25–30
S-300 EPC	90–80	30–33
N-440 FF	70	36
N-601 HMF	60–50	50–60
N-770 SRF	42–30	70–90
N-880 FT	20–15	150–200
N-990 MT	10–5	250–500
acetylene	64	43
Whiting		
Witco AA	0.55	3900
micronized	1.4	1500
Witcarb R-12	13	145
Witcarb R	32	50
Purecal V		40
Purecal M		1500
Atomite		1500
Calcene TM	15	100
Clay		
Catalpo		800
Dixie		1000
Silica		
Hi-Sil	110	25
Calcium silicate		
Silene EF	80	30

Another effect of fillers on the raw or uncured stock is to reduce the tendency of the compound to recover after being forced between mill or calender rolls or through an extruder die. This tendency is commonly called nerve and results from a relatively high elastic component in the compound. The use of increasing loadings of any dry filler suppresses nerve tendency. With increasing loadings, the coarser fillers reduce the nerve faster than finer fillers. Structure denotes the average aggregation of a filler. A filler with a small degree of aggregation is considered to have low structure, whereas a filler with a large degree of aggregation is considered to have high structure. The use of high structure fillers, and particularly if the structure remains during milling operations into the polymer, reduces the nerve of the compound to a much greater degree than the use of filler having a particulate structure. Thus with carbon blacks, extrusion shrinkage is a function of structure and amount of loading. Smaller particle-size black concomitant with higher structure produces compounds with the lowest extrusion shrinkage or die swell. Conversely, compounds containing thermal blacks with their large particle size and low structure have high shrinkage and die swell.

Another property of uncured rubber that is influenced by the type and amount of filler loading in the compound is scorch or incipient vulcanization during processing. A study of the effect of carbon black on scorch has been made (5). Carbon black affects the curing agents in the various polymers. In some cases the mechanism appears to be catalytic with the carbon black promoting the decomposition of the accelerators

Table 9.   Effect of Varying Surface Area and Structure Index on Mooney Viscosity[a]

Property	Grade of carbon black				
	N 990 (MT)	N 660 (GPF)	N 683 (APF)	N 347 (HAF-HS)	N 326 (HAF-LS)
surface area, m²/g	6	28	28.5	65.7	78
particle size index[b]	250	50	55	27	27
structure index[c]	65	54	45	42	59
Mooney viscosity (ML-4' at 100°C)					
50 parts of carbon black/100 parts rubber					
SBR	36	44	49	58	50
natural rubber	51	73	77	93	76
neoprene	45	78	99	126	90
nitrile	62	81	88	105	74
butyl	72	97	108	119	103
70 parts of carbon black/100 parts rubber					
SBR	40	61	69	80	73
natural rubber	52	97	113	137	98

[a] Ref. 4.
[b] Measured from electron photomicrographs.
[c] A low structure index value indicates a high structure black.

or intermediates present in the curing system. The surface properties probably affect the environment in which the reactions leading to cross-linking take place. There are a number of possible ways in which the surface can do this, eg, by providing acidity or basicity, influencing oxidation–reduction reactions, removing active species, providing moisture, etc. The pH values for several carbon blacks are given below:

Black	*pH*
S 300 (EPC)	4.5
N 990 (MT)	8.5
N 330 (HAF)	9.0
N 770 (SRF)	9.3
N 220 (ISAF)	9.3

The acidic channel blacks, which are no longer used in the United States, tend to retard the curing rate, whereas the alkaline furnace blacks tend to increase the curing rate. Scorch is greatly affected by the particle size of the carbon blacks. The largest particle-size blacks provide compounds with the greatest processing safety whereas the higher structure, small particle-size blacks produce compounds with shorter scorch times. Table 10 shows the effect of various carbon blacks on the Mooney scorch of a natural rubber stock.

In general, the effects produced in the vulcanizates by the addition of fillers to the various rubbers are similar. Tables 11–17 are comparisons of effects of the various types of carbon blacks in natural rubber, SBR, polybutadiene, Neoprene W, nitrile, butyl, and EPDM rubber compounds (6). Tables 18 and 19 are comparisons of the effects of various nonblack fillers in natural rubber and SBR (7).

Tensile Strength.   Generally the finer the particle size of a carbon black, the higher the tensile strength is. Structure of the carbon black also affects tensile strength. The role of structure reverses itself depending on whether the polymer is crystalline or amorphous. In natural rubber, butyl, and neoprene compounds, which are crystalline

Table 10. Mooney Scorch Times for Carbon Blacks in Natural Rubber[a,b], min

Grade of carbon black	Loadings		
	35 phr[c]	50 phr[c]	75 phr[c]
N 110 (SAF)	25	18	9
N 330 (HAF)	27	22	11
N 550 (FEF)	32	24	12
N 770 (SRF)	42	34	29
N 990 (MT)	>45	>45	>30
S 300 (EPC)	41	35	26

[a] Recipe, in parts by wt: natural rubber, 100; zinc oxide, 5; antioxidant, 1; stearic acid, 3; softener, 5; *N*-cyclohexyl-2-benzothiazolesulfenamide (CBS), accelerator, 0.4 (0.7 with S 300); sulfur, 2.5; carbon black, as indicated.

[b] Mooney scorch at 121°C.

[c] Phr = parts per hundred rubber.

Table 11. Effects of Carbon Blacks in Natural Rubber[a,b]

Carbon black	300% modulus, MPa[c]	Tensile strength, MPa[c]	Elongation at break, %	Hardness, Shore A	Rebound, %	Relative road wear
N 293 (CF)	12.6	28.3	550	63	63.8	
N 110 (SAF)	13.6	30.9	530	66	62.2	
N 220 (ISAF)	13.5	27.9	530	66	65.7	120
N 242 (ISAF-HS)	14.8	27.4	490	67	64.8	115
N 330 (HAF)	14.8	27.4	470	65	71.7	117
N 347 (HAF-HS)	16.2	26.6	460	67	70.4	100
S 300 (EPC)	12.4	28.6	550	63	67.5	105
N 440 (FF)	10.5	29.7	580	57	77.3	90
N 550 (FEF)	15.5	25.5	480	65	75.8	
N 660 (GPF)	12.9	25.0	500	61	79.7	
N 770 (SRF)	11.0	25.9	570	57	80.9	
N 880 (FT)	3.6	21.7	660	52	81.5	
N 990 (MT)	4.7	20.9	620	49	83.6	

[a] Ref. 6.

[b] Recipe (cured for 30 min at 145°C), in parts by wt: natural rubber, 100.0; carbon black, 50.0; stearic acid, 3.0; zinc oxide, 5.0; MBTS, 0.6; sulfur, 2.5.

[c] To convert MPa to psi, multiply by 145.

polymers, the small particle-size–low structure carbon blacks produce the highest tensile strengths. There is a gradual reduction in tensile strength as the loadings increase. The structure of the medium particle-size carbon blacks does not seem to have a significant effect on tensile strength in these crystalline polymers. With these medium particle-size blacks, the optimum loading is at a very low level and the tensile strength gradually decreases as the loadings increase. In SBR, nitrile, and EPDM rubbers, which are amorphous polymers, the carbon blacks have quite a different effect on tensile properties. Cured compounds containing small particle-size–high structure carbon blacks produce optimum tensile strength at lower loadings, whereas the tensile properties increase in vulcanizates containing low structure reinforcing blacks as loading is increased. Compounds with medium particle-size carbon blacks produce optimum tensile properties at higher loadings in these amorphous polymers. Thermal blacks give the lowest tensile strength.

Table 12.   Effects of Carbon Blacks in SBR[a]

Carbon black	300% modulus, MPa[c]	Tensile strength, MPa[c]	Elongation at break, %	Hardness, Shore A	Rebound, %	Relative die swell
N 293 (CF)	15.3	30.4	530	68	55.0	108
N 110 (SAF)	16.2	32.1	510	69	51.8	107
N 220 (ISAF)	16.0	30.0	520	68	53.5	104
N 242 (ISAF-HS)	17.1	30.0	490	69	54.5	102
N 330 (HAF)	15.7	28.6	500	67	58.0	109
N 347 (HAF-HS)	18.1	28.6	470	69	59.6	100
S 300 (EPC)	11.2	29.7	600	64	57.5	125
N 440 (FF)	10.0	26.2	560	58	61.3	126
N 550 (FEF)	15.5	23.5	530	64	64.8	100
N 660 (GPF)	11.9	22.1	560	62	66.8	110
N 770 (SRF)	9.7	21.2	600	60	67.8	117
N 880 (FT)	1.7	11.0	880	50	65.3	123
N 990 (MT)	2.8	11.0	750	48	67.7	120

[a] Ref. 6.

[b] Recipe (cured for 30 min at 145°C), in parts by wt: SBR 1500, 100.0; carbon black, 50.0; stearic acid, 1.5; zinc oxide, 5.0; MBTS, 2.0; sulfur, 2.0.

[c] To convert MPa to psi, multiply by 145.

Table 13.   Effects of Carbon Blacks in Polybutadiene[a,b]

Carbon black	300% modulus, MPa[c]	Tensile strength, MPa[c]	Elongation at break, %	Hardness, Shore A	Rebound, %
N 293 (CF)	8.1	16.9	500	57	59.6
N 110 (SAF)	8.3	17.3	490	58	58.9
N 220 (ISAF)	8.8	16.2	460	57	61.0
N 242 (ISAF-HS)	9.5	15.5	425	58	60.5
N 330 (HAF)	8.6	15.2	430	57	65.3
N 347 (HAF-HS)	9.8	14.7	380	59	65.1
S 300 (EPC)	6.0	15.5	510	53	63.5
N 440 (FF)	6.0	13.8	460	51	69.7
N 550 (FEF)	11.0	14.7	390	55	69.7
N 660 (GPF)	8.1	12.9	410	51	72.3
N 770 (SRF)	6.7	10.4	390	51	73.2
N 880 (FT)	2.2	5.7	600	42	75.8
N 990 (MT)	2.4	4.8	550	44	77.8

[a] Ref. 6.

[b] Recipe (cured for 45 min at 145°C), in parts by wt: Ameripol CB-880 (BF Goodrich), polybutadiene, 114.00; carbon black, 55.00; zinc oxide, 5.00; Flexamine (Uniroyal, Inc.), antioxidant, 1.00; CBS accelerator, 1.25; sulfur, 1.50.

[c] To convert MPa to psi, multiply by 145.

Among the nonblack fillers, the best tensile strengths are produced by the precipitated silicas, eg, HiSil. Calcium silicate (Silene EF) and chemically altered clay (Zeolex) are next, followed by zinc oxide, ultrafine carbonates, fine carbonates, and hard clay.

Table 14. Effects of Carbon Blacks in Neoprene W[a,b]

Carbon black	200% modulus, MPa[c]	Tensile strength, MPa[c]	Elongation at break, %	Hardness, Shore A	Rebound, %	ASTM #3 oil swell, vol %
N 293 (CF)	14.1	22.4	300	74	55.0	62.2
N 110 (SAF)	15.2	22.8	300	75	53.3	59.9
N 220 (ISAF)	14.5	23.5	300	73	55.1	62.0
N 242 (ISAF-HS)	15.9	22.8	260	75	56.0	59.8
N 330 (HAF)	14.8	22.4	270	71	62.1	57.0
N 347 (HAF-HS)	16.2	21.7	260	74	62.0	57.5
S 300 (EPC)	22.4	23.5	370	69	59.6	69.2
N 440 (FF)	10.2	23.1	370	66	68.2	68.4
N 550 (FEF)	14.1	21.1	280	69	66.9	64.0
N 660 (GPF)	11.7	20.4	320	66	68.6	65.0
N 770 (SRF)	8.3	21.4	370	62	72.5	64.0
N 880 (FT)	2.9	18.8	790	52	73.6	71.4
N 990 (MT)	3.1	17.3	750	54	73.6	73.4

[a] Ref. 6.
[b] Recipe (cured for 30 min at 153°C), in parts by wt: neoprene W, 100.0; carbon black, 50.0; stearic acid, 1.0; process oil, 12.0; antiozonant, 1.0; magnesium oxide, 4.0; NA-22 (ethylenethiourea), 0.5 (0.7 for S 300); zinc oxide, 5.0.
[c] To convert MPa to psi, multiply by 145.

Table 15. Effects of Carbon Blacks in Nitrile Rubber[a,b]

Carbon black	300% modulus, MPa[c]	Tensile strength, MPa[c]	Elongation at break, %	Hardness, Shore A	Rebound, %	ASTM #3 oil swell, vol %
N 293 (CF)	10.4	25.0	500	64	53.1	7.1
N 110 (SAF)	11.0	25.5	520	64	52.6	4.1
N 220 (ISAF)	11.0	25.5	490	64	54.1	4.7
N 242 (ISAF-HS)	12.4	25.2	490	64	54.0	3.7
N 330 (HAF)	11.9	22.8	470	62	57.0	3.6
N 347 (HAF-HS)	12.9	22.1	460	64	56.0	4.2
S 300 (EPC)	7.6	23.1	600	61	55.7	5.4
N 440 (FF)	7.6	21.2	630	57	60.8	7.0
N 550 (FEF)	11.4	19.8	510	61	60.2	4.8
N 660 (GPF)	8.6	18.5	560	56	62.5	3.9
N 770 (SRF)	7.2	17.9	640	56	63.8	5.4
N 880 (FT)	1.7	17.6	800	47	65.5	5.2
N 990 (MT)	1.9	15.5	840	49	66.3	4.8

[a] Ref. 6.
[b] Recipe (cured for 30 min at 153°C), in parts by wt: cold medium nitrile, 100.0; carbon black, 50.0; stearic acid, 1.0; zinc oxide, 3.0; dioctyl phthalate (DOP), 25.0; antiozonant, 1.0; MBTS, 1.5; sulfur, 1.5.
[c] To convert MPa to psi, multiply by 145.

Modulus. The modulus or force required to produce a given elongation in a vulcanizate is primarily a function of carbon black structure and loading. The higher structure carbon blacks produce the higher modulus. The modulus of rubber com-

Table 16.   Effects of Carbon Blacks in Butyl Rubber[a,b]

Carbon black	300% modulus, MPa[c]	Tensile strength, MPa[c]	Elongation at break, %	Hardness, Shore A	Rebound, %
N 293 (CF)	13.8	20.7	600	67	32.5
N 110 (SAF)	12.8	23.5	595	68	32.3
N 220 (ISAF)	13.8	20.5	590	65	34.4
N 242 (ISAF-HS)	14.5	19.7	550	68	34.2
N 330 (HAF)	13.6	19.0	580	63	36.8
N 347 (HAF-HS)	14.3	18.3	540	68	35.2
S 300 (EPC)	10.7	21.4	610	64	35.5
N 440 (FF)	10.0	20.7	580	61	36.0
N 550 (FEF)	12.4	16.2	545	64	35.8
N 660 (GPF)	10.5	14.3	550	61	36.9
N 770 (SRF)	8.6	14.0	595	56	40.0
N 880 (FT)	3.6	14.3	700	52	39.5
N 990 (MT)	4.0	15.2	700	50	39.8

[a] Ref. 6.

[b] Recipe (cured for 30 min at 153°C), in parts by wt: butyl 217, 100.0; carbon black, 50.0; Polyac (poly-*p*-dinitrosobenzene in wax, formerly by DuPont, now known as PolyDNB, by Lord Corporation), 0.5; stearic acid, 0.5; zinc oxide, 5.0; process oil, 3.0; MBTS, 1.0; TMTD, 1.0; sulfur, 2.0.

[c] To convert MPa to psi, multiply by 145.

Table 17.   Effects of Carbon Blacks in EPDM[a,b]

Carbon black	300% modulus, MPa[c]	Tensile strength, MPa[c]	Elongation at break, %	Hardness, Shore A	Rebound, %
N 293 (CF)	11.7	24.3	470	66	54.1
N 110 (SAF)	12.4	23.8	450	68	51.0
N 220 (ISAF)	13.1	22.1	430	68	53.3
N 242 (ISAF-HS)	14.5	22.1	410	70	52.4
N 330 (HAF)	14.5	22.1	400	65	55.0
N 347 (HAF-HS)	16.6	22.1	400	70	54.5
S 300 (EPC)	9.0	22.8	540	61	54.1
N 440 (FF)	10.4	21.1	480	58	63.9
N 550 (FEF)	14.5	17.6	360	64	62.8
N 660 (GPF)	11.4	15.2	430	60	65.0
N 770 (SRF)	9.3	13.6	450	57	67.8
N 880 (FT)	2.8	10.0	640	47	72.9
N 990 (MT)	3.8	9.3	500	48	74.2

[a] Ref. 6.

[b] Recipe (cured for 30 min at 160°C), in parts by wt: Royalene 300, 100.0; carbon black, 80.0; Circosol 42XH (Sun Oil), 40.0; zinc oxide, 5.0; stearic acid, 1.0; tetramethylthiuram monosulfide (TMTM), 1.8 (2.0 for S 300); MBT, 0.6 (0.75 for S 300); sulfur, 1.5.

[c] To convert MPa to psi, multiply by 145.

pounds is greatly increased by the use of increased loadings of reinforcing or semireinforcing carbon blacks. The particle size of carbon black has an apparent secondary effect on compounds. Thus the smaller particle size–low structure blacks produce a comparable modulus level to the medium particle-size blacks in all polymers except

Table 18. Effects of Nonblack Fillers in Natural Rubber[a,b]

Filler loading	Volume, parts	Mooney viscosity	Optimum cure (at 141°C), min	Modulus (at 300%), MPa[c]	Tensile strength, MPa[c]	Elongation, %	Hardness, Shore A	NBS abrasion (ASTM D 1630)	Rebound, %
pure gum		21	15	2.1	24.2	680	43		81.2
barytes	30	27	20	2.1	15.0	640	49	22.1	80.5
	50	24	20	2.1	11.5	610	56	19.4	
ground whiting	30	32	20	2.1	15.0	640	50	24.1	81.5
	50	43	15	1.9	10.7	590	57	21.6	
soft clay	30	35	25	7.2	17.9	500	49	44.4	81.5
	50	41	25	12.4	16.4	390	58	36.4	
hard clay	30	29	25	9.2	19.9	500	55	54.4	68.2
	50	31	25	14.9	18.4	370	64	43.8	
Purecal U	30	37	15	5.0	19.0	580	53	42.2	75.9
	50	46	15	6.1	14.9	540	57	31.8	
Calcene TM	30	32	15	3.7	19.2	640	48	50.0	70.5
	50	42	15	5.1	16.6	610	58	39.6	
zinc oxide	30	22	20	5.2	21.8	600	55	54.1	73.1
	50	23	20	5.5	15.4	550	64	48.1	
Zeolex	20	23	10	3.5	22.6	660	49	51.4	
	30	65	10	5.3	19.3	600	53	46.0	67.3
	40	62	10	7.9	18.0	540	64	39.4	
Silene EF[d]	20	47	10	4.0	24.8	670	52	59.7	
	30	51	10	5.2	20.1	610	61	48.9	67.3
	40	56	10	7.9	17.4	540	67	45.6	
HiSil 233[e]	20	64	10	5.6	27.2	640	58	57.2	77.5
	30	81	10	9.2	25.3	590	73	60.9	65.9
	40	89	10	13.0	23.9	500	78	79.6	53.9

[a] Ref. 7.
[b] Recipe, in parts by wt: smoked sheets, 100.00; zinc oxide, 5.00; filler, as indicated; nondiscoloring antioxidant, 1.00; MBTS, 1.00; TMTD, 0.10; sulfur, 2.75; stearic acid, 3.00.
[c] To convert MPa to psi, multiply by 145.
[d] Silene EF stocks contain 2.4, 3.8, and 5.0 parts of diethylene glycol, respectively.
[e] In addition to the 1 part of MBTS shown in recipe, the 20 vol HiSil 233 stock contains 0.25 parts TMTD and 0.5 parts triethanolamine. The 30 vol stock contains 0.5 parts TMTD and 2 parts triethanolamine. The 40 vol stock contains 0.5 parts TMTD and 3 parts triethanolamine.

Table 19. Effects of Nonblack Fillers in SBR [a,b]

Filler loading	Volume, parts	Mooney viscosity	Optimum cure (at 141°C), min	Modulus (at 300%), MPa[c]	Tensile strength, MPa[c]	Elongation, %	Hardness, Shore A	NBS abrasion, ASTM D 1630
pure gum		26	45	0.6	1.4	680	30	13.3
barytes	30	37	45	1.6	3.8	620	46	15.2
	50	48	45	2.3	3.8	530	54	15.3
ground whiting	30	40	45	1.2	4.0	620	48	16.5
	50	47	45	1.4	5.0	680	50	15.7
soft clay	30	42	90	2.1	9.0	1080	42	28.8
	50	51	90	2.8	7.0	1130	46	33.5
hard clay	30	41	90	2.8	11.4	890	50	34.2
	50	42	90	3.5	10.1	960	51	37.2
Calcene TM	30	45	45	1.6	12.6	860	45	26.6
	50	61	45	1.9	11.9	830	49	16.0
Purecal U	30	46	45	1.7	14.0	710	45	27.7
	50	65	45	2.5	15.2	680	52	27.3
zinc oxide	30	28	45	1.4	14.5	840	45	30.7
	50	39	45	2.1	14.9	760	54	25.6
Zeolex	20	50	30	2.1	14.0	680	47	34.1
	30	62	30	3.3	16.2	670	53	33.7
	40	79	30	4.2	14.0	620	56	32.1
Silene EF[d]	20	58	15	5.4	14.1	520	56	34.2
	30	72	15	5.3	15.9	570	60	42.0
	40	77	15	6.6	15.0	510	63	46.7
HiSil 233[e]	20	53	60	4.0	18.7	610	51	48.7
	30	81	30	5.8	23.5	610	56	45.9
	40	112	30	7.7	21.9	560	67	40.3

[a] Ref. 7.
[b] Recipe, in parts by wt: SBR 1502, 100.0; zinc oxide, 5.00; filler, as indicated; nondiscoloring antioxidant, 1.00; MBTS, 1.50; TMTD, 0.10; sulfur, 3.00; Cumar MH 2½, 15.00; stearic acid, 1.00.
[c] To convert MPa to psi, multiply by 145.
[d] Silene EF stocks contain 2.4, 3.8, and 5.0 parts of diethylene glycol, respectively.
[e] The HiSil 233 stocks contain 2.3, 3.5, and 4.8 parts of diethylene glycol, respectively. The 30 vol stock contains 1.2 parts MBTS and 0.15 parts TMTD instead of the combination in the recipe. The 20 and 40 vol stocks contain 1.5 parts MBTS and 0.1 parts TMTD.

EPDM. In butyl rubber the small particle-size blacks produce higher modulus at high loadings. With thermal blacks, increased loadings only slightly increase the modulus.

With the nonblack fillers, the stiffening types, eg, silicas, calcium silicate, Zeolex, or clay, produce high modulus compounds.

Elongation.   The elongation of a rubber compound is affected by the same factors that influence strength. The amount and nature of the reinforcing material governs the degree by which such reinforcement reduces elongation. Like modulus properties, elongation is basically a function of carbon black structure with lower elongations produced with the higher structure blacks. Elongation is reduced dramatically by increased loadings. At higher loadings, both the reinforcing and semireinforcing types have the same low elongations. Thermal black compounds generally have the highest elongations and are less affected by increases in loadings.

Elongation is reduced by fibrous materials, eg, cotton flock or asbestos. For high elongation, the medium particle-size precipitated calcium carbonates, ground limestone, or blanc fixe are the preferred nonblack fillers. Stiffening nonblack pigments should be avoided.

Hardness.   The hardness of a rubber compound depends upon the amount and type of filler used. With carbon blacks, hardness is controlled by structure, particle size, and loading with the higher structure blacks producing the highest hardness at equal particle size. Thermal blacks have the least effect of all the blacks on hardness.

Rod-shaped or platelike particles, which are oriented in parallel lines or planes during processing operations, produce greater hardness than spheres of the same material. Clays, laminar, and magnesium carbonate produce this effect. Sufficiently fine spherical particle materials, eg, silicas, Zeolex, or calcium silicate, also produce high hardness.

Resilience.   Pure gum compounds have the greatest amount of resilience or rebound. Thus the fillers having the least influence on rebound are the least reinforcing. Particle-size loading and structure of carbon black influence rebound. The resilience decreases as the carbon-black loadings are increased. The highest rebound values are obtained with the largest particle-size black. The rebound values decrease quicker with the small particle-size blacks as the loadings increase. Within each particle-size group, high structure produces a lowering of resilience. Whiting-loaded stocks have good resilience, whereas clay, zeolex, calcium silicate, and silicas have relatively poor resilience. Zinc oxide produces good resilience as well as good reinforcement.

Tear Resistance.   Resistance to tearing is greatly influenced by the particle size and particle shape of the fillers used in rubber compounding. The fine particle-size fillers, which are approximately spherical in shape, give better tear resistance than the needle-shaped or platelike particles. The relative tear-resistance ratings among the carbon blacks vary depending on the polymer used. The high structure blacks produce the best tear resistance at low loadings. The small particle-size blacks reach a peak and decline faster as loading increases. The low structure blacks gradually increase in tear resistance, which is retained at the higher loadings. The thermal blacks give a lower level of tear resistance than the reinforcing blacks. In neoprene, however, the thermal blacks have better tear resistance than the other blacks at high loadings, ie, 80 parts by wt.

Abrasion Resistance.   The best resistance to abrasion is produced by compounds containing small particle-size blacks at optimum loading. The furnace blacks have greater resistance to abrasion than the older channel blacks even though they have the same particle size. This is because the furnace blacks are made from oil instead of gas. These oil blacks have a greater tendency to form carbon gel, which leads to better abrasion resistance. In neoprene compounds, low structure blacks give better abrasion resistance. No other nonblack filler approaches the silicas in respect to abrasion resistance. The silicas come closest to the carbon blacks in this respect. Calcium silicate, Zeolex, zinc oxide, and clay are preferred in that order but considerably less than hydrated silica.

The superior tread-wear performance of modern tires results to a large degree from the abrasion resistance imparted to the tread compound by carbon black. The effect of structure, surface area, and concentration of the carbon black as well as of the process oil concentration in tread compounds used in bias tires under numerous conditions which influence wear rate or severity has been studied (8). Carbon blacks with high structure and surface area are substantially superior to the normal structure and surface area blacks at higher test severities. Increased oil content produces higher wear rates at higher general severities. At any given severity level, the wear rate passes through a minimum as the carbon black level is increased. At this minimum wear rate, the carbon-black level shifts to higher values as the general severity is raised. For maximum tread wear, the carbon black reinforcement system with any tread compound must be carefully adjusted to the anticipated severity level of the use.

Whether the influence of dispersion, structure, and surface area of carbon blacks used in radial tires is the same as in bias tires at variable test severities has been studied (9). In general, radial tires produce approximately twice the wear resistance of bias tires when they have the same tread compound and tread design. At all levels of severity, the 2:1 ratio of improved radial wear to bias wear was determined. In radial tires, there is less advantage in the use of high structure carbon blacks. The wear resistance improvement of carbon-blacks with increased surface area in radial tires is equivalent or better than that in bias tires. In all tires, good dispersion of the carbon black is important for good tread wear, but in radial tires the adverse effects of poor dispersion are reduced.

Carbon blacks are compared most frequently at equal volume loadings to assay their relative reinforcement potentials, as in Tables 11–17. Most rubber products are made to a definite hardness specification. Therefore, for practical applications, comparison of carbon black properties should be made at loadings which produce equivalent hardnesses. Table 20 shows such a comparison in natural rubber and SBR 1500. As the particle size of the carbon black decreases, the scorch resistance is reduced, the Mooney viscosity and ultimate elongation is increased, the tensile strength is greatly increased, and the compression set is very slightly increased.

For many reasons it is advantageous to manufacture certain rubber products in colors. For this purpose a variety of pigments is available to provide almost any desired shade of color. Crude natural rubber and the synthetics vary from very pale yellow to very dark brown. Pure gum stocks containing the minimum quantity of vulcanizing agents are translucent and vary in shade according to the color of the crude used. In mixes containing an accelerator with an excess of zinc oxide, the cured material loses its translucency and becomes opaque.

In developing colored stocks, the usual procedure is addition of a sufficient

Table 20. Carbon Blacks Compared at Equal Hardness[a]

Property	Carbon black						
	(MT) N-990	(SRF) N-770	(GPF) N-660	(FEF) N-550	(HAF) N-330	(ISAF) N-220	(SAF) N-110
surface area[b], m²/g	8	24	30	40	80	110	140
Natural rubber, 68 Shore A hardness[c]							
black, phr	135	77	70	62	60	55	54
Mooney scorch at 130°C	26.0	20.1	19.3	15.6	14.4	14.9	14.1
minimum Mooney viscosity at 130°C	18.0	21.5	21.5	26.5	32.5	34.0	37.5
45 min cure at 138°C, 300% modulus, MPa[d]	9.9	13.2	15.6	15.0	15.5	14.9	13.0
tensile strength, MPa[d]	16.3	22.1	22.6	25.4	28.0	30.6	31.8
ultimate elongation, %	500	500	470	460	515	550	585
compression set, % ASTM B	35.7	34.5	31.8	30.4	33.6	33.7	35.2
specific gravity	1.284	1.177	1.164	1.145	1.140	1.124	1.120
SBR-1500, 66 Shore A hardness[e]							
black, phr	145	84	68	59	59	54	50
Mooney scorch at 130°C	36.8	32.9	35.7	38.0	33.2	36.4	35.9
minimum Mooney viscosity at 130°C	26.0	25.5	25.0	26.5	27.5	29.0	29.5
45 min cure at 153°C, 300% modulus, MPa[d]	9.7	12.9	13.6	14.6	14.9	14.3	11.7
tensile strength, MPa[d]	11.3	18.2	17.9	22.3	25.9	28.4	30.0
ultimate elongation, %	480	490	485	490	510	520	575
compression set, % ASTM B	13.1	15.0	13.8	15.0	18.0	18.4	20.3
specific gravity	1.321	1.212	1.182	1.154	1.154	1.140	1.130

[a] Ref. 10.

[b] Measured by adsorption of nitrogen at cryogenic temperatures.

[c] Natural rubber compounding recipe, in parts by wt: #1 smoked sheets, 100; carbon black, as indicated; zinc oxide, 4; stearic acid, 3; pine tar, 4; Philrich 5 oil, 4; antioxidant, 2; sulfenamide accelerator, 0.5; sulfur, 2.5.

[d] To convert MPa to psi, multiply by 145.

[e] SBR-1500 compounding recipe, in parts by wt: SBR-1500, 100; carbon black, as indicated; zinc oxide, 4; stearic acid, 2; pine tar, 0; Philrich 5 oil, 10; antioxidant, 1; sulfenamide accelerator, 1; sulfur, 1.75.

quantity of a background pigment with a high hiding power, eg, one of the titanium pigments, and addition to this mixture of an organic dye to give the desired color. Shades that need not be bright are obtained by the use of inorganic pigments, eg, iron oxide, chromium oxide, antimony sulfide, cadmium selenide, and ultramarine blue (see Pigments, inorganic pigments). If pastel shades are not required, these materials can be added to stocks containing pigments with low opacity or hiding power, eg, whiting, and produce acceptable colors. The color pigments, whether organic or inorganic, are selected for their stability during cure and for their fastness in the service for which the article is intended.

A variety of ingredients other than those mentioned above are used in rubber for specific purposes. Among these are fibrous asbestos (qv) for the stiffening effect produced by fibers combined with the good heat resistance of the fiber; cotton or other textile fibers for the same purpose where extreme heat resistance is not required; graphite in a variety of particle sizes to provide reduced coefficient of friction; ground cork for compounds of low density; glue as a stiffener; litharge or other lead pigments where extremely high densities are required or for opacity to x rays; and stiffening resins, eg, poly(vinyl chloride), phenol–formaldehyde resins, polystyrene, or high styrene–low butadiene copolymer resins.

The reinforcing phenolic resins are available as thermosetting, phenolic, two-stage resins with hexamethylenetetramine (HEXA) or as a thermoplastic phenolic novalak base, which requires the addition of a cross-linking agent, eg, a methylene donor like HEXA, prior to vulcanization. The hardness and stiffness that these reinforcing phenolic resins produce is retained at high temperatures, which gives them an advantage over the high styrene–butadiene reinforcing resins. If during processing the mixing temperatures produce scorching, the thermoplastic type should be used and the HEXA added in a second pass mix with the curatives. The phenolic resins are used mainly with nitrile rubbers.

Organofunctional silanes have been used for many years to improve adhesion between glass fibers and various polymers. In general, they can be used to improve the compatibility of siliceous fillers with a variety of resins and polymers. However, many organofunctional silanes cannot enhance the reinforcement of silica-filled rubbers. The mercaptofunctional silanes, eg, mercaptopropyltrimethoxysilane (A-189, Union Carbide), give significant improvements in modulus, compression set, heat buildup, and abrasion resistance in SBR–silica-filled compounds. This improvement is also obtained in other rubbers filled with silica to various degrees depending on the type of rubber used. These silanes are capable of yielding compounds with the silica fillers which approach the reinforcement of carbon black compounds in synthetic rubber. The silane materials should be added to rubber with the silica well before the addition of zinc oxide at a concentration of 1–2 wt % based on the silica filler.

Softeners, Extenders, and Plasticizers

Softeners include a wide variety of oils, tars, resins, pitches, and synthetic organic materials and are used for a number of reasons, some of them having little or no relation to softness of either the raw, uncured stock or the vulcanizate. Some of these reasons are to decrease the viscosity and thereby improve the workability of the compound, to reduce mixing temperatures and power consumption, to increase tack and stickiness, to aid in the dispersion of fillers, to reduce mill and calender shrinkage, to provide lubrication so as to aid extrusion and molding, and to modify the physical properties of the vulcanized compound. A few of these reasons are exact opposites, so it is important to be able to pick the type of softener which gives the desired effect. Extenders are materials possessing plastic or rubberlike properties and can be used to replace a portion of the rubber usually with a processing advantage. Both softeners and extenders can be used simply as diluents. Thus the distinction between a softener or extender and a diluent is somewhat arbitrary. Certain materials may be used as both softeners and extenders on one hand and as extenders and diluents on the other.

Chemical plasticizers (qv), eg, pentachlorothiophenol (Renacit 7, Bayer), zinc salt of pentachlorothiophenol (Renacit 4, Bayer), 2,2'-dibenzamidodiphenyl disulfide (Pepton 22, American Cyanamid), and activated dithiobisbenzanilide (Pepton 44, American Cyanamid), are used in very low concentrations and their effect is chemical rather than physical. The chemical plasticizers or peptizers are generally used to lower the viscosity of the uncured compound. They function in the thermomechanical and thermo-oxidative breakdown of rubbers as oxidation catalysts at high temperatures but as radical acceptors at low temperatures. They have little effect on vulcanizate properties. This may be described in terms of the strong branching of the rubber molecules which occurs during the mastication process. This produces a more favorable rheological performance, but is not significant in terms of molecular weight or its distribution. The chemical plasticizers are more effective in natural rubber, SBR, and polyisoprene, but they are relatively ineffective in the other synthetic rubbers. About 0.25 parts are used for peptizing natural rubber, whereas 1.5 parts are needed for SBR. Bis(2,2'-benzothiazolyl) disulfide is used as a mild chemical peptizer for natural rubber for use in sponge compounds where it also serves as an accelerator. The thiuram disulfides, eg, TMTD and tetraethylthiuram disulfide (TETD), are used as plasticizing agents for the Neoprene GN types. The addition of certain compounding ingredients to the batch, eg, clay, carbon black, and sulfur, stops their plasticizing action.

The solubility of the softener in the rubber is one of its important characteristics. In general, the most effective softeners are those that are good solvents for the rubber. Nonsolvent or lubricating types are also used to facilitate certain manufacturing operations. Solvent softeners impart softness to the unvulcanized mix but do not reduce the nerve of the rubber. They also impart the maximum resilience and reduction in hardness to the vulcanizates. Less effective softeners from the standpoint of solubility produce smooth, low shrinkage processing compounds. Insoluble or only slightly soluble softeners reduce tackiness and stickiness of the compounded mix. Many of the latter group are soluble in the rubber at processing or vulcanizing temperatures but they bleed or bloom from the compound during storage at room temperature. Paraffin and castor oil (qv) are examples. The classes of softeners that are of the solvent type on one hand and of the lubricant type on the other are different for different rubbers. For example, a paraffin-based lubricating oil is a solvent-type softener for natural rubber, and SBR is a semilubrication type for neoprene and a lubricating type for the nitrile rubbers. Typical materials which are solvent-type softeners for these three types of rubbers are given below:

Natural rubber and SBR	*Neoprene*	*Nitriles*
all petroleum fractions	cycloparaffinic petroleum fraction	coal-tar fractions
pine tars and resins	coal-tar fractions	esters
coal-tar fractions	esters	

The physical characteristics of the softener have an important effect on the properties of the compound containing it. A low boiling-point softener that would evaporate during processing or curing or during the use of the article is undesirable. The melting point of the softener affects the low temperature brittle point of the vulcanizate in which it is used. The viscosity of the softener influences the stiffness of the cured compound; for example, a solid resin produces a harder stock than a very low viscosity oil. This very fluid oil gives compounds with maximum resilience and lowest hardness. Thus low molecular weight softeners reduce tensile strength and tear resistance of vulcanizates more than high molecular weight softeners.

Softeners may have an accelerating effect, a retarding effect, or no effect on the cure, depending on the pH of the material, its concentration, and the sensitivity of the vulcanizing system to pH variations. In butyl rubber, softeners with appreciable unsaturation must not be used or the cure of the butyl would be greatly retarded or completely stopped. The same is true in natural rubber or the butadiene polymers; unsaturated softeners affect the cure, and it is general practice to use higher ratios of vulcanizing agents when they are present in the compound.

For use in butyl rubber, mineral oils, paraffin wax and petrolatum are the preferred softeners. Paraffinic or slightly cycloparaffinic oils with viscosity–gravity constants less than 0.90 and viscosities of 40–80 SUs at 99°C are the best. Up to 10 parts paraffin wax is very effective in increasing extrusion speed and smoothness of butyl compounds, and up to 5 parts petrolatum improves processing. Petrolatum and wax are more effective than oil in smoothing butyl extrusions.

The development of SBR placed a greater demand on softeners for ease in processing than was the case with natural rubber, and petroleum oils fulfilled this softener function efficiently and inexpensively. The use of these petroleum oils greatly expanded with the adoption of oil-extended SBR for use in tire compounds and other applications. In this case, the oil, which is emulsified in water, is added to the SBR latex prior to coagulation and is coagulated with the rubber. The proportion of oil used is 25–50 parts per 100 parts of rubber. In the absence of the oil, the SBR polymer is of such high viscosity that it would be impossible to process by conventional techniques. The particular virtue of oil-extended SBR is that equal or better quality is obtained with the combination of a high molecular weight rubber and low cost oil than is obtained with a low viscosity rubber without oil. Thus, a fixed quantity of butadiene and styrene goes farther and appreciable savings is effected.

With SBR polymers, it has been common practice to use 5–20 parts of petroleum oils as processing aids (11). The extender and processing oils have been classified in ASTM D 2226 by the use of molecular analysis (see Table 21).

The highly aromatic oils give less Mooney viscosity change in SBR than the paraffinic types. The tensile strength increases as the aromaticity of the oil increases; this, however, is related to improvement in the dispersion of the fillers. The cyclo-

Table 21. ASTM D 2226 Classification of Extender and Processing Oils by Molecular Analysis

Oil type	Asphaltenes, % max	Polar compounds, % max	Saturates, % max	Former designation
101	0.75	25	20	highly aromatic
102	0.5	12	20.1–35	aromatic
103	0.3	6	35.1–65	cycloparaffinic
104	0.1	1	65 min	paraffinic

paraffinic oils have the lightest color and the least staining properties, so they are preferred for applications where color is important.

In polybutadiene polymers, the paraffinic and cycloparaffinic oils are preferred although the physical properties are only slightly affected. Heat buildup properties seem to be the most sensitive ones to the oil type used; the highly aromatic oils give the highest temperature rise. In blends where some other polymer predominates, the plasticizers or oils commonly used for that polymer are normally satisfactory.

In neoprene compounds, aromatic oils are very compatible, with over 100 parts or more being used in the highly loaded Neoprene WHV or WD compounds. Cycloparaffinic oils can be used up to 20 parts. The paraffinic oils tend to bloom even when only 5 parts are used. In order to obtain good low temperature properties in neoprene, dioctyl sebacate, butyl oleate, monomeric polyether, triethylene glycol caprylate–caproate, and trioctyl phosphate are used.

Nitrile rubber compounds contain softeners which aid processing and enhance particular properties, eg, low temperature flexibility. Monomeric esters are superior to other types of plasticizers for imparting low temperature properties to nitrile rubber. However, certain monomeric plasticizers are better than others in this respect. The adipates, sebacates, tributoxyethyl phosphate monomeric fatty acid ester (Synthetics L-1, Hercules Powder Company), methyl acetyl ricinoleate, di(butoxyethoxyethyl)-formal (TP-90B, Thiokol Chemical Corporation), di(butoxyethoxyethyl) adipate (TP-95, Thiokol Chemical Corporation), and triglycol ester of vegetable oil fatty acid (Plasticizer SC, Harwick Standard Chemical Company) are recommended for low temperature properties. Petroleum oils, polymeric polyesters, coal tar by-product resins, and chlorinated paraffins should not be used where low temperature properties are important.

The use of a monomeric ester plasticizer results in less swelling because the plasticizer is extracted and the rubber diluted. Compared to the monomeric ester plasticizers, the polymeric polyester and coal tar by-product plasticizers, eg, the coumarone–indene resins, result in increased volume change because they absorb oil. Table 22 illustrates these effects.

For increased tack with the inherently low tack nitrile rubbers, the following plasticizers in high concentrations are recommended: coumarone–indene resins, rosins, modified phenolics, tetrahydronaphthalene, and the low molecular weight monomeric esters, eg, dibutyl phthalate and dibutyl sebacate. Generally with nitrile rubber various

Table 22. **Effect of Type of Plasticizer on Fuel and Oil Resistance**[a]

| Plasticizer | Type of plasticizer | Fluid immersion, % vol change | | | | |
| | | ASTM oil no. | | | ASTM fuel | |
		1	2	3	A	B
none		+2	+8	+22	+7	+39
dioctyl phthalate	liquid ester	−16	−9	+2	−1	+16
tricresyl phosphate	liquid ester	−14	−7	+4	0	+19
coumarone–indene	coal tar resin	−6	+4	+16	+2	+22
Paraplex[b] G-25	polymeric ester	+1	+9	+24	+9	+44

[a] Base compound recipe, in parts by wt: medium nitrile rubber, 100.00; zinc oxide, 5.00; stearic acid, 1.00; sulfur, 1.50; Aminox (Uniroyal, Inc.), antioxidant, 1.50; N 770 (SRF) carbon black, 75.00; MBTS, 1.50; plasticizer (type indicated), 30.00.

[b] Trademark of Rohm and Haas Company.

plasticizer blends are used. In order to meet ASTM D 2000 and SAE J 200 BG series specifications, a three-way blend of Plasticizer SC, tributoxyethyl phosphate, and Paraplex G-25 plasticizers is recommended (12).

The EPDM polymers can be compounded with large loadings of petroleum oils. Up to 150 parts of oil have been used depending upon the carbon black loadings. The viscosity of the petroleum oil is more important than the chemical classification for most properties (13). The optimum viscosity of the oil seems to be 1700–2500 SUs at 38°C. Lower viscosity oils produce EPDM compounds with low tensile strength, whereas higher viscosity oils may cause processing problems. The paraffinic types generally give the best low temperature properties. When heat-resistant properties are necessary, a high molecular weight paraffinic oil is recommended. Generally cycloparaffinic oils are used for their balance of properties and low cost.

Extenders used in rubber compounding included factice and mineral rubber and were used in wire compounds. Factice is a class of extenders prepared by the reaction of certain unsaturated vegetable oils with sulfur. Factice is used to increase the tolerance for liquid plasticizers. Most mineral rubbers or rubber substitutes are simply blown asphalts. The original product was gilsonite (qv).

Pine products, eg, terpenes, rosin, and pine tars, are used as softeners in natural rubber and SBR. Pine tar aids in filler dispersion and imparts tack to the compounds. It is slightly retarding to the rate of cure.

The coumarone–indene resins improve processing; however, they have very little effect on hardness, rebound, and set properties. They aid in the dispersion of fillers and sulfur. Other materials that function as tackifiers are the alkyl phenol–formaldehyde nonheat-reactive resins, eg, Arofene resins (Ashland) and CRJ-418 and SP1068 (Schenectady Chemicals); and hydrocarbon resins, eg, Betaprene (Reichhold Chemicals), Escorez resins (Exxon Co.), Nevillac (Neville Chemical), Polyvel resins (Velsicol Chemical), and Wingtack (Goodyear).

A processing agent is a chemical product or mixture of materials that, when added to a rubber compound, improves processibility without adversely affecting physical or performance properties. These materials produce noticeable improvements in mixing by reducing temperatures and stickiness, in processing by increasing calender and extrusion rates, and in flow properties in molding operations. Materials of this type are the Struktols (Strucktol Company of America), Dynamar Brand Polymer Processing Additives (3M Company), Vanfres (R. T. Vanderbilt), and Processing Agent 748 (Uniroyal, Inc.). AC Polyethylenes (Allied Chemical) and Epolenes (C. P. Hall, Harwick Chemical) are low molecular weight polyethylenes and perform as processing aids. N,N'-ethylenebis(stearamide) (Acrawax C, Glyco Chemicals) functions as a detackifier and mold-releasing agent.

Vulcanization Agents and Auxiliary Materials

Natural Rubber.   Sulfur is the almost universal vulcanizing agent for rubber. The usual proportion used for obtaining soft rubber vulcanizates is ca 1.75–2.75 parts per 100 parts rubber. Lower ratios, down to 0.5 parts, are frequently used with a suitable upward adjustment of the accelerator. In general, the aging properties of the vulcanizate improve as the proportion of sulfur is decreased. Higher ratios than three parts may be used to obtain high modulus cures which very rapidly retract after stretching. Ratios of 30–60 parts give hard-rubber compounds. At 12–16 parts, compounds have relatively poor physical properties and poor aging characteristics. The good tensile strength properties of low sulfur (0.5–8.0 parts) compounds are attributed to the ability of the vulcanizates to crystallize readily on stretching. The crystallites thus formed function as reinforcing agents. Higher sulfur ratios interfere with crystallization and thus this reinforcement is not obtained.

When elemental sulfur is used as the vulcanizing agent, certain auxiliary materials must also be added to obtain desirable properties. The most important of these materials is the organic accelerator. It has long been recognized that the processing safety, the rate of vulcanization, and physical properties of sulfur-vulcanized rubber greatly depend on the choice of accelerator. The accelerator has a profound influence on the nature of the sulfur cross-link, which largely determines the thermal stability, flex-cracking, and aging resistance of the vulcanizate.

With the use of organic accelerators, scorch of rubber compounds plays an important part in the preparation, storage, and further processing of the compounds. How various types of accelerators affect scorch properties as measured by the Mooney viscometer in both natural rubber and N 330 (HAF) black compounds is shown in Table 23. Styrene–butadiene copolymer and polybutadiene compounds are also in-

Table 23.   Comparison of Mooney Scorch Times of Various Accelerators in Gum- and Carbon-Black-Loaded Compounds, min

Accelerator	Natural rubber (smoked sheets)[a]			BR[b]				SBR-1500[b]	
	Gum		Black	Gum		Black		Gum	Black
	121°C	138°C	121°C	121°C	138°C	121°C	138°C	138°C	138°C
N,N-diisopropyl-2-benzothia-zolesulfenamide (DIBS)	62.7	19.0	21.3		41.8		14.0	39.5	26.1
N-tert-butyl-benzothiazole-sulfenamide (TBBS)	>60	34.6	22.8		37.2		9.1	35.9	22.1
2-benzothiazyl-N,N-diethyl-thiocarbamyl sulfide (EA)	26.1	10.6	14.8		33.7		16.8	24.8	18.2
N-cyclohexyl-2-benzothiazole-sulfenamide (CBS)	>60	28.4	19.5		28.1		9.1	37.8	19.7
benzothiazyl disulfide (MBTS)	72.8	23.8	11.8		53.1		7.9	40.7	12.2
mercaptobenzothiazole (MBT)	11.8		8.8		37.9		7.9	23.5	14.9
tetramethylthiuram monosulfide (TMTM)	25.2		11.5		17.9		8.2		
tetraethylthiuram disulfide (TETD)	18.0		8.9		9.1	11.4	4.6		
zinc dibutyldithiocarbamate (ZBDC)	4.9		4.0		17.1		7.2		
zinc dimethyldithiocarbamate (ZMDC)	6.2		3.7		14.3	13.0	4.8		
condensation product of butyr-aldehyde and aniline (BA)	19.3		5.1	4.9	2.2	7.4	2.9		
diphenylguanidine (DPG)	16.0		9.6		20.0		14.0	14.2	10.1
tetramethylthiuram disulfide (TMTD)	13.2		6.4		7.2	7.1	3.5		

[a] Natural-rubber recipe, in parts by wt: rubber, 100; N 330 (HAF) black, 0 (gum) or 50 (black); zinc oxide, 5.00; stearic acid, 3.00; sulfur, 3.00; accelerator, 1.00.

[b] cis-1,4-Polybutadiene and SBR-1500 recipe, in parts by wt: rubber, 100; N 330 (HAF) black, 0 (gum) or 50 (black); zinc oxide, 5.00; stearic acid, 3.00; sulfur, 2.00; accelerator, 1.25.

cluded for comparative purposes. This comparison was based on work described in ref. 5. The use of carbon black activates most of the accelerator systems with the various polymers. The sulfenamide accelerators give the greatest processing safety. The dithiocarbamates, aldehyde–amines, and the thiuram disulfides are the most scorch-prone accelerators.

The extent of delayed action or length of scorch time required for a particular product depends on the processing procedure through which the material must go during manufacture. If all the operations are at relatively low temperatures and the intervals between operations are short, fast scorch accelerators, eg, zinc dimethyldithiocarbamate (ZMDC) or TMTD, might be tolerated. However, for high speed, high temperature operations, the maximum delayed action as exhibited by CBS or *N,N*-diisopropyl-2-benzothiazole sulfenamide (DIBS) cannot be too much. The scorch time and the cure rate are temperature-dependent; both can be changed by varying the accelerator concentration, and both are influenced by the other materials present in the compound, eg, the amount of sulfur and the kind and amount of filler that is used.

Accelerators are frequently used in combination in order to produce a faster cure than can be obtained by either material separately. The most common combinations consist of an acidic accelerator, eg, MBT, or one that becomes acidic during vulcanization, eg, benzothiazyl disulfide (MBTS), combined with a basic type, eg, diphenylguanidine (DPG) or thiuram disulfide.

Occasionally retarders are used to lengthen the scorch time and to slow the cure rate of excessively fast accelerator combinations. Until the development of prevulcanization inhibitors (PVIs), the most popular retarders were phthalic anhydride, benzoic acid, salicylic acid, and *N*-nitrosodiphenylamine (NDPA). The latter is somewhat discoloring and staining and was generally used with the sulfenamide accelerators in tire compounds. Its manufacture has been discontinued. The other retarders are still used mainly in footwear and mechanical goods compounds. Since they delay scorch time and slow the rate of cure, they are used in injection molding where they provide better mold flow as compared to the PVI types. The PVI types generally prolong scorch time but have essentially no effect on rate of cure or extent of cross-linking at temperatures above 160°C. The most popular PVI type is *N*-cyclohexylthiophthalimide (Santogard PVI, Monsanto). This material is nondiscoloring and nonstaining unlike NDPA (14).

The following are some typical accelerator–sulfur ratios used for natural rubber compounds:

In natural-rubber tire-tread compounds, 0.5–0.7 parts of a sulfenamide accelerator, eg, CBS or TBBS, with 2.75–1.75 parts of sulfur are used. If higher than normal curing temperatures are used, ie, 177°C, a combination of 2.0 parts of TBBS with 1.0 part of sulfur and 1.0 part of a sulfur donor, eg, dithiodimorpholine (DTDM) (Naugex SD-1, Uniroyal; Sulfasan R, Monsanto) should be used.

In black or tan heel compounds, 0.5 parts MBTS and 1.0 part DPG with 3.0 parts sulfur are used.

In hard-rubber compounds, 2.0 parts BA and 45.0 parts sulfur are used.

In black mechanicals, 0.2 parts TMTD, ZMDC, TETD, or TMTM (tetramethylthiuram monosulfide) with 3.0 parts of sulfur are used. Mechanical goods are those used in engineering, eg, gaskets, oil seals, rubber rollers and roll coverings, bridge pads, antivibration units, and various automotive parts.

In superaging black mechanical stocks, 3.0 parts TMTD or 3.7 parts TETD and no sulfur are used. Superaging means the rubber compound must pass specifications for heat resistance of over 121°C.

For low temperature cures (12–24 h at room temperature or 1–2 h at 82–100°C), 2.0 parts dibutyl xanthogen disulfide (CPB, Uniroyal), 2.0 parts dibenzylamine (DBA), and 4.0 parts of sulfur are used. This system is used for tank linings.

For wire-insulation compounds, 2.0 parts MBTS, 0.7 parts TMTM, and 1.0 part sulfur are used.

The usual dosage of zinc oxide is 5 parts per 100 parts rubber, although 1.0–2.0 parts are sufficient. Impurities present in the zinc oxide, principally lead, cadmium, and sulfur, may affect the curing rate. For special purposes, litharge, magnesium oxide, or lime are used in place of zinc oxide. Litharge is used in black wire compounds.

Fatty acids are normally present to 1–2 wt % in natural rubber. For the alde-hyde–amines, guanidines, and thiurams, additional fatty acid is not required. For other accelerators additional fatty acid, generally stearic or lauric, is added. For unloaded gum stocks 1–2 parts are adequate, but for carbon black stocks 3 or more parts are needed.

Natural rubber can also be vulcanized with a number of oxidizing agents including trinitrobenzene, p-quinone dioxime (GMF, Uniroyal; QDO, Lord Corporation), di-benzoyl-p-quinone dioxime (Dibenzo GMF, Uniroyal; DBQDO, Lord Corporation), and chloranil (Vulklor, Uniroyal). None of these materials yields vulcanizates com-parable to hard rubber. Also none of the cures of this type is of much practical im-portance in natural rubber because of high cost, a too rapid cure rate, inferior prop-erties, staining, or other disadvantages.

The elements selenium and tellurium also can vulcanize rubber but cannot pro-duce hard rubber. Because of their high cost, toxicity, and the relatively minor ad-vantages gained by their use, these materials have not been used widely.

Certain phenol–formaldehyde resins are capable of vulcanizing rubber. These resins are made from substituted phenols, which when treated with formaldehyde are incapable of forming a three-dimensional structure. So far as is known, there are no commercial applications of this method of vulcanization.

Practical vulcanization by organic peroxides, eg, dicumyl peroxide, has been demonstrated in a wide variety of applications (15–18). The dicumyl peroxide cure offers freedom from color opacity and resistance to ultraviolet light discoloration in white and transparent vulcanizates. In black compounds, dicumyl peroxide produces excellent compression set, low temperature flexibility, and aging resistance with nearly all rubber polymers except butyl. At 149°C, the optimum cure time is 45 min; at 177°C it is ca 7 min. In continuous-vulcanization wire curing at 204–260°C, the peroxide cures in seconds. Dicumyl peroxide does not function in hot-air curing. Furnace and thermal blacks increase the amount of dicumyl peroxide according to the amount of surface area present. HiSil requires the same amount of peroxide as N 330 (HAF) black. Di-cumyl peroxide has the disadvantage of producing an objectionable odor to the cured stock.

A resin–dicumyl peroxide–GMF (p-quinone dioxime) system is compared with a conventional MBTS–sulfur system for high temperature curing properties in natural rubber N 550 (FEF) black in Table 24 (19). Figure 1 shows the effect of curing tem-perature on tensile strength. The nonsulfur vulcanizing systems are vastly superior to the conventional MBTS–sulfur system, which is important in injection molding.

Table 24.   Recipes Used for Figure 1 Compounds[a]

Curing agent	Sulfur	Resin[b]	YDO[c]	Peroxide	GMF
#1X smoked sheets	100	100	100	100	100
N 550 FEF black	40	40	40	40	40
#8 oil	4	4	4	4	4
Laurex[d]	3	3	3	3	3
BLE-25	1.5			1.5	1.5
zinc oxide	2	2	2	2	2
resole resin[b,e]		10			
YDO[c]			6		
MBTS	0.5				2.0
dicumyl peroxide (90 wt %)				5.0	
sulfur	3.0				
stannous chloride dihydrate			2		
GMF					3.0

[a] Curing conditions: 45 and 90 min at 144°C; 2 and 4 min at 204°C; ½ and 1 min at 260°C.

[b] Resin cure (14).

[c] YDO = 2,2'-methylene-bis-4-chloro-6-methylolphenol.

[d] Zinc salt of lauric acid (Uniroyal).

[e] Commercial *tert*-butylphenyl resole resin.

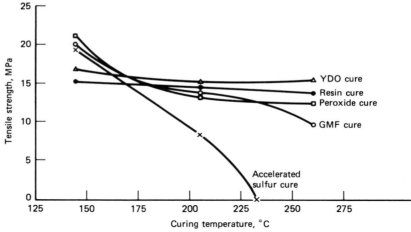

Figure 1.   Effect of type of curing system on the tensile strength of HEVEA carcass compounds. To convert MPa to psi, multiply by 145.

   Certain organic sulfur compounds release sulfur during heating; thus sulfur becomes available to produce vulcanization. Some of these materials are the alkyl phenol sulfides (Vultac 2,3,4,5,6,7, Pennwalt), dimorpholinyl disulfide (Naugex SD-1, Uniroyal; Sulfasan R, Monsanto), TMTD, TETD, butylxanthogen disulfide, and dipentamethylenethiuram hexasulfide (Tetrone A, DuPont; Sulfads, R. T. Vanderbilt). Materials of this type are frequently used when it is desired to have the minimum free sulfur in the vulcanizate or when excellent aging properties are desired.

   The development of efficient vulcanization (EV) systems has been described (20). In these systems, by the efficient utilization of sulfur for cross-linking, a network is obtained and yields natural rubber vulcanizates with the maximum resistance to thermal oxidation and good unaged physical properties. This network contains only monosulfidic cross-links with a minimal main-chain modification in the form of cyclic monosulfides. Examples of these EV accelerator systems are: 0.8 parts TMTD–0.73 parts MBTS–0.35 parts sulfur; 1.2 parts TMTD–1.8 parts CBS–0.25 parts sulfur. These EV systems have the advantages over the sulfurless 3.0 TMTD system of better processing safety and being relatively bloom-free. The EV systems generally have poorer flex life and poorer abrasion resistance than the conventional curing systems.

A greater number of curing systems with accelerator-to-sulfur ratios of 0.6 parts–2.5 parts are classified as semi-EV systems. These semi-EV systems produce networks intermediate in structure and thermal stability between EV and conventional systems. The use of a sulfur donor instead of and at the same level as conventional sulfur without changing the accelerator level is another way of producing a semi-EV system. Thus better properties with respect to reversion at high curing temperatures and aging resistance are obtained with the 0.6 parts CBS–0.6 parts DTDM–1.5 parts sulfur system than with the conventional 0.6 parts CBS–2.5 parts sulfur system. These semi-EV systems are comparable to the conventional curing systems in flexing and abrasion resistance, which makes them more suitable for tire compounding than the EV systems.

The use of soluble EV systems has produced an improvement in properties over the older EV systems. Thus a combination of 1.2 parts MBS (4-morpholinyl-2-benzothiazolesulfenamide)–0.5 parts tetrabutylthiuram disulfide (TBTD) (Butyl Tuads, R. T. Vanderbilt)–0.5 parts sulfur and the use of zinc-2-ethylhexanoate (Octoate Z, R. T. Vanderbilt) as a replacement for the normal stearic acid in a natural rubber compound overcome the problem of localized overcuring, which creates flaws that result in poorer physical properties in the older EV systems (21).

N-oxydiethylenethiocarbamyl-N'-oxydiethylenesulfenamide (OTOS) (Cure-rite 18, B. F. Goodrich) is an interesting thiocarbamylsulfenamide accelerator for use in semi-EV and conventional cure systems in natural rubber. It is a delayed-action accelerator, which becomes very active at curing temperatures >149°C. It is used in combination with MBS to give high modulus and reversion resistance to natural rubber compounds (22). Another class of accelerators, the triazines, has given good results in natural rubber. Bis-(2-ethylamino-4-diethylamino-triazine-6-yl)-disulfide (Triacet 20, Baydeg) provides processing safety similar to CBS, and it produces a higher cross-linking efficiency with good resistance to reversion.

Urethane cross-linking agents give excellent resistance to heat and compression set and to reversion in natural rubber. These agents were marketed as Novor by Hughson Chemical, but are now being marketed by MRPRA in England. A diurethane cross-linking agent, the adduct of nitrosophenol and a diisocyanate, dissociates during the natural-rubber vulcanization to allow nitroso addition, oxime formation, and reduction in the presence of ZMDC. The diisocyanate reaction with the resulting amine groups produces mainly urea–urea cross-links which combine the flexing resistance of polysulfide cross-links and the thermal stability of peroxide-generated carbon–carbon cross-links. Because of the small proportion of urethane–urethane cross-links, the hydrolytic stability is excellent. Applications for the diurethane cure system are in thick rubber articles requiring long cures, eg, heavy engine mounts and solid industrial tires (23).

The Peachy process can be used to effect vulcanization. This involves the alternate exposure of rubber to hydrogen sulfide and sulfur dioxide at room temperature.

Vulcanization is practically instantaneous at room temperature. This process, although of considerable theoretical interest, is of no commercial interest. Sulfur monochloride, S_2Cl_2, was used for effecting cold vulcanization of thin rubber products, eg, toy balloons or proofed fabric.

Styrene–Butadiene Polymers. Styrene–butadiene rubber is vulcanized with sulfur and can be vulcanized with dicumyl peroxide, p-quinone dioxime, trinitrobenzene, chloranil, and others. There are, however, certain fundamental differences between SBR and natural rubber which must be realized when compounding them. The normal sulfur ratio with SBR is 1.50–2.0 parts for carbon-black compounds and 3.0–4.0 parts for nonblack compounds. Styrene–butadiene rubber can vulcanize to hard rubber with 40–60 parts sulfur. Much more accelerator is required for SBR than for natural rubber. For example, a natural-rubber tread stock cures in 45 min at 138°C when accelerated with 0.9 parts per 100 parts rubber of MBT. A similar SBR stock requires 1.5–3.0 parts of MBT to attain a reasonable curing rate. The slower curing rate of these synthetics is attributed to the lowered reactivity of the molecule resulting from the replacement of the side methyl group in a polyisoprene with a hydrogen in what is essentially polybutadiene. Similar to natural rubber, SBR shows better resistance to aging, particularly at high temperatures when lower sulfur ratios are used. An additional improvement can be obtained by the use of a sulfur donor, eg, TMTD, as the source of sulfur, as in EV systems. The amounts of sulfur available for vulcanization from various sulfur donors are: TMTD, 13.3%; CPB, 21.4%; dimorpholinyl disulfide, 29.0%; dipentamethylene thiuram hexasulfide, 25.0%; alkylphenol polysulfide, 28.0%.

When used as primary accelerators, the guanidines and aldehyde amines are considerably less effective in SBR than they are with natural rubber. They are used mainly as activators for the thiazoles and sulfenamide accelerators.

The MBT, MBTS, CBS, TBBS, and DIBS accelerators are persistent, which means they continue to develop higher modulus and lower elongations in SBR during extended cure. The thiuram disulfides or dithiocarbamates are nonpersistent accelerators in SBR showing little change in physical properties after the optimum cure. The activating strength of the metal dialkyldithiocarbamates are in decreasing order of effectiveness: copper, bismuth, lead, and zinc.

The popular use of the basic type of accelerators as activators or secondary accelerators is derived primarily from efforts to modify the curing rate and maintain desirable physical properties. The persistence of thiazole or sulfenamide acceleration is lessened by reducing their concentration and adding an activator. Common activators are DPG, N,N'-di-o-tolylguanidine (DOTG), treated ureas (eg, BIK), TMTM, TMTD, and the dithiocarbamates.

The use of nonblack fillers, eg, HiSil, require modifications of both accelerators and activators in normal sulfur curing systems. For SBR and BR polymers, maximum reinforcement by HiSil can be obtained only in the absence of zinc oxide and stearic acid (24). The acceleration system for the silica soling stock is 1.2 parts MBTS–1.5 parts DOTG–0.25 parts TMTM–2.0 parts Carbowax 4000 (polyethylene glycol from Union Carbide Corp.)–2.5 parts sulfur.

Some typical accelerations for SBR compounds are tire treads, 1.2 parts TBBS–1.75 parts sulfur or 0.9 parts CBS–0.15 parts TMTD or 0.3 parts BIK–1.5 parts sulfur; retreads, 1.0 part TBBS–0.2 parts TMTM–2.0 parts insoluble sulfur; wire, 1.5 parts MBTS–0.3 parts TMTM–0.3 parts DPG–1.0 part sulfur or 2.0 parts MBTS–2.0 parts bismuth dimethyldithiocarbamate (BiMDC, Bismate, R. T. Vanderbilt)–1.0

part sulfur; mechanicals, 1.5 parts MBTS–0.3 parts DPG-2.0 parts sulfur or 0.5 parts MBTS–0.3 parts copper dimethyldithiocarbamate (CuMDC), Cumate, R. T. Vanderbilt)–2.0 parts sulfur.

The prevention of sulfur bloom in uncured rubber compounds is greatly aided by the use of insoluble sulfur. Sulfur occurs in a variety of allotropic forms which differ in such properties as molecular weight, density, crystalline form, and solubility. Insoluble sulfur, so-called because of its insolubility in carbon disulfide or rubber, is a polymeric form of sulfur in the $(1-3) \times 10^5$ molecular weight range. Commonly used regular ground sulfur is in the stable rhombic form, which is almost completely soluble in rubber at high temperatures. Because of its solubility, regular sulfur frequently migrates or blooms to the surface of the uncured rubber compound because of continued solution and recrystallization during hot-processing operations and storage. The sulfur bloom produces a loss of building tack, which is required for such articles as tire retreads, carcass and belt compounds, roll coverings, etc. Sulfur bloom problems are overcome by the use of insoluble sulfur if certain precautions are taken. The mixing temperature of the compound must be controlled to $\leq 100°C$ to prevent its conversion to the soluble form. A recent study has shown that amines from sulfenamide accelerators and HEXA have a detrimental effect on insoluble sulfur stored in preblended cure packages and in mixed compounds (25). The effect causes it to convert to the stable rhombic form and produce bloom.

With the rosin-acid-emulsified SBR rubbers, which are generally the cold types, a fatty acid is usually added in compounding.

Polyisoprene. The one manufacturer of polyisoprene rubbers in the United States uses an aluminum trialkyl–titanium tetrachloride catalyst, which gives a 96% *cis*-content polymer. Polyisoprene, on account of its chemical similarity to natural rubber (98% cis content), generally responds the same way to vulcanizing agents and accelerators. There are some differences, however, resulting from the inherent softness of polyisoprene. This inherent softness, plus the lower temperatures generated during mixing, result in dispersion problems with the high melting point accelerators. Polyisoprene reacts normally to variations in accelerator level but it is quite sensitive to overacceleration, which results in loss of physical properties. Sulfur reacts the same with polyisoprene as with other polymers. The level of sulfur is quite sensitive in gum compounds, which require ca 1.75–2.0 parts. With carbon black, whiting and clay loadings with ca 2.25–3.25 parts sulfur are required. The low melting point accelerators are favored with CBS and the other sulfenamide accelerators. They are used as primaries at 0.6–0.8 parts. As activators, TETD, TMTM, BIK, and liquid types, eg, Trimene base, TBTD, and triethanolamine, are used at 0.2–0.4 parts.

Polyisoprene differs from natural rubber in that it is free of fatty acid. Therefore, 2.0–3.0 parts of stearic acid or 1.0–2.0 parts zinc laurate (Laurex) are added during compounding. Zinc oxide is used in the normal concentrations of 3.0–5.0 parts.

Butyl Rubber. The base butyl polymer has a direct bearing on the physical properties and cure rate of the compound. Cure state is directly related to mole-percent unsaturation, ie, as the number of isoprene monomeric units per 100 isobutylene units increases, the sites for cross-linking increase and higher modulus or tighter cures are possible. Vulcanization is faster and heat resistance improves. The percent unsaturation varies from 0.6–1.0% to 1.1–1.4% to 1.5–2.0% to 2.1–2.5%.

Butyl rubber can be vulcanized by sulfur and by certain oxidizing agents that vulcanize the other polymers. Less than 1.0 wt % of sulfur can be combined with butyl if all the unsaturation is satisfied. An appreciable excess is usually used, ie, 1.0–2.0 parts. With sulfur, the organic accelerator is almost without exception a thiuram or dithiocarbamate type. Usually 0.5–1.0 parts MBT or MBTS is used as a modifier. For butyl inner tubes, the standard acceleration is 1.0 part TMTD–0.5 parts MBT–1.5 parts sulfur. For extrusions, 1.0 part zinc dibenzyldithiocarbamate (Arazate, Uniroyal)–1.5 parts ZBDC–0.5 parts tellurium diethyldithiocarbamate (TEDC)–2.0 parts sulfur is recommended (17). For heat-resistant molded compounds, a 2.0 parts TMTD–2.0 parts DTDM system is used.

Nonsulfur cures of butyl are commonly used commercially. Various dioxime compounds in the presence of an oxidizing agent act as rapid and efficient vulcanizing agents for butyl compounds. The most commonly used are *p*-quinone dioxime and its benzoic acid ester dibenzoyl-*p*-quinone dioxime. Because of their chemical structure, these materials are said to produce a quinoid type of cure. Wire and cable manufacturers have discovered important advantages of using quinoids in butyl insulation. The two compounds are used to obtain vulcanizates with superior electrical, heat-aging, ozone-resistant, and modulus properties. *p*-Quinone dioxime is normally used at 2.0 parts; the cures are faster and have a greater tendency to scorch than those of dibenzoyl-*p*-quinone dioxime. Six parts of the latter are used in butyl rubber. Bis(2,2′-benzothiazolyl) disulfide is highly recommended as an activator for *p*-quinone dioxime, whereas red lead is very effective with dibenzoyl-*p*-quinone dioxime. Other typical activators include various metal oxides, chloranil, and 1,3-dichloro-5,5-dimethylhydantoin. In some instances, several activators can be used in combination. Sulfur acts as a mild retarder and sometimes improves initial physical properties. Acidic materials initiate scorch in *p*-quinone dioxime compounds. Alkaline materials, eg, amines, function as retarders for *p*-quinone dioxime. With dibenzoyl-*p*-quinone dioxime, acidic materials act as retarders. Typical wire accelerations are 6.0 parts dibenzoyl-*p*-quinone dioxime–4.0 parts MBTS–10.0 parts red lead–1.0 part sulfur or 2.0 parts *p*-quinone dioxime–4.0 parts MBTS–6.0 parts red lead–1.0 part sulfur. These materials must first be oxidized to dinitrosobenzene to become effective as vulcanizing agents.

A method has been developed for vulcanizing butyl rubber with monomers or preferably polymers obtained from the condensation of formaldehyde with 4-alkyl phenols (26). The methylol resin curing system is considerably slower than the sulfur system, but the resin system has much greater stability towards prolonged vulcanization conditions. Because of its extreme heat resistance, the system is used in tire curing bladders. The heat-activated dimethylolphenol resin (Amberol ST137, Rohm & Haas Co.; SP1045, Schenectady Chemicals, Inc.) is slow curing and is generally used with a catalyst, eg, stannous chloride, ferric chloride, or halide-bearing elastomers, eg, neoprene, chlorosulfonated polyethylene, and halogenated butyl polymers (21,27–29). Zinc oxide is detrimental to the resin cure when metallic chloride activation is used, but it is needed when organic halogen activators are used. Generally, ca 12 parts of the methylol resin are used.

Halogenated Butyl Rubbers. Chlorobutyl and bromobutyl rubbers, which include either allylic chlorine or bromine and double bonds, can be cured by a number of systems. These include the normal systems for butyl rubbers, eg, the sulfur–organic accelerators, sulfur donor, *p*-quinone dioxime, and the methylol resins, and by some systems related to neoprene, eg, 5.0 parts zinc oxide–1.0 part magnesium oxide–1.5 parts ethylene thiourea (NA-22) and 3.0 parts zinc oxide–1.0 part TMTD–1.0 part

MBTS. A 0.8 parts Chloranil–10.0 parts zinc oxide system has produced excellent heat resistance (30).

For blends with natural rubber, SBR and EPDM are compounded with 3.0 zinc oxide–1.0 TBBS–1.0 Vultac #5 (Pennwalt)–0.5 sulfur.

Chlorosulfonated Polyethylene. Chlorosulfonated polyethylene (Hypalon) can be cross-linked by a variety of curing systems (31). Systems based on zinc oxide–magnesia as acid acceptors are most useful. If water resistance is required, litharge or organic lead bases, eg, dibasic lead phthalate or tribasic lead maleate, are used. If the curing is carried out with the metallic oxides alone, ionic cross-links form and compression set properties are poor. To improve set properties, some covalent cross-links must form through the use of 2.0 parts dipentamethylenethiuram hexasulfide (Tetrone A, DuPont; Sulfads, R. T. Vanderbilt) or 2.0 parts TMTD–1.0 part sulfur. Epoxy resins have been used as acid acceptors to produce good cures in conjunction with organic accelerators and sulfur. A 4.0 parts magnesia–3.0 parts pentaerythritol system is satisfactory for nondiscoloring compounds.

Polybutadiene. The preferred vulcanizing agent for BR is sulfur, but the well-known nonsulfur curing agents can also be used (32). The sulfur level should be kept as low as possible consistent with acceptable heat buildup in tread stocks. Low sulfur content improves tear resistance which minimizes cutting and chipping. In 100 wt % BR compounds, the sulfur level is 1.4–1.5 parts whereas in 65 wt % SBR–35 wt % BR compounds, 1.75 parts of sulfur are required. If natural rubber is used in place of SBR, the sulfur level is increased to 2.0 parts. In carcass compounds the sulfur level is increased to 2.0–2.5 parts to obtain higher resilience. The sulfenamide accelerators are preferred at 0.8 parts with 1.75 parts sulfur or 0.6 parts sulfenamide and 2.0 parts sulfur (33). Typical secondary accelerators like the dithiocarbamates, thiuram disulfides, DPG, or BIK are used at 0.1–0.2 parts with 0.7 parts sulfenamide. Two-to-three parts fatty acid are used with BR for activation and improved processing. The normal 3–5 parts of zinc oxide are required.

Nitrile Rubber. Nitrile rubber is vulcanized with sulfur with 1.0–1.75 parts in carbon black compounds and 2.0–3.0 parts sulfur for nonblack compounds. These polymers also form hard rubber with 30–50 parts sulfur. Nitrile rubber is similar to SBR requiring higher levels of accelerator than natural rubber.

One of the causes of reduced tensile properties in nitrile rubber is poor sulfur dispersion. The sulfur should be added early in the mix and special sulfurs should be used. These special sulfurs are surface-treated to promote more rapid blending during mixing, reduce dustiness, and prevent reagglomeration of the sulfur particles. Spider-brand sulfur (C. P. Hall, Stauffer Chemical Company) is treated with ca 2.5 wt % magnesium carbonate and is generally recommended for nitrile rubber. Tire-brand sulfur, which is treated with 2.5–10.0 wt % oil, is not recommended as the oil treatment produces repellency by the nitrile rubber.

One and one-half parts MBTS and 1.5 parts sulfur has been widely used as a general purpose curative; 0.2 parts TMTM–1.5 parts sulfur usually gives slightly safer processing than the MBTS–sulfur system and is considerably lower in cost. The 0.2 parts TMTM–1.5 parts sulfur system develops better compression set characteristics. The system 0.6 parts TMTM–1.0 part sulfur is considered the best general-purpose curative for nitrile rubbers. It is nonblooming and satisfies the 70-h-at-100°C heat agings and compression-set test requirements. A 2.0 parts CBS–1.5 parts TETD–1.5 parts TMTD–0.2 parts sulfur system is recommended for severe heat aging and high temperature compression-set requirements. For injection molding, the accelerator system 1.5 parts CBS–2.0 parts TMTD–0.75 parts sulfur is recommended.

Nitrile rubber can also be vulcanized by peroxides, eg, dicumyl peroxide (DiCup, Hercules, Inc., Varox DCP-T, R. T. Vanderbilt) lead peroxide, and 2,5-bis-(*tert*-butylperoxy)-2,5-dimethylhexane (Varox DBPH, R. T. Vanderbilt). These peroxide-cured nitrile rubbers have good low temperature properties, low compression set, and good aging resistance at elevated temperatures. The peroxides give shiny, bloom-free vulcanizates which do not stain silver or discolor in the presence of lead. 1–2 parts of active peroxide are required for curing.

p-Quinone dioxime, dimethylolphenol compounds, and chloranil also function as vulcanizing agents. Nitrile rubber can be vulcanized without curatives by heating for one or two hours in the absence of air at 200°C (34).

For 149°C heat resistance in nitrile rubbers, the following cadmium–magnesium curing system is recommended: 5.0 parts magnesium oxide–2.5 parts cadmium diethyldithiocarbamate (CdEDC) (Ethyl Cadmate, R. T. Vanderbilt)–1.0 part MBTS–1.0 part TMTD–5.0 parts cadmium oxide; the latter is used instead of zinc oxide.

When made with soap as the emulsifier, the nitrile rubbers are treated during their manufacture to remove the fatty acid to which the soap is converted during coagulation. The effect of the omission of fatty acid is very slight. Some improvement in set properties is obtained by the addition of 0.5–1.0 part of fatty acid, but other properties are not noticeably affected.

High concentrations of fatty acids in nitrile rubbers are likely to cause processing problems because of blooming from the uncured stock. They also bloom from the vulcanizate. Approximately 1–3 parts zinc oxide are used as an activator with the organic accelerators.

Chloroprene Rubbers.   The general-purpose neoprenes are classified in three groups, the G, W, and T types (1,35). The oxides of zinc and magnesium are vulcanizing agents but also serve other functions. Both oxides function as acid acceptors in the event that any hydrogen chloride is formed during processing or during service. The magnesium oxide also functions as a scorch retarder during processing. The grade of magnesium oxide is very important. It normally precipitates, ie, is not ground, and is calcined after precipitation. It also has a high surface area-to-volume ratio. Exposure of magnesium oxide to moisture and the air can cause loss of activity. Scorchguard O (Wyrough & Loser) helps to overcome this problem (36). Zinc oxide has little effect.

Generally, 4.0 parts of magnesia and 5.0 parts of zinc oxide are used in the neoprene compounds. If water resistance is a problem as in some wire compounds, 10–20 parts of red lead are used in place of the magnesia–zinc oxide system. Litharge has greater scorching tendencies than red lead.

Carbon blacks of all types accelerate the curing rate of neoprene, but variations in the pH of the blacks have only a slight effect.

Sodium acetate, MBT, MBTS, TMTD, and TETD are retarders. Sulfur accelerates the cure and stiffens the neoprene GN vulcanizates. It has little effect on scorch but harms heat resistance. Ethylene thiourea (ETU), the di-*o*-tolylguanidine salt of dicatechol borate (Permalux, DuPont), and *p*,*p*′-diaminodiphenylmethane (Tonox, Uniroyal) have been used as accelerators in the G types.

The W and T types require an organic accelerator in addition to the metallic oxides for practical cures. The accelerator for optimum heat and compression-set resistance is 0.5–1.0 parts ETU. Ethylenethiourea should be used in the form of an elastomeric dispersion, eg, ETU Poly-Dispersion END-75 (Wyrough and Loser Inc.), to prevent ingestion or inhalation of the compound. For maximum processing safety, 0.5 to 1.0 parts TMTM–0.5 to 1.0 part di-*o*-tolylguanidine (DOTG)–1.0 part sulfur is used. For very fast, high temperature cures, a 2.0 parts *N,N'*-diethylthiourea (DETU)–1.0 part MBTS–1.0 part DOTG system is used. Adequate ventilation should be used with the high temperature curing equipment to prevent eye irritation from breakdown products of DETU.

Two recent nonthiourea neoprene accelerators have been introduced to the rubber industry as replacements for ETU: dimethylammonium hydrogen isophthalate (Vanax CPA, R. T. Vanderbilt) and 3-methyl-thiazolidine-2-thione (Vulkacit CRV, Mobay) (37–38). One and one-half parts Vanax CPA or 0.5 parts Vulkacit CRV produce better processing safety at 121°C than 0.5 parts ETU and equivalent curing characteristics at 160°C.

It has been theorized that bifunctional neoprene accelerators become cross-links between chloroprene polymer chains as a result of being bisalkylated by the polymer chain at the active chlorine position (39). It also has been proposed that the vulcanizing mechanism with thiourea accelerators involves an intermediate complex of tertiary allylic chlorine with the thiourea (40). This complex reacts with zinc oxide to form a complex zinc salt, which decomposes and releases a molecule of urea corresponding to the starting thiourea. A zinc chloride mercaptide remains. The urea reacts with the second molecule of tertiary allylic chloride, which results in a sulfur bridge between the two polymer chains.

Ethylene–Propylene Terpolymers. Ethylene–propylene rubber can be vulcanized successfully by a variety of systems. The various third monomers which provide the unsaturation in the molecule allow EPDM to be vulcanized by the broad range of accelerator sulfur systems, *p*-quinone dioxime–dibenzoyl-*p*-quinone dioxime combinations, and resin systems. Vulcanization by peroxides and radiation is effective, though unrelated to molecular unsaturation. Ethylene–propylene terpolymer resistance to ozone, weather, and polar liquids and its excellent low temperature flexibility are independent of the vulcanizing system used. However, the vulcanizing system does affect vulcanizate stress–strain properties, heat buildup, compression set, and heat resistance. With more EPDM rubbers, the cure rate is slower than with SBR; the ultrafast curing EPDM rubbers are about the same as SBR. The general cure system has been a combination of 0.5 parts thiazole–1.5 parts dithiocarbamate or 1.5 parts thiuram–1.5 parts sulfur. Bloom has been a problem on molded parts but not on steam-cure articles. With normal sulfur levels (1.5–2.0 parts), the level of dithiocarbamates and thiurams accelerators must be controlled. Zinc dimethyldithiocarbamate, ZEDC, TMTD, TETD, TMTM, and Royalec 133 (Uniroyal) must be used at levels below 0.8 parts to prevent blooming. For nonblooming applications, ZBDC and DTDM are used at 2.0 parts. Generally for fast cures, a 1.5 parts MBTS–0.75 parts TMTD–0.75 parts dipentamethylenethiuram hexasulfide (DPTH)–1.5 parts sulfur system is used. In the ENB-type EPDM polymers, sulfenamide accelerators activated by DPG or a BIK sulfur system are used. For high temperature curing, the following systems are satisfactory: 3.0 parts MBTS–0.7 parts TMTD–1.5 parts ZBDC–0.7 parts DTDM–0.3 parts sulfur or 0.7 parts TMTD–0.7 parts TETD–0.7 parts DPTH–2.0 parts DTDM–0.3 parts sulfur. Dibenzoyl-*p*-quinone dioxime–red lead gives satisfactory cures and good heat resistance. Dibenzoyl-*p*-quinone dioxime also is useful as a coagent with the peroxide cures. The peroxide curing of EPDM and EPM (the copolymer of ethylene and propylene) is important in the vulcanization of wire insulation and jacket compounds. Various coagents for the peroxide cure have been studied (41–42). Ethylene dimethacrylate (SR 206, Sartomer Resin, Inc.) has been quite popular. Bromomethylmethylol phenolic resins plus stannous chloride give satisfactory cures. This resin system has no advantage over the sulfur-donor system and is generally poorer than the *p*-quinone dioxime cures and the peroxide system in heat protection at elevated temperatures.

Bonding Agents

In certain rubber articles designed to withstand considerable stresses in use, the rubber is reinforced with plies of comparatively inextensible textile, glass, or steel materials. Thus rubber tires, hose and belts, etc, are commonly reinforced with filamentary textiles, glass, or steel in the form of yarns, cords, or fabric. In such articles, it is important that the plies of reinforcing material be firmly adhered to the rubber and remain effectively adhered after the article has been subjected to repeated and varying strains in use. Thus the article's durability and its ability to perform under increasingly severe operating conditions is directly linked to the adhesion of the ply-reinforcing material to its adjacent rubber surface.

A technique for obtaining adhesion is the dry adhesive system in which the bonding agents are mixed directly into the rubber compound. Bonding agents used in the dry adhesive system are as follows: 2.5 parts resorcinol (dihydric phenol) and 1.6 parts HEXA; 2.0 parts Bonding Agent R-6 (Uniroyal) (a resorcinol donor) and 1.0 part Bonding Agent M-3 (Uniroyal) (a methylene donor); 2.5 parts hexamethoxymethylmelamine (Cohedur A, Mobay; Cyrez 963 resin, American Cyanamid) and 3.0 parts of the condensation product of resorcinol and formaldehyde which is deficient in formaldehyde (Penacolite Resin B-18, Koppers; SRF-1501, Schenectady Chemicals).

Cyrez BN (American Cyanamid) (2-naphthol) has been recommended as a resorcinol replacement in the bonding agent systems.

The HRH system was pioneered in the United States by PPG Industries and in Europe by DeGussa (43). HiSil or a siliceous filler with a high specific surface area, resorcinol, and HEXA are the ingredients of the system. The system produces strong adhesion between rubber compounds and fabrics or metals without the need for separate adhesive treatments. At least 15.0 parts HiSil are needed with 2.5 parts resorcinol and 1.6 parts HEXA. It is believed that the strong adhesion results because the resorcinol–formaldehyde resin forms hydrogen bonds with the substrate. Additional hydrogen-bondable sites are furnished by the silica present at the rubber surface. This permits hydrogen bonding in both directions from the resin to the fiber and to the rubber. Two parts Bonding Agent R-6 and 1.0 part Bonding Agent M-3 can replace the 2.5 parts resorcinol–1.6 parts HEXA portion of the HRH system without the loss of adhesion properties; but the former produces the added advantages of better processing safety and less chance for health hazards.

In studies in natural-rubber carcass compounds, the use of either the HRH or the R-6–M-3–HiSil systems with greige or untreated rayon or nylon cords produces the same H adhesion values (ASTM D 2229-73) as are obtained from commercially treated cords (44). With untreated polyester cords, however, no response is obtained.

The advent of the radial tire in the United States has greatly increased research and development efforts in the field of steel cord adhesion. Historically, brass plating has been the work horse in obtaining adhesion between rubber and metal. Today in conjunction with the use of steel cord, brass is still the most popular adhesive.

In order to obtain good adhesion, the brass plate must first be adherent to the steel and this generally is fairly easy to achieve. Secondly, the brass plate must be of optimum composition and thickness. The optimum brass composition is 66–70 wt % copper. No important relationship has been determined between adhesion and brass coating weight as long as a certain minimum thickness is exceeded; generally the coating thickness is 5–8 μm. Thirdly, the brass plate must be free from surface contaminations that would affect wettability or surface activity. Since most brass-plated steel cord is drawn to size after plating, it is important that the lubricant used is compatible with the rubber compound. Steel cord should be stored in desiccated containers whenever possible. In addition, the rubber compound must be compatible with the brass plate with respect to reaction kinetics (45–46).

The type of rubber and the vulcanizing systems have the greatest effect on steel cord adhesion. Natural rubber produces the highest adhesion followed by SBR, BR, NBR, and EPDM rubbers in that order. The sulfur level and accelerator types are very important factors, especially in natural rubber compounds. Two chemical reactions proceed simultaneously, both competing for sulfur. They are the copper–sulfur reaction and the rubber–sulfur reaction. It has been established that an excess of copper sulfide at the metal surface is detrimental to adhesion (47). The EV or sulfurless curing system with TMTD gives poor adhesion because of the attack of the thiuram disulfide on the copper to form CuMDC. Generally with natural rubber compounds, the more delayed-action sulfenamide accelerators give the best adhesion.

The bonding agents previously recommended for the dry adhesive systems produce increased adhesion to brass-plated steel wire. The bonding agents are generally used in combination with cobalt salts for the highest values. Cobalt naphthenate, cobalt stearate, cobalt abietate, and Manobond C (Manchem, Inc.) are the most widely used. The amount of cobalt used in the rubber compound is 0.1–0.3 wt % (48).

A bonding agent system of 2.0 parts Bonding Agent R-6 and 1.0 part tetrachlorobenzoquinone (Vulklor, Uniroyal) imparts excellent adhesion to brass-plated steel wire. The adhesion continues after aging and especially after oxygen and steam agings (49–50).

Antioxidants and Stabilizers

Natural rubber is different from SBR and other synthetic rubbers of the diene types in its need for protection. Latex as taken from the rubber tree contains natural antioxidants, proteins, and complex phenols, which protect it from deterioration during the coagulation and drying steps. These natural protectants are destroyed during vulcanization and thus synthetic protectants must be added to insure an adequate service life of cured natural rubber products. On the other hand, synthetic rubbers of the diene types are quite unstable when freshly prepared and require synthetic protectants both during the flocculation and drying of the polymers and sometimes after vulcanization. The former protectants are generally referred to as stabilizers and the latter are called antioxidants or antiozonants (see Antioxidants and antiozonants).

The nondiscoloring types of SBR generally contain ca 1.25 parts of an alkylated aryl phosphite (Polygard, Uniroyal, Inc.) or a substituted phenolic type (Wingstay S or T, Goodyear Tire & Rubber Co.). The staining types of SBR, which have been compounded with 1.25 parts of BLE-25 (Uniroyal, Inc.), N-phenyl-β-naphthylamine (PBNA), or Agerite Stalite (R. T. Vanderbilt Co.) in the past, have been mainly converted to a combination antiozonant–stabilizer. These antiozonant–stabilizers are used at 0.5–0.7 parts and are of the p-phenylenediamine type, eg, Flexzone 7L or 11L (Uniroyal, Inc.) or Wingstay 200 (Goodyear Tire & Rubber Co.). Their stabilizer advantage is the carry over of antiozonant activity in vulcanizates which is very important, especially in tire applications. The nondiscoloring grades of butyl rubber are compounded with ZBDC or octylated diphenylamine (Agerite Stalite or Naugard H, Uniroyal, Inc.), whereas the staining grades are compounded with PBNA. The solution polymers, eg, BR, polyisoprene, or EPDM, are stabilized with phenolics or phosphite types or combination of them at 0.5 parts. Nitrile rubbers are compounded at higher stabilizer levels (1.0 part) and in general with alkylated phenolics (BHT, Naugawhite; Uniroyal, Inc.) or a phosphite type (Polygard) or a combination of the two. The very slightly staining grades contain an octylated diphenylamine type, eg, Agerite Stalite or Naugard H. Neoprenes, especially the G types, are stabilized with a thiuram disulfide whereas the staining grades are stabilized with PBNA.

Polymers differ greatly in the way they deteriorate. The molecular changes that take place in degraded polymers can be described in three ways: chain scission resulting in a reduction in chain length and average molecular weight, cross-linking resulting in a three-dimensional structure and higher molecular weight, and chemical alteration of the molecule by the introduction of new chemical groups. Natural rubber, polyisoprene, and butyl rubbers degrade predominantly by chain scission resulting in a weak, softened stock often showing surface tackiness. Chemical analysis shows the presence of aldehyde, ketone, alcohol, and ether groups resulting from oxidative attack and mostly at alpha hydrogens and double bonds.

Styrene–butadiene copolymer, neoprene, EPDM, polybutadiene, acrylonitrile rubbers, and polyethylene degrade by cross-linking, giving boardy and brittle compounds with poor flexibility and elongation. An increase in the oxygen content of these polymers also occurs. Oil-extended synthetic rubbers may produce chain scission in an oxidative attack rather than cross-linking which occurs in the nonoil-extended rubbers. This chain scission can be controlled to a large degree by the use of a proper stabilizing system.

The extent of softening or hardening depends on many factors, some of them external and some internal to the rubber compound. The external factors are oxygen, heat, ozone, light, weather, fatigue, and atomic radiation. These external factors reduce the service life of the rubber compounds and, because of their nature, are not controllable. Since all rubberlike materials, whether natural or synthetic, cured or uncured, contain a certain amount of chemical unsaturation, they are subject to chemical attack by oxygen. Generally, when 1–2 wt % oxygen is combined with most polymers, the product is no longer useful. By speeding up the effect of oxygen, metallic salts are some of the most powerful catalysts of oxidation. Manganese, copper, iron, nickel, and cobalt are the worst; however, vulcanization helps to diminish their harmful effects.

Although some antioxidants, eg, the p-phenylenediamine derivatives, are active against the metallic-catalyzed oxidation of rubber, generally the standard antioxidants do not give protection against heavy metal ions. Since the activity of the metals depends on their being in an ionic form, it is possible to protect stocks by incorporating substances that react with ionic metals to give stable coordination complexes. Substances of this type are used in other industries as sequestering agents. The sequestering agents are normally used in combination with 1.0 part of the standard type of antioxidants.

In practice, the factor of heat results in a combination of cross-linking reactions and in increased rate of oxidation. The amount of oxygen necessary to cause a given reduction in tensile strength decreases markedly with increasing temperature. Thus, as the products are used at higher operating temperatures, much less combined oxygen causes deterioration.

Oxygen and heat generally are the main causes of deterioration. Usually 1–2.0 parts of the diarylamines, the reaction products of diarylamines and ketones, or the reaction products of primary aromatic amines and ketones are used in carbon black-loaded compounds. In nonblack compounds for heat or oxygen protection, 2.0 parts of the alkylated monophenolics or 1.0–2.0 parts of alkylated diphenolic or phenolic sulfide antioxidants are required.

Phenolic antioxidants are generally weaker than the amine antioxidants in aging protection. The phenolic types offer little flexing protection. Their effectiveness is drastically reduced in the presence of carbon black. The main use of the phenolic antioxidants is in compounds requiring a low degree of staining or discoloration.

The amount and type of antioxidant required for protection against heat up to 100°C is similar to the previous recommendations for oxidative protection. However, for heat protection against 121°C and higher temperatures, the amount of antioxidant should be increased to 2–4 parts and the volatility of the antioxidant should be very low (51). The curing system should be the sulfurless or EV type. In neoprene compounds, the use of 4.0 parts of dioctylated diphenylamine antioxidant (Octamine) plus 0.5–1.0 *p*-(*p*-toluenesulfonylamido)diphenylamine (Aranox) gives excellent heat protection (52).

When stretch vulcanizates are exposed to an outdoor atmosphere, a series of cracks may appear, the lengths of the cracks being perpendicular to the applied stress. This type of aging is caused by small traces of ozone (qv) in the air. After World War II, the switch from the use of natural rubber to SBR accentuated cracking in tires. These better-wearing SBR tires tended to form fewer but deeper cracks than the natural-rubber tires, which gave many more shallow cracks. This cracking was especially true in military tires that were stored from World War II to the Korean War. The discontinuance of the wrapping of passenger tires with paper also accentuated the ozone cracking problem. The study of smog and the cracking of rubber in the Los Angeles, Calif. area in the early 1950s pointed towards ozone as the main cause of the cracking. Prior to that, only the wire industry worried about cracking from ozone. The addition of ozone-aging requirements in purchase specifications for rubber parts has increased the awareness of the problem. The increased registration of automobiles with subsequent air pollutants from fuel exhaust has greatly increased the ozone concentration.

Ozone is formed in the stratosphere by the action of ultraviolet light from the sun on atmospheric oxygen. It is brought into the earth's atmosphere by wind currents. Thus, the amount of ozone in the air varies with geographical location and with the season. The general concentration of ozone in air is 0–6 parts per hundred million (10^8). The oxidant formation in the Los Angeles atmosphere is explained by the theory that ozone is formed by nitrogen dioxide photolysis. Nitrogen dioxide is formed by interaction between nitric oxide and hydrocarbon from fuel exhaust. The most active atmospheric ozone-forming hydrocarbons are olefins, alcohols, and aldehydes. Neither atmospheric gases, eg, sulfuric dioxide, nitrogen oxides, and chlorine, nor ultraviolet light cause the cracking associated with ozone. The nitrogen oxides do produce a discoloration problem with various antioxidants. Ozone absorption depends upon elongation, which is in contrast to oxygen absorption. Stretched rubber cracks as rapidly at night as it does in the daytime and very often more rapidly in the shade than in the sunlight. On outdoor exposure, black stocks are more vulnerable to ozone cracking than light-colored stocks because the latter more readily form a protective oxidized surface skin. However, once light-colored stocks begin to crack, they fail in a very short time.

For every rubber compound there is a critical elongation or stress at which the article is most vulnerable to ozone attack. This critical point varies, depending on the base polymer, amount, and type of filler loading, etc, although it is usually 10–50% elongation. As the elongation is increased above the critical point, cracks become progressively more numerous but smaller and more shallow. Below the critical point, cracks form more slowly but are usually deeper and, in time, are completely disruptive.

Processing and molding has a large influence on ozone cracking. Rubber, when molded under strain, cracks faster than when cured without strain. The greater the strain, the faster the rate of cracking. Aging of the uncured stock before molding also decreases cracking. Cracking is produced or accelerated by grit, foreign matter, impurities, eg, various metallic salts, and poor dispersion.

Since exposure cracking requires the presence of both strain and ozone, two methods can be used to improve the resistance of the stock. The rubber may undergo high rates of stress decay by being compounded with such materials as bitumens or the rubber can be compounded with materials that prevent ozone attack. Inasmuch as the attack of ozone is a surface phenomenon, the use of a wax to bloom from the rubber to form an impermeable surface coating is useful in preventing attack.

Once the surface coating is broken, the ozone will channel into the break and a very deep crack rapidly results. For this reason, it is essential to select the wax carefully. A blend of waxes is recommended to ensure adequate protection in both summer and winter and according to geographical region. Generally, between 2.0–10.0 parts of wax or wax blend are used depending on filler and plasticizer loading.

The use of a chemical to combat the attack of ozone in the same way as antioxidants combat oxidation has great merit. These chemicals, which are called antiozonants, do a good job in this respect and offer dynamic as well as static protection. These antiozonants are generally *para*-phenylenediamine derivatives. The main member is the diphenyl derivative (JZF, DPPD) which is widely used as a constituent in antiflex cracking mixtures but, because its solubility in rubber is limited, it cannot be used as an antiozonant.

The substitution of tolyl and xylyl groups for phenyl groups in the diaryl *p*-phenylenediamine class of antiozonants improves the solubility to the extent that these derivatives, eg, Wingstay 100 and 200 (Goodyear Tire & Rubber Company), can be used in rubber as moderately active antiozonants and retain the antiflex cracking ability of the diphenyl body. They have a slower migration rate than the dialkyl *p*-phenylenediamines and the alkyl–aryl *p*-phenylenediamines.

If one or both of the phenyl groups are replaced by alkyl groups, the products become very good antiozonants. The unsymmetrical bodies or N-alkyl-N'-phenyl-p-phenylenediamine derivatives are very active antiozonants, especially under dynamic conditions. The alkyl group varies from 3–8 carbons and they are represented by Flexzone 3C, 6H, 7L, 7F, and 11L (Uniroyal); Santoflex IP, 13, 13F, and 134 (Monsanto); Wingstay 300 (Goodyear); UOP-588, 688 (Universal Oil Products), Antiozite 67, 67F (R. T. Vanderbilt); and Vulkanox 4010NA and 4020 (Mobay). The most widely used antiozonant is the one containing the 6-carbon alkyl group.

The dialkyl p-phenylenes are very active antiozonants, especially under static conditions. The alkyl groups in these symmetrical compounds contain 7–8 carbons and are represented by Flexzone 4L, 8L, and 18F (Uniroyal); Santoflex 77, 17, and 217 (Monsanto); UOP-788, 88, 288, and 26 (Universal Oil Products); and Vulkanox 4030 (Mobay). These symmetrical molecules reduce the processing safety of rubber compounds and are more susceptible to heat and oxygen aging than the unsymmetrical antiozonants.

Blends of the unsymmetrical and symmetrical p-phenylenediamine antiozonants are available from rubber chemical suppliers (Flexzone 9L, 10L, L2L, and 15L from Uniroyal; Santoflex 734 from Monsanto; and UOP 256 from Universal Oil Products). These blends combine the best properties of each class and simplify inventory and storage problems.

The p-phenylenediamine derivatives produce stock discoloration and stain by contact, which restrict their use to black compounds. There are some differences among them in degree of migratory stain. These p-phenylenediamine derivatives are used in tire compounds at 1.0–4.0 parts and generally they are used with 1.0–3.0 parts of blended waxes, which function as migratory aids.

Styrene–butadiene rubber, polybutadiene, nitrile, polyisoprene, and natural rubber are quite susceptible to ozone attack, and the above p-phenylenediamines and 1,2-dihydro-2,2,4-trimethyl-6-ethoxyquinoline (Santoflex AW, Monsanto) give adequate ozone protection. Nickel dibutyldithiocarbamate (NBC) functions as an antiozonant at 2.0 parts in SBR, BR, nitrile, and neoprene. Nickel dibutyldithiocarbamate can be a prooxidant in natural rubber and polyisoprene. Poly(vinyl chloride) resins are used to improve the ozone cracking resistance of nitrile rubbers (Paracril OZO, Uniroyal; Chemivic, Goodyear; and Hycar Polyblends, BF Goodrich).

Hypalon, butyl, neoprene, and ethylene–propylene co- and terpolymers are the most resistant rubbers to ozone attack. The resistance of the Hypalon, butyl, and ethylene–propylene polymers results from their low unsaturation, whereas neoprene resistance results from its chlorine atoms making the carbon-to-carbon double bond less reactive.

With higher operating temperatures, butyl and neoprene lose their ozone resistance somewhat and must be compounded with antiozonants and waxes for protection. With neoprene, higher amounts, ie, 4.0–5.0 parts, of the standard amine antioxidants in conjunction with wax impart ozone resistance (53). High amounts of the phenolics tend to reduce the ozone resistance of neoprene.

Frosting is the product of an oxidation process produced by ozone under static conditions (54). A freshly cured shiny surface may become dull and frosted within a few minutes to several hours because of the formation of minute cracks. Frosting has a bloomlike appearance and it usually develops during hot humid weather. It can be prevented by compounding with certain waxes, antiozonants, and antioxidants.

Weather consists of a combination of many things, eg, visible light, water, heat, wind, and barometric pressure, and chemicals, dust, smoke, pollen, and microorganisms also may be present in the atmosphere. The combination of these many constituents constantly changes in direction and intensity at unpredictable and varying rates. This variability makes it difficult to predict the weathering resistance of any rubber compound by artificial methods. It also indicates that several methods of protection against weathering are probably needed.

Light promotes the action of oxygen on the surface of the rubber compounds. This photooxidation process leads ultimately to discoloration and loss in dielectric and mechanical strength. The effect of light on rubber and plastic parts varies with wavelength. Ultraviolet, ie, short wavelength, rays have the greatest effect. Since light-catalyzed oxidation is predominantly a surface effect, thin samples deteriorate more quickly than thick ones. This is contrary to experience in oxygen bombs, where the rate of oxidation is substantially independent of the thickness of the samples.

A depletion of antioxidant takes place also in the presence of light. Therefore, antioxidants added for sunlight protection should be used in greater amounts, ie, at 3.0 parts, than are customarily used for resistance to thermal oxidation. Because of the nondiscoloring or nonstaining factor generally involved, phenolic antioxidants are used alone or in combination with phosphite antioxidants.

The more important fatigue failure is the dynamic type, more popularly called flex cracking. The cause of this failure is twofold: stress breaking of chains and cross-links and, more important, oxidation accelerated by heat buildup in flexing. As with ozone cracking, flex cracking can be considered in two parts: initiation of cracks and crack growth. To date, the prevention of crack growth is a compounding problem, and no chemical commercially available is effective against it. The prevention of crack initiation, however, is a problem that can be helped by both antioxidants and compounding. Generally, the following suggestions should help extend flex life: use fine particle size, well-dispersed fillers, and avoid excessive volume loadings; obtain an optimum cure avoiding both over- and undercures, and avoid low sulfur; use a good flex-cracking antioxidant, eg, N,N'-diphenyl-p-phenylenediamine or mixtures containing it or the p-phenylenediamine antiozonants; choose the proper polymer (butyl is best, followed by neoprene at low temperatures and natural rubber at moderate temperatures); avoid large amounts of blooming waxes; avoid excessive heat buildup; design for minimum strain range.

In its effect on physical properties, radiation damage is very similar to heat aging of vulcanizates. Loss of tensile strength and increase in modulus after radiation treatment parallels similar changes for SBR stocks that have been heat-aged. A consequence of radiation damage is the formation of volatile products, hydrogen, or hydrocarbon gases. These products cause voids or blisters in the stock at high radiation doses. Apparently it makes no difference whether the rubber is exposed to high radiation dosages for a short time or low doses for a long time; the damage incurred is of the same type and approximately the same extent. Of the chemicals investigated by Goodrich as antirads, the best class was the N,N'-disubstituted p-phenylenediamines, particularly N-cyclohexyl-N'-phenyl-p-phenylenediamine (Flexzone 6H).

Different types of antioxidants often function by different mechanisms, and a given antioxidant may react in more than one way. Thus, a material that acts as an antioxidant under one set of conditions may become a pro-oxidant in another situation. It has been established that amine antioxidants act both by reacting with free-radicals and by decomposing peroxides. Phenolic antioxidants, on the other hand, react primarily as free-radical traps or chain terminators. Phosphites function by reacting with free peroxides.

Because of the different mechanisms of protection afforded by various types of antioxidants, synergistic rather than simple additive effects can be obtained. Combinations of amine and phenolic types, phenolic and phosphite types, or combinations of two chemicals of the same type can be advantageous in rubber compounds. Inhibitor regenerators, eg, the thiodipropionates, are useful in high temperature aging resistance when used in combination with phenolic antioxidants. The protective agents should be added to the rubber compounds at the beginning of the mixing cycle to be most effective.

Reclaimed Rubber and Ground Scrap

There are two basic factors that determine the type of reclaim. The first and most important is the type of scrap from which the reclaim is made. The second is the process by which the scrap is reclaimed (see Recycling, rubber). The most important properties of reclaimed rubber for compounding purposes are its viscosity and elasticity, its rubber hydrocarbon content, its freedom from lumps, and several less tangible properties, eg, tack, dryness, or stickiness and mushiness.

The choice of the relative proportions of new and reclaimed rubber to be used in a compound depends completely upon its intended use. The use of reclaim in tire treads always results in a sacrifice of quality, eg, abrasion loss, whereas in carcass and bead insulations, reclaim produces no sacrifice in quality. In radial sidewalls, reclaim is not used. Butyl reclaim is used in inner tubes and tire inner-liner compounds. Reclaim also is used in some mats, mechanical goods, solvent cements, and water dispersions. These water dispersions, which are artificial latex made from reclaim, are used as adhesives or fiber binders.

The reinforcing effect of fillers in reclaimed rubber follows the same order of effectiveness as in new rubber. The same amount and same type of softeners are used in reclaim as in new rubber. Reclaimed rubber needs less zinc oxide than new rubber. The same accelerators and vulcanizing agents are used with reclaimed rubber as with new rubber. Reclaimed rubber cures faster than SBR, so the compounder can either take advantage of the faster cure rate or reduce the amount of the accelerator and have a cost savings with an equivalent state of cure. The alkaline reclaims cure faster than the neutral types. The sulfur ratios are usually 2.5–3.5 parts per 100 parts of reclaimed rubber hydrocarbon (RHC), which are the ratios normally used for natural rubber. If synthetic rubber is blended with reclaim, then the sulfur ratio is normally the weighted average of the two ratios required for the respective rubbers. The use of reclaim in tires has been greatly reduced over the past 10–15 yr, which has resulted in a large reduction in the chemical reclaiming capacity.

The use of grinding to reclaim rubber scrap has been increasing. The conventional grinding methods produce coarse, ie, 600-μm, powders. The coarse powders generally reduce the physical properties below satisfactory limits. The use of cryogenic grinding of cured rubber scrap for reuse in molded parts has been advocated (55). Use of liquid nitrogen produces fine particles because the rubbers are pulverized at temperatures below the rubber compound's glass-transition or embrittlement temperature. The use of cryogenically ground neoprene and nitrile scrap produces little or no change in hardness, gives stable physical properties, and has a minimum effect on cure (56) (see Cryogenics).

Conditions of Cure

The combination of vulcanizing agents, accelerator, activator, type of polymer, kind and amount of filler loading, and other accessory materials chosen for a particular compound determines its curing rate. The curing rate is measured by the rapidity with which the physical properties of the rubber compound develops with the time of heating. The method of determining the curing rate is to use a cure meter to obtain a torque-vs-curing time curve. The old method has been to cure test sheets for various time intervals at the desired vulcanizing temperature, test the stress–strain properties of the resulting vulcanizates, and plot the tensile strength and the force required to produce a particular elongation, for example, 300% modulus vs the time of cure. In Figure 2 three cure-meter curves are shown illustrating the varying types of behavior encountered. The time required to attain the optimum cure is a measure of the rate of cure. Optimum cure is the state of vulcanization at which a desired property value or combination of property values is obtained. It is generally considered as the time required to attain 90% of full torque or modulus development. Cure time is the time in minutes to reach the optimum cure. Any cure shorter than the optimum cure is an

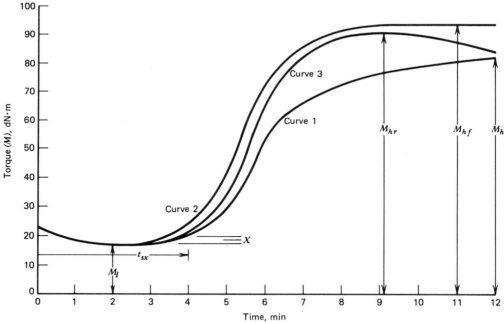

Figure 2. Curing curves. M = torque in dN·m (to convert dN·m to lbf·in., divide by 1.130); M_l = minimum torque, M_h = highest torque attained during specified period of time when no plateau or maximum torque is obtained (Curve 1), M_{hf} = maximum torque where curve plateaus (Curve 2), M_{hr} = maximum torque of reverting curve (Curve 3). T = temperature. t = time; t_{sx} = scorch time in minutes to increase x units above M_1, tX = cure time in minutes to $X\%$ of maximum torque, $t'X$ = cure time in minutes to $X\%$ of torque increase, ie, minutes to $M_1 + X(M_h - M_l)/100$ torque. The most commonly used values of X are 50 and 90.

undercure, and any cure longer than the optimum is an overcure. Extreme overcures may be of two types. One type is illustrated by curve 3 in Figure 2 by the drop in torque beyond the max. This happens with most natural rubber, polyisoprene, or butyl compounds, and this type of overcure is called reversion. The physical properties of a reverted stock are lower hardness, modulus, and tensile strength and higher elongation. In the other type of overcure, the stock continues to harden, the torque or modulus rises, and the tensile strength and elongation fall. This type is characteristic of SBR and most other synthetic rubbers. Curve 2 in Figure 2 represents a plateau or flat curing curve. Curve 1 represents a creeping torque or modulus curve, which has not reached optimum cure in the alloted test time.

The cure to be used on a particular product depends on the following factors:

The size and shape of the product. By all methods of curing except that of radiofrequency, heat is supplied to the product at the surfaces. Rubber is a poor conductor of heat and thick articles may require long periods of time during vulcanization before the temperature of the center regions of the article is the same as the impressed temperature. This time lag or incubation time is a function of the size and shape of the article and the thermal diffusivity of the rubber. Factors affecting the problem of obtaining equivalent cures in specimens of various shapes have been summarized (57). A nomograph has been developed that permits the estimation of incubation time for various shaped rubber articles. The nomograph is accurate at 138°C with a ca 5% error for temperatures of 138°C ± 33°C. The most favorable shape for low incubation time is a sphere, and the most unfavorable shape is a thick flat slab of relatively large length and width. The second factor is the minimum dimension. In the case of the sphere this dimension is the diameter, and for the large flat slab it is the thickness. Values for thermal diffusivities are given and all are quite low, eg, 0.0097–0.0845 cm^2/min for natural rubber tread stocks to 0.0539 cm^2/min for butyl. The thermal diffusivity values increase with increased pigment loading and decrease with increasing temperature.

The temperature of cure. The effect of temperature on the vulcanization rate over the range of temperature from room temperature to the highest temperatures employed commercially can be represented by an Arrhenius-type plot relating the logarithm of the time required, ie, the reciprocal of the rate, to the reciprocal of the absolute temperature. The curves closely approach straight lines and correspond to an apparent activation energy of ca 84–105 (kg·J)/mol (ca 20–25 (kg·cal)/mol). The published data on this point are not altogether reliable, probably because of the difficulties involved in determining the cure rates. Most investigators agree that some changes in the effect of temperature on cure rate result from changes in the curing system. As a rough approximation and for convenience in transposing cures from one temperature to another it is generally assumed that the cure rate is doubled for a 10°C increase in the curing temperature. At 125–160°C, ie, the range in which most rubber products are vulcanized, this is a fairly reliable approximation.

In making a change from one curing temperature to another, any incubation time which is included in the curing time of the article must be subtracted before the new curing time is calculated. Thus if the curing time of an article is 65 min at 138°C, including an incubation time of 15 min, and it is desired to find the equivalent curing time at 149°C, the incubation time is first subtracted giving 50 min at 138°C. The time at 149°C that is equivalent to 50 min at 138°C is 23 min. To this is then added the 15 min incubation time giving a final cure of 38 min at 149°C. The reason for this procedure is that the incubation time is independent of the temperature.

The temperature of cure selected for a particular product is determined by many factors, including the following:

For mold cures it is economically advantageous to have the fastest possible mold turnover. Thus the temperature chosen is usually the highest possible temperature consistent with the production of goods of satisfactory quality.

The size of the product frequently determines the highest temperature that can be used. A product having a large volume of rubber has a more uniform cure throughout its cross section if a long, low temperature cure is used rather than a short, high temperature cure. Frequently for products of this type, a step-up cure is used in which a period of time at a low temperature is followed by another period at a higher temperature.

The maximum temperature is sometimes limited by the pressure which the equipment safely withstands, ie, for steam cures.

The maximum temperature is frequently limited by the occurrence of manufacturing defects at high curing temperatures.

For colored stocks, lower temperatures are generally employed to obtain brighter colors.

In hard rubber products the danger of a violent exothermic reaction limits the maximum temperature.

In bladder curing of tires, the maximum cure temperature not only depends upon the thickness of the tire, but it is also limited by its effect on the service life of the curing bladders.

In air curing, the heat capacity of the air is so low that it is more important to provide adequate circulation of the air than to raise its temperature.

For products that are cured without pressure or in saturated steam without any additional pressure on the rubber other than the steam pressure; or in steam cures in which the product is cloth-wrapped to provide additional pressure, the maximum temperature that can be employed is limited by the tendency of most mixtures to blow and become porous if cured at too high a temperature.

In injection molding, liquid-curing-medium (LCM) process and continuous vulcanization (CV) wire curing at 204–260°C, the curing systems used and thermal diffusivity factors become very important.

Some of the reasons why the compounder does not run a full cure in the center region of the article are as follows: the article may perform adequately in service when cured for a shorter time even though some degree of undercure exists in the thickest portions; the reduced curing time results in cost savings; and the amount of overcure at the surface is reduced.

Hard Rubber

Hard rubber, also called ebonite or vulcanite, is the hard tough composition produced by prolonged vulcanization of rubber with a large proportion of sulfur. The ASTM classifies hard rubber as compositions having a combined sulfur-to-rubber hydrocarbon ratio in excess of 15%. The polymers used in compounding hard rubber are new or reclaimed natural rubber, polyisoprene, SBR, and nitrile rubber. Butyl, neoprene, EPDM, and Thiokol cannot be vulcanized to hard rubbers. Sulfur is the only vulcanizing agent for hard rubber.

Hard rubber has a unique combination of properties, which has caused it to be adopted for many varied uses. However, it is being challenged by many of the synthetic plastic resins. The properties of hard rubber are desirable physical properties, eg, toughness combined with an adequate degree of hardness; resistance to many chemicals; excellent electrical insulating properties; ease of machining; and attractive physical appearance, especially its high polish. There are some disadvantages associated with hard rubber. The material requires very long cures which are troublesome and costly. This is in contrast to the very short curing or molding cycles used in the plastic industry. The product is also subject to discoloration and loss of electrical properties after light and air exposure and during service at high temperatures, ie, >70°C.

When a mixture of either natural or synthetic rubber and sulfur is heated, the sulfur combines with the rubber in such a manner to render it unextractable with acetone. With natural rubber, this reaction results in one sulfur atom adding for each double bond of isoprene and, at saturation, $(C_5H_8S)_n$ forms, which corresponds to 32 wt % combined sulfur, ie, to a vulcanization coefficient of 47 (= 32 × 100/68). Synthetic rubbers with butadiene as one of their constituents, upon reaction with sulfur to saturation, form $(C_4H_8S)_n$, which corresponds to 37.3 wt % combined sulfur, ie, to a vulcanization coefficient of 59.5. Sulfur present in excess of the theoretical max required for saturation of the double bonds may react by long heating through substitution of the hydrogen atoms in the hydrocarbon molecule. In practice, sulfur ratios both lower and higher than those calculated above are employed to produce specific effects. The only physical property improved by low sulfur, ie, 28 wt %, is impact strength. High sulfur, ie, ≥35 wt %, improves the reduction of dielectric loss and the resistance to organic liquids and oils. Ca 40 wt % sulfur is the optimum content for swelling resistance with natural rubber. The only property for which there is a marked optimum sulfur content is resistance to heat distortion, which in natural rubber is ca 35 wt % sulfur. Mechanical strength, as measured by cross-breaking, or tensile strength of hard rubber prepared from natural rubber are little affected by the sulfur ratio, although a 35–40 wt % sulfur content gives the best strength. However, hard rubber from SBR and nitrile polymers gives the best strength with low sulfur content (25–35 wt %). Properties that are not affected by varying sulfur content are electrical breakdown strength, volume resistivity, and resistance to light.

At least some of the reactions involved in the formation of hard rubber are strongly exothermic. The heat is liberated at an early stage in the reaction, which indicates that substitution or dehydrogenation takes place in the rubber molecules. The temperature rise resulting from exothermic reaction depends mostly on the curing temperature and the thickness of the sample. When excessive substitution takes place in the rubber molecule, a considerable amount of hydrogen sulfide is given off, and this usually destroys the cured sample by making it porous. Differential thermal analysis (dta) indicates that the heat formed in vulcanizing a 68:32 natural rubber–sulfur mix is 1.35 kJ/g (323 cal/g) of rubber. The time required for the beginning of the exothermic reaction is considerably longer for SBR than for natural rubber; but the reaction, when it does occur, produces a much higher temperature rise in SBR. Accelerators and metallic oxides do not materially affect the amount of heat evolution. Their beneficial effects mainly result from a spreading out of the exothermic reaction resulting in a lower initiation temperature, a decrease in the rate of temperature rise, and to a small extent a decrease of the amount of heat evolution.

The curing times for hard rubber are quite long compared to those required for soft rubbers. A cure of 3–5 h at 70°C is required for a simple rubber–sulfur mix. In general, the thicker the product to be cured, the lower the impressed temperature must be to prevent the development of excessive temperatures in the center of the material with consequential blowing and porosity. The most effective way to make a hard rubber stock that can be cured quickly without overheating is to reduce the proportion of new rubber plus sulfur in the mix, since the reaction between these two produces the heat. The effectiveness of any diluent depends on the volume fraction of diluent in the total mix. If mineral fillers are used, there is a limit to the permissible dilution because such fillers make the hard rubber weak and brittle. Hard-rubber dust has much less weakening effect and is therefore the best filler for this purpose. This dust can be obtained by grinding hard rubber scrap, or it can be manufactured specifically for this purpose. Besides reducing the magnitude of the exothermic effect, hard-rubber dust improves processing and reduces shrinkage.

Shrinkage is a problem in hard rubber. A reduction in volume occurs, which for an unfilled composition amounts to 5.5%. This change in volume is frequently objectionable since it makes the control of dimensions difficult, and in molded products it produces shrinkage marks on the product.

Organic accelerators similar to those used in soft rubbers are effective in curing hard rubber, and they accelerate the final hard-rubber phase of the rubber–sulfur reaction. The degree of acceleration is much less than in soft rubber and the different accelerator types do not produce the same relative efficiencies. Neither zinc oxide nor magnesium oxide function as accelerator–activators in the rubber–sulfur reaction. The most effective accelerators for maximum safe cure rate are butyraldehyde–aniline (Beutene) and diphenylguanidine (DPG). Zinc oxide reduces the maximum safe rate when added to an accelerated mix, whereas magnesium oxide produces an improvement. Beutene plus magnesium oxide is the most effective combination for reducing the cure time without risk of overheating. Calcium oxide is used to accelerate the cure. Accelerators do not alter the relationship between combined sulfur and physical properties as they do in soft rubber; their function is simply to increase the rate of rubber–sulfur reaction and, perhaps as a consequence of this, enable more of the added sulfur to be combined.

As in soft rubber compounding, a variety of pigments and softeners is used. With the exception of the beneficial effects that pigments have in raising the softening temperature, their use does not enhance the physical properties of the hard rubber, but they are useful in aiding processing and in reducing costs. Carbon blacks are generally avoided because they produce brittleness and adversely affect electrical properties. Other pigments also reduce toughness but not as markedly as the carbon blacks. The most popular pigments are clay, asbestos, whitings, barytes, and related materials. The mineral fillers reduce the rate of absorption and the ultimate absorption of organic liquids.

Many of the softeners used in soft-rubber compounding cannot be used in hard-rubber compounding, because the reduction in solubility of the softener in the compound resulting from the high state of cure causes the softener to exude from the product. Frequently linseed and soybean oils are used; they react with sulfur during cure to give solid materials that do not bloom. Other materials used as processing aids include factices, mineral rubber, resins, petroleum oils, petrolatum, and waxes.

Hard rubber is generally cured by open steam, water, and press molding. In open-steam curing, the loss of sulfur by volatilization from the surface layers tends to reduce the degree of cure. However the stock receives a longer effective cure because of the quicker heating and the greater retention of exothermic heat. Open-steam curing gives a hard rubber which absorbs water more readily and has greater dielectric loss. This method of cure is used for products that do not require a lustrous surface, eg, tank linings, drums, battery cases, tubes, and rods. Curing in water, provided the hard rubber is properly protected from water contact by metal foil, gives results in all respects comparable with those of press curing except for a somewhat slightly lower cross-breaking strength. This method is used for combs, trays, and high grade sheet material. Generally, if a large number of piece parts are needed, it is necessary to resort to a press-molding operation. Since the molding cycle must be minimized, an accelerated rubber compound must be used. Magneto parts, pipe stems, tool handles, and syringes are produced by this method.

Natural rubber and SBR are the principal polymers used for making hard-rubber products, although nitrile and polyisoprene are also used to some extent. There are only minor differences between the properties of the various types of hard rubber obtained from these polymers. Styrene–butadiene and nitrile rubbers are generally inferior to natural and polyisoprene hard rubbers in toughness and impact strength, but they have the advantage of higher softening temperatures. The nitrile rubbers, eg, Paracril, are excellent in this respect and also in their resistance to solvents. Polyisoprene produces hard rubber with greater resistance to water absorption and has superior electrical insulation properties to natural rubber.

The physical properties of greatest interest in the development of hard-rubber material are the following: tensile strength and elongation, Shore D hardness, impact strength, distortion temperature, flexural strength, electrical properties, and stability on aging in air at elevated temperatures and on aging in light. Some typical applications for hard rubber include: withstanding corrosive chemicals in pipes and fittings, pumps, and tank linings; as a hard-rubber base for rolls and industrial solid tires; providing a means of adhering soft rubber to a metal core; forming sheet, rod, and tube for electrical insulating purposes; as storage-battery cases, dentures, and grinding wheels in which hard rubber is used to bond the abrasive particles; bowling balls, combs, fittings for drug sundries, and cellular gasoline floats.

Cellular Rubber

A cellular product is a two-phase system consisting of a gas as the dispersed phase in a continuous phase of solid material, eg, rubber. The properties of such a cellular product are generally determined by: the physical properties of the rubber phase in which the gas is dispersed, the density of the material, the type of cell structure, whether open or closed cells, and the nature of the dispersed gas. In the first cellular-rubber products, the rubber was expanded by means of a gas, eg, nitrogen, nitrous oxide, or carbon dioxide, introduced under high pressure into a partially cured compound. Completion of the cure at higher temperatures and subsequent release of pressure produced the expanded product. This process requires special equipment and considerable capital.

Chemical blowing agents are being more widely used then ever before in the manufacture of cellular rubber parts for consumer, automotive, and industrial applications, as well as in cellular plastics. There are two distinct classes of blowing agents:

the inorganic and the organic types. Inorganic blowing agents, eg, sodium bicarbonate, ammonium bicarbonate, and various forms of ammonium nitrite have been used. They generally release carbon dioxide gas upon decomposition. Sodium bicarbonate is the most popular. It is activated by fatty acids, eg, stearic acid. These inorganic blowing agents are difficult to disperse in rubber and may leave undesirable residues in the sponge. They should be finely ground. The inorganic blowing agents produce open-cell sponge.

The organic types are nitrogen compounds, which are stable at normal storage and mixing temperatures, but undergo controllable gas evolution at reasonably well-defined decomposition temperatures. Reviews of organic materials that have been suggested or used as blowing agents are given in refs. 59–61. Organic blowing agents have the advantage of being more varied and hence are more versatile than the inorganic types. They cover a wide range of chemical compositions, physical properties, and decomposition characteristics. Their varying decomposition temperatures are satisfactory for every known industrial use. Table 25 is a list of the commercial blowing agents. The higher temperature decomposing materials, eg, Celogen AZ or RA, can have their decomposition rates considerably accelerated at lower temperatures by promoters or activators, eg, zinc oxide, lead salts, BIK, glycols, etc (61–62). This

Table 25.   Blowing Agents for Rubber and Plastics

Chemical name	Trade name	Manufacturer	Decomposition temperature, °C	Amount of gas, cm³/g
diazoaminobenzene	Porofor DB	Mobay Chemical (Bayer)	100	115
benzenesulfonyl hydrazide	Geniron BSH	Whiffen & Sons	95–105	120
	Porofor BSH	Mobay Chemical (Bayer)		
p-toluenesulfonyl hydrazide	Celogen TSH	Uniroyal, Inc.	110	115
p,p'-oxybis(benzenesulfonyl hydrazide)	Celogen OT	Uniroyal, Inc.	164	
	Nitropore OBSH	Olin		
azodicarbonamide	Celogen AZ	Uniroyal, Inc.	196–205	225
	Kempore	Olin		
	Azocel	Fairmount Chemical Company		
	Genitron AC	Whiffen & Sons		
	Porofor ADC	Mobay Chemical (Bayer)		
	Ficel AC	Vulcanox		
activated azodicarbonamide	Celogen 754	Uniroyal, Inc.	165	200
	Porofor ADC-K	Mobay Chemical (Bayer)		
modified azodicarbonamide	Celogen AZNP, AZVR	Uniroyal, Inc.		
	Ficel EP-A, B, C, D	Vulcanox		
	Kempore FF, MC	Olin		
dinitrosopentamethylene-tetramine	Vulcacel BN 94	ICI	195	265
	Opex	Olin		
	Porofor DNO/N	Mobay Chemical (Bayer)		
p-toluenesulfonyl semicarbazide	Celogen RA	Uniroyal, Inc.	235	145

phenomenon enables the sponge manufacturer to control the decomposition rate to fit the requirements of the rubber formulations.

Cellular-rubber products prepared from chemical blowing agents are becoming increasingly popular for technical and economic reasons. Their special features include heat and sound insulation properties, lightness and comfort in apparel, shock absorbency, improved textures and consumer appeal, and low weight-per-volume costs.

It is possible by the use of chemical blowing agents to produce a variety of cellular-rubber products varying from those with low densities and low hardness to those with moderately high densities and high hardness, depending upon the final application of the component part. Semiblown resin shoe-soling material requires a high degree of hardness, ie, ca 90 Shore A with sp gr ca 0.9, whereas rug-underlay sponge has a compression resistance at 25% deflection of 21 kPa (3 psi) at sp gr ca 0.5. Cellular materials can be compounded for oil, ozone, and low temperature resistance as well as for meeting applicable government regulations on burning characteristics of materials used in specified applications.

The type of cell structure in cellular materials is a chief factor in determining their properties and uses. Open-cell cellular material made from a dry polymer or polymers is called a sponge, but when it is made from a latex or latices it is called a foam. Open-cell cellular material consists of a network of interconnecting cells which are formed by the expansion of decomposed gases or whipped-in air which develops sufficient internal pressure to rupture the polymeric cell walls. Open-cell sponges and foams usually have good resilience and compression-set properties but have poor mechanical strength and insulation properties and high water absorption. Closed-cell, expanded cellular material consists of a predominantly noninterconnecting network of cells formed by the entrapment of gas as discrete bubbles in a matrix of rubber or plastics. The closed cells are produced by the proper cure–blow balance, so that sufficient gel or cure has developed to encapsulate the nitrogen gas as the decomposition of the blowing agent begins. Closed-cell, expanded cellular materials have excellent thermal insulation and sound-absorption properties. The water-vapor transmission and water-absorption properties of the closed-cell materials are very low.

Blowing. Blowing can be compared to baking a cake with yeast; the blowing agent decomposes and causes the compound to expand and this is followed by curing. An increase in volume by several hundred percent can be obtained, depending upon the amount and type of blowing agent used and several other components in the base compound. Compounds used to produce cellular rubber by this process usually require the composition of the cellular compound to be very soft. The low plasticity of such compounds is essential to reduce the nerve of the rubber and to enable easy expansion of the nitrogen or carbon dioxide liberated from the blowing agent. Cellular materials prepared by this process can be blown in mold cavities by filling 10–50% of the volume of the mold with uncured compound and allowing the compound to fill the mold when heated. The compounds can also be extruded or calendered to any desired shape or profile and simultaneously expanded and cured in any one of several heat-transfer media, eg, hot-air ovens, between heated platens of a press, liquid-curing media, a fluidized bed, or a microwave unit. A product with either open or closed cells or a combination of both can be obtained, depending upon many factors including type and amount of blowing agent used, rate of cure, degree of expansion, etc. Cellular products prepared by the free-expansion method include sponge carpet underlay, direct-molded cellular soling, extruded and molded automobile sealing strips, pipe insulation, and sheet materials for flotation and insulation.

Expansion. In the expansion process, the uncured compound containing blowing agent is placed in a mold. This method differs from the blowing process; instead of the uncured compound filling the mold only partially, enough compound is placed in the mold to fill it completely with 5% excess to seal the mold. The blowing agent decomposes during the cure, which takes place under considerable pressure; under high pressure the nitrogen which is formed partly dissolves in the rubber. The press is opened before the cure of the rubber is complete. Owing to the high internal gas pressure, the precured compound expands and a material with very small closed cells is obtained.

A second curing operation, known as a postcure, is usually used to complete the cure and stabilize dimensions. Organic blowing agents are used almost exclusively for cellular rubbers prepared by this process and consequently a product with predominantly closed-cell structure results. Molded closed-cell sheet material prepared by the expansion process is usually split into sheets of convenient thickness from which finished articles, eg, gaskets, shoe soles, etc, can be dried out.

Latex

The compounding technique for latex differs from that of dry rubber and is fundamentally much simpler. Rubber latex is a colloidal aqueous emulsion of an elastomer. Natural latex is the milky exudation of certain trees and plants of which the only one of great commercial importance is the *Hevea brasiliensis* tree. Synthetic latices are mainly prepared by polymerizing low molecular weight monomers in a free-radical emulsion polymerization system. Artificial latices are water dispersions of reclaimed rubber, butyl rubber, *cis*-polyisoprene, *cis*-polybutadiene, or ethylene–propylene terpolymers.

There are certain applications where solid rubber or a rubber solution cannot be used, eg, in the preparation of foam sponge, in paper coating, or in saturation of a porous material. Less expensive and simpler equipment is required. Since the compounding of latex is done in the liquid state, milling or Banbury mixing is obviated. Very high application rates may be obtained, resulting in less expensive processes and, in many cases, the advantage of continuous operations. Power requirements are less than for nonaqueous applications. The use of latices avoids the hazards and additional expense associated with solutions requiring toxic and flammable solvents.

For dry-rubber compounding, the latex is coagulated, dried, and used in the solid dry state. Latex compounding, on the other hand, is done in the liquid state, and the compounded liquid latex is used directly to produce commercial articles without being first converted to a solid. For such use, the latex is usually concentrated to ca 60 wt % solids by one of several methods, the most common of which are centrifuging and creaming. As with dry rubber, most uses of latex require modification of the latex by addition of vulcanizing agents and other ingredients. The only exceptions are a few adhesives in which latex is used without modification by compounding.

The general principles of latex compounding are similar to those of dry-rubber compounding. The absence of mastication in the processing of latices saves time and power, and the resultant vulcanizate is inherently resistant to aging. A disadvantage of latex is the difficulty of obtaining any measure of reinforcement through the use of small particle-size inorganic fillers. This failure of reinforcement is related to the lack of mastication. Another disadvantage is the difficult drying of latex deposits. This difficulty with the usual accompanying shrinkage rules out the use of latexes in the production of solid articles of thick sections.

In order to obtain a homogeneous and stable latex compound, it is necessary that nonwater-soluble additives be dispersed or emulsified in water to about the particle size range of the latex particles. Thus the compounding ingredients must be capable of being readily dispersed. Therefore, water-soluble ingredients are advantageous, since they can be either added directly to the latex or used in the form of an easily prepared water solution.

Dispersions to be added to latex must have good storage stability, ie, the dispersions must not settle or change in pH. Highly basic systems tend to absorb carbon dioxide from the air, which lowers the pH and increases the electrolyte content as carbonate. The dispersions must be compatible with the latex. The pH should be similar to that of the latex, eg, pH 8.5–11 for alkaline latex and pH 3.5 for acid latex. Addition of low pH material to high pH latex or vice versa generally results in mutual precipitation and coagulation of the suspended particles. Latex particles are negatively charged. Since all the charges are of the same sign, the charges repel the particles from each other and prevent coagulation. Therefore, materials in water that produce Zn^{2+}, Mg^{2+}, or Ca^{2+} ions must be avoided in a latex compound designed for stability. However, these same materials under controlled conditions may be useful in effecting coagulation.

There are several limitations on pigmentation of latex in contrast to that of dry rubber. As indicated, materials giving polyvalent positive ions in water solution, including a number of the loading materials commonly used in dry rubber such as calcium carbonate, are avoided or at least limited to low amounts. Further, reinforcing pigments, eg, the carbon black used in dry rubber compounding, have little if any reinforcing effect in latex, probably because of the much less intimate contact obtainable between carbon particles and rubber particles in latex as contrasted with that obtained by the action of a rubber mill on a dry mix. Carbon black is used in small amounts for coloring. Heavy pigments, eg, lead compounds, are also usually avoided in latex compounding because they settle rapidly from latex mixes. If heavy pigments are used, the viscosity must be adjusted to prevent settling.

Offsetting the limitations or difficulties in latex compounding is possible. Since the mixing of latex compounds is accomplished by stirring the ingredients into the latex in the form of water solutions or dispersions, processing aids, eg, softeners and plasticizers, which are commonly used in dry rubber compounds to reduce nerve and to facilitate mill mixing or subsequent operations such as calendering or tubing, are not necessary and are not used because most of them are dark or cause discoloration. Similarly, because mill mixing is not required and the frictional heat developed with milling of dry compounds is not present, scorching is not a problem in latex compounding and retarders are not used. It is possible to process latex compounds designed for fast cures at low temperatures by use of ultra-accelerators and very fast curing combinations. Some latex compounds can be cured simply by allowing the dried film or article to remain at room temperature for a sufficient length of time. However, although there is no danger of scorch resulting from frictional heat during milling as with dry rubber, such fast-curing latex compounds do exhibit changes, referred to as cureup, in the tank when kept in tanks for long periods, particularly at high room temperatures and with mechanical agitation. Such mixes tend to become viscous and grainy; deposits, films, or articles made from them exhibit poor wet strength and may crack during formation of the deposit or article.

It is possible to further increase cures by preparation of prevulcanized compounds in which, by heating the latex with the curing ingredients, the latex particles in the liquid latex compound are vulcanized so that, after formation of a deposit or article, it is necessary only to dry at low oven or room temperatures to obtain products with properties similar to those of a conventionally postcured article. Although the physical properties obtainable in such an article are not quite as good as those of an article prepared from a conventional compound and then cured, they are adequate for most uses.

Pure gum latex compounds have generally higher tensile strengths than pure gum dry-rubber compounds because the latter lose some strength in mastication and milling. In the case of loaded compounds, however, it is possible to obtain superior tensile, tear, and abrasion properties in dry-rubber carbon black-reinforced compounds, eg, tire-tread compounds, whereas there are no satisfactory reinforcing fillers for latex. Sulfur is the almost universal vulcanizing agent for all polymers, whether in latex or dry form. Less sulfur is usually required for latex than in the corresponding dry mix. By increasing the sulfur concentration to 30–50 parts in latex compounds, satisfactory hard-rubber articles, particularly hard rubber–metal coatings, are obtained.

As in the case for many dry-rubber compounds, the addition of zinc oxide increases the rate of vulcanization and generally improves the physical properties and aging properties. Care is required in the use of zinc oxide in latex because at relatively low pH, Zn^{2+} ions form and tend to destabilize the latex compound. At relatively high pH caused by ammonia, complex zinc ions form which also destabilize latex compounds. It is therefore necessary to adjust the ammonia content and to add stabilizers, eg, fixed alkali, casein, soaps, and surface-active agents, to prevent viscosity increase and coagulation by the presence of the zinc oxide. Zinc oxide is used an an activator in amounts of 1.0–5.0 parts and in larger amounts as a filler. When used as a filler, zinc oxide imparts stiffness. The maximum thickening effect of zinc oxide in latex is at 0.7 wt % zinc oxide. The thickening is minimized with low ammonia and caustic potash content. Other metallic oxides, eg, litharge and magnesium oxide, which are used in dry compounds, are not suitable for latex compounding because of the instability they produce. Lead compounds discolor light-colored stocks.

Fillers, eg, clays and whiting, are used to reduce cost or to provide special properties. Fillers do not reinforce rubber deposited from latex. They are used to increase viscosity to give the latex compound suitable working properties for spreading.

The more active accelerators, particularly the dithiocarbamates, are used more widely in latex compounding than in dry-rubber compounding because scorching during processing is not a problem, many articles from latex compounds are light in color and it is desirable to keep vulcanizing temperatures at or below 104°C to prevent darkening, and in many processes it is very advantageous economically to reduce the time and temperature of vulcanization as much as possible. Dithiocarbamates are particularly suited for prevulcanization of latex in the liquid state. Zinc dibutyldithiocarbamate (ZBDC) is a much faster accelerator in latex compounds than zinc dimethyldithiocarbamate (ZMDC).

Sodium dibutyldithiocarbamate (Butyl namate, R. T. Vanderbilt) is a water-soluble accelerator that can be added directly to latex. Zinc dibenzyldithiocarbamate (Arazate) is a low temperature ultra-accelerator that combines a moderate rate of cure with an exceptional resistance to precure. These properties are advantageously utilized in formulating adhesives, dipping, and other compounds that require frequent replenishing with fresh material. When used in natural latex containing 0.3 wt % or more ammonia, no tendency to precure exists. However as soon as the ammonia evolves, vulcanization progresses at a rapid rate. The zinc salt of 2-mercaptobenzothiazole (ZMBT) is generally used in combination with the dithiocarbamates. It imparts good aging characteristics. Tetramethylthiuram disulfide (TMTD) is very effective in retarding heat and sunlight deterioration of low sulfur, low modulus, dipped latex films.

The use of fatty acids in latex compounds follows the same principles as in dry rubber and, although natural-rubber latex contains some fatty acids or their soaps, it is sometimes necessary to add more usually in the form of a soap. Antioxidants are used in latex compounding in a similar manner as in dry-rubber compounding. Because of the light color of many latex articles, the nondiscoloring and nonstaining phenolic antioxidants are of particular interest. Also, since many latex articles such as gloves are thin and parts of them are under some stress, they are particularly susceptible to cracking from ozone. Waxes, which form a protective coating by rising to the surface, are used in latex compounds as with dry-rubber compounds. Antiozonants can be used but they are very discoloring and staining.

Reclaimed rubber, which is very widely used in dry-rubber compounding, has little use in latex compounding. A dispersion or artificial latex must be made by a rather expensive process of milling in dispersing agents, eg, soaps or casein, and water. Some reclaim dispersion is used in latex compounds for such things as spread goods, adhesives, or fiber binders to reduce cost. However, for most latex compounds it is not desirable because of the poor physical properties that it imparts and the resultant darkening of the compound.

Synthetic. The main types of elastomeric polymers commercially available in latex form from emulsion polymerization systems are butadiene–styrene, butadiene–acrylonitrile, butadiene–styrene–acrylonitrile and chloroprene (neoprene). There are also a number of specialty latices that contain polymers that are basically variations of the above polymers, eg, those in which a third monomer has been added to provide a polymer that performs some specific function. The most important of these are products containing either a basic, eg, vinylpyridine, or an acidic monomer, eg, methacrylic acid. These latices are specifically designed for tire-cord solutioning, paper coating, and carpet backsizing respectively.

The basic constituents of all commercial emulsion-polymerization recipes are monomers, emulsifiers, and polymerization initiators. Other common components are modifiers, inorganic salts and free alkali, and shortstops. The function of these different components and the mechanism of emulsion polymerization have been described (63–64).

Most synthetic latices contain 5–10 wt % nonrubbers, of which more than half is an emulsifier or mixture of emulsifiers. One reason for this relatively high emulsifier concentration as compared with natural latex is that emulsifier micelles containing solubilized monomer play a principal role in the polymerization process. A high emulsifier concentration is usually necessary to achieve a sufficiently rapid rate of polymerization. Secondly, a considerable fraction of the surface of the polymer particles must be covered by adsorbed soap or equivalent stabilizer to prevent flocculation of the latex during manufacture or subsequent use. The emulsifier levels for synthetic latex is high also because of the small particle size relative to that of natural latex.

The most commonly used emulsifiers are the sodium, potassium, or ammonium salts of oleic acid, stearic acid, rosin acids, or disproportionated rosin acids, either singly or in mixture. An alkylsulfate or alkylarenesulfonate can also be used or be present as a stabilizer. A useful stabilizer of this class is the condensation products of formaldehyde with the sodium salt of β-naphthalenesulfonic acid. All these primary emulsifiers and stabilizers are anionic and, on adsorption, they confer a negative charge to the polymer particles. Latices stabilized with cationic or nonionic surfactants have been developed for special applications. Despite the high concentration of emulsifiers in most synthetic latices, only a very small proportion is present in the aqueous phase; nearly all of it is adsorbed on the polymer particles, as noted above.

Neutral or alkaline salts, eg, KCl, K_2SO_4, K_2CO_3, or Na_3PO_4, are often present in synthetic latices in quantities of about one percent or less based on the weight of the rubber. During emulsion polymerization they help to control the viscosity of the latex and, in the case of alkaline salts, the pH of the system. Many polymerizations are carried out at a high pH, requiring the use of fixed alkali, eg, KOH or NaOH. Very small amounts of ferrous salts can be employed as a component of the initiator system, in which case a sequestering agent, eg, ethylenediaminetetraacetic acid may be included to complex the iron. Water-soluble shortstops, eg, potassium dithiocarbamate, also may be included in very small amounts (ca 0.1 parts).

For applications, eg, foam rubber, high solids (>60%) latices are required. In the direct polymerization process, the polymerization conditions are adjusted to favor the production of relatively large average particle-size latices by lowering the initial emulsifier and electrolyte concentration and the water level in the recipe and by controlling the initiation step to produce fewer particles. Emulsifier and electrolyte are added in increments as the polymerization progresses to control latex stability. A latex of ca 35–40 wt % solids is obtained and is concentrated by evaporation to 60–65 wt % solids.

In the second process, a small particle-size latex is prepared and is then treated so that a limited and controlled degree of particle agglomeration occurs. The agglomerated latex is then concentrated as before but, because of the particle-size distribution obtained, the solids may be raised to ca 70 wt %. Two methods exist for agglomeration of latices, ie, chemical agglomeration and freeze agglomeration (65–66).

Styrene–Butadiene and Related Latices. The SBR latices can be classified into the hot and cold types, as determined by polymerization temperatures, and subdivided into low and medium solids and high solids. A summary of commercial SBR latices is given in Table 26. Hot SBR latices are polymerized at 49–66°C and are mainly supplied at medium solids content, ca 42–50 wt %. Only two high solids (ca 60–70 wt %) types, namely 2003 and 2004, are available and are used mainly for special applications such as the bases for graft polymerizations for high impact platics. Type 2006 is supplied at ca 27 wt % solids for the chewing-gum industry. There are a wide variety of high styrene resin latices for many diverse applications from carpet backing to paint and paper coating.

There has been a marked trend towards concentration of the higher styrene (>40%) polymers in the hot latices and the lower styrene (mostly 20–30% bound sty-

Table 26. Styrene–Butadiene Latices

Type	Monomer ratio: styrene/butadiene	Mooney viscosity (ML-4)	Solids, wt %	pH	Emulsifier[a]
Elastomeric, hot					
2000	46/54	70	39–44	10–11	RA
2001	46/54	30	39–44	11	RA
2004	0/100		58	10	RA–FA
2006	25/75	50	27	10	FA
J-9049	46/54	70	49	10	RA
J-8146	46/54	70	59	11	RA
Elastomeric, cold					
2105	30/70	140	62	10–11	RA–FA
2107	44/56	140	62	10–11	RA–FA
2108	25/75	>140	40	11	FA
2113	44/56	130	48	10–11	RA–FA
J-9428	25/75	135	69–72	10–11	RA–FA
5352	25/75	135	69–72	10–11	FA
High styrene latices, hot					
2711	59/41		53	10–11	RA
2714	82/18		54	10–11	RA

[a] RA = rosin acid; FA = fatty acid.

rene) types in the cold-latex series. This is a reflection of the fact that lowering the polymerization temperature of high styrene copolymers produces little or no gain in the physical properties of the copolymer.

Another difference between hot and cold elastomeric SBR latices is that the hot types are carried to ≥90% conversion and are not normally shortstopped. The cold latices are usually shortstopped at ca 60–80% conversion. Again the desired physical properties of the contained copolymer are responsible for these differences. Cold latices are used in applications where the modulus, eg, in foam, or retention of physical properties at high filler loadings, eg, in fabric backing, are required. The cold latices are generally supplied at a higher solids concentration than the hot series because of the above uses.

Styrene–butadiene latexes generally are quite stable mechanically because of the presence of relatively large amounts of emulsifying and stabilizing agents, and they therefore require addition of less stabilizer in compounding. The applications of SBR latex are classified in Table 27. This classification indicates the scope of the industry and illustrates the large number of diverse applications in which synthetic latices are employed. The latex types most suitable for particular applications are also listed.

Butadiene–Acrylonitrile Latices. Nitrile latices are copolymers of butadiene and acrylonitrile or products in which these copolymerized monomers are the main constituents (see Elastomers, synthetic, nitrile rubber). The latices differ mainly in respect to comonomer ratio and the type and concentration of stabilizer. They can be classified as medium and high acrylonitrile (ACN) types. The latter contain 35–40 wt % ACN in the copolymer and the medium types contain 25–35 wt %. Carboxylic acid-modified nitrile rubbers and butadiene–styrene–acrylonitrile terpolymers are also available as latices.

Though nitrile latices are usually thought of as special-purpose products, they

Table 27.   Applications of SBR Latices

General uses	Specific applications	Types
paper	saturation	2000, 2001, 2740
	beater addition	2000, J-9049
	coating	Dow 512R, 630, 636, 620
carpets	back sizes	2000, 2105, 2714, modified SBR
pile fabrics	coating back of pile fabrics	2000, modified SBR
fabric adhesives	fabric tie coats	2000, 2107, 2113
tires	cord dipping	2000, 2108, J-9049, Pyratex, Gen-Tac
footwear	combining fabrics	2000, 2002, 2107, 2113
asbestos fiber	brake linings, gaskets, linoleum base	2000, 2758, J-9049
jute	shoe insoles, etc	2000
jute and sisal	upholstery pads, rug underlay	2000, J-9049
foam sponge	cushioning, rug underlay	2105, J-9428, 5352
batteries	separator sheets	2000
chewing gum		2006, J-8146
paint	pigment binder	Dow 512-K, 762-W

are used in a wide variety of applications. Their use appears to depend more on the polarity and thermoplasticity of the polymer than on the oil resistance, although the latter is important in some applications. In the principal uses, eg, paper saturation and adhesives, small particle size and correct surface tension is desirable for rapid penetration and rapid setup or drying.

Films deposited from nitrile latices can be vulcanized with sulfur and ultra-accelerators, but for most uses the polymer is used in the unvulcanized form. The main uses of nitrile latices are paper saturation, paper-beater addition, textile and leather finishing, adhesives, and nonwoven fabric binders. A recent development in nitrile latex technology is the use of high solids nitrile latices, eg, Nitrex J-6849 and Polysar 845, to prepare oil-resistant foams for lubricants in heavy-duty bearings, such as railroad-car journal boxes. For general latex use and particularly in the manufacture of dipped articles, the nitrile latices suffer from the same disadvantages as the SBR latices.

Neoprenes. Of the synthetic latices, the type that can be processed most nearly like natural rubber latex and which is most generally adaptable to rubber-latex uses is the neoprene (chloroprene) type (see Elastomers, synthetic, neoprene). Neoprene latices exhibit high gum tensile strength, oil and solvent resistance, good aging properties, and flame resistance. There are several principal types of neoprene latex, some of which are available at both high and intermediate solids content (51). Differences in composition between the main types include the microstructure of the polymer, eg, sol or gel polymers; the type of stabilizer; and the total solids content (see Table 28).

As in dry-rubber compounding, neoprenes do not require sulfur or an accelerator for vulcanization, but zinc oxide is generally used to accelerate the cure and to function as an acid acceptor. Unlike dry-neoprene compounding, no magnesia is used because of its destabilization effect on the latex. Organic accelerators are used to improve the physical properties of neoprene latex films but their effect is not generally as great as when used with other polymers The most effective accelerators in neoprene latex are thiocarbanilide used either alone or in combination with DPG and combinations of tetraethylthiuram disulfide and sodium dibutyldithiocarbamate. Sulfur should be used in compounds of latex types 450A, 735A, 635, 750, 650, and 950 if a high state of cure is required, but it should not be used if color retention or heat resistance is more important than state of cure.

Two parts of a good antioxidant is recommended for almost all neoprene latex compounds. The most commonly used staining antioxidants are Agerite White (R. T. Vanderbilt Company) and Aminox (Uniroyal, Inc.). The substituted or hindered phenols, eg, Naugawhite, Uniroyal; Antioxidant 2246, American Cyanamid, are used where a minimum of discoloration and staining is required.

The tendency for light-colored neoprene vulcanizates made from latex to discolor when exposed to light is markedly retarded by compounding with triglycerides of long-chain fatty acids in combination with nondiscoloring antioxidants (67). Resistance to discoloration increases with increasing amounts of safflower oil up to 15 parts. Fillers are added to neoprene latex to modify the film's physical properties, to lower compound cost, to improve sunlight and water resistance, and to adjust the processing properties of the liquid. Ten to twenty parts of fillers reinforce neoprene latex films to some degree and improve tear strength. Generally, hard clays are used. Mineral and light process oils are used as plasticizers and to improve the hand in neoprene latex films, whereas ester plasticizers are used to improve low temperature properties.

Since the viscosity of neoprene latex at a given solids content is less than that of natural rubber latex, thickeners are generally needed with the former. Methyl cellulose and the water-soluble salts of polyacrylic acid are two most commonly used thickeners. Natural and synthetic gums also are used.

As in dry compounding, acid acceptors must be incorporated in neoprene latices, because of the wide use of the latices in coating cotton fabrics and metals. The hydrochloric acid which forms during service has a particularly destructive effect on coated cotton fabrics that are not adequately protected. High zinc oxide concentration (15 parts) and the use of 0.4 parts N-phenyl-N'-(p-toluenesulfonyl)-p-phenylenediamine (Aranox, Uniroyal) as an antioxidant provide adequate protection.

Vinylpyridines. The vinylpyridine latices were developed specifically for use in adhering rubber stocks to fibers, particularly nylon (68). In general, the polymers are high diene types containing 10–15 wt % copolymerized 2-vinylpyridine and an approximately equal amount of styrene. These latices are supplied at ca 41 wt % solids and are emulsified with anionic soaps. There are three commercially available vinylpyridine latices, namely, Gen-Tac (General Tire and Rubber Co.), Hycar 2518 (BF Goodrich), and Polysar 781 (Polymer Corp.).

Carboxylated. One advance in latex technology has been the introduction of latices in which the polymer phases contain functional groups (69). The functional groups are derived from the use of unsaturated monomers containing carboxy groups in the polymerization system. Carboxylated styrene–butadiene latices have been used increasingly in carpet-backing applications because of their self-curing feature, ie, the use of sulfur and accelerators is obviated, resulting in lower cost compounds but more particularly in very low odor compounds. Presumably carboxylated latices obtain their strength from the polar carboxyl groups and the fairly high styrene content. Metal oxides and melamine–formaldehyde resins can be used as curing systems. In carpet-backing applications, zinc oxide should be omitted from formulations since it can reduce tuft-retention values.

Table 28. Neoprene Latices[a]

Neoprene latex type	Comonomer	Emulsifiers	Emulsifier class	Chlorine content, wt %	pH at 25°C (typical values)	Standard solids, wt %	Distinguishing features	Primary applications
101	methacrylic acid	poly(vinyl alcohol)	nonionic	36	7.0	46	high strength, carboxylated polymer with outstanding mechanical and chemical stability	adhesives, bonded batts, coatings, saturants
102	methacrylic acid	poly(vinyl alcohol)	nonionic	36	7.0	46	medium strength, carboxylated polymer with outstanding mechanical and chemical stability	adhesives, bonded batts, coatings, saturants
357		potassium salts of disproportionated resin acids and fatty acid and polymerized potassium salts of alkylnaphthalenesulfonic acid	anionic	38	12.5	61	designed specifically for foam	foam
400	2,3-dichloro-1,3-butadiene	potassium salts of disproportionated resin acids	anionic	48	12.5	50	maximum chlorine content; rapid crystallizing; outstanding ozone and weather resistance	adhesives, bonded batts, coatings
450A	acrylonitrile	potassium salts of disproportionated resin acids and fatty acid	anionic	31	9.5	41	maximum hot-oil and plasticizer resistance	asbestos binder, back coatings
571	sulfur	sodium salts of resin acids	anionic	37.5	12.0	50	high strength cured films combined with low permanent set	dipped goods, elasticized aluminous cement, coatings, cord adhesives
572	sulfur	sodium salts of resin acids	anionic	38	12.0	50	rapid crystallizing; rapid coalescence under pressure	wet laminating adhesives

426

601A		sodium salts of resin acids	anionic	38	12.0	60	creamed Latex 842A	mastics
635		sodium salts of disproportionated resin acids	anionic	38	11.0	60	creamed Latex 735A	mastics
650	2,3-dichloro-1,3-butadiene	potassium salts of disproportionated resin acids	anionic	40	12.0	60	creamed Latex 750	mastics
671		potassium salts of disproportionated resin acids and fatty acid and polymerized potassium salts of alkylnaphthalenesulfonic acid	anionic	38	12.0	59	wet-gel strength; high solids with low viscosity; 671A has a lower viscosity than 671	dipped goods, laminating adhesives, mastics, bonded batts, impregnated paper, extruded thread, contact bond adhesives
671A					12.5	59		
735A		sodium salts of disproportionated resin acids	anionic	38.5	11.5	45	designed for wet-end addition to fibrous slurries	fiber binder
750	2,3-dichloro-1,2-butadiene	potassium salts of disproportionated resin acids	anionic	40	12.5	50	high wet-gel strength; low modulus; crystallization resistance	dipped goods, cord adhesives, contact bond and construction adhesives
842A		sodium salts of resin acids	anionic	37.5	12.0	50	medium strength cured films having a slow crystallization rate	adhesives, dipped goods, saturants, coatings, bonded batts
950		quaternary ammonium salt having one long-chain alkyl group	cationic	38	9.0	50	cationic colloidal system	elasticized portland cement, sealants and coatings

[a] Ref. 51.

427

Types of Latex Compounds. For comparison with dry-rubber compounds, some examples of various latex compounds and the physical properties of their vulcanizates are given in Table 29. Recipes of natural rubber latex compounds, including one without antioxidant, and data on tensile strength and elongation of sheets made from them, both before and after accelerated laboratory aging, are also listed. The effects of curing ingredients, accelerator, and antioxidant also are listed. Table 30 includes similar data for an SBR latex compound. Table 31 includes data for a typical neoprene compound. A phenolic antioxidant was used in all cases.

Table 29. Properties of a Natural Rubber Compound[a]

Property	No antioxidant	Antioxidant, 1.0 part (dry)
Unaged, cured in air at 121°C		
3 min cure 700% modulus, MPa[b]	8.2	9.7
tensile strength, MPa[b]	33.3	32.4
elongation, %	>970	>910
15 min cure 700% modulus, MPa[b]	5.5	9.1
tensile strength, MPa[b]	34.3	33.5
elongation, %	>990	>920
After air-oven aging for 7 d at 70°C		
3 min cure 700% modulus, MPa[b]	1.9	2.1
tensile strength, MPa[b]	35.6	37.1
elongation, %	920	880
15 min cure 700% modulus, MPa[b]	1.6	1.3
tensile strength, MPa[b]	32.8	34.2
elongation, %	980	920
After oxygen-bomb aging for 96 h at 70°C		
3 min cure 700% modulus, MPa[b]	1.9	3.3
tensile strength, MPa[b]	27.3	33.0
elongation, %	900	810
15 min cure 700% modulus, MPa[b]	1.5	2.8
tensile strength, MPa[b]	20.5	30.5
elongation, %	940	880
After oxygen-bomb aging for 7 d at 70°C		
3 min cure 70% modulus, MPa[b]	2.9	4.4
tensile strength, MPa[b]	20.4	30.1
elongation, %	870	820
15 min cure 700% modulus, MPa[b]	1.5	2.5
tensile strength, MPa[b]	11.0	16.8
elongation, %	920	840

[a] Dry basis natural-rubber compound recipe, in part by wt: natural latex (NC 356), 100.0; potassium hydroxide, 0.5; Nacconal 90F (an alkylarenesulfonate produced by Allied Chemical Company), 1.0; zinc oxide, 3.0; sulfur, 1.0; ZMBT, 1.0; zinc diethyldithiocarbamate (ZEDC) (trade names: Ethazate, Uniroyal, Inc.; Ethyl Zimate, R. T. Vanderbilt), 0.3; antioxidant, as indicated. Wet-basis natural-rubber compound recipe, in parts by wt: natural latex (NC 356), 167.9; potassium hydroxide, 2.5; Nacconal 90F, 5.0; zinc oxide, 5.45; sulfur, 1.65; ZMBT, 2.0; ZEDC, 2.0; antioxidant, as indicated. All films were poured from freshly mixed compounds, dried overnight in place, then lifted and dried one hour in air at 50°C before curing.

[b] To convert MPa to psi, multiply by 145.

Table 30.   Properties of an SBR Compound[a]

Property		No antioxidant	Antioxidant, 1.0 part (dry)
Unaged, air cured at 121°C			
10 min cure	500% modulus, MPa[a]	2.8	2.6
	tensile strength, MPa[a]	18.4	15.2
	elongation, %	790	790
30 min cure	500% modulus, MPa[b]	3.3	2.9
	tensile strength, MPa[b]	17.3	13.3
	elongation, %	750	710
After air-oven aging for 16 h at 130°C			
10 min cure	500% modulus, MPa[b]	4.0	3.7
	tensile strength, % retained	27.4	37.1
	elongation, % retained	58.5	76.0
30 min cure	500% modulus, MPa[b]	4.6	3.7
	tensile strength, % retained	33.6	41.5
	elongation, % retained	74.8	81.8
After oxygen-bomb aging for 96 h at 70°C			
10 min cure	500% modulus, MPa[b]	4.3	4.8
	tensile strength, % retained	63.4	106.0
	elongation, % retained	82.3	82.5
30 min cure	500% modulus, MPa[b]	3.7	3.5
	tensile strength, % retained	67.0	89.2
	elongation, % retained	89.5	97.2

[a] Dry basis SBR compound recipe, in parts by wt: SBR latex (type 2000), 100.0; Triton X-200 (sodium salt of alkylaryl polyether sulfonate, Rohm & Haas Company), 1.0; zinc oxide, 3.0; sulfur, 1.0; ZMBT, 1.0; ZEDC 0.3; antioxidant, as indicated. Wet basis SBR compound recipe, in parts by wt: SBR latex (type 2000), 265.5; Triton X-200, 5.0; zinc oxide, 5.45; sulfur, 1.65; ZMBT, 2.0; ZEDC, 0.6; antioxidant, as indicated.

[b] To convert MPa to psi, multiply by 145.

Selection of Rubbers for Specific Applications

The selection of the type of rubber to be used in a particular product depends on the technical requirements of the product, the properties attainable by compounding, and economic factors. By far the largest volume of rubber is used in the manufacture of pneumatic tires for passenger cars, trucks, airplanes, and farm machinery. The polymers used for these applications are natural rubber, SBR, polybutadiene, polyisoprene, halogenated butyl, and some EPDM rubber. Butyl, EPDM, and natural rubber are also used for inner tubes. The essential requirements for the tread of a typical passenger tire include: resistance to abrasion and to the closely related phenomena of cutting and chipping; reduced rolling resistance; resistance to cracking and to crack growth; adequate flexibility at the lowest temperature encountered in service; a sufficiently high coefficient of friction between the tread and the road to minimize slipping and skidding; adequate adhesion of the tread to the carcass; sufficient stability of the material with time so that excessive deterioration does not occur in the normal life of the tire; and a moderate hysteresis so that excessive temperatures do not develop in service. When compounded with 70 parts of N-339 (HAF-HS) or N 234 (ISAF-HS) carbon black, vulcanized with a suitable sulfenamide accelerator system, and adequately stabilized with antiozonants, the various grades of SBR,

Table 31. Properties of a Neoprene Compound[a]

Property		No antioxidant	Antioxidant, 2.0 parts (dry)
Unaged, after air cured at 125°C			
30 min cure	500% modulus, MPa[b]	2.6	2.8
	tensile strength, MPa[b]	25.5	24.2
	elongation, %	1010	960
60 min cure	500% modulus, MPa[b]	3.2	3.1
	tensile strength, MPa[b]	28.1	29.2
	elongation, %	900	920
After air-oven aging for 20 h at 130°C			
30 min cure	500% modulus, MPa[b]		7.5
	tensile strength, MPa[b]	5.2	19.2
	elongation, %	210	670
60 min cure	500% modulus, MPa[b]		9.2
	tensile strength, MPa[b]	5.7	16.0
	elongation, %	150	570
After oxygen-bomb aging for 7 d at 70°C			
30 min cure	500% modulus, MPa[a]	failed	2.6
	tensile strength, MPa[a]	failed	29.3
	elongation, %	failed	860
60 min cure	500% modulus, MPa[b]	failed	2.2
	tensile strength, MPa[b]	failed	28.3
	elongation, %	failed	850

[a] Dry basis neoprene compound recipe, in parts by wt: neoprene latex (type 571), 100; Aquarex MDL (sodium alkyl sulfate, DuPont), 0.45; zinc oxide, 10.0. Wet basis neoprene compound recipe, in parts by wt: neoprene latex (type 571), 200.0; Aquarex MDL, 1.8; zinc oxide, 18.2.

[b] To convert MPa to psi, multiply by 145.

polybutadiene, and natural rubber provide the above requirements. Generally, SBR and 25–35 wt % polybutadiene are used in passenger tires.

Neoprene is not used because it is too costly, has unfavorable low temperature properties, and does not adhere well to a carcass made from another rubber. It is used in some coverstrips. The nitrile rubbers are too costly and they also have poor low temperature flexibility. The silicones, polyacrylate rubbers, and polysulfide polymers would not be considered for a variety of reasons, including high cost and inadequate abrasion resistance. Butyl does not meet the abrasion and hysteresis requirements. The other problems with butyl in tires has been adhesion to fabric and the complete segregation of butyl compounds from other types of rubber.

The choice between SBR and natural rubber had been largely one of economics, although at various times it has been determined by government policy. Cold SBR performs somewhat better with respect to abrasion resistance than does natural rubber. However, cold SBR is somewhat poorer in heat buildup and is appreciably poorer in cut-growth resistance. The use of oil extension and carbon black masterbatching has placed SBR far ahead of natural rubber in passenger-tire compounding. The use of polybutadiene in tires is based on its excellent abrasion resistance, high resiliency, and excellent high and low temperature service properties. Polybutadiene also contributes to reducing groove-cracking problems in tire treads. It is used at up to 25 and 35 wt % in blends with SBR. There have been several factors which have slowed its use of higher levels. Mixing and processing become problematic when polybutadiene is used in levels greater than 50 wt %. Wet-skid resistance becomes poorer at higher levels as does cutting and chipping. However, the use of high molecular weight, oil-extended polybutadiene plus higher levels of carbon black and oil is overcoming these problems (70). For heavy-service truck and bus tires, natural rubber is preferred over SBR because of its lower hysteresis, better properties at the elevated temperatures encountered in such service, and better crack-growth resistance. Polybutadiene is used in blends with natural rubber to improve abrasion resistance and rebound.

Ethylene–propylene rubbers are used in tire compounds, mainly in the coverstrip and white sidewall portions. The substitution of EPDM, and Royalene 301T, in particular, for part of the unsaturated polymers in these compouds improves the static and dynamic cracking resistance. As the amount of EPDM in these compounds is increased, there is a loss in physical properties and in cured adhesion. Approximately 25% can be substituted without affecting the cured adhesion. Halogenated butyl rubbers are also used in the cover-strip, sidewall, and inner-liner portions of tires. An all-EPDM tire has been manufactured (71). These EPDM tires are somewhat poorer in wear than the regular SBR-BR blends and have inferior wet-skid resistance. The EPDM tires are equal to the SBR–BR tires with respect to comfort, handling, and noise.

For inner tubes, butyl rubber is used for tube sizes up to ca 12:00 because of its low air permeability. Natural rubber is used in the larger sizes because of processing advantages. Inner tubes for airplane tires are made from natural rubber because the high temperatures developed in braking cause excessive growth of butyl tubes; the growth of inner tubes is an increase in size accompanied by a decrease in thickness and results from flexing and exposure to elevated temperatures. Because the inner tube is confined in the tire casing, the size increase can only be accommodated by wrinkling or folding, which contribute to early failure. Styrene–butadiene rubber is not satisfactory for this use because of its poorer physical properties when compounded for this type of service, which requires a soft, highly extensible stock with good tear resistance. Chlorobutyl tubes cured with a Vulklor–zinc oxide system gives improved heat resistance in wheel tests (30). Some EPDM also is used with butyl in inner tubes.

For camelback and retreading operations, SBR–oil–carbon black masterbatches of SBR 1600 and 1800 series are used. The SBR 1600 series contains primarily masterbatches for truck treads and premium tires, with 1606, 1608, and 1609 being the most popular masterbatches. For lower price and lower quality, the 1800 series is used, with 1824 and 1848 being the most popular.

In the thousands of other products made from rubber, the choice of polymer depends, as in the case of tires, on the properties that can be developed by compounding, on the requirements of the service, and on the cost of the polymer. In products that require a soft, flexible, resilient, highly extensible rubber, eg, for stationers' bands or golf-ball thread, natural rubber and polyisoprene are preferred. Another choice for such applications is neoprene. For solid golf balls, the excellent rebound characteristics of polybutadiene make it the choice of compounders. For products of low-to-moderate quality, eg, soles, heels, garden hose, and mats, SBR and natural rubber are competitive and can be used with high proportions of reclaimed rubber to reduce cost. For applications requiring excellent ozone resistance, EPDM, butyl, neoprene, chlorosulfonated polyethylene, polysulfides, nitrile–vinyl blends (Paracril Ozo), chlorobutyl rubber (CIIR), acrylics, polyurethanes, silicones, fluo-

Table 32. Qualitative Comparisons of Rubber[a]

	NR	SBR	IR	BR	IIR	EPDM	CR	NBR	AUEU	OT and EOT	MQ VMQ PMQ PVMQ	ACM	FDM
Physical properties													
electrical resistance	B	B	B	B	B	B	C	X	C	C	A	C	B
flame resistance	X	X	X	X	X	X	B	X	C	X	C	X	B
gas permeation	C	C	C	C	A	C	B	A	B	A	C	C	
heat resistance	C	C	C	C	B	B	B	B	B	B	A	B	A
cold resistance	B	B	B	B	C	B	C	C	B	C	A	X	C
Mechanical properties													
tensile strength (at 7 MPa[b])													
pure gum	>21	<7	>21	<7	>10	<7	>21	<7	>28	<7	<7	>7	>14
reinforced	>21	>14	>21	>14	>14	>21	>21	>21	>28	>7	>7	>14	>14
hardness, Shore A	30	40	30	40	40	30	40	40	55	40	40	40	60
compression set	90	90	90	90	80	85	95	95	100	85	85	90	90
rebound, cold	A	A	A	A	B	B	A	A	B	X	A	B	A
rebound, hot	A	B	A	A	X	B	B	B	X	C	A	X	B
tear resistance	A	B	A	A	A	B	B	B	B	C	A		A
abrasion resistance	A	C	A	C	A	B	C	C	A	X	X	C	C
	A	A	A	A	B	A	B	A	A	X	X	C	B

Chemical stability

sunlight aging	C	C	C	C	A	A	A	C	A	A	A	A	A
oxidation resistance	B	B	B	B	A	A	A	A	A	A	A	A	A
ozone resistance	X	X	X	X	B	A	B	X	A	A	A	A	A
aliphatic hydrocarbon	X	X	X	X	X	X	B	A	A	A	X	B	A
aromatic hydrocarbon	X	X	X	X	X	X	C	B	B	A	X	X	A
chlorinated solvents	X	X	X	X	X	X	X	C	C	B	X	X	B
oxygenated solvents	B	B	B	B	B	B	C	C	B	A	C	B	X
petroleum, crude	X	X	X	X	X	X	B	A	B	A	X	B	A
natural gas	X	X	X	X	X	X	B	A	C	A	X	B	A
gasoline, fuel oil	X	X	X	X	X	X	B	A	A	A	C	A	A
lubricating oils	C	C	C	C	A	A	B	A	A	A	C	A	A
animal, vegetable oils	C	B	C	B	A	B	B	B	A	A	A	A	A
acids, dilute	B	B	B	B	A	B	A	C	B	B	A	X	A
acids, concentrated	C	C	C	C	A	B	B	C	X	C	B	X	B
sodium hydroxides	B	C	B	C	A	B	B	B	C	C	A	X	B
water swell resistance	B	A	B	A	A	B	B	A	C	A	A	C	A
specific gravity	0.92	0.93	0.92	0.94	0.91	0.86	1.23	1.00	1.05	1.27–1.34	0.95	1.10	1.4

[a] Ratings: A = excellent, B = good, C = fair, X = poor.
[b] To convert MPa to psi, multiply by 145.

433

roelastomers, and chlorohydrin rubbers can be used, the choice depending on what other properties are required by the service. Even the nonozone-resistant polymers, eg, SBR, BR, nitrile, natural rubber, etc, can be made resistant to ordinary ozone attack by the use of the proper blend of antiozonant and wax.

When oil resistance is a desired quality for special grades of hose, packing, heels and soles, conveyor belts, diaphragms, and many other products, it is necessary to use polymers other than natural rubber or SBR. The choice of polymer depends upon the kind of oil involved, the degree of oil resistance desired, and whatever other physical properties may be required by the service. The polymers to be considered are the neoprenes, the nitrile rubbers, the acrylate rubbers, polysulfides, polyurethanes, fluoroelastomers, and epichlorohydrin rubbers. It is unfortunate that good oil resistance and poor low temperature flexibility go together, so that in many applications it is necessary to compromise between the two properties.

For extreme heat resistance, the silicone rubbers are the best by far, but unfortunately most of their other physical properties are relatively poor. The next best heat-resistant polymers are the fluorinated rubbers followed by the acrylates, EPDM, butyl, chlorosulfonated polyethylene, neoprene, and nitrile rubbers. For low temperature flexibility, certain silicones are the best followed by EPDM, polybutadiene, natural rubber, and SBR.

All the properties mentioned above can be modified, sometimes over wide ranges, by the compounding technique and in nearly all applications more than one property is of importance in the service. For example, a hose line used in an aircraft for conducting lubricating oil might be required to be flexible at the lowest temperature encountered on the ground, eg, $-54°C$, but, while in flight, it would be required to contain the oil at a temperature of $149°C$. Whereas silicone rubbers would fulfill part of the requirements for this hose line because of their good low temperature flexibility and good high temperature resistance, they have inadequate oil resistance to perform satisfactorily. The best choice for this application is a nitrile rubber polymer with as low an acrylonitrile content as can be tolerated to meet the oil-resisting requirement and one that is compounded to give the maximum heat resistance.

In Table 32 a general rating of the various polymers is shown with respect to certain important properties of their vulcanizates. This is only an approximate rating since compounding techniques vary each property, and for most of the polymers various grades are available with varying properties. Also in some cases it may not be possible to combine the best performance in two or more properties in a single compound. For example, the excellent abrasion resistance indicated for natural rubber is obtained in a carbon-black-reinforced compound, whereas the best hysteresis properties are obtained in a gum stock.

Classification.   The Technical Committee on Automotive Rubber, sponsored jointly by ASTM and SAE, has designed the classification system ASTM D 2000–SAE J 200. The purpose of this system is to provide guidance to the engineer in the selection of practical, commercially available elastomeric materials and a method of specifying these materials by a simple, so-called line call-out designation.

All rubber compositions are classified according to two basic characteristics: heat resistance and oil resistance. First, a letter from A through J is assigned to characterize heat resistance and this letter designates type. A second letter, from A through K, indicates relative oil resistance and the class of the material. Both type and class letters are assigned on a basis of increasing performance. For example, AA indicates a material with low heat and oil resistance, whereas JK indicates a material with very high temperature resistance and highest oil resistance. Nitrile rubbers are classified in BF, BG, BK, and CH tables indicating their wide practical range of heat and oil resistance. The letter designation is always followed by three digits, which indicate required hardness and minimum tensile strengths. When requirements are needed that either supersede or are supplemental to basic requirements, they are indicated by grade numbers and by means of suffix letters and numbers. Suffix letters have the same meaning as in ASTM D 735. For example, the letter B denotes a compression-set requirement, the letter E denotes a fluid-resistance requirement, etc. Two numbers then follow the suffix letters. The first number indicates the test method and the second number denotes the temperature of the test. The prefix letter M indicates that the classification system is based on SI units. Call outs not prefixed by the letter M refer to an earlier classification system based on inch–pound units. This system was published in ASTM editions up to 1980.

Table 33 shows a sample compound illustrating the mechanics of the system. Table 34 is a list of the rubbers normally used to meet the material designation.

Fabrication

The recipe prescribing definite proportions of rubber, fillers, softeners, vulcanizing agents, and certain accessory materials which must be compounded or weighed, mixed, formed, and vulcanized. In some instances certain materials require pretreatment before compounding. For example, it may be necessary to dry fillers to remove excess moisture or to pass them through screens to ensure adequate fineness and the removal of lumps. Also the polymer in the case of natural rubber may require some processing,

Table 33. Sample Nitrile Compound Meeting ASTM D 2000–SAE J 200: Grade M2BG 407 B14E14E34E51F17L14[a]

Property	Specification requirement	Laboratory data (cured for 20 min at 155°C)
hardness, durometer type A	40 ± 5	43
tensile strength, MPa[b]	7 min	12.3
ultimate elongation, %	450 min	980
Heat-aged (D 573) for 70 h at 100°C		
tensile strength change, %	±30 max	−3.4
elongation change, %	−50 max	−22.0
hardness change, points	±15 max	+2.0
Oil immersion (D 471)		
(ASTM Oil #3) for 70 h at 100°C		
volume change, %	±40 max	+8.2
Compression set (D 395 Method B) for 22 h at 100°C		
% set	50 max	21.3[c]

[a] Recipe, in parts by wt: Paracril BJLT (medium nitrile), 100.0; zinc oxide, 5.0; Aminox (antioxidant), 1.5; N 880 (FT) black, 50.0; tributoxyethyl phosphate, 7.5; plasticizer SC, 7.5; Paraplex G-25, 10.0; stearic acid, 1.0; TMTM, 1.5; sulfur, 0.75; total parts, 184.75. The specific gravity of this nitrile compound is 1.17 g/cm³; Mooney viscosity (ML-4) at 100°C, 20; Mooney scorch (MS) at 121°C, >45 min.

[b] To convert MPa to psi, multiply by 145.

[c] 1.27 cm thick solid specimen cured for 30 min at 155°C.

Table 34. Polymers Used to Meet ASTM D 2000 Grades

ASTM D 2000–SAE J 200 material designation, type and class	Type of polymer normally used
AA	natural rubber, reclaimed rubber, SBR, butyl, polyisoprene, EP polybutadiene
AK	polysulfides
BA	high temperature SBR, butyl, and EPDM
BC	chloroprene polymers (neoprene)
BE	chloroprene polymers (neoprene)
BF	NBR polymers
BG	NBR polymers, urethanes
BK	polysulfides (Thiokol), NBR
CA	EPDM
CE	chlorosulfonate polyethylene (Hypalon)
CH	NBR polymers, epichlorohydrin polymers
DA	EPDM
DF	polyacrylic (butyl–acrylate type)
DH	polyacrylic polymers
FC	silicones (high strength)
FE	silicones
FK	fluorinated silicones
GE	silicones
HK	fluorinated elastomers (Viton, Fluorel, etc)

eg, washing or straining to remove foreign material, or it may require mastication to make the rubber more adaptable to succeeding processing steps.

Mastication

Probably the most important of these preliminary steps is the mastication of the rubber. Crude natural rubber as received from the plantation is high in viscosity, and for most uses it is advantageous to lower its viscosity prior to mixing.

The combination of heat and work on crude natural rubber produces a physical and chemical change that is measured in relative softness or plasticity. The amount of softness is prescribed by both the finished article use and the operation through which the compound will pass in its subsequent operations. Very high masticated or soft rubber is used in friction compounds, sponge stocks, and rubber solutions or cements. Medium soft rubber is used in calendering compounds, and lightly masticated rubber is used for stiff compounds.

Studies of mastication indicate that maximum softening effect on natural rubber is obtained by mastication either at <54°C or >132°C. Very little breakdown is obtained at 88–104°C. The breakdown of rubber is accomplished either on a roll mill, in an internal mixer, or in a screw plasticator, eg, the Gordon Plasticator. The roll mill functions in the low temperature region, whereas the plasticator and the Banbury operate at 149–177°C.

The roll mill consists of two parallel, horizontal rolls rotating in opposite directions so that the rubber or compound fed to them is pulled through the nip. The actual effect of mastication or mixing takes place in the nip or wedge formed by the two rolls. The remaining surface of the roll is used for heating or cooling the compound by contact and as a means of transportation for returning the compound to the nip or bite for additional mixing. Roll surface speeds are generally 0.51 m/s (100 ft/min) and the back roll rotates faster than the front roll by a friction ratio of 1.1:1 to 1.5:1. Cold or hot water, steam, or hot oil can be circulated through the hollow rolls so as to control the desired operating temperature. The nip width can be varied by the operator, since the front roll position is adjustable. The crude rubber, which may have been thermally softened in a hot room at 60°C, is thrown on the rolls and allowed to pass through the nip several times. After several passes, the rubber forms a sufficiently coherent band and clings to the slow roll and is fed back into the nip. Cross-blending is done by the operator by cutting the blanket on the roll and manipulating it end-to-end so as to create a complete interchange of rubber. At the end of the mastication, the rubber is cut in large sheets and cooled prior to piling on skids.

Compared with mixing rolls, an internal mixer, eg, a Banbury, gives the desired degree of mastication much more rapidly. Basically, the Banbury mixer consists of a completely enclosed mixing chamber in which two spiral-shaped rotors operate, a hopper to receive the rubber or compound for mixing, and a door for discharging the batch. The two rotors keep the stock in constant circulation by revolving in opposite directions and at slightly different speeds. Water or steam can be circulated through the hollow rotors and through the mixing-chamber walls. The rubber is charged directly into the mixing chamber, the ram is lowered to exert pressure on the batch, and mastication proceeds for the required time. At the end of the mastication period, the rubber is discharged onto a two-roll mill where the rubber is sheeted.

In masticating rubber in an internal mixer or a plasticator, a chemical plasticizer may be added to hasten the process. These plasticizers function best at the elevated temperatures developed in these two processes and therefore are not used for the low temperature, open-mill operation. In all three of these mastication operations, it is generally advantageous to blend several lots of rubber from different shipments of the same grade or from grades of similar quality to minimize differences that occur for a variety of reasons.

Viscosity-stabilized rubber is technically specified and marketed under the Standard Malaysian Rubber (SMR) scheme in two forms: SMR-5-CV and SMR-5-LV, which are subgrades of SMR-5. The CV and LV forms are stabilized in the same manner but the LV differs from the CV in that it contains an added 4 parts phr of light, nonstaining mineral oil. The latex is treated with a small amount of hydroxylamine salt before coagulation, which deactivates the cross-linking sites in the rubber that occur during subsequent drying. The viscosity of the stabilized dried rubber is similar to that of the fresh latex; therefore, it depends upon the clonal makeup of the latex. About six Mooney viscosity ranges are available. Thus SMR-5-CV-50 has a viscosity range of 45–55, and SMR-5-CV65 has a viscosity range of 60–70. These SMR-5-CV or LV rubbers also eliminate or greatly reduce the need for mastication.

Styrene–butadiene rubber is normally made to the viscosity range desired by the rubber manufacturer so additional breakdown is not needed. The 50°C hot-polymerized SBR responds to plastication like natural rubber, whereas the 5°C or cold SBR polymers are very difficult to soften. Butyl, neoprene, nitrile, polybutadiene, and EPDM polymers are not normally premasticated. Polyisoprene responds to mastication.

Mixing.   The mixing operation is one of the most important stages through which the composition must pass. The processing steps subsequent to mixing depend on an adequate and uniform mix, and the quality of the final product is directly affected by the kind of mixing. The primary objectives in mixing are to: attain a uniform blend of all the constituents of the mix, thus each portion of the resulting batch is of uniform composition; attain an adequate dispersion of the fillers, ie, avoid lumps or agglomerates of the fillers; and produce consecutive batches which are uniform both in degree of dispersion and viscosity (see Mixing and blending).

Roll mills and internal mixers, eg, the Banbury mixer, Intermix, and Bolling mixer, are used for batch operations. The rubber manufacturer has a variety of mills available to him which vary in size from the small laboratory mill to the large 305-cm mill, with the 107-cm, 152-cm, and 213-cm mills being the popular production sizes. The 213-cm mill can handle batch sizes of 68–136 kg. The selection of these mills for a particular batch is influenced by volume and, therefore, batch size and by the care with which the batch is to be handled. Very sensitive batches must be mixed in much smaller sizes and on a mill where the mill operator can control the compound. During the period that dry fillers are being worked into the batch, a continuous sheet must be maintained on the mill roll, and for this reason practically all of the work is performed on the slow roll. Ingredients of the same softness blend more readily than do hard and soft ones; thus in the preparation of the rubber portion of the batch, the sequence of ingredient addition is important. Fillers tend to work into the surface of a sheet on a mill, and in highly loaded batches it is necessary to stop the addition of fillers until the matrix is blended to form a homogeneous sheet. In batches containing high loadings of plasticizer, it is desirable to add the plasticizer concurrently with the fillers to prevent the stock from breaking up and falling off the mill roll. In batches where dispersion of a high degree of carbon black or other reinforcing agents is required, it is better to work in the dry fillers before adding the liquid plasticizers.

In the Intermix mixer, the great bulk of the shearing action occurs between the rotors rather than between the rotors and the chamber wall. In the operation of the Bolling mixer, as the ram pushes the preweighed charge of material down into the mixing chamber, the material is forced between helically fluted rotors. Shearing action occurs mostly between the rotors and the chamber walls, as in the Banbury. A spiral-blow arrangement inside the shell of the chamber wall is designed for effective circulation of steam so as to convey heat around the shell through baffles cast into the shell liner. In addition, separate channels running through the shell liner convey cooling water.

The continuous mixers offer the following advantages: time savings since there are no separate loading and discharging operations, labor savings, lower capital investment because production rates are generally higher than with batch mixers, and probably a more consistent mix overall than can be obtained in batch operations. A disadvantage of the continuous mixers is the relatively long mixing of a given compound. These mixers are generally fed by continuous weighing devices, which are usually not as accurate as batch weighing units and sometimes cannot be used. The material feed also must be in such a form as to enable it to be accurately weighed and to be free-flowing. The shear rate in continuous mixers cannot be controlled as closely as it can in batch units through a combination of ram pressure and rotor speed.

One such type of continuous mixer is the Shearmix extruder now known as the Transfermix (72–73). It is used also to replace mills in the handling of masterbatch and final mix compounds. This arrangement permits a reduction of cycle time in the Banbury, allows increased productive capacity of existing Banburys, and improves dispersion and processing safety. Other continuous mixers are the Farrel continuous mixer (FCM), Ko-Kneader, Plastificator, and Kneadermaster (74).

For each compound, a mixing procedure is developed to fit the requirements of the compound and the mixing equipment. For one that is simple and for which experience has given a good indication of the procedure to be used, it is possible to predict within fairly narrow limits what the optimum procedure should be. For more complicated compounds, some study involving a number of trial mixes might be required to design the most satisfactory mixing procedure.

Typical mixing procedures for an open 213-cm mill and a No. 27 Banbury for a natural rubber compound are described below. The natural rubber recipe, in parts by wt, is: premasticated smoked sheets, 100; zinc oxide, 5; Flexzone 7F (antioxidant), 2; N339 (HAF-HS) carbon black, 50; stearic acid, 2.5; petroleum softener, 5; CBS accelerator, 0.6; sulfur, 2.5.

Mill Mixing. For a 213-cm mill with rolls 61 cm in diameter, the batch size is 91 kg. The mill is set to a 6.4 mm nip width and cooling water is circulated through the rolls during mixing. The natural rubber is added to the mill and after several passes it sheets out well enough to cling to the front roll and to be fed back into the nip continuously. The rubber is then cut back and forth twice to ensure blending and to allow the rubber that is in the bank to go through the nip. The zinc oxide, Flexzone 7F antiozonant, stearic acid, and CBS accelerator are then added. The rubber is then cut back and forth twice to distribute these materials through the batch. By this time, ca 8 min has elapsed.

The mill is opened slightly and the N 339 carbon black is added slowly to the batch. This results in a high black concentration in that portion of the rubber near the center of the mill. To prevent excessive loading at the center, strips of rubber are cut from the ends of the rolls several times during this operation and are passed into the bank. When ca 90% of the black has been added, the oil and remaining black are added slowly and alternately to the batch. When all the carbon black is in and no free black is visible, the rubber is cut back and forth twice. This black and oil incorporation takes ca 11 min.

Since the total batch volume has been appreciably increased by the black and oil addition, the bank is now larger than at the start of the mixing. The mill again is opened until a rolling bank is obtained. The sulfur is added and worked in, which takes ca 3 min. The batch is then cut back and forth six times each way to ensure thorough blending. Next the batch is cut by the mill operator in sheets ca 0.9 m × 2.4 m, and these sheets are either dipped in water containing soapstone in suspension and then hung to dry and cool or they are thrown on screen-covered racks to cool, after which they are dusted with soapstone and piled on a skid. This operation requires ca 8 min. The total time for mixing the batch is 30 min.

Banbury Mixing. For a No. 27 Banbury mix, the batch size is ca 454 kg. Water is circulated through the rotors and jacket. The timing clock is set at 0 time, and the prebroken rubber, N 339 black, zinc oxides, CBS accelerator, Flexzone 7F antiozonant, and stearic acid are added to the Banbury. These powders have been automatically preweighed in low melting polyethylene bags. The ram is lowered and the mixing proceeds for 3 min. The ram is raised and the petroleum oil is automatically dispensed into the throat of the Banbury. The ram is lowered and the mixing is continued for another 2 min. The batch is then dumped, a procedure which requires ca 0.5 min, completing a total time of 5.5 min.

For handling the batch from the Banbury, a variety of arrangements can be used, including mills that automatically blend the compound and batch it to a conveyor system which cools the compound and then piles it on skids. One arrangement consists of five 213-cm mills. Two mills are side-by-side with one directly under the Banbury. The batch is dropped on the mill and, when half of the batch is blended, it is conveyed to its sister mill so that the whole batch is blended. The batch is then passed through a cooling arrangement where the batch temperature is decreased to ca 66°C. It is then conveyed to another 213-cm mill where the sulfur is added to the batch in the form of a sulfur masterbatch. The batch is then conveyed to the fourth 213-cm mill where additional blending and cooling takes place. Finally, at the fifth mill the batch is sheeted off for further use in the tubing or calendering operation.

Different manufacturers differ in their preferences as to the order of adding ingredients to the mix and therefore not all would agree that the previously described procedures are the best. The mixing procedure also varies greatly for different compounds and for different types of equipment.

The attainment of the first objective of mixing, that of uniform blending of all materials, is effected on the roll mill by cutting the batch back and forth. In the Banbury, the blending is accomplished by the action of the rotors plus subsequent sheeting out. The second objective is that a good dispersion of the filler is obtained by subjecting the rubber and filler to high shearing forces. In the open mill, this is done by the differential in surface speed of the rolls and by the shearing forces imposed between the moving rolls and the stationary bank, whereas in the Banbury it is done by smearing the rubber and filler between the rotor and the case. The attainment of a good dispersion, particularly of the finer blacks, is probably the most difficult of these objectives to attain. Generally, the best dispersions of carbon blacks are obtained when the filler is added to rubber that is relatively tough and viscous and yet not so tough as to cause crumbling in an internal mixer or to produce roughness or holes in the sheet on a roll mill. On a roll mill, the band of rubber should be free from holes, the black should be added evenly and near the center of the mill, the bank size should be small, and rolling and cutting of the rubber should not be done with dry black visible in the bank. Finally, resinous softeners should not be added with the black, either on a roll mill or in a Banbury.

The third objective, that of attaining uniformity of viscosity from batch-to-batch of the same composition, is promoted primarily by controlling those factors that influence the breakdown of the rubber that occurs during mixing. These factors are time, temperature, and in the case of roll mills the setting of the nip width.

The concept of unit work input and power integrators has been introduced as a means of scaling up the mixing of rubber compounds from the laboratory to the factory and as a method of control for factory mixing (75–77). It has been shown how the viscoelastic properties of the rubber mix depend on the mixing energy and how the power integrator charts can be used to identify various stages of the mixing process (78). Various criteria for determining when to terminate the rubber-mixing process have been reviewed (79). A recording chart indicating instantaneous power and integrated power or work done is required as are the normal control criteria of time-and-temperature charts.

A study of the mechanics of mixing carbon black in rubber by determining the influence of agglomerates in the mix on the final cured properties has been carried out (80). It is suggested that the mixing starts with the formation of soft agglomerates which contain rubber and black in about the proportion of the void ratio as determined by oil absorption. In later stages of mixing, these agglomerates are torn apart to smaller and smaller sizes and finally are dispersed. If these first-formed agglomerates are left intact, as is the case in a very short mixing cycle, the tensile and abrasion properties noticeably deteriorate. The Mooney viscosities and moduli for short mixing cycles can be at least twice as high as the final viscosity and modulus after all of the black is 100% dispersed. Hard particles added intentionally in an amount of 10–20% of the total carbon black become very harmful at an average diameter of 1 μm. The effect of these added hard particles on tensile and abrasion properties is much greater than that of soft agglomerates formed in short mixing cycles.

There is a tendency for zinc oxide and sulfur to reagglomerate along with wax in oil-extended SBR and natural-rubber tires (81). Poor groove-cracking resistance results from high concentrations of these relatively brittle reagglomeration areas. Reagglomeration in oil-extended SBR can be minimized by the use of a larger particle size or high structure carbon blacks and through improved dispersion of noncarbon ingredients.

Chemical compatibility has been studied as a factor in the dispersion of fillers in rubber (82). When acidic materials, eg, fatty acid, acidic accelerators, or retarders, are added simultaneously with basic activators, eg, litharge, magnesia, and zinc oxide, or with some basic fillers, eg, finely divided calcium and magnesium carbonates, filler agglomeration occurs. High temperature mixing (ca 163°C) usually gives better dispersion than low temperature mixing (ca 104°C). There are several methods available for measuring dispersion of fillers after mixing, ie, the microtomed-section method and the torn-surface method (83–85).

Forming. For the fabrication of most rubber products, the mixed compound must be formed in some way to prepare it for vulcanization. In some instances, the batch sheets as obtained from the mixing mills may not require processing other than to cut disks or rectangular pieces from the sheets, which are then suitable for charging into a mold. In most cases, however, the mixed compound must be processed into a form suitable for further fabrication. The most important of these processes are calendering and extrusion. Of less importance is the making of cements.

Calendering. In general, calendering is comprised of three separate operations: the formation of sheeted rubber, coating of fabric or other material, and the frictioning of fabric. The simplest calender consists of three parallel rolls arranged one over the other. The openings between the rolls are adjustable, and the rolls are hollow and can be heated or cooled and operated at varying speeds. Also the speeds of the top and bottom rolls can be made the same as the middle roll or slower. The calender operator must learn how to make the compound adhere to or follow one or the other rolls of the calender. In general, rubber sticks to a fast roll before it sticks to a slow roll, to a hot roll before a cold roll, and to a rough roll before a smooth roll.

Sheeted compound is generally produced on a three-roll calender with the middle and top rolls as the forming rolls and the bottom roll as the carrier or doubling roll. Preciseness in gauge is important in sheet compound. Generally, the formation of calender sheet is limited in gauge because of trapped air or blisters. In order to minimize blister formation, the calender operator reduces the gauge of the sheet and plies to final gauge. Special devices, eg, a bank bar, are used to squeeze the air out. There are many different arrangements for handling the finished sheet from the calender, with a belt windup or on drums, depending upon the type of product being made. In sheet calendering, the control of the roll temperature is very important. Cold checking, which is a rough surface seen on a calendered sheet when the temperature of the calender rolls is too low or when the sheet is chilled too suddenly, and calender effect, which is a grain in the finished sheet, are aggravated by calendering the compound at lower than optimum temperatures. The top temperature limit is determined by the scorch safety of the compound and by blister formation. The compound should be warmed on a mill for all types of calendering. A consistent rolling bank of compound is needed. The calender is generally run at odd speed, ie, with the rolls at different speeds, with the top roll running two-thirds as fast as the middle and bottom rolls which operate at the same speed. A liner that is permeable to air is used with sheeting compounds.

Coating operations are performed on either three- or four-roll calenders. The three-roll calender applies the coat to one side of the fabric and the four-roll calender coats both sides of the fabric. The top roll of the three-roll calender or the bottom and offset rolls of the four-roll calender are run at odd speed to the center roll, usually two-thirds as fast. The roll crown of the calender rolls in coating operations is very important. The Z calender (see part C of Fig. 3) ensures more precise control of the calendering operation. The rolls of this calender are in the Z arrangement, but instead of the pairs being horizontal or vertical, the rolls are inclined at an angle to facilitate feeding and working around the calender.

Friction compounds are rubber coatings produced by frictioning on a calender that forces the rubber among and into the interstices between the fibers as a result of the calender speed and roll pressure. In frictioning of fabric, the center roll is hotter and operates faster than the top or bottom rolls. Friction compounds are normally compounded to be soft and sticky. This result is obtained by the use of tackifying oils and extra softening of the base rubber. The fabric to be impregnated should be kept hot. These friction compounds are used in ply building in belts, tires, and footwear.

Extrusion. An extruder consists essentially of a power-driven screw rotating in a stationary cylinder. A hopper for feeding the compound is located near the driven end of the screw and a head is provided on the cylinder at the end of the screw. A variety of dies shaped to produce the desired cross section can be adapted to the head. The

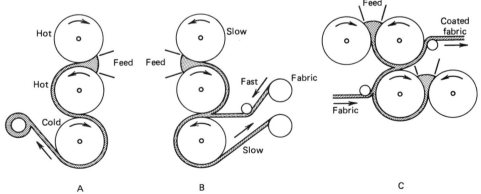

Figure 3. Arrangement of calender rolls. A, preparation of sheeting; B and C, preparation of coated fabric.

cylinder is jacketed for temperature control and, for larger size screws, the center section is bored for the same purpose.

The extruder can be warm-fed or cold-fed. Extruders handling cold feed, ie, at room temperature, are made differently than those handling warm feed. Cold-feed extruders are longer and stronger than the conventional type and the ratio of barrel length to tube diameter is ca 10:1. The conventional extruder usually has a 5:1 ratio. The compression ratio of the conventional extruder is usually two and can be developed by several methods. One method is to construct the screw with flights of decreasing pitch. The other method is to keep the pitch of the flights constant and increase the root diameter of the screw, which decreases the depth of the flights towards the discharge end. With cold-feed extruders, the extra length is needed to break down the rubber compound sufficiently for smooth extrusions. The screw flights are also shallower for effective compound softening and cooling.

The extruder can be fed by hand or by a force-feed system consisting of two feed rollers at the top of the feed box to force the compound into the screw and minimize air blisters in the extrusion. The newer extruders contain vacuum systems to eliminate trapped air and moisture.

An extruder may be too large or too small for a specific extrusion. If the extruder is too large, the rubber compound remains in the barrel too long and may scorch because of the unnecessary work done on it. If the extruder is too small and must be operated at very high screw speeds to meet the production rate, scorch also becomes a problem. Small extruders must be watched carefully to prevent them from running out of compound. Since cold-feed extruders perform work on the compound in addition to conveying and forming it, the stock temperature increases. Therefore, compounds to be cold-fed are more highly plasticized than normal stocks in order to prevent excessive heat generation and scorch in the extruder.

Because of the considerable elastic component of compounded rubbers in the unvulcanized state, appreciable expansion occurs when the material is forced through a die. Thus the opening in the die is much smaller than the corresponding dimensions of the extruded part. Since the swell at the die varies with the compound of the material being extruded, the die must be tailored to fit the properties of the particular compound to be used.

If the compound is intended to be used as insulation or a jacket on a wire or as a cover on a previously prepared hose carcass, a side-delivery head is used. In this case, the wire or hose carcass is fed through the head in a direction perpendicular to the axis of the extruder screw. The head is designed so that the compound is deflected 90° and completely surrounds the wire or hose carcass. A suitable die on the exit side controls the dimension of the part as it issues from the head.

Rubber Cements. By making a solution of the mixed composition in an organic solvent, certain forming operations are readily accomplished. For example, if it is desired to apply a coat of the material on fabric of a thickness less than can be applied by calendering, it can be done by spreading a layer of cement of suitable consistency. The cement or dough can be made of any desired consistency of deposit a film of the required thickness. Usually multiple layers are applied. Products include hospital sheeting, balloon fabric, life-raft fabric, and printing blankets.

In addition to these applications, cements are widely used in fabrication operations where it is necessary to build a composite structure of several elements that must be held together by the adhesive properties of the cement until vulcanization takes place. Some products, eg, nipples and gloves, are still made by the process of dipping a form of the desired shape in the cement, then drying the solvent, thus depositing a film. By successive dips, the required thickness is built. The product is then vulcanized on the form and subsequently stripped. Cements are made by mixing suitable proportions of the stock and a solvent in churns. The proportion of solvent is determined by its solvent power for the stock, the consistency desired in the final cement, and the character of the stock.

Assembly. Following the forming of the composition by calendering, extrusion or deposition from a cement, some articles can be vulcanized directly without further treatment. For example, sheeted material suitable for gasketing, sheet rubber from which thread may be cut after cure, extruded shapes such as tubing, and irregularly shaped window strips need no further treatment before vulcanization. For other products, assembly steps, which vary from simple to complex, may be required.

A typical tire-building operation for a four-ply bias passenger tire requires the following materials: An extruded section of tread having a suitable contour. Plies consisting of pick cord fabric or weftless cord fabric coated on both sides with rubber by calendering and cut from a long roll at an angle of ca 40° to the direction of the roll length; these bias-cut strips are spliced forming a continuous band of the required width. Two-bead assemblies built from wire insulated with a hard or semihard rubber compound; the bead assembly is wrapped with one or more layers of frictioned square, woven fabric.

The tire is built on a collapsible, rotatable drum. One layer of the bias-cut cord is applied and rolled down in contact with the drum, and the ends are spliced with an overlap equivalent to the width of about two cords. A second layer is applied with the cords running at ca 90° to those of the first ply. The previously assembled beads are then applied and the two-cord plies on the drum are folded over the bead assembly. Two additional plies are then applied as before with the cord direction alternating, and their edges are then folded over the bead assembly. The extruded tread section, which has been cut to the correct length and made into a band by cementing the ends together, is then applied and rolled down. The tire-building drum is collapsed and the uncured tire is removed. An uncured bias tire or a belted-bias tire appears like a barrel with both ends open. In radial tires before the belts are applied, the green tire is generally expanded from a cylindrical to a toroidal shape. The steel or glass belts and the tread are then applied to this toroidal shape. The tire is now ready for vulcanization.

During these operations, the property that is depended upon to hold the component parts in place is the tack of the rubber. Tack in rubber technology is the property of a rubber or compounded stock that causes two layers of stock which have been pressed together to adhere so firmly that they cannot be pulled apart at any of the areas of contact. Under these same conditions, sticky surfaces separate readily. A compound that is sticky tends to stick to other surfaces with which it may come in contact, such as mill rolls and calender rolls, whereas a tacky compound does not. In general the synthetic rubbers are less tacky than natural rubber, and SBR and BR in particular lack tackiness. Those parts of a tire that are made from SBR and BR are usually cemented with a natural-rubber cement or have sufficient added tackifiers to provide the necessary tack for building operations.

A braided hose structure is prepared by extruding the material for the inner tube in lengths of ca 152 m and coiling them on a circular platform. The textile reinforcement is then applied by a braider, which may apply one or two layers of cords. If two layers are applied, an insulating strip of rubber in the form of a calendered sheet is drawn through the double-deck braider and is applied over the first braid and under the second. The cover is then applied on a side delivery extruder and the cover is dusted with mica or a similar material to provide a suitable finish to the product.

Vulcanization. Vulcanization is an irreversible process during which a rubber compound, through a change in its chemical structure, eg, cross-linking, becomes less plastic and more resistant to swelling by organic liquids. The result is that elastic properties are conferred, improved, or extended over a wide range of temperature. The term vulcanization was originally employed to denote the process of heating rubber with sulfur, but has been extended to include any process with any combination of materials which produces this effect. Vulcanization can be carried out under numerous conditions.

Mold Curing. Compression molding is the oldest method for making molded parts. The uncured rubber is formed to the approximate shape and placed in the individual cavities. As the mold is closed under pressure, the compound conforms to the shape of the cavity and the excess material is forced into a flash groove. The compression molds may be either the simple flash type, the positive or plunger type, or the semipositive type. The plunger mold ensures that the full force of the press is applied directly on the rubber during molding and curing. This is in contrast to the flash mold, where both the metal and the rubber bear the press force. The advantage of the plunger mold is in molding very soft and very hard compounds. The disadvantages are that the amount of compound placed in the mold must be accurately controlled and the mold has poorer heat transfer. The semipositive mold incorporates some of the advantages of both the flash and positive molds.

The rubber overflow or flash must be removed from the part in most cases by hand or machines by punch-press-operated dies. When possible the rubber parts are tumbled in dry ice by means of machines that are similar to cement mixers. The thin rubber flash becomes brittle whereas the main body of the part is not cooled sufficiently and remains flexible. Thus the thinner, frozen flash breaks off in tumbling and the heavier main part is not harmed. Blasting frozen rubber parts with finer shot also removes flash. Rubbers that are freeze-resistant are not used in this dry-ice process. The freeze-resistant rubbers require the use of liquid nitrogen ($-196°C$) for deflashing. One of the disadvantages of compression molding is that the flash tends to vary towards the thick side, making removal by tumbling difficult or impossible. The cost of preparing individual preforms and the placing of each in the mold cavity is another disadvantage.

Transfer molding involves the transfer of the uncured rubber stock from one place to another within the mold. The compound is placed in a recess called the pot or transfer cavity. The pot is fitted with a ram or piston which is inserted over the compound. The force of the press when applied to the ram plus the heat from the mold causes the compound to soften and to flow through the sprues, runners, and gates into the previously empty molding cavities where the compound is cured in the desired form. Transfer molding permits closer dimensional control and generally reduces flash so that parts can be easily finished. Complex shapes and articles containing metal inserts are generally manufactured by this method. Small parts can be made more economically by this method because of the labor cost savings on preforms and finishing. The disadvantages of this method are that the formulations and control of compounds are much more critical than with compression molds. Compounds must flow well, knit properly, and cure quickly. A small amount of compound must be used to fill the transfer pot and thus is wasted as is the material in the runners. Therefore, this molding method requires careful cost calculations for scrap loss when making items requiring premium-priced polymers, eg, the fluororubbers, silicones, and polyacrylates. Transfer molds are more expensive and in general require more maintenance than compression molds (see Plastics processing).

Molded articles are usually produced from some kind of a press that is an arrangement of heated platens which close on the mold and hold it shut by hydraulic pressure. Generally the crosshead, which carries the top platen, is fixed and remains stationary. The moving table or lower platen is forced upward by means of hydraulic pressure. Around the ram are a gland and a packing ring which permit the ram to move freely up and down without loss of hydraulic fluid. A press not only must be capable of transmitting a great enough force to produce adequate molding pressure but the induced strains must be limited to minimize distortion of the platens. The more critical the molding tolerance, the less deflection permitted at the platen. This makes the ram-to-platen area relation very important. The minimum platen distortion results if the platen area is no larger than the horizontal cross-sectional area of the ram. A platen pressure of 4.1 MPa (600 psi) is the minimum limit for most molding operations and generally pressures of 8.3 MPa (1200 psi) are used. Platens can be heated by steam, oil, or electricity and the specified temperature is maintained preferably by automatic control. However, there are timing devices that automatically open the press at the end of curing cycle. With some presses the molds are fixed to the platens and, when automatically opened, the platens tilt apart. The advantages of these presses are ease of loading and ease stripping the articles from the mold plus reduced cooling of the mold during these operations. Some presses have been designed for automatic bumping, which lowers trapped air defects. The gooseneck type of press permits full access to the platens on three sides and are used for V belts.

Molds must be designed to withstand pressures of 4.1–69.0 MPa (600–10,000 psi) with a minimum of distortion. Therefore steel is generally used. However, other materials can be used eg, hard rubber or plastic molds can be used for many operations before being discarded. The surface finish on mold cavities is reflected in the surface of rubber molded against it. The higher the polish, the smoother the surface of the cured article. Chrome plating gives a smooth surface, and Teflon coating reduces sticking. Interchangeable mold-cavity linings are used in the tire industry and in molded goods to some extent. The linings are machined aluminum or die cast from soft zinc alloys and are placed inside a steel jacket, which supports the loading required for molding.

When a rubber article is removed from the mold and cooled to room temperature, the dimension of the article decreases as a result of thermal contraction. The difference between the dimensions of the article and the mold cavity in which it was molded is the mold shrinkage. Shrinkage becomes a molding problem only when the mold has not been designed to compensate for it correctly or when a required change in the compound greatly affects its coefficient of thermal expansion. Shrinkage is generally minimized by curing at lower temperatures. It varies with the type of rubber: neoprene gives very low shrinkage followed by natural rubber, SBR, nitrile rubber, and EPDM.

A more complicated molding operation is that of the pneumatic tire, in which a steel mold shapes the exterior surface of the tire from bead to bead, and the pressure during cure is supplied through a flexible bag which acts as a diaphragm by applying pressure internally to force the uncured tire against the mold surface. The diaphragm, which is an integral part of the press, is made of a resin-cured butyl stock which has extremely good heat resistance (86). Steam or hot circulating water is introduced to the inside of the diaphragm and, since the volume circulated is relatively large and temperature can be maintained, the greater percentage of the tire cure comes from the inside of the diaphragm and, since the volume circulated is relatively large and temperature can be maintained, the greater percentage of the tire cure comes from the inside. The tire is placed in the press just as it comes off the tire-building drum operations. A postinflation step may take place after curing; the tire is mounted on a rim and allowed to cool while inflated to reduce internal stresses.

Injection Cures. Injection molding is the same as transfer molding with the exception of injecting the compound into the cavities. There are three different types of injection-molding machines: in one machine a ram forces the compound through runners into the cavities, another uses a screw, and the third involves a ram and a reciprocating screw.

In the ram machine, the rubber compound is held in a cylinder and is thermally softened. During the injection cycle, the ram exerts sufficient pressure to force the rubber into the cavities. The problems with the ram type are scorch and flow. The compound must thermally soften to a point where it can be moved but must not set up or scorch during the resonance time. This requires a delicate balance of the acceleration system, viscosity control, and volume of compound moved.

In the second type of injection machine, the screw masticates the rubber compound resulting in viscosity reduction. This mastication is helped some by heating but generally less heat is required than with the ram machine. Problems with the screw machine are the mechanics of controlling the amount of mastication and injection characteristics, eg, pressure, volume, and type of injection.

In the reciprocating screw machine, the screw masticates the rubber compound. During screw rotation, the compound is fed to the front of the screw and this accumulation forces the screw to reciprocate to the rear. After enough compound has accumulated in front of the screw, the screw is stopped. Next the screw is forced to reciprocate forward by the hydraulic injection cylinder and the compound is forced into the mold. After the part is filled, a controlled amount of pressure is maintained in the stock to avoid any loss of the compound back into the cylinder before curing has started. The screw then starts to rotate again and preheats the required amount for the next shot. The mold can then be opened and the part removed; then the mold closes for the next shot. The amount of work put into the rubber compound can be controlled as can the ram pressures. Thus, the operator can make minor adjustments during each run.

In order to make injection molding profitable, very short cycles are required, generally 30–90 s. This requires curing temperatures of ca 204°C. Parts must be readily removed from the molds to keep the heat loss and cycle time to a minimum. All the advantages mentioned for transfer molding apply to injection molding. Efficiency is greatest for a large volume of small items with relatively thin walls and complicated shapes.

Steam Cures. In steam cures, the article to be vulcanized is placed in an autoclave and is subjected to saturated steam at some specified pressure until the cure is complete. The form of the product is provided during the preparation for cure and little, if any, change of shape occurs during the cure. Some products, eg, extruded shapes, are merely placed in pans, put in the autoclave, and vulcanized. Sometimes if the shape is such that it would tend to collapse during cure, a support of some kind is provided. Hose constructions consisting of tube, reinforcement, and cover built on a mandrel are cloth-wrapped and cured in open steam. The purpose of the cloth wrap is to provide pressure on the article during the cure. Tubing without reinforcement also is cured in this way. Tank linings and rubber-lined and rubber-covered equipment are usually cured in open steam.

A special case is the continuous vulcanization (CV) of insulated wire, in which the wire is rubber-covered by being passed through a cross-heat tuber and then run directly into a tube containing steam at 1.4–1.8 MPa (200–260 psi). This tube may be 30.5–61 m long, but at the speed the wire is traveling the cure must be completed in ca 15 s. The CV machines can be either the horizontal, catenary, or vertical type. The vertical CV machines are used for large cables.

Hydraulic Cures. Water is sometimes used in place of steam when it is considered to be advantageous to vulcanize under higher pressure than the pressure of saturated steam at the desired curing temperature. Thus the pressure of saturated steam corresponding to a temperature of 148°C is 345 kPa (50 psi). If pressures higher than this are desired, the water temperature is still maintained at 148°C. Boiling water at atmospheric pressure is also used for vulcanizing rubber linings in tanks that are too large to be vulcanized in a steam autoclave.

Air Cures. Vulcanization in air, either at atmospheric pressure or at elevated pressures, is commonly used for proofed goods and boots and shoes. Air is used instead of steam in some cases to avoid moisture during cure or to avoid staining or water spotting of the products. Air is a less satisfactory medium for the transfer of heat to a product than steam or water because of its lower thermal capacity and poor heat transfer to rubber. For this reason, air vulcanizers are generally designed to provide a rapid rate of air circulation. Blistering of hot-air-cured goods is lessened if the air in the vulcanizer is under pressure of 100–276 kPa (14.5–40 psi). Ammonia gas is sometimes used to produce a glossy surface on the footwear. Continuous vulcanization of unsupported solid and cellular extrusions is being carried out in hot air ovens.

Special and Combination Cures. A special case involving a combined mold and hydraulic cure is that of hose cured in lead. In this process, the prepared construction is surrounded with a lead wall applied by extrusion of lead through a die. The inside surface of the lead casing is given the desired design (usually ribbed) by the die. Long lengths of the leaded hose are wound on drums, and fittings are applied to the two ends to permit circulation of heated water under pressure. The drum is then placed in a steam autoclave and steam is applied externally. Thus the lead casing is in effect a mold, and pressure to force the hose firmly against the mold is provided by hot water under pressure. After the cure the lead jacket is split and removed. Combinations of air and steam are occasionally used to overcome the poor thermal characteristics of the air cure; by this combination, condensation of water with consequent staining and water spotting of the product can be avoided.

Continuous curing is also used in the belting or floor-matting industry. A continuous length of belt is made without splices or overlapping cures by using a Roto-cure.

This press consists of a rotating vulcanizing drum ca 1.5 m in diameter. A steel belt of highly polished, thin, band-saw steel is pressed tightly against this curing drum around a large portion of its circumference. The compound is fed into the space between the curing drum and the belt and is vulcanized as it passes around to where the belt is pulled away from the drum.

Sponge rug underlay and sheet sponge is made by passing a compound containing a chemical blowing agent between two mesh belts or cloth belts that travel between two long platens. The blowing agent decomposes before the cure is too advanced and expansion takes place to the desired thickness as governed by the spacing between the platens. The cure is complete by the time the sponge has passed to the end of the platens whereupon it is cooled and rolled up.

Cold curing has been used for the vulcanization of dipped goods produced from rubber cements and proofing made by the spreading method. The coated materials were dipped into a carbon disulfide solution of sulfur chloride or suspended in sulfur monochloride vapors to cure the thin films. This method is becoming obsolete.

The liquid-curing-medium (LCM) process is a continuous curing method for the production of solid and cellular extruded goods. The uncured compound passes directly from the extruder into a liquid heat-transfer medium which operates at 204–316°C. The compound is conveyed through the tank by either a belt, a series of driven rollers, or a combination of an organic liquid and molten metal. Hollow or sponge extrusion cannot be conveyed by the liquid–metal system since the extrusions would float on the liquid. The length of the tank depends on the bath temperature, the compound's curing rate, and the extrusion rate. The heat-transfer media can be molten metals, eg, a eutectic mixture of 58 wt % bismuth and 42 wt % tin; heat-stable organic liquids, eg, poly(alkylene glycol); or Hitec, a eutectic salt mixture. Porosity is one of the main problems, especially in soft compounds. Vacuum extruders should be used and volatile materials must be avoided. The use of calcium oxide dispersions is helpful to eliminate moisture from the compounds. Other fluid-bed techniques, both horizontal and vertical, are being used.

The principle of fluidizing small, solid particle masses by introducing a fluid into them is well established in the chemical-process industries. This principle is applied for the continuous vulcanization of rubber extrusions, and air is generally used as the fluidizing medium, although dry steam or nitrogen can be used for some polymers and small glass spheres, ie, ballotini, are used as the fluid base. Under the proper conditions, the mass of these small spheres behaves as a liquid through which the extrusions are moved and vulcanized. The fluidized bed is a very good heat-transfer medium and its buoyancy helps prevent deformation of the extrudate. Both horizontal and vertical beds are used with the horizontal type being more popular.

A method involving vulcanization by ultrahigh frequency (uhf) energy or microwaves is being developed. In all the methods previously described, the heat for vulcanization is supplied to the surface of the article, and the center portions of a thick article are heated by conduction from the surface. Since rubber is a poor conductor, thick articles need to be heated for long periods of time in order to attain the impressed temperature. This usually means that the outside layers receive too much cure. In microwave curing, the heat is generated throughout the mass so that the time ordinarily required for heating the mass by conduction is eliminated.

Compounding for microwave curing and the advantages of U.S. microwave curing equipment over European equipment is discussed in refs. 87–88. Polar compounds, eg, nitrile and neoprene, are highly receptive to microwave energy and heat very rapidly. Nonpolar compounds, eg, natural rubber, polyisoprene, SBR, EPDM, polybutadiene, and butyl, are not receptive but when the latter are compounded with polar rubbers or with carbon black or various other loadings, they can be heated effectively. There are two areas in the rubber industry where microwave curing is being used increasingly: in continuous vulcanization and in preheating prior to compression or transfer molding. Microwaves would enhance the curing of thick cross-sectional profiles but would be only marginally efficient in curing sheeting because of its large surface area-to-cross sectional area ratio (see Microwave technology).

Latex Mixes. The first step in latex fabrication is to bring the compounding ingredients into solution or dispersion form. In the case of water-soluble materials, it is necessary merely to dissolve in water, but most ingredients to be used are not water soluble, and it is necessary to emulsify the liquid ingredients and disperse the solid materials in water (see Latex technology).

Preparation of Dispersions. The general procedure for preparing dispersion is to make a coarse slurry of the powder with water that contains small amounts of dispersing agent and stabilizer, and then grind the slurry in a suitable mill to give the desired particle size. The water used should be as pure as possible. The dispersing agent is an additive that causes the dispersed particles to repel one another, thereby tending to keep the particles suspended (see Dispersants). The action is attributed to providing or increasing the magnitude of the electrostatic charge on the particles. Typical dispersing agents include the condensation product of sodium beta-naphthalenesulfonate and formaldehyde and an alkali metal salt of sulfonated lignin. The amount of dispersing agent required is specific and is determined experimentally. Factors to be considered in determining dosage are the type and particle size of the latex, the nature and desired particle size of the material being dispersed, and the nature of the dispersing agent itself. It has been found that surprisingly large differences in dispersion result from very minor differences in the concentration of the dispersing agent (89). This study showed the effect of a sodium naphthalene–formaldehyde sulfonate dispersing agent on the viscosity of zinc oxide dispersions.

XX-78 zinc oxide

dispersing agent, wt %	0.6	0.8	1.0	1.25	1.5	2.0	2.75
apparent viscosity, mPa·s (= cP)	975	190	112	88	118	215	740

The particle size in many synthetic latexes is much smaller than in natural latex, implying that a much greater surface area is exposed. This greater surface area may rob the added dispersion of some of the dispersing agent and create a degree of instability. Therefore, more dispersing agent should be used in a dispersion of SBR latex than for the same dispersion when it is added to natural latex. Sometimes it is necessary to add a wetting agent with the dispersing agent to produce a satisfactory dispersion. Usually less than 1 wt % wetting agent is used, since it tends to induce foaming during the dispersing operation.

Two classes of grinding equipment are used to prepare dispersions. The first, colloid mills, do not effect any ultimate particle-size reduction but does break down aggregates of fine particles. Colloid mills are used for such powders as clay, precipitated whiting, zinc oxide, etc. The second class of grinding equipment reduces ultimate particle size as well as breaks up agglomerates. Solids, eg, sulfur, antioxidants, and accelerators, are generally ground on this equipment, which is represented by ball and pebble mills, ultrasonic mills, and attrition mills (see Size reduction). Ultrasonic mills produce dispersions by causing the slurry to impinge against a stainless-steel blade at high velocity. These mills are fast and effective but have the disadvantage of being easily shut down by coarse slurry particles or large particles of foreign matter. An attrition mill functions by means of a reciprocating paddle, which violently agitates a mixture of the slurry, and a grinding medium usually consisting of pebbles. A pump also accelerates the grinding by circulating the slurry. These mills rapidly produce high concentrations of fine dispersions.

Typical recipes and directions for an antioxidant dispersion and an ultra-accelerator dispersion are given below.

The antioxidant dispersion recipe and directions are for Aminox (Uniroyal, Inc.), a low temperature reaction product of diphenylamine and acetone.

Material, parts by wt

A.	water	68.0
B.	water	22.8
	ammonia (28 wt % NH_3)	1.0
	Blancol (GAF Corporation)	4.0
	Dowicide A (Dow Chemical Company)	0.2
	casein	2.0
C.	bentonite	2.0
D.	Aminox	100.0
	Total	200.0
	total solids, wt %	54.2
	active solids, wt %	50

Procedure. Add A to ball mill. Compound B separately and add to mill. Add C and D to mill. Ball mill for 4 d, cooling with water to avoid sintering of the Aminox.

The ultra-accelerator dispersion recipe and directions are for Methazate (Uniroyal, Inc.), which is zinc dimethyldithiocarbamate.

Material, parts by wt

A.	water	70.0
B.	ammonia (28 wt % NH_3)	1.0
	Blancol (GAF Corporation)	4.0
	Dowicide A (Dow Chemical Company)	2.0
	casein	2.0
	water	22.8
C.	Methazate	100.0
	Total	200.0
	total solids, wt %	53
	active solids, wt %	50

Procedure. Add A to ball mill. Compound B and add to ball mill. Add C to ball mill. Ball mill for 48 h.

Preparation of Emulsions. An emulsion is a system in which a liquid is colloidally dispersed in another liquid (see Emulsions). The general method for preparing an oil-in-water emulsion is first to make a coarse suspension of oil droplets in water and then subject this suspension to a refining process which may involve intense shearing and impact such as is delivered in colloid or ultrasonic mills or in a homogenizer. A homogenizer is a machine that forces the rough emulsion through a fine orifice under high pressure. An emulsion can be simply made by adding the material to a soap solution. Where this is not effective, the soap is prepared directly in the machine. Here the fatty acid or anionic part, ie, stearic, oleic, or rosin acid, is dissolved in the material to be emulsified, and the cationic part of the soap, ie, potassium hydroxide or amine, is dissolved in water. Thus, when the oil or material phase is poured into the water phase with rapid stirring, the soap forms and an emulsion is obtained. Oleic acid is a convenient fatty acid to use since it is a liquid which is readily miscible with oil. Potassium hydroxide is preferred over sodium hydroxide because of its greater solubility. The addition of 2–10 wt % castor oil or a petroleum lubricating oil to the oil phase improves the mechanical stability of the emulsion. A slightly viscous emulsion is more stable during shelf storage than a water-thin emulsion. The use of fine particle-size insoluble material stabilizes some emulsions. A recipe and directions for an emulsion containing a liquid, nondiscoloring antioxidant, Naugawhite, is given below.

Material, parts by wt	Dry	Wet
water (hot)		19.0
Nopco 1444B (highly sulfonated castor oil produced by Nopco Chemical Company)	5.4	6.0
Naugawhite (75% active)	75.0	75.0
Total	80.4	100.0

Procedure.   Add Nopco 1444B to hot water with high speed stirring. Add Naugawhite slowly, allowing a few minutes between additions. After all the Naugawhite has been stirred in, continue stirring for 15 min.

A recipe and directions for a mineral-oil emulsion is given below.

Material, parts by wt		
A.	mineral oil	70.0
	oleic acid	1.5
B.	potassium hydroxide	1.5
	water	27.0
	Total	100.0

Procedure.   Add A to B using an agitator such as the Eppenbach Homo-mixer. Put the emulsion through a homogenizer to obtain a very small particle size and a high emulsion stability.

Mixing.   The mixing of latex compounds is a simple operation and involves the following: weighing the proper amounts of the various solutions, emulsions, or dispersions, based on the recipe and the concentration of the various dispersions; then stirring these materials into the latex, usually in a large tank equipped with a mechanical agitator. The ingredients can be stirred together by hand in a drum by means of a paddle. However, in some cases, eg, compounding latex for foam sponge, some of the ingredients must be added just prior to or during the foaming operation, and in these cases complicated automatic proportioning equipment is sometimes used.

Forming.   For the production of useful articles from latex compounds, it is necessary to convert them into solids of the desired form. A variety of methods are used to accomplish this. The straight-dip method is the simplest of any method used in making articles from latex. It is used for the manufacturing of very thin-walled dipped goods from which water can readily and quickly be removed by evaporation. The forms are dipped directly into the latex and are slowly removed. To ensure a uniform film thickness, the former is slowly rotated after dipping while the still liquid film is drying. The films are dried at room temperature or in warm air at 49–60°C. Thicker articles can be made by a tedious multiple-dipping process with drying between dips, but this process is not practical for most commercial articles. Latex deposits may be 0.127–0.25 mm thick per dip, depending on the viscosity of the latex compound.

The use of porous formers in the dipping process or porous molds prepared from plaster of Paris or unglazed porcelain whose pore size is smaller than the smallest rubber particles have been widely adapted in the latex industry. With the porous formers, the rubber particles are filtered on the surface of the formers. The rubber latex coagulates because of its high concentration to form a film. When plaster of Paris is used, the calcium ions contained in the plaster cause coagulation. With porous molds, the latex compound is poured through a funnel-shaped opening into the mold. The latex compound is allowed to stay in the mold until a deposit of the desired thickness has developed on the mold wall. The mold is empty of excess compound and is placed in an oven to dry at 60°C for one hour. The interior rubber surfaces are dusted with talc to prevent their sticking when being removed from the mold. The article after stripping from the plaster mold may be returned to the 60°C oven for 30 min or a shorter period at a higher temperature to facilitate drying. With some articles, to prevent pour lines, the mold is rotated on all planes for 15–30 min to give the latex an opportunity to flow to all extremities of the mold interior before setting. This technique is used for dolls and squeeze toys.

Electrodeposition. An interesting process that was used commercially for a time but has been dropped in favor of simpler methods is electrodeposition (90). The rubber particles in latex are negatively charged; therefore, when an electric current is passed through the latex, the rubber particles move towards the anode. When the particles arrive at the anode they are neutralized and are deposited on the anode as coagulated rubber. Nearly all of the common vulcanizing agents, when properly dispersed, are also negatively charged and are likewise deposited with the rubber on the anode. The chief drawback of this process is oxygen evolution at the anode which produces oxidation of the product and porosity.

Coagulant Deposition. The same results as those attained by electrodeposition can be achieved much more simply and economically by the use of chemical coagulants, eg, solutions of polyvalent metal salts, in place of the electric current. Two different forms of this general method were patented (91–92). In the American Anode process, the impermeable former is first dipped into a coagulant solution consisting usually of calcium chloride or a mixture of calcium chloride and calcium nitrate in a solvent, preferably alcohol. The former is withdrawn and the solvent evaporates leaving a very viscous film which is not distributed when the coagulant-coated former is dipped into the latex compound. The thickness of the film deposit depends upon the dwell time in the latex and upon the stability of the latex towards the coagulant. When the film has attained the desired thickness, it is washed on the former in hot water (60–71°C) for about an hour to remove the coagulant and other water-soluble ingredients. The film is then dried in air at room temperature or in a 66°C oven. The article is cured in the normal manner.

The Uniroyal process differs principally from the Anode process in that the first dip is in the latex compound rather than in the coagulating solution. The initial film of rubber is very thin and acts as a carrier for the coagulant absorbed by it. Volatile acids, eg, formic, acetic, or lactic acid, and cyclohexylamine acetate in solution in alcohol or acetone or in both are the generally used coagulants.

For neoprene latex and natural-latex dipping, the suitable coagulants are

	A	B	C
calcium nitrate, parts by wt	25	25	25
denatured ethyl alcohol, parts by wt	75	74.75	70.75
nonionic surfactant, parts by wt		0.25	0.25
release agent, parts by wt			4.0
Total	100	100	100

Examples of nonionic surfactant are: Emulphor ON (Antara Chemicals), the Igepals (Antara Chemicals), and Triton X-100 (Rohm & Haas Co.). Release agents are fine, insoluble, powders suspended in the coagulant and deposited on the form with the coagulant. Talc, clay, and diatomaceous earth can be used. Snow-Floss (Manville Corp.) is also suitable. Of the three coagulant formulations shown, Type A may be considered general-purpose and is suitable for most dipping applications. Type B containing the surfactant is used under conditions where the first type does not wet the forms evenly and thus gives uneven rubber deposits. Type C is used when the rubber tends to stick because of the shape of the form, making removal of the goods difficult. The release agent provides a layer of lubricant between the rubber and the form permitting easier stripping.

In dipping processes generally, but particularly with the Anode process, it is desirable to use tanks which circulate the mix to keep the surface of the mix clean and free from scum or lumps, and to keep the mix composition uniform. This creates a stability problem since agitation tends to destabilize and eventually coagulate mixes. Therefore, circulating tanks should be designed to minimize friction or shear action on the mix, and the compound should be stabilized for mechanical stability.

Kaysam Process. The Kaysam process, which has been used to some extent commercially, is somewhat similar to the roto-casting process with porous molds. However, in this process, the compound containing a gelling agent is introduced into metal molds and rotated until gelling takes place, usually with the application of heat to accelerate gelling. The gelling agents are electrolytes with a weak coagulating effect, eg, solutions of ammonium salts and sodium silicofluoride with large amounts of zinc oxide or zinc carbonate. The Kaysam process is based on the facts that latex sets to a gel with no change in volume, thereby taking the shape of the mold, and that, as the gel dries, it shrinks uniformly and without distortion. Thus, by using a mold larger than the required size in an amount calculated from predetermined shrinkage, a satisfactory article can be made. Adequate washing of Kaysam moldings is essential for good aging resistance.

Heat-Sensitizing Process. Another process that has been used to a limited extent involves sensitizing the compound so that it coagulates when heated to a certain temperature and then using heated forms of molds to build the articles to the desired thickness. Ammonia-preserved latex is used in this process in contrast to the Kaysam process. Polyether, polythioether, or poly(vinyl methyl ether) can be used as the heat-sensitizing agents.

Thread. Although manufacture of latex thread makes use of the general vulcanization principles and methods, it is unique with respect to the details of their application. The most widely used method is extrusion of latex compound through fine orifices into a coagulant bath which gels the thread, followed by mechanical handling of the thread through toughening, washing, drying, and curing operations. The coagulant bath is usually dilute acetic acid.

Latex Foam. The flexible-foam market, especially in the furniture and transportation industries, has been taken over by the flexible urethane foams because of their wide range of physical properties. The high resilience (hr) type especially shows superior properties and lower energy demand over the conventional molded hot-cure foam. Latex foam, however, still constitutes an important application for synthetic and natural latexes. Latex foam is one of the most interesting latex applications and probably the most challenging from the viewpoint of completely understanding all the phenomena that occur during manufacturing.

Many different processes are patented for preparing this product but there are only two of commercial interest for preparing molded cushioning stock: the Dunlop, which is the most widely used, and the Talalay processes. Some producers have developed variations, which are combinations of the two processes.

The basic aspects of the Dunlop process are whipping the latex to a froth by the mechanical incorporation of air into the latex, setting the frothed structure with a coagulant or gelling agent, and vulcanizing the rubber so that the foamed structure becomes permanent. The proper coagulation of the latex to give a stable foam, commonly referred to as gelation, is the key to the successful application. The gelling agent is one that can be mixed into the frothed latex and then remain dormant long enough to allow the froth to be poured into molds before exerting its gelling effect. Such a material is sodium silicofluoride Na_2SiF_6. The actual coagulating system does not involve this compound only; zinc oxide is believed to take an active part also. The pH of the aqueous phase begins to fall as soon as the Na_2SiF_6 is introduced as a result of the progressive formation of hydrofluoric acid. The ZnO is hydrolyzed at ca pH 10. At pH 8–9, the zinc forms insoluble soaps with the fatty acid latex stabilizers. The latex particles thus are sufficiently destabilized by removal of the fatty acids that they coalesce and form a gel. The pH continues to fall and ultimately reaches an equilibrium value between pH 7 and 8, and the bubbles in the system break to form an interconnecting cellular structure (93). The $Si(OH)_4$ may form in extremely fine particles, exposing a tremendous surface which could adsorb stabilizer and further act to induce gelation. Some latices may be so stable that gelation does not occur before the foam collapses. Then so-called secondary gelling agents or sensitizers can be employed to destabilize the system sufficiently to cause it to gel before foam collapse. Cationic soaps, salts of N,N'-diphenylguanidine (DPG), and amine compounds known as Trimene Base are employed for this purpose; presumably they allow gellation to occur at a higher pH.

The latex may consist entirely of natural latex or synthetic SBR latex or it may be a mixture of natural and an SBR latex. In the Dunlop process, natural-rubber foams shrink more than do SBR foams during washing and drying. The load-bearing capacity of the foams at a given density falls significantly as SBR is used in place of natural rubber.

After the compound is prepared, it is whipped either on a batch or continuous basis. The Oakes continuous mixer is the standard piece of equipment in the industry.

Then the gelling agents, ie, Na_2SiF_6 and ZnO, are added, the foam is poured into molds and cured, and the product is stripped from the mold and cured.

In the Talalay process, the froth is produced by chemical rather than mechanical means. Hydrogen peroxide and an enzymic decomposition catalyst are mixed into the latex and the mixture is placed in the mold. Decomposition of the peroxide by the added enzyme results in the liberation of oxygen which causes the latex mix to foam and fill the mold. The foam is rapidly chilled and CO_2 is introduced to gel the latex. The gelled foam is then handled in a manner similar to that used in the Dunlop process.

For supported flat stock foam, ie, foam backing on various fabrics such as carpets, scatter mats, upholstery fabrics, etc, either ammonium acetate or ammonium sulfate is employed in combination with zinc oxide as the gelation agent. The froth is prepared with an Oakes machine, the gelling agent is added at the machine, and the foam is applied to the fabric by spreading directly on the fabric or by spreading on a belt and transferring the wet gel to the fabric. Gelling is carried out at elevated temperatures, usually with the aid of infrared lamps. To prevent uneven shrinkage, the fabric is carried through the high temperature zone and drying ovens on tenter frames. For this application the foam is poured in narrow thicknesses, from 3.2 mm to a max of 12.7 mm.

Back Sizes. Backsizing defines the process of applying a film-forming adhesive material to the underside of carpets, rugs, or upholstery fabrics to achieve the following: anchor or lock the pile yarns to the basic weave or backing fabric, produce a firmer hand, increase weight, overcome raveling of cut edges, impart anti-skid properties, improve laying characteristics, which are especially important in wall-to-wall carpeting, and improve dimensional stability. In the cases of tufted carpet and pile fabric, the pile anchoring or tuft-locking function of the back size is critically important. A latex that accepts a high loading of filler for low costs and that at the same time develops the desired anchorage is required. The medium styrene–butadiene latices and particularly the so-called self-curing carboxy-modified SBR latices are most widely used.

The desired hand and drape is obtained by the proper selection of fillers and filler-loading ratio. Filler-loading ratio is the largest single factor in determining raw-material costs. The higher the loading, of course, the lower the cost of the compound. A high ratio of binders to filler results in strong backing of tufts or pile, improved dimensional stability, and excellent drape and nonskid properties. The common fillers are whiting and soft clays plus some titanium oxide for opacity. The stability of the size to degradation and discoloration under the influence of light and heat is important.

Paper Applications. In beater addition, the latex is mixed with the beaten paper pulp either by addition at the beater or to the stock chest at the wet end of the paper machine. In either case the pH of the pulp is reduced to 4.0–4.5, usually by the addition of a solution of alum to the pulp–latex mixture which has been thoroughly agitated. The latex, which for this application must be based on an anionic emulsifier, coagulates as the pH drops. The latex solids separate in intimate association with the pulp fibers. The pulp is then screened and the paper web is formed in the conventional way. A latex for this purpose must possess the proper balance between mechanical and chemical stability.

At low latex solids-to-pulp ratios, ie, ca 10–20 parts per hundred, latex is added to the beaten pulp to give a paper web with superior web strength, elongation, bursting strength, internal bond, and tear strength. The nitrile latices and medium styrene–butadiene are commonly used as beater additions. In a similar manner, latex can be deposited on asbestos fibers. Such compositions are used as gaskets, linoleum bases, etc.

Latex solids can be incorporated into a paper web either on or off the paper machine. In on-machine saturation, the extracted and sometimes partially dried web is passed through the saturation bath as a step in the papermaking. In off-machine saturation, the finished, dry web is saturated and usually by a converter.

In both on- or off-machine saturation, the web is passed through a bath of the latex. The pickup is controlled by the combination of bath strength, web speed, moisture content and squeeze roll pressure. The pickup usually covers the same range as for beater addition. The very heavy applications of resinous solids to paper are generally made by saturation. Again the nitrile and medium styrene–butadiene latices are used for saturation and for the same reasons as in the case of beater addition, ie, improved tear, tensile, and elongation.

At high saturation levels, ie, 50–100 parts per hundred, with a near thermoplastic elastomer, a leatherlike material results which can be embossed. The paper web used in pressure-sensitive tape is prepared by latex saturation in order to give it sufficient internal strength to release without delamination.

Since the latex solids in the saturation process are deposited in the structure of the paper web by drying, the colloidal system is not as critical as with beater addition. Nonionic and amphoteric surface-active materials can be effectively used in the latices. A low surface tension and small particle size are desirable features.

Both functional and decorative coatings are applied to paper from latices. The aqueous dispersions can be used on conventional paper converting machinery which usually cannot handle hot melts and solvent coatings. The lack of fire hazard because of the absence of solvents is an added advantage for the latex system.

If the coating is to be functional, the latex solids are generally film-forming and are deposited on the surface of the web in a continuous film. If the function of the film is to provide oil or grease resistance, a medium-to-high nitrile latex, or a plasticized poly(vinyl chloride) or poly(vinylidine chloride) latex may be used. The decorative coating applied to paper from latex may be a thin film of thermoplastic resin which, after application to the web, is cast against a mirror-finished heated roll. A high gloss paper stock results from this technique. The saturated vinyl and acrylic latices are used. More commonly the latex functions as the adhesive for a highly clay-loaded coating. Latices have displaced to a large extent natural materials, eg, casein and starch, which have long been used as fillers–binders for coatings applied to paper and paperboard. The latices are superior to the natural materials because of low viscosity combined with heavy coating weights. This combination of properties is necessary to the operation of the new high-speed paper-coating machines. Further, the latices provide higher gloss, better ink holdout, more flexibility, and better water resistance.

Adhesives. Both natural and synthetic latices are used in the preparation of combining adhesives for various applications. Latex adhesives are used for bonding sheets of fabric, paper, or leather, either to themselves or other materials. Essentially there are two types of latex adhesives: dry combining and wet combining. In the case of the dry-combining type, both materials to be combined are individually coated and dried. While still tacky, they are pressed together as they pass through a set of rollers. Natural latex is generally used for this application because it has a greater amount of tack than synthetic latices. Wet combining involves the application of the adhesive to only one of the two materials being combined. After the adhesive has been applied to one material and is still wet, the second material is laminated to it as both pass over drying rolls. Styrene–butadiene latices are generally used for wet combining. Nitrile and neoprene latices are used when oil resistance is required. The most important single outlet, which accounts for two thirds of the use of natural-rubber latex adhesives, is in the footwear industry, eg, for soling, lasting, and laminating (94).

Tire-Cord Solutions.   The latex solutioning of tire cord prior to building it into a carcass is a fabric adhesive application (see Tire cord). The adhesive bonds to the carcass rubber and to the textile must be stronger than the cohesive strength of the adhesive mass. In the case of rayon cord, a mixture of medium styrene–butadiene and butadiene–styrene–vinyl pyridine latices is generally applied. Nylon requires an adhesive with such strong functional groups that vinyl pyridine latices are almost universally used. Polyester tire cord is generally inert and more reactive chemicals must be added to the resorcinol–formaldehyde latex (RFL) system to promote adhesion to rubber. A predip solutioning of the polyester cord is made using 4,4′-methylene-bis(phenylcarbanilate) (Bonding Agent P-1, Uniroyal; Hylene MP, DuPont) and an epoxy resin curative (NER010, Nagase Trading Co.). The pretreated polyester cord goes through a second dip consisting of a RFL treatment where the latex is a vinyl-pyridine. In most instances, the tire-cord solutioning bath includes a preformed, water-soluble resin of the resorcinol–formaldehyde type in mixture with the latex.

Vulcanization.   Vulcanization of articles made from latex compounds generally follows the same principles and methods as those used in vulcanizing articles made from dry-rubber compounds. However there are a number of important differences. The method of curing articles in a mold by means of a press is not used with latex articles. In latex technology, molding refers to the forming of a hollow article by deposition from latex in a female form which is shaped to impart the desired contours. After drying of the deposit, moderately high temperatures can be employed for vulcanization but without the application of mechanical pressure. Hot-air, steam, and hot-water cures are widely used in the latex industry, and recently electronic curing has been introduced. Fast, low temperature curing compounds are usually used so that cures at 104°C or even lower temperatures are common. This is not true for hard-rubber latex compounds, which require curing temperatures of 149°C and are usually steam-cured. Synthetic latices usually require higher curing temperatures for normal soft compounds. For example, neoprene is usually vulcanized at a minimum of 121°C and often higher.

Articles made from suitably designed latex compounds can be cured by allowing them to remain at room temperature for long periods of time. This technique is seldom used, except for latex adhesives, because the properties of articles so made are not as good as those articles cured conventionally at higher temperatures and because the process takes a long time.

However, prevulcanized latex compounds, which are cured in the liquid state, are becoming more popular in industry, and articles made from them are merely dried or at most given a short cure at a low temperature in an oven. Prevulcanized natural rubber latices generally are used in making dipped goods or in products produced by casting or molding. Because the tear strength of prevulcanized latex films is inferior to those of post-vulcanized deposits, prevulcanization is restricted to products of fairly heavy-gauge sections. Styrene–butadiene, nitrile, and butyl latices have been prevulcanized.

Test Methods

Because of the long-range elasticity of soft-rubber vulcanizates and the various special conditions under which they are degraded in use, special test methods have been developed which in many respects are unlike those used for metals, wood, and hard plastics. The American Society for Testing Materials through its Committee D 11 on Rubber and Rubberlike Materials is constantly developing and promulgating new and improved methods for testing rubber and rubber products.

Unvulcanized Materials. Tests on raw or compounded but unvulcanized materials are chiefly concerned with rheological properties, that is, the response to the forces imposed during the operations of mixing, extrusion, calendering, and curing. Generally, some of these tests are used for control purposes at various stages in the factory operations to determine batch-to-batch uniformity of the material being processed to ensure that subsequent processing and curing steps are carried out uniformly. A combination of a number of these tests is usually necessary in the laboratory to evaluate the processing and curing characteristics of a new polymer or a new compound prior to producing it. The viscometers are based on one or more of three principles, namely, compression, rotation, or extrusion (see Rheological measurements).

An instrument used to measure plasticity and elastic recovery of elastomers and compounds is a compression-type viscometer. The parallel-plate plastometer (ASTM D 926-78), better known as the Williams Plastometer, was the first of this type to be used extensively. In this instrument, a sample initially in the form of a small pellet is pressed between two smooth plates under a given load. The thickness of the sample in millimeters after a given compression time, usually 3 and 10 min, is generally taken as a measure of its plasticity when multiplied by 100. A higher value denotes a stiffer, more viscous material. The elastic recovery is measured by the height of the sample a minute after removal from the plastometer. The most frequently used test temperatures are 70°C and 100°C.

One of the most popular and also the earliest rheological instruments of the rotational type is the shearing disk viscometer, generally referred to as the Mooney viscometer (ASTM D 1646-80). A high Mooney reading is indicative of a stiff, more viscous compound. The Mooney viscometer has been adapted for measurement of elastic recovery. A modified shearing-disk Mooney viscometer with rotor speeds of 0.05–77 rpm measures flow characteristics of elastomeric and thermoplastic materials at processing temperatures (95). This instrument is also very widely used to determine the curing characteristics of compounded stocks, most commonly for the determination of the scorch properties of the material. Curing characteristics can be determined from the viscosity–time curve or from the recorded values. The Mooney viscosity first decreases, then remains nearly constant for a period, and finally starts to increase at a rapid rate. The scorch time is the point on the viscosity curve 5 Mooney units above the minimum viscosity, and the time required for the viscosity to rise an additional 30 units is a measure of the cure rate. The test temperature is generally 125–135°C. Optimum cure can only be obtained from a mathematical extrapolation of the Mooney cure curve.

The need for an instrument to make measurements of some cure-dependent property continuously while cure is taking place and at the curing temperature so that a curing curve can be produced with good precision has been satisfied with the cure meters, eg, the oscillating disk cure meter (ASTM D 2084-79). The Plasti-Corder is another instrument which can be used to measure the rheological properties of rubber and plastics and it is very versatile (96). The Plasti-Corder can be used to measure compound or polymer flow characteristics over a wide range of shear rates and temperatures. The relative power requirements for different stocks can be predicted by the instrument. The extrusion performance of rubber compounds can be predicted from the Brabender curves. Generally, as the trace bandwidth increases, the extrusion quality decreases.

The extrudability of unvulcanized elastomeric compounds (ASTM D 2320-78) can be determined using ASTM Extrusion Die No. 1, Garvey type, which has a triangular shape. Systems for rating extrusions are described, and formulas and their preparation are given for compounds of known extrusion characteristics to allow each laboratory to evaluate its own technique. Since extrusion machines differ from laboratory to laboratory, these methods outline techniques to minimize differences in testing approaches between tubers.

Capillary rheometry is one of the few techniques available that covers the total range of shear rates involved in rubber processing. In one industrial method, the output rate of the extrudate is determined by applying constant pressure on the piston. The corresponding pressure is measured when a constant rate of piston movement is applied to give a fixed extrusion rate. The Monsanto Processability Tester (MPT) is designed with an advanced constant-rate capillary rheometer with die swell and relaxation measuring capability.

Vulcanizates.   There are two types of tests for vulcanizates. In the first type, the tests are either of specimens especially molded for the purpose or of specimens cut from a finished product. In the second type the tests are of the product itself, either in actual service or in machines designed to simulate or exaggerate conditions. The following discussion applies only to tests of the first type.

Tension.   The stress–strain test in tension is the most widely used test in the rubber industry. It is extremely useful for analyzing compound development, aiding in manufacturing control, and determining a compound's susceptibility to natural and artificial aging. Tensile strength and ultimate elongation values, however, have little significance for design or application engineers, since they cannot be used in design calculations and they bear little relation to the ability of a rubber part to perform its function. Tensile strength and elongation properties serve as an index to the general quality of a rubber part. Rubber compounds less than 6.9 MPa (1000 psi) in tensile strength are usually poor in most mechanical properties and those with tensile strengths over 20.7 MPa (3000 psi) are usually good in most mechanical properties. In the middle range, which is applied to most rubber products, correlation is at best haphazard between tensile strength and such properties as flex life, compression set, abrasion resistance, and resilience. In the standard test the specimen has a dumbell shape and is cut from a sheet with die C as described in ASTM D 412-80 and is ca 2.0 ± 0.2 mm thick. The test is conducted at room temperature and the jaws which grip the tab ends of the dumbbell specimens are separated at the rate of 8.5 mm/s. By means of suitable devices, the load required to elongate the specimen is recorded for each 100% extension of the restricted portion of the dumbbell, and both the elongation at break and the tensile strength at break are recorded. The load required to elongate to a specified elongation, eg, 300%, is referred to as the modulus of the material.

The values of stress at each increment of elongation and at break are calculated on the basis of the original, unstressed cross section, rather than the actual cross section at the time the measurements are made. If it is assumed that no change in volume occurs during stretching, it follows that for a conventional tensile strength value of 20.7 MPa (3000 psi) and an ultimate elongation of 900%, the actual stress at break on the cross-sectional area at the instant of breaking is

$$20.7((900/100) + 1) = 207 \text{ MPa } (30{,}000 \text{ psi})$$

Nonstandard tests are frequently made with specimens larger or smaller than die C or thicker or thinner than 2.0 ± 0.2 mm. The rate of jaw separation can be varied as can the testing temperature. All these conditions influence the results obtained.

In the United States much of the work of tensile testing is done on a Scott tensile-testing machine which has a pendulum dynamometer. These machines are considered accurate only at 15–85% of their maximum rated capacity. The Instron and Accr-O-Meter include strain gauges in their weighing system; thus they are useful in studying the low strain portion of stress–strain curves.

Hardness. Hardness (qv) is the relative resistance of the surface to indentation by an indenter of specified dimensions under a specific load. The objective of a hardness test is to measure the elastic modulus of the rubber compound under conditions of small strain. This property is one which is closely related to product performance, since most rubber products in use are subjected to relatively small strains. The ASTM hardness testing methods are ASTM D 2240-75, D 1415-68 (1975), and D 531-78. ASTM D 531 (Pusey and Jones Indentation) is used mainly for roll compounds.

The assumption that hardness is a close measure of stiffness may be problematic when testing rubber products such as motor mounts. There is a stress–strain relationship between hardness and stiffness, but it is established for two entirely different kinds of deformation. Hardness is derived from small deformations at the surface, whereas stiffness measurements are derived from gross deformations of the entire mass. Because of this difference hardness is not a reliable measure of stiffness.

Set, Creep, and Hysteresis. No rubber vulcanizate is perfectly elastic, and a great many tests are employed to measure the extent to which a material fails to be perfectly elastic. These can be grouped into static or long-time tests and dynamic or short-time tests. Among the static tests are tests of permanent set in terms of tension or compression, creep, and stress–relaxation. Perhaps the most common of these is the permanent-set test in compression (ASTM D 395-78), in which a specimen 12.7 mm thick and 28.7 mm in diameter is compressed between flat plates and compressed for a specified time at the desired test temperature, after which the compressing force is released and the specimen allowed to recover for a specified period of time. The height of the specimen is then measured and the permanent, unrecovered height is noted. This type of test is useful in developing materials or predicting the performance of a product which is utilized in compressive strain.

Permanent set in tension (ASTM D 412-80) is the permanent deformation caused by tensional forces. The tension-set tests are seldom used in practice except in the wire industry.

Creep is the increase of deformation with time under constant stress. Creep is important in motor mounts, since it influences the space relationships between various parts of equipment. It is difficult to predict creep for a given application without conducting simulated service tests because several factors influence creep, especially the amount of strain, temperature, and changes in these two resulting from vibration. The higher the initial strain, the higher the creep; also the higher the temperature, the higher the creep. The degree of creep depends on the type of strain. Creep is greater under tension strain than under equal shear or compression strain. Creep is also greater under dynamic loading than under static loading.

Stress relaxation of a cured rubber is the loss in stress with time at a constant deformation. A method of measuring stress relaxation in compression is described in ASTM D 1390-76. Stress relaxation is an important characteristic of a rubber gasket in its ability to maintain a seal.

The dynamic group of tests includes rebound tests and free-vibration tests either at resonance or at a frequency avoiding resonance. The objective in these tests is generally to determine the hysteresis or energy lost under the particular conditions employed, although the determination of the dynamic stiffness of the material is also important. One of the most widely used of these tests is the free-vibration test with the Yerzley oscillograph (ASTM D 945-79), whereby the specimen is vibrated either in compression or in shear and a damping curve is obtained from which the more important properties can be calculated. Another widely used method is the determination of hysteresis by means of the Goodrich Flexometer (method A of ASTM D 623-78). A cylindrical specimen 17.8 mm in diameter and 2.5 mm tall is vibrated at 30 Hz under controlled conditions of load, stroke, and ambient temperature. The temperature rise at the base of the specimen is measured and this is considered a measure of the hysteresis defect of the material under the particular conditions employed. Method B of ASTM D 623-78 describes the Firestone flexometer, which also vibrates the specimen at a constant amplitude. Method C of ASTM D 623 describes the St. Joe flexometer, which vibrates the specimen at either a constant load or a constant amplitude.

Impact resilience is determined in accordance with ASTM D 1054-79, also known as the Goodyear-Healy method. A free-falling pendulum hammer is dropped against a specimen. The resilience is the height to which it rebounds, expressed as the percentage of the height from which it was dropped.

Cracking and Crack Growth.   The flexing resistance of a rubber compound is its ability to withstand fatigue resulting from repeated distortion by extension, bending, or compression. This flexing fatigue may result in several different types of failure. The most important fatigue failure is popularly called flex cracking. The cause of this failure is twofold: stress breaking of rubber chains and cross-links and, more important, oxidation accelerated by heat buildup in flexing. This type of cracking occurs in tires, shoe soling, and belting. Both flex cracking and ozone cracking can be considered in two parts: initiation of cracks and crack growth.

ASTM D 430-73 methods B and C describe procedures by which the initiation of cracks and their subsequent growth can be measured. With some materials the initiation of cracks, as measured in method B, is slow and erratic, but once initiated they grow quite rapidly. To measure the growth of initiated cracks, a specimen as used in the initiation test is cut or pierced with a sharp tool at the base of the groove, and the rate of growth of this cut is measured as a function of the number of flexures (ASTM D 813-59) (1976). Method C, involving a specially molded, grooved specimen is used outdoors and in an ozone box to determine crack initiation and growth.

The Ross flexing machine is operated according to ASTM D 1052-55 (1976). In this method, which is particularly applicable to shoe-soling compounds, a specimen is bent freely over a rod through a 90° angle at the rate of 1.7 Hz. Prior to the test, the sample is pierced with an awl. Since flex life is a function of tear resistance, the thickness of the specimens should have close tolerances and, in comparison tests, should have the similar modulus characteristics. A test to evaluate crack growth in vulcanizates under a variety of environmental conditions has been described (97). Also, an improved flexing machine, called the TexUs flexing machine, has been developed (98). The TexUs flexing test is also described in ASTM D 3629-78.

For the study of ozone cracking, test method ASTM D 1149-78 has been developed. The material is exposed in a strained condition to air containing a known concentration of ozone at some specified temperature, usually 40°C or 49°C. Ozone exposure ratings can be made according to the method stated in ASTM D 1171-68 (1974).

ASTM D 3395-75 describes two methods for estimating the cracking resistance of vulcanized rubber compounds when exposed in an atmosphere containing ozone under dynamic conditions. These two methods, which consist of a tensile-elongation flexing test and a belt flexing test over pulleys, can be used to differentiate between degrees of ozone resistance of rubber compounds. Generally, these accelerated ozone tests are correlated with outdoor tests (ASTM D 518-61 (1974)) in compound-development work. The ozone box tests are excellent for control work.

Low Temperature Tests. Testing of rubber and rubber products at low temperatures has assumed increasing importance as rubber products for use in airplanes and in military vehicles must function satisfactory at very low temperatures. As the temperature is lowered, all rubbers become stiffer and harder and eventually become glass-hard and brittle.

The Gehman torsional test (ASTM D 1053-79) is perhaps the most widely used method to determine the change in stiffness with a lowering of the temperature. The brittleness temperature is generally defined as the highest temperature in a series of test temperatures at which a specimen fractures on sudden impact. For a particular material, this temperature depends on the geometry of the specimen and apparatus and on the speed of impact. The brittleness temperature of a rubber may vary widely, from 0 to −80°C, depending on the polymer used and how it is compounded. One brittleness test is described in ASTM D 2137-75.

Another method (ASTM D 1329-79) known as the T-R method is an adaptation of the T 50 test which involves stretching a specimen ca 250%, freezing it, releasing the clamp from one end, slowly raising the temperature of the bath, and noting the decrease in length or retraction of the sample as a function of temperature. The larger the differences between the temperature at which retractions of 10 and 70% occur, the greater is the crystallizing tendency of the materials. TR 70 correlates with low temperature compression set; TR 10 correlates with brittle points in vulcanizates of similar types of rubber.

ASTM D 797-64 (1976) involves measuring the deflection of a beam composed of the rubber stock. The beam is loaded in the middle and from the load deflection observed and the specimen dimensions, the Young's modulus can be calculated for each test temperature. Glass transition, which is relatively independent of time of exposure to low temperature and crystallization or other effects that are strongly dependent on time of low temperature exposure, can be determined by two procedures as described in ASTM D 797-64 (1976).

Accelerated Aging.   To determine the stability of vulcanizates to the effects of oxidation, a variety of accelerated aging tests is used. The most common of these is the Geer oven test (ASTM D 573-78), in which dumbbells cut from the material to be tested are exposed to circulating hot air in a suitable oven for various time periods.

No precise correlation exists between the effects of exposure to this accelerated test and the effects of shelf aging, ie, aging at room temperature in the presence of air and in the absence of other deteriorating influences. However, it is generally assumed that when two materials are being compared in the oven test and one gives results better than the other, the same relative difference would also be observed were the tests run for a long period of time under shelf-aging conditions.

Other accelerated aging tests for stability against normal oxidation include the oxygen pressure method (ASTM D 572-73), in which the test specimens are subjected to oxygen under pressure at an elevated temperature; the test-tube method (ASTM D 865-62 (1974)), which is similar to the oven method but which eliminates the possibility of contamination of one stock by another; the tubular-oven method (ASTM D 1870-68 (1978)), which involves exposing the specimens to a carefully controlled flow of air at an elevated temperature and at atmospheric pressure; and the air-pressure test (ASTM D 454-53 (1976)), in which the specimens are aged in air at 554 kPa (80.3 psi) and at 127°C.

Another deteriorating factor for some rubber products is exposure to various liquids, eg, fuels, oils, chemicals, and water. All rubber compounds are swollen by certain chemicals, and this swelling represents a deteriorating effect, since the physical properties generally decrease in proportion to the degree of swell. Procedures for conducting tests to study these effects are described in ASTM D 471-79. Method B is a quick approximate test for determining swell only, and method A is more accurate and permits the determination of certain physical properties after immersion. ASTM D 1460-60 (1974) is a method designed to give limited information on the change in length of a specimen while immersed in a liquid by observation of the test operator through a transparent wall of the container. This method is especially applicable to liquids that are so volatile that they must be maintained under pressure during the immersion period.

In order to study the effect of light and weathering on rubber samples, ASTM D 750-68 (1974) was developed. This method uses a carbon-arc light source and cyclic water spraying of the sample for various time periods. ASTM D 925-76 is used to check the contact, migration, and diffusion staining of rubber compounds with organic finishes.

Abrasion.   None of the abrasion tests has been uniformly successful, probably because abrasion resistance is not a fundamental property, but is one to which various other properties contribute in varying degrees depending on the nature of the service. The most important laboratory abrasion tests are those designed to simulate tire-tread wear. The more important abrasion test methods used involve the Williams abrader (formerly ASTM D 394-59 (1965)) also called the DuPont and Grasselli abrader, the Goodyear angle abrader, the Lambourn abrader, and the Pico abrader.

The Williams, Goodyear, and Lambourn abrasion test correctly evaluate different carbon blacks in the same polymer in order of excellence. However, although in the proper order, they do not rank quantitatively with road tests. These tests do not permit comparisons of abrasion resistance in different polymers. The Pico abrader (ASTM

D 2228-69 (1976)) gives good agreement with both carbon blacks and different polymers. The National Bureau of Standards abrasion test (ASTM D 1630-61 (1976)) is used for measuring the abrasion resistance of sole and heel compounds.

Tear. Tear resistance is the force per unit of thickness required to either propagate a nick or cut or initiate tearing in a direction normal to the direction of the applied stress. ASTM D 624-73 describes a tear test involving three independent specimen shapes. The disadvantage of this method is that the compound modulus has a direct influence on the measured tear strength. Based on theoretical work which showed that a characteristic tear strength can be measured independent of test-piece geometry, a modified trousers test was proposed involving fabric reinforcement of the legs and molding of a longitudinal groove of a specified cross-sectional geometry into the test piece (99). This modified trousers test correlates very well with the hand tear test but overcomes the qualitative nature of the latter.

Miscellaneous. Other commonly used tests include chemical analyses to identify the polymer used in a product or to determine the kind and amount of loading fillers, softeners, vulcanizing agents, antioxidants, antiozonants, etc (see ASTM D 297-79, D 1416-76, D 3677-78, D 3156-73, D 3566-77, and D-3452-78). Adhesion tests for multicomponent systems are important in the development of tire carcass, belting and hose, and other such compounds. Adhesion of rubber compounds to textiles and metal can be studied according to ASTM D 413-80, D 429-73, D 1871-68 (1975), D 2229-73 (H cord test), D 2138-72 (U cord test), and the diaphragm test methods (44,100).

BIBLIOGRAPHY

"Rubber Compounding and Fabrication" in *ECT* 1st ed., Vol. 11, pp. 892–946, by A. E. Juve and E. G. Partridge, B. F. Goodrich; "Rubber Compounding and Fabrication" in *ECT* 2nd ed., Vol. 17, pp. 543–645, by R. R. Barnhart, Uniroyal Chemical.

1. P. R. Johnson, *Rubber Chem. Technol.* **49**(3), (July–Aug. 1976).
2. B. M. Walker, *Elastomerics* **113**(4), (April 1981).
3. D. Parkinson, *Reinforcement of Rubbers*, The Institution of the Rubber Industry, London, 1957.
4. E. B. Hansen and F. C. Church, *Technical Report RG127*, Cabot Corporation, Boston, Mass., 1968.
5. M. L. Studebaker, *Fourth Annual TLARGI Foundation Conference, Los Angeles, Calif., June 5–6, 1963.*
6. D. F. Walker and C. P. Louthan, *Technical Report RG119*, Cabot Corporation, Boston, Mass., 1965.
7. R. F. Wolf, *Columbia Southern Chemical Corporation Laboratory Data*, Nov. 7, 1956, Barberton, Ohio.
8. A. G. Veith and V. E. Chirico, *Rubber Chem. Technol.* **52**(4), 748 (Sept.–Oct. 1979).
9. C. R. Wilder, J. R. Haws, and W. T. Cooper, *Rubber Chem. Technol.* **54**(2), 427 (May–June 1981).
10. *Rubber Chemicals Sales Bulletin No. 38*, Phillips Chemical Company, Akron, Ohio, 1962.
11. W. J. Stout and R. L. Eaton, *Rubber Age* **99**(12), 82 (Dec. 1967).
12. *Paracril Nitrile Rubber Bulletin No. 510-B10*, Uniroyal Chemical, Naugatuck, Conn., 1965.
13. W. G. Whitehouse, J. M. Mitchell, and R. R. Barnhart, *Rubber World* **150**(2), 39 (1964).
14. J. J. Luecken and H. L. Paris, *Rubber World* **117,** 47 (Nov. 1977).
15. L. M. Hobbs, R. G. Craig, and C. W. Burkhart, *Rubber World* **136,** 675 (1957).
16. C. H. Lufter, *Rubber World* **133,** 511 (1956).
17. Z. T. Ossefort, R. F. Shaw, and E. W. Bergstrom, *Rubber World* **135,** 867 (1957); **136,** 65 (1957).
18. L. O. Amberg and W. D. Willis, *Proceedings of the International Rubber Conference*, Nov. 1959, Washington, D.C.
19. F. B. Smith, *Rubber Chem. Technol.* **34**(2), 571 (April 1961).

20. T. D. Skinner and A. A. Watson, *Rubber Age* **99**(11), 76 (Nov. 1967); **99**(12), 69 (Dec. 1967).

21. D. J. Elliot and B. K. Kidd, *Prog. Rubber Technol.* **37**, 83 (1973–1974).

22. J. F. Krymowski and R. D. Taylor, *Rubber Chem. Technol.* **50**, 671 (1977).

23. S. A. Westley, *Rubber Chem. Technol.* **49**(5), (Nov.–Dec. 1976).

24. N. L. Hewitt, *Technical Service Bulletin*, PPG Industries, Pittsburgh, Pa., Jan. 1965.

25. M. A. Fath and D. A. Lederer, *Rubber World* **181**, 40 (Dec. 1979).

26. P. O. Tawney, J. R. Little, and P. Viohl, *Rubber Age* **83**, 101 (1958).

27. U.S. Pat. 2,726,224 (Dec. 6, 1955), C. Peterson and H. J. Batts (to U.S. Rubber Company (Uniroyal)).

28. U.S. Pat. 2,734,877 (Feb. 14, 1956), H. J. Batts and T. G. Delange (to U.S. Rubber Company (Uniroyal)).

29. U.S. Pat. 2,734,039 (Feb. 7, 1956), L. C. Peterson and H. J. Batts (to U.S. Rubber Company (Uniroyal)).

30. R. H. Dudley and A. J. Wallace, *Rubber World* **150**(4), 87 (1964).

31. J. T. Maynard and P. R. Johnson, *Rubber Chem. Technol.* **36**(4), 963 (Oct.–Nov. 1963).

32. D. V. Sarback, *Rubber World* **155**(1), 81 (Oct. 1966); **155**(2), 71 (Nov. 1966).

33. H. E. Railsback, W. T. Cooper, and N. A. Stumpe, *Rubber Chem. Technol.* **32**(1), 308 (Jan.–March 1959).

34. H. Luttropp, *Rubber Chem. Technol.* **31**, 132 (1958).

35. R. M. Murray and D. C. Thompson, *The Neoprenes*, E. I. du Pont de Nemours & Co., Inc., Wilmington, Del., 1963.

36. *Scorchguard O Bulletin*, Wyrough & Loser, Trenton, N.J., 1967.

37. H. C. Beadle and D. W. Gorman, *Elastomerics* **111**, 39 (Sept. 1980).

38. U. Eholzer, T. Kempermann, and W. Warrach, *Rubber Plast. News*, 48 (May 25, 1981).

39. P. Kovacic, *Ind. Eng. Chem.* **47**(5), 1090 (May 1955).

40. R. Pariser, *Kunststoffe* **50**, 623 (1960).

41. L. P. Lenas, *Ind. Eng. Chem. Prod. Res. Dev.* **2**, 202 (Sept. 1963).

42. J. A. Cornell and co-workers, A.C.S. Rubber Division, May 9, 1963, Toronto, Canada; *Rubber World* **152**(1), 66 (April 1965).

43. M. P. Wagner, *Rubber Chem. Technol.* **49**(3), (July–Aug. 1976).

44. W. Lyons, M. Nelson, and C. Conrad, *India Rubber World* **114**, 213 (1946).

45. W. J. van Ooij, *Rubber Chem. Technol.* **52**(3), 605 (July–Aug. 1979).

46. W. J. van Ooij, W. E. Weening, and P. F. Murray, *Rubber Chem. Technol.* **54**(2), 227 (May–June 1981).

47. S. Buchan, *Rubber to Metal Bonding*, Crosby Lockwood & Son, London, 1959; S. Buchan and W. D. Rae, *Trans. IRI* **19**, 25 (1943).

48. E. J. Weaver, *Rubber Plast. News*, 22 (July 10, 1978).

49. R. R. Barnhart and P. H. McKinstry, *J. Elast. Plast.* **10**, 216 (1978).

50. U.S. Pat. 3,728,192 (April 17, 1973), R. W. Kindle, R. R. Barnhart, and P. T. Paul (to Uniroyal).

51. R. B. Spacht and co-workers, *Rubber Chem. Technol.* **37**(1), 210 (Jan.–March 1964).

52. R. O. Becker, *Rubber Chem. Technol.* **37**(1), 76 (Jan.–March 1964); *Rubber World* **149**, 49 (1964).

53. D. C. Thompson, R. H. Baker, and R. W. Brownlow, *Ind. Eng. Chem.* **44**, 850 (April 1952).

54. W. F. Tuley, *Ind. Eng. Chem.* **31**, 714 (1939).

55. D. J. Zolin, N. B. Frable, and J. F. Gentilcore, *ACS Rubber Division Meeting*, Oct. 4–7, 1977, Paper No. 52, Cleveland, Ohio.

56. L. E. Peterson, J. T. Moriarty, and W. C. Bryant, *ACS Rubber Division Meeting*, Oct. 4–7, 1977, Paper No. 53, Cleveland, Ohio.

57. F. S. Conant, J. F. Svetlik, and A. E. Juve, *Rubber Chem. Technol.* **31**(3), 562 (July–Sept. 1958).

58. R. A. Reed, *Brit. Plast.* **33**(10), 169, 1960.

59. H. R. Lasman, *Modern Plastics Encyclopedia*, McGraw-Hill, Inc., New York, 1967, pp. 394–402.

60. R. L. Heck, *Modern Plastics Encyclopedia*, McGraw-Hill, Inc., New York, 1981.

61. B. A. Hunter, F. B. Root, and G. Morrisey, *J. Cell. Plast.* **3**(6), 268 (June 1967).

62. U.S. Pat. 3,152,176 (Oct. 6, 1964), B. A. Hunter (to U.S. Rubber Company (Uniroyal)); U.S. Pat. 3,235,519 (Feb. 15, 1966), B. A. Hunter (to U.S. Rubber Company (Uniroyal)).

63. G. S. Whitby, ed., *Synthetic Rubber*, John Wiley & Sons, Inc., New York, 1954.

64. F. Bovey and co-workers, *Emulsion Polymerization*, Vol. 10 of *High Polymers*, Interscience Publishers, New York, 1955.

65. L. H. Howland and co-workers, *Rubber Plast. Age* **42**, 868 (1961).

66. L. Talalay, *Rubber Chem. Technol.* **36**(3), 581 (July–Sept. 1963).
67. R. O. Becker and K. L. Seligman, *Rubber Chem. Technol.* **34**(3), 856 (July–Sept. 1961).
68. E. L. Borg, *Rubber World* **137,** 723 (1958).
69. H. P. Brown, *Rubber Chem. Technol.* **30**(5), 1382 (Dec. 1957).
70. D. V. Sarback, R. W. Hallman, and M. A. Cavicchia, *Rubber Age* **98**(11), 67 (Nov. 1966).
71. R. W. Kindle and S. van der Burg, *Rubber Age* **98**(4), 65 (1966).
72. U.S. Pat. reg. 26147 (Jan. 24, 1967); U.S. Pat. 2,744,287 (May 8, 1956), C. M. Parshall and P. Geyer (to U.S. Rubber Company (Uniroyal)).
73. S. E. Perlberg, *Rubber World* **156**(3), 71 (June 1967).
74. L. L. Scheiner, *Plast. Technol.* **13**(6), 39 (June 1967).
75. P. R. Van Buskirk, S. B. Turetsky, and P. F. Gunberg, *Rubber Chem. Technol.* **48,** 577 (1975); **49**(1), (1976).
76. *Ibid.*, **49**(1), (1976).
77. S. W. Newell, J. P. Porter, and H. I. Jacobs, *Rubber Chem. Technol.* **50,** 1099 (1977).
78. F. S. Myers and S. W. Newell, *Rubber Chem. Technol.* **51**(2), 180 (May–June 1978).
79. P. S. Johnson, *Blue Ridge Rubber Group Meeting*, Roanoke, Va., Feb. 26, 1981.
80. B. B. Boonstra and A. I. Medalia, *Rubber Age* **92,** 89 (March, 1963); **92,** 82 (April, 1963), reprinted in *Rubber Chem. Technol.* **36,** 115 (1963).
81. W. M. Hess and K. A. Burgess, *Rubber World* **147**(4), 57 (1963).
82. C. A. Carlton, *Rubber World* **143,** 59 (March 1961).
83. A. I. Medalia and D. F. Walker, *Evaluating Dispersion of Carbon Black in Rubber*, Tech. Report RG124, Cabot Corporation, Boston, Mass., 1966.
84. I. Medalia, *Rubber Age* **97**(4), 82 (1965).
85. C. W. Sweitzer, W. Hess, and J. E. Callan, *Rubber World* **138**(6), 869 (1958); **139**(1), 74 (1958).
86. U.S. Pat. 2,701,895 (Feb. 15, 1955), P. O. Tawney and V. R. Little (to U.S. Rubber Company (Uniroyal)).
87. C. W. Otterstedt and R. S. Auda, *ACS Rubber Division Meeting*, May 1979, Atlanta, Ga.
88. B. Krieger and R. D. Allen, *ACS Rubber Division Meeting*, Oct. 7–10, 1980, Detroit, Mich.
89. H. C. Jones and C. A. Klaman, *Rubber Age* **70,** 325 (1951).
90. U.S. Pat. 1,548,689 (Aug. 4, 1925), P. Klein (to Anode Rubber Company, Ltd.).
91. U.S. Pat. 1,996,051 (April 2, 1935), D. F. Twiss (to American Anode, Inc.).
92. U.S. Pat. 1,719,633 (July 2, 1929), M. C. Teague (to U.S. Rubber Company (Uniroyal)).
93. F. A. Murphy, *Trans. Inst. Rubber Ind.* **31,** 105 (1955).
94. P. E. Hurley, *Adhes. Age* (Feb. 1981).
95. W. E. Wolstenholme, *Rubber Chem. Technol.* **38**(4), 769 (1965).
96. A. D. Varenelli, *Wire Wire Prod.* (Oct. 1964).
97. J. R. Beatty and A. E. Juve, *Rubber Chem. Technol.* **38**(4), 719 (Nov. 1965).
98. R. W. Wu and D. I. Sapper, *Rubber Chem. Technol.* **38**(4), 730 (Nov. 1965).
99. A. G. Veith, *Rubber Chem. Technol.* **38**(4), 700 (Nov. 1965).
100. C. E. Taylor, *Rubber Chem. Technol.* **38**(4), 791 (Nov. 1965).

General References

Ref. 63 is also a general reference.
G. Alliger and I. J. Sjothun, eds., *Vulcanization of Elastomers*, Reinhold Publishing Company, New York, 1964.
Annual Book of ASTM Standards, Pt. 37, American Society for Testing and Materials, Philadelphia, Pa., 1980.
J. M. Ball, *Manual of Reclaim Rubber*, Rubber Reclaimers Association, Inc., New York, 1956.
D. C. Blackley, *High Polymer Latices*, Vols. 1 and 2, Palmerton Publishing Company, Inc., New York, 1966.
A. S. Craig, *Rubber Technology*, Oliver & Boyd, London, 1963.
A. J. DiMaggio, ed., *Basic Rubber Technology*, Philadelphia Rubber Group, Villanova University, Philadelphia, Pa., 1955.
F. R. Eirich, ed., *Science and Technology of Rubber*, Academic Press, Inc., New York, 1978.
L. H. Howland and R. W. Brown, *Rubber Chem. Technol.* **34**(5), 1501 (Dec. 1961).
D. S. LeBeau, *Rubber Chem. Technol.* **40,** 217 (1967).

E. W. Madge, *Latex Foam Rubber*, Interscience Publishers, a division of John Wiley & Sons, Inc., New York, 1962.

M. Morton, ed., *Introduction to Rubber Technology*, Reinhold Publishing Corporation, New York, 1959.

M. Morton, ed., *Rubber Technology*, 2nd ed., Van Nostrand Reinhold Company, New York, 1973.

W. J. S. Naunton, *The Applied Science of Rubber*, Edward Arnold Ltd., London, 1961.

R. C. Noble, *Latex in Industry*, 2nd ed., by *Rubber Age*, Palmerton Publishing Co., New York, 1953.

W. M. Saltman, ed., *The Stereo Rubbers*, John Wiley & Sons, Inc., New York, 1977.

J. R. Scott, *Ebonite*, Maclaren & Sons, London, 1958.

The Synthetic Rubber Manual, 8th ed., International Institute of Synthetic Rubber Producers, Inc., Houston, Texas, 1980.

R. O. Babbit, ed., *Vanderbilt Rubber Handbook*, R. T. Vanderbilt Company, Norwalk, Conn., 1978.

Vanderbilt News **25**(5), (Oct. 1959).

ROBERT R. BARNHART
Uniroyal Chemical

RUBBER, NATURAL

Natural rubber [*9006-04-6*] (NR) (*cis*-1–4-polyisoprene [*9003-31-0*]) occurs in over 200 species of plants, including dandelions and goldenrod (1–2). Only two species are commercially significant: the *Hevea brasiliensis* tree and the guayule bush *Parthenium argentatum*. Guayule once was an important source of rubber and provided 10% of the world's supply in 1910. Today, the *Hevea* tree accounts for over 99% of the world's natural-rubber supply, but researchers are again studying guayule as a potential source of rubber.

Columbus, during his second voyage to the New World in 1496, may have been the first European to see natural rubber when he observed the natives of Haiti playing with a ball made by smoke-drying the latex of the *Hevea* tree. One of the first written reports of natural rubber was a description of a rubbery material which the natives of the Amazon called Cahutchu or weeping wood. In 1770 the English chemist, Joseph Priestley, gave rubber its name when he noted that the gum from the *Hevea* tree could be used to rub out pencil marks.

Three events that occurred in the first half of the nineteenth century were of prime importance in the development of the rubber industry (3). Thomas Hancock, an Englishman, invented the masticator in 1820. This allowed rubber to be softened so that it could easily be dissolved in solvents. Solutions of rubber were used in 1823 to make waterproof garments, but the garments were not very satisfactory as the rubber softened in the sunlight and became brittle in the cold. These two events preceded the discovery of vulcanization. In 1839, Charles Goodyear discovered that rubber and sulfur, combined in the presence of heat, gave a product vastly different than the raw rubber. Goodyear's patent on the vulcanization of rubber was not issued until 1844. In the meantime, Thomas Hancock had also discovered vulcanization and had secured the British rights to the process. Whether Goodyear or Hancock made the discovery first is not known, but Goodyear is generally credited with the discovery.

The year 1876 signaled the beginning of the plantation era. Seventy thousand rubber seeds gathered from the Amazon valley were planted in Kew Gardens, London, and ca 2800 of the seeds germinated. In 1877, ca 1920 rubber plants were shipped to Ceylon (Sri Lanka) and planted in the botanical gardens. A few plants were sent from Ceylon to Malaya and from there, in 1882, seeds and plants were sent to Singapore, India, and Java. By 1888 there were ca 1000 trees growing in Malaya (4). In 1888, the first practical pneumatic tire was developed and eventually led to an unprecedented demand for rubber (4). At about the same time, it was demonstrated that rubber trees could be grown in cleared areas and in regular order and could be tapped at intervals of a few days, thus leading to an efficient and economical recovery of the latex.

Agriculture

Producing rubber trees grow to heights of 15–20 m and require 200–250 cm/yr of rainfall. They thrive at altitudes up to 300 m (5). Modern practice calls for the planting of trees at the rate of 400–500/ha (ca 160–200 per acre). Trees are commonly planted 3–4 m apart in parallel rows 6 m apart. Experiments are being conducted to determine the feasibility of higher planting densities. It takes up to 7 yr for a rubber tree to grow to maturity. The time of immaturity in the field can be reduced to 5 yr by growing the trees in plastic polybags in nurseries for 1–1.5 yr before transplanting to the field.

Rubber trees can be propagated either by sexual methods, ie, normal or controlled crossing and planting the seeds derived from the crosses, or by vegetative methods, eg, bud grafting. In normal practice, the trees are propagated by grafting buds from a mother tree onto the stem of a seedling. All trees derived by this method from a single mother tree either directly or indirectly are called a clone (6). Today, there are less than 100 clones that are commercially grown, but many times that number have been grown experimentally. The mother tree is chosen for the best combination of a variety of properties, including yield, resistance to disease, resistance to wind damage, bark characteristics, and various growth characteristics. Yields of dry rubber have been increased from the original levels of less than 500 kg/ha (ca 440 lb/acre) to present commercial levels of 2000–2500 kg/ha (ca 1785–2225 lb/acre). Experimental yields of 3000–4000 kg/ha (ca 2680–3570 lb/acre) have been achieved and theoretical considerations indicate yields of 9000 kg/ha (8025 lb/acre) may be possible (4).

Rubber trees are subject to a variety of diseases but the most devastating is the South American Leaf Blight caused by the fungus *Microcyclus ulei* (7). The disease, which is mainly confined to Central and South Americas, has prevented the expansion of rubber production in this area. Recently, thermal fogging with a combination of fungicides has shown promise in controlling this disease (see Fungicides, agricultural). If this disease can be controlled, South America may someday become a large producer of rubber.

Tapping

Latex is harvested from the tree by a process called tapping, which has been described as a controlled wounding of the tree (8). A specially designed tapping knife is used to remove shavings of bark from the surface of a groove made into the tree to a depth about 1 mm from the cambium. The groove is made from left to right at an angle of 30° to the horizontal across half the tree. This type of cut is the most common and is called a half spiral. A typical tapping intensity calls for tapping the half spiral on alternate days.

Techniques to increase yield and to reduce labor are constantly being researched. Two techniques in use are yield stimulation and puncture tapping. Ethylene gas is the active ingredient in commercially available stimulants, eg, Ethrel. The use of stimulants can increase the yield or maintain the same yield with a reduction in tapping intensity and thereby conserves bark. Puncture tapping is a recent innovation where trees are tapped by piercing the bark of the tree with sharp needles instead of in the conventional manner. The use of puncture tapping in conjunction with stimulation can give yields comparable to conventional tapping and enables the tapper to tap more trees in a given day. Early puncture tapping is also being investigated on trees 3–4 yr old. Since trees are planted in the field at 1.5–2 yr of age, early puncture tapping could reduce the unproductive period in the field to 1–2 yr (9).

With the conventional tapping method, a tapper begins tapping as soon as there is light enough to see. The tapper collects the cup lump, that is latex that has entered the cup and coagulated since the previous latex collection, and tree lace, which is a thin layer of coagulated rubber on the tapping cut. The tree is tapped by using the tapping knife to carefully remove 2–3 mm of bark from the cut. The latex exudes onto the surface of the cut and flows down the cut into the collection cup. In order to prevent the latex from coagulating before collection, a small amount of a preservative, eg, ammonia, sodium sulfite, formaldehyde, or boric acid, is often added to the cup. Later in the morning, the tapper returns to collect the latex and takes it to a centrally located bulking station where it is preserved further, usually with the addition of ammonia gas, before being transported to the factory.

Two types of raw material are brought to a processing factory: field latex and raw coagulum. Raw coagulum can be classified as either field or small-holder. Field coagulum is rubber that coagulates naturally in the field, ie, cup lump and tree lace. Small-holder coagulum is rubber coagulated either naturally or chemically by the small farmer who is unable to bring the rubber to the processing plant as latex. These raw materials are the basis for all types and grades of natural rubber.

Field Latex

Field latex is freshly tapped natural-rubber latex. The latex as it comes from the tree has a pH of 6.5–7.0, a density of 0.98 g/cm^3, and a surface free energy of 4.0–4.5-μJ/cm^2 (0.96–1.1 μcal/cm^2) (10). The total solids of field latex is 30–40 wt % depending on the clone, weather, tapping frequency, and various other factors. The dry rubber content is usually ca 3 wt % less than the total solids. The 3 wt % nonrubber portion is made up mainly of proteins, resins, and sugars.

The solids in freshly tapped latex are distributed through three phases: rubber, aqueous, and lutoid. The rubber phase typically contains 96 wt % rubber hydrocarbon, 1 wt % proteins, and 3 wt % lipids. There are also trace amounts of metals, mainly magnesium, potassium, and copper. The principal proteins in the rubber phase surround the rubber particle, and the main protein present is α-globulin. The lipids associated with the rubber particle are comprised of sterols, sterol esters, fats, waxes, and phospholipids. The phospholipids are adsorbed on the surface of the rubber particle and may be the means by which the proteins are linked to the rubber particle. The aqueous phase, which sometimes is referred to as C-serum, is a dilute aqueous solution with a density slightly over $1.0 \ g/cm^3$. Among the many compounds it contains are carbohydrates, proteins, amino acids, enzymes, and nitrogenous bases (11–12). The lutoid phase was first discovered in field latex in 1948 (13). Fresh field latex was centrifuged at low speeds and separated into fractions. The top fraction contained most of the rubber particles and the bottom yellow fraction contained aggregates that were distinctly different from the rubber particles. These aggregates were called lutoids because it was thought they caused the yellow color. The structure of the lutoids was described in detail in 1964 (14). Lutoids are made up mainly of water and small amounts of soluble proteins, insoluble proteins, and phospholipids. The yellow color of the bottom fraction is caused by Frey Wyssling particles, which are spherical, nonrubber particles. Their color results from the presence of carotenoid pigments, and it is these pigments that give natural rubber its yellow color whether Frey Wyssling particles are present or not. The intensity of the yellow varies from clone to clone.

Grades of Natural Rubber

Visually Graded (Conventional) Rubber. Visually graded natural rubber is often referred to as conventional rubber, since visual inspection is the oldest method of grading rubber. Thirty-five standard international grades of natural rubber of eight types produced only from the latex of *Hevea brasiliensis* are described in ref. 15. The type refers to the method of processing used in the preparation of the rubber. The grade refers to selected subdivisions made within each type of rubber with reference to quality. The rubber within each type is graded by visual inspection. Conventional rubber is graded by drawing samples from 10% of the bales within a lot and visually comparing them to the standard for that grade. In addition to the description of each grade given in ref. 15, master international samples have been prepared for 29 of the 35 grades. Official international samples have been copied from the master samples and supplied to the endorsing organization in 17 countries. From these international samples, the endorsing organizations prepare and distribute duplicate international samples to the rubber industry. The sample books contain carefully selected pieces of rubber indicating the quality that is acceptable in each grade. Rubber conforms to a specific grade when the average quality of the drawn samples is equivalent to the average quality represented by all the pieces in the sample book. Rubber that in part is inferior to any piece of the international sample can not be averaged with superior-quality rubber to make up a lot of a specific grade (see Production). The following general prohibitions are applicable to all grades as stated in ref. 15:

> Wet, bleached, undercured, and virgin rubber and rubber that is not completely visually dry at the time of buyer's inspection is not acceptable, except slightly undercured rubber as specified for no. 5RSS (ribber smoked sheet).
> Skim rubber made of skim latex shall not be used in whole or in part in the production of any grade.

The eight types of visually graded rubber, as classified by the source of raw material used in their manufacture, are latex grades: ribbed smoked sheet (RSS), white and pale crepes, and pure blanket crepes; and remilled grades: estate brown crepes, compo crepes, thin brown crepes or remills, thick brown crepes or ambers, and flat bark crepes. Latex grades are manufactured directly from field latex and remilled grades are produced from estate or small-holder coagulum. Compo crepes, thick blanket crepes (ambers), and pure smoked blanket crepes are not being produced in significant quantities.

Technically Specified Rubber. Technically specified natural-rubber (TSR) was introduced by the Malaysians in 1965 under the Standard Malaysian Rubber (SMR) scheme. Other rubber-producing countries soon followed with their own versions of TSR: Indonesia (SIR), Singapore (SSR), and Thailand (TTR). The introduction of TSR brought innovations in processing, packaging, and quality control to the natural rubber industry.

In 1980, TSR accounted for ca 40% of worldwide natural rubber production. The quality of TSR is determined by a series of chemical and physical tests rather than by visual inspection. Malaysia is the largest producer of TSR. Technical specifications for SMR rubber, which were revised in 1979, are shown in Table 1; SMR rubbers can also be classified as either latex or remilled types.

All versions of TSR are analyzed with the same basic tests to determine quality, but there are small differences in the test limits. Some countries do not include all the SMR grades in their schemes nor do some specify exactly the same raw materials for each grade.

A complete description of the test methods for determining SMR properties is given in ref. 16. These test methods, with slight variations, are used by most producers of TSR rubber. A sampling intensity of 10% of the bales is generally used, and each sample is tested individually for all properties. The dirt level and the type of raw material are the primary criteria for determining the grade. The dirt level is measured by dissolving a small sample of the rubber in mineral turpentine and filtering the solution through a 44-μm (325 mesh) sieve. The dirt levels for the TSR latex grades compare favorably with those for pale crepes and the premium grades of smoked sheet. The levels for the 10, 20, and 50 grades are comparable to those for brown crepes and the lower sheet grades.

The specification for Wallace plasticity for all grades of TSR is 30 minimum. In practice, the plasticity decreases from 45–55 for L and 10 grades to 30–45 for the 20 and 50 grades. The plasticity retention index (PRI) is determined by comparing the original Wallace plasticity P_o with the plasticity after aging the rubber for 30 min P_{30} in a 140°C oven, PRI = $(P_{30}/P_o) \times 100$. The plasticity retention index is a measure of the oxidizability of natural rubber and is an indication of the quality of the raw rubber, its processing behavior, and the properties of the vulcanizate (17).

Physical Properties

Natural rubber crystallizes at below 20°C because of its stereoregular molecular structure. The rate of crystallization varies with temperature and the type of rubber. At −26°C, a bale of rubber becomes hard after only a few hours, but at 8°C it takes about one month before crystallization is evident. Pale crepe crystallizes 2–5 times faster than TSR crumb rubbers at −26°C (see White and Pale Crepes; Technically Specified Rubber) (18). The molecular weight and plasticity of natural rubber vary over a wide range depending on the clone and sample history. The Mooney viscosity, Wallace plasticity, M_n, and M_w for the total solids film of 12 different clones have been

Table 1. Standard Malaysian Rubber (SMR) Specifications

Grade	Latex grades			SMR 5	SMR 10	Remilled grades		
	SMR CV[a]	SMR L[a]	SMR WF[a]			SMR 20	SMR 50	SMR GP[b]
dirt retained on 44 μm aperture, % wt (max)	0.03	0.03	0.03	0.05	0.10	0.20	0.50	0.10
ash content, wt % (max)	0.50	0.50	0.50	0.60	0.75	1.00	1.50	0.75
volatile matter, wt % (max)	0.80	0.80	0.80	0.80	0.80	0.80	0.80	0.80
Wallace plasticity (P_0, min)		30	30	30	30	30	30	
plasticity retention index (PRI), % (min)	60	60	60	60	50	40	30	50
nitrogen content, wt % (max)	0.60	0.60	0.60	0.60	0.60	0.60	0.60	0.60
color limit (Lovibond scale)		6.0						
Mooney viscosity ($ML_{(1+4)}$) at 100°C	60 ± 5[c]							58–72

[a] Cure information is provided in the form of a rheograph.
[b] This grade is produced from a combination of latex and field coagulum.
[c] CV_{50} and CV_{70} are available with viscosity limits of 50 ± 5 and 70 ±5, respectively.

473

compared (19). The number-average molecular weight was $(2.55-27) \times 10^5$ and the initial Wallace plasticity was 32–75. Mooney viscosity and Wallace plasticity begin to increase immediately after tapping because of a slow cross-linking process. This autocross-linking is caused by a very small concentration, $(2-6) \times 10^{-3}$ mol/kg of rubber, of aldehyde groups on the rubber molecule (18). The cross-linking is most rapid for the first couple of months after tapping but continues at a reduced rate for much longer periods. Over 4–5 yr, the Mooney of RSS1 may increase from the low 60s to over 100. Rubber prepared by intentional coagulation of field latex appears to storage-harden more than rubber made from field coagulum (see Field Latex). This difference probably results from the fact that field-grade rubbers undergo storage hardening during autocoagulation before the initial viscosity measurement is taken rather than any enhanced tendency of latex-grade rubber to storage-harden (20). The viscosity of natural rubber can be stabilized with the addition of small amounts of hydroxyl-amine hydrochloride or semicarbazide hydrochloride (21).

The physical properties of natural rubber vary slightly depending on the non-rubber content, degree of crystallinity, amount of storage hardening, and particular clone or blend of clones used as the raw material. Some average physical properties are listed in Table 2.

Chemical Properties

Commercial grades of *Hevea* natural rubber contain 93–95 wt % *cis*-1-4-poly-isoprene. The remaining nonrubber portion is made up of moisture (0.30–1.0 wt %), acetone extract (1.5–4.5 wt %), protein (2.0–3.0 wt %), and ash (0.2–0.5 wt %) (23). Natural rubber is soluble in most aliphatic, aromatic, and chlorinated solvents, but its high molecular weight makes it difficult to dissolve. The molecular weight is generally reduced by mastication on a mill or in a Banbury mixer prior to dissolving the

Table 2. Physical Properties of Natural Rubber[a]

Property	Value
specific gravity	
at 0°C	0.950
at 20°C	0.934
refractive index	
at 20°C, RSS	1.5195
at 20°C, pale crepe	1.5218
coefficient of cubical expansion, °C^{-1}	0.00062
heat of combustion, J/g[b]	44,129
specific heat	0.502
thermal conductivity, W/(m·K)[c]	0.13
dielectric constant	2.37
power factor (at 1000 cycles)	0.15–0.20
volume resistivity, Ω·cm	10^{15}
dielectric strength, V/mm	3,937
cohesive energy density, J/cm^3 [b]	266.5
glass-transition temperature, °C	−72

[a] Ref. 22.
[b] To convert J to cal, divide by 4.184.
[c] To convert W/(m·K) to (Btu·in.)/(h·ft^2·°F), divide by 0.1441.

rubber. A chemical peptizing agent can be added during mastication to reduce the molecular weight further and to enhance the solubility of the rubber.

The most important chemical reaction that natural rubber undergoes is vulcanization. Unvulcanized or raw rubber is soft, easily deformed, tacky when hot, and hard and rigid when cold. When natural rubber is masticated and combined with sulfur in the presence of heat, it is converted to an elastic material which is rubber as most people know it. The vulcanization of rubber in the presence of heat and sulfur alone is a very slow reaction, but the rate can easily be increased through the use of rubber accelerators and activators (see Rubber chemicals) (24–25).

The double bonds in natural rubber undergo many of the classical chemical reactions. Rubber can be hydrogenated with a platinum catalyst or heat and pressure to a nonelastic crystalline material. Natural rubber undergoes substitution and addition reactions with chlorine to yield chlorinated rubber as a fine white powder (26). Fully chlorinated natural rubber, ie, 65 wt % chlorine, is inert and is used in chemically resistant paints and in adhesives (qv) (see Paint). Hydrogen chloride adds to the double bonds in natural rubber to form the rubber hydrochloride (28–33 wt % chlorine), which is highly crystalline (27–28). *Para*-Toluenesulfenyl chloride reacts quantitatively with the double bonds of natural rubber in toluene solution. When purified natural rubber is used, a block polymer forms and is made up of blocks of unmodified rubber and blocks of p-toluenesulfenyl chloride-modified rubber (29).

Natural rubber treated with a p-nitrosoaniline or p-nitrosophenol yields a hydroxylamine as the primary product (30). The hydroxylamine can undergo further reaction to give a p-phenylenediamine or a p-aminophenol. These groups are excellent antioxidants and are chemically bound to the rubber molecule and are nonextractable (see Antioxidants and antiozonants). Unfortunately, colored by-products have prevented commercialization of this material.

Production

Conventional Rubber. *Ribbed Smoked Sheet.* Ribbed smoked sheets (RSS) are made from intentionally coagulated field latex. The latex is strained through a 250-μm (60-mesh) screen into factory bulking tanks where the dry rubber content (DRC) is determined, and the latex is diluted with clean water to ca 15 wt % solids. The latex is strained again as it is transferred into coagulation tanks ca 3 m long, 1 m wide, and 40 cm deep. The sides are slotted every 4 cm for aluminum partitions. A sufficient amount of a 5 wt % formic acid solution is added to reduce the pH to 4.8. After the acid is stirred into the latex and the surface foam is skimmed, the aluminum partitions are inserted into the tank before the latex coagulates. When the coagulum is firm enough, it is flooded with water to prevent oxidation. The common practice is to allow the rubber to coagulate overnight. The gelatinous coagulum slabs are then transferred to a sheeting battery along water-filled troughs. The sheeting battery is made up of a series of contrarotating rollers that compress the coagulum into a sheet 2–3 mm thick. The coagulum is kept immersed in water between each set of rollers to prevent oxidation and to wash the nonrubbers from the surface of the sheet. The final set of rollers is embossed with the characteristic ribbed pattern, which is impressed in the soft sheet. The sheets are dried 4–7 d in large smokehouses. The smokehouses may be heated by oil or wood, but enough wood is always burned to smoke the rubber since smoke acts as a fungicide and prevents mold growth during drying and storage. The smokehouses are divided into sections with the temperatures at 40–55°C. The rubber remains 1–2 d in each section before being moved to the next hotter section. Temperatures higher than 55°C cause the finished sheet to blister and have a tacky surface. The dried, golden-brown sheets are taken to the inspection room where each one is graded by visual inspection on a light table in relation to international samples.

Six grades of ribbed smoked sheet from RSS1X to RSS1 through RSS5 are described in ref. 15, but RSS1X and RSS5 are not produced in large quantities. Official samples have not been established for RSS1X. The sheets are graded on the basis of color, clarity, and the absence of bark, blisters, and holes. There are almost no visual defects or contamination present in RSS1. A well-run estate produces better than 95% RSS1 with the balance of sheet production downgraded to RSS2. Small holders produce sheet rubber in the same manner, but the latex collection, equipment, and processing are less efficient, so the rubber tends to have more visual defects and contamination. Some small holders are able to maintain high quality and produce RSS1 and RSS2, but the majority produce RSS3 and RSS4. The conventional method of packaging RSS is to press the sheets into blocks weighing 113.5 kg max and measuring 0.142 m^3 (5 ft^3). The minimum specified weight is 101.7 kg (15). The bales are wrapped with rubber of equal or higher quality, and this wrapper sheet is completely coated with a talc solution to prevent adhesion during transit. Recently, some producers have begun pressing the sheets into blocks weighing 33–35 kg, wrapping the blocks in polyethylene film, and packing 30 blocks in a wooden pallet crate.

White and Pale Crepes. White and pale crepes are produced from freshly coagulated field latex that is carefully selected and preserved. Field latex is selected from clones that produce rubber with a low carotene content. Sodium bisulfite is added, usually in the factory bulking tanks, to prevent enzyme darkening. The latex is strained several times during the processing to ensure clean rubber. A fractional coagulation process may be used to reduce the carotene level. In this process, the field latex is diluted to 25 wt % DRC with clean water and is transferred to the coagulation tanks where a dilute solution of acetic acid is added. In a few hours, a yellow fraction that is rich in carotenoid pigments coagulates and floats to the surface. This fraction, which represents 10–20 wt % of the latex, is skimmed from the top and the remaining latex is diluted to ca 20 wt % DRC. A bleach, usually xylyl or tolyl mercaptan, is then added to the latex. The mercaptan reacts with and neutralizes most of the remaining carotenoid pigments. Formic acid is added to reduce the pH to ca 4.5 and aluminum partitions are inserted to produce slabs of coagulum 10 cm thick. As soon as the coagulum is firm enough, it is flooded with water to prevent oxidation. The coagulum is ready for milling after 3–5 h. The slabs are processed through a series of mills under a water spray. The water keeps the rubber from warming and darkening, and it washes away the serum and any contaminants. The serum is high in nonrubbers and can cause yellow streaks in the crepe if not washed off thoroughly. The slabs are converted to thick blankets on a series of macerator mills with diamond grooves ca 3 mm deep. The mill rolls rotate at different speeds so that the rubber passing through the rolls undergoes a shearing and masticating action which exposes new surfaces to be washed. All rubber-processing mills operate according to this same basic principle. The thick blankets are then passed through a series of intermediate creping mills with grooves ca 1 mm deep. Finally, the rubber is milled on a series of smooth finishing mills until a smooth crepe, 1–2 mm thick, is obtained. The smooth crepes are hung in multistoried drying sheds for 6–7 d at 31–34°C. Higher temperatures or exposure to light can cause the crepe to darken. Production of high quality crepe may require 25 or more mill passes. An ample supply of clean water (60 L/kg dry rubber) is also required. Water with suspended solids or a high iron content causes discoloration and dullness of the crepe. The premium grades of crepe are generally packed in bales weighing $33\frac{1}{3}$ kg or 50 kg. The bales are wrapped in polyethylene film and are packaged either in 1 metric ton pallet crates or in individual paper bags, which are shipped without further packaging. The lower grades are usually packaged in 100-kg bales wrapped with crepe of equal or higher quality and talc-coated to prevent sticking during transit.

White and pale crepes are graded on the basis of color and uniformity of color. Sour or foul odors, dust, specks, sand, or other foreign matter are not allowed in any of the grades. There are two grades of white thin crepes: 1 and 1X. The no. 1 thin white crepe can have a very slight variation in shade but the 1X cannot. White crepes are often produced for manufacture into sole crepe. Pale crepes may be either thick or thin and are graded 1X, 1, 2, or 3. The lower grades permit more yellowness and less uniformity of color.

Estate and Thin Brown Crepes. Fresh cup lump and other estate-generated, high grade scrap make up the raw material for estate brown crepes; wet slab, unsmoked sheet, cup lump, and other high grade scrap from estates or small holdings are used to produce thin brown crepes. If tree-bark scrap is used, it must be cleaned to separate the rubber from the bark. The raw material is milled with water spray on a series of grooved wash mills, which thoroughly clean and blend the rubber. The types of mills and the number of mill passes are determined by the raw material quality and the grade desired. The crepes are dried for ca three weeks in large open-air drying sheds. After drying, the crepes are banded with three metal bands to form bales weighing ca 100 kg and they are coated with talc solution. Estate brown crepes are produced in thick or thin form in grades 1X, 2X, and 3X. Thin brown crepes are graded 1, 2, 3, and 4. The lower grades are darker and more variable in color and have more contamination.

Flat-Bark Crepes. Flat-bark crepes are produced from all types of scrap natural rubber including earth scrap. The processing and packaging is similar to that for the various brown crepes. The two grades of flat-bark crepes are standard flat-bark crepe and hard flat-bark crepe.

Technically Specified Rubber. Technically specified rubber (TSR) is produced from either intentionally coagulated field latex or estate and small-holder coagulum by means of a crumbing process. The coagulum is milled into a blanket on a series of wash mills, which clean and homogenize the rubber. The rubber can be converted to crumb by a variety of milling or granulating equipment with or without chemicals. The wet crumb is placed in open metal baskets and is dried either continuously or batchwise in an air-circulating oven at ca 100°C. The warm dried crumb is compressed into bales by a hydraulic press and the bales are wrapped in strippable or thin (0.040 mm), low melting (108°C max) polyethylene film. The bales measure ca 67 × 33 × 20 cm and 30 bales are packaged to a polyethylene-lined wooden pallet crate. Shrink-wrapping of the 30-bale unit to the pallet base with heavy polyethylene film is a recent innovation, which greatly reduces the possibility of contamination from wood.

The Heveacrumb process for manufacturing TSR was invented in 1964, and the first Standard Malaysian rubber (SMR) rubber was produced by this process (31–32). Castor oil (qv) (0.7 wt %) was added to the field latex prior to coagulation. During milling of the coagulum, the castor oil acted as a crumbing agent converting the blanket to a fine crumb. Some consumers were concerned about the possible use of excess castor oil which could bloom to the surface in certain vulcanizates. Because of this concern and with the development of improved mechanical granulating equipment, the amount of castor oil was drastically reduced. Castor oil is used mainly as a partitioning agent to prevent the crumbs from agglomerating after being mechanically converted. Rubber that is produced with even small amounts of castor oil is commonly referred to as Heveacrumb rubber. If the rubber is produced by strictly mechanical processing without any castor oil or other incompatible oil, it is referred to as comminuted rubber. Comminuted rubber is preferred by some customers, especially those manufacturing adhesives or using rubber in rubber-to-metal bonding. In practice, it is extremely difficult to show any difference between modern Heveacrumb rubber and comminuted rubber. The SMR scheme is the most complicated of the TSR schemes and is made up of the largest number of grades.

SMR CV. Natural rubber undergoes irreversible chemical cross-linking on storage, which results in an increase in Mooney viscosity (33). SMR CV is a viscosity-stabilized rubber made from field latex that produces a rubber with the desired initial viscosity. The initial viscosity is maintained by the addition of ca 0.15 wt % hydroxylamine hydrochloride or neutral hydroxylamine sulfate, which inhibits storage hardening by reacting with the aldehyde groups on the rubber molecule which are the sites for cross-linking (34). Subsequent work has shown that a lighter color CV rubber can be produced with semicarbazide hydrochloride (21). SMR CV is produced in three viscosity ranges (Table 1) with CV 60 being the most common grade.

SMR L. SMR L is a light-amber grade produced from field latex, which is chosen for its light color. The rubber must be processed quickly, usually the same day, and sodium bisulfite may be added to the latex in the bulking tank to prevent enzyme darkening. There is no bleaching or partial fractionation of the latex to lighten the rubber. The color is determined by comparing the color of a thin pressed sheet of rubber with a series of colored transparent disks from almost clear to distinct yellow.

SMR WF. SMR WF is SMR L-type rubber that is too dark to meet the color specification. Any SMR CV that does not meet the viscosity specification does not qualify as SMR WF.

SMR 10, 20, and 50. The SMR grades 10, 20, and 50 are produced from estate and small-holder coagulum and have replaced much of the old brown crepe production.

SMR 5. SMR 5 is produced from sheet raw material. Sheet raw material is typically ribbed smoked sheet, air-dried sheet, or partially dried but unsmoked sheet obtained from small holders. The sheet raw material is milled, crumbed, and dried in the same manner as described for the other SMR rubbers. Pale crepes and ribbed smoked sheets prepared in the conventional manner may be pressed into SMR-size bales without crumbling, tested to the SMR 5 specification, and designated SMR 5 (pale crepe) and SMR 5 (RSS).

SMR GP. SMR GP is a general-purpose grade introduced in 1979 (35). It is a blend of deliberately coagulated latex rubber and field coagulum according to the following formulation:

Raw material	Wt %
SMR-factory/group-processing center latex and/or sheet material	60
Field coagulum	40

At least 20% of the first item must be processed latex or sheet material of SMR 5 quality from an SMR factory or group processing center. The balance can be unsmoked or smoked sheets from either small holders or estates. The rubber is viscosity-controlled to a Mooney of 65 ± 7. This grade was developed to meet demand for large volume, consistent-processing rubber, but it has not had much acceptance among consumers.

Other Types of *Hevea* Rubber

Superior-Processing (SP). Superior-processing rubber is a mixture of cross-linked and unmodified natural rubber. It is produced in forms corresponding to conventional types of rubber, ie, RSS, pale crepes, air-dried sheets, and brown crepes, each of which contain 20 wt % vulcanized and 80 wt % unvulcanized rubber. Superior-processing rubbers are also produced in masterbatch form from blends of prevulcanized and regular latex. The two most common masterbatches are PA80 and PA57: PA80 contains 80 parts of cross-linked rubber and 20 parts unmodified rubber, and PA57 is an oil-extended version of PA80 and is designed for easy dispersion. It contains 57 wt % vulcanized rubber, 14 wt % unmodified rubber, and 29 wt % of a light colored non-staining oil.

These rubbers are used as processing aids in mixes with regular grades of natural rubber to give lower die swell, smoother and faster extrusions, and improved calendering. In 1979, Malaysia produced 4536 metric tons of SP rubbers (36).

Technically Classified (TC). Technically classified rubbers are either ribbed smoked sheet or air-dried sheets, which are classified according to cure rate. There are three classifications: red, yellow, and blue circle, which correspond to slow, medium, and fast cure rates. The cure rate is determined from strain data using the ACS No. 1 test recipe (2). The production of TC rubber in Malaysia for 1979 was 5684 t (37).

Air-Dried Sheet (ADS). Air-dried sheet is a light-colored sheet prepared in the same manner as ribbed smoke sheet but dried in the absence of smoke.

Skim. The skim latex produced as a by-product from concentrating latex by centrifuging can be coagulated to produce a very clean rubber with a high nonrubber content. The nonrubber portion is rich in protein, which causes the rubber to be fast curing and to have a strong odor. The skim latex can be treated to reduce protein content and metal content, which is also high. The quality of skim rubber varies significantly among producers and the rubber is usually sold based on a producer's specification or to sample. Skim rubber is used in sponge rug underlay and adhesives, although the strong odor has caused some problems for adhesive manufacturers.

Deproteinized Natural Rubber (DPNR). Deproteinized natural rubber (DPNR) has very low nitrogen (0.07 wt %) and ash (0.06 wt %) levels and is produced by treating field latex with an enzyme that breaks down the proteins to water-soluble products, which are washed from the rubber during processing (38). The latex is diluted prior to coagulation to reduce further the nonrubber content. Deproteinized natural rubber has a low affinity for water compared to conventional natural rubber. When water from the air is absorbed by natural rubber compounds, the scorch safety is reduced, the modulus varies, and electrical properties are impaired. These problems are eliminated with DPNR. Its vulcanizates also exhibit greater resilience and reduced creep than NR.

Oil-Extended Natural Rubber (OENR).   Oil-extended natural rubber (OENR) is produced and marketed in various grades by several Malaysian producers (39). The raw-rubber portion may be either a latex or remilled-type rubber. Aromatic or non-staining cycloparaffinic oils are used at 10, 25, and 30 wt %. The rubber is packaged in the same manner as SMR rubbers. Oil-extended natural rubber with a high oil content crystallizes slowly and does not need premastication. Vulcanizate properties are similar to those for natural-rubber vulcanizates if the oil is added during the mixing.

Heveaplus MG.   Heveaplus MG is produced by polymerizing methyl methacrylate in natural rubber latex. The result is a graft polymer in which the methyl methacrylate has been grafted onto the natural-rubber molecule. The amount of methyl methacrylate can be varied up to 50 wt %. Heveaplus MG 49 (49 wt % methyl methacrylate) can be used in hard, rigid, molding applications where high hardness, good mold flow, and high temperature hardness retention is required (40). This polymer also can be used in adhesives either as a latex, solution, or dry rubber film (41).

Epoxidized Natural Rubber.   Natural-rubber field latex or concentrate can be epoxidized with peracetic acid and hydrogen peroxide (42–43). The glass-transition temperature increases linearly from −72°C for straight NR to 5°C for 100 mol % epoxidized NR. Epoxidized NR can be vulcanized with standard sulfur curing systems or with peroxide curing agents. The vulcanizates have high damping characteristics with low stress-relaxation rates and low heat buildup. Epoxidation also significantly increases the resistance of natural rubber to swelling by hydrocarbons and may be a replacement for nitrile rubber in certain application. It is anticipated that epoxidized natural rubbers will be commercially available in 1982 (see Epoxidation).

Economic Aspects

When official statistics on natural-rubber production were first compiled in 1899, world production totaled 4 t. Over the next 40 yr the supply, demand, and price of natural rubber varied greatly; but almost all the rubber consumed was natural. During World War II, the Japanese cut off almost 90% of the world's natural-rubber supply, which led to a rapid development and expansion of synthetic-rubber plants in the United States. Over 7×10^8 was invested in ca 87 plants. Synthetic-rubber production grew from 4.2×10^4 metric tons in 1940 to ca 9×10^5 t in 1945 (45). Synthetic rubber increased its share of the market through the early 1970s because of inexpensive petroleum-derived raw materials oil and the slow expansion of natural-rubber production. From 1945 through 1974 the demand for all rubbers grew at ca 7%/yr, whereas the natural-rubber supply grew at 2–3%/yr (45). The rapid rise of oil prices, which began in 1973, led to renewed interest in natural rubber by producers and consumers.

In 1980, Malaysia, Indonesia, and Thailand accounted for almost 80% of the world's natural-rubber production (see Table 3). World consumption of natural rubber for 1980 was 3.79×10^6 t, which represented 30.5% of the total rubber market. The United States continued to be the single largest consumer with 7.32×10^5 t or 19.0% of world production (see Table 4).

The world demand for natural rubber is expected to grow 4.3%/yr from 1980 to 1985 and 5.5%/yr during 1985–1990 (48). Natural-rubber production has been projected to grow at a rate of only 4%/yr. If these projections hold true, natural rubber's share of the world elastomer market will fall from over 30% in 1980 to 26% in 1990 (48).

Table 3. World Natural-Rubber Production, Thousand Metric Tons[a]

	1960	1965	1970	1975	1976	1977	1978	1979	1980
Malaysia	785	949	1269	1459	1612	1588	1607	1600	1552
Indonesia	620	717	815	823	848	835	903	905	1020
					(estd)	(estd)	(estd)	(estd)	(estd)
Thailand[b]	171	216	290	335	412	431	467	531	501
Sri Lanka	99	118	159	149	152	146	156	153	133
India	25	49	90	136	148	151	133	147	155
Africa (estd)	144	159	210	215	204	209	203	193	177
others (estd)	173	172	270	198	209	265	286	331	277
Total[c]	2017	2380	3103	3315	3585	3625	3755	3860	3815

[a] Refs. 46 and 47.
[b] Exports plus consumption plus changes in stocks.
[c] Including allowances for apparent discrepancies in officially reported statistics.

Table 4. World Consumption of Natural Rubber, Thousand Metric Tons[a]

	1970	1975	1976	1977	1978	1979	1980
U.S.	568	666	687	802	761	732	585
UK	195	171	168	172	139	138	131
France	158	156	167	164	163	177	188
FRG	201	197	195	176	185	185	180
Italy	113	118	135	128	113	128	132
Netherlands	22	23	22	21	19	20	20
other Western Europe[b]	180	213	200	220	215	225	220
Eastern Europe[c]	465	475	460	425	425	430	440
People's Republic of China[c]	208	225	240	280	300	355	340
Australia	40	50	50	41	41	45	42
Brazil	37	59	66	71	72	76	81
Canada	51	73	85	90	89	94	80
India	86	129	133	143	158	168	171
Japan	283	285	302	320	355	390	427
other	383	528	595	662	693	707	753
Total[d]	2990	3368	3505	3715	3728	3870	3790[e]

[a] Ref. 47.
[b] Consumption for most countries estimated by correcting net imports to allow for working stocks at 1.5 mo consumption.
[c] Provisional.
[d] Including allowances for discrepancies in officially reported statistics.
[e] Including year-end adjustment.

The International Natural Rubber Agreement developed under the auspices of United Nations Conference on Trade and Development (UNCTAD) became provisionally effective as of October 1980 (49). The agreement is a cooperative effort of producing and consuming countries whose objective is the stabilization of the price of natural rubber and, as a result, encouragement of production expansion. The agreement calls for a buffer stock of 5.5×10^6 t of rubber which is supposed to maintain the price of rubber at ca $0.77–1.17/kg fob Malaysia or Singapore. These prices are a composite of RSS1, RSS3, and TSR20 fob prices as determined in four markets: London, Singapore, Kuala Lumpur, and New York. The consuming countries hope that the floor price offers sufficient incentive to the small farmer to replant and expand. The small farmer is the key to increasing rubber production and accounted in 1980 for over 60% of world production. It takes a minimum of 5 yr after a tree is planted in the field for the owner to begin recovering the investment. In order to justify this investment of time and money, the farmer must have assurance of a reasonable return. In the past, natural-rubber prices fluctuated widely, as shown in Table 5, thus discouraging investment by the small farmer. The price stabilization agreement coupled with incentives from the governments of producing countries should encourage an expansion of natural-rubber production by both large estates and small holders.

Natural Rubber vs Synthetic. Natural rubber's share of the world elastomer market has declined from well over 90% in the early 1940s to ca 30% in 1980. Synthetic-rubber usage increased during this period because of relatively inexpensive raw materials and the slow expansion of natural-rubber production. The decline in the percentage use of natural rubber caused many experts in the 1960s to predict that it would disappear from the market. This viewpoint was changed by the rapid rise in oil prices and a recognition of the excellent general properties of natural rubber. Natural rubber cannot compete with specialty elastomers that are designed for particular service requirements, eg, heat or solvent resistance. However, its excellent processability and physical properties, eg, tensile, elongation, resilience, and abrasion resistance, and its relatively low cost make it one of the best general-purpose rubbers.

When the compounder selects an elastomer for a specific application, the most important service condition is considered first. A general point-by-point comparison of polymers determines which one gives the best combination of properties. The compounder then tries to enhance the most critical properties and to balance the rest. The general comparisons in Table 6 are based on a number of compounds designed for various applications. Natural and synthetic rubbers are compared in refs. 50 and 51 (see Elastomers, synthetic).

Latex Concentrate

Natural-rubber latex concentrate is made by concentrating field latex with a dry-rubber content of 30–40 wt % to a minimum of 60 wt % dry rubber. The concentrate must be carefully preserved so that the quality is maintained during transit and storage. Natural latex is used in a variety of applications: dipped goods, adhesives, latex thread, carpet backing, foam, and various miscellaneous uses, eg, reconstituted

Table 5.   Actual Spot Prices for RSSI fob New York, $/kg

Year	High	Low	Year	High	Low
1913	2.48	1.31	1960	1.01	0.63
1920	1.21	0.39	1965	0.61	0.52
1925	2.31	0.78	1970	0.57	0.40
1932	0.10	0.06	1975	0.63	0.69
1935	0.29	0.25	1976	0.94	0.73
1940	0.66	0.41	1977	0.98	0.86
1945	0.50	0.50	1978	1.29	0.96
1950	1.61	0.40	1979	1.50	1.20
1955	1.07	0.68	1980	1.83	1.49

Table 6. Relative Properties of Elastomers [a]

	Natural rubber	Synthetic polyisoprene	Styrene–butadiene rubber	Polybutadiene	Ethylene–propylene rubber	Butyl rubber	Chloroprene rubber	Nitrile rubber
Physical properties								
tensile	E	E	E	F	E	F	E	G
elongation	E	E	G	G	E	E	E	G
resilience	E	E	G	E	G	P	E	G
compression set	G	G	G	G	G	F	G	G
rebound cold	E	E	G	E	G	P	G	G
rebound hot	E	E	G	E	G	P	G	G
adhesion to metals	E	E	E	G	G	G	E	E
adhesion to fabrics	E	E	G	G	F	G	E	G
Resistance properties								
abrasion	E	E	E	E	G	F	G	G
tear	G	G	G	G	G	G	G	G
ozone	F	F	F	F	E	E	E	F
gas permeation	F	F	F	F	G	E	G	G
Fluid-resistance properties								
aliphatic hydrocarbons	P	P	P	P	P	P	G	E
aromatic hydrocarbons	P	P	P	P	P	P	F	G
acids, dilute	E	G	G	G	E	E	E	G
acids, concentrated	G	G	G	G	G	E	G	G
alkalies	E	E	G	G	G	E	E	G
water absorption	G	G	G	G	G	G	G	G

[a] Code: E, excellent; G, good; F, fair; P, poor. Rating applies only to the individual property for which it is indicated and may be available only in compounds specifically designed for the particular service.

483

leather board and rubberized hair for upholstery. Table 7 shows the estimated 1980 product distribution of latex imported into the United States. Use of latex in dipped goods and adhesives is increasing whereas its use in carpet backing and in foam is declining because of lower cost synthetic latex and polyurethane foam, respectively (52) (see Latex technology; Urethane polymers).

In 1980, ca 7% of the natural rubber produced was in the form of latex concentrate. The production of latex concentrate is limited to large, well-run estates, since small farmers generally do not have the volume or equipment necessary to manufacture high quality latex economically. Malaysia, Indonesia, and Liberia accounted for all but a fraction of the latex exported in 1980, with Malaysia being the main producer (see Table 8).

Table 9 shows a history of natural-rubber latex imports. The United States is the largest single importer of latex; its 1980 imports accounted for ca 16% of that year's world imports.

Production. *Preservation and Quality.* Freshly tapped field latex has a pH of 6.5–7.0. As soon as the latex comes in contact with air, changes in its chemical composition begin to take place. Enzymes and bacteria act on the latex to lower the pH and, unless preventive measures are taken, the latex coagulates in a short time. Ammonia is the most common preservative used to keep the latex in a stable colloidal state. It is common practice to add small amounts of ammonia to the cup prior to tapping and additional ammonia gas at the field collection stations. A number of other secondary preservatives can be used in conjunction with ammonia, but ammonia is difficult to replace as the primary preservative. The lutoids and Frey Wyssling particles

Table 7.   U.S. Natural Latex Market

	%
dipped products	46
adhesives	18
carpet industry	10
thread	10
foam	7
miscellaneous	9
Total	*100*

Table 8.   Net Exports of Natural Rubber Latex, t [a]

	1970	1975	1976	1977	1978	1979	1980
Malaysia	172,594	189,838	195,389	203,855	203,153	205,628	200,077
Indonesia	28,620	25,513	38,013	36,055	36,894	27,169	43,896
Liberia	41,455	32,092	28,310	25,997	26,500 (estd)	26,250 (estd)	24,000 (estd)
Sri Lanka		43	90				10
Vietnam	1,085						
Cambodia	7,000 (estd)						
Total	*250,754* (estd)	*247,486*	*261,802*	*265,907*	*266,547* (estd)	*259,047* (estd)	*267,983* (estd)

[a] Ref. 53.

Table 9.   Net Imports of Natural Rubber Latex, t [a]

	1970	1975	1976	1977	1978	1979	1980
U.S.	66,609	43,805	56,185	57,081	52,381	44,958	42,499
UK	20,835	15,011	15,415	17,806	14,338	19,647	14,111
USSR (estd)	18,500	42,750	31,250	36,500	37,750	39,750	52,500
France	11,692	10,305	10,670	8,887	8,989	8,059	8,753
FRG	28,631	19,277	19,336	18,648	16,877	16,711	15,175
Japan	20,586	15,126	18,804	19,015	18,264	17,593	19,134
Italy	19,476	14,126	16,603	17,357	18,385	21,150	18,458
Eastern Europe (estd)	10,500	10,250	9,750	12,000	12,250	10,250	11,500
other (estd)	53,171	66,350	73,987	78,706	79,016	85,132	77,870
Total (estd)	*250,000*	*237,000*	*252,000*	*266,000*	*258,250*	*263,250*	*260,000*

[a] Ref. 54.

dissolve in the serum when field latex is ammoniated so that ammoniated latex is a two-phase system of rubber particles and serum. Ammonia inhibits bacterial development, acts as a alkaline buffer and thereby raises the pH, lowers the viscosity, and neutralizes free acid formed in the latex. Ammonia also promotes hydrolysis of the proteins and phospholipids in the latex. The proteins are hydrolyzed to polypeptides and amino acids; the phospholipids to glycerol, fatty-acid anions, phosphate anions, and organic bases (55). The fatty-acid anions are absorbed on the surface of the rubber particle, thereby contributing to the stability of the latex. The removal of proteins through hydrolysis destabilizes the latex. The rubber particles in ammoniated latex are stabilized by both fatty-acid anions and protein anions, whereas raw field latex is stabilized by protein anions only.

The quality of natural-latex concentrate is measured by its mechanical stability (MST), potassium hydroxide number (KOH no.), and volatile fatty acid number (VFA no.) (56). The mechanical stability of latex is measured by subjecting it to shear on a machine resembling a milkshake mixer and measuring the time until the rubber starts to coagulate. Ammonia-preserved latex concentrate increases in stability under aerobic conditions (57). The stability of concentrate that is stored under aerobic conditions decreases (58). Other researchers have shown that MST, KOH no., and VFA no. increase during storage (59). The MST of a freshly prepared, ammonia-preserved concentrate is significantly affected by temperature regardless of other storage conditions. Older, mature latices are less affected.

The volatile fatty acids, ie, formic, acetic, and propionic, are formed by the action of bacteria on the serum carbohydrates. The ammonia salts of these fatty acids reduce the mechanical and chemical stability of the latex. The stability of ammonia-preserved latex that has been in storage for several months or longer is determined by the balance between protein hydrolysis and VFA formation, which destabilize the latex, and lipid hydrolysis which enhances stability.

The KOH no. is the number of grams of KOH required to neutralize the acids present in 100 g of latex solids. High values for KOH no. have often been regarded as an indication of poor preservation and therefore of poor quality latex. A recent study has shown that although latices with high VFA nos. have high KOH nos., the reverse is not necessarily true (56). The KOH no. by itself is not a reliable means for determining latex quality.

Centrifuged Latex. The most common method for concentrating field latex is to pass it through a centrifuge after adequate stabilization (60). Field latex is ammoniated at the field collecting station to ca 0.4 wt % of the total wet weight. At the factory, the latex is strained into bulking tanks where additional ammonia may be added prior to centrifuging. Certain field latices have relatively high magnesium contents. Di-ammonium phosphate is added to the latex and magnesium ammonium phosphate settles as a sludge. The de Laval centrifuge is the most common type in use today. It separates the latex into a rubber-rich concentrate and a rubber-poor serum or skim portion. The centrifuge is regulated to give the desired total solids content and an economic concentrate-to-skim ratio. Approximately 90 wt % of the total solids of the field latex is in the concentrate with the remaining solids in the skim. Table 10 shows the difference in composition of skim and concentrate produced by centrifuging.

After centrifuging, the ammonia level in the concentrate is adjusted to 1.6 wt % minimum based on the water phase. This type of latex is referred to as ASTM type 1 or high ammonia natural latex (see Table 11). Lauric acid or ammonium laurate is added to the concentrate to increase mechanical stability so the latex can be pumped and transported without danger of coagulum formation. The amount added is carefully determined, as MST generally increases during storage. Too high an MST could cause difficulties in destabilization at the consumer's factory.

Creamed Latex. Creaming is another method for concentrating *Hevea* latex. Ammoniated latex creams slowly on its own, but the rate is not sufficient to achieve the desired solids level. Creaming agents are used to hasten the process. They are generally hydrophilic colloids, which swell in water and form viscous solutions at very low concentrations. Many creaming agents have been used, but the salts of alginic acid, which are obtained from seaweed, are preferred since they leave no solid residue. Ammonium alginate is the most common agent in use. Alginates vary in efficiency depending on age, manufacturer, and the particular lot. The selection and control of alginate quality and quantity are important factors to the success of a creaming operation (see Gums).

The mechanism by which the creaming agent works is not completely understood. One theory is that Brownian movement is slowed by the increase in viscosity and that a loose agglomeration of rubber particles results. The specific gravity of the agglomerated particles is less than that of the serum and they gradually move upward. This cannot be the total explanation since alginates or similar compounds are necessary for efficient creaming. The polar groups of the alginate molecule may provide points where the molecule can be absorbed to the rubber particle. The probability that por-

Table 10. Constituents of Concentrated Latex and Skim[a]

Constituent	Concentrated latex	Skim
total solids, wt %	61.5	11.0
dry rubber (by coagulation), wt %	60.0	7.1
acetone extract of the rubber, wt %	3.2	10.3
ash, wt % of the rubber	0.37	2.2
water-soluble matter, wt % of the rubber	0.23	35.6
nitrogen, wt % of the rubber	0.36	2.2

[a] Ref. 61.

Table 11. Detail Requirements for Types 1, 2, 3, and 4 Concentrated Latex[a]

	Type 1	Type 2	Type 3	Type 4
total solids, wt % min	61.5	64.0	61.5	64.0
dry rubber content (DRC)[b], wt % min	60.0	62.0	60.0	62.0
total solids minus dry rubber, wt % max	2.0	2.0	2.0	2.0
total alkalinity[c], wt %	1.6 min	1.6 min	1.0 max	1.0 max
sludge, wt % (max) of wet mass	0.10	0.10	0.10	0.10
coagulum, wt % (max) of wet mass	0.050	0.050	0.050	0.050
KOH no., max[d]	0.80	0.80	0.80	0.80
mechanical stability, min	540	540	540	540
copper, wt % (max) of total solids	0.0008	0.0008	0.0008	0.0008
manganese, wt % (max) of total solids	0.0008	0.0008	0.0008	0.0008
color on visual inspection	no pronounced blue or gray[e]			
odor	no putrefactive odor			

[a] Ref. 63.
[b] Dry rubber content is the acid coagulable portion of latex after washing and drying.
[c] Calculated as ammonia; expressed as a percentage of water in the latex.
[d] It is accepted that KOH values for boric-acid-preserved latices are higher than normal latices and are equivalent to the amount of boric acid in the latex.
[e] Blue or gray color usually denotes iron contamination caused by improper storage in containers.

487

tions of a single alginate molecule become absorbed to more than one rubber particle is high since the molecular weights of commercially available alginates are 21,000–70,000 and with the higher molecular weight salts are more efficient (62).

In the creaming process, field latex is preserved with ca 1.25 wt % ammonia. This is sufficient for long-term preservation and is necessary since creaming takes several weeks. The processing is batchwise rather than continuous as in the centrifuge process. The size of the batches is determined by available tank space. After the addition of ammonia, diammonium phosphate is added if necessary to reduce the magnesium level of the latex. The latex is now ready for the addition of creaming agent and soap. The amount of creaming agent required is determined by experimentation. The goal is to obtain a concentrate with the required dry rubber content (DRC) and a serum low in rubber as well as a reduction in the tendency of the latex to cream further after shipment. High levels of alginate increase recovery, but the DRC of the concentrate falls once the optimum concentration is reached. The soap increases DRC but also reduces recovery. Amounts are chosen for the best economic and technical results (63).

The properties of cream latex are similar to those of centrifuged latex, but the solids and viscosity are generally higher for the former. High ammonia-creamed latex is an ASTM type 2 natural latex (see Table 11). Creamed latex is used in the United States almost exclusively for the production of rubber thread because of its superior filterability.

Low Ammonia Latices. Both centrifuged and creamed natural latices are produced in low ammonia versions, and the centrifuged ASTM type 3 is the most common (see Table 11). Low ammonia preservation systems typically have ammonia levels of 0.15–0.25 wt %, based on the total wet weight, with a secondary preservative. There are three commercially significant secondary preservative systems: sodium pentachlorophenate (SPP) (64), tetramethylthiuram disulfide (TMTD) and zinc oxide, and sodium dimethyldithiocarbamate (SDC) and zinc oxide.

The low ammonia latices are preferred by some consumers because of their lower odor and the elimination of a costly and time-consuming deammoniation step. These latices account for only a small portion of the latex produced, but this may change. In 1976, OSHA proposed changing the present 50 ppm time-weighted-average exposure limit to a 50 ppm ceiling limit. Although no action has been taken on this proposal to date, it has caused many latex users to examine the amount of ammonia used in their plants and some have switched to low ammonia latex. Another factor that may lead to increased usage is the discovery that formaldehyde is a potential carcinogen. Many users deammoniate their high ammonia latex with formaldehyde and are considering low ammonia latex as an alternative.

Balata and Gutta-Percha. Natural rubber (polyisoprene) exists in nature as both the cis and trans isomer. The rubber from the *Hevea* tree occurs only in the cis form whereas both balata and gutta-percha occur as the trans isomer. Gutta-percha is obtained from trees of the family *Sapotaceae*, which are native to Malaysia, Borneo, and Sumatra. Gutta trees are grown on plantations, and the rubber is recovered by maceration and hot-water treatment of the leaves. Balata is harvested from wild bushes and trees which grow primarily in the Surinana and Guiana regions on the northeastern coast of South America. Raw balata is very high in dirt and resin and must be purified by a solvent-extraction process. Refined balata contains less than 2 wt % resin and nonrubber.

Trans-Polyisoprene is a tough horny material at room temperature but is soft and tacky at 100°C. It is insoluble in most aliphatic hydrocarbons, but is soluble in aromatic hydrocarbons and most chlorinated hydrocarbons. *Trans*-Polyisoprene is used in golf-ball covers, sheeting, tubing, and submarine cables. In 1979, 3174 t of balata and 536 t of gutta-percha were imported into the United States.

Guayule. Guayule (*Parthenium argentatum*) is a member of the sunflower family. The shrub grows to ca 0.5–1 m tall in its natural habitat of northern Mexico and the Big Bend area of Texas. It thrives at altitudes of 1200–2300 m with yearly rainfalls of 23–59 cm. In the early 1900s several companies were established in northern Mexico to extract rubber from wild guayule. By 1910, guayule accounted for 10% of the world's natural-rubber production with Mexico being the principal producer. In 1942, the Emergency Rubber Project (ERP) was enacted by the U.S. Government and 130 km² (32,000 acres) of guayule was planted in California and various parts of the southwest United States (65). With the end of World War II and the advent of synthetic rubber, only ca 15% of this guayule was harvested. The remainder was destroyed and the land was returned to the original owners. In 1976, the Mexican Government built a pilot plant to extract rubber from wild guayule. The objective was to reduce Mexico's dependence on imported natural rubber and to create jobs for northern Mexico.

There are no significant structural differences between guayule and *Hevea* rubber. Guayule is more linear and is practically gel-free, which should result in better processing. It lacks the nonrubber impurities of *Hevea*, some of which are natural accelerators and antioxidants. In addition, it contains a resin that must be removed as it contributes to a slower and lower state of cure. After minor compounding changes, guayule vulcanizates compare favorably with *Hevea*. Passenger, truck, and off-the-road tires have been produced with guayule rubber in all components where natural rubber is required. In order to produce guayule at competitive prices, a great deal of work must be done in terms of agronomics and by-product utilization. The best methods of planting, fertilizing, weed control, water control, and harvesting must be developed. For each metric ton of rubber 2 t of wood fiber, 0.5 t of resins, and 1 t of leaves are also produced. The wood fiber may be used in lower quality paper and cardboard, whereas the leaves contain ca 2.5 wt % of a hard wax (65).

The chemical analysis of the resin portion has been reported but the relative amounts of the various components have not been determined (66). The volatile essential oils in guayule resin are α-pinene, dipentene, and cadinene. The nonvolatiles are carotenoids, partheniols, and unidentified dihydroxy diterpenes, palmitic acid, stearic acid, linoleic acid, linolenic acid, *trans*-cinnamic acid, hard wax (mp 76°C), parthenyl cinnamate, and a shellaclike drying resin.

BIBLIOGRAPHY

"Rubber, Natural" in *ECT* 1st ed., Vol. 11, pp. 810–826, by T. H. Rogers, Jr., The Goodyear Tire & Rubber Company, Inc.; "Rubber, Natural" in *ECT* 2nd ed., Vol. 17, pp. 660–684, by T. H. Rogers, The Goodyear Tire & Rubber Company, Inc.

1. H. Brown, *Rubber: Its Sources, Cultivation, and Preparation*, John Murray, London, 1918.
2. G. Martin, *I.R.I. Trans.* **19**, 38 (1943).
3. S. T. Semegen and Cheong Sai Fah, *The Vanderbilt Rubber Handbook*, R. T. Vanderbilt & Co., Inc., Norwalk, Conn., 1978, pp. 18–41.
4. P. E. Hurley, *Nat. Rubber News*, 3 (Aug. 1980).

5. A. T. Edgar, *Manual of Rubber Planting*, The Incorporated Society of Planters, Kuala Lumpur, Malaysia, 1958, Chapt. 3.
6. *Ibid.*, Chapt. 4.
7. *Ibid.*, Chapt. 9.
8. *Ibid.*, Chapt. 11.
9. J. J. Riedl, remarks presented to IISRP, Mexico City, April 1980.
10. D. C. Blackley, *High Polymer Latices*, Vol. 1, Palmerton Publishing Co., Inc., New York, 1966, p. 214.
11. G. F. J. Moir and S. J. Tata, *J. Rubber Res. Inst. Malays.* **16,** 155 (1960).
12. Ng Tet Soli, *Proc. Nat. Rub. Res. Conf.*, Kuala Lumpur, Malaysia, 809 (1960).
13. L. N. S. Homans and G. E. van Gils, *Proc. Rub. Tech. Conf.*, London, 292 (1948).
14. P. B. Dickenson, *N.R.P.R.A. Silver Jubilee Conf.*, *Cambridge*, Maclaren, London, Eng., 1964.
15. *The International Standards of Quality and Packing for Natural Rubber Grades*, *The Green Book*, *The International Rubber Quality and Packing Conferences*, Office of the Secretariat, The Rubber Manufacturers Assn., Inc., Washington, D.C., effective date, Jan. 1, 1979.
16. *SMR Bulletin No. 7*, Rubber Research Institute of Malaysia (available through the Malaysian Rubber Bureau, Hudson, Ohio).
17. L. Bateman and B. C. Sekhar, *J. Rubber Res. Inst. Malays.* **19**(3), 133 (1966).
18. *NR Technol.* **9**(3), 55 (1979).
19. S. Nair, *J. Rubber Res. Inst. Malays.* **23**(1), 76 (1970).
20. C. M. Bristow, *NR Technol.* **5**(1), 1 (1974).
21. Ong Chong Oon, *J. Rubber Res. Inst. Malays.* **24**(III), 160 (1976).
22. S. T. Semegen, "Rubber," in M. Morton, ed., *Rubber Technology*, 2nd ed., Van Nostrand Reinhold Company, New York, 1973, p. 162.
23. *Ibid.*, 22, p. 161.
24. *Ibid.*, Chapt. 2.
25. Ref. 3, Chapts. 3, 6, and 9.
26. C. Kraus and W. B. Reynolds, *J. Am. Chem. Soc.* **72,** 5621 (1950).
27. S. D. Gehman, J. E. Field, and R. P. Dinsmore, *Proc. First Rubber Tech. Conf.* 961 (1938).
28. C. W. Bunn and E. V. Garner, *J. Chem. Soc.*, 654 (1942).
29. Wong Ah Kiew and Seow Pin Kwong, *J. Rubber Res. Inst. Malays.* **28**(1), 1 (1980).
30. G. T. Knight and B. Pepper, *Tetrahedron* **27,** 6201 (1971).
31. D. J. Graham and J. E. Morris, *RRIM Planters Conference, Preprint No. 11*, Rubber Research Institute of Malaysia, July 1966.
32. J. E. Morris, *J. Rubber Res. Inst. Malays.* **22**(1), 39 (1966).
33. G. M. Bristow, *NR Technol.* **5**(1), 1 (1974).
34. B. C. Sekhar, *Proc. Nat. Rub. Res. Conf., Kuala Lumpur* **3,** 512 (1960).
35. *SMR Bulletin No. 9*, Rubber Research Institute of Malaysia, 1978.
36. *Monthly Statistical Bulletin*, Malaysian Rubber Producers' Council, Jan. 1980.
37. *Ibid.*, Jan. 1979–Jan. 1980.
38. *NR Technol.* **7**(1), 21 (1976).
39. *NR Technol.* **8**(2), 39 (1977).
40. *NR Technol.* **8**(4), 69 (1977).
41. Ref. 22, p. 166.
42. D. Barnard, *The Chemical Modification of Natural Rubber*, paper presented at Saltillo Conference, Saltillo, Mexico, 1950.
43. I. R. Gelling and J. F. Smith, *Controlled Viscoelasticity by Natural Rubber Modification*, MRPRA, Malaysian Rubber Producers' Research Association, Reprint #877, 1980.
44. P. Radhakrishman, *Commodities*, 23 (Aug. 1975).
45. P. W. Allen, *Plast. Rubber Internat.* **4**(4), 161 (1979).
46. *Rubber Statistical Bulletin* **24**(3), 3 (1969).
47. *Rubber Statistical Bulletin* **35**(9), 6, 7, 13 (1981).
48. *World Bank Report*, No. 814/78, June 1978.
49. *Nat. Rubber News*, 8 (Nov. 1980).
50. P. J. Larsen, *Machine Design*, Penton/IPC Inc., Cleveland, Ohio, Jan. 25, 1979.
51. *Handbook of Molded and Extruded Rubber*, Goodyear Tire & Rubber Company, Akron, Ohio, 1969, pp. 8 and 9.
52. B. G. Lewis, *Rubber World* **180**(6), 36 (1979).

53. Ref. 47, p. 22.
54. Ref. 47, p. 22–24.
55. Ref. 18, p. 230.
56. H. C. Chin and M. M. Singh, *Plastics and Rubber: Materials and Applications*, Plastics and Rubber Institute, Stevenage Herts, England, Nov. 1979, pp. 164–169.
57. M. C. Gavack and E. M. Bevilaqua, *Ind. Eng. Chem.* **43,** 475 (1951).
58. H. M. Collier, *Trans. I.R.I.* **31,** 166 (1955).
59. *Plant. Bull. Rubber Res. Inst. Malays.* **113,** 102 (1971).
60. Brit. Pat. 319,410 (Nov. 13, 1929), (to Dunlop Research Co.).
61. D. F. Twiss, *Trans. Inst. Rubber Ind.* **7,** 280 (1931).
62. K. E. Bristol and J. D. Strong, *Creaming Studies of Natural Rubber Latex*, *Report 3*, Goodyear Tire & Rubber Co., Akron, Ohio, 1952.
63. I. H. Duckworth, *Plant. Bull. Rubber Res. Inst. Malays.* **74,** 111 (1964).
64. U.S. Pat. 2,254,267 (Sept. 2, 1941), T. S. Carswell (to Monsanto Chemical Company).
65. *Guayule: An Alternative Source of Natural Rubber*, National Academy of Sciences, Washington, D.C., 1977.
66. *U.S. Natural Rubber Research Station Report*, *1953*, U.S. Natural Rubber Research Station, Bureau of Agriculture and Industrial Chemistry, Salinas, California.

DAVID R. ST. CYR
Goodyear Tire & Rubber Company, Inc.

RUBBER, SYNTHETIC.   See Elastomers, synthetic.

RUBBERY ACRYLONITRILE POLYMERS.   See Acrylonitrile polymers; Elastomers, synthetic, nitrile rubber.

RUBIDIUM AND RUBIDIUM COMPOUNDS

Rubidium [7440-17-7] (Rb) is an alkali metal in Group IA of the periodic table. Its chemical and physical properties generally lie between those of potassium and cesium (see Cesium and cesium compounds; Potassium; Potassium compounds). Rubidium is the sixteenth most prevalent element in the earth's crust (1). Despite its abundance, it usually is widely dispersed and is not found as a principal constituent in any mineral. At present, most of the rubidium production is obtained from lepidolite containing 2–4% rubidium oxide [18088-11-4]. Lepidolite is found in Zimbabwe and at Bernic Lake, Canada.

Rubidium was discovered in 1861 by Bunsen and Kirchoff by means of an optical spectroscope. It was named rubidium for the prominent red lines in its spectrum, the Latin word *rubidus* meaning darkest red. Bunsen prepared free rubidium during the same year by an electrolytic method. After cesium, rubidium is the second most electropositive and alkaline element. The two isotopes of natural rubidium are ^{85}Rb [13982-12-1] (72.15%) and ^{87}Rb [13982-13-3] (27.85%).

Physical Properties

Rudibium is a soft, ductile, silvery-white metal and is the fourth lightest metallic element. Table 1 lists certain physical properties (1–10).

Chemical Properties

The reactions of rubidium are very similar to those of cesium and potassium, but rubidium is slightly more reactive than potassium. Rubidium burns with a violet flame in the presence of air and reacts violently with water, liberating hydrogen which spontaneously explodes if oxygen or air are present. Rubidium forms a mixture of four oxides: the yellow monoxide [12509-27-2], Rb_2O; the dark brown peroxide [23611-30-5], Rb_2O_2; the black trioxide [12137-26-7], Rb_2O_3; and the dark orange superoxide [12137-25-6], RbO_2 (2). Rubidium is the second strongest Lewis base.

Rubidium metal alloys with the other alkali metals, the alkaline-earth metals, antimony, bismuth, gold, and mercury. Rubidium forms double halide salts with antimony, bismuth, cadmium, cobalt, copper, iron, lead, manganese, mercury, nickel, thorium, and zinc. These complexes are generally water insoluble and not hygroscopic.

The soluble rubidium compounds are acetate, bromide, carbonate, chloride, chromate, fluoride, formate, hydroxide, iodide, nitrate, and sulfate. These compounds are generally hygroscopic.

Manufacture and Processing

Rubidium is found widely dispersed in potassium minerals and salt brines. Lepidolite [1317-64-2], a lithium mica with the composition $KRbLi(OH,F)Al_2Si_3O_{10}$, contains up to 3.5% Rb_2O and is the principal source of rubidium. An ore that is basically pollucite, $Cs_2O.Al_2O_3.4SiO_2$, contains up to 1.5% RbO_2, and some rubidium

Table 1.   Properties of Rubidium

Property	Value	Refs.
atomic weight	85.47	
melting point, °C	39.0	2–3
boiling point, °C	689	4–5
density, g/cm^3		
solid, 18°C	1.522	2
liquid, 39°C	1.472	2
viscosity at 39°C, mPa·s (= cP)	0.6713	2
surface tension at 39°C, mN/m (= dyn/cm)	75	2
vapor pressure, kPa		
(Antoine equation)		4
liquid, 427–1093°C $\log_{10} P^a =$	-3891.8/T (K) $+ 6.0494$	
heat of fusion, J/g^b	25.69	6
heat of vaporization, J/g^b	887	1
specific heat J/(kg·K)^b		
solid	331.37	3
liquid	368.19	4
vapor	241.83	7
thermal conductivity, W/(m·K), liquid	29.3	1
electron work function, eV	2.09	1
electron affinity, eV	0.486	10
neutron-absorption cross-section, thermal, m^2 ^{c}	7.3×10^{-29}	1
ionization potential, V	4.159	8
ionic radius, nM	0.148	9

^{a} To convert $\log_{10} P_{kPa}$ to $\log_{10} P_{mm\ Hg}$, add 0.8751 to the constant.

^{b} To convert J to cal, divide by 4.184.

^{c} To convert m^2 to barn, multiply by 10^{28}.

is produced as a by-product of cesium manufacture from this source. The traditional processes used to recover rubidium from these sources involve extraction of mixed alkali alums from the ore. The ore is subjected to a prolonged sulfuric acid leach to form the alkali alums. The alum solution is filtered from the residue, which is washed with water. Calcination of the ore before leaching increases the yield. The other alkali alums are separated by fractional recrystallizations. The purified rubidium alum is converted to rubidium hydroxide by neutralization to precipitate the aluminum; subsequent treatment with barium hydroxide precipitates the sulfate. The chlorostannate method requires a partial separation of potassium (9). The remaining dissolved carbonates are converted to chlorides, and the solution is treated with enough stannic chloride to precipitate cesium chlorostannate, which is less soluble than its rubidium counterpart. The cesium-free chloride solution is treated with an excess of stannic chloride to precipitate rubidium chlorostannate [17362-92-4]. The purified rubidium chlorostannate may be decomposed to separate the rubidium and tin chlorides by pyrolytic, electrolytic, or chemical methods. Rubidium compounds can also be separated from other alkali-metal compounds by solvent extraction and ion exchange.

Pure rubidium metal is obtained by reducing pollucite or lepidolite ore with an active metal, followed by vacuum distillation (11). Another method is to reduce pure rubidium compounds thermochemically according to the following reactions (12):

$$2\,RbCl + Ca \rightarrow CaCl_2 + 2\,Rb$$

$$2\,RbOH + Mg \rightarrow Mg(OH)_2 + 2\,Rb$$

$$Rb_2CO_3 + 3\,Mg \rightarrow 3\,MgO + C + 2\,Rb$$

Packaging, Shipping, and Storage

Rubidium metal and its compounds are labeled as follows (13):

Rubidium metal	Flammable solid and dangerous when wet
Rubidium hydroxide	Corrosive
Rubidium nitrate	Oxidizer
Rubidium oxide	Flammable solid
Rubidium perchlorate (solid)	Oxidizer

Additional information and details covering the regulations for packaging and shipping of dangerous materials may be found in ref. 14.

Special containers are used for rubidium metal and oxide (15). For quantities up to 100 grams, a borosilicate ampul may be used; for larger quantities, stainless-steel containers are used to facilitate safe handling. Both are hermetically sealed. The environment in the capsules may be vacuum or an inert-gas atmosphere. The metal may also be handled and stored safely in a dry saturated hydrocarbon liquid within a suitable container.

For shipping, the ampuls are protected by wrapping in aluminum foil. The wrapped ampul or stainless-steel container is packed in a metal can and surrounded by a material, eg, expanded vermiculite, as protection against fire and mechanical shock. Rubidium compounds must also be handled and transported with special care. Although not as reactive as the metal, the compounds must be shipped and stored in sealed polyethylene containers, and hygroscopic salts must be stored in a dry place. The containers are packed in a metal can surrounded by expanded vermiculite or plastic foam.

Economic Aspects

The supply and demand for rubidium compounds has grown steadily in recent years. The U.S. demand in 1979 was ca 1040 kg of contained rubidium (9). Total world demand in 1979 was estimated at approximately twice that of the United States. The demand for rubidium metal is small compared to the demand for rubidium compounds.

Primary producers of rubidium in the United States are KBI division of Cabot Corporation at Revere, Pennsylvania, and MSA Research Corporation at Callery, Pennsylvania. Rubidium is also produced in the Federal Republic of Germany and probably in the USSR.

Rubidium metal is offered in two purity grades, 99.5% min and 99.8% max. According to KBI division of Cabot Corporation, the price for metal has not changed since 1965, whereas compound prices have increased (see Table 2). Higher purity and optically pure compounds are available at higher prices.

Table 2. Prices of Rubidium Metal and Compounds, $/kg

Substance	Grade 99.5%	99.8%
metal (1965–1980)	661	827
compounds		
1965	64–106	150–172
1980	141–157	227–243

Specifications and Analytical Methods

Tables 3 and 4 give typical specifications for rubidium metal and rubidium compounds, respectively. The hydroxide is supplied as 50% solution; other forms are available as well.

Analyses of alkali impurities in rubidium compounds are determined by atomic absorption. Other metallic impurities and sodium and potassium impurities are determined by emission spectrograph. For analysis, rubidium metal is converted to a compound such as rubidium chloride.

Health and Safety Factors

Few toxicity data are available regarding the response of humans to exposure to rubidium or its compounds. Physiological experiments indicate exchangeability of rubidium for potassium in blood, plasma, and tissue. Although medical and toxicological literature indicate a very low degree of toxicity (16), nontoxicity cannot be assumed. Localized ventilation of equipment and the use of approved dust respirators are recommended when handling dry rubidium salts.

Animal skin tests show that a 5% RbOH solution is nonirritating to intact skin;

Table 3. Analysis and Specifications of Rubidium Metal, Wt %[a]

Element	Technical grade Typical	Specification	High purity grade Typical	Specification, max[b]
lithium	0.001		<0.001	0.005
sodium	0.100		0.005	0.02
potassium	0.100		0.010	0.10
cesium	0.200		0.050	0.10
aluminum	0.003		<0.0001	0.002
calcium	0.001		<0.0001	0.0005
chromium	0.001		<0.0001	0.0005
copper	0.001		<0.0001	0.0005
iron	0.002		<0.0005	0.002
lead	0.001		<0.0001	0.0005
magnesium	0.001		<0.0001	0.0005
manganese	0.001		<0.0001	0.0005
nickel	0.001		<0.0001	0.0005
silicon	0.002		<0.0001	0.001
rubidium (by difference)	99.6	99.5 min	99.93	99.8 min

[a] Ref. 15.
[b] Unless otherwise stated.

Table 4.   Analysis and Specifications of Rubidium Compounds, Wt %[a]

Assay	Technical grade		High purity grade	
	Typical	Specification, max[b]	Typical	Specification, max[b]
total alkali (Cs, K, Li, Na)	0.65	0.90		
cesium and potassium combined			0.15	0.20
lithium			<0.001	0.005
sodium			0.005	0.010
silicon	0.01	0.02	<0.001	0.010
aluminum		0.01	0.005	0.010
barium		0.01	0.005	0.010
calcium		0.01	0.001	0.005
chromium		0.01	<0.001	0.001
copper		0.01	<0.0005	0.0005
iron		0.01	<0.001	0.001
lead		0.01	<0.0005	0.0005
magnesium		0.01	<0.001	0.001
manganese		0.01	<0.0005	0.0005
nickel		0.01	<0.0005	0.0005
insolubles	0.01	0.01	<0.01	0.010
rubidium compound[d]	99.2	99.0[c]	99.82	99.8[c]
particle size				
acetate, fluoride, formate, oxide	<2.5 cm		<2.5 cm	
bromide, carbonate, chloride, chromate, iodide, nitrate, perchlorate, sulfate	<0.64		<840 μm (20 mesh)	

[a] Ref. 15.
[b] Unless otherwise stated.
[c] Minimum.
[d] By difference.

however, it is a mild irritant to abraded skin (17). A 5% rubidium iodide solution is not irritating for either intact or abraded skin. Impervious gloves are recommended in order to avoid direct skin contact during handling of all rubidium salts.

Animal tests show that a 5% solution of RbOH is extremely irritating and corrosive to the eyes. Tests with a 1% RbOH solution and with 5% solution of RbI show no ocular reaction. Eye contact with rubidium salts should be avoided through the use of a face shield.

Rubidium metal and its compounds are similar in reactivity to sodium and potassium and many of their compounds. The metal reacts violently with water, ice, steam, and flammable vapors; it also can ignite spontaneously in the presence of oxygen and tarnishes rapidly when exposed to air. Compounds generally are not as reactive as the metal. However, they must be handled, transported, and stored with care. Small amounts of rubidium metal may be destroyed by reaction with dry isopropyl alcohol. However, this should be done with caution in a well-ventilated area to prevent a build-up of the potentially explosive hydrogen that evolves during this reaction.

Some compounds, such as rubidium chromate and rubidium iodide, emit toxic vapors when heated. Other compounds, eg, rubidium hydroxide, are caustic; compounds such as rubidium perchlorate are irritants. The levels of toxicity of the metal and many of its compounds are not known. Because of the high reactivity of rubidium and its compounds and their tendency to be irritants or toxic, these materials are classified as hazardous.

Rubidium Compounds

Rubidium acid salts are usually prepared from rubidium carbonate or hydroxide and the appropriate acid in aqueous solution, followed by precipitation of the crystals or evaporation to dryness. Rubidium sulfate is also prepared by the addition of a hot solution of barium hydroxide to a boiling solution of rubidium alum until all the aluminum is precipitated. The pH of the solution is 7.6 when the reaction is complete. Aluminum hydroxide and barium sulfate are removed by filtration, and rubidium sulfate is obtained by concentration and crystallization from the filtrate.

Rubidium aluminum sulfate dodecahydrate (alum), $RbAl(SO_4)_2.12H_2O$, is formed by sulfuric acid leaching of lepidolite ore. Rubidium alum is more soluble than cesium alum and less soluble than the other alkali alums. Fractional crystallization of Rb alum removes K, Na, and Li values, but concentrates the cesium value.

Rubidium hydroxide, RbOH, is prepared by the reaction of rubidium sulfate and barium hydroxide in solution. The insoluble barium sulfate is removed by filtration. The solution of rubidium hydroxide can be evaporated partially in pure nickel or silver containers. Rubidium hydroxide is usually supplied as a 50% aqueous solution.

Rubidium carbonate, Rb_2CO_3, is formed readily by bubbling CO_2 through a solution of rubidium hydroxide, followed by evaporation to dryness in a fluorocarbon container.

Other rubidium compounds can be formed in the laboratory by means of anion-exchange techniques.

Table 5 lists some properties of certain rubidium compounds.

Uses

The principal use for rubidium metal and its compounds is in research (1,18). Properties of rubidium and its compounds are so similar to those of cesium and its compounds that, in many cases, rubidium and cesium may be used interchangeably. The latter, however, is more electropositive and, therefore, usually preferred.

In early magnetohydrodynamics (MHD) and thermionic experiments, rubidium

Table 5.   Properties of Rubidium Compounds

Compound	CAS Registry No.	Formula	Solubility, g/100 cm³ °C		Mp, °C	Bp, °C
			Hot water	Cold water		
acetate	[563-67-7]	$RbC_2H_3O_2$	$86^{44.7}$		246	
aluminum sulfate dodeca- hydrate	[7488-54-2]	$RbAl(SO_4)_2.12H_2O$	43^{80}	1.3^0	99	
bromide	[7789-39-1]	RbBr	$205^{113.5}$	98^5	682	
chloride	[7791-11-9]	RbCl	139^{100}	77^0	715	1390
fluoride	[13446-74-7]	RbF		130.6^{18}	760	1410
iodide	[7790-29-6]	RbI	163^{25}	152^{17}	642	1300
sulfate	[7488-54-2]	Rb_2SO_4	82^{100}	36^0	1073	
chromate	[13446-72-5]	$RbCrO_4$	95.7^{60}	62^0		
nitrate	[13126-12-0]	$RbNO_3$	452^{100}	19.5^0		
carbonate	[584-09-8]	Rb_2CO_3	450^{20}		837	740 (dec)
hydroxide	[1310-82-3]	RbOH		180^{15}	301	
perchlorate	[13510-42-4]	$RbClO_4$	100^{18}	0.5^0		

oxide was used (see Coal, coal conversion processes, magnetohydrodynamics). When it was used as the seeding material, the conductivity of the plasma improved and the ionization temperature was lowered (see Plasma technology, Supplement).

In biological research, rubidium salts, eg, RbCl, are used in conjunction with or instead of cesium compounds as a density-gradient medium for ultracentrifugal separation of deoxyribonucleic acid (DNA), viruses, and other large particles (9) (see Centrifugal separation).

Rubidium salts are used in pharmaceuticals as soporifics, sedatives, and in the treatment of epilepsy (see Hypnotics, sedatives, and anticonvulsants). Rubidium iodide has been used as a substitute for potassium iodide in the treatment of goiters. In another medical use, radioactive rubidium is used as a tag element to trace the flow of blood in the body (19). In order to identify their products in liability and counterfeit cases, manufacturers add small amounts of rubidium as a chemical tag (9) (see Radioactive tracers).

Cathodes made of rubidium telluride [12210-70-7] are used as radiation detectors for wavelengths in the 200–5000 nm range. Small additions of Rb_2CO_3 reduce the conductivity of glass and improve its stability and durability. Therefore, Rb_2CO_3 is used in the production of special glasses and in fiber optics (see Fiber optics). The carbonate is also used for synthetic fiber production.

Rubidium-87 emits beta particles and decomposes to strontium. The age of some rocks and minerals can be measured by the determination of the ratio of the rubidium isotope to the strontium isotope (see Isotopes).

An early commercial application for rubidium was as a getter of oxygen in vacuum tubes (9). This application has diminished in importance with the advent of solid-state electronics. Rubidium has also been used in photoelectric cells.

Rubidium compounds act as catalysts in some organic reactions; they are used as crown ether- and cryptate-complexing agents for the cations (20–21) (see Chelating agents).

BIBLIOGRAPHY

"Rubidium" under "Alkali Metals" in *ECT* 1st ed., Vol. 1, pp. 451–453, by Elizabeth H. Burkey, Jean A. Morrow, and Muriel S. Andrew, E. I. du Pont de Nemours & Co., Inc.; "Rubidium Compounds" in *ECT* 1st ed., Vol. 2, pp. 946–949, by J. J. Kennedy, Maywood Chemical Works; "Rubidium and Rubidium Compounds" in *ECT* 1st ed., Suppl. 2, pp. 734–736, by J. N. Hinyard, American Potash & Chemical Corporation; "Rubidium and Rubidium Compounds" in *ECT* 2nd ed., Vol. 17, pp. 684–693, by Robert E. Davis, American Potash & Chemical Corporation.

1. C. A. Hampel, "Rubidium and Cesium" in *Rare Metals Handbook*, 2nd ed., Reinhold Publishing Corp., New York, 1961, pp. 434–440.
2. J. W. Mellor, *Comprehensive Treatise on Inorganic and Theoretical Chemistry*, Vol. 2, Suppl. 3, John Wiley & Sons, Inc., New York, 1963, pp. 2136–2293, 2488–2505.
3. "Rubidium" in *Gmelins Handbuch der anorganischen Chemie*, Vol. 24, 8th ed., Verlag Chemie, Berlin, FRG, 1955.
4. F. Tepper, A. Murchison, J. Zelenak, and F. Rochlich, *Thermophysical Properties of Rubidium*, Report No. ASD-TDR-63-133, Feb. 1963.
5. F. M. Perelman, *Rubidium and Cesium*, 1st English edition, The Macmillan Company, New York, 1965.
6. K. K. Kelley, *U.S. Bur. Mines Bull. 601*, Washington, D.C., 1962.
7. W. D. Weatherford, Jr., J. C. Tyler, and P. M. Ku, "Properties of Inorganic Energy-Conversion and Heat-Transfer Fluids for Space Applications," *WADD Tech. Report 61-96*, Southwest Research Institute (Nov. 1961).

8. E. N. Simons, *Guide to Uncommon Metals*, Hart Publishing Co., New York, 1967, pp. 161–163.
9. R. J. Bascle, *U.S. Bur. Mines Bull. 671*, Washington, D.C., 1980.
10. T. A. Patterson, H. Hotop, A. Kasdan, D. W. Norcross, and W. G. Lineberger, *Phys. Rev. Lett.* **39,** 189 (1974).
11. U.S. Pat. 3,207,598 (Sept. 21, 1965), C. E. Berthold (to San Antonio Chemicals, Inc.).
12. U.S. Pat. 2,668,778 (Feb. 9, 1954), E. A. Tuft.
13. N. I. Sax, *Dangerous Properties of Industrial Materials*, 5th ed., Reinhold Publishing Corp., New York, 1979.
14. *Explosives and Combustibles*, Interstate Commerce Commission Bureau of Explosives, Tariff No. 10, Chapt. 39.
15. KBI Division of Cabot Corporation Product Data, *Metals and Alloys—Rubidium*, File No. 320-PD-1, Reading, Pa.
16. U.S. Department of Health, Education and Welfare, NIOSH, *Health Hazard Evaluation Report 71-72*, July 1973.
17. G. T. Johnson, T. R. Lewis, and W. D. Wagner, *Toxicol. Appl. Pharmacol.* **32,** 239 (1975).
18. *The Economics of Cesium and Rubidium*, Roskill Information Services, Ltd., SWIP #R2, London, UK, 1975.
19. G. L. Brownell, "New Imaging Systems" in *Nuclear Medicine Technical Progress Report*, July 1, 1974–April 1, 1975, Massachusetts General Hospital, Boston, Mass., Contract: AT(11-1) 3333.
20. J. J. Christensen, D. J. Etough, and R. M. Izatt, *Chem. Rev.* **74,** 351 (1974).
21. R. M. Izatt and J. J. Christensen, *Progress in Macrocyclic Chemistry*, Vol. I, John Wiley & Sons, Inc., New York, 1979.

FREDERICK B. WHITE, JR.
W. G. LIDMAN
Kawecki Berylco Industries, Inc.

RUTHENIUM.   See Platinum-group metals.

RUTHERFORDIUM.   See Actinides and transactinides.

S

SABADILLA, SABADINE, SABALINE. See Insect control technology.

SACCHARIN. See Sweeteners.

SAFETY. See Plant safety; Materials reliability.

SALICYLIC ACID AND RELATED COMPOUNDS

Compounds of the general structure

where the hydroxy is ortho [69-72-7], meta [99-06-9], or para [99-96-7] are commonly known as the monohydroxybenzoic acids. Of the three acids, the ortho isomer, salicylic acid, is by far the most important. The main importance of salicylic acid and its derivatives lies in their antipyretic and analgesic actions (see Analgesics, antipyretics, and anti-inflammatory agents). Natural salicylic acid, which exists mainly as the glucosides of methyl salicylate [119-38-6] and salicyl alcohol [90-01-7], is widely distributed in the roots, bark, leaves, and fruits of various plants and trees. As such, their use as preparations for ancient remedies is probably as old as herbal therapy. Hippocrates recommended the juice of poplar trees as treatment for eye diseases. Salicyl alcohol glucosides (salicin) [138-52-3] occur in *Populus balsamifera* (poplar) and *Salix helix* (willow) trees. Methyl salicylate glucosides occur in *Betula* (birch) and *Fagus* (beech) trees. A more familiar source of methyl salicylate is the leaves of *Gaultheria procumbens* (wintergreen) (see also Hydroxy carboxylic acids).

Free salicylic acid occurs in nature only in very small amounts. It has been isolated from the roots, plants, blossoms, and fruit of *Spiraea ulmaria*, from which its original name, *acidium spiricum*, was derived. Salicylic acid as well as salicylates occur in tulips, hyacinths, and violets, and in common fruits, eg, oranges, apples, plums, and grapes, which explains the presence of salicylic acid in most wines (1–2).

Physical Properties.   Salicylic acid is obtained as white crystals, fine needles, or fluffy white crystalline powder. It is stable in air and may discolor gradually in sunlight. The synthetic form is white and odorless. When prepared from natural methyl salicylate, it may have a lightly yellow or pink tint and a faint, wintergreenlike odor. *m*-Hydroxybenzoic acid crystallizes from water in the form of white needles and from alcohol as platelets or rhombic prisms. *p*-Hydroxybenzoic acid crystallizes in the form of monoclinic prisms. Various physical properties of hydroxybenzoic acids are listed in Tables 1–4.

Reactions.   The hydroxybenzoic acids have both the hydroxyl and the carboxyl moieties and, as such, participate in chemical reactions characteristic of each. In addition, they can undergo electrophilic ring substitution. Reactions characteristic of the carboxyl group include decarboxylation; reduction to alcohols; and the formation of salts, acyl halides, amides, and esters. Reactions characteristic of the phenolic hydroxyl group include the formation of salts, esters, and ethers. Reactions involving

Table 1.   Physical Properties of Hydroxybenzoic Acids

Property	Value (isomer)		
	Ortho	Meta	Para
molecular weight	138.12	138.12	138.12
melting point, °C	159	201.5–203	214.5–215.5
boiling point, °C	211 sub		
density	1.443_4^{20}	1.473_{25}^{25}	1.497_{20}^{20}
refractive index	1.565		
flash point (Tag closed-cup), °C	157		
K_a (acid dissociation) at 25°C	1.05×10^{-3}	8.3×10^{-5}	2.6×10^{-5}
heat of combustion, mJ/mol[a]	3.026	3.038	3.035
heat of sublimation, kJ/mol[a]	95.14		116.1

[a] To convert J to cal, divide by 4.184.

Table 2.   Solubilities of the Hydroxybenzoic Acids in Water, Wt %[a]

Temperature, °C	Isomer		
	Ortho	Meta	Para
0	0.12	0.35	0.25
10	0.14	0.55	0.50
20	0.20	0.85	0.81
30	0.30	1.35	0.81
40	0.42	2.0	1.23
50	0.64	3.0	2.3
60	0.90	4.3	4.2
70	1.37	7.0	7.0
80	2.21	11.0	12.0

[a] Ref. 3.

Table 3.   Solubilities of the Hydroxybenzoic Acids in Nonaqueous Solvents, Wt %[a]

| Solvent | Isomer | | |
	Ortho	Meta	Para
acetone at 23°C	396	327	285
benzene at 25°C	0.775	0.010	0.0035
1-butanol	$28.8_{38°C}$	$20.7_{36.5°C}$	$19.5_{32.5°C}$
ethanol (99 wt %)	$40.6_{41°C}$	$39.6_{65°C}$	$38.75_{67°C}$
n-heptane	$2.09_{92.2°C}$	$2.0_{197°C}$	$1.5_{197°C}$
methanol at 15°C	39.87	40.38	36.22
carbon tetrachloride at 25°C	0.262		
chloroform (satd in H_2O) at 25°C	1.84		
ethanol (abs) at 21°C	34.87		
1-propanol at 21°C	27.36		

[a] Ref. 3.

Table 4.   Saturated Vapor Pressure (p) of o- and p-Hydroxybenzoic Acids[a]

Temperature, °C	o-Hydroxybenzoic acid p, Pa[b]	p-Hydroxybenzoic acid p, Pa[b]
95	30.9	
100	48.7	
105	70.1	
110	104	
115	153	
120	220	
125	322	3.03
130		4.59
135	649	6.94
140		10.6
145		16.1
150		23.9
155		34.6
159		47.3

[a] Ref. 4.
[b] To convert Pa to mm Hg, divide by 133.3.

ring substitution include nitration, sulfonation, halogenation, alkylation, and acylation. The following reactions are illustrated only with salicylic acid; however, these reactions are characteristic of all the hydroxybenzoic acids.

Salicylic Acid

Reactions.   Carboxyl.   Typical decarboxylation by simple heating of a free acid occurs with only a few types of acids. However, decarboxylation of salicylic acid takes place readily because of the presence of the hydroxyl group, which is electron donating (see Fig. 1). Upon slow heating, salicylic acid decomposes to phenol and carbon dioxide; when heated rapidly, it sublimes.

Generally, the carboxyl group is not readily reduced. Lithium aluminum hydride is one of the few reagents that can reduce an acid to an alcohol. The scheme involves the formation of an alkoxide, which is hydrolyzed to the alcohol. Commercially, the alternative to direct reduction involves esterification of the acid followed by reduction of the ester.

Salicylic acid dissolves in aqueous sodium carbonate or sodium bicarbonate to

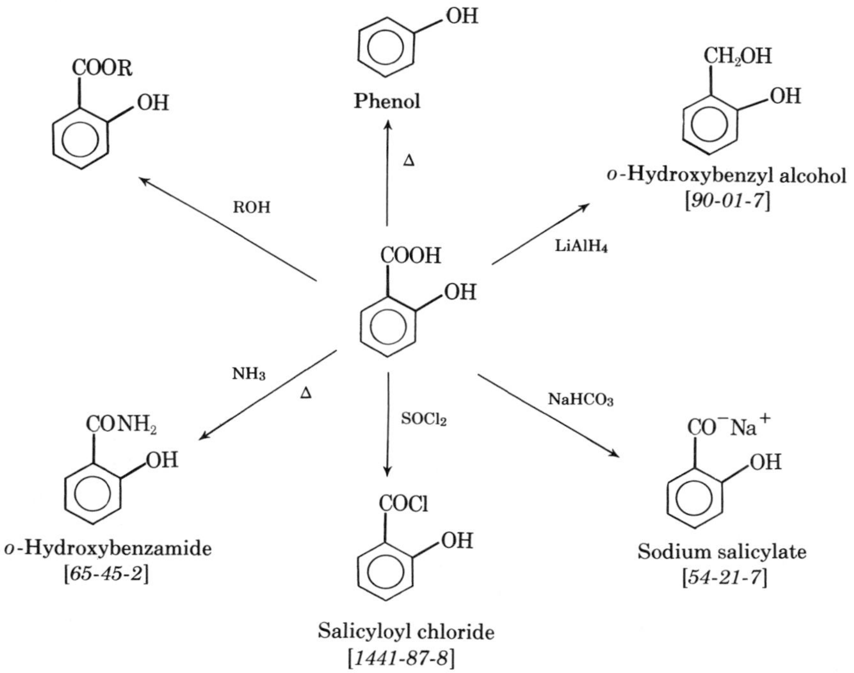

Figure 1. Reactions of the carboxyl group of salicylic acid.

form sodium salicylate. However, if salicylic acid dissolves in the presence of alkali metals or caustic alkalies, eg, excess sodium hydroxide, the disodium salt forms.

Salicylic acid can be converted to salicyloyl chloride by reaction with thionyl chloride in boiling benzene. However, the formation of acyl halides can be complicated by the presence of the phenolic hydroxyl. For example, the reaction with phosphorus tri- and pentachlorides is not restricted to the formation of the acid chloride. Further interaction of the phosphorus halide and the phenolic hydroxyl results in the formation of the phosphoric or phosphorous esters.

The formation of amides can be accomplished by the dehydration of the ammonium salt of salicylic acid. The more common method for amides is the reaction of the ester, acyl halide, or anhydride with an amine or ammonia. Each step is fast and essentially irreversible.

Esterification is frequently carried out by direct reaction of the carboxylic acid with an alcohol in the presence of a small amount of mineral acid, usually concentrated sulfuric or hydrochloric acid. The ester of commercial importance is methyl salicylate. Direct esterification has the advantage of being a single-step synthesis; its disadvantage is the reversibility of the reaction. The equilibrium can be shifted to the right if either raw material is used in large excess, or by selective removal of one of the products. One less frequently employed technique is the transformation of the acid to the acid chloride followed by alcoholysis; each step is essentially irreversible. Another method is the reaction of the alkali salt, eg, sodium salicylate, with an alkyl or an arylalkyl halide.

Hydroxyl. The hydroxyl group is alkylated readily by the sodium salt and an alkyl halide (Williamson ether synthesis) (see Fig. 2). Normally, only *O*-alkylation is ob-

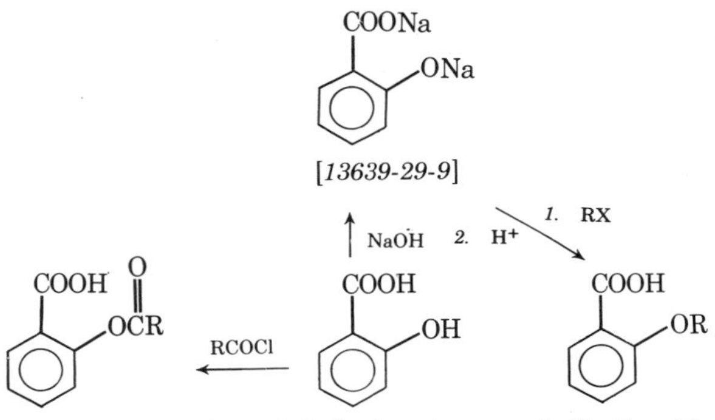

Figure 2. Reactions of the hydroxyl group of salicylic acid.

served. However, phenolate ions are ambident nucleophiles and, as such and under certain conditions, as with the use of alkyl halides, the problem of C- versus O-alkylation can occur. Either reaction can be made essentially exclusive by the proper choice of reaction conditions. For example, polar solvents favor formation of the ether, whereas nonpolar solvents favor ring substitution (see Alkylation).

Esters of the phenolic hydroxyl are obtained easily by the Schotten-Baumann reaction. The reaction in many cases involves an acid chloride as the acylating agent. However, acylation can also be achieved by reaction with an acid anhydride. The single most important commercial reaction of this type is the acetylation of salicylic acid with acetic anhydride to yield acetylsalicylic acid [50-78-2] (aspirin).

Ring Substitution. In the introduction of a third group into a disubstituted benzene, the position the group takes depends on the groups present (see Fig. 3). In the case of salicylic acid, the hydroxyl directs ortho and para and the carboxyl directs meta substitution. It is generally accepted that if both an ortho-para and a meta director are competing for the orientation of a third group, the ortho-para director prevails since, unlike the meta director, it activates the ring. Specifically, the hydroxyl group is electron-donating which, on the basis of resonance considerations, increases the

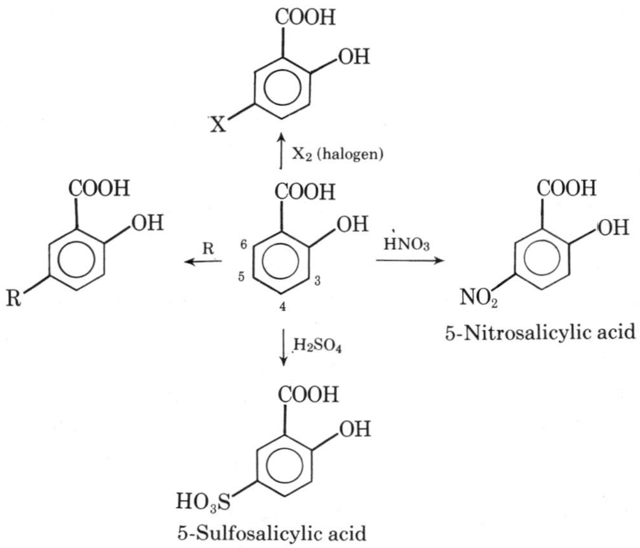

Figure 3. Ring-substitution reactions of salicylic acid. X = halogen.

electron density in the 3 and 5 positions. The electron-withdrawal nature of the carboxyl group decreases the electron density around the 4 and 6 positions, which further enhances the electron density of the 3 and 5 positions. As a rule, direct substitution occurs more easily in the less sterically hindered 5 position, but most often small amounts of the 3 substituted and 3,5-disubstituted product also form. High yields of the 3-substituted salicylic acid usually can only be prepared indirectly.

Direct halogenation of salicylic acid is generally carried out in glacial acetic acid. As expected, the main product is the 5-halo-salicylic acid with small quantities of the 3-halo- and 3,5-dihalosalicylic acids.

Reaction with cold nitric acid results primarily in the formation of 5-nitrosalicylic acid [96-97-9]. However, reaction with fuming nitric acid results in decarboxylation as well as the formation of 2,4,6-trinitrophenol [88-89-1] (picric acid). Sulfonation with chlorosulfonic acid at 160°C yields 5-sulfosalicylic acid [56507-30-3]. At higher temperatures (180°C) and with an excess of chlorosulfonic acid, 3,5-disulfosalicylic acid forms. Sulfonation with liquid sulfur trioxide in tetrachloroethylene leads to a nearly quantitative yield of 5-sulfosalicylic acid (5).

Because salicylic acid contains the deactivating meta-directing carboxyl group, Friedel-Crafts reactions (qv) are generally inhibited. This effect is somewhat offset by the presence of the activating hydroxyl group. Salicylic acid also reacts with isobutyl or t-butyl alcohol in 80 wt % sulfuric acid at 75°C to yield 5-t-butylsalicylic acid [16094-31-8]. In the case of isobutyl alcohol, the intermediate carbonium ion rearranges to $(CH_3)_3C^+$.

Miscellaneous. The Reimer-Tiemann reaction of salicylic acid with chloroform and alkali results in the 3- and 5-formyl derivatives.

If the reaction is carried out with carbon tetrachloride, the corresponding dicarboxylic acids form.

2-hydroxy-1,3-benzenedicarboxylic acid [606-19-2]   4-hydroxy-1,3-benzenedicarboxylic acid [636-46-4]

Alkylation involving formaldehyde in the presence of hydrogen chloride is known as chloromethylation. The reagent may be a mixture of formalin and hydrochloric acid, paraformaldehyde and hydrochloric acid, a chloromethyl ether, or a formal. Zinc chloride is commonly employed as a catalyst, although many others can be used. Chloromethylation of salicylic acid yields primarily the 5-substituted product.

The reaction of salicylic acid with formaldehyde with catalytic amounts of acid results in the condensation product methylene-5,5'-disalicylic acid [122-25-8].

COOH OH + CH$_2$O + HCl $\xrightarrow{\text{ZnCl}_2}$ ClH$_2$C— COOH OH + H$_2$O

5-chloromethylsalicylic acid
[10192-87-7]

2 COOH OH + CH$_2$O $\longrightarrow$ HO— COOH —CH$_2$— COOH OH + H$_2$O

Salicylic acid, upon reaction with amyl alcohol and sodium, reduces to a ring-opened aliphatic dicarboxylic acid, ie, pimelic acid. The reaction proceeds through the intermediate cyclohexanone-2-carboxylic acid.

COOH OH $\xrightarrow[\text{Na}]{\text{CH}_3(\text{CH}_2)_4\text{OH}}$ COOH O $\longrightarrow$ HOOC(CH$_2$)$_5$COOH

During certain substitution reactions, the carboxyl group is often replaced by the entering group. An example is fuming nitric acid, which results in the formation of trinitrophenol. Another is the bromination of salicylic acid in aqueous solution to yield the tribromophenol derivative.

COOH OH + 3 Br$_2$ $\longrightarrow$ Br— Br Br OH + 3 HBr + CO$_2$

[25376-38-9]

Salicylic acid couples with diazonium salts in the expected manner. With diazotized aniline, ie, benzenediazonium chloride, the primary product is 5-phenylazosalicylic acid [3147-53-3].

COOH OH + N$_2$Cl $\longrightarrow$ N=N COOH OH + HCl

The close proximity of the carboxyl and the hydroxyl groups can be used for heterocyclic synthesis, as in the preparation of hydroxyxanthones (6).

Manufacture. R. Piria is generally credited with the first laboratory synthesis of salicylic acid (7–8). In 1859, a synthetic method of preparing salicylic acid was discovered by treating phenol with carbon dioxide in the presence of metallic sodium (9). However, the only commercial means of manufacturing salicylic acid until 1874 was the saponification of methyl salicylate obtained from the leaves of wintergreen or the bark of sweet birch. The first technically suitable commercial process was introduced in 1874 and involved the reaction of dry sodium phenate with carbon dioxide under pressure at high temperatures (180–200°C) (10).

There were limitations, however; not only was the reaction reversible, but the best possible yield of salicylic acid was 50%. At lower temperatures (120–140°C) and under pressures of 500–700 kPa (5–7 atm), the absorption of carbon dioxide forms the intermediate phenyl carbonate almost quantitatively (11–12). The sodium phenyl carbonate rearranges to give predominately the ortho isomer, sodium salicylate.

The Schmitt modification, which involves lower reaction temperatures and reaction times, significantly increases the yields. Modern methods of commercial manufacture still employ the basic Kolbe-Schmitt reaction (13).

Phenol and a slightly greater than equal molar amount of hot aqueous caustic are mixed. The resulting solution is heated to ca 130°C and evaporated to dryness, initially at atmospheric pressure and, in the final stages, under vacuum. The dry sodium phenate is cooled to ca 100°C, at which time a known excess of dry carbon dioxide at 500–600 kPa (5–6 atm) is introduced to the carbonator. Air is excluded to minimize oxidation and the formation of colored compounds. The resulting mixture is agitated and heated to 150–170°C for several hours. The pressure is reduced, and any regenerated phenol is recovered by vacuum distillation.

The crude sodium salicylate is cooled and then dissolved in water. The solution is generally purified in one of two ways. One technique consists of treating the solution with activated carbon containing zinc dust; zinc acts as a reducing agent upon color bodies. Another technique is the crystallization of the hexahydrate. The sodium salicylate solution is acidified, generally with sulfuric acid, to precipitate technical-grade salicylic acid.

The technical-grade salicylic acid can be further purified by recrystallization or sublimation. The products obtained from sublimation are salicylic acid USP and much smaller quantities of a slightly colored, technical sublimed grade. During sublimation and recovery, the risk of dust explosions is minimized by circulating an inert gas through the reaction chambers. The overall yield of salicylic acid USP is greater than 80% based on an 85–90% yield of technical-grade salicylic acid and 95% recovery in sublimation.

A number of specific variations of the basic Kolbe-Schmitt reaction are given in the literature. The Wacker process involves carbonation of sodium phenate in excess phenol (13). Carbonation of phenate salts in alcoholic solutions have also been described (14). The carbonation of dried sodium phenate followed by gaseous hydrogen chloride neutralization–sublimation has recently been patented (15–16). The following are alternative, although not commercially practiced, syntheses. Salicylic acid can be produced from derivatives of carboxylic acids, eg, the reaction of amino acids with nitrous acid, the fusion of 2-sulfo- and 2-halobenzoic acids with alkali, and the oxidation of the creosols, o-hydroxybenzyl alcohol, and o-hydroxybenzaldehyde.

Economic Aspects. Technical- and USP-grade salicylic acid are produced in the United States by The Dow Chemical Company, Hilton-Davis Company, Monsanto Company, and Tenneco Chemicals, Inc. U.S. production and sales data for salicylic acid, as reported by the U.S. International Trade Commission, are shown in Table 5. The production figures shown are for technical-grade salicylic acid and include the quantities used in the manufacture of USP grade. The difference between the production and sales figures reflect in-plant consumption for the manufacture of aspirin and other derivatives, eg, methyl salicylate.

In May 1981, the prices per kilogram of salicylic acids were as follows: salicylic acid technical grade, $2.71; salicylic acid sublimed technical grade, $2.80; and salicylic acid USP, $2.82.

Specifications and Analysis. Salicylic acid USP complies with the following standards of purity as established by the *U.S. Pharmocopeia XX* (17): salicylic acid contains not less than 99.5 wt % and not more than 101.0 wt % of $C_7H_6O_3$, calculated on the dried basis. Other USP specifications include a melting range of 158–161°C, loss on drying over silica gel for 3 h of not more than 0.5%, residue on ignition of not more than 0.05%, chloride content of not more than 140 ppm, sulfate content of not more than 200 ppm, and a heavy-metals content of not more than 20 ppm. Table 6 presents typical specification values for the technical and technical sublimed grades.

Health and Safety Factors (Toxicology). *Handling.* In the laboratory, salicylic acid should be handled so that its irritant properties do not become a problem. Avoidance of skin or eye contact and general good hygiene and good housekeeping should be sufficient. If large quantities are to be handled, protective clothing, ie, long sleeves

Table 5.   U.S. Production and Sales of Salicylic Acid[a]

Year	Production, 1000 metric tons	Sales		
		Quantity, 1000 t	Value, 10^6 $	Unit value, $/kg
1969	18.1	3.7	2.6	0.71
1970	17.6	3.4	3.1	0.91
1971	18.8	4.2	3.8	0.91
1972	21.4	4.1	3.7	0.91
1973	18.8	6.4	5.6	0.88
1974	19.8	5.7	8.3	1.46
1975	13.9	3.0	5.4	1.81
1976	14.2	1.8	3.3	1.81
1977	20.5	2.6	4.8	1.85
1978	21.4	2.9	5.5	1.90
1979	18.3	2.3	4.4	1.90

[a] From U.S. International Trade Commission monthly reports, 1980.

Table 6. Typical Salicylic Acid Specifications

Type	Melting range, °C	Assay, %	Ash, wt %	Water, wt %
technical	157–161	98.5	1.0	0.4
technical sublimed	157–161	99.5	0.1	0.1

and gauntlets, and NIOSH-approved dust respirators should be used. Dust concentrations as low as 9 g/m^2 can ignite; therefore, "No Smoking" signs should be posted in or near the work area. Eye fountains and safety showers should also be in the vicinity of the work area, especially if large quantities are being handled.

Toxicity and First Aid. The single-dose oral toxicity of salicylic acid is moderate. The LD$_{50}$ in the rat is 400–800 mg/kg. Safety glasses with side shields should be worn when handling salicylic acid because eye contact with the chemical may produce irritation and marked pain. In severe cases, corneal injury has occurred; however, such lesions heal with time. If salicylic acid or its dusts should contact the eyes, they should be washed promptly with plenty of water for at least 15 min and medical attention should be sought. Salicylic acid may be absorbed through the skin, but the hazard is considered to be very low. Single short exposures are not expected to cause significant irritation, but prolonged or repeated exposures may cause marked irritation or a skin rash. If the acid contacts the skin, it should be washed off promptly with soap and water. Contaminated clothing should be washed before reuse. If a rash appears, the patient should seek medical attention. Breathing of dusts generated in handling salicylic acid should be avoided. If the product is handled with reasonable care and dusts are controlled by ventilation, there is little likelihood of injury resulting from inhalation. If dusts are created, handlers should wear NIOSH-approved dust respirators and their clothing should be washed before reuse.

Uses. Approximately 60% of the salicylic acid produced in the United States is consumed in the manufacture of aspirin: this statistic has remained relatively constant for at least the last ten years. Approximately 10% of the salicylic acid produced is consumed in various applications, eg, foundry and phenolic resins, rubber retarders, dyestuffs, and other miscellaneous uses. The remaining 30% is used in the manufacture of its salts and esters for a variety of applications.

There are many foundry-resin systems in use. Salicylic acid is a small component only in the Shell process. It is used as a cross-linking agent in the phenol–formaldehyde resin used as a sand core and mold binder and imparts higher tensile strength. More recent developments have demonstrated that higher concentrations of salicylic acid than previously used further improve cold and hot tensile strength and reduce cure and machine processing time (18). The continuing interest in energy and environmental considerations has led to the low energy processes, which typically do not use salicylic acid. Their growth has been somewhat limited because of the large capital expenditures required; however, the economics is expected to shift as the cost of energy increases. Therefore, a zero or small negative growth for salicylic acid is predicted in foundry-resin applications. Salicylic acid has also been used in other phenolic resin applications, ie, binders for grinding wheels, fiber glass, and brake linings (qv) (see Phenolic resins).

Some salicylic acid is used as an intermediate in the manufacture of dyes. Spe-

Table 7. Physical Properties of Salicylic Acid Salts

Salt	CAS Registry No.	Formula	Mol wt	Crystalline form
basic bismuth salicylate	[5798-98-1]	$Bi(C_7H_5O_3)_3 \cdot Bi_2O_3$	877.3	white microcrystal
calcium salicylate dihydrate	[824-35-1]	$Ca(C_7H_5O_3)_2 \cdot 2H_2O$	350.34	white octahedral
lithium salicylate	[552-38-5]	$LiC_7H_5O_3$	144.05	white powder
magnesium salicylate tetrahydrate	[18917-89-0]	$Mg(C_7H_5O_3)_2 \cdot 4H_2O$	370.60	slightly red crystalline powder
potassium salicylate	[578-36-9]	$KC_7H_5O_3$	176.22	white powder
sodium salicylate	[54-21-7]	$NaC_7H_5O_3$	160.11	white crystalline powder

cifically, it is a coupling agent for azo dyes (qv) and a chelating agent in chromium dyes (see Chelating agents). It had been used in benzidine dyes until benzidine was determined to be carcinogenic.

Salicylic acid USP is also used medicinally as an antiseptic, disinfectant, antifungal, and keratolytic agent. It is applied externally as a dusting powder, lotion, or ointment for the treatment of dandruff, eczema, psoriasis, and parasitic skin diseases. To destroy warts or corns, it is applied in a collodion or as a plaster. It is anticipated that the future market growth of salicylic acid will depend largely on the use of pharmaceuticals derived from it (see Disinfectants and antiseptics; Chemotherapeutics, antibacterial and antimycotic).

Salts. Sodium salicylate is by far the most important salt of salicylic acid. The salt can be obtained directly from the Kolbe-Schmitt carboxylation or by the reaction of salicylic acid with either aqueous sodium bicarbonate or sodium carbonate. The resulting mixture is heated until effervescence stops and is then filtered and evaporated to dryness at low temperatures. Generally, the solution must be kept slightly acidic so that a white product is obtained; if the mixture is basic, a colored product results, owing to complex reactions. Other salts of salicylic acid may be prepared in a similar manner by use of the appropriate carbonates.

A large number of salts of salicylic acid have been prepared and evaluated for therapeutic or other commercial use. The physical properties of the salicylic acid salts most frequently reviewed in modern literature are listed in Table 7.

Sodium salicylate has analgesic, anti-inflammatory, and antipyretic actions (see Analgesics, antipyretics, and anti-inflammatory agents). The principal use of sodium salicylate USP is in the treatment of acute rheumatic fever. The USP product contains 99.5–100.5% $NaC_7H_5O_3$ calculated on the anhydrous basis. Sodium salicylate has also been used for the treatment of other conditions, eg, rheumatoid arthritis; however, because the salts have an unpleasant effect on the stomach, their use has been limited. In general, aspirin is preferred because it is a more effective analgesic. The May 1981 price was $6.94/kg.

Magnesium salicylate is an analgesic–anti-inflammatory agent and is claimed to have exceptional ability to relieve backaches. To a much lesser extent it is also used for the symptomatic relief of arthritis.

Basic bismuth salicylate (oxysalicylate, subsalicylate) is employed as an anti-

Mp, °C	Solubility, g/100 cm^3		
	Cold water	Hot water	Other
-2 H$_2$O, 120	4^{25}	sl sol	sol in acid and alkaline; insol in CH$_3$OH and ether
			sl sol in CH$_3$OH
	133.3		50 in CH$_3$OH
	sl sol		sl sol in CH$_3$OH
440, decomposes under vacuum	sl sol		sl sol in CH$_3$OH
	125^{25}	decomposes	17^{15} in CH$_3$OH

diarrheal agent. It is taken orally in combination with other ingredients for protective, antacid action as well as antidiarrheal and antiseptic effects. It can also be used to impart a pearly surface to cellulose-base, polystyrene, and phenol–formaldehyde resins.

Other salts of salicylic acid that are of interest are aluminum [18921-11-4], ammonium [528-94-9], calcium, lead [15748-73-9], lithium, mercury [5970-32-1], potassium, and strontium [526-26-1]. Most of these salicylates have been used in medicinal applications, but only a few were described in the USP and the *National Formulary*. Their applications ranged from the treatment of catarrhal infections of the nose and pharynx to the treatment of syphilis and gout. The salts in general also have a number of nonmedicinal applications. For example, lead salicylate has been used experimentally in paints to improve their resistance to moisture, light, and heat. It also has been used as a stabilizer in vinyl plastics, where it acts both as a uv light absorber and as an antioxidant (see Uv absorbers).

Commercially, the salts of salicylic acid account for ca 5% of the salicylic acid produced.

Esters. The esters of salicylic acid account for ca 25% of the salicylic acid produced. Typically, the esters are commercially produced by esterification of technical-grade salicylic acid with the appropriate alcohol. A strong mineral acid, eg, sulfuric, is employed as a catalyst. After completion of the esterification, the excess alcohol is distilled and recovered. The crude product is further purified, generally by distillation. Other techniques reported for the manufacture of salicylate esters include the transesterification of methyl salicylate and the reaction of sodium salicylate and the appropriate alkyl halide.

The main commercial applications for the esters are as flavor and fragrance agents and as pharmaceuticals (see Flavors and spices). The single ester of principal commercial importance is methyl salicylate. A number of salicylate esters of commercial interest and their physical properties are listed in Table 8. Annual U.S. production of selected salicylate esters as reported to the U.S. International Trade Commission are listed in Table 9.

Methyl salicylate NF can be obtained from natural sources by the maceration and subsequent steam distillation of wintergreen leaves or sweet birch bark. Generally, the commercial product is produced synthetically by the esterification of salicylic acid

Table 8. Physical Properties of Salicylic Acid Esters

Ester	CAS Registry No.	Structure	Mol wt	Mp, °C	Bp, °CkPa[a]	Density, g/cm³	Refractive index, n_D	Water	Methanol	Ether	Acetone	Benzene	Other
amyl salicylate	[2050-08-0]	$R = (CH_2)_4CH_3$	208.24		265	1.1799^{30}_{4}	1.5805^{20}	δ	sl sol	sl sol			
benzyl salicylate	[118-58-1]	CH_2– (benzyl)	228.25	25	320	1.0535^{20}_{4}	1.5080^{20}	insol	v	sl sol			chloroform
isoamyl salicylate	[87-20-7]	$(CH_2)_3(CH_3)_2$	208.24		$276{-}277^{99}$	1.0639^{20}_{4}	1.5807^{20}	insol	sl sol	sl sol			
isobutyl salicylate	[87-19-4]	$CH_2CH(CH_3)_2$	194.23	5.9	260–262	1.0729^{20}_{4}	1.5065^{20}	insol	sl sol				
isopropyl salicylate	[607-85-2]	$CH(CH_3)_2$	180.21		240–242			insol	∞	∞			
menthyl salicylate	[52253-93-7]	(menthyl structure)	276.38		190^{2}	1.0467^{20}	1.5198^{26}	insol	∞	v	v	∞	
normenthyl salicylate	[118-56-9]	(normenthyl structure)	262.38	4.1									
methyl salicylate	[119-36-8]	CH_3	152.16	−8.6	223.3	1.1738^{20}_{4}	1.5369^{20}	δ	v	v			
phenyl salicylate	[118-55-8]	(phenyl ring)	214.22	43	$173^{1.6}$	1.2614^{20}_{4}		insol	v	sl sol	v	v	CCl$_4$
salicyl salicylate	[552-94-3]	(ring with COOH)	258.23	149				insol	v	sl sol	v		CH$_3$OH

Structure general form: salicylate ester, benzene ring bearing $-OH$ (ortho) and $-C(=O)OR$.

[a] To convert kPa to mm Hg, multiply by 7.5.

[b] δ = partially soluble; v = very soluble; ∞ = infinitely soluble.

512

Table 9. U.S. Production of Selected Salicylate Esters, Metric Tons[a]

Year	Ester			
	Benzyl	Isobutyl	Isopentyl [607-85-2]	Methyl
1969	205		326	3152
1970	109	20[a]	348	2448
1971	194	9[a]	275	2280
1972	190	8[a]	183	2644
1973	257	8[a]	345	3081
1974	371	7[a]	423	3181
1975	261[b]	3[a]	265	2330
1976	669	6[a]	427	
1977		5[a]	477	
1978			380	
1979			411	

[a] From U.S. International Trade Commission monthly reports, 1980.

[b] Reported sales; production data were not given.

with methanol. Whatever the source, it must be declared on the label. The product must assay not less than 98.0% and not more than 100.5%. Methyl salicylate is by far the most important commercial derivative, other than aspirin, of salicylic acid. As a pharmaceutical, it is used in liniments and ointments for the relief of pain in the lumbar and sciatic regions, and for rheumatic conditions. As a flavor and fragrance agent, it is used in confectionary, dentifrices (qv), cosmetics (qv), and in perfumes (qv). Other miscellaneous applications for methyl salicylate are as a dye carrier (qv), uv light stabilizer in acrylic resins, and chemical intermediate. The May 1981 price was $3.62/kg.

Phenyl salicylate (salol) generally is manufactured by heating salicylic acid and phenol in the presence of phosphorus oxychloride for 4–5 h at 110–115°C. The molten product is separated, mixed with water, dried, and distilled under vacuum. Another process involves the transesterification of a salicylate, eg, methyl salicylate, with phenol in the presence of an alkali or alkaline-earth phenate. Medicinally, phenyl salicylate can be used as an enteric coating for pills and capsules and it was formerly used as an intestinal antiseptic. The main applications of phenyl salicylate are related to its ability to absorb uv light over the wavelengths 290–325 nm. As a very effective uv light absorber, phenyl salicylate has been incorporated in waxes (qv), polishes, and polymers, eg, acrylics, celluloses, polyesters, polyethylene, and polypropylene. It can also be incorporated into suntan lotions to block the uv rays of the sun. It is also used in the manufacture of adhesives (qv) and lacquers and as a plasticizer for ethyl cellulose (19). The May 1981 price was $5.18/kg for the technical grade and $6.06/kg for a purified grade.

Benzyl salicylate can be prepared by the reaction of benzyl chloride with an alkali salt of salicylic acid at 130–140°C or the transesterification of methyl salicylate with benzyl alcohol. It is used as a fixative and solvent for nitro musks. Benzyl salicylate is also effective in absorbing uv light and can be used in protective sunscreen lotions. The May 1981 price was $5.27/kg.

Menthyl salicylate and, to a greater extent, normenthyl salicylate are also widely employed as sunscreen agents in suntan preparations. The May 1981 price was $6.72/kg.

Isoamyl salicylate is perhaps the most important ester of salicylic acid for perfumery purposes. It generally is manufactured by the esterification of salicylic acid with isopentyl alcohol. It has a characteristic flowery aroma. It is used in many floral applications and is particularly useful in soap fragrances. The May 1981 price was $4.23/kg. Other salicylates of commercial interest as flavor and fragrance agents include isopropyl, isobutyl, phenethyl [87-22-9], and 2-ethylhexyl [118-60-5] salicylates.

Salicyl salicylate (salicylsalicylic acid) is prepared by the action of phosphorus trichloride, phosphorus oxychloride, or thionyl chloride on salicylic acid at low temperatures in benzene, toluene, ether, or pyridine. The crude product is recrystallized rapidly from ethyl alcohol to avoid hydrolysis and esterification. It is used as an analgesic and antipyretic and in the treatment of acute and chronic rheumatism, acute rheumatic fever, and arthritis. It is considered to be better than aspirin for certain conditions because it is only slowly hydrolyzed in the intestines and does not induce perspiration or gastric disturbances. Owing to the slowness of its hydrolysis (two molecules of salicylic acid per molecule of the ester), the action of salicylsalicylic acid is less prompt but more persistent than that of other salicylates. Other salicylates of commercial interest as medicinals include ethylene glycol monosalicylate [87-28-5], dipropylene glycol monomethyl ether salicylate, choline salicylate [2016-36-6], bornyl salicylate [560-88-3], and *p*-acetamidophenyl salicylate [118-57-0].

Other Derivatives. The derivatives of salicylic acid have been used in a wide variety of applications; however, the primary emphasis has been in the development of medicinal agents. Table 10 lists the common derivatives of salicylic acid and their physical properties.

p-Aminosalicylic acid USP and its salts are used in the treatment or prophylaxis of tuberculosis. *p*-Aminosalicylic acid can be prepared by the carboxylation of *m*-aminophenol (20). Aminosalicylic acid USP assays not less than 98.5% and not more than 100.5%, calculated on the anhydrous basis. *p*-Aminosalicylic acid is more of an irritant and nauseant than its salts and, therefore, is less tolerated. The calcium [133-15-3], potassium [133-09-5], and sodium [133-10-8] salts and the ethyl [6069-17-2] and phenyl [133-11-9] esters of *p*-aminosalicylic acid are more frequently administered as antitubercular agents than the free acid.

Methylene-5,5-disalicylic acid is produced by heating two parts salicylic acid with 1–1.5 parts of 30–40 wt % formaldehyde in the presence of an acid catalyst. The resulting product is a mixture of isomers, primarily the 5,5′-isomer and small amounts of low molecular weight polymers. Methylene-5,5-disalicylic acid is used in the production of printing inks and dyestuffs (see Dyes). It has been suggested for the formulation of alkyd resins (qv) and modified phenolics for paints and varnishes to give rapid-drying coatings with superior gloss, hardness, and outdoor durability. It has also been used as a lubricating-oil additive where it reportedly contributes detergency properties. It has been used as an intermediate in the production of bacitracin methylenedisalicylate, which is used as a supplement in feed to promote growth and as a medicament in swine, poultry, and cattle (see Pet and livestock feeds). The May 1981 price was $6.17/kg.

Salicylamide is prepared by the reaction of methyl salicylate with ammonia. Salicylamide has analgesic, anti-inflammatory, and antipyretic properties similar to but milder than those of aspirin. It was a widely used ingredient in a number of over-the-counter analgesics and antipyretics. Salicylamide is unlike other salicylates in that it causes sedation and central nervous system depression. Salicylamide is not

Table 10. Physical Properties of Salicylic Acid Derivatives

Derivative	CAS Registry No.	Mol wt	mp, °C	Bp, °CkPa[a]	Density	Solubility[b]					
						Water	Methanol	Ether	Acetone	Benzene	Other
p-aminosalicylic acid	[65-49-6]	153.14	150–151 dec			sol	sol	sol	sol	insol	
methylene-5,5-disalicylic acid	[27496-82-8]	228.26	243–244			δ	sol	sol	sol	δ	
salicylamide	[65-45-2]	137.13	142	181.5^{14}		δ	sol	δ			
salicylanilide	[87-17-2]	213.24	135.8–136.2		1.175_4^{140}	sol	δ	δ		δ	δ in chloroform
5-sulfosalicylic acid	[97-05-2]	218.18	120 (anhydrous)			v	v	v			

[a] To convert kPa to mm Hg, multiply by 7.5.
[b] δ = partially soluble; v = very soluble.

hydrolyzed to a salicylate; thus, its effectiveness depends on the entire molecule. Moreover, its use is expected to decline as a result of a recent Federal Register report and other studies indicating that it may not be an effective analgesic, pending further investigation (21). Salicylamide has also been useful for protection against mildew and fungus in a variety of soaps, salves, lotions, and oils. The May 1981 price was $6.39/kg.

Salicylanilide is prepared by heating salicylic acid and aniline in the presence of phosphorus trichloride (22). It is used as an intermediate in the production of other chemicals and as a slimicide, fungicide, and medicament. As an active fungicide and antimildew agent, it is used in cotton fabrics, cordage, paints, and lacquers. It is also reported to be one of the best fungistatic agents for plastics. As a medicament, salicylanilide is the active ingredient in creams used to treat fungus infections of the scalp. 3,4',5-Tribromosalicylanilide is an antimicrobial with reported applications in soaps, shampoos, textiles, melamine–formaldehyde and polyethylene plastics, synthetic fibers, paints, adhesives, and paper (see also Industrial antimicrobial agents).

5-Sulfosalicylic acid is prepared by heating 10 parts of salicylic acid with 50 parts of concentrated sulfuric acid, by chlorosulfonation of salicylic acid and subsequent hydrolysis of the acid chloride, or by sulfonation with liquid sulfur trioxide in tetrachloroethylene (5). It is used as an intermediate in the production of dyestuffs, grease additives, catalysts, and surfactants (see Surfactants and detersive systems). It is also useful as a colorimetric reagent for ferric iron and as a reagent for albumin.

m-Hydroxybenzoic Acid

Of the three hydroxybenzoic acids, the meta isomer is of little commercial importance. It offers no outstanding points of chemical interest and is used industrially in small quantities in a limited number of applications.

Reactions. *m*-Hydroxybenzoic acid has both the hydroxyl and the carboxylic acid moieties and participates in reactions characteristic of each (see Salicylic Acid). *m*-Hydroxybenzoic acid is quite thermally stable. Upon heating to 300°C, it turns brown and does not decarboxylate. The reduction of *m*-hydroxybenzoic acid affords a variety of products, depending on the catalyst and conditions employed. Catalytic reduction over platinum black or platinum oxide in alkaline solution gives 3-hydroxycyclohexanecarboxylic acid [22267-35-2]. Reduction of a warm aqueous solution over platinum oxide or over colloidal platinum yields cyclohexanecarboxylic acid. *m*-Hydroxybenzaldehyde can be prepared by reducing *m*-hydroxybenzoic acid with sodium amalgam. Finally, reduction over Raney nickel gives cyclohexanol.

Nitration of *m*-hydroxybenzoic acid with fuming nitric acid in the presence of sulfuric acid and acetic anhydride gives a mixture of the 2-nitro [602-00-6] and 4-nitro [619-14-7] substitution products. Bromination and iodination yield the 4-halogenated derivatives (4-Br [14348-38-0], and 4-I [58123-77-6]). When *m*-hydroxybenzoic acid is treated with formalin in the presence of hydrochloric acid, 4-hydroxyphthalide [13161-32-5] is obtained as shown:

Unlike salicylic acid, *m*-hydroxybenzoic acid does not undergo the Friedel-Crafts reaction. It can be converted in 80% yield to *m*-aminophenol by the Schmidt reaction, which involves treating the acid with hydrazoic acid in trichloroethylene in the presence of sulfuric acid at 40°C (23).

Manufacture. *m*-Hydroxybenzoic acid was first obtained by the action of nitrous acid on *m*-aminobenzoic acid (24). It is more conveniently prepared by the sulfonation of benzoic acid with fuming sulfuric acid. The resulting *m*-sulfobenzoic acid is mixed with salt and fused with caustic soda at 210–220°C. The fusion melt is dissolved in water and acidified with hydrochloric acid to precipitate the crude product. Final purification generally is achieved by recrystallization from water with activated charcoal.

Uses. *m*-Hydroxybenzoic acid is reported as an intermediate in the manufacture of germicides, preservatives, pharmaceuticals, and plasticizers (qv). In the production of pharmaceuticals, the *m*-hydroxybenzoic acid ester of tropine, ie, (*m*-hydroxybenzoyl)tropeine [52418-07-2], is claimed to cause dilation of the pupils (mydriatic effect), whereas the sodium salt [81256-75-9] is a cholagogic agent promoting the discharge of bile. Esters and metal salts of *m*-hydroxybenzoic acid have been used as germicides and preservatives in foods and meats. Ethers of alkyl esters and carbonates of glycol esters of the acid have been patented as plasticizers for vinyl and cellulosic plastics. It is useful in the manufacture of B-stage epoxy resins (qv) having long shelf-lives and short curing times (25). Condensation with formaldehyde gives the best ion exchanger of all of the isomeric hydroxybenzoic acids (see Ion exchange).

(*m*-hydroxybenzoyl)tropeine

p-Hydroxybenzoic Acid.

p-Hydroxybenzoic acid is of significant commercial importance. The most familiar application is in the preparation of several of its esters, which are used as preservatives.

Reactions. In its chemical behavior, *p*-hydroxybenzoic acid undergoes the typical reactions of the carboxyl and hydroxyl moieties. When heated above its melting point, it decomposes almost completely into phenol and carbon dioxide. It reacts with electrophilic reagents in the predicted manner and does not undergo the Friedel-Crafts reaction. Nitration, halogenation, and sulfonation afford the 3-substituted products. Heating *p*-hydroxybenzoic acid with 8 *N* nitric acid results in a 95% yield of picric acid. In a similar fashion, treatment with chlorine water yields 2,4,6-trichlorophenol (26).

Manufacture.   Several methods have been described for the preparation of *p*-hydroxybenzoic acid. The commercial technique is similar to that of salicylic acid, ie, Kolbe-Schmitt carboxylation of phenol. The modification includes the use of potassium hydroxide in place of caustic (27). The dried potassium phenate is heated under pressure 2.7 kPa (20 atm) or more with dry carbon dioxide at 180–250°C. The potassium salt [*16782-08-4*] of *p*-hydroxybenzoic acid forms almost quantitatively and can be converted to the free acid by use of a mineral acid.

Other reported syntheses include the Reimer-Tiemann reaction, in which carbon tetrachloride is condensed with phenol in the presence of potassium hydroxide. A mixture of the ortho and para isomers is obtained; the para isomer predominates. *p*-Hydroxybenzoic acid can be synthesized from phenol, carbon monoxide, and an alkali carbonate (28). It also can be obtained by heating alkali salts of *p*-cresol at high temperatures (260–270°C) over metallic oxides, eg, lead dioxide, manganese dioxide, iron oxide, or copper oxide, or with mixed alkali and a copper catalyst (29). Heating potassium salicylate at 240°C for 1–1.5 h results in a 70–80% yield of *p*-hydroxybenzoic acid (30). When the dipotassium salt of salicylic acid is heated in an atmosphere of carbon dioxide, an almost complete conversion to *p*-hydroxybenzoic acid results. The *p*-aminobenzoic acid can be converted to the diazo acid with nitrous acid followed by hydrolysis. Finally, the sulfo- and halogenobenzoic acids can be fused with alkali.

Uses.   There are many polymer and plastic applications for *p*-hydroxybenzoic acid. It has been converted to epoxy resins by the reaction with epichlorohydrin, used as a modifier for ethylene–propadiene copolymers, and employed in copolyether esters as fibers for use in radial tire cord (qv) (31–33).

One of the more recent applications is the development of a linear *p*-hydroxybenzoic acid polymer (34). The polymer displays long-term stability in air at over 325°C. In addition, the polymer has a self-lubricating character combined with the highest reported elastic modulus, thermal conductivity, electrical insulating character, and solvent resistance of any available polymer.

p-Hydroxybenzoic acid is used in the manufacture of its methyl, ethyl, *n*-propyl, *n*-butyl, and benzyl esters. These esters have been used as preservatives for food, pharmaceuticals, and cosmetics for many years. The physical properties of the esters are listed in Table 11. For details of the determination of assay and other specifications, consult ref. 17. These esters or parabens are effective bacteriostatic and fungistatic agents against a wide variety of microorganisms.

Although the parabens are only slightly soluble in water, conversion to their sodium salts permits their incorporation into products without the use of alcohol or heat. Furthermore, since they are more soluble in oils and organic solvents than in water, they tend to migrate to the oil or organic phase of emulsion products. As a result, combinations of the esters are used to attain a satisfactory oil–water distribution and are even more effective than an equal weight of any one ester used alone.

The esters of *p*-hydroxybenzoic acid are available in purified and technical grades. Although the technical grades also have a minimum assay of 99.0%, they do not meet the color and odor standards of the purified product. The NF grades are used in pharmaceuticals, cosmetics, parenteral solutions, and other substances that are applied to the skin or taken internally. The technical grades are used industrially for preserving glues, pastes, gum solutions, and polishes. The methyl and propyl esters are also approved as food-grade by the FDA.

The May 1981 prices for the methyl, ethyl, *n*-propyl, and *n*-butyl parabens were

Table 11. Physical Properties p-Hydroxybenzoic Acid Esters (Parabens)

Property	Methyl p-hydroxy-benzoate [99-76-3]	Ethyl p-hydroxy-benzoate [120-47-8]	n-Propyl p-hydroxy-benzoate [94-13-3]	Butyl p-hydroxy-benzoate [94-26-8]	Benzyl p-hydroxy-benzoate [94-18-8]
mp, °C	125–128	116–119	95–98	68–72	108–113
assay (min), %	99.0	99.0	99.0	99.0	99.0
solubility (at 25°C), g/100 g solvent					
water	0.25	0.17	0.05	0.02	0.006
water (at 80°C)	2	0.86	0.30	0.15	0.09
methanol	59	115	124	220	79
ethanol	52	70	95	210	72
propylene glycol	22	25	26	110	13
peanut oil	0.5	1	1.4	5	0.5
acetone	64	84	105	240	102
benzene	0.7	1.65	3	40	2.6
ether	23	43	50	150	42
carbon tetrachloride	0.1	0.9	0.8	1	0.08

$9.48/kg, $10.69/kg, $10.14/kg, and $15.21/kg, respectively. Production data for methyl and n-propyl parabens, as reported by the U.S. International Trade Commission, are shown in Table 12.

Acetylsalicylic Acid (Aspirin)

Acetylsalicylic acid [530-75-6] (o-acetoxybenzoic acid) was first synthesized in 1853 from the reaction of acetyl chloride with sodium salicylate. As a medicament, acetylsalicylic acid was introduced to Germany in 1899 and into the United States in 1900. The first U.S. patent (35) for the manufacture of acetylsalicylic acid expired in 1917. Since that time, the growth of production and consumption has been astonishing. Aspirin is a registered trademark in many nations. However, in the United States and in general, it is accepted as a generic name.

Table 12. U.S. Production of Parabens, Metric Tons

Year	Ester Methyl	n-Propyl
1969	344	114
1970	381	142
1971	346	111
1972	416	177
1973	492	162
1974	476	
1975	221	
1976	362	87
1977		104
1978	385[a]	
1979		

[a] Sales figures; no production data were reported

Aspirin normally occurs in the form of white, tabular or needlelike crystals, or as a crystalline powder. It melts at 135–137°C and decomposes at 140°C. The solubility of aspirin is ca 1 g/300 mL of water at 25°C, ca 1 g/5 mL of ethanol at 25°C, 29 g/100 g of acetone at 20°C, and 1 g/17 mL of chloroform at 25°C.

Manufacture. Aspirin is manufactured by the acetylation of salicylic acid with acetic anhydride (36–37).

Salicylic acid and acetic anhydride are introduced into a glass-lined or stainless-steel reactor. The reactor temperature is kept below 98°C for 2–3 h; careful control of the temperature-time cycle is important. The resulting solution is pumped through a filter to remove extraneous solids and then into a crystallizer. The temperature is reduced to ca 0°C during crystallization. The resulting suspension is centrifuged. The crystals are thoroughly washed, dried to ca 0.5 wt % moisture, and then separated by particle size. In some processes, the crystals from the centrifuge are recrystallized prior to washing.

Various processes involve acetic acid or hydrocarbons as solvents for either acetylation or washing. Normal operation involves the recovery or recycle of acetic acid, any solvent, and the mother liquor. Other methods of preparing aspirin, but which are not of commercial significance, involve acetyl chloride and salicylic acid, salicylic acid and acetic anhydride with sulfuric acid as the catalyst, salicylic acid and ketene, and sodium salicylate with acetyl chloride or acetic anhydride.

Production and Economic Aspects. Aspirin is produced in the United States by The Dow Chemical Company, Monsanto Company, Morton-Norwich Products, Inc., and Sterling Drug, Inc. Production data for aspirin, as reported by the U.S. International Trade Commission, are shown in Table 13. In 1976, production increased only moderately. Many factors have contributed to this condition over the last decade, including fewer exports and, to a significant extent, the competition from nonaspirin

Table 13.   U.S. Production of Aspirin[a]

Year	Production, metric tons
1969	16,907
1970	15,953
1971	14,364
1972	15,879
1973	14,584
1974	15,100
1975	11,537
1976	12,828
1977	14,250
1978	14,627
1979	14,475

[a] From U.S. International Trade Commission monthly reports, 1980.

substitute products. The U.S. aspirin market is generally considered to be mature, and only population increases and new uses will affect its production. The May 1981 price was $3.66/kg.

Specifications. Acetylsalicylic acid USP complies with the following standards of purity as established by the USP (17): aspirin contains not less than 99.5% and not more than 100.5% of $C_9H_8O_4$, calculated on the dried basis. Other specifications include a loss on drying of not more than 0.5 wt %, a residue on ignition of not more than 0.05 wt %, a chloride content not more than 140 ppm, a sulfate content not more than 400 ppm, and a heavy-metal content not more than 10 ppm.

Uses. Aspirin has analgesic, anti-inflammatory, and antipyretic actions. It is used for the relief of less severe types of pain, eg, headache, neuritis, acute and chronic rheumatoid arthritis, myalgias, and toothache. Aspirin can be purchased in a variety of forms: powder, tablet, capsule, time-release, plain, buffered, in combination with other ingredients, extra-strength, and numerous special formulations. It is available in pharmacies, supermarkets, and gasoline stations. It is by far the most widely used drug. Even with the advent of new therapeutic agents, it maintains the preeminent position of all over-the-counter medications.

Aspirin appears to produce many of its effects by interfering with the body's production of prostaglandins (qv, Supplement Vol.), which are hormonelike compounds occurring in all body tissues and many fluids. Prostaglandins apparently play an important role in numerous tasks, eg, elevation of body temperatures, sensitization of pain receptors, and inflammation. Aspirin's interference in such functions helps explain some of its long-standing therapeutic benefits. As more is learned about aspirin and prostaglandins, new uses for aspirin have been suggested; the most newsworthy has been its use as an antithrombotic. Aspirin inhibits platelet cyclooxygenase, an enzyme that mediates certain prostaglandins that promote blood clotting. Such interference with clot formation appears to suggest that aspirin may help reduce heart attacks and strokes. Aspirin has been shown to reduce significantly transient ischemic attacks or strokes, particularly in men who had prior cerebral ischemia (38–39). This effect was sex-dependent; no significant reduction was observed among women. The results of these studies were statistically significant and, as such, the FDA's Advisory Committee on Peripheral and Central Nervous System Drugs has recommended that therapy for transient ischemic attacks in males be added to the list of approved uses for aspirin.

Although there is proof that aspirin may be effective in preventing strokes in many patients with cerebral transient ischemic attacks caused by platelet clots, such proof in heart attacks is not conclusive. Data of early controlled studies designed to evaluate the efficacy of aspirin in the secondary prevention of recurrent infarction indicate a trend favoring the use of aspirin. However, the results are not considered significant because of the small size of the study groups. More recent studies are of substantially larger populations, but produce conflicting data (40–41). However, the results of each test also are considered statistically insignificant.

Although the results of these studies are not conclusive, they still seem to indicate aspirin's possible prevention of recurrent infarction and to warrant further investigation. Additional studies are underway to test this hypothesis further and to corroborate results relative to dosage, sex dependencies, time when administration begins, etc.

Salicyl Alcohol

Salicyl alcohol [90-01-7] (saligenin, o-hydroxybenzyl alcohol), $C_6H_4(OH)CH_2OH$, crystallizes from water in the form of needles or white rhombic crystals. It occurs in nature as the bitter glycoside salicin, which is isolated from the bark of *Salix helix*, *S. pentandra*, *S. praecos*, some other species of willow trees, and the bark of a number of species of poplar trees, eg, *Populus balsamifera*, *P. candicans*, and *P. nigra*.

The alcohol has the following properties: mp, 86°C; d_{25}^{13}, 1.161 g/cm³; heat of combustion, 3.542 MJ/mol (846.6 kcal/mol); solubility in 100 mL water at 22°C, 6.7 g; very soluble in alcohol and ether; sublimes readily.

Reactions.   Saligenin undergoes the typical reactions of phenols and benzyl alcohol. When heated above 100°C, it transforms into a pale-yellow resinous material. Amorphous condensation products are obtained when saligenin is treated with acetic anhydride, phosphorus pentachloride, or mineral acids. Upon boiling with dilute acids, saligenin is converted into a resinous body, saliretin, $C_{14}H_{14}O_3$, which is a polymerized aldehyde of saligenin. Condensation reactions of saligenin with itself in the absence of any catalysts and in the presence of bases have also been studied.

Oxidation of saligenin with chromic acid or silver oxide yields salicylaldehyde as the first product. Further oxidation results in the formation of salicylic acid, which is also obtained when saligenin is heated with sodium hydroxide at 200–240°C. Chlorination of an aqueous solution of the alcohol gives 2,4,6-trichlorophenol and bromination in an alkaline medium yields 2,4,6-tribromophenol and tribromosaligenin. When saligenin is heated with one mole of resorcinol in the presence of anhydrous zinc chloride, 3-hydroxyxanthene forms.

Manufacture.   The hydrolysis of the naturally occurring β-glycoside (salicin) with hydrochloric or sulfuric acid affords saligenin and glucose.

Numerous methods for the synthesis of salicyl alcohol have appeared in the literature. These involve the reduction of salicylaldehyde or of salicylic acid and its derivatives. The alcohol can be prepared in almost theoretical yield by the reduction of salicylaldehyde with sodium amalgam, sodium borohydride, or lithium aluminum hydride; by catalytic hydrogenation over platinum black or Raney nickel; or by hydrogenation over platinum and ferrous chloride in alcohol. The electrolytic reduction of salicylaldehyde in sodium bicarbonate solution at a mercury cathode with carbon dioxide passed into the mixture also yields saligenin. It is formed by the electrolytic reduction at lead electrodes of salicylic acid in aqueous alcoholic solution or sodium salicylate in the presence of boric acid and sodium sulfate. Salicylamide in aqueous alcohol solution acidified with acetic acid is reduced to salicyl alcohol by sodium amalgam in 63% yield. Salicyl alcohol forms along the p-hydroxybenzyl alcohol by the action of formaldehyde on phenol in the presence of sodium hydroxide or calcium oxide. High yields of salicyl alcohol from phenol and formaldehyde in the presence of a molar equivalent of ether additives have been reported (42). Phenyl metaborate prepared from phenol and boric acid yields salicyl alcohol after treatment with formaldehyde and hydrolysis (43).

Uses. Saligenin has been used medicinally as an antipyretic and tonic. It possesses marked local anesthetic powers and has been tried clinically and with good results in concentrations of 4–10 wt %. It is practically nontoxic. Saligenin's taste is pungent at first and then numbing. It has also been used in the synthesis of anion-exchange resins and in the stabilization of rosin and rosin size against crystallization. Unsymmetrical diphenylolmethanes prepared from salicyl alcohol and substituted phenols at 160–170°C in the presence of alkaline catalysts have been claimed as resin components and resin-hardening agents (44).

Thiosalicylic Acid

Thiosalicylic acid [147-93-3] (o-mercaptobenzoic acid) is a sulfur-yellow solid that softens at 158°C; has a mp of 164°C; sublimes; is slightly soluble in hot water and freely soluble in glacial acetic acid and alcohol; and yields dithiosalicylic acid [527-89-9] upon exposure to air.

Reactions. Thiosalicylic acid reacts with ethylmercuric chloride in alcohol and in the presence of sodium hydroxide to yield sodium ethylmercurithiosalicylate [54-64-8] (thimerosal; Merthiolate, Eli Lilly and Co.) (45).

Manufacture. Thiosalicylic acid is prepared by heating o-chlorobenzoic acid with an alkaline hydrosulfide in the presence of copper sulfate and sodium hydroxide (46).

Uses. Thiosalicylic acid has been used as an anthelmintic (see Chemotherapeutics, anthelmintic), as a bactericide, and as a fungicide. It also has been used as a rust remover, a corrosion inhibitor for steel, and a polymerization inhibitor (see Corrosion and corrosion inhibitors). In photography (qv), it has application in print-out emulsions and as an activator for photographic emulsions.

BIBLIOGRAPHY

"Salicylic Acid, Salicylaldehyde, and Salicyl Alcohol" in *ECT* 1st ed., Vol. 12, pp. 45–66, by R. T. Gottesman, Heyden Chemical Corporation; "Salicylic Acid" in *ECT* 2nd ed., Vol. 17, pp. 720–743, R. T. Gottesman and D. Chin, Tenneco Chemicals, Inc.

1. M. Gross and L. A. Greenberg, *The Salicylates—A Critical Bibliographic Review*, Hillhouse Press, New Haven, Conn., 1948.
2. M. J. H. Smith and P. Smith, *The Salicylates—A Critical Bibliographic Review*, Interscience Publishers, a division of John Wiley & Sons, Inc., New York, 1966.
3. H. Stephan and T. Stephen, *Solubilities of Inorganic and Organic Compounds*, Vol. 1, Parts 1 and 2, Macmillan Inc., New York, 1963.
4. J. Timmermans, *Physico-Chemical Constants of Pure Organic Compounds*, Vol. 2, Elsevier, North-Holland, Inc., New York, 1965.
5. E. E. Gilbert, B. Veldhuis, E. J. Carlson, and S. L. Giolito, *Ind. Eng. Chem.* **45**, 2067 (1953).
6. P. Grover, G. Shah, and R. Shah, *J. Chem. Soc.*, 3982 (1955).
7. R. Piria, *Ann. Chim. Phys.* **69**, 281 (1838).

8. R. Piria, *Prakt. Liebigs Ann.* **30,** 151 (1839).

9. H. Kolbe and E. Lautemman, *Prakt. Liebigs. Ann.* **115,** 157 (1860).

10. H. Kolbe, *J. Prakt. Chem.* **21,** 443 (1880).

11. Schmitt, Ger. Pat. 29,939.

12. Schmitt, Ger. Pat. 38,742.

13. A. S. Lindsey and H. Jeskey, *Chem. Rev.* **57,** 583 (1957).

14. U.S. Pat. 2,824,892 (Feb. 25, 1958), L. B. Barkley (to Monsanto Chemical Co.).

15. U.S. Pat. 4,137,258 (Jan. 30, 1979), E. R. Moore, D. C. McDonald, R. Hoffman, and R. L. Briggs (to The Dow Chemical Co.).

16. U.S. Pat. 4,171,453 (Oct. 16, 1979), E. R. Moore, D. C. McDonald, J. Willner, and R. L. Briggs (to The Dow Chemical Company).

17. *The United States Pharmacopeia XX (USP XX–NF XV)*, The United States Pharmacopeial Convention, Inc., Rockville, Md., 1980.

18. U.S. Pat. 4,090,991 (May 23, 1978), J. G. Smillie (to Aurora Metal Corp.).

19. U.S. Pat. 2,392,361 (Jan. 8, 1946), E. C. Britton and E. Monroe (to The Dow Chemical Co.).

20. U.S. Pat. 2,644,011 (June 30, 1953), R. P. Parker and J. M. Smith, Jr. (to American Cyanamid Co.).

21. *Fed. Regist.* **42**(131), 35345 (July 1977).

22. U.S. Pat. 2,763,683 (Sept. 18, 1956), F. L. Beman and E. C. Britton (to The Dow Chemical Co.).

23. L. H. Briggs and J. W. Lyttleton, *J. Chem. Soc.*, 421 (1943).

24. Gerland, *Ann.* **86,** 143 (1853); **91,** 189 (1854).

25. Brit. Pat. 1,017,699 (Jan. 19, 1966), M. P. Rainton (to CIBA A.R.L. Ltd.).

26. K. V. Sarkanen and C. W. Dence, *J. Org. Chem.* **25,** 715 (1960).

27. Can. Pat. 843457 (June 2, 1970), R. Ueno and T. Mayazaki (to Ueno Pharmaceutical Research Laboratory).

28. Jpn. Pat. 31,333 (Dec. 15, 1969), H. Yasuhara, T. Nogi, and I. Takahashi (to Toyo Rayon Co., Ltd.); H. Yasuhara and T. Nogi, *Chem. Ind.* **77** (1969).

29. Belg. Pat. 665,631 (Dec. 20, 1965), W. W. Kaeding and E. J. Strojny (to The Dow Chemical Co.).

30. C. A. Buehler and W. E. Cate in A. H. Blatt, ed., *Organic Syntheses*, Collective Vol. 2, John Wiley & Sons, Inc., New York, 1943, p. 341.

31. U.S. Pat. 3,061,171 (Dec. 4, 1962), M. Hoppe (to Inventa A.G.).

32. U.S. Pat. 3,274,158 (Sept. 20, 1966), S. Tocker (to E. I. du Pont de Nemours & Co., Inc.).

33. F. Nagasawa and M. Nuiya, *Jpn. Chem. Q.* (Eng.) **3**(2), 46 (1967).

34. J. Economy, B. E. Nowak, and S. G. Cottis, *Polym. Prep.* **2**(1), 332 (1970); *Sample J.* **6**(6), 21 (1970).

35. U.S. Pat. 644,077 (Feb. 27, 1900), F. Hoffman (to Farbenfabriken Bayer and Heyden).

36. U.S. Pat. 2,987,539 (June 6, 1961), W. C. Stoesser and W. R. Surine (to The Dow Chemical Co.).

37. U.S. Pat. 3,235,583 (Feb. 15, 1966), R. T. Edmunds (to Norwich Pharmacal Co.).

38. W. S. Fields and co-workers, *Stroke* **8,** 301 (1977).

39. The Canadian Cooperative Study Group, *N. Engl. J. Med.* **229,** 53 (1978).

40. Aspirin Myocardial Infarction Study Research Group, *J. Am. Med. Assoc.* **243**(7), 661 (1980).

41. Persantine Aspirin Re-infarction Study Research Group, *Circulation* **62,** 449 (1980).

42. G. Casiraghi, G. Casnati, G. Ruglise, and G. Sartori, *Synthesis* **2,** 124 (1980).

43. U.S. Pat. 3,321,526 (May 23, 1967), P. A. R. Marchand and J. B. Grenet (to Rhône Poulenc S.A.).

44. U.S. Pat. 2,744,882 (May 8, 1956), H. L. Bender and co-workers (to Union Carbide Corp.).

45. Swiss Pat. 227,349 (Sept. 16, 1943), J. R. (to Geigy A.G.).

46. U.S. Pat. 1,672,615 (June 5, 1928), S. M. Kharasch.

STEPHEN H. ERICKSON
The Dow Chemical Company

SAMPLING

The chemical industry produces many gaseous, liquid, and solid materials, ranging from basic chemicals to functional specialties. In addition, many processes require the use of intermediates in the form of gases, liquids, and solids. These various materials are sampled for the purposes of process control, product quality control, environmental control, and occupational health control. Sampling of some materials can be hazardous, particularly those involving toxic, unstable, and pressurized substances, which require special safety precautions. However, with the exception of these special circumstances, the problems encountered in sampling materials in the chemical industry are principally those of proper choice and handling of a sampling device. Long experience with highly varied materials has produced many methods of sampling, only some of which are discussed here. Extensive coverage of sampling methods is reported in ref. 1. A summary of the current ASTM sampling standards is given in Table 1.

Definitions and Problems

Sampling is the operation of removing a portion from a bulk material for analysis in such a way that the portion is representative of the physical and chemical properties of that bulk material. From a statistical point of view, sampling is expected to provide analytical data from which some property of the chemical may be determined with known and controlled errors and at the lowest cost.

For pure liquids and gases, sampling is relatively easy, but it becomes difficult when particulates are involved, and almost all samples taken in the chemical industry contain solids in a subdivided form. Some raw materials, intermediates, or products are particulates; others contain them as contaminants. Some exist in natural deposits, eg, strata, in a heap on the ground; in storage bins, tanks, pipes, or ducts; in railcars, drums, bottles, and bales; or in other containers that may or may not be subdivided easily into representative units. Furthermore, many systems containing particulates tend to segregate during handling and storage, and this introduces a sampling error in the form of bias. In a heap of particulate material that exhibits segregation, the coarse particles collect near the bottom perimeter of the heap, whereas the fine particles concentrate in the center core. The coarse particles roll down the sloping surfaces of heaps, and the fine ones percolate through the larger particles at the apex and build a central core. With particulate systems that segregate in this manner, the act of pouring this material onto a stationary apex to make a heap, fill a bin, load a conveyor, etc, always results in this cross-sectional pattern. Any form of sampling from such a stationary heap is biased with respect to particle-size distribution and should be avoided.

Generally, little is known in advance concerning the degree of mixing of most sampled systems. Uniformity is rarely constant throughout bulk systems, and it is often nonrandom. During the production of thousands of metric tons of materials, size distribution, shape distribution, surface and bulk composition, density, moisture, etc, can vary; thus, in any bulk container, the product may be stratified into zones of variable properties. In gas and liquid systems, particulates segregate and concentrate in specific locations in the container as a result of sedimentation or flotation processes.

Table 1. ASTM Sampling Standards

Chemical	ASTM	Chemical	ASTM
aerospace components	F 304	hardened concrete in constructions	C 823
aerospace fluids	F 310, F 309, F 303, F 302, F 329, F 301, F 306	high gravity glycerol	D 1258
		hollow microspheres	D 2841
		hydraulic cement	C 183
		hydrocarbon fluids	D 3700
aggregate batches	C 702	industrial chemicals	E 300
aggregates	D 75	inorganic alkaline detergents	D 501
agricultural liming materials	C 602	ion-exchange materials	D 2687
alkylbenzenesulfonates	D 1568	lac resins	D 29
aluminum and aluminum alloys	E 716	leather and leather products	D 2813
aluminum powder and paste	D 480	lime and limestone	C 50
amphibole asbestos	D 3879	liquefied petroleum	D 1265
aniline	D 3456	liquid cyclic products	D 3437
asbestos–cement flat sheets	C 459	magnesium oxide, electrical grade	D 2755
asbestos fiber	D 2590		
atmospheric analysis		man-made staple fibers	D 3333
for analysis of gases and vapors	D 1605	manufactured gas	D 1247
to collect organic compounds	D 3686	metallographic specimens	E 3
bituminous materials	D 140	metal powders, finished lots	B 215
bituminous paving mixtures	D 3665	mica paper	D 1677
board, electrical insulating	D 3394	naphthalene, maleic anhydride, phthalic anhydride	D 3438
brick	C 67		
calcium chloride	D 345	natural gas	D 1145
carbon black	D 1900, D 1799	nonferrous metals and alloys	E 88, E 55
cellulosic pulps for electrical insulation	D 3376	paper	D 202
		paper, paperboard, fiberboard, and related products	D 585
chemical analysis of metals	E 173		
chlorine-containing bleaches	D 2022	particulate matter from stacks and flues, for emission testing	D 2928
coal	D 388, D 2234, D 2013, D 410		
		peat materials	D 2944
coke, for laboratory analysis	D 346	pesticides	F 725
concrete	C 140	petroleum and petroleum products	D 270
cotton fibers	D 1441		
creosote and creosote–coal tar solution	D 38	phenol	D 3852
		pinene	D 233
cresylic acid[a]	D 3852	pine oil	D 268
dipentene and related terpene solvents	D 268	pine tars and pine-tar oils	D 856
		plasticizers	D 1045
drying oils, fatty acids, and polymerized fatty acids	D 1466	plastics	D 1898
		preformed thermal insulation	C 390
electrical insulating liquids	D 923	pressurized gas, from aerospace systems	F 307, F 327, F 308
electrical insulating oils, for gas analysis	D 3613		
		pulverized coal	D 197
electrical insulating solids	D 3636	raw rubber	D 1485
electrodeposited metallic coatings	B 602	raw wool	D 1771
		rosin	D 509
ferroalloys	E 32	rubber	D 3896, D 3183
ferroalloys, for size	A 610	shellac varnish	D 1650
fiberboard	D 585	soaps and soap products	D 460
freshly mixed concrete, test	C 172	soil	
gas from transformer	D 3305, D 2759	by auger borings	D 1452
glass containers	C 224	by ring-lined barrel	D 3550
graphite electrodes	C 783	by split-barrel sampler	D 1586

[a] See also D 3538, Part 29.

Table 1 (*continued*)

Chemical	ASTM	Chemical	ASTM
soil		thermal insulating cements	C 163
by thin-walled tube	D 1587	turpentine	D 233
soil and rock for engineering	D 2113, D 420,	volatile solvents	D 268
purposes	E 311	water	D 3370, D 1192
standard sand	C 778	wool	D 1060, D 2525,
steam	D 1066		D 584, D 1234,
structural clay tile	C 67		D 2258

Figure 1 shows the particulate loading of a pipe containing gas and particulates and shows the nonuniformity induced by a disturbance, eg, a 90° bend (2). Figure 2 shows a profile of concentration gradients in a long, straight, horizontal pipe containing suspended solids. In the latter, segregation occurs as a result of particle mass, and certain impurities, eg, metal-rich particulates, occur near the bottom, and others, eg, oily flocculates, near the top (3). However, liquid-velocity disturbances and pipe roughness may affect the distribution.

Although these systems introduce sampling difficulties, they do obey one rule of good sampling, which is that all samples should be taken when the system is in motion. For example, powders should be sampled from the flowing stream forming a heap, and not from the heap after its formation. Also, a cut should be taken through the complete cross-section of a moving stream, and not from a part of it. With gas systems, this is not possible; with liquids, it can be done only at the outfall from a pipe. A third rule is that small quantities should be cut from the flowing stream frequently rather than large quantities taken infrequently. Obviously, the ideal place to sample would be where the sample is well-mixed and uniform; however, such places are rarely accessible.

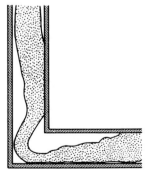

Figure 1. Particle-flow pattern near a 90° bend.

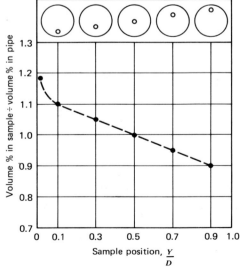

Figure 2. Particle-concentration profile of liquid flowing in a pipe.

The design engineer must be extremely familiar with the potential problems that any sampling operation can impose. Rarely are chemical plants well-designed in terms of sampling capabilities, and rarely do they permit all rules of sampling to be practiced. Therefore, most sampling involves some compromise. Decisions must be made regarding several factors, including the use to which the subsequent analysis is to be put; the test procedure to be employed and the quantity of material to be taken; how representative is the quantity and quality of the sample; minimization of chemical and physical changes during sampling; analytical facilities available, whether manual or automatic; skill and experience of the sampling personnel; sample and analysis use correlation; and cost-benefit relationship.

Careful consideration of the use to which the subsequent analysis is to be put can save time and effort in the initial sampling and possible resampling operations. For the collection of particulates from hot ducts, the decision to collect particles on a filter for analysis is generally acceptable if mass loadings are required. If size distribution and classification are necessary to provide data on chemical composition as a function of size, then an in-duct impactor should be used to classify the sample directly at the temperature and location of the sampling nozzle. Agglomeration and aggregation resulting from the formation of liquid and solid bridges upon cooling can transform a flowable particulate into a solid mass, which is not easily redispersed.

Representative sampling demands a knowledge of the chemistry and chemical reactivity of the species being sampled. If a sample is being drawn from a very hot, high pressure reactor in which the carrier gas is oxygen-free, the extracted sample, when cooled, stored, and analyzed, rarely fails to contact the atmosphere at some stage; this contact can induce hydrolysis and change of the particulate surface. If size analysis is needed, any atmospheric reaction may not affect the results, depending on the increase in agglomeration or aggregation, but it will significantly affect the surface chemical composition. Physically, the sample may be representative, but chemically, it is not. It is important that steps be taken to minimize such possible physical and chemical changes before the sampling operation begins.

In terms of quantity of sample, there are two aspects that must be considered: sample volume and statistical sample size. The sample volume is selected to permit the completion of all the required analytical procedures. The sample size is the necessary number of subsamples taken from the stream to characterize the lot. Sound statistical practices are not always feasible physically or economically in any industry, because cost and engineering difficulties are often prohibitive. In most sampling procedures, subsamples are taken at different levels and locations and form a composite sample. If some prior estimate of the standard deviation of a lot is available, then the sample size n required to build that composite sample is calculated by

$$n = \left(\frac{t\sigma}{E}\right)^2$$

where σ is the estimate of the standard deviation, E is the maximum allowable difference between the estimate to be made from the sample and the actual value, and t is the probability factor giving the selected level of confidence that the difference is greater than E (4). For a standard deviation of 0.187, the number of samples required to assure with 95% confidence that the average quality of a lot lies within ±0.15 of the mean of the determination is

$$n = \left(\frac{2 \times 0.187}{0.15}\right)^2 = 6.22 \text{ or } 7 \text{ samples}$$

If the standard deviation of the lot is not known, a sampling program of greater sample size is required to generate an estimate of the standard deviation for future sampling operations.

In some cases, sample size can be increased and sampling costs reduced by the use of automatic samplers. These offer a distinct reduction in labor costs but an increase in capital costs.

Finally, skill and experience of sampling designers and personnel cannot be overemphasized in chemical-plant sampling. Safety precautions are of the utmost importance, and the necessary steps must be taken to document the hazards involved in the operation and to ensure that the staff is well-informed, protected, and capable of performing the sampling operation. With the exception of bulk powders, most chemical-plant sampling is hazardous and difficult and must be designed with care. The following discussions are based on the assumptions that most of these decisions have been made and a satisfactory sampling procedure has been planned.

Gases

By far the largest proportion of gas-sampling operations in industry are conducted in the environmental field, and the sampling methods have been well researched and well documented (4). The preparation, precautions, and equipment requirements involved in the sampling of air-pollution sources are applicable to most other gaseous environments (see Air pollution control methods).

Before a source-analysis program is undertaken, it is important to decide which information is really required. In the chemical industry, most operations include both very simple and very complex emission problems. Incorrect sampling can cause the greatest single error in analysis and can often incur the largest fractional cost. Sampling sites must be selected with care, as the choice of the sampling point can significantly influence accuracy and cost. For most purposes, measurement requires the determination of the temperature, concentration, and characteristics of the gas contaminants. It requires the mass rates of emission of each contaminant and, therefore, concentration and volumetric-flow data need to be taken.

Sampling Site. The location of the sampling site and the numbers of sampling points required are based on the need to obtain representative data. Any sampling point should be at least eight-stack or -duct diameters downstream and two diameters upstream from any disturbance (5). A disturbance is interpreted as a bend, expansion, contraction, valve, baffle, or visible flame. For rectangular ducts, it is generally sufficient to divide the cross-section area into twelve equal rectangular segments with a sampling point at the centroid of each segment. For circular ducts and stacks, if the sampling point is beyond the minimum eight diameters downstream and two diameters upstream from the nearest disturbance, the cross-section can be divided into twelve equal areas with traverse points located at the centroid of each segment on two perpendicular diameters. These configurations are shown in Figure 3 (5). Table 2 lists the number of stack diameters upstream or downstream of a disturbance with the required number of traverse points on each diameter for ten specific situations (5).

Measurement of Gas Velocity and Temperature. Stack-gas velocity is determined at each traverse point based on the gas density and a measurement of the average velocity head with a type S pitot tube (see Fig. 4) (5). The measured velocity pressure is the difference between the total pressure as measured against the gas flow and the

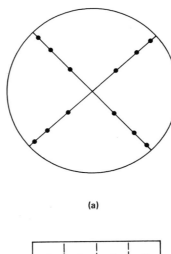

(a)

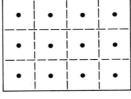

(b)

Figure 3.   Location of traverse points. (**a**) Cross section of circular stack showing location of traverse points on perpendicular diameters. (**b**) Cross section of rectangular stack divided into 12 equal areas, with traverse points at centroid of each area.

Table 2.   Number of Stack Diameters from Flow Disturbance [a]

Downstream	Upstream	No. of traverse points, diameter
>8	>2	6
7.3	1.8	8
6.7	1.7	10
6.0	1.5	12
5.3	1.3	14
4.7	1.2	16
4.0	1.0	18
3.3	0.8	20
2.6	0.6	22
2.0	0.5	24

[a] Ref. 5.

static pressure measured perpendicular to the gas flow. The type S pitot tube has the advantage of easy entry into a duct and a low incidence of plugging when large amounts of particulate are present. However, it does not measure velocity pressure directly but must be calibrated for the velocity being measured. Correction factors of 0.78–0.92 have been reported. Details of this procedure have been documented by the EPA (5). During sample and velocity traverses, the type S pitot tube is rarely used in isolation.

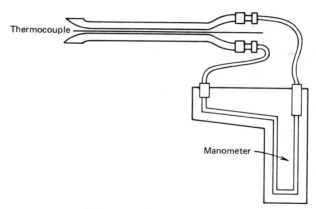

Figure 4.   Type S pitot tube.

It is necessary to measure stack- or duct-temperature profiles to determine variation in gas distribution. This is generally done at the same time as the velocity profiles. If the temperatures measured at each sampling or traverse point are the same, then a single gas-sampling point will suffice later; but if it varies by more than 5%, then samples must be drawn from the traverse points selected in the initial stack survey. For most purposes then, the pitot tube is usually combined with a thermocouple and sampling nozzle in a sampling assembly. As the presence of other components can significantly affect the correction coefficient applied to the type S pitot tube, the placement of various components is critical to minimize aerodynamic interference. These placements are shown in Figure 5 (5).

Sample Extraction.   Once the velocity and temperature profiles have been taken, the significant sampling points have been determined and gas samples can be withdrawn. In the sampling of noncondensable gases free of particulates, the gases are extracted from a duct by one of the following methods: a single-point grab sample, a single-point integrated sample, and a multipoint integrated sample. The latter method is applicable for the collection of CO_2, CO, O_2, excess air, and nitrogen from

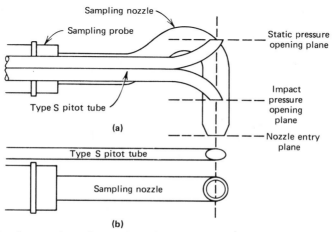

Figure 5.   Correct pitot tube-nozzle configuration. (**a**) Side view. (**b**) Bottom view.

any process in which other gases and compounds are not present in concentrations likely to affect the result. The sampling probe can be made of stainless steel, borosilicate, quartz glass, aluminum, copper, or Teflon. A one-way squeeze bulb is attached to the probe to extract the gas in a grab sample. For integrated sampling, the train shown in Figure 6 is recommended (5). A glass or Pyrex-wool filter is inserted in the probe tip to remove any unwanted particulates and to prevent blockage of the train and gas adsorption or reaction on cooling. A pump first sucks the gas through a cold trap to remove moisture and then through a rotameter to measure flow rate into a gas bag, where the sample is stored prior to analysis. In some cases, the moisture content is measured directly by passing the gas through a set of impingers in an ice bath inserted in the train (see Fig. 7) (5).

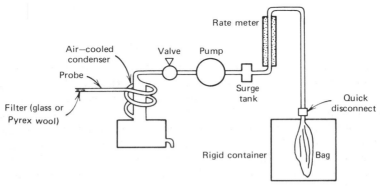

Figure 6.   Integrated gas-sampling train.

For other gases, specific trains are recommended. Sulfur compounds, eg, SO_2, SO_3, H_2S, and mercaptans, are generated during combustion, ore roasting, paper manufacture, and other industrial operations. In most instances, total sulfur is measured; Figure 8 shows the sampling train for SO_2 (5). Sulfur dioxide is highly reactive and is sampled through a heated or well-insulated probe and sample line to an impinger train. The probe contains a quartz or Pyrex-wool filter to remove particulates. In the impinger train, the midget bubbler contains 80 wt % isopropyl alcohol to remove SO_3, the first two midget impingers contain 3 wt % H_2O_2, and the third midget impinger is dry. The gas is drawn through a silica-gel drying tube. The SO_2 is estimated by titration with 0.01% N barium perchlorate with thorin as an indicator. Nitrogen oxide sampling is simpler, since the gas is drawn into an evacuated sample flask containing dilute sulfuric acid and hydrogen peroxide. The flask is shaken and allowed to stand for 16 h before the flask pressure is measured. The solution is made alkaline, and nitrogen oxides are determined by the phenoldisulfonic acid colorimetric test.

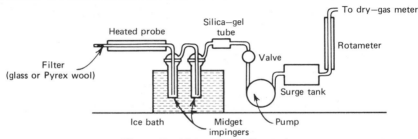

Figure 7.   Moisture sampling train.

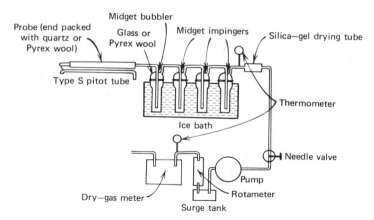

Figure 8.   SO$_2$ sampling train.

Sample Extraction When Particulates Are Present.   Different designs of probe and train may be required, depending on the reason for sampling particulates. The simplest sampling train is for the determination of mass loading only. Representative sampling of particulates is only obtained when the sampling velocity of the gas in the sample nozzle is the same as the gas velocity in the pipe, duct, or stack. This is termed isokinetic sampling. If the sampling velocity in the sampling nozzle exceeds that in the duct, there is a lower than actual concentration of particles collected and the size distribution is finer, because large particles of high inertia do not follow the streamlines into the nozzle but rather pass by. The opposite situation occurs when the sampling velocity is less than that in the duct. Thus, the sample contains a higher than actual concentration and the size distribution is coarser, because large particles enter the nozzle, whereas fine particulates follow the streamlines around it. This effect becomes significant for particle diameters in excess of 3–5 μm. Furthermore, the plane of the sample nozzle must be perpendicular to the gas flow or probe malalignment errors become significant.

For mass loadings, particulate matter is withdrawn isokinetically and is collected on a glass-fiber filter maintained at 120 $\pm$ 4°C. The appropriate sampling train is illustrated in Figure 9 (5). The sample nozzle is made of stainless steel with a sharp-pointed leading edge. The taper is on the outside of the nozzle to provide a constant internal-probe diameter. Because of alignment requirements, the probe is usually of the buttonhook or elbow design. In the impinger train, the first two impingers contain a known amount of water, the third is usually empty, and the last contains silica gel. Particulate matter present as solid or liquid at the sampling temperature is collected on the preweighed filter and is determined by gravimetric analysis. Organic condensable matter is collected in the water, extracted with chloroform and then ether, and weighed after evaporation to dryness. The water phase is also evaporated to dryness, and the residue is reported as inorganic condensable matter. Sometimes particulates are present in liquid form instead of solid. Typical examples are acid mists.

Figure 10 shows the sampling train for sulfuric acid-mist collection (5). The first impinger contains 80 wt % isopropyl alcohol, and the second and third contain 3 wt % H$_2$O$_2$. The first impinger and filter retain the acid mist and SO$_3$, and the next two retain the SO$_2$. After sampling, the filter is added to the contents of the first impinger and the total acid is titrated and reported as sulfuric acid.

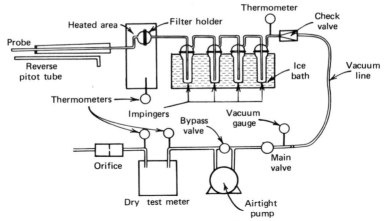

Figure 9.   Particulate sampling train.

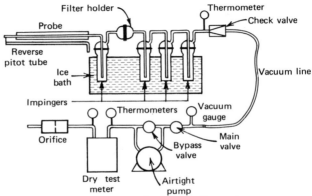

Figure 10.   Sulfuric acid-mist sampling train.

In many instances, mass loadings are simply not sufficient. For example, size distributions are often needed to correlate respirable fractions in health physics and other operations. Under the conditions of high temperature followed by condensation and solidification or cooling, it is rarely possible to redisperse a filtered particulate into its original size distribution as found in the duct because of agglomeration and aggregation. In these cases, size distributions are measured directly by use of an in-stack impactor. A typical example is the Anderson stack sampler (6). All conditions that are necessary in mass-loading measurements are required in the use of this sampler, ie, isokinetic sampling and nozzle alignment. The sampler classifies the particulates drawn into the housing according to their aerodynamic diameter and deposits the different sizes on collection plates. These can be weighed or chemically analyzed to permit calculation of aerodynamic size distributions and determination of chemical composition as a function of size.

For acid mists, the Brink impactor is often used (see Fig. 11) (7). The mist is first drawn through a cyclone to remove all particles greater than 3 μm. A five-stage impactor is used to classify and collect all mist particles of 0.3–3.0 μm dia (7).

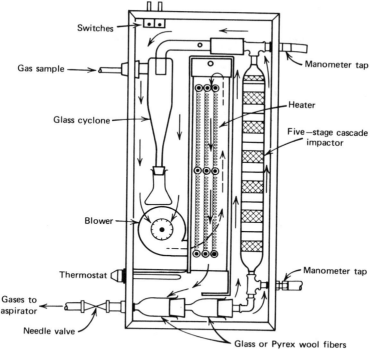

Switches

Gas sample

Glass cyclone

Blower

Thermostat

Gases to aspirator

Needle valve

Manometer tap

Heater

Five—stage cascade impactor

Manometer tap

Glass or Pyrex wool fibers

Figure 11.   Brink impactor.

Liquids

Unlike gases, liquids are contained and stored in a wider variety of containers, which require specific designs of samplers. In the chemical industry, liquids are sampled from process vessels, tanks, tank trucks, tank cars, ships, barges, pipelines, transfer lines, drums, carboys, cans, bottles, open lagoons, settling ponds, sewers, and open, flowing streams and rivers. For simplicity, the procedures for sampling from liquids have been subdivided into three classes: tanks and similar containers, pipelines, and open streams and lagoons. Liquid systems, eg, gas systems, are difficult to sample representatively when particulates are present. Conditions of isokinetic sampling, probe alignment, and sampling location are important to the design of the sampling procedure; however, with liquid systems, they are not as critical as with gases. Viscosity tends to dampen the effects of sudden changes in flow direction, and particles do not separate readily from streamlines unless the particle masses and velocities are large.

Tanks. Sampling methods for tanks, trucks, tank cars, barges, etc, usually involve the use of fixed sample taps, thief samplers, or bottle samplers. Tanks are often equipped with stationary taps that are attached to pipes extending 0.6–0.9 m inside the tank. Usually, three taps, one located in each third of the tank height, are sufficient. A delivery tube that is long enough to reach the bottom of the sample container to allow submerged filling is attached to the tap. Separate samples from each tap are taken or a composite sample is obtained by attaching the delivery tube to each tap in succession and filling the bottle one-third each time. When particulates are present, as in slurries or suspensions, samples can only be drawn from well-stirred tanks. In those cases where particulates are present as contaminants and the tanks have no provision for stirring or mixing, the tap-sampling method rarely makes possible the collection of a representative particulate sample.

A more flexible sampling system involves the use of thief tubes and bottle devices. If the tank is designed so that at least half of the cross-section area of the liquid surface can be exposed for sampling, samples are withdrawn in a regular grid pattern similar to the one used in gas sampling. The tank is subdivided into twelve equal areas, and the thief sampler is inserted at the centroid of each area. Sample thief tubes for liquids consist of long tubes containing one or more compartments along their lengths; they are isolated from the liquid by valves. At a specified depth, in the case of single-compartment thief tubes and, upon contact with the tank bottom, in multicomponent thief tubes, the valves are opened mechanically, the compartments filled, and the valves closed before the thief is removed from the liquid. Each compartment is analyzed independently or the several compartments are combined into a composite sample. Another approach is to use a bottle sampler, which is a heavy-metal, perforated screen cage surrounding a one-liter bottle, which can be lowered into the liquid by a string or chain. The metal casing weights the bottle to ensure that it sinks in the liquid. The bottle is stoppered and is opened by pulling out the stopper with a second length of string. Separate samples are taken at different depths by filling the bottle completely and pulling it to the surface. In some instances, the sampler is mounted on a rigid rod fitted with a mechanical bottle opener, which enables depth sampling to be varied and a composite sample to be withdrawn.

In many cases when sampling is conducted on tank cars, trucks, barges, etc, access is limited to a single, fixed filling hole or vent, and only the one port is available for thief sampling. This limits the statistical accuracy of the sampling procedure but is better than no sampling at all. For drums, carboys, bottles, and cans, thief tubes are simpler. The thief usually consists of an open tube, which is inserted to a measured depth in the container. The upper open end is then closed by the presence of a finger or thumb, and a sample is withdrawn and deposited in a sample container by simply removing the finger. Sometimes, aspirator bulbs or peristatic pumps are used to withdraw larger samples.

Pipes.   Samples are withdrawn from closed-pipe cross-sections and from the outfalls of open pipes; the sampling methods are different for each. In the chemical industry, many pipe samples are taken from organic liquids, which may be toxic, highly reactive, and flammable. The quantity of sample to be extracted is important, because there is no sense in exposing a worker to large volumes of hazardous material if only $5\ \mu L$ is needed (8). Furthermore, sample disposal after analysis is an added problem. Adequate ventilation must be provided for personnel, as vapors from many liquids may be more hazardous than the liquid. Splash guards are also necessary when sampling toxic and corrosive liquids. For these applications, in-line samplers are generally employed. These trap and isolate a predetermined, precise volume of liquid from the line and deliver it to the sample bottle. They can be installed on either side of the suction or discharge side of pumps; typical devices include sampling plugs, multiport valves, and pneumatic samplers (8).

The sampling plug is usually inserted in a bypass line, as shown in Figure 12. The plug has both sample and vent connections. When open, the liquid pumped through the bypass line passes through the plug and can be returned to the line. When closed,

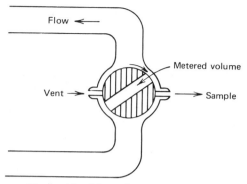

Figure 12. Metered sampling plug.

a small but constant volume of liquid is trapped in the plug and, on further rotation of the tap, is drained into the sample bottle. The tap is simultaneously vented (8). If air or oxygen causes a problem in the process, the valve can be vented with nitrogen and then closed, and the liquid in the bypass line is returned to the process. Sampling of the main stream is usually performed with a sharp-edged probe facing directly into the liquid flow at some preset point. Rarely in pressurized systems can sampling sites be easily changed or statistically designed with regard to location. With such systems, care has to be taken when discharging plug samplers because the pressurized liquid can vent through the valve connections (8). Sometimes a safety relief valve is inserted in the bypass to prevent blockage of the pump discharge (8).

Multiport valves can be used to obtain larger sample sizes (see Fig. 13) (8). During normal operation, the liquid flows through both valves in a bypass line. The sample is extracted by turning the two valves at the same time, first to isolate the sample and then to let it discharge into the sample bottle. This is best achieved by gear-and-linkage devices, which mechanically turn both valves simultaneously. Failure to do this simultaneously can result in the full line pressure being vented through the sample line (8). This type of sampler, although offering advantages in sample volume, presents potential hazards with high pressure systems.

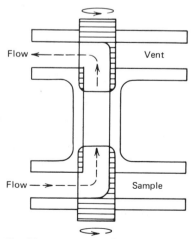

Figure 13. Two-port metered sampling valve (closed).

Air–electric samplers can be installed directly in the pipe wall. One liquid sampler is operated by a solenoid valve that activates an air cylinder. A shaft is moved by this cylinder in and out of the pipe, and samples are expelled into a container below the sampler. The unit is machined to a fine tolerance between shaft and housing and contains Teflon seals. Sample volumes of 2–30 mL are possible.

For low pressure pipelines that have ports open to the atmosphere, eg, sewers or closed effluent culverts, samplers are designed to sample through manholes. In a typical system, a vacuum lifts liquid through a suction line to the sampling chamber. When filled, the chamber is automatically closed to the vacuum, the pump shuts off, and the sample drains into the sample jar. A secondary float prevents any liquid from reaching the pump. The suction line then drains by gravity back to the source.

A more permanent installation is a chain-driven sampler (9). It is used extensively in steel and paper mills. The cup, which is attached to the chain which is positioned perpendicular to the flow, travels down through the liquid flow and returns to the upper sprocket, where the sample is discharged into a container. Flow-proportional timers can be installed to change the rate of sampling with flow rate.

For low pressure pipelines and enclosed troughs, cutter-type samplers can be installed. These samplers contain a movable cutter connected to a flexible hose through which the sample is extracted. Such devices meet all three rules of good sampling, since they sample a moving stream by taking small samples frequently from the whole stream cross-section. The cutter must have a large enough opening, ie, with a diameter equal to or greater than three times the diameter of the largest particle, and it must be hydraulically correct for the flow. Sample overflow must not occur for representative sampling. With these devices, the cutter is located in a housing directly in the pipeline and is driven across the whole stream by a motor-driven traverse. Two designs for sloping troughs and vertical pipes are shown in Figure 14. Cutter samplers can also be placed at the outfalls of pipes or on weir dischargers (10). Alternatively, a rotary cutter (Vezin sampler) can be used for flow rates up to 3.8 L/s (10) (see Solids).

Open Streams and Lagoons. Open-stream discharges are encountered in wastewater plants in industry (see Wastes, industrial; Water, water pollution). Settling ponds and lagoons are a part of the wastewater-treatment plants. Details on the monitoring of industrial wastewater are given in ref. 11. A successful water-pollution-abatement program is based upon information obtained by sampling (11).

Obtaining a representative sample should be a principal concern in any abatement program. Waste flows vary widely in magnitude and composition. Often, they differ significantly during the weekends. This difference is not always observed as an improvement in quality. General maintenance of settling lagoons is often performed during weekends when the plant is not at full production, and this can result in unusually high suspended-solids loadings in the discharge stream over short periods of time. Boilers are often cleaned out at such times, which can give rise to slugs of very highly alkaline water. Sampling must, therefore, be carried out each day on a 24-h basis, unless the discharge is known to be free of such variations. Several samples should be taken during a 24-h period, the number depending on the variability of the discharge. Ideally, the sample should be taken from a place where the flow is well-mixed, eg, an outfall of well-mixed tanks, but regulations are often based on the conditions of the final outfall of the effluent to a stream or river, where the flow and composition can vary widely. Consequently, the flow rate must be measured so that the total waste being discharged can be calculated, and samples must be taken at high frequency in

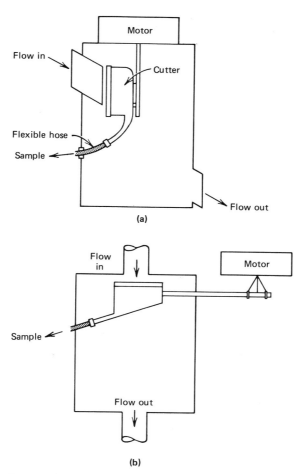

Figure 14. (a) Sloping trough cutter. (b) Vertical pipe cutter. Courtesy of Denver Equipment Company.

proportion to the flow. Methods of measuring flow rates are reported in ref. 11. Such samples are often composited to represent the conditions over an 8-h shift or any other suitable time frame.

Many final discharges are mixed effluents, which may contain oil and particulates that tend to separate near the surface and near the pipe bottom, respectively. Furthermore, the conditions promoting solids suspension, eg, turbulence, also entrain air into the sample, which can significantly change the dissolved oxygen content and affect COD (chemical oxygen demand) and BOD (biochemical oxygen demand) measurements. Plant-induced turbulence prior to discharge can be used, but sampler-induced turbulence, eg, vigorously shaking a grab sample prior to pouring it into a sample bottle to suspend the particle, is not a good practice. Avoidance of contamination and confusion in labeling, etc, should also be eliminated by careful operator procedures. Reference 11 provides detailed instructions for the correct sampling methods.

Preserving the sample is more critical in wastewater sampling than in gas or powder sampling. Biological and chemical changes can take place fairly rapidly, and analysis must be performed as quickly as possible. For unstable samples, the use of composite samples is not good practice. Instead, special grab samples should be taken and submitted to the laboratory for analysis within one hour of sampling. Tables 3 and 4 show recommended storage and preservation conditions for various analytical parameters (12–13).

Table 3. Recommended Storage Procedure[a]

	Sample storage	
Analysis	Refrigeration at 4°C	Frozen
total solids	ok	ok
suspended solids	up to several days	no
volatile suspended solids	up to several days	no
chemical oxygen demand (COD)	up to several days	ok
biochemical oxygen demand (BOD)	up to one day in composite sampling systems	lag develops, must use fresh sewage seed

[a] Ref. 12.

Table 4. Sample Preservation[a]

Parameter	Preservative	Maximum holding period
acidity–alkalinity	refrigeration at 4°C	24 h
biochemical oxygen demand (BOD)	refrigeration at 4°C	6 h
calcium	none required	7 d
chemical oxygen demand (COD)	2 mL H_2SO_4/L	7 d
chloride	none required	7 d
color	refrigerated at 4°C	24 h
cyanide	NaOH to pH 10	24 h
dissolved oxygen	determine at site	no holding
fluoride	none required	7 d
hardness	none required	7 d
metals, total	5 mL HNO_3/L	6 mo
metals, dissolved	filtrate: 3 mL 1:1::HNO_3:H_2O/L	6 mo
nitrogen, ammonia	40 mg $HgCl_2$/L at 4°C	7 d
nitrogen, Kjeldahl	40 mg $HgCl_2$/L at 4°C	unstable
nitrogen, nitrate–nitrite	40 mg $HgCl_2$/L at 4°C	7 d
oil and grease	2 mL H_2SO_4/L at 4°C	24 d
organic carbon	2 mL H_2SO_4/L (pH 2)	7 d
pH	determine at site	no holding
phenolics	1.0 g $CuSO_4$/L + H_3PO_4 to pH 4.0 at 4°C	24 h
phosphorus	40 mg $HgCl_2$/L at 4°C	7 d
solids	none available	7 d
specific conductance	none required	7 d
sulfate	refrigeration at 4°C	7 d
sulfide	2 mL zinc acetate/L	7 d
threshold odor	refrigeration at 4°C	7 d
turbidity	none available	7 d

[a] Ref. 13.

Grab samples are good for noncontinuous flows, but they cannot be sampled proportional to the flow. They are reliable when the discharge composition is constant and are essential to pinpoint the times of isolated high pH or high solids content. Composite samples can be taken proportional to flow or time, but such samples provide averages of the conditions of isolated high values of pH and suspended matter to the point where they may not be obvious from the final analysis. The difficulties of deciding upon grab or composite sampling are not the only factors of concern. Another is whether to sample the discharge manually or automatically. Manual sampling involves high labor costs but low capital costs. It also permits the sampler to observe unusual conditions and to determine the source of a contaminant quickly, eg, suspended solids. Automatic sampling involves higher capital costs but is beneficial for high frequency and round-the-clock sampling programs. It does not enable unusual conditions to be observed but can be instrumented with alarms to warn of any unusual characteristic.

Bacteriological sampling is performed by manual techniques because of careful sterilization requirements. Samples are taken in wide-mouthed, sterile, glass-stoppered bottles that are wrapped in paper prior to sterilization in an autoclave at 138 kPa (20 psi) or in an oven at 170°C. The bottle is unwrapped and the lower portion of it is held in the hand. The sample is taken with the bottle mouth in the direction of the flow or by pushing the bottle mouth away from the hand. The stopper must be protected from contamination, the bottle only partially filled, and the sample stored at 4°C after sampling. If bacteriological samples are withdrawn from a tap, the water should run for five minutes and then be shut off; the tap should be sterilized by flaming before the bottle is filled.

In instances where free chlorine is present, eg, in drinking water, it is measured on-site, and a crystal of sodium thiosulfate is added to the bottle prior to sterilization to convert free chlorine to chloride.

Radioactive samples require other, special techniques, some of which are discussed in ref. 11 (see also Radioactive tracers).

Equipment. Manual sampling is performed with one-liter, wide-mouthed bottles so that sampling can be performed quickly. A long-handled, wide-mouthed scoop is often used for less accessible sampling points. Weighted bottles or specially designed samplers that open at any required depth under water are also used. Hand-operated pumps are used for less accessible locations. A wide range of automatic samplers is commercially available (14–15). No single sampler exists that is universally applicable with equal efficacy. Over 40 manufacturers supply automatic liquid-sampling equipment, and comparative data show that there are marked differences in results obtained with different types of equipment. The sampling method is more site-dependent than any design attribute (14).

When the flow is nearly constant, a nonproportional sampler can be used. The sample is simply drawn from the waste stream at a continuous flow rate. The sampling lines should be as short as possible and free of sharp bends, which can lead to particle separation by settling. Proportional samplers are designed to collect either definite volumes at irregular time intervals or variable volumes at equally spaced time intervals. Both types depend upon the flow rate. Examples of some of these devices are the vacuum and chain-driven wastewater samplers. Other types, which have cups mounted on motor-driven wheels, vacuum suction samplers, and peristaltic pump samplers, are described in refs. 14 and 15.

Such samplers must be designed and constructed to withstand the chemical composition extremes present in the individual discharges. Corrosion-resistant fabrication must be used for equipment in contact with many chemical-industry discharges (see Corrosion and corrosion inhibitors).

Solids

Solids occur in several forms in the chemical industry. Raw materials from natural deposits are compacted in the ground, and sampling is performed during the exploration stages. This type of material is typified by minerals and fossil fuels. Before use, such materials must be mixed, crushed, ground into particulate form, cleaned, and stored in a heap on the ground or in a silo, bin, or hopper. Products in particulate form are usually stored in drums, railcars, ships, barges, cans, bags, boxes, etc. During manufacture, they are transported by conveyors, pipes, and chutes and are packaged with the use of free-flowing streams, pneumatic conveyors, spouts, and gravity chutes. Sampling may be required before, during, or immediately after any one of these operations, and different methods are used for most of them.

Natural Deposits. Natural deposits, eg, minerals and fossil fuels, are located by drilling operations. An auger, eg, a screw or worm, is turned into the earth and pulled out, and material is scraped from the auger for analysis. Alternatively, samples can be taken by hollow-core drills which, when withdrawn, enclose a core of earth that is representative of the strata through which the drill has passed. Such core samples are used in geological surveys for fossil fuels. As the drill drives deeper into the strata, each core is extracted and placed in a shallow box and coded so that a complete cross-section of the geological strata can be reconstructed. From this, the relative thicknesses of coal and mineral seams can be directly measured.

Stored Nonflowing Materials. Nonflowing materials are comprised of very fine cohesive powders, sticky materials, moist materials, and fibrous solids. Such materials are stored in boxes, bags, bales, and similar small containers; these are usually sampled by a manual or power-driven thief. For this purpose, a split-tube thief, which is a tube with a slot running its entire length with a sharp cutting edge at the lower end of the tube, is used. The thief is inserted in the center of the container and is rotated to cut out a core of material. The thief is withdrawn and the material is scraped from it for analysis.

Stored Free-Flowing Materials. Free-flowing particulate solids should be sampled in accordance with the three rules of sampling. Natural deposits obviously cannot be sampled this way, and nonflowing materials do not exhibit segregation; therefore, they do not require as much care. Batch nonuniformities could exist in nonflowing materials, but the requirement that they flow to meet the first sampling rule is impossible because of the nature of their properties. Free-flowing materials, however, can be made to meet all three rules (see Flowing Product Streams). For free-flowing materials stored in small hoppers, drums, cans, boxes, and bags, static sampling by the use of thief samplers is quite common. The typical sampler consists of two tubes, one fitting snugly inside the other. The tip of the outer tube is sharply pointed, and holes are cut in both tubes to mate at a specific angle of rotation. The outer tube is rotated until the holes are closed, and the thief is firmly inserted into the powder bed. At the desired depth, the inner tube is rotated until the holes mate and the thief is opened. Powder flows into the inner tube cavities, after which the inner tube is again rotated to the closed position. The thief is withdrawn and reopened to ensure that powder flows out into sampling containers. Inner tubes are compartmentalized to ensure that depth profiling can be conducted. Alternatively, a composite sample can be formed. Such thief tubes are used extensively to examine corn and associated products. The thief sampler is a deluxe version of a scoop sampler. Samples can be extracted from the same containers with a shovel, scoop, or spatula, and depth profiles can be taken by scoop sampling while the container is being systematically emptied. For larger storage bins, automatic auger samplers are used to sample free-flowing materials. A solenoid-controlled air cylinder opens and closes an aperture in the sampling tube, and the sample is drawn to the discharge point by a motor-driven auger.

In industry, heaps are scoop-sampled by means of thief tubes and shovels, but small heaps are more often sampled by coning and quartering (see Fig. 15). The heap is first flattened slightly at the top and is then separated into four equal segments with a sharp-edged board or shovel. The four segments are pulled apart so that the heap is broken into four separate piles. Two opposite quadrants are then recombined into a new heap when shovels of particulate are added onto a new heap apex. The remaining two quadrants are discarded. The process is then continuously repeated until a quantity of sample equivalent to the analytical needs remains. This method is based on the assumption that the heap is perfectly symmetrical with respect to size distribution and segregation and that each recombined heap meets the same specifications. However, in practice, heaps are not always formed by the flow of material directly on the apex. A worker shoveling material into a heap may stand a couple of meters away from it. Consequently, heaps are formed nonuniformly and, since segregation is influenced by the angle of the sloping heap, segregation bias can result. Careful coning and quartering of this heap biases the sample.

Flowing Streams.   All free-flowing particle systems are transported at some time in their manufacture as flowing streams. Hoppers are emptied by screw or belt conveyors. Solids are transferred to bagging operations by pneumatic conveyors, and most solids pass through transfer points in gravity-flow pipes and chutes. Even small samples in boxes, bags, cans, bottles, etc, can be made to flow by emptying them into volumetric feeders, and sampling is carried out on the resultant moving stream.

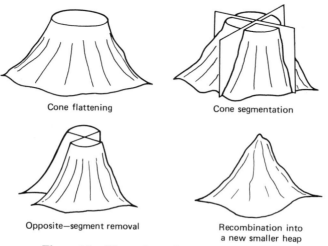

Cone flattening   Cone segmentation

Opposite—segment removal   Recombination into
                                  a new smaller heap

Figure 15.   The coning and quartering method.

The sampling from pneumatic conveyors parallels gas sampling, with the exception that solids loadings can be as high as 50 kilograms of solids per kilogram of gas. Commercially available samplers extract particles directly from a transport line. These samples form a fixed position, since they are mounted directly on the pneumatic conveyor pipe. In a device manufactured by the Quality Control Equipment Company (QCEC), frequent samples are extracted from the product stream by the projection of a sample tube into the flow (9). The particles impact on the tube and fill the open cavity. The tube is then withdrawn, and an internal screw discharges the collected material. In another model, the sample is extracted by compressed air. Gustafson, Inc. markets a similar device (16). The product is sampled when a solenoid-controlled air cylinder moves a snorkel in and out of the product stream. Pressure between the conveying line and the sample container is equalized after sampling to allow the discharge of the sample by gravity.

QCEC and Gustafson, Inc. also market slide-gate samplers for screw or drag conveyors. With these conveyors, sampling probes cannot be inserted into the lines; therefore, sliding gates are positioned on the bottom of drag housings or on the bottom or sides of screw conveyors. In the QCEC version, an air cylinder opens and closes the discharge gate, permitting powder to fall through the discharge tube into a sample container.

The belt-conveyor sampler manufactured by the Bristol Engineering Company has only one moving part in contact with the material (17). A rotary arm scoop passes across the belt to sample the material and moves in the same direction but slightly faster than the belt. An electric brake stops the arm near the top of the arc, and discharges the sample into a hopper at the side of the conveyor. In another method of belt-conveyor sampling, a cross-cutter sampler is placed at the discharge or transfer point of the belt. Conveyor belts, unlike pneumatic conveyors, often carry large-diameter particulates, eg, 5–15 cm. Consequently, such cutters must be adjustable in width; the minimum opening should be 2.5–3 times wider than the largest particle. The Denver Equipment Company manufactures the largest selection of cutter samplers, one of which, the Type-C cutter, is shown in Figure 16. The cutter blades enter the flow stream perpendicular to the particle trajectory, and the cutter discharge chute is adequately sloped to permit free flow into a flexible hose. Various designs are available for high and low tonnages. These are available with abrasion-resistant plates and, to prevent contamination, baffle plates attached to the cutter body (10).

With large tonnages, samples taken from conveyors can represent large quantities of material. Often, a sample cutter is used as a primary sampler, and the extracted sample is further cut into a convenient quantity by a secondary sampling device. Most primary cutters extract samples and discharge them by flexible hoses, pipes, gravity chutes, or other, similar, enclosed discharges. Secondary samplers may also include cutters or, for finer industrial powders, slotted sampling tubes, depending on the size of the particulate.

Secondary enclosed cutter samplers similar to the ones just described are marketed by the Denver Equipment Company, QCEC, and Gustafson, Inc. for tube, pipe, and chute use. An alternative design is the radial cutter or Vezin sampler shown in Figure 16. These samplers vary in size from a 15-cm laboratory size to a 152-cm commercial unit (10). For free-flowing powdered solids, rotating-tube samplers are used. These designs fit permanently in a gravity chute, pipe, or duct and extract samples continuously by gravity or by auger. The Gustafson Model D contains an auger sampler

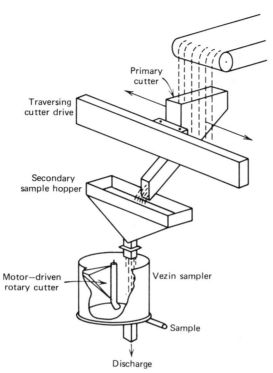

Figure 16. Schematic of a primary and secondary sampling system based on Denver Equipment Company's type C and Vezin samplers.

in a sampling tube. The tube rotates at 10 rpm and the auger at 72 rpm. The machine permits continuous extraction of a sample, which fills the tube when the opening faces the product flow.

Sampling Efficiency

Sampling of bulk solids from grinding circuits or chemical plants represents tons of material per day. A primary sampler generally removes 10–100 kg as a gross sample, which is further subdivided into 1–10-kg quantities by a secondary device. This secondary or laboratory sample is submitted for analysis. The sampling might be analyzed in its present form or it might be crushed prior to examination, depending on the type of analysis needed. Samples are required from this analytical sample in gram and milligram quantities. Because of the convenient size and reasonable quantities of test material, more has been reported on the efficiency of these devices than on commercial units (18–19). However, the findings are important to sampling design on a small and large scale (20).

In one study, sugar and sand mixtures were used and several techniques were compared, ie, coning and quartering, scoop sampling, and chute, table, and rotary riffling (19). A chute riffler consists of a hopper subdivided into several compartments that are combined to discharge equally in two directions. The sample is emptied into the riffler and is collected in two bins, one on each side of the device. A sample from one side is poured through repeatedly until the quantity for laboratory purposes remains in one bin. A table riffler consists of a flat, sloping surface upon which are triangular dividers that continuously subdivide a powder as it flows down over the surface. Samples are collected in cups at the bottom edge of the table, where the powder dischargers. The rotary riffler, or spinning riffler as described in ref. 15, consists of a hopper and vibratory feeder, from which powder is made to flow in a constant stream. The powder falls onto a rotating circular tray subdivided into various compartments. These compartments are sometimes small hoppers with discharge valves, as in large laboratory units, or boxes or test tubes, as in small laboratory units. According to one study, there should be 35 presentations of each sample compartment to the flow stream for accurate sampling, but at a speed of rotation low enough to prevent losses of fine particles resulting from cyclonic air flow (19). The results of the study described in ref. 19 are shown in Table 5. Rotary sample dividing is by far the best analytical sampling method to use, as it extracts samples from a moving stream, extracts samples from the full width of the stream, extracts 35 small samples frequently from the whole stream, and all of the sample passes through the device. In fact, it simply follows all the rules for good sampling discussed at the beginning of this section. Figure 17 shows the Ladal sampler (19) and Figure 18 shows a microsampler manufactured by Quan-

Table 5.   Reliability of Sampling Methods[a]

Method	Standard deviation, %	Maximum sampling error (estd), %
cone and quartering	5.76	19.2
scope sampling	6.31	21.0
table sampling	2.11	7.0
chute riffling	1.10	3.7
rotary riffling	0.27	0.9
random variation	0.09	0.3

[a] Ref. 19.

Figure 17.   The Ladal spinning riffler.

Figure 18.   The Quantachrome rotary microsplitter.

tachrome, Syosset, N.Y., based on similar principles. Microsamples for microscopy can also be extracted by the use of one, narrow, raised sample slit on a rotating disk plate. Other analytical rotary sample dividers are marketed by the Alfred Fritsch Company, Geos Corp., and Leeds and Northrup.

BIBLIOGRAPHY

"Sampling" in *ECT* 1st ed., Vol. 12, pp. 84–91, by H. W. Eckweiler, U.S. Bureau of Customs, and pp. 91–95, by C. L. Dunn, Hercules, Inc. "Sampling" in *ECT* 2nd ed., Vol. 17, pp. 744–762, by Charles A. Bicking, The Carborundum Company.

1. *ASTM Standards Index*, American Society for Testing and Materials, Philadelphia, Pa., 1981.
2. T. Allen, *Particle Size Measurement*, 2nd ed., Chapman & Hall, Ltd., London, UK, 1975.
3. J. H. Rushton and J. G. Hillestad, *Paper presented at 24th Midyear Meeting of the American Petroleum Institute*, Preprint No. 52, May 1964.
4. ASTM E 122-5 Standard Designation, American Society for Testing and Materials, Philadelphia, Pa., 1958.
5. *U.S. EPA Regulations on Standards of Performance for New Stationary Sources*, 40 CFR 60, Appendix A, Reference Methods, Washington, D.C., 1980.
6. "Anderson Stack Sampler," manufactured by 2000 Inc., Salt Lake City, Utah.
7. J. A. Brink, *Ind. Eng. Chem.* **50,** 645 (1958).
8. B. G. Lovelace, *Chem. Eng. Proc.* **51,** (Nov. 1979).
9. "E Sampler," manufactured by Quality Control Equipment Co., Des Moines, Iowa.

10. D. K. Fields, *Soc. Mining Eng.*, 1486 (Oct. 1979).
11. *Handbook for Monitoring Industrial Wastewater*, U.S. EPA, Washington, D.C., Aug. 1973.
12. *Water Pollut. Control Res. Ser.*, 12020 EJD (Mar. 1971).
13. *Methods for Chemical Analysis of Water and Wastes*, 16020 EPA, National Environmental Research Center, Washington, D.C., July 1971.
14. *An Assessment of Automatic Sewer Flow Samplers*, U.S. Department of Commerce, PB-25, 987, Washington, D.C., 1975.
15. *A Survey of Commercially Available Automatic Wastewater Samplers*, U.S. EPA, 600/4-76-051, Washington, D.C., 1976.
16. "RX Sampler," manufactured by Gustafson, Inc., Hopkins, Minn.
17. *DS3 Sampler Bulletin 101*, Bistrol Engineering Co., Yorkville, Ill.
18. B. H. Kaye, PhD Thesis, London University, UK, 1961.
19. T. Allen and A. A. Khan, *Chem. Eng.* **238,** CE 108 (1970).
20. G. J. Hahn, *Chemtech* **12,** 286 (1982).

REG DAVIES
E. I. du Pont de Nemours & Co., Inc.

SCANDIUM. See Rare-earth elements.

SCREEN PRINTING. See Dyes, application and evaluation; Printing processes; Reprography.

SCREENS, SCREENING. See Size classification; Size measurement of particles.

SCRUBBERS. See Absorption; Gas cleaning.

SEALANTS

The Bureau of Census of the U.S. Department of Commerce includes sealants with adhesives (qv) in its SIC 2891 classification. According to this source, the adhesive and sealants industry employed almost 16,000 workers in 1977, and the total value of shipments was ca 2×10^9. The annual sales of caulking compounds and sealants was ca 400×10^6. The annual consumption of adhesives and sealants in the United States is 4×10^6 metric tons, of which ca 2×10^6 is sealants (1).

Although there is some overlap in application, adhesives and sealants usually have different functions. The former are selected for their ability to bind two materials, whereas the latter are selected as load-bearing elastic jointing materials that exclude dust, dirt, moisture, and chemicals that contain a liquid or gas. Sealants are also used to reduce noise and vibrations and to insulate or serve as space fillers (see Insulation, acoustic; Chemical grouts).

The term caulking compound is derived from the verb to caulk, which describes the forcing of tarred oakum between the planks of a ship. The original caulking compounds were nonload-bearing materials based on mixtures of bitumens and asbestos (qv). The sealant is a more modern term and describes compositions based on synthetic resins as well as asphaltic and oil-based caulking compounds (2–4). The formulations for several hundred commercially available adhesives and sealants have been described (5).

These composites require little if any energy for conversion to solids and usually conserve energy when applied. The principal resins used as sealants are polysulfides, silicones, polyurethanes, acrylics, neoprene [31727-55-6], butyl rubber [9006-49-9], and sulfochlorinated polyethylene [68037-39-8] (6).

Polysulfides

Poly(ethylene sulfide) [24936-67-2] was one of the first synthetic elastomers and was produced by the condensation of α,ω-dichloroalkanes with sodium polysulfide (7) (see Polymers containing sulfur). The principal polysulfide is prepared from bis-(chloroethyl) formal. The utility of these solvent-resistant solid elastomers was increased when they were reduced to lower molecular weight liquid mercaptans by heating with sodium sulfite and sodium hydrosulfite. The liquid products can be oxidized by metallic oxides, eg, lead dioxide, or by epoxy resins *in situ* at ambient temperatures to produce flexible caulking compounds (8). It is customary to add fillers (qv), eg, carbon black (qv); calcium carbonate or clay; plasticizers (qv), eg, dibutyl phthalate; retarders, eg, stearic acid; accelerators, eg, amino compounds; and adhesion promotors, eg, epoxy resins (qv) (9) (see Rubber chemicals). Polysulfide sealants are commercially designated as LP-2, LP-32, LP-3, LP-33, etc. These liquid prepolymers have different average molecular weights and different cross-link densities.

A typical formulation for a two-package polysulfide sealant in parts by weight is as follows:

Composition	Package 1	Package 2
polysulfide	100	
liquid epoxy resin		90
solid epoxy resin		60
titanium dioxide	100	
carbon black	5	
urea–formaldehyde resin		20
methyl ethyl ketone	45	
plasticizer	15	

These materials, like most two-package sealants, caulking compounds, and cements, are thoroughly mixed before application by trowel, knife, or caulking gun. The prepolymer hardens in place at ambient temperatures to a semielastic solid.

Polysulfide sealants are the most widely used sealants for airplane fuel tanks, curtain-wall construction, glazing, marine decks, binders for solid-rocket propellants, and sealants for joints in airport runways, highways, and canals (10). The principal advantages of polysulfide sealants are ease of application, good adhesion, good resistance to weathering and solvents, negligible shrinkage, and good moisture-vapor transmission (MVT) (11).

Silicones

Silicone sealants based on polydimethylsiloxanes were developed in the early 1940s, and the two-package, room-temperature-vulcanizing (RTV), elastomeric silicone sealants with silanol ends were introduced in the 1950s (12). One-package silicone sealants, which cure by exposure to moisture in the air, were introduced in the 1960s (13). The first one-package silicone sealants were characterized by a tacky surface. This deficiency was eliminated by the addition of metal salts of carboxylic acids. These one-component RTV systems account for the main share of the silicone-sealant market. These products are used in adhesives, encapsulation, impregnation, moldmaking, and sealants.

Silicone sealants are based on mixtures of fillers, eg, silica, silicone polymers, cross-linking components, and catalysts. The polymer has a siloxane backbone, ie, —Si—O—Si, with alkyl and alkoxy or acetoxy pendant groups. The latter are readily hydrolyzed to silanol groups (SiOH), which form larger chains by condensation and loss of alcohol or acetic acid.

Trifunctional silanes, eg, trimethoxylmethylsilane or triacetoxymethylsilane, are used as cross-linking agents. Metal carboxylates such as stannous octanoate and titanates, eg, tetraisopropyl titanate, are used as catalysts (14).

The components of the two-package silicone sealants or caulking compositions are mixed thoroughly just before use, and the mixtures are usually applied with a caulking gun or knife. The one-component systems are cured by exposure in moist air. The cured products are used as flexible space fillers or sealants.

Silicone sealants that were introduced in 1967 have low shrinkage characteristics and can be applied and used over a wide temperature range (−40–65°C). The fluorosilicones (polyfluoroalkylsiloxanes), which retain their flexibility at even lower temperatures (−50°C), are used as sealants for aircraft, automobiles, and light bulbs as well as for sealing oil-well heads in Alaskan oil fields (15).

Silicone sealants with cyanoalkyl pendant groups have superior solvent resistance and are used as oil rings on fuel lines of jet aircraft (16–17). Room-temperature-vulcanizing silicones have been used for formed-in-place gaskets and potting compounds (18–19) (see Silicon compounds, silicones).

Butyl Rubber

Polyisobutylene [9003-27-4] produced by the low temperature cationic polymerization of isobutylene has been used to a small extent as a one-package sealant since the 1930s. However, butyl rubber, which is a copolymer of isobutylene and isoprene, is more readily available and can be readily cross-linked by agents, eg, p-quinone dioxime and an oxidizer (see Elastomers, synthetic).

Commercial butyl sealant compositions usually contain fillers, such as carbon black and zinc oxide, or silica; tackifiers, such as pentaerythritol esters of rosin; and solvents, eg, cyclohexane. A typical butyl rubber caulking compound has the following composition:

Composition	Parts by weight
butyl rubber	175
mineral spirits	270
petroleum resins	34
pentaerythritol esters of rosin	8
bentone clay derivative	23
finely divided silica	364
fiber	91
titanium dioxide	45

Both polyisobutylene and butyl rubber sealants harden by evaporation of the solvent. Butyl rubber compositions may be used as hot-melt sealants. These compositions may be admixed with ethyl acrylate copolymers and used as a gunnable hot melt for sealing dual-pane windows. Butyl rubber sealants are used for joint-and-panel sealing as well as for glazing. Butyl rubber is also available as a tape sealant, which is pressed into the cavities to be filled by sealants.

Polyurethanes

Polyurethanes, which were developed in the 1940s for use as fibers, elastomers, plastic foams, and adhesives, are also useful as sealants. These sealants are usually two-package systems, but one-package systems are also available. The two-package systems are based on a diisocyanate, eg, tolyl diisocyanate (TDI), and a poly(ethylene glycol). These systems also contain antioxidants, pigments, fillers, alkylenetriol cross-linking agents, and organometallic catalysts, eg, dibutyltin dilaurate (see Urethane polymers).

These two-component systems are used on a large scale in reaction-injection molding (RIM), in which the components are allowed to react in a large mold (20). Castable polyurethane sealants are characterized by a wide range of hardness and flexibility. They have good sag resistance, good thixotropy, and good adhesion to the substrate (21–22). These sealant compositions are also used as dental fillings (23) (see Dental materials). The surfaces to be sealed must be dry to prevent foam formation from the reaction of water and the diisocyanates. Polyurethane sealants also are being used in the assembly of several makes of automobiles.

These compositions, like most sealants, are not energy-intensive. Their hardness and flexibility may be controlled by the selection of appropriate diols and fillers. The life expectancy of polyurethane sealants is ca 10 yr, in contrast to a life expectancy of 15 yr for polysulfides and butyl sealants (24). Polyurethane sealants have been used successfully in lip-type launch seals in submarine missile systems (25). Over 75,000 metric tons of polyurethane-type sealants are used annually.

Acrylics

Emulsions and solutions of poly(methyl methacrylate) or its copolymers have been used as caulking materials and sealants. The solvent-type acrylics are similar to high solids (90 wt %), acrylic coatings. A typical formula for a water-based acrylic sealant is as follows:

Composition	Parts by weight
acrylic latex (50 wt % solids)	41.9
wetting agent	1.7
plasticizer	9.5
ethylene glycol	32
calcined China clay	42.7
titanium dioxide	1.7
mineral spirits	0.3

Two-package systems consist of acrylic monomers, fillers, and initiators. They polymerize at RT and are also used as polymer concrete (26). These compositions have been used for repairing potholes and cracks in concrete highways and for bone cement (27). One-package acrylic sealants harden by evaporation of water and solvent. The two-package systems are essentially solvent-free but precautions must be taken to avoid inhalation of acrylic monomers. Automated equipment is available for mixing the components of both systems (see Acrylic ester polymers).

Polychloroprene

Polychloroprene (neoprene) was commercialized by E. I. du Pont in the 1930s. The preferred sealant composition is a two-package system consisting of the polymers in one container and accelerators, eg, tertiary amine, in the other (28). A typical polychloroprene sealant has the following composition:

Composition	Parts by weight
Neoprene AG	110
magnesia	4
zinc oxide	5
heat-reactive resin	45
methyl ethyl ketone, naphtha, or toluene	600

The components are mixed in an intensive mixer. Some two-package systems contain litharge (PbO) and cure at ambient temperatures. The one-package systems set by evaporation of solvent and cure at elevated temperatures. Appropriate precautions must be taken to avoid inhalation of the flammable solvent used in the one-package system (see Elastomers, synthetic).

Chlorosulfonated Polyethylene

Hypalon is the trade name of the product obtained by chlorination and sulfonation of polyethylene, and it is used as a sealant (28). It is customary to blend at least two polymers with different degrees of substitution (DS) with a plasticizer, eg, chlorinated naphthalene; fillers, eg, carbon black; and curing agents, eg, litharge. Compounded mixtures of sulfochlorinated polyethylene and neoprene are also used as sealants for sealing masonry. A typical formulation for a sealant of this type is as follows:

Composition	Parts by weight
blend of sulfochlorinated polyethylene	17.5
chlorinated paraffins	17.5
asbestos and other fillers	20.0
titanium dioxide	14.0
curing agents, eg, litharge	7.5
dibutyl phthalate plasticizer	19.0
solvents, eg, isopropyl alcohol, etc	4.5

These compositions are cured by cross-linking at ambient temperatures to produce flexible sealants with good adhesion (see Elastomers, synthetic).

Bitumens

Bitumens or asphaltic materials have been used as sealants for many centuries (see Asphalt). Bituminous hot-melt compositions usually contain scrap rubber or neoprene, which acts as a flexibilizing agent. These dark-colored sealants are used for joints in highways and buildings. Their tendency to cold-flow is reduced by the incorporation of epoxy resins. Sealants produced from asphalt emulsions have been used to prevent the escape of radon gas from storage bins of uranium-mill tailings.

Latex Caulking Compositions

Filled polymeric emulsions, which are related to latex paints, have been used as tub caulks and spackling compounds. A typical caulking composition consists of 45 wt % poly(vinyl acetate), filler, plasticizer, and water. Since these systems set by evaporation, their shrinkage is greater than hot-melt or prepolymer sealants (see Paint; Latex technology).

Oil-Based Caulks and Sealants

Oil-based caulks or putty have been used for centuries as glazing sealants. These 100% solids compositions are much more rigid than other sealants, but the rigidity is often overcome by the addition of elastomers, eg, neoprene. A typical formula for an oil-based caulking composition is as follows:

Composition	Parts by weight
bodied vegetable oil	100
kettle-bodied vegetable oil	70
polyisobutylene	100
cobalt carboxylate	0.20
calcium carbonate	483
asbestos (long fiber)	23
asbestos (short fiber)	27
titanium dioxide	17

Hot-Melt Sealants

In addition to butyl rubber and bitumens, several other polymers are used as hot-melt sealants. Among these are the copolymers of ethylene and vinyl acetate (EVA), atactic polypropylene, and mixtures of paraffin wax and polyolefins (see Olefin polymers). Over 250,000 t of hot-melt sealants is used annually. Silica- or carbon-black-filled sulfur cements are also used as hot-melt sealants. Because they must be heated before use, hot-melt sealants are more energy-intensive than many room-temperature-curing sealants.

Poly(vinyl Chloride) Plastisols

Poly(vinyl chloride) (PVC) plastisols have been used widely as gap fillers and sealants in automobiles. These plastisols are dispersions of finely divided PVC in liquid plasticizers, eg, dioctyl phthalate (DOP). These liquid sealants set to form flexible compositions when heated, ie, they are activated at 125–200°C. Filled plastisols or plastigels may be used like putty. Plastigels have less tendency to leak through large crevices. However, the liquid plasticizer may bleed from either of these PVC sealants and cause softening of painted surfaces (see Vinyl polymers).

Polyesters

Fibrous-glass-reinforced unsaturated polyesters have been used as structural materials since the early 1940s. Silica- and calcium carbonate-filled polyester grouting compositions called cultured marble, plastic concrete, and Vitroplast were developed in the early 1950s (29). These strong, lightweight plastic cements have been used for making sewer pipe (30). Plastic concrete, trade name Polysil, is used for making utility poles, bathroom fixtures, and manholes (31). Polyester compositions have been used to fill crevices and for automobile body repair.

A typical two-package, unsaturated, polyester sealant consists of a filler, eg, silica, calcium carbonate, or alumina trihydrate (ATH), and an initiator, eg, benzoyl peroxide (BPO). The liquid compound consists of an ethylene glycol maleate dissolved in styrene monomer and a tertiary amine, eg, N,N-dimethylaniline. Since the styrene monomer is flammable and has a relatively high vapor pressure, adequate ventilation must be provided to protect the applicators.

Epoxy Resins

Epoxy resins (qv) were first synthesized by the reactions of bisphenol A and epichlorohydrin in the early 1940s by Castans in Switzerland (32). Most of the annual production of 400,000 t of epoxy resin is used as *in situ* polymerized plastics, which are used as adhesives (6.5%), flooring (6.0%), protective coatings (45%), and laminates (16.6%) (33) (see also Embedding).

It is customary to cure or cross-link epoxy prepolymers with polyamines, eg, diethylenetriamine. Epoxy resins have been used to seal bridge surfaces and to repair concrete structures. Filled epoxy resins have been used as jointing materials for brick and as monolithic heavy-duty coatings. These compositions are resistant to many nonoxidizing acids, salts, and alkalines (34–35).

A typical formulation for a corrosion-resistant epoxy resin sealant is as follows:

Composition	Parts by weight
liquid epoxy resin prepolymer	100
liquid poly(ethylene sulfide)	50
dimethylaminopropylamine	10

Phenolic Resins

In situ cured silica-filled phenolic resin composites were introduced under the trade name Asplit in Germany in the 1930s (36) (see Phenolic resins). The two-package system consists of a silica or carbon filler containing an acid setting agent, eg, p-toluenesulfonic acid (PTA), and an A-stage resole phenolic resin. The filler and liquid resin must be thoroughly mixed and the mixture must be applied to bricks or tile while it is a fluid. Phenolic grouting compositions and other resinous sealants that are cured by strong acids cannot be used as grouts for concrete or other alkaline substrates.

Urea Resins

The disadvantage of the dark color associated with the phenolic resin cement was overcome by the use of urea–formaldehyde prepolymers. Kaolin-filled, *in situ* cast urea resinous composites had been used for sealing cracks in underground rocks and for the fabrication of irrigation pipes. The acid setting agents used with urea resins are usually inorganic acids, which are less apt to cause dermatitis than the p-toluenesulfonic acid (PTA) used with phenolic resin compositions. However, these resin compositions can be used for filling voids if the surfaces of the substrate are coated with a protective coating (see Amino resins).

Furan Resin

The lack of resistance of phenolic resin cements to alkali was overcome by the development of comparable cements based on furan resins (37). These two-package resinous cements have been used extensively as mortars, grouts, and setting beds for brick and tile (38). Most of these high solids composites are essentially solvent-free. These jet-black materials are chemurgic products (see Chemurgy). The furfuryl alcohol used to produce furan resins is obtained by the reduction of furfural. Furfural is obtained by acid-degradation of pentosans, eg, those present in corn cobs and oat hulls (see Furan derivatives).

Economic Aspects

The use of sealants is growing at an annual rate of 6%, and the total sales should exceed 10^9 in 1985 (1). The annual U.S. sales for various uses are shown in Table 1.

There are over 350 compounders and producers of sealants in the United States. Over 60% of the total value of sealants used is supplied by 15 producers. General Electric, Dow Corning, and Thiokol are the only vertically integrated producers of sealants; General Electric, DAP, and Dow Corning are the leading sealant suppliers.

The modern automobile contains 40 kg of adhesives, sealants, and sound deadeners. Poly(vinyl chloride) and asphalt have been the principal products used for these applications, but asphalt is being replaced by more complex polymers. Considerable quantities of sound deadeners are being used in the floor and rear-seat areas of diesel automobiles. The emphasis on the reduction of energy costs has resulted in an increase in the use of insulated glass in residential construction. Polysulfides account for 90% of the sealants used in this application, but these products are being displaced by butyl rubber hot melts.

Health and Safety Factors

Although the death rate of sealant workers is no greater than that of other workers, appropriate precautions must be taken to assure the health and safety of those working with the production or application of sealants. Care should be taken to prevent exposure to solvents, plasticizers, and fillers used with polysulfide sealants. The solvents should not be allowed to evaporate in a closed area, and potable water or foods should

Table 1.   Sales Volume for Sealants, 10^6

Use	1979	1985
building construction	135	283
do-it-yourself market	95	201
insulating glass	56	122
glazing	42	88
membranes	37	84

not contact the set sealants for long periods. Adequate precaution should be taken during production and application if asbestiform fillers are used in the formulation. Polysulfide sealants produce hydrogen sulfide when heated and, therefore, should not be used at elevated temperatures.

Silicone sealants are usually considered to be nontoxic. However, the acetoxysilanes produce acetic acid in the setting reaction, and precaution must be taken to counteract any deleterious effects of this weak acid.

Butyl rubber melt sealants are nontoxic, but room-temperature-setting butyl rubber sealants contain solvents that are volatile and flammable. Thus, provisions must be made for exhausting these solvents when the sealants are applied in enclosed areas.

The diisocyanates present in two-package polyurethane systems are toxic, and adequate ventilation must be provided. The cured polyurethanes are considered to be nontoxic unless heated above their decomposition temperature.

Sealants based on acrylic emulsions are nontoxic. Monomers and solvents in the two-package systems are toxic and flammable and, therefore, adequate precautions should be taken to remove the fumes from working areas. Additional precautions should be taken when acrylic sealant systems containing asbestos fillers are used.

Sealants based on chlorosulfonated polyethylene may contain chlorinated aromatic and phthalate plasticizers, asbestos fillers, volatile solvents, and lead curing agents. Appropriate precautions must be taken to avoid contact with the asbestos filler or plasticizers during application of these sealants. Solvents must be exhausted from the working area and foodstuffs should not be placed in contact with the cured sealant.

Hot-melt asphaltic compositions should not be allowed to come in contact with the skin. The solvent in bituminous mastics must be exhausted from the working area. Both the solvent and the set bitumens are flammable.

Since many oil-based caulks contain asbestos, appropriate precautions must be taken during production and application. Since most hot melts are flammable, adequate precaution must be taken to prevent them from igniting and from coming in contact with the skin of the applicator.

Initiators or catalysts used to cure polyester resins are unstable compounds that may explode by impact. Hence, initiators, eg, benzoyl peroxide, should be used in small quantities and preferably in a liquid medium. Styrene is said to be carcinogenic; it is also flammable. Therefore, adequate ventilation must be used to maintain concentrations of styrene monomer in air of less than 10 ppm.

Because many epoxy resin curing agents are vesicants, the skin of the applicator should be protected to prevent dermatitis. Since the fumes of these amines may cause asthma, precautions should be taken to prevent the inhalation of these fumes. These problems can be reduced by automation and good housekeeping.

Formaldehyde which evolves from phenolic and urea resins is said to be carcinogenic. The acid curing agent can cause dermatitis if it comes in contact with the skin. If the process is not automated, adequate precaution should be taken to prevent inhalation of formaldehyde fumes and to prevent contact between the applicator's skin and the acid curing agent. The acid setting agent in furan cements can also cause dermatitis if the applicator's skin is not protected by salves or clothing.

BIBLIOGRAPHY

1. *Sealants*, Skeist Laboratories, Inc., Livingston, N.J.
2. A. Damusis, *Sealants*, Reinhold Publishing Co., New York, 1967.
3. R. B. Seymour, ed., *Plastic Mortars, Sealants, and Caulking Compounds*, ACS Symposium Series 113, Washington, D.C., 1979.
4. H. H. Buchter, *Industrial Sealing Technology*, John Wiley & Sons, Inc., New York, 1979.
5. E. W. Fleck, *Adhesive and Sealant Compounds and Their Formulations*, Noyes Data Corp., Park Ridge, N.J., 1978.
6. S. S. Maharajan and N. D. Ghatge, *Paintindia* **29**(6), 37 (1979).
7. Brit. Pat. 302,220 (Dec. 13, 1927), J. C. Patrick and N. M. Mnookin (to Thiokol Corp.).
8. U.S. Pat. 2,466,963 (Apr. 12, 1949), J. C. Patrick and H. R. Ferguson (to Thiokol Corp.).
9. E. R. Bertozzi, *Rubber Chem. Technol.* **41**, 114 (1968).
10. U.S. Pat. 2,910,922 (Nov. 3, 1959), F. P. Horning (to Thiokol Corp.).
11. M. A. Schuman and A. D. Yazujian in ref. 3, Chapt. 11.
12. U.S. Pat. 2,571,039 (Oct. 9, 1951), J. F. Hyde (to Dow Corning, Inc.).
13. U.S. Pat. 3,105,061 (Sept. 24, 1963), L. B. Bruner (to Dow Corning, Inc.).
14. J. M. Klosowski and G. A. L. Gent in ref. 3, Chapt. 10.
15. O. R. Pierce and Y. K. Kim, *J. Elastoplast.* **3**, 82 (1971).
16. U.S. Pat. 3,513,508 (Sept. 29, 1970), G. K. Goldman and L. Morris (to Dow Corning, Inc.).
17. W. J. Ratelle, *Mach. Des.* **49**, 133 (1971).
18. T. J. Gair and W. W. Wadsworth, *SAE Tech. Pap. 720129* (1972).
19. A. Grey, *Elastomerics* **109**, 23 (1977).
20. W. M. Haines, *Elastomerics* **110**(9), 26 (1978).
21. G. Eagle, *Plast. Eng.* **36**(7), 29 (1980).
22. L. J. Lee, *Rubber Chem. Technol.* **53**, 542 (1980).
23. G. G. Stecher, *New Dental Materials*, Noyes Data Corp., Park Ridge, N.J., 1980.
24. R. B. Seymour in ref. 3, Chapt. 8.
25. J. P. Meier, G. F. Rudd, A. J. Molna, V. D. Jerson, and M. A. Mendelsohm, *Resins for Aerospace*, ACS Symposium Series 132, Washington, D.C., 1980, Chapts. 14–15.
26. R. Martino, *Mod. Plast.* **57**(3), 46 (1980).
27. S. W. Shalaby, *Polymer News* **5**, 176 (1979).
28. L. S. Bake in I. Skeist, ed., *Handbook of Adhesives*, R. E. Kreiger Publishing Co., Huntington, N.Y., 1973, Chapt. 20.
29. R. B. Seymour in ref. 3, Chapt. 5.
30. B. J. Schrock and G. Rangel, *Paper presented at 30th Anniv. Tech. Conf. Reinforced Plastics/Composites Institute SPI*, 1975.
31. G. Miller, *Plast. Des. Process.* **19**(1), 30 (1979).
32. I. Skeist, *Epoxy Resins*, Van Nostrand Reinhold Co., New York, 1958.
33. J. M. Sosa in ref. 3, Chapt. 3.
34. R. B. Seymour and R. H. Steiner, *Chem. Eng. Progr.* **49**, 220 (1953).
35. F. T. Watson, *Symposium on Engineering Applications of Epoxy Resins at 57th Annual Meeting of AICHE*, Boston, Mass., Dec. 6, 1964.
36. R. B. Seymour in ref. 3, Chapt. 1.
37. U.S. Pat. 2,226,049 (Dec. 26, 1944), C. R. Payne and R. B. Seymour (to Atlas Mineral and Chemical Products Co.).
38. R. H. Leitheiser, M. E. Londrigan, and C. A. Rude in ref. 3, Chapt. 2.

RAYMOND B. SEYMOUR
The University of Southern Mississippi

SEASONINGS.   See Flavors and spices.

SEAWEED COLLOIDS.   See Gums.

SEDIMENTATION

Sedimentation is defined as "the separation of a suspension into a supernatant clear fluid and a rather dense slurry containing a higher concentration of solid" (1). This definition is too broad since it does not specify the external field of acceleration that causes the separation: gravitational, centrifugal, magnetic, or electrostatic. There is also a possible ambiguity whether the suspension is gaseous or liquid. In this article, the subject is restricted to the most common definition of sedimentation, ie, the gravitational settling of solids in liquids (see also Centrifugal separation; Gravity concentration; Magnetic separation).

The uses of sedimentation in industry fall into the following categories: solid–liquid separation (2); solid–solid separation; particle-size measurement by sedimentation (3); and other operations such as mass transfer, washing, etc.

In solid–liquid separation, the solids are removed from the liquid either because the solids or the liquid are valuable or because they have to be separated before disposal. If the primary purpose is to produce the solids in a highly concentrated slurry, the process is called thickening. If the purpose is to clarify the liquid, the process is called clarification (see Denaturing, Supplement Volume). Usually, the feed concentration to a thickener is higher than that to a clarifier. Some types of equipment, if correctly designed and operated, can accomplish both clarification and thickening in one stage (see also Extraction, liquid–solid).

In solid–solid separation, the solids are separated into fractions according to size, density, shape, or other particle property (see Size reduction).

Sedimentation is also used for size separation, ie, classification of solids. It is one of the simplest ways to remove the coarse or dense solids from a feed suspension. Successive decantation in a batch system produces closely controlled size fractions of the product. Generally, however, particle classification by sedimentation does not give sharp separation (see Size measurement).

In particle-size measurement, gravity sedimentation at low solids concentrations (<0.5% by volume) is used to determine particle-size distributions of equivalent Stokes diameters in the range from 2 to 80 μm. Particle size is deduced from the height and time of fall using Stokes' law, whereas the corresponding fractions are measured gravimetrically, by light, or by x-rays. Some commercial instruments measure particles coarser than 80 μm by sedimentation when Stokes' law cannot be applied.

In addition to the above-mentioned applications, sedimentation is also used for other purposes. Relative motion of particles and liquid increases the mass-transfer coefficient. This motion is particularly used in solvent extraction in immiscible liquid–liquid systems (see Extraction, liquid–liquid extraction). An important commercial use of sedimentation is in continuous countercurrent washing, where a series of continuous thickeners is used in a countercurrent mode in conjunction with reslurrying to remove mother liquor or to wash soluble substances from the solids.

Most applications of sedimentation are, however, in straight solid–liquid separation, and the following account of principles and equipment is from that point of view.

Principles

Gravity Settling of a Single Particle.   If a particle moves relative to the fluid in which it is suspended, the force opposing the motion is known as the drag force. Knowledge of the magnitude of this force is essential if the particle motion is to be studied. Conventionally, the drag force F_D is expressed according to Newton:

$$F_D = C_D \cdot A \cdot \frac{\rho v^2}{2} \tag{1}$$

where v is the particle–fluid relative velocity, ρ is the fluid density, A is the area of the particle projected in the direction of the motion, and C_D is a coefficient of proportionality known as the drag coefficient. Newton assumed that the drag force is due to the inertia of the fluid and that C_D would then be constant.

Dimensional analysis shows that C_D is generally a function of the particle Reynolds number:

$$Re_p = \frac{v \cdot x \cdot \rho}{\mu} \tag{2}$$

where x is particle size; the form of the function depends on the regime of the flow. This relationship for rigid spherical particles is shown in Figure 1. At low Reynolds numbers, under laminar flow conditions when viscous forces prevail, C_D can be determined theoretically from Navier-Stokes equations and the solution is known as Stokes' law

$$F_D = 3 \pi \mu v x \tag{3}$$

This is an approximation that gives the best results for $Re_p \rightarrow 0$; the upper limit of its validity depends on the error that can be accepted. The usually quoted limit for the Stokes region of $Re_p = 0.2$ is based on the error of ca 2%.

Elimination of F_D between equations 1, 2, and 3 gives another form of Stokes' law

$$C_D = \frac{24}{Re_p} \qquad (Re_p < 0.2) \tag{4}$$

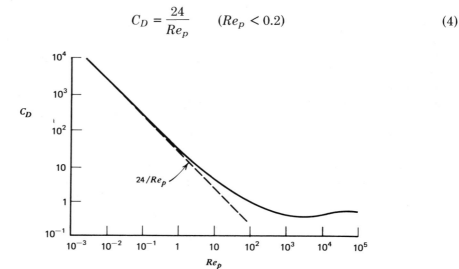

Figure 1.   Drag coefficient versus particle Reynolds number for spherical particles.

as shown in Figure 1 as a straight line. For Reynolds numbers >1000, the flow is fully turbulent with inertial forces prevailing, and C_D becomes constant and equal to 0.44 (the Newton region). The region in between $Re_p = 0.2$ and 1000 is known as the transition region and C_D is either described in a graph or by one or more empirical equations.

In solid–fluid separation, the fine particles are most difficult to separate, ie, Re_p is low, almost inevitably below 0.2, owing to low values of x and v; therefore, only the Stokes region has to be considered.

A single particle settling in a gravity field is subjected primarily to drag force, gravity force, and buoyancy, which have to be in equilibrium with the inertial force:

$$m \frac{dv}{dt} = (m \cdot g) - (m \cdot g) \frac{\rho}{\rho_s} - F_D \tag{5}$$

$$\text{inertial} \quad \text{gravity} \quad \text{buoyancy} \quad \text{drag}$$
$$\text{force} \qquad \text{force} \qquad \qquad \text{force}$$

(assuming positive downward forces) where m is particle mass, g is gravity acceleration, ρ_s is particle density, and t is time. Equation 5 can be solved, assuming Stokes' law:

$$v(t) = \frac{gx^2(\rho_s - \rho)}{18 \, \mu} \left[1 - \exp\left(-\frac{t \, 18 \, \mu}{x^2 \rho_s} \right) \right] \tag{6}$$

This relationship is exponential with respect to time t and with increasing time quickly approaches

$$v_g = \frac{gx^2(\rho_s - \rho)}{18 \, \mu} \tag{7}$$

This is known as the terminal settling velocity under gravity.

The terminal velocity in the case of fine particles is approached so fast that in practical engineering calculations the settling is taken as a constant-velocity motion and the acceleration period is neglected. Equation 7 can also be applied to nonspherical particles if the particle size x is the equivalent Stokes diameter as determined by sedimentation or elutriation methods of particle–size measurement.

Settling of Suspensions. As the concentration of the suspension increases, particles get closer together and interfere with each other. If the particles are not distributed uniformly, the overall effect is a net increase in settling velocity since the return flow caused by volume displacement predominates in particle-sparse regions. This is the now well-known effect of cluster formation which is significant only in nearly monosized suspensions. With most practical widely dispersed suspensions, clusters do not survive long enough to affect the settling behavior and, as the return flow is more uniformly distributed, the settling rate steadily declines with increasing concentration. This phenomenon is referred to as hindered settling and can be theoretically approached from three premises (4): as a Stokes'-law correction by introduction of a multiplying factor; by adopting effective fluid properties for the suspension different from those of the pure fluid; and by determination of bed expansion with a modified Carman-Kozeny equation. These three approaches yield essentially identical results:

$$\frac{v_p}{v_g} = \epsilon^2 f(\epsilon) \tag{8}$$

where v_p is the hindered settling velocity of a particle, v_g is the terminal settling velocity of a single particle as calculated from Stokes' law (eq. 7), ϵ is volume fraction of the fluid (voidage), and $f(\epsilon)$ is a voidage function, which for Newtonian fluids has different forms, depending on the theoretical approach adopted. The differences between the available expressions for $f(\epsilon)$ are not great and are frequently within the experimental accuracy. The most important forms are as follows: from the Carman-Kozeny equation (5)

$$f(\epsilon) = \frac{\epsilon}{10\,(1 - \epsilon)} \tag{9}$$

from Brinkman's theory (6), applied to Einstein's viscosity equation (7)

$$f(\epsilon) = \epsilon^{2.5} \tag{10}$$

and from the well-known Richardson and Zaki equation (8)

$$f(\epsilon) = \epsilon^{2.65} \tag{11}$$

For irregular or nonrigid particles, eg, flocs, the Einstein constant (2.5) and the Richardson and Zaki exponent (2.65) can be considerably larger than for spheres.

Strictly speaking, the above correlations apply only to the cases where flocculation is absent, such as with coarse mineral suspensions. Suspensions of fine particles, because of the very high specific surface of the particles, often flocculate and therefore show different behavior. With increasing concentration C of such suspensions, at a particular concentration C_1, an interface is observed that becomes sharper at $C > C_1$. The slurry is then said to be in the zone-settling region. The particles below the interface, if their size range is not more than 6:1, settle *en masse*, that is, all at the same velocity irrespective of their size. There are two possible reasons for this: either the flocs become similar in size and settle at the same velocity, or they are joined and fall as a web. Interestingly enough, the settling rates of the interface, and of the solids below it, of many practical suspensions can still be described by the Richardson and Zaki equation but the value of v_g has to be determined by extrapolation of the experimental log-linear plot for $\epsilon = 1$. The value of this intercept has in fact been used for indirect size measurement of the flocs. The slope of the plot determines the exponent.

The concentration C_1 at which zone-settling is first observed depends very much on the material and its state of flocculation and no guidance can be given. Addition of flocculation or dispersing agents drastically changes this concentration and only experimental evaluation can yield its value (see Flocculating agents; Dispersants). If the concentration is increased still higher, a point is reached when the flocs become significantly supported mechanically from underneath as well as hydraulically and the suspension is then known to be in compression or compression settling. The solids in compression continue to consolidate. The consolidation rate depends not only on the concentration but also on the structure of the solids, which in turn depends on the pressure and flow conditions. This is a very complex problem closely related to cake filtration and expression (see Filtration). At intermediate concentrations between those of zone settling and fully established uniform compression, a phenomenon is sometimes observed called channeling which particularly occurs in slowly raked large-scale thickeners. Under those conditions, a coarser structure of pores becomes interconnected in the form of channels.

Most authors who studied the consolidation process of solids in compression used the basic model of porous medium with point contacts which yields a general equation of the mass-and-momentum balances. This must be supplemented by a model describing filtration and deformation properties. Probably the best model to-date uses two parameters to define characteristic behavior of suspensions (9). This model can be potentially applied to sedimentation, thickening, cake filtration, and expression.

Coagulation and Flocculation. Both these classical pretreatment methods are used to increase the effective particle size, thereby improving sedimentation settling rates. Although the two terms, coagulation and flocculation, are often used interchangeably, coagulation is sometimes defined as agglomeration of the primary particles into particles up to 1 mm in diameter. Flocculation on the other hand, not only agglomerates particles but also interconnects them by means of long-chain molecules of the flocculating agent into giant loose flocs up to 1 cm in size. The term flocculation is used here to include coagulation as defined. Chemical agents create favorable conditions for flocculation by neutralization of surface charges and thus reduce interparticle repulsion. Mineral coagulants are in the form of electrolytes, such as alums or lime, whereas flocculation agents (qv) are mostly synthetic polyelectrolytes of high molecular weight. Development of the latter group in the past decade has resulted in a remarkable improvement in sedimentation equipment. Such agents are relatively expensive and the correct dosage has to be carefully optimized. Overdosage is not only uneconomic but may inhibit the flocculation process and cause operating problems. As surface charges are also affected by pH, its control is essential in pretreatment.

Although reduction or elimination of the repulsion barrier is a necessary prerequisite of successful flocculation, the actual flocculation in such a destabilized suspension is effected by particle–particle collisions. Depending on the mechanism that induces the collisions, the flocculation process may be either perikinetic or orthokinetic.

Perikinetic flocculation is the first stage of flocculation, induced by the Brownian motion. It is a second-order process that quickly diminishes with time and therefore is largely completed in a few seconds. The higher the initial concentration of the solids, the faster is the flocculation.

The well known D.L.V.O. theory of colloid stability (10) attributes the state of flocculation to the balance between the van der Waals attractive forces and the repulsive electric double-layer forces at the liquid–solid interface. The potential at the double layer, called zeta potential, is measured indirectly by electrophoretic mobility or streaming potential. The bridging flocculation by which polymer molecules are adsorbed on more than one particle is due to charge effects, van der Waals forces, or hydrogen bonding (see Colloids, Supplement Volume).

The flocculation rate is determined from the Smoluchowski rate law which states that the rate is proportional to the square of the particle concentration, inversely proportional to the fluid viscosity, and independent of particle size.

Orthokinetic flocculation is flocculation induced by the motion of the liquid obtained, for example, by paddle stirring or any other means that produce shear within the suspension. Orthokinetic flocculation leads to exponential growth which is a function of shear rate and particle concentration. Large-scale one-pass clarifiers used in water installations employ orthokinetic flocculators before introducing the suspension into the settling tank (see Water, municipal water treatment).

The scale-up of orthokinetic flocculators, which are generally in the form of paddle devices, is based on the product of mean velocity gradient and time, for a constant-volume concentration of the flocculating particles.

Another type of flocculation is due to particle–particle collision caused by differential settlement. This effect is quite pronounced in full-size plants where large rapidly falling particles capture small particles that settle more slowly.

A third type of flocculation has recently been introduced, so-called mechanical syneresis (11). It is defined as the process of shrinkage and densification of loose and bulky flocs through uneven application of local fluctuating mechanical forces which leads to exudation of the liquid from the floc. This is achieved by slowly stirring the blanket zones with rotating paddles in sludge-blanket clarifiers. Pellet-like flocs can be produced by this process which allows higher overflow rates than those obtained with conventional blanket clarifiers. So far, the process has been applied successfully only to a few suspensions in water treatment.

Settling Tests

For the simplest case of particulate clarification where no flocculation takes place during settling (either the flocculation process is completed before entering the settling tank or the suspension is entirely nonflocculant), the basic test is the so-called short-tube procedure (12). A sample settles in and is decanted from a large measuring cylinder in order to evaluate the settling rate, ie, the specific overflow rate that produces a satisfactory clarity of the overflow. The long-tube procedure is designed for systems where flocculation (or deflocculation) takes place during settling and thus the settling-tank's performance depends not only on the specific overflow rate but also on the residence time in the tank. Tests are conducted in a vertical tube that is as long as the expected depth of the clarifier, under the ideal assumption that a vertical element of a suspension which has been clarified maintains its shape as it moves across the tank.

When the overflow clarity is independent of overflow rate, depending only on detention time (as in the case for high solids removal from a flocculating suspension), the required time is determined by simple laboratory testing of residual solid concentrations in the supernatant versus detention time under the conditions of mild shear. This determination is sometimes called the second-order test procedure because the flocculation process follows a second-order reaction rate.

The design of the sludge-blanket clarifiers used primarily in the water industry is based on the jar test and a simple measurement of the blanket expansion and settling rate (12).

Different versions of the jar test exist, but essentially it consists of a bank of stirred beakers that is used as a series flocculator to optimize the flocculant addition that produces the maximum floc-settling rate. Visual floc-size evaluation is usually included.

The critical settling flux essential for evaluating the requirement of the zone-settling-layer area of a gravity thickener is measured either with the Coe and Clevenger method (13) or the simpler Talmage and Fitch procedure (12). The former consists of a series of settling tests in a measuring cylinder where the initial settling rates of a visible interface within the settling suspension are measured for different initial solids concentrations. The Talmage and Fitch procedure is simpler because it requires only one test at any concentration, providing it is in the zone-settling regime. Theoretically, the two methods should give an identical critical settling flux and therefore identical pool areas, but this is not so in practice; usually the Coe and Clevenger method leads to underdesign of the thickener area whereas the Talmage and Fitch procedure leads to overdesign. With highly flocculant slurries, the area requirement of the compression layer may exceed that of the zone-settling layer; the compression zone also has a depth requirement. No laboratory tests for the latter existed until recently, when a multiple batch upflow test for compression-zone evaluation was described (9). This test is by no means generally accepted and its reliability remains to be demonstrated.

Design Methods

The simplest case of a gravity-settling tank without coagulation or flocculation in clarification applications, ie, when removing small amounts of solids, is based on the identical principle as laminar settling chambers for cleaning gases (14) (see Gas cleaning). The grade-efficiency curve $G(x)$ is given by:

$$G(x) = \frac{v_g A}{Q} \qquad (12)$$

where v_g is the terminal settling velocity of a particle of size x (see eq. 7), Q is the feed flow rate which equal to overflow rate, and A is the plan area of the tank. The above equation was derived assuming laminar flow in the tank and no end effects. The tank area is the only design parameter affecting the theoretical separational performance irrespective of the shape or depth of the pool. The above equation can be expressed in terms of the more conventional dimensionless groups as

$$Stk\ Fr = G(x) \qquad (13)$$

if Stokes' law is assumed for the particle-settling velocity and, therefore, the Stokes' number is defined as

$$Stk = \frac{x^2 \Delta \rho}{18\,\mu} \cdot \frac{Q}{A \cdot H} \qquad (14)$$

and Froude number

$$Fr = \frac{HgA^2}{Q^2} \qquad (15)$$

where H is a characteristic dimension of the pool, eg, the height, which in equation 13 cancels; $\Delta \rho$ is particle–fluid density difference, μ is liquid viscosity, and g the acceleration of gravity.

Scale-up can be calculated with the help of equation 12, based on a simple specific overflow-rate model: for the same performance, the flow rate is proportional to area, or vice-versa. To measure the grade-efficiency curve of a settling tank is difficult, particularly on a large scale, because of the long residence time involved during which the feed solids must remain constant. It is therefore more practical to measure the specific overflow rate (or overflow volume flux) Q/A that gives satisfactory overflow clarity from simple settling tests (see the short-tube procedure above). If the clarification is not completely particulate, ie, when flocculation takes place and has not been completed before the suspension enters the settling tank, the overflow clarity depends not only on the overflow rate but also on the detention time. The scale-up under such conditions is based on the long-tube procedure mentioned above. Sometimes the effect of the detention time is so strong that the overflow rate can be ignored and the scale-up is based on the detention procedure (see above). For the whole range of coagulation clarifiers used predominantly in the water industry, the scale-up is usually based on the overflow rate determined from jar tests primarily designed to select the best flocculant and determine the settling rates that give a clear supernatant.

Probably more relevant to the chemical industry is the scale-up of thickeners. Thickeners are basically gravity-settling tanks that, apart from producing a clear overflow, are designed to have a thick underflow with as little water content as possible. The feed into a thickener is generally more concentrated than the feed into a clarifier and quite often exhibits the zone settling behavior discussed above because of the application of flocculants.

An operating thickener has basically three layers: the topmost clarification layer, the zone-settling layer and the compression layer at the bottom. Each of the three layers requires a certain area. Ideally, the largest of these governs the design of the thickener. In most cases, it is the function of the clarification layer to prevent those particles that have escaped from the zone-settling layer or from the feed from leaving with the overflow. This function is frequently less important than the thickening function and thus the thickener area is usually chosen on the basis of the zone-settling or compression-layer requirement.

The conventional design and scale-up of thickeners operated with mineral or certain metallurgical slurries is based on the area requirement of the zone-settling layer and assumes that the compression zone only imposes a solids-detention (hence depth) demand and has no independent demand on area, with the exception of the empirical three-foot (1 m) rule, which if applicable, introduces an area demand (12). It is the basic principle of this method that the solids on their way downward from the feed layer to the underflow continuously increase in concentration from that of the feed to that of the underflow (usually determined by a time-retention test). The total-solids flux (mass flow rate of solids per unit area), which different layers in the thickener are capable of accommodating, usually goes through a minimum between the feed zone and the underflow. This critical solids flux G_c determines the minimum design area of the thickener. A thickener of this or greater area should be in no danger of overflowing the solids through backing up in the thickener of the zone-settling layer which should be so small as to introduce no demand on the depth allocated to this layer in the thickener. The area of the thickener A is calculated from the critical solids flux G_c, the feed flow rate Q, and concentration C_f using a simple mass balance

$$G_c \cdot A = Q \cdot C_f \tag{16}$$

which assumes a complete separation of all feed solids.

Thus the design and scale-up of the thickeners in the above-mentioned category centers around the determination of the critical solids flux G_c. This value cannot be estimated from the primary properties of the particulate system because of the unpredictable effect of flocculation. It must be obtained experimentally from large-scale or pilot-scale thickeners. However, with a few exceptions, this method is impracticable because of the scale and cost of such experiments. If the settling velocity of the solids is assumed to be a function of concentration only, ie, $v = v(c)$, then this function should be unique for a given suspension and should be the same in batch-settling and continuous operations. This is the basis of the conventional Coe and Clevenger, and

Talmage and Fitch methods which only differ in the way in which G_c is determined from experimental tests. The latter method requires simpler tests. According to Coe and Clevenger, if the function of $v = v(c)$ is known, the critical flux G_c corresponds to the minimum on the total flux G curve which is

$$G = v\left(\frac{1}{C} - \frac{1}{C_v}\right) \qquad (17)$$

where the underflow concentration C_v is determined by time-detention tests. The thickness of the compression zone is also determined by time-detention tests, assuming that the solids concentration reached in the compression layer in a batch test after a given time is the same as in the compression layer of a thickener.

Equipment

Sedimentation equipment can be divided into batch-operated settling tanks and continuously operated thickeners or clarifiers. The operation of the former is very simple and their use has recently diminished. They are still used, however, when small quantities of liquids are to be treated, for example in the cleaning and reclamation of lubricating oil (see Recycling, oil). Most sedimentation processes are operated in continuous units.

Clarifiers. The largest user of clarifiers is probably the water-treatment industry. The conventional one-pass clarifier uses horizontal flow in circular or rectangular vessels. Figure 2 shows a schematic diagram of a rectangular basin with feed at one end and overflow at the other. The feed is preflocculated in an orthokinetic (paddle) flocculator which often forms an integral part of the clarifier. Settled solids are pushed to a discharge trench by paddles or blades on a chain mechanism or suspended from a travelling bridge (see Water, industrial water treatment).

Circular basin clarifiers are most commonly fed through a centrally located feed well; the overflow is led into a trough around the periphery of the basin. The bottom gently slopes to the center and the settled solids are pushed down the slope by a number of motor-driven scraper blades that revolve slowly around a vertical center shaft. This design closely resembles the conventional thickener described below. Like thickeners, circular clarifiers can be stacked in multitray arrangements to save space. Some juice clarifiers are arranged in this way.

Circular raking mechanisms are sometimes also used in square basins with horizontal flow across the basin. Such designs have to incorporate supplementary rake arms that reach into the corners of the square vessel (see Fig. 3).

The conventional one-pass clarifier is designed for the lowest specific overflow rate (flow per unit area of liquid surface), which is usually 1–3 m/h depending on the

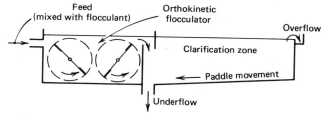

Figure 2. Schematic diagram of a rectangular basin clarifier.

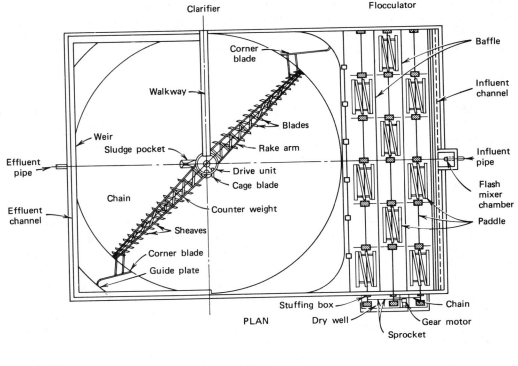

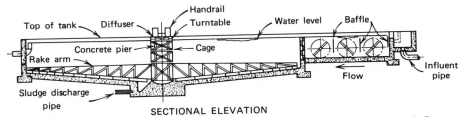

Figure 3. The Dorrco Flocculator-Squarex clarifier combination (cross-flow arrangement). Courtesy of Dorr-Oliver Inc.

degree of flocculation. These clarifiers can be started and stopped without difficulty.

New designs incorporate some vertical flow and combine flocculation, gravity, and inertial clarification, and solids recirculation. Because such units achieve higher overflow rates, they are referred to as rapid settling or high rate clarifiers. Figure 4 gives an example in this category.

Flocculation is accelerated and higher overflow rates are achieved by external or internal recirculation of settled solids into the feed which leads to the collection of fine particles by interception. Addition of conditioned fine sand to the feed induces separation by differential sedimentation, and sometimes increases overflow rates to 6–8 m/h. Design and operation of recirculation systems can be complicated. Problems are avoided by using a sludge-blanket clarifier, in which feed enters below a blanket of accumulated and flocculated solids which become fluidized in the zone-settling regime

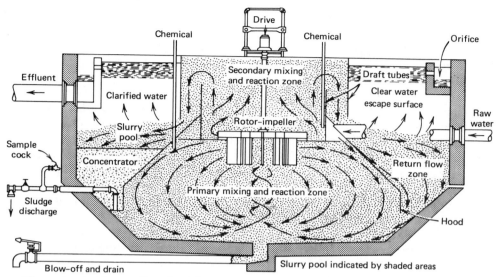

Figure 4. Accelerator. Functional diagram. Courtesy of Infilco Fuller Company/General American Transportation Corporation.

by the upflowing feed. Feed solids are trapped in the blanket. The solids content of the blanket continuously increases and part must be bled off in order to maintain the mass balance.

Sludge-blanket clarifiers are available in flat-bottom, trough-bottom, and hopper-bottom types. The hopper-bottom vertical-flow clarifier shown in Figure 5 achieves rise rates of 1–6 m/h in wastewater applications. It is a 3–4 m deep, 60° triangular or circular hopper tank, with feed introduced through a downward-pointing inlet at the bottom of the tank. The flocculated feed suspension is clarified by passage through the blanket and overflows into decanting troughs that are usually 1–1.5 m above the blanket, to allow for blanket-level variations with feed flow rate. The blanket can be continuously bled off through a submerged weir-type regulator and then thickened in a conical concentrator, or the clarifier can be periodically shut down to allow settlement bleeding.

Sludge-blanket clarifiers are difficult to start up (the first blanket must be established) and large-scale units require extensive excavation. Sizes range from 600

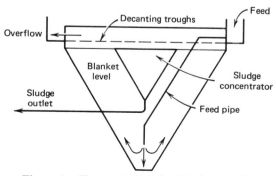

Figure 5. Hopper-tank sludge-blanket clarifier.

× 600 mm to 50 × 50 m. Precipitation and crystallization can be carried out in similar hopper-designed units, with overflow rates of 80 m/h or higher.

Thickeners

The most common thickener is the circular-basin type shown in Figure 6. After treatment with flocculant, the feed stream enters the central feed well which dissipates the stream's kinetic energy and disperses it gently into the thickener. The feed finds its height in the basin where its density matches the density of the inside suspension and spreads out at that level. Solids concentration increases downward in an operating thickener giving stability to the process. The settling solids and some liquid move downward. The amount of the latter depends on the underflow withdrawal rate. Most of the liquid moves upward and into the overflow which is collected in a trough around the periphery of the basin.

A typical thickener has three operating layers: clarification, zone settling, and compression. Frequently, the feed is contained in the zone- settling layer which theoretically eliminates the need for the clarification zone because the particles would not escape through the interface. In practice, however, the clarification zone provides a buffer for fluctuations in the feed and the sludge level.

The most important design dimensions of a thickener are pool area and depth. The pool area is chosen to be the largest of the three layer requirements. In most cases, only the zone-settling and compression-layer requirements need to be considered. However, if the clarity of the overflow is critical, the clarification zone may need the largest area. As to the pool depth, only the compression layer has a depth requirement because the concentration of the solids in the underflow is largely determined by the time detention and sometimes by the static pressure. Thickness of the other two layers is governed only by practical considerations.

Thickeners are widely used, particularly in the mineral-processing industry and in wastewater treatment. Typical applications include thickening of alumina red mud, alumina hydrate, coal tailings, copper middlings and concentrate, magnesium hydroxide, china clay (kaolin), phosphate slimes, potash slimes, pulp-mill wastes, and gas-washing effluents. In hydrometallurgical installations, thickeners are employed for the separation of dissolved components from leached residues in countercurrent washing configurations, eg, in the copper, uranium, alumina, and precious-metals production (see also Flotation).

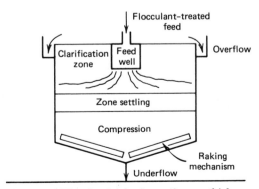

Figure 6.   The circular-basin continuous thickener.

The conventional thickeners are constructed of steel (up to ca 25 m diameter) or concrete (up to 200 m diameter); the floor is usually sloped toward the underflow discharge in the center. Large thickeners have earth bottoms. Raking mechanisms turning slowly around the center column consolidate the solids in the compression layer and facilitate the discharge of solids. Smaller basins can be covered for conservation of heat or to prevent freezing.

The center-drive mechanism and feed launder are usually supported by a walkway that extends across one half or the whole diameter of the basin. Devices with drive mechanism and rakes supported by a truss across the diameter of the thickener are referred to as bridge machines. The bridge thickeners usually do not exceed 25–45 m dia. In thickeners with larger diameters, the drive mechanism is supported by a central column or pier and the rakes are driven and supported by a drive cage. The sediment is discharged into an annular trench around the bottom of the column.

In even larger caisson thickeners, the central column is sufficiently large to accommodate a discharge pump at the bottom. The discharge passes through the column and along the access walkway above the basin. Caisson thickeners can be built with diameters of up to 200 m, often with an earth bottom.

The rake arms are driven by fixed connections or dragged by cables or chains suspended from a drive arm that is rigidly connected to the drive mechanism. The rake arms are connected to the bottom of the central column by a special arm hinge that allows both horizontal and vertical movements. This arrangement lifts the rakes automatically if the torque becomes excessive. The drive arm can be attached below the suspension level or, if scaling is a problem, above the basin.

The traction thickener includes a traction mechanism where the movement of the rake is supplied by a single long arm pivoted around the center column and driven by a trolley that moves on a peripheral rail around the basin. Such units have diameters of 60–130 m.

The thickeners described above are the most widely used conventional types which, in metallurgical and mineral-processing applications, give solids fluxes (mass flow rate of solids per unit area) in the range of 0.011–0.022 kg/(m²·s) (15).

Improved application of flocculating agents resulted in several new, high capacity thickeners capable of handling fluxes up to 0.19–0.38 kg/(m²·s). A good example is the deep-cone thickener developed by the National Coal Board (NCB, UK) (16–17). It is based on the deep-cone vessel used for the processing of coal and metallurgical ores since the turn of the century (see Coal). As can be seen in Figure 7, the vessel is equipped with a slow-turning stirring mechanism (2 rpm) which enhances flocculation in the upper part and acts as a rake in the lower section. The unit is used for densification of froth-flotation tailings at overflow rates from 6.5–10 m/h; the final discharge contains 25–35 wt % moisture. Other commercial deep-cone thickeners are of particular advantage (as is the NCB unit) where the final underflow density is increased by the large static head above the discharge point, eg, with flocculated clays.

There are two U.S. thickeners based on feeding the slurry under the settling-solids interface (sludge blanket) in a way similar to the sludge-blanket clarifiers described above. They also offer other features designed to accelerate flocculation and thus increase the capacity.

In the Eimco Hi-Capacity thickener the feed is introduced from the top through a hollow shaft which incorporates flocculant addition and a mixing device. The feed is then directed into the established sludge blanket under the sludge line which par-

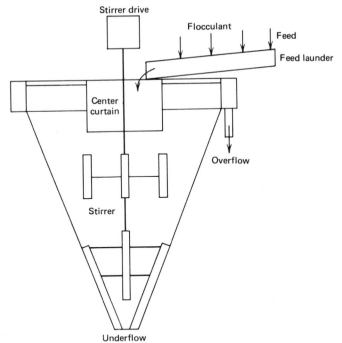

Figure 7.   The NCB deep-cone thickener.

tially submerges a set of inclined settling plates. The settled solids are moved by a conventional raking mechanism at the bottom of the basin.

The principle of another high capacity unit, the Enviro-Clear thickener is shown schematically in Figure 8. Here the flocculated feed enters vertically from the bottom and is directed horizontally at a controlled velocity into the sludge blanket by an impingement plate. A number of other arrangements are also available. The feed can, for example, enters from the top through a center well surrounding the rake drive shaft. A glass window allows visual observation of the sludge line. Available unit sizes range from 4–18 m with typical overflow rates from 2.4–14.4 m/h. The applications include sugar and paper production, and mineral processing (see Sugar; Paper; Extractive metallurgy).

Stacking of sedimentation units in vertical arrangements increases the capacity per unit area. Multiple compartment or tray thickeners consist of two or more conventional thickener compartments up to 35 m in diameter stacked vertically. Each compartment has a set of raking arms operating from a rotating central shaft common to the whole stack. The compartments are used either in series or parallel.

A more recent development in this category is the Swedish Lamella thickener (see Fig. 9). It consists of a number of inclined plates stacked closely together. The flocculated feed enters the stack from the side feed box. The flow moves upward between the plates while the solids settle onto the plate surfaces and slide down into the sludge hopper underneath where they are further consolidated by vibration or raking. Even distribution of the flow through the plates is assisted by flow-distribution orifices placed in the overflow exit. In theory, the effective settling area is the sum of the horizontal projected areas of all plates. In practice, only ca 50% of the area is utilized.

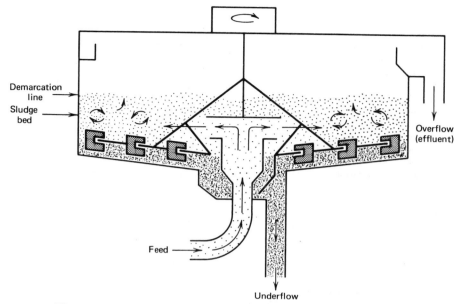

Figure 8.   The Enviro-Clear thickener. Courtesy of Amstar Corporation.

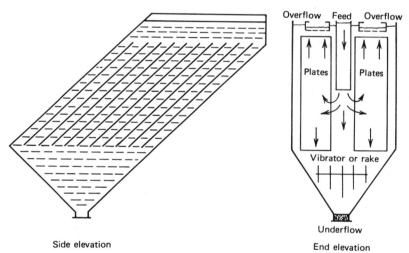

Figure 9.   The Lamella thickener.

When treating sticky sludges, the whole Lamella pack can be vibrated intermittently or continuously to assist the sliding motion of the solids down the plates. In some instances, the plates are corrugated instead of flat or they are replaced by tube bundles in the tube settler; these tubes are square and have a cross-section of ca 50×50 mm^2 and are 950 mm long. The Lamella thickener and the tube settler are used in the treatment of coal, gas-scrubber effluents, fly ash, leach solutions, and many other applications. In coal concentration, the typical overflow rates range from 1.7–2.9 m/h.

574   SEDIMENTATION

Nomenclature

A	= area of the particle projected in the direction of motion; plan area of a settling tank or thickener
C	= solids concentration
C_D	= drag coefficient
C_f	= feed solids concentration
C_v	= underflow concentration
$F(\epsilon)$	= voidage function
F_D	= drag force
Fr	= Froude Number
G	= total flux
g	= gravity acceleration
$G(x)$	= grade efficiency
G_c	= critical solids flux
H	= height
m	= particle mass
Re_p	= particle Reynolds number
Stk	= Stokes number
t	= time
Q	= flow rate
v	= particle–fluid relative velocity
$v(c)$	= velocity as a function of concentration
v_g	= terminal settling velocity
v_p	= hindered settling velocity of a particle
$v(t)$	= velocity as a function of time
x	= particle size
Δ_p	= particle–fluid density difference
ϵ	= voidage
μ	= liquid viscosity
ρ	= liquid density (fluid density)
ρ_s	= solids density; particle density

BIBLIOGRAPHY

"Sedimentation" in *ECT* 1st ed., Vol. 12, pp. 126–145, by W. A. Lutz and D. C. Gillespie, The Dorr Company; "Sedimentation" in *ECT* 2nd ed., Vol. 17, pp. 785–808, by F. L. Bosqui, Consultant.

1. G. G. Brown and Associates, *Unit Operations*, Seventh Printing, John Wiley & Sons, Inc., New York, 1960.
2. L. Svarovsky, ed., *Solid-Liquid Separation*, Butterworths, London, 1977.
3. T. Allen, *Particle Size Measurement*, *IIIrd Ed.*, Chapman and Hall, London, 1981.
4. P. H. T. Uhlherr, *Hindered Settling in Non-Newtonian Fluids—an Appraisal*, CHER 75-1, Department of Chemical Engineering, Monash University, Melbourne, Australia, Jan. 1975.
5. P. C. Carman, *Flow of Gases through Porous Media*, Butterworths, London, 1956.
6. H. C. Brinkman, *Appl. Sci. Res.* **A1,** 27 (1947); **A1,** 81 (1948); **A2,** 190 (1949).
7. A. Einstein, *Ann. Phys. Leipzig* **19,** 289 (1906); **34,** 591 (1911).
8. J. F. Richardson and W. N. Zaki, *Chem. Eng. Sci.* **3,** 65 (1954).
9. P. Kos, *paper presented at The Second World Filtration Congress 1979*, Olympia, London, Sept. 18–20, 1979, pp. 595–603.
10. Ref. 2, Chapt. 4.
11. T. Ide and K. Katoka, *paper presented at The Second World Filtration Congress 1979*, Olympia, London, Sept. 18–20, 1979, pp. 377–385; *Filtr. Sep.*, 52 (March/April 1980).
12. D. B. Purchas, ed., *Solid-Liquid Separation Equipment Scale-up*, Upland Press Limited, London, 1977.
13. H. S. Coe and C. H. Clevenger, *Trans. Inst. Min. Eng. (London)* **55,** 356 (1916).
14. L. Svarovsky, *Solid-Gas Separation*, Elsevier, Amsterdam, 1981.

15. J. M. Keane, *World Min.*, 44 (Nov. 1979).
16. J. Abbott and co-workers, *paper presented at the International Coal Preparation Congress*, Paris, 1972, paper 20E VI.
17. J. Abbott, *Filtr. Sep.* **16**(4), 376 (July–Aug. 1979).

LADISLAV SVAROVSKY
University of Bradford

SELENIUM AND SELENIUM COMPOUNDS

Selenium

Selenium [7782-49-2] (Se, at no. 34, at wt 78.96) is between sulfur and tellurium in group VIA and between arsenic and bromine in period 4 of the periodic table. Its outer electronic configuration is $3d^{10}4s^24p^4$, and its three inner shells are completely filled (1–4). Strikingly similar to sulfur in most of its chemistry, its important oxidation states are −2,0, +2,+4, and +6. As far as is known, the +2 state does not occur in nature. Selenium exists in various allotropic modifications and forms many inorganic and organic compounds, which are like those of sulfur and some of which are isomorphous. Selenium was discovered in 1817 and its name was derived from *selene* (moon, in Greek).

Physical Properties. The six stable isotopes of selenium are ^{74}Se [13981-33-4], ^{76}Se [13981-32-3], ^{77}Se [14681-72-2], ^{78}Se [14833-16-0], ^{80}Se [14681-54-0], and ^{82}Se [14687-58-2], and they occur naturally in the approximate abundances of 0.87, 9.02, 7.58, 23.52, 49.82, and 9.19%, respectively. A number of artificial radioactive isotopes have been prepared by neutron activation. One of these, the gamma-emitting ^{75}Se [14265-71-5] (half-life 120.4 d) is a diagnostic tool in medicine.

Solid selenium has several allotropic forms, including an amorphous one resembling plastic sulfur. The stable form at ordinary temperatures, ie, gray or tetragonal selenium, is the most dense form and is semimetallic in appearance. It crystallizes in a hexagonal lattice with $a = 0.4366$ nm and $c = 0.4954$ nm. The electrical conductivity, which makes gray selenium useful in photoelectrical and photochemical applications, is low in the dark but increases several hundredfold on exposure to light. Crystalline red selenium exists in two monoclinic forms, which are obtained by evaporation of carbon disulfide extracts of amorphous red selenium. The α-monoclinic form ($a = 0.9054$ nm, $b = 0.9083$ nm, $c = 1.1060$ nm, and $\beta = 90.81°$) has a unit cell formed of four Se_8 ring molecules. The β-monoclinic form ($a = 1.285$ nm, $b = 0.807$ nm, $c = 0.931$ nm, and $\beta = 93.13°$) is made up of puckered Se_8 [13494-81-0] rings. Amorphous selenium can be black vitreous and is formed by rapid cooling of liquid selenium; the red amorphous and red colloidal forms are involved in reduction reactions. Heating and catalysts transform all the amorphous forms and both monoclinic crystalline forms to gray selenium. Liquid selenium is black in the bulk and brownish red in thin films. The liquid probably contains chains and rings of variable numbers of atoms. Selenium vapor is also complex in nature. The most important species are Se_2 [12185-17-0], Se_5 [12597-28-3], Se_6 [12597-30-7, 20721-13-5], Se_7 [12597-32-9], and Se_8. Se_2 is an important species at 900°C, whereas at 2000°C the vapor is mainly monatomic. The saturated vapor pressure is given by the equation below (5):

$$\log P_{kPa} = 7.2355 - 5010.7/T$$

(to convert $\log P_{kPa}$ to $\log P_{mm\ Hg}$, subtract 0.8751 from the constant). The saturated vapor pressure increases from 133 Pa (1 mm Hg) at 343.7°C to 101.3 kPa (1 atm) at 685.4°C, the boiling point of selenium.

Some physical constants for selenium are given in Table 1. More extensive data and many sources are given in refs. 1–2, 5–7. For a selenium atom, the covalent radius is ca 0.115 nm, the electron affinity for two electrons is ca −4.6 eV (energy absorbed), and the first ionization potential is 0.75 eV.

Chemical Properties. The chemical properties of selenium are intermediate between those of sulfur and tellurium. Selenium reacts with active metals and gains electrons to form ionic compounds containing the selenide ion Se^{2-}. Selenium forms covalent compounds with most other substances. The oxidation states in elemental

Table 1.   Physical Constants of Selenium[a]

Property	Value
melting point, °C	217
boiling point, °C	ca 685
heat of fusion (trigonal liquid), J/mol[b]	6.224
heat of evaporation, J/mol[b]	95.90
heat of combustion (at 298 K), J/mol[b]	−236.8
heat capacity, J/(g·K)[b]	
trigonal	24.52
vitreous	25.627
liquid	29.288
thermal conductivity, W/(m·K)	248.1
thermal expansion coefficient	3.24×10^{-5}–7.5×10^{-5}, depending on the form
viscosity, mPa·s (= cP)	
at 220°C	221
at 360°C	70
density, g/cm³	
trigonal at 298 K	4.819
monoclinic	4.4
liquid at 490 K	3.975
vitreous	4.285
standard reduction potential, V	
$Se + 2\,e \rightarrow Se^{2-}$	−0.78
$Se + 2\,H + 2\,e \rightarrow H_2Se$ (aq)	−0.36
surface tension (liquid), mN/m (= dyn/cm)	
at 220°C	105.5
at 310°C	95.2
electronegativity (Pauling scale)	2.4

[a] Refs. 2 and 6.
[b] To convert J to cal, divide by 4.184.

and combined forms are as follows (5): Na_2Se [*1313-85-5*], −2; Na_2Se_2 [*39775-49-0*], −1; Se_8, 0; Se_2Cl_2 [*10025-68-0, 21317-32-8*], +1; $SeCl_2$ [*14457-70-6*], +2; Na_2SeO_3 [*10102-18-8*], +4; and Na_2SeO_4 [*13410-01-0*], +6.

Selenium combines with metals and many nonmetals directly or hydrochemically. The selenides resemble sulfides in appearance, composition, and properties. Selenium forms halides by reacting vigorously with fluorine and chlorine, and less so with interhalogen compounds and bromine; selenium does not react with iodine. It does not react with pure hydrogen fluoride or hydrogen chloride but decomposes hydrogen iodide to liberate iodine and form hydrogen selenide [*7783-07-5*]. Selenium combines with oxygen yielding a number of oxides, the most stable being selenium dioxide [*7446-08-4*]. Under proper conditions, selenium forms selenides with hydrogen, carbon, nitrogen, phosphorus, and sulfur (see Inorganic Compounds). Crystalline selenium does not react with water, even at 150°C.

Selenium remains unaffected by dilute sulfuric acid or hydrochloric acid, but it dissolves in a nitric–hydrochloric acid mixture, strong nitric acid, and concentrated sulfuric acid. It is oxidized by ozone and solutions of alkali-metal dichromates, permanganates, and chlorates and calcium hypochlorite. Selenium dissolves in strong alkaline solutions yielding selenides and selenites. It forms selenocyanates, MSeCN, with alkali-metal cyanides as well as many inorganic and organic derivatives of the corresponding acid, HSeCN [*13103-11-2*]. Selenium also dissolves in alkali-metal sulfites forming selenosulfates, M_2SSeO_3 and, because tellurium does not undergo this reaction, this method can be used to separate the two elements. Selenium mixes in all proportions with sulfur and tellurium forming a continuous series of solid solutions and alloys.

In many reactions, selenium is an oxidant as well as a reductant. Strong oxidants convert selenium dioxide and its derivatives to the hexavalent state. Although the products are oxidants, they are less active and are difficult to reduce. Selenium salts resemble the corresponding sulfur and tellurium salts in behavior.

Selenium also forms a large number of organic compounds. Of special interest is the oxidizing and reducing action of selenium and its compounds on many organic compounds.

Chemical reactions are described in detail in refs. 1–3, 5, 7–11. Organic reactions are reviewed in refs. 12–13. The organic chemistry and biochemistry of selenium is discussed in refs. 14–17.

Occurrence. The occurrence of selenium is discussed in refs. 19–21. At 0.68 atom per 10,000 atoms Si, selenium is the thirtieth element of cosmic abundance. It ranks about seventieth in the order of crustal abundance, the average amount in the crustal rocks being 0.05 ppm. Selenium is widely dispersed in igneous rocks probably as selenide minerals; in volcanic sulfur deposits in sulfides, where it is isomorphous with sulfur; in hydrothermal deposits, where it is associated isothermally with silver, gold, antimony, and mercury; and in massive sulfide and porphyry copper deposits, where it appears in large quantities but only in small concentrations. In such sedimentary rocks as sandstones, carbonaceous siltstones, phosphorite rocks, and limestones, it is syngenetic, ie, it was introduced during deposition, probably by adsorption on precipitated ferric hydroxide. Like vanadium, phosphorus, arsenic, and antimony, selenium has been reported in sedimentary iron ores in amounts larger than its average abundance in the crust. The estimated selenium content in oceans is only ca 0.090 $\mu g/L$, a small fraction of the selenium transported into the sea by weathering and erosion

(21). The principal species in the seas is SeO_4^{2-}. Adsorption by some marine organisms also contributes to the removal of selenium from seawater. Thus selenium often, though not always, occurs in the pyrite and marcasite of sedimentary formations as well as in soils derived from them. It is particularly concentrated in soils of the drier regions, eg, the North American great plains from Mexico to the prairie provinces of Canada and westward to the Pacific Ocean, but especially in Wyoming and South Dakota, and in localities in Colombia, Ireland, Israel, South Africa, and the People's Republic of China. It is also found in Argentina, Venezuela, Spain, Bulgaria, Algeria, Morocco, Australia, New Zealand, and some regions of the USSR, as shown by analyses of locally grown crops. Selenium occurs in coals in 0.5–12 ppm concentrations, probably associated with pyrite and marcasite (4). Crude oil from different locations contains variable amounts of selenium, usually less than 0.5 ppm.

Biosphere. Selenium occurs in alkaline soils chiefly as selenates which, when water-soluble, are readily available to plants. In acid soils, selenides and to some degree elemental selenium are prevalent but little is available to plants. The upper soils can be enriched by certain native plants that accumulate high selenium concentrations from the seleniferous soils and earth formations. These plants, which require selenium for their growth, include about 24 species and varieties of *Astragalus* (milk vetch), *Xylorhyza machaeranthera* (woody aster), *Oonopsis haplopappus* (goldenweed), and *Stanleya* (princes plume). Other species, eg, *Aster*, *Atriplex*, and *Grindelia* (gum weed), absorb moderately large quantities when growing in soils with high available selenium concentrations. These so-called secondary selenium absorbers and the primary selenium indicators aid in locating seleniferous regions. Some selenium-tolerant plants may contain up to 1.5 wt % selenium.

Certain of the indicator plants have an offensive odor, which varies with the selenium concentration. Other vegetable matter grown on seleniferous soils may have a sufficiently high selenium content to be toxic when ingested by animals or humans (see Health and Safety Factors). Apart from its appearance in seleniferous plants, selenium has been considered as a variable contaminant. The current view is that selenium is a necessary microconstituent in the living organism and a needed micronutrient for humans as well as for animals.

Data on selenium in waters are limited. The drinking-water content is usually less than 1 μg/L, somewhat higher in several wells in seleniferous areas, and markedly higher in some river waters where irrigation drainage from seleniferous soils contains up to 2680 μg/L.

Deposits. Selenium forms natural compounds with 16 other elements. It is a main constituent of 39 mineral species and a minor component of 37 others, chiefly sulfides. The minerals are finely disseminated and do not form a selenium ore. Because there are no deposits that could be mined for selenium alone, there are no reserves. Nevertheless, the 1980 world reserves, chiefly in nonferrous metal sulfide deposits, are ca 112,000 metric tons and resources are ca 291,000 t (22). The principal resources of the world are in the base metal sulfide deposits that are mined primarily for copper, zinc, nickel, and silver (and, to a lesser extent, lead and mercury), where selenium recovery is incidental.

Manufacture. *Recovery.* Practically all selenium is obtained as a by-product of precious-metal recovery from electrolytic copper refinery slimes. A minor amount is recovered from sludges and dusts of sulfuric acid manufacture and of the pulp and paper industry where iron sulfides are burned to sulfur dioxide. There is a small but vigorous industry of secondary selenium recovery from rectifier and electrophotographic plates and other materials.

The composition of the slimes varies widely among the refineries because of variation in the composition of ore smelted to anodes. The range of slimes composition is Cu, 3–77 wt %; Ag+Au, 1–31 wt %; Se, 1.5–21 wt %; Te, 0.1–10 wt %; Pb, 0.3–33 wt %; and As+Sb+Bi, <1–27 wt % (23). Decopperized slimes from several USSR copper refineries contain Cu, 1–4.2 wt %; Se, 7–13.8 wt %; Te, 0.8–2.8 wt %; Pb, 6–25.4 wt %; and As+Sb+Bi, 9.4–13.6 wt % (24).

Selenium occurs in the slimes mainly as CuAgSe [12040-91-4], Ag_2Se [1302-09-6], and $Cu_{2-x}Se_x$ [20405-64-5] where $x < 1$. The primary purpose of slimes treatment, ie, the recovery of precious metals, is carried out in a silver refinery. Because of the complexity of slimes composition and the need of separating at least five valuable components, ie, copper, silver, gold, selenium, tellurium, and in some cases metals of the platinum group, in a more-or-less pure state, treatment flow sheets are complicated. As with analytical determination, treatment methods usually involve conversion of selenium to a water-soluble form, followed by reduction to the elemental state. The most important methods of achieving this are smelting with soda ash, roasting with soda ash, direct oxidation, and roasting with sulfuric acid.

Soda Smelting. Soda smelting is a pyrometallurgical method. The raw slimes are first decopperized by sparging with air, ie, aeration in a modified flotation cell, or oxidative pressure leaching with dilute sulfuric acid, or by an oxidizing roast followed by sulfuric acid leach. In all cases, the copper is dissolved as copper sulfate. Copper is removed because in large concentration it makes the subsequent smelting of slimes difficult. The decopperized slimes are then mixed with soda and silica and are smelted in a small reverberatory or doré furnace. The first slags are mainly siliceous and low in selenium (ca 1 wt %), and the bulk of impurities eg, iron, arsenic, antimony, and lead, is eliminated. The molten charge is then blown with air to oxidize and volatilize as much selenium as possible. Niter may be added to decompose copper selenide. The volatilized oxides of selenium and other elements are caught in the flue and the scrubber-Cottrell systems. The charge is rabbled, ie, stirred, with additional soda ash to remove almost all the remaining selenium as a soda slag. The subsequent treatment of the furnace charge for precious-metal recovery is described elsewhere (25) (see Copper; Gold and gold compounds; Silver and silver alloys). Carbonaceous materials may be added to the molten slag to control the formation of difficult-to-reduce sodium selenate (26). The soda slag is water-leached and filtered to yield an alkaline liquor containing sodium selenite, sodium selenate, and sodium tellurite. The liquor is neutralized to pH 5.5–6.5 with raw sulfuric acid or acidic solutions from the scrubber-Cottrell system. This precipitates tellurium as tellurous acid, which is separated and treated for the recovery of tellurium. The filtrate and the remaining scrubber solutions are acidified and treated with sulfur dioxide to precipitate selenium as a red amorphous sludge. The sludge is then boiled with steam to coagulate any colloidal selenium and to transform the red precipitate into the almost black metallic modification. The latter is washed, dried, and pulverized. In some refineries, it is shipped as commercial refined selenium, but it is generally preferable to carry out a separate refining operation.

The selenium recovery is ca 80% as a result of losses in slags, flue dusts, and flue gas. Precipitation of tellurium upon neutralization is not complete, and some tellurium appears in the selenium as an impurity. Nevertheless, the process is simple and economical and has been employed with modifications at several refineries, including International Nickel Company and U.S. Metals Refining Company (AMAX, Inc.).

Soda Roasting.   A modification of the pyrometallurgical method is the roasting or baking of raw or decopperized slimes mixed with soda ash (27). The temperature must be kept below the sintering point to ensure access of air, which is essential for thorough oxidation of the selenides. The selenides are converted to sodium selenite or sodium selenate according to the following equations:

$$Se \text{ (selenide)} + Na_2CO_3 + O_2 \rightarrow Na_2SeO_3 + CO_2$$

$$Se \text{ (selenide)} + Na_2CO_3 + 1\frac{1}{2} O_2 \rightarrow Na_2SeO_4 + CO_2$$

Increasing the temperature, the time, and the amount of soda in the charge promotes the second reaction. The calcine is leached with water, and the selenium is recovered from the leach liquor by either of two methods. In one method, the solution is neutralized to separate tellurium as H_2TeO_3 and then treated with sulfur dioxide, which causes precipitation of selenium from selenite (28). The selenate ion is reduced to the selenite state with the addition of hydrochloric acid or sodium chloride and sulfuric acid,

$$SeO_4^{2-} + 2 HCl \rightarrow SeO_3^{2-} + Cl_2 + H_2O$$

Ferrous sulfate may be added to accelerate the reaction, as shown:

$$3 H_2SeO_4 + 6 HCl + 6 FeSO_4 \rightarrow 3 H_2SeO_3 + 2 FeCl_3 + 2 Fe_2(SO_4)_3 + 3 H_2O$$

The second method was developed by Boliden Aktiebolag in Ronnskar, Sweden, for slimes of negligible tellurium content. The leach liquor is evaporated to dryness, and the sodium selenite and sodium selenate mixture is reduced with coke to selenide (29). The latter is redissolved and blown with air to liberate elemental selenium, which is separated, and to regenerate sodium hydroxide. Selenium recovery may be $\geq 95\%$. Because any tellurium present in the slimes is similarly affected, soda roasting is usually not feasible with high tellurium slimes, where recovery of high purity selenium is desirable.

Direct Oxidation.   In one variation of direct oxidation, the slimes were air-roasted to oxidize the metal selenides to selenites and leached with caustic soda. The sodium selenite solution was treated as in the soda smelting and roasting methods. The high dust loss necessitated dust collection and recirculation, and the roasting temperature required close control to prevent sintering. A variation involves roasting the slimes in a fluidized bed or pelletizing and roasting the slimes in a stationary or a traveling bed or in a rotary kiln. The purpose is to volatilize selenium as the dioxide, which is collected dry or wet. A similar process, developed at Leningrad Mining Institute, is used in the USSR in a modified form (30–31). Air is passed through granulated raw or decopperized slimes in a semicontinuously operating shaft furnace at 750–800°C. Selenium dioxide is absorbed in sodium carbonate solution. The recovery from slimes is 91–92%.

In a wet method, the slimes are leached with hot sodium hydroxide solution under oxygen pressure. Selenium extraction is 95% or higher. This method has not been used commercially because an appreciable amount of selenium is oxidized to the selenate state (32). A similar oxidative caustic pressure leach is claimed to dissolve 96% of the slimes selenium, of which more than 90% is in the selenite state (33).

Sulfuric Acid Roasting. Concentrated sulfuric acid is added to the slime before or during roasting. At Canadian Copper Refiners Ltd. (now Noranda Mines Ltd., CCR Division) in Montreal East, Canada, the slimes–acid mix was heated to sulfatize the copper and then roasted to volatilize selenium as the dioxide for recovery in a wet-scrubber-Cottrell system (34). This method is no longer in use; however, variations of it are practiced or have been practiced in many refineries throughout the world. Because of atmospheric pollution with sulfur dioxide, the tendency is to replace acid roasting with wet sulfatization of copper followed by selenium removal without the use of acid.

The slimes at Montreal East are decopperized by oxidative pressure leaching with sulfuric acid solution. The dissolved tellurium is cemented on metallic copper for subsequent recovery. The decopperized slimes are pelletized and roasted to remove and recover one half of the selenium as the dioxide. Most of the remaining selenium is recovered during doré furnace smelting (35–36).

Wet decopperization methods include the time-honored but slow and inefficient air sparging of slimes suspended in discarded copper electrolyte, where the iron sulfate impurity promotes the reaction, and the use of Mekhanobr flotation machines provided with mechanical impellers in the USSR, which leave 2–2.5% residual copper but dissolve arsenic, antimony, and bismuth, but not Cu_2S or $CuAgSe$ (37).

Miscellaneous. Where copper and lead are treated in adjoining plants and especially where the selenium content is low, it may be expedient to smelt decopperized slimes with lead slimes or other lead materials. The lead bath is blown with air or is cupeled, and the selenium is volatilized and recovered from flue dust and fume by leaching. Variations of this method are used or have been used by Métallurgie Hoboken-Overpelt N.V. in Oolen, Belgium and Cerro Corporation in Oroya, Peru. There are also methods for selenium recovery from sulfuric acid plant sludge and from pyrite, zinc, lead, and other metallurgical fumes and dusts.

Secondary selenium is recovered mechanically from rectifier and electrophotographic copier plates and electronic-industry scrap by hammer milling; shot blasting; use of high pressure water jets (38); or dissolving in aqueous sodium sulfite, sodium sulfide, polysulfide, fused caustic soda, fused caustic soda–sodium chloride mixtures, and other solvents. The recovery with nitric acid can be low and the fumes of nitrogen oxides are objectionable. The use of dialkyl dicarboxylates or trimethylhexamethylenediamine in poly(ethylene glycol) solution and of lauryl mercaptan has been suggested to remove or loosen the selenium layers (38,40). Protective coatings on rectifiers must be removed or sufficiently corroded to prevent interference with recovery. The accompanying tin, bismuth, and cadmium from the counterelectrode and the metallic aluminum substrate are usually recovered.

Recovery from Silver-Refinery and Other Solutions. The selenium-bearing solutions may be alkaline or acidic. The solutions are alkaline when soda slag, roasted or baked slimes–soda mixtures, and other alkaline materials are leached with water. The selenium occurs as sodium selenide Na_2Se, sodium polyselenide [37285-90-8] Na_2Se_x, sodium selenite Na_2SeO_3, and sodium selenate Na_2SeO_4. Such solutions may contain tellurium, lead, and other impurities. The acidic solutions are formed when selenium, which is volatilized as the dioxide, is recovered by water-leaching the flue dust and by scrubbing the fumes with water. Such solutions contain selenium as selenious acid H_2SeO_3, as well as some sulfuric acid. For separation of the selenium, the alkaline selenide solution is blown with air with or without carbon dioxide. The alkaline selenite solution is acidified with sulfuric acid and is treated with sulfur dioxide. The reaction vessel is water-cooled to <32°C to precipitate the selenium as an amorphous sludge, which is readily discharged from the vessel. At higher temperatures, the sludge may form a cokelike or even plastic mass which adheres to the walls. However, precipitation with vigorous agitation at >85°C, preferably at >100°C, yields finely divided black crystalline selenium.

Selenic acid is not readily affected by sulfur dioxide. When present in low concentration (0.5 g/L), its recovery may not be economical. If the sodium selenate concentration is high, the solution is evaporated to dryness and reduced by heating with carbon or fuel oil; or the solution is acidified, sodium chloride and ferrous sulfate are added, and selenium precipitates upon addition of sulfur dioxide. The use of chlorides is usually avoided if the product is to be refined to a high purity grade, because chlorine is a persistent impurity which affects the electrical properties of purified selenium.

Purification. Single-stage distillation of selenium at atmospheric pressure and at 680°C from cast iron or iron alloy retorts expels most of such occluded impurities as sulfur dioxide, water, organic substances, halogens, sulfuric acid, and mercury, and leaves the bulk of the nonvolatile impurities, eg, metals and tellurium, in the residue. Distillation can be carried out at 1.5–3.75 kPa (0.2–0.5 mm Hg) and 310–400°C. The vapor is condensed at 100–250°C. Distillation from a quartz retort upgrades the quality of purified selenium.

In a cyclic method, which is favored in the USSR, selenium is dissolved in hot sodium sulfite solution

$$Se_2 + Na_2SO_3 \rightarrow Na_2SeSO_3$$

sodium selenosulfate [25468-09-1]

and tellurium and many other impurities remain undissolved. The solution is filtered and cooled to reverse the reaction and to deposit solid selenium. The mother liquor is recycled to the dissolution step. Combined selenium is not affected.

At Kennecott's Utah refinery, the crude selenium was dissolved in aqueous sodium sulfite, and the solution was filtered and acidified with sulfuric acid:

$$Na_2SeSO_3 + H_2SO_4 \rightarrow Na_2SO_4 + Se + SO_2 + H_2O$$

The liberated selenium was distilled several times, and the first and last fractions were discarded. The selenium was then shotted and sold as the high purity grade. Unlike the cyclic method, this method involved consumption of reagents and evolution of sulfur dioxide.

A less expensive commercial purification method is oxidation with air or oxygen to selenium dioxide, which is then dissolved in water. Purified selenium precipitates upon addition of sulfur dioxide and is washed, dried, distilled, and shotted.

Melting a mixture of crude selenium with ammonium nitrate at 210–230°C oxidizes and causes slagging of the impurities (41).

$$Te + 2 NH_4NO_3 \rightarrow TeO_2 + N_2 + 4 H_2O$$

The refined selenium yield is 98.5–99%, volatilization loss is ca 0.5%. The slag containing 40–70 wt % Se is reworked. This method has been used in the USSR since 1957. A treatment with hydrochloric acid is claimed to purify selenium (42).

High purity selenium is made in batches, and each batch is sampled and analyzed spectrographically. The ultrahigh purity grade is prepared on a laboratory scale only. Selenium of high purity in the semiconductor sense is not readily available because zone refining is not feasible. At practical zone-refining speeds, crystallization does not occur and impurities do not segregate. However, a controlled differential thermal treatment of selenium in a long vertical glass tube has been described (43). The treatment time is several weeks to several months. This method is similar to zone refining (qv).

The methods of manufacturing commercial selenium products are summarized in Table 2.

Economic Aspects. Commercial production of selenium in the United States was established in ca 1910 when ca 4500 kg was recovered. Until then, selenium was imported from Germany at $55/kg. Demand was at first small but increased gradually until selenium became widely used and, in many cases, an essential industrial material. All primary selenium producers are electrolytic copper refiners. Asarco (formerly American Smelting & Refining Co.), AMAX, Inc. (formerly American Metal Climax, Inc.), and Kennecott Copper Corporation are the U.S. producers. The Anaconda Company and Phelps Dodge Refining Corporation produce electrolytic slimes but ship them elsewhere for treatment. Most of the selenium producers are listed in Table 3 by country and location.

Japan, Canada, and the United States accounted for 75% of the 1980 estimated production by noncommunist countries of 1450 metric tons (see Table 4). Production in the United States was low because of a strike in the copper industry and because of antipollution regulations which resulted in the closing of several smelters and refineries and the export of concentrates to Japan. Production in communist-block countries, including the People's Republic of China, is not known but is estimated at 3.50 t. Secondary selenium recovery from electronic, xerographic, and other scrap and residues was 50–55 t. The estimated consumption by different industries was as follows: glass and pigments, 50%; metallurgy, 5%; xerography, 25–30%; antidandruff preparations, feed additives, catalysts, and minor uses, 15–20%.

Because selenium is difficult to replace in most uses, the price has only a moderate effect on consumption and depends on demand and supply. Price has tended to change rapidly in periods of surplus or short supply. Merchant prices are very volatile. In September 1981, the price of refined selenium was $15.44/kg and the high purity grade was $33.63/kg. Quotations for commercial ferroselenium, nickel–selenium, sodium selenite, and sodium selenate are based on the selenium content. U.S. dealer prices for commercial-grade selenium ranged from $7.30 to $8.80 in the first quarter of 1982. The supply of selenium is expected to be ample. Import of selenium metal and scrap, selenium dioxide, and selenium salts into the United States is duty-free. Export is not limited by regulations but is subject to license.

To stimulate interest and to widen the base of uses, the principal selenium producers have organized the Selenium–Tellurium Development Association, Inc. in Darien, Connecticut, U.S. The Association sponsors research at universities and, through various institutions, symposia, and publications, and it distributes a periodic newsletter and monthly abstracts on selenium and tellurium.

Specifications and Standards. There are no official specifications, apart from those of the U.S. Government for purchases under the national stockpile plan. Some producers publish their own standards, and some users specify a screen analysis and maximum content of certain impurities. Three grades of selenium are available. The commercial or refined grade contains >99.5 wt % selenium and impurities, usually <0.2 wt % tellurium, <0.1 wt % iron, <0.005 wt % lead, and <0.005 wt % copper. It is

Table 2.   Commercial Selenium Products

Product	CAS Registry No.	Formula	Method	Grade	Se content, theoretical %	Ref.[a]
ferroselenium (iron–selenium)	[1310-32-3]	FeSe	melting a mixture of powdered iron and powdered selenium	crushed lump	58.6	
nickel–selenium	[1314-05-02]	NiSe	melting a mixture of powdered nickel and powdered selenium	crushed lump	57.2	
selenium dioxide	[7446-08-4]	SeO_2	oxidizing selenium with nitric acid and evaporating; oxidizing selenium with air or oxygen	commercial and high purity	71.1	
sodium selenite	[10102-18-8]	Na_2SeO_3	neutralizing selenious acid with soda ash		45.7	
sodium selenate	[13410-01-0]	Na_2SeO_4	neutralizing selenic acid with soda ash		41.8	
			electrolyzing sodium selenite solution; product contains no selenite			44
selenic acid	[7783-08-6]	H_2SeO_4	oxidizing selenious acid with hydrogen peroxide		54.5	
			electrolyzing selenious acid; product contains no selenious acid			45
selenium oxychloride	[7791-23-3]	$SeOCl_2$	chlorinating a mixture of selenium and selenium dioxide dry or suspended in carbon tetrachloride		47.6	46
cadmium sulfoselenide	[11112-63-3]	CdSSe	calcining a mixture of cadmium sulfide and selenium		35.3	47
selenium diethyldithio-carbamate (Selenac)	[5456-28-7]	$[(C_2H_5)_2N\overset{\displaystyle S}{\overset{\|}{C}}S]_4Se$	reaction of diethyl-amine, carbon disulfide, and selenium dioxide in alcoholic solution			48

[a] The patents cited are basic; improvements have been made, but little has been published.

sold mainly as 74-μm (200-mesh) powder as well as in smaller quantities of coarser size and as lump. The high purity grade contains 99.99 wt % selenium, although this figure is obtained by difference and not by a direct determination. It is sold usually in shotted form and in small amounts as a powder. Some users have their own specifications which reflect a particular application of selenium. Impurities, eg, mercury, tellurium, iron, arsenic, and nonferrous metals, that are harmful in electronic and

Table 3. Selenium Producers

Country	Producer
Australia	Electrolytic Smelting and Refining Company of Australia Pty., Port Kembla
Belgium	Métallurgie Hoboken-Overpelt S.A., Oolen
Bulgaria	Damyanov Combine, Plovdiv
Canada	Inco Ltd., Copper Cliff, Ontario
	Noranda Mines Ltd., CCR Division, Montreal East, Quebec
Chile	Empresa Nacional de Mineria (ENAMI)
People's Republic of China	Shen-Yan
Finland	Outokumpu Oy, Pori
Federal Republic of Germany	Norddeutsche Affinerie, Hamburg
German Democratic Republic	Mansfeld Kombinat
Japan	Asia Bussan Zairyo Company
	Mitsubishi Metal Corporation
	Mitsui Mining and Smelting Company
	Nippon Mining Company
	Shinko Chemical Company
	Sumitomo Metal Mining Company
	Toho Zinc Company
Mexico	Cobre de Mexico S.A.
Peru	Centromin-Peru
Romania	Baja Mare
Sweden	Boliden Metal AB, Rönnskar
United Kingdom	Johnson Matthey Chemicals Ltd.
United States	Amax Copper, Inc., Carteret, N.J.
	Asarco, Inc., Amarillo, Texas
	Kennecott Corporation
USSR (copper refineries)	Alaverd
	Almalyk
	Balkhash
	Kyshtym
	Norilsk
	Pyshma
Yugoslavia	Rudarsko-Topionicarski Basen, Bor
Zambia	Roan Consolidated Mines Ltd.

electrostatic uses must be <1–2 ppm each. A somewhat higher concentration of inert contaminants, eg, sodium, magnesium, calcium, aluminum, and silicon, can be tolerated. The concentration of halogens, sulfur, and oxygen must be low, depending on the use. Ultrahigh purity selenium is claimed to contain 99.999–99.9999 wt % selenium and is commercially available.

The USSR specifications of different grades of technical selenium are given in Table 5. The USSR specifications of several grades of pure selenium are given in ref. 49.

Analytical Methods. Comprehensive accounts of the various gravimetric, volumetric, polarographic, spectrophotometric, and neutron activation analytical methods have been published (1–2,5,18,20,50–52). The sampling and analysis of biological materials and organic compounds is treated in refs. 53–54. Many analytical methods depend on conversion of the selenium in the sample to selenious acid and then reduction to elemental selenium. The volatility of selenium in steam complicates the

Table 4.   1980 Refined Selenium Production by Country[a]

Country	Production, t
Japan	500 estd
Canada	395
United States	231
USSR	186 estd
Mexico	80
Sweden	67
Belgium	60 estd
Yugoslavia	52

[a] Ref. 22.

Table 5.   GOST 10298-69 Specifications for Technical Selenium[a]

Grade	Se content, min wt %	Impurity content, max wt %						
		Fe	Cu	Pb	Hg	Te	As	S
ST0	99.4	0.005	0.003	0.003	0.001	0.05	0.003	0.005
ST1	99.0	0.015	0.005	0.005	0.005	0.1	0.005	0.02
ST2	97.5	0.5	0.05	0.05	0.05	0.5	0.05	0.5

[a] Ref. 41.

analysis, especially when hydrochloric acid and hydrobromic acid vapors are present.

In many cases, the samples are treated with fuming or concentrated nitric acid or with a sulfuric–nitric acid mixture. Addition of potassium chlorate assists in dissolving any carbonaceous matter. Perchloric acid is an effective oxidant, but its use is hazardous and requires great care. Organic selenium compounds and siliceous materials, eg, rock, ore, and concentrate, are fused with mixtures of sodium carbonate and alkaline oxidants, eg, sodium peroxide and potassium nitrate, or with potassium persulfate. With volatile compounds, this fusion is carried out in a bomb. Oxidizing fusion usually converts selenium into Se(VI) rather than Se(IV). When the sample contains less than 0.1 wt % selenium or if interfering substances are present, selenium may be preconcentrated by distillation from a bromine–hydrobromic acid mixture. A number of substances can readily reduce selenious and selenic acid solutions to an elemental selenium precipitate. This precipitation separates selenium from most elements and serves as a basis for gravimetric determination. Sulfur dioxide is the most convenient and widely used precipitant. Hydroxylamine hydrochloride may be used alone or with sulfur dioxide if Se(VI) is present. Separation from tellurium is effected by holding the hydrochloric acid concentration at 6–9.5 N during reduction.

Gravimetric determination by weighing as Se in a tared Gooch crucible after drying at 105°C can be used if there is a minimum 5-mg selenium content. Volumetric methods include the iodimetric, thiosulfate, and permanganate methods. In one of the several iodimetric methods, the dissolved selenium is reduced to the Se(IV) state, and a moderate excess of potassium iodide is added to liberate iodine, which is titrated with a standard thiosulfate solution.

$$H_2SeO_3 + 4\ HCl + 4\ KI \rightarrow Se + 2\ I_2 + 4\ KCl + 3\ H_2O$$

In the thiosulfate method, selenious acid is treated with an excess of standard sodium thiosulfate solution as follows:

$$H_2SeO_3 + 4\,Na_2S_2O_3 + 4\,HCl \rightarrow Na_2SeS_4O_6 + Na_2S_4O_6 + 4\,NaCl + 3\,H_2O$$

The excess $Na_2S_2O_3$ is back-titrated with a standard iodine solution. The permanganate method is based on the oxidation of Se(IV) to Se(VI).

In pure solutions containing up to 10 $\mu g/mL$, selenium precipitates upon addition of sulfur dioxide or hydrazine and it can be estimated colorimetrically or nephelometrically. Color reactions with alkaloids, eg, codeine phosphate, in 95 wt % sulfuric acid are useful for microdeterminations, especially in organic materials. The highly selective and sensitive 3,3'-diaminobenzidine is used widely in spectrophotometric and spectronephelometric determination of selenium traces.

Polarography is used to a lesser degree than it has been in the past. A differential potentiometric method is highly accurate (55). Spectrographic determination requires a vacuum spectrograph and short-wavelength, ultraviolet-sensitive plates; the detection limit is 0.001 wt %. Atomic absorption spectrometry is sensitive to 1 $\mu g/mL$ of solution. Neutron-activation analysis is used to determine nanogram amounts.

Ion-exchange (qv), solvent-extraction, chromatographic (paper, thin-layer, and gas), ring-oven, spectrofluorimetric, radiochemical, and isotope-dilution techniques and methods are also used. Spectrophotometry with an added graphite furnace is very widely used for selenium analysis.

Accurate analysis for selenium is difficult. Loss of selenium may occur on preparing the sample or during the determination. Liquid occlusions in the precipitate may cause an error in the gravimetric method. Most methods require close control of determination conditions.

Health and Safety Factors. Commercial elemental selenium is relatively inert and can be handled without special precautions. This applies also to the stable metallic selenides, ie, those of copper, nickel, lead, but other selenium compounds should be treated with care. These include the reactive selenides; the gaseous, volatile, and soluble compounds; and particularly hydrogen selenide, the halides and oxyhalides, and the organics. Some of these can enter the body through the lungs or the skin, especially if the tissue is damaged, and may affect the body organs. Because of the rare incidence of occupational poisoning, the pathology of selenium in humans has not been studied adequately. Nevertheless, many selenium compounds are believed to cause damage to some body organs when absorbed through the skin or when inhaled or ingested.

Selenium plays a dual role in a living organism, depending on the compound and the amount absorbed. Controlled small doses of some compounds are used in medicine and as a diet supplement in agriculture, eg, for poultry and stock (see Mineral nutrients); larger amounts can be toxic.

Contact with elemental selenium does not injure the skin, but some of its compounds may cause dermatitis. Selenium dioxide and selenious acid attack the skin and may cause local irritation and dermatitis, and the effects resemble those produced by hydrofluoric acid. Selenium oxyhalides are extremely vesicant and, by hydrolysis to selenious acid and halogen acid, cause slowly healing burns. Hydrogen selenide affects the mucous membranes of the upper respiratory tract and the eyes.

Industrial precautions include the common-sense measures of good housekeeping, proper ventilation, personal cleanliness, frequent change of clothing; and provision of dust masks where needed, gloves, and either safety glasses or chemical goggles. Food should be consumed in a clean room separate from the handling area and after washing hands and face. A shower should be taken at the end of a working day. Calamine lotion, calamine ointment, and various creams have been used to protect the skin. Operators in one selenium recovery plant prefer a 10 wt % sodium thiosulfate solution applied to the skin. Skin that has been in contact with selenium compounds and their solutions should be washed copiously with water and treated with thiosulfate solution. Selenium and its commercial products are not a fire hazard. The toxicity of seleniferous plants ingested by domestic animals and instances of poisoning among humans consuming the products of these animals and the food grown on highly seleniferous soils have been discussed (20). Measures have been taken to cease grazing and cultivation on such soils. The aspects of selenium in human and animal biology and in medicine are treated in refs. 54, 56–59. The toxicology is discussed in refs. 60–63.

Inorganic Compounds

Selenium forms inorganic and organic compounds similar to those of sulfur and tellurium. The oxidation states are −2, 0, +4, and +6. The most important inorganic compounds are the selenides, halides, oxides, and oxyacids. Detailed description of the compounds, techniques and methods of preparation, and references to original work are given in refs. 1–3, 5, 7–11, 64–68. Some important physical properties of selenium compounds are listed in Table 6.

Selenides. Selenium forms compounds with most elements. Binary compounds of selenium with 58 metals, 8 nonmetals and alloys with 3 other elements have been described (68). Most of the selenides can be prepared by direct reaction, which varies from very vigorous with alkali metals to sluggish and requiring high temperature with hydrogen.

Table 6.   Physical Properties of Some Inorganic Selenium Compounds[a]

Compound	CAS Registry No.	Formula	Mp, °C	Bp, °C	Heat of formation (at 25°C), kJ/mol[b]
hydrogen selenide	[7783-07-5]	H_2Se	−65.7	−41.3	85.7
carbon selenide	[506-80-9]	CSe_2	−40–45	125	155.2
selenium chloride	[10025-68-0]	Se_2Cl_2	−85	127 (dec)	−43.7
selenium bromide	[7789-52-8]	Se_2Br_2		227 (dec)	
selenium dichloride	[14457-70-6]	$SeCl_2$			−40.6
selenium tetrafluoride	[13465-66-2]	SeF_4	−13.2	101	
selenium tetrachloride	[10026-03-6]	$SeCl_4$	305	196 (sub)	−188.3
selenium tetrabromide	[7789-65-3]	$SeBr_4$	75 (dec)		
selenium hexafluoride	[7783-79-1]	SeF_6	−34.6	−46.6 (sub)	−1029.3
selenium dioxide	[7446-08-4]	SeO_2	340	315 (sub)	−238.5
selenium trioxide	[13768-86-0]	SeO_3	118		−184
selenious acid	[7783-00-8]	H_2SeO_3			−531.3
selenic acid	[7783-08-6]	H_2SeO_4	60		−538.0
selenium oxyfluorides	[7783-43-9]	$SeOF_2$	15	125–126	
	[14984-81-7]	SeO_2F_2	−99.5	−8.4	
	[27218-12-8]	$SeOF_6$	−54	−29	
selenium oxychloride	[7791-23-3]	$SeOCl_2$	10.8	177.6	
selenium oxybromide	[7789-51-7]	$SeOBr_2$	41.7	217	

[a] Refs. 1, 2, and 6.
[b] To convert J to cal, divide by 4.184.

Hydrogen Selenide. The only important hydride of selenium is hydrogen selenide, H_2Se, although there is some evidence for H_2Se. Deuterium selenide [13536-95-3], D_2Se, has been prepared and its properties studied. Hydrogen selenide may be prepared by the action of acids or water on some metal selenides, usually aluminum selenide or iron selenide, or by passing hydrogen and selenium vapor over pumice at 177°C resulting in a ca 58% yield. Thermodynamically, hydrogen selenide is unstable at room temperature, but the rate of decomposition is very slow. The gas is colorless, flammable, highly toxic, and has an offensive odor. It irritates the mucous membranes, nose, throat, and eyes. Dry oxygen has no effect on hydrogen selenide. Moist air decomposes it, liberating elemental selenium. It burns in air with excess oxygen to SeO_2 and, with insufficient oxygen, to Se. Hydrogen selenide is a strong reductant. The aqueous solution is weakly acidic, although it is stronger than aqueous acetic acid.

Binary Selenides. Binary selenides are formed by heating selenium with most elements, reduction of selenites or selenates with carbon or hydrogen, and double decomposition of heavy-metal salts in aqueous solution or suspension with a soluble selenide salt, eg, Na_2Se or $(NH_4)_2Se$ [66455-76-3]. Atmospheric oxygen oxidizes the selenides more rapidly than the corresponding sulfides and slower than the tellurides. Selenides of the alkali, alkaline earth metals, and lanthanum elements are water-soluble and are readily hydrolyzed. Heavy metal selenides are insoluble in water. Polyselenides form when selenium reacts with alkali metals dissolved in liquid ammonia. Metal hydrogen selenides of the *M* HSe type are known. Some heavy metal selenides show important and useful electric, photoelectric, photo-optical, and semiconductor properties. Ferroselenium and nickel selenide are made by melting a mixture of selenium and metal powder.

The carbon selenides CSe_2 [506-80-9], COSe [1603-84-5], and CSSe [5951-19-9] are not very stable, especially when exposed to light. Nitrogen selenide [12033-88-4], Se_4N_4, is an orange-red amorphous powder and is unstable. It detonates easily when scratched or even heated to 200°C. Phosphorus selenides, P_4Se_3 [1314-86-9], and the readily flammable P_2Se [12137-67-6], are known. Sulfur and selenium form homogeneous solutions in the liquid state, and these form a series of solid solutions on crystallization. Brick-red SeS [7446-34-6] has been prepared by precipitation. Selenium and tellurium form a continuous series of solid solutions or alloys, which probably contain different simple and mixed chains of various lengths.

Halides and Oxyhalides. ***Halides.*** Selenium combines directly with fluorine, chlorine, and bromine, but not iodine, and forms the monohalides, Se_2X_2; the dihalides, SeX_2; the tetrahalides, SeX_4; and the hexafluoride, SeF_6. The compounds are covalent and volatile. The stability decreases from the fluorides to the iodides. Although SeF_6 is stable in an electric-arc discharge, SeI_2 [81256-76-0] and SeI_4 [13465-68-4] exist in solution only. Selenium halides and sulfur halides have similar properties. The tetrahalides are more stable than the dihalides. When heated, the dihalides decompose into selenium and a tetrahalide. The chlorides and the bromides are easily hydrolyzed and the hexafluoride hydrolyzes very slowly. The hydrolysis products are selenious acid and halogen acid.

Selenium tetrafluoride, SeF_4, is a colorless liquid and fumes in air. It is a powerful oxidant. It attacks silicon, phosphorus, arsenic, antimony, and bismuth; dissolves sulfur, selenium, bromine, and iodine; reacts violently with water; and attacks glass. Selenium tetrachloride, $SeCl_4$, is a white or pale yellow crystalline substance. It forms addition compounds with metal and nonmetal chlorides, ammonia, thiourea, and amines. It is a strong chlorinating substance; it is readily hydrolyzed to selenious acid and is reduced to the monochloride by sulfur and selenium. Selenium monochloride, Se_2Cl_2, is a brownish red oily liquid. It is decomposed by a number of substances and reacts with many metals and nonmetals: with some it reacts violently and even explosively. Selenium tetrabromide, $SeBr_4$, is orange-red, crystalline, hygroscopic, and readily hydrolyzed. In carbon disulfide solution and with ammonia, it forms the highly explosive Se_4N_4. Selenium monobromide is a dark-red (almost black), pungent-smelling, oily liquid. It is hygroscopic and is readily hydrolyzed by water and alkalies and forms addition compounds with amines and heterocyclic bases.

Oxyhalides. Selenium oxyhalides, $SeOX_2$, are prepared by the addition of a halogen to preferably a dry mixture of selenium and selenium dioxide or to a carbon tetrachloride suspension. Selenium oxyfluoride is conveniently prepared by treating selenium oxychloride with silver fluoride. The product is a colorless liquid with a pungent odor. It reacts with water, glass, silicon, and very violently with red phosphorus, and it forms addition compounds with hydrofluoric acid. Selenium oxychloride is also a colorless liquid with a pungent odor, is readily hydrolyzed, and fumes in moist air. It is a strong chlorinating agent and oxidant; it reacts with many metals, nonmetals, and organic substances and at times it does so explosively. It dissolves many metal chlorides and chalcogenides, and it forms solvates and addition compounds. When mixed with sulfur trioxide, it dissolves alumina, chromic oxide, and rare metal oxides. Selenium oxychloride has been called the universal solvent. It has a high dielectric constant and has been used extensively as an ionizing solvent. Selenium oxybromide is an orange solid and is very easily hydrolyzed, decomposing in air at 50°C. It is a brominating agent with solvent properties similar to those of the oxychloride.

Oxides, Acids, and Salts. Oxides. Selenium monoxide [*12640-89-0*] has only been detected spectroscopically as the gas. Selenium dioxide, SeO_2, is prepared by burning selenium in a current of air or oxygen and, optionally, by passing it over a catalyst or by oxidation with nitric acid to selenious acid followed by evaporation to dryness and dehydration by heating. The compound is white and crystalline with a tetragonal structure and is yellowish green as the vapor. It dissolves easily in water forming selenious acid and less readily in acetone, alcohols, glacial acetic acid, and dioxane. Selenium dioxide is less stable than sulfur dioxide or tellurium dioxide. Like sulfur dioxide it is an acid anhydride, but unlike SO_2 it is a strong oxidant and is reduced to elemental selenium by sulfur dioxide, hydrogen, hydrogen sulfide, ammonia, phosphorus, carbon, and even organic dust particles in moist air. The reaction with sulfur dioxide does not occur without the presence of water vapor. Selenium dioxide strongly catalyzes the oxidation of nitrogen compounds in Kjeldahl digestions. The use of selenium dioxide as an oxidant in organic chemistry has been reviewed (69). In the gaseous state it is reduced by ammonia and by hydrocarbons with emission of light. This gaseous reaction with ammonia was used at one time in the production of high purity selenium. Selenium dioxide is oxidized by fluorine to SeO_2F_2 and by hydrogen peroxide and potassium permanganate to form selenates.

Selenium trioxide is white, crystalline, and hygroscopic. It can be prepared by the action of sulfur trioxide on potassium selenate or of phosphorus pentoxide on selenic acid. It forms selenic acid when dissolved in water. The pure trioxide is soluble in a number of organic solvents. A solution in liquid sulfur dioxide is a selenonating agent. Selenium trioxide is a strong oxidant and reacts with many inorganic and organic substances. It is stable in very dry atmospheres at room temperature and, on heating, it decomposes first to selenium pentoxide [*12293-89-9*] and then to selenium dioxide.

Acids and Salts. The important oxyacids are selenious acid, H_2SeO_3, and selenic acid, H_2SeO_4. Selenious acid is formed by wet oxidation of the element or of a solution of selenium dioxide in water. It is colorless, crystalline, and highly soluble in water, forming a weakly acidic solution and two series of salts: selenites and hydrogen selenites. Selenious acid is an oxidant. It is easily reduced by nascent hydrogen, hydrogen sulfide, sulfur dioxide, and hydrochloric, hydrobromic, and hydroiodic acids. It is also readily oxidized to selenic acid by fluorine, bromine, chlorine, chloric acid, ozone, hydrogen peroxide, or permanganate; the oxidation by chlorine or bromine is reversible. Electrolysis of selenious acid yields selenic acid free of selenious acid. The selenate and hydrogen selenate salts are similar in their properties to the sulfates and hydrogen sulfates. Selenic acid and selenates are better oxidants than sulfuric acid or sulfates.

Some of the other selenium oxyacids are permonoselenic acid [*81256-77-1*], H_2SeO_5, perdiselenic acid [*81256-78-2*], $H_2Se_2O_8$, and pyroselenic acid [*14998-61-9*], $H_2Se_2O_7$. Selenosulfuric acid, H_2SeSO_3, is not known but its alkali metal salts have been prepared.

Other inorganic selenium compounds include sodium selenocyanate [*4768-87-0*], NaSeCN, which is prepared by melting together selenium and sodium cyanide; selenocyanogen [*27151-67-3*], (SeCN), (SeCN)$_2$; sodium selenosulfate [*25468-09-1*], Na_2SeSO_3, which is prepared by dissolving selenium in aqueous sodium sulfite (acidification decomposes this compound); and selenate alums, eg, $Al_2(SeO_4)_3 \cdot K_2SeO_4$ [*13530-59-1*].

Organic Compounds

The chemical properties of organosulfur, organoselenium, and organotellurium compounds are markedly similar. Because bond stability decreases with increasing atomic number of the element, thermal stability and stability on exposure to light of all wavelengths decrease and oxidation susceptibility increases. Thus, selenium compounds often turn red when exposed to light or air. These factors as well as the unpleasant odor of many compounds and difficulties in analytical determination in the early 1900s hindered the development of organoselenium chemistry. In the last 50 yr, interest has grown appreciably. A large and steadily increasing number of selenium analogues of organosulfur have been prepared and their properties and possible uses have been studied. The compounds range from the simple COSe, CSSe, and CSe_2 to complex heterocyclic compounds, selenium-containing coordination compounds, and selenium-containing polymers. Some of the more important compounds are given in Table 7.

Examples of the more simple ring compounds include selenophene [*288-05-1*] (**1**), selenolo[2,3-*b*]selenophene [*250-85-1*] (**2**), seleno[3,2-*b*]selenophene [*251-49-0*] (**3**), seleno[3,4-*b*]selenophene [*250-71-5*] (**4**), benzo[*b*]selenophene [*272-30-0*] (**5**), selenanthrene [*262-30-6*] (**6**), selenazole [*288-52-8*] (**7**), selenetane [*287-28-5*] (**8**), and cyclic ethers 1,4-diselenane [*1538-41-6*] (**9**) and 1,4-selenoxane [*5368-46-7*] (**10**).

The organoselenium compounds and their chemistry and methods of preparation are treated in refs. 3, 6, 8, 12–13, 15, 70. The selenium-containing polymers discussed in ref. 71 are of interest because of their semiconducting and plastic properties. The

Table 7.   Types of Organoselenium Compounds

Type	Formula[a]
selenols (selenomercaptans)	RSeH
selenides	RSeR
diselenides	RSeSeR
triselenides	RSeSeSeR
seleninic acids	RSeOH
Derivatives:	
selenamides	$RC(Se)NH_2$
selenocyanates	$RSeCN^2$
seleninic acids	RSeO(OH)
selenonic acids	$RSeO_2(OH)$
halides	R_3SeX
dihalides	R_2SeX_2
trihalides	$RSeX_3$
selenoxides	R_2SeO
selenones	R_2SeO_2
selenonium compounds	$R_3Se^+X^-$
selenoaldehydes	RCHSe
selenoketones, cyclic selenoketones	RC(Se)R
sulfenoselenides	RSSeR
selenourea [630-10-4]	$(NH_2)_2CSe$
Derivatives:	$(RNH)_2CSe$
selenocarboxylic acids, esters, and N derivatives	
selenacycloalkanes	$\overline{CH_2(CH_2)_x Se}$
selenophenols	C_6H_5SeH
selenosemicarbazides	$H_2NC(Se)NHNHR$
selenium-containing N heterocycles	
selenium-containing carbohydrates (selenogluconides, selenosugars)	
selenium-containing dyes (cyanine and noncyanine)	

[a] R = aliphatic or aromatic group.

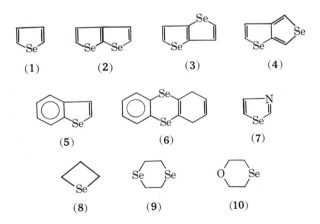

biologically important compounds include selenoaminocarboxylic acids and derivatives, selenium-containing peptides, and selenium derivatives of pyrimidines, purines,
cholines, steroids, coenzyme A, and other compounds (6,15). The biochemical and

medical aspects and uses of organoselenium compounds are also discussed in refs. 14–15, 54, 58.

Uses

Electrical and Optical. The electrical and optical uses are based on the unique photoelectric and semiconducting properties of selenium (see Photomultiplier tubes; Photovoltaic cells). The electrical conductivity of selenium is low in the dark, but increases by about three orders of magnitude on exposure to light. As semiconductors, selenium and some of its compounds and alloys possess asymmetric conduction, ie, they pass electric current more freely in what is called the forward or conducting direction than in the reverse or nonconducting direction (see Semiconductors). This phenomenon arises at a boundary surface of thin layers or films of the semiconductor in contact with various metals. Bulk gray selenium by itself has no such asymmetric properties.

As early as 1876 it was observed that conduction in selenium was nonlinear; the current increased less than proportionately as the applied voltage was increased. In 1883 a selenium photovoltaic cell, which had many features common to the modern selenium rectifier, was described, and applications of selenium in rectifiers was described in 1909. The first commercial unit was developed in 1924.

Photoelectric Cell. The photoelectric cell consists of a layer of selenium on a metal plate substrate coated with a translucent film of gold or another metal (see Film deposition techniques). The photoconductivity response of the cell is linear, rapid, and steady for small and medium illumination intensities. The cell is more efficient than other photoelectric devices but suffers greater fatigue when exposed to high intensities of illumination. It is more sensitive in the spectral range of the human eye than potassium, cesium, and cuprous oxide cells.

Selenium cells are used in photometers, colorimeters, electric eyes, pyrometers, and other light-sensitive instruments. The commercial photographic exposure meter may be a selenium or, more frequently, a cadmium selenide cell connected to a deflecting galvanometer. Selenium photoelectric cells, which have only limited application and have never required significant quantities of selenium, are of greater historical than commercial interest. They have been almost entirely superceded by cells of the photovoltaic type.

Photovoltaic Cell. A photovoltaic cell is basically a rectifier with a transparent counterelectrode, which usually is of cadmium oxide. The cell consists of an iron or brass plate on which is deposited a 100-μm layer of polycrystalline selenium and then a 1-μm layer of cadmium, which forms an intermediate layer of cadmium selenide with selenium. The cell is, therefore, a CdSe–Se heterojunction illuminated through the transparent cadmium oxide electrode.

Rectifier. A selenium rectifier is similar in construction to the photovoltaic cell. In manufacture the metallic base plate or disk, which is of steel, stainless steel, or aluminum, is sandblasted or etched, nickel-plated, coated by a 50–60-μm layer of selenium, and sprayed with low melting cadmium alloy as a counterelectrode. The nickel selenide interface, which is less than 1 μm thick, gives a good ohmic contact for selenium. The thin cadmium selenide interface, formed between the cadmium compound and the selenium, provides a barrier layer between the counterelectrode and the selenium. The barrier layer is important for the function of selenium rectifiers and photovoltaic cells.

Rectifier manufacturers require selenium of high purity and prefer to dope it themselves with additives. Sodium, tellurium, and the halides generally accelerate crystallization, whereas arsenic and phosphorus retard it. Sulfur and mercury adversely affect the electrical characteristics. The manufacture of selenium rectifiers is proprietary; most of the available information is in patent literature.

The nonlinear and thermal properties of selenium rectifiers have led to their use as voltage-sensitive resistors or varistors and temperature-sensitive resistors or thermistors. Individual selenium cells are connected in series or stacks. The stacks are then assembled in either series or parallel to rectifiers of a wide variety of sizes and capacities, ranging from a fraction of a watt to hundreds of kilowatts. Selenium rectifiers have been used in mechanical equipment with pulsating dc; excitation of synchronous motors; electrochemical processes, in which they can deliver 40 kA at 6 V; electrostatic precipitation, in which they rectify small currents at several thousand volts; welding equipment; battery chargers; voltage regulators; etc. The efficiency of selenium rectifiers is high, ca 85%, at full load and over a wide range of load. Fan cooling allows a selenium rectifier to carry a current several times its rated capacity.

In the 1940s, selenium rectifiers were favored over thermionic tubes because of their almost instantaneous operation, longer life, lower operating temperature and power consumption, and lower cost. Compared to cuprous oxide rectifiers, selenium rectifiers are lighter and more compact and are characterized by a higher operating voltage and permissible voltage overload and long operating life. By the 1950s, their use reached such industrial and military importance as to dislocate the U.S. selenium market and bring about government control of distribution. In ca 1952, silicon and germanium became competitive with selenium and displaced it to a large degree. Nevertheless, partly owing to the development of thin-film devices, the consumption of selenium in the rectifier field continues to be significant.

Maximum operating temperatures are 150°C for the selenium, 200°C for the silicon, and 85°C for the germanium rectifiers. Selenium rectifiers are somewhat larger than silicon rectifiers of equal capacity, and the former resist nuclear-radiation damage better. They also withstand voltage overloads better than do germanium rectifiers.

Xerography. Xerography is the most highly developed form of electrophotography or photocopying and is a primary application of the photoconductive properties of selenium. The process employs reusable photoreceptors. The photoreceptor is a 50 μm-thick film of vitreous selenium on an aluminum substrate. The film is sensitized by electrostatic charging with a 10^5 V/cm field and is exposed to a light-and-dark image pattern. The surface potential is decreased in the light-struck areas to form a latent image. The latter is developed by an oppositely charged powder, ie, toner particles, which adhere only to the charged areas and form a visible image. The toner particles are then transferred to charged bond paper, and the image is made permanent by fusing or melting the toner particles into the paper surface. The photoreceptor is made ready for another print cycle by removing any residual toner and flooding with light to discharge the surface potential (72) (see Electrophotography; Reprography).

Manufacturers of xerographic equipment require selenium and selenium alloys of high purity. This field is still growing and it provides an outlet for ca 25% of selenium production and generates important amounts of scrap for retreatment. Attempts have been made to replace the selenium in xerography with organoselenium compounds in a polymeric matrix (73) and other semiconductor materials.

Miscellaneous. Selenium and selenium compounds are used in some phosphors and vidicon targets. An electrolyte for lead storage batteries comprised of selenic acid, ferrous sulfate, sodium chloride, and manganous sulfate in small concentrations has been patented (74). Selenium is used in laser technology (see Lasers). Cadmium selenide [1306-24-7] and copper indium selenide [12018-95-0] are promising materials for solar (photovoltaic) cells.

Metallurgical. *Ferrous Metals.* Selenium in small concentrations decreases the surface tension of molten steel (qv) more than S, N, O, P, and C, but probably less than Te. It retards nitrogen absorption and improves the impact resistance of the steel. In low Mn cast iron, selenium promotes carbide formation, but when Mn is present at higher levels, it does not suppress graphite nucleation. In small amounts selenium is a mild deoxidizer and grain-refining agent in steels, especially those intended for carburizing and direct hot rolling without further treatment. Selenium alters the filamentary shape of MnS in hot-rolled steel to a globular shape and thus improves the transverse properties. It permits more severe cold forging than does sulfur and improves machinability. The forging steels contain 0.18–0.22 wt % selenium.

Elemental selenium, nickel–selenium, and iron selenide in 0.005–0.02 wt % additions decrease the pinhole porosity in high alloy steels. Adding 0.15–0.30 wt % selenium to carbon and alloy steels improves the machinability and the production rate of parts and improves surface finish. The low carbon, free-machining selenium-containing steels may also contain S, Pb, Ni, Bi, and Te. On a weight basis, selenium provides better machinability than does sulfur and has a less deleterious effect on corrosion resistance, hot shortness, cold working, and strength of finished products. The increase in production rate over selenium-free metal is 16–60% for steels and 25–50% for stainless steels. Addition of 0.1 wt % selenium to silicon electric transformer steels improves the magnetic permeability and decreases the power-core loss. Selenium is required in several varieties of military steels of the 18-8 type. The retention of selenium in ferrous metals is 67–90% depending on the addition method.

Copper and Copper Alloys. Selenium is highly surface-active in liquid copper metals. Selenium additions (0.25–1 wt % as copper selenide) improve the machinability more than do sulfur, tellurium, or bismuth, have little effect on the strength, and decrease the ductility and conductivity only slightly. Unlike lead, selenium does not cause hot or cold shortness, whereas bismuth additions cause embrittlement and prevent cold working. Study of the microstructure of selenium-containing copper shows the presence of Cu–Cu$_2$Se eutectic, which serves to break up the chips during machining. Selenium does not impair the soldering or brazing properties of copper metals but impairs their weldability.

Nickel–Iron and Cobalt–Iron Alloys. Selenium improves the machinability of Ni–Fe and Co–Fe alloys, which are used for electrical applications. Unlike sulfur and tellurium, it does not cause hot brittleness. Addition of 0.4–0.5 wt % selenium enhances the coercive force of cobalt–iron–titanium alloys in permanent magnets.

Lead and Lead Alloys. Selenium increases the recrystallization temperature of low antimony (0.5–3.5 wt % Sb) alloys and improves their casting and mechanical properties. Grids made of selenium-containing low antimony alloys are less susceptible to gas evolution and self-discharge in low maintenance, automotive, lead storage batteries (75) (see Batteries).

Chromium Plating. Sodium selenate or selenic acid have been added to chromium-plating baths since 1959 with a marked improvement in corrosion protection from pitting, blistering, and rusting, especially of plated automobile articles exposed to salt in the North American snow belt. The optimum concentration of the selenate ion is 0.015–0.018 g/L in a bath containing 250 g/L chromic acid and 2.5 g/L sulfuric acid at a cathodic current density ca 21 A/m^2 (225 A/ft^2) and 43–44°C. The selenate ion causes ca 600 uniformly distributed microcracks per linear centimeter in the plate. The network of fine intersecting cracks distributes the galvanic action over the surface and thus offers corrosion protection of the base metal. Such chromium plate has a dull luster with a decreased glare. A similar microporous, decorative, corrosion-protective plate has been developed recently (76) (see Electroplating).

Miscellaneous. Addition of selenium to magnesium and magnesium alloys is claimed to improve corrosion resistance by seawater. However, the technique of making this addition requires great care. Selenium dioxide has been added to baths used in etching, formation of decorative colors on galvanized iron and nonferrous metals and alloys, and corrosion-protective coatings on magnesium alloys. Inorganic and organic selenium compounds are brighteners in electroplating. Selenious acid improves the current efficiency in electrowinning of manganese. Diffusion coating of ferrous metals with selenium in a molten salt bath improves resistance to wear and to seizure of contact surfaces. In the manufacture of friction brake facings by powder metallurgy, the addition of selenium to copper oxide followed by reduction, mixing with nonmetallic particles, pressing, and sintering improves the adhesion of the metallic to the nonmetallic particles and increases the pressed green strength by 30%.

Glass and Ceramics. The use of selenium in decolorizing glass was introduced into the United States in 1915. Until then, manganese dioxide from the USSR had been employed in conjunction with arsenic oxide. As manganese dioxide of uniform quality was then difficult to obtain owing to wartime conditions, it was successfully and completely replaced by selenium. Selenium is generally used in the elemental form, although sodium and barium selenites and frits containing them have also been used. The amount required depends mainly upon the iron content of the glass batch; additions of 0.01–0.015 kg/t of glass batch have been reported. Volatilization loss is high, in some cases 80% or more, although efforts have been made to improve the degree of utilization and to decrease the iron content of the batch.

The glass industry consumes important amounts of selenium, chiefly to decolorize container glass and, to a smaller degree, table, lighting, and industrial glass. Some selenium is used to produce ruby glass and glasses of different colors. In flat plate glass, selenium is used to produce bronze or smoky colors and to block solar heat transmission, especially in office buildings.

Selenium ruby glass has been used in the United States since 1895 for religious articles, vases, and signal-light lenses. The color can be obtained by adding cadmium sulfide and selenium to the glass batch and melting in a reducing atmosphere. The resulting faint-yellow glass is reheated to develop the ruby color. A typical high grade signal-lens glass contains 2 wt % selenium, 1 wt % cadmium sulfide, 1 wt % arsenious oxide, and 0.5 wt % carbon. Varying the proportions of the additives can effect a final color from intense yellow-orange to ruby red. Selenium rubies give very good transmission in the red part of the spectrum with a sharp cutoff of the other colors. They are specified in railway, marine, airport, and other signal lenses. Selenium produces a variety of colors with other additives, eg, yellow with oxides of arsenic, antimony, and bismuth; amber with ferrous oxalate and silicon; topaz with bismuth trioxide and silicon; emerald green with chromium oxide and silicon; and black with cobalt or iron oxides.

Selenium is also used by the ceramic industry in the compounding of colored glazes for chinaware, porcelain, and pottery; enamels for cast iron; and printing inks for glass and other containers. Selenium-containing glass acts as an infrared filter in laser windows. A selenium-arsenic glass was developed for use in infrared photography.

Pigments. Cadmium sulfide selenide [12626-36-7] pigments vary from orange to red to deep maroon, depending on the selenium content. They are characterized by their long life, brilliancy, and stability to heat, sunlight, and chemical action. Originally developed for use in ceramics and paints, they are used most importantly in plastics, especially those cured at high temperatures. The olive-drab zinc–chromium selenate pigments have corrosion-resistant properties (see Corrosion and corrosion inhibitors). An intensely golden-yellow pigment based on zinc sulfide contains copper–indium selenide sulfide (77) (see Pigments).

Rubber. The rubber industry consumes finely ground metallic selenium and Selenac (selenium diethyl dithiocarbamate). Both are used with natural rubber and with styrene–butadiene rubber (SBR) to increase the rate of vulcanization and to improve the aging and mechanical properties of sulfurless and low sulfur stocks. Selenac is also used as an accelerator in butyl rubber and as an activator for other types of accelerators, eg, thiazoles (see Rubber chemicals). Selenium compounds are useful as antioxidants (qv), uv stabilizers (qv), bonding agents, carbon-black activators, and polymerization additives. Selenac improves the adhesion of polyester fibers to rubber.

Lubricants. Selenium and its compounds are added to lubricating oils and greases for normal and extreme pressure service (see Lubrication and lubricants). Dialkyl-selenides are oxidation inhibitors in lubricants. Barium, calcium, and zinc salts of seleninic acids ($RSe(O)OH$), where R is an alkyl or alkaryl radical with 10–40 carbon atoms, improve the detergent properties of lubricating oils. Dry, powdered lubricants containing diselenides of tungsten, niobium, tantalum, and molybdenum are manufactured for use at elevated temperatures in an ultrahigh vacuum. Under such conditions, they possess lower outgassing properties, lower coefficients of friction, and higher stability than graphite and the corresponding metal sulfides. Some self-lubricating iron alloys contain selenium.

Organic Chemistry and Pharmaceuticals. Selenium dioxide is an important oxidizing agent and catalyst in the synthesis of organic chemical and drug products (78) (see Hormones; Vitamins). It has been used in the manufacture of cortisone and nicotinic acid (niacin). Selenium and its compounds are useful in a wide variety of organic reactions, eg, oxidation, ammoxidation, reduction, hydrogenation, dehydrogenation, addition, condensation, isomerization, cracking, halogenation, dehalogenation, and polymer treatment. Dehydrogenation of hydroaromatic compounds $\geq 250°C$ with the evolution of hydrogen selenide has been valuable in elucidating the structure of some natural products. Selenium is often used as a catalyst in Kjeldahl determinations to hasten the digestion of nitrogenous materials. Some organoselenium compounds act as miticides. Others have neuroleptic and antidepressant properties. An antibiotic selenomycin is known. An organoselenium compound is claimed to possess psychotropic activity (39) (see Psychopharmacological agents).

Medicine and Nutrition. A stabilized buffered suspension of selenium sulfide has been marketed for many years as Selsun (Abbott Laboratories) for control of seborrheic dermatitis of the scalp. A similar sulfur or selenium sulfide shampoo, containing a metallic cation complex, has been prepared (79). Topical application of selenium sulfide controls dermatitis, pruritis, and mange in dogs (see Cosmetics).

In conjunction with vitamin E, selenium is very effective in the prevention of muscular dystrophy in animals. Sodium selenite and sodium selenate are administered to prevent exudative diathesis in chicks (a condition in which fluids leak out of the tissues), white muscle disease in sheep, and infertility in ewes. It lessens the incidence of pneumonia in lambs and of premature, weak, and still-born calves, controls *hepatosis dietetica* in pigs, and decreases muscular inflammation in horses. White muscle disease, widespread in sheep of the selenium-deficient areas of New Zealand and the United States, is insignificant in high selenium soil areas. The supplementation of animal feeds with selenium was approved by the FDA in 1974 (see Mineral nutrients).

Inorganic and organic selenium compounds, including selenoproteins, have been used to treat selenium deficiency in humans and warm-blooded animals, control stress and high blood pressure in poultry and animals, combat aging, and improve the healing of surgically incised, lacerated, and burned tissue (see Memory enhancing agents and antiaging drugs). Some organoselenium compounds are useful in immunoassays. Tablets containing several micrograms sodium selenite are sold over-the-counter in the United States.

Miscellaneous. Sodium selenite is used in photography (qv) to produce prints with warm brown tones; sodium selenosulfate gives similar results. Derivatives of selenazole [288-52-8], benzoselenazole [273-91-6], and naphthoselenazole and selenium-containing dyes are used in supersensitive photographic emulsions. Selenocarbazides are used in photographic bleach fixing baths. Some photographic antifoggants contain selenium.

Sodium selenate has been used on a small scale in commercial greenhouses, primarily for growing carnations and chrysanthemums. It is transformed by the plants into volatile selenides, which repel red spiders, mites, thrips, and aphids. Sodium selenite is not intended for crops which could ultimately be used as food for humans or domestic animals.

Selenium and selenium compounds are also used in electroless nickel-plating baths, delay-action blasting caps, lithium batteries, xeroradiography, cyanine- and noncyanine-type dyes, thin-film field effect transistors (FET), thin-film lasers, and fire-resistant functional fluids in aeronautics (see Electroless plating).

BIBLIOGRAPHY

"Selenium and Selenium Compounds" in *ECT* 1st ed., Vol. 12, pp. 145–163; "Selenium" in *ECT* 2nd ed., Vol. 17, pp. 809–833, by E. M. Elkin, Canadian Copper Refiners Ltd., and J. L. Margrave, University of Wisconsin.

1. K. W. Bagnall, *The Chemistry of Selenium, Tellurium, and Polonium*, 2nd ed., Elsevier Publishing Company, Amsterdam, The Netherlands, 1966.
2. M. Schmidt, W. Siebert, and K. W. Bagnall, *The Chemistry of Sulphur, Selenium, Tellurium, and Polonium*, Pergamon Press Ltd., Oxford, UK, 1973.
3. N. V. Sidgwick, *The Chemical Elements and Their Compounds*, Oxford University Press, London, 1950, pp. 948–994.

4. S. P. Bapu, ed., *Trace Elements in Fuel, Advances in Chemistry*, Series 141, American Chemical Society, Washington, D.C., 1975.

5. A. A. Kudryavtsev, *The Chemistry and Technology of Selenium and Tellurium*, Collets (Publishers) Ltd., London and Wellingborough, UK, 1974.

6. R. A. Zingaro and W. C. Cooper, *Selenium*, Van Nostrand Reinhold Company, New York, 1974.

7. *Gmelins Handbuch der Anorganischen Chemie*, 8th ed., Verlag Chemie, Weinheim/Bergstrasse, System-Nummer 10, Teil A, Lfg. 1, 1942, Lfg. 2, 1950, Lfg. 3, 1953; Teil B, 1949.

8. J. N. Friend, ed., *Textbook of Inorganic Chemistry*, Griffin, London, Vol. 7, Pt. 2, 1931; Vol. 9, Pt. 4, 1937.

9. J. W. Mellor, *Comprehensive Treatise on Inorganic and Theoretical Chemistry*, Vol. 10, Longmans, Green & Co., Inc., New York, 1930, pp. 693–932.

10. D. M. Yost and H. Russell, Jr., *Systematic Inorganic Chemistry of the Fifth-and-Sixth-Group of Nonmetallic Elements*, Prentice-Hall, Inc., Englewood Cliffs, N.J., 1944.

11. P. Pascal, *Nouveau Traité de Chimie Minérale*, Vol. XIII, Part 2, Masson & Cie., Paris, 1960, pp. 1651–1912.

12. K. J. Irgolic and M. V. Kudchadker in ref. 6, p. 408.

13. E. P. Painter, *Chem. Rev.* **28**, 179 (1941).

14. H. E. Ganther in ref. 6, p. 546.

15. D. L. Klayman and W. H. H. Gunther, eds., *Organic Selenium Compounds: Their Chemistry and Biology*, John Wiley & Sons, Inc., New York, 1973.

16. Y. Okamoto and W. H. H. Gunther, eds., *Ann. N.Y. Acad. Sci.* **192**, (April 17, 1972).

17. *2nd Int. Symp. on Org. Selenium and Tellurium Chemistry, including Biochemistry*, Lund, Sweden, Aug. 18–22, 1975, published in *Chem. Scr.* **8A**, (1975).

18. N. D. Sindeeva, *Mineralogy and Types of Deposits of Selenium and Tellurium*, Interscience Publishers, a division of John Wiley & Sons, Inc., New York, 1964.

19. A. M. Lansche, *Selenium and Tellurium—A Materials Survey*, U.S. Bureau of Mines Information Circular 8340, Government Printing Office, Washington, D.C., 1967.

20. I. Rosenfeld and O. A. Beath, *Selenium-Geobotany, Biochemistry, Toxicity, and Nutrition*, Academic Press, Inc., New York, 1964.

21. B. Mason, *Principles of Geochemistry*, 3rd ed., John Wiley & Sons, Inc., New York, 1966.

22. J. R. Loebenstein in *Mineral Facts and Problems*, U.S. Bureau of Mines Bulletin 671, U.S. Govt. Printing Office, Washington, D.C., 1980; *Proceedings of the Symposium on Industrial Uses of Selenium and Tellurium*, Toronto, Canada, Oct. 1980, pp. 6–17.

23. P. H. Jennings and J. C. Yannopoulos in ref. 6.

24. L. A. Soshnikova and M. M. Kupchenko, *Pererabotka Medeelektrolitnykh Shlamov* (*Treatment of Electrolytic Copper Refinery Slimes*), Metallurgiya, Moscow, USSR, 1978.

25. J. H. Schloen and E. M. Elkin in A. Butts, ed., *Copper*, Reinhold Publishing Corporation, New York, 1954, Chapt. 11; *Trans. Am. Inst. Min. Metall. Pet. Eng.* **188**, 764 (1950).

26. U.S. Pat. 2,863,731 (Dec. 9, 1958), C. B. Porter, A. L. Labbe, Jr., and K. N. Pike (to American Smelting & Refining Co.).

27. Eur. Pat. Appl. 21,744 (Jan. 7, 1981), G. S. Victorovich, R. Sridhar, and M. C. E. Bell (to Inco Ltd.).

28. U.S. Pat. 2,378,824 (June 10, 1945), J. O. Betterton and Y. E. Lebedeff (to American Smelting & Refining Co.).

29. Brit. Pat. 440,004 (Dec. 18, 1935), (to Bolidens Gruvaktiebolaget); U.S. Pat. 2,010,870 (Aug. 13, 1935), A. R. Lindblad (to Bolidens Gruvaktiebolaget).

30. U.S.S.R. Pat. 112,634 (Aug. 15, 1958), T. N. Greiver.

31. T. N. Greiver, I. G. Zaitseva, and V. M. Kosover, *Selen i Tellur* (*Selenium and Tellurium*), Metallurgia, Moscow, USSR, 1977, pp. 164–200.

32. U.S. Pats. 2,835,558 (May 20, 1958) and U.S. Pat. 2,990,248 (June 27, 1961), L. E. Vaaler (to Diamond Alkali Company); U.S. Pat. 3,127,244 (May 31, 1964), E. M. Elkin and P. R. Tremblay (to Canadian Copper Refiners Ltd.); B. H. Morrison in M. E. Wadsworth and F. T. Davis, eds., *Unit Processes in Hydrometallurgy*, AIME Metallurgical Society Conferences, Vol. 24, Gordon & Breach Science Publishers, Inc., New York, 1964, pp. 227–249.

33. U.S. Pat. 4,229,270 (Oct. 21, 1980), K. N. Subramanian, M. C. E. Bell, J. A. Thomas, and N. C. Nissen (to International Nickel Company, Inc.).

34. U.S. Pat. 2,322,348 (June 22, 1943), C. W. Clark (to Canadian Copper Refiners Ltd.); G. Bridgstock, E. M. Elkin, and S. S. Forbes, *Trans. Can. Inst. Min. Metall.* **63**, 523 (1960).

35. U.S. Pat. 4,047,939 (Sept. 13, 1977); Can. Pat. 1,091,035 (Dec. 9, 1980), B. H. Morrison (to Noranda Mines Ltd.).

36. R. Bresee, D. Vleeschhouwer, and J. Thiriar, *Proceedings of the Symposium on Industrial Uses of Selenium and Tellurium, Toronto, Canada, Oct. 1980*, pp. 31–49.

37. Ref. 24, pp. 36–45.

38. U.S. Pat. 4,047,973 (Sept. 13, 1977), J. K. Williams (to Xerox Corporation).

39. USSR Pat. 551,327 (May 23, 1977), I. N. Azerbaev and co-workers.

40. U.S. Pat. 4,192,692 (Mar. 11, 1980), H. Herrmann (to Hoechst A.G.).

41. Ref. 24, p. 107.

42. Jpn. Kokai 80 158,112 (Dec. 9, 1980), (to Sumitomo Metal Mining Company, Ltd.).

43. Can. Pat. 1,003,190 (Jan. 11, 1977), S. T. Henriksson (to Boliden AB).

44. U.S. Pat. 2,486,464 (Nov. 1, 1949), C. W. Clark and J. H. Schloen (to Canadian Copper Refiners Ltd.).

45. U.S. Pat. 2,583,799 (Jan. 29, 1952), J. H. Schloen and L. V. Franchetto (to Canadian Copper Refiners Ltd.).

46. U.S. Pat. 1,382,920, U.S. Pat. 1,382,921, and U.S. Pat. 1,382,922 (June 28, 1921), V. Lenher; *J. Am. Chem. Soc.* **42**, 2498 (1920).

47. U.S. Pat. 2,189,480 (Feb. 6, 1940), and U.S. Pat. 2,196,380 (Apr. 9, 1940), C. W. Hewlett (to General Electric Company); U.S. Pat. 2,300,196 (Oct. 27, 1942), A. R. Bozarth (to Harshaw Chemical Company); U.S. Pat. 2,306,109 (Dec. 22, 1942), R. H. Long (to Harshaw Chemical Company).

48. U.S. Pat. 2,347,128 (Apr. 18, 1944), W. F. Russel (to R. T. Vanderbilt Company); U.S. Pat. 2,351,985 (June 20, 1944), J. Loeffler (to Alien Property Custodian).

49. D. M. Chizhikov and V. P. Shchastlivyi, *Selenium and Selenides*, Collets (Publishers) Ltd., London and Wellingborough, 1968, p. 124.

50. T. E. Green and M. Turley in I. M. Kolthoff and P. J. Elving, eds., *Treatise on Analytical Chemistry*, Vol. 7, Pt. 2, Interscience Publishers, a division of John Wiley & Sons, Inc., New York, 1961, pp. 137–205.

51. H. W. Furman, ed., *Standard Methods of Chemical Analysis*, 6th ed., Vol. 1, Pt. B.D., Van Nostrand Company, Inc., Princeton, N.J., 1962.

52. A. I. Vogel, *A Text-Book of Quantitative Inorganic Analysis*, 3rd ed., Longmans, Green & Co., London, 1962.

53. J. F. Alicino and J. A. Kowald in ref. 15, pp. 1049–1081.

54. U.S. National Research Council, *Medical and Biologic Effects of Environmental Pollutants—Selenium*, U.S. National Academy of Sciences, Washington, D.C., 1976.

55. S. Barabas and P. W. Bennett, *Anal. Chem.* **35**(2), 135 (1963).

56. D. L. Klayman in ref. 15, pp. 629–814.

57. H. L. Cannon and H. W. Lakin, *Trace Metals In The Environment*, N.T.I.S. PB-274,428, U.S. Dept. of Commerce, Washington, D.C., and Research Triangle Park, N.C., Dec. 1976.

58. D. Hausknecht and R. Ziskind, *Health Effects of Selenium*, N.T.I.S. PB-250,568, Electric Power Research Institute, Palo Alto, Calif., Jan. 1976.

59. W. Krause and P. Oehme, *Dtsch. Gesundheitswes.* **34**(37), (1979).

60. J. Glover, O. Levander, J. Parizek, and V. Vouk in L. Friberg and co-eds., *Handbook of the Toxicology of Metals*, Elsevier/North Holland Biomedical Press, The Netherlands, pp. 555–577.

61. C. G. Wilber, *Clin. Toxicol.* **17**(2), 171 (1980).

62. M.-T. Lo and E. Sandi, *J. Environ. Pathol. Toxicol.* **4**, 193 (1980).

63. *Proceedings of the Symposium on Industrial Uses of Selenium and Tellurium, Toronto, Canada, Oct. 1980*, Selenium–Tellurium Development Association, Darien, Conn.

64. R. C. Brasted, *Comprehensive Inorganic Chemistry*, Vol. 8 of *Sulfur, Selenium, Tellurium, Polonium, and Oxygen*, D. Van Nostrand Company, Inc., Princeton, N.J., 1961.

65. *Inorganic Syntheses*, McGraw-Hill Book Company, Inc., New York, published at intervals since 1939.

66. J. Brauer, ed., *Handbuch der Präparativen Anorganischen Chemie*, Thieme, Leipzig, Germany, 1925–1941 (photolithoprinted by Edwards Bros., Ann Arbor, Mich.).

67. P. J. Durrant and B. Durrant, *Introduction to Advanced Inorganic Chemistry*, John Wiley & Sons, Inc., New York, 1962, p. 864.

68. D. M. Chizhikov and V. P. Shchastlivyi, *Selenium and Selenides*, Collets (Publishers) Ltd., London and Wellingborough, 1968.

69. G. R. Waitkins and C. W. Clark, *Chem. Rev.* **36**, 235 (1945).

70. V. Krishnan and R. A. Zingaro in ref. 6, p. 337.
71. L. Martillaro and M. Russo in ref. 15, p. 816.
72. G. Lucovsky and M. D. Tabak in ref. 6, pp. 788–807.
73. U.S. Pat. 4,047,947 (Sept. 13, 1977), J. Y. C. Chu and W. H. H. Gunther (to Xerox Corporation); Can. Pat. 1,064,969 (Oct. 23, 1979), W. H. H. Gunther (to Xerox Corporation).
74. U.S. Pat. 4,245,015 (Jan. 13, 1981), D. Burke.
75. U.S. Pat. 3,801,310 (Apr. 2, 1974), S. C. Nijhawan (to Varta A.G.).
76. U.S. Pat. 4,007,099 (Feb. 8, 1977), S. H. L. Wu (to Harshaw Chemical Company).
77. Ger. Offen. 3,005,221 (Aug. 21, 1980), G. P. Kinstle (Ferro Corporation).
78. V. Kollonitsch and C. H. Kline, *Ind. Eng. Chem.* **55**(12), 18 (1963).
79. U.S. Pat. 4,089,945 (May 16, 1978), R. E. Brinkman and R. L. Vogenthaler (to The Procter & Gamble Co.).

E. M. ELKIN
Noranda Mines Limited

SEMICONDUCTORS

THEORY AND APPLICATION

Semiconductors are a class of materials exhibiting electrical conductivities intermediate between metals and insulators. The pivotal point in the history of semiconductors is the invention of the transistor in 1947. Historical developments in semiconductor technology are described in refs. 1–13. Numerous important inventions have appeared in the last 25 years, including the laser in 1957, the superconducting junction in 1962, the III–V microwave oscillator in 1963, the floating-gate memory in 1967, magnetic bubble memory in 1969, and the charge-coupled device in 1970 (see also Semiconductors, fabrication and characterization; Integrated circuits; Light-emitting diodes and semiconducting lasers).

Since 1960, the number of components on a single integrated circuit has roughly doubled each year, a phenomenon known as Moore's law (14–15). An integrated circuit made in 1981 may contain over 10^5 independent electronic components. The cost reductions resulting from large-scale integration have radically influenced market demand. The price of a large computing system, for example, decreased by a factor of 1000 from 1960 to 1980. Small and medium-sized computers have made possible further cost reductions for specific applications. The number of computers in the United States increased from 1000 in the middle 1950s to 2.2×10^5 in 1976. At the same time, the number of computer professionals increased from 10^5 to 2.5×10^6. According to one estimate, semiconductor-device sales have grown from 0.5×10^9 in 1961 to $2.4

$\times 10^9$ in the United States and 5.4×10^9 worldwide in 1976 (13). A 15–20% annual growth rate through 1985 and a worldwide sales volume of 10×10^9 in 1982 have been predicted (16) (see Computers).

Solid-State Theory

Intrinsic Properties. Metals, insulators, and semiconductors display a wide variety of electrical and optical properties which are elegantly described by the band theory of solids (17). The origin of energy bands is often explained according to two complementary approaches; chemical and physical (18–19). The hydrogenic atom is the basis of the chemical approach. Basic concepts of quantum mechanics, which are quantified in Schrödinger's wave equation, predict the existence of discrete energy levels around the hydrogenic nucleus, which may be occupied by electrons. If two identical nuclei are placed near each other, their energy levels are perturbed. An electron situated between the nuclei is attracted to both and, therefore, experiences a lower potential energy. The energy level associated with the unperturbed electron undergoes splitting. The lowered energy level corresponds to an electron having a high probability of residing between the nuclei. Two electrons of opposite spin may occupy this bonding level, which results in a net lowering of the electron energy of the system. If this energy reduction exceeds the gain in potential energy as a result of coulombic repulsion between the two nuclei, a covalent chemical bond forms. The level that increases in energy, called the antibonding level, corresponds to an electron that is unlikely to be found between the nuclei.

As more nuclei are added to this system, additional energy-level splitting occurs. For a large atomic array, many closely spaced levels form bands of allowed electron energies. For example, in the case of silicon (see Fig. 1) (17), the silicon atom contains

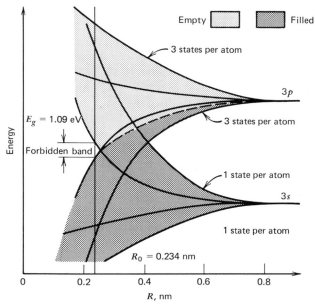

Figure 1. Atomic levels broaden into energy bands as the distance R between atoms in a tetrahedrally coordinated silicon lattice is reduced (17).

two electrons in the $3s$ level and two in the $3p$ level. As the silicon atoms are brought together into the diamond configuration, the $3s$ and $3p$ levels begin to broaden and the gap between them decreases. Eventually, the gap disappears, and electron states take on a mixed s and p character, ie, hybridization. Further decrease in atomic spacing results in the appearance of two new energy bands separated by a gap E_g. The lower valence band consists of four states per atom associated with the four bonds to the nearest-neighbor atoms. Since four electrons per atom may occupy these states, a significant lowering of the total electronic energy results in the formation of covalent bonds. Electrons in the valence band have bonding character, ie, they tend to reside in regions of low potential energy, whereas electrons in the upper conduction band have antibonding character. Figure 1 illustrates that a decrease in lattice spacing, such as might be induced by applied pressure, results in an increase of the energy gap E_g. Similarly, thermal expansion of the lattice decreases the energy gap. These effects are exploited in pressure- and temperature-sensitive semiconductor transducers (20).

The chemical view of energy bands describes electron states as superpositions of atomic orbitals. This approach is cumbersome for electron states near the band-gap edge, which are of primary importance in electronic devices, since these states do not interact strongly with the atomic nuclei. The physical view describes electron states in a crystalline solid as superpositions of free-electronlike states or plane waves; boundary conditions are defined by the periodicity of the lattice. Such plane waves are identified by an inverse wavelength or wave number k and a direction of motion. An electron state with the periodicity of the lattice is described by a standing-wave solution corresponding to zero kinetic energy. The symmetry of the lattice constrains this solution to be either in phase or out of phase with the nuclei. The in-phase solution corresponds to electrons residing in regions of low potential near the nuclei. The out-of-phase solution has an antibonding character associated with electrons in high potential energy regions. The difference in energy between these two states is the energy gap E_g, which decreases as the internuclear spacing increases. The group IVA elements in the periodic table, ie, carbon, silicon, germanium, and tin, display successively larger internuclear spacings and, hence, successively smaller band gaps. Values of the energy gap for several semiconductors are given in Table 1 (21).

With the approaches described above, it is possible to construct an energy E vs wave number k diagram for a solid, as shown in Figure 2a. The wave number k is sometimes called the crystal momentum, since it is connected through de Broglie's relation $(p = \hbar k)$ to the momentum p of the electron. The energy-vs-momentum diagram for a free particle is parabolic $(E = p^2/2m)$. The parabolic curves in the E-vs-k diagram indicate that electrons behave as free particles near the band-gap edges, although their effective mass may be very different from the rest mass m of the electron as a result of lattice-diffraction effects. Effective masses m^* are an important device parameter and are also tabulated in Table 1. The E-vs-k diagram of Figure 2 describes electron states for a single direction of motion in the crystal. A more complete description of the solid is obtained by constructing these diagrams for the high symmetry directions. Three such diagrams are presented in Figure 3 for germanium, silicon, and gallium arsenide (21).

The density of states, or number of states available for occupation within a given energy increment, is important to the understanding of material properties. The chemical approach illustrates that energy levels or bands contain a fixed number of

Table 1. Properties of Important Semiconductors

Semiconductor		CAS Registry No.	Band gap, eV		Mobility (at 300 K), cm²/(V·s)[a]		Band[c]	Effective mass[b], m*/m		Relative dielectric, ϵ_s/ϵ_0
			300 K	0 K	Electrons	Holes		Electrons	Holes	
Element	C	[7440-44-0]	5.47	5.48	1800	1200	I	0.2	0.25	5.7
	Ge	[7440-56-4]	0.66	0.74	3900	1900	I	1.64[d]	0.04[e]	16.0
								0.082[f]	0.28[g]	
	Si	[7440-21-3]	1.12	1.17	1500	450	1	0.98[d]	0.16[d]	11.9
								0.19[f]	0.49[g]	
	Sn	[7440-31-5]		0.082	1400	1200	D			
IV–IV	α-SiC	[409-21-2]	2.996	3.03	400	50	I	0.60	1.00	10.0
III–V	AlSb	[25152-52-7]	1.58	1.68	200	420	I	0.12	0.98	14.4
	BN	[10043-11-5]	ca 7.5				I			7.1
	BP	[20205-91-8]	2.0							
	GaN	[25617-97-4]	3.36	3.50	380			0.19	0.60	12.2
	GaSb	[12064-03-8]	0.72	0.81	5000	850	D	0.042	0.40	15.7
	GaAs	[1303-00-0]	1.42	1.52	8500	400	D	0.067	0.082	13.1
	GaP	[12063-98-8]	2.26	2.34	110	75	I	0.82	0.60	11.1
	InSb	[1312-41-0]	0.17	0.23	80,000	1250	D	0.0145	0.40	17.7
	InAs	[1303-11-3]	0.36	0.42	33,000	460	D	0.023	0.40	14.6
	InP	[22398-80-7]	1.35	1.42	4600	150	D	0.077	0.64	12.4
II–VI	CdS	[1306-23-6]	2.42	2.56	340	50	D	0.21	0.80	5.4
	CdSe	[1306-24-7]	1.70	1.85	800		D	0.13	0.45	10.0
	CdTe	[1306-25-8]	1.56		1050	100	D			10.2
	ZnO	[1314-13-2]	3.35	3.42	200	180	D	0.27		9.0
	ZnS	[1314-98-3]	3.68	3.84	165	5	D	0.40		5.2
IV–VI	PbS	[1314-87-0]	0.41	0.286	600	700	I	0.25	0.25	17.0
	PbTe	[1314-91-6]	0.31	0.19	6000	4000	I	0.17	0.20	30.0

[a] The values are for drift mobilities obtained in the purest and most perfect materials available to date.
[b] m^* = effective mass, m = rest mass.
[c] I = indirect, D = direct.
[d] Longitudinal effective mass.
[e] Light-hole effective mass.
[f] Transverse effective mass.
[g] Heavy-hole effective mass.

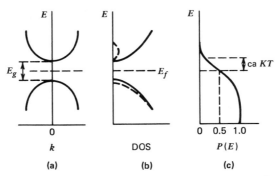

Figure 2. (a) Energy E vs crystal momentum k for free-particlelike conduction- and valence-band electrons. (b) The corresponding density of available electron states DOS. (c) The Fermi-Dirac function, ie, the probability $P(E)$ that a state is occupied. The product of $P(E)$ and DOS yields the energy distribution of electrons shown by the dashed lines in (b).

electron states in direct proportion to the number of atoms in the crystal. The narrow lower-lying energy bands, which correspond to atomiclike states, display a rather flat E-vs-k relationship. Electrons in these states are localized near the nuclei, the density of states and effective mass at these energies are large. The broad valence and conduction bands have relatively low densities and low effective masses. The density of states is related to the slope of the E-vs-k diagram (see Fig. 2). The density of states is low near the band edges, where electrons have little kinetic energy.

The probability that an energy state is occupied by an electron is governed by Fermi-Dirac statistics. According to the Pauli principle, no more than two electrons may occupy a single energy state. It is useful to define a reference level, called the Fermi energy E_F, at which states have a 50% probability of being occupied. At a temperature of absolute zero, the total electronic energy is minimal. All states below the Fermi level are occupied and all above are empty. As the temperature increases, electrons are excited above the Fermi level. The probability of occupation is given by the following formula:

$$P(E) = \frac{1}{1 + \exp\left|\dfrac{E - E_F}{KT}\right|} \tag{1}$$

where K is Boltzmann's constant. When the energy E is distant from E_F ($|E - E_F| \gg KT$), this function falls off in a simple exponential fashion, as would the classical Boltzmann distribution function for particles. This is so because the Pauli principle is no longer important in sparsely populated regions where two electrons are not likely to occupy the same state. The Fermi-Dirac distribution function is illustrated in Figure 2**c**.

The actual energy distribution of electrons in the solid is obtained by multiplying the occupation probability by the density of states; this is shown in Figure 2**b** by the dashed regions. For each electron in the conduction band, there is a corresponding electron vacancy or hole in the valence band. The Fermi level in this case is located at the midpoint of the band ("midgap"). In the presence of an electric field, the hole behaves as a singly-charged positive particle with an effective mass equal to that of the missing valence electron. The conductivity of the semiconductor, ie, its ability to carry current under an applied electric field, depends on the number of conduction electrons and holes. At absolute zero, the Pauli principle dictates that no current can flow because there are no empty states in the valence band into which an electron can move. As the temperature increases, conduction electrons are released into a sea of empty states; their movements are unrestricted by the Pauli principle. Similarly, because holes are empty states, they have freedom of movement in the valence band. The conductivity of the semiconductor increases with temperature because of the creation of additional charge carriers. From Fermi-Dirac statistics, the concentration of conduction electrons or holes n_i varies as $\exp(-E_g/2\,KT)$. In an insulator, $E_g > 3$ eV and n_i is very small. A metal displays a large low temperature conductivity, which decreases as the temperature rises. The metal is characterized by a partially filled energy band. Electrons near the Fermi level may move into unoccupied states that are also located near the Fermi level. As the temperature increases, thermal agitation of the lattice inhibits the flow of electrons.

The energy-band theory helps to explain the optical properties of materials. Photons with energy hν greater than the semiconductor gap energy E_g may excite electrons from the valence to the conduction band. Lower energy photons cannot excite electrons into the forbidden gap, ie, the region in which electrons cannot exist; therefore, they pass through the material. Visible photon energies are in the red-to-blue range, ie, 1–3 eV. Thus, insulators are frequently transparent ($E_g > 3$ eV) and metals ($E_g = 0$) are opaque. Coloration of semiconducting and insulating crystals is generally caused by the presence of energy levels within the gap.

The germanium, silicon, and gallium arsenide E-vs-k diagram of Figure 3 illustrate the difference between an indirect-gap and a direct-gap semiconductor (21). In gallium arsenide, the valence-band maximum and conduction-band minimum lie at the same value of k. Electron-hole recombination may occur directly between these points without the participation of a phonon (lattice vibration) to conserve momentum. Such band-to-band recombination is associated with the emission of a photon with energy $h\nu = E_g$. In germanium and silicon, the valence-band maximum and conduction-band minimum are not aligned, and electron-hole recombination occurs nonradiatively by means of phonons and traps. Thus, the direct-gap compound semiconductors are used in light-emitting applications.

Extrinsic Properties. Device materials are bounded by interfacial regions and contain defects or purposely introduced impurities. These nonidealities tend to introduce energy states into the band gap and may significantly alter the number of available charge carriers and, thus, the conductivity. The important case of impurities in silicon illustrates these effects.

The most common impurities or dopants introduced into the silicon crystal are the group V elements phosphorus, arsenic, and antimony, and the group III element boron. These impurities are substitutional, ie, they occupy normal Si-atom positions in the crystal lattice. The group V elements contain five valence electrons, four of which participate in covalent bonding with the surrounding Si atoms. The fifth electron is

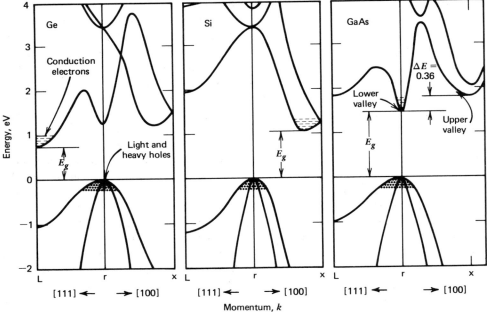

Figure 3. Detailed E-vs-k diagrams for germanium, silicon (indirect gaps), and gallium arsenide (direct gap) along the [111] and [110] crystalline directions (21).

weakly bound to the impurity atom at 0 K and, at higher temperatures, is excited into the conduction band. For this reason, the group V impurities are called electron donors. The small excitation energy is indicated by placing the donor energy level near the conduction-band edge (see Fig. 4) (22). At 0 K, the Fermi level is between the conduction band and the donor level. As the temperature rises, the Fermi level drops gradually toward midgap as the impurities are ionized. When the silicon is doped predominantly with electron donors, it is called n-type. In n-type material, the electrons are dubbed majority carriers and the holes are minority carriers. The group III element boron contains three valence electrons that participate in covalent bonding. At 0 K, the fourth bond is half filled, leaving the boron atom in a neutral charge state. Little energy is required to excite a valence electron into this fourth bond, and its introduction creates a hole in the valence band. Boron is an electron acceptor. The acceptor level is located near the valence-band edge (see Fig. 4), and the Fermi level lies between the two at 0 K. When the temperature rises, the Fermi level moves toward midgap as the boron atoms become ionized. The acceptor-doped material is termed p-type, and holes are the majority carriers.

The impurities give rise to shallow levels. Shallow levels are located within ca 0.1 eV of the band edges so that, at RT, the majority of impurity atoms contribute charge carriers to the conduction or valence bands. In this manner, the conductivity can be increased by orders of magnitude over the intrinsic RT value. Other impurities frequently give rise to deep levels located throughout the energy gap. Deep levels can act as electron or hole traps, thereby inhibiting conductivity. Gold, for example, gives rise to four energy levels in the band gap of germanium, corresponding to four possible charge states (23). The gold atom can act as a multiple acceptor or as a donor by giving up its lone valence electron to the conduction band. An empty lattice site or vacancy also gives rise to deep levels (24). The energy levels associated with atoms surrounding the vacant site become atomiclike and, therefore, move into the gap. Associated with each atom is a dangling bond directed toward the empty lattice position. These atoms undergo a Jahn-Teller spatial distortion and cause the filled electron levels to drop lower in the gap. The magnitude of the distortion depends on the charge state of the vacancy. Grosser crystalline defects, eg, stacking faults or grain boundaries, similarly give rise to deep levels.

Group IIIA–VA compounds are doped in an analagous manner to silicon and

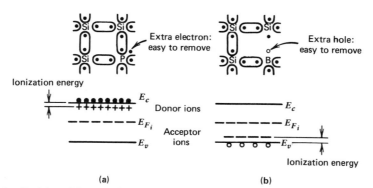

Figure 4. Position of donor and acceptor levels in the silicon band gap. (**a**) n-Type semiconductor. (**b**) p-Type semiconductor. E_c = conduction-band edge, E_{F_i} = intrinsic Fermi energy, E_v = valence-band edge (22).

germanium (23). Groups IIA and VIA elements generally enter the lattice substitutionally. The group IIA atoms replace group IIIA atoms in the lattice to form acceptors, whereas group VIA atoms replace group VA atoms to form donors. The compound semiconductor may act to compensate added donor- or acceptor-type impurities through the formation of vacancies of the opposite type. Such effects become more severe in the larger band-gap, ie, more ionic, materials. In general, the compound semiconductors are difficult to purify and must be doped heavily to overshadow the effect of remnant impurities. Exceptionally heavy doping of compound or elemental semiconductors may result in the formation of impurity bands because of the spatial interaction of the impurity states, which narrows the energy gap. A large impurity concentration may also result in stress variations throughout the lattice, again causing the energy gap to narrow.

Another important class of defects occurs at the interface between the semiconductor and vacuum, insulator, or metal. The behavior at the vacuum interface is similar to that of the isolated vacancy. Dangling bonds extend from the surface atoms into vacuum. Atomiclike states associated with these bonds might be expected at midgap, which implies that the half-filled bonds have an equal probability of emptying or filling. However, a Jahn-Teller distortion or surface reconstruction occurs at the semiconductor surface so as to minimize the electronic energy. Level splitting results in filled states dropping below midgap and empty states rising above it. In gallium arsenide and other compound semiconductors, the states may move out of the gap entirely, although such an ideal vacuum interface is difficult to form or maintain (25).

The presence of a large number of surface states can completely overshadow any doping effects in the surface region. It is useful in electronic devices to passivate the surface, ie, to remove the great majority of surface states. In silicon, this is easily accomplished by thermally oxidizing the surface to form a protective insulating SiO_2 layer (26). Electrons in the interfacial Si—O bonds are strongly bound and do not participate in conduction processes in the silicon bulk. The number of surface states is reduced by four orders of magnitude. The remaining states are thought to result from isolated silicon and oxygen dangling bonds at the interface. Silicon has a great advantage over other semiconductors, since it can be passivated easily and efficiently. Metal coverage of the semiconductor surface also tends to remove the native interfacial states, although electrons in the metal conduction band can communicate easily with the semiconductor levels. Recent evidence indicates that, in compound semiconductors, the remaining interfacial states are simple defects in the semiconductor near the interface (27).

Equilibrium and Steady State. Chemical equilibrium requires that the Fermi level be at a fixed energy throughout the system of interest; that is, all states at a given energy have equal occupation probability. This rule facilitates mathematical computation, although a more intuitive approach to equilibrium and steady-state behavior is taken here.

Equilibrium is maintained in the semiconductor by a balance between the generation and recombination of electrons and holes (22). The generation rate G of electron–hole pairs is a function of temperature only. The electron–hole recombination rate R is proportional to the concentration of electrons n and of holes p ($R = \alpha np$). Since the two rates balance in equilibrium ($R = G$), the np product must be a function of temperature, not of the doping level, and it therefore may be set equal to the intrinsic product n_i^2:

$$np = n_i^2 \tag{2}$$

Equation 2 is known as the law of mass action. n_i^2 was shown earlier to vary as exp $(-E_g/KT)$. In silicon, $n_i =$ ca 10^{11} cm^{-3} at RT. Typically, the number of dopant atoms, and hence the majority-carrier concentration, is 10^{15}–10^{20} cm^{-3}.

If the semiconductor is exposed to light, the steady-state generation rate is increased by photon absorption, and the np product must similarly increase. For weak illumination, the minority-carrier concentration may increase by orders of magnitude with only a small percentage change in the majority-carrier concentration. When the light source is removed, the carrier concentrations decay to their equilibrium values. The rate of decay is estimated easily. For example, when the light is removed from an illuminated n-type semiconductor for which the recombination rate is approximately $\alpha n(p + \Delta p)$, the generation rate immediately returns to αnp. The difference in these two rates $\alpha n \Delta p$ corresponds to the initial rate at which the excess minority carriers are consumed. In a time $\tau = 1/\alpha n$, the majority of the excess carriers Δp will have recombined; τ is the minority-carrier lifetime. It should be noted that the recombination rate $\alpha n \Delta p$ becomes simply $\Delta p/\tau$.

The lifetime estimate $\tau = 1/\alpha n$ assumes band-to-band recombination of electrons and holes. Typically, the recombination and generation rates are greatly accelerated by the presence of deep levels or traps in the band gap. Such traps may capture or emit electrons and holes from the conduction and valence bands. These events are more probable than band-to-band transitions since the required energy jumps are smaller. To be effective, the trap must capture an electron and subsequently capture a hole before the electron is reemitted into the conduction band or vice versa. Trap levels near midgap are most effective in supporting this recombination process. A straightforward analysis yields the following application for the trap-dominated minority-carrier lifetime:

$$\tau = \frac{1}{\sigma v_{th} N_t} \tag{3}$$

σ is the effective cross-sectional area for minority-carrier interaction with the trap, N_t is the trap concentration, and v_{th} is the thermal velocity of the minority carrier as determined from the thermodynamic relation

$$\tfrac{1}{2} m v_{th}^2 = \tfrac{3}{2} kT \tag{4}$$

Surface and interface states can also mediate the recombination process and may drastically reduce the carrier lifetimes.

Transport. Charge carriers may move through a semiconductor device by random thermal motion or with the assistance of an electric field. These two modes of transport are called diffusion and drift, respectively (28). The magnitude of either process depends on a parameter called the mobility, which is a measure of the ease with which a charge carrier moves through the lattice. The mobility μ enters the relation $v = \mu E$, where v is the average velocity of the charge carrier in the direction of the electric field $E (v \ll v_{th})$. The carrier is accelerated by the field but undergoes collisions with impurities or phonons, ie, lattice vibrations, which result in a loss of momentum. Equating the average carrier momentum $m^* v$ to the applied impulse between collisions $qE\bar{t}$, where $\bar{t}$ is the mean time between collisions, yields the following relation for the mobility:

$$\mu = \frac{q\bar{t}}{m^*} \tag{5}$$

Equation 5 illustrates the importance of the effective mass m^* in device performance. The hole mobility is smaller than the electron mobility in most semiconductors because of the effective mass difference. The collision time $\bar{t}$ is controlled by impurity scattering at low temperatures and lattice scattering at high temperatures: At very low temperatures, the low thermal velocity of carriers enables them to interact strongly with ionized impurities. As the temperatures increase, $\bar{t}$ and hence μ increase. A maximum mobility is reached (at ca 100 K in heavily doped Si) and μ then begins to decrease with temperature as a result of the increased probability of phonon interaction. Once again, surface or interface irregularities can decrease $\bar{t}$ significantly.

The current density J in the semiconductor, which results from drift and diffusion, is given by the following formula:

$$J = \sigma E - qD \frac{\partial n(x)}{\partial x} \tag{6}$$

The conductivity σ is proportional to the carrier concentration n, the electron charge q, and the mobility:

$$\sigma = nq\mu \tag{7}$$

The diffusion term indicates that in the presence of a concentration gradient $\partial n/\partial x$, random flow of charges lessens or smooths the gradient. The rate at which this smoothing occurs is proportional to the thermal velocity of the carriers and to the scattering length $l (l \sim v_{th}\bar{t})$. It is not difficult to show that

$$D = \tfrac{1}{3} l v_{th} = \tfrac{1}{3} l^2/\bar{t} \tag{8}$$

The diffusion constant would be expected to depend on the carrier mobility, or ease of movement, and this is demonstrated by the Einstein relation:

$$D = \frac{kT}{q} \mu \tag{9}$$

This relation is easily verified by substituting equations 4, 5, and 8.

The physical meaning of the diffusion constant D may be illustrated by considering the problem of the random walk (29). It can be shown that the mean net distance R traveled by a randomly moving particle is given by the relation $R^2 = Nl^2$, where N is the number of random steps taken by the particle and l is the step length. The number of steps N occurring in a time t is simply $t/\bar{t}$, so that

$$R^2 = Nl^2 = \frac{tl^2}{\bar{t}} = 3 Dt \tag{10}$$

Thus, the parameter $\sqrt{Dt}$ has the units of length and corresponds roughly to the net distance traveled by the particle in time t. In particular, with $t = \tau$, the parameter $L = \sqrt{D\tau}$ describes the mean distance traveled by a charge carrier prior to recombination. The carrier diffusion length L plays a key role in the description of diode behavior (see Device Physics).

High Field Effects. The phenomena of velocity saturation, avalanching, and tunneling become important in the presence of high electric fields (23). The simple proportionality $v = \mu E$ no longer holds. Instead the velocity varies nonlinearly with electric field, as shown in Figure 5 for electrons in silicon and gallium arsenide. At high fields ($E > $ ca 10^4 V/cm), the velocity is insensitive to variations of E and is said to be satu-

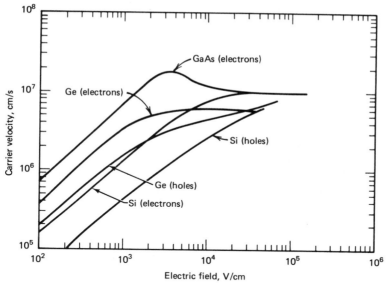

Figure 5.   Electron and hole velocities vs electric field for germanium, silicon, and gallium arsenide at 300 K (21).

rated. A discussion of acoustic and optical phonons is necessary to explain this behavior.

Lattice vibrations or phonons cause periodic distortions in the width of the energy gap (19). The modulating potential of the conduction- and valence-band edges results in the scattering of electrons and holes. The low energy or acoustic phonons, with energies of ca KT, are associated with long wavelength lattice distortions. Atoms and their neighbors vibrate in concert so that, although displaced from their original sites, their relative spacing is only slightly affected. A higher energy vibrational mode exists, in which neighboring atoms move in opposing directions. These vibrations typically occur at optical frequencies and are associated with the optical phonons. The minimum energies required to excite these phonons are listed in Table 1 for some of the semiconductors and are 0.03–0.06 eV. At high electric fields, carriers may acquire sufficient kinetic energy before scattering to excite an optical phonon. The scattering probability increases with phonon energy; thus, optical phonon emission is more likely than acoustic phonon emission once the optical threshold energy is reached. Since the majority of carriers do not surpass this threshold energy, the average carrier velocity is limited.

The unusual behavior of the gallium arsenide velocity-field curve results from the band structure illustrated in Figure 3. The conduction band consists of a central valley with minimum energy at $k = 0$, and a satellite valley with the minimum located higher in energy. The central valley has a large curvature, which is associated with a low effective mass and a low density of states. The relatively flat satellite valley has a correspondingly high effective mass and density of states. A high field may raise electrons in the central valley above the minimum energy of the satellite valley. These electrons are likely to scatter into and remain in the satellite valley because of its high density of states. Their mobility is then reduced by the effective mass increase. The decrease of velocity with increasing field at $E = $ ca 10^4 V/cm results from this electron transfer and is the basis of the Gunn-effect devices (see Applications).

Avalanching and tunneling effects become important as the electric field increases above ca 10^5 V/cm. These effects are illustrated in the band diagrams of Figure 6. The band diagram is a plot of the conduction- and valence-band edges as a function of position. The slope of the band edges or potential drop per unit distance indicates the applied E field. The kinetic energy of a carrier is indicated by its vertical distance from the band edge. Carriers are accelerated horizontally in the diagram, electrons to the right and holes to the left, so that their total energy remains constant, with the kinetic energy increasing at the expense of the potential energy. Scattering causes the carriers to drop toward the band edges. At very high fields, a small fraction of carriers may acquire kinetic energies on the order of the band gap E_g. These carriers may then excite an electron from the valence to the conduction band. This phenomenon is known as impact ionization. The electron and hole created in this manner travel in opposite directions under the influence of the field and may create other electron-hole pairs. At sufficiently high fields, a feedback or avalanche effect occurs. Electrons and holes continuously traverse the high field region of the semiconductor and thereby create electron-hole pairs that also participate in the process, which results in a large increase in current.

A horizontal constant-energy line drawn through the high field band diagram of Figure 6b intersects the conduction- and valence-band edges. As the distance between these points of intersection decreases below ca 10 nm with increasing E, the probability that a valence electron tunnels through the gap and appears in the conduction band increases dramatically. This tunneling mechanism, also called Zener breakdown, results in an increased current. The field required to produce tunneling in silicon is ca 10^6 V/cm. The field required to induce avalanching depends on the spatial extent of the high field region. If it is very thin, carriers may be accelerated into lower field regions before impact ionization occurs. The critical field for avalanche breakdown also depends to some extent on the temperature. Higher temperatures increase the phonon emission probability and therefore reduce the impact-ionization rate.

Several important devices are based on the avalanche and Zener breakdown. These phenomena also place fundamental limits on the operating voltages of diodes and transistors. The velocity-saturation effect, in general, limits the operating speed of these devices.

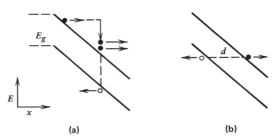

Figure 6.   Band diagrams for (**a**) impact ionization and (**b**) tunneling. The slope of band edges is proportional to the electric field. In (**a**), an energetic electron causes the formation of an electron–hole pair, which imparts ionization. In (**b**), a valence electron tunnels through the forbidden gap when $d \gtrsim 10$ nm; ○ = hole, ● = electron.

Noise. All discussions to this point have been based on the most probable distributions of charge carriers. However, statistical fluctuations of these distributions in time and space result in corresponding fluctuations in the device current or voltage. Such fluctuations or noise place a limitation on the minimum signal with which the device can operate unambiguously (30).

Three types of noise affect semiconductor devices. Thermal noise is caused by the random motion of charges and is proportional to temperature. Shot noise is related to random fluctuations of carriers passing a semiconductor junction and is proportional to device current. Shot noise generally dominates over thermal noise in semiconductor devices. Flicker noise or 1/f noise is most troublesome at low frequencies, as compared to thermal and shot noise, which generate white or frequency-independent power spectra. The universal nature of 1/f noise in various materials is poorly understood. In surface field-effect devices, a strong correlation has been observed between 1/f noise and surface-state or trap charging and discharging. Avalanche multiplication adds significantly to the shot noise of a device. Thus, in general, a low noise device has low trap densities and is operated at low temperatures below the avalanche threshold field.

Device Physics

The ideal rectifying diode is a two-terminal device that allows current to flow in one direction only. The transistor is a three-terminal device that may act as an amplifier or a switch.

p-n Junction. In the case of two silicon slabs, one p-type and the other n-type, that are joined to form an abrupt interface, at the instant they are joined there is an enormous difference in electron and hole concentrations on either side of the interface. Electrons therefore diffuse from the n-type material into the p-type material and vice versa for the holes. The majority-carrier concentrations are much reduced in the neighborhood of the interface, whereas the local concentration of immobile impurity ions is roughly equal to the original majority-carrier concentration. In the interfacial region, the ions greatly outnumber the mobile charges. The charge imbalance at the interface results in an electric field, which sweeps carriers in a direction opposite to the diffusive flow. In equilibrium, a balance is struck between the diffusion and drift currents. The high field region is called the depletion region, since it is depleted of mobile charges. The interfacial region and equilibrium band diagram are illustrated in Figure 7 (28). The Fermi levels are aligned, and the potential developed across the interface, called the built-in potential V_o, is somewhat less than the band gap (ca 1 V for silicon).

When a voltage bias is applied to the p-n diode, the voltage is dropped primarily across the interfacial region. The depletion region width and, thus, the charge imbalance, are altered. The voltage drop occurring outside the depletion region is proportional to the induced diode current. This drop is small because of the high conductivity of the doped semiconductor. The reverse- and forward-bias cases are illustrated in Figure 7.

The minority-carrier concentrations outside the depletion region are defined as follows: p_p and n_p refer to the hole and electron concentrations on the p side, and p_n and n_n refer to the n-side concentrations. The minority-carrier concentration n_p is roughly equal to the fraction of majority carriers γn_n with sufficient energy to sur-

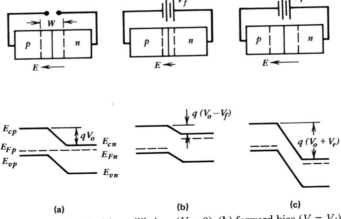

Figure 7.   The *p-n* junction in (a) equilibrium ($V = 0$), (b) forward bias ($V = V_f$) and (c) reverse bias ($V = -V_r$). The applied voltage appears primarily across the depletion width W. E = energy. Courtesy of Prentice-Hall (28).

mount the interfacial potential barrier $q(V_o - V)$. It is assumed that generation and recombination in the depletion region is negligible. From Fermi-Dirac statistics, γ is simply $\exp(-q(V_o - V)/KT)$. At $V = 0$, $n_{po} = n_n \exp(-qV_o/KT)$, so that

$$\frac{n_n}{n_{po}} = \frac{p_n}{p_{po}} = \exp(-qV/KT) \tag{11}$$

In forward bias, injected minority carriers travel a short distance (ca L) from the depletion-region edge before recombining with majority carriers. As majority carriers are consumed by recombination, additional majority carriers are provided by the external diode contacts to maintain charge neutrality. The magnitude of the diode current is, therefore, determined by the recombination rate of excess minority carriers. The excess concentrations are denoted by $\Delta n = n_p - n_{po}$ and $\Delta p = p_n - p_{no}$. Within a diffusion length L_n of the depletion edge, an amount of charge $qL_n\Delta n$ per unit area recombines in a time τ_n. The diode current density J is thus given by

$$J = \frac{qL_n\Delta n}{\tau_n} + q\frac{L_p\Delta p}{\tau_p} = q\left|\frac{L_n n_{po}}{\tau_n} + \frac{L_p p_{no}}{\tau_p}\right||e^{qV/KT} - 1| \tag{12}$$

The minority-carrier concentrations are easily determined from the law of mass action ($np = n_i^2$) with the majority-carrier concentration equal to the dopant concentration. Frequently n_{po} and p_{no} differ greatly, and the term associated with the heavily doped side can be neglected.

Equation 12 is also valid for the reverse-biased diode. The exponential term of equation 11 drops to zero, since majority carriers can no longer surmount the large potential barrier. The diode current is now caused by minority carriers generated within a distance L of the depletion region. Once they reach the depletion-region edge, they are swept across the junction by the field. Far away from the depletion region, the generation rate is equal to the recombination rate n_{po}/τ_n or p_{no}/τ_p. The recombination rate is negligible within a diffusion length L of the depletion edge, since there are few minority carriers, whereas the generation rate is unchanged. The reverse current is given by the product of the diffusion length and generation rate, as indicated in equation 12 with $V \ll 0$. The reverse saturation or leakage current is relatively in-sensitive to the reverse bias. The I–V characteristic of the p-n junction is illustrated in Figure 8. The diffusion length $L = \sqrt{D\tau}$. From the Einstein relation (see eq. 9), the diode current varies as $\sqrt{\mu/\tau}$.

In order to change the voltage across the p-n diode, the depletion-region width and minority-carrier distributions near the interface must be altered. The amount of charge dQ that is added to or subtracted from the interfacial region to effect a given change in voltage dV is a measure of the maximum switching speed of the diode. Thus, the capacitance $C = |dQ/dV|$ must be minimized for optimum device speed. The p-n-diode capacitance consists of two components: the junction capacitance, which dominates in reverse bias, and the charge-storage capacitance, which tends to dominate in forward bias. The junction capacitance results from charge movements associated with the changing depletion-region width and is inversely proportional to that width. The charge-storage capacitance results from the charge variations caused by generation and recombination, which are necessary to alter the excess minority-carrier distri-butions outside the depletion region. This capacitance is proportional to the carrier lifetime τ. Gold centers are occasionally introduced into the p-n junction to reduce τ and the capacitance, although the leakage current is simultaneously increased.

Avalanche or tunnel breakdown occurs in the p-n junction for sufficiently large reverse bias. The mode of breakdown depends on the strength of the electric field in the depletion region and the width of this region. As the doping on either side of the junction is increased, the depletion-region width at a given bias decreases and the electric field increases. The avalanche-breakdown voltage is lowered by the width effect, although the actual critical field at which breakdown occurs increases. For doping levels greater than ca 10^{18} cm^{-3} in silicon, the critical field for avalanche breakdown exceeds the threshold of tunnel breakdown, so that the latter mechanism dominates. Tunnel breakdown occurs at reverse biases of just a few volts (21).

Bipolar Transistor. The bipolar transistor, so named because both majority and minority carriers participate in its operation, consists of two back-to-back p-n junctions located in close proximity. The central region between the junctions is termed the base, and the outer regions are called the emitter and collector. In typical operation, small changes in current supplied to the base cause proportionately large changes in the emitter and collector current. The transistor can be fabricated in the n-p-n or p-n-p configuration; only the latter is discussed below.

In the p-n-p transistor, with the emitter-base junction forward-biased ($V_{BE} <$

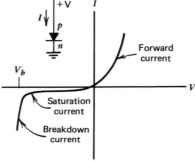

Figure 8.   Typical current-voltage characteristic of a p-n junction diode. V_b = breakdown voltage (28).

0) and the collector-base junction reverse-biased ($V_{CB} > 0$), the width of the base W is much less than the diffusion length L_p, so that the holes injected into the base by the emitter have a high probability of reaching the collector junction before recombining. Once at the collector junction, they are swept into the collector by the junction field. The emitter and collector currents are nearly equal, and their magnitudes are controlled by the much smaller base current. The collector current is relatively independent of V_{CB}, since carriers reaching the collector junction are swept through regardless of the magnitude of the junction field. The voltage at the forward-biased emitter junction V_{BE} varies between 0 and 0.6 V, ie, one diode drop, corresponding to large changes in emitter and collector currents. This voltage is controlled by the base current. The presence of excess holes in the base (injected by the emitter) acts to diminish the forward bias of the emitter junction. The base contact supplies electrons that maintain charge neutrality in the base region and thereby fixes the emitter-junction voltage. The base current consists primarily of electrons that recombine in the base and of electrons that are injected into the emitter. The latter component is greatly reduced by heavily doping the emitter. This causes the first term in equation 12 to become negligible. The recombination current is much less than the emitter current, since only a small fraction of injected holes recombines with electrons in the base.

The ratio of the collector current to the base current β is a measure of the amplifying power of the transistor (28). It is estimated in the following manner: the average time spent in the base region by a hole, called the transit time τ_t, is much less than the lifetime of a hole or electron τ_p, since $W \ll L_p$. The transit time is defined by the relation $W = \sqrt{2 D_p \tau_t}$. (The factor of two arises from the nature of the charge distribution in the base.) The probability that the hole will recombine in the base, thereby contributing to the base current, is τ_t/τ_p, whereas the probability that it will be swept into the collector is near unity. Therefore,

$$\beta = \frac{i_c}{i_\beta} = \frac{\tau_p}{\tau_t} = \frac{2 D_p \tau_p}{W^2} = \frac{2 L_p^2}{W^2} \tag{13}$$

A simple common-emitter circuit is illustrated in Figure 9 with the transistor-characteristic curves. The load line that connects points S and C is a plot of the relation $-V_{CE} + i_c(5\ k\Omega) = 40$ V, which is dictated by the external circuitry. The signal voltage e_s alternates between +10 V and −10 V, resulting in a base current of 0.1 mA when the emitter junction is forward-biased, and no current when this junction is reverse-biased. The transistor behaves as a switch. When no base current flows, it is in the cutoff or off state. The base current of 0.1 mA is sufficient to drive the transistor into the saturation mode, for which i_c drops below βi_B. In this mode, the collector junction becomes slightly forward-biased and the amount of stored charge in the base increases greatly. V_{CE} is near zero and the transistor is on.

The operation speed of the transistor is governed by the same processes controlling the p-n junction, namely, the movement of charge into and out of the depletion regions, and the generation–recombination of stored charge. The latter effect may be reduced by keeping the transistor unsaturated and by gold-doping to reduce the carrier lifetime. However, gold-doping also reduces β, as evidenced in equation 13. If the generation-recombination effects are overcome, the basic speed limitation of the device is the transit time τ_t of equation 13. Optimum speed is then obtained by minimizing the base width and maximizing D_p or the hole mobility–temperature product. The minimum

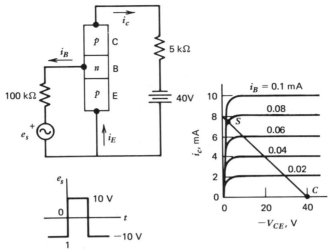

Figure 9. Characteristic curves for a common emitter circuit. The load line is defined by the relation $V_{CE} + i_c(5\,k\Omega) = 40$ V. The transistor switches between saturation S and cutoff C. C = collector, B = base, E = emitter, i_c = collector current, i_E = emitter current, i_B = base current, e_s = supply voltage, t = time, V_{CE} = voltage between collector and emitter.

base width is determined by the depletion-region widths. The base must be doped heavily enough to prevent the depletion regions from contacting each other. This heavy doping increases the electron injection into the emitter and also reduces the avalanche-breakdown voltages of the junctions. Contact of the depletion regions is called punchthrough and results in loss of current control by the base, since injected carriers are swept through the entire base region. With effort, τ_t can be reduced to ca 10^{-11} s.

Field-Effect Transistor. The field-effect transistor (FET) is a unipolar device, ie, its operation depends on the motion of one type of charge carrier. All FETs contain a channel, in which charge flows through the device. The channel is constricted or blocked by the application of an electric field, which alters the current flow. The dominant device of this type is the metal-oxide-semiconductor FET (MOSFET).

The band diagrams for three bias states of a metal-SiO_2-p-type silicon capacitor are illustrated in Figure 10 (22). The metal electrode is termed the gate, and the voltage at this electrode is V_G with respect to the grounded silicon substrate. Most of the gate voltage appears across the thin oxide; if the oxide thickness is increased, a proportionately greater fraction of V_G appears across it. In the accumulation case, $V_G < 0$, majority carriers or holes are attracted to the Si–SiO_2 interface by the gate charge. When V_G rises above zero, majority carriers are repelled from the interface, which results in the formation of a depletion region, as in the case of the p-n junction. As V_G increases, the depletion region grows; the charge on the gate is balanced by the ionized impurity charge in the depletion region. For sufficiently large V_G, a stable population of minority carriers can be supported at the Si–SiO_2 interface. These interfacial electrons form an inversion layer if their numbers exceed the original hole concentration. The surface layer then behaves as a thin n-type region. No further voltage drop can occur in the semiconductor, since additional gate charge is compensated by the addition of electrons to the inversion layer. The inversion-layer charge is supplied by generation current in the depletion region. However, in the MOSFET, this charge

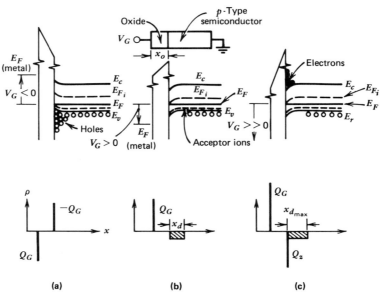

Figure 10.   Charge distributions in the voltage-biased MOS capacitor. (a) Accumulation of majority carriers near surface. (b) Depletion of majority carriers from surface. (c) Inversion or accumulation of minority carriers near surface (22). V_G = gate voltage; Q_G = gate charge; x_o, x_d, and $x_{d\ max}$ = depletion widths; ρ = charge density.

is supplied by a *p-n* junction, and the inversion layer acts as the channel of the device.

The minimum gate voltage necessary to create an inversion layer is called the threshold voltage V_T. Increasing the substrate doping also increases V_T. This results from a larger electric field in the depletion region, which must be supported by a larger gate voltage. Reducing the oxide thickness decreases V_T. The threshold voltage is also altered by the presence of surface states, oxide charges, and work-function differences. Surface state and oxide charge must be balanced by gate charge, thus accounting for a threshold change. Metal-oxide-semiconductor FETs are typically produced on [100] Si surfaces, since fewer surface states appear at the [100] Si–SiO$_2$ interface than at other Si surfaces (26). The [100] surface has the fewest broken bonds per unit area. Sodium is the primary source of charges in the oxide. The oxide-charge problem has been greatly reduced by recent improvements in fabrication techniques. The work function is the energy required to excite an electron from the Fermi level to the vacuum level. When two dissimilar materials are joined, a charge readjustment occurs at the interface so as to bring the Fermi levels into alignment. The voltage drop at the interface is equal to the work-function difference and must again be compensated by the gate voltage of the MOSFET. Threshold control is one of the main problems in MOS technology.

In the MOSFET structure illustrated in Figure 11, two $n+$ regions have been added to the MOS capacitor to act as a source and a drain, respectively, for inversion-layer charge. Typically, the source region and substrate are fixed at zero potential and the gate and drain voltages, V_G and V_D, respectively, are varied. For example, when V_D is slightly positive and $V_G \gg V_T$, which causes an inversion layer to form under the gate, inversion-layer charge is now supplied by the source and, under the

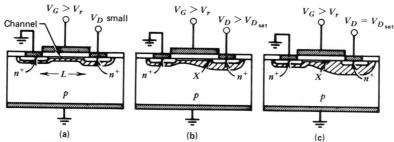

Figure 11.   MOS n-channel transistor with channel length L and width Z. (a) Linear region. (b) Onset of saturation, pinchoff X occurs at drain. (c) Saturation, pinchoff position X moves away from drain as dashed depletion region grows (22).

influence of the drain bias, drifts toward the drain where it is collected. The drain current I_D is proportional to the amount of charge Q in the inversion layer, which can contribute to the current flow. The inversion charge per unit area Q is equal to $C_o(V_G - V_T)$, where C_o is the oxide capacitance and $(V_G - V_T)$ is approximately the voltage decrease across the oxide. The oxide capacitance is ϵ_o/W_{ox} where ϵ_o is the oxide dielectric permittivity and W_{ox} is the oxide thickness. The charge Q moves at an average velocity $v = \mu E = \mu V_D/L$, where μ is the surface mobility and L is the distance between source and drain or channel length. The drain current for a MOSFET with gate area $A = ZL$ is

$$I = AQ\left|\frac{v}{L}\right| = \mu \frac{Z}{L}\frac{\epsilon_o}{W_{ox}}|V_G - V_T|\,V_D \tag{14}$$

L/v is the device transit time (26).

As V_D increases, the depletion width at the drain junction grows (see Fig. 11). The larger depletion region under the channel edge near the drain accommodates more charge. Less charge is needed in this portion of the inversion layer to balance the gate charge. The surface potential at the drain edge of the channel is V_D. If $V_G - V_D < V_T$, an inversion layer can no longer be maintained. Thus, when the drain voltage reaches the value $V_G - V_T$, the inversion layer is pinched off at the drain edge. The charge contained in the inversion layer gradually increases toward the source edge of the channel, corresponding to a decreasing depletion width and surface potential. The total inversion-layer charge is roughly half the amount, ie, $Q/2$, calculated previously. As V_D increases above $V_G - V_T$, the pinch-off position moves only slightly away from the drain junction. The pinched-off portion of the channel contains no stable population of mobile charge and can therefore accommodate a large potential drop. Consequently, the total charge contained in the inversion layer varies only slightly from the value $Q/2$ when $V_D > V_G - V_T$, and the drain current is fixed at its saturated value. By substituting $Q/2$ for Q and $V_G - V_T$ for V_D in equation 14, the saturation current is obtained:

$$I_{D_{sat}} = \frac{\mu}{2}\frac{Z}{L}\frac{\epsilon_o}{W_{ox}}|V_G - V_T|^2 \tag{15}$$

If V_D is large, the velocity may reach its saturated value v_{sat} rather than μE, in which case $I_{D_{sat}}$ becomes

$$I_{D_{sat}} = \frac{Z}{2}\frac{\epsilon_o}{W_{ox}}v_{sat}|V_G - V_T| \tag{16}$$

The prefactors in equations 15 and 16 are a figure of merit for the MOSFET, as β was for the bipolar transistor. $I_{D_{sat}}$ is optimized by increasing μ, perhaps by lowering the temperature; reducing the oxide width W_{ox} and channel length L; and minimizing V_T. The oxide thickness is limited by its breakdown voltage and by processing capabilities. The channel length is limited by lithographic resolution and by a punchthrough condition similar to the bipolar transistor; ie, the depletion regions of the source and drain, with $V_G = 0$ and V_D being large, must not come into contact, since a large drain current may result that cannot be controlled by the gate. The depletion-region widths may be reduced by increasing the substrate doping. A thinner oxide is then needed to reduce V_T. The heavy substrate doping also increases the junction capacitances.

The operating speed of the MOSFET is affected by the gate capacitance and the two junction capacitances. The junction capacitances may be reduced by decreasing the junction areas and the substrate doping. If the junction capacitances become negligible, the speed is limited by the transit time L/v_{sat}. (The switching speed is determined by the rate at which charge is applied to the gate. When like transistors are driving each other, the transit time is a useful measure of switching speed.) This transit time becomes comparable to the bipolar transit time from equation 13, when L drops below 5 μm. A more detailed comparison of bipolar and MOS capabilities is presented in the discussion on integrated circuits (qv).

The MOSFET may be enhancement mode or depletion mode, depending on the sign of the threshold voltage. If $V_T > 0$, the device is normally off (enhancement mode) and a positive gate voltage is applied to turn the device on. If $V_T < 0$, the device is normally on and a negative gate voltage is applied to turn it off. When V_G is close to V_T, the channel is only weakly inverted, the I–V characteristic departs somewhat from the previous equations, and the device is said to be in the subthreshold region. This region of operation can be important in low voltage operation (31).

Schottky Diodes and Ohmic Contacts. The metal-semiconductor junction may be rectifying, in which case it is called a Schottky barrier, or it may be ohmic (21). The ohmic junction permits current to flow freely under forward or reverse bias and is used to provide contacts to the devices discussed previously. The analysis of the metal-semiconductor junction is quite similar to that of the p-n junction. In equilibrium, the metal and semiconductor Fermi levels must align. The resultant band-bending at the interface is equal in magnitude to the difference between the metal work function $q\phi_M$ and the semiconductor work function $q\phi_s$. The work function is the energy required for an electron at the Fermi level to escape into the vacuum. A similar quantity, the electron affinity $q\chi$, is defined as the energy required to raise an electron from the conduction-band edge to the vacuum level. Equilibrium band diagrams for a metal–n-type semiconductor junction are illustrated in Figure 12. In the first case, $q\phi_M < q\chi$; there is no barrier to electron flow at the junction; thus, the contact is ohmic. This situation is seldom encountered in practice. Figure 12b shows the formation of a normal Schottky barrier with height $q\phi_{Bo} = q\phi_M - q\chi$.

Charge redistribution at the interface results in the formation of a depletion region in the n-type semiconductor. In equilibrium, the flow of majority carriers or electrons that surmounts the barrier $q\phi_{Bo}$ and enters the metal is balanced by thermal emission of metal electrons over the barrier $q\phi_M - q\chi$. If a bias voltage V is applied, $q\phi_B = q\phi_{Bo} - V$, and the barrier to metal electrons $q\phi_M - q\chi$ remain unchanged. The number of majority carriers varies as $\exp(-qV/KT)$, as in the p-n junction. The Schottky diode current for a thin depletion region is approximately

$$I = \frac{4\pi g m^*}{h^3} T^2 \exp|-\phi_{Bo}/KT| \; |\exp(-qV/KT) - 1| \qquad (17)$$

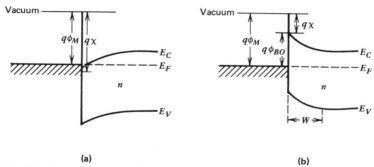

Figure 12. Schottky-barrier band diagrams. (a) A rare situation where the metal work function $Q\phi_M$ is less than the semiconductor electron affinity $q\chi$, resulting in an ohmic contact. (b) Normal Schottky barrier with barrier height $q\phi_{Bo}$; when the depletion width W drops below 10 nm, an ohmic tunneling contact forms.

The Schottky-barrier current is controlled by the rate at which majority carriers can be thermally emitted over the Schottky barrier. The injected electrons equilibrate rapidly with the metal. This is in marked contrast to the p-n junction, where the rate-limiting step is the recombination of injected carriers. The $\exp|-q\phi_{Bo}/KT|$ dependence is analogous to the $\exp|-qV_o/KT|$ dependence for the p-n junction, which is contained implicitly in the minority-carrier concentrations in equation 12. Since the Schottky diode is not limited by charge-storage effects, and with $\phi_{Bo} < V_o$, the Schottky diode can carry significantly more current at a given bias than the p-n junction; it is inherently a much faster device.

In a heavily doped semiconductor, the depletion-region width is thin. If a reverse bias is applied, the conduction-band edge drops below the metal Fermi level at some distance x_o from the junction. As the reverse bias increases, this distance decreases. If $x_o <$ ca 10 nm, tunneling of electrons from the metal Fermi level into the conduction band may occur. The resultant tunneling current is quite large and is limited by the external circuit. The junction permits current flow in either direction and is therefore ohmic. Dopant concentrations of ca 10^{19} cm^{-3} are typically used to form ohmic contacts on silicon.

The presence of interfacial states alters somewhat the previous description of Schottky-barrier behavior. A large number of such states pins the Fermi level at the semiconductor surface. Pinning occurs because a small movement of the Fermi level charges many surface states, thereby creating a sheet of surface charge. In this case, the total band-bending or Schottky-barrier height on the semiconductor is no longer affected by the metal work function. In general, the Fermi levels of the more covalent semiconductors, ie, Si, Ge, GaAs, and InP, are pinned, whereas the Fermi levels of the more ionic materials, ie, SiO_2, ZnS, and Al_2O_3, are not (21).

Applications

Devices. Many of the currently available two- and three-terminal devices are simple extensions of the four basic device structures discussed previously (28). The capacitive and resistive properties of the p-n junction, for example, are exploited in the varactor, varistor, and avalanche diode. The varactor has an a-c capacitance that varies with the width of the reverse-biased depletion region and is frequently used in tuning circuits. The varistor is based on the bias-dependent a-c resistance. The avalanche or Zener diode has a doping profile tailored to cause avalanche breakdown at a specific reverse-bias voltage. This device provides voltage regulation or protection by constraining the voltage at a given circuit node to be less than or equal to the diode breakdown voltage.

The conducting channel of the junction FET (JFET) is constricted by the depletion regions of two parallel p-n junctions. The JFET I–V characteristics are similar to those of the MOSFET. The JFET can accomodate a larger current density, since charge flow is not restricted to a thin planar region; however, the leakage current at the gate junctions increases the gate current that must be supplied and also acts as a source of noise. In the silicon-on-sapphire (SOS) MOSFET, the source and drain regions extend through the thin (ca 1 μm) Si region to the insulating sapphire. In this manner, the p-n-junction area, and hence the junction capacitance, are greatly reduced. The resulting speed increase is somewhat offset by the effect of the floating Si substrate, ie, the substrate voltage can vary, and by the relatively poor quality of the Si film (32). The SOS MOSFET is less susceptible to radiation effects because of the decreased junction area. External radiation forms traps in the junction depletion regions, which results in excess leakage current. Radiation-tolerant devices are described as radiation-hard and are useful in nuclear and extraterrestrial applications. A device similar to the JFET can be fashioned on the SOS structure. The silicon substrate acts as a channel, in which current is constricted between the gate depletion region and the sapphire. This MOSFET is said to operate in the deep-depletion mode. The final FET to be considered is of the deep-depletion variety. In the metal-semiconductor FET (MESFET), the undoped, semi-insulating, semiconductor substrate acts as one boundary of the conducting channel. The channel is again constricted by the gate depletion region. The gate, however, is a reverse-biased or slightly forward-biased (V < 0.3 V) Schottky diode. The MESFET has a reduced punchthrough limitation, and the gate length can therefore be quite small, ie, <1 μm. The short gate length and insulating substrate offer significant operating-speed advantages. The gallium arsenide MESFET is common in microwave applications. Virtually all GaAs MESFETs are n-channel devices, since the electron mobility is much larger than the hole mobility. Integration of MESFET devices is somewhat complicated. Gallium arsenide MOSFETs are as yet impractical because of the lack of a suitable insulator (33).

Optoelectronics is a main segment of the semiconductor industry. Optoelectronic devices include the light-emitting diode (LED), semiconductor laser, and various photodetectors (qv), including the solar or photovoltaic cell (qv) (see Light-emitting diodes and semiconductor lasers). Light emission in each device results from the radiative recombination of electrons and holes, as described earlier. Emitting devices are usually fabricated in direct-gap materials with band-gap energies E_g near the photon energy of interest. Gallium arsenide, for example, emits in the infrared, whereas gallium phosphide emits in the green. The mixed compound $GaAs_{1-x}P_x$ has a band-gap intermediate between GaAs and GaP, depending on the relative As and P concentrations, and is frequently used to obtain red and yellow light emission. An efficient blue emitter has not yet been fabricated, although silicon carbide shows some promise in this area (21). In the LED, recombination occurs in the charge-storage region of a forward-biased p-n junction. In the semiconductor laser, the charge-storage region may also be the intermediate layer of a three-layer heterostructure (21). A heterostructure consists of layers of mixed compound semiconductors with similar lattice spacing but differing energy gaps. If a thin layer of low band-gap material is placed between two larger band-gap materials, injected carriers may be confined to this layer. A portion of the laser light output is fed back to the charge-storage region, where it further stimulates the emission of radiation.

Photodetectors rely on light-induced carrier generation (21). The simplest such device is a photoconductor, which is merely a semiconductor slab. The carrier concentration is increased under illumination, which increases the conductivity of the slab. The photoconductor speed of response is limited by the carrier transit time through the length of the device and by the recombination lifetime. Low lifetime material is frequently employed in fast detectors. The transit-time limitation is somewhat eased in the p-n junction or Schottky-barrier photodiode. Minority carriers generated in the diode depletion region or within a diffusion length of the depletion edges are swept through the junction, thereby increasing the drift-current component of the diode. The response time is then limited by the recombination rate, the transit time through the depletion region, and the time required for carriers to diffuse to the depletion edge. The efficiency is limited by the amount of light reaching the junction, which is located near the surface of the device. The efficiency is improved in a two-layer heterostructure photodiode. The band gap of the exposed layer is greater than the photon energy of interest; thus, photons proceed directly to the junction region, where they are absorbed by the lower band-gap material.

The solar cell is an important example of the photodiode (21). Under illumination, the cell develops a slight forward bias to offset the light-related current flow. This bias can be used to power an external load. Most solar cells are fabricated in silicon or gallium arsenide. The latter, because of its larger band gap, intercepts less of the solar spectrum but is able to develop a larger photovoltage and is therefore somewhat more efficient than silicon. The GaAs cell is also able to operate efficiently at high temperatures and, therefore, can be used in applications where sunlight is focused onto a small cell area. The Si cell is less expensive to fabricate and is used in unfocused applications, eg, a satellite power panel. Light emitters and detectors are used primarily in alphanumeric displays and in the emerging field of integrated optics.

The final class of discrete devices to be considered is those that display a negative-resistance characteristic. These devices are employed in switching, microwave amplification, and microwave generation (see Microwave technology). The I–V characteristic of one such device is illustrated in Figure 13. A load line, such as is shown in Figure 9, may intersect this curve in one, two, or three points. In the latter case, the device has two stable states, ie, on and off, and one metastable state in the negative-resistance region, ie, where the slope of the I–V curve is negative. This corresponds to the switching mode of operation. A common switching device used in high power applications is the p-n-p-n diode (28). As the applied voltage is increased, this device remains in a current-blocking state because of the reverse-biased, central n-p junction. When a critical voltage V_c is reached, all three junctions become forward-biased and the device reverts to a conducting state. If a third lead is connected to one of the base regions, the critical voltage V_c can be controlled externally. This three-terminal device is called a semiconductor-controlled rectifier (SCR).

The tunnel diode, which is illustrated in Figure 13, is formed by joining degenerately doped $n+$ and $p+$ regions. The heavy doping results in the formation of impurity bands and a very narrow equilibrium depletion region. At RT, the Fermi levels lie outside the energy gap (see Fig. 13b). When a small forward bias is applied, con-

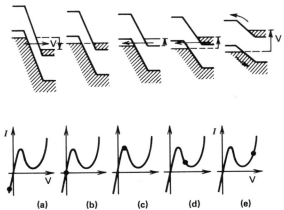

Figure 13. Tunnel-diode current vs voltage. (**a**) Reverse tunnel current. (**b**) Unbiased. (**c**) Forward tunnel current. (**d**) Tunneling ceases, since no available states lie opposite the conduction-band edge. (**e**) Normal diode conduction (21).

duction-band electrons may tunnel across the depletion region to empty-valence-band states. For sufficiently large forward bias, filled-conduction-band states no longer lie opposite empty-valence-band states of equal energy, tunneling ceases, and the diode reverts to normal p-n junction behavior. The tunnel diode can operate as a very fast microwave switch but has not been suitable for integrated applications because of its high power consumption (21).

The negative conductance of the GaAs Gunn diode arises from the unusual shape of the velocity-field curve in Figure 5. Over a certain range of electric field, the velocity decreases with increasing field as a result of the transfer of electrons into the low mobility satellite valley. If an electric field in this range is applied across a slab of gallium arsenide, charge-density variations develop. Electrons congregate into charge domains, and there is a resultant reduction of the field outside the domain regions. Each domain drifts across the GaAs slab and is collected at a contact; other domains may then form. The collection of such charge pulses allows the generation of microwave power. The generated frequency depends on the transit time through the slab. Other modes of operation that make use of the Gunn effect are also possible. The impact-avalanche-transit-time (IMPATT) diode also displays negative conductivity at microwave frequencies. Typically, this diode has an $n+$–p–i (intrinsic)–$n+$ structure, and it is also known as the Read diode. A d-c voltage is applied and biases the $n + p$ junction near avalanche breakdown. An applied microwave voltage brings the diode in and out of the avalanche mode at the microwave frequency. The avalanche-generated charges must then traverse the length of the intrinsic region. The combined avalanche and transit-time delays result in the collected a-c current being out of phase with the applied a-c voltage at an appropriate resonant frequency. Amplification is possible at this frequency. These negative-conductance microwave devices are employed in various communications, spectroscopic, and astronomical applications (28).

Integrated Circuits. A large number of transistors, diodes, resistors, and capacitors can be fabricated simultaneously and interconnected on a single semiconductor chip to form a monolithic integrated circuit (IC). The costs of packaging, interconnection, and system design are greatly reduced in electronic systems incorporating ICs. The fabrication cost is comparable to that of a discrete device, whereas the cost of designing a complex circuit may be divided among the number of such circuits produced. The small size and high packing density of individual components on the IC facilitates low power and high speed operation. The vast majority of integrated circuits is fabricated in silicon because of its ease of p- and n-type doping and the passivating nature of SiO_2. A small number of GaAs ICs have been fabricated in the laboratory (33).

Integrated circuits are classified according to whether their function is linear, ie, analogue, or digital (34). Linear ICs operate on and produce signals of varying amplitude. Typical linear devices include amplifiers, filters, phase-locked loops, digital-to-analogue (D/A) converters, and analogue-to-digital (A/D) converters and are used in such applications as signal measurement, process control, and communications. The signal levels in digital ICs are quantized and typically assume two states, which are termed on and off or high and low. The low state is usually fixed at ground potential and the high state is a small positive voltage. Digital ICs are employed in information processing and storage and in switching applications. Among the more important digital devices are the microprocessor and the semiconductor memory. Digital ICs are produced in far greater numbers than their linear counterparts and provide the main impetus for achieving higher operating speeds and lower power dissipation. The evolution of digital integrated circuits has been governed by economic factors. The IC cost must be minimized subject to its speed and power requirements. Several IC families have been developed and offer compromises between cost, speed, and power. The future potential of these families is a significant focus of research in the electronics industry.

The cost of an integrated circuit has historically decreased as the number of individual devices on the IC has increased. The four generations of integrated-circuit complexity, ie, small-, medium-, large-, and very-large-scale integration, SSI, MSI, LSI, and VLSI, respectively, correspond roughly to 10^2, 10^3, 10^4, and 10^5 transistors per semiconductor chip. The complexity is limited by the size of the chip and the individual device dimensions. If the chip is too large, the probability that a flaw exists within the chip confines becomes excessively large. As device dimensions shrink, the susceptibility to such flaws increases. The percentage of fabricated ICs that meet the desired performance criteria is termed the process yield. Yields well below 50% are tolerated for certain complex circuits. The largest chip size currently in production is ca 1 cm^2. Reducing device dimensions also improves the operating speed, since the speed depends on junction areas, transit times, and interconnection lengths. Although device capacitances shrink with device size, the rate at which they are charged or discharged depends on the device current which, in turn, depends on the supply voltage, the device geometry, and the circuit design. Power dissipation is proportional to the supply voltage V_s and the average device current and, hence, to V_s^2. Consequently, the operating speed may be increased at the expense of greater power dissipation. In very fast circuits, special cooling or heat-sinking procedures are often required. In VLSI circuits, a fundamental limitation to IC complexity is imposed by the finite thermal conductivity of silicon. A standard silicon chip can dissipate little more than one watt of power at RT. Individual transistors must, therefore, dissipate less than one milliwatt of power (35).

The basic building block of the integrated circuit is the switch or inverter illustrated in Figure 14a. Each IC family employs a different inverter configuration. The active transistor, an n-channel MOSFET in this case, is called the driver. The boxed

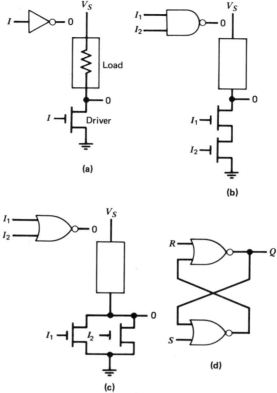

Figure 14. Circuit building blocks. (**a**) Inverter or switch consists of a load (resistive element) and a driver, in this case a MOSFET (metal-oxide-semiconductor field-effect transistor). (**b**) NAND ("not and") function produced by two MOSFETs in series. If I_1 and I_2 are raised to V_s, ϕ drops to ground. (**c**) Parallel NOR ("not or") configuration, in which a high voltage V_S at either I_1 or I_2 drops ϕ to ground. (**d**) Two NOR gates are connected to form a flip-flop; the output Q sits stably at high or low voltage and may be switched by pulsing R or S inputs to high voltage.

device, which may be a resistor or a transistor configured to behave resistively, is termed the load. A positive voltage applied to the inverter input turns the MOSFET on, forcing the output voltage to ground. If the input is grounded, the MOSFET turns off and the output is pulled up to the supply voltage through the load device. Logic operations can be performed by forming series and parallel combinations of driver transistors. A NAND ("not and") gate is illustrated in Figure 14**b**. The output sits at the supply voltage, unless both driver transistors are turned on by positive input voltages, in which case the output drops to ground. The NOR ("not or") gate output of Figure 14**c** is at ground unless both inputs are grounded. The output is then increased to the supply voltage by the load. Multi-input logic gates can be connected sequentially to form adders, multipliers, or other circuits in which the output voltages depend in a logical manner on the sequence of input voltages. Two NOR gates or NAND gates can feed into each other, as shown in Figure 14**d**, to form a flip-flop. The flip-flop inputs are normally low, whereas the output may sit at high or low voltage. If the S (set) input is briefly raised to the high voltage level, the output is fixed at the high voltage. If the R (reset) input is pulsed to the high voltage level, the output switches to and remains at the low voltage level. The flip-flop is the heart of the static semiconductor memory (34).

Some of the more popular bipolar logic families are transistor-transistor logic (TTL), integrated-injection logic (I²L), and emitter-coupled logic (ECL) (34). A cross-sectional view of a typical bipolar IC structure is illustrated in Figure 15a. The transistor is fabricated in a p-type epitaxial Si layer that has been grown on an n-type Si substrate. The n-type diffusions, which reach through to the substrate regions on the outside of the device, isolate the p-collector region from other devices. These isolation regions form large-area, reverse-biased diodes that consume much chip area. The base and emitter are formed by two successive n and p diffusions. More than one emitter may be introduced into a single collector region to form a multi-input transistor. A resistor may be formed by contacting both ends of a long, thin, n-type diffusion placed in the p-type epitaxial layer. Transistor-transistor logic consists of a reasonably straightforward combination of such transistors and resistors. Because of its ease of design, low cost, and moderately high speed, TTL was, until the late 1970s, the dominant IC technology. Integrated-injection logic (sometimes called merged transistor logic or MTL) was developed in 1972 as a new bipolar technology. In I²L, the TTL load resistors, which consume area and power, are replaced by merged, complementary transistors; p-n-p is complementary to n-p-n. Integrated-injection logic also requires fewer isolation regions and fewer metal interconnects. Power consumption is reduced and packing density increases to a level competitive with MOS technologies. The basic I²L structure is illustrated in Figure 16b. The emitter of the load transistor, called the injector, injects charge into the base of the driver transistor, thereby controlling whether it is on or off. Both I²L and TTL are saturating logic families, ie, the driver transistors are driven into saturation during the switching cycle and the resultant excess charge stored in the base significantly reduces the switching speed. Saturation may be prevented through additional processing to form a Schottky clamp. A Schottky diode inserted between collector and base prevents this junction from becoming forward-biased into saturation. Power consumption also increases. The most popular nonsaturating logic family is emitter-coupled logic (ECL). Conventional bipolar transistors and resistors are configured to prevent saturation at the expense of greatly increased power dissipation. Subnanosecond switching cycles are obtained, although elaborate cooling procedures are often required. In 1980, the world's fastest computers employed ECL logic.

The base width in bipolar technologies is controlled by the vertical emitter and

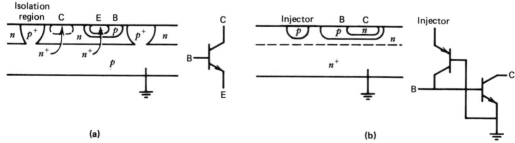

Figure 15.   (a) Standard bipolar transistor structure with collector C, emitter E, and base B regions shown. Reverse-based isolation regions isolate the collector from adjacent devices. (b) Integrated-injection logic: two merged complementary transistors form a compact inverter. The injector, base, and collector regions are shown.

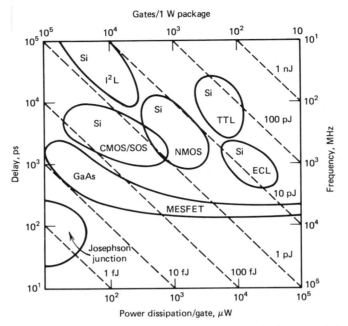

Figure 16. Comparison in 1980 of commercial Si technologies and laboratory GaAs and Josephson devices. The delay is the average switching time of a transistor or gate in a moderately complex circuit and is plotted against the average power dissipation per transistor in such a circuit. Opposing axes show the maximum switching frequency and the number of gates that can be accommodated on a chip with a one-watt power-dissipation limit, which is typical for silicon. Dashed lines are for energy dissipation.

base diffusions and is typically ca 1 μm. In MOS technologies, the channel length is controlled by lateral lithographic capability and, until 1980, was 5 μm or more in commercial devices. Consequently, metal-oxide-semiconductor families had significantly slower operating speeds and were employed in applications requiring high packing density or low power operation. With the approach of submicrometer lithographic capabilities, MOS devices should operate at speeds as fast or faster than bipolar devices (36). The two most popular MOS families are n-channel MOS (NMOS) and complementary MOS (CMOS). The former is preferred over p-channel MOS because of its higher electron mobility. The load device in NMOS is generally an enhancement- or depletion-mode MOSFET. The gate of the enhancement-mode device is either tied to the power supply or to a separate high supply that prevents the load from saturating. The gate of the depletion-mode device, which is normally on, is tied to the output. Depletion-load NMOS is faster than enhancement mode but requires additional processing. Isolation is not required in NMOS and the packing density is therefore excellent. In CMOS, the load device is a p-channel MOSFET, with its gate tied to the n-channel gate. This configuration is inherently faster than NMOS and dissipates no static power in the idle, ie, nonswitching, state since one transistor is always nonconducting. CMOS is ideal for low power or battery-driven circuits where individual IC transistors are idle much of the time. Much additional processing is required to create the p-channel FET, control n- and p-channel thresholds, and prevent peculiar problems, eg, parasitic bipolar action (34). Metal-oxide-semiconductor ICs fabricated by the silicon-on-sapphire technology have shown some improvement in speed and good radiation immunity at the expense of considerable additional processing (31).

The semiconductor memory is a periodic array of storage cells, each of which stores a binary digit: 1 or 0 (37). Memories are classified as volatile or nonvolatile, depending on whether stored information is lost or retained when chip power is removed. Two types of memory, the random-access memory (RAM) and read-only memory (ROM) are manufactured. The RAM essentially serves as a computer scratch pad. Information may be encoded or sensed at a rapid rate. The static RAM maintains its stored information indefinitely so long as chip power is on. The static RAM cell is a flip-flop (see Fig. 14**d**) and has one or two additional transistors that allow the cell to be selected for reading or writing. The dynamic RAM contains one or two transistor cells that store information capacitively for short periods of time; the cells must be periodically refreshed. Dynamic RAMs exhibit the ultimate in VLSI packing densities ($>10^5$ cells per chip). The ROM contains permanent or semipermanent stored information. It can be programmed during manufacture simply by removing transistors from the appropriate array points, but most often is electrically programmed by the user. The metal-nitride-oxide semiconductor (MNOS) ROM stores charge at interface states between the nitride and oxide layers. These states may be charged by application of a high voltage to induce avalanching or tunneling through the thin oxide. In the floating-gate memory, an isolated gate is used for charged storage. In either case, the charge can be stored for long periods of time, depending on the thickness of the intervening insulator layer. These memories are nonvolatile, since charge is stored when the power supply is off.

The charge-coupled device (CCD) is a unique array structure in which charge is stored beneath closely spaced MOS gates (38). Charge may be transferred from one gate to the next by the properly timed application of gate voltages. Reading and writing is accomplished by sequential motion of charge packets through lines of the CCD array. Charge packets may also be produced by light absorption under the gates. Charge-coupled-device photodetector arrays are the central element of portable television cameras. They have recently been fabricated in HgCdTe, which is a low band-gap material, for use as an ir imager (39).

Outlook

Some newly emerging, alternative electronic technologies include superconductors, integrated optics, magnetic bubbles, display devices, and amorphous semiconductors. All of these technologies are in early or intermediate stages of development, but show considerable promise.

Superconductors. The superconducting computer is expected to have an operating-cycle time of ca 1 ns, which is an order of magnitude faster than any conventional computer (40). In one nanosecond, signals moving at the speed of light travel only 15 cm. In order to keep signal paths small, the logic circuitry of the superconducting computer must be contained in a box roughly 5 cm on a side. Since millions (10^6) of devices are to be confined in this region, power dissipation per device must be very small to assure thermally stable operation. The basic switching device in the superconducting computer is the Josephson junction. This junction consists of two superconducting metals separated by a thin (<5 nm) insulator. At liquid-helium temperatures, ie, ca 3 K, superconducting current, which dissipates no device power, may pass through the thin insulator by means of the tunneling mechanism. Application of a magnetic field or sufficient current switches the junction into a resistive mode.

Switching can occur in a few picoseconds and, combined with the low power dissipation, yields a speed power product of ca 10^{-17} J (2.4×10^{-18} cal), which is perhaps an order of magnitude better than any conventional transistor envisioned (see Fig. 16) (see Superconducting materials).

Considerable technological hurdles must be overcome; in particular, the insulator thickness must be well controlled and the operating voltages, which are a few millivolts, must be well regulated. If the superconducting technology proves practical, it might well be restricted to large, ultrafast computing systems because of cooling and cost requirements. Thus, superconductors are not expected to infringe seriously on the semiconductor market in the forseeable future, although the superconducting computer may prove of great importance to the meteorological service, the aerospace and military industries, and the scientific community in general.

Integrated Optics. Optical glass fibers can transmit large amounts of information with low attenuation and, hence, are of great interest to the telecommunications industry. The sending and receiving circuits for such fibers will contain semiconductor lasers or LEDs, detectors, wave guides, lenses, couplers, and other optical devices. Integrated optics or photonics seeks to integrate these devices on monolithic substrates (41). Photonic systems will undoubtedly proliferate in the 1980s and may be employed in high traffic telephone areas and as links between high speed computers. It is hoped that new, as yet unspecified, photonic devices will emerge to compete with electronic devices in conventional applications (see Fiber optics).

Magnetic Bubbles. Large computer systems utilize a hierarchy of memory types that compromise access speed for high density and low cost. In the past, disk and tape storage have been used to supplement the semiconductor memory. Magnetic bubbles show great promise as a source of durable, nonvolatile, high density, low cost memory (42). The bubble is a cylindrical magnetic domain existing in thin films of certain magnetic materials. The structure of the magnetic-bubble memory is similar to that of the CCD. Bubbles are shifted sequentially to adjacent positions in the memory array and are read in serial fashion. The energy required to shift a bubble one position is small compared to the switching energy of a semiconductor memory element. The reading speed is slow, ie, in the kHz range, and the cost is high, but magnetic-bubble memories have already been used in the telecommunications industry and should increase in popularity as they are developed (see Magnetic materials, thin film).

Displays. Flat-panel visual displays are an active area of research aimed at a lucrative market (43). Such displays may eventually displace the bulky and fragile cathode-ray tube (CRT) in television and computer applications, despite continual advances in CRT technology. A flat-panel display consists of an array of closely spaced elements. Each element must be separately contacted and may draw considerable power. The amount of wiring required for a large matrix is prodigious. The individual elements may be light-emitting, as in the LED, plasma panel, cathodoluminescent device, and electroluminescent device, or nonemitting, as in liquid crystals (qv), electrophoretics, and electrochromics (44) (see Chromogenic materials). Light-emitting diodes draw much current and are not likely to be useful in large matrixes. The nonemitters, which transmit or reflect light, draw considerably less power and have the advantage of being viewable under various ambient light conditions. Flat-panel displays are small and very expensive. Only limited application is foreseen in the coming decade.

Amorphous Semiconductors. Amorphous semiconductors lack long-range crystalline order (see Semiconductors, amorphous). Amorphous hydrogenated silicon is under intensive investigation as a possible low cost solar-cell material. Efficient cells have been fabricated, but their performance tends to degrade after modest exposure to sunlight (45). The chalcogenide memory switch is based on thermally induced reversible transitions between the crystalline and amorphous state. Amorphous switches are in a very early stage of development (46).

The technologies just described pose no serious threat to the overall dominance of semiconductor electronics in the next few decades.

Limitations. As base widths and channel lengths shrink to the submicrometer level, high field effects become increasingly deleterious. If the operating voltage drops near the band gap (1.1 V), avalanching multiplication cannot occur, and tunneling becomes the main limitation (21). Avalanching in the MOS channel causes charge injection into the gate oxide and thereby affects the threshold voltage (47).

Heavy doping is required to prevent punchthrough. In the bipolar device, the emitter injection efficiency is reduced. In the MOS device, a thin-gate oxide, which is difficult to fabricate, is required to maintain a low threshold voltage. High current density in the bipolar device may also lead to formation of current channels and local hot spots.

The silicon power-dissipation limit requires that speed-power products be substantially reduced in future generations of VLSI circuits. Since power dissipation is a function of the square of the operating voltage, it is desirable to reduce this voltage. In bipolar devices, the minimum operating voltage must be on the order of built-in diode potential, ie, ca 0.6 V. In MOS devices, threshold-voltage variations are a principal limitation. Fabrication nonuniformities, eg, statistical fluctuations in doping, must also be considered.

The high packing density, low static-power consumption, and ease of designability of MOS ICs suggest that MOS technologies will soon dominate in high speed applications. The electron saturation velocity intrinsically limits MOS device speed. Reduction of load capacitances by shrinking metal line widths is effective to a point. As line spacings approach 1 μm, however, interline capacitance becomes appreciable (36). Thin-metal lines become resistive and are also more susceptible to electromigration, ie, local thinning and melting induced by current flow (48). Fabrication of MOS devices on dielectrically isolated substrates can significantly reduce junction capacitances and thereby improve speed. Mobilities may be adversely affected, as in SOS MOSFETs, and higher operating voltages are needed to optimize speed. Recent work on the laser recrystallization of silicon on SiO_2 shows some promise in this area (49).

Significant speed improvement and higher packing density may be achieved through low temperature operation (50). At 100 K, the low field mobility in heavily doped silicon is near its maximum at roughly double the RT value. The saturation velocity also increases somewhat and may be achieved with a lower applied voltage. Thermal conductivity is improved by nearly an order of magnitude and allows greater power dissipation and, hence, greater packing density. Thermal noise is also reduced. Carrier freeze-out, which reduces the effective doping, is a minor nuisance. Temperatures of 100 K can be achieved by liquid-nitrogen cooling and may be practical in systems of moderate size.

Gallium arsenide has a significantly higher low field electron mobility and saturation velocity than silicon. Gallium arsenide MESFET technology is therefore viewed as a potential competitor in high speed applications (32). To date, the potential 2–5 factor of improvement in speed has been offset by difficulties in large-scale integration. Low hole mobility also precludes the use of complementary circuits.

The device potential of lower band-gap compound semiconductors is being investigated. One of these, ie, HgCdTe, is already used in ir detectors. Others are used in semiconductor lasers. A fascinating new entry to the list of potential materials is the superlattice, which consists of very thin alternating layers of lattice-matched compound semiconductors grown by molecular-beam epitaxy (51). Transfer of electrons into the lower band-gap, high mobility layers can result in large effective mobilities.

Future high speed devices may also make use of an effect known as ballistic transport (52). Ballistic transport occurs when the critical device dimension is on the order of the electron-scattering length. This distance is approximately 100 nm in gallium arsenide but only 10 nm in silicon. Electrons that are field-accelerated over this distance may substantially overshoot the saturation velocity.

As packing densities increase, device design becomes increasingly complex. Fault detection, for example, is a difficult problem with an integrated circuit containing 10^6 devices. Circuit and system design and programming fall under the label of software, as opposed to the technological aspects discussed previously, which fall into the hardware category. Although hardware is entering a mature stage of development, software is in a very early developmental stage (15). A curious bootstrapping phenomenon is occurring, in which VLSI-circuit hardware requires advanced software in its implementation and, in turn, provides the means for creating more complex software systems. The number of software employees in the semiconductor industry has mushroomed throughout the 1970s and early 1980s. Attempts to place software on a firmer scientific foundation should help to stabilize this growth. Advances in computer architecture will ease human–machine interaction and thereby expand the semiconductor market. Thus, although integrated-circuit complexity may cease its exponential growth in the coming decade, continuing advances in software should extend the rapid pace of progress in electronics into the following decades.

BIBLIOGRAPHY

"Semiconductors (Theory)" in *ECT* 2nd ed., Vol. 17, pp. 834–861, by P. J. Dean, Bell Telephone Laboratories, Inc.

1. E. Braun and S. MacDonald, *Revolution in Miniature—The History and Impact of Semiconductor Electronics*, Cambridge University Press, New York, 1978.
2. H. K. Henisch, *Metal Rectifiers*, Clarendon Press, Oxford, UK, 1949.
3. G. F. J. Tyne, *Saga of the Vacuum Tube*, Howard W. Sams & Co., Inc., Indianapolis, Ind., 1977.
4. G. L. Pearson and W. H. Brattain, *Proc. IRE*, 1794 (1955).
5. A. H. Wilson, *Proc. R. Soc. London* **A133,** 458 (1931).
6. J. Bardeen, *Phys. Rev.* **21,** 717 (1947).
7. W. Shockley, *IEEE Trans. Electron Devices* **ED-23,** 597 (1976).
8. W. Shockley, *Electrons and Holes in Semiconductors with Applications to Transistor Electronics*, D. Van Nostrand Co., Inc., Princeton, N.J., 1950.
9. D. M. Chapin, C. S. Fuller, and G. L. Pearson, *J. Appl. Phys.* **25,** 676 (1954).
10. B. R. Nag, *Electron Transport in Compound Semiconductors*, Springer Verlag, Berlin, FRG, 1980.
11. L. Esaki, *Proc. IEEE* **62,** 825 (1974).
12. D. Kahng. *A Historical Perspective on the Development of MOS Transistors and Related Devices*, *IEEE Trans. Electron Devices* **ED-23,** 655 (1976).

13. A. L. Robinson, *Science* **195,** 1179 (1977) (special issue devoted to the electronics revolution).
14. *Sci. Am.*, (Sept. 1977) (special issue on semiconductor electronics).
15. C. L. Seitz, ed., *Proceedings of CalTech Conference on Very Large Scale Integration*, CalTech Computer Science Department, 1979.
16. *Chem. Week*, 15 (Nov. 21, 1979).
17. R. L. Sproull, *Modern Physics*, 2nd ed., John Wiley & Sons, Inc., New York, 1963.
18. C. Kittel, *Introduction to Solid-State Physics*, 5th ed., John Wiley & Sons, Inc., New York, 1980.
19. J. P. McKelvey, *Solid State and Semiconductor Physics*, Harper & Row, New York, 1966.
20. H. V. Malmstadt, C. G. Enke, and S. R. Crouch, *Electronic Analog Measurements and Transducers*, W. A. Benjamin, Inc., Menlo Park, Calif., 1973.
21. S. M. Sze, *Physics of Semiconductor Devices*, 2nd ed., John Wiley & Sons, Inc., New York, 1981.
22. A. S. Grove, *Physics and Technology of Semiconductor Devices*, John Wiley & Sons, Inc., New York, 1967.
23. S. Wang, *Solid State Electronics*, McGraw-Hill, Inc., New York, 1966.
24. R. B. Fair in D. Kahng, ed., "Physics and Chemistry of Impurity Diffusion and Oxidation of Silicon," *Silicon Integrated Circuits*, Part B, Academic Press, Inc., New York, 1981.
25. W. E. Spicer and P. E. Gregory, *CRC Crit. Rev. Sol. St. Sci.* **5,** 231 (1975).
26. B. E. Deal, *J. Electrochem. Soc.* **121,** 198C (1974).
27. W. E. Spicer, J. Lindau, P. Skeath, and C. V. Su, *J. Vac. Sci. Technol.* **17,** 1019 (1980).
28. B. G. Streetman, *Solid-State Electronic Devices*, Prentice-Hall, Inc., Englewood Cliffs, N.J., 1972.
29. B. Tuck, *Introduction to Diffusion in Semiconductors*, Peter Perigrinus Ltd., 1974.
30. B. M. Oliver, *Proc. IEEE* **53,** 436 (1965).
31. J. R. Brews in D. Kahng, ed., "Physics of the MOS Transistor," *Silicon Intergrated Circuits*, Part A, Academic Press, Inc., New York, 1981.
32. A. C. Ipri, "The Properties of Silicon-on-Sapphire Substrates, Devices, and Integrated Circuits," in ref. 31.
33. R. C. Eden, B. M. Welch, R. Zucca, and S. I. Long, *IEEE Trans. Electron. Devices* **ED-26,** 299 (1979).
34. D. J. Hamilton and W. G. Howard, *Basic Integrated Circuit Engineering*, McGraw-Hill Book Co., New York, 1975.
35. B. Hoeneisen and C. A. Mead, *Solid State Electron.* **15,** 819, 891 (1972).
36. J. A. Cooper, Jr., *Proc. IEEE* **69,** 226 (1981).
37. D. A. Hodges, ed., *Semiconductor Memories*, IEEE Press, New York, 1972.
38. G. F. Amelio, *Sci. Am.*, 22 (Feb. 1974).
39. R. A. Chapman, M. A. Kinch, A. Simmons, S. R. Borello, H. B. Morris, J. S. Wrobel, and D. D. Buss, *Appl. Phys. Lett.* **32,** 434 (1978).
40. J. Matisoo, *Sci. Am.*, 50 (May 1980).
41. P. K. Tien and J. A. Giordmaine, *Bell Lab Rec.*, 371 (Dec. 1980).
42. T. H. O'Dell, *Magnetic Bubbles*, John Wiley & Sons, Inc., New York, 1974.
43. L. E. Tannas, Jr. and W. F. Goede, *IEEE Spectrum*, 26 (July 1978).
44. J. I. Pankove, ed., *Display Devices*, Springer Verlag, Berlin, UK, 1980.
45. R. Williams and R. S. Crandall, *RCA Rev.* **40,** 371 (1979).
46. R. Dalven, *Introduction to Applied Solid State Physics*, Plenum Publishing Corp., New York, 1980.
47. R. B. Fair and R. C. Sun, *IEEE Trans. Electron. Devices* **ED-28,** 83 (1981).
48. S. Vaidya, T. T. Sheng, and A. K. Sinha, *Appl. Phys. Lett.* **36,** 464 (1980).
49. J. F. Gibbons, *Jpn. J. Appl. Phys. Suppl.* **19-1,** 121 (1979).
50. S. K. Tewksbury, *IEEE Trans. Electron. Devices* **ED-28,** 1519 (1981).
51. R. Dingle, A. C. Gossard, and H. L. Stormer, *Bell Lab Rec.*, 274 (Sept. 1980).
52. K. Hess, *IEEE Trans. Electron. Devices* **ED-28,** 937 (1981).

S. A. SCHWARZ

Bell Laboratories, Inc.

FABRICATION AND CHARACTERIZATION

Semiconductor technology is dominated by silicon integrated circuits (ICs) that are used in the consumer, communications, entertainment, and computer industries (see Semiconductors, theory and applications; Integrated circuits). An understanding of silicon IC fabrication provides the basic knowledge necessary for understanding other fields of semiconductor-device fabrication, with the possible exception of compound semiconductors (1–2) (see Light-emitting diodes and semiconductor lasers). Therefore, the main topics treated here are the Si and GaAs technologies.

Early Semiconductor Devices

Point-Contact Detectors. The semiconductor device first used on a significant scale in ca 1905 was the point-contact, also called whisker detector (3). The sharp point of a metal wire was placed onto a semiconductor, eg, galena (PbS), copper oxide, selenium, germanium, or silicon. The earliest example was the wire-galena rectifier made by E. Braun in 1874 (4). Subsequent landmarks were the use of metal–germanium detectors in 1925 and of metal–silicon contacts for radar detection in the 1940s. Much of the work on germanium in the 1940s is summarized in ref. 5.

Germanium Transistors. The important properties of Ge were characterized in the 1940s and include relatively low melting temperature, high carrier mobility, and long minority carrier diffusion lengths (6).

The transistor was invented in 1947, and the results were first reported in 1948 (7–10). The emitter and collector contact wires were phosphor bronze, and the base contact was soldered. By 1951 this type of transistor was commercially manufactured by Western Electric as the type A transistor (8,11). n-Type germanium was used initially because the commercially purchased high purity Ge was generally n-type and initial attempts at carrier injection into p-type material were unsuccessful (12). Many materials were used for the emitter contact, but those containing copper, eg, BeCu, gave the best results (8).

The first grown junctions were reported in 1951 (13). The junction transistor was described in 1949, and commercial manufacture of not only diodes but also transistors with grown junctions immediately followed (8,11,14–15). Crystals of mostly (111) orientations were used because of the ease of locating the (111) plane and of growing germanium boules having [111] axes. The (hkl) crystallographic notation is described in ref. 16. The high electron mobility in Ge was important because methods of fabricating thin base regions had not been perfected. This thickness limits the cut-off frequency; with higher mobility, thicker base regions can be tolerated. Initially, npn transistors were made because the available high grade n-type germanium facilitated the initiation of crystal growth with the first n layer. Therefore, the choice of npn rather than pnp was one of convenience. These new npn transistors were not interchangeable with the older type A transistors which behaved like pnp transistors. The manufacture of these transistors is described in refs. 8 and 11.

Devices made in the early 1950s were the point contacts and grown junctions, ie, rectifiers, zeners, and photodetectors; the type A transistor (n-type Ge with two point contacts); the filament transistor (n-type Ge filament with one point-contact emitter and two ohmic contacts); the *npn* grown-junction transistor; and the *pnpn* or the *pn* hook transistor (8,11,17). The point-contact transistors were outmoded by the more reliable grown-junction transistors. By 1955, the alloyed-junction transistor was readily available (15,18). The surface-barrier transistor was first reported in 1953 (17,19).

Semiconductor Technology

The advantages of solid-state devices over electron tubes and other devices motivated intense research and development efforts in solid-state technology (15). The semiconductor devices were less expensive, more reliable, smaller, and faster, and they consumed less power and provided greater flexibility in design. New devices and methods of making them began to proliferate, and several electronics firms began to supply them in commercial quantities (1,15,17,20–21).

Thermal diffusion was introduced in 1955 as a method of semiconductor doping (22–25). By this time, the advantages of silicon were well recognized. Compared to germanium, the silicon band gap is larger, and the melting point is higher, which means lower leakage currents, higher temperature operation, and higher reliability (1,26). The greatest advantage of silicon is the existence of amorphous SiO_2, which produces a nearly ideal dielectric–silicon interface. The discovery that SiO_2 can be used as a mask for dopant diffusions led to the development of planar technology (24,27). Other important innovations were the mesa structure (22–23,28), the epitaxial process (29), the junction field-effect transistor (JFET) (1952) (30–31), the integrated circuit (1958) (32), and the metal oxide semiconductor (MOS) transistor (1960) (33–36). In the 1970s, ion implantation (qv) began to be used widely, and beam-lead technology provided devices with excellent reliability and electrical performance (37). The evolution of silicon semiconductor technology is reviewed in ref. 38.

Silicon Technology

Discrete, Monolithic, Hybrid, Integrated, Power, and High Voltage Devices. Semiconductor devices made in the 1960s generally were of three types: discrete devices, monolithic circuits, and hybrid circuits. A discrete device contained just one element, eg, a diode, transistor, light-emitting diode (LED), etc. Monolithic circuits were made on a single chip of silicon by planar technology and contained many elements. Hybrid circuits were made by attaching monolithic circuits to substrates, which were usually ceramic and contained interconnects, resistors, and capacitors. Integrated circuits (ICs) were made by building up the circuit elements, eg, transistors, resistors, and interconnects simultaneously by use of planar, thin film, or thick film technologies. Monolithic and hybrid ICs were the most important types. Improvements in fabrication technology led from the first integrated circuits, ie, small-scale integration (SSI) in 1960 with tens of transistors, to medium-scale integration (MSI) in 1965 with ca 100 transistors, to the large-scale integration (LSI) in the 1970s with ca 10^4 transistors, to the present very large-scale integration (VLSI) in 1980 for incorporating more than 10^5 transistors on a chip.

In parallel with these advances for low-power devices were the developments for high-power applications; these were the power transistor, ie, the *npn* transistor and its derivatives, and the thyristor, ie, the *pnpn* and its derivatives (39–41).

Commercial manufacture of high voltage ICs began in 1980 (41–43). These ICs can replace the costly and clumsy mechanical relays used in communications switching networks, because they have sufficient bandwidth, power, and voltage-blocking capability.

Integrated Circuits: Unipolar and Bipolar. Bipolar transistors are the direct descendants of junction transistors; unipolars are field-effect devices, eg, field-effect transistors (FETs) and MOS transistors. Most bipolars are current-driven devices and have low input impedance. Most unipolars are voltage-driven devices; junction field-effect transistors (JFETs) generally have higher input impedance than bipolars, and MOS devices have the highest impedance. High input impedance is advantageous for large fanouts (number of devices that can be driven by a given transistor) and low power dissipation.

Metal oxide semiconductor ICs have advantages over bipolars because of the former's lower power consumption and higher circuit density. Early development of MOS technology was delayed by problems with trapped impurities and charges in the gate oxide. Less than 1 ppm of certain contaminants can cause more than a 5-V shift in the threshold voltage or turn-on voltage of MOS transistors. A new technology based on clean-room operation needed to be devised. Fabrication of the first MOS device of Kahng and Atalla succeeded because of the availability of high pressure SiO_2 (36,44). The first commercial MOS devices appeared in the late 1960s; these were p-MOS on (111) Si with aluminum gate metallization (45). The p-MOS was successfully fabricated because the trapped charges in the gate oxides are positive and cause n-MOS devices to be permanently on, whereas they only raise the threshold voltages of p-MOS devices. (111) Silicon was the first substrate used because it was easiest to grow [111] oriented boules of Si by the Czochralski method, and it was also easiest to cut and polish (111) oriented wafers. However it was soon established that the oxide-charge density was lowest for (100) Si, and most manufacturers switched to this orientation to produce lower threshold devices. Aluminum was used for the gate metal because it could be vacuum-evaporated at comparatively low temperatures for depositing thin films to provide gate and interconnect metallizations. In the late 1960s, the silicon gate was introduced (46). This was an important advance because silicon is a refractory material and a high temperature source and drain diffusion can be performed after gate lithography, permitting fabrication of self-aligned gates with lower parasitic capacitance. The lower capacitance results in higher speed. n-Metal oxide semiconductor ICs were successfully fabricated in early 1970 by maintaining clean facilities with deionized water with resistivities in excess of 10 MΩ·cm, and by purging the oxidation furnaces of alkali metals by flushing with chlorine-containing gases, eg, HCl. In addition, gettering of impurities by use of wafer backside stress or damage or by use of SiO_2 containing 4–9 wt % phosphorus with high temperature (ca 1100°C) treatment and proper annealing of the gate structure, eg, at 450°C in hydrogen, contributed to successful n-MOS devices (47–48). An n-MOS transistor is faster than a p-MOS transistor because electron mobility in silicon is about three times higher than hole mobility. In early 1970, speed was further increased by the use of n-MOS depletion-mode active loads. The main efforts in the late 1970s have been in producing smaller devices and in devising higher conductivity interconnect metallizations, eg, metal silicides (49). With smaller devices, the advantages of higher yield, lower power, higher speed, greater reliability, and lower cost could all be gained simultaneously. In the middle 1970s, complementary MOSs (CMOSs) began to be accepted for low power applications (1). Currently, n-MOS technology is the most important IC technology because it is the simplest, provides the greatest speed, and has been scaled down in size. p-Metal oxide semiconductors and, therefore, CMOSs are difficult to scale down owing to lack of manufacturers' experience with gate metals appropriate to p-MOS at small dimensions. It is not clear what the most popular IC technology succeeding n-MOS will be. Technologies, eg, CMOS and dielectric-isolated structures have certain advantages over n-MOS and are being actively researched. The extent of this research is demonstrated by the fact that in 1980, well over 50 new transistor devices were patented.

Organization of IC Fabrication Plants. The IC facility is composed of several interacting groups centered around the device-processing line. These are the design, the mask, the development, and the electrical testing groups, and representatives of product users.

Design. The design group designs the circuit and chip layout using computer simulation as the main working medium. A typical design begins with computer-aided design (CAD), in which each circuit element, ie, transistor, capacitor, conductor line, etc, is computer-modeled in precise detail (50). Then the circuits are computer-simulated. These simulations are extremely accurate and take tremendous amounts of computation. For large chips, eg, microprocessors, complete simulation is not possible, and they are fabricated after partial simulation. The finished product of the designers is the layout, which is given to the mask makers. After initial manufacture, the chips are tested and the necessary modifications are made. This process is iterated until the chip meets specifications.

Designing is broken into numerous levels and the simplified output of one level is used as the input for the next level of simulation. For example, one of the first CAD programs simulates the basic device fabrication processes, eg, diffusion, implantation, and oxidation. After experimental verification, results from these programs are used to model devices, eg, n-MOS transistors, that are made by these same processes. Outputs from these device models are used to simulate logic elements, eg, memory cells and flip-flop circuits. The resulting outputs are in turn the inputs for logic simulations, of which there are several types, eg, input–output relations, timing simulations, and programs for fault detection.

Designers work closely with the processing-line personnel and the development group to agree on a set of design rules, ie, the limits of technology that the processing line can attain. They include the minimum pitch, ie, the distance between two parallel lines, minimum gate oxide thickness, maximum metallization conductivity, die or chip dimensions, etc.

Masking. The mask group takes the layouts from the designers and produces masks for use in manufacturing. The chromium masks are the most popular ones; the layout patterns are transferred to the chromium film on glass plates by lithography, and the chromium is etched to form the transparent regions of the masks. At first, photolithography was exclusively used, but by 1975, electron-beam writing had become the most important method for generating masks. Mask making is very specialized, and only the largest semiconductor-device manufacturers can afford their own mask facilities. Companies that specialize in mask generation often provide faster and more economical service than the limited facilities at most device-fabrication centers.

Main Processing. The main processing line typically costs ca 10^7, exclusive of the building and utility supplies, eg, the deionized water and air-filtration systems; the latter two can total another 10^7. Such a line is capable of beginning 100–500 wafers per day, each wafer containing 20 to more than 200 devices. It takes 2 wk–2 mo to process each lot of wafers, ie, each set of 10–100 wafers that are processed together. In the 1970s, silicon wafers 75 mm in diameter were in standard use; since 1980, many facilities have switched to the use of 100 mm wafers. Wafers 125 mm in diameter are now available, and there is little question that some manufacturers will be switching to this larger size in the early 1980s.

The facilities must be ventilated with clean air because any dust on the wafers creates defects which destroy the devices. Many dust particles are larger than the device features, and typical room air contains more than 10^6 particles/m³. The process line, or sometimes just the critical areas within the line where the wafers are exposed to room air, must contain less than 10^3 particles/m³, and particle sizes must not be larger than 0.5–5 μm. This quality air is achieved by multistage filtration.

An ample supply of deionized water is also required, and usually it must meet the following standards: resistivity, 18 MΩ·cm at 25°C; particles, <1000/L and 0.2 μm; bacteria (includes all microorganisms), <25 colonies/100 cm³; total carbon, <0.2 ppm; total residue, <0.5 ppm; and sodium, <1 ppb. This water is prepared from a local water supply in the following stages: clarification by settling and flocculation; first filtration with sand, charcoal, or both; primary deionization by electrodialysis or reverse osmosis; two-bed deionization; storage of recirculation water; sterilization by ultraviolet irradiation; mixed-bed polishing deionization; and ultrafiltration through pores <10 nm in diameter. The terms two-bed and mixed-bed refers to beds of resins designed to remove anions and cations; the mixed-bed deionizer provides purer water than the two-bed type because both kinds of ions are removed together.

The first four stages apply to new incoming water. Most of the water flow occurs in the recirculating system comprised of storage, sterilization, and polishing deionization. The final filtration is performed at the point of use.

The processing line communicates mostly with the development group regarding new technology and troubleshooting. The line must inform the design group of the capabilities and limitations of the facility, and it obtains device performance data from the testing group.

Development. The development group studies new technologies and tries them out on a pilot line before transferring them to the manufacturing line. It is well-equipped for analytical work and is experienced in materials science and device diagnostics; it also assists the line in problem solving. The more common diagnostic tools used by the research group are described below.

The optical microscope is used to examine at magnifications of (10–600) ×. Almost all samples are optically inspected before other, more powerful techniques are applied (51).

The scanning electron microscope (SEM) is used for examining features to 10 nm (51).

An ellipsometer is used to determine the thicknesses of optically transparent films from the change in polarization of reflected light (52). Ideally an ellipsometer gives a thickness resolution of ca 0.1 nm.

Stylus technique involves the measurement of film thickness by determining the height of a step in the film with a stylus. Its resolution is ca 20 nm (51).

An electron microprobe is used to determine chemical composition (51). An electron beam excites characteristic x rays, which are analyzed to determine chemical composition. Modern instruments are computerized and can provide quantitative analyses.

The simplest type of stress-measuring device is the optically levered laser instrument in which laser light is reflected from the surface. The curvature of this surface is determined by the change in the reflection angle as the sample is translated, and the stress is calculated from this curvature.

Many electrical properties must also be measured, eg, film resistivity, various capacitances, MOS properties of dielectrics, contact resistances, *pn*-junction properties, etc. For more complex analyses, the more powerful techniques can be applied, eg, transmission electron microscopy; Auger electron spectroscopy; x-ray photoelectron spectroscopy; Rutherford backscattering spectroscopy; neutron activation analysis; optical-emission and absorption spectroscopies; x-ray and electron diffraction; and mass spectrometry (51).

Electrical Testing. Testing for functional operation, for meeting the designers' specifications and for failure modes are the main tasks of this group. Testing for reliability is also an important function, although this responsibility is shared with the research group. Stress aging is the main procedure in reliability studies. Stress refers to the application of usually extreme values of current, voltage, or humidity during device operation. Accelerated aging is accomplished by raising the temperature.

User Representatives. Users often are not represented by any group in the device-manufacturing organization, except at the largest IC-manufacturing centers. In the case of discrete devices, the design, research, or fabrication group may be responsible for market research.

For integrated-circuits manufacture, representatives of systems personnel, ie, those in charge of designing complete systems such as computers, provide information for the IC designers. They specify the type of operation, cycle times, operating voltage, temperature range of operation, etc. As ICs become more powerful, the chip designers become responsible for more of the jobs formerly performed by systems engineers, by incorporating systems capabilities into single chips.

Basic Processing Operations. The basic processing steps in IC production are cleaning; crystal growth, eg, of Si and compound semiconductors (2,53–55); oxidation (atmospheric and high pressure); lithography (photolithography, electron-beam, and x-ray); doping by diffusion, implantation, and molecular-beam epitaxy (MBE); deposition (electron-gun, sputter, chemical vapor deposition (CVD), low pressure chemical vapor deposition (LPCVD), and plasma); etching (wet chemical and dry or plasma) (56); and electrical testing.

Wafer cleaning is one of the most important procedures in IC production. Scrubbing is an indispensable part of many cleaning operations and is commonly performed with brushes. Water jets, sometimes combined with ultrasonic agitation, have also been used; however, the electrostatic charging caused by a jet can lead to arcing and destruction of the devices. This problem can be alleviated by doping the water to make it conductive. In brush scrubbing, the brushes should not touch the wafer; if they do, the sharp edges of some features on the wafer can slice bits of brush off, creating more debris than is cleaned away. The brush surface should be hydrophilic, so as to retain a film of water between the brush and the wafer. Hydrophobic surfaces are easier to clean because they retain no water film, which tends to hold particles. Hydrophilic surfaces are easier to dry after cleaning because the surface water film can be reduced to only several monolayers by spin drying, and then it can be dried by evaporation without any residue remaining. Hydrophobic wafers tend to retain water in a few isolated drops, which can leave residues when the wafers are dried. The adsorbed surface water remaining after spin drying and evaporation is removed by baking.

Buried-Collector, Bipolar, Integrated-Circuit Processing.   A flow diagram for bipolar IC fabrication is shown in Figure 1, and a diagram of a finished bipolar transistor is displayed in Figure 2. The starting material is p-type Si (111) with a conductivity of 10 Ω·cm. Wafers cut 3° from (111) orientation are often preferred because this misorientation results in better epitaxial films. Process 1 (see Fig. 1) is the collector-implant process; it starts with a cleaning operation, which includes scrubbing with a brush in a mild detergent and several chemical oxidation–etch sequences. A steam oxide is grown at 1100°C for use as a mask to define the buried collector. The first mask,

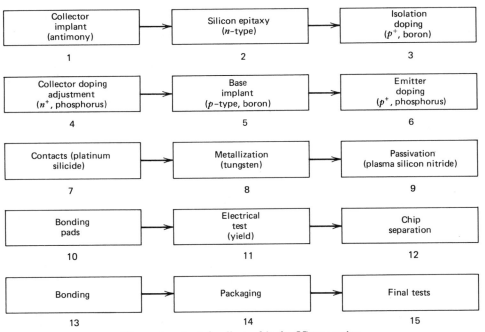

Figure 1.   Buried-collector bipolar IC processing.

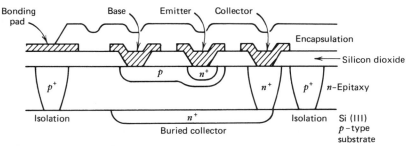

Figure 2.   Buried-collector bipolar transistor.

photomask 1, is used for the lithography. There are different resist application, baking, exposure, and development operations that depend on whether positive, negative, or a special brand of photoresist is chosen. Typically, the resist is spun on, then it is prebaked to remove solvents from it. After exposure, development, and inspection, the wafers are baked again to promote resist adhesion to the wafer. Without this bake, the etch solution would penetrate underneath the photoresist at the edge of a pattern, resulting in undercutting. The collector pattern is etched into the oxide using buffered HF (BHF). The resist is then stripped. The procedure starting from the initial cleaning and oxide growth and ending with the final resist strip is referred to as patterning or pattern transfer, and the particular patterning sequence just described is called collector definition. Antimony is then implanted at 150 keV to dope the collector region as n-type. Antimony is chosen because it is a large atom and cannot readily diffuse outward from the collector region during the following high temperature operations. At 150 keV, the Sb atoms are unable to penetrate the masking oxide and, therefore, implantation into Si occurs only in the patterned areas where the oxide was etched. A heat or drive-in treatment at 1250°C in argon gas with 10 wt % O_2 activates the implanted Sb atoms, which are initially at interstitial sites, by allowing them to settle into substitutional sites. All surface oxide is then etched in preparation for the next operation.

In process 2 an epitaxial layer is grown by chemical vapor deposition (CVD) and forms the buried collector structure. The thickness and doping level of the epitaxial layer depends on the particular device code being processed, which must be specified at this point (see Film deposition techniques).

In process 3 the individual transistor areas are isolated from each other by an isolation definition, which is similar to the collector definition of process 1, except that a different mask, photomask 2, is used. Boron in BBr_3 gas diffuses into the isolation areas at 1100°C. Boron tribromide is obtained by bubbling argon through liquid BBr_3 held at a precisely controlled temperature to give the desired BBr_3 vaporization rate. During this isolation predeposition, a boron glass forms over the wafer. This glass has a different etch rate from pure SiO_2; therefore, it is purified by oxidizing in steam at 900°C. The resulting oxide can be etched in BHF very reproducibly. After boron predeposition, a heat treatment normally follows. However, this is not required in this case because there are later heat treatments at sufficiently high temperatures to activate the boron. In process 1, a high temperature annealing was necessary, because there would be no subsequent heat treatment at the high temperature of 1250°C required to activate the antimony atoms.

In process 4 the doping profile in the collector region is adjusted by starting with another collector definition procedure with photomask 3. Phosphorus is doped into the collector areas by diffusion at 1040°C with PBr_3 gas. Argon or nitrogen is bubbled through liquid PBr_3 to obtain the desired amount of PBr_3 for doping. The surface oxide is etched with BHF. This collector doping adjustment improves device properties, eg, the breakdown voltage.

In process 5 and after base definition using photomask 4, boron is implanted at 30 keV. Implantation is needed to attain a precise doping profile. The surface is then cleaned in ozone at 300°C to ensure that organic contaminants deposited during implantation are removed. A heat treatment to activate the boron atoms is accomplished at 1150°C in an N_2 atmosphere with 10 wt % O_2.

Process 6 starts with emitter definition using photomask 5. Emitter predeposition is performed by thermal diffusion at 1000°C with PBr_3. The critical parameter is the diffusion time, which is adjusted with monitors to attain a precise base width. Monitors are sacrificial wafers processed with the device lot for various purposes; in this instance, they are used for performing trial experiments to determine the exact diffusion conditions for the real device wafers. After predeposition, all surface oxide is stripped in 100:1 H_2O:HF. At this point, many electrical tests can be performed, eg, the measurement of sheet resistances of monitors.

Process 7 first involves contact definition with photomask 6. The contact windows are then tested to see if all oxide has been removed by spraying a mist of water onto the window. Silicon dioxide is hydrophilic and silicon is hydrophobic; therefore, water forms small droplets if the windows are open; if not, the water spreads uniformly across the window. The wafers are etched in 100:1 H_2O:HF solution to remove any oxide that may have grown during air exposure. A 40 nm thick layer of platinum is immediately deposited by sputtering and is sintered at 650°C to form platinum silicide. Platinum outside the contact windows does not form silicide because that part of the platinum film is over SiO_2. This unreacted Pt is etched with aqua regia, which does not dissolve platinum silicide. This process silicides only the contact windows.

Metallization refers to establishing metallic interconnection lines between circuit elements. Process 8 involves a simple tungsten metallization. Tungsten is sputter-deposited and then patterned with photomask 7. Tungsten can be etched in $K_3Fe(CN)_6$ that has been buffered to pH 9 with $KOH-H_3BO_3$. The devices are then ready for electrical testing. Processing beyond this point depends on the device code and its applications (see Passivation, Separation, Bonding, and Packaging).

_n_-Metal Oxide Semiconductor IC Processing. The instructions for IC processing consist of two parts: a general one, which describes the overall process flow, and a set of detailed instructions for each step. The detailed instructions are not presented because many of the instructions apply only to particular machines in use at a facility and some of the steps are proprietary. The following type of processing (called HMOS) was used at Intel in the 1970s (57). A diagram of the finished device is shown in Figure 3, and the processing sequence is outlined in Figure 4.

In process 1 (Fig. 4) a thermal oxide is first grown, and then silicon nitride is deposited by LPCVD and patterned with photomask 1 by a dry etching technique. The oxide layer serves as the etch stop for the nitride, and it also prevents contamination of the Si substrate surface with nitrogen. The slope of the field-oxide edge can be controlled by adjusting the thickness of the oxide.

After the nitride is patterned, boron is implanted over the entire wafer at 70 kV. The nitride pattern is the mask for both the boron implantation and field oxidation. Thus, when the field oxide is grown under steam at 920°C, this extra boron doping

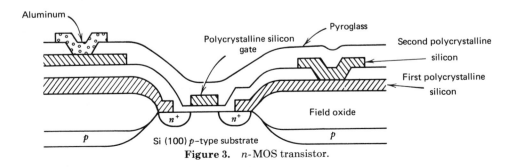

Figure 3. _n_-MOS transistor.

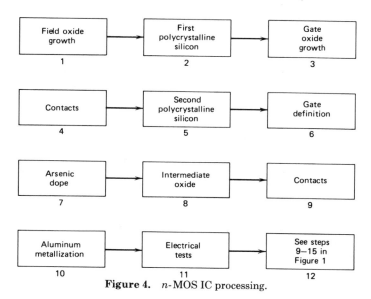

Figure 4. *n*-MOS IC processing.

occurs only under the field oxide. The boron increases the *p*-type doping to discourage transistor action under the field oxide when large voltages are applied to conductors over the field oxide. The nitride mask is removed with a H_3PO_4-based etch, which does not attack SiO_2.

In process 2, the first polycrystalline silicon is deposited, doped, and patterned. Low pressure chemical vapor deposition is used because hundreds of wafers can be processed simultaneously with uniform film thickness and contaminant distribution. This uniformity can be maintained even at low deposition temperatures near 600°C. Films deposited at these temperatures are nearly amorphous; they also are smooth and provide very reproducible products. Films grown at higher temperatures consist of crystallites; as these crystallites grow in size during high temperature processing, the film surface becomes bumpier, leading to many processing problems.

The polycrystalline silicon is doped with phosphorus as in bipolar processing (process 4, Fig. 1). An oxide is first grown to pattern the polycrystalline silicon. In critical applications, this oxide growth step can be a multistep process, eg, wet oxidation followed by oxide stripping to remove a surface layer containing excess phosphorus, followed by dry oxidation. The last dry oxide provides a film with a low phosphorus concentration and no adsorbed water, which improves photoresist adhesion. The polycrystalline silicon is then patterned using photomask 2 and, at the same time, the gate areas are exposed during etching.

Process 3: The growth of gate oxide is the most critical step in the entire processing and, therefore, requires extra care. One practice is to grow a sacrificial oxide which is etched prior to gate oxidation, thus removing a controlled amount of surface silicon and associated contaminants. This improves the gate breakdown characteristics and provides uniform threshold behavior. The gate oxide is then grown by a two-step wet–dry process. The wet oxidation allows thermal oxidation at the lowest possible temperature (850°C) to discourage undesirable side effects, eg, migration of dopants and formation of oxidation-induced defects in the silicon. In the following dry oxidation at 1075°C, the quality of the oxide, which is poor when grown in steam, is improved.

In process 4 the contacts for the second polycrystalline silicon layer are patterned with photomask 3. In processes 5, 6, and 7 the second polycrystalline silicon is deposited by LPCVD and patterned with photomask 4. This second film, which is also the gate metal, and the source and drain regions are simultaneously doped with arsenic-doped SiO_2 slurry, which is spun on. The arsenic is driven into the silicon by heating to 1075°C. After removal of the arsenic glass by etching in a 5:1 H_2O:HF solution, the exposed silicon areas are oxidized to seal the entire structure. It is especially critical to reoxidize the edges of the gate metal properly, because this is often a weak point for gate breakdown.

In process 8 a thick intermediate oxide is then deposited by CVD to insulate the underlying conducting elements from subsequent aluminum metallization. After deposition of such thick CVD films, there is excessive material growth in certain locations at the wafer edges and this must be removed. The oxide is then baked at 920°C to remove moisture. Process 9 involves the photolithography and etching of the contact windows using photomask 5. A gettering operation is performed to remove impurities, eg, sodium, prior to covering the wafers with aluminum.

In process 10 a special pre-aluminum-etch procedure is necessary to ensure that the windows are open just prior to Al deposition by electron-gun evaporation which provides an extremely pure aluminum film. The aluminum can be doped with copper to facilitate etching and to increase its resistance to electromigration failure and with silicon to reduce spiking at the contacts. Many manufacturers use sputter deposition of Al instead of electron-gun evaporation because the sputter deposition method is simpler, although the resulting film is not as pure.

An alloying operation removes most of the radiation damage introduced during aluminum deposition. This damage is particularly critical in the gate oxide. The alloying also stabilizes the metallurgy of the aluminum contacts to silicon.

Following the electrical tests of process 11, special-case wafers may be used for various purposes, eg, data gathering or experimentation. The remaining processing steps for n-MOS and bipolar devices are similar and are therefore described together.

Passivation, Separation, Bonding, and Packaging. A passivating dielectric is first deposited to protect the wafers from handling and contamination during the ensuing processing (process 9, Fig. 1). Some manufacturers use phosphorus-doped SiO_2 because it is a barrier to sodium. This phosphorus glass is sometimes capped with undoped glass because the doped glass is highly hygroscopic. Other manufacturers use silicon nitride because it is impervious to both sodium and moisture.

The bonding-pad pattern is then etched through the capping layers by means of a lithography step. Such a procedure applies to the n-MOS ICs just described. An example of a bonding-pad metallization that results in bipolar beam-leaded devices (37) starts with the deposition of Ti, TiN, and Pt (process 10, Fig. 1). Gold is then plated over the platinum. The titanium promotes adhesion, and the TiN and Pt combine to form a diffusion barrier. The gold layer serves as a corrosion-resistant cover over the pads and as a soft metal suitable for bonding. Patterning this stack of different metals by chemical etching would be quite difficult. Thus a combination of physical and chemical methods is used. First a mask is used to delineate the negative of the beam-lead pattern on the wafer, thereby exposing only the gold areas that are to become the beam leads. Nickel is plated onto the gold to serve as a mask for patterning of the gold. The photoresist is stripped, and the gold is patterned by backsputtering.

Nickel is a good sputtering mask because, under certain sputtering conditions, nickel sputters much more slowly than gold. The backsputtering automatically stops at the titanium layer because it, like nickel, also backsputters slowly. Finally, the titanium is etched with ethylenediaminetetraacetic acid (EDTA). The different layers deposited on the backside of the wafers are removed by abrasion and Au or other metal back contacts are then deposited.

The wafers are inspected optically, then they are sorted and tested electrically to identify the functional chips (process 11, Fig. 1). Nonworking chips are marked with ink dots. The ratio of functional chips to total chips fabricated is the yield.

For low-level integrated circuits, ie, medium-scale integration (MSI) and below, the chips can be separated (process 12, Fig. 1) by scribing and cleaving. For higher level ICs, the chips are separated by either laser scribing or dicing with diamond-impregnated wheels with the wafers mounted on a holder. Two common methods are wax mounting or deposition of a magnetic film, eg, permalloy, on the back of the wafers and the use of a magnetic chuck. In both laser scribing and mechanical dicing, it is not necessary to cut through the entire thickness of the wafer to cause chip separation. Dicing is preferred over laser scribing because it is faster and cleaner.

In the bonding operation (process 13, Fig. 1), electrical connections are established between the bonding pads on the IC chip and the pins on the chip support. The support can be a metal-lead frame, a chip carrier, or a header, ie, a sealed container. Bonding can be accomplished by wire bonding or the use of beam leads, plated bumps, or bumps of conducting epoxy. The most popular methods are wire bonding and tape bonding, both of which involve thermocompression and ultrasonic techniques. The conducting elements generally contain gold. Wires can be attached by ball bonding or wedge bonding. In ball bonding, the wire is first cut, then a ball is formed at the cut end by melting in a flame, and the ball is bonded onto the IC bonding pad by thermocompression and ultrasonic techniques. In wedge bonding, the wire is compressed under a wedge-shaped anvil. Excess wire is cut off by pulling the wire while the anvil is still down which causes the excess wire to break at the bond. For tape bonding, the conductor pattern is plated onto a plastic tape. The conductors terminate in a pattern that matches the pattern of bonding pads on the IC chip. The bonding is accomplished by compressing the conductors onto the bonding pads by use of an anvil whose shape also corresponds to the pattern of the bonding pads. Tape bonding is the most readily automated of the bonding techniques. The entire assembly is then molded in plastic, is covered with some encapsulant, eg, room-temperature-vulcanizing (RTV) rubber, or is hermetically sealed (process 14, Fig. 1) (see Electrical connectors; Embedding).

The finished product may be in molded plastic cases, sealed containers, dual-in-line packages (dips), chip carriers (rectangular chip holders with solder bumps), etc. The products can then be subjected to final electrical testing. For critical applications, the ICs are operated under well-controlled conditions to weed out devices with early failures (process 15, Fig. 1).

Compound Semiconductor Devices

Compound semiconductors provide a variety of materials properties that permit the fabrication of devices that can not be made using silicon. The main difficulties with compound semiconductor processing arise from the complexity of processing materials containing at least two elements and from the lack of a nearly ideal dielectric–semiconductor system, eg, SiO_2–Si. However, most of these difficulties originate from manufacturers' inexperience, rather than any fundamental limitations. Because compound-semiconductor technology is less advanced than silicon technology, applications have been developed mostly in areas where silicon cannot compete, eg, microwave technology (qv) and photonics. The greatest difference between silicon technology and compound-semiconductor technology is the dominance of MOS and bipolar devices for silicon and the almost complete absence of these devices for compound semiconductors.

Many compound semiconductors are being studied but only a few have attained commercial importance (1–2,53–55,58). These are GaP, GaAsP, and AlGaAs for light-emitting diode (LED) applications, GaAs for field-effect transistors (FETs), and GaAs and AlGaAs for lasers. Indium phosphide and indium gallium arsenic phosphide are being actively researched and are expected to become commercially important before 1985. The subscripts indicating the exact composition, eg, the x in $Al_x Ga_{1-x} As$, have been omitted for simplicity.

The most important field in compound-semiconductor technology is crystal growth. The methods used are Czochralski, Bridgeman, CVD, vapor-phase epitaxy (VPE), liquid-phase epitaxy (LPE), and molecular-beam epitaxy (MBE) (2,53,55,58).

Gallium Phosphide Light-Emitting Diode Processing. The procedure for LED manufacture is outlined in Figure 5. The process is for the fabrication of red (ZnO-doped) or green (N-doped) GaP LEDs; the finished device is illustrated in Figure 6.

A boule of n-type GaP is first grown (process 1, Fig. 5) by the liquid-encapsulated Czochralski (LEC) method, in which liquid GaP is encapsulated by liquid B_2O_3 and kept under pressure at ca 5 MPa (ca 50 atm) to prevent the loss of phosphorus at the growth temperature, ca 1465°C (53,58). The boules are then sliced into wafers, cleaned, polished, and cleaned again (process 2) in preparation for p-n-junction growth.

Then n and p layers are grown by LPE (process 3). In this method, the growth solutions are melted inside a hollow that is machined into graphite boats placed over sliding graphite substrate holders (53,58). The substrates are slid into position under the growth solutions for film growth and the process is terminated by pulling the substrates out. Although this technique produces excellent device quality material, it also yields films with rough surfaces. There are other methods of junction formation,

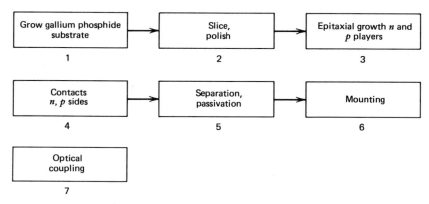

Figure 5. Gallium phosphide LED processing.

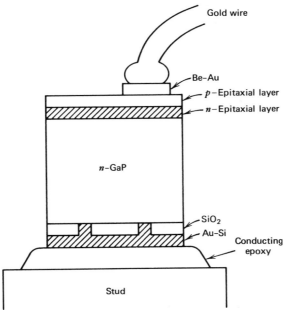

Figure 6. Gallium phosphide LED.

eg, diffusion or growth by VPE, CVD, MBE, and tipping (53,58). The doping in the epitaxial films, whether by ZnO or N, determines the color, red or green, respectively, of the emitted light.

Before contacts are deposited on the back, ie, reflecting, side of the diode, a layer of SiO_2 is deposited and the contact pattern is defined photolithographically and then etched (process 4, Fig. 5) into the SiO_2. Thus, the subsequently deposited metal, eg, Au–Si, layer makes contact at only a fraction of the total area. The reason for the use of the SiO_2 is the light-absorbing nature of the contact areas; the areas protected by SiO_2 reflect light, thus increasing light output. By a judicious choice of the contact area, a compromise is struck between sufficient contact area to provide the required electrical current and sufficient reflective area to minimize light loss. On the p side, through which the light escapes, a small dot of contact film, eg, Be–Au, is deposited through a shadow mask leaving enough area for light transmission.

The wafers are then coated with a film, eg, silicon nitride, silicon dioxide, or photoresist, to protect them from contaminants that deposit during laser cutting or dicing with a diamond-impregnated wheel, and the chips are separated (process 5). After separation, the damaged layers are etched, and critical areas that were exposed during chip separation can be passivated.

The devices are then mounted (process 6) on studs by means of low melting point alloys or conducting epoxy. Wires are attached by thermocompression bonding to the contact film on the p side. Finally, the diode is encapsulated in an optical enclosure, eg, a lens or light guide (process 7).

Gallium Arsenide Power Field-Effect-Transistor Fabrication. The procedure for making FETs is summarized in Figure 7, and a sketch of the finished device is shown in Figure 8.

It is difficult to grow large substrate GaAs crystals of sufficient quality for direct

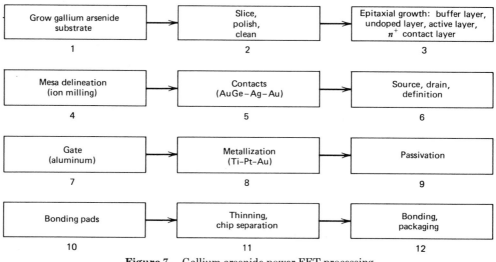

Figure 7.   Gallium arsenide power FET processing.

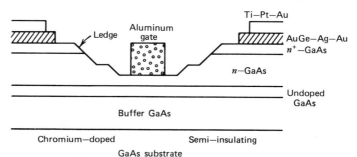

Figure 8.   Gallium arsenide power FET.

device fabrication. Therefore, an inexpensive substrate is first grown by Bridgeman or Czochralski methods (process 1, Fig. 7). This material is chromium-doped and semi-insulating. The boule is then sliced, polished, and cleaned in preparation for epitaxial growth (process 2).

A high purity epitaxial film of a semi-insulating, chromium-doped, buffer layer several micrometers thick is grown (process 3) to isolate the active layer from the defects and contaminants in the substrate. Then a 700 nm thick undoped layer is grown to discourage chromium diffusion. Next is the active layer, which is sulfur-doped (ca 5×10^{16} atoms/cm^3) and ca 700 nm thick. The final layer is the n^+ contact layer, which is ca 200 nm thick and heavily doped with S or Si (10^{18} atoms/cm^3). All these films can be grown by CVD with $AsCl_3$–Ga–H_2, using CrO_2Cl_2 for Cr doping.

The FETs are made in mesa structures delineated (process 4) by ion milling the n^+ and n layers using the photoresist as the mask. After photoresist stripping, the contact metals (AuGe–Ag–Au) are deposited and patterned by the lift-off technique (process 5). This technique obviates the need for chemically etching these complex layers of metals which are difficult to etch without attacking GaAs. The patterns are first delineated in the photoresist, then the metals, ie, AuGe–Ag–Au, are deposited. When the photoresist is subsequently stripped, the unwanted metal over the photoresist is removed or lifted-off with it. The lift-off procedure is illustrated in Figure 9 for the gate structure of Figure 8. Each layer in the contact metallization serves a specific purpose. The AuGe eutectic is used because its melting temperature is sufficiently low for ease of processing but is not too low to affect device reliability. Germanium is an n-type dopant, and gold readily alloys with gallium at low temperatures, thus providing strong adhesion. Alloying is accomplished by increasing the temperature to 550°C. The silver layer prevents balling of the AuGe when it melts during alloying. The final gold layer prevents oxidation of the surface. The alloying is performed in an oxygen-free atmosphere. This complicated scheme is required to produce low resistivity, ohmic contacts. The source and drain areas where the n^+ layer must be removed for gate deposition are delineated by photolithography (process 6). The n^+ layer is made to extend a short distance from the source and the drain contacts towards the gate. This extension or ledge (see Fig. 8) increases the device breakdown voltage (2).

The gate pattern, which can be as narrow as 1 μm, is first patterned in the photoresist over GaAs (process 7), in preparation for another lift-off procedure. Then the GaAs is chemically etched. This creates a groove in the GaAs that is wider than the intended gate length because the etch undercuts the photoresist (Fig. 8). The depth of the groove is carefully controlled, since it determines the final device operating parameters. The aluminum is then evaporated to form the gate at the bottom of the groove, and the unwanted aluminum over the photoresist is removed by the lift-off technique when the photoresist is stripped.

For metallization (process 8), the photoresist is patterned, then Ti–Pt–Au (200 nm, 200 nm, 1000 nm thick, respectively) layers are electron-gun evaporated. Such thick layers are difficult to remove by the conventional lift-off technique. Therefore a second lithography is performed with a negative pattern so that areas to be metallized are covered with photoresist but the areas where the metals are to be removed are exposed. The gold and platinum are then removed by ion milling. This procedure works because titanium is an excellent ion-milling etch stop under certain milling conditions. The patterning procedure is completed by stripping both layers of photoresist. The reasoning behind the use of the Ti–Pt–Au metallization is as follows: the Ti layer promotes adhesion; the Pt layer prevents oxidation of the Ti layer and, with the Ti layer, acts as a diffusion barrier against Au, Ga, and As; and the Au layer is the soft metal to which wires can be bonded.

The final steps are passivation, opening of windows for bonding pads, thinning, chip separation, and bonding, as outlined in processes 9–12 in Figure 7. Thinning the substrate to ca 40 μm decreases the thermal impedance, which is important for power devices. Thinning can be accomplished with a chemical etch, eg, 5 H_2SO_4:1 H_2O_2:1 H_2O, with the wafer wax mounted on silicon. Approximately 25 μm of electroless

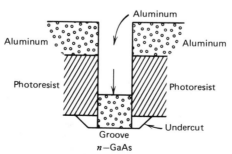

Figure 9. Gallium arsenide power-FET processing: after aluminum-gate deposition, before lift-off.

gold is plated on the backside and the device is bonded backside-down on a preform plated with AuSn. Wires are then bonded to the source, drain, and gate contacts (see Electroless plating).

Solid-State Lasers. The most important step in semiconductor laser processing is crystal growth (59). Impurities and defects in the crystal provide paths for carrier recombination, and nucleation and growth mechanisms for more defects that can extinguish the lasing action or change the laser properties. The most successful crystal-growth method has been liquid-phase epitaxy because of its simplicity. Molecular-beam epitaxy, the various CVD techniques, and other methods show more promise than LPE and they are becoming increasingly popular. The main problems with LPE are the difficulty of controlling the film thickness, surface morphology, and doping and compositional variations (see Lasers).

The device described below is called an oxide stripe laser because the lasing area is defined by a stripe that is etched into a surface oxide layer. It is a heterostructure laser because it is fabricated from heteroepitaxial material; ie, epitaxial layers of different chemical compositions. A diagram of the finished device is shown in Figure 10. Devices of this type were made at the California Institute of Technology in ca 1975 (60). The processing for commercial lasers today differs from that described below as the technology has been developing rapidly. Most of the new developments are proprietary. The following procedures, however, do include all principal fabrication steps.

The following materials are required for GaAs heterostructure-laser processing: n^+ GaAs substrate, 400 μm thick, carrier concentration 2×10^{18} atoms/cm^3 Te-doped, etch-pit density <1000 etch pits/cm^2; a dummy substrate with the same characteristics as the n^+ GaAs substrate; 99.9999% pure Al wire; 99.9999% pure Sn n-type dopant; 99.9999% pure Ge p-type dopant; high purity polycrystalline GaAs for saturating LPE growth solutions; and 99.9999% pure Ga for making LPE growth solutions.

The starting GaAs substrate is prepared in the (100) orientation because this is the nonpolar surface which is normal to cleavage planes (011) that are used at the end of processing to form the mirror faces.

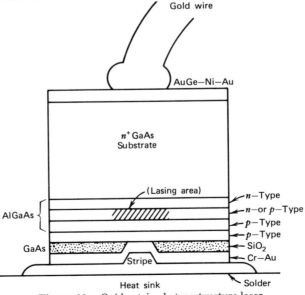

Figure 10.   Oxide-stripe heterostructure laser.

The growth furnace is first prepared for LPE growth. Oxygen must be eliminated from the system because it causes defect formation. This is done by flushing the furnace with N_2 for two hours to remove air and then flushing with H_2 for two hours or until the O_2 content is less than 0.1 ppm.

The substrates and growth materials are then carefully cleaned. Organic contaminants are removed from the polycrystalline GaAs by treatment with acetone and methanol. It is then etched in a 5-wt % bromine–methanol solution at room temperature for 10 min. The gallium arsenide is then rinsed with methanol and deionized water and is blown dry. Cleaning of the GaAs substrates involves removal of organic contaminants in acetone and methanol, removal of dust and lapping residue by scrubbing, rinsing in deionized water followed by blow drying, removal of surface residue with concentrated H_2SO_4, rinsing in deionized water followed by blow drying, etching in a 4:1:1 concentration of H_2SO_4:30 wt % H_2O_2:H_2O for 1 min, and finally rinsing in deionized water followed by blow drying. The aluminum wire and tin are etched in concentrated HCl for 10 s, rinsed in deionized water, and blown dry. Organic contaminants are removed from the germanium with concentrated H_2SO_4 and the germanium is rinsed in deionized water and blown dry. The GaAs substrate should have either a high quality commercial surface polish, or it should be polished in a 5-wt % bromine–methanol solution over a soft pad.

Immediately after cleaning, all materials are loaded into growth boats in the proportions listed in Table 1. There should be excess GaAs floating on the Ga solutions so that it is not completely dissolved in the epitaxial sequence (see Table 1). This excess GaAs assures As saturation of the solutions at all times. The GaAs wafers are inserted and covered with graphite so that the wafer surface does not deteriorate before growth. The solutions are flushed with N_2 for 5 min to remove air, then with H_2 for 24 h to remove O_2, and then are baked at 800°C for 2 h to equilibrate the solutions.

Epitaxial growth starts at 800°C and the cooling rate is 0.2°C/min. After cooling by 3°C, the dummy substrate is slid under the first growth solution just ahead of the true substrate on a single slider. The first epitaxial solution is equilibrated by growth on the dummy substrate while the furnace is cooled another 3°C. In the next sliding operation, the dummy moves under the second solution and simultaneously the true substrate moves under the first solution. Thus, equilibrium is established in the second solution while growth occurs on the first. This sliding operation is repeated for each growth layer. The cooling proceeds by 6°C, 0.16°C, 4°C and 1°C for growth of four layers. The product of this growth procedure is an epitaxial $Al_x Ga_{1-x} As$ film with the layer sequence shown in Table 2.

The thinnest layer is the active layer where lasing occurs. This layer must be lightly doped but can be either *n*- or *p*-type. The substrate and fourth layer contain no aluminum; this facilitates bonding by eliminating the possibility of aluminum oxide formation.

Table 1. Compositions of Growth Solutions

Solution	Ga, g	Al, mg	Dopant	Polycrystalline GaAs, mg
1	4	4	120 mg Sn	200
2	4	0.5	0	300
3	4	4	2 mg Ge	200
4	4	0	2 mg Ge	300

Table 2.   Layer Sequence for an Epitaxial $Al_x Ga_{1-x}As$ Film

Layer	Thickness, μm	x	Carrier concentration, atoms/cm^3
1	2	0.4	10^{17} n-type
2	0.2	0.1	10^{16} n- or p-type
3	2	0.4	10^{17} p-type
4	1	0	10^{17} p-type

After growth, the wafers are cleaned by scrubbing in a CH_3OH–HCl solution to remove excess Ga; then SiO_2 is deposited by CVD, the stripes are patterned, and Cr–Au contacts are evaporated on the p side. The zinc diffusion preceding this contact metallization is performed in a sealed ampul of ca 3 cm^3 with 5 mg of $ZnAs_2$ at 610°C for 20 min. After wafer thinning to 100 μm, the n contact is made by depositing 50-nm AuGe–50-nm Ni–300-nm Au. This metallization is very similar to the AuGe–Ag–Au described in GaAs FET fabrication.

After alloying at 450°C for 30 s, the chips are cleaved; the cleavage faces become the end mirrors for the lasers. Because of the intense radiation, mirror damage occurs during use and thus limits the device life. The mirrors can be passivated with half-wavelength Al_2O_3 facet coatings, which are electron-gun evaporated, to reduce early failures. The half-wavelength optical thickness of the coating assures that the waves reflected from the two surfaces of the film are in phase, thus preserving the mirror quality.

The lasers are bonded with indium solder onto gold-plated copper heat sinks, with Indalloy 4 flux. As shown in Figure 10, the laser is bonded with the lasing side closest to the heat sink. The devices are next cleaned in water. Gold wires are then ball-bonded, and the lasers are ready for testing.

Economic Aspects

Estimates of semiconductor device sales and capital investments vary considerably depending on the source. However, most published data agree within ca 50%. The most common sources of these estimates are the periodic reports of semiconductor-based organizations (61).

Some figures for 1980 are as follows: Total worldwide semiconductor sales in 1980 were ca 15×10^9, of which ca 6×10^9 was from discrete devices, and most of the rest was related to ICs. Memory devices accounted for 4×10^9; the IC memory market was the fastest growing one because of a large demand and decreasing production costs. Integrated-circuit memory benefits from cost-saving advances in IC technology because the circuit is mostly a repetition of one basic memory cell and is most easily adapted to new technologies. Total compound semiconductor sales were ca 0.7×10^9. The two top producers were the United States and Japan, with sales of ca 7×10^9 and 5×10^9, respectively. Investments in semiconductor processing equipment were 10–15% of sales. Total worldwide employment associated with U.S. semiconductor firms was ca 250,000 workers.

BIBLIOGRAPHY

"Semiconductors, Fabrication" in *ECT* 2nd ed., Vol. 17, pp. 862–883, by J. R. Carruthers, Bell Telephone Laboratories, Inc.

1. S. M. Sze, *Physics of Semiconductor Devices*, John Wiley & Sons, Inc., New York, 1981.

2. J. V. DiLorenzo, ed., *GaAs FET, Principles and Technology*, Artech House Books, Dedham, Mass., 1982.

3. G. L. Pearson and W. H. Brattain, *Proc. Inst. Radio Eng.* **43**, 1794 (1955).

4. E. Braun and S. MacDonald, *Revolution in Miniature*, Cambridge University Press, London, 1978.

5. H. C. Torrey and C. A. Whitmer, *Crystal Rectifiers*, Clarendon Press, Oxford, England, 1949.

6. K. Lark-Horowitz, *Electr. Eng.* **68**, 1047 (1949).

7. W. Shockley, *Electrons and Holes in Semiconductors*, D. Van Nostrand, Princeton, N.J., 1950.

8. H. E. Bridgers, J. H. Scaff, and J. N. Shive, eds., *Transistor Technology*, Vol. 1, D. Van Nostrand Company, 1958.

9. J. Bardeen and W. H. Brattain, *Phys. Rev.* **74**, 230 (1948).

10. *Ibid.*, **75**, 1208 (1949).

11. F. J. Biondi, ed., *Transistor Technology*, Vol. II, D. Van Nostrand Company, 1958.

12. R. M. Ryder and W. G. Pfann, Bell Laboratories, private communications, 1981.

13. G. K. Teal, M. Sparks, and E. Buehler, *Phys. Rev.* **81**, 637 (1951).

14. W. Shockley, *Bell Syst. Tech. J.* **28**, 453 (1949).

15. A. B. Phillips, *Transistor Engineering*, McGraw-Hill, Inc., New York, 1962.

16. C. Kittel, *Introduction to Solid State Physics*, John Wiley & Sons, Inc., New York, 1968.

17. E. J. M. Kendall, *Transistors*, Pergamon Press, Inc., Elmsford, N.Y., 1969.

18. R. N. Hall and W. C. Dunlop, *Phys. Rev.* **80**, 467 (1950).

19. W. E. Bradly and co-workers, *Proc. Inst. Radio Eng.* **41**, 1702 (1953).

20. J. Evans, *Fundamental Principles of Transistors*, D. Van Nostrand Company, Princeton, N.J., 1962.

21. W. W. Gartner, *Transistors—Principles, Design, and Applications*, D. Van Nostrand Company, Princeton, N.J., 1960.

22. M. Tannenbaum and D. E. Thomas, *Bell Syst. Tech. J.* **35**, 1 (1956).

23. C. A. Lee in ref. 22, p. 23.

24. C. J. Frosch and L. Derrick, *J. Electrochem. Soc.* **104**, 547 (1957).

25. F. M. Smits, *Proc. Inst. Radio Eng.* **46**, 1049 (1958).

26. A. S. Grove, *Physics and Technology of Semiconductor Devices*, John Wiley & Sons, Inc., 1967.

27. J. A. Hoerni, *Institute of Radio Engineers Electron Devices Meeting*, Washington, D.C., 1960.

28. C. H. Knowles, *Electron. Ind.* (Aug. 1958).

29. H. C. Theurer, J. J. Kleimack, H. H. Loar, and H. Christensen, *Proc. Inst. Radio Eng.* **48**, 1642 (1960).

30. W. Shockley, *Proc. Inst. Radio Eng.* **40**, 1365 (1952).

31. G. C. Dacey and I. M. Ross, *Proc. Inst. Radio Eng.* **41**, 970 (1953).

32. R. N. Noyce, *Sci. Am.* **237**, 63 (Sept. 1977).

33. U.S. Pat. 1,745,175 (Jan. 28, 1930), J. E. Lilienfeld.

34. Brit. Pat. 439,457 (Dec. 6, 1935), O. Heil.

35. W. Shockley and G. L. Pearson, *Phys. Rev.* **74**, 232 (1948).

36. D. Kahng and M. M. Atalla, *Institute of Radio Engineers Solid State Device Research Conference*, Carnegie Institute of Technology, Pittsburgh, Pa., 1960.

37. M. P. Lepselter, *Bell Syst. Tech. J.* **45**, 233 (1966).

38. B. E. Deal and J. M. Early, *J. Electrochem. Soc.* **126**(1), 20C (1979).

39. S. K. Ghandhi, *Semiconductor Power Devices*, John Wiley & Sons, Inc., New York, 1977.

40. A. Blecher, *Thyristor Physics*, Springer-Verlag, Inc., New York, 1976.

41. B. J. Baliga in D. Kahng, ed., *Applied Solid State Science*, Suppl. 2B, Academic Press, Inc., New York, 1981, p. 109.

42. T. Kariya, S. Okuhara, M. Tokunaga, and T. Kamei, *Hitachi Rev.* **29**(3), 131 (1980).

43. P. W. Shakle, A. R. Hartman, B. T. Murphy, R. S. Scott, R. Lieberman, and M. Robinson, *Proc. IEEE* 292 (March 1981).

44. J. R. Ligenza, *J. Phys. Chem.* **65**, 2011 (1961).

45. R. M. Jecmen, C. H. Hui, A. V. Ebel, V. Kynett, and R. J. Smith, *Electronics* 1 (Sept. 13, 1979).

46. J. C. Sarace and co-workers, *Solid State Electron.* **11**, 653 (1968).

47. G. A. Rozgonyi, P. M. Petroff, and M. H. Read, *J. Electrochem. Soc.* **122**, 1725 (1975).

48. R. L. Meek, T. E. Seidel, and A. G. Cullis in ref. 46, p. 786.

49. S. P. Murarka, *J. Vac. Sci. Technol.* **17**, 775 (1980).

50. J. C. Bowers, G. Wzobrist, C. Lors, T. Rodby, and J. E. O'Reilly, *Wright-Patterson Air Force Base Technical Report AFAPL TR-76-73*, Dayton, Ohio, April, 1976.

51. P. F. Kane and G. B. Larrabee, eds., *Characterization of Solid Surfaces*, Plenum Publishing Corporation, New York, 1974.
52. R. M. A. Azzam and N. M. Bashara, *Ellipsometry and Polarized Light*, North-Holland, Amsterdam, The Netherlands, 1977.
53. A. A. Bergh and P. J. Dean, *Light Emitting Diodes*, Oxford, London, 1976.
54. H. Kressel and J. K. Butler, *Semiconductor Lasers and Heterojunction LED's*, Academic Press, Inc., New York, 1977.
55. H. C. Casey and M. B. Panish, *Heterostructure Lasers*, Pts. A and B, Academic Press, Inc., New York, 1978.
56. B. Chapman, *Glow Discharge Processes*, John Wiley & Sons, Inc., New York, 1980.
57. G. Moore and E. Flath, private communication, Intel Corp., 1981.
58. E. G. Bylander, *Materials for Semiconductor Functions*, Hayden Book Company, Inc., New York, 1971.
59. R. W. Dixon, *Bell Syst. Tech. J.* **59**, 669 (1980).
60. D. Wilt, private communication, Bell Laboratories, 1981.
61. Periodic reports of the Semiconductor Industry Association, Cupertino, Calif.

CHUAN C. CHANG
Bell Laboratories

AMORPHOUS

Although the modern theory of solids is intimately connected with crystalline materials, the vast majority of solids that is encountered in everyday experience is amorphous. Technically, an amorphous material exhibits no correlation between two atoms located more than ca 5 nm apart, so that the long-range periodicity of crystals is entirely absent. Nevertheless, amorphous solids ordinarily possess a great deal of short-range order, particularly correlations between nearest-neighboring atoms, which arises from their chemical interactions.

Since, historically, solid-state theory was restricted to periodic crystals, it was believed for many years that amorphous semiconductors did not exist. Consequently, it was a surprise when it was reported in the 1950s that many chalcogenide glasses, ie, amorphous alloys containing significant concentrations of one or more of the chalcogen elements and prepared by rapid cooling of the liquid, were semiconducting (1). The field of amorphous semiconductors attracted little scientific interest before 1968, when the existence of reversible switching effects in thin films of chalcogenide glasses was reported and a wide array of potential applications was proposed (2). This work spurred intensive investigations into the electronic properties of chalcogenides and amorphous analogues of the conventional, tetrahedrally coordinated, crystalline semiconductors, eg, silicon [7440-21-3] and germanium [7440-56-4]. Many amorphous-based semiconductor devices, from computer memories and television pickup tubes to solar cells, are commercially available.

Classification

The classification scheme that is most useful for the analysis of electronic behavior is based on predominant atomic coordination. As is the case for crystalline solids, semiconducting behavior is generally associated with covalent $s-p$ bonding, and the maximum coordination number under ordinary circumstances is four. When sp^3 bonding predominates, the solid is tetrahedral. Examples are amorphous silicon-based alloys, amorphous germanium, and III–V alloys, eg, amorphous GaAs [1303-00-0]. When p^3 bonding predominates because of the presence of large concentrations of Group VA atoms, ie, pnictogens, the amorphous solids are usually called pnictide glasses. When p^2 bonding predominates as a result of the presence of large concentrations of Group VIA atoms or chalcogens, the amorphous solids are called chalcogenide glasses. Although other amorphous semiconductors exist, these three groups have been by far the most intensively studied.

Structure

Before any analysis of the physical properties of an amorphous solid can be undertaken, a knowledge of its structure is essential. Because of the lack of long-range periodicity in amorphous materials, it is much more difficult to determine their structure than those of crystalline solids. In addition, the preparation methods necessary to obtain the material in the amorphous phase require the consideration of several issues that are not usually contemplated when crystals are investigated. First, there is the question of composition. It is not always evident what is actually in the solid. If the material has been prepared by cooling from the liquid, there are usually not many compositional uncertainties. However, many of the most important amorphous semiconductors cannot be fabricated in this way but rather are deposited as thin films directly from the vapor phase, and even relatively easily formed glassy solids are often prepared in thin-film form for convenience or for commercial purposes. In such cases, the composition must be carefully determined, and preferential deposition and unsuspected impurities are the rule rather than the exception. The composition can be a sensitive function not only of the preparation technique but also of the many deposition parameters and of the geometry of the system. The development of modern techniques, eg, secondary-ion mass spectroscopy (SIMS), has been an immense aid in determining the composition, but most important is the necessity of such a determination.

Once the composition is known, two other investigations are essential. First, it is important to determine the homogeneity of the material. Identification of microscopic phase separation over regions greater than 100 nm is relatively straightforward, but inhomogeneities on a smaller scale may be much more subtle. Aggregation of one of the species can be important on a scale of only ca 1 nm. The second consideration is of possible anisotropy. It is evident that all thin films have two interface regions, ie, one near the substrate and the other contacting a solid, liquid, or gas, and these interfaces probably are significantly different from the bulk. In addition to these lateral effects, anisotropy perpendicular to the substrate, eg, columnar growth, is often observed.

Amorphous solids do not exhibit the long-range order of crystals, and this results in a freedom from periodic constraints that allows a wide range of potential compositions and structures. In this sense, an amorphous solid can be considered as a giant chemical molecule. Since small molecules can have many different structures with very different properties, it is clear that amorphous solids with the same exact composition are far from unique. Under ordinary conditions, a great deal of short-range order is expected, ie, each atom is most likely to have a local coordination consistent with its electronic structure. Thus, neutral atoms in Groups I–IV should have 1–4 nearest neighbors, respectively, whereas neutral atoms in Groups V, VI, and VII should have coordination numbers of 3, 2, and 1, respectively, in accordance with the 8-N rule of chemical bonding. This arises because any deviation from the optimal coordination would ordinarily cause a large increase in total energy and thus would be suppressed if at all possible.

It is important to consider the ideal structure of an amorphous semiconductor. In amorphous alloys different chemical bonds are possible, and the strongest bonds are favored. Each type of bond has an optimal length, and the large values of the bond-stretching frequencies indicate the presence of strong forces which tend to suppress significant bond-length deviations. Although considerably weaker, the bond-bending frequencies are usually sufficiently large that there are at most small bond-angle distortions under ordinary circumstances. The forces tending to impose third-neighbor, ie, dihedral-angle, constraints are generally very much weaker. It has been recently suggested that the optimal average coordination for an amorphous alloy is the one in which the number of constraints introduced by fixed bond lengths and bond angles is equal to the number of degrees of freedom, namely three (3). If the average coordination number is m, then the average number of bonds per atom is $m/2$. Since the average number of bond angles is $(m(m-1))/2$, then the criterion suggests that the optimal coordination for glass forming, m_o, is given by

$$\tfrac{1}{2} m_o + \tfrac{1}{2} m_o (m_o - 1) = 3$$

or   (1)

$$m_o = \sqrt{6} \simeq 2.4$$

This is an oversimplification, particularly if some of the bonds are predominately ionic, in which case the bond-bending frequencies are relatively small. Nevertheless, most glass-forming alloys have average coordinations between two and three, and amorphous solids with higher coordinations tend to be overconstrained. For such materials, there is competition between formation of the optimal number of bonds and relief of the concomitant strains thereby introduced.

The total energy of amorphous alloys containing one or more elements from Groups I–III with one or more from Groups V–VI can be reduced by the formation of dative bonds. This tends to increase the average coordination at the expense of additional strains.

Therefore, the lowest energy structure of a particular amorphous semiconductor is that which maximizes the number of chemical bonds consistent with the given composition, maximizes the strength of the possible chemical bonds, optimizes the bond lengths, and optimizes the bond angles. If the average coordination is ≤2.4, these criteria can be satisfied at least in principle without undue strain. However, if the average coordination is >2.4, some strains are introduced.

In reality, the ideal structure is never attained even for glasses with low average coordinations. This is because of the preparation techniques which ordinarily require rapid thermal quenching to prevent crystallization. Even when bulk glasses are made by cooling from the liquid phase, significant atomic motion does not occur below a well-defined glass-transition temperature T_g in the vicinity of which the viscosity rapidly increases with decreasing temperature from values characteristic of a liquid to those of a solid. Since T_g is a finite temperature, thermodynamics suggest that the equilibrium phase is not that one with the lowest energy; rather it is the one with the lowest free energy. For example, if a particular defect, ie, nonoptimally coordinated, configuration with a relatively low creation energy ΔE_d exists, then a minimum concentration of these defects, which is given by

$$N_d = N_o \exp\left[-\Delta E_d / k T_g\right] \tag{2}$$

where N_o is the total concentration of atoms in the glass, is thermodynamically quenched during preparation. Less careful fabrication techniques can increase the defect concentrations well beyond that given by equation 2. Defect configurations ordinarily control the electrical properties of amorphous semiconductors.

Electronic. As indicated, the quantum theory of solids as originally formulated was entirely based on crystalline periodicity. In the absence of such periodicity, the problem of electronic structure becomes mathematically much more complex, and additional approximations are essential. The electronic structure of amorphous semiconductors is generally regarded as at least qualitatively understood.

The main results of the quantum theory of crystalline solids are the electronic states lie in densely packed regions of energy called bands, which are separated by regions called gaps; the electronic density of states $g(E)$ decreases near a band edge E_o as

$$g(E) = A|E - E_o|^{1/2} \tag{3}$$

where A is a constant; all states within a band are extended, ie, the occupying electron has equal probability of being located near any equivalent atom in the periodic solid; and localized states can exist only within gaps and result only from the presence of defects or surfaces. The extended nature of the vast majority of electronic states in crystals results in high carrier mobilities and ultimately in metallic behavior. The gaps are responsible for the existence of insulators and semiconductors. The presence of localized states within the gap is a prerequisite for semiconductor devices. Since amorphous metals exist, extended states must also be present in noncrystalline solids. Also, since amorphous insulators exist, gaps must be present even in the absence of periodicity. However, equation 3 is not valid for noncrystalline materials (4). Instead of sharp band edges, the breakdown of long-range periodicity introduces a tailing of states from the bands into the gaps; these states are appropriately called band tails. Theoretical calculations ordinarily indicate a gaussian band tail, although exponential tails have also been determined (5). The necessary presence of these tails effectively eliminates the band gap and emphasizes the question of how an amorphous semiconductor can exist at all (6).

This problem was avoided with the proposal that a critical density of states exists above which all electronic states can be thought of as extended in the sense that they exhibit a finite mobility for free carriers even at absolute-zero temperature (7). The energies at which points this critical density is reached are called the mobility edges in analogy to the band edges of crystalline solids. The band structure is shown in Figure 1(**a**). There has been a great deal of theoretical effort recently to make the important concept of mobility edges rigorous, although some controversy remains (8).

One group assumed the existence of band tails and mobility edges and suggested that, for multicomponent amorphous solids, the valence and conduction band tails could overlap in the center of the gap, thus yielding a finite density of states at the Fermi energy $g(\epsilon_f)$ (9). This group also assumed that all the atoms locally satisfied their chemical valence requirements, thus precluding the existence of sharp bumps within the mobility gap, as occurs in doped crystalline semiconductors. Their model, often called the CFO model (see Fig. 1(**b**)), provides a useful structure for the analysis of transport and optical data. However, it is important to ask how the assumptions of the model might be tested by such experiments.

Extensive band tails should be evident in ordinary optical-absorption experiments. In crystalline semiconductors, the material is essentially transparent to light of frequency below that of the energy gap, but the optical-absorption coefficient rapidly increases with photon frequency above the gap. It might be expected that if extensive band tails exist, the optical-absorption coefficient should begin to increase with photon frequencies well below the mobility gap, and this increase should be much less sharp than in the case of crystals with direct edges.

The most obvious manifestation of a sharp mobility edge would be an electrical conductivity that varies as

$$\sigma = N_c e \mu_e \exp\left[-(E_c - \epsilon_f)kT\right] + N_v e \mu_h \exp\left[-(\epsilon_f - E_v)/kT\right] \tag{4}$$

where N_c and N_v are the effective densities of states in the conduction and valence bands, μ_e and μ_h are the electron and hole mobilities beyond the mobility edges, and E_c and E_v are the positions of the conduction- and valence-band mobility edges, respectively. Equation 4 would yield a log σ vs T^{-1} plot that is either a single straight line or two intersecting straight lines. On the other hand, a diffuse mobility edge should yield a concave upward curve on a plot of log σ vs T^{-1}, reflecting the fact that at lower temperatures the conductivity is dominated by low mobility carriers much closer to the Fermi energy.

If the suggestion that a finite $g(\epsilon_f)$ exists is correct, the Fermi energy would then appear to be pinned, in the sense that conductivity would be insensitive to substitutional doping or to the injection of excess carriers. In addition, the redistribution of electrons, which should leave high energy valence-band states above ϵ_f (see Fig. 1), should create large densities of unpaired spins; these would be expected to be observable in both electron paramagnetic resonance (epr) or magnetic-susceptibility experiments.

It was pointed out that for a sufficiently large $g(\epsilon_f)$, phonon-assisted hopping among localized states near ϵ_f might dominate bandlike conduction beyond the mobility edges at sufficiently low temperatures (10). It was further suggested that at very low temperatures, when very few energetic phonons are present, such hopping conduction would take place preferentially to farther rather than to nearest-neighboring localized states in order to reduce the energy that must be obtained from the phonons. For this mechanism of variable-range hopping conduction, the conductivity varies as

$$\sigma = \sigma_o \exp\left[-(T_o/T)^{1/4}\right] \tag{5}$$

where σ_o and T_o are constants. When equation 5 is obeyed, the plot log σ vs T^{-1} is concave upward, but the plot log σ vs $T^{-1/4}$ is linear.

The possibility of variable-range hopping somewhat muddles the test for the

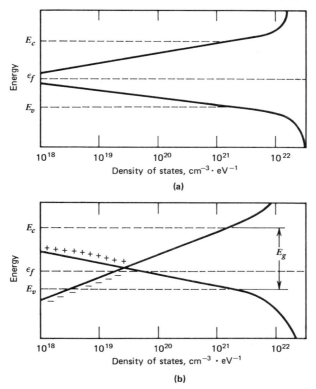

Figure 1. Mott-CFO model for the density of states of covalent amorphous semiconductors: (**a**) elemental amorphous semiconductors, eg, a–Si; (**b**) multicomponent amorphous alloys. E_v and E_c represent the valence and conduction-band mobility edges, respectively; E_g is the mobility gap; and ϵ_f indicates the position of the Fermi energy.

existence of sharp mobility edges. A linear plot of log σ vs T^{-1} could indicate either bandlike conduction beyond a sharp mobility edge or phonon-assisted hopping between nearest-neighboring localized sites. Alternatively, a concave upward plot of log σ vs T^{-1} could reflect either the absence of a sharp mobility edge or the predominance of variable-range hopping conduction.

In principle, field-effect measurements provide an excellent tool for determining the density of states near the Fermi energy. This technique involves inducing either excess negative or excess positive charge in the semiconductor. The resultant increase in the electron concentration should then cause ϵ_f to shift upwards or downwards, respectively. The extent of this shift can be determined from the corresponding change in conductivity and it can then be related directly to $g(E)$. In principle, the field-effect technique can also be used to determine if $g(E)$ is smooth or exhibits structure, the latter indicating a significant lack of local valence satisfaction.

Thus, the model reported in ref. 9 suggests that semiconductors, eg, amorphous Si, Ge, or Se [7782-49-2], should exhibit a large field effect, the possibility of substitutional doping, a small density of unpaired spins, no variable-range hopping, and a sharp absorption edge. In contrast, a multicomponent amorphous alloy should be characterized by a small field effect, no substitutional doping, a large unpaired-spin density, variable-range hopping, and a diffuse absorption edge.

When experiments were carried out, the results were surprising. For example, pure amorphous Si or Ge deposited on a room-temperature substrate exhibits none of the predicted properties: there is no observable field effect, no doping, a large unpaired-spin density, variable-range hopping conduction, and a diffuse absorption edge. Many of these properties, however, are not intrinsic and anneal at temperatures below the crystallization point. Figure 2 indicates that the diffuse absorption edge in deposited a(amorphous)-Ge films evolves upon annealing into an edge no more diffuse than in the crystallized film (11). The strong correlation between the unpaired-spin density, the electrical conductivity, and the optical edge of a-Si as functions of annealing are evident in Figure 3 (12).

All of these properties arise because of the overconstrained nature of the fourfold-coordinated tetrahedral network. Strains introduced upon deposition lead to distorted bond angles and well-defined chemical defects, eg, dangling bonds. These defects can be reduced by depositing the film on a high temperature substrate or by annealing it after deposition, but such procedures are limited by the crystallization temperature. The behavior predicted by the model given in ref. 9 is never observed.

Alternatively, when amorphous silicon is hydrogenated during deposition or afterwards, the expected behavior is observed in each case. This was originally discovered accidentally when silane gas, SiH_4, was decomposed by a radio-frequency glow discharge in an attempt to obtain pure amorphous silicon (13). The resulting material, particularly when deposited on a substrate at ca 350°C, exhibits a field effect indicative of a very low (ca 10^{17} cm^{-3} eV^{-1}) density of states near ϵ_f, as is evident from Figure

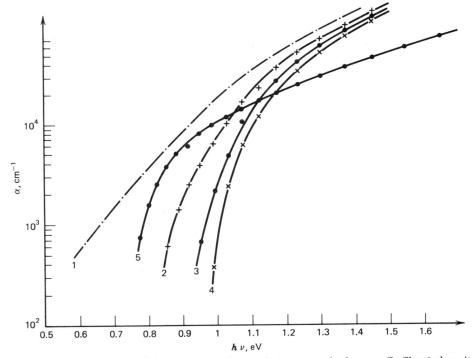

Figure 2.   Absorption coefficient α as a function of photon energy $h\nu$ for an a–Ge film: 1, deposited; 2, annealed at 200°C; 3, annealed at 300°C; 4, annealed at 400°C; 5, crystallized (11).

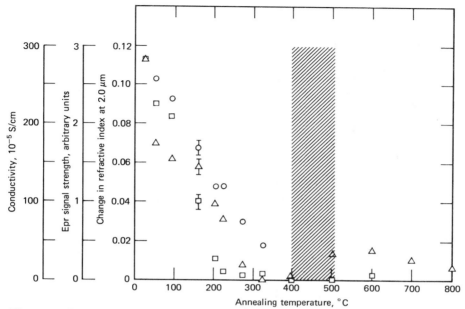

Figure 3. Correlation between room-temperature electrical conductivity, epr signal strength, and change in refractive index for a–Si films as a function of annealing (S/cm = $\Omega^{-1}\cdot$cm^{-1}). The shaded area represents the region in which the films crystallize. O = optical; □ = electrical, △ = spin (12).

4 (14). The density of defects is sufficiently low that p and n doping of the films are easily achieved by mixing small concentrations of either B_2H_6 or PH_3 gas with the SiH_4 prior to decomposition (15). The results are shown in Figure 5. All silane-decomposed films contain 5–30 wt % residual hydrogen, which bonds only monovalently (ignoring the possibilities of three-center and hydrogen bonding), thus decreasing the average coordination number and vastly reducing the strains (16). The reason for the optimal substrate temperature near 350°C is the thermodynamic requirement that hydrogen effuses from the network at high temperatures. Thus, the minimum defect density represents a balance between the beneficial effects of annealing and the deleterious effects of hydrogen effusion, as is evident from the temperature dependence of the unpaired-spin density shown in Figure 6 (16).

Other alloys besides a(amorphous)-silicon–hydrogen alloy (Si–H) exhibit excellent electronic properties. One particularly promising material is a-Si–F–H, which is fabricated by the glow-discharge decomposition of SiF_4–H_2 mixtures (17). This alloy appears to have lower defect densities than a-Si–H, most likely because of the extra flexibility of the primarily ionic Si—F bond. The Si—F bond is also stronger than the Si—H bond, resulting in a harder material which retains its properties to higher annealing temperatures. Curiously, no high quality a-Ge alloys have been reported. Decomposition of GeH_4 always yields a-Ge—H films with relatively high defect densities (18). The probable origin of this is that the relative weaknesses of the Ge—H bond requires the effusion of H before the films are sufficiently annealed.

Despite these wide variations in behavior, the properties of tetrahedral amorphous semiconductors can be understood by extending one or the other of the models sketched in Figure 1. However, the same is not the case for chalcogenide alloys. In almost all of these materials, the Fermi energy appears to be very strongly pinned, as

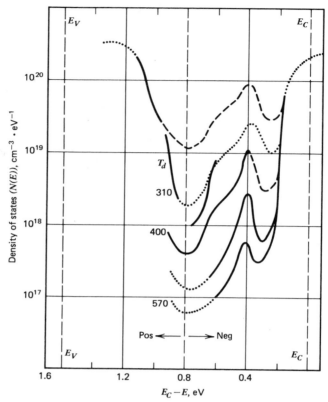

Figure 4.   Density of states as a function of energy for evaporated films of a–Si and glow-discharge films of a–Si:H deposited at various substrate temperatures T_d in degrees Kelvin. The position of the Fermi energy is indicated by the arrow. The full lines represent the results deduced from field-effect measurements (14).

evidenced by the absence of extrinsic conduction, doping, or an observable field effect under ordinary conditions. The extremely linear relation between log σ and T^{-1} over many decades in conductivity for many different amorphous chalcogenides is shown in Figure 7 (19); no extrinsic region is evident down to conductivities of 10^{14} S/cm (= $\Omega^{-1}\cdot cm^{-1}$). On the other hand, no variable-range hopping or any unpaired-spin density has been observed except under the most nonequilibrium conditions, and the absorption edge is no more diffuse in multicomponent glasses than in a-Se or a-Te, as is clear from Figure 8. Relatively large field effects have been observed in multicomponent glasses containing Te and As, but these are transient and decay with time (20). Thus, there is evidence both for and against a large density of states at the Fermi energy, and the experimental results appear to be contradictory.

The resolution of these and other problems that have emerged with regard to the electronic structure of amorphous semiconductors lies in the recognition that well-defined chemical defects exist in these materials. As indicated previously, these defects control the transport properties of the semiconductors.

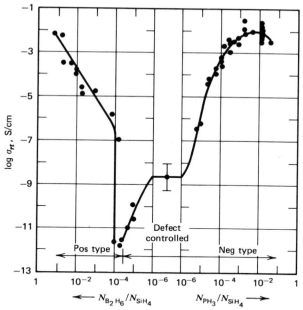

Figure 5. Room-temperature electrical conductivity as a function of phosphorus and boron doping in a–Si:H films (15).

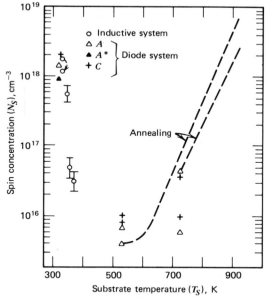

Figure 6. Unpaired-spin concentration as a function of substrate temperature and annealing for a–Si:H films (16).

Defects

Although many of the defects common in crystalline solids, eg, vacancies, interstitials, dislocations, and grain boundaries, are not likely to occur in the absence of

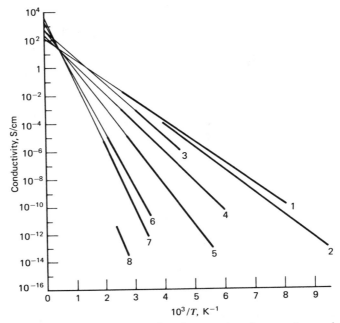

Figure 7. Electrical conductivity as a function of temperature for several amorphous chalcogenide alloys (S/cm = $\Omega^{-1} \cdot cm^{-1}$) (19).

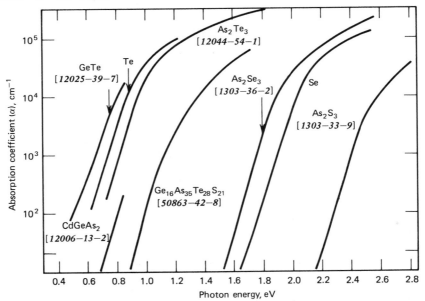

Figure 8. Absorption coefficient α as a function of photon energy for several amorphous chalcogenide glasses (19).

periodicity, chemical defects, eg, undercoordinated or overcoordinated atoms and wrong bonds, are possible. There is a great deal evidence for their existence in all classes of amorphous semiconductors (21). One of the simplest examples is the apparent substitutional doping of amorphous silicon alloys (15). If phosphorus and boron do enter these materials in tetrahedral sites, such centers then have the wrong chemical coordination and represent defects. In the model described in ref. 9, the valence requirements of each atom are locally satisfied, and this type of doping would then be impossible. Even in undoped samples of a-Si—H, the existence of defects can be inferred unambiguously. For example, it is clear from the results in Figure 6 that the spin density decreases with increasing substrate temperature until ca 350°C. However, as samples are annealed above 350°C, hydrogen effuses and the spin density sharply increases. When hydrogen effuses, defect centers must form. The spin density is observed by epr measurements, which indicates a common center for all spins. Clearly, samples prepared at low substrate temperatures must contain some of the same defect centers as those created when hydrogen effuses. It can be concluded that well-defined defects are present in tetrahedrally coordinated amorphous solids.

There is also a great deal of evidence for defects in amorphous chalcogenides. Perhaps the simplest demonstration is the recent photoluminescence investigation of crystalline and amorphous SiO_2 [7631-86-9] (22). Amorphous SiO_2 has an absorption edge at ca 10 eV. However, the photoluminescence excitation spectrum peaks near 7.6 eV and is clearly separated from the continuum, as is evident from Figure 9. Also shown in Figure 9 are the photoluminescence, excitation, and absorption spectra for amorphous As_2S_3 [1303-33-9] and amorphous SiO_2. The similarities in data for two such different glasses are strong evidence in favor of a similar mechanism for the luminescence.

The photoluminescence (PL) and excitation (PLE) spectra for both neutron-irradiated crystalline and amorphous SiO_2 are essentially the same (23). However, whereas the temperature dependence of the luminescence intensities of unirradiated a-As_2S_3 and a-SiO_2 are identical when the temperature is scaled to the glass-transition temperature, irradiated a-SiO_2 has a luminescence 20 times stronger. Irradiated samples of crystalline SiO_2 have a strong absorption peak at 7.6 eV, which can result

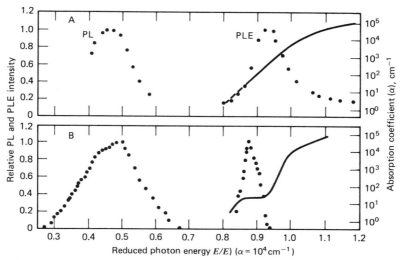

Figure 9. Photoluminescence (PL), excitation (PLE), and absorption spectra for a–As_2Se_3 and a–SiO_2, normalized to the energy at which the absorption coefficient α is 10^4 cm^{-1}. A, for As_2S_3, $E(\alpha = 10^4$ cm$^{-1}) = 2.55$ eV; B, for SiO_2, $E(\alpha = 10^4$ cm$^{-1}) = 8.6$ eV (22).

from nothing but a defect center. After irradiation, the excitation spectrum still peaks near 7.6 eV and the luminescence still peaks at 4.5–5.0 eV, but the luminescence is greatly enhanced. Irradiation simply increases the density of the defects that are responsible for the luminescence. In addition, the similarities in the scaled temperature quenching of the photoluminescence and the scaled energy dependence of luminescence and excitation spectra for several different glasses suggest a common mechanism for the entire process. Thus, it can be concluded that well-defined defects similar to those in the corresponding crystals are responsible for photoluminescence in chalcogenide glasses.

The recognition that well-defined defects exist in a-Si alloys provides a simple explanation of the diverse electronic properties of pure and hydrogenated amorphous silicon. Pure a-Si has an average coordination near four, and the resulting strains lead to high defect concentrations. These defects yield localized states in the gap, which pin the Fermi energy, produce large spin densities, and lead to the predominance of variable-range hopping conduction. However, a-Si–H has a much lower average coordination, which relieves the vast majority of the strains and removes almost all of the localized states from the gap. This is clear from the field-effect results shown in Figure 4. The structure evident in the data from Figure 4 is usually assumed to result from dangling bonds on Si atoms, although the situation is undoubtedly much more complex (24).

The resolution of the anomalous behavior in amorphous chalcogenides is much more subtle. These materials are not overconstrained and strain-produced defects are not expected. However, a particularly low energy defect, which has been called a valence-alternation pair (VAP), exists in these materials (25). A typical VAP, sketched in Figure 10, consists of a positively charged threefold-coordinated chalcogen and a negatively charged singly coordinated chalcogen. In the notation given in ref. 25, such a VAP is represented by C_3^+–C_1^-. The creation energy of this VAP can be es-

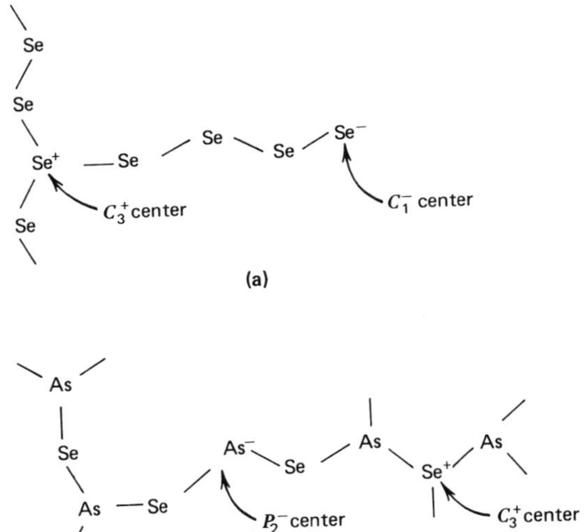

Figure 10.   Examples of valence-alternation pairs (VAPs): (**a**) a–Se; (**b**) a–As$_2$Se$_3$. C = chalcogen, P = pnictogens.

timated to be approximately equal to a quantity U, the additional Coulomb repulsion that results from the presence of the extra electron on the C_1^- site. U represents the difference between the ionization potential of the chalcogen atom and its electron affinity screened by the dielectric response of the material. Crude estimates for chalcogen atoms suggest that U is ca 0.5–1.0 eV; therefore, equation 2 predicts that there are at least 10^{16}–10^{20} cm^{-3} VAPs in chalcogenide glasses under ordinary conditions.

The reason for the low creation energy of a VAP is that the total number of bonds in a C_3^+–C_1^- pair is four, exactly the same as in a pair of chalcogen atoms in their ground states (C_2°). Furthermore, both C_3^+ and C_1^- centers have low energies: a positively charged chalcogen is isoelectronic with a pnictogen, and therefore it is optimally threefold-coordinated; a negatively charged chalcogen is isoelectronic with a halogen, and thus is optimally singly coordinated.

In addition to their low creation energy, VAPs also have a very unusual property which resolves the question of why the Fermi energy is so strongly pinned in chalcogenides, whereas there is no measurable unpaired-spin density and variable-range hopping conduction is not observed in any temperature range. The origin of this is the fact that the two defect centers, C_1 and C_3 have coordination numbers separated by two. This provides the possibility of converting a C_3 to a C_1 center simply by the breaking of a bond, since breaking of a two-electron bond lowers the total coordination of the atoms in a network by two. It is this possibility, and the fact that a positively charged C_3 center and a negatively charged C_1 center have a lower total energy than two neutral centers with either the C_1 or C_3 configuration, that leads to Fermi-energy pinning. The Fermi energy is the average energy necessary to add a few electrons to the material. However, if VAPs are present, adding two electrons simply converts a C_3^+ to a C_1^- center, independent of the charge state of the material, until no more C_3^+ centers are left. This occurs in two steps. The first electron is attracted to the C_3^+ center converting it to C_3°. The second electron then moves to any one of the three nearest-neighboring sites of the C_3°, and the bond between that site and the C_3° breaks. This returns the C_3° center to a C_2° ground state, leaving the neighbor as a C_1^- center. Since C_3° contains an antibonding electron but C_1^- does not, the second electron enters with a lower energy than the first. This means that excess electrons tend to enter the material in relatively localized pairs; since each pair has the same total energy, ϵ_f is pinned. The presence of VAPs also explains a great deal of other unusual characteristics of chalcogenide glasses including the complete absence of spins, ie, neither a C_3^+ nor a C_1^- center contains any unpaired spins, no observed variable-range hopping (two electrons or holes must hop simultaneously to interconvert C_3^+ and C_1^- centers, and the resulting electrostatic relaxations require too much energy), and the details of the photoconductivity, photoluminescence, and photostructural effects (24–25).

Under the assumption that heteropolar bonding predominates in binary alloys, the lowest energy VAPs in V–VI alloys are expected to be C_3^+–P_2^- pairs, where P represents a pnictogen atom; similarly, in IV–VI alloys the lowest energy VAP is a C_3^+–T_3^- pair, where T represents a tathogen (Group IVA) atom (21). However, in both of these cases, the electronegativity differences are such that they have larger creation energies than a C_3^+–C_3^- pair in a pure chalcogen material and thus appear with considerably lower concentrations. In addtion the C_3^+–T_3^- VAP only pins the Fermi energy for excess positive charge because additional bond formation is impossible for tathogens. Although a P_4^+–T_3^- pair has a low creation energy in IV–V alloys, it is not a VAP because of steric constraints (24). Although a P_4° center can in principle convert into a T_3° center by breaking a bond, the resulting relaxation would require both the T and the P atoms to move towards each other after the bond breaks, which is a highly energetically unfavorable motion. In pure pnictide materials, the creation energy of a P_4^+–P_2^- VAP would be expected to be large relative to those discussed previously because of the additional s–p promotion necessary to form the P_4^+ center, and there are some steric constraints retarding interconvertibility. In general, pnictogens exhibit behavior intermediate between chalcogenides and tetrahedral solids. In pure tathogens, VAPs are impossible because four is the maximal coordination with only s and p orbitals. However, charged defect centers are still likely (24).

When an amorphous solid is not overconstrained, it would be expected that the lowest energy defect predominates. However, there are several complications. First, it is possible that minimizing the free energy at T_g requires the presence of states in which the oppositely charged centers are located near each other, ie, intimate pairs (IVAPs), in which case a distribution of such pairs with various separations are present in the glass (24). Second, in nonstoichiometric alloys, homopolar bonds must exist, and these can yield localized gap states. Third, many other defects besides single overcoordination or undercoordination can exist. In particular, tathogens often bond with twofold coordination (T_2°), thereby saving the s–p promotion energy, and these might be expected in tetrahedral alloys (26). Other unusual configurations, eg, three-center bonds, could also be present (27). None of these defects would ordinarily be expected to exist in sufficiently large concentrations to be observable in direct structural studies, but they could be of the utmost importance in the transport behavior.

Doping

The fabrication of conventional semiconductor devices requires the ability to decouple the electrical activation energy from the optical gap (see Semiconductors, fabrication and characterization). In crystallization semiconductors, this is accomplished by substitutional doping, in which small concentrations of an impurity atom with a different valence from the host are introduced into the crystal. The periodic constraints then force a defect configuration, which leads to a large change in ϵ_f. This would appear to be impossible in an amorphous semiconductor, in which no crystalline constraints exist. However, a-Si—H and a-Si—F—H alloys can be routinely doped with either phosphorus or boron. It is thus important to determine why P and B enter the amorphous network tetrahedrally instead of in their ground-state (P_3° or B_3°) configurations. One possibility is that phosphorus or boron act as nucleation centers for the growth of doped microcrystalline Si, and there is evidence that this occurs in some very heavily doped samples (28). However, there is no direct structural evidence for any crystallites in lightly or moderately doped a-Si alloys.

Both a-Si—H and a-Si—F—H have sufficiently low defect concentrations that doping would be straightforward if P_4 and B_4 centers could form. It is fairly certain that the residual defects in undoped material include T_3 centers, which are necessary to relieve the remaining strains. If this is the case, doping can be understood in a simple way (24). For example, since the energy of a P_4^+–T_3^- pair is lower than that of a P_3°–T_3° pair, because of the extra bond, the formation of tetrahedrally coordinated phosphorus can be energetically favorable. For moderately heavy doping, the presence of twofold-coordinated Si may be essential (24). Boron doping can be explained analogously by the creation of T_3^+–B_4^- pairs (24). However, the unusual chemistry of boron complicates the situation considerably (27).

Chemical Modification

When low energy VAPs exist, the Fermi energy is effectively pinned and ordinary doping is impossible. Even if an impurity atom could be positioned in a nonoptimal chemical environment, the resulting release of excess electrons, for example, would only convert some of the C_3^+ centers to C_1^- without moving ϵ_f. To overcome this problem, one researcher introduced an ingenious technique for decoupling ϵ_f, even in the presence of large VAP densities, thereby making possible the use of chalcogenides in conventional semiconductor devices (29).

The technique of chemical modification requires the use of large concentrations of one or more modifying elements to overcome the large VAP density and the introduction of the modifier in a nonequilibrium manner to preclude its incorporation in the network. A simple example is the postdiffusion of indium into a chalcogenide glass. Because indium has an odd valence, it cannot enter the formed matrix in a fully bonded position. A low energy state for indium is one in which it gives up an electron and forms two ordinary and two dative bonds (27). For moderate indium concentrations, the excess electrons in pairs convert C_3^+ to C_1^- centers, and the conductivity is unaffected. However eventually the C_3^+ centers are completely depleted and ϵ_f begins to increase, thus reducing the electrical activation energy. Typical results are shown in Figure 11.

Alkali atoms and transition metals can also be used as chemical modifiers, and cosputtering and ion-implantation (qv) techniques can be effective under certain conditions (29). Pure materials and alloys from Groups III–VI are amenable to chemical modification. Because of the nonequilibrium preparation methods, the chemistry of the phenomena can be extremely complicated (27).

Transient Effects

Virtually all of the present electronic uses of amorphous semiconductors involve nonequilibrium processes introduced by the application of light or applied electric fields. Under these conditions, the nature and concentration of the defects are of the utmost importance, as they ordinarily completely control the kinetics of the return to equilibrium.

When VAPs are present, the positively and negatively charged centers represent traps with large cross sections for electrons and holes, respectively. However, because of the interconvertibility of the charge centers, equilibration requires atomic relaxations, eg, bond breaking. The complete return to equilibrium requires at least three steps. For example in a chalcogen these are trapping of an electron by the positively charged center

$$C_3^+ + e^- \rightarrow C_3^\circ \tag{6}$$

a relaxation involving, in this case, bond breaking,

$$C_3^\circ \rightarrow C_1^\circ \tag{7}$$

and trapping of a second electron by the neutral center,

$$C_1^\circ + e^- \rightarrow C_1^- \tag{8}$$

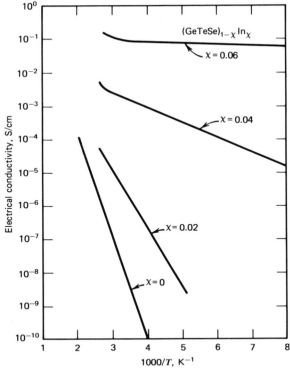

Figure 11.   Electrical conductivity as a function of temperature for pure a–GeTeSe and films chemically modified with In (S/cm = $\Omega^{-1}\cdot$cm^{-1}) (30).

In field-effect observations, for example, excess positive or negative charge is induced in the semiconductor by applying a voltage to a gate electrode. If a negative charge is induced, processes 6–8 ensure that ϵ_f is pinned and the field effect is unobservable. However, this pinning can manifest itself only if the time scale of the experiment is long compared to the longest of the three processes. This can be the resolution of the question of why field effects are routinely observed in AsTe glasses (20). If any one of the rates controlling processes 6–8 is long, a bottleneck can retard equilibrium. The most likely origin of such a bottleneck is process 7, since probably both C_3° and C_1° are local minima in the potential energy of the network; thus, a barrier should exist between the two defect centers, and an appropriate Boltzmann factor is included in the equilibration kinetics. For very short times, C_3^+ centers trap electrons and C_1^- centers trap holes, but the two neutral centers do not interconvert. Thus, the apparent density of states is that of a doped, compensated semiconductor. The Fermi energy is not pinned in this time scale, and a transient field effect should be observable. However, this field effect should decay with time with the time constant yielding the height of the potential barrier. Such transient effects have been observed in TeAs [12044-33-6] glasses and have been quantitatively analyzed (20). The relaxation time for process 7 is given by

$$\tau \simeq \tau_{ph} \exp\left[\Delta E/kT\right] \tag{9}$$

where ΔE is the height of the barrier and τ_{ph} is a phonon time of ca 10^{-12} s. Thus, at room temperature with a ΔE of 0.5 eV, τ is ca 100 μs. However, for a ΔE of 1.0 eV, τ

is several hours. These transient effects can persist for very long times, and this is the case for glasses in the TeAs system (20).

Despite the sharp contrasts in steady-state behavior between amorphous silicon alloys and chalcogenides, many of the unusual transient phenomena observed in chalcogenide glasses have also been observed in a-Si–H. These include dispersive transport (31), photoluminescence fatigue (32), light-induced creation of unpaired spins (33), and apparent photostructural changes (34). These and other recent observations provide evidence that charged centers also exist in a-Si–H (24). However, since unpaired spins always occur in the dark in a-Si–H, it is very likely that only intimate pairs are stable. The most probable candidate is a T_3^+–T_3^- pair, because each center is the lowest energy bonding configuration for its charge state, T^+ and T^- being isoelectronic to Groups III and V atoms, respectively. However, since a T_3^+ center is in its lowest energy state with a 120° (sp^2) bond angle but a T_3^- center requires a bond angle of about 95° (primarily p^3 bonding), the two neutral centers that result from electron and hole capture, respectively, are interconvertible but not identical. Therefore, a potential barrier can retard the kinetics of equilibration just as in chalcogenides, and this can account for the many similarities in transient phenomena between these two classes of materials which otherwise exhibit very different behavior.

A metastable state can be induced in a a-Si–H after exposure to intense light for long times (34). Whereas deposited films exhibit room-temperature conductivity of ca 10^{-6} S/cm (= $\Omega^{-1}\cdot cm^{-1}$) and activation energy of 0.57 eV, after the film is exposed to light, the conductivity decreases to ca 10^{-10} S/cm and the activation energy increases to 0.87 eV. The original state can be restored by annealing the film for two hours at 150°C. The relaxation time for recovering the deposited state has an activation energy of ca 1.5 eV, so that the room-temperature relaxation time is >25,000 yr.

The existence of charged defect centers provides a natural explanation of such effects, since electrons can be trapped at the positive centers to form a metastable state (35). If the potential barrier retarding equilibration is 1.5 eV, the very long time relaxations can be understood.

Health and Safety Factors

The following compounds, which are involved in amorphous-semiconductor technology, should be handled with great care. Silane is spontaneously flammable in air and is water-reactive. Silane is also irritating to the respiratory tract. Silicon tetrafluoride gives off HF and thus is highly toxic; TLV of SiF_4 <1 ppm. Phosphine is flammable in air and causes lung damage, convulsions, coma, and death. Exposure to 500 ppm PH_3 after 30 min is dangerous; TLV = ca 0.3 ppm. Diborane is flammable in air, irritates lung tissue, and causes edema; TLV = 0.1 ppm. Arsine is toxic to the gastrointestinal tract and causes chronic kidney and liver degeneration, shock, and death; LD_{50} = ca 20 ppm for 30 min, TLV = 0.05 ppm. Selenium causes liver damage; however, quantitative toxicity information is not available.

Uses

The applications of amorphous semiconductors arise primarily from their low preparation costs and their resistance to deterioration under severe conditions rather than from their unique physical properties. The main application has been as the photoreceptor in the electrophotographic process, ie, xerography (see Electrophotography). It is the excellent photoconductivity, the low dark conductivity, and the resistance to corrosion under conditions of high electric field and humidity that have led to this enormously successful use of amorphous chalcogenides. Other important devices based upon the photoconductive properties of amorphous semiconductors include the use of chalcogenide alloys in television pickup tubes, ie, vidicons.

A second area of application results from the reversible structural changes induced by either an external electric field or light (2). Electrically alterable and programmable read-only-memories and an array of imaging films have been commercially available for several years, and a film for updatable microfiche applications has been developed.

Future applications are particularly promising in many areas. The reversible switching phenomena exhibited by chalcogenides in high electric fields should result in novel control functions not possible with conventional materials (2). A wide array of related computer applications are also being developed (see Computers).

Amorphous silicon-based alloys are certain to have significant uses as thin-film semiconductor devices. In particular, they are extremely promising as thin-film transistors (TFTs) because of their low cost and high frequency response. In addition, the high ratio of photoconductivity to dark conductivity in high quality, amorphous silicon alloy films has led to their application in electrophotographic copying machines, because of the low cost and high frequency response of the films.

Perhaps the most important application in the near future will be as solar cells (see Photovoltaic cells). Solar photoelectric energy conversion has been available for over 25 yr, but it has never been of much use outside the space program because of its enormous cost. However, new interest has been stirred by the decrease in the supply and concomitant sharp increase in the price of the world's fossil-fuel reserves. Amorphous silicon-based alloys can be deposited inexpensively over very wide areas in thin-film form without requiring any expensive crystal growth or diffusion processes. The first high efficiency, amorphous silicon solar cell was reported in 1976 shortly after the demonstration that these alloys could be doped (36). Since then, there has been a great deal of activity in this area and solar-energy conversion efficiencies of ca 8% have been achieved (37–42) (see Solar energy). A device with an efficiency of more than 9% has been reported (43).

BIBLIOGRAPHY

1. B. T. Kolomiets, *Phys. Status Solidi* **7**, 359, 713 (1964).
2. S. R. Ovshinsky, *Phys. Rev. Lett.* **21**, 1450 (1968).
3. J. C. Phillips, *J. Non-Cryst. Solids* **34**, 153 (1979).
4. I. M. Lifshitz, *Usp. Mat. Nauk* **7**, 170 (1952).
5. T. P. Eggarter and M. H. Cohen, *Phys. Rev. Lett.* **25**, 807 (1970).
6. J. M. Ziman, *J. Non-Cryst. Solids* **4**, 426 (1970).
7. N. F. Mott, *Adv. Phys.* **16**, 49 (1967).
8. D. Weaire in F. Yonezawa, ed., *Fundamental Physics of Amorphous Semiconductors*, Springer-Verlag, Berlin, 1981, p. 155.
9. M. H. Cohen, H. Fritzsche, and S. R. Ovshinsky, *Phys. Rev. Lett.* **22**, 1065 (1969).
10. N. F. Mott, *Philos. Mag.* **16**, 49 (1967).
11. M. L. Theye, *Mat. Res. Bull.* **6**, 103 (1971).
12. M. H. Brodsky, R. S. Title, K. Weiser, and G. D. Pettit, *Phys. Rev. B* **1**, 2632 (1970).

13. R. C. Chittick, J. H. Alexander, and H. F. Stirling, *J. Electrochem. Soc.* **116**, 77 (1969).

14. A. Madan, P. G. LeComber, and W. E. Spear, *J. Non-Cryst. Solids* **20**, 239 (1976).

15. W. E. Spear and P. G. LeComber, *Philos. Mag.* **33**, 935 (1976).

16. H. Fritzsche, C. C. Tsai, and P. Persans, *Solid State Technol.* **21**, 55 (1978).

17. S. R. Ovshinsky and A. Madan, *Nature* **276**, 482 (1978); A. Madan, S. R. Ovshinsky, and E. Benn, *Philos. Mag.* **40**, 259 (1979).

18. W. Paul in ref. 8, p. 72.

19. N. F. Mott and E. A. Davis, *Electronic Processes in Non-Crystalline Materials*, 2nd ed., Clarendon Press, Oxford, England, 1979.

20. R. C. Frye and D. Adler, *Phys. Rev. Lett.* **46**, 1027 (1981).

21. D. Adler, *Solar Cells* **2**, 199 (1980).

22. C. M. Gee and M. Kastner, *Phys. Rev. Lett.* **42**, 1765 (1979).

23. C. M. Gee and M. Kastner, *J. Non-Cryst. Solids* **35–36**, 807 (1980).

24. D. Adler, *J. de Physique* **42**, C4-3 (1981).

25. M. Kastner, D. Adler, and H. Fritzsche, *Phys. Rev. Lett.* **37**, 1504 (1976).

26. D. Adler, *Phys. Rev. Lett.* **41**, 1755 (1978).

27. S. R. Ovshinsky and D. Adler, *Contemp. Phys.* **19**, 109 (1978).

28. A. Matsuda, S. Yamasaki, K. Nakagawa, H. Okushi, K. Tanaka, S. Iizima, M. Matsumura, and H. Yamamoto, *Jpn. J. Appl. Phys.* **19**, L 305 (1980).

29. S. R. Ovshinsky in W. E. Spear, ed., *Amorphous and Liquid Semiconductors*, C.I.C.L., University of Edinburgh, Scotland, 1977, p. 519.

30. S. R. Ovshinsky, unpublished data, 1977.

31. J. M. Hvam and M. H. Brodsky, *Phys. Rev. Lett.* **46**, 371 (1981).

32. J. I. Pankove and J. E. Berkeyheiser, *Appl. Phys. Lett.* **37**, 705 (1980).

33. J. C. Knights, D. K. Biegelsen, and I. Solomon, *Solid State Commun.* **22**, 133 (1977).

34. D. L. Staebler and C. R. Wronski, *J. Appl. Phys.* **51**, 3262 (1980).

35. R. C. Frye and D. Adler, *Phys. Rev. B* **24**, 5812 (1981).

36. D. E. Carlson and C. R. Wronski, *Appl. Phys. Lett.* **28**, 671 (1976).

37. D. Adler, *Kinam*, in press.

38. J. I. B. Wilson, J. McGill, and S. Kinmond, *Nature* **272**, 152 (1978).

39. W. E. Spear, R. A. Gibson, P. G. Le Comber, and A. J. Snell, *J. Non-Cryst. Solids* **35–36**, 725 (1980).

40. Y. Hamakawa, *Appl. Phys. Lett.* **39**, 237 (1981).

41. Y. Hamakawa, *Kinam*, in press.

42. S. R. Ovshinsky, *J. de Physique* **42**, C4-1095 (1981).

43. S. R. Ovshinsky, unpublished data, 1982.

DAVID ADLER
Massachusetts Institute of Technology

ORGANIC

Many families of organic compounds have thermally activated conductivity (1–8); however, actual working organic-semiconductor junction devices, eg, diodes, crude transistors, photovoltaics, etc, were not reported until 1978 (9) (see Photovoltaic cells; Polymers, conductive). Other typical, low conductivity, high dielectric-constant semiconductor components, eg, electrolytic capacitors, were prepared in the mid-1960s but have only limited application (10). On the other hand, poly(vinylcarbazole–trinitrofluorenone) [9020-74-0; 39613-12-2] (PVK–TNF) has wide application in electrophotographic copiers (see Electrophotography). The commercial uses of organic semiconductors in terms of their electrical properties are extremely rare, and this article is concerned largely with their theoretical and experimental aspects. Examples of common materials used in the manufacture of organic semiconductors and metals are listed in Table 1.

Fundamental research on organic conductors has led from anthracene (resistivity of ca 10^{22} Ω·cm) in the early part of this century to the tetracyanoquinodimethane (TCNQ) salts (semiconductors, resistivities of ca 10^{-1} Ω·cm) in the 1960s (2) to tetrathiafulvalenium (TTF) salts (organic metals, resistivities of 10^{-2}–10^{-3} Ω·cm) in the 1970s (3) to tetramethyltetraselenafulvalenium (TMTSF) salts (metals, resistivities of 10^{-5} Ω·cm and superconductivity (11) at 1.3 K) in the 1980s (see Fig. 1). Thus, in only two decades, there has been a revolution in terms of electrical conductivity, with a progression from ca 10 S/cm (= Ω^{-1} cm^{-1}) to infinite conductivity. Fundamental research on organic conductors has also led to profound contributions to theoretical solid-state physics, particularly with respect to low dimensional solid-state transitions. Theories on hypothetical one-dimensional metals were corroborated by the pseudo-one-dimensional organic solids, eg, TTF–TCNQ (3). The intimate relationship between Fermi-surface instabilities, eg, charge-density waves (CDWs) and metal-insulator phase transitions in low dimensional solids, were elucidated by studies of organic metals (5,10). Other collective modes, eg, spin-density waves (SDWs), occur (12–13) (see Superconducting materials).

Theory

The theory of the conductivity of these materials is a fast-growing field marked by many points of contention (1–7). The more exotic transport mechanisms that have been suggested have played an important role in stimulating chemists to become involved in the synthesis of these materials. The excitonic superconductivity model provides the best example of this generation of interest. It has been suggested that the dipole fields set up by electronic transitions could replace the phonon field in supplying the necessary electron-pair coupling mechanism for the transition to a superconducting state, presumably at higher than usual temperatures (14). The verity of this mechanism has yet to be demonstrated.

In a general sense, the features that distinguish organic metals, semiconductors, and insulators are just those that are operative in the case of the standard inorganic materials, ie, the conduction properties of all these materials depend on the nature of the electronic energy bands in these solids. The organics are somewhat atypical in that electronic transport is often associated with a preferred direction, ie, they exhibit anisotropic behavior, and this considerably complicates their properties.

Table 1. Some Common Materials Used in the Manufacture of Organic Semiconductors and Metals

Compound	CAS Registry No.	Structure
Dyes		
crystal violet	[548-62-9]	
rhodamine B	[81-88-9]	
malachite green	[569-64-2]	
violanthrene	[81-31-2]	
Polymers		
polyvinylcarbazole (PVK)	[25067-59-8]	
trans-polyacetylene	[25067-58-7]	
polythiazyl, poly(sulfur nitride)	[56422-03-8]	
poly(*p*-phenylene sulfide)	[25212-74-2]	
poly-*p*-phenylene	[25190-62-9]	

Table 1 (*continued*)

Compound	CAS Registry No.	Structure
Donors		
2,2′,5,5′-tetrathiafulvalene (TTF)	[31366-25-3]	
N,N,N′,N′-tetramethyl-p-phenylenediamine (TMPD)	[100-22-1]	
dehydrotetrathionaphthazarin (TTN)	[35753-06-1]	
5,6,11,12-tetrathiotetracene (TTT)	[193-44-2]	
5,6;11,12-tetraselenotetracene (TSeT)	[193-45-3]	
5,6;11,12-tetratellurotetracene	[64479-92-1]	
3,3′,4,4′-tetramethyl-2,2′,5,5′-tetraselena-fulvalene (TMTSF)	[55259-49-9]	
hexamethylenetetraselenafulvalene (HMTSF)	[56366-76-8]	
phthalocyanines (PC), M = Cu	[147-14-8]	
perylene (PER)	[198-55-0]	
N-methylacridinium	[26456-05-3]	

Table 1 (*continued*)

Compound	CAS Registry No.	Structure
N-methylphenazinium (NMP)	[7432-06-6]	
Acceptors		
7,7,8,8-tetracyanoquinodimethane (TCNQ)	[1518-16-7]	
tetracyanoethylene (TCNE)	[670-54-2]	
9,9,10,10-tetracyano-2,6-naphthylidene-dimethane (TNAP)	[6251-01-0]	
trinitrofluorenone (TNF)	[129-79-3]	

Some representative energy-band schemes are shown in Figure 2. They signify the allowed energy levels resulting from the orbitals, which are delocalized throughout the crystal. Their bandwidth W provides a measure of the strength of the interaction between molecules, eg, charge-transfer salts, or fragments, eg, polymers, in adjacent unit cells. For example, the orbitals associated with electron transport in the charge-transfer salt TTF X and the polymer $(CH)_x$ are shown in Figure 3. In both cases, these conduction molecular orbitals are primarily composed of $p\pi$ atomic orbitals, and they differ only in the main direction of conductivity with respect to the axes of the p orbitals. Nevertheless, because of the very different spacings between unit cells along the chain or stacking axis, the bandwidths W in polymers $[(CH)_x, (SN)_x, \text{ca } 10 \text{ eV}]$ are much greater than those in charge-transfer salts $[\text{TTF X} \leq 1 \text{ eV}]$.

The formation of energy bands does not automatically imply metallic properties. The additional prerequisite is the occupancy of these energy bands; if they are either completely filled or entirely vacant, the material is an insulator. A metal is associated with partly filled energy bands. If one or two bands are slightly filled or slightly vacant, the material is a semimetal or a semiconductor. In a semimetal, one band is partly filled and another is partly empty at all temperatures, but in an undoped semiconductor, all the energy bands are either vacant or full at zero kelvin. Thus, the chief characteristic of a semiconductor is the requirement of thermally activated conductivity (the energy gap E_g in Fig. 2), and this is shown by the increase in the resistivity of such materials with a decrease in temperature. The conductivity of metals shows the opposite temperature dependence, since the scattering mechanisms are less effective at low temperatures.

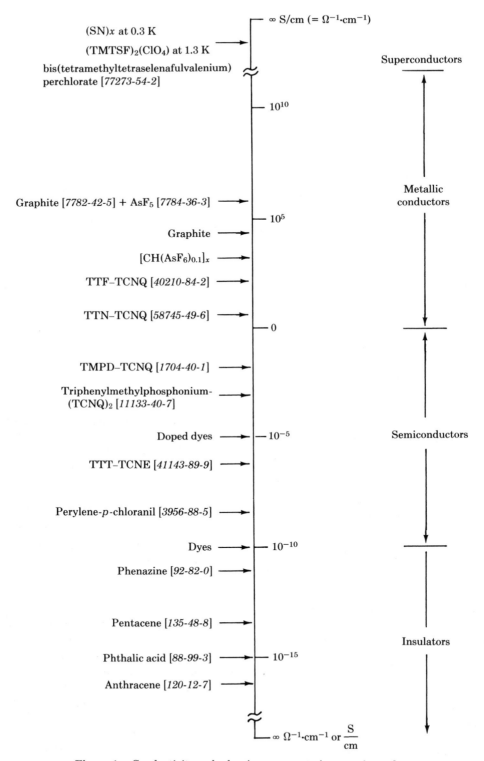

Figure 1. Conductivity scale showing representative organic conductors.

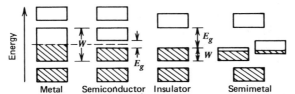

Figure 2. Schematic representation of energy bands in metals, semiconductors, insulators, and semimetals. The open boxes represent the density of vacant electronic states whereas those with cross-hatching denote occupied levels. The band width is represented by W and E_g is used to denote the energy between the valence (cross-hatched) and conduction (open) bands. The dashed lines represent the Fermi level.

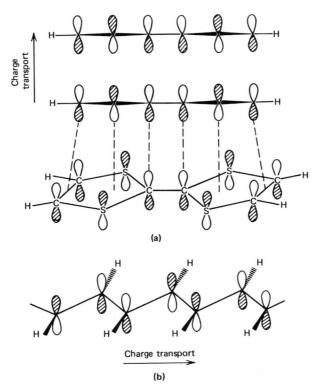

Figure 3. (a) Conduction molecular orbitals in TTF. (b) Conduction molecular orbitals in $(CH)_x$

There is an additional complication often associated with the organic metals that arises from the anisotropy in their electronic properties. One-dimensional metals are susceptible to structural changes, which are often associated with a transition to a semiconducting ground state as a result of Fermi-surface instabilities. Charge-density-wave (CDW) formation provides the most prevalent example of the ramifications of such an instability and is characterized by the occurrence of an electron-density distribution that has a different periodicity than the lattice. The Peierls CDW transition refers particularly to the adjustment in nuclear configuration accompanying the development of a lower symmetry electronic wave function, which sets up a superlattice in the crystal so that the unit cell becomes larger along the unique axis. The periodicity of this distortion is determined by the nature of the partially filled energy bands, in particular the Fermi-wave vector, and is readily determined from the degree of band filling. A half-filled band is the simplest example, and it is also the most favorable energetically for such a transition, as it is commensurate with the lattice, which results in a dimerization or doubling of the unit cell. The half-filled-band situation is exemplified by $(CH)_x$ (see Fig. 3b), which would be a one-dimensional metal (see Fig. 2) if it possessed equal C—C bond lengths and a uniform electron-density distribution, as in benzene. The polymer $(CH)_x$ exhibits alternating C—C single and double bonds (pinned Peierls CDW) and is best described as a semiconductor (see Fig. 2). Examples of Peierls transitions are known for various degrees of band filling; some are associated with superlattice formation over many lattice constants. The most important effect of this transition on organic metals is the gap that is opened up in the band at the Fermi energy, which causes the occupied and vacant orbitals to separate, resulting in the formation of a semiconductor.

Apart from influencing the occurrence of the Peierls transition, the degree of band filling exerts a powerful influence on the conductivity of one-dimensional organic metals by coulombic forces (charge repulsion), which become significant as electronic charge is transported through the lattice. This is depicted in Figure 4 for the case of TCNQ salts. On a molecular level, charge transport is associated with an intramolecular coulomb repulsion in the case of the half-filled band, but this reduces to the much smaller intermolecular Coulomb repulsion in the quarter-filled-band situation, thereby facilitating conductivity. Where the on-site, ie, intramolecular, Coulomb repulsion becomes large compared to the bandwidth, the conduction process may be inhibited by this effect, which is usually referred to as Mott-Hubbard localization. This phenomenon also is related to the spin-density-wave (SDW) transition, which occurs when the orientation of the spins on the adjacent sites becomes correlated, which leads to low dimensional (anti)ferromagnetic behavior. Thus, the occurrence of mixed valence (variation in band filling) plays a central role in determining the transport properties of one-dimensional organic conductors.

Charge-Transfer Salts. Charge-transfer salts are composed of regular arrays, ie, stacks or columns, of individual molecules. They are conductors because one or both of the components possesses unpaired electrons that can participate in the formation of a partially filled energy band. The unpaired electrons are generated by electron transfer between the constituents, so that one component is termed an electron donor and the other an electron acceptor, and the whole is usually referred to as a charge-transfer salt (see Table 1). The salts need not be stoichiometric, and the amount of charge transferred per molecule on the average can be less than unity, which results

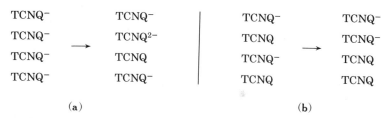

Figure 4. Molecular representation of the effect of band-filling on conductivity. (a) Intramolecular Coulomb repulsions in TCNQ half-filled band. (b) Intermolecular Coulomb repulsions in TCNQ quarter-filled band.

in mixed valence. The transport properties may be associated with the cationic and anionic stacks of molecules.

Many of the one-dimensional charge-transfer salts undergo Peierls transitions to semiconducting ground states. Repulsive intermolecular forces limit the approach of the molecules within a stack; thus, the interactions are small and the bandwidths narrow, usually less than 1 eV. As a result, some of these compounds are susceptible to Coulomb localization because the charge repulsion involved in the conduction process may exceed the bandwidth, which leads to a gap in the orbital energies and a semiconducting ground state, ie, Mott-Hubbard transition. The Peierls transitions are peculiar to materials with highly anisotropic band structures and in particular to those that are one-dimensional.

Polymers. The principal difference between the charge-transfer salts and the polymers is the magnitude of the bandwidth. Whereas the former are narrow-band materials (≤ 1 eV), the bandwidths of the polymers are of the order of those exhibited by inorganic metals and semiconductors (ca 10 eV). As a result, these materials are more susceptible to Peierls CDW transitions than the charge-transfer salts but are less prone to Coulomb localization.

The polymer $(CH)_x$ is a good example of CDW formation, although there is no temperature regime in which the CDW is absent in $(CH)_x$. This is the classic case of the half-filled band with a doubling of the unit cell; as expected, the pristine material is a poor conductor. Doping this polymer with donors or acceptors generates carriers, ie, electrons and holes, respectively, and greatly increases the conductivity. Polythiazyl does not undergo a Peierls transition and remains metallic to very low temperatures, where it eventually becomes superconducting. Unlike $(CH)_x$, $(SN)_x$ has strong interchain interactions and is sufficiently three-dimensional in its electronic structure to avoid CDW formation.

Physical Properties

The physical properties of the organic conductors, which set them apart from normal solids, arise from their anisotropic electronic structure. The electrical properties are the focus of interest and provide the most useful materials classification (see Fig. 1).

Electrical. *Charge-Transfer Salts. Class 1 (Semi-) Conductors.* Class 1 conductors are characterized by room-temperature resistivities of $1-10^{10}$ $\Omega\cdot$cm and a monotonic decrease in conductivity as the temperature decreases (12,15). The interpretation of this behavior is in accord with that of the usual inorganic semiconductors, eg, silicon. The activation energy for conduction in these semiconductors originates in either of two distinct effects. In the amorphous or disordered solids, a band picture is not appropriate and the electrons or holes migrate by hopping. Therefore, thermal energy must be supplied in the form of lattice vibrations in these solids in order to facilitate the movement of the localized charge carriers.

In the case of crystalline compounds, eg, K TCNQ [12153-61-6], a band model may be applied, and activated conductivity is brought about by a band gap that results from the Coulomb localization of a Mott-Hubbard transition. The energy gap in these materials is sufficiently large to classify them as semiconductors, because the degree of band filling is commensurate with the lattice and the transition to an insulating ground state is strongly favored.

Class 2 (Intermediate) Conductors.   Class 2 conductors exhibit electrical properties that fall between those expected of semiconductors and metals, and they are often referred to as semimetals. As such, they show room-temperature resistivities of $1-10^{-2}$ $\Omega \cdot$cm, and their conductivities gradually increase to a broad maximum as the temperature decreases, after which they decline rapidly. Some of these salts are borderline cases, and different classification schemes place them in one of the other categories. Thus, although there is general agreement that quinolinium $(TCNQ)_2$ [*12261-34-6, 30419-88-6*] is an intermediate conductor, NMP–TCNQ [*34504-21-7*] is sometimes placed with class 3 conductors.

In these compounds, the degree of band filling is such that the various insulating transitions are no longer commensurate with the lattice and, therefore, are not as strongly favored as in the case of the class 1 conductors. This is brought about by nonstoichiometry, as in the complex salts, eg, quinolinium $(TCNQ)_2$, or by back-charge transfer, as in NMP–TCNQ, by which each of the donor-acceptor components bears less than unit charge. An increase in mobility of the charge carriers may be reponsible for the initial small increase in conductivity as the temperature is lowered in these materials.

Class 3 (Metallic) Conductors.   The crystals of class 3 conductors are typified by room-temperature resistivities of less than 10^{-2} $\Omega \cdot$cm and a steady increase in conductivity as the temperature decreases. These compounds, therefore, show distinctive metallic behavior; however, in most cases, they undergo a low temperature phase transition, and their conductivities decrease rapidly below this point. Such behavior is exemplified by TTF–TCNQ, which shows a very sharp minimum in resistivity (10^{-4} $\Omega \cdot$cm) at 59 K.

These materials are characterized by incommensurate degrees of band filling, and this considerably reduces the driving force of the insulating transition. For example, in TTF–TCNQ, on the average 59% of an electron per molecule is transferred from TTF to TCNQ. There are a number of explanations for the initial increase in conductivity as the temperature is lowered in these compounds (1–7), including the increased carrier mobility mechanism, which is operative in normal metals (15). Among other models, the high conductivity near the phase transition has been attributed to various nonlinear collective modes, eg, sliding charge-density waves (16).

Only those systems with some three-dimensionality in their band structures, ie, strong electronic interchain interactions, remain metallic to the lowest accessible temperatures, ie, millikelvins. Structure studies show that such interactions occur between the cationic and anionic stacks in HMTSF–TCNQ [*56366-77-9*], and this compound remains highly conducting to 0.02 K (17). Interchain coupling also occurs in $(TMTSF)_2ClO_4$ between the cationic stacks, and this material not only remains metallic at low temperatures but becomes superconducting at 1.3 K (11).

Polymers.   The polymers differ from the charge-transfer salts in that their bandwidths are about an order of magnitude broader, and as a result, the insulating transitions are of the CDW type. This accounts for the poor conductivity of the amorphous polymer $(CH)_x$ which, in the pure state, has a room temperature resistivity of 10^9 $\Omega \cdot$cm. Doping the material with electron donors, eg, alkali metals, which results in the addition of electrons, or electron acceptors, eg, halogens or Lewis acids, which create holes, greatly increases the conductivity (18). The doped polymer $[CH(AsF_6)_y]_x$ (y = ca 0.1) has a room-temperature resistivity of 2×10^{-3} $\Omega \cdot$cm that gradually increases on cooling, and the material exhibits an insulating ground state at very low temperatures. This behavior is typical of the one-dimensional amorphous polymers.

Two features distinguish polythiazyl from the polymers discussed previously: crystallinity and well-developed interchain interactions. Although this material is formally described as a linear chain polymer, the three-dimensionality in the band structure is sufficient to inhibit the occurrence of a CDW transition. The low room-temperature resistivity of the polymer along the chain axis (5×10^{-4} Ω·cm) shows a metallic decrease with a decrease in temperature, and the material eventually becomes superconducting at 0.3 K (19). The conductivity perpendicular to the chain axis is limited by the nature of the interfiber contacts and has an essentially temperature-independent value of 0.1 Ω·cm, with the exception of the superconducting temperature regime where the change of state extends isotropically over all three directions. Thus, polythiazyl provides the first example of a superconducting polymer and the first occurrence of superconductivity in the S, N region of the periodic table. The material can be doped with halogens, but this merely increases the room-temperature conductivity by about an order of magnitude and leaves the superconducting-transition temperature essentially unchanged.

Magnetic. *Charge-Transfer Salts.* *Class 1 (Semi-) Conductors.* The class 1 compounds are semiconductors by virtue of the band gap that is opened by Coulomb localization, since the charge-transport process involves the intramolecular repulsion energy of two electrons or holes, ie, the half-filled-band case (see Fig. 4). The localized spins on adjacent molecules couple antiferromagnetically, and the paramagnetic susceptibility decreases strongly as the temperature decreases (see Magnetic materials).

Most of the alkali-metal TCNQ complex salts exhibit an anomaly in magnetic susceptibility, which is thought to result from a Peierls SDW transition. Although the TCNQ stacks are uniform above the transition, they are dimerized at temperatures below this point. As the conductivity is activated both above and below the transition, this effect is apparently of magnetic origin.

Class 2 (Intermediate) Conductors. The intermediate conductors are characterized by incomplete charge transfer; for example, in NMP–TCNQ, about 10% of the NMP and TCNQ molecules are thought to be neutral as a result of back-charge transfer. Thus, the carrier density in this material is low and the unpaired spins are well-separated. As a result, the spin–spin exchange term is small, and there is a large antiferromagnetic contribution to the susceptibility at low temperature.

Class 3 (Metallic) Conductors. The magnetic susceptibilities of the highly conducting charge-transfer salts show complex behavior, and no single model accounts for the temperature dependence of the magnetism at all temperatures. In TTF–TCNQ, three distinct regimes may be distinguished. Above 300 K, the magnetism is essentially temperature-independent and may be viewed as arising from Curie behavior or from an enhanced Pauli susceptibility. Below 300 K, the susceptibility decreases and, at temperatures above 60 K, this has been attributed to the formation of a pseudogap either as a result of CDW fluctuations or rehybridization effects, ie, interchain coupling. At 60 K, there is a break in the magnetic susceptibility data and, at temperatures below this point, the magnetism rapidly approaches the core diamagnetism of the constituent molecules. Apparently, the CDW becomes pinned to the lattice at this point and a real gap is produced in the density of states. Nevertheless, the magnitude of the gap is small, and magnetic excitation is not completely quenched until ca 20 K.

Polymers. In polythiazyl, the magnetic properties are characteristic of a normal semimetal with low carrier density. When the core diamagnetism is subtracted, the magnetic susceptibility is positive, small, and essentially temperature-independent; these features are characteristic of the Pauli susceptibility of a normal semimetal.

In pure *trans*-polyacetylene, there is a small concentration of highly mobile free spins that form as defects during preparation from the cis compound, which does not contain mobile, unpaired spins. At very low doping levels, it is possible to quench these spins and obtain an essentially diamagnetic material. As doping proceeds, a temperature-independent Pauli-type susceptibility develops.

Optical. **Charge-Transfer Salts.** *Class 1 (Semi-) Conductors.* The class 1 salts only exhibit one important electronic transition, apart from intramolecular excitonic features. In K TCNQ, this is a charge-transfer transition (energy, ca 1 eV) between adjacent pairs of $TCNQ^-$ molecules in the stack, ie, $2\,TCNQ^- \rightarrow TCNQ + TCNQ^{2-}$, and is a reflection of the electronic structure of these compounds: a half-filled band with a single unpaired electron per TCNQ molecule.

Class 2 (Intermediate) Conductors. Class 2 conductors are characterized by incomplete charge transfer; thus, in a compound such as triethylammonium $(TCNQ)_2$ [12261-33-5], there are neutral and negatively charged TCNQ molecules within the stacks. Consequently, in addition to the absorptions that occur in class 1 salts, there is an additional low energy charge-transfer band (ca 0.5 eV) caused by excitation of an electron from an anionic $TCNQ^-$ molecule to a neighboring neutral TCNQ molecule, and it is usually referred to as a mixed-valence transition.

Class 3 (Metallic) Conductors. The most significant feature of the optical properties of the highly conducting charge-transfer salts, eg, TTF–TCNQ, is the metallic plasma edge in the optical reflectivity when light is polarized parallel to the conducting axis and the absence of this structure when light is polarized perpendicular to the same direction. This behavior emphasizes the anisotropic band structure of these materials. The reflectance spectra do not change appreciably at temperatures below the phase transition, ie, 59 K in TTF–TCNQ. The electronic transitions that are present in the class 1 and 2 materials also occur in the absorption spectra of the class 3 salts.

Polymers. With light polarized parallel to the highly conducting axis, the optical reflectivity of polythiazyl exhibits a well-defined, metallic plasma edge. The data obtained with light of perpendicular polarization do not lead to a straightforward interpretation, although it is thought that the perpendicular reflectivity is best interpreted in terms of a strongly damped plasma edge.

Pure *trans*-polyacetylene shows a band edge at 1.4 eV with a steep rise to the peak at 2.0 eV, which is attributed to transitions from the valence band to the conduction band. Absorption also grows at ca 0.7 eV with light doping (<1%), with a concomitant strong decrease in intensity of the interband absorption. It has been suggested that the new feature is associated with transitions from the conduction band to positively charged solitons, which produce the midgap states (20). At doping levels beyond 5%, polyacetylene films show no transmission in the infrared region, which implies a continuous or metallic excitation spectrum.

Chemical Properties

Structure. **Single-Stack Donor.** *Tetrathiafulvalene (TTF) Halides and Pseudohalides.* In the case of TTF halides and pseudohalides, the materials adopt the simplest one-dimensional structure with the TTF molecules in stacks, where each molecule lies directly above the other, and the stacking axis is therefore perpendicular to the TTF planes (see Fig. 3a). The counterions or halides are in the channels generated by the TTF stacks, and it is assumed that the halide plays no part in the transport properties. A wide range of stoichiometries occurs in these salts (21); in TTF X_n, n exhibits a number of (fractional) values, which usually are 0.57–1. This variability directly affects the electronic population of the conduction band.

Tetramethyltetrathiafulvalene and Tetramethyltetraselenafulvalene. Although the planes of the molecules TMTTF and TMTSF are perpendicular to the stacking axis, they are slipped relative to one another and, as a result, there are two donor molecules in the unit cell. The molecules within neighboring stacks are parallel to one another rather than perpendicular, as observed in the TTF halides. As a result, (TMTSF)$_2$Y salts are thought to possess considerable two-dimensionality resulting from selenium–selenium interactions between adjacent stacks.

Single-Stack Acceptor. Virtually all investigations and applications of single-stack acceptors are concerned with TCNQ as an acceptor. Whereas TCNQ universally forms slipped-stack salts, the intermolecular distances in the stacks are not always uniform as in Cs$_2$(TCNQ)$_3$ [11119-04-3], where there are two different sets of TCNQ spacings and distinct triads of TCNQ molecules occur.

Double-Stack Donor-Acceptor. *Mixed Stacks.* In mixed stacks, the donor (D) and acceptor (A) molecules are interleaved within a single type of stack. A good example is provided by TMPD–TCNQ, in which the molecules form a slipped stack with uniform spacings. As a result of the —DADA— structure, these materials are necessarily poor conductors.

Segregated Stacks. In the case of segregated stacks, the donors and acceptors form independent slipped stacks, each one of which plays some part in the transport properties. The degree of charge transfer plays an important role in determining the electronic properties of these materials, and mixed-valence compounds are often associated with high conductivity.

Polymers. Because of their mode of preparation (see Synthesis and Manufacture), most polymers are amorphous or microcrystalline at best. The two outstanding exceptions to this behavior are provided by graphite and polythiazyl. Graphite consists of infinite sheets of repeating, condensed, benzene rings (see Carbon and artificial graphite). Polythiazyl may be viewed as strongly interacting sulfur–nitrogen chains. Polyacetylene consists of long chains of repeating ethylene units in a cis or trans configuration, ie, infinite polyene, although the material is polycrystalline. Thermal isomerization of the cis to the trans form occurs readily. Catalyst residues are not completely removed in preparations.

Electrochemistry of Donors and Acceptors. Charge transfer is an essential feature for conductivity in donor–acceptor salts and, thus, electrochemical oxidation and reduction potentials provide a useful index for evaluating likely combinations. Consider the following redox reaction

$$D + A \rightarrow D^{\rho+} + A^{\rho-} \tag{1}$$

For solids, this reaction may not occur at all (molecular crystal, $\rho = 0$), may go to completion (ionic crystal, $\rho = 1$), or may proceed partially (mixed-valence salt, $0 < \rho < 1$). In the first case, the crystal is an insulator; in the second case, it is probably a Mott-Hubbard semiconductor. Only in the third case is metallic behavior a possibility. Solution electrochemical redox potentials provide a very good guide to solid-state behavior and degree of charge transfer (value of ρ in eq. 1). In principle, the gas-phase ionization potentials and electron affinities should provide the same sort of information, but these quantities are much harder to obtain, and they seem to be less useful in predicting solid-state properties.

Donors. Empirically, it has been shown that sulfur- and selenium-based donors produce organic metals if their first oxidation potential ($E_{ox}^{1/2}$) in solution

$$D \rightarrow D^+ + e \quad E_{ox}^{1/2} \text{ (SCE)} \tag{2}$$

does not exceed 0.6 V, where $E_{ox}^{1/2}$ (SCE) is the half-cell oxidation potential measured against the standard calomel electrode (SCE). The hydrocarbon perylene (PER), for which $E_{ox}^{1/2} = 1$ V, can be converted to the salt (PER)$_2$PF$_6$ [75506-31-9], which shows a resistivity of 10^{-2}–10^{-3} Ω·cm at RT. As a result of the high oxidation potential of perylene, such salts are not stable to the atmosphere. The same applies to the polymers (CH)$_x$, (SN)$_x$, and graphite when they have been intercalated with dopants.

Acceptors. Tetracyanoquinodimethanide is the predominant progenitor of organic semiconductors and metals. Only materials based on TCNQ as an acceptor have possible commercial application (see Uses). Although there is a large family of cyanocarbons, eg, TCNE, that is potentially useful because of high electron affinity, only TCNQ and TNAP have been the basis for organic semiconductors and metals. It has been possible to vary the reduction potential of TCNQ by substitution, and derivatives with $E_{red}^{1/2}$ of -0.02 to 0.65 V have been prepared. The first reduction potential of an acceptor is given by

$$A + e \rightarrow A^- \quad E_{red}^{1/2} \text{ (SCE)} \tag{3}$$

Quinones are the oldest family of organic acceptors, but they have little use in the chemistry of organic semiconductors and metals, even though they exhibit a wide range of reduction potentials.

Stability. *Thermal Stability.* The organic semiconductors of technological importance are thermally labile when compared to inorganic systems. However, for some of their applications, eg, photocopying, the materials need not be subjected to large temperature extremes. The highest temperature to which an organic compound can be exposed in the absence of air is ca 500°C because of the nature of carbon–carbon, carbon–hydrogen, and carbon–heteroatom bonds. This temperature is reduced by ca 200°C when the materials are exposed to prolonged heating in the atmosphere. For example, TCNQ decomposes at its melting point of 287°C and TTF melts at 119°C without apparent decomposition, but becomes coated with sulfoxides derived from oxidation of one or more of its sulfur atoms (22). Polyacetylene [(CH)$_x$] can be heated to 200°C in the absence of air without apparent decomposition (23). However, it is degraded at or below room temperature by atmospheric oxygen and moisture regardless of the degree of doping.

Photostability. No in-depth studies on the photostability of TCNQ in the solid state have been reported. In solution, TCNQ rapidly decomposes upon uv irradiation. Tetrathiafulvalene is unstable to uv-visible radiation in the presence of mild acceptors (CCl$_4$ forms a weak complex that is photolabile) or oxygen, both in solution and in the solid state (22). However, PVK–TNF is quite stable to radiation, and it is this stability that makes it so useful in its photocopying applications.

Atmospheric Stability. No quantitative studies exist regarding the atmospheric degradation of TTF or TCNQ or any of the other known components of organic conductors. As mentioned above, TTF can be oxidized to a sulfoxide under mild conditions, whereas TCNQ is degraded by atmospheric moisture and light, particularly in solution (10).

Solubility. Since the sample-preparation technology of organic materials in the majority of cases depends on their formation from solution, it is important that they be soluble in the more common solvents. Most nitrogen- and sulfur-based donors are soluble in commonly employed solvents, with the exception of TTN and TTT, for which the solubilities are less than 1 mg/100 mL CH_3CN. Of the selenium-based donors, HMTSF and TSeT are the least soluble. The latter is slightly soluble in tetrahydrofuran (THF).

In general, the larger and more symmetrical molecules are the least soluble. Symmetrical methylation has an adverse effect on the solubility of TTF. Lack of solubility is not completely deleterious, since most of the insoluble donors can be cosublimed with acceptors to form charge-transfer salts as the sublimate. This technique has been used for the preparation of the TTT iodides (24).

Synthesis and Manufacture

The compounds TTF, TCNQ, PVK, TNF, polyvinylpyridine [25014-15-7], poly(phenylene sulfide), phthalocyanines, $(CH)_x$, and many dyes are commercially available. The preparative literature of TTF derivatives has been reviewed (25). None of the selenafulvalenes are commercially available.

Organic Metals. Organic metals are prepared by combining donors and acceptors or by electrolysis (see Crystal Growth).

Donors. Some donors may be very easily prepared from the elements. For example, TTT is synthesized by heating tetracene with sulfur in refluxing dimethylformamide (26), and TSeT is prepared from dichlorotetracene and selenium metal in Dowtherm as the solvent (27). However, tetraselenafulvalene [54489-01-9] (TSF), TMTSF, HMTSF, and related compounds are prepared by complex synthetic schemes, as shown below for the preparation of TMTSF (28):

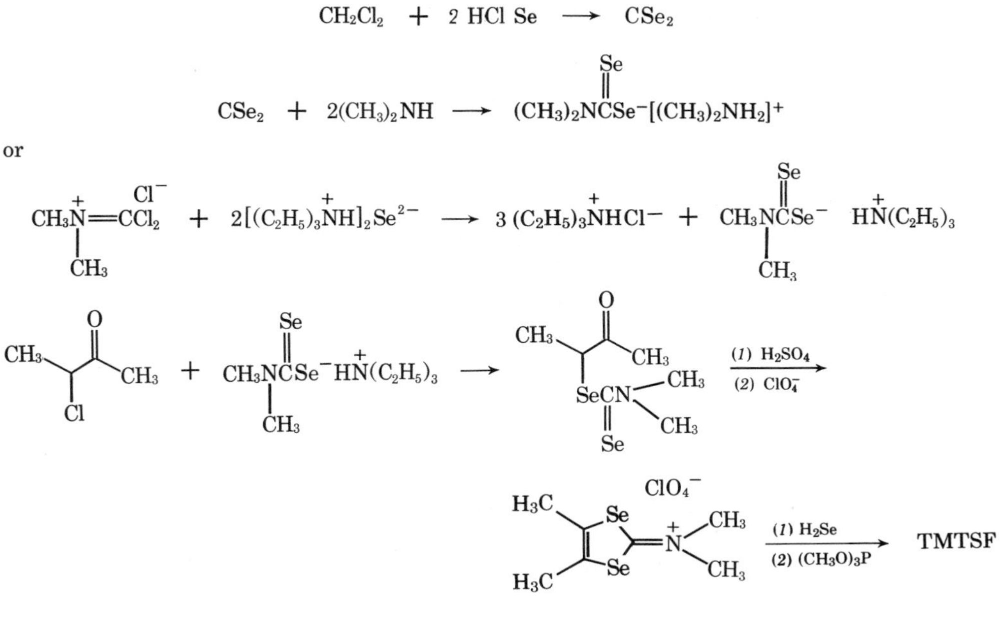

$$CH_2Cl_2 \ + \ 2 \ HCl \ Se \ \longrightarrow \ CSe_2$$

$$CSe_2 \ + \ 2(CH_3)_2NH \ \longrightarrow \ (CH_3)_2N\overset{\overset{\text{Se}}{\|}}{C}Se^-[(CH_3)_2NH_2]^+$$

or

$$CH_3\overset{+}{N}=CCl_2 \ \overset{Cl^-}{\underset{\underset{CH_3}{|}}{}} \ + \ 2[(C_2H_5)_3\overset{+}{N}H]_2Se^{2-} \ \longrightarrow \ 3\,(C_2H_5)_3\overset{+}{N}HCl^- \ + \ CH_3\overset{\overset{\text{Se}}{\|}}{N}Se^- \ \overset{+}{H}N(C_2H_5)_3$$

TMTSF

Tetramethyltetraselenafulvalene that is produced in this way is impure and must be recrystallized and then gradient-sublimed twice onto Teflon.

Acceptors. The more esoteric acceptors are prepared by complex synthetic schemes in fair yields (27% in three steps) as, for example, TNAP (29).

Organic Semiconductors. Charge-transfer organic semiconductors are prepared by the combination of a donor and acceptor in solution. If a polymer happens to be a good donor, eg, $(CH)_x$, then it is exposed either to the gaseous acceptor or to the dopant, eg, AsF_5, I_2, etc, or is electrolyzed in the presence of the dopant-derived anion or cation. Most TCNQ-based semiconductors are prepared by *in situ* reduction of the acceptor by iodide salts, eg, CsI, poly(vinylpyridinium iodide), etc, and the resulting iodine is washed away with a nonpolar organic solvent. Once the charge-transfer complex has been formed, it is usually not purified further but is used directly.

Crystal Growth. With a few rare exceptions, all organic semiconductor and metal crystal growths are carried out in solution at ca RT. There are three main methods: diffusion, metathesis, and electrolysis.

Diffusion. Since the electron-transfer step is very fast and the charge-transfer complexes are usually quite insoluble, formation of microcrystalline precipitates is prevented by slowing down the approach of the constituent molecules. This is usually accomplished by separating solutions of the donor and the acceptor in different compartments, which are connected directly by an intervening pure solvent that has been passed through alumina and which, in some cases, is doubly distilled, ie, aceto-nitrile, chlorobenzene, tetrahydrofuran, from sodium, and methylene chloride (30). This can be done with an H cell (see Fig. 5).

Further resistance to diffusion can be achieved by separating the donor solution, the acceptor solution, and the pure solvent with medium-porosity glass frits (31). The crystals of the complex, which may be as long as 2 cm, appear in the general area between the donor and acceptor compartments. In the case of TTF–TCNQ, the optimum temperature is 30°C. Higher temperatures are required for other systems (30). Another method to slow down diffusion is to increase the viscosity of the medium with thickening agents, eg, poly(methyl methacrylate), polystyrene, etc (30). This method does not have universal application and works best for TTF–TCNQ.

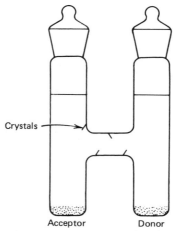

Crystals

Acceptor   Donor

Figure 5.   Crystal growth of charge-transfer compounds in H-shaped diffusion apparatus.

Metathesis. There are families of organic metals that can only be formed by oxidizing the donor to a radical cation or reducing the acceptor to a radical anion and then exchanging the counterion for the desired salt. For example, tetrathiafulvalenium selenocyanate [62152-48-1] $(TTF)_{12}(SeCN)_7$ and NMP–TCNQ have been prepared by first synthesizing the individual components (32). Tetrathiafulvalene is oxidized with hydrogen peroxide and fluoroboric acid or with triphenylmethyl fluoroborate to tetrathiafulvalenium tetrafluoroborate [60665-67-0] $(TTF)_3(BF_4)_2$. Lithium (TCNQ) [1283-90-5] is obtained by reduction of TCNQ with LiI. Methylation of phenazine with dimethyl sulfate gives *N*-methylphenazinium methosulfate [299-11-6] $(NMP^+CH_3SO_4^-)$. Both $(TTF)_3(BF_4)_2$ and Li TCNQ are relatively soluble in most of the solvents mentioned above. Reactions 4 and 5 below are typical metathetical reactions and work well because the desired product is less soluble than the starting materials.

$$(TTF)_3(BF_4)_2 + (C_4H_9)_4N^+SeCN^- \xrightarrow{\text{slow-cooling}} (TTF)_{12}(SeCN)_7\downarrow \qquad (4)$$

$$LiTCNQ + NMP^+CH_3SO_4^- \rightarrow NMP.TCNQ\downarrow \qquad (5)$$

Very slow diffusion does not work for reaction 4; instead, slow cooling of a homogeneous solution of the salts in acetonitrile affords the best crystals.

Electrolysis. Some donor-acceptor salts have been grown electrolytically with moderate success (33). More successful have been electrolyses of donors in the presence of inorganic anions. For example, electrolytic oxidation of TTF in the presence of tetrabutylammonium selenocyanate as supporting electrolyte produces the desired salt as crystals growing from the platinum anode. The usual cell is shown in Figure 6.

The electrolysis is usually at constant current, although constant potential electrolyses appear to give the same results. The initial potential is usually in the vicinity of the first oxidation wave of the donor. Salts that are grown by this method only are $(TMTSF)_2X$ (X = PF_6^- [73261-24-2], AsF_6^- [73731-75-6], SbF_6^- [73731-77-8], ClO_4^- [77273-54-2], and NO_3^- [73731-81-4]) (11).

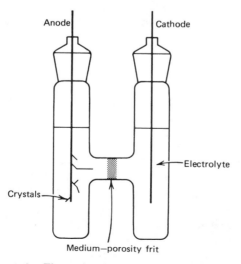

Figure 6.   Electrochemical crystal-growth apparatus.

Miscellaneous. Other methods, eg, vapor-phase crystal growth, are usually used if all else fails. For example, TTT and TSeT are so insoluble in polar solvents that solution techniques are difficult. The iodide is grown by coevaporation of TTT and iodine at high temperature under vacuum (24,26). When the opposite case is true and the complex is too soluble in polar solvents, then a precipitation method can be used whereby the solution of the complex is placed in an open vessel within a closed container, at the bottom of which is a nonpolar, volatile solvent. As the nonpolar solvent evaporates and dissolves in the polar solution, it displaces the complex in the form of crystals (30).

Since there are many possible stoichiometries and phases in these solids, it is important to know in detail which method of crystallization was employed for the preparation of a particular salt. For example, TTF–TCNQ grown in the vapor phase has the same stoichiometry as in the solution-grown material, but the stacks are along the crystallographic a direction rather than the b direction (30).

Polymers. The usual polymer-fabrication techniques apply to PVK, polyvinylpyridine, poly(phenylene sulfide), etc, but not to $(CH)_x$. The latter is prepared from acetylene and a large excess of a Ziegler-Natta catalyst in toluene at low temperature. The only modifications that can be carried out after polymerization are stretching, by a factor of ca 3, and isomerization. The $(CH)_x$, when initially formed at low temperature, is all cis; heating to 200°C converts it to trans. The trans isomer is not as stretchable. Conversion of cis to trans also occurs upon exposure to acceptors at low temperature. Upon exposure to air, both isomers lose their elasticity.

Polypyrrole [30604-81-0] and related 5-membered polyheterocycles are also not fabricable in the usual way. Polypyrrole and its congeners are prepared by electrolysis of a pyrrole derivative or another 5-membered heterocycle, eg, thiophene, at a platinum anode. One ion of the electrolyte is usually incorporated as the counterion, and the material is formed in doped form. Once deposited, the polymer is mechanically removed from the electrode as a film.

Poly(p-phenylene) is produced by exposure of benzene to copper halide at 35–50°C or exposure of biphenyl to a Lewis acid. This polymer can be fabricated prior to oxidation or reduction but, upon doping, its physical properties deteriorate. It is claimed that the doped polymer is more stable to atmospheric exposure than $(CH)_x$ (34).

Polyacetylene can be oxidized or reduced electrolytically. For example, if polyacetylene is used as a cathode in propylene carbonate with lithium perchlorate as the electrolyte, lithium is incorporated, and the polyacetylene is reduced.

Health and Safety Factors

Since most of the molecules employed in the preparation of organic semiconductors and metals exhibit facile oxidation and reduction processes, they should probably be used with care, because they could have profound effects on the biochemical electron-transport processes. Tetrathiafulvalene is toxic when administered intraperitoneally to mice of 200–400 mg/kg body weight, and it is not active against L-1210 lymphoid leukemia and P-388 lymphocytic leukemia (35).

Organoselenium Compounds. The toxicity of hydrogen selenide and inorganic selenides is well known. However, organoselenium compounds in small quantities are metabolically active and necessary. No toxicity studies have been performed on organoselenium donors, semiconductors, or metals. When organoselenium compounds are ingested, they are usually metabolized to methyl selenides, which are volatile and have a characteristic garlic odor (36). As a consequence, most bodily excretions, eg, breath, perspiration, etc, take on that odor.

Organotellurium Compounds. Hydrogen telluride, sodium telluride, and tellurium hexafluoride are among the most lethal compounds known to man (36). The toxicity of the former and the latter results from their volatility. The first step in physiological action apparently is reduction to tellurium metal. Eventually, this element is slowly excreted as the volatile and malodorous methyl tellurides. Concentrations as small as parts per million (10^6) of tellurium, when ingested, make their presence known by the odor of tellurium breath. The latter can apparently be controlled by administration of ascorbic acid (36). None of the very few known organotellurium donors have been tested for physiological activity.

Uses

The practical applications of organic semiconductors are still largely experimental, and their inroads into commercial uses are extremely limited. However, a large technical literature suggests possible applications for organic semiconductors. Skepticism has arisen about such applications because of the relatively poor stability and the unpredictable behavior typical of organic semiconductors. Certain aspects of organic semiconductor use are described elsewhere (see Dyes, sensitizing; Photodetectors; Liquid crystals). The following uses of organic semiconducting substances, although not of commercial interest at the moment, suggest research directions that might be taken.

Photovoltaic Devices. Many organic semiconductors display photovoltaic responses when used in appropriate cell structures (see Photovoltaic cells). The organic material is dissolved in a suitable solvent, and films are cast on transparent conducting surfaces, eg, SnO and CdS n-type semiconductors, by solvent evaporation. The semiconducting substance can be confined if it is first dispersed in a gel, eg, agar. A second electrode is pressed against the gel surface, and the cell is cemented together. A typical design is shown in Figure 7. Finished devices contain films ca 0.1-mm thick, with enough dye so that photoactive dye aggregates can form and, presumably, contact each other to make the device usable. Thus, when the organic semiconductor (p-type) is illuminated, an electron–hole pair is produced, and migration of the electron and the hole to the electrodes produces a photovoltage. The dye crystal violet has been utilized in a configuration resulting in an efficiency of ca 0.05%, which is not at all comparable

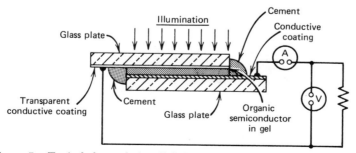

Figure 7. Typical photovoltaic-cell design based on organic semiconductor.

with the much higher (by a factor of ca 250) efficiencies of inorganic-based systems (37).

The essential structure can be modified if, instead of a p-type transparent electrode, a combination of a p-type organic and an n-type organic is used. One arrangement includes phthalocyanine or merocyanine dyes as the p-type material with malachite green, Rhodamine B, or TMPD compounds as the n-type material (38–40). Such devices can be used as photoelectric cells that develop photovoltages and photocurrents. Their efficiencies are also well below those in the inorganic semiconductor-based solar cells. Also, the question of their long-term chemical stability suggests that they may never be competitive, even if their efficiencies were increased substantially.

Thermometers. The electrical resistance of organic semiconductors exhibits a wide temperature-dependence range. This characteristic can be employed in temperature-measurement devices. A recent Japanese patent discloses a planar temperature sensor, which is made by forming a pair of electrodes on a plastic substrate (41). The organic semiconductor is a tetraalkylammonium TCNQ salt and is deposited as a film between the electrodes to form the device; the temperature is measured from the resistivity. Other sensing components, eg, thermistors, have also been suggested (42) (see also Temperature measurement).

Switches and Diodes. Voltage-controlled negative resistance is exhibited by numerous inorganics, including the oxides of nickel, silicon, and aluminum, when fabricated in a metal-oxide-metal arrangement. The behavior of these devices suffers from stability problems and irreproducibility (43). Amorphous glasses containing up to four elements show switching behavior (43). Stable and reproducible, bistable, current-controlled threshold switches have been fashioned from copper or silver TCNQ or TNAP polycrystalline anion-radical salts (43). This type of switch passes current only when an applied field in excess of a threshold value is reached, similar to the operation of silicon-controlled rectifiers (SCR); SCRs, however, carry much greater currents. A schematic diagram of a typical device construction is shown in Figure 8. The anion-radical salt is prepared by merely dipping the substrate into a saturated solution of the neutral compound. Oxidation–reduction occurs according to

$$Cu \cdot + TCNQ \rightleftharpoons Cu^+[TCNQ]^- \tag{6}$$

When the desired thickness is obtained, the substrate is removed from the solution, an electrode is sputtered onto the polycrystalline semiconducting layer, and wire leads are connected. When an applied field exceeds the threshold voltage, rapid off-on

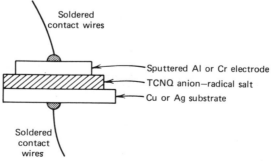

Figure 8.   Bistable switch based on organic species.

switching occurs. Switching times of ca 15 ns are obtained. In addition, the device remains in the low impedance state even after the initial applied field is removed, thus acting as a memory switch. Further work on the silver and copper TNAP systems relates the switching phenomena to a redox-induced phase transition in the polycrystalline radical-ion salts (44). The ease of manufacture of such switches is encouraging; however, their utility will ultimately depend on their cost and their long-term reliability and resistance to wear, about which no information is available. Switching phenomena in amorphous semiconductor diodes of tetracene–lead phthalocyanine and polystyrene films have also been observed (43).

Reproduction and Resist Materials. It has been shown that the conducting organic charge-transfer compounds derived from TTF and TTT with halogen acceptors are useful as electron-beam resists (45). The lithographic process is based on a reverse charge-transfer mechanism. When an electron beam of sufficient current density interacts with the conducting charge-transfer compound, a reverse reaction occurs and volatile free-halogen and neutral-donor D molecules form, as in

$$(D)_{1-y}(D^+)_y X_y^- \xrightarrow{e\text{-beam}} D + X_2\uparrow \tag{7}$$

where X = halogen, and $0 < y < 1$. This process is accompanied by a loss of conductivity in the exposed regions and the introduction of differential solubility between the exposed and unexposed areas resulting from cross-linking of the neutral donor with high current density. Negative images form upon development with 0.5 μm resolution. Positive images can also result if the current density is kept low and the exposed area, ie, neutral donor, is removed by a nonpolar solvent.

The properties of organic semiconductors that make them suitable for use in reproduction are their low inherent dark conductivity and their photoconductivity. One such process involves a photoelectrophoretic system (46–47). Semiconducting dye pigments are dispersed in an insulating medium and migrate from a carrier electrode during illumination and under the influence of an electric field. Thus, when a counter electrode is rolled over the illuminated dye medium, a subtractive color image remains on the carrier electrode.

Another imaging sequence makes use of the memory property of certain organic photoconductors that permits them to remain conductive for some time after the illumination source is removed. Positive or negative charge can be transferred to a dielectric layer and developed with appropriate powders, depending on the sign of the potential applied to the photoconductor. A single original exposure can result in several copies of equivalent quality (48). Among the materials suggested for use as photoconductors exhibiting memory effects are poly(N-vinylcarbazole)s containing cation radicals (49).

Optical printing of conductive images composed of TTF halides and related compounds has been reported (50). This method requires an organic π-donor, eg, TTF or TTT, dissolved in a halogenated hydrocarbon, such as CCl_4, to be deposited on a suitable substrate, eg, paper, which is then selectively exposed to actinic radiation. The mechanism is thought to involve photochemical oxidation of the donor and subsequent formation of the highly colored, conducting halide salt. For example,

$$TTF + CCl_4 \xrightarrow{h\nu} (TTF^+)(CCl_4^-) \tag{8}$$

$$(TTF^+)(CCl_4^-) \rightarrow (TTF)Cl_{0.77} + \text{other products} \tag{9}$$

The resolution obtained is ca 10–20 lines/mm, and the resistivity of the image is 10^4–10^5 Ω·cm. Because of the volatility and toxicity of CCl_4, systems of this type may never be of practical value.

Batteries.   Nonaqueous electrolytic primary cells have been demonstrated for a number of TCNQ-based systems (see Batteries and electric cells, primary). A cell based on tetraalkylammonium $(TCNQ)_2$ dissolved in acetonitrile has an output of 0.11 V at RT (51). In solid-state electrochemical cells, power density could be increased by using barium, calcium, or magnesium as the anode metal and a conducting iodine or TCNQ charge-transfer (CT) salts as the cathode (52–53). When magnesium is the anode material, a cell voltage of 1.5 V is achieved. Such a cell is constructed, for example, by placing phenothiazine-I_2 [25724-19-0], which is a CT complex, between electrodes of magnesium and platinum. Electrical-energy production is thought to involve oxidation of the anode metal and reduction of iodine. Cells that utilize charge-transfer complexes for the cathode, eg, phenazine-I_3 [25724-19-0], have also been reported (54). Such a cell based on a lithium anode and a poly(vinyl pyridine)–iodine complex cathode is being employed in cardiac pacemakers (55–56). The energy density, 120 W·h/kg (186 Btu/lb), is four times that of an equivalent lead storage battery.

Electrolytic Capacitors and Rectifiers.   Solid electrolytic capacitors that are based on MnO_2 have drawbacks that include relatively unstable frequency characteristics, low ratio of breakdown-to-formation voltage, and high leakage current. Substitution in this device with an organic semiconductor (see Fig. 9) has been carried to ameliorate most of these defects (57). Compounds that have performed well include N-n-propylquinolinium$(TCNQ)_2$ [12771-99-2] and N-methylacridinium$(TCNQ)_2$ [1927-47-1]. Their efficiency seems to be related to their ability to enhance passivation of the metal core (58). There is improvement of the low temperature and high frequency characteristics over MnO_2-based capacitors.

Electrochromic Devices.   The application of a voltage to some organic solids between electrodes leads to oxidation or reduction, depending on the material used and the current direction, with a concomitant change in color. Typical electrochromic displays are prepared, as shown in Figure 10. The organic electrochromic is lightly

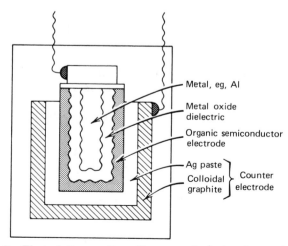

Metal, eg, Al

Metal oxide dielectric

Organic semiconductor electrode

Ag paste
Colloidal graphite } Counter electrode

Figure 9.   Electrolytic capacitor with an organic semiconductor electrode.

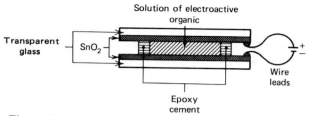

Figure 10.   An organic-based electrochromic display device.

colored or colorless before the cell is switched on. When current is permitted to flow through an easily reducible substance, a dark color caused by the formation of radical ions appears at the cathode. When the current polarity is reversed, the dark color fades and the image is erased (see Chromogenic materials).

One such system uses viologen materials as the electrochromic with hydroquinones to complete the redox system (59). The reactions that occur are as follows:

$$R—N^+ \langle \rangle —\langle \rangle N^+—R \ + \ e \ \longrightarrow \ R—N \langle \rangle —\langle \rangle N^+—R \tag{10}$$

$$\longrightarrow \qquad + \ 2\,e \ + \ 2\,H^+ \tag{11}$$

2,3,5,6-tetrachlorodihydrobenzoquinone   tetrachloro-*p*-benzoquinone

Similar devices have been described with the use of a variety of organics, including pyrazoline compounds (60–61), conjugatively linked tetrathiafulvalenes (62), polymer-bound π donors (63), and lutetium diphthalocyanine [*12369-74-3*] (64–65).

Miscellaneous.   Related to the photoelectric effects of *p-n* junctions are the photovoltaic effects resulting from metal–semiconductor or metal–gas–semiconductor interfaces. The directions of the photoelectric currents produced in such systems are wavelength-dependent (66). It has been proposed that this type of sensitivity can be exploited where organic semiconductors become the targets in vidicon-television pickup tubes. Exploratory work has been reported in medical ir vidicons and television cameras (67–68). It is not clear that such devices were ever really prepared. Also, as the references are over ten years old and nothing more has appeared in the literature on organic semiconductor vidicon technology, this may not be a fruitful field of investigation. Although the use of organic materials for vidicon tubes remains merely a literature footnote, such devices based on photoconductors, silicon, eg, charge-coupled devices (CCD) or antimony trisulfide, are in general use and have excellent properties.

For some organic semiconductors, the data suggests that the extent of various types of catalytic activity can be correlated with solid-state conductivity and with the type of conductance, ie, *n* or *p*-type (69–71). Such conclusions may be misleading because they may merely reflect chemical reactivities of families of semiconductors.

Thus, although the electrical properties may be incidental, there are a large number of examples of uses for organic catalysts that are also semiconductors. For example, selectivity of catalytic activity has been reported in dye photoconductors used to photooxidize isopropyl alcohol (72). Rather than the *p* or *n* nature of the dye, the activity differences may well result from differences in dye triplet energies, the chemical reactivity, or both. Para-to-ortho hydrogen conversion, which occurs with paramagnetic species, has been accomplished with phthalocyanines or electron donor–acceptor complexes, eg, tetracene or violanthrene halides (73–74). The polymerization of vinyl monomers has been shown to be initiated by the TCNQ salt of *N,N*-dimethylaniline-*N*-oxide [874-52-2] (75). None of the above systems is in commercial use.

Another group of devices that has generated some recent research activity is that based on organic semiconductor thin films. For example, preliminary investigations on the prototype of a light-modulated metal-oxide-semiconductor (MOS) transistor have been reported (76). An evaporated-dye film is used to photosensitize charge injection into an insulator when it is placed between the metal gate and oxide of the MOS transistor. The spectral sensitivity of the device can thus be chosen. Organic *n-p-n* photodiodes of the type ZnO/merocyanine/rhodamine B have been investigated (77). Variation of the bias voltage permits changing of the spectral sensitivity in the visible spectrum. The transient time of the photoresponse is ca 5 ms with a bias voltage of +1 V under 150 kW/cm^2 (36 kcal/(cm^2·s)) of 580-nm irradiation.

Aromatic organic dianhydride films, eg, 3,4:9,10-perylenetetracarboxylic dianhydride [128-69-8], are converted to black, electrically conducting (1 S/cm or 1/(Ω·cm)) patterns on exposure to electron-beam doses of >0.05 C/cm^2 (78). Fine lines ca 100 nm wide are suggested to have some potential application in miniature device structures.

BIBLIOGRAPHY

1. D. Fox, ed., *Physics and Chemistry of the Organic Solid State*, Wiley-Interscience, New York, 1965.
2. F. Gutmann and L. E. Lyons, *Organic Semiconductors*, John Wiley & Sons, Inc., New York, 1967.
3. H. J. Keller, ed., *Chemistry and Physics of One-Dimensional Metals*, Plenum Press, Inc., New York, 1977.
4. J. S. Miller in A. J. Epstein, ed., *Synthesis and Properties of Low-Dimensional Materials*, *Annals*, Vol. 313, N.Y. Acad. Sci., New York, 1978.
5. J. T. Devreese and R. P. Evrard in V. E. van Doren, ed., *Highly Conducting One-Dimensional Solids*, Plenum Press, Inc., New York, 1979.
6. W. E. Hatfield, ed., *Molecular Metals*, Plenum Press, Inc., New York, 1979.
7. L. Alcacer, ed., *The Physics and Chemistry of Low-Dimensional Solids*, R. Deidel Publishing Co., Dordrecht, Holland, 1980.
8. J. Kommandeur in ref. 1, Vol. 2, p. 1.
9. A. J. Heeger and A. G. MacDiarmid in ref. 7, p. 353.
10. S. Flandrois and L. Alcacer in ref. 7, p. 403.
11. K. Bechgaard, K. Cameiro, M. Olsen, F. B. Rasmussen, and K. S. Jacobsen, *Phys. Rev. Lett.* **46,** 852 (1981).
12. J. B. Torrance in ref. 4, p. 210; J. B. Torrance, *Acc. Chem. Res.* **12,** 79 (1979).
13. W. M. Walsh, Jr., F. Wudl, G. A. Thomas, D. J. Nalewajek, J. J. Hauser, P. A. Lee, and T. O. Poehler, *Phys. Rev. Lett.* **45,** 829 (1980).
14. W. A. Little, *Phys. Rev.* **134A,** 1416 (1964).
15. A. J. Epstein, E. M. Conwell, and J. S. Miller in ref. 4, p. 183.
16. A. J. Heeger in ref. 5, p. 69.
17. A. N. Bloch, T. F. Carruthers, T. O. Poehler, and D. O. Cowan in ref. 3, p. 47.

18. A. G. MacDiarmid and A. H. Heeger, *Synth. Met.* **1,** 101 (1980).
19. G. B. Street and R. L. Greene, *IBM J. Res. Dev.* **21,** 99 (1977).
20. N. Suzuki, M. Ozaki, S. Etemad, A. J. Heeger, and A. G. MacDiarmid, *Phys. Rev. Lett.* **45,** 1209 (1980).
21. B. A. Scott, S. J. La Placa, J. B. Torrance, B. D. Silverman, and B. Welber in ref. 4, p. 369.
22. K. Bechgaard in ref. 6, p. 1.
23. A. G. MacDiarmid in ref. 6, p. 161.
24. L. C. Isett and E. A. Perez-Alberne, *Solid State Commun.* **21,** 433 (1977).
25. M. Narita and C. U. Pittman, *Synthesis,* 489 (1976).
26. U.S. Pat. 3,723,417 (Mar. 27, 1973), E. A. Perez-Alberne (to Eastman Kodak).
27. C. Marschalk, *Bull. Soc. Chim. France,* 800 (1952).
28. K. Bechgaard, D. O. Cowan, and A. N. Bloch, *Chem. Commun.,* 937 (1974); F. Wudl and D. Nalewajek, *Chem. Commun.,* 866 (1980).
29. D. J. Sandman and A. J. Garito, *J. Org. Chem.* **39,** 1165 (1974).
30. J. R. Andersen, E. M. Engler, and K. Bechgaard, *Ann. N.Y. Acad. Sci.* **313,** 293 (1978); A. Diaz, *Chem. Scr.* **17,** 145 (1981).
31. M. L. Kaplan, *J. Cryst. Growth* **33,** 161 (1976).
32. F. Wudl, *Am. Chem. Soc.* **97,** 1962 (1975).
33. R. C. Wheland and J. L. Gillson, *J. Am. Chem. Soc.* **98,** 3916 (1976).
34. R. H. Baughman, *Synth. Met.* **1,** 307 (1979).
35. Drug Evaluation Branch, National Cancer Institute, Bethesda, Md., 1978.
36. W. C. Cooper, ed., *Tellurium,* Van Nostrand Reinhold Co., New York, 1971, p. 313.
37. U.S. Pat. 3,900,945 (Aug. 26, 1975), R. E. Kay and R. Walwick (to Philco-Ford Corp.).
38. H. Meier, *J. Phys. Chem.* **69,** 719 (1965).
39. D. Kearns and M. Calvin, *J. Chem. Phys.* **29,** 950 (1959).
40. U.S. Pat. 3,057,947 (Dec. 31, 1962), M. Calvin and D. R. Kearns.
41. Jpn. Kokai Tokkyo Koho 80,109,310 (Feb. 16, 1979), (to Matsushita Electric Indust. Co., Ltd.).
42. I. M. Paushkin, A. F. Lumin, V. A. Leksandrov, S. S. Oganesov, and V. B. Markovich, *Isz. Vyssh. Ucheb. Zaved. Fiz.* **12,** 90 (1969).
43. R. S. Potember, T. O. Poehler, D. O. Cowan, and A. N. Bloch in ref. 7, p. 419.
44. R. S. Potember, T. O. Poehler, A. Rappa, D. O. Cowan, and A. N. Bloch, *J. Am. Chem. Soc.* **102,** 3659 (1980).
45. Y. Tomkiewicz, E. M. Engler, J. D. Kuptsis, R. G. Schad, and V. V. Patel in ref. 7, p. 413.
46. U.S. Pat. 3,442,781 (Jan. 6, 1966), L. Weinberger (to Xerox Corp.).
47. U.S. Pat. 3,445,227 (May 20, 1969), L. Weinberger (to Xerox Corp.).
48. J. H. Dessauer and H. E. Clark, *Xerography and Related Processes,* Focal Press, London, UK, 1965.
49. H. Block, M. A. Cowd, and S. M. Walker, *Polymer* **18,** 781 (1977).
50. U.S. Pat. 4,036,648 (July 19, 1977), E. M. Engler, F. B. Kaufman, and B. A. Scott (to I.B.M. Corp.).
51. U.S. Pat. 3,110,630 (Nov. 12, 1963), W. R. Wolfe (to E. I. du Pont de Nemours & Co., Inc.).
52. F. Guttman, A. M. Hermann, and A. Rembaum, *J. Electrochem. Soc.* **114,** 323 (1967).
53. *Ibid.,* **115,** 359 (1968).
54. M. Pampallona, A. Ricci, B. Scrosati, and C. A. Vincent, *J. Appl. Electrochem.* **6,** 269 (1976).
55. A. M. Hermann and E. Luksha, *J. Card. Pulm. Tech.* **6,** 15 (1978).
56. Jpn. Kokai 49-56132 (May 31, 1974), T. Wada.
57. S. Yoshimura and M. Murakami in ref. 4, p. 269.
58. S. Yoshimura and M. Murakami, *Bull. Chem. Soc. Jpn.* **50,** 3153 (1977).
59. U.S. Pat. 3,806,229 (Apr. 23, 1974), C. J. Schoot and J. J. Ponjee (to U.S. Philips Corp.).
60. U.S. Pat. 4,093,358 (June 6, 1978), M. D. Schattuck and G. T. Sincerbox (to I.B.M. Corp.).
61. U.S. Pat. 4,090,782 (May 23, 1978), K. E. Bredfeldt, B. Champ, and K. J. Fowler (to I.B.M. Corp.).
62. U.S. Pat. 4,249,013 (Feb. 3, 1981), R. C. Haddon, M. L. Kaplan, and F. Wudl (to Bell Laboratories).
63. F. B. Kaufman, A. H. Schroeder, E. M. Engler, and V. V. Patel, *Appl. Phys. Lett.* **36,** 422 (1980).
64. G. A. Corker, B. Grant, and N. J. Clecak, *J. Electrochem. Soc.* **126,** 1339 (1979).
65. M. M. Nicholson and F. A. Pizzarello, *J. Electrochem. Soc.* **126,** 1490 (1979).
66. H. Meier and W. Albrecht, *Ber. Bunsenges. Phys. Chem.* **73,** 86 (1969).
67. P. Delius, *Brit. J. Photogr.* **111,** 278 (1964).
68. U.S.S.R. Pat. 335,740 (Apr. 13, 1970), G. A. Morozov, I. V. Antonova-A-Fanaseva, Yu. A. Popov, and I. Ya. Markova.

69. W. Hauke, *Z. Chem.* **9**, 1 (1969).
70. M. M. Sakharov and O. A. Golovina, *Probl. Kinet. Katal.* **15**, 94 (1973).
71. W. Hanke and W. Karsch, *Mbr. Dtsch. Akad. Wiss. Berlin* **9**, 323 (1967).
72. H. Inoue, S. Hayashi, and E. Imoto, *Bull. Chem. Soc. Jpn.* **37**, 326 (1964).
73. E. K. Rideal, *J. Res. Inst. Catal. Hokkaido Univ.* **16**, 45 (1968).
74. M. Tsuda, T. Kondow, H. Inokuchi, and H. Suzuki, *J. Catalysis* **11**, 81 (1968).
75. T. Sato, M. Yoshioka, and J. Otsu, *Makromol. Chem.* **177**, 2009 (1976).
76. B. W. Flynn, J. Muvor, and A. E. Owen, *Solid-State Electron Dev.* **2**, 94 (1978).
77. K. Kudo and T. Moriizumi, *Appl. Phys. Lett.* **39**, 609 (1981).
78. P. H. Schmidt, D. C. Joy, M. L. Kaplan, and W. L. Feldmann, *Appl. Phys. Lett.* **40**, 93 (1982).

General References

Ref. 2 is a general reference.
H. Meier, *Organic Semiconductors*, Verlag Chemie, Weinheim, 1974.
K. Masuda and M. Silver, eds., *Energy and Charge Transfer in Organic Semiconductors*, Plenum Press, Inc., New York, 1974.
J. J. Brophy and J. J. Buttrey, eds., *Organic Semiconductors*, Macmillan Inc., New York, 1962.
H. Kallmann and M. Silver, eds., *Electrical Conductivity in Organic Solids*, Interscience Publishers, New York, 1961.

ROBERT C. HADDON
MARTIN L. KAPLAN
FRED WUDL
Bell Laboratories

SEPARATION SYSTEMS SYNTHESIS

The separation of chemical species (components) is required in most chemical processes. Often such equipment participates in most process operations and the selection, arrangement, interconnection, and energy integration of the separation operations is a complex task.

An example of the synthesis of a separator process is the recovery of butenes in a butadiene process (1) (see Butadiene; Butylenes). Specifications for a typical system are shown in Figure 1. The feed is a C_4 concentrate from the catalytic dehydrogenation of *n*-butane, which is to be separated into four fractions: a propane-rich stream containing 99% of the propane entering the separation system; an *n*-butane-rich stream that contains 96% of the entering *n*-butane, which is recycled; a stream containing a mixture of the three butenes, at a 95% recovery, which is sent to a butenes-dehydrogenation reactor to produce butadienes; and an *n*-pentane-rich stream containing 98% of the entering *n*-pentane (see also Hydrocarbons, C_1–C_6).

An economical separation process for this problem, based on the availability of relatively inexpensive energy and similar to the system described in ref. 1, is given in Figure 2. In the first separation, 1-butene and propane are removed from *n*-butane

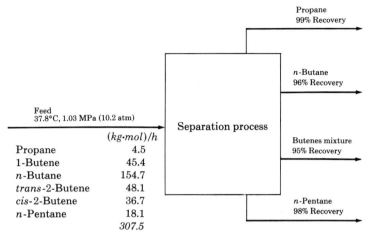

Figure 1. Specification for butenes recovery system.

and the heavier components in a 100-tray distillation (qv) column (shown as two 50-tray columns), C-1. This separation is difficult because the relative volatility between 1-butene and n-butane is only about 1.2 and, therefore, a reflux ratio of 25:1 is needed in conjunction with 100 trays. In the second step, propane and 1-butene, which have a relative volatility of about 2.2, are easily separated by distillation in C-2 with 25 trays and a reflux ratio of about 12:1 on a small quantity of distillate. The pentane in the bottoms from C-1 is easily removed by distillation in the deoiler, C-3, with 20 trays and a reflux ratio of about 1:1. The fourth step is essentially impossible by distillation, because the relative volatility between n-butane and the lower-boiling 2-butene isomer is only about 1.03. However, in the presence of appreciable quantities of 96% furfural in water, this relative volatility is increased to about 1.17. Thus, extractive distillation in C-4 removes the n-butane (see Azeotropic and extractive distillation, Vol. 3 and Supplement). Furfural solvent is recovered for recycle in the stripper, C-5.

   Although the scheme in Figure 2 is practical, a needless separation of the three butenes is performed and 1-butene must be remixed with *cis*- and *trans*-2-butenes. Furthermore, reflux ratios are significantly above the minimum, tray requirements are approximately twice the minimum, and no attempt is made to conserve energy by integrating heat-transfer requirements (see Heat-transfer technology).

   The total steam requirement for the five separation operations is very substantial. A more efficient process, dictated by the high cost of energy, would reduce reflux and, therefore, reboiler-heat-duty requirements, increase the number of trays, and integrate condensers and reboilers as well as other heat-exchange equipment where possible, even though process control would become more complex and column-operating pressures would have to be altered. For example, by lowering the operating pressure of C-1 and slightly raising the operating pressure of C-4, part of the heat for E-1 could be provided by condenser E-8.

   Although the process of Figure 2 utilizes distillation-type operations exclusively, a large number of other types of separation operations are available and should be considered. Nevertheless, distillation-type operations are usually preferred.

   A low-cost adiabatic technique for superpurification of volatile liquids was pre-

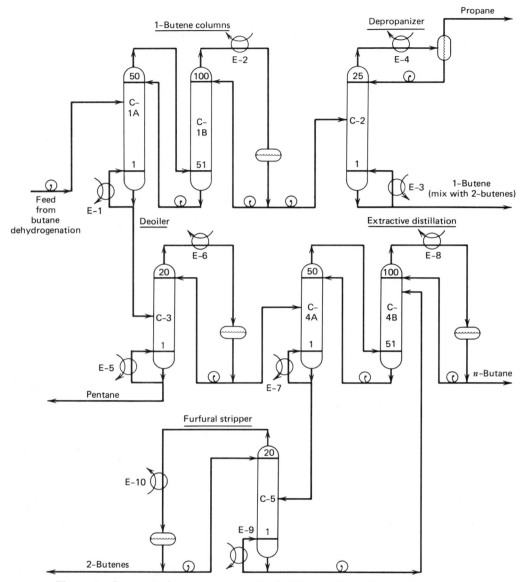

Figure 2.   Process for butenes recovery. C = distillation column; E = heat exchanger.

sented at the 2nd World Congress of Chemical Engineering. In this method, which is said to be particularly useful for closely boiling liquids, a thin layer of liquid feed enters a horizontal vacuum chamber having a pressure slightly lower than the vapor pressure of the mixture. The impurity begins to vaporize, which causes the desired product to crystallize; it exits to a melt chamber for removal from the system. The vapor, which contains some product and essentially all of the impurities, is condensed to provide the vacuum for the system. Final purities over 99.99% are reported to be obtainable from 90% pure feed (2).

Types and Characteristics of Separation Operations

The feed to a separation process generally consists of a single phase, ie, vapor, liquid, or solid. However, in some cases a feed may be comprised of two or more coexisting phases. If so, phase separation by mechanical means, to the extent practical, should precede the species-separation process. For example, a slurry of crystals and solvent might first be centrifuged to recover most of the liquid solvent, and the remainder is recovered from the wet cake by a species-separation operation (see Extraction, liquid–liquid).

The large number of available species-separation techniques may be classified into three categories: interphase mass-transfer separation, intraphase mass-transfer separation, and chemical-reaction separation operations (see also Mass transfer).

In interphase mass-transfer separations, another phase is created by an energy-separating agent (ESA), a mass-separating agent (MSA), or both. The ESA may be heat transfer, work transfer, or both. Alternatively, the new phase may be created by reducing the pressure across a valve. An MSA may be a selective solvent or adsorbent or, by its presence, may alter species-separation indexes, eg, in extractive distillation. Regardless of the nature of the separating agent, a potential must exist for the species in the feed to partition between the two or more phases. This potential is governed by equilibrium thermodynamics. The rate of approach to the limiting equilibrium composition is controlled by interphase mass transfer, which can be enhanced by intimate mixing of the phases. Following contact, the phases are generally separated by mechanical means.

Even when an MSA is used, an additional operation involving an ESA is generally required to recover the MSA for recycle. Therefore, most interphase mass-transfer operations are energy-intensive. Because of the steadily mounting energy costs in recent years, interest has greatly increased in separation operations that do not create or introduce a new phase. Such operations are referred to here as intraphase mass-transfer separation operations. They are based on the application of some barrier or imposed force field. The degree of separation is governed by differing transport rates across the barrier or through the force field of the species involved. Large-scale applications of selective membrane permeation and chromatographic separation operations are appearing in industries where interphase separation operations have previously been used almost exclusively (see Membrane technology).

Table 1 includes commonly used continuous-separation operations based on inter- and intraphase mass transfer (3–4).

Selection of Criteria

The development of a separation process involves the selection of separation operations, including any necessary ESAs or MSAs, types of equipment, and the sequencing of equipment for multicomponent separations.

If the feed has only two components, a single operation may suffice, provided that an ESA is used with either an interphase or intraphase mass-transfer separation operation. If an MSA is needed, at least one recovery operation is required, unless the MSA and associated material are expendable.

If the feed contains many components, the desired products may be essentially the pure species, or one or more multicomponent products. In the latter case, it is

Table 1. Separation Operations[a]

Interphase mass-transfer operations

Separation operation	Kirk-Othmer article[b]	Initial or feed phase	Developed or added phase	Separating agent(s)	Principle	Industrial application	Kirk-Othmer article[b]
flash vaporization	Distillation, Vol. 7, p. 860	liquid	vapor	pressure reduction	difference in volatility	recovery of water from seawater	Water, supply and desalination, Vol. 24
partial vaporization and evaporation	Evaporation, Vol. 9, p. 472	liquid	vapor	heat transfer	difference in volatility	evaporation of water from a solution with urea	Urea and urea derivatives, Vol. 23
partial condensation	Carbon monoxide, Vol. 4, p. 785	vapor	liquid	heat transfer	difference in volatility	recovery of H_2 and N_2 from ammonia	Ammonia, Vol. 2, pp. 494–496
distillation	Distillation, Vol. 7, p. 849	vapor or liquid	vapor and liquid	heat transfer or work transfer	difference in volatility	stabilization of natural gasoline to remove isobutane and lighter hydrocarbons	Petroleum, refinery processes, Vol. 17, pp. 199–201
extractive distillation	Azeotropic and extractive distillation, Vol. 3, p. 352	vapor or liquid	vapor and liquid	liquid solvent and heat transfer	difference in volatility as altered by solvent	separation of toluene from close-boiling nonaromatic compounds; phenol used as solvent	Toluene, Vol. 23
azeotropic distillation	Azeotropic and extractive distillation, Vol. 3, p. 352	vapor or liquid	vapor and liquid	liquid entrainer and heat transfer	difference in volatility as altered by solvent to cause formation of azeotrope	separation of acetic acid from water; n-butyl acetate used as an entrainer to form an azeotrope with water	Azeotropic and extractive distillation, Vol. 3, pp. 365–367
absorption	Absorption, Vol. 1, p. 52	vapor	liquid	relatively nonvolatile liquid absorbent	difference in volatility as evidenced by solubility	separation of carbon dioxide from combustion gases by absorption with	Carbon dioxide, Vol. 4, pp. 730–731

stripping	Absorption, Vol. 1, p. 52	liquid	vapor	relatively noncondensable stripping vapor	difference in volatilities	aqueous solutions of an ethanolamine steam stripping of naphtha, kerosene, and gas-oil side cuts from a crude-distillation unit	Petroleum, refinery processes, Vol. 17, p. 217
reboiled absorption	Adsorptive separation, Vol. 1, pp. 565–575	vapor or liquid	vapor and liquid	relatively nonvolatile liquid absorbent and heat transfer	difference in volatilities	removal of ethane and lighter hydrocarbons from the overhead of a catalytic cracker main fractionator	Petroleum, refinery processes, Vol. 17, p. 206
reboiled stripping	Absorption, Vol. 1, p. 52	liquid	vapor	heat transfer	difference in volatilities	removal of light ends from a naphtha cut	Petroleum, refinery processes, Vol. 17, p. 199
refluxed stripping (steam distillation)	Distillation, Vol. 7, p. 881	vapor or liquid	vapor and liquid	stripping vapor and heat transfer	difference in volatilities	distillation of reduced crude oil under vacuum with use of stripping steam	Petroleum, refinery process, Vol. 17, p. 199
liquid–liquid extraction	Extraction, liquid–liquid, Vol. 9, p. 672	liquid I	liquid II	liquid solvent	difference in solubilities in the two liquid phases	deasphalting of a reduced crude oil; propane used as a solvent	Asphalt, Vol. 3, pp. 297–298
two-solvent liquid–liquid extraction	Extraction, liquid–liquid, Vol. 9, p. 672	liquid I	liquid II	2 liquid solvents	difference in solubilities in the two liquid phases	separation of paraffins from aromatic and cycloparaffinic compounds; propane and cresylic acid used as solvents	Petroleum, refinery processes, Vol. 17, p. 227
crystallization	Crystallization, Vol. 7, p. 243	liquid	solid and sometimes vapor	heat transfer	difference in freezing tendency	crystallization of p-xylene from m-xylene	Xylenes and ethylbenzenes, Vol. 24

703

Table 1 (*continued*)

Separation operation	Kirk-Othmer article[b]	Initial or feed phase	Developed or added phase	Separating agent(s)	Principle	Industrial application	Kirk-Othmer article[b]
adsorption	Adsorptive separation, Vol. 1, p. 531	vapor or liquid	solid	solid adsorbent	difference in adsorption tendency	selective adsorption of n-paraffins from petroleum fractions	Adsorptive separation, Vol. 1, pp. 550–551
drying	Drying, Vol. 8, p. 75	liquid and sometimes solid	vapor and solid	gas or heat transfer	difference in volatility	removal of water from poly(vinyl chloride) with hot air	Vinyl polymers, vinyl chloride, Vol. 23
desublimation		vapor	solid	heat transfer	difference in tendency to condense as a solid	removal of phthalic anhydride from light gases and organic compounds by condensation to solids	Phthalic acids and other benzene polycarboxylic acids, Vol. 17, pp. 738–754
leaching (liquid–solid extraction)	Extraction, liquid–solid, Vol. 9, p. 721	solid	liquid	liquid solvent	difference in solubility	aqueous leaching of slime to recover copper sulfate	Copper, Vol. 6, pp. 844–845
ion exchange	Ion exchange, Vol. 13, p. 678	liquid	solid	solid ion-exchange resin	interchange of ions between solid and liquid	demineralization of water using both cation- and anion-exchange resins	Ion exchange, Vol. 13, p. 700
Interphase mass-transfer operations							
pressure diffusion	Diffusion separation methods, Vol. 7, p. 695	gas		centrifugal force	induced pressure gradient	separation of isotopes of uranium	Diffusion separation methods, Vol. 7, p. 696
gaseous diffusion	Diffusion separation methods, Vol. 7, p. 659	gas		porous barrier	forced flow through barrier	separation of isotopes of uranium	Diffusion separation methods, Vol. 7, pp. 639–641
reverse osmosis	Reverse osmosis,	liquid		membrane	pressure gradient	desalinization of	Reverse osmosis,

	Vol. 20, pp. 230–248					Vol. 20, p. 239
selective-membrane permeation	Membrane technology, Vol. 15, pp. 113–137	gas or liquid	membrane	forced flow through semipermeable membrane to overcome osmotic pressure	removal of hydrogen; water	Hydrogen, Vol. 12, p. 970
dialysis	Dialysis, Vol. 7, p. 564	liquid	membrane	difference in diffusion rate	recovery of caustic from hemicellulose	Dialysis, Vol. 7, p. 572
foam fractionation	Foams, Vol. 11, pp. 127–141	liquid	foam interface	selective concentration of species at interface	enzyme and dye separations	Foams, Vol. 11, pp. 127–141
chromatographic separations	Analytical methods, Vol. 1, pp. 599–600	gas or liquid	solids	selective concentration in and on solids	mixed vapor–solvent recovery	Analytical methods, Vol. 1, pp. 599–602
zone melting	Zone refining, Vol. 24	solid and liquid	temperature gradient	difference in freezing tendency	germanium purification	Zone refining, Vol. 24
thermal diffusion	Diffusion separation methods, Vol. 7, p. 681	gas or liquid	temperature gradient	induced concentration gradient	separation of uranium isotopes in the liquid phase	Diffusion separation methods, Vol. 7, p. 681
electrolysis	Deuterium and tritium, Vol. 7, p. 539	liquid	electric field	different rates of discharge of ions	separation of hydrogen and deuterium	Deuterium and tritium, Vol. 7, p. 539
electrodialysis	Electrodialysis, Vol. 8, p. 726	liquid	electric field and charged membranes	tendency of membrane to pass only anions	production of potable water from brackish water	Electrodialysis, Vol. 8, p. 733
electrodecantation	Electrodecantation, Vol. 8, p. 721	liquid and colloids	electric field	different ionic mobilities of colloids	concentration of rubber latex	Electrodecantation, Vol. 8, p. 724

a Refs. 3–4.
b Third edition.

preferable not to separate components that must be blended later to form desired products. However, many exceptions exist to this rule. For example, in the operation shown in Figure 1, a six-component mixture is separated into four products, one of which contains 1-butene and *cis-* and *trans-*2-butene.

In the process shown in Figure 2, separating 1-butene from 2-butene cannot be avoided and, thus, a final blending operation is required. Nevertheless, except for possible improvement by energy integration, the process shown in Figure 2 is the most economical one known. For multicomponent-separation processes, each separation operation generally separates between two components, and the minimum number of operations is one less than the number of products. However, there are a growing number of exceptions to this rule, and cases are described below where a single separation operation may produce only a partial separation of three or more products.

First, the phase condition of the feed has to be considered. If the feed is a vapor or is readily converted to a vapor, as shown in Table 1, operations such as partial condensation, various distillation techniques, absorption, desublimation, and adsorption may be applicable. If the feed is a liquid or is readily converted to a liquid, flash or partial vaporization, evaporation, various distillation techniques, stripping, liquid–liquid extraction, crystallization, and adsorption are applied. Feeds consisting of wet solids are generally separated by drying, whereas feeds consisting of dry solids are leached. However, for the separation of liquid or vapor feeds, several alternatives are usually available. Generally, intraphase mass-transfer separation is only considered when a suitable interphase mass-transfer operation cannot be employed.

The selection of a particular separation operation is based on the nature of the feed components and any MSA used, as well as temperature, pressure, and phases. In general, interphase mass-transfer separation operations are readily converted into efficient countercurrent cascades that produce very sharp separations. Thus, unless only a moderate degree of separation is required, certain intraphase mass-transfer separation operations may not be feasible.

The reasons for a separation must be considered next. They usually include purification of a species or group of species, removal of undesirable constituents, and recovery of constituents for subsequent processing or removal. In the case of purification, an MSA may prevent exposure to high temperatures that may cause decomposition. Removal of undesirable species containing a modest amount of desirable species may be economically acceptable. Likewise, in the recovery of constituents for recycle, a high degree of separation may not be necessary.

For the interphase mass-transfer separation operations listed in Table 1, the separation factor (SF) is based on the concept of approaching thermodynamic equilibrium between two or more phases. If components 1 and 2 (referred to as the key components) of the mixture are separated and the two phases are labeled I and II, the separation factor may be defined as

$$\text{SF} = \frac{C_1^{\text{I}}/C_2^{\text{I}}}{C_1^{\text{II}}/C_2^{\text{II}}} \qquad (1)$$

where C is some composition unit such as mol fraction, mass fraction, concentration, etc. Most commonly, the composition is expressed in terms of mol fractions x. Thus

$$\text{SF} = \frac{x_1^{\text{I}}/x_2^{\text{I}}}{x_1^{\text{II}}/x_2^{\text{II}}} \qquad (2)$$

At thermodynamic equilibrium, the fugacity f of each species i is the same in each phase. Thus,

$$f_i^{I} = f_i^{II} \tag{3}$$

which can be expressed in terms of the mixture fugacity coefficient, ϕ, pure-component fugacity coefficients, v, or activity coefficients, γ, as

$$x_i^{I}\phi_i^{I} = x_i^{II}\phi_i^{II} = \gamma_i^{I}x_i^{I}v_i^{I} = \gamma_i^{II}x_i^{II}v_i^{II} \tag{4}$$

From equation 4, several equivalent expressions can be derived for determining SF:

$$SF = \frac{\phi_1^{II}/\phi_2^{II}}{\phi_1^{I}/\phi_2^{I}} = \frac{\gamma_1^{II}/\gamma_2^{II}}{\gamma_1^{I}/\gamma_2^{I}} = \frac{\gamma_1^{II}v_1^{II}/\gamma_2^{II}v_2^{II}}{\phi_1^{I}/\phi_2^{I}} \tag{5}$$

Values of ϕ and v are generally determined by pressure–volume–temperature equations of state; these values also depend on composition (5). Values of γ are estimated from free-energy models and depend mainly on temperature and composition. However, ϕ, v, and γ are also related to melting point, normal boiling point, vapor pressure, solubility, and the Henry's law constant. These physical-properties values are usually found in tables. They are related to fundamental molecular properties, such as weight, volume, shape, dipole moment, and polarizability, and to other, more fundamental characteristic properties, such as critical temperature, critical pressure, acentric factor, and solubility parameter (see also BTX processes).

In general, components 1 and 2 are designated in such a manner that SF > 1.0. Consequently, the larger the value of SF, the more feasible the particular separation operation. However, in order to obtain a desirable SF value, it is best to avoid extreme conditions of temperature, ie, which may require refrigeration or damage heat-sensitive materials; pressure, ie, which may require gas compression or vacuum; and MSA concentration, ie, which may require expensive means to recover the MSA. Operations employing an ESA are economically feasible at a lower value of SF than are those employing an MSA. In particular, provided that vapor and liquid phases are readily formed, distillation should always be considered as a possible separation operation. A multicomponent mixture that forms nearly ideal liquid solutions is a good example. If the pressure is low, the ideal gas law holds. Then, the SF for vapor–liquid separation operations employing an ESA such as partial evaporation, partial condensation, or distillation is referred to as the relative volatility, α; for this case, it is given in terms of vapor pressures, $P°$, because in equation 5: $\gamma^{II} = 1$, $\phi^{I} = 1$, and $v^{II} = P°/P$.

$$SF = \alpha_{1,2} = P_1°/P_2° \tag{6}$$

where II is the liquid phase.

For vapor–liquid separation operations, eg, extractive distillation, that use an MSA that causes the formation of a nonideal liquid solution,

$$SF = \alpha_{1,2} = \gamma_1^{II}P_1°/\gamma_2^{II}P_2° \tag{7}$$

if the pressure is low and II is the liquid phase.

If the MSA is used to create two liquid phases, such as in liquid–liquid extraction, the SF is referred to as the relative selectivity β where

$$\beta_{1,2} = \frac{\gamma_1^{II}/\gamma_2^{II}}{\gamma_1^{I}/\gamma_2^{I}} \tag{8}$$

In general, MSAs for extractive distillation and liquid–liquid extraction are selected according to their ease of recovery for recycle and to achieve relatively large values of SF. Such MSAs are often polar organic compounds, eg, phenol (for recovery of heptane from toluene), furfural (for recovery of n-butane from 2-butenes), acetone (for recovery of acetylene from ethylene), acetonitrile (for recovery of butadiene from butenes), and tetraethylene glycol (for recovery of aromatic from nonaromatic compounds).

In some cases, the MSA is selected in such a way that it forms one or more homogeneous or heterogeneous azeotropes with the components in the feed. For example, the addition of n-butyl acetate to a mixture of acetic acid and water results in a heterogeneous minimum-boiling azeotrope of the acetate with water. The azeotrope is taken overhead, the acetate and water layers are separated, and the acetate is recirculated.

Single-stage operations, eg, partial vaporization or partial condensation with the use of an ESA, are preferred if SF is very large. For example, if SF = 10,000, a mixture containing equimolar parts of components 1 and 2 could be partially vaporized to give a vapor containing 99.01 mol % of component 1 and a liquid containing 99.01 mol % of component 2.

At lower values of SF, perhaps as low as 1.05, multiple-stage distillation is often suitable. However, as illustrated in Figure 3, extractive distillation or liquid–liquid extraction may be preferred if the SF can be suitably enhanced (6). If SF = 2 for distillation, it must be above 3.5 for extractive distillation to be an acceptable alternative, and above 20 for liquid–liquid extraction.

Unless values of SF are ca 10 or above, multistage absorption and stripping operations cannot achieve sharp separation between two components. Nevertheless, these operations are used widely for preliminary or partial separations where the separation of one key component may be sharp, but only a partial separation of the other key component may be necessary. The degree of sharpness of separation is indicated by the recovery factor

$$\Phi_i^I = \frac{\text{mass (or moles) of } i \text{ to separator product I}}{\text{mass (or moles) of } i \text{ in the feed}} \tag{9}$$

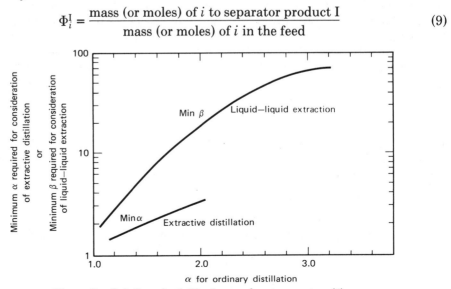

Figure 3.   Relative selectivities for equal-cost separators (4).

where most of i is removed in product I rather than in product II, etc.

Separations of solid phases may be necessary when one or more of the components is not readily melted, sublimed, or vaporized. Such operations may even be preferred when boiling points are close but melting points are far apart, as is the case with many isomer pairs. The classic example of such an operation is the separation of m-xylene from p-xylene, whose normal boiling points differ only by 0.8°C, but whose melting points differ by 64°C. With an SF of only 1.02, as determined from equation 6, distillation to produce relative pure products from an equimolar mixture of the two isomers would require about 1000 stages and a reflux ratio of more than 100. For crystallization, the SF is nearly infinity because essentially pure p-xylene is produced. However, the mother liquor contains at least 13 mol % p-xylene in m-xylene corresponding to the limiting eutectic composition. For nearly perfect separation between the two xylene isomers, an extractive crystallization could be conducted (7). Other operations for separating these two isomers are discussed in ref. 3.

If only small amounts of one component are to be removed from large amounts of one or more other components, changing the phase of the latter components should be avoided. In such a case, the minor component is best removed by selective adsorption.

For intraphase mass-transfer separation operations, SF may still be defined by equations 1 or 2. However, it is governed by relative rates of mass transfer rather than by equilibrium considerations. These rates may often be influenced by factors other than those associated with ordinary diffusion processes, including membrane transport, centrifugal-force fields, electric and magnetic fields, and pressure and temperature gradients. In most cases, the applicability of intraphase-type separation operations must be established experimentally. They are only considered if a feasible interphase-type separation operation is unavailable. A recent development of interest is the Prism selective-permeation operation (8), which uses hollow fibers for separating hydrogen, helium, carbon dioxide, and water vapor from gases containing oxygen, nitrogen, carbon monoxide, and light hydrocarbons (see Hollow-fiber membranes). A typical application is shown in Figure 4, where the SF between hydrogen and methane is 40. The recovery of hydrogen in the permeate stream is 60% whereas the recovery of methane in the nonpermeate stream is 96%.

Equipment Selection

In general, equipment selection is based on stage- or mass-transfer efficiency, pilot-plant tests, scale-up feasibility, investment and operating cost, and ease of maintenance (qv) (see Pilot plants; Economic evaluation; Process development).

For distillation, a myriad of choices is available to the design engineer. The stages may consist of trays or packing; the latter may be random or stacked. Trays can be designed with and without "downcomers" and can be of the perforated, bubble-cap, or valve-cap, etc, type.

For liquid–liquid extraction, an even greater variety of equipment is available, including multiple mixer-settler units or single countercurrent-flow contactors with or without mechanical agitation. In some cases, a centrifugal extractor may be the best choice (see Centrifugal separation).

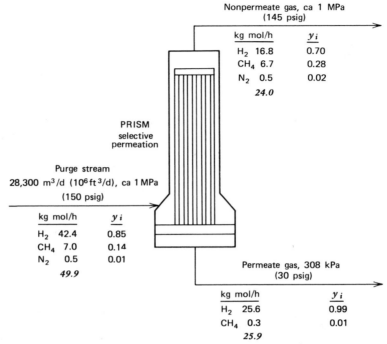

Figure 4.   Prism selective-permeation operation.

Synthesis of Separation Sequences

In most industrial separations, multicomponent mixtures are separated into more than two products. Although one separator of complex design can often be devised to produce all the desired products, a sequence of two-product separators is more commonly used. For example, a mixture of benzene, toluene, and biphenyl can be conveniently separated by distillation.

A single complex distillation operation, shown in Fig. 5a, is compared with the sequence of two simple distillation operations in Figure 5b giving the same products.

Even if a sequence is used, not all its separators may give products. For example, Figure 6 shows two distillation schemes for the separation of a mixture of ethylbenzene, p-xylene, m-xylene, and o-xylene into three products consisting of nearly pure ethylbenzene, a mixture of p- and m-xylene, and nearly pure o-xylene (see Xylenes and ethylbenzene). Necessary heat exchangers and pumps are not shown. In Figure 6a, two columns are used in sequence. The first column removes ethylbenzene as an overhead product, whereas the second removes o-xylene from the mixture of p- and m-xylene. In Figure 6b, the same products are obtained by interlinking two distillation columns. In the first column, a partial separation is made by taking ethylbenzene and a portion of p- and m-xylene overhead, and the o-xylene and other portion of p- and m-xylene as bottoms. The second column produces the three products as well as reflux and boil-up for the first column.

Consider the synthesis of a separation process with a single multicomponent feed of C components that is to be separated by (C-1) single-feed, two-product distillation

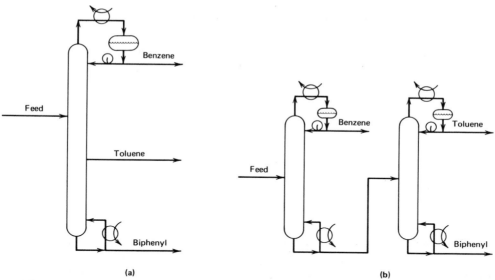

Figure 5. Distillation schemes for separation of a benzene–toluene–biphenyl mixture. **(a)** Single complex operation. **(b)** Sequence of two simple operations.

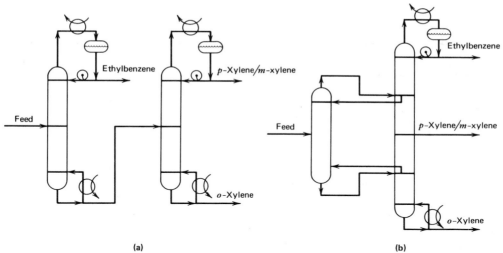

Figure 6. Distillation schemes for separation of ethylbenzene and xylene isomers into three products. **(a)** Simple sequence. **(b)** Interlinked sequence.

columns into C essentially pure products. No interlinking such as in Figure 6**b** is to be employed. Only sequences of the type of Figure 6**a** are to be considered under the assumption that the order of volatility of the C components does not change as the sequence proceeds. The equation for the number of different sequences N_s that can be developed is as follows (9).

$$N_s^1 = \frac{[2(C\text{-}1)]!}{C!(C\text{-}1)!} \tag{10}$$

Results for feeds containing up to 10 components are given in Table 2. As shown, the number of sequences grows rapidly as C increases.

Table 2.   Sequences for Separation by Simple Distillation

Number of components, C	Number of separators in sequence	Number of sequences, N
2	1	1
3	2	2
4	3	5
5	4	14
6	5	42
7	6	132
8	7	429
9	8	1430
10	9	4862

The five possible sequences for a four-component feed are shown in Figure 7. The first, where all but one product are distillates, is often referred to as the direct sequence and is widely used in industry. If all products but one are bottom products, the sequence is referred to as indirect.

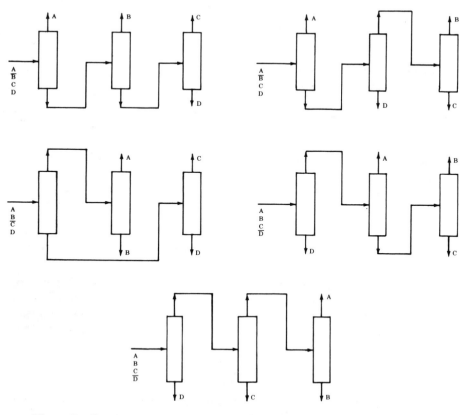

Figure 7.   Sequences for separation by simple distillation of four-component feed.

The combinatorial problem is increased greatly when more than one separation technique is considered, as indicated in the following equation (9):

$$N_s^T = T^{C-1} N_s^1 \tag{11}$$

where T is the number of different types of separators. When a separator-type utilizes an MSA, equation 11 may be applied if the MSA is recovered for recycle in the separator following the separator into which it is introduced, and if these two separators are counted together as one type. For example, if $C = 3$ and simple distillation, extractive distillation with solvent I, extractive distillation with solvent II, and liquid–liquid extraction with solvent III are to be considered, then the application of equations 10 and 11 gives 32 possible sequences.

Equations 10 and 11 can be applied to even more complex problems, such as those of Figure 1, where the feed contains six components but only four products are specified. If two separation methods, ie, simple distillation and extractive distillation with aqueous furfural, are considered, and only the four desired products are produced without blending, then $N_s^2 = 40$. If five products, including the 2-butene isomers, are produced with the two butene streams being blended at the end of the sequence, $N_s^2 = 224$. If six products are produced with the three butene streams being blended at the end of the sequence, then $N^2 = 1344$. Therefore, the total number of sequences possible is 1608.

The number of possible sequences can often be greatly reduced by excluding certain separations. For example, for the separation process shown in Figure 1, *cis*- and *trans*-2-butene remain adjacent to each other in volatility order for simple distillation and extractive distillation with aqueous furfural. Therefore, sequences that include their separation are not necessary. This reduces the total number of possible sequences to 264. Further reduction can be made by excluding other separations. For example, the relative volatility between *cis*-2-butene and *n*-pentane for simple distillation is approximately 2.5; accordingly, extractive distillation for this separation probably need not be considered, as indicated in Figure 3. In this manner, the total number of sequences for consideration can be reduced to 227 (10).

To synthesize the optimal or near-optimal sequences, all feasible sequences may be examined or synthesis techniques may be applied to find the best sequences with the least effort. These techniques include heuristic, evolutionary, and algorithmic methods.

Heuristic Methods. Efficient column sequences can be devised by heuristic methods. A list of useful heuristics is given below (11):

Heuristics for sequences of simple distillation columns
1. Single-component or multicomponent products are removed one by one as distillates (the direct sequence).
2. Splits are sequenced in the order of decreasing relative volatility, and the most difficult separation is saved for the last operation.
3. Splits are sequenced to remove components in the order of increasing molar percentage in the process feed.
4. Splits are favored that give equimolar amounts of distillate and bottoms.
5. Splits are sequenced to leave last those separations that give high purity products.
6. Thermally unstable, corrosive, or otherwise chemically reactive components are removed early in the sequence.

All but the sixth heuristic are reasonably consistent with the overall goal of minimizing the total boil-up and, thus, reboiler energy requirement, in the columns forming the sequence. However, when applying the heuristics, they are often in conflict with each other, and the optimum sequence is not obvious. Alternative sequences should then be developed.

The more general heuristics given below can be applied to sequences involving simple distillation as well as other separation operations.

Heuristics for general sequences of separation operations

1. Simple distillation is favored provided that
   a. relative volatility between the two key components is >1.05.
   b. reboiler duty is not excessive.
   c. tower pressure does not cause the mixture to approach the critical temperature.
   d. overhead can be at least partially condensed to provide reflux without excessive refrigeration requirements.
   e. bottoms temperature is not so high that decomposition occurs.
   f. azeotropes do not prevent the desired separation.
   g. tray-pressure drop is tolerable, particularly if operation is under vacuum.
2. An MSA should be removed in the separator following the one into which it is introduced.
3. When multicomponent products are specified, sequences are favored that produce these products directly or with a minimum of blending unless relative volatilities are appreciably lower than for a sequence that requires additional separators and blending.
4. In developing a sequence starting from the process feed, the cheapest alternative should be chosen for the next separator.
5. An MSA should not be recovered with the help of another MSA.
6. Ambient processing conditions are favored over extreme conditions.

The application of these heuristics can lead directly to the development of sequences (see Fig. 2) to solve the kind of problems illustrated in Figure 1.

In some cases, complex distillation configurations should be considered. Some guidance is available from a study of ternary mixtures in which eight alternative sequences of one to three columns were considered, seven of which are shown in Figure 8 (12). The configurations include two interlinked cases (III and IV), five cases that include the use of sidestreams (III, IV, V, VI, and VII), and one case (V) involving a column with two feeds. As shown in Figure 9, optimal regions for the various configurations depend on process-feed composition and on an ease-of-separation index (ESI), which is defined as the relative volatility ratio α_{AB}/α_{BC}. The results of Figure 9 can be extended to multicomponent-separation problems involving more than three components, if difficult ternary separations are performed last.

Evolutionary Methods. To conduct an evolutionary synthesis (13), an initial flow sheet must be available or developed; rules must be made for systematic changes to the flow sheet to create neighboring flow sheets; an effective strategy must be employed to apply these rules; and a means must be selected to compare the original flow sheet to any of its neighbors. The following procedure is suggested (14):

Step 1. Six ordered heuristics generate an initial flow sheet. Only if a heuristic does not apply should an attempt be made to apply the heuristic following it.

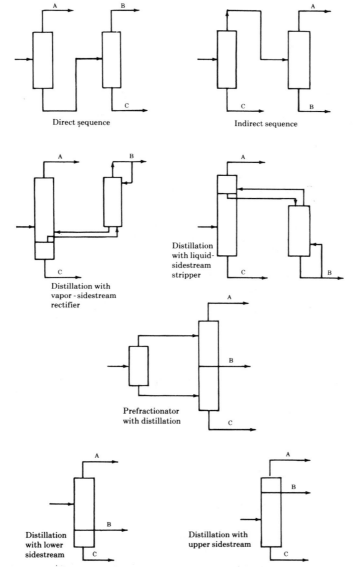

Figure 8. Configurations for ternary distillation.

Heuristic 1: A preliminary screening of separation methods is conducted to determine feasible methods and separating agents. Uneconomical splits are eliminated.

Heuristic 2: For each feasible separation method, an ordered list of components is developed according to the separation index, eg, K-value. From the separation indexes, relative indexes, eg, relative volatility, are computed between adjacent pairs of components in the ordered list. When these adjacent relative volatilities vary widely in the process feed, the splits are sequenced in the order of decreasing adjacent relative volatility.

Heuristic 3: The splits are sequenced to remove components in the order of de-

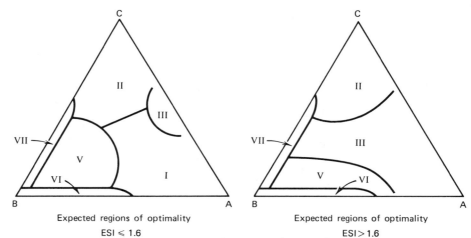

Figure 9.   Regions of optimality for ternary distillation configurations (12).

creasing molar percentage in the process feed when that percentage varies widely, but relative volatility does not vary widely.

Heuristic 4: When neither Heuristic 2 nor 3 appears applicable, the direct sequence can be used to remove the components one by one as overhead products. When applied to simple distillation, pressure tends to be highest in the first separator and is reduced in each subsequent separator.

Heuristic 5: A mass-separating agent is removed in the next separator.

Heuristic 6: When one or more multicomponent products are specified, sequences are favored that give desired products directly or with a minimum of blending, unless adjacent relative volatilities are appreciably lower than for a sequence that needs additional separators and blending.

Step 2. From the initial flow sheet, each possible neighbor is generated and all neighbors are retained that would have been generated had the ranking of Heuristics 2, 3, 4, and 6 been relaxed (owing to the difficulty in determining their applicability for a particular problem). Each neighbor retained is listed in an order that reflects the heuristic responsible for its retention.

Step 3. Each neighboring flow sheet is evaluated economically and in sequence in order to find a better flow sheet, in which case Step 2 is repeated. Otherwise, Step 4 follows.

Step 4. Before accepting the apparent best flow sheet, all remaining neighbors are accepted that were not retained in Step 2; the best flow sheet is kept.

In this strategy (14), neighboring flow sheets are generated by interchanging splits. Thus, in Figure 10, sequence (**b**) is obtained from sequence (**a**) by interchanging splits C/D and E/F. Sequence (**c**) is obtained from sequence (**b**) by interchanging A/B and C/D.

Compared to the heuristic method, the evolutionary method requires that each flow sheet given serious consideration be designed and economically evaluated with respect to investment and operating cost. However, the number of such flow sheets evaluated is generally small. For example, for the problem in Figure 1, only three of 227 possible sequences have to be evaluated. The study in ref. 13 ignored the fourth heuristic in the list of heuristics for sequences of simple distillation columns given

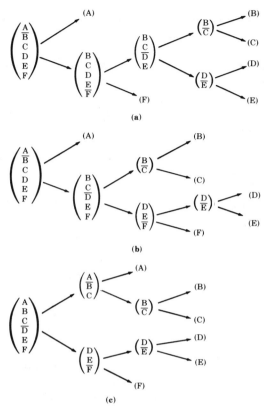

Figure 10. Evolutionary development of optimal distillation sequence (14). (**a**) Initial sequence; cost = $1,234,000/yr. (**b**) Result of first interchange; cost = $1,213,000/yr. (**c**) Final flow sheet; cost = $1,153,000/yr.

above. That heuristic may be more important than the first, and possibly the second, heuristic in that list (15).

Algorithmic Methods. Although good sequences are usually developed with either the heuristic or the evolutionary method, neither necessarily leads to the optimal result, which can be obtained with an algorithmic method. The first such method developed for the synthesis of separation sequences utilized list processing in combination with dynamic programming (10). The latter applies the principle of optimality (16). Simply stated, an optimal policy requires that optimal decisions be made whatever the preliminary situation or decisions may have been. This method is best applied by excluding the obviously unfeasible or uneconomical splits indicated in Heuristic 1 of Step 1 of the evolutionary method discussed above. In addition, Heuristic 2 of Step 1 is applied, but only to develop ordered component lists and feasible split points. Dynamic programming is then applied by use of a cost encompassing both capital investment and operating costs that is determined from a design for each separator. The cost of each separation is presumed to be independent both of the nature of the upstream separations and the accompanying splits of less-important components and of the pressure or phase condition of the feed to that separation. An example of this method is if components A, B, C, and D are to be separated by simple distillation, where the order of components is that of decreasing volatility. No splits are excluded, and lists of all possible splits and annual costs of corresponding separators are as given in Table 3. These costs are for separators operating at the optimal pressure with the optimal reflux rate and the optimal amount of feed preheat. All splits are relatively sharp and, therefore, material balances for each separation are essentially independent of location of the separator in any particular sequence. Dynamic programming proceeds backward from the two-component separator feeds, (AB), (BC), and (CD), each of which, as seen in Table 3, is unique. For the three-component feed (BCD), two partial sequences are possible (see Table 4) starting with B/CD or BC/D. The first partial sequence is the least expensive and is marked with an asterisk. For (ABC), the second partial sequence is best and is marked with an asterisk. For the four-component feed, ie, the process feed, three partial sequences equivalent to the full sequence are shown where, for example, *(B/CD) represents the optimal partial sequence already determined earlier in Table 4 for that three-component feed. The optimal sequence at $689,000/yr is

$$
\begin{pmatrix} A \\ \underline{B} \\ C \\ D \end{pmatrix} <
\begin{aligned}
\left(\frac{A}{B}\right) &< \begin{aligned} (A) \\ (B) \end{aligned} \\
\left(\frac{C}{D}\right) &< \begin{aligned} (C) \\ (D) \end{aligned}
\end{aligned}
$$

For this example from Table 2, five sequences are possible, but only three are developed in Table 4 to find the optimal one.

An alternative algorithmic strategy sometimes reduces the number of separators that must be designed and evaluated economically. It is based on list processing and an ordered-branch search (17). Unlike dynamic programming, this algorithm starts at the process-feed end and uses forward branching and backtracking in conjunction with the cheapest-separator-next heuristic (18) to find the optimal sequence. The procedure is best illustrated by reference to the complete search space shown in Figure 11 for the separation problem of Table 3. An initial sequence is developed that makes use of the cheapest-separator-next heuristic through the design and economic evaluation of all possible next separators leading from a square. Starting from the process feed (ABCD), the cheapest separator is A/BCD, based on a cost of $85,000/yr (see Table 3). This leads to product A and intermediate feed BCD, for which the cheapest separator is B/CD, at a cost of $247,000/yr. This leads to product B and intermediate feed CD at a total cost thus far of $85,000 + $247,000 or $332,000/yr. For CD, only one separation step is available, at a cost of $420,000/yr. Products C and D are obtained, which completes the sequence at a total cost of $752,000/yr. This sequence is the initial

Table 3. Costs for Dynamic Programming Example Given in Figure 11

Component group	Separation	Separator no.	Cost, $/yr
ABCD	A/BCD	1	85,000
	AB/CD	2	254,000
	ABC/D	3	510,000
ABC	A/BC	8	59,000
	AB/C	9	197,000
BCD	B/CD	4	247,000
	BC/D	5	500,000
AB	A/B	6	15,000
BC	B/C	10	190,000
CD	C/D	7	420,000

Table 4. Partial Sequences Developed by Dynamic Programming

Feed group	Partial sequence	Cost of partial sequence, $/yr
BCD		667,000
		690,000
ABC		249,000
		212,000
ABCD		752,000
		689,000
		722,000

upper limit. The optimal sequence is either this sequence or lower-cost sequences, which is next determined by backtracking and branching to develop additional sequences as summarized in Table 5. If the total cost of a subsequent sequence is lower, it becomes the new upper limit. However, each additional sequence is developed only to the point where the upper limit is exceeded. In Table 5, the second sequence is obtained by backtracking from separator 7 to box CD. Since only one alternative exists for this box, another backtracking step leads to separator 4 and hence to box BCD,

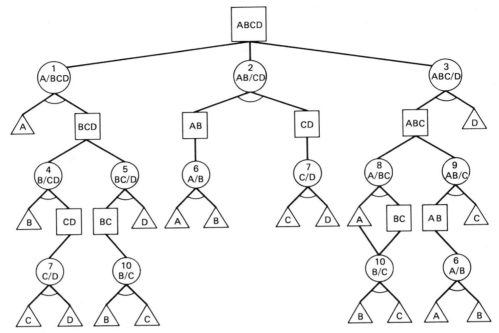

Figure 11. Search space for ordered-branch search method. O, separator; □, process feed or intermediate product; △, final product. The numbers in the separators are described in Table 3.

Table 5. Sequences Developed by Ordered-Branch Search Method

Sequence[a] by separator numbers of Table 3	Total cost, $/yr
1-4-7	752,000[b]
1-5-10	775,000[c]
2-6-7	689,000
3-8-10	759,000
(3-9)	(707,000)

[a] Or partial sequence.
[b] Initial upper limit, cheapest-separator-next heuristic.
[c] New upper limit.

which has a second alternative, separator 5. The sum of the costs of partial sequence 1–5 is $585,000/yr, which is still less than the upper limit of $752,000/yr. Therefore, a forward-branching step is taken from separator 5 to box BC that leads only to separator 10, which completes a second sequence 1-5-10 at a higher cost of $775,000. The remaining searches are done in a similar manner. From Table 5, the optimal sequence is 2-6-7, which is the same result as obtained by dynamic programming. In Table 5, the last sequence is only developed partially, because with just separators 3 and 9 (and not 6), the cost already exceeds the upper limit. In this example, the ordered-branch search method has little apparent advantage over dynamic programming. However, the former offers significant advantages for more complex cases involving more products or more separation methods.

Energy Integration

Before the energy crisis in the mid 1970s, energy (heat and work) integration in separation sequences was not common except at cryogenic or high temperature conditions. Rarely were condensers and reboilers coupled because of control problems. Today, consideration of energy integration is common at all levels, with and without phase change, and control problems are solved by digital process control.

When the feed to a distillation sequence is a subcooled liquid or the distillate temperature is significantly lower than the bottoms temperature, heat integration can be employed to preheat the feed with the products and reduce the reboiler duty of the first column. An application of this technique is shown in Figure 12 (4).

An even more complex heat-integration scheme (see Fig. 13) includes matching of condensers and reboilers of different columns. Column pressures are adjusted such that the condenser temperature of the third column is higher than the reboiler temperatures of the first two columns. Since the condensing duty for the overhead vapor of the third column is greater than the combined reboiler duties for the first two columns, an auxiliary condenser is required for the third column.

For sequences involving simple distillation steps, condenser and reboiler matching can greatly increase the number of possible equipment arrangements. For example, fourteen different nonheat-integrated sequences for simple distillation columns are possible (see Table 2). When energy integration of condensers and reboilers is allowed, the number of possible arrangements is more than an order of magnitude greater, with the exact number depending upon the particular problem and on whether a condenser or reboiler is allowed to match against more than one reboiler or condenser. Algorithms for synthesizing optimal heat-integrated sequences are given in refs. 19–21.

When condenser temperatures are lower than reboiler temperatures, heat can be pumped from the condenser to the reboiler. Techniques for this operation are shown in Figure 14 (22). In Figure 14**a**, the heat is transferred indirectly by an external re-

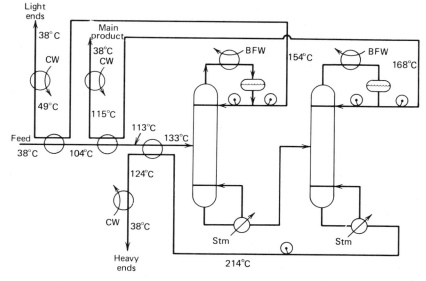

Figure 12.   A direct sequence with heat integration (4). CW = cold water; BFW = boiler feedwater; Stm = steam.

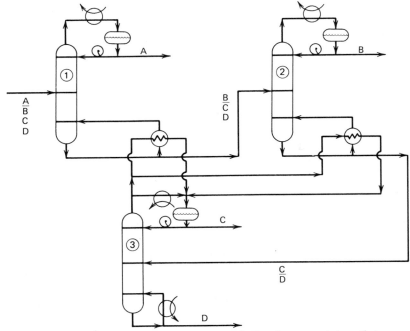

Figure 13. Matching of condenser and reboilers for energy integration.

frigerant. In **14b**, the overhead vapor is compressed (vapor recompression) to achieve a higher temperature in order to transfer the heat directly. In **14c**, the bottoms are flashed (reboiler flashing) to lower the temperature in order to transfer heat directly. All three schemes require an expansion valve and a compressor permitting heat rejected in the condenser to be used in the reboiler. Although not shown, an auxiliary condenser or reboiler is usually necessary because the two heat duties are generally not the same. The heat-pump scheme can be extended to sequences of distillation columns, thereby permitting operation of the columns at any desirable pressures. More advanced heat-pump concepts are discussed in refs. 23–25.

Proposals for energy conservation in distillation are summarized as follows (26):

1. The temperature driving force in the reboiler is reduced to permit use of lower pressure steam.
2. The temperature driving force in the condenser is reduced to permit heat recovery at a higher temperature.
3. Intermediate condensers and reboilers are utilized to increase the condensation temperature and decrease the reboiling temperature. However, reflux and boiling requirements may be higher.
4. The number of trays is increased or trays are replaced by high efficiency packing (retrofit).
5. When the reflux requirement is high or the temperature difference between distillate and bottoms is small, a heat pump is used. Alternatively, the operation is converted to a double-effect system operating at two different pressures, wherein overhead vapor from the first effect at higher pressure is condensed in the reboiler of the second effect at lower pressure.

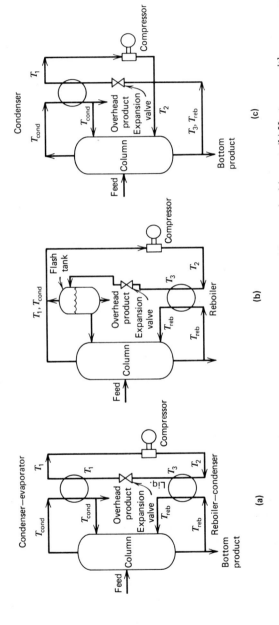

Figure 14. Use of heat pumps with single distillation operation (22). (**a**) Heat pump with external refrigerant. (**b**) Heat pump with vapor recompression. (**c**) Heat pump with reboiler liquid flashing.

6. With a high distillate-to-bottoms ratio, eg, 4:1, partially vaporized feed reduces the heat requirement of the reboiler.
7. Minimizing tray pressure drop reduces bottoms temperature and permits use of lower pressure steam in the reboiler.
8. The location of the feed-entry tray should be optimized.
9. Maximum insulation should be provided.
10. Sensible heat is recovered from distillate and bottoms products.

Nomenclature

C	= composition factor (mole fraction, mass fraction, composition, etc)
ESA	= energy-separating agent
ESI	= ease of separation index α_{AB}/α_{BC}
f	= fugacity
K-value	= ratio of vapor-to-liquid mole fractions for a given component
MSA	= mass-separating agent
N_s	= number of sequences
P°	= vapor pressure
SF	= separation factor
T	= number of different types of separators
x	= mole fraction
ϕ	= mixture fugacity coefficient
υ	= pure component fugacity coefficient
γ	= activity coefficient
α	= relative volatility
β	= relative selectivity

BIBLIOGRAPHY

1. C. K. Buell and R. G. Boatright, *Ind. Eng. Chem.* **39,** 695 (1947).
2. *Chem. Eng.*, 35 (Oct. 19, 1981).
3. C. J. King, *Separation Processes*, 2nd ed., McGraw-Hill Book Co., New York, 1980.
4. E. J. Henley and J. D. Seader, *Equilibrium-Stage Separation Operations in Chemical Engineering*, John Wiley & Sons, Inc., New York, 1981.
5. R. C. Reid, J. M. Prausnitz, and T. K. Sherwood, *The Properties of Gases and Liquids*, 3rd ed., McGraw-Hill Book Co., New York, 1977.
6. M. Souders, *Chem. Eng. Prog.* **60**(2), 75 (1964).
7. R. A. Findlay and J. A. Weedman in K. A. Kobe and J. J. McKetta, eds., *Advances in Petroleum Chemistry and Refining*, Vol. 1, Interscience Publishers, New York, 1958.
8. E. A. Maciula, *Hydrocarbon Process.* **59**(5), 115 (1980).
9. R. W. Thompson and C. J. King, *Synthesis of Separation Schemes*, Technical Report No. LBL-614, Lawrence Berkeley Laboratory, Berkeley, Calif., 1972.
10. J. E. Hendry and R. R. Hughes, *Chem. Eng. Prog.* **68**(6), 71 (1972).
11. N. Nishida, G. Stephanopoulos, and A. W. Westerberg, *A Review of Process Synthesis*, Report DRC-06-17-80, Design Research Center, Carnegie-Mellon University, Pittsburgh, Pa., 1980.
12. D. W. Tedder and D. F. Rudd, *AIChE J.* **24,** 303 (1978).
13. G. Stephanopoulos and A. W. Westerberg, *Chem. Eng. Sci.* **31,** 195 (1976).
14. J. D. Seader and A. W. Westerberg, *AIChE J.* **23,** 951 (1977).
15. A. Gomez, *Synthesis of Separation Processes*, Ph.D. thesis in chemical engineering, University of Utah, Salt Lake City, June 1981.
16. R. Bellman, *Dynamic Programming*, Princeton University Press, Princeton, N.J., 1957.
17. F. R. Rodrigo and J. D. Seader, *AIChE J.* **21,** 885 (1975).
18. R. W. Thompson and C. J. King, *AIChE J.* **18,** 941 (1972).
19. R. N. S. Rathore, K. A. Van Wormer, and G. J. Powers, *AIChE J.* **20,** 491 (1974).
20. D. C. Faith III and M. Morari, *Comput. Chem. Eng.* **3,** 269 (1979).

21. N. Nishida, G. Stephanopoulos, and A. W. Westerberg, *AIChE J.* **17**, 321 (1981).
22. H. R. Null, *Chem. Eng. Prog.* **72**(7), 58 (1976).
23. D. C. Freshwater, *Brit. Chem. Eng.* **6**, 388 (1961).
24. J. R. Flower and R. Jackson, *Trans. Inst. Chem. Eng.* **42**, T249 (1964).
25. R. S. H. Mah, J. J. Nicholas, and R. B. Wodnik, *AIChE J.* **23**, 651 (1977).
26. R. M. Stephenson and T. F. Anderson, *Chem. Eng. Prog.* **76**(8), 68 (1980).

General References

N. Doukas and W. L. Luyben, "Economics of Alternative Distillation Configurations for Separation of Ternary Mixtures," *Ind. Eng. Chem. Process Des. Dev.* **17**, 272 (1978).

D. C. Freshwater and B. D. Henry, "Optimal Configuration of Multicomponent Distillation Systems," *Chem. Eng. London* **301**, 533 (1975).

D. C. Freshwater and E. Ziogou, "Reduction Energy Requirements in Unit Operations," *Chem. Eng. J.* **11**, 215 (1976).

A. Gomez and J. D. Seader, "Separation Sequence Synthesis by a Predictor Ordered Search," *AIChE J.* **22**, 970 (1976).

V. D. Harbert, "Which Tower Goes Where?" *Pet. Refiner* **36**(3), 169 (1957).

"Distillation Processes and Apparatus," U.S. Pat. 4,025,398 (May 24, 1977), G. G. Haselden.

D. L. Heaven, "Optimum Sequencing of Distillation Columns in Multicomponent Fractionation," M.S. thesis in chemical engineering, University of California, Berkeley, 1969.

J. E. Hendry, D. F. Rudd, and J. D. Seader, "Synthesis in the Design of Chemical Processes," *AIChE J.* **19**, 1 (1973).

B. D. Henry, "Economies Possible in the Separation of Multicomponent Mixtures by Distillation," *Inst. Chem. Eng. Symp. Ser.* **54**, 75 (1978).

V. Hlavacek, "Synthesis in the Design of Chemical Processes," *Comput. Chem. Eng.* **2**, 67 (1978).

R. Krishna, "A Thermodynamic Approach to the Choice of Alternatives to Distillation," *Inst. Chem. Eng. Symp. Ser.* **54**, 185 (1978).

F. J. Lockhart, "Multicolumn Distillation of Natural Gasoline," *Pet. Refiner* **26**(8), 104 (1947).

V. P. Maikov, G. G. Vikkov, and A. V. Gallstov, "Optimum Design of Multicolumn Fractionating Plants from the Thermoeconomic Standpoint," *Int. Chem. Eng.* **12**, 426 (1972).

R. Nath and R. L. Motard, "Evolutionary Synthesis of Separation Processes," *AIChE J.* **27**, 578 (1981).

H. Nishimura and Y. Hiraizumi, "Optimal Systems Pattern for Multicomponent Distillation Systems," *Int. Chem. Eng.* **11**, 188 (1971).

R. B. Petlyuk, V. M. Platonov, and D. M. Slavinskii, "Thermodynamically Optimal Method of Separating Multicomponent Mixtures," *Int. Chem. Eng.* **5**, 555 (1965).

W. C. Petterson and T. A. Wells, "Energy-Saving Schemes in Distillation," *Chem. Eng.* **84**(20), 78 (1977).

G. J. Powers, "Heuristic Synthesis in Process Development," *Chem. Eng. Prog.* **68**(8), 88 (1972).

R. N. S. Rathore and G. J. Powers, "A Forward Branching Scheme for the Synthesis of Energy Recovery Systems," *Ind. Eng. Chem. Process Des. Dev.* **14**, 175 (1975).

V. Rod and J. Marek, "Separation Sequences in Multicomponent Rectification," *Collect. Czech. Chem. Comm.* **24**, 3240 (1959).

J. J. Siirola, "Progress Toward the Synthesis of Heat-Integrated Distillation Schemes," *paper presented at 85th National Meeting of AIChE, Philadelphia, Pa.*, 1978.

D. F. Rudd, G. J. Powers, and J. J. Siirola, *Process Synthesis*, Prentice-Hall, Inc., Englewood Cliffs, N.J., 1973.

G. Stephanopoulos and A. W. Westerberg, "Synthesis of Optimal Process Flow Sheets by an Infeasible Decomposition Technique in the Presence of Functional Non-Convexities," *Can. J. Chem. Eng.* **53**, 551 (1975).

W. J. Stupin and F. J. Lockhart, "Thermally Coupled Distillation: A Case Study," *Chem. Eng. Prog.* **68**(10), 71 (1972).

B. D. Tyreus and W. L. Luyben, "Two Towers Cheaper Than One?," *Hydrocarbon Process.* **54**(7), 93 (1975).

T. Umeda, T. Harada, and K. Shiroko, "A Thermodynamic Approach to the Synthesis of Heat Integration Systems in Chemical Processes," *Comput. Chem. Eng.* **3**, 273 (1979).

T. Umeda, K. Niida, and K. Shiroko, "A Thermodynamic Apporach to Heat Integration in Distillation Systems," *AIChE J.* **25**, 423 (1979).

A. W. Westerberg and G. Stephanopoulos, "Studies in Process Synthesis—I," *Chem. Eng. Sci.* **30**, 963 (1975).

J. D. SEADER
University of Utah

SHALE OIL.   See Oil shale.

SHAPE-MEMORY ALLOYS

The shape-memory effect is based on the continuous appearance and disappearance of martensite with falling and rising temperatures. This thermoelastic behavior is the result of transformation from a phase stable at elevated temperature to the martensite phase. A specimen in the martensite condition may be deformed in what appears to be a plastic manner but is actually deforming as a result of the growth and shrinkage of self-accommodating martensite plates. When the specimen is heated to the temperature of the parent phase, a complete recovery of the deformation takes place. Complete recovery in this process is limited by the fact that strain must not exceed a critical value which ranges from 3–4% for copper memory-effect alloys to 6–8% for the Ni–Ti system. A number of other characteristics associated with shape memory are referred to as pseudoelasticity or superelasticity, two-way shape-memory effect, martensite-to-martensite transformations, and rubberlike behavior.

Martensite is a metastable phase that forms when a phase stable at elevated temperature, such as austenite in steel, is cooled at a certain rate, thereby suppressing the formation of phases that are diffusion-controlled (see Steel). A characteristic of martensite is the relationship between the parent-phase crystal and the martensite. For steel, the austenitic phase is fcc. When it transforms to martensite, the orderly shift to a bct (body-centered tetragonal) structure takes place (see Fig. 1). The structure stable at elevated temperature varies with the alloy system, and the martensite varies from a simple bcc to a more complex structure which may have as many as 18 atom layers to define the unit cell.

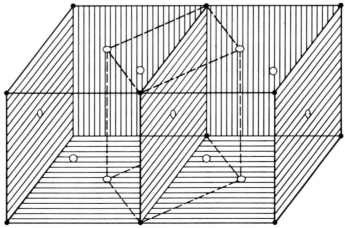

Figure 1.   The original fcc structure (0) changes to the bct with (111) $\gamma \parallel$ (110) α and the [110] $\gamma \parallel$ [111] α.

The temperature at which martensite starts to form on cooling is referred to as the M_s and the temperature at which the elevated temperature phase has been completely transformed is the M_f. On heating a martensitic specimen, the temperature at which the reaction reverses to the elevated temperature phase is designated A_s; the reaction is completed at a higher temperature designated A_f (see Fig. 2).

Thermoelastic behavior was first discussed in 1938 in a study of a Cu–Zn alloy which showed that martensite could be made to appear and disappear with a change in temperature (1). In a later study in the USSR, the phase relationships in brass between the high temperature β phase and martensite were examined (2). The length changes that occur on martensite transformations under load in the Fe–Ni system were determined (3). A later investigation of the shape-memory effect in Au–Cd alloys demonstrated that useful force could be generated in this type of transformation. These findings led to research on the practical applications for the shape-memory effect, and investigation of the martensite transformation kinetics in the Ni–Ti system (4). This system is commonly referred to as Nitinol (Nickel–Titanium Naval Ordnance Laboratory). The essential features of the martensite transformation are common to all alloy systems exhibiting the memory effect, whether they be 2H, 3R, 9R, or 18R structures (5–6). In some alloy systems, the various martensites are both internally faulted or internally twinned and may possess different crystal structures. However, in all cases studied to date, an initial parent phase transforms to self-accommodating martensite plates that are characterized by six plate groups, each consisting of four variants. Because of the self-accommodating character of the transformation, the average shape deformation in a particular plate group is effectively zero.

Table 1 lists the martensite alloy systems that have been investigated with respect to thermoelasticity, pseudoelasticity, shape-memory effect, and two-way shape-memory effect. The bcc (β) phase is the dominant parent because of the comparatively large thermodynamic difference between the martensite transformation from a bcc parent to 3R, 2H, 18R, or 9R, and from an fcc parent to bcc or bct. The former is typical for nonferrous alloys, whereas the latter accounts for the hardening of steel.

For an excellent general review of martensite kinetics, microstructure, and thermodynamics as they relate to shape memory, see ref. 7 and the First International Conference on Shape Memory (8).

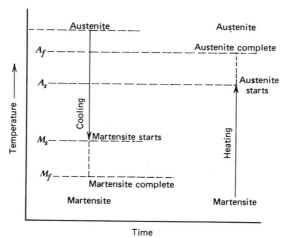

Time

Figure 2. Martensite phases as a function of temperature. A_f = reverse reaction is finished; A_s = reverse reaction starts; M_s = martensite starts to form; and M_f = martensite is completely transformed.

Table 1.   Alloys Exhibiting Martensitic Effects

Thermoelastic	Pseudoelastic	Shape memory	Two-way shape memory
Ag–Cd	Ag–Cd	Ag–Cd	Cu–Al
Au–Cu–Zn	Au–Cd	Au–Cd	Cu–Zn–Al
Cu–Al–Ni	Au–Cu–Zn	Au–Cu–Zn	In–Tl
Cu–Zn	Cu–Al–Mn	Cu–Al	Ti–Ni
Cu–Zn[a]	Cu–Al–Ni	Cu–Al–Ni	
Fe–Pt	Cu–Au–Zn	Cu–Zn	
Ti–Ni	Cu–Zn	Cu–Zn–Al	
	Cu–Zn–Al	Cu–Zn–Ga	
	Cu–Zn–Sn	Cu–Zn–Si	
	Cu–Zn–X	Cu–Zn–Sn	
	Fe_3Be	Fe–Pt	
	Fe_3Pt	Fe–Ni	
	In–Tl	In–Cd	
	Ni–Ti	In–Tl	
	Ti–Ni	Ni–Al	
		Ni–Ti	
		304 stainless steel	
		Ti–Nb	
		Ti–Ni	

[a] With ternary additions of Ni, Ag, Au, Cd, In, Ga, Si, Ge, Sn, and Sb.

The behavior of shape-memory-effect (SME) alloys is exactly opposite to that of normal metals in the following essential feature: as the temperature rises above A_s and martensite is increasingly converted to the β phase, the modulus of elasticity increases. This change is spectacular for Cu–Al–Zn SME alloys where the Young's modulus changes by a factor of 50 from 0.4 GPa (58,000 psi) at the A_s to 20.7 GPa (3 $\times 10^6$ psi) near the A_f. Obviously, a spring tension or torsion device develops a greater force with increasing temperature and this force can be either static or it can be used to produce a controlled motion. The force that can be developed by an SME device is as much as 200 times the force that could be developed by a bimetallic element of the same size or volume.

The Crystallographic Nature of Shape Memory

The martensitic memory or marmem effect occurs in alloys where both the parent and the martensite are ordered and exhibit crystallographically reversible, thermoelastic martensite transformations (7,9–10).

It has been demonstrated in one of the simplest systems, Cu–Zn, that a single orientation of the bcc β phase transforms on cooling below the M_f to self-accommodating variants of martensite. The habit planes for the transformation are symmetrically disposed around the (110) family of planes, of which there are six in the cubic system. Habit planes are the planes in the parent phase from which transformation takes place. Thus, for the plane (011) the four variants are (2 11 12), (2 12 11), ($\bar{2}$ 12 11), and ($\bar{2}$ 11 12); this set is termed a plate group. The six (110) planes offer a total of 24 martensite variants. When the martensite group is deformed, deformation proceeds by a gradual conversion of four variants to a single martensite plate rather than by grain-boundary sliding or slip. The surviving orientation depends on whether the strain is tensile or compressive and whether the growth or shrinkage is such as to minimize strain-energy accumulation. Upon heating, the reverse transformation from martensite to the β phase occurs between A_s and A_f and the single crystal (plate) of martensite transforms to a single β crystal with the original orientation of the parent phase. In the case of Cu–Zn and other systems with relatively complex ordered martensite structures (9R or 18R), the reverse transformation is crystallographically restricted. Thus, although there are many variants that can develop on transformation from parent to martensite, only a single-parent orientation is possible in the reverse, or shape-recovery, transformation.

When a martensite group is deformed to coalesce into a single orientation, the dominant mechanism is twinning. Each twin is actually an alternative variant of the martensite crystal. Thus, for the four variants that cluster about the (110) habit, each orientation is a twin of another, and by this degenerate variant-twin relationship, a group of martensite plates formed from a single parent crystal can, on deformation, coalesce to a single crystal (single variant) of martensite.

The parent phase is usually ordered B_2 or DO_3 symmetry, although initially a transformation can take place from disordered to an ordered or superlattice structure. This ordered structure transforms to one of the four martensite crystal forms 2H, 3R, 9R, or 18R that exhibit shape memory. These designations refer to the sequence of stacking of planes to form the ordered structure. A sequence could be *abab*, *abcabc* or *abcbcacab* or similar variations. The repetition of these planes to define a unit cell is then 2, 3, and 9, respectively.

Typical alloys that transform to these various martensites are

> 2H, Cu–Al–Ni and Ag–Cd
> 3R, Ni–Al
> 9R, Cu–Zn
> 18R, Cu–Zn–Al.

A distinction exists in the habit and deformation characteristics of the 2H and 3R types and the 9R and 18R martensites. The former are internally twinned and deformation occurs by a detwinning of a variant plate. The latter are internally faulted and deformation proceeds by variant-to-variant coalescence followed by group-to-group coalescence. Although these structural differences exist, the self-accommodating habit-plane grouping with respect to an (011) plane is common to all systems exhibiting the marmem effect. The 9R martensite is derived from a B_2 parent, whereas the 18R transforms from a DO_3 superlattice. The difference in the stacking of (110) planes is because of the requirement for an invariant plane strain that involves a restricted stacking of close-packed planes. In order to obtain the required invariant plane-strain condition, both the 9R and 18R martensite contain stacking faults to provide the necessary accommodation. The sequence of β to β_2 to orthorhombic 9R for a Cu–Zn alloy is shown in Figure 1.

As noted previously, the atoms in each plane are displaced relative to those above and below and from a repetitive sequence of a nine-plane group. These complex atomic displacements take place by a combined process of shuffling and shear to arrive at the 9R structure. The 3R martensite twin plane is identical to the 9R and 18R fault plane. In the case of 2H martensites, no such twin-fault correspondence exists, and the twin is derived from a different parent (110) plane.

When sufficient strain takes place during deformation of a martensitic structure, six plate groups coalesce to a single orientation where the surviving variant from the original 24 allows the greatest extension in strain direction. This process varies with the nature of the strain, tensile, compression, or bending properties (Fig. 2). Once a group becomes a single variant, it can change its orientation to that of an adjacent group by twinning.

Although some of the crystallographic details differ for different memory-alloy systems, the essential processes described above are common. Some systems such as Ni–Ti (Nitinol) and Cu–Zn–Al have premartensite transformations with ordering, but in the final analysis the sequence of β to martensite cluster to single-variant martensite to β parent is the same. Schematically, the memory effect is summarized in Figure 3. A stress–strain, temperature-strain diagram is shown in Figure 4. Assuming an initial martensite structure, as stress increases a point A is reached where the martensite starts to coalesce to form a single variant. When the stress is released at B, a residual strain ϵ is left. As the temperature is raised, transformation of the martensite starts at A_s and continues with increasing temperature to A_f at which point all the strain is recovered.

Simple shape-memory behavior can be extended to two-way memory. In order to exhibit two-way memory where the part spontaneously changes from one shape to another on cooling or heating, a conditioning of the martensite must take place. This

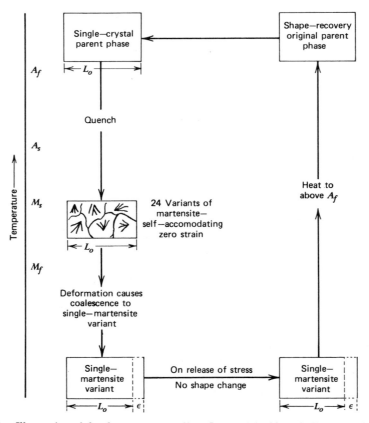

Figure 3.   Illustration of the shape-memory effect. L_O = original length. For other definitions, see Figure 2.

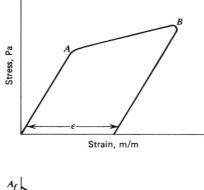

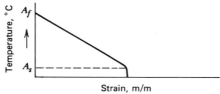

Figure 4. Stress–strain curve demonstrating the shape-memory effect. ϵ = residual strain.

conditioning is brought about by limiting the number of martensite variants that form upon cooling through the applications of an external stress during transformation. The stress favors the initial formation of selected variants, similar to the fact that stressing a martensite group causes the selective growth and shrinkage of suitably oriented plates. This limit imposed upon the number of variants formed reduces the self-accommodating feature of the usual transformation and increases the residual stress. By repeating the process a number of times, the restricted variant group and its associated internal stress spontaneously revert to β on heating and then to a singular martensite group on cooling.

Owing to the fact that the internal boundaries of martensite–martensite groups are relatively mobile, a cyclic stress causes a back-and-forth motion of these boundaries which produces a high internal friction. This behavior is the basis for using shape-memory alloys as vibration-damping materials.

The other property peculiar to marmem alloys is the ability under certain conditions to exhibit superelastic behavior. Although in one sense, the 3–8% apparently recoverable strain of the memory effect is truly an extended or pseudoelastic behavior, an even further elastic range is possible. When many of the martensitic alloys are deformed well beyond the point of the initial single-coalesced martensite stage, a stress-induced martensite–martensite transformation can occur. In this mode of deformation strain is reversible through stress release and not by a temperature-induced phase change, and recoverable strains as high as 17% have been observed.

In the shape-memory process, the temperature is controlled at which the effect takes place. The alloy composition determines the working temperature range of a device, whether one-way or two-way. In the case of Nitinol, the composition is usually ca 50:50. The M_s is extremely sensitive to variations in the Ni–Ti ratio. This sensitivity is reduced by the addition of copper to the Ni–Ti binary system. The other alloy system that has been exploited for a very broad range of applications is the Cu–Zn–Al ternary. The M_s of this alloy can be varied from 170° to -105°C, although a range of -100° to 60°C is more acceptable for long-term stability. This range of M_s temperature falls within the ternary field of Cu–(13–27%) Zn–(4.5–9%)–Al. Additions of Sn, Mn, Si, and Ni to this ternary system have been investigated for special devices.

Economic Aspects

For a period, the United States Naval Surface Weapons Center was the sole supplier of Nitinol, usually as small-diameter wire; the supply was limited to a few compositions.

In 1981, Nitinol Devices (11) offered 0.457-mm wire with a zero-force transition temperature of 38°C for $1.95/m or $1760/kg.

The Raychem Corp. (12) produces Nitinol, but sells it only in the form of finished devices. This company also produces SME brass under the trade name of Betalloy, sold only as a finished actuator.

The price of SME brass, typically the Cu–Zn–Al family of alloys, is not published since it is sold only as a finished and heat-treated device. A typical small spring sells in the range of $0.50–1.50.

At this time, only Delta Memorial Metal Ltd. (13), UK, and Proteus Metal (14), Brussels, Belgium, manufacture SME brass actuators. Both companies market finished, heat-treated components ready for installation.

Applications

The first shape-memory alloy to be placed in commercial use was Nitinol employed as a high reliability coupling. In this device, known as a Cryofit connector (12), a small tubular piece is machined to an initial inside diameter ca 4% smaller than the outside diameter of the tube to be joined. The tube is heated to the β phase at elevated temperature and quenched to yield martensite. It is then expanded to a diameter sufficient to allow it to fit over the tubes to be joined, usually ca 4%. Upon heating, the ca 8% strain is recovered, producing a very high compressive-force joint of superior reliability. An M_s of ca -120°C was chosen because these connectors were first used for aircraft hydraulic couplings which experience a service temperature range from -100°C to perhaps 45°C. This application was so successful that it was extended to the coupling of piping systems on ships where the pipe diameter ranged from 1.9–5 cm (see Piping systems). In spite of the high cost of Nitinol for both alloy shapes and components, the labor savings over other joining systems, particularly welding (qv), has made the process economically attractive. The fact that once expanded to the open position the coupling must be stored at cryogenic temperature has not prevented their use for undersea oil-line connections. Undersea pipes are usually joined by welding in a hyperbaric chamber, a difficult and costly operation. Using Cryofit units, pipes up to 15 cm dia have been joined at depths of >100 m.

Another use for the Ni–Ti memory alloys is in high reliability strip-chart pen-recorder drives (15). The usual D'Arsonval movement was replaced with a Nitinol wire which, when it expanded or contracted, moved the recording pen through a system of simple pivots. The wire is heated by a high frequency induction heater whose frequency, and therefore heat-coupling efficiency, is proportional to the input signal. Since this is a two-way memory effect, such wire must go through a training period to develop a repeatable shape recovery. These highly reliable and shock resistant recorders have been very successful.

Medical Applications. The alloy Nitinol resulted from a successful research program intended to create a high strength metal resistant to corrosion by seawater. With these properties, it is not surprising that it has interesting applications in memory devices for medical implants requiring passive chemical behavior (16) (see Prosthetic and biomedical devices).

An early medical device exploits the superelastic behavior of Nitinol rather than its memory effect. This device, the restraining wire arch in an orthodontic brace attached to teeth being straightened, provides a constant force which produces the desired effect more quickly than conventional braces. The wire is deformed in a martensive condition to a point where martensite–martensite reaction occurs. Because of their wide elastic range, these wires are superior to the usual stainless-steel wires, which because of stress relaxation, require constant readjustment.

Exploiting the true memory effect has been the aim of a medical research team specializing in cardiovascular problems (see Blood, coagulants and anticoagulants; Cardiovascular agents). Addressing the prevention of blood clots, a group at Beth Israel Hospital in Boston has successfully demonstrated on animals a system for screening blood clots large enough to be hazardous (17). A Nitinol wire bundle in the martensitic state, with a composition selected to give an A_s slightly above body temperature, is deformed into an umbrella-like screen. The screen is collapsed into a straight strand of wires and inserted by means of a catheter into the vena cava where, upon warming, it opens to form a clot screen. This device may offer an effective method to guard against the aftereffects of operations on legs or lower trunk or in cases of phlebitis where clots are a common occurrence. This system avoids complicated and sometimes high risk surgery and the side-effect problems associated with anticoagulant drugs.

In the area of orthopedic surgery, Nitinol applications have been highly successful. Their range is as diverse as bone-fracture fixation, hip implants, and the reduction of bone deformaties and structural abnormalities such as scoliosis.

A most common orthopedic problem is the accurate fixation of fractures, particularly multiple breaks common in sports. Screws and plates are generally employed but the alignment is not always exact and the fracture interfaces are not always in close contact. With a shape-memory plate that shrinks on heating, a fracture can be aligned by screwing the plate to bridge the fracture and, when it is warmed by body heat or artificially heated by diathermy, the contractions pull the surfaces into tight contact ensuring rapid calcification and healing. Devices of this type are under investigation in the FRG (18), as well as in the U.S. and the UK.

Although hip replacements are now a common procedure, they still present problems of fixing the new ball joint rigidly in the femur. A long spike is usually driven into the hollow of the bone and cemented with acrylic or similar organic adhesive. The occurrence of hairline fractures is common, and misalignment may require a second operation. A new device employs a beta-conditioned hollow tube which is split lengthwise and bent to the shape of a C. After quenching to martensite, it is closed. Upon insertion, the tube is warmed by the body heat and expands to the original C configuration, providing a tight fit, which, since the grip is along the entire length, minimizes the chance of fracture. The ball joint is then attached.

Scoliosis, a lateral curving of the spine, is a disfiguring and disabling ailment. The cure is long and difficult and requires the repeated and gradual straightening of a Harrington rod to force the spine into alignment. These rods have been replaced by a memory alloy which, when initially attached to the spine, has a matching curvature.

When heated externally, the rod attempts to straighten, applying a constant controlled force that gradually brings the spine into normal alignment (19).

Active exploration of memory-effect devices is underway for application in automatic drug dispensing, an artificial heart muscle, prosthetic devices for arthritic joints, and systems to generate artificial peristalsis action (see Pharmaceuticals, controlled release).

Mechanical Devices. As the Ni–Ti alloys have dominated the biomedical field, so have the Cu–Zn–Al family of alloys taken a commanding lead in the field of thermally actuated mechanical devices. These include circuit breakers and electric relays, thermostatic valves, window openers, climate-control devices for buildings and cars, safety devices, aerospace-antenna systems, and various fastener and connector devices.

Since the Cu–Zn–Al alloys are readily hot-worked, they have been generally used in the form of extruded wire. Springs in torsion, tension, or compression have been developed as well as tubes and rods in torsion. Flat strip or sheet can also be produced although not as readily because the cold workability of these alloys is limited.

The automobile industry was among the first to develop SME actuators. An automobile contains ca 25 devices which could exploit SME, including thermostatic fluid and air controls, air-conditioner valves, exhaust emission controls, carburetor valves, and automotive fans. Such devices are now in the development stage in many countries.

An interesting application now in road tests is an automotive-engine fan whose speed is proportional to the cooling demand. An SME spring controls the relative slip of a clutch that engages the belt drive and fan. A considerable amount of fuel is saved by this system. Exhaust-emission control is a worldwide concern and although various catalytic devices provide a reasonably effective reduction in toxic emission, climate changes, particularly temperature, reduce their efficiency because of a shift in the fuel–air mixture. A carburetor jet with an SME alloy orifice that changes diameter in response to fuel temperature maintains a constant mixture and, as a result, stabilizes the exhaust chemistry (see Exhaust control, automotive).

A concept to use the car-passenger heater as part of the engine cooling system allows a reduction in the size of the main engine radiator. When the heat generated by the interior heater is not needed, louvers controlled by SME devices conduct the heat to the outside.

In the area of home-climate installation, SME valves controlling water circulation in hydronic heating systems offer the advantage of simplicity and economy over the conventional combination of electric thermostat, relay, and electrically operated valves (see Air conditioning). An SME valve can be placed in each room to give individual temperature control. In solar heating, passive wall-heat absorbers, so-called Tromb walls, require the opening and closing of shutters to control the insolation. The SME actuators developed for the control of greenhouse windows offer reliable, automatic, and powerful operation (see Solar energy).

A broad variety of thermal-overload devices are candidates for SME actuation, ranging from small relay systems to large circuit breakers. Protection against thermal overload is required on industrial electrical machinery to improve their fail-safe operation. For example, SME spring elements, because of their high force and small size, are incorporated in these devices. In addition, copper-based SME alloys are used as couplings for specialized tube connections, eg, the connection between copper and aluminum tube ends used in air-conditioning systems. The SME couple has a polymer inner liner that acts as a sealant as well as a barrier to possible galvanic corrosion (12).

As new forms of energy generation are needed, solar collectors, both passive and active, have been developed and the sun-induced temperature difference between the ocean surface waters and the ocean depths offers a particularly intriguing possibility. Shape-memory engines offer the potential of capturing this vast source of low temperature energy. No matter what the system, the scale would be enormous; however, active support is being given to such low temperature energy converters where the source is ocean thermal difference, thermoclines in high hydroelectric power-dam lakes, and effluents from conventional fossil and nuclear-fuel generator plants (20).

Theoretical studies of shape-memory engines (21) have compared the thermal efficiency of a shape-memory device operating as solid-state engine or solid-state heat-pumping device with more conventional Carnot fluid systems. The first SME engine demonstrating a useful power output with a modest temperature difference was a rotary engine in which Nitinol elements operated in a bending mode and provided the motive force as they dipped alternately in hot and cold water (22). Although relatively inefficient, it showed the possibilities inherent in solid-state engines.

In order to calculate the efficiency of a solid-state SME engine, a temperature-entropy diagram can be constructed similar to a Carnot cycle. The Gibbs free-energy change in the transformation of martensite to the parent phase is the driving force for such a system. Because of the hysteresis in the memory-effect transformation, the efficiency of a solid-state engine is always below that of a Carnot cycle. The sensible heat involved in the cycle, as compared with the latent heat of transformation must, for optimum efficiency, be kept low. This restriction is more difficult to achieve in a solid-state engine than in a fluid engine. Estimates of the efficiency of these engines range from 3–5%. These calculations are difficult because of the problem of accurately estimating entropy and free energy change for the martensite–parent and parent–martensite under stress.

Model engines built to date indicate that a solid-state engine can yield an efficiency at low temperature differences that is about equal to a gas- or fluid-phase engine and also offer a considerable advantage in size.

Another possible application for SME materials is the storage of large quantities of thermal energy by cycling between M_f and A_f. In effect, in heating from M_f to A_f the energy stored is the latent heat of transformation $T\Delta S$ plus the specific heat C_p. When the specimen is cooled from A_f to M_f, this energy is released. Since the temperature difference can, for a selected alloy, be on the order of 20°C, a large amount of energy can be stored in a relatively small volume.

A final area of interest in SME motors is in solar-driven irrigation pumps for use in underdeveloped countries. Efficiency is not as important as the ability to pump water at modest rates from modest depths with no fuel requirement and at a low capital cost. Several excellent designs have been demonstrated under a United Nations Development Program. Assuming 6 h of pumping per day and a peak pumping capacity twice the average, the required capacity is 5–10 L/s from a depth of 5 m. This corresponds to an average pumping power of 125–250 W/hm^2 of irrigated land (23).

BIBLIOGRAPHY

1. A. B. Greninger and V. G. Mooradian, *Trans. Metall. Soc. AIME* **12B,** 337 (1938).
2. G. V. Kurdyumov and L. G. Khandros, *Dokl. Akad. Nauk SSSR* **66,** 211 (1949).
3. E. Scheil, *Z. Anorg. Allg. Chem.* **207,** 21 (1932).
4. W. J. Buehler, J. V. Gilfrich, and R. C. Wiley, *J. Appl. Phys.* **34,** 1475 (1963).
5. R. J. Wasilewski, *Metall. Trans.* **2,** 2973 (1971).
6. A. Nagasawa, *J. Phys. Soc. Jpn.* **31,** 136 (1971).
7. L. Delaey, R. V. Krishman, H. Tas, and H. Warlimont, *J. Mater.* **9,** 1521 (1974).
8. J. Perkins, ed., *Proceedings of the First International Conference on Shape Memory*, Plenum Press, Toronto, Canada, 1975.
9. T. A. Shroeder, I. Cornelis, and C. M. Wayman, *Metall. Trans.* **7A,** 535 (1976).
10. T. A. Shroeder and C. M. Wayman, *Acta Metall.* **25,** 1375 (1977).
11. Nitinol Devices, Escondido, Calif.
12. Raychem Corporation, Menlo Park, Calif.
13. Delta Memorial Metal Ltd., Ipswich, UK.
14. Proteus Metal, Brussells, Belgium.
15. Foxboro Instrument Company, Foxboro, Mass.
16. L. S. Castleman, S. M. Motzkin, F. P. Alicandri, V. L. Bonawit, and A. A. Johnson, *J. Biomed. Mater. Res.* **10,** 695 (1976.
17. H. C. Ling and R. Kaplan, *Mater. Sci. Eng.* **48,** 241 (1981).
18. G. Bensmann, F. Baumgart, and J. Hartwig, *Metall* (*Berlin*) **35,** 312 (1981).
19. F. Baumgart, G. Bensmann, J. Haasters, J. Nolker, and K. F. Schlegal, *Arch. Orthop. Traumatic Surg.* **91,** 67 (1978).
20. W. S. Ginell, J. L. McNichols, and J. S. Cory, *ASME Publ. 78-ENA-7* (1978).
21. L. Delaey and G. Delepeleire, *Scr. Metall.* **10,** 959 (1976).
22. R. Banks, P. Hernandez, and D. Norgren, *Report NSF/RANN/SE/AG-550/FR72/2* (*UCID 3739*), National Science Foundation, Washington, D.C., July 1975.
23. *Testing and Demonstration of Small-Scale Solar-Powered Pumping Stations*, *Project GLO/78/004*, United Nations Development Program, Washington, D.C., Dec. 1979.

L. McDonald Schetky
International Copper Research Association

SHELLAC

Shellac is the purified product of the hardened resinous secretion (lac) of an insect that is parasitic on selected trees and bushes of India, Burma, and Thailand. This tiny insect, *Kerria lacca* (formerly *Laccifer lacca*) of the family Coccoidea secretes the lac as a protective covering for its larva. Lac is the only known commercial resin of animal origin. Its outstanding properties and versatility as a resin are attested to by its continued widespread use in industry (see also Resins, natural).

Shellac is a hard, tough, nontoxic, amorphous resin that produces films of good water resistance and exceptional gloss. Its chemical formula has eluded chemists even after many years of investigation, but it is generally believed to be a mixture of two resins secreted simultaneously by the lac insect. These resins are composed of aliphatic polyhydroxy acids present in the form of lactones, lactides, and intermolecular esters. Associated with the secreted resin are a water-soluble dye, laccaic acid, a water-insoluble dye, erythrolaccin, and a wax, also produced by the insect.

The value of shellac as a protective and decorative coating probably evolved from the use of the blood-red dye laccaic acid extracted from the resin (1–2). Lac is believed to have been used in India for several thousand years. Records of the great Mogul ruler of India, Akbar, tell of mixing lac and pigments to decorate public buildings in 1590. Jan Huyghen van Linschooten, sent to India by the King of Portugal, wrote in 1596 of the use of lac as a protective and decorative coating. An analysis of the first use of lac and its early history, suggests that its name may have originated in South China (3).

The world's main lac-growing area extends from Central India, northeast to Assam, and then south to Burma, Thailand, Laos, Cambodia, and Vietnam. Before and shortly after World War II, India practically had a monopoly of lac, since most of the raw lac grown outside of India was shipped to Calcutta for processing and export. Since about 1950, Thailand has become a large competitor in exports of seed lac (see below).

Most lac is cultivated part-time by the peasant farmer or ryotat on the farmer's own or rented trees, using cheap family labor. Branches are cut from the host trees that contain the young insects just before swarming. These branches, called brood lac, are either tied to the trees that are to serve as the new hosts or placed in baskets attached to these trees.

After the young larvae emerge from the brood lac, they swarm up the new branches until they find young twigs where they secure themselves in large colonies by piercing the thin bark with their proboscises. Then they begin sucking the sap and secreting the resin. The insect also secretes a wax filament to prevent the pores from becoming blocked. As the insect grows, it continues to secrete the lac resin.

In India, the two main strains of the lac insect are the Kusmi and Rangeeni; each has two life cycles a year. With two strains and two life cycles, four lac crops are harvested, as shown below.

Insect strain	Host tree	Harvesting time	Crop
Rangeeni	Palas, Ber, Ghont	April	baisakhi
		November	katki

Insect strain	Host tree	Harvesting time	Crop
Kusmi	*Kusum*	July	*jethwi*
		December	*aghani*

Physical Properties

In consideration of the properties of shellac, it must be remembered that it is a natural product of animal origin and differs somewhat from one source to another. Moreover, some of the earlier property studies were conducted before the effects of temperature and humidity were recognized. Hence, some of the results recorded are questionable. Average property values, as reported in the literature, are given in Table 1.

Shellac is nonconductive after it has been subjected to an electric arc. This non-tracking property is found in shellac varnish film, shellac moldings, and synthetic-resin moldings containing shellac.

Shellac is stable to ultraviolet radiation. It also retains its electrical insulating properties under the influence of ultraviolet radiation.

Solubility. The best solvents for shellac are the lower alcohols, methyl and ethyl, followed by amyl alcohols, glycols, and glycol ethers (4). Shellac dissolves in acetone if a small amount of a polar solvent, such as water or alcohol, is present.

Shellac is insoluble in esters, ethers (other than glycol ethers), hydrocarbons, chlorinated solvents, and water. It can be readily dispersed in water with the aid of soda ash, borax, ammonia, morpholine, or triethanolamine.

Chemistry

Shellac contains ca 67.9% carbon, 9.1% hydrogen, and 23.0% oxygen, corresponding to an empirical formula of C_4H_6O. A number of investigators (5–6) have studied the molecular weight of shellac and concluded it to be ca 1000. Based on this calculation, the average shellac molecule has the formula $C_{60}H_{90}O_{15}$. It contains one free acid group, three ester linkages, five hydroxyl groups, and possibly a free or potential aldehyde group, as indicated by acid value, hydroxyl value, saponification value, and carbonyl value. It has an ionization constant K_a of 1.8×10^{-5} (7).

Shellac does not contain any of the constituents found in other natural resins, such as oxidized polyterpenic acids, aromatic acids, phenolic compounds, and other resinous compounds; instead it is composed of hydroxyl fatty acid derivatives that are totally absent in other natural resins. Since lac is of animal origin, whereas the other resins are of plant origin, this fact is not surprising.

Shellac is believed to be a combination of hard and soft resins secreted by the insect at different rates (8). The former is secreted at a regular rate and the latter at an irregular rate. Extraction with ether gave 30% of a soft ether-soluble resin and 70% of a hard ether-insoluble resin (9).

Chemical properties of orange and bleached shellacs are given in Table 2 as well as the properties of hard- and soft-resin components separated by solvent extraction.

Table 1. Physical Properties of Shellac

Property	Value
melting point, °C	77–90
softening range, t_f–t_g °C	30.5–56.5
specific gravity at 15.5°C	1.110–1.217
molecular weight	964–1,100
energy of activation of viscous flow E_0 of molten lac, kJ[a]	120.–159.
refractive index, n_D^{20}	1.514–1.524
temperature coefficient of refractive index at 40–50°C	$(1.12$–$2.1) \times 10^{-4}$
hardness on copper	
Shore	60–61
Brinell	18.1–19.1
Vickers	16.2–17.0
scratch (1-mm ball)	4.5–5.5
abrasion resistance, sand[b], mm^{-1}	2.3–2.8
ultimate tensile strength at 20°C, MPa[c]	13
modulus of elasticity, MPa[c]	
by sound transmission at 20°C	1094
by beam method at 15–20°C	1324
adhesion, MPa[c]	
to glass	7.6
brass	17.2–22.8
copper	22.8
steel	22.1
an optically plane surface	44.1
heat of fusion, J/g[a]	52.72
specific heat at 10–40°C, J/(g·°C)[a]	1.51–1.59
thermal conductivity, mW/(cm·K)	
at 35°C	2.42
at 63°C	2.09
thermal expansion (cubical), $\Delta V/V$ at 46°C = $(\Delta t) + \beta(\Delta t)^2$	
at −80 to 46°C, α, per °C	2.73×10^{-4}
β per (°C)^2	0.39×10^6
at 46–200°C, α, per °C	13.10×10^{-4}
β, per (°C)^2	0.62×10^{-6}
fluidity, (Pa·s)$^{-1}$ [d], at 57°C	1×10^{-8}
at 70°C	200×10^{-8}
polymerization time[e] at 150°C, min	10–120
flow	
method A at 125°C, s	55–700
method B at 100°C, mm	10–100
dielectric constant (K) at 20°C	3.23–4.61
volume resistivity, Ω·cm	
at 20°C	1.8×10^{16}
at 30°C	1.2×10^{16}
surface resistivity, Ω	
at 20% rh	2.2×10^{14}
at 40% rh	1.1×10^{14}
dielectric strength at 20°C, V/cm	200×10^3–480×10^3
surface flashover strength at 60% rh, V/cm	6.2×10^3
power factor, tan θ, at 20°C	$(4.4$–$7.2) \times 10^{-3}$
loss factor, K tan θ, at 20°C	$(1.52$–$2.8) \times 10^{-2}$
permittivity, %	2.3–3.8
magnetic susceptibility, m^3/mol (cgs)	3.77×10^{-12} (0.30×10^{-6})
sound transmission at 20°C, m/s	970

[a] To convert J to cal, divide by 4.184.
[b] ASTM D 968-76.
[c] To convert MPa to psi, multiply by 145.
[d] (Pa·s)$^{-1}$ = rhe.
[e] ASTM D 411-73.

Table 2.   Chemical Properties of Shellac

Property	Orange, mol wt 1006	Bleached, mol wt 949	Hard resin, mol wt 1900–2000	Soft resin, mol wt 513–556
acid value, mg KOH/g	68–79	73–91	55–60	103–110
saponification value, mg KOH/g	220–232	185–260	218–225	207–229
ester value, mg KOH/g	155–167	103–155	163–165	104–119
hydroxyl value, mg KOH/g	250–280	230–260	116–117	235–240
iodine value, mg I/g				
Wijs (1 h)	1.3–1.6	0.7–1.0	1.1–1.3	5.0–5.5
carbonyl value				
sodium sulfite method	7.8–27.5		17.6	17.3

Aleuritic acid, more fully explored than any other constituent, was identified as the optically inactive 9,10,16-trihydroxypalmitic acid, mp 101.5°C (10). Several isomers of aleuritic acid are also present with the same or lower melting points, such as 89–90°C and 97–97.5°C (11). Aleuritic acid is readily obtained by saponifying dewaxed shellac with a 20% sodium hydroxide solution. After ca six days at room temperature, the sodium aleuritate can be filtered. Yields as high as 43% have been reported (12). Aleuritic acid can be converted to civetone and dihydrocivetone (13). Ambrettolite, the natural constituent of ambrette seed, is also made of aleuritic acid (see Perfumes). The preparation of α-dicarboxylic acids and ω-hydroxy acids from aleuritic acid for the synthesis of perfume compounds has been reported (14).

Another acid isolated from saponified shellac, so-called shellolic acid, $C_{15}H_{20}O_6$, melts at 199.5–201°C and decomposes at 202–203°C. This dihydroxy dibasic acid reacts readily with active hydrogen to give dihydroshellolic acid (mp 157–158°C).

Since a number of investigators have had considerable difficulty in obtaining shellolic acid, the question arises whether it is present in shellac as such, or whether the hydrogen chloride used as a catalyst in preparing the crystalline dimethyl ester, may not cause some structural or isomeric change.

In studies by acid degradation and synthesis (15–16), shellolic acid was identified as a sesquiterpene of the form given in structure (1) with the α-cedrene skeleton, where R is COOH.

R = ◂COOH, shellolic acid
R = - - -CHO, (+)-jalaric acid
R = ◂CH₂OH, (+)-lakshollic acid

(1)

Even under mild conditions of saponification the alkali-labile groups of shellac undergo some change, and therefore, the acids separated may not be the primary products of hydrolysis (17). The changes may be owing to aldol condensation, Cannizzaro reaction, hydrogen disproportionation, or dehydrogenation.

A pure aldehydic acid, $C_{15}H_{20}O_5$, mp 178–180°C, was isolated and called (+)-jalaric acid (18–19). Its ir spectrum shows the presence of hydroxy-α,β-unsaturated carboxyl and aldehyde groups, and a conjugated trisubstituted ethylenic linkage. When oxidized at room temperature with alkaline silver oxide, (+)-jalaric acid yields two dicarboxylic acids, epimers of shellolic acid. This behavior of jalaric acid has been attributed to epimerization under the alkaline conditions of silver oxide oxidation. It was also deduced that (+)-jalaric acid is the compound present in shellac and that shellolic acid is derived from it. From these studies and ^1H nmr it was concluded that the configuration of (+)-jalaric acid has the form given in structure (1), where R is CHO. In addition, lakshollic acid (mp 183°C) and its epimer (mp 203°C) were isolated (20). These are trihydroxymonobasic acids with structure (1), where R is CH_2OH. A monohydroxy fatty acid (mp 54–55°C) was also isolated (21). It was named butolic acid and was later identified as 6-hydroxytetradecanoic acid (22). The isolation and characterization of a number of minor constituent acids of lac resin was also reported (23), including myristic, palmitic, and stearic acids, and the corresponding unsaturated acids, 6-ketomyristic acid, 6-hydroxypalmitic acid, 6-hydroxy-cis-palmitoleic acid, and threo-9,10-dihydroxy-myristic and palmitic acids.

A series of papers on the chemistry of the lac resin structure have appeared in refs. 19 and 20. A more plausible structure for "pure lac resin" has been proposed from glc and 220 MHz ^1H nmr spectra (24). The structure is based on a 1:1 aleuritic acid: terpenic acid ratio.

Threo-Aleuritic acid having no protection of the hydroxy groups was esterified (20). Results indicated that greater stability and preferential ester formation with mixed stearic–carbonic anhydride at C-13 rather than C-10.

Of the many acids reported to occur in shellac, only the presence of (±)-aleuritic acid (2) and its isomers (−)-R-butolic acid (3), shellolic acid and the aldehyde acid, jalaric acid, has been substantiated. The exact percentage of these acids has not been established beyond doubt. To arrive at a complete structural picture, more work is necessary. The lac resins are not a single chemical entity, but a solid solution of several different molecular complexities formed by aliphatic hydroxy and terpenic acids (see Terpenoids).

$$OH$$
$$|$$
$$HOCH_2(CH_2)_5CHCH(CH_2)_7COOH$$
$$|$$
$$OH$$

(±)-aleuritic acid

(2)

$$OH$$
$$|$$
$$CH_3(CH_2)_7CH(CH_2)_4COOH$$

(−)-R-butolic acid

(3)

Laccaic Acid and Erythrolaccin. Laccaic acid (4), the water-soluble dye associated with the crude lac (11), is a mixture of several components, one of which contains nitrogen (25–26). The components are anthraquinone derivatives. Other investigators, however, believe that the dye remains associated with a protein and belongs to the monochrome group of pigments (27).

Erythrolaccin (5), the water-insoluble dye, $C_{15}H_{10}O_6$, melts with decomposition at 314°C (28).

Lac Wax. The lac insect secretes the wax in the form of thin white filaments along with the lac resin. These filaments are embedded in the resin and thus form an essential, although minor, constituent between 3.5, and 5.5% associated with the resin. A melting point of 72–82°C, acid value of 12.0–24.3, and saponification value of 79–126 have been reported (29), including the following composition:

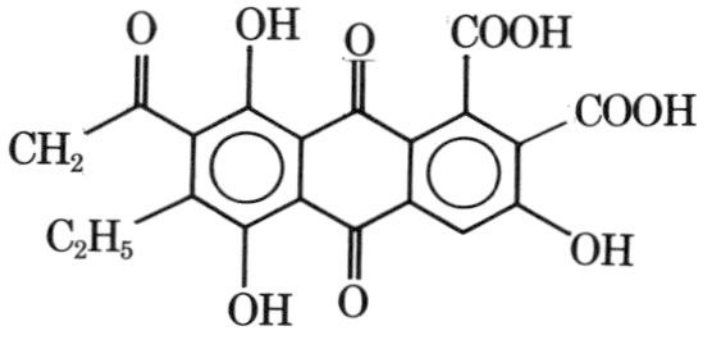

laccaic acid

(4)

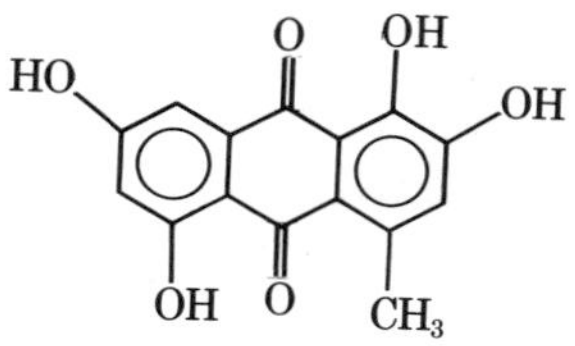

erythrolaccin

(5)

Constituent	Wt %
esters of wax acids	80–82
free wax acids	10–14
free wax alcohols	1
hydrocarbons	2.6
lac resins	2.4

However, the wax is composed of 1.8 wt % hydrocarbon, 77.2 wt % fatty alcohols, and 21 wt % acids (30).

Results obtained in analyzing lac wax have so far been conflicting, and although the overall compositions from different sources may be similar, the compounds present and the proportions in which they occur seem to depend on the origin of the lac.

Polymerization. When heated above their melting points, lacs behave as thermoplastic materials for short periods of time. After a given interval, which varies with the temperature, polymerization begins. The melt increases in viscosity and suddenly passes through a rubbery stage to turn into a brittle, horny product that is insoluble in alcohol. Only water is split off; after ca 70% of the material has reached the alcohol-insoluble stage, the reaction slows down and never reaches a complete end point. It seems that ca 5–10% of the constituents remain alcohol-soluble; hence, shellac is not a true thermosetting resin.

Polymerization does not change the saponification value or the aleuritic acid content of lac. The splitting off of water during heat polymerization is reversible; that is, when polymerized shellac is autoclaved under pressure in the presence of water, it is converted back into an alcohol-soluble resin with properties similar to those of the original shellac (31).

Esterification. When dissolved in low molecular weight alcohols such as methanol or ethanol, shellacs tend to esterify slowly, even in the absence of a catalyst. Shellac solutions that are stored for prolonged periods become slow drying as esterification proceeds and produce tacky films. Bleached shellac tends to esterify faster than orange shellac because residual chlorine lowers the pH to 3.5–4.0. Alcoholic solutions of orange shellac with a pH of 5.0–5.5 are more stable.

Shellac esters, because of their adhesion and flexibility, are excellent resin-plasticizers for cellulose lacquers, and large volumes are used commercially for this purpose. These are prepared by heating shellac in presence of an acid catalyst (see Plasticizers).

Processing

After the life cycle of the insect has been completed, the lac is ready for harvest. The ryotat cuts the coated twigs and scrapes off the encrustation or chops up the twigs into small pieces, called stick lac, which he takes to the market. There it is bought and transferred to refining centers.

The stick lac is scraped to remove the lac or crushed to separate the lac resin from the sticks. At this stage the crushed lac contains a mixture of resin, insect remains, dye, twigs, and other impurities. The large sticks are removed by screening and the material, known as bueli, is washed with water to remove the water-soluble dye and small sticks and insect bodies. After a few washings, the crushed lac is dried; this product is known as seed lac.

The shellacs of commerce are made from seed lac and are grouped under three processes: handmade, machine-made, and bleached.

Handmade Shellac. This indigenous process is used by small native factories and has changed little over the years. The seed lac is melted and squeezed through a long, thin canvas bag; impurities such as twigs and sand remain in the bag. The molten lac is then stretched into thin sheets by a native worker using his hands, feet, and teeth. The sheets are cooled and broken up into flakes. Alternatively, the melted lac is cast into button-shaped cakes instead of being drawn into sheets; the product is called button lac.

Grades produced by this process are T. N. (truly native), superfine, lemon No. 1 and No. 2, button lac, etc.

Machine-Made Shellac. These shellacs are produced either by a heat process or a solvent process. In the heat process, seed lac is melted on steam-heated grids. The molten lac is forced through a filter cloth or fine wire screen with the aid of hydraulic pressure. The filtered material is dropped on rollers, where it is squeezed out and removed in thin sheets.

The solvent process produces three types of shellac: wax-containing, dewaxed, and, dewaxed-decolorized. For wax-containing grades, the raw seed lac and solvent (usually ethyl alcohol) are heated in a tank under reflux for one or two hours and then filtered to remove the undissolved impurities. The filtrate is fed into a series of evaporators to concentrate the material to a viscous melt. This melt is dropped onto rollers, where it is squeezed and removed in flake form.

Dewaxed shellacs are made by dissolving seed lac in alcohol at a temperature depending on the concentration (ie, slightly elevated temperature for more dilute alcohol). The solutions are first passed through filter presses to remove the wax and then concentrated into a viscous melt, and finally removed in flake form.

Dewaxed decolorized shellacs are produced by the same process except that after dewaxing, the solutions are treated with activated carbon to remove the darker materials. By using light-colored seed lac and varying the amount of carbon and contact time, grades of shellac are produced varying in color from very light yellow to a dark orange.

Bleached Shellacs. These come in two grades: regular bleached shellac, containing the natural shellac wax, and refined wax-free bleached shellac. Seed lac is dissolved in aqueous sodium carbonate at high temperature. The solution is then centrifuged or passed through a fine screen to remove insoluble residue along with sand, wood, or other impurities. For regular bleached shellac, the centrifuged solution is treated with a dilute solution of sodium hypochlorite. The solution is acidified with sulfuric acid to precipitate the resin. The precipitated shellac is filtered, washed free of acid, and dried to form small granules. For refined wax-free bleached shellac, the treated solution is dewaxed by high speed centrifugation or filter pressing. The effluent is bleached and precipitated by the same method as regular shellac.

Economic Aspects

India held a monopoly on the raw-lac trade ever since lac became an industrial commodity in western countries. Before World War II, practically all the lac collected outside India, in the region from Burma to Vietnam, was exported as stick lac to India, where it was refined and exported with Indian lac. Since World War II, Thailand has progressively increased its production and now exports substantial amounts of lac, in the form of seed lac, to the United States and Japan. In 1961, Thailand's export of seed lac exceeded that of India.

Very little lac is consumed in the producing countries of India and Thailand, and about 95% is exported to the highly developed industrial countries in the Western Hemisphere, Asia, and Europe. The United States has always been the largest importer and consumer of lac (see Table 3); the United Kingdom, the USSR, Japan, and the People's Republic of China also import large amounts.

The United States uses large quantities of seed lacs for the manufacture of bleached shellac and flake shellac for industrial uses. Japan imports mostly seed lac. The United Kingdom imports mostly handmade and machine-made shellac and some seedlac. The USSR and the People's Republic of China import only handmade shellac from India. Refuse lac (Kiree), a by-product of handmade shellac, is used for solvent-processed machine-made shellacs in the FRG.

Total exports of lac from India and Thailand reached a peak in 1956, when about 50,000 metric tons was exported; since then exports have leveled off to ca 10,000 t/yr. Accurate export figures for each crop cannot be obtained because part of the crop may be carried from one year to the next, depending on demand. Production of lac, in both India and Thailand, fluctuates from year to year, depending on growing conditions. The 1979–1980 Thailand lac crop was a total failure because of adverse weather conditions, and of an expected 4000 t, only 1000 was produced. Because of this crop failure,

Table 3.   U.S. Imports of Lac[a]

Year	Metric tons
1970	5921
1973	4624
1975	4928
1977	5250
1979	4680

[a] Ref. 32.

seed lac price rose from an average of $0.66/kg to $2.86/kg. Customers had to exert pressure on India to meet their needs. As a result, the Indian Government established quotas for seed lac exports. Since India had sufficient crop to make up for Thailand's failure, there was no world shortage, but consumers had to pay higher prices.

Specifications, Standards, Analytical Methods

A number of commercially available grades of shellac are produced from different sources of raw lac, which vary in coloring matter, wax, and impurities (present in small amounts). Analytical values vary to some degree because of test methods and origin of samples.

Since the United States and United Kingdom are the largest importing countries of shellac, they established grade standards. In the United Kingdom grading of lac has been governed by the London Shellac Trade Association (LSTA), and in the United States by the Shellac Importers' Association (USSIA), the American Bleached Shellac Manufacturers' Association (ABSMA), and the ASTM specifications are based on consumer requirements and the grades available in India and Thailand. The Indian Standards Institute (ISI) standardized seven grades of seed lac and six grades each for handmade and machine-made shellac that are mainly based on the color index percentage of matter insoluble in hot alcohol (33). Recently, the ISO has become interested in shellac and has issued seed lac specifications (34). As of July 1, 1980, pharmaceutical glaze (35) and shellac (36) were listed in the USP/NF; both specifications and test methods are given. Shellac is also listed in the *Food Chemicals Codex*, 3rd ed. (37).

The practice of adding rosin to shellac goes back to the days when shellac buttons were used in lac turning, a method of applying a decorative coating on wooden furniture and toys by frictional heating. Addition of other natural resins similar to rosin facilitated the melting.

The addition of orpiment (arsenic trisulfide) can also be traced to the lac-turning industry; as little as 0.5% gave a light-yellow color.

Today shellac is not longer adulterated with rosin and orpiment, and standards are not based on color alone but also on various specifications. However, it is difficult to establish absolute standards because shellac has such widely divergent applications in a great number of industries with specialized requirements.

Rosin is detected by a fast and simple method using the Halphen-Hicks reaction of phenol and bromine. For the determination of arsenic, the Gutzeit test is used.

Most analytical test methods for such constants as acid value, iodine value, and saponification value, are based on standard procedures as outlined in *Official Methods of Analysis, Standards, Specifications, and General Information* (38).

Emphasis in recent years has been shifted to physical properties such as color, flow, and heat behavior for which tests have been developed by consumers to meet their special requirements.

Uses

There are no statistics available for the amounts of shellac used by the different lac-consuming industries. The phonograph-record industry was the leading consumer of shellac until the World War II. Vinyl copolymer resins have largely replaced shellac in this field except in less developed countries (see Recording disks).

The electrical-insulating industry still consumes large quantities of shellac to make shellac-bonded paper for transformer and high tension terminals. Because of its nontracking properties under arcing, shellac is excellent for this purpose (see Insulation, electric).

Shellac varnishes are used extensively in wood finishing. The varnishes are quick-drying and tend to seal off sap left in the wood. As a top coating for floors, shellac has excellent wear properties. When modified with urea–formaldehyde resins and catalyzed with acid (39), shellac provides a protective coating that is extremely tough, has excellent water and solvent resistance, and is fast-drying (see Coatings, resistant). Shellac varnish mixed with oils (French polish) and rubbed on wood produces a deep-toned finish with excellent gloss depth.

Although shellac has been replaced by polystyrene and acrylic emulsions as the principal resin in high gloss no-rub floor polish, it still is used to some extent as a leveling resin in some formulations. Straight polystyrene emulsions do not have the wear properties suitable for floor polishes until modified by shellac, which is incorporated as the emulsifier during emulsion polymerization of the styrene monomer (40) (see Styrene plastics).

Pharmaceutical tablets and pellets are coated with shellac to protect against moisture and to seal off active ingredients. Enteric coatings protect the active ingredients from stomach acids until the pills reach the intestine, which is alkaline. Here the shellac coating dissolves to release the medicaments to the body (see Pharmaceuticals).

Shellac has been used in printing inks for a number of years, with substantial recent growth in this field due to new packaging technology. Both alcoholic and aqueous alkali varnishes are used in inks for flexographic and gravure printing. The grade and formulation depend on the stock to be printed (41) (see Inks).

Long-established industrial applications, such as the coating of candies, shellac-bonded grinding wheels, and hat stiffening, still use large quantities of shellac. Refs. 42 and 43 give formulations and application data on industrial uses.

BIBLIOGRAPHY

"Shellac" in *ECT* 1st ed., Vol. 12, pp. 243–260, by Wm. Howlett Gardner, Polytechnic Institute of Brooklyn; "Shellac" in *ECT* 2nd ed., Vol. 18, pp. 21–32, by James W. Martin, William Zinsser & Co.

1. S. Krishnaswami, *A Monograph on Lac*, Indian Lac Research Institute, Nankum, Ranchi, Bihar, India, 1962, Chapt. 1.
2. K. N. Dave, *Lac and the Lac Insect in the Atharva-Veda*, L. Chandra, Nagpur, India, 1950.
3. *Shellac*, Angelo Brothers Ltd., Cassipore, Calcutta, India, 1965, pp. 6–10.
4. W. H. Gardner and W. F. Whitmore, *Ind. Eng. Chem.* **21,** 226 (1929); W. H. Gardner and H. Weinberger, *Ind. Eng. Chem.* **30,** 451 (1938).
5. W. H. Gardner, W. F. Whitmore, and H. J. Harris, *Ind. Eng. Chem.* **25,** 696 (1933).
6. S. Basu, *J. Indian Chem. Soc.* **25,** 103 (1948).
7. N. R. Kamath and S. P. Potnis, *J. Sci. Ind. Res. (India)* **15B,** 437 (1955).
8. M. Venugopalan, *Indian Lac Res. Inst. Bull.* **3,** (1929).
9. C. D. Harries and W. Nagel, *Ber.* **55B,** 3833 (1922).
10. W. Nagel, *Ber.* **60B,** 605 (1927).
11. P. K. Bose, Y. Sankaranarayanan, and C. S. Sen Gupta, *Chemistry of Lac*, Indian Lac Research Institute, Nankum, Ranchi, Bikar, India, 1963, Chapt. IV.
12. B. S. Gidvani, *J. Chem. Soc.* 306 (1944).
13. H. Hunsdiecker, *Ber.* **76B,** 142 (1943).

14. Indian Pat. 65543 (1958), H. H. Mathur and S. O. Bhattacharya; G. B. V. Subramanian, U. Majumdar, R. Nuzhat, V. K. Mahajan, and K. N. Ganesh, *J. Chem. Soc. Perkin I*, 2167 (1979); S. C. Sengupta, S. C. Agarwal, and N. Prasad, *J. Oil Colour. Chem. Assoc.* **62**, 85 (1979).

15. B. Yates and G. F. Field, *J. Am. Chem. Soc.* **82**, 5764 (1960).

16. W. Carruthers, J. W. Cook, N. A. Glen, and F. D. Gunstone, *J. Chem. Soc.*, 5251 (1961).

17. N. R. Kamath and V. B. Mainkar, *J. Sci. Ind. Res.* **14B**, 555 (1955).

18. M. S. Wadis, V. V. Mhaskar, and Sukh Dev, *Tetrahedron Lett.* (8), 513 (1963).

19. N. R. Kamath and S. P. Potnis, *Intern. Congr. Pure Appl. Chem.*, *XIV*, *Congress Handbook*, 1955, p. 186.

20. R. G. Khurana, M. S. Wadis, V. V. Mhaskar, and Sukh Dev, *Tetrahedron Lett.*, 1537 (1964); Sukh Dev, V. V. Mhaskar, R. G. Khurana, and M. W. Wadia, *Tetrahedron* **25**, 3841 (1969); Sukh Dev, A. B. Upadhye, and A. N. Singh, *Tetrahedron* **30**, 867 (1974).

21. S. C. Sengupta and P. K. Bose, *J. Sci. Ind. Res.* **11B**, 458 (1952); Sukh Dev, V. V. Mhaskar, A. B. Upadhye, and M. W. Wadia, *Tetrahedron* **26**, 4177 (1970).

22. W. W. Christie, F. D. Gunstone, and H. G. Prentice, *J. Chem. Soc.*, 5768 (1963).

23. W. W. Christie, F. D. Gunstone, H. G. Prentice, and S. C. Sengupta, *J. Chem. Soc. Suppl.*, 5833 (1964).

24. Sukh Dev, A. B. Upadhye, and A. N. Singh, *Tetrahedron* **30**, 3689 (1974).

25. R. Burwood, G. Read, K. Schofield, and D. E. Wright, *J. Chem. Soc.*, 6067 (1965).

26. E. D. Pandhare, A. V. Rama Rao, R. Srinivasan, and K. Venkataraman, *Tetrahedron Suppl.* 8(1), 229 (1966).

27. H. Singh, T. R. Seshadri, and G. B. V. Subramanian, *Tetrahedron Lett.* (10), 1101 (1966).

28. P. Yates, A. C. Mackay, L. M. Punde, and M. Amin, *Chem. Ind.* (*London*), 1991 (1964).

29. A. H. Warth, *The Chemistry and Technology of Waxes*, 2nd ed., Reinhold Publishing Corp., New York, 1956, pp. 112–113.

30. E. Faurot-Bouchet and G. Michel, *J. Am. Oil Chem. Soc.* **41**, 418 (1964).

31. M. Rangaswami and R. W. Aldis, *Indian Lac. Res. Inst. Bull.* **14**, (1933).

32. *U.S. Imports for Consumption*, *TSUSA Commodity by Country of Origin*, FT 246/calender year, TSUSA items 188.1020, 188.1040, and 188.1060, U.S. Department of Commerce, Bureau of the Census, Washington, D.C.

33. *Indian Standard Specification for Seedlac IS:15-1973*, Indian Standards Institution, Manak Bhavan, New Delhi, India.

34. *Seedlac—Specifications Ref. #ISO 55-1977 (E)*, International Organization for Standardization Case Postale 56, Geneva, Switzerland.

35. "Pharmaceutical Glaze," *USP XX/NF XV*, Supplement 1, The U.S. Pharmacopeia XX/The National Formulary XV, Rockville, Maryland, 1980, pp. 86, 168.

36. "Shellac" *USP XX/NF XV*, Addendum a to Suppl. 1, The U.S. Pharmacopeia XX/The National Formulary XV, Rockville, Md., 1980, p. 172.

37. "Shellac," The *Food Chemicals Codex*, 3rd ed., National Academy of Sciences, National Academy Press, Washington, D.C., 1981.

38. *Official Methods of Analysis, Standards, Specifications, and General Information*, American Bleached Shellac Manufacturers' Association and United States Shellac Importers, Association, New York, 1957.

39. Brit. Pat. 963,608 (Nov. 4, 1964), D. Lovering (to William Zinsser & Co.).

40. U.S. Pat. 2,961,420 (Nov. 22, 1960), R. J. Frey, Jr., and M. Roth (to Monsanto Co.).

41. *Shellacs for Flexographic Inks*, PD#53-2-1, William Zinsser & Co., New York, 1963.

42. J. W. Martin in *Treatise on Coatings*, *Film-Forming Compositions*, Vol. 1, Pt. II, Marcel Dekker, Inc., New York, 1972, p. 442.

43. G. S. Misra and S. C. Sengupta in N. M. Bikales, ed:, *Encyclopedia of Polymer Science and Technology*, Vol. 12, Interscience Publishers, a division of John Wiley & Sons, Inc., New York, 1970, p. 419.

JAMES MARTIN
William Zinsser & Co.

SHORTENINGS AND OTHER FOOD FATS.   See Fats and fatty oils; Vegetable oils.

SHRIMP MEAL.   See Pet and livestock feeds; Aquaculture.

SIDERITE.   See Pigments.

SIENNA, BURNT.   See Pigments.

SIEVES.   See Size measurement of particles.

SIGNALING SMOKES.   See Chemicals in war.

SILANES; SILANOLS.   See Silicon compounds.

SILICA

INTRODUCTION

The term silica denotes the compound silicon dioxide [7631-86-9], SiO_2. In technological usage, this designation includes various, primarily amorphous, forms of the parent compound which are hydrated or hydroxylated to a greater or lesser degree, eg, types of colloidal silica and silica gel.

Silicon dioxide is the most common binary compound of silicon and oxygen, the two elements of greatest terrestrial abundance. It constitutes ca 60 wt % of the earth's crust, occurring either alone or combined with other oxides in the silicates. It is thus a ubiquitous chemical substance and, owing to its rich chemistry, is of great geological importance. Commercially it is the source of elemental silicon and is used in large quantities as a constituent of building materials. In its various amorphous forms it is used as a desiccant, adsorbent, reinforcing agent, filler, and catalyst component (see Drying agents; Fillers). It has numerous specialized applications, eg, piezoelectric crystals, in vitreous-silica optical elements and glassware. Silica is a basic material of the glass, ceramic, and refractory industries and an important raw material for the production of soluble silicates, silicon and its alloys, silicon carbide, silicon-based chemicals, and the silicones (see Carbides; Ceramics; Glass; Refractories).

Structure and Bonding

Silicon shares with the other elements of group IV A of the periodic system the property of forming an oxide of formula MO_2. All these oxides show acidic properties, most distinctly in the case of CO_2 and SiO_2. Basic character becomes more pronounced for the heavier members of the group; eg, SnO_2 is distinctly amphoteric. Like the dioxides of germanium, tin, and lead, silica is a solid of high melting point, although it differs from the former in having in its common forms a three-dimensional lattice based on four-coordinate silicon, whereas the heavier analogues possess the more ionic structures of the rutile type (except for a high-temperature form of GeO_2 which has the four-coordinate structure). Carbon dioxide is markedly different. It is a gas which condenses at $-78.5°C$ at ca 10 kPa (1 atm) to a molecular solid with a linear OCO structure. The principal reason for this structural difference between CO_2 and SiO_2 is the ability of carbon, along with other elements of the second period, to form strong π bonds using valence-shell p orbitals. Thus CO_2 assumes a molecular structure characterized by carbon–oxygen double bonds. For elements of the third and lower periods, larger internuclear separations (bond distances) make p_π–p_π overlap considerably less favorable than for the lighter elements. The π bonding is accordingly diminished, and the possibility of association through formation of additional σ bonds is correspondingly enhanced. Consequently, in numerous examples of formally similar compounds of the second and third periods, the lighter compounds are monomeric molecular species, whereas the heavier analogues of the same stoichiometry have oligomeric or polymeric structures. Among these are the analogous compounds boron trichloride, BCl_3, and aluminum chloride, Al_2Cl_6 (g); carbon disulfide, CS_2, and poly(silicon disulfide), $(SiS_2)_x$; nitrogen, N_2, and phosphorus, P_4; nitric acid, HNO_3, and poly(metaphosphoric acid), $(HPO_3)_x$; and nitrogen oxide (2:3), N_2O_3, and phosphorus oxide (4:6), P_4O_6. For the same reason, the silicones, $(R_2SiO)_x$, form polymeric structures in preference to the monomeric structures of the analogous ketones, R_2CO.

The basic structural unit of most of the forms of silica and of the silicate minerals is a tetrahedral arrangement of four oxygen atoms surrounding a central silicon atom. Although the Si—O bond clearly possesses considerable covalent character, these materials may be treated to a good approximation according to the empirical rules dealing with the stability of ionic crystals formulated by Pauling (1). This approach treats the silica structure as an aggregation of Si^{4+} cations (ionic radius 41 pm) and O^{2-} anions (ionic radius 140 pm). The ratio of cation:anion radius (0.29) is in the range for which tetrahedral coordination of O about Si is predicted from geometric arguments based on anion–anion and cation–anion contact in coordination polyhedra (2). The SiO_2 stoichiometry requires that on the average each oxygen must be shared by silicons in two tetrahedra, whereas in accordance with the electrostatic valency rule a single oxygen cannot be shared between more than two tetrahedra. Sharing of corners is the common mode of linkage of the coordination polyhedra; sharing of edges is rarely encountered and sharing of faces never occurs because of the decrease in stability that would result from the close approach of the silicon cations.

Structurally, silica represents a limiting case in which an infinite three-dimensional network is formed by the sharing of all oxygen atoms of a given tetrahedron with neighboring groupings. The possibility of linking tetrahedra with some corners remaining unshared gives rise to a wide range of structural possibilities, some of which are encountered in the silicates. In structures for which all corners of the tetrahedra are not shared, each unshared oxygen atom contributes to the anionic groups thus formed a formal negative charge, which is satisfied by the presence of other cations in the silicate structure. The various ways in which the SiO_4 tetrahedra may be linked form a basis for a structural classification of the silicates and of silica itself (3). A more extensive treatment of silicate structures is given in ref. 2.

Five structural types, summarized in Figure 1, include the following: (a) Structures containing discrete SiO_4^{4-} tetrahedra. These mononuclear anions, variously called silicate, orthosilicate, or tetraoxosilicate(IV) ions (4), are found in the mineral olivine, $(Mg,Fe)_2SiO_4$, and zircon, $ZrSiO_4$. (b) Structures containing discrete $Si_2O_7^{6-}$ polyhedra. In these disilicate (formerly called pyrosilicate) anions, two SiO_4 tetrahedra share a single corner. Thortveitite, $Sc_2Si_2O_7$, and a series of disilicates of the $4f$ (lanthanide) elements, $M_2Si_2O_7$, are examples of this relatively rare silicate type (see Rare-earth elements). (c) Structures composed of tetrahedra sharing two oxygen atoms give rise to discrete cyclic anions such as $Si_3O_9^{6-}$ (cyclotrisilicate) as in benitoite, $BaTiSi_3O_9$, or $Si_6O_{18}^{12-}$ as in beryl. Alternatively, single-chain structures may be formed as in the pyroxenes. Different spatial configurations of the single chains are possible, resulting in structures with different numbers of tetrahedra in the spatially repeating unit. (d) Double-chain structures in which the SiO_4 tetrahedra are topologically inequivalent in the sense that some silicon atoms share two, and some share three oxygen atoms. They may be considered as arising from the lateral linking of single chains as in the amphiboles, in which the repeating unit is $Si_4O_{11}^{6-}$. The composition of such double chains depends on the spatial configuration of the corresponding single chains. (e) Sheet structures in which three oxygen atoms of each SiO_4 tetrahedron are shared. A sheet structure consisting of linked rings formed by six tetrahedra is characteristic of the micas. A different sheet structure containing alternate four- and eight-membered rings is found in the mineral apophyllite.

The structures in which SiO_4 tetrahedra share all four oxygen atoms lead to the principal forms of silica. Replacement of some silicon atoms by aluminum gives a negatively charged framework of composition $Al_xSi_yO_{2(x+y)}$ in which positive ions are accommodated in holes in the structure. Examples of these framework silicates include the feldspars, zeolites, and ultramarines (see Clays; Molecular sieves).

Although the treatment of silica and the silicates in terms of Pauling's rules suggests an ionic model, it should be emphasized that the silicon–oxygen bond possesses appreciable covalency. Pauling estimated an Si—O bond distance of 163 pm (observed value 162 pm) from the sum of the covalent radii modified by the Schomaker-Stevenson correction for the electronegativity difference between Si and O and a shortening correction for double bonding $(d_\pi—p_\pi)$ between oxygen and silicon (1,5). On the basis of this model, the residual charge on the silicon atom in SiO_4^{4-} arising from partial ionic character is reduced by π bonding to a value in accord with the electroneutrality principle. Pauling's estimate for this residual charge is approximately +1 electronic units. From the viewpoint of a totally covalent bonding model, the silicon atom forms bonds to oxygen utilizing four equivalent sp^3 hybrid orbitals with possible participation of $p_\pi—d_\pi$ bonding. The resulting tetrahedral geometry, which follows

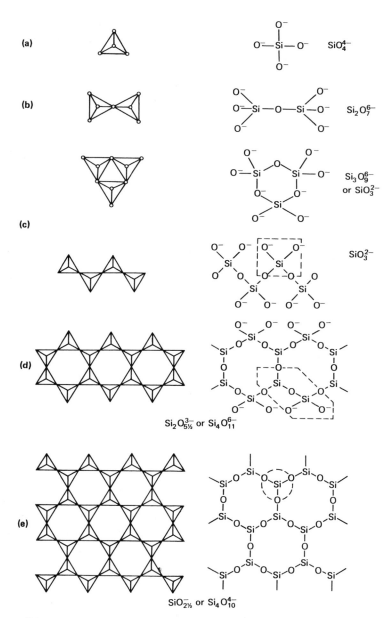

Figure 1. Schematic representation of modes of linkage of SiO$_4$ tetrahedra in basic silicate structures (3). The Si atoms in (**d**) and (**e**), which appear to be joined to only three O atoms, are joined to a fourth also, which is above the plane of the diagram.

from the assumption of sp^3 hybridization, is that found universally in the uncomplexed molecular compounds of silicon. Since the bonding in the silicates is not dissimilar to that in other silicon compounds, it is not surprising that the various silicate structures have analogues in molecular silicon–oxygen chemistry, where oxygen bridging between silicon atoms is commonly encountered. Thus, the silicon–oxygen ring system of the cyclosilicate anions is found in the organocyclosiloxanes, whereas the single-chain silicates are formally similar to linear silicone polymers, in which the unshared oxygen atoms are replaced by univalent organic groups. The organosiloxanes include, however, some structural types that have no silicate analogues, such as discrete, three-dimensional polyhedra based on sharing of three oxygens per silicon atom (6). The relationship between the organopolysiloxanes and the silicates may be invoked to relate some structural and chemical properties of silicates to those of their organic analogues (7).

Properties

At ordinary temperatures silica is chemically resistant to many common reagents. However, it undergoes a wide variety of chemical transformations under appropriate conditions, particularly at high temperatures or when volatile products escape from the reaction. Reactivity is strongly dependent upon the form, pretreatment, and state of subdivision of the particular sample investigated. Finely divided amorphous silica is in many circumstances considerably more reactive than crystalline silica in bulk. The reactivity of high surface area amorphous silica is conditioned by the presence of surface hydroxyl (silanol) groups, half of which may be retained even after heating to 400°C (8). A large amount of information on the reactivity of these surface groups has been developed in the course of surface-chemistry investigations (9–10). Silanol groups can be expected to be present on the surface of any silica sample and to exert an influence as reactive sites.

Surveys of the chemical properties of silica are given in refs. 11 and 12, both of which include numerous references to the earlier primary literature.

Common aqueous acids do not attack silica, except for hydrofluoric acid which forms fluorosilicate anions, eg, SiF_6^{2-}. The rate at which the various forms of silica are dissolved by aqueous HF decreases with increasing density, ρ, in the sequence: vitreous silica ($\rho = 2.2$ g/cm^3) < tridymite [15468-32-3] ($\rho = 2.22$ g/cm^3) $\simeq$ cristobalite [14464-46-1] ($\rho = 2.33$ g/cm^3) < quartz [14808-60-7] ($\rho = 2.65$ g/cm^3). Coesite [13778-38-6] ($\rho = 3.01$ g/cm^3) is practically insoluble in aqueous HF. Stishovite [13778-37-5] ($\rho = 4.35$ g/cm^3) is even less soluble, perhaps owing to the presence in the latter of a filled octahedral coordination shell about the silicon atom. Low quartz is attacked on the plane perpendicular to the optic axis at a rate about a hundred times greater than that on the prism faces, possibly reflecting the different surface structures on different crystal planes (13). Phosphoric acid attacks vitreous silica at elevated temperatures, forming a crystalline silicophosphate. The solubility of silica is greater in dilute than in more concentrated aqueous phosphoric acid (14). Quartz and vitreous silica are affected only slightly by aqueous alkali at room temperature. The attack is faster at higher temperatures. Precipitated amorphous silica is more reactive than vitreous silica which in turn is more reactive than quartz.

Silica is reduced to silicon at 1300–1400°C by hydrogen, carbon, and a variety of metallic elements. Gaseous silicon monoxide is also formed. At pressures of ≥40 MPa (400 atm), in the presence of aluminum and aluminum halides, silica can be converted to silane in high yields by reaction with hydrogen (15). Silicon itself is not hydrogenated under these conditions. The formation of silicon by reduction of silica with carbon is important in the technical preparation of the element and its alloys and in the preparation of silicon carbide in the electric furnace. Reduction with lithium and sodium occurs at 200–250°C, with the formation of metal oxide and silicate. At 800–900°C, silica is reduced by calcium, magnesium, and aluminum. Other metals reported to reduce silica to the element include manganese, iron, niobium, uranium, lanthanum, cerium, and neodymium (16).

Of the halogens, only fluorine attacks silica readily, forming SiF_4 and O_2. A number of halogen compounds of the nonmetals and metalloids react more or less readily with silica forming volatile silicon halogen compounds (see Table 1). The formation of $SiCl_4$ by direct chlorination of mixtures of silica and carbon is of some technical importance.

The acidic character of silica is shown by its reaction with a large number of basic oxides to form silicates. The phase relations of numerous oxide systems involving silica are summarized in ref. 23. Reactions of silica at elevated temperatures with alkali and alkaline earth carbonates result in the displacement of the more volatile acid, CO_2, and the formation of the corresponding silicates. Similar reactions occur with a number of nitrates and sulfates. Treatment of silica at high temperature with sulfides gives thiosilicates or silicon disulfide, SiS_2.

The reactions of silica with organic and organometallic compounds result in compounds containing Si—C and Si—O—C bonds. Treatment of silica with alkyl or aryl Grignard reagents, followed by hydrolysis, gives organocyclosiloxanes in high yields (24). Small amounts of low molecular weight silicon compounds can be obtained by reaction of methanol or sodium methoxide with silica over a period of days or weeks (25). Other studies indicate that fracture of silica, eg, in grinding, produces active sites that react with alcohols to form surface esters and with olefins to yield oligomerized species bonded to the surface through Si—O bonds that cannot be hydrolyzed (26).

An important aspect of silica chemistry concerns the silica–water system. The interaction of the various forms of silica with water has geological significance and is applied in steam-power engineering where the volatilization of silica and its deposition on turbine blades may occur, in the production of synthetic quartz crystals by hydrothermal processes, and in the preparation of commercially important soluble silicates, colloidal silica, and silica gel.

Reliable determination of the solubility of silica in water has been complicated by the effects of impurities and of surface layers that may affect attainment of equilibrium. Solubility behavior of silica is discussed in refs. 9 and 27. Reported values for the solubility of quartz at room temperature are in the range 6–11 ppm (as SiO_2). Typical values for massive amorphous silica at room temperature are around 70 ppm

Table 1. Reactions of Halogen Compounds with Silica

Halogen compound	Products	Conditions	Reference
HF	SiF_4		
FNO	SiF_4, N_2O_3	slowly at 150°C	17
$SeOF_2$	SiF_4, SeO_2	quantitatively	18
BrF_3	SiF_4, O_2, Br_2	quantitatively	19
BF_3	SiF_4, $(BOF)_3$	thermally	20
	SiF_4, cyclic $(SiOF_2)_n$, $(BOF)_3$, B_2OF_4, F_2BOSiF_3	microwave discharge	21
CF_3CF_3	SiF_4, CO, CO_2	at 800°C	22
BCl_3, S_2Cl_2, PCl_3	$SiCl_4$	elevated temperatures	

and for other amorphous silicas in the range 100–130 ppm. Solubility increases with temperature, approaching a maximum ca 200°C. It appears to be at a minimum ca pH 7 and increases markedly above pH 9 (9).

Results obtained at high temperatures indicate that the solubilities of the crystalline modifications of silica are in the order tridymite > cristobalite > quartz, an order that parallels to some extent the chemical reactivity of these forms. Lower values for solubility of crystalline as compared to amorphous silica are consistent with the free-energy differences between them.

The solution in equilibrium with amorphous silica at ordinary temperatures contains monomeric monosilicic acid, $Si(OH)_4$. The acid is dibasic, dissociating in two steps (28):

$$Si(OH)_4 + H_2O \rightarrow SiO(OH)_3^- + H_3O^+ \qquad pK_1 = 9.8 \ (20°C)$$

$$SiO(OH)_3^- + H_2O \rightarrow SiO_2(OH)_2^{2-} + H_3O^+ \qquad pK_2 = 11.8 \ (20°C)$$

The possibility of six-coordinate silicon species in aqueous solution has been suggested by several workers; Raman studies have indicated, however, that monosilicic acid in solution contains a tetracoordinate silicon species (29).

Solutions of monosilicic acid may also be obtained by careful hydrolysis of tetrahalo-, tetraalkoxy-, or tetraacyloxysilanes, by electrolysis or acidification of alkali silicate solutions, or by ion exchange. By operating under carefully controlled conditions at low temperature and pH, solutions may be obtained that remain supersaturated with respect to amorphous silica for hours at temperatures ca 0°C. Eventually, however, polymerization reactions involving the formation of siloxane linkages occur, leading ultimately to the formation of colloidal particles and further aggregation or gel formation.

The basic features of silica polymerization and its technical relevance are summarized in ref. 9. A general polymerization scheme is shown in Figure 2. When a solution of $Si(OH)_4$ is formed (as by acidification of a solution of a soluble silicate) at a concentration greater than the solubility of amorphous silica (100–200 ppm), the monomer polymerizes to form dimers and higher molecular weight species. The mechanism is ionic, and the rate is proportional to OH^- concentration above pH 2 and to H^+ concentration below pH 2. Several mechanisms may be involved, since the reaction is reported to be third order below pH 2 and second order at higher pH. The polymerization occurs in such a way as to maximize formation of siloxane linkages (Si—O—Si), forming particles with internal siloxane linkages and external SiOH groups. Above pH 7, stabilized particles (sols) grow to radii of ca 100 nm. At lower pH or if salts are present to neutralize the charge on the growing particles, aggregation of particles occurs with the formation of chains and, ultimately, three-dimensional gel networks. Control of these processes by adjustment of pH and addition of coagulants is the basis of the technology for formation of the amorphous silicas described below.

Silica dissolves in water at high temperatures and pressures. For amorphous silica up to 200°C, the solubility in liquid water is given (30):

$$c = 0.382(13.6 + t) \times 10^{-3}$$

where c is the concentration of dissolved silica in wt % and t is the temperature in °C. The solubility of quartz under pressure in $H_2O(l)$ passes through a maximum at ca 330°C, at which temperature the saturated solution contains 0.07 wt % silica, and

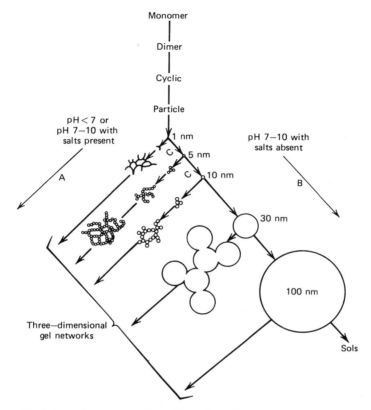

Figure 2.   Schematic representation of the polymerization of monosilicic acid (9).

declines rapidly as the temperature approaches the critical point (31). Vaporization of silica in steam is of potential importance to the degradation of refractories, eg, in coal-gasifier environments (32). The volatility of silica in steam is discussed in ref. 33. Solubility of quartz in steam may be as high as 61.8 wt % at 963 MPa (9500 atm) and 1050°C (34–37). At high pressures and temperatures, dimeric, trimeric, and higher oligomeric forms of H_2O become significant molecular constituents of steam, and thus probable reactants in the SiO_2–H_2O system. Vapor species including $Si(OH)_4$, $Si_2O(OH)_6$, and $[SiO(OH)_2]_x$ appear to be involved in these processes, with the oligomers more important at higher steam densities (33).

Forms of Silica

Crystalline Silica. Silica exists in a variety of polymorphic crystalline forms (23,38–40), in amorphous modifications, and as a liquid. The literature on the crystalline modifications is to some degree controversial. According to the conventional view of the polymorphism of silica, there are three main forms at atmospheric pressure: quartz, stable below ca 870°C; tridymite, stable from ca 870–1470°C; and cristobalite, stable from ca 1470°C to the melting point at ca 1723°C. In all these forms, the structures are based on SiO_4 tetrahedra linked in such a way that every oxygen atom is shared between two silicon atoms. The structures, however, are quite different in detail.

At the temperature limits of their stability ranges, these forms interconvert. The transformations involve a change in the secondary (non-nearest-neighbor) coordination and require the breaking and reformation of Si—O bonds. The transformation processes, known as reconstructive polymorphic transformations (41), are slow, as shown by the fact that the high temperature polymorphs can persist outside their normal stability range. The transformations are aided by or may require the presence of impurities or added mineralizers such as alkali metal oxides. Indeed, it has been suggested that tridymite cannot be formed at all in the absence of impurities, and some modern texts assert that pure SiO_2 occurs in only two forms, ie, quartz and cristobalite (42).

In addition to the reconstructive transformations, each of the main forms of silica undergoes one or more transformations of a different sort, the so-called high–low, displacive, or martensitic transformations. These involve relatively small structural rearrangements such as minor rotations of the tetrahedra without bond-breaking. In general they are facile and reversible. For example, low or α-quartz, the form stable at room temperature, transforms displacively to high or β-quartz at 574°C. This transformation is important in ceramic technology, since it involves a significant volume change that can lead to cracking of ceramic bodies containing large amounts of silica. The use of α and β to denote the forms of quartz is not uniform in the literature. Some authors use β to denote the low form. Cristobalite undergoes a similar transition at 270°C. The displacive transitions of tridymite are the subject of an extensive literature (38,43–44).

The melting point of high cristobalite is 1723 ± 5°C (International Temperature Scale of 1948) (45). Because of the slowness of the tridymite–cristobalite conversion, it is possible to observe the melting point of the metastable form at 1680°C. Similarly, quartz melts at a temperature lower than that of either cristobalite or tridymite, probably ca 1470°C, but the rate of fusion is comparable to the rate at which cristobalite is formed (46).

The transformations among the principal crystalline forms of silica may thus be represented in simplified form (41):

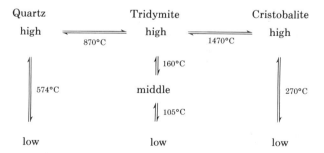

The vertical directions represent the facile, displacive polymorphic transitions, whereas the horizontal directions represent the sluggish reconstructive transitions. Some thermodynamic properties of these forms are given in Tables 2 and 3 (47).

The stable phases by definition show no changes in properties with time at constant temperature and pressure. Metastable or unstable phases may persist essentially indefinitely at temperatures and pressures outside their ranges of stability. Thus, tridymite and cristobalite may be investigated at temperatures within the stability

Table 2. Thermodynamic Properties of Quartz, Cristobalite, and Liquid SiO_2 [a]

Property	Low quartz	Low cristobalite	High cristobalite	Liquid SiO_2
heat of formation, $\Delta H^{\circ}_{f\,298.15\,K}$, kJ/mol [b]	-910.9 ± 2	-908.3 ± 2	-905.5	-902.7
entropy, $S^{\circ}_{298.15\,K}$, J/(mol·K) [b]	41.5 ± 1	43.40 ± 0.13	50.05	47.93 ± 1
heat capacity, $C^{\circ}_{p\,298.15\,K}$, J/(mol·K) [b]	44.59	44.95	26.58	44.18

[a] Ref. 47.
[b] To convert J to cal, divide by 4.184.

Table 3. Heat of Transformation of Various Forms of SiO_2 [a]

Transformation	Temperature, K	ΔH°, kJ/mol [b]
low quartz → high quartz	847 ± 1.5	0.73 ± 0.2
high quartz → high cristobalite	1079 ± 250	2.01 ± 0.6
low cristobalite → high cristobalite	5.3 ± 3	1.34 ± 0.3
high quartz → liquid	1696 ± 50	7.7 ± 0.8
high cristobalite → liquid	1996 ± 5	9.6 ± 2

[a] Ref. 47.
[b] To convert J to cal, divide by 4.184.

range of quartz and may undergo transitions of the high–low type at these temperatures. Such thermodynamically unstable but kinetically stable phases are said to be thermally stranded. Under particular conditions, a metastable or unstable phase may convert either to the phase which is stable under those conditions, or into some other phase which is lower in free energy than the first but also unstable or metastable. Pure quartz, for example, when heated to 867–1470°C, usually converts to a disordered cristobalite rather than to tridymite.

Another representation of the stability relations of the silica minerals is shown by Figure 3. This diagram, which is essentially that developed in the classical studies early in this century (48), illustrates the relationship of vapor pressure to temperature on the basis that vapor pressure increases with temperature, and that the form with the lowest vapor pressure is the most stable form. The actual values of the vapor pressures are largely unknown, and therefore, the ordinate must be considered to be only an indication of relative stabilities. The diagram does not show all the various forms of tridymite which now appear to have been identified.

In addition to the three principal polymorphs of silica, three high pressure phases have been prepared: keatite [17679-64-0], coesite, and stishovite. The pressure–temperature diagram in Figure 4 shows the approximate stability relationships of coesite, quartz, tridymite, and cristobalite. A number of other phases, eg, silica O, silica X, silicalite, and a cubic form derived from the mineral melanophlogite have been identified (9) along with a structurally unique fibrous form, Silica W.

Quartz. The atomic arrangement in high quartz (β-quartz) consists of linked tetrahedra-forming helixes which, in a given crystal, are either right- or left-handed (49). The hexagonal unit cell contains three SiO_2 units with $a_0 = 0.501$ nm and $c_0 = 0.547$ nm at ca 600°C (50), space group $P6_22$. The Si—O distance is 0.162 nm. Density at 600°C is ca 2.53 g/cm³. The structure of low quartz (α-quartz) is closely similar, but somewhat less regular (Fig. 5). The unit cell has dimensions $a_0 = 0.4913$ nm and c_0

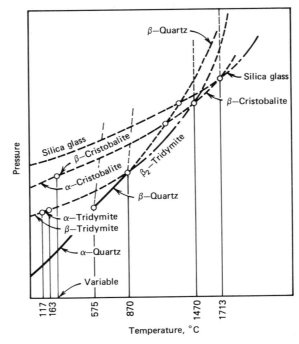

Figure 3. Stability relations of the silica minerals.

= 0.5405 nm with three formula units in the hexagonal unit cell (space group P3$_2$) and two slightly different Si—O distances 0.1597 and 0.1617 nm in the tetrahedra (2). Density at 0°C is 2.65 g/cm^3. In the high–low inversion the atoms in the structure are only slightly displaced. Quartz is birefringent (n_0 = 1.5442, n_E = 1.5533, Na$_D$ line) and optically active; individual crystals are either dextro- or levorotatory with α = 21.71°/mm for the sodium D line. The sense of rotation is unchanged on passing from the low to the high form.

The most common form of silica is low quartz, which by virtue of its piezoelectric and other properties is of considerable commercial importance (see Silica, synthetic quartz). Crystallographic data cited are representative values from the literature. Ref. 51 gives a detailed compilation of crystal structure data.

The high–low thermal inversion of quartz is accompanied by discontinuities in many physical properties, including density, indexes of refraction and birefringence, optical rotatory power, dielectric constant, and thermal expansion coefficients. It is not established, however, that all these changes are precisely coincident with the inversion point, ie, the temperature at which a crystal undergoes an abrupt, reversible change in dimensions and structure. The most common method of observing the transition is probably differential thermal analysis, by which the absorption or evolution of heat accompanying the change can be readily detected. The accepted value of the inversion temperature is 574 ± 1.5°C under atmospheric pressure, measured with rising temperature. The inversion temperature of natural-quartz samples may vary by 40°C, and a variation of as much as 160°C has been found for synthetic samples (52). This variation is believed to be associated with the formation of solid solutions with impurities. Substitution of small amounts of germanium for silicon raises the

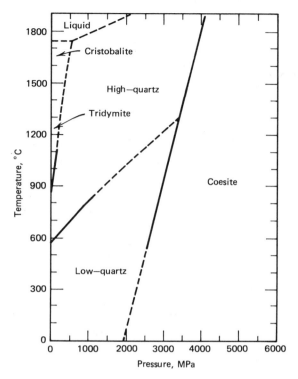

Figure 4. Pressure–temperature diagram for the more familiar SiO₂ polymorphs (40). To convert MPa to atm, divide by 0.101.

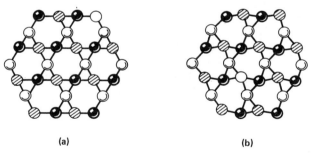

Figure 5. Schematic illustration of structural relationship between (**a**) high-temperature and (**b**) low-temperature forms of quartz (41).

inversion temperature, whereas introduction of interstitial lithium (with substitution of aluminum preserving charge balance) lowers it.

The pressure dependence of the high–low inversion was investigated up to 260 MPa (ca 2570 atm) (53) and up to 1 GPa (ca 9900 atm) (54). The change in the inversion temperature was found in the latter study to obey the equation:

$$\Delta T = 1.6 + 2.871 \times 10^{-2}\, p - 4.284 \times 10^{-7}\, p^2$$

where p is the pressure expressed in 10^5 Pa (bars).

Tridymite. Tridymite is reported to be the silica form stable from 870–1470°C at atmospheric pressure (41). Owing to the sluggishness of the reconstructive tridymite–quartz conversion, which requires mineralizers such as sodium tungstate, alkali metal oxide, or the action of water under pressure, tridymite may persist as a metastable phase below 870°C. It occurs in volcanic rocks and stony meteorites.

The two high–low inversions in tridymite were observed at 117 and 163°C (48). The three forms were designated α, β_1, and β_2 in order of increasing temperature, and named low tridymite, lower-high tridymite, and upper-high tridymite (43). Further work has, however, revealed a considerably more complex situation (38). Several varieties of tridymite have been suggested, including a monotropic tridymite M which is transformed to stable tridymite S by a transition of the reconstructive type and possibly a highly disordered tridymite U (44). Furthermore, six modifications of tridymite S are recognized, denoted S-I to S-VI in order of rising temperatures, with high–low inversions at 64, 117, 163, 210, 475°C, and three modifications of tridymite M, with inversions at 117 and 163°C (38).

The structure of tridymite is more open than that of quartz and is similar to that of cristobalite (Fig. 6). The high temperature form, probably S-IV, has a hexagonal unit cell containing four SiO_2 units with $a_0 = 0.503$ nm and $c_0 = 0.822$ nm > 200°C, space group P6$_3$/mmc. The Si—O distance is 0.152 nm. Density at 200°C is ca 2.22 g/cm^3.

The existence of tridymite as a distinct phase of pure crystalline silica has been questioned (39,55–60). According to this view, the only true crystalline phases of pure silica at atmospheric pressure are quartz and a highly ordered three-layer cristobalite, with a transition temperature variously estimated as 806 ± 250°C to ca 1050°C (47,57). Tridymites are considered to be defect structures with two-layer sequences predominating. The stability of tridymite as found in natural samples and in fired silica bricks has been attributed to the presence of foreign ions. This view is, however, disputed by others who cite evidence of the formation of tridymite from very pure silicon and water and of the conversion of tridymite M, but not tridymite S, to cristobalite below 1470°C (44). It has been suggested that the phase relations of silica are determined by the purity of the system (39). However, the assumption of existence of tridymite phases is well established in the technical literature pertinent to practical work.

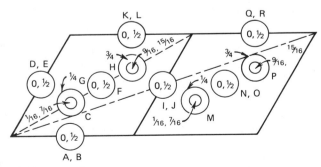

Figure 6.   A projection on its base of the atomic contents of two of the hexagonal units of high, or β-, tridymite (50).

Cristobalite. Cristobalite is the high-temperature solid form of silica, stable from 1470–1723°C. It is capable of metastable existence and is often obtained below 1470°C. It occurs naturally in some volcanic rocks. Pure, well-crystallized cristobalite has a high–low inversion at ca 270°C, but the inversion temperature is variable and may occur as low as 170°C (61). The inversion temperature and the sharpness of the transition are dependent on the history of the sample and may be associated with the presence of foreign substances and with crystal perfection (61). The structure of high cristobalite is cubic, with $a_0 = 0.716$ nm at 290°C in the eight-molecule unit cell (Fig. 7), space group $P2_3$. The oxygen atoms of the SiO_4 tetrahedra of tridymite and cristobalite have the relationship of hexagonal close-packing to cubic close-packing, ie, the difference between the structures is like the difference in wurtzite and zinc blende structures. The idealized tridymite structure thus involves two-layer sequences of SiO_4 tetrahedra, and the idealized cristobalite structures, three-layer sequences (57).

Keatite. Keatite has been prepared (62) by the crystallization of amorphous precipitated silica in a hydrothermal bomb from dilute alkali hydroxide or carbonate solutions at 380–585°C and of 35–120 MPa (345–1180 atm). The structure (63) is tetragonal with twelve SiO_2 units in the unit cell; $a_0 = 0.745$ nm and $c_0 = 0.8604$ nm, space group $P4_2$. Keatite has a negative volumetric expansion coefficient from 20–550°C. It is unchanged by heating at 1100°C, but is transformed completely to cristobalite in 3 h at 1620°C.

Coesite. Coesite, the second most dense phase of silica (3.01 g/cm^3), was first prepared in the laboratory by heating a mixture of sodium metasilicate and diammonium hydrogen phosphate or another mineralizer at 500–800°C at 1.5–3.5 GPa (14,800–34,540 atm). Coesite has also been prepared by oxidation of silicon with silver carbonate under pressure (64). The structure is monoclinic, with $a_0 = b_0 = 0.717$ nm, $c_0 = 1.238$ nm, $\gamma = 120°$, and sixteen SiO_2 units per unit cell, space group $C2/c$. Coesite persists as a basically stranded phase at atmospheric pressure and is a stable phase at high pressure. Density values are 2.93–3.01 g/cm^3 (40). The existence of a true equilibrium between coesite and quartz has been demonstrated. Pressure–temperature diagrams for the coesite–quartz equilibrium are summarized in reference 23. Coesite has been found in nature in the Meteor Crater in Arizona.

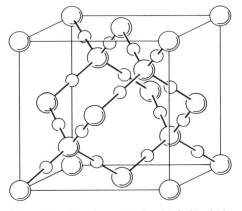

Figure 7.   Structure of high cristobalite (41).

Stishovite. Stishovite was first prepared (65) in the laboratory in 1961 at 1200–1400°C and pressures >16 GPa (158,000 atm). It was subsequently discovered, along with natural coesite, in the Arizona meteor crater. These minerals have been suggested to be geological indicators of meteorite impact structures. Stishovite is the densest known phase of silica (4.35 g/cm^3). The structure (space group P4$_2$/nmn) is like that of rutile, TiO$_2$, and the silicon atom is octahedrally coordinated by six oxygens (66); four Si—O distances in the octahedra are 0.176 nm; two are 0.181 nm.

Silica W. This phase is the lightest silica phase (1.97 g/cm^3). It was prepared by the disproportionation of silicon monoxide (67). It forms microcrystalline fibers and differs radically from all other silica phases because the (SiO$_4$) tetrahedra share edges rather than corners. The structure consists of parallel chains and is analogous to that of SiS$_2$ and SiSe$_2$. Silica W reacts rapidly with traces of water vapor, transforming to amorphous silica with silica tetrahedra sharing corners rather than edges.

Microcrystalline Silicas. Various microcrystalline (cryptocrystalline) materials such as flint, chert, and diatomaceous earth are found in nature (see Diatomite). They may arise from amorphous silica, often of biogenic origin, which undergoes compaction and microcrystallization over geologic time.

Noncrystalline Silica. The noncrystalline forms of silica include bulk vitreous silica and a variety of other amorphous types.

Vitreous silica (silica glass) is essentially a supercooled liquid formed by fusion and subsequent cooling of crystalline silica. It is found in nature in fulgerites, ie, fused bodies resulting from lightning striking sand.

Liquid silica is highly viscous, and supercooling to the glassy form occurs readily. In practice, vitreous silica is prepared by fusion of crystalline quartz or quartz sand. The primary source of high quality silica sand used in the United States is Brazil, though commonly available domestic quartz sand is available for applications not requiring highest purity. Vitreous silica is also made by flame or plasma hydrolysis of silicon tetrachloride, by thermal decomposition of silicate esters, or by sputtering of SiO$_2$. Glasses prepared by flame-fusion processes may contain significant amounts (>1000 ppm) of hydroxyl impurity which affect optical transmission as well as thermal and mechanical properties.

The structure of vitreous silica is a continuous random network of SiO$_4$ tetrahedra, linked through sharing of corners. It differs from crystalline silica in having a broader distribution of Si—O—Si bond angles and a more random distribution of one tetrahedron with respect to another (41). The density is 2.2 g/cm^3.

The properties of high quality vitreous silica which determine its uses include high chemical resistance, low coefficient of thermal expansion (5.5 × 10^{-7}°C), high thermal shock resistance, high electrical resistivity, and high optical transmission, especially in the ultraviolet. Bulk vitreous silica is difficult to work because of the absence of network-modifying ions present in common glass formulations. An extensive review of the properties and structure of vitreous silica is given in ref. 68.

Amorphous silica exists also in a variety of forms which are composed of small particles, possibly aggregated. Commonly encountered products include colloidal silica, silica gels, precipitated silica, and fumed or pyrogenic silica (9,69).

Amorphous silicas are characterized by small ultimate particle size and high specific surface area. Their surfaces may be substantially anhydrous or may contain silanol (SiOH) groups. They are frequently viewed as condensation polymers of silicic acid, Si(OH)$_4$ (see Properties, above).

Colloidal silicas (silica sols) are stable dispersions of amorphous silica particles in water. Commercial products contain silica particles with diameters of ca 3–100 nm, specific surface areas of 50–270 m^2/g, with silica contents of 15–50 wt %. They contain small amounts (<1 wt %) of stabilizers, most commonly sodium ions.

Silica gels contain three-dimensional networks of aggregated silica particles of colloidal dimensions. As formed, the pores are filled with the medium (usually water) in which the gel is prepared. Simple removal of the liquid results in extensive shrinkage owing to surface-tension forces. Silica gels dried this way are termed xerogels. If the water in the pores is replaced by alcohol and the gel is heated under pressure above the critical temperature of the liquid, resulting in the disappearance of the liquid–vapor interface, surface-tension effects are absent and a very voluminous dry silica gel (aerogel) is obtained.

Precipitated silicas are powders obtained by coagulation of silica particles from an aqueous medium under the influence of high salt concentrations or other coagulants.

Fumed silicas (aerosils, pyrogenic silica) are produced by vapor-phase processes, generally by the vapor-phase hydrolysis of silicon tetrahalides. Other methods include vaporization of SiO_2, vaporization and oxidation of Si, and high temperature oxidation and hydrolysis of silicon compounds such as silicate esters.

Vaporization. Silica vaporizes principally by dissociation to gaseous SiO and O_2; these are the predominant vapor species, with some contribution from atomic oxygen and gaseous SiO_2 (70–71). The total vapor pressure over the liquid at the melting point is in the range 1–10 Pa (10^{-4}–10^{-5} atm). The boiling point of silica is estimated as 2797 ± 75°C (72). Experimental values ranging from 2230–3500°C have been reported. The heat of vaporization of SiO_2 at the melting point is given as 560 kJ/mol (134 kcal/mol), whereas the heat of the reaction

$$SiO_2(l) \rightarrow SiO + \tfrac{1}{2} O_2$$

is 750 kJ/mol (179 kcal/mol).

Health and Safety Factors

An overview of silica in biological systems is given in ref. 9. Silica is chemically inert and, in bulk, does not appear to have significant toxic effects. Food-grade silica is permitted as an additive in food for human and animal consumption. The principal hazard has been associated with inhalation of dust over long periods, particularly of crystalline silica, which is the main cause of a disabling pulmonary disease known as silicosis (73). Federal standards are in effect for work-place exposures (74).

Uses

The diversity of silica forms and their properties leads to a broad range of applications.

Silica is the basic raw material of the glass industry. The vitreous silica structure forms the basis of commercial glass compositions, whose properties are modified by the addition of other metal oxides (see Glass). Silica is a main constituent of ceramics; for example, refractory silica brick containing small amounts of Al_2O_3 is used as roof brick for open-hearth furnaces at temperatures >1600°C (see Ceramics; Refractories).

In space technology, fused silica is used in windows for the Apollo spacecraft and in the thermal protection tiles on the Columbia space-shuttle orbiter (75).

Quartz. Because of its piezoelectric properties, synthetic quartz is used for frequency control in electrical oscillators and filters and in electromechanical transducers.

Vitreous Silica. Because of its chemical and thermal resistance, vitreous silica is used in laboratory glassware (up to 1000°C), in furnaces and radiant heaters, and as lamp envelopes. Silica fibers are used in precision instruments, eg, balances and thermal-expansion apparatus. Thin films of vitreous silica are applied to dielectric components of integrated surfaces. Because of optical transparency, particularly in the ultraviolet region, vitreous silica is used for prisms, cells, windows, and other optical components. The low thermal expansion has made vitreous silica a material of choice for astronomical telescope-mirror blanks.

Amorphous Silica. Colloidal silicas are used as binders and stiffeners, for modifying frictional properties of waxes and fibers, modifying adhesion between surfaces, reinforcing polymers, as polishing agents, and as viscosity agents. Silica gels are used to modify adhesives and the viscosity and thixotropy of liquids; as adsorbents, drying agents, catalyst supports; and for other related purposes (9).

BIBLIOGRAPHY

"Silica and Silicate Minerals" treated in *ECT* 1st ed., under "Silica and Inorganic Silicates," Vol. 12, pp. 268–303, by George W. Morey, Geophysical Laboratory, Carnegie Institution of Washington; "Silica (Introduction)" in *ECT* 2nd ed., Vol. 18, pp. 46–61, by Thomas D. Coyle, National Bureau of Standards.

1. L. Pauling, *The Nature of the Chemical Bond*, 3rd ed., Cornell University Press, Ithaca, N.Y., 1960.
2. A. F. Wells, *Crystal Chemistry*, 4th ed., Oxford University Press, London, 1975, Chapt. 23.
3. W. E. Addison, *Structural Principles in Inorganic Compounds*, John Wiley & Sons, Inc., N.Y., 1963, p. 141.
4. International Union of Pure and Applied Chemistry (IUPAC), *Nomenclature of Inorganic Chemistry*, 2nd ed., Butterworths, London, 1970.
5. L. Pauling, *J. Phys. Chem.* **56,** 361 (1952).
6. A. J. Barry, W. H. Daudt, J. J. Domicone, and J. Gilkey, *J. Am. Chem. Soc.* **77,** 4248 (1955).
7. W. Noll, *Angew. Chem. Intern. Ed. Engl.* **2,** 73 (1963).
8. G. J. Young, *J. Colloid Sci.* **13,** 67 (1958).
9. R. K. Iler, *The Chemistry of Silica*, John Wiley & Sons, Inc., New York, 1979.
10. H. P. Boehm, *Adv. Catal.* **16,** 179 (1966).
11. R. Calas, P. Pascal, and J. Wyart, *Nouveau Traite de Chimie Minerale*, Vol. 8, Pt. 2, Masson et Cie, Paris, France, 1965.
12. *Gmelins Handbuch der Anorganischen Chemie*, Vol. 15, Pt. B, Verlag Chemie G.m.b.H., Weinheim, FRG, 1959.
13. F. M. Ernsberger, *J. Phys. Chem. Solids* **13,** 347 (1960).
14. V. N. Sveshnikova and E. P. Damlova, *Zh. Neorg. Khim.* **2,** 928 (1957).
15. H. L. Jackson, F. D. Marsh, and E. L. Muetterties, *Inorg. Chem.* **2,** 43 (1963).
16. F. Trombe and M. Foëx, *Compt. Rend.* **216,** 268 (1943).
17. O. Ruff, W. Menzel, and W. Neumann, *Z. Anorg. Chem.* **208,** 293 (1932).
18. E. B. R. Prideaux and C. B. Cox, *J. Chem. Soc.* 739 (1928).
19. H. J. Emeleus and A. A. Woolf, *J. Chem. Soc.* 164 (1950).
20. P. Baumgarten and W. Bruns, *Chem. Ber.* **74B,** 1232 (1941).
21. F. E. Brinckman and G. Gordon, *Proceedings of the International Symposium on Decomposition of Organometallic Compounds to Refractory Ceramics, Metals, and Metal Alloys*, Nov. 1967, Dayton, Ohio.
22. L. White and O. K. Rice, *J. Am. Chem. Soc.* **69,** 267 (1947).

23. E. M. Levin, C. R. Robbins, and H. F. McMurdie, *Phase Diagrams for Ceramists*, American Ceramic Society, Columbus, Ohio, 1964, and supplements 1969, 1975, 1981.
24. Ger. Pat. 1,028,784 (April 24, 1958), H. Kautsky.
25. E. Daubach, *Z. Naturforsch.* **8B**, 58 (1953).
26. R. E. Benson and J. E. Castle, *J. Phys. Chem.* **62**, 840 (1958).
27. W. Stober, *Adv. Chem. Ser.* **67**, 161 (1967).
28. S. A. Greenberg, *J. Chem. Ed.* **36**, 218 (1959).
29. D. Fortnum and J. O. Edwards, *J. Inorg. Nucl. Chem.* **2**, 264 (1956).
30. A. S. Berezhnoi, *Silicon and Its Binary Systems*, Consultants Bureau, New York, 1960, p. 137.
31. G. C. Kennedy, *Econ. Geol.* **45**, 639 (1950).
32. M-C. Cheng and I. B. Cutler, *J. Am. Ceram. Soc.* **62**, 593 (1979).
33. J. W. Hastie, *High Temperature Vapors*, Academic Press, Inc., New York, 1975.
34. G. C. Kennedy, G. J. Wasserburg, H. C. Heard, and R. C. Newton, *Am. J. Sci.* **260**, 501 (1962).
35. A. I. Semenova and D. S. Tsiklis, *Zh. Fiz. Khim.* **44**, 205 (1970).
36. E. L. Brady, *J. Phys. Chem.* **57**, 706 (1953).
37. O. Glemser and H. G. Wendtlandt, *Adv. Inorg. Chem. Radiochem.* **5**, 215 (1963).
38. R. B. Sosman, *The Phases of Silica*, Rutgers University Press, New Brunswick, N.J., 1965.
39. N. A. Toropov, V. P. Barzakovskii, I. A. Bondai, and Yu. P. Udalov, *Handbook of Phase Diagrams of Silicate Systems*, Vol. II, Israel Program for Scientific Translations, Jerusalem, Israel, 1972.
40. C. Frondel, *Dana's System of Mineralogy*, Vol. 3, John Wiley & Sons, Inc., New York, 1962.
41. F. A. Cotton and G. Wilkinson, *Advanced Inorganic Chemistry*, 3rd ed., Wiley-Interscience, New York, 1976.
42. W. D. Kingery, H. K. Bowen, and D. R. Uhlmann, *Introduction to Ceramics*, 2nd ed., John Wiley & Sons, Inc., N.Y., 1976.
43. R. B. Sosman, *The Properties of Silica*, American Chemical Society Monograph 37, Reinhold Publishing Corporation, New York, 1927.
44. V. G. Hill and R. Roy, *Trans. Brit. Ceram. Soc.* **57**, 496 (1958).
45. S. J. Schneider, *Compilation of the Melting Points of the Metal Oxides*, National Bureau of Standards Monograph 68, Washington, D.C., 1963.
46. Ref. 38, Chapt. 7.
47. Joint Army-Navy-Air Force (JANAF), *Thermochemical Tables*, 2nd ed., NSRDS-NBS 37, 1971.
48. C. N. Fenner, *Am. J. Sci.* **36**, 331 (1913).
49. W. G. Moffatt, G. W. Pearsall, and J. Wulff, *The Structure and Properties of Materials*, Vol. 1, John Wiley & Sons, Inc., New York, 1964.
50. R. W. G. Wyckoff, *Crystal Structures*, Vol. 1, 2nd ed., Interscience Publishers, a division of John Wiley & Sons, Inc., New York, 1963.
51. *Crystal Data, Determinative Tables*, Vol. 2, J. D. H. Donnay and H. M. Ondik, eds., *Inorganic Compounds*, 1973; Vol. 4, H. M. Ondik and A. D. Mighell, eds., *Inorganic Compounds*, 1978, National Bureau of Standards/Joint Committee on Powder Diffraction, Washington, D.C.
52. M. L. Keith and O. F. Tuttle, *Am. J. Sci.* **250A**, 203 (1952).
53. R. E. Gibson, *J. Phys. Chem.* **32**, 1197 (1928).
54. H. S. Yoder, *Trans. Am. Geophys. Union* **31**, 827 (1950).
55. Ref. 30, p. 117.
56. O. W. Flörke, *Silikattechn.* **12**, 304 (1961).
57. O. W. Flörke, *Ber. Deut. Keram. Ges.* **32**, 369 (1955).
58. Y. E. Budnikov and Yu E. Pivinskii, *Russ. Chem. Rev.* 210 (1967).
59. W. F. Ford, *The Effect of Heat on Ceramics*, MacLaren & Sons, London, 1967, pp. 82–83.
60. R. Wollast, *Proceedings of the 8th Conference on the Silicate Industry*, Akadémiai Kiado, Budapest, Hungary, 1966.
61. O. W. Flörke, *Ber. Deut. Keram. Ges.* **33**, 319 (1956).
62. P. O. Keat, *Science* **120**, 328 (1951).
63. J. Shropshire, P. P. Keat, and P. A. Vaughan, *Z. Kryst.* **112**, 409 (1959).
64. L. Coes, *Science* **118**, 131 (1953).
65. S. M. Stishov and S. V. Popova, *Geokhimiya*, 837 (1961).
66. S. M. Stishov and N. V. Belov, *Dokl. Akad. Nauk* **143**, 951 (1962).
67. A. Weiss and W. Weiss, *Z. Anorg. Allgem. Chem.* **276**, 95 (1954).
68. R. Brückner, *J. Non-Cryst. Solids* **5**, 123, 177 (1970).
69. B. Alexander, *Silica and Me*, Anchor Books, Doubleday & Co., Inc., Garden City, N.Y., 1967.

70. R. F. Porter, W. A. Chupka, and M. C. Inghram, *J. Chem. Phys.* **23,** 216 (1955).
71. L. Brewer and D. F. Mastick, *J. Chem. Phys.* **19,** 834 (1951).
72. H. L. Schick, *Chem. Rev.* **60,** 331 (1960).
73. N. I. Sax, *Dangerous Properties of Industrial Materials*, 5th ed., Van Nostrand Reinhold Company, New York, 1979.
74. M. Sittig, *Handbook of Toxic and Hazardous Materials*, Noyes Publications, Park Ridge, N.J., 1981.
75. L. J. Korb, C. A. Morant, R. M. Calland, and C. S. Thatcher, *Ceram. Bull.* **60,** 1188 (1981).

T. D. COYLE
National Bureau of Standards

AMORPHOUS SILICA

The word amorphous, when used to describe silica, denotes a lack of crystal structure, as defined by x-ray diffraction. Some short-range organization may be present and is indicated by electron diffraction studies but this ordering gives no sharp x-ray diffraction pattern. Silica [7631-86-9], SiO_2, can be either hydrated (up to ca 14%) or anhydrous. The chemical bonding in amorphous silica is of several types, including siloxane (—Si—O—Si—), silanol (—Si—O—H), and at the surface, silane (—Si—H) or organic silicon (—Si—O—R or —Si—C—R).

Early interest in amorphous silica was purely academic. It was reported in 1640 that amorphous silica in the presence of excess alkali became a liquid, and that subsequent neutralization of the liquid with acid caused precipitation of silica. In 1861 it was found that a silica sol formed a hydrogel if the sol was not stabilized by dialysis to remove electrolytes. Interest in silica gel increased during World War I when it was considered for use as an adsorbent in gas masks. A method of acid gelation of alkali silicate was perfected in 1919, followed shortly thereafter by commercial production of silica gel for use as an adsorbent and desiccant (see Drying agents). The use of precipitated silicas or sols was limited at this time by the tendency of silica sols that contained more than 10% SiO_2 to solidify to a gel. An ion-exchange process for removing sodium from a solution of sodium silicate was patented in 1941, and a gel-washing process followed by autoclaving patented in 1945 allowed stable sols to be prepared that contained 15–20% silica. Further refinements resulted in formation of silica particles of uniform size and high purity; commercial production of colloidal silica in the United States began before 1950. Amorphous silica products are now manufactured by ca 30 companies in the United States, The Federal Republic of Germany, the United Kingdom, France, Japan, Finland, and Sweden (see under Economic Aspects).

Amorphous silica can be broadly divided into three categories: vitreous silica or glass made by fusing quartz; silica M made by irradiating either amorphous or crystalline silica with high speed neutrons; and microamorphous silica. Silica M is a dense form of amorphous silica; it is thermally unstable and converts to quartz at 930°C after 16 h. Microamorphous silica includes sols, gels, powders, and porous glasses, all of which are composed of ultimate particles or structural units <1 μm in diameter. These silicas have high surface areas, generally >3 m²/g. Microamorphous silica can be further divided into microparticulate silica, microscopic sheets and fibers, and hydrated amorphous silica (1). The microparticulate silicas are the most important group commercially and include pyrogenic silicas and silicas precipitated from aqueous solution. Pyrogenic silicas are formed at high temperature by condensation of SiO_2 from the vapor phase, or at lower temperature by chemical reaction in the vapor phase followed by condensation. Silica formed in aqueous solution can occur as sols, gels, or particles (see Fig. 1). A gel has a three-dimensional, continuous structure, whereas a sol is a stable dispersion of fine particles. Macroscopic particles are formed by aggregation of smaller particles from either a gel or sol. Microscopic sheets of amorphous silica are prepared by oxidation and hydrolysis of gaseous $SiCl_4$ or SiF_4, followed by polymerization of silicic acid in water, by freezing or coagulation of silica sols, or by hydrolysis of $HSiCl_3$ in ether followed by evaporation of the ether. Amorphous silica fibers are prepared by drying thin films of sols, oxidation of SiO, or chilling fibrous, very high surface-area silica. Hydrated amorphous silica is prepared by polymerization of silicic acid in water at low temperature in slightly acidic solution. The water content of the resulting solid silica can be as high as 14%, which corresponds to one mole of water for two moles of silica. This water is stable to 60°C and is not lost by evaporation at room temperature.

Properties of silica sols, silica gels, precipitated silica, and pyrogenic silica are given in Table 1.

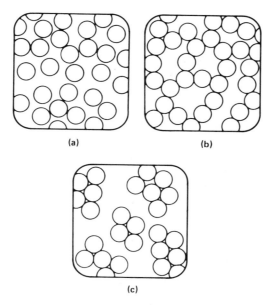

(a)

(b)

(c)

Figure 1. Ultimate particles in common forms of colloidal silica. Aggregation is actually in three dimensions but is represented here only in two. (**a**), sol; (**b**), gel; (**c**), powder (2).

Table 1. Properties of Different Forms of Amorphous Silica

Property	Silica sols	Dry silica gels	Silica precipitated from solution	Pyrogenic silica
SiO_2, %	10–50	96.5–99.6	80–90	99.7–99.9
CaO, %	na	na	0.1–4	na
Na_2O, %	0.1–0.8	0–1	0–1.5	na
wt loss, %				
at 105°C	50–80	na	5–7	0.5–2.5
at 1200°C	50–90	2–17.5	10–14	0.5–2.5
ultimate particle size, nm	5–100	1–100	10–25	1–100
aggregate particle size, μm		3–25	1–10	2–3
surface area, m^2/g	50–700	200–700	45–700	15–400
pH, aqueous suspension	3–5, 8–11	2.3–7.4	4–9	3.5–8
apparent or bulk density, g/cm^3	1.2–1.4	0.1–0.8	0.03–0.3	0.03–0.12
true density, g/cm^3	2.2–2.3	2.22	2.0–2.1	2.16
refractive index, n_D	1.35–1.45	1.35–1.45	1.45	1.45
oil absorption, g/g		0.9–3.15	1–3	0.5–2.8

Heterogeneous Reactions

Dissolution. Amorphous silica dissolves (or depolymerizes) in water according to the following equation:

$$SiO_2(s) + 2\,H_2O(l) \longrightarrow \qquad H_4SiO_4(aq)$$

$$\text{orthosilicic acid, monosilicic acid}$$

$H_4SiO_4(aq)$ can also be expressed as $Si(OH)_4(aq)$ or $H_2SiO_3(aq)$. The solubility of amorphous silica in water at 25°C is in the range of 80–130 ppm SiO_2 (1.4–2.2 mmol/kg). This range of solubility is probably due to changes in the silica surface during dissolution. An area with a small positive radius of curvature dissolves most extensively, and an area with a small negative radius of curvature dissolves least extensively. The result is a flat surface of lower area and lower interfacial energy and hence lower solubility. The solubility of bulk silica at 25°C is ca 70 ppm SiO_2 (1.2 mmol/kg). Because of this surface chemistry, particle size affects solubility; particles with a diameter of less than ca 4 nm have a progressively greater solubility than larger particles (3). Several solubility studies have observed a decrease in solubility with time as the silica surface becomes more uniform, which makes determination of initial solubility difficult.

The solubility of amorphous silica in water increases with increasing temperature and increasing pressure (4). The solubility at 101.3 kPa (1 atm) and 0°C is ca 50 ppm SiO_2 (0.8 mmol/kg), increasing to 100 ppm (1.7 mmol/kg) at 25°C, and 750 ppm (12.5 mmol/kg) at 100°C. The solubility increases by ca 50% as pressure increases from 101.3 kPa to 101 MPa (1–1000 atm); the experiments were conducted in the 0–25°C range.

The solubility of amorphous silica in neutral dilute aqueous salt solutions is only slightly less than in pure water. The solubility in aqueous solutions at concentrations above ca 0.5 M decreases and is probably related to the amount of water available for hydration. Amorphous silica is essentially insoluble in methanol; solubility in methanol–water mixtures decreases with increasing percentage of methanol. Small amounts of impurities (especially aluminum or iron) in solid silica cause a decrease in solubility. Acid cleaning of impure silica to remove metal ions increases the solubility. Natural biogenic silica often has an organic coating that inhibits dissolution; it can be removed by acid washing. Solubility is affected by the crystal structure of the solid phase: amorphous silica has the highest solubility and quartz the lowest (ca 6 ppm SiO_2 or 0.1 mmol/kg at 25°C). Hence, theoretically, amorphous silica should dissolve and quartz should precipitate. However, at the earth's surface this is a very slow process occurring over geologic time (4–6).

Below pH 9 the solubility of amorphous silica is independent of pH; above pH 9 the solubility increases because of increased ionization of silicic acid (silica hydrate). The reactions for silicic acid dissociation at 25°C and 101 kPa are

$$H_2SiO_3(aq) \rightarrow H^+(aq) + HSiO_3^-(aq) \qquad K_1 = 1.6 \times 10^{-10}$$
$$\text{metasilicic acid, monosilicic acid}$$

$$HSiO_3^-(aq) \rightarrow H^+(aq) + SiO_3^{2-}(aq) \qquad K_2 = 7.4 \times 10^{-13}$$

Silicic acid is a very weak acid. At pH 7, only 0.2% of the dissolved silica occurs as silicate ion; at pH >9, the dimers $Si_2O_2(OH)_5^-$ and $Si_2O_3(OH)_4^{2-}$ also become important species in addition to $H_3SiO_4^-$, and above pH 11, $H_2SiO_4^{2-}$ and $Si_2O_4(OH)_3^{3-}$. At pH 10, undissociated silicic acid is less than 10%. The point of zero charge for the silica surface is approximately pH = 2 (7).

The dissolution of amorphous silica follows first-order kinetic behavior according to the equation

$$dc/dt = k_2S(c_e - c)$$

where c = concentration at time t, k_2 = rate constant for deposition, S = surface area of the solid silica per unit volume of solution, and c_e = the equilibrium concentration of silica. In neutral solution

$$\log[(c_e - c)/c_e] = -k_2St$$

Small particles dissolve more rapidly than large ones, and hydrated amorphous silica dissolves more rapidly than anhydrous amorphous silica. The dissolution is catalyzed by hydroxyl; fluoride can act as a catalyst at low pH. Dissolved salts increase the rate of dissolution in neutral solution (8).

Molecular Precipitation. Amorphous silica is precipitated from a supersaturated solution. Supersaturation is obtained by concentrating an undersaturated solution, cooling a hot saturated solution, or generating $Si(OH)_4$ by hydrolysis of a silica ester, SiH_4, SiS_2, $SiCl_4$, or Si. Monomeric silica can be deposited from supersaturated solution onto an oxide surface containing hydroxyl groups. Certain living organisms are able to precipitate solid silica even from undersaturated solutions (see below). Molecular silica is precipitated by such a process which is essentially the reverse of dissolution. These condensation reactions are catalyzed by hydroxyl ion and are accelerated by the presence of dissolved salts (9).

Polymerization. Dissolved silica undergoes polymerization to give discrete particles which associate to give chains and networks. The three stages of reaction are polymerization of monomers to form particles, growth of particles, and linking particles to give chains and then networks. In basic solution, the particles grow in size and decrease in number. In acid solution or in the presence of flocculating salts, particles aggregate into three dimensional networks and form gels (see Fig. 2). Solid hydrated polymeric amorphous silica is sometimes called silicic acid.

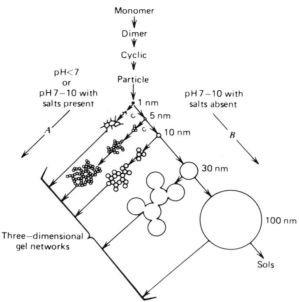

Figure 2. Polymerization behavior of silica. In basic solution (*B*), particles in sol grow in size with decrease in numbers; in acid solution or in presence of flocculating salts (*A*), particles aggregate into three-dimensional networks and form gels (10).

At concentrations below the solubility of amorphous silica, monosilicic acid, H_4SiO_4, is in true solution in water. Polymerization occurs only when solubility is exceeded and there is no solid phase present on which the silica can be deposited. The rate of polymerization is proportional to the concentration of hydroxyl ion at pH >2 and to the concentration of hydrogen ion at pH <2. Bonds are formed through collisions. Because of the tendency to maximize Si—O—Si bonds and minimize Si—O—H bonds, polymers condense to form ring structures early in the polymerization process. These ring structures become nuclei on which particle growth occurs. Because of the solubility difference, small particles dissolve, whereas larger ones grow (Ostwald ripening). The presence of dissolved salts neutralizes surface charges and thus enhances aggregation. Increasing temperature either increases the growth of sol particles or causes gel formation. The rate of aggregation also increases with dissolved silica concentration in solution. As polymerization occurs, the solution still contains dissolved silica at a concentration equivalent to the solubility of amorphous silica (11–14).

Organic Reactions. Silicon forms chelate-type bonds with some oxygen- and nitrogen-containing organic compounds; silicon is then hexacoordinated. Silicon may also occur as a chelate in humic compounds. Polymerized silicic acid can attach to certain polar organic compounds like ethers, amides, alcohols, ketones, and amines through hydrogen bonds, and sometimes coacervates are formed. Silica surfaces and poly(silicic acid) also undergo esterification with alcohols to form products having the Si—O—C—R structure. Esterification of a silica surface renders the surface hydrophobic. Organosilicon compounds of the type R_ySiX_z (in which R is a hydrocarbon or other organic group, X is a halogen or hydroxyl ion, and $y + z = 4$) react with a silica surface to give a product where surface bonds are Si—C—H rather than Si—O—H. This type of reaction is used to bond fiber glass to an organic resin. Certain water-soluble organic polymers can be adsorbed onto silica surfaces either by hydrogen bonding or by electrostatic attraction (15–16) (see Silicon compounds).

Silica Sols and Colloidal Silica

Properties. A silica sol is a stable dispersion of discrete, colloid-size particles of amorphous silica in aqueous solutions. Silica sols do not gel or settle even after several years of storage. Sols may contain up to 50% silica, and particle sizes of up to 300 nm are possible, although particles larger than 70 nm slowly settle.

In the absence of a suitable solid phase for deposition and in supersaturated solutions of pH 7–10, silicic acid polymerizes to form discrete particles. Electrostatic repulsion of the particles prevents aggregation if the concentration of electrolyte is below ca 0.1–0.2 N. The particle size attained depends on temperature, ie, it decreases with decreasing temperature. Particles of 4–8 nm are obtained at 50–100°C, whereas particles of up to 150 nm form at 350°C in an autoclave. However, the size of the particles obtained in an autoclave is limited by the conversion of amorphous silica to quartz at high temperature. Particle size influences the stability of the sol, since particles smaller than 7 nm tend to grow spontaneously in storage which may affect the sol properties. However, sols can be stabilized by addition of sufficient alkali (17–18).

The stability of a silica sol depends on several factors (see Fig. 3). The pH must be above 7 to maintain the negative charges on the silica particles that prevent aggregation. This surface charge is neutralized by soluble salts that ionize and form a double layer around the silica surface, which then allows aggregation; therefore, sols are only stable at low salt concentrations. In the low pH region sols are metastable, and gelling and aggregation are catalyzed by even very small amounts of fluoride ion. In this low pH region, water-miscible organic solvents like alcohol retard gelling. Gelling is, however, more rapid at higher temperature. The higher the silica content of a sol, the more likely it is to gel. Surface characteristics of the silica particles control sol stability (20).

Silica sols can be destabilized by aggregation, particle growth plus settling, gelling, or crystallization. Aggregation occurs by coagulation in which particles collide or by

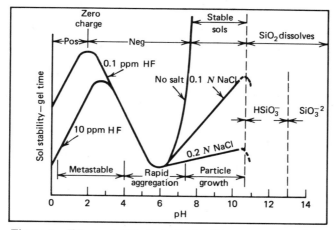

Figure 3. Effects of pH in the colloidal silica–water system (19).

flocculation in which particles become linked by bridges of flocculating agent. Aggregation or gelling is prevented by maintaining a charge on the silica particles by keeping the pH at 9–10 and maintaining a low salt content. This charge separates the particles by electrostatic repulsion. Aggregation is also prevented at neutral or lower pH by an adsorbed layer of inert material on silica surfaces; this prevents direct contact of silanol groups (steric hindrance). Aggregation as well as particle growth is minimized by maintaining low silica concentration, low temperature, and a pH of 9–10. Destabilization by crystallization of silica rarely occurs, since this is a very slow process at ordinary temperatures (21).

Preparation. To produce sols that are stable at relatively high concentration, particles must be grown to a certain size in weakly alkaline aqueous dispersion. The following general methods are discussed in more detail in ref. 22.

Addition of acid to a dilute solution of sodium silicate to lower the pH to 8–10 produces a sol at sodium concentrations below ca 0.3 N at elevated temperature to promote rapid particle growth. Rapid mixing prevents formation of low pH regions where gelation might occur.

Sodium ions are removed from a solution of sodium silicate by electrodialysis or with an ion-exchange resin which forms active silica. The latter is stored at 82–100°C and more active silica is added as the sol particles grow. The sol is concentrated by evaporation or ultrafiltration.

A silica sol can be made from a freshly formed silica gel by washing out the salt and peptizing at elevated temperature and pressure.

Very pure but dilute sols are produced by the hydrolysis of silicon tetrachloride or organosilicates, eg, ethyl silicate plus isopropyl alcohol. Such sols generally gel quickly. •

After pulverized silicon metal is cleaned with hydrofluoric acid to remove the oxide coating, it reacts with water in an alkaline medium to produce hydrogen gas and colloidal silica. The latter can be stabilized as a sol with alkali.

Oxidation at high temperature of ethyl silicate or silicon tetrachloride or vaporization of silica in the presence of a reducing agent gives pyrogenic silica powder. The particles are partly coalesced and have a low concentration of surface silanol groups. They are therefore difficult to disperse in water to form sols without addition of a wetting or other dispersing agent or a treatment like grinding. The resultant sols contain aggregates.

Silica sols are purified with the aid of an ion-exchange resin, dialysis, electrodialysis, or washing. Washing is usually effected by centrifugation, flocculation, ultrafiltration or electrodecanting, and finally redilution and reconcentration. A preservative, eg, formaldehyde, is sometimes added to prevent the growth of microorganisms. Sols are concentrated by evaporation using forced-circulation evaporators. If the sol becomes too concentrated, a hard layer of silica is deposited on the equipment walls, especially on heat-exchanger surfaces, and evaporation must be carefully monitored. Sols with particle size >30 nm are concentrated by centrifugation. Ultrafiltration, where the sol collects near the filter but does not pass through, can also be used to remove water and salts (see Ultrafiltration). Sols that are low in soluble salts are concentrated by electrodecantation (23).

Characterization. A silica sol is characterized by its chemical composition, its physical properties, and by the characteristics of its particles. The most important chemical constituents are silica, stabilizers, carbon (as carbon dioxide and organic carbon), soluble salts, total solids, nonsiliceous ash, and metals (especially aluminum or iron). Physical characteristics include pH, density, viscosity, turbidity, refractive index, light-scattering properties, and sedimentation rate by ultracentrifugation. The particles are characterized by size and size distribution, porosity, degree of aggregation, specific surface area, and rate of dissolution. Particle size is measured by electron microscopy, light-scattering, absorbance of visible light, low-angle x-ray scattering, centrifugation, rate of reaction with molybdic acid, or surface area measurement of dried particles (24) (see Analytical methods).

Modifications and Alterations. Coagulation, flocculation, or gelling and drying of silica sols produces amorphous silica powders. Gellation is rapid at pH 4–6. It is effected by increasing the silica content, eg, by evaporation. Further drying of the resulting gel yields colloidal aggregates. Coagulation of a sol is prevented by surface charge and surface hydration; conversely, neutralizing the surface charge by lowering the pH or adding salt causes coagulation. Addition of salts also causes surface adsorption of cations which reduce surface hydration. Polyvalent cations like Al^{3+} are strongly adsorbed by silica surfaces and neutralize the surface charge. Hydrated metal cations act as bridging compounds or flocculating agents for silica particles (see Flocculating agents). Flocculating agents used with silica sols are cationic surfactants such as octadecyltrimethylammonium bromide or organic polymers such as polyethyleneimine or polyacrylamide (21).

Sol particles are extremely small, and particulate silica obtained from sols is composed of aggregates or porous particles that have a much higher specific surface area than estimated from apparent size. Aggregate particles are also called secondary particles or clusters. Once silica particles have been formed as a sol, the surface can be modified by the attachment of different atoms or groups to obtain specific properties. Addition of aluminate ions gives an aluminosilicate surface that is much less soluble than the silica surface and much more stable toward gelling even at pH 4–6 and in the presence of salts. Addition of polyvalent metal oxides reverses the surface charge of silica and prevents bridging by siloxane bonds. Organic groups can be added to silica surfaces in the form of organic ions through formation of Si—O—C bonds as in esters or Si—C bonds as in organosilicon compounds. These surface alterations make silica easily dispersible in organic solvents including hydrocarbons. Some of these processes may cause the silica surface to become hydrophobic. Even a small change in particle surface characteristics may change the chemical behavior of the particle. Thus a silica particle covered with alumina behaves as an alumina particle, and a silica particle covered with a hydrocarbon coating acts as a large hydrocarbon molecule (25–29).

Silica Gel

Silica gels are classified into three types. Regular-density gel is made by gelling in an acid medium, which gives very small particles with high surface area (750–800 m^2/g). The average pore diameter is 2.2–2.6 nm, and the pore volume is 0.37–0.40 mL/g. Regular-density gel contains ca 6 wt % water as surface hydroxyl groups, which imparts a high capacity for water adsorption and adsorption of other polar molecules (see Drying agents). Regular-density gel exhibits a high selectivity for polar molecules and a large percentage of small pores. Intermediate-density silica has a lower surface area (300–350 m^2/g) but larger pore volume (0.9–1.1 mL/g). The average pore size is 12–16 nm in diameter, and the particles are larger than those of regular-density gel. Because of the large pore size, intermediate-density gel has a high capacity for water adsorption at high humidities. Intermediate-density gel is often used as a fine particle powder because particle size and porosity can be controlled. Low-density silica gel (eg, some aerogels) has a lower surface area (100–200 m^2/g), larger average pore diameter (18–22 nm), and larger pore volume (1.4–2.0 mL/g) than the other types. It is usually prepared as a very fine powder of extremely low density; shrinkage of the gel during drying is minimized.

Properties.   Silica gel is a coherent, rigid, continuous three-dimensional network of spherical particles of colloidal silica. A hydrogel is a gel in which the pores are filled with water, whereas a xerogel is a gel from which the liquid medium has been removed, causing the structure to collapse, thus decreasing porosity. If the liquid medium is removed in order to prevent shrinkage and collapse of the gel structure, an aerogel is formed. Porous glass is similar to a silica gel. Silica powder can be made by grinding or micronizing xerogels, which decreases the size of the gel fragments but leaves the ultimate gel structure unchanged. Gels and powders are characterized by the density, size, and shape of the particles, particle distribution, and by aggregate strength or coalescence (12–14,30–34).

When silica is used as an adsorbent, the pore structure determines the gel-adsorption capacity. Pores are characterized by specific surface area, specific pore volume (total volume of pores per gram of solid), average pore diameter, pore-size distribution, and the degree to which entrance to larger pores is restricted by small pores. These parameters are derived from gas- or vapor-adsorption isotherms, mercury-penetration studies, low-angle x-ray scattering, electron microscopy, gas permeability, ion or molecule exclusion studies, or measurement of volume of imbibed liquid (35).

Surfaces can be categorized as fully hydroxylated in which the surface consists solely of silanol (Si—O—H) groups, as in a siloxane (Si—O—Si), or as an organic surface. Silanol surfaces are formed by drying silica gels or precipitates from water below 150°C. These surfaces are readily wetted by water. Hydroxylated surfaces heated from 300–1000°C progressively develop a siloxane surface by dehydration; pyrogenic silicas also have siloxane surfaces. The behavior of particles with organic surfaces depends on the coating material. The particles may become dispersible in water, oil, or other organic solvents. If fluorocarbons are the surface group, the silica becomes both hydrophobic and oleophobic. The nature of the surface can be determined by measuring the heat of nitrogen adsorption, dye adsorption, infrared adsorption, or chemical analysis (14,18,26–28,36–37).

Preparation.   Silica gels can be prepared by several methods (32,38). Most commonly, a sodium silicate solution is acidified to a pH less than 10 or 11; the gel time varies as shown in Figure 3.

The bulk-set process consists of the following steps: A silica hydrosol is prepared by mixing sodium silicate with a strong mineral acid. After it is allowed to set to a rigid mass, it is broken up mechanically. The silica concentration, temperature, and pH affect gelling time and final gel characteristics such as density, strength, hardness, surface area, and pore volume. The hydrogel particles are then washed free of electrolytes, and the washed gel is dried and activated. The rate of drying affects final gel properties.

In the slurry process, sodium silicate solution and acid are mixed, either batch or semicontinuously, to produce a gelatinous precipitate. This hydrogel is washed and dried, often by spray-drying. Sodium silicate is the cheapest source of silica although natural sources can also be used. Certain clays yield relatively pure silica when leached with acid. Gels made from soluble silicates usually have very small particle size and the salts must be washed away before use. Gels can be made directly from salt-free colloidal silica, which provides a larger ultimate particle size and hence greater stability, along with low salt content. Surface characteristics may also be different, such as lower specific surface area and larger pore diameter.

Hydrolysis of pure silicon compounds like ethyl silicate, silicon tetrachloride, and other volatile hydrolyzable silicic esters is a third method of preparing gels. This method is more expensive, but produces very dense gels of high purity and very small pore size.

For some applications, silica gel is converted to pelletized or granular form by extruding pulverized gel with a binder or shaping the hydrogel during drying. Silica can be gelled in spherical form by spray-drying, or by spraying droplets into an immiscible liquid (emulsion polymerization). Freezing of a silica sol produces silica-gel particles of nonspherical shapes.

Characterization. The properties of a finished gel are determined by the size of the primary particles at the moment they aggregate into the gel network; the concentration of the primary particles in solution and thus the compactness of the gel network; the pH, salt concentration, temperature, and time during which the gel is aged while wet; mechanical pressure or shear forces applied to the gel before or during drying; temperature, pressure, pH, salt content, and surface tension of the liquid medium as it is being evaporated from the pores of the gel; and temperature, time, and type of atmosphere in which the gel is heated after being dried (39).

Modification. Once a gel structure is formed, it can be modified in the wet state to strengthen the structure or enlarge the pore size and reduce surface area (40). Gel reinforcement can be carried out in several ways: active or low molecular weight silica can be added to a broken-up gel in order to deposit it at a uniform rate, active silica can be added to a sol as the gel is growing which causes strong gel bridges to form between particles, or the wet gel can be heat-aged to increase coalescence of the particles. In this aging process silica is dissolved from smaller particles and deposited at the points of contact between larger particles, which causes strengthening. Washing can also be an aging step. Aging a wet gel increases interparticle bonding, which leads to less shrinkage of the gel during drying. The drying procedure affects the gel characteristics. Low density gels are made by minimizing shrinkage during drying by first reinforcing the gel. The wet gel is aged and the water is replaced with a liquid of lower surface tension, eg, alcohol. The gel is then heated to a temperature above the critical point of the liquid, thereby releasing the liquid as a vapor (aerogel process) (41).

Sintering a dried silica surface in air or in a vacuum causes shrinkage which decreases the surface area, whereas sintering in steam also increases pore size. Micropores are obtained by heating a hydrated gel at 1000°C for 10 h. The presence of impurities like aluminum tends to minimize changes caused by heating; however, at some temperature >1000°C, and in the presence of impurities, silica gel is converted to cristobalite or to nonporous silica glass. Gels can be made with extremely small pores. Such gels include impervious silica, porous glass, and silica used as an adsorbent for certain specific materials which are determined by the surface composition and pore size of the silica gel.

Precipitated Silica

Precipitated silica (also called particulate silica) is composed of aggregates of ultimate particles of colloidal size that have not become linked in a massive gel network during the preparation process. Precipitated silicas are either formed from the vapor phase (fumed or pyrogenic silicas) or by precipitation from solution. Precipitated silica powders have a more open structure with higher pore volume than dried pulverized gels.

Silica can be precipitated from a sodium silicate solution using a lower concentration than in gel preparation. In the absence of a coagulant other than the sodium salt that is being formed, silica is precipitated from a hot sol at pH 9–10 when the concentration of sodium ion exceeds 0.3 N. Precipitation proceeds in several steps including nucleation of particles, growth of particles to desired size, coagulation to form aggregates by control of pH and sodium ion concentration, and reinforcement of the aggregate to the desired degree without further nucleation (42). Coagulating agents include sodium, calcium, or other polyvalent metal cations; ammonium ions; certain organic compounds; and fluoride ion. Reinforcement is carried out by adding active silica (silica that quickly dissociates to give monomeric silica) to the suspension of particles under alkaline conditions above 60°C. Silica is precipitated from silica sols by adjusting pH and salt concentration; care must be taken to prevent gelling (see Figs. 2 and 3). Silica is also precipitated by adding aqueous ammonium hydroxide to ethyl silicate, $(C_2H_5O)_4Si$, in alcohol.

Pyrogenic or fumed silica is prepared differently. Silica (usually sand) can be vaporized at ca 2000°C. On cooling, anhydrous amorphous silica particles form. In the presence of a reducing agent, eg, coke, silica sublimes at ca 1500°C to produce SiO, which can then be oxidized to produce particulate SiO_2. Oxidation of silicon tetrachloride vapor at high temperature produces SiO_2 and Cl_2. Alternatively, $SiCl_4$ can be burned with methane or H_2 to produce SiO_2, H_2O, and HCl; the latter process is an important commercial method. Silicon ester vapors can be oxidized and hydrolyzed to produce particulate silica of high purity though at high cost. SiF_4, a by-product of the phosphate fertilizer industry, can also be used to produce silica by hydrolysis of the vapor at 1600–2200°C. HF is a product which can then react with sand to produce more SiF_4 (43–44).

The physical and chemical properties of precipitated silicas vary according to the manufacturing process. Ultimate and aggregate particle size in silicas precipitated from solution can be varied by reinforcement and control of suspension pH, temperature, and salt content. The particle size in pyrogenic silicas is controlled by combustion conditions. The surface area, as determined by nitrogen adsorption, is a function of particle size. Pyrogenic silicas tend to be less dense and more pure than silicas precipitated from solution since the latter contain coagulating agents. Pyrogenic silicas are much less hydrated and are sometimes completely anhydrous, whereas precipitated silicas may contain up to 10% water as surface hydroxyl groups that remain after drying at 150°C.

Some important chemical and physical properties of silica sols, silica gels, silica precipitated from aqueous solution, and pyrogenic silica are given in Table 1.

Naturally Occurring Amorphous Silica

Biogenic Silica. Several aquatic organisms, including diatoms, radiolarians, sponges, and silicoflagellates, secrete solid amorphous silica in the form of shells, skeletons, spines, or plates. These organisms extract silica from very dilute solutions (0.1 ppm SiO_2, or 2 mmol/kg). These organisms are widespread, occurring both in marine and freshwater environments. They form a significant part of marine sediments in the equatorial Pacific (radiolarians) and in high latitude areas of all oceans (diatoms). However, most silica dissolves in water before becoming incorporated into the sediments. Dissolution is inhibited by incorporation of small amounts of metals and by the presence of an organic membrane coating over the silica (45) (see Diatomite).

Opal. Biogenic silica is sometimes called opal-a. With time opal-a becomes more structured, altering to opal-CT and then to opal-C, which is well crystallized cristobalite (46). These structural differences can be detected by x-ray diffraction or infrared spectroscopy. The gem opal is a cryptocrystalline (very fine-grained crystals) form of cristobalite with submicroscopic pores that contain water; the amount of water varies and can constitute several wt % (47).

Diatomaceous Earth. Diatomaceous earth [7631-86-9], also called kieselguhr or diatomite, is a loosely coherent chalk-like sediment made up of fragments and shells of diatoms (one-celled algae). It is used as an absorbent, filler, insulating material, and polishing agent. The particles are very fine and have high surface area; silica content may be as high as 94%. Initially, silica in diatomites was amorphous, but many of these deposits are millions (10^6) of years old, and the silica may now be present as cryptocrystalline quartz (47).

Chert. Chert is a diagenetic rock made up of microcrystalline or cryptocrystalline quartz with or without opaline silica. Most chert is white, tan, or gray and often contains other minerals. It may occur as nodules in limestone. Because of its hardness, chert is not very useful although it is sometimes substituted for sand in ceramics. Varieties of chert include flint, jasper, fossiliferous chert, oolite chert, novaculite, porcelanite, and tripoli. Most cherts have a biochemical origin and were originally composed of diatoms, radiolarian shells, or sponge spicules. These siliceous materials are transformed by dissolution, precipitation, recrystallization, and compaction over time to give a less soluble, denser material of lower surface area, which is called chert if granular, chalcedony if fibrous, or flint if dark gray. Some chert may originate from chemical precipitation. In these cherts, no traces of siliceous fossils remain (47).

Tripoli, a form of chert, is utilized as abrasive, filler, and extender. It is microcrystalline quartz (<10 μm dia) although it is sold as amorphous silica (48). It is derived from calcite- or dolomite-bearing chert from which the carbonates have been leached, leaving a very pure silica. Tripoli is white or gray, soft, porous, and friable, and is mined extensively from Devonian deposits in Southern Illinois. Tripoli contaminated with iron oxide is red and has little commercial value (47) (see Abrasives; Fillers).

Amorphous Silica of Volcanic Origin. During the Roman Empire a colloidally subdivided high surface-area amorphous silica was mined at Pozzuoli, Italy, and on the Greek island of Santorini. This material is an alteration product of volcanic ash and when mixed with lime and sand gave an extremely impervious cement used to line cisterns (48).

Geothermally Deposited Silica. Amorphous silica is sometimes precipitated from the hot supersaturated waters of hot springs and geysers. This precipitation reduces the supersaturation caused by cooling hot subterranean solutions under pressure in equilibrium with quartz which has a solubility of 200 ppm SiO_2 or 3 mmol/kg at 200°C. The resulting precipitate is called siliceous sinter, whereas the precipitate deposited from geysers is called geyserite. Siliceous sinter often occurs as incrustations around springs or geysers, and sometimes along with calcareous sinter. Pure siliceous sinter is white; it can be porous and loose or dense and compact. Its precipitation is a potential problem in the development of geothermal power (47,49) (see Geothermal energy).

Silicification of Biogenic Materials. Fossils preserved by silicification often match the original perfectly. Amorphous silica was initially deposited and, over geological time, converted to chert with a crystal structure small enough to prevent disruption of the initial fossil structure. The source of the silica may be dissolution from volcanic ash or geothermal water rich in dissolved silica (50).

Economic Aspects

The only commercially available natural amorphous silica is diatomaceous earth, and much of this material is not truly amorphous. A microcrystalline quartz from a tripoli deposit is sold as amorphous silica and is the least expensive form of amorphous silica, costing less than half as much as industrially prepared amorphous silica. Precipitated silicas and gels made from sodium silicate and acid are more expensive, followed by colloidal silica sols also made from sodium silicate. The pyrogenic silicas are still more expensive, and specialty silicas (hydrophobic or with other surface alterations or very high purity) are the highest in price. The price of amorphous silica precipitated from sodium silicate solution was fairly stable from 1950 to 1970 and ranged between 11 and 88¢/kg. In 1975 the price had increased to $0.22–1.10/kg, and by 1980 to $0.55–1.54/kg. The average price for silica precipitated from sodium silicate solution increased from 18¢/kg in 1970 to 59¢/kg in 1980. The U.S. amorphous silica industry had sales of ca $150 × 10^6 in 1981. A list of producers is given in Table 2.

Health and Safety Factors

Amorphous silicas do not usually cause silicosis. This disease may be caused by inhalation of fine crystalline particles (0.5–5.0 μm) which do not dissolve readily in body fluids (51–53).

Amorphous silica is nontoxic when ingested and is used as an anticaking agent (up to 2%) in food and medicines (see Food additives). Dissolved silica occurs in well water, and is commonly ingested and then excreted. Because of the presence of silica in most foods, no need for the addition of silica to human diets has been established. Trace amounts of silica are probably necessary for birds and mammals (52,54–58).

The only health hazard reported is so-called transient dermatitis, which causes skin dehydration and loss of skin oils (59).

Uses

Amorphous silica, depending on its form and purity, is used mainly as a filler and reinforcing material in rubber; to improve ink retention on paper; as a pigment and

Table 2. Producers of Amorphous Silica

Producer	Country	Producer	Country
Bayer A.G.	Federal Republic of Germany	Ketjen NV	Netherlands
Davison Chemical Division, W. R. Grace Corporation	United States	Monsanto	United Kingdom, United States
Degussa	Federal Republic of Germany, United States	Nalco Chemical Company	United States
DuPont	United States	Nisson Chemical Industries	Japan
Glidden Pigments	United States	Nyacol	United States
Godfrey L. Cabot Corporation	United States	Nynas-Petroleum	Sweden
Hoesch	Federal Republic of Germany	Pechiney	France
J. M. Huber	United States	PQ (Philadelphia Quartz)	United States
Illinois Mineral Company	United States	PPG Industries	United States
		Si-France	France

filler in paints and coatings; as an abrasive, adsorbent, and catalyst base; and in electrical insulation. Amorphous silica is obtained on a large scale as a by-product in the silicon and ferrosilicon industry, but has little use in this form.

Silica in colloidal and sol form is used in preparing silica gels; as a stiffening and binding agent; to increase friction; to provide antisticking, antistatic, and antisoiling effects; for hydrophilic surfaces; to alter adhesion properties; in electrical conducting and insulating film; for polymer cross-linking and reinforcement; to polish silicon wafers; to modify surfactant and viscosity properties; and in ceramics (qv), refractories (qv), and photographic film.

Silica gel is used as a desiccant; as an adsorbent; as a catalyst base; to increase viscosity and thixotropy; for surfactant and optical effects; as a source of reactive silica; for cloud seeding; in chromatographic column packing; as an anticaking agent; and in paper coating.

Precipitated silica is used as a filler for paper and rubber; as a carrier and diluent for agricultural chemicals; as an anticaking agent; to control viscosity and thickness; and as molecular-sieve material (see Molecular sieves).

Pyrogenic or fumed silica is used as a thixotropic agent in polyester-glass reinforced plastics; as a reinforcing and thickening agent in rubber, plastics, silicone, and epoxy resins; and as a thickening and gelling agent.

Hydrophobic silica is used as defoaming agent and as a carrier in aerosols (see Defoamers).

The common commercial applications of the various forms of amorphous silica are discussed in refs. 32, 34, 60–62.

BIBLIOGRAPHY

1. R. K. Iler, *The Chemistry of Silica*, John Wiley & Sons, Inc., New York, 1979, pp. 21–28.
2. *Ibid.*, p. 23.
3. *Ibid.*, pp. 46–58.
4. J. V. Walther and H. C. Helgeson, *Am. J. Sci.*, 277 (1977).

5. W. Stöber in R. F. Gould, ed., *Advances in Chemistry Series 67*, ACS, Washington, D.C., 1967, p. 161.

6. Ref. 1, pp. 30–62.

7. Ref. 1, pp. 40–49.

8. Ref. 1, pp. 62–76.

9. Ref. 1, pp. 83–94.

10. Ref. 1, p. 174.

11. Ref. 1, pp. 172–311.

12. R. K. Iler in E. Matijevic, ed., *Surface and Colloid Science*, John Wiley & Sons, Inc., New York, 1973, p. 6.

13. R. K. Iler, *J. Colloid Interface Sci.* **75**(1), 138 (1980).

14. Z. Z. Vysotskii, V. I. Galinskaya, V. I. Kolychev, V. V. Strelko, and D. N. Strazhesko in D. N. Strazhesko, ed., *Adsorption and Adsorbents No. 1*, John Wiley & Sons, Inc., New York, 1973, p. 72.

15. Ref. 1, pp. 155–158, 288–300, 689–709.

16. L. Bokasanyi, O. Liardon, and E. Kovats, *Adv. Colloid Interface Sci.* **6**, 95 (1976).

17. Ref. 1, pp. 312–330.

18. D. H. Napper and R. J. Hunter, *Med. Tech. Publ. Int. Rev. Sci.: Phys. Chem. Ser. One London* **7**, 241 (1971).

19. Ref. 1, p. 367.

20. Ref. 1, pp. 313–328.

21. Ref. 1, pp. 364–407.

22. Ref. 1, pp. 331–336.

23. Ref. 1, pp. 337–344.

24. Ref. 1, pp. 344–364.

25. Ref. 1, pp. 407–415.

26. D. Barby in G. D. Parfitt and K. S. W. Sing, eds., *Characterization of Powder Surfaces*, Academic Press, Inc., New York, 1976, p. 353.

27. L. C. F. Blackman and R. Harrop, *J. Appl. Chem.* **18**, 37 (1968).

28. J. A. Hockey, *Chem. Ind. London*, 57 (1965).

29. A. Snoeyink and W. Weber in J. F. Danielli, M. D. Rosenberg, and D. A. Cadenhead, eds., *Progress in Surface and Membrane Science*, Vol. 5, Academic Press, Inc., New York, 1972, p. 63.

30. Ref. 1, pp. 462–478.

31. S. Kondo, *Hyomen* **10**(6), 321 (1972).

32. I. E. Neimark and R. Y. Sheinfain, *Silica Gel: Preparation Properties and Uses*, Nauk Dumka, Kiev, USSR, 1973.

33. C. Okkerse in B. G. Linsen, ed., *Physical and Chemical Aspects of Adsorbents and Catalysts*, Academic Press, Inc., New York, 1970, p. 214.

34. H. Teicher in T. C. Patton, ed., *Pigment Handbook*, Vol. 1, John Wiley & Sons, Inc., New York, 1973.

35. Ref. 1, pp. 478–505.

36. Ref. 1, pp. 505–554.

37. A. V. Kiselev, *Trans. Faraday Soc. Disc.* **52**, 14 (1971).

38. Ref. 1, pp. 510–554.

39. Ref. 1, pp. 516–528.

40. Ref. 1, pp. 528–533.

41. Ref. 1, pp. 533–554.

42. Ref. 1, pp. 554–564.

43. Ref. 1, pp. 565–568.

44. G. D. Ulrich, *Combust. Sci. Technol.* **4**(2), 47 (1971).

45. R. Wollast in E. D. Goldberg, ed., *The Sea*, Vol. 5, Wiley-Interscience, New York, 1974, pp. 359–392.

46. J. B. Jones and E. R. Sequit, *J. Geol. Soc. Aust.* **18**(1), 57 (1971).

47. R. V. Dietrich and B. J. Skinner, *Rocks and Rock Minerals*, John Wiley & Sons, Inc., New York, 1979.

48. Ref. 1, p. 569.

49. R. O. Fournier in *Proceedings of the Symposium on Hydrogeochemistry and Biogeochemistry*, International Association of Geochemistry and Cosmochemistry, Clarke Company, Washington, D.C., 1973, pp. 122–139.

50. Ref. 1, pp. 88–91.
51. Ref. 1, pp. 769–782.
52. G. Bendz and I. Lindquist, eds., *Biochemistry of Silicon and Related Problems*, Plenum Press, New York, 1977.
53. W. Stöber, *Arch. Environ. Health* **16,** 706 (1968).
54. Ref. 1, pp. 753–769.
55. A. G. Heppleston in E. Kulonen, ed., *Biol. Fibroblast, Sigrid Juselius Found. Symp. 4th*, 1972, Academic Press, Inc., New York, 1973, pp. 529–537.
56. R. F. Leo and E. S. Barghoorn, *Acta Cient. Venez.* **27,** 231 (1976).
57. K. Stalder and W. Stöber, *Nature* **207**(4999), 874 (1965).
58. M. G. Voronkov, G. I. Zelchan, and E. Y. Lukevitz, *Silizium Und Leben*, transl. by K. Ruhlman, Akademie-Verlag, East Berlin, German Democratic Republic, 1975.
59. PPG Bulletin A-628-85, PPG Industries, Inc., Pittsburgh, Pa., 1976.
60. Ref. 1, pp. 415–439, 578–598.
61. M. P. Wagner, *Rubber Chem. Technol.* **49**(3), 703 (1976).
62. W. Kress, *Double Liaison* **18**(186), 63 (1971).

General References

Ref. 1 is also a general reference; it includes several thousand references regarding silica-related subjects.

Ref. 52 is also a general reference.

S. Aston, ed., *Silica Geochemistry and Biogeochemistry*, Academic Press, London, in press.

R. A. Berner, *Principles of Chemical Sedimentology*, McGraw-Hill, Inc., New York, 1971.

J. G. Falcone, ed., *Soluble Silicates*, American Chemical Society Symposium Series, Washington, D.C., in press.

G. V. Kukolev, *Chemistry of Silicon and Physical Chemistry of the Silicates*, Vols. 1–3, transl. from Russian by E. H. Murch, National Lending Library of Science and Technology, Boston Spa, UK, 1971.

R. B. Sosman, *The Phases of Silica*, Rutgers University Press, New Brunswick, N.J., 1965.

J. G. Vail, *Soluble Silicates*, American Chemical Society Monograph Series Number 116, Vols. 1–2, Reinhold, New York, 1952.

JOAN D. WILLEY
University of North Carolina at Wilmington

VITREOUS SILICA

Vitreous silica [60676-86-0] is a glass composed essentially of SiO_2. It has been the subject of considerable study for two reasons. First, it is a material with many unique and useful properties, eg, low thermal expansion, high thermal shock resistance, high ultraviolet transparency, good refractory qualities, dielectric properties, and chemical inertness. A second reason is the simplicity of its chemical constitution. It is one of the relatively few binary oxide glasses and consequently has been investigated by countless chemists, physicists, spectroscopists, and materials scientists. However, vitreous silica is actually a very complicated material whose properties vary with, among other things, raw material, method of manufacture, and thermal history (see also Glass).

The question arises as to why, if vitreous silica has such outstanding properties, it is not used even more extensively. The answer is the high cost of manufacture as compared to most glasses caused by very high viscosity, small temperature coefficient of viscosity, and volatility at forming temperatures. This means that even at 2000°C the melt is very stiff and difficult to shape, particularly by mass production methods, although there are significant research and development efforts aimed at circumventing some of these obstacles.

Vitreous silica is either transparent or nontransparent. The nontransparent fused material contains a large number of microscopic bubbles that create a milky appearance caused by the scattering of light. This material, sometimes called translucent fused silica, is more economical to produce than the transparent type and is often used where optical properties are not important. Another nontransparent type is opaque and is formed by sintering powdered vitreous silica.

Probably the earliest record of vitreous silica was in a communication from Marcet, a physician, to a Dr. Thomson, dated July, 1813. By directing a current of oxygen through the flame of an alcohol lamp, Marcet melted wires of platinum and iron and small needles of quartz.

The properties of vitreous silica were first described in 1839 by a French scientist (1). The glass that he produced had remarkable strength and an elasticity that resembled that of iron; it was not broken up when rapidly cooled from the molten state. As a result of these excellent properties, this glass could be put to many uses including small springs, torsion threads, and high temperature forceps. Furthermore, the sandstone near Paris could be fused into a glass which, instead of being transparent, formed silky white threads and pearl-like droplets. These droplets served as very good imitation pearls because of their high strength and luster.

Splintering during melting creates air lines (see under Manufacture), but if the crystals are heated to a red heat and quenched in water with fracturing into tiny crystals, they can be fused without splintering. This discovery in 1900 has given rise to the clear-fused-quartz industry (2–4).

Methods for the manufacture of translucent vitreous silica by fusion of sand surrounding a graphite rod through which a current is passed and subsequent manipulation of the hot plastic material were patented around the turn of the century (5–8), followed by a patent for feeding powdered quartz crystal into an arc (9).

Vitreous silica occurs sometimes as the result of natural phenomena (10). A bolt of lightning can fuse quartz sand at a temperature high enough to give glass tubes called fulgurites. The best-known deposit, the mineral lechatelierite, is in the Libyan desert.

Deposits of natural fused quartz have also been found near Canyon Diablo, Ariz., and in small meteorite craters in Australia and Arabia. Their origin is uncertain, but the most widely accepted theory is that pressure created at the moment of meteorite impact caused adiabatic instantaneous heating of sandstone well above the quartz melting point.

In an interesting modification of this theory (11), it has been proposed that impact pressure creates stishovite, a crystalline form of SiO_2 with Si in sixfold coordination, which changes to a glass with sixfold-coordinated Si, followed by formation of the observed fourfold-coordinated silica glass. The process can take place in a matter of 14 min at 400°C and in 1–2 min at 800°C.

Until the mid-1970s, the primary source for natural quartz crystal for fusion was Brazil, which also supplied electronic-grade crystals for the U.S. electronic industry. Dramatic price increases for Brazilian imports, in addition to the increased capacity of U.S. synthetic quartz crystal production, has reduced quartz imports from Brazil. Some optical-grade quartz is still imported for fusing. The Brazilian product, called lasca, has been largely replaced by beneficiated quartz deposits and sands from North America (North Carolina, Arkansas, North Dakota, Mexico, and Canada). This quartz is normally processed by acid washing, flotation, and the usual procedures. The average particle size is smaller, and lower temperatures are required for melting the beneficiated minerals than for lumps of lasca. High purity beneficiated materials currently available are Iota Quartz and Quintus Quartz, supplied by International Minerals and Chemical Corporation.

The nomenclature of vitreous silica has been confusing, ambiguous, and often incorrect (12–14).

Vitreous silica is the most unambiguous general term which covers the entire field of noncrystalline silica.

Silica glass, although technically correct, can be misinterpreted as covering any glass composition containing silica.

Fused silica should apply to any form of vitreous silica manufactured by fusion. However, it has been used by some to denote all vitreous silica not produced by quartz fusion and by others for only the translucent vitreous silica.

Synthetic fused silica is a term used to describe the material formed by vapor-phase hydrolysis.

Fused quartz is the material formed by direct melting of quartz crystals. This term is used by some to denote all forms of vitreous silica and by others to specify all transparent vitreous silica.

Quartz glass is the same as fused quartz but is ambiguous since quartz is crystalline and glass is vitreous.

Quartz is a crystalline form of silica. This much-abused designation for vitreous silica is incorrect.

The words transparent, nontransparent, or translucent sometimes precede the above terms.

Structure

A theory of the structure of vitreous silica was first formulated in 1930 (15). Because of the coincidence of the maxima of the x-ray diffraction peaks for vitreous silica and cristobalite, it was concluded that vitreous silica consists of units of cristobalite crystallites. These crystallites are ca 1.5–2.0 nm in size. This theory was supported by the USSR school of glass science (16–19).

New theories are continually being proposed, indicating the lack of certainty of the actual structure. In general, they refer to the original random network theory which describes vitreous silica as follows (20): It is built up of tetrahedra with oxygen atoms at the corners surrounding a silicon atom in the center. The tetrahedra share only corners in such a way that an oxygen atom is linked to two silicon atoms. The oxygen–silicon–oxygen bond angle varies throughout the network in contrast to the crystalline forms. The extended three-dimensional network lacks symmetry and periodicity.

X-ray, neutron, and electron diffraction studies, carried out by numerous investigators, have created several areas of agreement: The x-ray diffraction pattern is typical of an amorphous material, having broad diffuse rings with no indication of crystallites of significant size. The radial distribution functions indicate that the separation between bonded silicon and oxygen atoms is in the 0.159–0.162 nm range. The oxygen-to-oxygen distances similarly calculated are 0.260–0.265 nm, whereas Si–Si distances are 0.305–0.322 nm. The —SiOSi— angles range from 120–180°, with an apparent max ca 150°. The —OSiO— (tetrahedral) angle appears normal (109.5°).

The infrared and Raman spectra (21) and the x-ray and neutron diffraction patterns of vitreous silica were compared with the patterns observed for the various crystalline polymorphs of silica. Thus, in addition to cristobalite, tridymite and β-quartz have been offered as structural models. Vibrational frequencies observed for vitreous silica are in agreement with the frequencies calculated for the β-quartz model (22–23). As with many other properties of vitreous silica, the thermal history of the sample should be taken into account. The Raman spectra of Type III-vitreous silica change when the samples are stabilized at 1000°C instead of 1300°C (24).

Physical models have been built to scale, and computer models have been generated using the known near-neighbor structural relationships established to date. Some of these models have been successfully used in predicting intrinsic properties, eg, density, and have allowed comparison of observed coordination spheres with the radial distribution functions obtained by diffraction experiments. The existence of crystal-related order beyond distances of 0.7–1.0 nm is still uncertain.

Theoretical calculations based on the random-network theory offer partial explanations for such properties as thermal capacity, thermal expansion, compressibility, elasticity, viscous flow, and stress–optical effects (25–26). These calculations are in reasonably good agreement with experimental evidence.

Devitrification

Devitrification of vitreous silica at atmospheric pressure occurs as cristobalite formation from 1000°C to the cristobalite liquidus at 1723°C with a maximum growth rate at ca 1600°C. Crystals form and grow from nuclei found predominantly at the glass surface, but internal crystallization, though rarely seen, may also occur (27). In all observations to date, the crystallization process has been the result of heterogeneous nucleation, regardless of whether it began internally or externally.

The general effect of temperature on devitrification rate for crystals growing inward from the glass surface is shown in Figure 1. In a study of the growth rate–temperature relation for internally nucleated cristobalite crystals in a fused quartz of low hydroxyl content at 1350–1620°C, the observed growth rates were linear with time and were the lowest measured for vitreous silica. For example, the growth rate at 1350°C was 6.5×10^{-4} μm/min at 1620°C. Because the growth was internal and free from surface contamination, the rates seem to be very near the intrinsic rates for the material.

The devitrification rate is extremely sensitive to both surface and bulk impurities, especially alkali. Fingerprints can be developed by heating a piece of vitreous silica that has been touched, causing an accelerated crystal growth rate in the area of finger contact. As little as 0.32 wt % soda, added as a bulk impurity, increases the maximum devitrification rate 20–30 times and lowers the temperature of the maximum to ca 1400°C (29). The addition of small amounts of alumina to fused silica also increases the devitrification rate, reaching a maximum when the Al^{3+}:Si^{4+} ratio is ca 0.225×10^{-2} (30).

The water content and stoichiometry of the glass also affect the devitrification rate. A high hydroxyl content in the glass as well as water vapor and oxygen in the atmosphere enhance devitrification, whereas oxygen deficiency of the glass and neutral or reducing atmospheres inhibit devitrification (28,31–34). The oxygen affects the rate by diffusing through the cristobalite layer to the glass–crystal interface and oxidizing the glass, thereby bringing it closer to stoichiometry (31–33). The water vapor and hydroxyl content show a similar effect, probably by dissociating at the elevated temperatures to give free oxygen and by weakening the glass structure through the

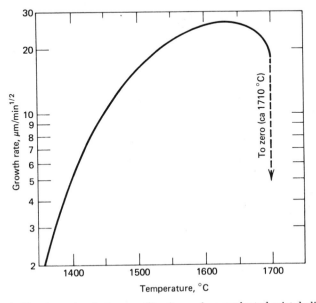

Figure 1. Devitrification rate of vitreous silica for surface-nucleated cristobalite as a function of temperature (28). The rate of growth is proportional to the square root of the time. Measurements were made on General Electric fused silica, heat-treated in air.

formation of silicon–hydroxyl bonds (31–32). The effect of stoichiometry on the devitrification rate is the basis for the development of a vitreous silica resistant to crystallization formed by the addition of silicon to silica (34).

The effect of pressure on devitrification was investigated with Spectrosil (Thermal Syndicate Ltd.) and G.E. Type 204 (General Electric Company) fused silicas (35). At pressures above 2.5 GPa ($<25 \times 10^3$ atm), devitrification occured at temperatures as low as 500°C and at 4 GPa ($<40 \times 10^3$ atm), as low as 450°C. Heat treatments at these temperatures lasted for 70 h. Though the temperatures and pressures were in the stability field of coesite, both coesite and quartz were observed. Both the devitrification rate and the formation of the stable phase (coesite) were enhanced by the presence of water. In the 1000–1700°C region at 500–4000 MPa ($<5,000$–40,000 atm), α- and β-quartz were the primary phases. Crystal-growth rates increased with pressure and with hydroxyl content for a given stoichiometry. Under the influence of heat treatments in oxidizing or reducing atmospheres, crystal growth rates also increase as the vitreous silica approaches ideal stoichiometry. A growth-rate expression was obtained (36).

The formation of cristobalite from vitreous silica occurs at temperatures as low as 400°C when the pressure is equal to 35 MPa (<350 atm) and the glass is immersed in weak NaOH solutions (37). In stronger NaOH solutions, quartz is formed. The formation of the crystalline phases is due to the hydrolysis of the anions present. No crystallization occurs with HF, H_2SO_4, and H_3PO_4 in $KHSO_4$ solutions or in pure water.

Physical Properties

Vitreous silica has many exceptional properties, a number of which are abnormal when compared to other glasses and even other solids (38). The following anomalous properties are more or less interdependent: The expansion coefficient is negative below ca -80 to -100°C and is positive and very small above these temperatures. The elastic moduli increase with increasing temperatures above ca -190°C. Young's modulus increases linearly with applied longitudinal stress at -196°C, whereas that of soda glass decreases (39). The compressibility increases as the pressure increases in the low to moderate range, ie, to ca 3 GPa ($<30,000$ atm). The equilibrium density decreases with heat treatment in the transformation range, in contrast to that of most glasses. The bulk modulus shows a negative pressure dependence. There is a divergence from diffusion-controlled permeation of hydrogen at elevated temperatures. The temperature coefficient of sound velocity is positive over the range 0–800°C (40).

Efforts have been made to reconcile some of these anomalies with the supposed glass structure (41). In addition, discontinuities in property-vs-temperature relationships are often associated with structural rearrangements observed in the various crystalline phases of SiO_2 (42). The abnormalities in acoustic loss, pressure and temperature dependence of compressibility, thermal expansion, and specific heat have been related to a bimodal distribution of —SiOSi— bond angles (43). The large amount of free volume in the vitreous material permits reorientation of the silica tetrahedra by changing —SiOSi— angles (keeping Si—O distances constant) or by favoring the lower angles in the bimodal distribution.

Physical properties are dependent on thermal history, as discussed below.

Thermal Expansion. Most manufacturers' literature (44–48) quotes a linear expansion coefficient within the 0–300°C range of ca 5.4 × 10⁻⁷/°C to 5.6 × 10⁻⁷/°C. The effect of thermal history on low-temperature expansion of Homosil (Heraeus-Schott Quarzschmelze GmbH) and Osram's vitreous silicas is shown in Figure 2 (49). The 1000, 1300, and 1720°C curves are for samples held at these temperatures until equilibrium density was achieved, then quenched in water. The effect of temperature on linear expansion of vitreous silica is compared with that of typical soda–lime and borosilicate glasses in Figure 3. Vitreous silica annealed at 1100°C has been designated NBS Standard Reference Material 739. Its expansion coefficient (α/K) may be calculated for 300–700 K from the following expression (50):

$$\alpha \times 10^6 \, K = -0.8218 + 7.606 \times 10^{-3} \, T - 1.266 \times 10^{-5} \, T^2 + 6.487 \times 10^{-9} \, T^3$$

where T = absolute temperature. Very precise measurements of the dimensional stability of low expansion materials indicate that vitreous silica, eg, Corning 7940 and Homosil, displays a length change at 25°C of ca 0.5 ppb per day (51).

Viscosity. The viscosity of vitreous silica in the transformation range depends primarily on thermal history and impurities (especially hydroxyl content) (52) (see Table 1). IR Vitreosil with a fictive temperature of 1400°C has a viscosity of 1 TPa·s (= 10¹³ P) at ca 1700°C, whereas a 1000°C fictive material has a viscosity of 1 TPa·s (= 10¹³ P) at ca 1320°C (53). Hydroxyl-containing (>0.1 wt %) vitreous silicas, eg, Corning Code 7940, Suprasil, Spectrosil, and Dynasil, have annealing points of 1087 ± 50°C (45–48). Hydroxyl-free (<0.001 wt %) high purity fused-quartz glasses, eg, IR Vitreosil, some of the General Electric types, and various Amersil grades, have softening, annealing, and strain points up to ca 100°C higher than the hydroxyl-containing types.

A complete viscosity–temperature curve is shown in Figure 4 for Amersil commercial-grade fused quartz (54–55).

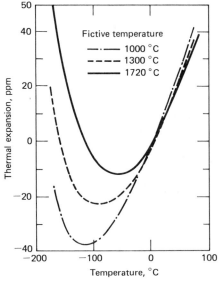

Figure 2. Effect of thermal history on low temperature thermal expansion (dimensionless) of vitreous silica (49).

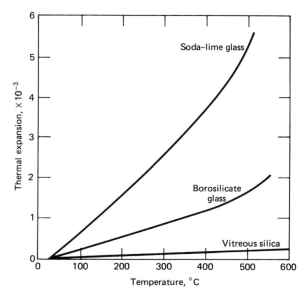

Figure 3. Comparison of thermal expansion (dimensionless) of vitreous silica with that of other glasses.

Table 1. Viscosity Data[a]

	QSI[b] Transparent	Translucent	GE 204	Vitreosil[c] IR	OG	TAFQ[d] Spectrosil	GCW[e] 7940	Dynasil
type	I	I	II	II	I	III	III[c]	III	III
OH, ppm	na	na	low	3	400	high	1200	1000	600–1000
softening point, °C	1670	1650	1813	1582	1597	1580	1594	1585	1600 ± 25
annealing point, °C	1140	1100	1213	1190	1108	1070	1082	1075	1100 ± 20
strain point, °C	1070	1040	1107	1108	1015	1000	987	990	1000 ± 20

[a] Manufacturers' literature, except where otherwise stated.
[b] Quartz Syndicate, Inc.
[c] Ref. 53; IR = infrared; OG = optical grade.
[d] Thermal American Fused Quartz.
[e] Corning Glass Works.

Very high temperature data have been obtained by x radiography with a falling tungsten ball in a closed tungsten crucible (56). Viscosity values range from 930 Pa·s (= 9300 P) at 2085°C to 860 Pa·s (= 8600 P) at 3210°C.

Mechanical Properties. The Young's modulus of vitreous silica at 25°C is 73 GPa ($<1.06 \times 10^7$ psi), the shear modulus is 31 GPa ($<4.5 \times 10^6$ psi), and the Poisson's ratio is 0.17. There may be small differences of ca ±1 GPa (<147,000 psi), depending on the type. The relationship of elastic moduli with temperature is approximately parabolic; increasing from ca 25°C to a maximum at ca 1050–1200°C. The Young's modulus at the max is 11% higher than the value at 25°C; the shear modulus increases by ca 9% (57).

Strength. The theoretical ultimate strength of vitreous silica has been calculated to be 1.6 GPa (<235,000 psi) (58); higher values of 2.3 GPa (<338,000 psi) have also been obtained (59–60). Values close to theoretical have been obtained experimentally

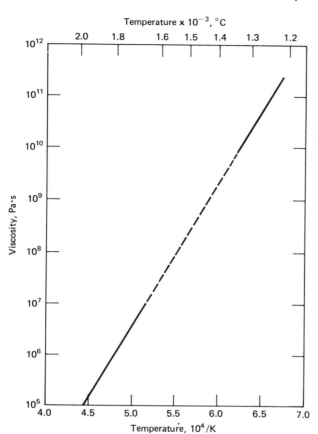

Figure 4. Viscosity–temperature curve for Amersil commercial-grade silica (54–55). 10 Pa·s = 1 P.

using carefully prepared flame-drawn rods with diameters as large as 1 mm and testing them under both bend and elongation conditions. For example, tests gave values in excess of 1.3 GPa (191,000 psi) at −196°C (61–62). Thus, vitreous silica is one of the strongest known bulk materials and attains nearly 60% of the estimated theoretical strength of 2.3 GPa (<338,000 psi).

The failure of this material to attain the theoretical value may be due to intrinsic surface flaws (58,61,63–64). However, its strength is also impaired by temperature, atmosphere, and mechanical abrasion. For the effect of temperature on flame-drawn fiber strength, see Figure 5. The strength of fused silica rods decreased in the presence of saturated vapors of alcohols, benzene, acetone, and water (65). The largest decrease was caused by water vapor, ie, almost 50% from a strength of 91 MPa (<13,400 psi).

The critical fracture energy γ_c is 4.3–4.4 J/m^2 (18–18.4 cal/m^2) in dry nitrogen gas at 27°C, and 3.7 J/m^2 (15.5 cal/m^2) in air at 40% rh (66–67). In aqueous solutions, pH effects crack growth (68). In high vacuum only abrupt failure with no subcritical crack growth was observed ≤775°C. The fracture mechanics of vitreous silica and other glasses has been reviewed (69).

Mechanical abrasion is a primary cause of decreased strength and as a result strength of manufactured materials is considerably lower than the theoretical values.

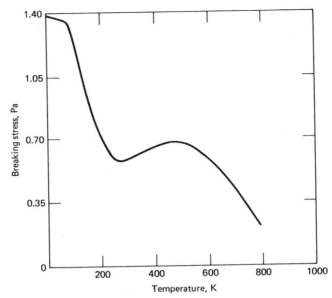

Figure 5.   Effect of temperature on vitreous silica fiber strength (64). To convert Pa to mm Hg, multiply by 0.0075.

For example, transparent vitreous silica has an impact strength of 83 MPa (<1220 psi), whereas nontransparent impact strength is 81 MPa (<1190 psi) (70). The modulus of rupture of a transparent rod abraded by sand blasting with a sand of 150–230 μm (<65 to >100 mesh) grit for 5 s is 49 MPa (<720 psi) at 25°C (44). This value increases to ca 65 MPa (<955 psi) at 550°C and 74 MPa (<1090 psi) at 900°C. At ambient temperature the strength of a flame-drawn fiber is decreased from ca 4.7 GPa (<690 psi) to 344 MPa (<50,600 psi) by merely placing a finger upon it (71).

The Knoop indentation hardness of vitreous silica is in the 473–593 kg/mm^2 range (72). The diamond pyramidal (Vickers) hardness is in the 600–750 kg/mm^2 range. The Vickers hardness for fused quartz decreases with increasing temperature but suddenly decreases at ca 70°C. In addition, a small positive discontinuity occurs at ca 570°C which may be due to a memory of quartz structure (73). A max at 570°C is attributed to the presence of small amounts of quartz microcrystals (74).

Density.   The density depends on thermal history and type of the sample. The density of transparent Vitreosil is 2.21 g/cm^3, of translucent Vitreosil 2.07–2.15 g/cm^3, and of Corning 7940 2.202 g/cm^3 (44–45).

Differences exist between the densities of natural (Types I and II) and synthetic (Type III) vitreous silica, with maxima occurring at 1550 and 1460°C, respectively (75–76). These differences may be attributed to hydroxyl content. Relationships have been demonstrated between density, hydroxyl content, and fictive temperature (77).

The degree of densification under pressure depends on time, temperature, and pressure. In general, the density increases with an increase in these variables, but the results obtained by various investigators on transparent glass are widely divergent (78–80). The wide range may reflect differences in the amount of shear stress applied by the various investigators, and densification may be primarily the result of structural rearrangement under shear (18). A density increase of almost 19% was obtained when the glass was subjected to a pressure of 8 GPa (<80,000 atm) at 575°C for 2 min. The samples remained completely amorphous, though the densities and refractive indexes were similar to those of cristobalite and in some cases approached those of α-quartz (79,81). However, crystallization takes place when the glass is subjected to similar pressures, eg, 8 GPa (<80,000 atm) for a very long time at temperatures as low as 450°C. Densification of vitreous silica by as much as 3% has also been attained by fast neutron irradiation (82) (see under Radiation Effects). Both kinds of densification can be reversed by annealing (83).

Electrical Properties. The d-c conductivity of vitreous silica is due primarily to impurities, and, particularly, alkali ions. The high resistivities are therefore the result of a low concentration of conducting species. Actually, the vitreous silica network is relatively open, and small differences in sodium content on the order of a few ppm significantly affect resistivity (84). The log resistivity–reciprocal temperature curves for a high hydroxyl–low alkali and a low hydroxyl–higher alkali vitreous silica are shown in Figure 6 (85). A small but significant change in slope of the hydroxyl-containing Corning Code 7940 curve is observed at ca 225°C. Several theories for this anomaly have been proposed but none has been confirmed (42).

The dielectric properties for two types of vitreous silica are given in Table 2 (85).

Ultrasonic Properties. Vitreous silica of high purity, such as the synthetic type, has an unusually low attenuation of high frequency ultrasonic waves (see Ultrasonics). The loss A is a linear function of frequency f up to the 30–40 MHz region and can be

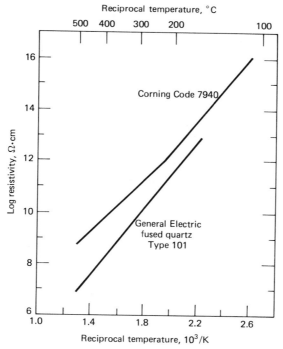

Figure 6. Relationship of d-c resistivity to temperature of vitreous silicas. Corning Code 7940 has a high hydroxyl–low alkali content; G.E. Type 101 has a low hydroxyl–high alkali content.

Table 2. Dielectric Properties of Vitreous Silica[a]

| Property | Type III | | GE Type 101[c] |
	Corning Code 7940[b]	Dynasil	
dielectric constant			
at 25°C, 1 kHz	3.95	3.83	3.95
1 MHz	3.95	3.83	3.93
8600 MHz	3.78	3.82	
at 300°C, 1 kHz	3.98	3.86	4.00
1 MHz	3.96	3.85	3.96
8600 MHz	3.78	3.85	
loss tangent			
at 25°C, 1 kHz	0.00006	0.000002	0.0003
1 MHz	0.00002	0.000015	0.0002
8600 MHz	0.00018	0.00012	
at 300°C, 1 kHz	0.0058	0.0093	0.089
1 MHz	0.0002	0.0001	0.0008
8600 MHz	0.00018	0.00012	

[a] Ref. 85.

[b] Sodium content <0.001 wt %; water content >0.1 wt %.

[c] Sodium content >0.001 wt %; water content <0.005 wt %.

expressed as $A = Bf$, where $B = 0.26$ (dB·MHz)/m for shear waves and 0.16 (dB·MHz)/m for compressional waves (86).

The ultrasonic relaxation loss may be due to a thermally activated structural relaxation associated with a shifting of bridging oxygen atoms between two equilibrium positions (87).

The velocity v of ultrasonic waves in an infinite medium is given by

$$v = (M/d)^{1/2}$$

where M = appropriate elastic modulus and density $d = 2.20$ g/cm^3.

With a shear modulus of 31.1 GPa (4.5×10^6 psi) vitreous silica has a plane-shearwave velocity of 3.76×10^5 cm/s.

Shear velocity and acoustic absorption have been studied as a function of OH content and fictive temperature for four different vitreous silica samples (88). All showed a shear-velocity minimum at 80 K; its magnitude was influenced by OH content. The samples having the highest OH content and lowest fictive temperature display the lowest losses.

Gaseous Diffusion. The permeability K, diffusivity D, and solubility S of a gas in a solid for a given temperature are related as follows (89–93):

$$K = K_o \exp(-E_k/RT)$$

$$D = D_o \exp(-E_d/RT)$$

$$S = S_o \exp(-E_s/RT)$$

where K_o, D_o, and S_o = constants (temperature dependence neglected), E_k, E_d, and E_s = activation energies for various processes, R = gas constant, and T = absolute temperature.

When the gas pressure outside the specimen is 101 kPa (1 atm) and any two of the above values (K, D, or S) are known for a given temperature, then the third may be obtained from the following simple relation:

$$K = DS$$

The important gas-diffusion parameters for vitreous silica, as determined experimentally by various investigators, are summarized in Table 3.

Physical and chemical solubility for inert gases (eg, He and Ne) vs interacting gases (eg, H_2) are distinguished in ref. 100. Hydrogen and D_2 are exchanged with hydroxyl groups in the glass above 300°C (101) or upon γ irradiation (102), leading to a universal expression for hydrogen solubility from 0 to 800°C up to 142 MPa (<1420 atm).

In addition to the technical importance of gas solubility and diffusivity data, additional information can be inferred about the structure of vitreous silica from the physical transport properties of inert gases in the glass (103).

There are apparently two different mechanisms involved in the oxidation of vitreous silica, depending on whether the material contains hydroxyl groups or not (96). For hydroxyl-containing material, the rapid oxidation probably occurs by the diffusion and removal of hydrogen, according to the following reaction:

$$-Si^{3+}HO-Si^{4+} \rightarrow -Si^{4+}-O-Si^{4+}- + \tfrac{1}{2}H_2$$

Conversely, in hydroxyl-free vitreous silica, the oxidation is much slower and is controlled by the diffusion of oxygen through the solid according to the following reaction:

$$-Si^{3+}Si^{3+}- + \tfrac{1}{2}O_2 \rightarrow -Si^{4+}-O-Si^{4+}-$$

The diffusion parameters for the diffusion of ^{22}Na in Infrasil fused quartz have also been determined (94). Between 1000 and 573°C, they are $D_o = 3.44 \times 10^{-2}$ cm^2/s, and $E_d = 88.3$ kJ/mol (21.1 kcal/mol); $D_o = 0.398$ cm^2/s and $E_d = 108$ kJ/mol (25.8 kcal/mol) between 573 and 250°C; and $D_o = 2.13$ cm^2/s and $E_d = 118$ kJ/mol (28.3 kcal/mol) between 250 and 170°C.

Thermal Properties. The mean heat capacity (0–900°C) can be calculated from the expression:

$$C_p, \text{J/(kg·°C)} = 700 + 0.79\,t - 5.23 \times 10^{-4}\,t^2$$

where t = temperature in °C. This gives a value of 0.71 J/(g·°C) [0.17 cal/(g·°C)] at 25°C (70).

Thermal conductivity data on clear transparent vitreous silica collected from a number of different sources are shown in Table 4. The difference between various types, including low and high hydroxyl content, is small. The thermal conductivity of the translucent form may be as much as 1% lower than that of the clear type when the heat flow is parallel to the striations.

Thermal diffusivity at 25°C is 9×10^{-3} cm^2/s; at 400°C it is 8–10% lower; it increases rapidly at higher temperatures (105).

Optical Properties. Optical transmission is influenced by the raw material and the method of manufacture. Hence, it offers a good method of characterization. The ultraviolet cutoff for ultrapure material of 1 cm thickness is slightly lower than 160 nm. Various impurities such as ferric ion, which is highly absorbing, or network defects caused by reducing conditions move this cutoff to longer wavelengths.

Table 3. Gaseous Diffusion through Vitreous Silica

Gas	Sample type	K_o, mol/(cm·s)	E_k, kJ/mol[a]	D_o, cm²/s	E_d, kJ/mol[a]	S_o, mol/cm³	E_s, kJ/mol[a]	Temperature, °C	Refs.
helium	fused quartz			3.04×10^{-4}	23.3 ± 0.25	1.99×10^{17}	-0.285 ± 0.25	24–300	90
helium	fused quartz			7.40×10^{-4}	27.7 ± 0.17	1.28×10^{17}	-4.90 ± 0.50	300–1034	90
helium	G.E. fused quartz		20.5					−80–600	89
neon-20	G.E. 204 fused	$1.25 \pm 0.17 \times 10^{13}$	39.4 ± 0.59	$2.21 \pm 0.12 \times 10^{-4}$	47.6 ± 0.33	$5.59 \pm 0.84 \times 10^{16}$	-8.2 ± 0.6	440–985	91
neon-22	not identified	$1.20 \pm 0.17 \times 10^{13}$	39.3 ± 0.63	$2.08 \pm 0.17 \times 10^{-4}$	47.4 ± 0.54	$5.75 \pm 0.92 \times 10^{16}$	-8.2 ± 0.7	440–985	91
hydrogen	G.E. 204 fused quartz and Suprasil	5.30×10^{13}	37.7	5.65×10^{-4}	43.5	9.4×10^{18}	−5.8	300–1000	92, 94
hydrogen	O.G. Vitreosil			9.5×10^{-4}				800–1050	95
deuterium	G.E. 204 fused quartz and Suprasil	5.12×10^{13}	39.2	5.10×10^{-4}	66.1	1.0×10^{17}	−4.6	300–1000	92, 94
deuterium	Corning 7943	1.9×10^{10}		1.5×10^{-7}		1.2×10^{17}		985	93
oxygen	Amersil			2×10^{-9}	121			850–1250	96
oxygen	IR Vitreosil			2.7×10^{-4}	113			800–1050	95
oxygen	not identified		92		113			950–1080	97
oxygen	not identified				298 ± 22			925–1225	98
oxygen	fused quartz			1.51×10^{-2}	234			900–1200	99

[a] To convert J to cal, divide by 4.184.

794

Table 4. Thermal Conductivity of Vitreous Silica[a]

Temperature, K	Thermal conductivity, W/(m·K)	Temperature, K	Thermal conductivity, W/(m·K)
100	0.674	400	1.51
200	1.126	500	1.62
300	1.37	600	1.74

[a] Ref. 104.

An absorption band at 242 nm is characteristic of many reduced vitreous silicas, such as IR Vitreosil, General Electric fused-quartz types, and Amersil grades, Homosil and Optosil. This band has been attributed to germanium impurity present in a partially reduced state (106). Reduced silicon Si^{2+} has also been postulated as causing the absorption (107). The band could also be the result of a reduced center Si^{3+} associated with a network substituent like Al^{3+} (108). Another suggestion is absorption by a trapped electron or hole in the vicinity of an impurity or associated vacancy, which could be insensitive to the specific impurity (109). All explanations have some merit but none is completely satisfactory.

Heraeus fused quartz shows fluorescence at 280 and 390 nm with 253.7 nm excitation (110). Corning Code 7940 vitreous silica, which is very low in impurities such as alumina, does not have the 242 nm absorption band required to initiate fluorescence.

Water incorporated in vitreous silica as —OH has a strong fundamental absorption band at 2.73 μm with overtones at 1.38 μm, 945 nm, and 585 nm. The Si—O vibration has two strong fundamental absorption bands that affect the transmission of vitreous silica. The fundamental vibration at 8.83 μm has a strong overtone at 4.45 μm, the practical infrared cutoff for vitreous silica of usable thickness. Another strong Si—O absorption at 12.41 μm combines with the strong —OH absorption at 2.73 μm to produce a peak at 2.22 μm. This peak occurs only in glass with high hydroxyl content (111).

Vitreous silicas have transmission curves of one of the types shown in Figure 7. The curves represent only the general shapes and should not be used for exact transmittance values. The types of vitreous silica and their corresponding curve designations are also listed. The glasses corresponding to curve A were prepared by vapor-phase hydrolysis and those corresponding to curve C were melted in a dry atmosphere, probably vacuum. Curve B only approximately represents the designated glasses in Figure 7. Glasses melted electrically from raw material with a relatively high hydroxyl content may have a minimum at 2.73 μ up to 20% higher than that shown. Glasses melted in a gas–oxygen atmosphere may have a minimum up to 20% lower than that shown. The peak at 2.22 μ becomes apparent as the minimum at 2.73 μm drops below 20%.

The spectral normal emissivity of Corning Code 7940 vitreous silica, as computed from measurements of transmittance and reflectance at 25°C, is shown in Figure 8 (112). The total normal emissivity of Code 7940 vitreous silica is shown in Figure 9 (113).

The refractive indexes of three synthetic vitreous silicas (Corning Code 7940, Dynasil high purity synthetic fused silica, and G. E. Type 151) for 60 wavelengths from 0.21 to 3.71 μ at 20°C (114) were compared with computed indexes using the following

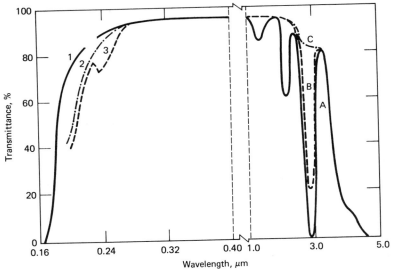

Figure 7.   Transmission curves for vitreous silica, 1 cm thick.

Manufacturer	Type	Uv curve	Ir curve
Amersil, Inc. (Heraeaus)	Optosil, Homosil	3	B
	Suprasil-W	1	C
	Ultrasil	2	B
	Infrasil	2	C
	Suprasil	1	A
Corning Glass Works	Code 7940	1	A
	Code 7943	3	C
Dynasil Corporation of America	Dynasil	1	A
General Electric	Types 204, 214	3	C
	Type 124 boule	2	C
Quartz et Silice	Pursil 453	3	C
	Pursil Ultra	2	C
	Tetrasil	1	A
Thermal American Fused	IR Vitreosil	3	C
Quartz Company, Thermal	Vetreosil 055, 066, 077	3	B
Syndicate Ltd.	Spectrosil A&B	1	A
	Spectrosil WF	1	C
Westdeutsche Quarzschmelze GmbH	Synsil	1	A

four-term Sellmeier dispersion equation (115):

$$n^2 - 1 = \sum (\lambda^2 \times a_{2i-i})/(\lambda^2 - a_{2i}^2)$$

where n = refractive index, λ = corresponding wavelength, a = parameter, and i = index number of the equation term.

With systematic errors removed from computed data, agreements with an average deviation of 4.3×10^{-6} were obtained. The refractive index at 25°C is 1.53429 at 0.21386 μm; 1.46313 at 0.48613 μm; 1.45841 at 0.58926 μm; 1.45637 at 0.65627 μm; and 1.39936 at 3.7067 μm. Reciprocal relative dispersion is 67.8. Refractive index changes with temperature ca $>1.00 \times 10^{-5}$/°C from 0 to 100°C (112).

The stress optical coefficient is 3.5×10^{-12} Pa^{-1}.

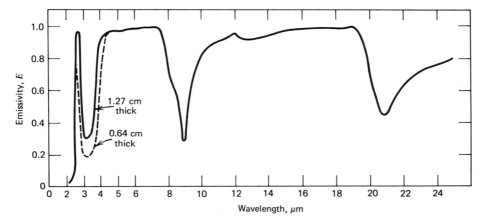

Figure 8.   Spectral normal emissivity of vitreous silica (112).

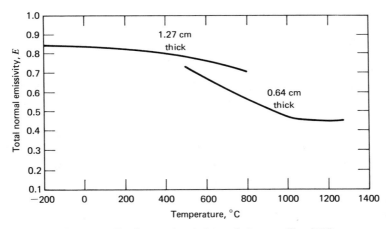

Figure 9.   Total normal emissivity of vitreous silica (113).

Radiation Effects

Depending upon the type of radiation, significant structural changes are generally due to high energy particle radiation, eg, a neutron stream, whereas electronic changes are usually the result of ionizing radiation, such as electron beams, x rays, γ rays, or protons. Damage by ionizing radiation manifests itself in the formation of optical absorption centers (also called color centers or defect centers). Though the predominant effect of neutron radiation is structural, electronic changes can also occur and, if the ionizing radiation is of very high intensity, structural defects may be produced.

Irradiation by fast neutrons causes a densification of vitreous silica that reaches a maximum value of 2.26 g/cm^3 (ie, an increase of ca 3%) after a dose of ca 1×10^{20} neutrons per square centimeter. Doses of up to 2×10^{20} n/cm^2 do not further affect this density value (116). Quartz, tridymite, and cristobalite attain the same density after heavy neutron irradiation, which means a density decrease of 14.7% for quartz and 0.26% for cristobalite (117). The resulting glasslike material is the same in each case, and shows no x-ray diffraction pattern but has identical density, thermal expansion (118), and elastic properties (119). Other properties are also affected, ie, the heat capacity is lower than that of vitreous silica (120), the thermal conductivity increases by a factor of 2 (121), and the refractive index n_D increases to 1.4690 (122). This new phase is called amorphous silica M after metamict, a word used earlier to designate minerals disordered by radiation in the geological past (123).

A number of mechanisms have been proposed by which this common irradiated state is obtained. The most widely accepted is the thermal-spike theory. It considers the heat generated in the wake of a fast particle passing through a solid as being sufficient to cause severe structural disturbances which are then frozen in by rapid cooling. Many property changes can be explained with this theory (122).

The structure of silica M is still in doubt. It is generally agreed that it is amorphous, though it exhibits a higher degree of order than vitreous silica. An electron diffraction pattern resembling that of α-quartz can be obtained from very heavily irradiated vitreous silica (124). No crystallinity was observed in samples given doses below 8.6×10^{19} nvt (integrated flux, where n = number of neutrons, v = velocity, t = time of exposure). Crystallinity so induced shows a large number of lattice defects and escapes detection by x-ray diffraction methods. This α-quartz model is further supported by the fact that amorphous silica M changes to polycrystalline quartz when heated at 940°C for 16 h (123). Confirmation of these results is required before this theory can be fully accepted. Radiation compaction of vitreous silica has been reviewed (125), and the effects of electron irradiation have been described (126–128).

Irradiation by both x rays and γ rays produces absorption centers in vitreous silica, although structural damage is negligible. These absorption centers show up primarily in the visible and ultraviolet spectral regions. Little change is seen in the infrared region, although a Raman line at 606 cm^{-1} has been ascribed to nonbridging oxygen defects (129). The types and number of absorption centers produced depend upon the purity of the vitreous silica and the total radiation dosage received. The rate of coloration of high purity vitreous silica (ie, very low concentration of metallic impurities, such as type III, Table 1) is very slow, and doses of ≥ 10 Gy (10^7 rad) are required to obtain appreciable absorption. For comparable absorption intensities in less-pure material (ie, types I and II, Table 1), radiation ca two orders of magnitude lower is required (130). Extensive research has been conducted in the field of radiation-induced absorption centers in vitreous silica but only the principal centers and their probable causes are discussed here. Excellent reviews of past work, including a thorough compilation of all absorption centers studied, are available (131–133).

A typical absorption curve for vitreous silica containing metallic impurities after x-ray irradiation is shown in Fig. 10 (134). The sample was a Heraeus fused quartz (the main impurity was aluminum), and the radiation dose was 10^4 Gy (10^6 rad). The primary absorption centers are at 550, 300, and between 220 and 215 nm. The 550 nm band is due to a center consisting of an interstitial alkali cation associated with a network substituent of lower valency than silicon (eg, aluminum) (131). Only alkalies contribute to the coloration at 550 nm; lithium is more effective than sodium, and sodium more effective than potassium. Pure silica doped with aluminum alone shows virtually no coloration after irradiation (135). The intensity of the band is determined by the component that is present in lower concentration. The presence of hydrogen does not appear to contribute to the 550 nm color-center production (135).

The absorption band at 300 nm may also be associated with alkali ions, possibly

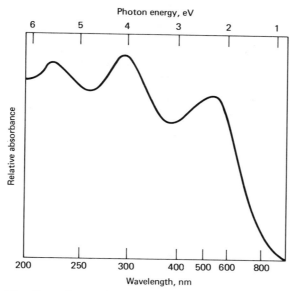

Figure 10.   Absorption spectrum of irradiated impure vitreous silica (134).

the result of a trapped electron stabilized by an alkali ion. The band shifts to longer wavelengths with heavier alkali ions, and growth rates for the band show a definite dependence on the type of alkali (131,135).

The 215 nm band may be intrinsic to silica, since it can be produced in Corning Code 7940 glass by long-term x-ray irradiation (136). This band is now generally assigned to an E' center, which may also be observed in irradiated α-quartz. Structurally, the E' center is assumed to be a pyramidal SiO_3 unit with an unpaired electron in the dangling sp^3 orbital of Si (133).

A typical absorption curve obtained for a metal-free vitreous silica after a large dose of γ rays (26 Gy or 2600 rad) is shown in Figure 11 (137). The main band is at ca 215 nm, with three smaller bands at 230, 260, and 280 nm. The 230 nm band may be due to an electron trapped at a silicon atom with an incomplete oxygen bond (131).

The band at 280 nm has been shown to be a germanium center (133). Nonbonding oxygens with a negative charge can acts as hole-traps in irradiated silica. These oxygen-associated-hole centers (OHCs) have been associated with an absorption at 163 nm. Dry OHCs, observed in vitreous silicas with low hydroxyl content, are peroxy-radical defects (138).

Chemical Properties

Stoichiometric vitreous silica contains two atoms of oxygen for every one of silicon, but it is extremely doubtful if such a material really exists. In general, small amounts of impurities derived from the starting materials are present; water is incorporated in the structure as —OH.

In the presence of water vapor at high temperature, the following reaction can take place:

$$\text{—SiOSi— + H}_2\text{O} \rightarrow 2\text{ —SiOH}$$

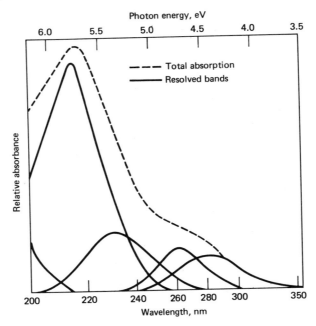

Figure 11.   Absorption spectrum of irradiated high purity vitreous silica. Measurements made on Corning Code 7940 glass (137).

Reduction to a certain degree also takes place, depending on the method of manufacture. In the presence of carbon the following reaction may occur:

$$SiO_2 + C \rightarrow SiO + CO$$

With pure carbon this reduction takes place at temperatures as low as 1200°C (70). A similar reduction with tungsten, tantalum, or molybdenum occurs at 1300–1400°C in vacuum (139). This type of reaction is very important in quartz fusion since SiO_2 is not volatile even over 1800°C, but SiO and Si have appreciable vapor pressures.

The resistance of nontransparent vitreous silica to chemical attack is slightly less than the resistance of transparent vitreous silica. This difference is primarily due to the higher surface area of the former caused by the presence of a large number of bubbles. Most data in the literature are on the transparent material.

Metals do not generally react below 1000°C or their melting point, whichever is lower. Exceptions are magnesium, alkali metals, and aluminum; the latter readily reduces silica at 700–800°C. Alkali-metal vapors attack at temperatures as low as 200°C. Sodium-vapor attack involves a diffusion of sodium into the glass, followed by a reduction of the silica:

$$4\,Na + 5\,SiO_2 \rightarrow 2\,Na_2Si_2O_5 + Si$$

This reaction is accompanied by a blackening of the vitreous silica and a flaking of the surface. As evidence for the diffusion step, the sodium absorption by the much denser quartz is 12 mg/(1000 h·cm²) at 350°C, whereas vitreous silica absorbs 23 mg/(1000 h·cm²) at 286°C (140). Molten sodium is much less reactive.

Fused basic salts and basic oxides react at elevated temperatures. Reaction with alkaline earth oxides takes place at ca 900°C. Halides may tend to dissolve vitreous silica at high temperatures; fluorides are the most reactive. Fused borates react appreciably (70).

Dry halogen gases do not react below 300°C. Reaction with hydrogen is very slight at 1000°C, but at much higher temperatures reduction takes place. The formation of —SiOH and SiH groups was demonstrated by infrared and Raman spectroscopy (141), in addition to some SiO formation:

$$2 \, SiO_2 + H_2 \rightleftharpoons {\geqslant}SiOH + {\geqslant}SiH$$
$$\text{(trace)}$$
$$3 \, SiO_2 + H_2 \rightleftharpoons 2 \, {\geqslant}SiOH + SiO$$

Hydrogen fluoride attacks vitreous silica readily:

$$SiO_2 + 4 \, HF \rightarrow 2 \, H_2O + SiF_4{\uparrow}$$

In the presence of an excess of hydrofluoric acid, the volatile reaction product may also be H_2SiF_6. The effects of two hydrofluoric acid solutions of different concentrations on various silica phases are shown in Table 5 (142). Phosphoric acid causes some attack at ca 150°C, whereas most other acids do not react. In 5% HCl at 95°C, surface erosion is <0.05 μm in 24 h (44). Similar results are obtained with H_2SO_4.

Table 5.   Dissolving Rates of Silica Phases[a]

Silica phase[b]	Silica dissolved, %	
	In HF, 5%, 1/2 h	In HF, 1%, 1 h
quartz	30.1	5.2
tridymite	76.3	20.3
cristobalite	74.3	25.8
vitreous silica	96.6	52.9

[a] Ref. 142.
[b] Samples of uniform particle size, ca 40 μm in dia.

Attack of dilute basic solutions is very slight at room temperature. In 5% NaOH at 95°C, surface erosion is ca 10 μm in 24 h; however, crazing may occur (44). At higher caustic concentrations and temperatures, the reaction rate increases significantly. With 45 wt % NaOH at 200°C, dissolution proceeds at 0.54 mm/h (143); this is used to advantage in the investment-casting industry. Typical durability test results are given in Table 6.

Vitreous silica does not react with water or steam at moderate temperatures and pressures. At ca 200–400°C and ca 1–30 MPa (<10–300 atm) the solubility S in g

Table 6.   Durability of Vitreous Silica at 95°C

Test solution	Duration of test, h	Weight loss, mg/cm²	Depth of attack, μm
5% HCl	24	<0.01	<0.05
5% NaOH	6	0.7	3.2
0.02 N Na$_2$CO$_3$	6	0.02	0.09
5% H$_2$SO$_4$	24	<0.01	<0.05
H$_2$O	24	<0.01	<0.05

$SiO_2/kg\ H_2O$ can be expressed as follows:

$$\log S = 2 \log d - \frac{2679}{T} + 4.972$$

where d = density of the vapor phase, and T = absolute temperature.

The solution occurs as a bimolecular, heterogeneous gas reaction (144):

$$SiO_2 + 2\ H(OH) \rightarrow [Si(OH)_4]$$

Other solubility data are given in Table 7.

Manufacture

A summary of the important manufacturing methods, manufacturers, and product names and forms is given in Figure 12.

Translucent Vitreous Silica. Translucent vitreous silica is produced by fusion of high purity quartz sand crystals. Sand is placed around a strong, conducting graphite rod through which a current is passed. The fusion product thus formed is plastic and can be blown into molds, drawn into tubing, or shaped by rolling or pressing. Separation from the graphite rod is facilitated by gaseous products formed by interfacial reaction. Since the outside is sandy, the product is known as sand-surface ware. A mat finish is obtained by mechanical buffing; a glazed surface is produced by quickly fusing the outside surface with an electric carbon arc or flame. The product is called glazed ware (see also Ceramics). Drawn tubing with the glazed surface is called satin tubing (6).

The Rotosil process employed by Heraeus and Heraeus-Amersil is used for the production of tubular or cylindrical shapes. It permits greater uniformity and dimensional control than the older process. The quartz is washed in hydrofluoric.acid and distilled water, to remove impurities. Impurities impair product performance and accelerate crystallization during manufacture. The sand is then placed in a rotating horizontal steel tube where it is held at the circumference by centrifugal force. A carbon arc is passed slowly down the center of the drum to fuse the sand. A modification, using open rotating molds on a vertical axis, allows formation of crucibles, beakers, and bowls (145).

Transparent Vitreous Silica. Clear, transparent, bubble-free vitreous silica may be obtained by melting natural quartz minerals or synthetic silica, or by flame or plasma vapor-deposition methods. Generally, four types are recognized and identified (75,146): Type I is obtained by electric melting of mineral quartz (sand, chunks, or powder), usually in vacuum. Type II is made by flame fusion of quartz. Type III is made by vapor-phase hydrolysis of pure silicon compounds carried out in a flame. Type IV is made by oxidation of pure silicon compounds which are subsequently fused elec-

Table 7.   Solubility of Silica in Water, g/10^6 g H_2O

Temperature, °C	Pressure, MPa[a]	Vitreous silica	Clear vitreous silica	Quartz
400	13.8	36	31	5.2
500	34.5	346		
500	103	4179		

[a] To convert MPa to atm, divide by 0.101.

trically or by means of a plasma. Types I and II are fused quartz, whereas Types III and IV are fused silica.

Fused Quartz. On heating, quartz undergoes an inversion from low to high at 573°C in which volume and elastic constants are rapidly changed. If heating is rapid and nonuniform, large differences in temperature develop within a large crystal. These conditions can cause fracture. If this phenomenon occurs during the melting process, gaseous inclusions in the form of air lines are present in the product (splintering). This problem is overcome by washing the crystals in hydrofluoric acid and distilled water to remove impurities. The crystals are dried and then heated to ca 800°C and plunged into distilled water to facilitate subsequent crushing. This procedure not only solves splintering problems, but also gives small particles with a larger surface area that are especially advantageous for vacuum melting, since they melt faster and expel gaseous impurities more completely than the raw quartz crystals.

The melting point of quartz is <1450°C, probably ca 1400°C. It is difficult to determine, since between the melting point of quartz and that of cristobalite at 1723°C, both glass and cristobalite are formed when quartz is heated in air for longer than 15 min. Quartz microcrystals are present even after heating at 1860°C for 30 min; however, crystallinity disappears at 1900°C on heating for 1 h (147). Fusion is generally carried out in an oxy–hydrogen flame or in molybdenum or carbon crucibles in an inert or reducing atmosphere. Electric melting procedures use a vacuum to minimize the gas content in the pore space before fusion. This procedure is sometimes followed by the application of pressure to reduce the remaining bubbles.

Although fused-quartz production is usually carried out above the melting point of cristobalite, a technique was developed for producing vitreous silica from quartz below 1723°C (148). For example, powdered quartz is pressed into a given shape in a mold and sintered at 1500°C for 8 min. Then the temperature is raised to 1650°C for 2 min. Owing to the rapid heating, no cristobalite is present and the product is vitreous. The sintering process takes no more than 15 min in the 1400–1710°C range.

In the Heraeus process, quartz crystals are fed through an oxy–hydrogen flame onto a rotating fused-quartz tube and withdrawn slowly from the burner as clear fused quartz is built up (149). In order to obtain a high purity material, particularly for lamp envelopes, chlorine or a chlorine compound is introduced into the system above 1600°C and preferably between 1800 and 2200°C. Any impurities that form volatile chlorides, eg, sodium, magnesium, calcium, barium, aluminum, copper, zinc, titanium, or iron, are removed (150). Alternatively, an alkali salt or alkaline earth salt may be added to the granular quartz batch, followed by stirring at 800–1700°C before fusing (151).

In the Osram process, quartz is fed into the top of a resistance-heated tubular furnace and melted in a molybdenum crucible protected by an inert or reducing gas. Tubing is drawn from the bottom (152).

In a similar process, purified sand is melted in an induction-heated furnace in a hydrogen–helium atmosphere and drawn into tubing or rod at the bottom (153).

Tubing may also be obtained by redrawing a hollow, vitreous silica cylinder formed under vacuum in an electrically heated rotating graphite container (154).

A similar hollow cylinder for redrawing is obtained by flame-atomizing quartz powder onto a thin fused-quartz tube on a refractory metal mandrel (155).

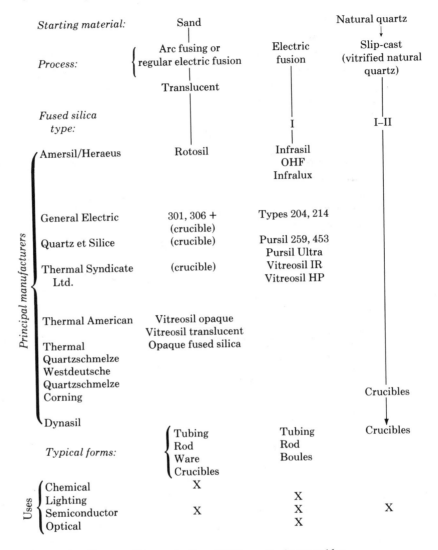

Figure 12.   Vitreous silica production. OWG = optical waveguides.

Fused Synthetic Silica.   Synthetic silica may be prepared from a variety of silicon compounds by oxidation or hydrolysis in a flame or plasma where it is fused to form silica glass. Synthetic material obtained as flocculent, microparticulate powder is called fumed silica (see also Silica, amorphous). Because of the possibility of purifying the starting material, products of very high purity may be obtained. This technology has been effectively utilized in a variety of processes for the manufacture of fiber-optic waveguides (see Fiber optics). Similar processes are also capable of producing very pure vitreous silica of good optical quality.

In a process called vapor-phase hydrolysis, oxygen is passed through a volatile silicon compound, such as silicon tetrachloride. In preparing the chloride, moisture

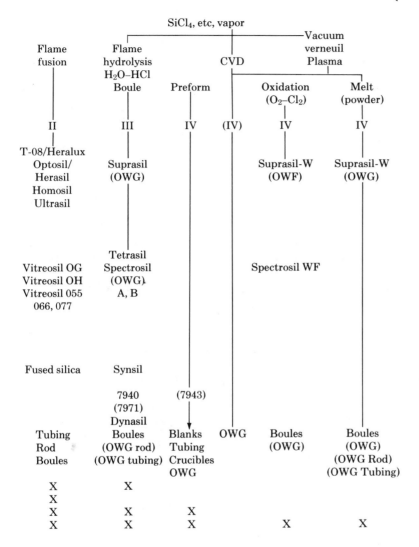

Figure 12. (*continued*)

must be excluded to prevent hydrolysis. With care, dry anhydrous silicon tetrachloride may be handled in metal vessels and tubing without contamination (see also Silicon compounds). The mixture of oxygen and silicon chloride is fed together with natural gas into a burner. (European manufacturers prefer hydrogen.) Hydrolysis takes place in the flame, producing fused vitreous silica. It can also be formed by deposition on a substrate of preheated sand and consolidated to clear glass by the heat from the burners (156). This product contains a high percentage of water.

Hydroxyl-free vitreous silica may also be produced by vapor-phase hydrolysis. The particles (ca 0.1 μm dia) formed are not consolidated in the burner flame but deposited on an air-cooled aluminum mandrel. As the mandrel rotates and moves laterally through the flame, layers of these particles are built up since the particles adhere tightly to one another. The deposited shape is then removed from the aluminum and placed over a graphite mandrel which is about one third the size of the aluminum mandrel. The particles consolidate in ca 30 min to a clear glass at 1450–1500°C in an electrically heated furnace containing hydrogen or a hydrogen–nitrogen mixture. The glass assumes the shape of the graphite mandrel.

Another vapor-phase method uses an oxygen plasma (157–158). A volatile, hydrogen-free silicon compound, such as silicon tetrachloride, is introduced into a hydrogen-free stream of oxygen which is passed through an induction-coupled plasma torch. The resulting silica is deposited on a refractory substrate and fired to a clear transparent glass. Low hydroxyl concentration is obtained by adding chloride, generally in the 200 ppm range. A CO_2 laser has also been used as a heat source (159) (see Lasers).

Fused silica materials obtained by the various vapor phase methods are formed in the shape of rod, tubing, and boules, and may range in size up to hundreds of kilograms. Any of these may be subsequently reworked to make a variety of useful products. Crucibles have been made from both tubing and massive pieces. Optical elements are cut from the boules and polished, and often finely annealed to control refractive index and reduce internal stress.

Impurities.   The maximum content of impurities of the different types of vitreous silica is given in Table 8. The impurities present depend mainly on the treatment and purity of the quartz crystals. Manufacturers' literature should be consulted for material designation.

Sintering and other Processes.   Since fabrication of large shapes is very difficult, sintering techniques have been developed to obtain large pieces with good dimensional control.

For certain refractory applications, sand of moderate purity is fused by an electric arc into an ingot, which is subsequently crushed and ground to a size suitable for injection-molding or slurry-casting. The fused refractory materials are utilized as bricks in glass-melting furnaces, shaped refractory articles for glass-delivery systems, pouring tubes and nozzles for the steel industry, and coke-oven doors and cast parts. Fused silica is particularly suitable for repairs in hot glass and coke furnaces, since its low thermal expansion renders it immune to thermal shock in contrast to many other typical refractories.

In a typical hot-pressing technique, vitreous silica is crushed, washed, and screened to ca 5 μm size. The powder is pressed into shapes at 6.9 MPa–17.3 MPa (1000–2500 psi) and 1100–1200°C up to 90 min (160–161).

A slip-casting method uses a water slip of powdered silica which is poured into a plaster mold. After removal from the mold, the body is dried slowly and then fired at ca 1100–1200°C for 1–4 h (162).

Transparent vitreous silica crucibles are obtained by firing slip-cast precursors in a vacuum or a helium atmosphere (163–164) or in helium followed by argon (165). The crucible is usually supported, eg, on a graphite form. A process requiring no supporting form uses an array of burners to fuse only one longitudinal section at a time (166), whereas another uses a frozen shaped slurry which is cast in a nonmelting granular refractory material before firing (167).

Fumed silica has also been used as a silica source. The finely divided silicon oxide

Table 8. Maximum Impurity Concentrations in Various Types of Vitreous Silica[a]

Impurity, ppm	Translucent vitreous silica	Transparent vitreous silica[b]		
		Type I	Type II	Type III[c]
aluminum	500	74	68	<0.25
antimony	na	0.3	0.1	0.1
arsenic	na	na	na	<0.02
boron	9	4	3	0.1
calcium	200	16	0.4	<0.1
chromium	na	0.1	na	0.03
cobalt	na	na	na	0.0001
copper	na	1	1	<1
gallium	na	na	na	<0.02
gold	na	na	na	<0.1
hydroxyl	na	60	450	1200
iron	77	7	1.5	<0.2
lithium	3	7	1	na
magnesium	150	4	na	na
manganese	na	1	0.2	<0.02
mercury	na	na	na	<0.1
phosphorus	na	0.01	0.005	<0.001
potassium	37	6	<1	0.1
sodium	60	9	5	<0.1
titanium	120	3	2	na
uranium	na	na	0.0006	na
zinc	na	na	na	<0.1
zirconium	15	3	na	na

[a] Data taken from the literature, private communications, and various manufacturers' publications. Courtesy of Thermal Syndicate Ltd.

[b] The types referred to are those described in the text. The raw material is not specified.

[c] The metallic impurities of Type IV vitreous silica are similar to those of Type III material, except for a negligible hydroxyl content.

is converted to a sol with a polar solvent, dried, calcined, wet-milled to a slip, cast, and fired as a crucible (168–169).

Other processes are intended to effect vitrification at lower temperatures, to increase purity, or both. For example, hydrolysis of silicon organometallic compounds, eg, tetraethyl orthosilicate, yields a silica–alcohol–water gel. The dried silica gel is subsequently fused (170–173). Sodium silicate solutions have been used to form silica sols by ion exchange; these are slowly dried and finally sintered at 1200°C under vacuum to a nonporous product (174). A similar process uses mixtures of alkali metal silicates and colloidal silica to yield solids of very high porosity with unusually uniform pore size. The porous bodies are then sintered to yield solid, transparent vitreous silica (175).

Vitreous silica tubing is also made from metallic silicon, carbon dioxide, and oxygen (176) (see Fiber optics).

Flame-working and Sealing. Flame-working of vitreous silica is difficult because of its extremely high viscosity and volatility. However, this drawback is compensated for by excellent resistance to thermal shock.

A gas–oxygen flame is satisfactory for most manipulations, whereas an oxy-hydrogen flame (1:2) provides somewhat more energy. An oxy–acetylene flame gives even more heat but with excessive volatilization.

When silica volatilizes, vapors condense on cooler areas to form a white bloom that can be removed by heat or dilute hydrofluoric acid. Because dilute hydrofluoric acid also attacks the substrate, a mild careful treatment is required. To minimize volatilization, the temperature should be as low as possible.

Annealing of flame-worked pieces is generally not necessary because of the low thermal expansion. However, massive pieces should be annealed at a temperature corresponding approximately to the annealing point of the specific type and for a time determined by size. Once the piece is uniformly heated throughout, stress is relieved in a matter of minutes.

Owing to recent developments in infrared lasers, vitreous silica up to 6 mm thick is cut and drilled with CO_2 gas lasers and numerically controlled equipment. Very thin kerf widths are obtained at higher speeds than with mechanical cutting and without induced stresses (177).

Occasionally, vitreous silica is attached to other glasses, eg, Corning Code 7740 borosilicate glass, by a graded seal. The silica is first sealed to Code 7230 glass (expansion coefficient $\alpha = 14 \times 10^{-7}/°C$, from 0–300°C) which in turn is sealed to Code 7240 glass ($\alpha = 21 \times 10^{-7}/°C$). The assembly is then sealed to Code 7740 glass ($\alpha = 32.5 \times 10^{-7}/°C$).

Sealing to metals is difficult since the thermal expansion of metals is much higher than that of vitreous silica. Nevertheless, such sealing is necessary, particularly in the case of mercury-vapor lamps where vacuum-tight lead connections are required. A molybdenum foil seal is used based on the principle that a very thin foil is able to contract during cooling without inducing stress in the glass. The molybdenum foil is connected to tungsten electrodes and placed in the tube which is flushed with a neutral gas. The tube is heated strongly on the outside with a gas–oxygen flame. It collapses around the foil and is mechanically pinched to make firm contact (178).

Economic Aspects

The first sales of fused silica were made in 1906 and consisted principally of pipes and muffles (179). The same year saw the incorporation of the Thermal Syndicate Ltd. Since then this industry has developed an annual business of $>$1 \times 10^8$ in 1980 in the United States alone.

Because of the wide diversity of uses and the proprietary nature of many products, economic data on vitreous silica are difficult to obtain. Certain estimates have been made, however (see Table 9). The annual U.S. consumption of quartz has been estimated at close to 45,000 metric tons, with prices ranging from \$0.50–2.00/kg.

Prices for individual products vary greatly with volume, purity, and optical quality. In 1980 the price for a 500-mL beaker was about \$45, whereas the price for a 13 cm $\times$ 13 cm $\times$ 6 mm plate ranged from \$40–340, depending on optical homogeneity, inclusions, and transmission. Commercial-grade tubing, 25–30-mm OD, sells for \$45–90/m and ultrapure 10-mm rod \$686/m, ie, \$3960/kg.

Table 9. Estimated Annual U.S. Sales of Vitreous Silica Products

Product	Sales, 10^6 \$
transparent tubing for lighting	20^a
optical applications[b]	12
hollow ware[c]	20
electronic applications	10–15

[a] Foreign sales are ca 2–3 times higher.
[b] Exclusive of fiber-optic waveguides.
[c] Tubing, chemical apparatus, crucibles.

The principal U.S. and foreign manufacturers of vitreous silica are given in Table 10.

In constant dollars, many items have decreased in cost over the past decade because of improved manufacturing methods, new raw material sources, and higher production volumes. However, many of these products were relatively unknown 10 years ago (eg, photomask substrates), whereas others have been practically displaced by new materials or technologies.

Table 10. Manufacturers and Suppliers of Vitreous Silica

Company	Location
United States	
Corning Glass Works	Corning, N.Y.
Dynasil Corporation of America	Berlin, N.J.
General Electric Company	Willoughby, Ohio
GTE-Sylvania[a]	Greenland, N.H.
Harbison-Walker Refractories	Pittsburgh, Pa.
Heraeus-Amersil, Inc.[b]	Sayreville, N.J.
Leco Corporation	St. Joseph, Mich.
Pyromatics, Inc.	Cleveland, Ohio
Quartz Scientific, Inc.	Fairport Harbor, Ohio
Sherwood Refractories, Inc.	Cleveland, Ohio
Thermal American Fused Quartz[c]	Montville, N.J.
Westinghouse[d]	Bloomfield, N.J.
Foreign	
Heraeus Quarzschmelze, GmbH	Hanau, FRG
Japan Electronic Metals Company Ltd.[e]	Tokyo, Japan
Nippon Seiki	Tokyo, Japan
Osram GmbH	Munich, FRG
N.V. Philips' Gloeilampenfabriken	Eindhoven, Netherlands
Quartz et Silice	Paris, France
Thermal Quarzschmelze GmbH	Wiesbaden-Biebrick, FRG
Thermal Syndicate	Wallsend, UK

[a] Internal for lamp envelopes.
[b] Subsidiary of Heraeus Quarzschmelze GmbH.
[c] Subsidiary of Thermal Syndicate, Ltd.
[d] Primarily internal.
[e] Subsidiary of Mitsubishi Metals.

Uses

Chemical Applications. With its excellent chemical durability, high purity, thermal shock resistance, and usefulness at high temperature, vitreous silica has a wide range of applications in chemical analysis and preparations. Tubing, rods, crucibles, dishes, boats, and other containers and special apparatus are available in both transparent and nontransparent varieties (45–48).

Because of its inertness, vitreous silica is used as a chromatographic substrate in the form of microparticles and open columns for high resolution gas chromatography (180) (see Analytical methods).

Thermal Applications. The protection of precious-metal thermocouples in high temperature pyrometry is an important application of vitreous silica. Although satin tubing is usually employed, transparent tubes are much superior for protecting couples when used in a reducing atmosphere (181).

Because of its electrical resistivity, impermeability, and low expansion, vitreous silica is used for gas-heated or electrically heated devices in various shapes, eg, as a tube or muffle. In its simplest form, an electric-resistance furnace consists of a vitreous siica tube or pipe on which the resistance element is wound (see Furnaces, electric). Because of its indifference to temperature gradients, a tubular furnace of vitreous silica may be made to operate at different temperatures at various portions of the tube, either by arrangement of the heating elements or by cooling sections of the tube with water. Vitreous silica pipes may be employed in vacuum-induction and gas-fired furnaces (182).

Radiant heaters employing resistance wire encased in vitreous-silica tubes have the capability to modify the emitted radiation and furnish a higher proportion of shorter wavelengths (1–2 μm) which constitute the desirable radiations (182). Immersion heaters for use with acid solutions are of similar construction. An overhead heating unit consisting of a resistance wire sealed inside a vitreous-silica container permits acid liquids to be concentrated or evaporated without ebullition or spattering. Some radiators have been developed for power requirements up to 3 kW, ca 64 cm long, for a large variety of applications that require high radiation efficiencies. They are used as energy sources in heat exchangers (see Heat-exchange technology), high duty enamel drying equipment, and copying devices (182).

Insulation with high thermal endurance has been made from vitreous-silica fibers (see Ablative materials). Such material forms the basis for the ca 30,000 insulating tiles, 125 mm thick, that protect the aluminum skin of the space shuttle. The tiles add a minimum of weight, since the density of the insulation is only 144 kg/m³ (9 lb/ft³), similar to balsa wood.

Mechanical Applications. The volume of vitreous silica used in fiber application is a very small part of the total consumption. However, some interesting and significant applications have been developed in the laboratory, particularly in the area of measurements.

The sorption balance employed 15 small vitreous silica springs enclosed in glass tubes (183). The fiber was 2 mm in dia and the coils 1.25 cm in dia and 5.99 cm long. A platinum bucket, which weighed 202 mg, contained up to 0.5 g charcoal; the balance could be read to 0.2 mg. The sorption of gases was determined over a wide range of temperatures and pressures. This type of balance has been used for density determinations, measurements of heat loss and evaporation, study of chemical reactions between gas and solid phases, and weight determination of wet tissues of plant and animal origin. The advantages of vitreous silica are its resistance to corrosion, ease of cleaning and sterilization in an enclosed tube, and absence of damping from internal friction.

Because of its low and regular thermal expansion, vitreous silica is employed in apparatus used to measure the thermal expansion of solids. The NBS has published a detailed account of the different methods used for this purpose (184). The most common form of dilatometer utilizes a vitreous-silica tube closed at the bottom and containing the test sample. A movable rod of vitreous silica, resting on the sample, actuates a dial indicator resting on the top of the rod. The assembly containing the sample is placed in a furnace, bath, or cooling chamber to attain the desired temperature.

Optical Applications. Because of its excellent ultraviolet transmission, resistance to radiation darkening, and physical and chemical stability, vitreous silica is the ideal refractive optical material. It is used for prisms, lenses, cells, windows, and other optical components where ultraviolet transmission is critical and for solar-cell covers that must resist radiation darkening (see Fiber optics; Solar energy).

A more recent application is in spacecraft windows. Every U.S. manned space vehicle has been equipped with windows made of high optical-quality vitreous silica (Corning Code 7940) for visual observation, photography, and television. Most recently, the space shuttle has utilized triple-layer windows that have outer and central panes of vitreous silica with a tempered aluminosilicate inner pane. The outer pane is thinner for thermal endurance, whereas the inner two panes are thick to supply strength.

The ability of vitreous silica to withstand dimensional change with changing temperature has made it ideally suited for mirror blanks of telescopes (185–187). These massive pieces range from 51 cm to 4 m in dia and are too heavy for spaceborne systems, but structures were developed with ca one-third the weight of the solid blank of comparable diameter and thickness. The solid surface plate is supported rigidly by an array of slitted struts or some other cellular structure of vitreous silica. Lightweight mirror blanks up to 2 m dia have been fabricated.

More recent mirrors have utilized various materials of exceedingly low expansion, partially replacing vitreous silica.

With the development of laser-fusion systems to shorter uv wavelengths for energy and military applications, vitreous silica has become an important component of laser optics (see Lasers). Very high purity materials have the required uv transmittance, relatively low refractive index, thermal endurance, and chemical inertness.

Lighting. An important application of clear fused quartz is as envelope material for mercury-vapor lamps (188) (see Mercury). In addition to resistance to deformation at operating temperatures and pressures, fused quartz offers ultraviolet transmission to permit color correction. Color is corrected by coating the inside of the outer envelope of the mercury-vapor lamp with phosphor (see Luminescent materials). Ultraviolet light from the arc passes through the fused-quartz envelope and excites the phosphor, producing a color nearer the red end of the spectrum (189). A recent improvement is the incorporation of metal halides in the lamp (190–191).

Incandescent tungsten–halogen (F, Br, I) cycle lamps first became available in 1959 for general use. In this lamp, tungsten evaporating from the filament deposits on the envelope wall where it reacts with halogen vapor to form a volatile halide. The tungsten halide diffuses back to the filament where it dissociates to halogen vapor and tungsten. This permits a higher temperature operation, which affords higher efficiency and longer life than the conventional incandescent lamp. The wall temperature must be at least 250°C and, if possible, ca 600°C. Fused quartz has been essential for this development since it is one of the few readily available transparent materials that can be used in lamp envelopes. Recent improvements include the internal deposition of a barrier (eg, aluminum fluoride, aluminosilicate) (192–193) and reflective layers (eg, titanium silicate, zirconia) (194–195). These protect the silica glass from vapor attack, slow the diffusion of deactivating impurities, and reflect heat onto the filament while transmitting visible light. These lamps are used in the illumination of airfields, sports arenas, buildings, and automobiles. They are also used in slide and film projectors and specialized optical instruments (196–197).

Electronic Applications. In electronic systems, such as radar and computers, signal delay is sometimes necessary. A transducer converts electrical signals to ultrasonic elastic waves, which pass through a connecting medium to another transducer where they are reconverted to electrical signals. Vitreous silica is an ideal connecting medium since it has excellent physical stability and low ultrasonic transmission losses; it transmits an ultrasonic signal almost 100,000 times slower than an electric signal in a wire. The vitreous silica delay line is in the form of a flat polyhedral plate. The facets of the edges of this plate are ground to a predetermined angle with great precision. The transducer is fastened to one of these facets. The plates may be designed for path lengths of 4 cm to nearly 12 m giving delay times for shear waves of 10–3000 μs (198).

By far the most significant electronic application, and perhaps the highest volume application for vitreous silica, is in semiconductors (see Semiconductors).

This association began with silicon-wafer technology, where vitreous silica was the principal high purity material for crucibles for silicon-crystal growth. Both opaque sintered and transparent vitreous silica from tubing or boules are used. Crucible sizes have been increasing in response to the demand for ever-larger silicon wafers.

Vitreous silica tubes are used for the subsequent high temperature diffusion, doping, and epitaxial growth treatments performed on wafers destined for large-scale integrated circuits. Once again, the high purity of certain grades of vitreous silica, along with its high thermal endurance, make it the material of choice for diffusion tubes, wafer holders, boats, and associated hardware. The thermal endurance of the tubes may be prolonged by causing some surface crystallization; the increased rigidity prevents sagging (47). Diffusion-resistant barrier coatings have also been used (199–200). In addition to silicon wafers, many other high temperature, high purity solid-state reactions are performed in vitreous-silica vessels, including crystal growth, zone refining, and the preparation of gallium and indium arsenides and germanium and silicon optical materials.

Thin films of vitreous silica have been used extensively in semiconductor technology. They serve as insulating layers between conductor stripes and a semiconductor surface in integrated circuits, and as a surface-passivation material in planar diodes, transistors, and injection lasers. They are also used for diffusion masking, as etchant surfaces, and for encapsulation and protection of completed electronic devices. They serve an important function in multilayer conductor–insulation technology where a variety of conducting paths are deposited in overlay patterns and insulating layers are required for separation.

Thin vitreous-silica films are usually formed by vapor deposition or r-f sputtering (see Film-deposition techniques). Vapor deposition is generally effected by the pyrolytic decomposition of tetraethoxysilane or another alkoxysilane. Silica has been most extensively used in r-f sputtering of dielectric films, which are of very high quality.

Large-scale and very large-scale integration (VLSI) of electronic circuits requires the photoreduction of complex conductor and insulator patterns, which are reproduced on semiconductor devices by various photoresist processes. Vitreous silica is frequently used for the photomask substrate, ie, the transparent substrate for the negative image mask that contains the basis for conduction or insulator patterns. Because of the extremely high dimensional tolerances of VLSI, vitreous silica is a natural choice for the substrate. Its uv transmission properties allow fast exposure of the photoresists, and its very low thermal expansion prevents pattern distortions due to temperature gradients encountered during processing. Perfectly flat and inclusion-free material is required (201).

Other Uses. Vitreous silica also has small-volume applications in the form of powders, fibers, wool, and chips. The powder may be used as an inert filler or as a coating material in the investment-casting industry for the lost-wax process. Chips are used as an inert substrate in certain chemical process applications. Vitreous silica wool or fiber is an excellent insulation or packing material.

Complex shapes rendered in extruded or sintered vitreous silica are used to produce precisely shaped holes, tunnels, cavities, and passages in investment-cast metal parts (see also Refractories). The vitreous silica shapes are chemically compatible with most alloys and dimensionally stable at casting temperatures; they give smooth internal surfaces to the cast pieces. The cooled casting is subsequently treated in hot caustic solutions to remove the silica core. Sandblasting may also be used. Internal features such as long holes and cooling passages for turbine rotors, which would be impossible to fabricate by drilling or machining, are obtained routinely (143).

BIBLIOGRAPHY

"Silica and Silicates (Vitreous Silica)" in *ECT* 1st ed., Vol. 12, pp. 335–344, by W. Winship, The Thermal Syndicate Ltd.; "Silica (Vitreous)" in *ECT* 2nd ed., Vol. 18, pp. 73–105, by William H. Dumbaugh and Peter C. Schultz, Corning Glass Works.

1. Gaudin, *C. R. Acad. Sci. Paris* **8**, 678, 711 (1839).
2. W. A. Shenstone and H. G. Lacell, *Nature* **62**(1592), 20 (1900).
3. W. A. Shenstone, *Nature* **61**(1588), 540 (1900).
4. *Ibid.*, **64**(1649), 126 (1901).
5. Ger. Pat. 113,817 (1899), (to Deutsche Gold-u.Silber-Scheideanstalt).
6. Brit. Pat. 10,670 (1904), J. F. Bottomley, R. S. Hutton, and R. A. S. Paget.
7. Brit. Pat. 10,437 (1904), J. F. Bottomley and R. A. S. Paget.
8. U.S. Pat. 778,286 (Dec. 27, 1904), E. Thomson.
9. Brit. Pat. 10,930 (1910), H. A. Kent and H. G. Lacell.
10. R. B. Sosman, *The Phases of Silica*, Rutgers University Press, New Brunswick, N.J., 1965, pp. 164, 165.
11. B. J. Skinner and J. J. Fahey, *J. Geophys. Res.* **68**(19), 5595 (1963).
12. Ref. 10, pp. 148–150.

13. J. S. Laufer, *J. Opt. Soc. Am.* **55,** 458 (1965).
14. G. Hetherington, *J. Brit. Ceram. Soc.* 3(4), 595 (1966).
15. J. T. Randall, H. P. Rooksby, and B. S. Cooper, *J. Soc. Glass Technol.* **14,** 219 (1930).
16. N. N. Valenkov and E. A. Porai-Koshits, *Nature* **137,** 273 (1936); *Z. Kristallogr.* **95,** 195 (1936).
17. K. S. Evstropyev in *Proceedings of a Conference on the Structure of Glass, Leningrad, 1953*, USSR Academy of Science Press, Moscow, USSR, 1953, transl. by Consultants Bureau, New York, 1958, pp. 9–15.
18. E. A. Porai-Koshits in ref. 17, pp. 25–35.
19. A. A. Lebedev, *Bull. Acad. Sci. USSR Phys. Ser.* (*English Transl.*) 4(4), 585 (1940).
20. W. H. Zachariasen, *J. Am. Chem. Soc.* **54,** 3841 (1932).
21. J. Wong and C. A. Angell, *Glass-Structure by Spectroscopy*, Marcel Decker, Inc., New York, 1976, pp. 436–442.
22. A. H. Narten, *J. Chem. Phys.* **56,** 1905 (1972).
23. J. B. Bates, *J. Chem. Phys.* **57,** 4042 (1972).
24. S. W. Barber in S. T. Pantelides, ed., *The Physics of SiO$_2$ and Its Interfaces*, Pergamon Press, New York, 1978, pp. 139–143.
25. H. T. Smyth, *Theory of Glass Structure*, Rutgers University, School of Ceramics, New Brunswick, N.J., 1961.
26. H. T. Smyth, *Theoretical and Experimental Properties of Simple Glasses*, Rutgers University, School of Ceramics, New Brunswick, N.J., 1966.
27. F. E. Wagstaff, *J. Am. Ceram. Soc.* 51(8), 449 (1968).
28. N. G. Ainslie, C. R. Morelock, and D. Turnbull, *Symposium on Nucleation and Crystallization in Glasses and Melts, Toronto, 1961*, American Ceramic Society, 1962, pp. 97–107.
29. H. Rawson, *Inorganic Glass-Forming Systems*, Academic Press, Inc., New York, 1967, pp. 48–61.
30. S. D. Brown and S. S. Kistler, *J. Am. Ceram. Soc.* 42(6), 263 (1959).
31. F. E. Wagstaff, S. D. Brown, and I. B. Cutler, *Phys. Chem. Glasses* 5(3), 76 (1964).
32. A. G. Boganov, V. S. Rudenko, and G. L. Bashina, *Izv. Akad. Nauk SSSR Neorg. Mater.* 2(2), 363 (1966).
33. F. E. Wagstaff and K. J. Richards, *J. Am. Ceram. Soc.* 48(7), 382 (1965).
34. U.S. Pat. 3,370,921 (Feb. 27, 1968), F. E. Wagstaff.
35. D. R. Uhlmann, J. F. Hays, and D. Turnbull, *Phys. Chem. Glasses* 7(5), 159 (1966).
36. V. J. Fratello, J. F. Hays, and D. Turnbull, *J. Appl. Phys.* 51(9), 4718 (1980).
37. R. G. Yalman and J. F. Corwin, *J. Phys. Chem.* **61,** 1432 (1957).
38. O. L. Anderson and G. J. Dienes in V. D. Frechette, ed., *Non-Crystalline Solids*, John Wiley & Sons, Inc., New York, 1960, pp. 449–486.
39. F. P. Mallinder and B. A. Procter, *Phys. Chem. Glasses* 5, 91 (1964).
40. G. W. Morey, *The Properties of Glass*, Reinhold Publishing Corporation, New York, 1954, pp. 319–320.
41. F. M. Ernsberger in *Elasticity and Strength in Glass*, Vol. 5 of D. R. Uhlmann and N. J. Kreidl, eds., *Glass: Science & Technology*, Academic Press, Inc., New York, 1980, pp. 9–12.
42. C. L. Babcock, *Silicate Glass Technology Methods*, John Wiley & Sons, Inc., New York, 1977, pp. 56–62.
43. M. R. Vukcevich, *J. Non-Cryst. Solids* **11,** 25 (1972).
44. *Fused Silica Code 7940*, Corning Glass Works, Corning, New York, 1978.
45. *TAFQ*, Thermal American Fused Quartz Company, Montville, N.J., 1980.
46. *Dynasil Fused Silica*, Dynasil Corporation of America, Berlin, N.J., 1976.
47. *Fused Quartz Tubing*, General Electric Company, Cleveland, Ohio, 1974.
48. *Optical Fused Quartz and Fused Silica*, Heraeus-Amersil, Inc., Sayreville, N.J., 1981.
49. R. Brückner, *Glastech. Ber.* 37(10), 459–475 (1964).
50. T. A. Hahn and R. K. Kirby in M. G. Graham and H. E. Hagy, eds., *Thermal Expansion—1971*, American Institute of Physics, New York, 1972, pp. 13–24.
51. J. W. Berthold and S. F. Jacobs, *Appl. Opt.* 15(10), 2344 (1976).
52. R. Brückner, *J. Non-Cryst. Solids* **5,** 177 (1971).
53. G. Hetherington, K. H. Jack, and J. C. Kennedy, *Phys. Chem. Glasses* **5,** 130 (1964).
54. E. H. Fontana and W. A. Plummer, *Phys. Chem. Glasses* **7,** 139 (1966).
55. R. Brückner, *Glastech. Ber.* 37(9), 413 (1964).
56. D. W. Bowen and R. W. Taylor, *Ceram. Bull.* 57(9), 818 (1978).
57. S. Spinner, *J. Am. Ceram. Soc.* **45,** 394 (1962).

58. A. Kelly, *Strong Solids*, Clarendon Press, Oxford, England, 1966, p. 5.
59. W. B. Hillig in J. D. Mackenzie, ed., *Modern Aspects of Vitreous State*, Vol. 2, Butterworth & Co., Washington, D.C., 1962, p. 190.
60. I. Náray-Szabó and J. Ladik, *Nature* **188,** 226 (1960).
61. J. G. Morley, P. A. Andrews, and I. Whitney, *Phys. Chem. Glasses* **5,** 1 (1964).
62. W. B. Hillig, *Symposium on the Strength of Glasses and the Means to Improve It*, Union Scientifique Continentale du Verre, Charleroix, Belgium, 1962, p. 206.
63. F. M. Ernsberger in J. E. Burke, ed., *Progress in Ceramic Science*, Vol. 3, Pergamon Press, Inc., New York, 1963, pp. 59–75.
64. J. G. Morley, *Proc. Roy. Soc. London Ser. A* **282,** 43 (1964).
65. M. L. Hammond and S. F. Ravity, *J. Am. Ceram. Soc.* **46**(7), 329 (1963).
66. S. M. Wiederhorn, *J. Am. Ceram. Soc.* **52,** 99 (1969).
67. J. J. Mecholsky and co-workers, *J. Am. Ceram. Soc.* **57,** 440 (1974).
68. S. M. Wiederhorn and H. Johnson, *J. Am. Ceram. Soc.* **56,** 192 (1973).
69. S. W. Freiman in ref. 41, pp. 9–12.
70. *Vitreosil*, Thermal American Fused Quartz Company, Montville, N.J., 1966.
71. W. B. Hillig, *J. Appl. Phys.* **32,** 741 (1961).
72. *Annual ASTM Standards*, Pt. 17, American Society for Testing and Materials, Philadelphia, Pa., 1978, p. 750.
73. J. H. Westbrook, *Phys. Chem. Glasses* **1,** 32 (1960).
74. J. D. Mackenzie, *J. Am. Ceram. Soc.* **43,** 615 (1960).
75. R. Brückner, *J. Non-Cryst. Solids* **5,** 123 (1970).
76. *Ibid.*, 281 (1971).
77. J. F. Shackelford and co-workers, *J. Am. Ceram. Soc.* **53**(7), 417 (1970).
78. P. W. Bridgman and I. Simon, *J. Appl. Phys.* **24**(4), 405 (1953).
79. R. Roy and H. M. Cohen, *Nature* **190,** 798 (1961).
80. E. B. Christiansen, S. S. Kistler, and W. B. Gogarty, *J. Am. Ceram. Soc.* **45**(4), 172 (1962).
81. J. D. Mackenzie, *J. Am. Ceram. Soc.* **46**(10), 461 (1963).
82. W. Primak, L. H. Fuchs, and D. Day, *J. Am. Ceram. Soc.* **38,** 135 (1955).
83. S.-Y. Hsich and co-workers, *J. Non-Cryst. Solids* **6,** 37 (1971).
84. A. E. Owen and R. W. Douglas, *J. Soc. Glass Technol.* **43,** 159T (1959).
85. Private communication, M. T. Splann and W. H. Barney, Corning Glass Works, 1968.
86. M. D. Fagan, *Proc. Natl. Electron. Conf.* **7,** 380 (1951).
87. O. L. Anderson and H. E. Bömmel, *J. Am. Ceram. Soc.* **38,** 125 (1955).
88. J. T. Krause, *J. Appl. Phys.* **42**(8), 3035 (1971).
89. F. J. Norton, *J. Am. Ceram. Soc.* **36**(3), 90 (1953).
90. D. E. Swets, R. W. Lee, and R. C. Frank, *J. Chem. Phys.* **34**(1), 17 (1961).
91. R. C. Frank, D. E. Swets, and R. W. Lee, *J. Chem. Phys.* **35**(4), 1451 (1961).
92. R. W. Lee, *J. Chem. Phys.* **38**(2), 448 (1963).
93. R. W. Lee and D. L. Fry, *Phys. Chem. Glasses* **7**(1), 19 (1966).
94. G. H. Frischat, *J. Am. Ceram. Soc.* **51**(9), 528 (1968).
95. E. L. Williams, *J. Am. Ceram. Soc.* **48**(4), 190 (1965).
96. G. Hetherington and K. H. Jack, *Phys. Chem. Glasses* **5**(5), 147 (1964).
97. F. J. Norton, *Nature* **191,** 701 (1961).
98. E. W. Sucov, *J. Am. Ceram. Soc.* **46**(1), 14 (1963).
99. R. Haul and G. Dümbgen, *Z. Elektrochem.* **66**(8–9), 636 (1962).
100. J. F. Shackelford and co-workers, *J. Appl. Phys.* **43**(4), 1619 (1972).
101. J. E. Shelby, *J. Appl. Phys.* **48**(8), 3387 (1977).
102. *Ibid.*, **50**(8), 5533 (1979).
103. J. F. Shackelford, *J. Non-Cryst. Solids* **42,** 165 (1980).
104. L. C. K. Carwile and H. J. Hoge, *U.S. Army Technical Report 67-7-PR*, U.S. Army Natick Labs, Natick, Mass., July 1966.
105. E. B. Shand, *Glass Engineering Handbook*, McGraw-Hill Book Company, New York, 1958, p. 30.
106. V. Garino-Canina, *Verres Refract.* **6,** 313 (1958).
107. H. Mohn, *60 Jahre Quarzglas-25 Jahre Hochvakuumtechnik*, W. C. Heraeus GmbH, Hanau, Germany, 1961, p. 114.
108. G. Hetherington, K. H. Jack, and M. W. Ramsay, *Phys. Chem. Glasses* **6**(1), 6 (1965).

109. W. H. Turner and H. A. Lee, *J. Chem. Phys.* **43**, 1428 (1965).
110. A. Kats and J. M. Stevels, *Philips Res. Rep.* **11**, 115 (1956).
111. R. V. Adams and R. W. Douglas, *J. Soc. Glass Technol.* **43**, 147T (1959).
112. C. J. Parker, Corning Glass Works, private communication, 1968.
113. W. A. Clayton, *Space Aeronaut.* 129 (June 1963).
114. I. H. Malitson, *J. Opt. Soc. Am.* **55**, 1205 (1965).
115. B. Brixner, *J. Opt. Soc. Am.* **57**, 674 (1967).
116. E. Lell, N. J. Kreidl, and J. R. Hensler in J. Burke, ed., *Progress in Ceramic Science*, Vol. 4, Pergamon Press, Inc., New York, 1966.
117. M. C. Wittels and F. A. Sherrill, *Phys. Rev.* **93**, 1117 (1954).
118. I. Simon, *J. Am. Ceram. Soc.* **41**, 116 (1958).
119. G. Mayer and M. Lecomte, *J. Phys. Radium* **21**, 846 (1960).
120. A. E. Clark and R. E. Strakna, *Phys. Chem. Glasses* **3**, 121 (1962).
121. A. F. Cohen, *J. Appl. Phys.* **29**, 591 (1958).
122. W. Primak, *Phys. Rev.* **110**, 1240 (1958).
123. Ref. 10, pp. 150, 173–175.
124. S. Weissman and K. Nakajima, *J. Appl. Phys.* **34**, 3152 (1963).
125. W. Primak, *Compacted States of Vitreous Silica*, Gordon & Breach Science Publications, New York, 1975, pp. 83–127.
126. C. B. Norris and E. P. EerNisse, *J. Appl. Phys.* **45**(9), 3876 (1974).
127. W. Primak, *J. Appl. Phys.* **49**(4), 2572 (1978).
128. T. A. Dellin and co-workers, *J. Appl. Phys.* **48**(3), 1131 (1977).
129. F. L. Galeener and co-workers in ref. 24, pp. 284–288.
130. G. W. Arnold, Jr., *J. Phys. Chem. Solids* **13**, 306 (1960).
131. Ref. 116, pp. 3–93.
132. P. H. Gaskell and D. W. Johnson, *J. Non-Cryst. Solids* **20**, 153, 171 (1976).
133. D. L. Griscom in ref. 24, pp. 232–252.
134. J. M. Stevels in V. D. Frechette, ed., *Non-Crystalline Solids*, John Wiley & Sons, Inc., New York, 1960, pp. 412–441.
135. E. Lell, *Phys. Chem. Glasses* **3**, 84 (1962).
136. A. J. Cohen, *J. Chem. Phys.* **23**, 765 (1955).
137. P. W. Levy, *Phys. Chem. Solids* **13**, 287 (1960).
138. E. J. Friebele and co-workers, *Phys. Rev. Lett.* **42**(20), 1346 (1979); M. Stapelbroek and co-workers, *J. Non-Cryst. Solids* **32**, 313 (1979).
139. A. deRudney, *Vacuum* **1**, 204 (1951).
140. C. A. Elyard and H. Rawson, *Advances in Glass Technology, Proceedings of the 6th International Congress on Glass, Washington, D.C., 1962*, Plenum Press, New York, pp. 270–286.
141. G. H. A. M. VanderSteen, Ph.D. Thesis, *Introduction and Removal of Hydroxyl Groups in Vitreous Silica*, Eindhoven, The Netherlands, 1976, pp. 64–70.
142. Ref. 10, p. 146.
143. J. J. Miller and D. L. Eppink, *Leaching of Preformed Ceramic Cores*, Sherwood Refractories, Inc., Cleveland, Ohio, 1977.
144. R. Moseback, *J. Geol.* **65**, 347 (1957).
145. J. Fortey, *BIOS Final Report 1202*, Biological Investigations of Space, 1947.
146. G. Hetherington, G. W. Stephenson, and J. A. Winterburn, *Electron. Eng.* **52**, (May 1969).
147. J. D. Mackenzie, *J. Am. Ceram. Soc.* **43**, 615 (1920).
148. U.S. Pat. 2,270,718 (Jan. 20, 1942), F. Skaupy and G. Weissenberg.
149. U.S. Pat. 2,904,713 (Sept. 15, 1959), W. H. Heraeus and H. Mohn (to Heraeus Quarzschmelze GmbH).
150. U.S. Pat. 3,128,166 (April 7, 1964), H. Mohn (to Heraeus Quarzschmelze GmbH).
151. U.S. Pat. 3,850,602 (Nov. 26, 1974), R. Brüning (to Heraeus-Schott Quarzschmelze GmbH).
152. U.S. Pat. 2,155,131 (April 18, 1939), W. Hanlein (to Patent-Treuhand-Gesellschaft für Elektrische Glühlampen GmbH).
153. U.S. Pat. 3,764,286 (Oct. 9, 1973), S. Antczak and co-workers (to General Electric Company).
154. U.S. Pat. 3,853,520 (Dec. 10, 1974), K. Rau.
155. U.S. Pat. 3,486,870 (Dec. 30, 1969), A. P. Vervaart and A. J. P. Ansems (to U.S. Philips Corporation).

156. U.S. Pat. 2,272,342 (Feb. 10, 1942), J. F. Hyde (to Corning Glass Works).
157. U.S. Pat. 3,275,408 (Sept. 27, 1966), J. A. Winterburn (to The Thermal Syndicate, Ltd.).
158. K. Nassau, T. C. Rich, and J. W. Shiever, *Appl. Opt.* **13**(4), 744 (1974).
159. S. Kobayashi and co-workers, *Appl. Opt.* **14**(12), 2817 (1975).
160. T. Vasilos, *J. Am. Ceram. Soc.* **43**, 517 (1960).
161. U.S. Pat. 3,116,137 (Dec. 31, 1963), T. Vasilos and R. Wagner (to Avco Corporation).
162. J. D. Fleming, *Am. Ceram. Soc. Bull.* **40**, 748 (1961).
163. U.S. Pat. 4,072,489 (Feb. 7, 1978), T. A. Loxley and co-workers (to Sherwood Refractories, Inc.).
164. U.S. Pat. 3,837,825 (Sept. 24, 1974), T. A. Loxley and co-workers (to Sherwood Refractories, Inc.).
165. U.S. Pat. 3,775,077 (Nov. 27, 1973), C. A. Nicastro and co-workers (to Corning Glass Works).
166. U.S. Pat. 3,620,702 (Nov. 16, 1971), A. DeKalb and A. G. Whitney (to Corning Glass Works).
167. Can. Pat. 915,889 (Dec. 5, 1972), F. J. Edwards and co-workers (to Thermal Syndicate, Ltd.).
168. U.S. Pat. 4,042,361 (Aug. 16, 1977), P. P. Bihuniak and co-workers (to Corning Glass Works).
169. U.S. Pat. 4,200,445 (Apr. 29, 1980), P. P. Bihuniak and co-workers (to Corning Glass Works).
170. U.S. Pat. 4,243,422 (Jan. 6, 1981), A. Lenz and co-workers (to Dynamit Nobel Aktienges.).
171. U.S. Pat. 4,098,595 (July 4, 1978), A. Lenz and co-workers (to Dynamit Nobel Aktienges.).
172. G. J. McCarthy, R. Roy, and J. M. McKay, *J. Am. Ceram. Soc.* **54**(12), 637 (1971).
173. G. J. McCarthy and R. Roy, *J. Am. Ceram. Soc.* **54**(12), 639 (1971).
174. U.S. Pat. 3,535,890 (Oct. 27, 1970), K. W. Hansen and H. P. Hood (to Corning Glass Works).
175. U.S. Pat. 4,059,658 (Nov. 22, 1977), R. D. Shoup and W. J. Wein (to Corning Glass Works).
176. U.S. Pat. 4,054,641 (Oct. 18, 1977), J. N. Carman (to J. S. Pennish).
177. F. Foster, Applied Laser Systems, Inc., Santa Clara, Calif., personal communication, 1981.
178. W. G. Houskeeper, *J. Am. Inst. Electr. Eng.* **42**, 954 (1923).
179. G. E. Stephenson, *J. Soc. Glass Technol.* **39**, 37T (1955).
180. S. G. Hurt and co-workers, *Chromatogr. Newsl.* **8**(2), 32 (1980).
181. R. A. Ragatz and O. A. Hougen, *Chem. Met. Engr.* **33**, 415 (1926).
182. Ref. 107, pp. 174–177.
183. J. W. McBain and A. M. Bakr, *J. Am. Ceram. Soc.* **48**, 690 (1926).
184. P. Hidnert and W. Sauder, *Natl. Bur. Std. (U.S.) Circ.* **486** (1950).
185. C. J. Parker, *Appl. Opt.* **7**, 740 (1968).
186. C. L. Rathmann, C. H. Mann, and M. E. Nordberg, *Appl. Opt.* **7**, 819 (1968).
187. Ref. 107, pp. 143–149.
188. E. W. Beggs, *Illum. Engr.* **42**, 435 (1947).
189. H. D. Frazer and W. S. Till, *Illum. Engr.* **47**, 207 (1952).
190. D. A. Larson, H. D. Fraser, W. V. Cushing, and M. C. Unglert, *Illum. Engr.* **58**, 434 (1963).
191. E. C. Martt, L. J. Smialek, and A. C. Green, *Illum. Engr.* **59**, 34 (1964).
192. J. R. Fitzpatrick and V. W. Goddard, *J. Illum. Engr. Soc.* **10**(2), 107 (1981).
193. Ger. Offen. 2,524,410 (Dec. 11, 1975), E. M. Clausen (to General Electric Company).
194. U.S. Pat. 3,988,628 (June 13, 1974), E. M. Clausen (to General Electric Company).
195. U.S. Pat. 3,879,625 (Oct. 9, 1973), C. I. McVey and O. M. Uy (to General Electric Company).
196. J. A. Moore and C. M. Jolly, *Gen. Electr. Corp. J.* **29**, 99 (1962).
197. T. M. Lemons and E. R. Meyer, *Illum. Engr.* **59**, 723, 728 (1964).
198. E. B. Shand, *Glass Engineering Handbook*, McGraw-Hill Book Company, New York, 1958, pp. 354, 355.
199. K. M. Eisele and R. Ruthardt, *J. Electrochem. Soc.: Solid State Sci. Technol.* **125**(7), 1188 (1978).
200. *T07 Stabilised Diffusion Tubes*, Heraeus-Amersil, Sayreville, N.J., 1980.
201. *Photomask Substrates for Microlithography*, Corning Glass Works, Corning, New York, 1980.

PAUL DANIELSON
Corning Glass Works

SYNTHETIC QUARTZ CRYSTALS

Silicon dioxide [7631-86-9], SiO_2, exists in both crystalline and glassy forms. In the former, the most common polymorph is α-quartz (low quartz). All commercial applications use α-quartz which is stable only below ca 573°C at atmospheric pressure. Some of the properties of α-quartz are listed in Table 1.

Quartz is mainly used in electronic applications for which it must be free of electrical and optical twinning, voids, inclusions of foreign minerals and liquids, and must be large enough for convenient processing. The principal source of electronic-grade natural quartz is Brazil, but today manufacturers use synthetic quartz for electronic devices. Until the discovery of economic processes for quartz synthesis in the 1950s, natural quartz was used for commercial applications. It is generally no longer used because its size, perfection, and properties vary. In addition, because natural crystals are generally irregular in shape, automated cutting is cumbersome and the yield is lower. Vitreous silica consists of small, less perfect crystals which are also used as a starting material for synthetic quartz growth. A typical recent price for such material was <$2.00/kg.

Table 1. Properties of α-Quartz[a]

Property	Value
structural	
crystal class, space group	32[b]
lattice const, nm	
a	0.491
b	0.540[c]
optical	
indexes of refraction, Na D line	
n_o	1.5442
n_E	1.5533[d]
optically active, Na D line	
α, °/mm	27.71
transmission, good from, nm	150–3000
electrical	
resistivity, Ω·cm	10^{15}
dielectric const	
ϵ_1^T	4.58
ϵ_3	4.70
piezoelectric coupling coefficient, %	10
piezoelectric const, FC/N[e]	
d_{11}	−23.12
d_{14}	727
mechanical	
hardness, Mohs	7
thermal conductivity, W/(m·K)	6.69–12.13
acoustic Q	0.1×10^6–3×10^6

[a] Many properties are directionally dependent; therefore, the values listed are indicative only.
[b] Trigonal trapezohedral class of the rhombohedral subsystem.
[c] Variable, depending on purity.
[d] Birefringent; $n_E - n_o = 0.0091$.
[e] To convert FC/N to stat C/dyn, divide by 333×10^8.

The principal regions where natural quartz of size and perfection suitable for electronic applications is found are in the states of Minas Gerais, Goiaz, and Bahia in Brazil. The quartz occurs in veins, pipes, and pockets in Precambrian sedimentary rocks. Natural quartz is expensive because scattered deposits are often mined on a desultory basis by individuals, and only a small yield of crystals large and perfect enough to be usable is obtained at a given site.

In the 1970s in an attempt to stimulate onshore production of synthetic quartz and piezoelectric devices, Brazil imposed an embargo on exports and ultimately raised the price several-fold for small quartz crystals used as the starting material for quartz growth. However, sources of suitably pure quartz were located in the United States and Canada, including vein and pegmatic deposits (1). The interaction of process variables with the form and purity of the starting material was studied and processes compatible with U.S. starting material from a variety of sources were developed making production relatively independent of imports (1).

At present, synthetic quartz is produced in the UK, France, Belgium, the USSR, The People's Republic of China, Japan, Brazil, Poland, and by a half dozen U.S. firms including Thermo Dynamics, Inc., Shawnee Mission, Kansas; Motorola Inc., Carlisle, Pa.; P. R. Hoffman Co., Carlisle, Pa.; Sawyer Research Products, Inc., Eastlake, Ohio; and Western Electric Co., Inc., North Andover, Mass.

The price for electronic-grade synthetic quartz is ca $100–200/kg, depending on size, quality, and orientation. World capacity for production of synthetic quartz easily exceeded 500 metric tons in 1980.

Synthesis

α-Quartz cannot be crystallized from its pure melt because viscous SiO_2 melts almost always form silica glass upon cooling. When crystals are formed, they are high temperature polymorphs of SiO_2 (cristobalite [14464-46-1] and tridymite [15468-32-3]) that do not easily transform to untwinned α-quartz. α-Quartz is soluble in a variety of molten salts but the melts are viscous and crystallization below the α-transition temperature is not practical. Silicon dioxide is insoluble in most aqueous solvents at ambient conditions except in HF solutions and α-quartz is not the stable solid phase in equilibrium with such solutions. No successful practical vapor-transport reaction for α-quartz growth has been discovered. However, α-quartz is stable and soluble in water at elevated temperatures and pressures (hydrothermal conditions). The solubility at a convenient temperature (400°C) and pressure (ca 17 MPa = 25,000 psi) is only a few tenths of a wt % and is not large enough for crystal growth. However, reactions of the type

$$2\,OH^- + SiO_2 \rightarrow SiO_3^{2-} + H_2O$$

take place in basic solutions under hydrothermal conditions with the result that the wt solubility of quartz is increased to several percent. α-Quartz is the stable solid phase in equilibrium with such solutions and large crystals can be grown. This is the basis of the processes used for commercial quartz growth.

The first known successful attempt to grow quartz crystals hydrothermally was reported in 1905 (2). Small crystals were formed in a thermal gradient from a sodium silicate solution. In Germany during World War II, quartz crystals were grown in an isothermal system using α-quartz seeds and vitreous silica nutrient (3). After the war, several laboratories began research programs aimed at practical quartz production (4–7). It soon became apparent that processes depending upon the supersaturation caused by the presence of silica glass were impractical. Metastable phases have a higher solubility than stable phases, but the supersaturation caused by their presence persists only as long as they are present. In the case of silica glass, its surface rather quickly devitrifies under hydrothermal conditions and growth on α-quartz seeds ceases. All successful quartz-growth processes depend upon the supersaturation produced by dissolving small particles of quartz nutrient in a hot region of the high pressure system and crystallizing it onto α-quartz seeds in a cooler part of the system. Thus, it is necessary to employ a solvent in which quartz is the stable solid phase with reasonable solubility and in which the dependence of solubility upon temperature produces an appropriate supersaturation (ΔS) with an appropriate temperature differential (ΔT) between the dissolving and the growth zones. All commercial processes (8–10) use either NaOH (4) or Na_2CO_3 (5). The dissolving mechanism is similar in both solvents since CO_3^{2-} hydrolyzes to OH^-. Sodium salts are required because insoluble sodium iron silicates form on the steel walls of the high pressure vessels when Na^+ and silicates are present. These compounds are relatively insoluble under hydrothermal conditions and form a protective coat on the walls allowing the use of unlined, relatively inexpensive vessels. The slope of the solubility vs temperature curve is greater in CO_3^{2-} than in OH^- solutions. Thus, with a given ΔT, the ΔS and hence the crystal-growth rates would be larger in CO_3^{2-} solutions. However, small changes in temperature (unavoidable with even the best process control) in CO_3^{2-} cause larger changes in ΔS than in OH^-. Thus loss of control in CO_3^{2-} with poor quality growth and spontaneous nucleation is more likely than in OH^-. The practical solution to this problem in CO_3^{2-} is to use a lower ΔT which produces a ΔS safely below the poor quality region. Thus in CO_3^{2-} processes the growth rates are usually lower. Indeed in OH^- the temperature control required is generally less than in CO_3^{2-}.

The OH^- process described below was developed by Bell Telephone Laboratories (8–9) and is used by the Western Electric Company and most other producers (11). It is usually operated at higher pressures than the CO_3^{2-} process and quartz is grown at faster rates; less precise temperature control is required.

Equipment. A typical commercial quartz-growing autoclave is illustrated in Figure 1. The material of construction for use at 17 MPa (25,000 psi) and 400°C can be a low carbon steel, such as 4140, or various types of low alloy steel. The closure is a so-called modified Bridgman closure. It is based upon the unsupported-area principle (12); that is, the pressure in the vessel is transmitted through the plunger to the steel surfaces which initially are nearly line contacts. Thus, the pressure in the seal surface greatly exceeds the pressure in the vessel, since most of the area of the plunger is unsupported. Hydrothermal equipment is further discussed in ref. 10.

After long periods of time at operating temperature, the brittle-ductile transition temperature in autoclave steels increases (13). At temperatures much above 200°C for the solutions and fills used in ordinary hydrothermal processes, pressures and hence stresses in autoclaves could cause failure of metal in the brittle state. Ordinarily, the brittle region is well below these temperatures but careful monitoring of the brittle-ductile transition is necessary for safe autoclave use over many years.

The baffle is a perforated metal disk which restricts convective circulation within the vessel and thus creates two isothermal regions, the dissolving and the growth zones. Therefore, all seed crystals experience the same ΔT and ΔS and grow uniformly, and

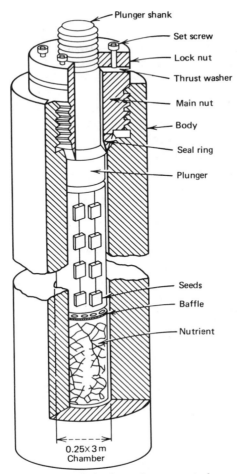

Figure 1. Hydrothermal quartz autoclave.

no growth takes place on the nutrient. The autoclave is heated by appropriately placed external resistance heaters; the temperature is measured and controlled from externally placed thermocouples and the pressure is measured by strain and Bourdon gauges.

Modern process facilities are controlled by computers which program temperature and ΔT upward during the start of a growth cycle, monitor and control pressure and temperature, and provide pressure and temperature overshoot alarms and overrides. Such systems also store data from previous runs for correlations with properties or for identical replication of past conditions.

Typical Run. In a typical run, small particle size α-quartz nutrient is added to a large autoclave, a 5% open area baffle is inserted and a frame holding many seed plates, whose principle surface is (0001) or an orientation close to that, is placed within the autoclave. The autoclave is filled to 82% of its free volume with 1.0 M NaOH and closed. The vessel is heated so that the nutrient zone is at a temperature of about 400°C and the growth zone at a temperature of about 350°C. Under these conditions, the pressure is ca 16.5 MPa (24,000 psi). The vessel is held at these conditions for several weeks and then cooled. The grown crystals are removed, rinsed with distilled water, and are ready for processing. A typical growth rate under such conditions would be ca 1.27 mm/d for growth on the basal (0001) planes. Depending on the particular piezoelectric application, the seed orientation is adjusted to produce a grown crystal that can most efficiently be cut into piezoelectric plates of the orientation dictated by the application. The three seed orientations generally used are basal, +5°X (rotated +5° from the basal seed) and seed plates rotated 3° off the minor rhombohedral face sometimes called r- or z-face (14). This latter orientation, in spite of the fact that its growth rate is ca half of basal growth, gives a grown crystal particularly useful for AT-cut piezoelectric resonators (15) and is often used (14).

Effect of Conditions on Rate.   It is essential for commercial growth to obtain a high rate with usable perfection at pressure and temperature conditions allowing economical equipment design. The dependence of rate on the process parameters has been studied (8,14). The results of these studies are important for production analysis and may be summarized as follows:

Growth rate depends upon crystallographic direction; (0001) is one of the fastest directions.

Growth rate is linear with ΔT (because ΔS is approximately linear with ΔT).

Growth rate has an Arrhenius equation dependence upon the temperature in the crystallization zone, as shown in the following equation:

$$R \propto e^{-\Delta E/R'T}$$

where R = rate in a particular crystallographic direction; E = energy of activation; T = absolute temperature; R' = gas constant.

Growth rate is increased by an increase in fill, ie, percent of free volume filled with solvent at room temperature.

Piezoelectrical quartz must be free of twins, bubbles, and particulate inclusions. If the seeds used in quartz synthesis are untwinned and the growth takes place at an appropriate rate and pressure–temperature conditions, synthetic quartz is free of these imperfections. If the rate is too high for the pressure-temperature conditions, the growth becomes limited by the rapidity with which silica can diffuse across a locally depleted zone close to the growing seed. The result is imperfect or crevice-flawed growth. Such growth contains many voids and liquid inclusions and is not usable for piezoelectric applications. It is caused by conditions analogous to those causing dendritic growth in the preparation of metal crystals. The success of commercial processes depends upon careful mapping of the pressure, temperature, and temperature differential conditions in order to find regions where crevice flawing does not occur at commercially useful rates.

Particular difficulty with cracking occurs in z-face-grown quartz. High strain in seeds leads to a strain build up in the grown quartz that exceeds the elastic limit which produces cracks in the as-grown material or leads to cracking during subsequent processing. Severe losses in yield can result (14). Careful inspection of seeds between crossed polarizers in an immersion oil of matching index of refraction reveals strain and allows high quality seeds to be selected so that a high yield of unstrained material is possible (14). Strains result from dislocations generated at particulate inclusions of Na–Fe silicates (16). Growth in the absence of Fe in silver cans contained in hydrothermal autoclaves greatly reduces inclusions and strain and produces strain-free seeds for subsequent use under normal conditions (16). The pressure in the can is balanced by filling the space between the can and the autoclave with H_2O to a height slightly above that of the NaOH solution used in the can. The higher fill is required to ensure that the can is under compression, a situation less prone to rupture. Careful choice of seeds and conditions using silver-lined vessels has even resulted in the growth of dislocation-free quartz in the laboratory (16).

An additional requirement for piezoelectrical quartz is a high acoustic or mechanical Q (13,17–19). The acoustic Q of a piezoelectric material is equal to the Q of the resistance-capacitance-inductance circuit, which is electrically equivalent to the piezoelectric resonator circuit. The higher the Q of a piezoelectric material, the more efficiently it converts mechanical to electrical energy. In low-Q materials, much energy is lost by thermal processes in this conversion. The acoustic Q of ordinary materials may be thought of as being higher the longer they ring when struck mechanically. Lead is not used to make bells because it has a low Q.

The acoustic Qs of quartz prepared under a variety of conditions are given in Table 2. The acoustic Q of natural quartz is $(1–2) \times 10^6$.

Infrared absorption studies have shown that Q correlates with an absorption at 3 μm associated with an OH stretching frequency (20). Indeed, infrared absorption provides a useful tool for Q evaluation in rapid production-quality control. Infrared and other studies show that Q degradation is caused by proton inclusion in the grown quartz.

The distribution constant for (OH) or a proton depends upon the presence of other ions since the proton enters the quartz lattice interstitially and is compensated by +3 or +2 ions such as Fe^{2+}, Fe^{3+}, and Al^{3+} substituted at Si^{4+} lattice sites. Iron originates in the autoclave and Al^{3+} is an impurity in the natural quartz starting material. Other +1 ions such as Na^+ or Li^+ can interstitially compensate for non +4 ions at the Si^{4+} site. Thus, Li^+ doping improves Q as well as quartz growth in a silver liner or can to exclude iron (21). Furthermore, the distribution constant for impurities depends on the quartz growth rate in a predictable manner and thus slower growth excludes protons (22). Systematic studies of distribution constant have been used to identify the commercial conditions used for high Q and high optical-transmission-quartz growth (22–24). Thus high Q quartz can be grown at such rates that large crystals (>5 cm grown material) are produced in 28–30 d. Since energy (electric power for the heaters) is one of the principal costs in quartz production (as much as 50% of total cost in slow growth rate processes), high growth rate processes will probably tend to dominate in the future.

Table 2. Acoustic Q of Synthetic Quartz Prepared under a Variety of Conditions

Synthetic quartz growth solution	Q
1 M NaOH	1×10^5
1 M NaOH + 0.2 M LiOH	$(2–3) \times 10^5$
1 M NaOH + 0.2 M LiNO$_2$	$(0.5–1) \times 10^6$
1 M NaOH + 0.2 M LiNO$_2$[a]	2×10^6

[a] Crystallization in silver-lined tube.

Health and Safety Factors

The principal consideration in quartz synthesis is the safe management of the high pressures required for growth. The autoclave must be designed on the basis of proper stress analysis and materials selection. Autoclaves are shielded either by armor plate or by placement in cement-lined pits in the ground. Rupture disks that release pressures of ca 35 MPa (5000 psi) above the operating conditions as well as pressure and temperature overrides and alarms are standard practice. Autoclave embrittlement over long service life must be monitored and autoclaves in danger of operating at pressures much above ambient in the brittle range are removed from service. The hydroxide solutions used in growth should be handled with standard safety measures, including eye shields for operators, safety showers, etc. Activities in high pressure areas should be limited to essential maintenance. The disposal of spent solutions presents no special pollution problems since, at the conclusion of growth, runs are not strongly basic and can be easily neutralized.

Uses

The principal use of α-quartz depends upon the fact that it is piezoelectric (see also Ferroelectrics). Crystals are used in electrical filters and oscillators for frequency control and timing and frequency-division multiplexing and demultiplexing. Quartz also has been used as an electromechanical transducer in ultrasonic generators and various pickups (see Ultrasonics). Quartz has important uses in consumer products such as electronic watches where a quartz resonator provides the timing function and in citizen's-band radios where α-quartz resonators establish broadcast frequencies. The piezoelectric properties and designs of filters and resonators are discussed in refs. 15 and 25.

Quartz also has modest but important uses in optical applications, primarily as prisms where its dispersion makes it useful in monochromators for spectrophotometers in the region of 0.16–3.5 μm. Specially prepared optical-quality synthetic quartz is required since ordinary synthetic quartz is usually not of good enough quality for such uses mainly because of scattering and absorption at 2.6 μm associated with (OH) in the lattice.

BIBLIOGRAPHY

"Synthetic Quartz Crystals" in *ECT* 1st ed., Vol. 12, pp. 331–335, by G. T. Kohman, Bell Telephone Laboratories; "Synthetic Quartz Crystals" in *ECT* 2nd ed., Vol. 18, pp. 105–111, by R. A. Laudise and A. A. Ballman, Bell Telephone Laboratories, Inc.

1. E. D. Kolb, K. Nassau, R. A. Laudise, E. E. Simpson, and K. M. Kroupa, *J. Cryst. Growth* **36,** 93 (1976).
2. G. R. Spezia, *Acad. Sci. Torino* **44,** 95 (1908).
3. R. Nacken, *U.S. Dept. Commerce, Office Technical Service Report*, PB-6498, Washington, D.C., 1945.
4. A. C. Walker and G. T. Kohman, *Trans. Am. Inst. Electr. Eng.* **67,** 565 (1948).
5. D. R. Hale, *Science* **108,** 393 (1948).
6. C. S. Brown, R. C. Kell, L. A. Thomas, N. Wooster, and W. A. Wooster, *Nature* **167,** 940 (1951).
7. R. A. Laudise, *J. Am. Chem. Soc.* **81,** 562 (1959).
8. R. A. Laudise in F. A. Cotton, ed., *Progress in Inorganic Chemistry*, Vol. III, John Wiley & Sons, Inc., New York, 1962, pp. 1–47.

9. A. A. Ballman and R. A. Laudise in J. J. Gilman, ed., *The Art and Science of Growing Crystals*, John Wiley & Sons, Inc., New York, 1963, pp. 231–251.

10. R. A. Laudise and J. W. Nielsen in F. Seitz and D. Turnbull, eds., *Solid State Physics*, Academic Press, Inc., New York, 1961, pp. 149–222.

11. R. A. Laudise and R. A. Sullivan, *Chem. Eng. Prog.* **55,** 55 (1959).

12. P. W. Bridgman, *Proc. Am. Acad. Arts Sci.* **49,** 625 (1914).

13. P. L. Key, E. D. Kolb, R. A. Laudise, and E. Bresnahan, *J. Cryst. Growth* **21,** 164 (1974).

14. R. L. Barns, E. D. Kolb, R. A. Laudise, E. E. Simpson, and K. M. Kroupa, *J. Cryst. Growth* **34,** 189 (1976).

15. R. A. Heising, *Quartz Crystals for Electrical Circuits*, D. Van Nostrand, Co., Inc., New York, 1946.

16. R. L. Barns, P. E. Freeland, E. D. Kolb, R. A. Laudise, and J. R. Patel, *J. Cryst. Growth* **43,** 676 (1978).

17. J. C. King, A. A. Ballman, and R. A. Laudise, *J. Phys. Chem. Solids* **23,** 1019 (1962).

18. A. A. Ballman, R. A. Laudise, and D. W. Rudd, *Appl. Phys. Lett.* **8,** 53 (1966).

19. A. A. Ballman, *Am. Mineral.* **46,** 439 (1961).

20. D. M. Dodd and D. B. Fraser, *J. Phys. Chem. Solids* **26,** 673 (1965).

21. R. A. Laudise, A. A. Ballman and J. C. King, *J. Phys. Chem. Solids* **26,** 1305 (1965).

22. N. Lias, E. Grudenski, E. D. Kolb, and R. A. Laudise, *J. Cryst. Growth* **18,** 1 (1973).

23. A. A. Ballman, D. M. Dodd, N. A. Keubler, R. A. Laudise, D. L. Wood, and D. W. Rudd, *J. Appl. Opt.* **7,** 1387 (1968).

24. E. D. Kolb, D. A. Pinnow, T. C. Rich, R. A. Laudise, N. Lisa, and E. Grudenski, *Mater. Res. Bull.* **7,** 397 (1972).

25. W. P. Mason, *Piezoelectric Crystals and Their Application to Ultrasonics*, D. Van Nostrand Co., Inc., New York, 1950.

R. A. Laudise
E. D. Kolb
Bell Laboratories

SILICA, BRICK.   See Refractories.

SILICATES.   See Silicon compounds.

SILICIDES.   See Silicon and silicon alloys.

SILICON AND SILICON ALLOYS

Pure silicon, 826
Metallurgical, 846

PURE SILICON

Silicon [*7440-21-3*] (from the Latin *silex*, *silicis* for flint) is the 14th element of the periodic series (at wt 28.083). It was first reported by Berzelius in 1817. Elemental silicon does not occur in nature; however, as a constituent of various minerals, eg, silica and the silicates, it accounts for ca 25% of the earth's crust. There are three stable isotopes that occur in nature as well as several that are artificially prepared and radioactive (see Table 1) (1). Silicon has a gray, metallic luster and may appear iridescent if it has a thin oxide covering. It is a brittle material with a hardness slightly less than that of quartz. Small single-crystal filaments are very strong and exhibit breaking strengths of up to ca 1.4 GPa (200,000 psi). In more massive pieces, substantially smaller values are observed because of difficulties in removing the effect of stress enhancement at small surface cracks. Below 800°C there is very little plastic flow. In thin sections, single crystals cleave along (111) planes, but in larger pieces, fracture is conchoidal.

Crystal Structure

At atmospheric pressure, silicon has a diamond cubic structure, ie, two interpenetrating face-centered cubes are displaced ¼, ¼, ¼ from each other. Additional data are given in Table 2 (1–3). When subjected to ca 15 GPa (<150,000 atm), the fcc structure is converted to a body-centered lattice (2). Vapor deposition below ca 500°C (the exact temperature depends on the deposition rate) produces amorphous silicon; upon reheating to a somewhat higher temperature, crystallization will occur. In the early literature, a stable hexagonal form was reported, but its existence has never been verified.

Under most circumstances, the slow-growing faces of silicon are (111). Thus, the equilibrium shape is octahedral. Very small crystallites grown from vapor are octa-

Table 1. Isotopes of Silicon[a]

Isotope	CAS Registry No.	Natural abundance, %	Half-life
^{25}Si	[*15759-89-4*]		0.23 s
^{26}Si	[*14932-60-6*]		2.1 s
^{27}Si	[*14276-59-6*]		4.2 s
^{28}Si	[*14276-58-5*]	92.23	
^{29}Si	[*14304-87-1*]	4.67	
^{30}Si	[*13981-69-6*]	3.10	
^{31}Si	[*14276-49-4*]		2.62 h
^{32}Si	[*15092-72-5*]		ca 650 yr

[a] From ref. 1.

Table 2.   Structure of Silicon

Characteristic	Type I[a] at 101 kPa (1 atm)	Type III[b] at 15 GPa[c]
crystal structure	diamond cubic	bcc
lattice spacing, nm	0.5430[d]	0.6636
atoms per unit cell	8	16
space group	Fd3m	Ia3

[a] From ref. 1.

[b] Metastable at ca 15 GPa, but persists at 101 kPa. From ref. 2.

[c] To convert GPa to psi, multiply 145,000.

[d] This value is dependent on the crystal purity. A 0.1 atom % boron, for example, causes a lattice contraction of about 0.03%. For more details see ref. 3.

hedral, but those larger than a few μm on a side have distorted shapes influenced by concentration gradients in the gas stream. Although crystals grown from melt have vestigial (111) facets, they are shaped primarily by thermal gradients in the melt.

Physical Properties

Values for various thermal and mechanical properties are given in Table 3 (1,4–7) (see also Figs. 1–3).

Table 3.   Thermal and Mechanical Properties

Property	Value	Ref.
at wt	28.085	1
atomic density (atoms/cm^3)[a]	5.0×10^{22}	
melting point, °C	1410	1
boiling point, °C	2355	1
vapor pressure, Pa[b]		4
800°C	1.33×10^{-8}	
1000°C	1.33×10^{-5}	
1500°C	2.66	
2000°C	80	
density, g/cm^3 at 25°C	2.329	1
critical temp, °C	4886	5
critical pressure, MPa[c]	53.6	5
hardness, Mohs/Knoop	6.5/950	6
elastic constants, GPa[d]		7
C_{11}	167.4	
C_{12}	65.23	
C_{44}	79.59	
heat of fusion, kJ/g[e]	1.8	1
heat of vaporization at mp, kJ/g[e]	16	5
volume contraction on melting, %	9.5	5

[a] Calculated.

[b] To convert Pa to mm Hg, multiply by 0.0075.

[c] To convert MPa to atm, divide by 0.101.

[d] To convert GPa to dyn/cm^2, multiply by 10^{10}.

[e] To convert J to cal, divide by 4.184.

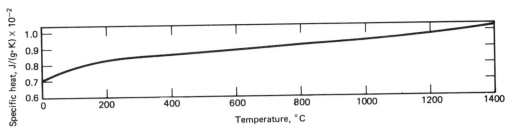

Figure 1. Specific heat of silicon. To convert J to cal, divide by 4.184.

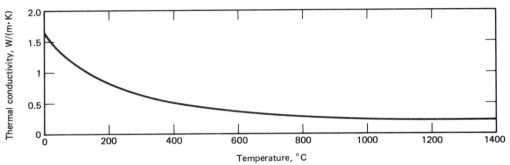

Figure 2. Thermal conductivity of silicon.

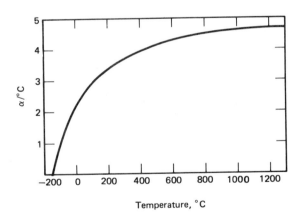

Figure 3. Thermal expansivity of silicon.

Unlike structural materials whose properties are generally given for polycrystalline samples, the properties of high purity silicon are almost always measured on single-crystal samples. Even so, a diversity of values is reported by various investigators. Single-crystal properties may also be anisotropic, depending both on the particular property and the crystal class. For cubic crystals, properties described by other than a zero or a second-rank tensor are anisotropic (11) and include, for example, hardness and elastic constants. Silicon is not a liquid semiconductor, and currently there are no applications in the liquid state. The liquid properties summarized in Table 4 are nevertheless of particular importance because they affect single-crystal growth, an operation through which essentially all semiconductor-grade silicon must pass (5,12–14).

Table 4. Properties of Liquid Silicon[a]

Property	At mp	At 1500°C	Ref.
thermal conductivity, W/(m·K)	41.84		5
dynamic viscosity, mPa·s (= cP)	0.88	0.7	5
kinematic viscosity, mm²/s (= cSt)	0.347[b]	0.28[b]	
surface tension, mN/m (= dyn/cm)	736	720	5
heat capacity, J/(kg·K)[c]	0.16	6.84	5
density, g/cm³	2.533	2.50	5
electrical resistivity, μΩ·cm	80	100	12
total optical emissivity	0.33	0.33	13
reflectivity at 633 nm, %	72	70	14

[a] Many of these values are extrapolated (see ref. 5 for more details).
[b] Calculated.
[c] To convert J to cal, divide by 4.184.

The optical transmissivity of silicon, as well as that of other semiconductors, has the general characteristic shown in Figure 4a. Because of the high index of refraction (see Fig. 5) (15), reflection losses are very high. When coated with a suitable low-reflecting coating, the maximum transmissivity may be much higher; absorption is low for wavelengths corresponding to energies less than the band-gap energy. The absorption is comprised of a background resulting from free-carrier absorption and various narrow absorption bands resulting from impurities (see Table 5) (16) and the

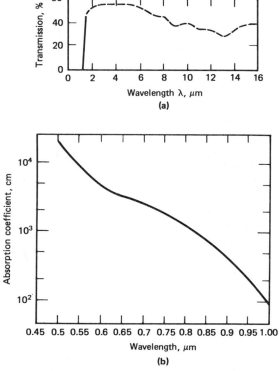

Figure 4. (a) Transmissivity vs wavelength. (b) Short-wavelength absorption coefficient.

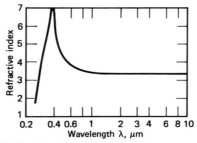

Figure 5.   Refractive index of silicon at room temperature (15).

Table 5.   Optical Absorption Bands of Trace Impurities Occurring in Silicon [a]

Source	Wavelengths, μm [b]	Temperature, K
oxygen	9.042	ca 298
carbon	16.53	ca 298
boron [c]	31.2	4
	32.2	
aluminum [c]	21.1	4
	22.5	
gallium [c]	21.2	4
	20.0	
phosphorus [c]	31.3	
arsenic [c]	31.6	
antimony [c]	33.9	
	31.0	

[a] From ref. 16.

[b] There are several bands from each source; the ones listed are the most intense.

[c] These impurities are ionized at room temperature. When the temperature is reduced to the point where they are no longer ionized, transitions may occur between their levels and the valence or conduction bands.

silicon lattice. There are several lattice absorption bands, the most intense is found at 16.13 nm. Figure 4**b** shows the very high absorption coefficient in the short-wave region (17). The high absorptivity is accompanied by free-carrier generation and forms the basis for photovoltaic solar cells. The low absorptivity at longer wavelength allows silicon to be used as an infrared optical material in the 1–8 μm range. Silicon surfaces take a good optical finish. The sawing, grinding, and polishing operations are similar to those for glass (18). The high index of refraction requires antireflective layers for almost all optical applications; otherwise about 30% of the incident light energy is lost at each reflecting surface. In the wavelength region where solar cells operate, the index, and hence the loss, is even higher.

Electrical Properties.   Silicon is a semiconductor with a band gap E_g of 1.12 eV at ca 25°C; E_g is the amount of energy required to raise an electron from the valence band to the conduction band. The electrical conductivity σ is given by

$$\sigma = q(\mu_n n + \mu_p p) \tag{1}$$

where q = electronic charge, μ = carrier mobility in cm^2/(V·s), n = density of free electrons present (electrons per unit of volume), and p = density of free holes. For absolutely pure silicon, the density of holes and electrons is equal and is given by

$$n_i = [1.5 \times 10^{33} \, T^3 \exp (1.21/\kappa T)]^{1/2} \tag{2}$$

where k = Boltzmann's constant and T = temperature in K. At 25°C, $n_i = 1.4 \times 10^{10}/\text{cm}^3$, which corresponds to a resistivity of approximately 23×10^4 $\Omega \cdot$cm. As can be seen from equation 2, carrier generation increases very rapidly with temperature, and at 700°C the resistivity drops to ca 0.1 $\Omega \cdot$cm. Many impurity elements, when incorporated into the silicon lattice during crystal growth or by diffusion, ionize at relatively low temperatures and provide either free electrons or holes. In particular, group IIIA elements (n-type dopant or donor) supply electrons and group VA elements (p-type dopant or acceptor) supply holes. Of these, boron, phosphorus, arsenic, and antimony are most commonly used. In very pure, high perfection silicon, minority-carrier lifetime may be in the millisecond range. Intentional doping with elements that reduce lifetime, eg, gold, allows that value to be reduced to a few nanoseconds in a controlled fashion (19).

The mobility μ refered to in equation 1 is given by

$$\frac{1}{\mu} = \frac{1}{\mu_L} + \frac{1}{\mu_I} \tag{3}$$

with simple theory predicting that lattice mobility μ_L should vary as $T^{-3/2}$, and the ionized impurity mobility μ_I as $T^{3/2}$. In addition, μ_I varies inversely with the ionized impurity concentration. The mobility-vs-temperature plot thus has a maximum that varies with impurity concentration. Combining the free-carrier generation with the impurity-induced carrier concentration (doping level) and the temperature dependence of mobility gives resistivity curves with the general characteristic shown in Figure 6.

Impurity concentration in semiconductors is commonly expressed in atoms of impurity per cm^3 of host material. In silicon devices, these concentrations vary from about 10^{14} to 10^{20} atoms per cm^3. The relation between impurity concentration and

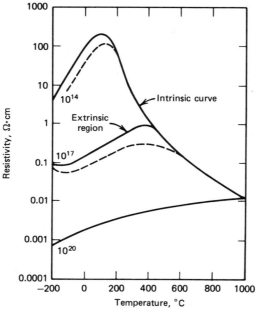

Figure 6.   Calculated resistivity vs temperature for various doping levels. To the left of each peak, the resistivity is determined by the impurity doping. —— = p-type; - - - = n-type.

resistivity at 25°C is given in Figure 7 (20). The difference between n- and p-type curves arises because of the differences between μ_n and μ_p.

In bulk, the resistivity is independent of crystallographic orientation because silicon is cubic. However, if the carriers are constrained to travel in a very thin sheet, eg, in an inversion layer, the mobility, and thus the resistivity, becomes anisotropic (21). The mobility is sensitive to both hydrostatic pressure and uniaxial tension and compression, which gives rise to a substantial piezoresistive effect. The resistivity gradually decreases as hydrostatic pressure is increased and then abruptly drops several orders of magnitude at ca 15 GPa (150,000 atm), where a phase transformation occurs and the silicon changes from a semiconductor to a metal (22). The longitudinal piezoresistive coefficient varies with the direction of stress, the impurity concentration, and the temperature. At ca 25°C, stress in a [100] direction, and resistivities of a few hundredths of an $\Omega \cdot$cm, the coefficient values are 500–600 m²/N (50–60 cm²/dyn, reciprocal stress). Other electrical data are given in Table 6 (23).

Radiation Effects

Gamma radiation produces free carriers much as does visible light (24). High energy electrons and protons produce deep-level defects that reduce minority-carrier lifetime according to the equation

$$\frac{1}{\tau_f} = \frac{1}{\tau_o} + k\phi \tag{4}$$

where τ_f is the lifetime after irradiation with a fluence ϕ, τ_o is the original lifetime, and k is a radiation-damage constant. Neutrons produce deep-level defects that not only degrade the lifetime (see eq. 4) but also increase resistivity through the removal of carriers. In the latter case, the number of carriers N remaining after irradiation is given by

$$N = N_o - K\phi \tag{5}$$

where N_o is the initial number of carriers and K is a different radiation-damage constant. The damage-constant values are dependent on the kind and energy of the ir-

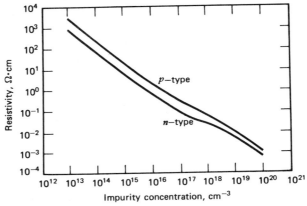

Figure 7. Impurity concentration vs resistivity at room temperature. These curves are for phosphorus and boron; other group III A and V A impurities behave similarly. Adapted with permission of the *Bell System Technical Journal*, 1960.

Table 6.   Electrical Properties of Silicon[a]

Property	Value
band gap, E_g at ca 25°C, eV	1.12
dE_g/dT, eV/K	-2.4×10^{-4}
dE_g/dP, (eV·cm²)/kg	-2.4×10^{-6}
effective mass[b]	
longitudinal	0.97
transverse	0.19
light hole	0.16
heavy hole	0.3
mobility, high purity, cm²/(V·s)	
electron	1500
holes	600
work function, V	4.8
maximum drift velocity, cm/s	ca 10^7
dielectric constant	11.8
impurity ionization energy (eV)	
aluminum	0.057[c]
boron	0.045[c]
gallium	0.065[c]
indium	0.16[c]
antimony	0.039[d]
arsenic	0.049[d]
phosphorus	0.044[d]
gold	$\begin{cases} 0.35^e \\ 0.54^f \end{cases}$
iron	0.40[e]
	0.55[d]
platinum	0.37[d]
thermoelectric power, mV/K	
0.1 Ω·cm n-type	-1
100 Ω·cm n-type	-1.7
0.1 Ω·cm p-type	$+1$
100 Ω·cm p-type	$+1.7$

[a] From ref. 23.
[b] Mass of electrons moving in lattice per mass of free electrons.
[c] Acceptor level above valence band.
[d] Donor level below conduction band.
[e] Donor level above valence bond.
[f] Acceptor level below conduction band.

radiation, the kinds of impurities in the silicon, and the temperature. Moreover, many of the defects anneal out at higher temperatures.

Neutrons also transform some of the silicon atoms to phosphorus through the reaction

$$^{30}\text{Si} + n \rightarrow {}^{31}\text{Si} + \gamma \rightarrow {}^{31}\text{P} + \beta \tag{6}$$

This reaction is occasionally used for doping crystals uniformly after they have been grown (transmutation doping) (25). Table 7 lists neutron cross-sections for various neutron energies (26–27).

Table 7. Neutron Cross-Section, 10^{-28} m^2 ^{a}

Isotope	Thermal^b	Energy level		
		30 MeV^c	45 MeV^c	55 MeV^c
^{28}Si	80			
^{29}Si	280			
^{30}Si	110			
natural Si	160	1961 ± 25	1866 ± 16	1688 ± 13

^{a} To convert m^2 to barn, multiply by 10^{28}.
^{b} From ref. 26.
^{c} Interpolated from data in ref. 27.

Chemical Properties

Silicon, carbon, germanium, tin, and lead comprise the group IVA elements of the periodic system (28–30). Silicon and carbon form silicon carbide which, although most widely known as an abrasive and heating element, is also a semiconductor. Germanium and silicon are isomorphous and thus are mutually soluble in all proportions. Tin and lead do not react with silicon; indeed, molten silicon is immiscible in both molten tin and molten lead. Otherwise, molten silicon is an excellent solvent, and no container material has been found that is not noticeably dissolved.

The elements in the adjacent columns (IIIA and VA) form compounds with silicon and also enter in small amounts into the lattice of a silicon crystal. These elements are commonly used to provide the p- and n-type carriers discussed under Electrical Properties. They may be introduced during the crystal-growing operation, or later by ion implantation (qv) or solid-state diffusion, or they may be residuals left from the initial purification steps.

Oxygen forms strong bonds with silicon. There are two oxides, numerous silicates, and almost endless variations of silicones. The bond energy is 25.8 kJ (108 kcal) (in silica) and the observed bond length is 0.163 nm. The nature of the silicon oxygen bond is of great importance to the semiconductor industry, since the silicon–oxygen bonding at the interface between the single crystal silicon and the protective oxide has a significant effect on device performance. At room temperature, silicon is covered with an oxide layer 2.0–3.0 nm thick. This oxide, as well as those intentionally grown in steam or oxygen, is amorphous. Occasionally, some contaminant causes small areas to crystallize, but those circumstances are normally avoided. For the manufacture of semiconductor devices, oxide thickness from a few hundred to several thousand nm is required. Oxide layers are formed at elevated temperatures, and sometimes at elevated pressures, in an atmosphere containing either oxygen or steam (31). Oxidation consumes silicon at a rate of about 0.4 times the rate of oxidation. The amount of oxidation is given by

$$X^2 + AX = B(t + \tau) \tag{7}$$

where X is the thickness of oxide, A and B are rate constants, τ is a constant involving the initial oxide thickness, and t is the oxidation time. Initially, X varies linearly with time, but as growth continues, the rate slows and X varies with $t^{-1/2}$. Figure 8 shows some typical thickness–time curves. The exact values depend on the temperature, the pressure, the oxidant, the crystal orientation of the surface being oxidized, and the amount and type of impurity in the silicon.

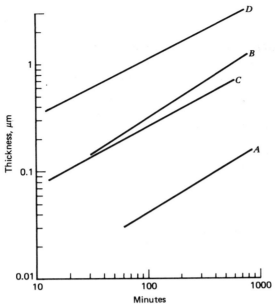

Figure 8. Typical oxidation rates at atmospheric pressure for single-crystal surfaces (111). In general, rates increase somewhat with impurity concentration. Over several atmospheres, the rate increases nearly linearly with pressure. $A = 900°C$ dry O_2; $B = 900°C$ steam; $C = 1200°C$ dry O_2; $D = 1200°C$ steam.

Oxygen can also dissolve in the silicon crystal lattice and form SiO_x complexes of varying composition that may radically affect the electrical properties of the host. The oxygen is usually unintentionally introduced during the crystal-growing operation (see below) in concentrations up to the solubility limit (ca 2.5×10^{18} atoms per cm^3). When oxygen-containing silicon is annealed in the 450°C range, the —OSi≡ complexes become donors and contribute to electrical conductivity. At higher temperatures, larger, electrically inactive clusters occur that sometimes cause crystal defects to form. Above ca 1200°C, the clusters begin to dissolve and the oxygen is redispersed (32).

At elevated temperatures, silicon reacts with the halogens as well as with their anhydrous acids to form $SiCl_4$, $SiBr_4$, SiI_4, and SiF_4. In addition, hydrogen bonding gives rise to a series of hydrides (SiH_4, Si_2H_6, etc) and halosilanes, eg, $SiHCl_3$, SiH_2Cl_2, SiH_3Cl, and numerous organosilanes. The halides and the monohydride have been used in the manufacture of high purity silicon, in the deposition of thin layers of polycrystalline silicon, and in the overgrowth of silicon layers on a single-crystal surface (epitaxy).

At ca 25°C, silicon appears relatively inert, partly because of the rapid formation in air of the protective oxide coating. Silicon is insoluble in acids but can be dissolved in a two-stage operation in which the surface is first oxidized and the oxide then removed (33). For this operation, an aqueous solution of HNO_3 and HF is generally used.

If a hydrated oxide forms, eg, with hydrazine or potassium hydroxide, it can be removed with water and a suitable complexing agent (34). Pyrocatechol and water are sometimes used with hydrazine or ethylenediamine, and isopropyl alcohol or water are used with KOH. The HNO_3–HF–H_2O system usually produces isotropic etching and is used for chemical polishing. The other systems are anisotropic and have very low etch rates on (111) crystal faces. The anisotropy can be very pronounced; if there are no ledges to promote etching, ratios of up to 400 between the (111) and other faces can be obtained. This feature is useful in producing grooves in silicon with closely controlled dimensions and has only limited application in semiconductor-device fabrication (35). Compositional variations, usually based on HF–HNO$_3$, are sensitive to the silicon resistivity and type; they etch very heavily doped regions much more rapidly than lightly doped ones. Electrolytic (anodic) etching of silicon occurs in HF solutions (36). With concentrated HF, the silicon is dissolved in the divalent state:

$$\text{Si} + 2\,\text{HF} \rightarrow \text{SiF}_2 + 2\,\text{H}^+ + 2\,e$$

In dilute solutions and higher anodic voltages, the silicon is dissolved in the tetravalent state:

$$\text{Si} + 4\,\text{H}_2\text{O} \rightarrow \text{Si(OH)}_4 + 4\,\text{H}^+ + 4\,e$$

Free radicals such as F$\cdot$ can be used to etch silicon at ambient temperature (37). The radicals are formed by dissociating compounds such as CF$_4$ in a plasma (see Plasma technology, Supplement Volume). Thus,

$$\text{CF}_4 + \text{Plasma} \rightarrow \text{CF}_3\cdot + \text{F}\cdot$$

$$4\,\text{F}\cdot + \text{Si} \rightarrow \text{SiF}_4$$

Plasma etching is used widely in the semiconductor industry to etch patterns in the thin layers of polycrystalline silicon used for interconnections in MOS ICs (metal oxide semiconductor integrated circuits).

Manufacture

Reduction. Semiconductor-grade silicon is prepared by careful purification of some easily reduced silicon compound followed by reduction with an equally pure reducing agent. In general, naturally occurring quartz is first converted to metallurgical-grade silicon by heating it with coke in an electric furnace. The low grade silicon is then converted to halide or halosilane which, in turn, is reduced with a high purity reagent.

In the earliest commercial process for high purity semiconductor-grade silicon, silicon tetrachloride was reduced with zinc. Although only silicon and ZnCl$_2$ were produced, the process had the disadvantages of requiring very high purity zinc, the piping of molten zinc, system clogging caused by the ZnCl$_2$, and the necessity of reclaiming the zinc. Today, most producers reduce trichlorosilane with hydrogen. Deposition occurs on a silicon filament, usually U shaped, heated to about 1150°C. At the end of the cycle, the U rod (1.2–2.4 m long) is ca 10 cm in diameter. It is tempting to write the equation for this reaction as

$$\text{SiHCl}_3 + \text{H}_2 \rightarrow \text{Si} + 3\,\text{HCl}$$

However, many other compounds coexist in a H–Cl–Si system, depending on the temperature and concentrations of hydrogen, chlorine, and silicon. Thus, the exit gases contain not only HCl and unreacted SiHCl$_3$, but also SiCl$_2$, SiCl$_4$, SiH$_2$Cl$_2$, and even some long-chain oily and explosive polymers (38). Gases and by-products are purified and recycled. A complete system, including provisions for making the trichlorosilane, is shown in Figure 9 (39).

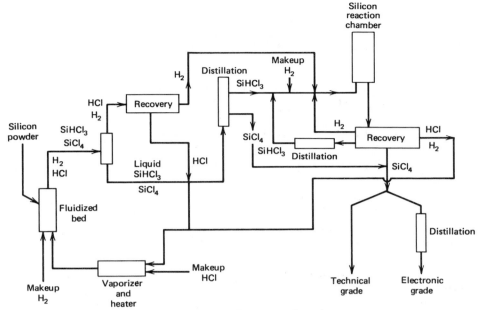

Figure 9.   Flow sheet of silicon manufacturing process.

Other reactions, such as those listed below, have been considered for the production of high purity silicon (28,40).

$$SiH_4 \xrightarrow[heat]{} Si + 2\,H_2$$

$$SiI_4 \xrightarrow[heat]{} Si + 2\,I_2$$

$$SiBr_4 + 2\,H_2 \xrightarrow[heat]{} Si + 4\,HBr + \ldots$$

$$SiF_4 + 4\,Na \xrightarrow[heat]{} Si + 4\,NaF_4$$

The choice is dictated by how well the reactants can be purified, by the resulting purity of the silicon, and by the economy of the process.

Electrolytic Deposition and Metallurgical Methods. Silicon is deposited electrolytically from molten mixtures such as K_2SiF_6–LiF–KF or SiO_2–Na_3AlF_6 and, with less success, from $SiCl_4$ or $SiHCl_3$ in an organic solvent (41). However, the purity obtained to date has not been sufficient to serve the semiconductor market.

Silicon can also be purified by various metallurgical methods. The earliest high performance silicon diodes were made from crushed silicon leached with acid to remove impurities that had segregated at the grain boundaries during freezing. This method was improved by directionally freezing silicon ingots, discarding the last part to freeze (where there was a high concentration of impurity), and then repeating the cycle (42). In zone refining (similar in concept to directional freezing), a thin zone is melted in a long rod and moved from one end to the other (see Zone refining). By repeating sweeps in the same direction, the impurities are moved toward one end. Leaching cannot provide a high quality final product but is sometimes used to upgrade very impure material. Directional freezing and zone refining are not very efficient, although they can in principle be used to reduce any concentration of impurities. When very high purities are required, eg, material that permits crystals from 200 to 10,000 Ω·cm to be grown, the best available chemically purified material is subsequently zone refined as required.

Crystal Growth. Because most devices require single-crystal semiconductors for best performance, the growth of single crystals of silicon has been studied extensively since the 1950s. A number of processes are available, depending on the desired properties (43). The most common and least expensive is the Czochralski process, first described in 1918, also called the Teal-Little method in honor of G. K. Teal and J. B. Little who first applied the process to semiconductor materials (44) (see Fig. 10). Molten silicon freezes gradually onto a properly oriented single-crystal seed that is both rotated and withdrawn from the melt. The freezing (growth) rate is controlled by a combination of melt temperature and radiation and conduction heat losses from the crystal. Crystals as small as 3 mm and as large as 20 cm in dia have been grown by this process. The most common size is between 75 and 125 mm in dia and approximately one meter long. Growth rates vary with the diameter but are usually in the 2.5 cm/h range.

The principal disadvantage of this method is the difficulty of maintaining the reactive molten silicon free of contaminants for hours at a time. These contaminants are difficult to control since they arise from the fused-silica container which is slowly dissolved by the silicon, from gaseous impurities in the inert atmosphere, and from vaporization of hot portions of the furnace. The crystal must be grown with a specified concentration of impurity (dopant) in order to provide the desired resistivity. This concentration is ordinarily controlled by the addition of the proper quantity of dopant to the melt before or during the growing operation. In general, the concentration of impurity actually incorporated into the crystal is less than the concentration in the melt and can be expressed as

$$N_{\text{crystal}} = kN_{\text{melt}} \tag{8}$$

where the N = concentrations and k = effective segregation coefficient. Thus, as growth continues and the melt volume decreases, the concentration in the melt in-

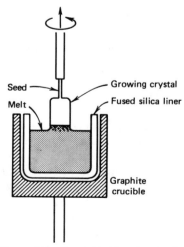

Figure 10.   Crystal pulling from melt.

creases. If the growth conditions are held constant over the length of the crystal, the concentration of impurity in the crystal is given by

$$N_{\text{crystal}} = N_o k (1 - l)^{k-1} \tag{9}$$

where N_o is the original concentration in the melt and l is the fraction of melt solidified (45). Table 8 gives values of k appropriate for the growth conditions normally encountered.

Table 8.   Segregation Coefficients, k

Element	k
B	0.80
Al	0.0020
Ga	0.0080
In	4×10^{-4}
P	0.35
As	0.3
Sb	0.023
Bi	7×10^{-4}
Cu	4×10^{-4}
Au	2.5×10^{-5}
Fe	8×10^{-6}
O	1.24
C	0.07

[a] From ref. 46.

In order to provide very high resistivity material, float-zoning is sometimes used (see Fig. 11). A polycrystalline rod along with a single-crystal seed at one end is held vertically, and an r-f-heated molten zone is caused to traverse its length, starting at the seed end. The length of the zone and the diameter of the rod are carefully chosen in such a way that surface tension and interaction between circulating currents in the

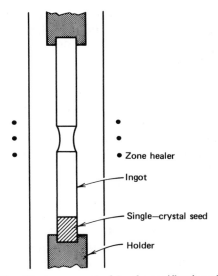

Figure 11.   Float-zone method for the purification of silicon.

melt and the field from the surrounding induction coil keep the molten zone in place. In this case, the molten silicon is in contact only with silicon, the volume of melt at any instant is very low, and the chamber can be kept cool. Repeated passes of the zone may be used to sweep increasing amounts of the residual impurities in the original rod to one end. The maximum diameter obtained by this method is about 10 cm.

Most semiconductor applications require thin, flat, single-crystal slices with damage-free surfaces for subsequent processing. Both the Teal-Little and float-zone methods generate long rods from which slices must be cut and subsequently lapped and polished. In attempts to circumvent these steps, various modifications of melt growth have been proposed in order to grow single-crystal ribbons directly. The initial work started in the early 1960s and concentrated on dendrites and on growth in a [211] direction through a shaped orifice (47). Neither the dendrites nor the dendrite webs that followed were of sufficient quality for most uses, and shaped crystals could not be grown thin enough. In addition, these methods could not compete economically with conventional crystal pulling. In the mid-1970s, the realization that reasonably efficient solar cells could be made from poor-quality crystalline material, combined with the search for very low cost cells, led to renewed studies of ribbon or sheet growth (48). Among the more promising approaches, in addition to webs, has been edge-defined growth, where a silicon carbide die is used with a cross-section equal to that of the crystal desired. The lower end is immersed in molten silicon in such a way that surface tension can draw molten silicon to the top of the die. The crystal is then pulled from that small reservoir. It can grow no larger in cross-section than the silicon supply, and the orientation is so chosen that the direction of growth is [211] and the sides of the ribbon are (111) faces.

Silicon can be grown from various fluxes and by a combination of electrolysis and flux at temperatures well below the melting point of pure silicon (41). The main disadvantages are the inclusion of the flux in the crystal and the poor quality obtained. Potential advantages are purification during electrolysis and a decrease of growth temperature.

Vapor-phase growth at temperatures below the melting point is now used widely for adding thin layers to slices. If this process results in a single-crystal layer, it is referred to as epitaxy. Vapor-phase growth allows rapid change of doping impurities, permitting the sequential growth of layers of different resistivity. The method of transport may be by direct evaporation (molecular-beam epitaxy) (49) or by chemical means. With chemical means, any of the reactions previously mentioned for manufacturing semiconductor-grade silicon may be employed. At present, the thermal decomposition of silane and the hydrogen reduction of $SiCl_4$, $SiHCl_3$, or SiH_2Cl_2 are the principal processes (50) (see Fig. 12).

For vapor-phase doping, an easily vaporizable compound is either thermally decomposed or reduced along with the silicon compound. Typical examples are diborane, phosphine, and boron tribromide. Deposition (growth) rates vary from fractions of micrometers per minute for molecular-beam epitaxy to about 1 μm/min for the various hydrogen reduction processes. In some cases, a layer of polycrystalline silicon is placed over an oxidized silicon surface. Furthermore, a small grain size and smooth surface are preferred. Under such circumstances, deposition takes place at 600–700°C; reduced pressure increases the deposition rate. Deposition at even lower temperatures, eg, by evaporation or a plasma-assisted SiH_4 decomposition, gives highly resistive amorphous silicon. However, when hydrogen is present during the deposition,

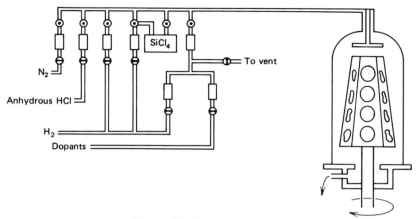

Figure 12. Vapor deposition.

some of the bonds are apparently terminated, and the material can be used in some semiconductor applications, eg, photovoltaic detectors and solar cells (see Photovoltaic cells).

Economic Aspects

The principal producers of silicon in the noncommunist world are given in Table 9. The 1980 price of silicon was ca $75/kg, whereas in 1955, one year after the commercial introduction of the silicon transistor, the price ranged between $650 and $2000/kg, depending on the quality.

Consumption, in metric tons, of semiconductor-grade silicon is given below, based on Bureau of Mines data and industry estimates.

1955	1960	1965	1975	1980
5–10	30	50	1500	2800

Metallurgical applications and the silicon industry consumed ca 5×10^6 t in 1980.

Table 9. Producers of Semiconductor-Grade Silicon, 1980

Company	Country
Great Western	United States
Hemlock	United States
Komatsu	Japan
Monsanto	United States
Motorola	United States
Osaka Titanium	Japan
Shinetsu	Japan
Smiel (Dynamit Nobel)	Italy
Texas Instruments	United States
Wacker	FRG

Analytical Methods

Wet Chemistry.   In most methods, silicon is first converted to silica or silicic acid. For organic materials, simple ashing is usually sufficient. For inorganic materials, a fluxing agent, eg, Na_2CO_3 or $Na_2Br_4O_7$, followed by an acid treatment to precipitate gelatinous silicic acid are required. A qualitative test is the yellow color of the silico-molybdate complex formed by the reaction of silicic acid with ammonium molybdate. Quantitative analysis by colorimetry is also possible at this point. Alternatively, the precipitate may be dried and the amount of SiO_2 determined by dissolving in HF.

Spectrography.   Silicon has very strong lines and can be easily detected in the 1–100 ppm range (51). The two intense lines occur at 251.6123 and 288.1578 nm.

X Ray.   For nondestructive analysis, both x-ray diffraction and x-ray fluorescence may be used (51). The latter has a sensitivity of about 30 ppm.

Determination of Impurities.   Trace impurities in silicon are determined by the following methods: optical spectroscopy (see Table 5); solid-mass spectroscopy; neutron-activation analysis; x-ray fluorescence; Auger spectroscopy combined with sputtering; or various indirect methods that require the measurement of some electrical property dependent on the impurity content (51–52) (see Analytical methods). Direct determinations may be difficult because the amounts to be detected are often parts per billion or less. Indirect measurements provide a more powerful tool for the detection of impurities. For example, if resistivity is used as a measure of impurity concentration, the presence of fractions of parts per billion of electrically active impurities is detectable, although the impurity itself cannot be identified. Hall measurements as a function of temperature can be used to determine activation energies of the impurities present. However, many have multiple energy levels or levels very close to each other. Thus, it is always difficult and sometimes impossible to establish the identify of the impurity.

The distribution of impurities over a surface can be measured by autoradiography, stepping over the surface with resistivity probes, or by scanning with an x-ray microprobe, secondary-ion mass spectrometer, or Auger spectrometer. Depth measurements can be made by combining the above with the repeated removal of thin layers, either by etching, or by sputtering, as in the secondary-ion mass spectrometer.

Health and Safety Factors

Elemental silicon is inert; in air, it is only classified as a nuisance particulate (53) with a TLV of 10 mg/m^3. However, crystalline-quartz dust is considered hazardous. Many silicon compounds are poisonous, and the silanes are highly explosive as well. Compounds such as $SiCl_4$, $SiHCl_3$, and $SiBr_4$ decompose in the presence of moisture, forming SiO_2 and the corresponding acid. If this decomposition takes place in the respiratory tract, the result is similar to breathing the acid fumes. The long-chain chlorinated polymers are unstable and detonate under mild shock.

Uses

Silicon is used widely in metallurgical applications, both as a constituent of various alloys and as an oxidizer in steelmaking (see Steel). It is also basic to the silicone industry, which is built around compounds with long oxygen–silicon chains (see Silicon compounds, silicones; Elastomers). However, silicon is best known as the material from which transistors and integrated circuits are made, although the quantities consumed are small compared to metallurgical uses. The consumption of only a few thousand tons a year is the basis of the 15×10^9-dollar-a-year semiconductor industry (see Semiconductors). Although the metallurgical silicon industry depends on product purity in the 95–99% range at best, impurities in semiconductor-grade silicon are measured in parts per billion (10^9). Semiconductor-silicon requirements differ in one other significant manner. Whereas a polycrystalline material is used in all the other main applications, the semiconductor applications require single-crystal silicon.

Although not originally recognized as such, the semiconductor properties of silicon have been exploited at least since 1906, when metal-silicon contacts were studied for r-f detection. During the early years of radio, a small metal wire (cat's whisker) contacting a piece of galena (naturally occurring PbS) was a common detector, but by the mid-1930s galena was replaced by silicon. During World War II, silicon-diode detectors made of polycrystalline material reached a high degree of complexity and were used widely in military radar (42). Shortly after the invention of the germanium transistor, in which single-crystal germanium was used, work began on the production of single crystals of silicon for use in silicon transistors (44).

The quantity of silicon required to make a single transistor has steadily decreased, particularly after the invention of the integrated circuit, which allowed many transistors to be interconnected on one piece of silicon. For example, in 1955, 50 g of high purity silicon might produce 50 salable transistors. By 1980, 50 g provided over one million (10^6) transistors in the form of integrated circuits. Even so, the demand for transistors has grown so phenomenally that the semiconductor-grade silicon requirement has risen substantially (see Economic Aspects).

Nomenclature

A, B	=	rate constants
E_g	=	band gap; amount of energy required to raise an electron from the valence band to the conduction band
k	=	radiation-damage constant; effective segregation coefficient
l	=	fraction of melt solidified
N	=	concentrations
n	=	density of free electrons present
n_i	=	intrinsic carrier concentration
N_o	=	original concentration
p	=	density of free holes
q	=	electron charge
T	=	temperature
t	=	oxidation time
X	=	oxide thickness
σ	=	electrical conductivity
μ	=	carrier mobility
τ_f	=	lifetime after irradiation with a flux
ϕ	=	flux
τ_o	=	original lifetime
τ	=	constant involving the initial oxide thickness
κ	=	Boltzmann's constant

BIBLIOGRAPHY

"Silicon (Pure)" in *ECT* 2nd ed., Vol. 18, pp. 111–125, by Walter R. Runyan, Texas Instruments, Inc.

1. R. C. Weast, ed., *Handbook of Chemistry and Physics*, 62nd ed., CRC Press Inc., Boca Roton, Fla., 1981.
2. J. S. Kasper and S. M. Richards, *Acta Cryst.* **17**, 752 (1964).
3. G. Celotti, D. Nobili, and P. Ostaja, *J. Mat. Sci.* **9**, 821 (1974) and references contained therein.
4. R. E. Honig, *RCA Rev.* **23**, 567 (1962).
5. C. Y. Yaws and co-workers, *Solid State Tech.* **24**, 87 (Jan. 1981).
6. A. A. Giardini, *Am. Mineral.* **43**, 957 (1958).
7. H. J. McSkimin, M. L. Bond, E. Buehler, and G. K. Teal, *Phys. Rev.* **83**, 1080(L) (1951).
8. Y. S. Touloukian and co-workers, *Thermophysical Properties of Matter*, Vol. 4, IFI/Plenum, New York, 1970.
9. Ref. 8, Vol. 1, 1970.
10. Ref. 8, Vol. 13, 1977.
11. J. F. Nye, *Physical Properties of Crystals*, Oxford University Press, Fairlawn, N.J., 1960.
12. V. M. Glazov and co-workers, *Liquid Semiconductors*, Plenum Press, Inc., New York, 1969.
13. S. N. Rea, *Large Area Czochralski Silicon*, Texas Instruments Inc. Monthly Technical Progress Report Report 03-76-31, May 1976 (ERDA/JPL 954475-76II).
14. M. O. Lampert, J. M. Koebel, and P. Siffert, *J. Appl. Phys.* **52**, 4975 (1981).
15. H. R. Philipp and E. A. Taft, *Phys. Rev.* **120**, 37 (1960); C. D. Salzberg and J. J. Villa, *J. Opt. Soc. Am.* **47**, 244 (1957).
16. J. I. Pankove, *Optical Processes in Semiconductors*, Dover Publications, Inc., New York, 1975; T. S. Moss, *Optical Properties of Semiconductors*, Academic Press, Inc., New York, 1959.
17. R. Braunstein, A. R. Moore, and F. Herman, *Phys. Rev.* **99**, 1151 (1955).
18. W. R. Runyan and S. B. Watelski, "Semiconductor Materials" in C. A. Harper, ed., *Handbook of Materials and Processes for Electronics*, McGraw-Hill Book Co., New York, 1970.
19. W. M. Bullis, *Solid State Elect.* **9**, 143 (1966).
20. J. C. Irvin, *Bell System Tech. J.* **41**, 387 (1962).
21. D. Coleman, R. T. Bate, and J. P. Mize, *J. Appl. Phys.* **39**, 1923 (1968).
22. S. Minomura and H. G. Drickamer, *J. Phys. Chem. Solids* **23**, 451 (1962).
23. S. M. Sze, *Physics of Semiconductor Devices*, 1st ed., John Wiley & Sons, Inc., New York, 1969.
24. N. J. Rudie, *Principles and Techniques of Radiation Hardening*, Vol. 2, 2nd ed., Western Periodicals Company, North Hollywood, Calif., 1980.
25. H. A. Herrmann and H. Herzer, *J. Electrochem. Soc.* **122**, 1568 (1975); J. M. Meese, *Second International Conference of Neutron Transmutation Doping in Semiconductors*, University of Missouri, Columbus, Mo., Plenum Press, Inc., New York, 1979.
26. E. G. Rochow in *Comprehensive Inorganic Chemistry*, Vol. 1, Pergamon Press, New York, 1973, p. 1331.
27. M. Auman and co-workers, *Phys. Rev. C* **5**, 1 (1972).
28. J. W. Mellor, *A Comprehensive Treatise on Inorganic and Theoretical Chemistry*, Vol. 4, Longmans, Green & Co., Inc., New York, 1957.
29. A. S. Berezhnoi, *Silicon and Its Binary Systems* (trans. from Russian), Consultants Bureau, New York, 1960.
30. E. A. V. Ebsworth, *Volatile Silicon Compounds*, Academic Press, Inc., New York, 1964.
31. R. B. Fair, *J. Electrochem. Soc.* **128**, 1360 (1981); R. R. Razouk, L. N. Lie, and B. E. Deal, *J. Electrochem. Soc.*, 2214 (1981).
32. V. Cazcarra, *Inst. Phys. Conf. Ser.*, (46), 303 (1979) and references therein.
33. B. Schwartz and H. Robbins, *J. Electrochem. Soc.* **123**, 1903 (1976) (see also the preceding three papers in the series).
34. W. K. Zwicker and S. K. Kurtz, "Anisotropic Etching of Silicon Using Electrochemical Displacement Reactions" in H. R. Huff and R. R. Burgess, eds., *Semiconductor Silicon 1973*, The Electrochemical Society Softbound Symposium Series, Princeton, N.J., 1973, p. 315.
35. K. E. Bean and W. R. Runyan, *J. Electrochem. Soc.* **124**, 5C (1977) and the included references.
36. R. Memming and G. Schwandt, *Surface Sci.* **4**, 109 (1966).
37. J. L. Vossen and W. Kern, eds., *Thin Film Processes*, Academic Press, Inc., New York, 1978.
38. M. Bawa, J. Truitt, and W. Haynes, *Texas Instruments Semiconductor Eng. J.* **1**(2), 47 (1980); M. Bawa, R. Goodman, and J. Truitt, *S.C. Eng. J.* **1**(3), 42 (1980).

39. L. D. Crossman and J. A. Baker, "Polysilicon Technology" in H. R. Huff and E. S. Sirtl, eds., *Semiconductor Silicon 1977*, The Electrochemical Society, Princeton, N.J., 1977.

40. C. H. Lewis, M. B. Giusto, and S. Johnson, *The Preparation of Transistor-Grade Silicon from Silane or Analogous Compounds*, Metal Hydrides Incorporated, Final Report, Contract AF19(604)-3463, 1959; G. Szekely, *J. Electrochem. Soc.* **104,** 663 (1957); R. C. Sangster, E. F. Maverick and M. L. Croutch, *ibid.*, p. 317; A. Sanjurjo, L. Nanis, K. Sancier, R. Bartlett, and V. Kapur, *J. Electrochem. Soc.* **128,** 179 (1981).

41. D. Elwell, *J. Cryst. Growth* **52,** 741 (1981).

42. H. C. Torrey and C. A. Whitmer, eds., *Crystal Rectifiers*, McGraw-Hill Book Co., New York, 1948.

43. R. A. Laudise, *The Growth of Single Crystals*, Prentice-Hall Inc., Englewood Cliffs, N.J., 1970.

44. G. K. Teal, *IEEE Trans. Elec. Dev.* **23,** 621 (1976).

45. W. G. Pfann, *Zone Melting*, 2nd ed., John Wiley & Sons, Inc., New York, 1958.

46. F. A. Trumbore, *BSTJ* **39,** 205 (1960).

47. For early work on semiconductor dendrites and crystal shaping by various other means, see papers in R. O. Grubel, ed., *Metallurgy of Elemental and Compound Semiconductors*, Interscience Publishers, New York, 1961.

48. The September 1980 special issue of the *J. Cryst. Growth* is completely dedicated to shaped crystal growth and covers the 1980 state of understanding and application.

49. P. E. Luscher, *Solid State Tech.*, (Dec. 1977).

50. C. H. L. Goodman, ed., *Crystal Growth, Theory and Techniques*, Vol. 2, Plenum Press, Inc., New York, 1978; E. Sirtl, "Thermodynamics and Silicon Vapor Deposition" in R. R. Haberecht and E. L. Kern, eds., *Semiconductor Silicon*, The Electrochemical Society, Princeton, N.J., 1969; See also other volumes in this series, ie, *Semiconductor Silicon 1973*, *Semiconductor Silicon 1977*, etc.

51. P. F. Kane and G. B. Larrabee, *Characterization of Semiconductor Materials*, McGraw-Hill Book Co., New York, 1970.

52. W. R. Runyan, *Semiconductor Measurements and Instrumentation*, McGraw-Hill Book Co., New York, 1975.

53. *Threshold Limit Value For Chemical Substances and Physical Agents in the Work Room Environment with Intended Changes for 1980*, American Conference of Governmental Industrial Hygienists, 1980.

General Reference

Ref. 31 is a general reference for silicon oxidation and can assist in locating more material on this subject.

WALTER RUNYAN
Texas Instruments, Inc.

METALLURGICAL

The most important and most widely used method for making metallurgical silicon and silicon alloys is the reduction of oxides or silicates with carbon in an electric furnace. In the reduction process, metal carbides usually form first because of lower temperature requirements. As silicon is formed, it displaces the carbon, because silicon alloys and silicides have a higher heat of formation than metal carbides (1).

Silicon

The commercial production of silicon [7440-21-3] in the form of binary and ternary ferroalloys began early in the twentieth century with the development of electric-arc and blast furnaces and the subsequent rise in iron and steel production (1) (see Steel).

Silicon is soluble in aluminum in the solid state to a maximum of 1.65 wt % at 577°C (2). It is soluble in silver, gold, and zinc at temperatures above their melting points. Phase diagrams of systems containing silicides are given in refs. 2–4.

Silicon metal with a purity of ca 98% Si is obtained by carbon reduction of nearly pure silica in an electric furnace. In 1979, about 134,000 metric tons of silicon metal was produced in the United States, an increase of ca 25,000 t over 1978 (5). Preliminary figures for 1980 indicate a 12% decline from 1979 (6). Silicon metal is used extensively by the nonferrous metal industry, mainly in the production of aluminum and copper alloys. In aluminum, silicon improves castability, reduces shrinkage and hot-cracking tendencies, and increases corrosion resistance, hardness, tensile strength, and wear resistance (7). Silicon metal added to copper produces silicon bronzes. The silicon improves fluidity, minimizes dross formation (7), and enhances corrosion resistance and strength (8).

Silicon Alloys and Silicides

Metal silicides form well-defined crystals with a bright metallic luster; they are usually hard and high-melting. The metals that form silicides and those that do not are given in Table 1 (2). Most of the former form several silicides; for example, iron forms Fe_5Si_3 [12023-77-7], FeSi [12022-95-6], $FeSi_2$ [12022-99-0], Fe_2Si_5 [12063-29-5], and others (see Fig. 1).

Ferrosilicons. Ferrosilicons for steelmaking and foundry uses have a silicon content of 14–95%. Ferrosilicons are made by melting low alloy steel scrap, usually machine shop turnings or short shovelings, in electric furnaces, or by the reduction of high purity silica rock or quartz with carbon.

Silvery pig iron containing <25% Si is magnetic, which is an advantage in handling. It is used primarily as a furnace block and is added in the form of controlled-weight piglets for initial deoxidation (7,9) (see Iron; Steel). It is also furnished in powder form for ore beneficiation (9). The low silicon content of silvery pig iron makes it unsuitable for ladle addition, because the large quantity normally required would cause excessive chilling.

In the United States, the preferred silicon alloy is 50% ferrosilicon, which indicates

Table 1. Metal Silicide Formation[a]

Metals that form silicides			Metals that do not form silicides	
barium	lithium	rhenium	aluminum	silver
beryllium	magnesium	strontium	antimony	sodium
boron	manganese	tantalum	arsenic	thorium
calcium	molybdenum	titanium	bismuth	zinc
chromium	nickel	tungsten	cadmium	
cobalt	niobium	uranium	gold	
copper	platinum	vanadium	mercury	
iron	rare earth	zirconium	ruthenium	
lead	elements			

[a] From ref. 2.

that there is an abundant and economical supply of steel scrap. It is the least expensive silicon alloy on a silicon-content basis (see Economic Aspects). Regular 50% ferrosilicon is used as a deoxidizer and alloying agent in the production of killed and semikilled steels. So-called killed steels are those that have been made quiet and free from bubbling, while molten, by the addition of a deoxidizer. This process minimizes reaction of carbon and oxygen during solidification. The melting point of 50% ferrosilicon at 1220°C is the lowest of the iron–silicon alloys, except the 1200°C eutectic for silvery pig iron containing about 22% Si (see Fig. 1). Iron foundries add 50% ferrosilicon to the cupola and subsequently to the ladle for the purpose of inoculation. Inoculation may be defined as a change in the physical properties of iron that is not explainable by a change in chemistry. In practice, inoculation consists of withholding from the base iron some portion of the desired silicon and adding it to the ladle. The effects of this method are decreased chilling tendency and reduced iron carbide formation without appreciable effect on pearlitic stability; increased randomness of graphite flakes and decrease in flake size; decreased section sensitivity; increased ratio of tensile strength to hardness; and improved machinability at all strength levels.

High purity grades of 50 and 75% ferrosilicon with a max aluminum content of 10% are sources of silicon for electrical steels containing less than 2.0% silicon, where the low residual aluminum content contributes to the attainment of desired electrical properties, eg, significant reduction of eddy currents. The formation of this alloy is endothermic and for this application is normally limited to heats (charges) of more than 90 metric tons, or where the temperature loss from the addition is of little importance (7).

The 65% grade of ferrosilicon decreased in commercial importance in the past decade. It is thermodynamically balanced with respect to heat effect on molten steel; ie, the exothermic effect developed by the dissolving silicon is equal to the heat required to raise the ferrosilicon to the temperature of the molten steel. The 75 and 90% ferrosilicon grades are used mainly for high alloy cast steels that require large additions of silicon. Of the 65–95% ferrosilicon grades consumed in the United States in 1979 at a rate of ca 140,000 metric tons, the 75% grade accounted for more than 130,000 t. The increased importance of 75% ferrosilicon is mainly owing to economic reasons. Rapidly rising U.S. energy costs stimulated imports of 75% ferrosilicon at prices competitive with the domestic product. Between 1978 and 1980, the difference between the prices of domestic and imported 75% ferrosilicon fell from ca 24 to 6.6 ¢/kg con-

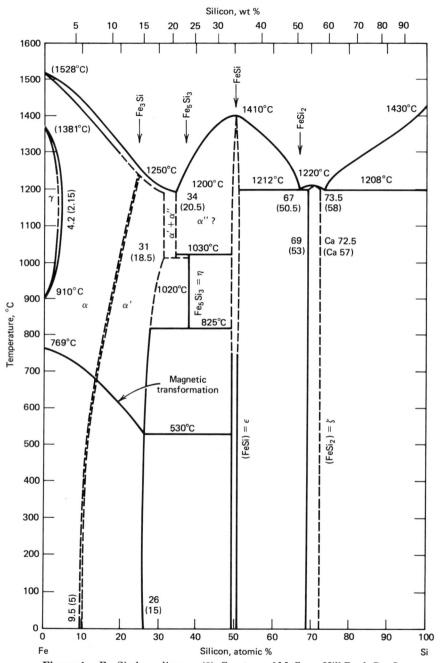

Figure 1. Fe–Si phase diagram (2). Courtesy of McGraw-Hill Book Co., Inc.

tained silicon (5). This reduction makes 75% ferrosilicon economically the most attractive grade between 65 and 95%, and even threatens to compete with 50% ferrosilicon as the most economical source of silicon.

Boron-Bearing Ferrosilicons. These alloys are made by either carbon reduction of boron and silicon oxides in an electric furnace or by silicon reduction of boron oxide ladle additions. Borosil (Ohio Ferroalloys Co.) contains 2–3% boron and is used to increase the hardness of steel. Only a few thousandths of one percent of boron in steel is needed for this purpose, replacing considerably greater amounts of more expensive alloys such as chromium (10). Boron-bearing 50% ferrosilicon with ca 0.10% boron is used for furnace or ladle additions in the production of malleable iron.

Calcium Silicons. Calcium silicon and calcium–barium–silicon are made in the submerged-arc electric furnace by carbon reduction of lime, silica rock, and barites. Calsibar (Elkem Metals) is used in place of calcium–manganese–silicon. Commercial calcium–silicon contains 28–32% calcium, 60–65% silicon, and 3% iron max. Calsibar contains 16–20% calcium, 14–18% barium, and 53–59% silicon. These alloys are used to deoxidize and degasify steel. They produce complex calcium silicate inclusions that are minimally harmful to physical properties and prevent the formation of alumina-type inclusions, a principal source of fatigue failure in highly stressed alloy steels. As a sulfide former, they promote random distribution of sulfides, thereby minimizing chain-type inclusions. In cast iron, they are used as an inoculant.

Ferrochrome–Silicon Alloys. These alloys have declined in importance since the introduction of the basic oxygen process in steelmaking. Like ferrosilicon alloys, they are made in an electric furnace; the main difference is that ferrosilicon production is virtually slag-free. The standard grades of ferrochrome–silicon still in use are 35 Cr–40 Si and 40 Cr–43 Si with maximum carbon specified as 0.06 and 0.05%, respectively (11). These alloys were originally developed for stainless steel. The silicon in the alloy reduces the metal oxides, principally chromium, permitting them to sink from the slag back into the metal bath while the chromium in the alloy dissolves in the metal.

Magnesium Ferrosilicons. Magnesium ferrosilicons in alloy combination with other elements such as nickel, calcium, barium, titanium, cerium, and rare earths are extensively used in the manufacture of ductile or compacted-graphite cast iron. Most widely used are nickel–iron–silicon–magnesium, iron–silicon–magnesium, and iron–silicon–magnesium–rare earths. The magnesium contents are between 5 and 10%. The content of rare-earth elements, if present, ranges from 0.1 to 1.0%.

Ductile cast iron is made by converting the flakes of graphite in gray iron into tiny balls or spherulites by the addition of one or more elements to the molten metal. Magnesium, calcium, cerium, barium, or other elements produce spherulitic graphite structures (12); magnesium and cerium (rare earths) are commercially important. Ductile iron with a compacted-graphite structure is a more recent modification (13). The addition of 1.50–1.95% nickel to nodular iron significantly increases the strength of the pearlitic matrix. Conversion of graphite flakes into spherulites increases tensile strength and notch hardness of cast iron; subsequent heat treatment produces desirable mechanical properties that cannot be obtained with iron-containing graphite in flakes.

Rare-Earth Silicides. Rare-earth silicides in the form of a ferroalloy that contain up to 33% rare earths are used increasingly by the iron and steel industry. (The term silicides, although no longer used for alloys of this type, is still in common use in this instance.) For nodular iron, addition of rare earths gives spheroidal rather than flaky graphite. Rare earths desulfurize and, combined with silicon, deoxidize and alloy steels; they may be substituted for misch metal.

Aluminum–Silicon Alloys. Aluminum–silicon alloys containing up to 80% aluminum, with the remainder mostly silicon, are usually made by secondary melting of aluminum with silicon metal. For better control of impurities and for economic reasons, most producers of aluminum prefer commercial silicon metal as their source of silicon; several make their own supply of silicon metal by carbon reduction. The production of aluminum–silicon alloys by direct carbon reduction of alumina and silica has been the aim of aluminum producers for many years. Such alloys usually contain aluminum and silicon in a 60:40 ratio, including 5% iron plus titanium, depending upon the source of clay or bauxite used. Removing the iron and titanium impurities greatly increases costs (14).

Silicon–Aluminum Alloys. Silicon–aluminum alloys contain up to 50% silicon and 10–20% aluminum; the remainder is mostly iron, manganese, calcium, or a combination of these. The base alloy (50% ferrosilicon) is melted in an electric furnace and aluminum is added to the ladle of molten metal. These alloys are preferred as a ladle addition in iron and steel production for final deoxidation and alloying. These alloys control the grain size of steel and reduce the harmful effect of nitrogen. Increasing energy costs have led to a decline in demand for these alloys and production is minimal.

Manganese–Silicon Alloys. These alloys usually contain various amounts of iron and carbon. They are made by carbon reduction of manganese ore or manganese slag in the presence of silica in an electric furnace. The lower silicon grades are referred to as silicomanganese. They contain ca 1.5–3% carbon, 18–20% silicon, and 65–68% manganese; the remainder is mainly iron. Silicomanganese is particularly suitable for introducing manganese into low carbon steel in which the added silicon is not objectionable. It is also used as a deoxidizer for open-hearth and electric-furnace steel, both acid and basic (7). The use of silicomanganese is often preferred to separate additions of manganese and silicon because of its low carbon content, which produces a higher purity steel. A low carbon grade of silicomanganese containing ca 30% silicon and 0.06% carbon max is made by a two-stage process. It is used extensively in the production of the 200 and 300 stainless-steel series.

Strontium–Silicon. This alloy contains 30–40% strontium and 45–55% silicon; the remainder is mainly iron. It is made in an electric furnace by carbon reduction of silica- and strontium-bearing ore. In cast irons, this alloy effectively reduces chill. It dissolves rapidly, produces little dross, and minimizes shrinkage defects (15).

Titanium–Silicon. This alloy is made by adding titanium scrap to molten metallurgical silicon or silicon alloys. Graphidox (Foote Mineral Co.) contains 50–55% silicon, 9–11% titanium, and 5–7% silicon; the remainder is mainly iron. It is made by carbon reduction of titanium ore, limestone, and quartz in a submerged-arc furnace. Titanium–silicon alloy is an efficient graphitizing inoculant for chill reduction in gray cast iron. It is also a supplementary deoxidizer for wrought and cast steels (16).

Vanadium–Silicon. This alloy is made by the reduction of vanadium oxides with silicon in an electric furnace. Application is essentially the same as that of the titanium alloys; that is, these vanadium alloys sometimes offer the most economical way of introducing vanadium into molten steel.

Zirconium–Silicon. This alloy is made in an electric furnace by carbon reduction of the oxides. It is used by the iron and steel industry as a deoxidizer and scavenger. Zirconium combines readily with oxygen, nitrogen, and sulfur, forming nonmetallic inclusions that either float out of the molten bath or are rendered minimally harmful. A special grade called SMZ (Elkem Metals Co.) contains 60–65% silicon, 5–7% zirconium; the remainder is mainly iron. It is used in the production of gray iron castings to control the depth of chill at edges and corners and in light sections.

Economic Aspects

Before the 1970s, the United States was dominant in the manufacture of metallurgical silicon and silicon alloys. However, since the reduction of silicon from its ores is energy-intensive, rapidly rising costs of energy, carbon reducing agents, and carbon electrodes have contributed to declining profits. The high cost of pollution control, more than 500×10^6 in the 1970s (17), tied up capital. Other countries, including the developing nations, became more competitive. Chromium ferroalloys and also manganese alloys were most affected, because these metals are not found in the United States. However, the United States probably has the largest resources of naturally occurring silica materials (8) and the world's greatest reserves of metallurgical coals. These facts are of prime importance in the continuing survival of U.S. production of metallurgical silicon and silicon ferroalloys.

The availability of inexpensive scrap steel contributes to low cost 50% ferrosilicon and silvery pig iron, a U.S. asset under pressure from higher energy costs. The plants of two main U.S. producers (Union Carbide and Airco) of silicon metal and silicon ferroalloys were recently acquired by Elkem of Norway and SKW of the Federal Republic of Germany.

In 1978, U.S. consumption of silicon alloys, including magnesium–ferrosilicon, in the manufacture of cast irons, both gray and ductile, was ca 90,000 metric tons (5). This fell to ca 65,000 t in 1979, reflecting a dramatic decline in usage by the automotive industry. The decline in magnesium–ferrosilicon consumption was partly a result of a new technology in which a granular magnesium–ferrosilicon is added to the mold rather than to larger-size melters. Improved magnesium recoveries and enhanced environmental standards combined with economic recession have had an extraordinary impact on the consumption of magnesium–ferrosilicon.

Estimated production in 1979 of silvery pig iron, ferrosilicon, silicon metal, and miscellaneous silicon alloys, excluding silicomanganese, was ca 950,000 t, an increase of ca 75,000 t over 1978 (5). Preliminary figures for 1980 indicate a 22% decline from 1979 (6).

Prices and composition for silicon and several of its alloys are given in Table 2 (11,18), and estimated capacities and production data are given in Table 3. A ten-year price index for silicon metal and ferrosilicon is given in Table 4 (19). The trend is a dramatic reflection of energy and environmental costs in an industry suffering from declining profits.

Health and Safety Factors

The details of toxicity associated with metallurgical silicon are unknown (20). Toxicities of metal silicides are variable and depend upon hazards identified with the other metallic elements (see Manganese and manganese alloys).

The principal health hazard that may be associated with silicon is caused by the oxide. Silica, the natural oxide of silicon, is both the raw material and main polluting

Table 2. Silicon Alloys, Consumption and 1981 Prices[a]

Product	Composition, % Si	Fe	Ca	Others	Price[b], $/kg
regular ferrosilicon	47–51	remainder			
50%	47–51	remainder			0.99[c]
75%	74–79	remainder			1.09[c]
silicon metal	98 min	1.0 max	0.07 max		1.41[c]
calcium silicon	60–65	1.5–3.0	28–32		1.80[d]
magnesium ferrosilicon	44–48		1.0	5–6 Mg, 0.30 Ce	1.12[d]
silicomanganese	18–20	remainder		65–68 Mn, 1.5–3 C	0.54[d]
SMZ alloy	60–65	remainder	3–4	5–7 Mn, 5–7 Zr	1.08[d]
Graphidox	50–55	remainder	5–7	4–11 Ti	1.17[d]

[a] From refs. 11 and 17.
[b] Bulk, fob producer.
[c] Per kg silicon.
[d] Per kg alloy.

Table 3. Estimated Capacities and Production of Metallurgical Silicon and Silicon Alloys, 10³ Metric Tons Alloy[a]

	Silvery pig iron[b] Capacity	Production	Ferrosilicon 45–50% Capacity	Production	75% Capacity	Production	Silicon metal Capacity	Production	Miscellaneous[c] Production
U.S.									
1979	105	56	610	511	220	161	185	134	89
1980	105	41	665	408	220	107	200	118	66
Other countries[d]									
1979			745	650	1775	1550	490	410	na
1980			790	640	1850	1560	540	460	na

[a] From refs. 5, 6, and 14.
[b] Produced only in the United States.
[c] Excluding silicomanganese; capacity included in ferrosilicon.
[d] Except USSR and Comecon.

effluent in the electric-furnace production of metallurgical silicon and silicon alloys. Silica in its crystalline form is the chief cause of a disabling pulmonary fibrosis, ie, as silicosis. Over a period of years, the breathing of air containing excessive amounts of crystalline silica can cause shortness of breath (20) (see Silica). Similar symptoms may be caused by prolonged exposure to fine metal dust generated in the sizing of silicon metal and silicon alloys. Workers engaged in these activities often are provided with dust masks and respirators.

The white smoke, commonly called silica fume, emanating from electric-furnace production of silicon or silicon alloys includes particles of dust containing about 90% silica. Unlike crystalline quartz, silica fume is amorphous. It formerly polluted the environment to a far greater extent than at present. By 1980, Federal and state laws governing the protection of the environment had secured near-total compliance with the belief that electric-furnace dust need not be spread over the countryside. Inside the plants, the atmosphere was also improved, and workers are not exposed to higher health risk than the general public.

Uses

In the iron and steel industry, silicon alloys (silicides) are used for alloying, deoxidizing, and reducing other alloying elements such as manganese, chromium, tungsten, and molybdenum. In the nonferrous metal industry, silicon metal is used primarily as an alloying agent for copper, aluminum, magnesium, and nickel. Substantial quantities of silicon metal are used in the production of silicones, which was greatly expanded in the 1970s. Silicon in the form of 75% ferrosilicon is used as a reducing agent in magnesium manufacture by the Pidgeon process. Hydrogen is obtained by the reaction of ferrosilicon and caustic soda. Metallurgical-grade silicon is used for the production of high purity silicon crystals required by the electronics industry and of monocrystalline and polycrystalline material for solar cells.

Ferrosilicons containing 14–95% silicon (11) are used extensively by the iron and steel industry, and silicon is present in most commercial grades of steel and cast iron. For these applications, silicon metal of better than 96–99% purity is too expensive (5). Most silicon metal is consumed in the manufacture of aluminum-alloy castings containing from 2 to 25% silicon (8). Ternary and quaternary ferroalloys, often proprietary, are also available for the simultaneous addition of silicon and other alloying elements, usually manganese and iron. Other elements added in this way include aluminum, boron, barium, calcium, chromium, magnesium, strontium, titanium, vanadium, zirconium, and the rare earths. These combinations are widely used in the iron foundry and steel industries (see Iron; Steel).

As an effective and economical deoxidizer, silicon is used to refine most grades of carbon and alloy steels. In importance to steelmaking, it is second only to manganese. As an alloying element in steels, it increases tensile strength and elastic limits, improves resistance to corrosion and high temperature oxidation, improves electrical characteristics, and decreases yield point. As a reducing agent, it reduces metal oxides from slag, thereby permitting the desired element, such as chromium, to sink and be recovered as an alloying agent. The wide range of properties available in cast iron is controlled primarily by carbon and silicon content (7). Silicon in cast iron reduces the stability of iron carbide and promotes the formation of graphitic carbon. As an effective graphitizer in cast iron, silicon softens the iron and improves its fluidity and machinability. For wear resistance, silicon in gray iron ranges from 0.50 to 1.50%. In ductile iron, silicon ranges from 1.50 to 3.00%. Increased percentages of silicon improve the corrosion and oxidation resistance of gray and ductile cast irons.

Table 4. Silicon Metal and Alloy Prices [a]

Metal or alloy	1970	1975	1980
ferrosilicon 50% [b]	0.33	0.71	0.92
ferrosilicon 75% [b]	0.42	0.80	1.02
silicon metal [b], 1% Fe	0.44	0.93	1.31
calcium silicon [c]	0.53	1.25	1.68
silicomanganese [c]	0.22	0.53	0.54

[a] Bulk, fob producer; from ref. 18.
[b] Dollar per kg contained silicon.
[c] Dollar per kg alloy.

BIBLIOGRAPHY

"Silicon and Silicon Alloys" in *ECT* 1st ed., Vol. 12, pp. 360–365, by J. C. Vignos, Ohio Ferro-Alloys Corporation; "Metallurgical Silicon and Silicides" in *ECT* 2nd ed., Vol. 18, pp. 125–132, by E. G. Simms, Ohio Ferro-Alloys Corporation.

1. H. Moissan, *The Electric Furnace*, 2nd ed., trans. by V. Lenker, Chemical Publishing Co., Easton, Pa., 1920.
2. M. Hansen, *Constitution of Binary Alloys*, 2nd ed., McGraw-Hill Book Co., New York, 1958.
3. F. A. Shunk, *Constitution of Binary Alloys, Second Supplement*, McGraw-Hill Book Co., New York, 1969.
4. *Metals Handbook*, Vol. 8, 8th ed., American Society for Metals, Metals Park, Ohio, 1973.
5. F. J. Schottman and P. H. Kuck, "Silicon" in *Minerals Yearbook, 1978–79*, Vol. 1, U.S. Bureau of Mines, Washington, D.C., 1981
6. "Silicon in December 1980," *Mineral Industry Surveys*, U.S. Bureau of Mines, Washington, D.C., 1981.
7. D. N. Matter, *Silicon: Its Alloys and Their Use*, Ohio Ferro-Alloys Corp., Canton, Ohio, 1961.
8. *Silicon and Ferro-Silicon Survey of World Production, Consumption and Prices*, Roskill Information Services Ltd., London, 1973.
9. *Pulverized Silvery Pig Iron*, Foote Minerals Co., Exton, Pa., 1980.
10. *The Steel-Making Process and the Function of Silicon in Steel*, Technical Bulletin, Ohio Ferro-Alloys Corp., Canton, Ohio, 1961.
11. *ASTM Specifications for Ferroalloys*, PCN #06-109078-01, American Society for Testing and Materials, Philadelphia, Pa., 1978.
12. *Metals Handbook*, Vol. 1, 8th ed., American Society for Metals, Metals Park, Ohio, 1961.
13. *A Foote Ferroalloys Progress Report*, Technical Bulletin, Foote Mineral Co., Exton, Pa., 1976.
14. R. F. Silver, unpublished work, Elkem Metals Company.
15. R. A. Clark and T. K. McCluhan, *Trans. Am. Foundrymen's Soc.* **74**, 394 (1967).
16. *Graphidox for Gray Cast Iron*, Foote Mineral Co., Exton, Pa., 1978.
17. *The Ferroalloys Association Statistical Yearbook, 1979*, The Ferroalloys Association, Washington, D.C., 1980.
18. *Am. Met. Mark.* **89**, 8 (May 14, 1981).
19. *Met. Week* **41**, 10 (Dec. 14, 1970); **46**, 4 (Dec. 15, 1975); **51**, 7 (Dec. 29, 1980).
20. N. I. Sax, *Dangerous Properties of Industrial Minerals*, 3rd ed., Reinhold Book Corp., New York, 1968.

RONALD F. SILVER
Elkem Metals Company

SILICON COMPOUNDS

SYNTHETIC INORGANIC SILICATES

Naturally occurring silicate minerals make up ca 90% of the earth's crust. These minerals are slightly soluble and are generally in dynamic chemical equilibrium with the mineral components of the aquasphere in the process of mineral breakdown and re-formation. Because of this slight solubility, concentrations of dissolved silica usually are 10–100 ppm. These minerals may be viewed as the natural analogues of the synthetic silicates, both soluble and insoluble (see Derivatives for discussion of insoluble silicates).

The soluble commercial silicates have the general formula

$$M_2O.mSiO_2.nH_2O$$

where M is an alkali metal and m and n are the number of moles of SiO_2 and H_2O relative to one mole of M_2O; m has been called either the ratio or modulus of the silicate. The most common soluble silicates are sodium silicates [1344-09-8]. Potassium silicate [1312-76-1] and lithium silicate [12627-14-4] are manufactured to a limited extent for use in special applications. Commercial forms of these materials are generally manufactured as a glass that dissolves in water to form viscous alkaline solutions. The values of m for commercial materials generally are 0.5–4.0. The most common form of soluble silicate, sometimes called waterglass, has an m value of 3.3.

Although the knowledge of soluble glass has been traced to antiquity, industrial development began in Germany in the early 19th century. Soluble sodium silicates were first produced in North America during the Civil War. They were used in laundry soaps as a replacement for rosin, which was scarce because of the war. Sodium carbonate (soda ash) and sand [14808-60-7] were fused in an open-hearth furnace to produce a glass which was then cooled, crushed, and dissolved.

These glasses, essentially the salts of a strong base and a weak acid, formed highly alkaline solutions, which made them a cost-effective functional additive builder for bar soaps (see Soap). Commercial availability of these glass solutions gradually led to other uses. The glassy nature of very concentrated solutions of sodium silicate led to the development of adhesive and binder applications after the turn of the century (see Adhesives). A third market for soluble silicates arose because these solutions contained silica, the basic building block for the insoluble silicas and silicates, in a highly reactive form. Seen as value-added sand, this material can be used to manufacture synthetic pigments and fillers (qv), silica gels and sols, and synthetic clays and zeolites (see Pigments; Molecular sieves). Soluble silicates also have miscellaneous uses in, for example, cements, coatings, bleaching, water treatment, soil stabilization, and ore beneficiation. These applications are generally based on the ability of silicates to form gels or to react with multivalent metal ions or oxide surfaces in solution. Sodium silicates or their derivatives are used in almost every large industry; they make up a versatile and stable commodity and are ranked in the top fifty commodity chemicals. The relative importance of these uses of silicates has shifted somewhat since World War II, when adhesives and soaps were the main markets. Currently, the primary uses of soluble silicates are as active sources of silica (40%), for detergency (32%), in paper and board adhesives (6%), and miscellaneous uses (22%). As the structure and chemistry of these solutions containing polymeric silicate species, ie, active SiO_2, are better understood, new and growing markets emerge, eg, enhanced oil recovery, which promise to generate continued market growth of silicates (see Petroleum, chemicals for enhanced recovery).

Structure of Soluble Silicates

The structure of silicate materials, including glasses, solutions, and crystalline forms, is of great interest to those concerned with understanding the chemical properties and applications of these materials. The 1970s have been marked by significant advances in the understanding of the glass and solution structures, particularly as a result of ^{29}Si Fourier-transform nuclear magnetic resonance spectroscopy (Ft-nmr). Because of these significant recent advances and the exhaustive coverage of the general physical properties of silicate materials (1–3), the following discussion of properties primarily addresses structure and property concepts important for understanding applications of these materials.

Silicate Glasses. Synthetic silicates and silica are made up of oligomers of the basic silicate building block SiO_4^{4-}. These orthosilicate monomers have a tetrahedral structure and may be viewed as regular tetrahedra. Silicate polygons can be used to construct more complex structures according to Pauling's rules, ie, Si—O—Si bonds are permitted at the polygon corners only. When these rules are followed, five general structural categories, which are outlined in Table 1, are possible. The Q^s structure notation refers to the connectivity of silicons (4). The value of the superscript s, which is representative of a given silicon atom, is the number of nearest-neighbor silicon atoms (see Fig. 1). Many specific modifications are available within these categories, especially in the framework structures. Connections of the monomer structure SiO_4^{4-} at all vertices leads to the completely condensed structure $(SiO_2)_n$, ie, quartz. This silica structure is not close-packed, owing to Pauling constraints; therefore, there are alternate holes in half of the tetrahedral positions. In the formation of the soluble sodium silicate glasses, Na^+ and O^{2-} are introduced into the quartz network and break Si—O—Si bonds, ie, bridging oxygens, to form Si—O$^-$ sites, ie, nonbridging oxygens. The Na^+ ions are thought to be distributed nonuniformly in interstices in the disordered network, and they produce regions rich in SiO_2 polymers and other regions rich in cations (5). The basic mineral-silicate structures are expected to exist in short-range order in a random, polymer order-disorder equilibrium model not unlike the flickering-cluster model of water structure. Thus, in the glass state, a complex distribution of silicate anions is envisioned as well as an equally complex distribution of alkali cations, presumably in the interstitial cages bounded on average by a number of oxygen atoms equal to the cation coordination number. Thus, the physical and chemical properties of these glasses are quite sensitive to the modulus of the glass and to the ion size or coordination number of the modifying cation (6). Melt or glass containing an SiO_2: Na_2O ratio of 1, ie, sodium silicate [15915-98-7], may be expected to possess a high

Table 1. A Summary of the Five Silicate Structural Categories

| Silicate type | Unit structure | Mineral example | | | Q^s structures[a] |
		Name	CAS Registry No.	Formula	
discrete, noncyclic					
orthosilicate	SiO_4^{4-}	zircon	[14940-68-2]	$ZrSiO_4$	Q^0
pyrosilicate	$Si_2O_7^{6-}$	thortveitite	[17442-06-7]	$Sc_2Si_2O_7$	Q^1Q^1
discrete, cyclic					
cyclic tetramer	$Si_3O_9^{6-}$	benitoite	[15491-35-7]	$BaTiSi_3O_9$	$(Q^2)_3$
cyclic hexamer	$Si_6O_{18}^{12-}$	beryl	[1302-52-9]	$Be_3Al_2Si_6O_{18}$	$(Q^2)_6$
infinite chains					
pyroxenes	$(SiO_3^{2-})_n$	diopside	[14483-19-3]	$CaMg(SiO_3)_2$	$(Q^2)_n$
amphiboles	$(Si_4O_{11}^{6-})_n$	tremolite	[14567-73-8]	$Ca_2Mg_5(Si_4O_{11})_2(OH)_2$	$(Q^3Q^2)_n$
sheets	$(Si_2O_5^{2-})_n$	talc	[14807-96-6]	$Mg_3[(OH)_2|Si_4O_{10}]$	$(Q^3)_n$
framework	$(SiO_2)_n$	silica, quartz	[7631-86-9]		$(Q^4)_n$

[a] End groups or surface silicons for more condensed structures are ignored.

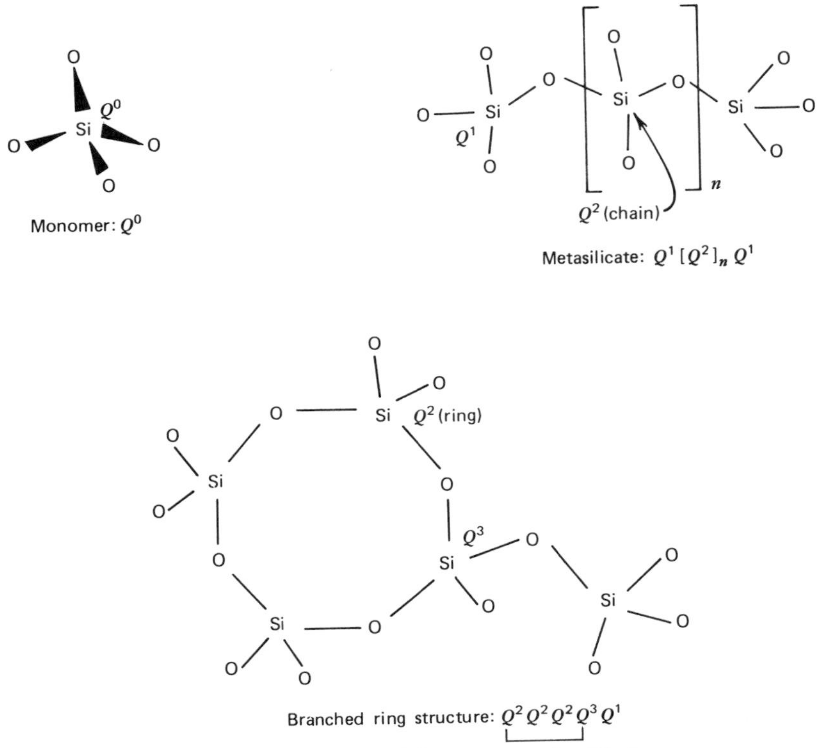

Figure 1.   Examples of Q^n structures for silicates.

proportion of $(SiO_3^{2-})_n$ chains. At $n = 2$, the presence of sheets might predominate; however, little if any convincing evidence has been shown for any clear predominance of these structures. The possible structures in melts of different ratios are discussed in refs. 7–9.

Silicates in Solutions.   The distribution of silicate species in solution has long been of interest because of the wide variations in the typical properties of these solutions when the ratio varies. Studies in the 1920s led to a dual description of silicate components (10). It was pointed out that the analysis of the sodium silicate must be clearly stated and that the terms sodium silicate or waterglass should not be used indiscriminately. Prior to 1928, sodium silicate solutions were thought to be composed of products of hydrolysis, colloidal silicic acid, hydroxide ion, and sodium ion. However, it was systematically demonstrated, primarily through the analysis of the colligative properties of solutions with various ratios of $SiO_2:Na_2O$, that the silicate solutions also contain what was called crystalloidal silica (10). It was assumed that this crystalloidal silica was analogous to the simple species then thought to be the components of the known crystalline sodium silicates, charged aggregates of these unit structures and silica (ionic micelles), or definite complex silicate ions.

Subsequent research in light scattering showed that stable silicate solutions do not contain very large particles; however, aggregation of particles was indicated in high ratio solutions at high solids levels (11). Polymers in the ratio range of commercial silicate solutions were studied by preparing and identifying trimethylsilyl derivatives (12). This work presented fairly strong evidence for the presence of a variety of silicate structures even in highly alkaline 1 M solutions at an m value or ratio of 0.5. The presence of highly polymerized species was indicated in the very low ratio solutions; however, these conclusions were weakened by the possibility that the derivatization method itself could induce polymerization as a result of HCl formation during the reaction (13).

Until the mid 1970s, these and other indirect methods of studying the silicate species in solution indicated that these solutions are a complex mixture of silicate anions with varying degrees of polymerization in a dynamic equilibrium. However, these methods are all indirect and only recently has a more direct method, ie, the use of Ft-nmr spectroscopy, been perfected. The nmr spectra for a range of silicate ratios were measured, and the various silicon centers could be identified and relative concentrations of various types of silicon could be estimated (4,14) (see Fig. 2). Others have used this method to develop more detailed analyses of the individual species of silicate solutions (15–16).

Species determination in soluble silicate solutions has been reviewed and, although it was determined that ^{29}Si nmr studies yield the greatest potential for understanding the quantitative makeup of equilibrium silicate solutions, the use of a careful trimethylsilylation technique offers the best results in the study of dynamic systems (17). The results, albeit semiquantitative, reveal that species equilibrium attainment is rapid in alkaline solution above a pH value of ca 10, unless the solutions contain particles of colloidal dimensions. In acidic solutions at ca pH 2, equilibrium is attained slowly. Polymerization of smaller species appears to occur in a sequential manner with a given-sized polymer species at first increasing its size and then disappearing, presumably because of its inclusion in higher-order polymers. Depolymerization appears rapid, since crystalline Na$_2$SiO$_3$ [6834-92-0, 15915-98-7] and Na$_2$H$_2$SiO$_4$.9H$_2$O [13517-24-3] yield equivalent species in solution very soon after dissolution.

Based on information to date, the following simplified representations of the polymeric species in silicate solutions have been developed, ie, for H$_{2x}$Si$_y$O$_{(2y + x)}$ or

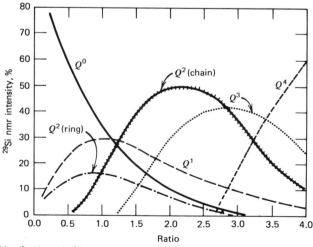

Figure 2. Distribution of silicon centers in soluble silicate solutions from ^{29}Si Fourier-transform-nmr spectroscopy (4,14).

any of its ionized forms, the following is true:

number of nonbridging oxygens ($\geqslant$SiO$^-$ or $\geqslant$SiOH) = $2\,x$
number of bridging oxygen atoms = $2\,y - x$
number of closures = $y - (x + 1)$

The mean number of shared corners (connectivity)

$$s = 2\,(2\,y - x)/y = \text{mean } s \text{ value on } Q \tag{1}$$

When the available information on the connectivity of silicate species in solution was matched to the number of silicon atoms in the species, the empirical relation

$$s = 7\,(y - 1)/(2\,y + 3) \tag{2}$$

was developed. It predicts that as the number of SiO_2 units in an oligomer increases, the connectivity tends towards 3.5. This result suggests that even very large particles possess a fair proportion of internally trapped, nonbridging oxygens or active sites on the surface of the particles.

Crystalline Soluble Silicates. The most common crystalline soluble silicates are of the metasilicate family $Na_2O.SiO_2.nH_2O$. The anhydrous sodium metasilicate Na_2SiO_3 contains SiO_2 chains, whereas the hydrates ($Na_2H_2SiO_4.XH_2O$, where $X = n - 1$) contain silicate monomer (18). The structure variations in the series with $X = 4$, sodium metasilicate pentahydrate [10213-79-3], 5, sodium metasilicate hexahydrate [35064-64-3], 7, sodium metasilicate octahydrate [27121-04-6], and 8, sodium metasilicate nonahydrate, appear to result primarily from the order and coordination of the hydrated sodium ion. However, some differences are observed in symmetry and interatomic differences in the $Si(OH)_2O_2^{2-}$ groups. Only the anhydrous and tetrahydrate ($X = 4$) forms are of general commercial importance.

In addition to the above metasilicates, there are five known mineral structures: natrosilite [56941-93-6], $Na_2Si_2O_5$, magadiite [12285-88-0], $Na_2Si_{14}O_{29}.11H_2O$, kenyaite [12285-95-9], $Na_2Si_{22}O_{45}.10H_2O$, makatite [27788-50-7], $Na_2Si_4O_9.5H_2O$, and kanemite [38785-33-0], $NaHSi_2O_5.3H_2O$ (19). These silicates appear to have layer structures and exhibit some degree of inner crystalline reactivity.

Dissolution

The dissolution of soluble silicates is of great commercial importance. The rate of solution is generally thought to depend on the glass ratio, concentration, temperature, pressure, and glass particle size. The dissolution of silicate glass involves a two-step mechanism (20):
Ion exchange

$$\geqslant\text{SiONa} + H_2O \rightleftharpoons \geqslant\text{SiOH} + Na^+ + OH^- \tag{3}$$

Network breakdown

$$\geqslant\text{SiOSi}\leqslant + OH^- \rightleftharpoons \geqslant\text{SiO}^- + HOSi\leqslant \tag{4}$$

Thus, the removal of silica from the glass follows the alkali. The rate of SiO_2 extraction increases as the pH value rises above 9. Alkali extraction occurs more readily at pH values below the pK_a of the glass-surface silanol groups. If a sufficient hydroxide activity is not generated in the ion-exchange step, the network breakdown does not occur. In this case, glass leaching results. The transition state in the network-breakdown process is thought to include five-coordinate silicon. The presence of Na^+ in the glass network is also thought to have a retarding effect on the overall rate of solution, most likely resulting from hindrance of the nucleophilic attack. The rate of SiO_2 removal from a potassium silicate glass depends on the rate of removal of alkali and not on the quantity removed (21). Smaller cations with higher charge densities produce less soluble silicate glasses. Thus, the solubility trend is $K^+ > Na^+ > Li^+$. The presence of multivalent metal ion impurities, eg, Al^{3+}, Ca^{2+}, or Fe^{3+}, in the alkali silicate reduces the dissolvability of the glass. There is a correlation between the ratio of free and total silanol groups and the dissolvability of a glass (22). This ratio varies roughly linearly, as different metal ions are added to Na-, K-, or Li-based 4.0-mol-ratio alkali-silicate glasses at the expense of SiO_2. The addition of water to glass depresses the ratio rapidly.

A novel tower dissolver operated at atmospheric pressure for dissolving glass has been described (23). The rate of solution of glass is independent of dissolving liquor concentration and rate of circulation. It is suggested that the principal factors in commercial silicate-glass dissolution are temperature, glass composition, and surface area. It is well recognized that the glass must be sized to avoid a phenomenon called a sticker, where the dissolving glass mass solidifies. Recent studies of the rate of dissolution of a 2.0-ratio glass into high solids solutions indicate that, for the commercial sizes, the rate of glass dissolution (expressed as kg dissolved/(kg glass·h)) is independent of the initial glass particle size (24). Also, under all conditions where the Na^+ ion exchange is suppressed, eg, by increased Na^+ activity in the dissolving liquor, only linear increases in percent solids in solution are observed with time. The rate of solids increase in solution appears to be inversely related to sodium ion activity.

Polymerization of Silicates in Solution

Polymerization and depolymerization of silicate anions has been and continues to be a subject of great interest to technologists in diverse areas, eg, gel and catalyst manufacture, detergency, geothermal energy (qv), and limnology (see Surfactants and detersive systems). The following is a summary of the current understanding; however, the reader is referred to refs. 25 and 26 for more detailed discussions.

The complex silanol condensation process may be represented empirically as

$$\gtrsim SiOH + \gtrsim SiO^- \underset{K_2}{\overset{K_1}{\rightleftharpoons}} \gtrsim SiOSi \lessgtr + OH^- \tag{5}$$

It appears that this condensation occurs most readily at a pH value equal to the pK_a of the participating silanol group. This empirical representation becomes less valid at pH values greater than ca 10, where the rate of the depolymerization reaction K_2 becomes significant, and at very low pH, where H^+ exerts a catalytic influence. The pK_a value of monosilicic acid is 9.91 ± 0.04, and it has been shown that the pK_a value decreases to 6.5 for high polymers (27–28). This result is consistent with another report of a pK_a of 6.8 ± 0.2 for surface silanol groups of silica gel (29). Thus, the acidity of silanol groups increases as the degree of polymerization of the anion increases. The exact relationship between the connectivity of the silanol silicon and acidity is not known.

Silicate polymerization in dilute solutions at pH values up to 10 is sensitive to pH and other factors that generally influence colloidal systems, eg, ionic strength, dielectric constant, and temperature. Larger particles grow at the expense of smaller particles, especially at higher pH values where the latter dissolve more readily. This results from the tendency of smaller particles to condense at the surface of the larger particles. For example, if colloidal silica particles are dispersed in a soluble silica solution, polymerization occurs primarily on the particle surfaces. Lower pH values and higher ionic strengths lead to the growth of smaller particles. If the concentration of SiO_2 is sufficiently high, ie, ca 1 wt %, interparticle aggregation and ultimately network formation, ie, gelation, occurs, yielding a continuous structure throughout the medium. This structure initially encompasses the whole system and appears uniform but, in time, further condensation occurs with gel shrinkage and water release, ie, syneresis.

In more concentrated solutions, network formation sufficient to produce a gel point occurs at a lower total conversion of low molecular weight polymers to larger ones (30). Therefore, the pH values for maximum gel rate are lower for more dilute silicates than for more concentrated solutions, given a constant initial degree of polymerization. Also, the second-order reaction mechanism predicts the general pH changes that are observed when polymerization occurs, a prediction based on the initial pH of the solution and knowledge of the acid dissociation constant. A quantitative description of the polymerization-depolymerization mechanism requires a more complete representation of the specific anions formed in solution and greater knowledge of the activity variations of the potential condensation sites on the silica in solution.

Because of their polymeric nature, compositionally equivalent solutions of soluble silicates may have observably different physical properties and chemical reactivities. This effect seems to be true for the very concentrated solutions sold commercially, apparently because of the viscous nature of these solutions, which prevents more rapid attainment of an equilibrium state.

Chemical Activity

Soluble Silicate Polymer–Metal Ion Interactions in Solution. Since the reaction of metal ions in solution with polymeric silicate species may be thought of as an ion-exchange process, it might be expected that the silicate species as ligands would exhibit a range of reactivities toward cations in solution (31). It has been shown that silica-gel surfaces form complexes with multivalent metal ions in a manner that indicates a correlation between the ligand properties of the surface OH groups on silica gel and metal-ion hydrolysis (32–33). For Cu^{2+}, Fe^{3+}, Cd^{2+}, and Pb^{2+},

$$pK_1^s = 0.09 + 0.62 \, p^*K_1 \tag{6}$$

K_1^s = stability constant of surface complex

$^*K_1$ = metal-ion hydrolysis constant

It has been suggested that metal-ion adsorption is initiated at a pH value corresponding to surface nucleation, which seems to relate to the reduction of cation-solvent interactions leading to conditions favorable to the adsorption of hydrated metal ion from solution (34). It is suggested in ref. 34 that metal-ion hydrolysis is required, whereas direct participation by unhydrated ions is proposed in ref. 33. Other studies suggest that cations adsorb onto silica-gel surfaces as a result of hydroxyl-ion adsorption, which drags an equivalent amount of cations to the surface (35).

Solution activities of Ca^{2+}, Mg^{2+}, and Cu^{2+} decrease at a given pH value to a greater extent in the presence of SiO_2 obtained from 2.0- and 3.8-ratio silicate solutions than they do with solutions of 0.5-ratio sodium orthosilicate (36). Thus, soluble silicates at very high degrees of association appear to interact with metal ions in solution in a manner analogous to silica gel and, as the degree of polymerization decreases, these silicates species exhibit decreased interaction with cations. These results are consistent with the following general observation: silica suspended in a solution of most polyvalent metal salts begins to adsorb metal ions when the pH is raised to within 1–2 pH units below the pH at which the polyvalent metal hydroxide precipitates (37).

It is likely that the increased acidity of the larger polymers leads to this reduction in metal-ion activity through the more facile development of active sites in these polymers. Thus, one might expect a range of instantaneous interaction constants between metal ions and soluble-silicate polymer sites that is a function of the silicate polymer size distribution. Also, the process of ion interaction with a silicate anion, which leads to a reduction in pH, usually produces larger anions which, in turn, leads to increased interaction. Therefore, the metal-ion distribution in an amorphous metal ion-silicate particle might be expected to be nonhomogeneous. It is not known whether this is the case, but it is clear that interactions of metal ions and silicate can be thought of as making up a complex process related to the hydrolysis of the metal ion, since the reactions are comparable. The products of high concentration reactions of soluble silicates with metal salts at ambient temperature have generally been viewed as complex metal-hydroxide and silica-gel mixtures because of the presence of OH^- and the similarity of hydroxide to polymer-silicate surface sites.

Effect of Soluble Silicates on Oxide–Water Interfaces. The adsorption of ions at clay, mineral, and rock surfaces is an important factor in many natural and industrial processes. Silicates are adsorbed on oxides to a far greater extent than would be predicted from their concentration (38). This adsorption maximum at a given pH value is insensitive to ionic strength, and the pH value of maximum adsorption occurs near the pK_a value of the orthosilicate species. A correlation is observed between the pH value at maximum adsorption of a series of weak acid anions and the pK_a values of the weak acids. This observation indicates that the presence of both dissociated and undissociated forms of these species is required. The adsorption of silicate species, however, is observed to be greater at lower pH values than the simple acid-base equilibria would predict. This was suspected to be caused by ion-surface interactions or ions already on the surface. It may well be that this deviation results from silica polymerization and varying silicate anion acidity. Similar behavior for the adsorption of aqueous silica onto a γ-Al_2O_3 surface has been observed (39). This suggests that lateral interactions involving adsorbed silicate ions are important. These interactions lead to an amorphous aluminosilicate on the surface of the γ-Al_2O_3. The addition of divalent metal ions tends to reduce silicate adsorption.

The addition of polymeric silicate anions to oxide mineral suspensions generally increases the magnitude of the negative surface charge of the particles in suspension. The influence of the soluble silicates on the surface charge of a model oxide–water system has been studied (40) (see Dispersants). When silicate solutions are used as a replacement for NaOH to adjust the pH values of suspensions of ground quartz in 0.1-mM $PbCl_2$, the pH value of maximum positive charge shifts to lower pH values and the pH-range of positive charge narrows. These effects are more pronounced for more polymeric silicates. A similar reduction of the influence of multivalent cations on the quartz surface occurs for 0.1-mM $FeCl_3$ and mixtures of Fe^{3+} and Pb^{2+} with and without the addition of 400 ppm Ca^{2+} and Mg^{2+} hardness. The influence of silicate polymers on iron oxide sol surfaces also has been studied; at a given pH value and silica concentration, the effectiveness of soluble silica in discharging and recharging the sol surface increases with the modulus of the original sodium silicate sample (41). Thus, it appears that soluble silicates not only specifically adsorb onto oxide surfaces, but they also play a significant role in maintaining a negative surface charge on oxide surfaces in the presence of cations that could reverse the surface charge.

General Characteristics

General characteristics that are relevant to various uses of soluble silicates are the pH behavior of solutions, rate of water loss from silicate films, and dried-film strength. The pH values of silicate solutions are a function of composition and concentration. They are quite alkaline, since they are solutions of a salt of a strong base and weak acid. The solutions exhibit up to twice the buffering action of other alkalies, eg, phosphate. An approximately linear relationship exists between the modulus of the sodium silicate in solution and the maximum solution pH value between $m = 2.0$ and 4.0.

$$pH = 13.4 - 0.69 \, m \qquad (7)$$

This relationship allows for an estimation of the modulus of a relatively pure concentrated solution. The rate of water loss from silica solutions and films is greater at higher values of m. If films or solutions are dried at a given temperature and humidity, a change in condition initiates further drying or rehydration, depending on the case; drying is not sufficient to obtain insolubility of a silicate film. Hydrated glass films made from silicates with higher m values or from those containing metal ions that decrease solubility rehydrate less rapidly. Finally, the dried-film characteristics have been known to depend on the value of the glass modulus. It is generally observed that higher m values produce more brittle films.

Manufacture and Processing

The soluble silicate glasses are manufactured usually in oil- or gas-fired open-hearth regenerative furnaces. The glass is obtained by reaction of quartz sand and sodium carbonate (soda ash) at a temperature sufficient to provide a reasonable quartz dissolution rate in the molten batch and a manageable melt viscosity. The liquidus diagram for sodium silicate glasses and lines of constant melt viscosity are shown in Figure 3. The reaction rate of quartz with Na_2CO_3 is controlled by silica diffusion and varies inversely with the square of the radius of the quartz particle (42). As Na_2CO_3 melts and envelops the sand grains, the slow process of quartz network breakdown and diffusion into the melt occurs. From thermogravimetric analysis data, it has been shown that the kinetics of the reaction between Na_2CO_3 and SiO_2 may be adapted to a modified Ginstling and Brounsthein model (43–44). The melts produced are very corrosive toward refractory materials and care is required in furnace design. Where electric power is available and costs are low, electric melting furnaces can be used satisfactorily (45).

The sand and soda ash required for the manufacture of the soluble silicate must be of high purity. Typically, a no. 1 grade of glass sand containing no more than 300 ppm iron and a medium density soda ash, which is obtained from mined Wyoming

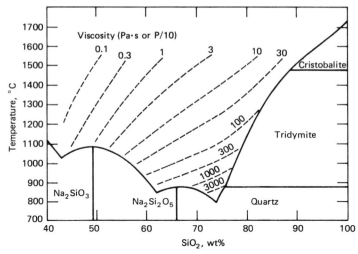

Figure 3.   The viscosity and liquidus curves for molten sodium silicates (1). Courtesy of PQ Corp.

trona ore, are used. The typical composition of commercially available 3.30-ratio sodium silicate glasses and the trace impurities normally observed in 3.3-ratio solutions are shown in Tables 2 and 3, respectively.

Once the glass is produced in the furnace, it can be drawn and formed into solid lumps or drawn directly into a rotary dissolver. The glass lumps can also be dissolved in pressure dissolvers. The concentrations and ratios of solutions are monitored during manufacture with the use of alkali-gravity-viscosity (AGV) charts (see Fig. 4). From these charts, ratio and solids content of a solution may be determined from an analysis of the density and Na_2O content. The lump glass is sold directly to those who have their own dissolving equipment, or it is ground to powders of various particle size distributions. Concentrated solutions of glass are generally slightly turbid because

Table 2.   Range of Composition of Typical Sodium Silicate Glasses (3.3 SiO_2:Na_2O) Manufactured at Different Plants

Assay	Wt %	
	Low	High
Na_2O	23.21	23.89
SiO_2	75.36	76.00
K_2O	0.00	0.10
Fe_2O_3	0.005	0.030
TiO_2	0.004	0.052
Al_2O_3	0.15	0.51
CaO	0.032	0.17
MgO	0.004	0.10
CdO	0.00012	0.0022
NiO	0.00008	0.0026
SO_3	0.008	0.19
CO_2	nil	0.23
Cl	0.025	0.12
ignition loss	0.03	0.36
ratio by wt SiO_2:Na_2O	3.154	3.249

Table 3. Range of Trace Impurities in Typical Sodium Silicate Solutions (3.3 $SiO_2:Na_2O$) Manufactured at Different Plants

Assay	Impurities content[a], ppm[b]	
	Low	High
F	6.7	9.5
Cl	130	1900
SO_4	<160	1700
N	0.1	44
As	<1	<1
Hg, ppb	<0.26	2.5
Pb	0.17	0.60
Cd, ppb	<10	21
Fe	36	120
Mg	4	26
Ca	<1	76
Al	50	220
P	<18	<18
V	<0.3	0.8
Cr	<0.3	1.0
Ni	<0.3	0.3
Co	<0.3	<0.3
Zn	<0.6	2.8
Cu	<0.6	1.1
Bi	<25	<25
Sr	<0.2	1.5
Ba	<0.2	2.8
Mn	0.1	1.8
Sn	<60	<60
Sb	<15	<15
Se	<20	<20

[a] Where < value is shown, the value indicates the detection limit of the method.
[b] Ppm except where noted otherwise.

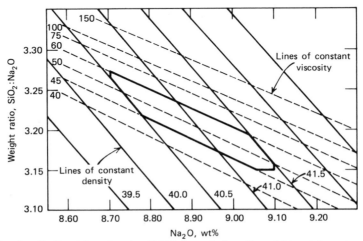

Figure 4. An alkali-gravity-viscosity (AGV) control chart for a 3.22-weight ratio sodium silicate solution. The area enclosed in the darkened line represents the range of properties for a typical product. Courtesy of PQ Corp.

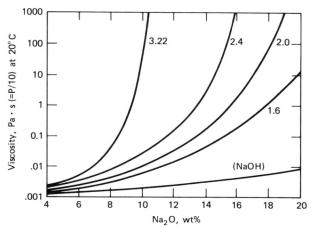

Figure 5.   Viscosities of sodium silicate solutions at various ratios vs percent solids. Courtesy of PQ Corp.

of the presence of either very fine unreacted sand particles or amorphous-particulate impurities containing metal ions and silica. This particulate can be removed readily by settling or, in most cases, filtration through diatomaceous earth, which yields brilliantly clear solutions. Silicate solutions up to a molar ratio of ca 2.65 can be manufactured in an autoclave at ca 160°C by dissolving finely ground sand in a NaOH solution. Higher ratios can be produced with amorphous silica. Higher ratio silicate solutions can also be adjusted to make lower ratio solutions. The choice of the process to make a given ratio is generally based on economics. The solids level of the solutions sold commercially is normally determined by the maximum reasonable viscosity that can be handled readily. Figure 5 shows how the viscosity of silicate solutions increases with increases in the solids level. The more siliceous ratios show rather abrupt increases in viscosity. This property is of value in certain adhesive applications. As a higher ratio solution evaporates, the viscosity continues to increase to the point where solid solutions form. These materials have commercial significance, since the rate of solution of soluble silicate solids is a function of particle size, ratio, and water content. These dried silicate solutions, or hydrous silicates containing ca 18 wt % water, are stable enough to be handled commercially, yet they dissolve significantly faster than their ground-glass counterparts. Thus, as shown in Table 4, powdered soluble silicate products can be made in a wide range of ratios and solution rates to meet the demands of diverse applications. The hydrous silicates are manufactured in drum dryers or spray towers.

Crystalline metasilicates are manufactured by processing high solids solutions of 1.0 ratio sodium silicate or by direct fusion of sand and soda ash followed by grinding and sizing. The corrosive nature of the latter reaction mixtures on refractories and the difficulty in effecting a complete reaction make this process less satisfactory for a high quality product. For this purpose, solutions at the metasilicate ratio are dried in a rotary moving bed or a fluid-bed dryer. The metasilicate solution in this process is sprayed onto the bed and forms beads that are screened as they exit the dryer. The fines are recycled to provide seed for further growth. A uniformly sized, readily soluble, anhydrous sodium metasilicate (ASM) product is thus made. The only readily available

Table 4.   Solution Rates of Amorphous Sodium and Potassium Silicate Powders (3 Parts Water + 1 Part Silicate Powder)

Silicate, wt % ratio	Particle (Tyler screen) sizing, μm (mesh)	Time needed to dissolve at 25°C			Time needed to dissolve at 50°C		
		50%	75%	100%	50%[a]	75%	100%
3.22 ratio sodium, anhydrous glass	230 (65)	60 h			15% in 30 min		
3.22 ratio sodium, hydrated (18.5 wt % H_2O)	149 (100)	19 min	45 min		54 s	76 s	100 s
2.00 ratio sodium, anhydrous glass	230 (65)	10 h	70 h		17 min	1 h	
2.00 ratio sodium, hydrated (18.5 wt % H_2O)	149 (100)	27 s	54 s		15 s	22 s	29 s
2.50 ratio potassium, anhydrous glass	230 (65)	60 min	7.5 h	48 h	12 min	45 min	

[a] Unless noted.

commercial hydrate of sodium metasilicate, ie, $Na_2H_2SiO_4 \cdot 4H_2O$, is manufactured by preparing a solution with this composition at a temperature above 72.2°C and allowing the mass to cool to yield the hydrate.

Finally, sodium orthosilicates are produced either by blending ASM and NaOH beads or by fusion and grinding similar to the direct manufacture of ASM. The relationships among these processes are shown in Figure 6.

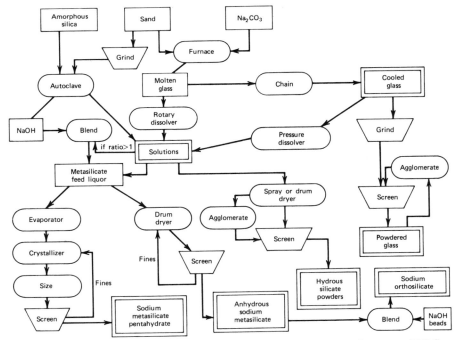

Figure 6.   Various manufacturing routes for commercial sodium silicates. Courtesy of PQ Corp.

Potassium silicates are manufactured in a similar manner by the reaction of K_2CO_3 and sand. However, crystalline products are not manufactured and the glass is generally sold as a flake. A 3.90-ratio potassium silicate flake glass can be dissolved readily in water initially at ca 88°C without pressure by incremental addition of glass. The exothermic heat of solution causes the water temperature to rise to the boiling point. Because of the insolubility of the lithium silicate [10102-24-6] glasses, these silicate solutions are generally prepared by dissolving silica gel in a LiOH solution or mixing a silica sol with LiOH.

Commercial Products

The average composition and relevant properties or characteristics of commercial soluble silicates are shown in Table 5. The largest volume of these materials is sold as liquids, which are differentiated by the ratio of SiO_2 to Na_2O and by specific gravity. The powdered forms are desired as an ingredient in dry blended products, to control silicate reactivity, for convenience in handling and storage, or to increase silicate solution concentrations above the ranges available in liquid form. Sodium orthosilicate [15859-24-2] and metasilicates are valued for their high alkalinity content as well as for the preceding reasons. The Federal specifications for commercial materials are listed below:

Commodity	Specification no.
sodium metasilicate	O-S-604 D
type I, pentahydrate	
type II, anhydrous	
sodium orthosilicate	P-S-651 E
sodium silicate solution	O-S-605 D

Handling and Storage.   The liquids are available in drums or bulk quantities. Bulk shipments are available by truck, rail, or tankers. Gravity or pump unloading may be utilized when transferring large volumes of silicate solution. Storage is recommended in closed but vented carbon-steel tanks with capacities of 45–57 m^3 (45,000–57,000 L) for rail and not less than 23 m^3 (23,000 L) for tank-truck deliveries. Powdered silicates are available in bags or drums, and a pressure-differential bulk unloading system is recommended for bulk quantities. When bulk storage is used, a minimum storage capacity for 28,000 kg is advisable, and the storage bins should have provisions for humidity control because of the hygroscopic nature of these materials. Although the packaging of most bagged products is designed to provide humidity protection for reasonable periods of time, additional protection should be provided if unusually high humidities or long storage times are expected.

Regulatory Status.   *Additive to Food and Potable Water.*   Sodium silicate is considered generally recognized as safe (GRAS) by the U.S. FDA if it migrates to food from cotton and cotton fabrics used in dry-food packaging or to food from paper and paperboard products (46). Sodium silicate is recognized by the FDA as a secondary

Table 5. Typical Commercial Sodium and Potassium Silicates

Commercial silicates	Wt ratio[a] $SiO_2:M_2O$	Modulus[a] $SiO_2:M_2O$	M_2O[a], wt %	SiO_2, wt %	Softening point[b], °C	Flow point[c], °C	H_2O, wt %	Baumé[d] (at 20°C)	d^{20}_{20}, g/cm³	Viscosity (at 20°C), Pa·s[e]	pH
anhydrous glasses											
sodium silicates	3.22	3.33	23.5	75.7	655	840					
	2.00	2.06	33.0	66.0	590	760					
potassium silicates	2.50	3.92	28.3	70.7	700	905					
hydrated amorphous powders											
sodium silicates	3.22	3.33	19.2	61.8			18.5				
	2.00	2.06	27.0	64.0			18.5				
solutions											
sodium silicates	1.60	1.65	19.7	31.5				58.5	1.68	7.00	12.8
	2.00	2.06	18.0	36.0				59.3	1.69	70.00	12.2
	2.50	2.58	10.6	26.5				42.0	1.41	0.06	11.7
	2.88	2.97	11.0	31.7				47.0	1.49	0.96	11.5
	3.22	3.32	8.9	28.7				41.0	1.39	0.18	11.3
	3.75	3.86	6.8	25.3				35.0	1.32	0.22	10.8
potassium silicates	2.50	3.93	8.3	20.8				29.8	1.259	0.04	11.30
	2.20	3.45	9.05	19.9				30.0	1.261	0.01	11.55
	2.10	3.30	12.5	26.3				40.0	1.381	1.05	11.70
	1.80	2.83	16.4	29.5				47.7	1.490	1.30	12.15
crystalline solids											
sodium orthosilicate		0.50	60.8	28.8			9.5				
anhydrous sodium metasilicate		1.00	51.0	47.1			2.0				
sodium metasilicate pentahydrate		1.00	29.3	26.4			42.0				
sodium sesquisilicate [1344-09-8]		0.67	36.7	24.1			38.1				

[a] M represents Na or K.
[b] Viscosity reaches 4 MPa·s (4 × 10⁷ P).
[c] Viscosity reaches 10 kPa·s (10⁵ P).
[d] To convert Be° to sp gr, divide 145 by (145 − Be°).
[e] To convert Pa·s to P, multiply by 10.

870

direct food additive when it is used in boiler water for food-contact steam (47). Sodium silicate also has FDA unpublished GRAS status as a corrosion preventative in potable water (see Corrosion and corrosion inhibitors). The safety of it in these and other food-related applications has been recently reviewed for the FDA (48). The use of sodium silicate in canned emergency drinking water at concentrations of up to 100 ppm is also described in a Federal specification (49).

Transportation and Disposal. Only highly alkaline forms of soluble silicates are regulated as DOT hazardous materials for transportation and, when discarded, they are classified as a Resource Conservation and Recovery Act (RCRA) hazardous waste. Typical members of this class are sodium silicate solutions with silica-to-alkali ratios <1.6 and solid sodium silicates with silica-to-alkali ratios <1.0. The recommended treatment and disposal method for soluble silicates is to neutralize them with aqueous acid (6 M HCl or equivalent) and to dispose of the solids, principally silica gel, in a landfill, according to local, state, and Federal regulations. The neutral liquid, which is a salt solution, can be flushed into sewer systems (50).

Consumer Products. The Consumer Product Safety Commission's (CPSC's) labeling criteria under the Federal Hazardous Substances Act is based on the results of biological testing. In the absence of specific biological test data, CPSC requires specific cautionary statements for consumer products containing certain types and amounts of sodium silicates (51).

Occupational Health and Safety. There are no specific OSHA exposure limits for sodium or potassium silicates (52). A prudent industrial exposure standard could range from the permissible exposure limit (PEL) for inert or nuisance particulates up to the PEL for sodium hydroxide, depending on the rate of solution and the degree of airborne material. Material safety data sheets are available from silicate producers and should be consulted for specific handling precautions, recommended personal protective equipment, and other important safety information for specific products (see Health, Safety, and Environmental Aspects).

Production and Economic Aspects

Worldwide silicate production is estimated to be near 3.0×10^6 metric tons per year (3). 1981 prices for typical silicate products are listed in Table 6 (53).

Table 6.   1981 Prices for Silicate Products, $/t

Product	Price	
	Bulk	Bags
sodium silicate glass (3.30 ratio)	220	375
sodium silicate glass (2.00 ratio)	280	500
anhydrous sodium metasilicate	450	480
sodium metasilicate pentahydrate	325	360
sodium orthosilicate (technical grade, anhydrous)		481

The price for a liquid product on a 100% solids basis is generally quite close to the price for the glass. International trade of these products, especially the liquids, is inhibited by the costs of transportation relative to actual product costs. The U.S. Bureau of Census production and shipment statistics indicate a modest increase, as shown in Table 7 over a 20-yr period. The Producer Price Index (PPI) shows the costs of the raw materials for the manufacture of soluble silicate, ie, the price of energy, sand, sodium carbonate, and industrial chemicals in general are rising faster than those of the soluble silicate products (see Table 8) (54).

The main producers of soluble silicates in the United States, in order of their estimated capacities, are The PQ Corporation, Diamond Shamrock, DuPont, PPG Industries, W. R. Grace, and Chemical Products. There are ca 30 soluble silicate plants in the United States. Production capabilities are more than adequate for current sodium silicate demands. However, costs of emissions control and rapidly rising raw materials costs combined with limited growth and the general excess capacity have introduced great pressure on profitability. This situation has fostered increased operating efficiencies at some existing plants and could well lead to additional shutdowns of less efficient ones (55).

Despite this somewhat negative picture, the market for these materials is expected to at least continue its modest growth and may experience enhanced growth. Enhanced growth for this environmentally acceptable material or its derivatives might be expected in any of a number of markets, eg, mineral and oil recovery, detergency, catalysis, and binder applications.

Analysis and Test Methods

Standard test methods for the soluble silicates are given by ASTM D 3400-75 and TAPPI (Tentative Standard T632ts632) and, most recently, by ANSI/AWWA (American Waterworks Association) Standard B 404-80 for aqueous silicate solutions and ASTM (ANSI/ASTM Standard D 501-67, Sections 46–59) for the determination of Na_2O, SiO_2, H_2O, and insoluble matter in the alkaline crystals. Atomic absorption and plasma-emission spectroscopy may be used effectively to determine the levels of trace impurities in soluble silicates. Silica may be determined by the usual gravimetric technique, but a satisfactory and rapid volumetric method of titration with $2 N$ HCl takes advantage of the following reaction with fluoride ion:

$$SiO_2 + 6 F^- + 4 H^+ \rightleftharpoons SiF_6^{2-} + 2 H_2O \tag{8}$$

Alkali is usually determined by acid titration. Recently, a thermometric titration method was reported that involves the reaction of the silicate solution first with HCl to measure Na_2O and then with HF to measure the precipitated SiO_2 (56). Lower concentrations of silica may be determined with the silicomolybdate colorimetric method. Carbon dioxide, density, and viscosity can be determined by standard methods; however, for density and viscosity, the industry uses the pragmatic measures of °Baumé and Stormer seconds, respectively. Crystallinity of silicates may be determined quantitatively by x-ray diffraction, and qualitative determinations by x-ray diffraction patterns may be identified by normal search methods in the Joint Committee for Powder Diffraction Standards (JCPDS) file.

Table 7. Production and Shipment Statistics for Sodium Silicates from 1959 to 1979, 10^3 t

Year	Sodium silicate solution[a]		Sodium metasilicate[b]		Sodium orthosilicate[c]
	Production	Shipment	Production	Shipment	Production
1959	466	365	169	122	35
1960	450	365	175	127	34
1961	476	387	180	131	33
1962	501	410	198	142	34
1963	500	397	216	142	34
1964	512	412	207	171	37
1965	533	414	228	186	38
1966	565	417	208	164	42
1967	556	398	199	157	38
1968	574	409	218	158	38
1969	596	462	215	165	33
1970	570	423	204	162	31
1971	577	421			
1972	600	445			
1973	655	485			
1974	699	507			
1975	657	501			
1976	677	519	171	129	33
1977	689	513	159	136	31
1978	722	523	155	107	28
1979	698	501			

[a] Soluble sodium silicate glass solid and liquid (anhydrous) (SIC no. 2819741). Excludes quantities consumed in the manufacture of meta, ortho, and sesquisilicates. Includes quantities consumed in the manufacture of glass powder, hydrated glasses, and precipitated products. Shipment figures include unspecified amounts shipped to other plants for use in manufacturing *meta-*, *ortho-* or sesquisilicates. Source: U.S. Department of Commerce.
[b] All sodium metasilicate products (SIC no. 2819743) on a 100 wt % sodium metasilicate pentahydrate basis.
[c] Estimated production data for 100 wt % sodium orthosilicate (SIC no. 2819745).

Table 8. Producer Price Index (PPI) of Soluble Silicate Raw Materials[a,b]

Commodity	PPI (December 1980)
sodium silicate	258.6
sodium metasilicate	225.7
sodium carbonate	335.4
energy (average)	1002.7
industrial chemicals	334.6

[a] Ref. 54.
[b] 1967, PPI = 100.

Health, Safety, and Environmental Aspects

The moderate-to-strong alkalinity of commercial soluble silicates is their primary hazard. Contact-exposure effects can range from irritation to corrosion, depending on the concentration of the soluble silicate, its silica-to-alkali ratio, the sensitivity of the tissue exposed, and the duration of exposure. Sodium silicate solutions of commercial concentration with a $SiO_2:Na_2O$ ratio <1.8 and sodium silicate powders with a $SiO_2:Na_2O$ ratio <2.4 are corrosive to the skin when tested according to the Federal Hazardous Substances Act (FHSA) protocol (57). If ingested or inhaled, soluble silicates are rapidly absorbed and eliminated in the urine (58–59). Trace quantities of silicon are essential to nutrition, but if normal dietary amounts are greatly exceeded, siliceous urinary calculi may result (48).

Compounds of silicon and oxygen are the primary constituents of the earth's land masses. Dissolved silica is a minor but ubiquitous constituent of the hydrosphere. The earth's biomass contains dissolved silica, amorphous silica in the solid phase, and silica that is bound to organic matter. Dissolved silica is supplied to the environment by chemical and biochemical weathering processes that involve the transfer of energy from biological systems to silicate minerals, ion substitution and chelation, and dissolution (60). The immense magnitude of the flux of the natural-silica cycle can be appreciated by considering the average silica weathering rate for watersheds: ca 2000 $kg/(km^2 \cdot yr)$ (61). When rapid algal growth occurs in inland waters, dissolved silica may become depleted resulting in the replacement of diatoms by species that accelerate eutrophication (62). Commercial soluble silicates have a higher degree of polymerization than natural dissolved silica because of higher concentrations; however, when diluted, they are thought to depolymerize rapidly to molecular species that are indistinguishable from natural dissolved silica (63).

Uses

Estimates of the 1978 uses of sodium silicates in the United States are listed in Table 9 (64).

Detergency. The largest single use for soluble silicates as a functional additive is in soaps and detergents (see Surfactants and detersive systems). These materials provide a constant pH value in the detergent system and aid in the saponification of oils and fats by means of their alkaline nature and buffering ability. They provide a crisp quality to a spray-dried detergent granule because of their glassy nature when dehydrated. They also enhance the effectiveness of surfactants, help maintain soil-particle dispersion, and prevent soft-metal, ie, tin, aluminum, copper, and brass, corrosion because the silicate polyanion forms complexes with metal ions and adsorbs at oxide surfaces. In dishwashing detergents, silicates provide protection for chinaware and metal utensils. Silicates are particularly effective in minimizing the negative effects of Mg^{2+} hardness and therefore work well in combination with soda ash and zeolites, which are more effective Ca^{2+}-ion-activity reducers (65).

Water Treatment. There are four main areas where silicates have historically been used in water treatment: to make activated silica sol, a stable, acid-polymerized silicate

Table 9.   Uses of Sodium Silicates in the United States in 1978[a]

Application	Percentage of U.S. market
detergents	30
silica source for the chemical industry	
silicate-based pigments	20
cracking catalysts	10
silica gels and sols	4
zeolites	1
calcium silicates	1
titania pigments	1
Total	*37*
other	
adhesives	4
cements	4
roofing granules	3
ore flotation	2
water treatment	2
textile bleaching	1
foundry binders	1
welding rods	1
miscellaneous	15
Total	*33*

[a] Ref. 64.

solution that is a coagulant aid in the alum coagulation of suspended matter in raw or wastewater streams (see Flocculating agents); for the prevention of corrosion of metal surfaces in contact with water; for stabilization of reduced iron and manganese in water supplies (see Chelating agents); and in boiler-water treatment (66) (see Water, industrial water treatment). The boundaries between these areas are indefinite and tend to merge into other fields that involve large quantities of water, eg, detergency, papermaking, bleaching, and oil recovery.

These applications appear to rely on the alkalinity of silicates and the ability of the polymeric silicate anion to form complexes with metal ions in solution and adsorb at charged interfaces. Silicates may be particularly effective in high hardness systems, where ions other than Mg^{2+} and Ca^{2+} are of concern, because of the relative insensitivity of silicates to Ca^{2+} compared with other inorganic sequestering agents.

Mineral Beneficiation.   Sodium silicates are used primarily in froth flotation as strong and selective depressants by increasing the hydrophilic nature of specific particles; however, small concentrations may activate the flotation of some calcium minerals (see Flotation). Soluble silicates are also excellent dispersants (qv) and are useful if fine slimes occur as a result of grinding to fine sizes (67). Soluble silicates adsorb onto mineral surfaces selectively at relatively low concentrations and improve separation of different particles by causing variable rates of settling. However, excess silicate causes nonselective depression of all minerals. Soluble silicates also can deactivate certain mineral surfaces to flotation by preventing collector, eg, oleic acid, adsorption. Higher modulus silicates are generally more active and appear to derive their values primarily from their ability to aid inexpensively in attenuation of particle separation through the development of appropriate hydrophobic and hydrophilic surfaces. The latter effect results from simple silicate anion adsorption, and the former is either inherent in the mineral or a result of collector, ie, surfactant, adsorption.

Silicates are also thought to be a cost-effective additive to wet-ore grinding processes, in which they reduce wear on steel grinding balls and rods (68). It is thought that 85% of the 4.5×10^5 t loss of steel per year worldwide is the result of corrosion (69).

Adhesives and Binders.   The applications of soluble silicates as binders or adhesives are numerous. Silicates are often used to bind sand molds in foundries by means of an appropriate acid either directly, as in the CO_2 process, or indirectly with an organic ester, which hydrolyzes and subsequently gels the silicate sand mass (70–71). Silicates are also used to stabilize soils in underground construction projects (72). Silicate solution mixed with appropriate reactants, ie, gelling agents, and water are pumped into the ground, where they gel and harden the soil. In a similar manner, silicates are used to divert or control water flows in oil fields or to seal defects in concrete pipes by direct impregnation (73) (see Chemical grouts).

As an adhesive, silicates are used extensively in spiral-tube winding, fiber drums, end sealing, laminating metal foil to paper, and in corrugated boxes. They are also used in the manufacture of refractory and acid-resistant mortars and cements. Potassium silicates are preferred when high refractoriness and a nonblooming surface are required. In gunning applications, the sodium silicates might be preferred because of their higher tack. One of the better-known examples of the use of soluble silicates as a binder or coating involves the binding with potassium silicates of phosphor to black-and-white television screens. Silicates are also used in pelletizing, granulating, and briquetting of fine particles and as a vehicle for water-based coatings (see Size enlargement). If a water- or humidity-resistant system is not required in any of the preceding applications, then simple air drying or heating is sufficient (74–75). However, when increased water insensitivity is required, the use of either a setting agent, eg, a reactive pigment or a heavy metal salt; post-treatment, eg, baking or curing with an acid wash; or glyoxal curing is required (76). The degree of water insensitivity depends on the setting agents used and the temperature and method of cure. Examples of some applications where high water insensitivity is required are welding-rod coatings, roofing granules, and zinc-rich corrosion-resistant paints for iron and steel.

Enhanced Oil Recovery.   A new and growing use of silicates has been developed in enhanced oil recovery. Silicates are used for reasons that are related by analogy to their uses in detergency and ore beneficiation. In a chemical flooding method called alkaline flooding, a sodium orthosilicate solution is superior to that of NaOH. The hydroxide ion promotes *in situ* formation of petroleum-acid-based surfactants, which improve oil flow from the porous substrate. The presence of silicates seems to enhance a surfactant's effectiveness, especially in high hardness reservoir brines (77–78). Studies of the addition of silicates, both before and during direct surfactant floods, ie, micellar-polymer flood preflushes and low tension waterflooding with a silicate sacrificial agent, indicate that silicates retard surfactant adsorption or retention by the substrate and improve flood sweep efficiency, possibly by selective-permeability reduction (79). The former effect is extremely significant because surfactant retention by the substrate not only reduces surfactant effectiveness, but may contribute to the holding of oil in the reservoir. In this manner, silicates are thought to improve the water-wetting nature of the reservoir. This application offers great promise for silicate market growth.

Other. *Bleach Stabilization.* Silicates are used in combination with H_2O_2 as a bleach stabilizer and presumably as a peroxide activator through the formation of an active peroxysilicate intermediate compound (80). The stabilizing action is a result of suppression of the negative effects of heavy metals (see Bleaching agents).

Deflocculation and Slurry Thinning. The activity of silicates as deflocculants, ie, agents that maintain high solids-slurry viscosities at increased solids levels, is used in several applications. These applications appear to be based on the ability of silicates to suppress the ordering that creates increased resistance to viscous flow within the various slurry systems. In these applications, laboratory trials are necessary, since the complexity of the systems precludes the general use of any universal deflocculant. Silicates are used in the thinning of limestone or clay slurries in wet-process manufacturing of cements, manufacture of bricks, clay refining, and petroleum drilling muds (see Petroleum).

Manufacture of Derivatives. The single largest use of the soluble silicates is as an active source of building-block silicate anions for the manufacture of silica-based products, eg, silica gels and sols, synthetic zeolites (see Molecular sieves) and clays, pigment-grade silicas, and heavy-metal silicates (see Derivatives).

Derivatives

Precipitated Silica, Silica Sols, and Gels. Solutions of soluble silicates can be modified by ion exchange or acid addition and post-treatment to make a range of products primarily distinguishable by their SiO_2 and H_2O content, pore volume and surface area, particle size and morphology, and residual salts content (see Silica). Silica sols are manufactured from diluted 3.3-ratio silicate solution by H^+/Na^+ ion exchange and reconcentration. This process produces a dispersion of colloidal-sized silica, which is used, for example, as antislip agents, investment castings, binders, and polishing solutions for silicon wafers used to make semiconductors (see Semiconductors, fabrication and characterization). Precipitated silicas, which are used widely as fillers in rubber, paints, and flow-control agents, can be made by neutralizing the interparticle forces holding polymeric silicate anions in solution, eg, by the addition of an organic compound or a sodium salt and mineral acid. Silica gels are prepared by acidifying concentrated 3.3-ratio sodium silicate solutions, which rapidly produces a gel network, ie, a hydrogel, after passing through a sol stage. Xerogels or aerogels can be manufactured based on the method of gelation, milling, washing, and drying. A detailed discussion of the various gels and their properties is given in ref. 81. These gels are used as desiccants, clear-gel-toothpaste thickeners and abrasives, beer clarifiers, catalyst carriers, and specialty fillers for plastic films and lacquers (see Fillers).

Synthetic Insoluble Silicates. The synthetic insoluble silicates, ie, the mineral-type compounds, are generally synthesized by hydrothermal techniques (82). Of greatest commercial importance today are the many varieties of zeolites [1318-02-1],

$$M_{x/n}[(AlO_2)_x(SiO_2)_y].zH_2O$$

where M is usually an alkali metal or alkaline earth ion. They are generally synthesized from gels made from active SiO_2 and Al_2O_3 sources containing appropriate modifiers. Synthetically manufactured zeolites are more cost-effective than natural ones, and thus almost all commercial applications rely on these materials (83). These crystalline aluminosilicates have as their basic structural units the SiO_4 tetrahedral network with partial replacement of silicon by aluminum atoms. The aluminosilicates have an open framework, which may vary as the Si:Al ratio and mobile charge-neutralizing cation vary. Thus, these materials, which are sometimes called molecular sieves, are used as ion exchangers, selective sorbing agents, and catalysts (see Ion exchange; Catalysis). Other uses for these materials may exist in treating radioactive waste, as plastics additives, and for the controlled release of chemicals (see Microencapsulation). Ultramarine pigments may be produced synthetically and apparently derive their color from S^{2-} in the pigment structure (84). The estimated worldwide market growth for these synthetic aluminosilicates indicates that the consumption of these materials has increased significantly as a result of the use of zeolite NaA in detergents as a partial replacement for phosphates. The present worldwide market is ca 4.5×10^5 metric tons (85).

Another insoluble silicate of commercial interest is synthetic magnesium silicate clay, hectorite [12173-47-6], which may be prepared from sodium silicate ($m = 3.3$), LiF, $MgSO_4$, and Na_2CO_3 (86). The synthetic material is free from the impurities associated with natural hectorite and has better suspending and thickening properties. This inorganic thickening agent produces translucent thixotropic gels that are neither tacky, gummy, nor stringy, and they can be used in antiperspirants, gel toothpastes, shampoos, makeup, paints, and household cleansing products. Although numerous other crystalline silicates can be prepared at higher temperatures and pressures, eg, asbestiform minerals (87), calcium silicates (88), and miscellaneous materials (89), the commercial applications and volumes are not significant in comparison to those of zeolites and hectorite, particularly the zeolites. In many cases, these materials are of interest in understanding the chemistry of cement formation and mineral development. However, there are several significant products and markets for amorphous insoluble silicates made by reactions of soluble salts of magnesium, aluminum, calcium, and others. These materials are useful as sorbents, ion exchangers, rubber reinforcers, paint hardeners, fillers, thixotropes, and pigments.

BIBLIOGRAPHY

"Soluble Silicates and Synthetic Insoluble Silicates" in *ECT* 1st ed., Vol. 12, pp. 303–330 and "Silicon Compounds, Synthetic Inorganic Silicates" in *ECT* 2nd ed., Vol. 18, pp. 134–165, by John H. Wills, Philadelphia Quartz Company.

1. J. G. Vail, *Soluble Silicates*, Reinhold Publishing Corp., New York, 1952.
2. H. H. Weldes and K. R. Lange, *Ind. Eng. Chem.* **61**, 29 (1969).
3. D. Barby, T. Griffiths, A. R. Jacques, and D. Pawson in R. Thompson, ed., *The Modern Inorganic Chemistry Industry*, The Chemical Society, London, UK, 1977.
4. Von G. Engelhardt and co-workers, *Z. Anorg. Allg. Chem.* **418**, 17 (1975).
5. B. D. Mosel and co-workers, *Phys. Chem. Glasses* **15**(6), 154 (1974).
6. S. P. Zhdanov, *Fizika i Khimya Stekla* **4**(5), 515 (1978).
7. H. P. Calhoun, C. R. Masson, and J. M. Jansen, *J. Chem. Soc. Chem. Comm.*, 576 (1980).
8. S. A. Brawer and W. B. White, *J. Non Cryst. Solids* **23**, 261 (1977).
9. S. Urnes and co-workers, *J. Non Cryst. Solids* **29**, 1 (1978).
10. R. W. Harman, *J. Phys. Chem.* **32**, 44 (1928).
11. R. V. Nauman and P. Debye, *J. Phys. Chem.* **55**, 1 (1951); **65**, 5 (1961).
12. C. W. Lentz, *Inorg. Chem.* **3**(4), 574 (1964).
13. L. S. Dent Glasser and S. K. Sharma, *Br. Polym. J.* **6**, 283 (1974).
14. H. C. Marsmann, *Z. Naturforsch.* **29B**(7–8), 495 (1974).
15. R. K. Harris and R. H. Newman, *J. Chem. Soc. Faraday Trans. 2* **79**(9), 1204 (1977).
16. R. K. Harris, private communication, 1980.

17. L. S. Dent Glasser and E. E. Lachowski, *J. Chem. Soc. Dalton*, 393, 399 (1980).
18. L. S. Dent Glasser and P. B. Jamieson, *Acta Crystallogr.* **B32**(3), 705 (1976).
19. K. Beneke and G. Lagaly, *Am. Mineral.* **62**, 763 (1977).
20. T. M. El-Shamy, J. Lewins, and R. W. Douglas, *Glass Technol.* **13**(3), 81 (1972).
21. R. W. Douglas and T. M. El-Shamy, *J. Am. Ceram. Soc.* **50**(1), 1 (1967).
22. C. Wu, *J. Am. Ceram. Soc.* **63**(7–8), 453 (1980).
23. F. R. Jorgensen, *J. Appl. Chem. Biotech.* **24**, 303 (1977).
24. R. W. Spencer and J. S. Falcone, unpublished data, The PQ Corporation, Lafayette Hill, Pa., 1980.
25. R. K. Iler, *The Chemistry of Silica*, John Wiley & Sons, Inc., New York, 1979, p. 172.
26. A. R. Marsh, G. Klein, and T. Vermeulen, *Polymerization Kinetics and Equilibria of Silicic Acid in Aqueous Systems*, National Technical Information Service, #LBL 4415, Springfield, Va., 1975.
27. R. Schwartz and W. D. Muller, *Z. Anorg. Allg. Chem.* **296**, 273 (1958).
28. V. N. Belyakov and co-workers, *Ukr. Khim. Zh. Russ. Ed.* **40**(3), 236 (1974).
29. P. W. Schindler and H. R. Kamber, *Helv. Chim. Acta* **51**, 1781 (1968).
30. K. R. Lange and R. W. Spencer, *Environ. Sci. Technol.* **2**(3), 212 (1969).
31. L. H. Allen, E. Matijevic, and L. Meites, *J. Inorg. Nucl. Chem.* **33**, 1293 (1971).
32. R. W. Maatman and co-workers, *J. Phys. Chem.* **68**(4), 757 (1964).
33. P. W. Schindler and co-workers, *J. Colloid Interface Sci.* **55**, 469 (1976).
34. R. O. James and T. W. Healy, *J. Colloid Interface Sci.* **40**, 42, 53, 65 (1972).
35. I. M. Kolthoff and V. A. Stenger, *J. Phys. Chem.* **36**, 2113 (1932); **38**, 249, 475 (1934).
36. J. S. Falcone, *paper presented at ACS Symposium of Soluble Silicates*, New York, Aug. 27, 1981.
37. Ref. 25, p. 667.
38. F. J. Hingston, R. J. Atkinson, A. M. Posner, and J. P. Quirk, *Nature* **215**, 1459 (1967).
39. C. P. Huang, *Earth Planet. Sci. Lett.* **27**, 265 (1975).
40. F. Tsai and J. S. Falcone, *paper presented at the ACS/CSJ Chemical Congress*, Honolulu, Hawaii, Apr. 6, 1979.
41. F. J. Hazel, *J. Phys. Chem.* **49**, 520 (1945).
42. C. Kröger, *Glass Ind.* **37**(3), 133 (1956).
43. W. R. Ott, *Ceramurgia Int.* **5**(1), 37 (1979).
44. A. M. Ginstling and B. I. Brounsthein, *J. Appl. Chem. USSR* **23**, 1327 (1950).
45. A. G. Pincus and G. M. Dicken, eds., *Electric Melting in the Glass Industry*, Magazine for Industry, Inc., New York, 1976.
46. 21 CFR §182.70–182.90, revised Apr. 1, 1979.
47. 21 CFR §173.310, revised Apr. 1, 1979.
48. Select Committee on GRAS Substances, *Evaluation of the Health Aspects of Certain Silicates as Food Ingredients*, SCOGS-61, Federation of American Societies for Experimental Biology, NTIS Pb 301-402/AS, Springfield, Va., 1979, p. 26.
49. Specification MIL-W-15117D, Amendment 3, Jan. 22, 1975.
50. *Sodium Silicate—Oil and Hazardous Materials—Technical Assistance Data System*, National Institute of Health, EPA, Washington, D.C., an on-line data base.
51. *Hazardous Labeling Guide*, CPSC 9010.125, Appendix 1, 1975, p. 45.
52. CFR §1910.1000, revised July 1, 1979.
53. *Chem. Mark. Rep.* **219**(10), 42 (1981).
54. *Current Industry Report*, Inorganic Chemical Section M28A, U.S. Department of Commerce, Bureau of Census, Washington, D.C., 1981.
55. *Chem. Purchasing* **55**(12), 37 (1979).
56. H. Strauss and R. Rutkowski, *Silikattechnik* **29**(11), 339 (1978).
57. W. L. Schleyer and J. G. Blumberg in ref. 36.
58. G. M. Berke and T. W. Osborne, *Food Cosmet. Toxicol.* **17**, 123 (1979).
59. R. Michon, *C.R. Acad. Sci.* **243**, 2194 (1956).
60. J. R. Boyle and G. K. Voigt, *Plant Soil* **38**(1), 191 (1973).
61. M. A. Soukup, *The Limnology of a Eutrophic Hardwater New England Lake with Major Emphasis on the Biogeochemistry of Dissolved Silica*, Xerox Univ. Microfilms, Ann Harbor, Mi., No. 75-27-527, 1975.
62. P. Kilham, *Limnol. Oceanogr.* **16**(1), 10 (1971).
63. J. L. O'Connor, *J. Phys. Chem.* **65**(1), 1 (1961).
64. L. S. Dent Glasser, *Chem. Br.* **18**, 33 (Jan. 1982).
65. T. C. Campbell, J. S. Falcone, and G. C. Schweiker, *HAPPI* **15**(3), 31 (1978).

66. S. Sussman, *Ind. Water Eng.* **3**(3), 23 (1976).

67. V. Klassen and V. A. Mokrousov, *An Introduction to the Theory of Flotation*, Butterworth, London, UK, 1963, pp. 320–335.

68. G. R. Hoey, W. Dingley, and A. W. Lui, *Can. Chem. Process.* **59**(5), 36 (1975).

69. J. L. Briggs, *Mater. Perform.* **12**(1), 20 (1974).

70. K. E. L. Nicholas, *The CO_2-Silicate Process in Foundries*, British Cast Iron Research Association (BCIRA), Alvechurch, UK, 1972.

71. M. Roberts, *Foundry Trade J.* **133**(2925), 783 (1972).

72. K. M. O'Connor, R. J. Krizek, and D. K. Atrmotzidis, *Proc. Am. Soc. Civ. Eng.* **104**(GT7), 939 (1978).

73. G. V. Topil'skii and co-workers, *J. Appl. Chem. USSR* **51**(3), 482 (1978).

74. L. S. Dent Glasser and C. K. Lee, *J. Appl. Chem. Biotechnol.* **21**(5), 127 (1971).

75. N. R. Horikawa, K. R. Lange, and W. L. Schleyer, *Adhes. Age* **10**(7), 30 (1967).

76. L. S. Dent Glasser, E. G. Grassick, and E. E. Lackowski, *J. Chem. Tech. Biotechnol.* **29,** 283 (1979).

77. T. C. Campbell, *paper presented at the 50th Annual California Regional Meeting of SPE*, Pasadena, Calif., Apr. 9–11, 1980.

78. L. W. Holm and S. D. Robertson, *J. Pet. Tech.* **30**(1), 161 (1978).

79. P. H. Krumrine, T. C. Campbell, and J. S. Falcone, *paper presented at the 2nd SPE/DOE Joint Symposium on EOR*, Tulsa, Ok., Apr. 5–8, 1981.

80. M. Ya. Kanter and co-workers, *J. Appl. Chem. USSR* **50**(4), 697 (1977).

81. D. Barby in G. D. Parfitt and K. S. W. Sing, eds., *Characteristics of Powdered Surfaces*, Academic Press, Inc., London, UK, 1976, p. 353.

82. R. M. Barrer, *Chem. Br.* **2**(9), 380 (1966).

83. P. H. Shimizu, *Soap/Cosmetics/Chemical Specialties* **53,** 33 (June 1977).

84. F. A. Cotton and G. Wilkinson, *Advanced Inorganic Chemistry*, Interscience Publishers, a division of John Wiley & Sons, Inc., 1966, pp. 468–474.

85. G. C. Schweiker, *J. Am. Oil Chem. Soc.* **55**(1), 36 (1978).

86. U.S. Pat. 3,586,478 (June 22, 1971), B. S. Neumann (to LaPorte Industries, Ltd.).

87. W. H. Bauer and P. P. Keat, *Cer. Age* **56**(5), 19 (1950).

88. E. Thilo, *Angew. Chem.* **63,** 201 (1951).

89. C. I. v. Nieuwenberg, *Rev. Gen. Sci. Appl. Brussels* **1**(3), 61 (1952).

General References

Ref. 25 is a general reference.

J. G. Vail, assisted by J. H. Wills, *Soluble Silicates*, 2 Vols., ACS Monograph No. 116, Reinhold Publishing Corp., New York, 1952.

W. Eitel, *Silicate Science*, Vols. 2, 3, and 4, Academic Press, Inc., New York, 1964.

R. K. Iler, *The Colloid Chemistry of Silica and Silicates*, Cornell University Press, Ithaca, N.Y., 1955.

R. Houwink and G. Solomon, eds., *Adhesion and Adhesives*, Vol. 1, Elsevier Publishing Co., New York, 1965, Chapt. 8.

G. W. Morey, *The Properties of Glass*, ACS Monograph No. 77, Reinhold Publishing Company, New York, 1938.

W. A. Weyl and E. C. Marboe, *The Constitution of Glasses: A Dynamic Interpretation*, Vols. 1 and 2, John Wiley & Sons, Inc., New York, 1963.

W. A. Weyl in F. R. Eirich, ed., *Rheology*, Academic Press, Inc., New York, 1960, Chapt. 8.

JAMES S. FALCONE, JR.
The PQ Corporation

SILICON HALIDES

The study of silicon halides began in the early 1800s (1). Since then essentially all of the monomeric silicon halides have been extensively studied and reported in the literature. These include mixed silicon halides and halohydrides. A large number of halogenated polysilanes have also been reported in the literature. Despite the extensive research in silicon halides, only two of these chemicals are produced on a large industrial scale (excluding organohalosilanes). These are tetrachlorosilane ($SiCl_4$) and trichlorosilane ($HSiCl_3$).

Physical Properties

The physical properties of silicon tetrahalides are listed in Table 1; those of the halohydrides are listed in Table 2. A more complex review of the physical properties of these chemicals is given in ref. 2. Detailed lists of properties of the colorless fuming liquids, silicon tetrachloride and trichlorosilane, are given in Table 3. A review of the physical and thermodynamic properties of silicon tetrachloride is given in ref. 3.

Several of the mixed silicon halides are formed simply by heating a mixture of the tetrahalosilanes at moderate temperature (3), eg,

$$SiCl_4 + SiBr_4 \rightleftharpoons SiBrCl_3 + SiBr_2Cl_2 + SiBr_3Cl$$

Table 1.   Silicon Tetrahalides

Compound	CAS Registry No.	Mp, °C	Bp, °C	Density, g/cm³ (°C)	Bond energy, kJ/mol[a]
SiF_4	[7783-61-1]		−90.3	1.66 (−95)	146
$SiCl_4$	[10026-04-7]	−68.8	56.8	1.48 (20)	381
$SiBr_4$	[7789-66-4]	5	155.0	2.81 (29)	310
SiI_4	[13465-84-4]	124	290.0		234

[a] To convert J to cal, divide by 4.184.

Table 2.   Properties of Silicon Halohydrides

Compound	CAS Registry No.	Mp, °C	Bp, °C	Density, g/cm³ (°C)
H_3SiF	[13537-33-2]		−99.0	
H_2SiF_2	[13824-36-7]	−122.0	−77.8	
$HSiF_3$	[13465-71-9]	−131.2	−97.5	
H_3SiCl	[13465-78-6]	−118.0	−30.4	1.145 (−113)
H_2SiCl_2	[4109-96-0]	−122.0	8.3	1.42 (−122)
$HSiCl_3$	[1025-78-2]	−128.2	31.8	1.3313 (25)
H_3SiBr	[13465-71-1]	−94.0	1.9	1.531 (20)
H_2SiBr_2	[13768-94-0]	−70.1	66.0	2.17 (0)
$HSiBr_3$	[7789-57-0]	−73.0	111.8	2.7 (17)
H_3SiI	[13598-42-0]	−57.0	45.4	2.035 (14.8)
H_2SiI_2	[13760-02-6]	−1.0	149.5	2.724 (20.5)
$HSiI_3$	[13465-72-0]	8.0	111.0[a]	3.314 (20)

[a] At 2.9 kPa (21.8 mm Hg).

Table 3.   Physical Properties of Silicon Tetrachloride and Trichlorosilane

Property	Value	
	$SiCl_4$	$SiHCl_3$
refractive index	1.4146	1.3983
d_4^{25}, g/cm^3	1.4736	1.3313
viscosity (at 25°C), mm^2/S (= cSt)	0.35	0.23
vapor pressure, kPa^a		
at −54°C		1.3
at −34.4°C	1.3	
at −26°C		8.0
at −16°C		13.3
at 5.0°C	13.3	
at 14.6°C		53.3
at 38.4°C	53.3	
heat of vaporization, J/g^b	167.4	195.4
heat of formation, kJ/mol^b		
liquid	640	
gas	610	
flash point (Cleveland open cup), °C		−28
autoignition temperature, °C		104
specific heat, J/g^b	0.20	0.96
coefficient of expansion, °C^{-1}	0.0011	0.0019
solubility	soluble in organic solvents; reacts with H_2O and alcohol	

^{a} To convert kPa to mm Hg, multiply by 7.5.

^{b} To convert J to cal, divide by 4.184.

Mixed halosilanes can also be prepared by heating a mixture of the appropriate halides of silicon and aluminum (5–6). Some physical properties of these mixed tetrahalosilanes are listed in Table 4. In addition to these, the properties of several halogenated polysilanes are listed in Table 5. A summary of the preparation and properties of the chlorinated polysilanes has been published (7).

Table 4.   Properties of Mixed Silicon Halides

Compounds	CAS Registry No.	Mp, °C	Bp, °C
$SiBr_3Cl$	[13465-76-4]	−208	128
$SiBr_2Cl_2$	[13465-75-3]	−45.5	104.4
$SiBrCl_3$	[13465-74-2]	−62.0	80
$SiClI_3$	[13932-03-1]	2.0	234–237
$SiCl_2I_2$	[13977-54-3]	<−60.0	172
$SiCl_3I$	[13465-85-5]	<−60.0	113–114
$SiBr_3I$	[13536-76-0]	14.0	192
$SiBr_2I_2$	[13550-39-5]	38.0	230–231
$SiBrI_3$	[13536-68-0]	53.0	255
$SiBr_3F$	[18356-67-7]	−82.5	83.8
$SiCl_3F$	[14965-52-7]	−120.8	12.2
$SiBr_2F_2$	[14188-35-3]	−66.9	13.7
$SiCl_2F_2$	[18356-71-3]	−139.7	−32.2
$SiClF_3$	[14049-36-6]	−142.0	−70.0
$SiBrF_3$	[14049-39-9]	−70.5	−41.7
$SiBrCl_2F$	[28054-58-2]	−112.3	35.5
$SiBr_2ClF$	[28054-61-7]	−99.2	59.5

Table 5. Properties of Some Halogenated Polysilanes

Compound	CAS Registry No.	Mp, °C	Bp, °C	Density, g/cm^3 (°C)
Si_2F_6	[13830-68-7]	−18.6	−19.1	
Si_2Cl_6	[13465-77-5]	−1.0	144.5	1.5624 (15)
Si_3Cl_8	[13596-23-1]	−67.0	216.0	1.61 (15)
Si_4Cl_{10}	[13763-19-4]	93.6	100.0	
Si_5Cl_{12}	[13596-24-2]	−80.0	130.0^a	
Si_6Cl_{14}	[13596-25-3]	319.0 (dec)		
Si_2Br_6	[13517-13-0]	95.0	240.0	
Si_3Br_8	[51804-32-9]	133.0		
Si_4Br_{10}	[81626-34-8]	185.0		
Si_2I_6	[13510-43-5]	250.0 (dec)		

^{a} At 2.9 kPa (21.8 mm Hg).

Chemical Properties

Silicon halides are typically tetravalent compounds. The silicon–halogen bond is very polar, thus the silicon is susceptible to nucleophilic attack. This in part accounts for their broad range of reactivity with various chemicals. Furthermore, their reactivity generally increases with the atomic weight of the halogen atom.

Halosilanes are very reactive with a variety of protic chemicals. They generally react violently with water forming silicon dioxide and the respective hydrohalogen. Other examples include reaction with alcohols and amines as follows:

$$SiCl_4 + 4\ CH_3CH_2OH \rightarrow Si(OCH_2CH_3)_4 + 4\ HCl$$

$$SiCl_4 + (excess)HN(CH_3)_2 \rightarrow Si[N(CH_3)_2]_4 + 4\ HN(CH_3)_2.HCl$$

Such substitution reactions typically are reversible and facilitated by the removal of the hydrohalogen being formed. This is commonly accomplished by adding a tertiary amine or excess amine reactant, which generally precipitates as the amine hydrohalide. The halosilanes are reduced to silicon with hydrogen at elevated temperatures. Tetrafluorosilane reacts with hydrogen only above 2000°C, whereas tetrachlorosilane and tetraiodosilane are reduced by hydrogen at 1000–1200°C (1,8–10). The silicon halohydrides and mixed halides are reduced at lower temperatures.

Silicon halides are stable to oxygen at room temperature but react at high temperature to form mixtures of oxyhalosilanes (11–12). Silicon halides are also reduced by a number of metals. One of the earliest methods of producing silicon was the reduction of a silicon halide with sodium or potassium (1). A similar reduction process with zinc was the first commercial process for producing semiconductor-grade silicon from silicon tetrachloride (13–14).

Magnesium does not form stable Grignard reagents with silicon halides, although some silicon halohydrides do react forming polysilanes (15).

$$2\ HSiBr_3 + 3\ Mg \rightarrow \frac{2}{x}(SiH)_x + 3\ MgBr_2 \cdot$$

As would be expected, all silicon halides are readily reduced by hydride ions or complex hydrides to silicon hydrides (16–18).

$$4 R_2AlH + SiX_4 \rightarrow SiH_4 + 4 R_2AlX$$

Historically, one of the most important reactions of silicon halides is with metal alkyls and metal alkyl halides. The Grignard reaction, eg, was the first commercial process for manufacturing organosilicon compounds, which were later converted to silicones (19).

$$SiCl_4 + n\ RMgX \rightarrow R_n SiCl_{4-n} + n\ MgXCl$$

The halohydrides are particularly useful intermediate chemicals because of their ability to add to alkenes as follows (19–20):

$$HSiCl_3 + RCH{=}CH_2 \rightarrow RCH_2CH_2SiCl_3$$

This reaction is catalyzed by uv radiation, peroxides, and some metal catalysts, eg, platinum. This reaction has led to the production of a broad range of alkyl and functional alkyl trihalosilanes. These alkylsilanes have important commercial value as monomers and are also used in the production of silicone fluids and resins.

Additional information on the chemistry of silicon halides is given in refs. 19, 21–24.

Manufacturing Processes

The silicon halides can be easily prepared by the reaction of silicon or silicon alloys with the respective halogens (24). Fluorine and silicon react at room temperature to produce silicon tetrafluoride. Chlorine reacts with silicon exothermally, but the mixture must be heated to several hundred degrees centigrade to initiate the reaction (25). Bromine and iodine react with silicon at red heat.

Hydrogen halides also react freely with elemental silicon at moderate temperatures to yield halosilanes (26–31).

$$2 Si + 7 HCl \rightarrow HSiCl_3 + SiCl_4 + 3 H_2$$

Although a mixture of silicon halides and hydrides is formed, the formation of silicon tetrachloride can be maximized by increasing the temperature (22). The direct reaction of silicon with hydrogen chloride is also the primary manufacturing procedure for producing trichlorosilane, $HSiCl_3$. In this case, the reactor-bed temperature is maintained so as to optimize the yield of trichlorosilane. Trichlorosilane can also be produced according to the following reaction in a fluid-bed reactor (30).

$$3 SiCl_4 + 2 H_2 + Si \underset{500°C}{\overset{Cu}{\rightleftharpoons}} 4 HSiCl_3$$

At equilibrium the vapors are predominantly hydrogen and silicon tetrachlorides. These, however, can be easily removed from the trichlorosilane and be recycled.

Another common commercial manufacturing procedure for silicon tetrachloride is the reaction of chlorine gas with silicon carbide

$$SiC + 2 Cl_2 \rightarrow SiCl_4 + C$$

However, this reaction is highly exothermic, thus it is difficult to control the reaction temperature (31).

The oldest method for producing $SiCl_4$ is the direct reaction of silica with chlorine in the presence of carbon as a reducing agent (24).

$$SiO_2 + 2\,C + 2\,Cl_2 \rightarrow SiCl_4 + 2\,CO$$

In one modification of this procedure, the starting material is pyrolyzed rice hulls in place of more conventional forms of silicon dioxide (31). Another unique process that is described in ref. 32 involves chlorination of a combination of SiC and SiO_2 with carbon in a fluid-bed reactor. The advantages of this process are that it is less energy-intensive and substantially free of lower silicon chlorides.

Silicon Tetrachloride. A substantial percentage of commercially available silicon tetrachloride is made as a by-product from industrial processes. The two primary sources are the production of metal halides, particularly $ZrCl_4$ and $TiCl_4$, and of semiconductor-grade silicon by thermal reduction of trichlorosilane.

$$2\,HSiCl_3 \xrightarrow{\;>1000°C\;} Si + SiCl_4 + 2\,HCl$$

In addition to these sources, $SiCl_4$ is manufactured and marketed by Dow Corning Corporation, Union Carbide Corporation, and Van DeMark Chemical Company, Inc. Two additional companies, ie, Cabot Corporation and Degussa Corporation, manufacture silicon tetrachloride to produce fumed silica (33). These companies generally consider their manufacturing processes proprietary; however, Van DeMark manufactures silicon tetrachloride by chlorination of silicon carbide.

Trichlorosilane. Trichlorosilane, like other organochlorosilanes, is produced exclusively by the direct reaction of hydrogen chloride gas with silicon metal in a fluid-bed reactor. As described earlier, this process produces both trichlorosilane and silicon tetrachloride. The silicon tetrachloride production, however, can be minimized by proper control of the reaction temperature (22). The domestic producers and suppliers of trichlorosilane are Dow Corning Corporation and Union Carbide Corporation.

Health and Safety Factors

Halosilane vapors react with moist air to produce the respective hydrohalogen acid mist. Federal standards have not been set for exposure to halosilanes, but it is generally believed that there is no serious risk if vapor concentrations are maintained below a level that produces an irritating concentration of acid mist. The exposure threshold limit value (TLV) for HCl is 5 ppm expressed as a ceiling limit, which means that no exposure above 5 ppm should be permitted. Because most people experience odor and some irritation at or below 5 ppm, HCl is considered to have good warning properties.

Liquid halosilanes react violently with water to produce the respective hydrohalogen. Contact with skin and eyes produces severe burns. Eye contact is particularly serious and may result in loss of sight.

Halosilanes should only be handled in areas that are equipped with adequate ventilation, eye wash facilities, and safety showers. It is recommended that personnel handling halosilanes wear rubber aprons and gloves and chemical safety goggles. Furthermore, all personnel handling halosilanes should be thoroughly trained in safe handling procedures, hazardous characteristics of halosilanes, and emergency procedures for all forseeable emergencies.

Uses

Silicon Tetrachloride.   Although there is a broad range of industrial applications for SiCl$_4$, the vast majority is used in the manufacture of fumed silica. This process is carried out by burning silicon tetrachloride in a mixture of hydrogen and oxygen (33). High-purity grades of silicon tetrachloride are also used to prepare silica, which is used in the manufacture of special glasses for the electronics industry and in the production of optical wave guides, ie, fiber optics (qv).

Silicon tetrachloride is also used to prepare silicate esters, eg, ethyl silicate:

$$SiCl_4 + 4\ CH_3CH_2OH \rightarrow Si(OCH_2CH_3)_4 + 4\ HCl$$

Silicate esters are used in the production of coatings and refractories. A broad range of purity grades of silicon tetrachloride is available to meet the requirements of these different applications.

Trichlorosilane.   There are essentially only two large industrial applications for trichlorosilane. These are the synthesis of organotrichlorosilanes (see Chemical Properties) and the production of semiconductor-grade silicon metal. In the production of semiconductor-grade silicon metal, the purified trichlorosilane is reduced in the presence of hydrogen at temperatures greater than 1000°C. Although a large number of silicon halides and silicon halohydrides can be reduced to semiconductor-grade silicone, trichlorosilane is most commonly used because of its favorable balance of manufacturing, purification, handling, and chemical reduction properties (1,12,30).

BIBLIOGRAPHY

"Silicon Halides" treated under "Silicon Compounds," in *ECT* 1st ed., Vol. 12, pp. 368–370 by E. G. Rochow, Harvard University; "Silicon Halides" treated under "Silicon Compounds" in *ECT* 2nd ed., Vol. 18, pp. 166–172, by A. R. Anderson, Anderson Development Company.

1. W. R. Runyan, *Silicon Semiconductor Technology*, Texas Instrument Electronics Series, McGraw-Hill Book Company, New York, 1965.
2. E. A. Ebsworth, *Volatile Silicon Compounds*, Pergamon Press Limited, Oxford, UK, 1963.
3. C. L. Yaws, G. Hsu, P. N. Shah, P. Lubwak, and P. M. Patel, *Solid State Technol.* **22**(2), 65 (1979).
4. H. H. Anderson, *J. Am. Chem. Soc.* **67**, 859 (1945).
5. M. Schmeisser and H. Jenkner, *Z. Naturforsch.* **7B**, 191 (1952).
6. K. Moedritzer, *Organomet. Chem. Rev.* **1**, 179 (1966).
7. E. F. Hengge, *Rev. Inorg. Chem.* **2**(2), 139 (1980).
8. N. C. Cook, J. K. Walfe, and J. D. Cobine, *paper presented at the 128th Meeting American Chemical Society*, Minneapolis, Minn., 1955.
9. R. Schwarz and H. Merckback, *Z. Anorg. Allgem. Chem.* **232**, 241 (1937).
10. R. B. Litton and H. C. Anderson, *J. Electrochem. Soc.* **101**, 287 (1954).
11. A. D. Gaunt, H. Mackle, and L. E. Sutton, *Trans. Faraday Soc.* **47**, 943 (1954).
12. D. W. S. Chambers and C. J. Wilkins, *J. Chem. Soc.*, 5088 (1960).
13. D. W. Lyon, C. M. Olson and E. D. Lewis, *J. Electrochem. Soc.* **96**, 359 (1949).
14. U.S. Pat. 2,773,745 (Dec. 11, 1956), K. H. Butler and C. M. Olson (to E. I. du Pont de Nemours & Co., Inc.).
15. G. Schott, W. Herman, and R. Hirschmann, *Angew. Chem.* **68**, 213 (1956).
16. A. E. Finholt, A. G. Bond, K. E. Wilzback, and H. I. Schlesinger, *J. Am. Chem. Soc.* **69**, 2692 (1947).
17. J. E. Baines and C. Eaborn, *J. Chem. Soc.*, 1436 (1956).
18. E. G. Rochow, *J. Am. Chem. Soc.* **67**, 963 (1945).
19. W. Noll, *Chemistry and Technology of Silicones*, Academic Press, New York, 1968.
20. U.S. Pat. 2,823,218 (May 12, 1955), J. L. Speier and D. E. Hook (the Dow Corning Corp.).

21. A. G. MacDiarmid, *Organometallic Compounds of the Group IV Elements; Volume 2, The Bond to Halogens and Halogenoids*, Marcel Dekker Inc., New York, 1972.
22. V. Bazant, V. Choalovsky, and J. Rathousky, *Organosilicon Compounds; Volume 1, Chemistry of Organosilicon Compounds*, Academic Press, New York, 1965.
23. W. Noll, *Chemistry and Technology of Silicone*, Academic Press, New York, 1968.
24. E. Hengge in V. Gutmann, ed., *Inorganic Silicon Halides in Halogen Chemistry*, Vol. 2, Academic Press, New York, 1967.
25. K. A. Andrianov, *Dokl. Akad. Nauk. SSSR* **28,** 66 (1940).
26. H. Buffard, F. Wohler, *Am. Chem.* **104,** 94 (1857).
27. S. Friedel and J. Crafts, *Am. Chem.* **147,** 355 (1863).
28. L. Gatterman, *Chem. Ber.* **22,** 186 (1889).
29. *Inorg. Synth.* **1,** 38 (1939).
30. *Feasibility of the Silane Process for Producing Semiconductor Grade Silicon*, Jet Propulsion Laboratory Contract 954334, June, 1979.
31. P. K. Basu, Ph.D. dissertation, *Development of a Process for the Manufacture of Silicone Tetrachloride from Rice Hulls*, University of California, Berkeley, Calif., 1972.
32. U.S. Pat. 3,173,758 (March 16, 1965), R. N. Secord (to Cabot Corp.).
33. L. J. White and G. L. Duffy, *Ind. Eng. Chem.* **3,** 235 (1959).

WARD COLLINS
Dow Corning Corp.

SILANES

Silanes are compounds containing a hydrogen–silicon bond. They also are referred to as silicon hydrides. Silane, SiH_4, is the simplest hydride and provides the basis of nomenclature for all silicon chemistry (1). Compounds are named as derivatives of silane with the substituents prefixed, eg, trichlorosilane [10025-78-2], $HSiCl_3$; disilane [1590-87-0], H_3SiSiH_3; methyldichlorosilane [75-54-7], $CH_3SiH(Cl_2)$; methylsilane [992-94-9], CH_3SiH_3; diethylsilane [542-91-6], $(C_2H_5)_2SiH_2$; and triethylsilane [617-86-7], $(C_2H_5)_3SiH$. Two or more substituents are listed alphabetically with substituted organic moieties being named first, followed by simple organic fragments. Alkoxy substituents are named next, followed by acyloxy, halogen, and pseudohalogen groups; for example, ethylmethylethoxysilane [68414-52-8], $C_2H_5(CH_3)SiH(OC_2H_5)$, and (3-chloropropyl)methylchlorosilane [33687-63-7], $ClCH_2CH_2CH_2SiH(CH_3)Cl$. Complete rules for nomenclature with examples are described in ref. 2. Organosilanes have also been referred to as organosilicon hydrides and organohydrosilanes. This broad classification is based on comparison of the electronegativities of silicon and hydrogen.

The classic work in the field was completed in the early 1900s and involved the study of silane and higher binary silanes $(Si_n H_{2n+2})$ by means of precision vacuum techniques (3).

Only a few of the thousands of silane compounds reported have any commercial significance. These include inorganic silanes, organic silanes, and polymeric siloxanes.

Despite the small number of compounds, a wide range of applications has developed, including high purity and electronic-grade silicon metal, epitaxial silicon deposition, selective reducing agents, monomers, and elastomer intermediates. Not least is the use of these materials as intermediates for production of other silanes and silicones.

Inorganic Silanes

The inorganic silanes of commercial importance include silane, dichlorosilane, and trichlorosilane. The last, trichlorosilane, is preponderant. It is not only the preferred intermediate for the first two, but it is also used in the production of high purity silicon metal and as an intermediate for silane adhesion promoters, coupling agents, silicone resin intermediates, and surface treatments. Other silanes that appear to have potential in solar electronics are monochlorosilane, disilane, and some silylmetal hydrides. Additionally, siloxene [27233-73-4], $(H_6Si_6O_3)_x$, an inorganic polymer containing silicon hydride bonds, is of interest as a catalyst.

Properties. There is a great temptation to compare silanes and chlorosilanes with simple hydrocarbons and chlorinated hydrocarbons. Boiling points, melting points, and dipole moments are comparable. Both silanes and hydrocarbons are colorless gases or liquids at room temperature. The similarity ends with the description of simple physical characteristics (4). Silane, chlorosilane, disilane, and disiloxane [13597-73-4] are pyrophoric, igniting immediately on contact with air. Liquid disilane explodes on contact with air. The chlorosilanes react with moist air liberating hydrogen chloride. Dichlorosilane hydrolyzes to a polymeric material that may ignite spontaneously. Even trichlorosilane is highly flammable. The ability of chlorosilanes to permeate or solvate materials of construction coupled with their hydrolysis to corrosive hydrogen chloride, formation of abrasive silica, and ability to act as reducing agents make them difficult-to-handle materials.

The simple inorganic silanes are similar to carbon in that they form stable, covalent, single bonds. Double bonds involving silicon and silicon–carbon are relatively unstable. Only recently have examples of isolable silicon double-bond-containing materials been reported (5–6). Most silane materials have a tetrahedral bonding geometry consistent with formation of sp^3 hybrid orbitals. Although in some cases participation of $3d$ orbitals in five- or six-coordinate silicon compounds, eg, SiF_6^{2-}, has been invoked, their degree of participation is debated. Silicon is more electropositive than both hydrogen and carbon. This generally leads to a more polar bond structure than occurs in carbon analogues. However, the expected inductive release of electrons from R_3Si does not occur. Disiloxanes, for example, are less basic than ethers. One explanation for this is that back-bonding from lone pairs or π-electron systems to the vacant $3d$ silicon orbitals occurs. Another factor which may contribute to the greater reactivity of silicon compared to carbon is its greater size.

The physical and thermodynamic properties of silane in the context of semiconductor applications have been reviewed in detail (7). Tabulations of properties of various silanes in the context of inorganic chemistry have also been published (8). Table 1 contains selected physical properties of inorganic silanes.

Thermal. Silanes have less thermal stability than hydrocarbon analogues. The C—H bond energy in methane is 414 kJ/mol (98.9 kcal/mol) compared to 364 kJ/mol (87 kcal/mol) for each Si—H bond in silane (9). Silane, however, is one of the most

Table 1. Properties of Inorganic Silanes

Property	SiH$_4$ [7803-62-5]	H$_3$SiCl [13465-78-6]	H$_2$SiCl$_2$ [4109-96-0]	HSiCl$_3$ [10025-78-2]	H$_3$SiSiH$_3$ [1590-87-0]	H$_3$SiOSiH$_3$ [13597-73-4]	(H$_3$Si)$_3$N [13862-16-3]
mp, °C	−185	−118	−122	−126.5	−132.5	−144	−105.6
bp, °C	−111.9	−30.4	8.2	31.9	−14.5	−15.2	52
ΔH vaporization, kJ/mol[a]	12.5	20.0	25.2	26.6	21.2	21.6	
ΔH fusion, kJ/mol[a]	0.67						
critical temperature, °C	−3.5		176		109		
critical pressure, MPa[b]	472		455				
ΔH formation, kJ/mol[a]	32.6						
dipole moment, C·m × 10^{-30} [c]			3.913	3.24		0.80	0
density, g/cm^3	0.68 at −185°C	1.145 at −113°C	1.22	1.34	0.69 at −15°C	0.881 at −15°C	0.895 at −106°C

[a] To convert kJ/mol to kcal/mol, divide by 4.184.
[b] To convert MPa to psi, multiply by 145.
[c] To convert C·m to debye, divide by 3.336 × 10^{-30}.

thermally stable inorganic silanes. It decomposes at 500°C in the absence of catalytic surfaces. Decomposition occurs at 300°C in glass vessels and at 180°C in the presence of charcoal (10). Disilanes and other members of the binary series are less stable. Halogen-substituted silanes are subject to disproportionation reactions at higher temperatures (11). The thermal decomposition of the silanes in the presence of hydrogen into silicon for production of ultrapure, semiconductor-grade silicon has become an important art and is known as the Seimens process (12). A variety of process parameters, which usually include the introduction of hydrogen, have been studied. Silane can be used to deposit silicon at temperatures <1000°C (13). Dichlorosilane deposits silicon at 1000–1150°C (14–15). Trichlorosilane is ordinarily a source for silicon deposition at >1150°C (16).

Reactions. *Oxidation.* All inorganic silicon hydrides are readily oxidized. Silane, disilane, and disiloxane are pyrophoric in air and form silicon dioxide and water as combustion products. In contrast to carbon analogues, soot from these materials is white. The activation energies of the reaction of silane with molecular and atomic oxygen have been reported (17–18). The oxidation reaction of dichlorosilane under low pressure has been used for the vapor deposition of silicon dioxide (19).

Water and Alcohols. Silanes do not react with pure water or slightly acidified water under normal conditions. A rapid reaction occurs, however, in basic solution with quantitative evolution of hydrogen (3). Alkali leached from glass is sufficient to lead to the hydrolysis of silanes. Complete basic hydrolysis

$$SiH_4 + 2\,KOH + H_2O \rightarrow K_2SiO_3 + 4\,H_2$$

$$Si_2H_6 + 4\,KOH + 2\,H_2O \rightarrow 2\,K_2SiO_3 + 7\,H_2$$

followed by the quantitative measurement of hydrogen formed is often used to determine the number of Si—H and Si—Si bonds present in a particular compound. One molecule of H_2 is liberated for each Si—H and Si—Si bond present. The total silicon content can be obtained from analysis of the resulting silicate solution.

Silane reacts with methanol at room temperature to produce methoxymonosilanes of types $Si(OCH_3)_4$ [78-10-4], $HSi(OCH_3)_3$, and $H_2Si(OCH_3)_2$ [5314-52-3], but not H_3SiOCH_3 [2171-96-2] (20). The reaction is catalyzed by copper metal. In the presence of alkoxide ions, SiH_4 reacts with various alcohols, except CH_3OH, to produce tetraalkoxysilanes and hydrogen (21).

Halogens, hydrogen halides, and other covalent halides. Most compounds containing Si—H bonds react very rapidly with the free halogens. An explosive reaction takes place when chlorine or bromine is allowed to react with SiH_4 at room temperature, presumably forming halogenated silane derivatives (3). At lower temperatures, the reactions are moderated considerably.

$$SiH_4 + Br_2 \xrightarrow{-80°C} SiH_3Br\ (+ SiH_2Br_2) + HBr$$

Halogen derivatives also form when the silanes are allowed to react with anhydrous hydrogen halides, ie, HCl, HBr, or HI, in the presence of an appropriate aluminum halide catalyst (22–23). The reactions are generally quite moderate and can be carried out at room temperature or slightly above, ie, 80–100°C.

$$SiH_4 + HX \xrightarrow{Al_2X_6} SiHX_3 + H_2$$

where X = Cl, Br, or I. Hydrogen bromide reacts more readily than HCl or HI. Increasing the temperature or the duration of reactions generally leads to the formation of more fully halogenated derivatives.

Metals and metal derivatives. Silane reacts with alkali metals (potassium has been the most commonly studied) dissolved in various solvents, forming as the chief product the silyl derivative of the metal, eg, $KSiH_3$ [13812-63-0] (24–27). When 1,2-dimethoxyethane or bis(2-methoxyethyl) ether are used as solvents, two competing reactions occur, ie,

$$SiH_4 + 2\,M \rightarrow SiH_3M + MH$$

$$SiH_4 + M \rightarrow SiH_3M + \tfrac{1}{2}\,H_2$$

With hexamethylphosphoramide as the solvent, only the second reaction occurs. Disilane also reacts with potassium in 1,2-dimethoxyethane to form $KSiH_3$, although SiH_4 and nonvolatile polysilanes are also produced (27–28). Pure crystalline $KSiH_3$ prepared from SiH_4 and potassium in 1,2-dimethoxyethane has been obtained by slow evaporation of the solvent. When liquid ammonia is used as the solvent, only a small fraction of SiH_4 is converted into metal salt; most of the SiH_4 undergoes ammonolysis (29).

Disilane undergoes disproportionation in 1,2-dimethoxyethane, forming SiH_4 and a solid material ($-SiH_2-$) when an alkali metal salt, eg, KH or LiCl, is present (30–31).

Silanes react with alkyllithium compounds, forming various alkylsilanes. Complete substitution is generally favored; however, less substituted products can be isolated by proper choice of solvent. All four methylsilanes, vinylsilane [7291-09-0], and divinylsilane [18142-56-8] have been isolated from the reaction of SiH_4 and the appropriate alkyllithium compound with propyl ether as the solvent (32). Methylsilane and ethyldisilane [7528-37-2] have been obtained in a similar reaction (33).

Electrical discharge, irradiation, and photolysis. Early reports of the decomposition of SiH_4 in an electrical discharge indicated that the main products were hydrogen, solid silicon subhydrides of composition $SiH_{1.2-1.7}$, and small quantities of higher silanes (34). However, more recent studies indicate that under certain conditions reasonably large quantities of higher silanes up to n- [7783-29-1] and iso-Si_4H_{10} [13597-87-0] with smaller amounts of various isomers of higher silanes up to Si_8H_{18} can be produced by this method (35–37). In addition, mixed-hydride derivatives can be prepared by subjecting mixtures of SiH_4 and certain other volatile hydrides to such a discharge. Thus, SiH_3GeH_3 [13768-63-3], SiH_3PH_2 [14616-47-8], and SiH_3AsH_2 [15455-99-9] have been prepared from SiH_4-GeH_4, SiH_4-PH_3, and SiH_4-AsH_3 mixtures, respectively (38–39). Although both disilylphosphine [14616-42-3], $(SiH_3)_2PH$, and disilanylphosphine [14616-42-3], $SiH_3SiH_2PH_2$, are obtained in the SiH_4-PH_3 system, they can be obtained free of each other in the discharge of $SiH_4-SiH_3PH_2$ and $Si_2H_6-PH_3$ (40). Methyldisilane [13498-43-6] can be isolated from the products of a $SiH_4-CH_3SiH_3$ or a $SiH_4-(CH_3)_2O$ discharge system (32,41). Hexachlorodisilane [13465-77-5] and hexafluorodisilane [13830-68-7] can be prepared from trichlorosilane and trifluorosilane [13465-71-9], respectively (42). Glow discharge also provides a method for a high deposition rate of silicon from disilane (43). Silicon has been produced from silane by Penning discharge (44).

Photolysis studies of SiH_4 have involved a sensitizer, eg, mercury vapor, since SiH_4 does not absorb ultraviolet (uv) radiation at wavelengths above 185 nm (5). The ^{31}P mercury-sensitized photolysis of SiH_4 leads to the formation of H_2, Si_2H_6, Si_3H_8

[*7783-26-8*], and polymeric solid silanes, where the quantum yields of H_2 and Si_2H_6 are ca 1.6 and 0.6, respectively (46–47). The photolysis of SiH_4 in the presence of GeH_4 or CH_3I produces SiH_3GeH_3 or CH_3SiH_3, respectively (48).

Manufacture and Processing.   Four fundamental methods of production of compounds containing a Si—H bond are noteworthy. Silicides of magnesium, aluminum, lithium, iron, and other metals react with acids or their ammonium salts to produce silane and higher binary silanes. This method is uniquely applicable to the inorganic silanes. It is also the historic method for production of the silicon hydrides. Under optimum conditions, ie, the addition of magnesium silicide to dilute phosphoric acid, a 23% conversion to volatile silanes is possible (3). The composition of the silane mixture is 40 wt % SiH_4, 30 wt % Si_2H_6, 15 wt % Si_3H_6, and 15 wt % higher silanes. The highest reported yields of silicon hydrides have been achieved by treatment of magnesium silicide with NH_4Br in liquid ammonia at $-33°C$ and $N_2H_4.2HCl$ in hydrazine (49–51). The formation of silane from magnesium silicide is thought to involve a step-wise series of reactions (52):

$$Mg_2Si \ + \ 2 \ H_2O \ \longrightarrow \ \begin{bmatrix} HOMg \\ \diagdown \\ \diagup \\ HOMg \end{bmatrix} SiH_2 \end{bmatrix}$$

$$\begin{bmatrix} HOMg \\ \diagdown \\ \diagup \\ HOMg \end{bmatrix} SiH_2 \end{bmatrix} \ + \ H_2O \ \longrightarrow \ SiH_4 \ + \ 2 \ Mg(OH)_2$$

The relatively low yields of silane provided by this method and the difficulties in subsequently purifying the materials led to the discontinuation of commercial processes based on this technology in the United States in the late 1960s and early 1970s. The method was generally abandoned in favor of methods involving reduction of silicon halides.

Treatment of calcium silicide with HCl–ethanol or glacial acetic acid yields the complex polymer called siloxene (53–54).

The reduction of chlorosilanes by lithium aluminum hydride, lithium hydride, and other metal hydrides offers the advantages of higher yield and purity and the flexibility in producing a range of silicon hydrides comparable to the range of silicon halides (55). The general reaction for this reduction is

$$R_nSiX_{4-n} + (4-n)MH \rightarrow R_nSiH_{4-n} + (4-n)MX$$

where X is halogen. The most versatile reagent is lithium aluminum hydride. The most convenient solvent is tetrahydrofuran followed by ethyl ether and bis(methoxyethyl) ether. The reduction occurs at room temperature. Limitations of the lithium aluminum hydride reagent system are the inability to produce partially reduced materials, eg, $SiCl_4$ is reduced completely to SiH_4 without formation of SiH_2Cl_2, and the catalytic rearrangement of susceptible bonds because of the formation of by-product aluminum chloride. Lithium hydride is perhaps the most useful of the other hydrides. Its principal limitation is its poor solubility which essentially limits reaction media to such solvents as dioxane and dibutyl ether.

The direct reduction of chlorosilanes with hydrogen at high temperatures have been inefficient processes and have not been used commercially (56–57). In a significant process innovation, it has been shown that the hydrogenation of silicon tetrachloride over Si—Cu at less than 2.45 MPa (500 psi) proceeds in good conversion (57–58). The reduction of silicon tetrachloride in a plasma has also been reported (59–60) (see Plasma technology, Supplement Volume). Other methods of reduction include electrolytic reduction in molten salt and catalytic reduction with aldehydes (61–63).

Disproportionation reactions of silicon hydrides occur readily in the presence of a variety of catalysts. For example,

$$4\ HSiCl_3 \xrightarrow{\text{catalyst}} SiH_4 + 3\ SiCl_4$$

The most common catalysts in order of decreasing reactivity are halides of aluminum, boron, zinc, and iron (64). Alkali metals and their alcoholates, amines, nitriles, and tetraalkylureas have been used (65–68). A new method of particular promise is the use of a resin-catalyst system (69). Trichlorosilane refluxes in a bed of anion-exchange resin containing tertiary amino or quaternary ammonium groups. Contact time can be used to control disproportionation to dichlorosilane, monochlorosilane, or silane.

Direct synthesis is the preparative method that ultimately accounts for most of the commercial silicon hydride production. Trichlorosilane is produced by the reaction of hydrogen chloride with silicon, ferrosilicon, or calcium silicide in the presence of a copper catalyst (70). High purity trichlorosilane is usually produced in a fluidized bed at 300–450°C. Standard purity is produced in a static bed at 400–900°C.

$$Si + 3\ HCl \rightarrow SiHCl_3 + H_2$$

Substantial amounts of silicon tetrachloride also form in the process. Other minor by-products are dichlorosilane and higher silicon halides.

Other methods for forming inorganic silicon hydrides are shown below (71–77):

$$Cl_3SiSiCl_3 + HCl \rightarrow HSiCl_3 + SiCl_4$$

$$Cl_3SiSiCl_3 + NH_3 \rightarrow HSiCl_3 + SiCl_3NH_2$$

$$SiCl_4 + HCHO \rightarrow H_3SiCl + H_2SiCl_2 + \text{other products}$$

$$3\ SiCl_4 + 4\ Al + 6\ H_2 \rightarrow 3\ SiH_4 + 4\ AlCl_3$$

$$HSiCl_3 + (diphos)_2CoH \xrightarrow{175°C,\ 900\ atm} H_2SiCl_2 + (diphos)_2CoCl$$

$$[diphos = (C_6H_5)_2PCH_2CH_2P(C_6H_5)_2]$$

Economic Aspects. Trichlorosilane is the only inorganic silicon hydride produced in large scale. U.S. production is ca 24,000–27,000 metric tons. Substantial quantities of the material are generated in the production of silicon tetrachloride streams used in fumed silica production, but these are never isolated. Most (80–85%) trichlorosilane is used in the production of electronic-grade silicon (Fig. 1). Seimens-type processes consume most of the material, but some is used for epitaxial deposition. Other uses for trichlorosilane include disproportionation to silicon hydrides and the conversion to organosilanes, eg, adhesion promoters.

The cost of trichlorosilane is $1.50–5.00/kg, depending on grade and container.

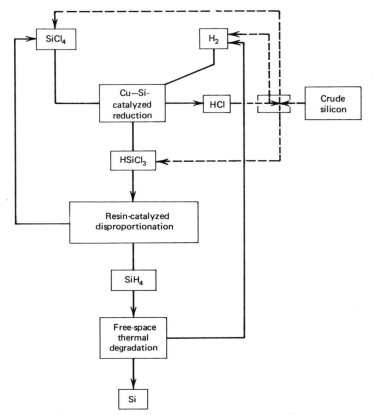

Figure 1.   Method for production of ultrapure silicon.

Trichlorosilane is produced in the United States by Union Carbide, Dow Corning, and Texas Instruments. General Electric has announced a start-up production date of 1983. Other worldwide producers are Dynamit-Nobel, Wacker, Chisso, and Shinetsu.

Silane production is <10 t/yr with prices of $150–500/kg. Union Carbide has announced a planned start-up date of September 1982, for a free-space silicon unit that will generate 100 t/yr (78). This marks the first electronic-grade silicon production that is substantially different from the Seimens approach. Producers of silane are Union Carbide, Liquid Carbonic, and Matheson.

Other specialty silanes used in electronic applications include dichlorosilane, disilane, tribromosilane [7789-57-3], and trisilylamine [13862-16-3].

Health and Safety Factors, Toxicology.   The acute hazards of silicon hydrides are of overwhelming importance when considering worker safety. Both the pyrophoric nature of compounds containing three or more atoms of hydrogen and the rapid production of HCl by the chlorine-containing compounds reduce long-term hazard considerations. At low concentrations, chlorosilanes affect nasal and pulmonary membranes. Only a minimal amount of toxicity information is available on these materials. The LD_{50} for trichlorosilane is 1050 mg/kg (79).

Organic Silanes

The organosilane of greatest commercial importance is methyldichlorosilane. Careful hydrolysis of this material with water affords polymethylhydrosiloxanes, which are used in the textile industry to waterproof and improve the wear resistance of fabrics. It is also used as a waterproofing agent in the leather industry, in the paper industry for sizing, in electronic applications, and in construction (80–82). Similar fluids based on ethyldichlorosilane [1789-58-8] have been developed for commercial application in the USSR, but they are not available elsewhere. Methyldichlorosilane is also used captively by silane and silicone producers in thermal condensation reactions to produce vinyl, phenyl, and cyanoalkyl precursors to silicone fluids. The addition reaction of methyldichlorosilane to fluorocarbon alkenes has enabled production of methyltrifluoropropyl silicone fluids, gums, and rubbers. Trialkoxysilanes, eg, triethoxysilane and trimethoxysilane, are being used to prepare a number of organic coupling agents utilized by the plastics industry as adhesion promotors (see Adhesives). Organosilanes containing one or more Si—H bonds have excellent reducing capabilities (83).

Physical Properties. The physical properties of organosilanes, as with inorganic silicon hydrides, are determined largely by the properties of the silicon atom (see Table 2). Silicon is larger and less electronegative than both carbon and hydrogen. The polarity of the Si—H bond is therefore opposite to that of the C—H bond. This difference in polarity imparts hydride character to the Si—H bonds of organosilanes. This difference in electronegativities is not as great as in ionic hydrides, eg, LiH, NaH, and CaH_2, and the Si—H bond is still largely (98%) covalent. The size of the silicon atom is greater than carbon, and this increase in atomic volume enables nucleophilic attack on the silicon to occur more readily than on carbon. Electrophilic attack on hydrogen bonded to silicon is also facilitated by the small steric constraints of hydrogen and the increased bond length for Si—H. In addition to the increased bond length, the Si—H bond energy is considerably lower than C—H and is reflected in the thermal stabilities of such bonds. Organohydrosilanes begin to decompose at 440–460°C through homolytic cleavage of the Si—H bond and subsequent radical formation. The length and strength of silane bonds are also manifested by their vibrational spectra. Vibration frequencies of silanes (2100–2260 cm^{-1}) are lower than those of C—H (2700–3000 cm^{-1}). The presence of inductive groups, eg, halogen, vinyl, phenyl, or fluoroalkyl, on silicon increases the vibrational frequency of the Si—H bond. Conversely, the presence of alkyl substituents generally decreases the vibrational frequency because of the donation of electron density.

Unlike carbon, the silicon atom may utilize vacant $3d$ orbitals to expand its valence beyond four to five or six and form new bonds with electron donors; this is shown by isolated amine complexes. The stability of the organosilane–amine complexes vary over a wide range and depend on the nature of the donor and acceptor (2). A comparison of some physical properties of carbon and silicon and their bonds to hydrogen is given in Table 3.

Chemical Properties. Organohydrosilanes undergo a wide variety of chemical conversions. The Si—H bond of organohydrosilanes reacts with elements of most groups of the periodic system, especially Groups VIA and VIIA. There are no known reactions where the Si–H bond is replaced by stable bonds of silicon with elements of Groups IIA, IIIA, and VIII (53).

Table 2. Properties of Commercial Organosilanes, Siloxanes, and Silazanes [a]

Compound	CAS Registry No.	Mol wt	Bp, °C$_{kPa}$ [b]	Mp, °C	d_4^{20}	n_D^{20}
CH_3SiH_3	[992-94-9]	46.1	-57	-157	0.6277[c]	
$(CH_3)_3SiSiH_3$	[18365-32-7]	104.3	80–81			
$C_2H_5SiH_3$	[2814-79-1]	60.2	-14	-180	0.6396[d]	
$C_4H_9SiH_3$	[1600-29-9]	88.2	56	-138	0.6764	1.3922
$C_6H_5SiH_3$	[694-53-1]	108.2	120		0.8681	1.5125
$C_6H_{13}SiH_3$	[1072-14-6]	116.3	114–115	-98	0.7182	1.4129
$C_{18}H_{37}SiH_3$	[18623-11-5]	284.6	195–196$_2$			
$(CH_3)_2SiH_2$	[1111-74-6]	60.2	-20	-150	0.6377[e]	
$CH_3(C_6H_5)SiH_2$	[766-08-5]	122.2	139		0.889	1.506
$(C_2H_5)_2SiH_2$	[542-91-6]	88.2	56	-134	0.6832	1.3920
$(C_2H_5O)_2SiH_2$	[18165-68-9]	120.2	90–92			
$(C_6H_5)_2SiH_2$	[775-12-2]	184.3	100–101$_4$		0.9964	1.5756
$CH_3SiH(Cl)_2$	[75-54-7]	115.0	41–42	-93	1.105	1.422
$CH_3SiH(OCH_3)_2$	[16881-77-9]	106.2	61			
$CH_3SiH(OC_2H_5)_2$	[2031-62-1]	134.3	94.5		0.829[f]	1.372[f]
$CH_3SiH(Cl)N(CH_3)_2$	[18209-60-4]	123.7	85–87			
$CH_3SiH[N(CH_3)_2]_3$	[3768-57-8]	176.4	112–113			
$CH_3SiH(O\overset{O}{\overset{\|}{C}}CH_3)_2$	[3435-15-2]	162.2	83–84$_6$		1.1081	1.4022
$CH_3SiH(C_2H_5)Cl$	[6374-21-6]	108.6	67–68		0.8816	1.4020
$(CH_3)_2SiH(Cl)$	[1066-35-9]	94.6	36	-111	0.851	111
$(CH_3)_2SiH(C_2H_5)$	[758-21-4]	88.2	46		0.6681	1.3783
$(CH_3)_2SiH(OC_2H_5)$	[14857-34-2]	104.2	54		0.7572	1.3683
$(CH_3)_2SiH[N(C_2H_5)_2]$	[13686-66-3]	131.3	109–110			
$(CH_3)_3SiH$	[993-07-7]	74.2	67	-136	0.6375	
$(CH_3O)_3SiH$	[2487-90-3]	122.2	86–87			1.3687
$C_2H_5SiH(Cl_2)$	[1789-58-8]	129.1	75	-107	1.0926	1.4148
$(C_2H_5)_2SiH(Cl)$	[1609-19-4]	122.7	100		0.889	1.4152
$(C_2H_5)_2SiH(CH_3)$	[760-32-7]	102.3	77–78		0.7054	1.3984
$(C_2H_5)_3SiH$	[617-86-7]	116.3	107–108	-157	0.7318	1.4119
$(C_2H_5O)_3SiH$	[998-30-1]	164.3	131.5	-170	0.875	1.337
$(2\text{-}ClC_2H_4O)_3SiH$	[10138-79-1]	267.6	117–118$_{0.3}$		1.2886	1.4577
$(C_3H_7)_3SiH$	[998-29-8]	158.4	173		0.7580	1.4272
$(C_6H_5)_2SiH(Cl)$	[1631-83-0]	218.8	143$_{1.3}$		1.118	1.581
$C_6H_5SiH(Cl_2)$	[1631-84-1]	177.1	65–66$_{1.3}$		1.212	1.526
$C_6H_5SiH(CH_3)Cl$	[1631-82-9]	156.7	113$_{13}$		1.054	1.571
$C_6H_5SiH(CH_3)_2$	[766-77-8]	136.3	156–157		0.8891	1.4995
$C_6H_5(H_2C{=}CH)SiH(CH_3)$	[17878-39-6]	148.3	56–57$_{0.9}$		0.891	1.5115
$(C_6H_5)_2SiH(CH_3)$	[776-76-1]	198.3	266–267		0.9945	1.5717
$(C_6H_5)_3SiH$	[789-25-3]	260.4	160–165$_{0.4}$	42.44		
$(C_6H_{13})_3SiH$	[2929-52-4]	284.6	160–161		0.7992	1.448
$(C_8H_{17})_3SiH$	[18765-09-8]	368.8	163–165$_{0.02}$		0.821	1.454
$[(CH_3)_2SiH]_2O$	[3277-26-7]	134.3	70–71		0.757	1.370
$[(CH_3)_2SiH]_2NH$	[15933-59-2]	133.3	99–100			
$(CH_3)_3SiOSiH(CH_3)_2$	[1438-82-0]	148.4	85–86		0.7578	1.3740
$[(CH_3)_3SiO]_2SiH(CH_3)$	[1873-88-7]	222.5	141–142			1.3815
$[CH_3SiHO]_4$	[2370-88-9]	240.5	134–135	-69	0.9912	1.3870
$[SiH(CH_3)_2]_2$	[814-98-2]	118.3	118.3	84–86		

[a] Refs. 84 and 85.
[b] To convert kPa to mm Hg, multiply by 7.5.
[c] At $-58°C$.
[d] At $-14°C$.
[e] At $-20°C$.
[f] At $25°C$.

Table 3. Comparison of Some Physical Properties of Carbon and Silicon and of Their Bonds to Hydrogen [a]

Element	Electronegativity	Covalent radius, nm	Coordination no.	Bond with hydrogen	
				Bond length, nm	Bond energy, kJ/mol [b]
silicon	1.8	0.117	4	0.149	335–365
hydrogen	2.1	0.030		0.074	432
carbon	2.5	0.077	4	0.106–0.110	368–506

[a] Ref. 55.

[b] To convert J to cal, divide by 4.184.

Reactions. *Oxidation.* Unlike the majority of inorganic silanes that are oxidized spontaneously on contact with oxygen and air, the simplest organosilane, ie, methylsilane, is stable to air and does not spontaneously inflame. It has been reported to explode in the presence of mercury and oxygen (86). The larger-chain alkylsilanes are more stable but ignite spontaneously when vaporized in oxygen under pressure. Phenyl and cyclohexylsilane can be distilled open to the atmosphere. Trialkyl- and triarylsilanes are more stable and have been distilled at as much as 325°C without decomposition (87). These differences between organosilanes and inorganic silanes are attributed to the electronic and steric effects of the organic substituents. The oxidizability of Si—H bonds is much greater compared to that of C—H bonds. This is manifested by the ease of oxidation of organohydrosilanes with metal oxides; the most noteworthy is mercuric oxide. Reactions of alkylsilanes with mercuric oxide in inert solvents, eg, toluene, afford nearly quantitative yields of silanols or disilanols.

$$R_2SiH_2 + 2\,HgO \xrightarrow[\text{-80 to 20°C}]{\text{toluene}} R_2Si(OH)_2 + 2\,Hg$$

Oxidation has also been cited as occurring in the cure of polymethylhydrosiloxane [9004-73-3] (PMHS) on cellulose acetate fibers. Investigation of the cured, cross-linked silicone shows no evidence of the Si—H bond. The same compound under an atmosphere of nitrogen does not cure and retains the Si—H bonds (82).

Hydrolysis. As with inorganic silanes, no reaction occurs between organohydrosilanes and water. The presence of acidic or alkaline catalysts, however, greatly accelerates the reaction according to the following scheme:

$$\geqslant\!SiH + H_2O \xrightarrow{\text{catalyst}} \geqslant\!SiOH + H_2$$

The ease of hydrolysis depends on the pH and is more rapid in alkaline than acidic conditions. Organosilanes are most stable in weakly acidic media. The rate and completeness of hydrolysis reactions is conveniently monitored by measurement of the rate and amount of hydrogen evolved. Alkaline hydrolysis is more widely used and has been more extensively studied. Alkaline catalysts may be organic, eg, triethylamine, pyridine, etc, or inorganic, eg, KOH, NaOH, NH_4OH, etc. The rate and completeness of these reactions depend on the nature and amount of catalyst, solvent, reagent concentration, temperature, and organic substituents of the organosilane. The substituents influence rate through steric and inductive effects. Such effects determine the ease of formation of the pentacoordinate silicon complex believed to be the transition state (88–89). The mechanism is thought to proceed through rapid attack of the base on the organohydrosilane with formation of a pentacoordinate silicon via bonding through $3d$ orbitals. Slow decomposition of this complex through interaction with a solvent proton affords the product.

$$\geq\!\!\text{SiH} + \text{OH}^- \xrightarrow{\text{fast}} \left[\geq\!\!\text{Si}\!<\!\!\begin{array}{c}\text{H}\\\text{OH}\end{array}\right]^- + \text{H-solvent} \xrightarrow{\text{slow}} \geq\!\!\text{SiOH} + \text{H}_2 + \text{solvent}^-$$

Steric and inductive effects determine the rate of formation of the pentacovalent silicon reaction complex. In alkaline hydrolysis, replacement of a hydrogen by alkyl groups, which have lower electronegativity and greater steric requirements, leads to slower hydrolysis rates. Replacement of alkyl groups with bulkier alkyl substituents has the same effect as demonstrated by the following series of decreasing reaction rates:

$$\text{C}_4\text{H}_9\text{SiH}_3 > \text{CH}_3(i\text{-C}_3\text{H}_7)\text{SiH}_2 > (\text{C}_2\text{H}_5)_2\text{SiH}_2 > \text{CH}_3(\text{C}_2\text{H}_5)_2\text{SiH} > (\text{C}_2\text{H}_5)_3\text{SiH} > (\text{C}_3\text{H}_7)_3\text{SiH}$$
$$[18165\text{-}88\text{-}3]$$

Mechanistically the rate-determining step is nucleophilic attack of the hydroxide ion on the more positive silicon atom in the Si—H bond. This attack has been related to the Lewis acid strength of the silane to act as an acceptor for a given attacking base (85). Similar inductive and steric effects apply for acid hydrolysis of organosilanes (90).

Alcohols, phenols, silanols, and carboxylic acids. The catalyzed reaction of organosilanes with hydroxyl-containing organic compounds affords organoalkoxy- and organoaryloxysilanes, usually in high yields. Alkali-metal oxides, hydrogen halides, and metal halides are most often used as catalysts. Of the metal halides, ZnCl_2 and SnCl_2 are the most commonly used.

$$\geq\!\!\text{SiH} + \text{HOR} \xrightarrow{\text{catalyst}} \geq\!\!\text{SiOR} + \text{H}_2$$

where R = alkyl or aryl. The reaction of organosilanes with alkali-metal alkoxides proceeds vigorously and in high yields. Aryloxy derivatives of silanes can be produced from phenols by catalysis with colloidal nickel. Yields of up to 90% have been obtained for the reactions of trialkylsilanes with phenols in the presence of this catalyst (53). Other metals, eg, copper, cobalt, chromium, zinc, and tin, also exhibit a catalytic effect. Reactivity increases for selected alcohols in the order $\text{CH}_3\text{OH} < \text{C}_2\text{H}_5\text{OH} < n\text{-C}_3\text{H}_7\text{OH} < n\text{-C}_4\text{H}_9\text{OH}$ (91). This suggests that the alkoxide is the active nucleophile.

$$\text{C}_6\text{H}_5\text{SiH}_3 + 3\,\text{ROH} \xrightarrow{\text{Cu}} \text{C}_6\text{H}_5\text{Si(OR)}_3 + 3\,\text{H}_2$$

Organic amines, eg, pyridine and piperidine, have also been used successfully as catalysts in the reactions of organosilanes with alcohols and silanols. The reactions of organosilanes with organosilanols lead to formation of siloxane bonds. Nickel, zinc, and tin also exhibit a catalytic effect.

$$\geq\!\!\text{SiH} + \text{HOSi}\!\leq \xrightarrow{\text{catalyst}} \geq\!\!\text{SiOSi}\!\leq + \text{H}_2$$

The reaction is of practical importance in the vulcanization of silicone rubbers (see Rubber compounding). Linear hydroxy-terminated polydimethylsiloxanes are conveniently cross-linked by reaction with methyldiethoxysilane or triethoxysilane [998-30-1]. Catalysts are amines, carboxylic acid salts of metals (Zn, Sn, Pb, Fe, Ba, Ca), and organotin compounds. Hydroxy-terminated polysiloxanes react with Si—H-containing polysiloxanes to produce cross-linked materials; metal salts of carboxylic acids are used as catalysts. The liberated hydrogen can act as a blowing agent to produce silicone foam rubbers (82).

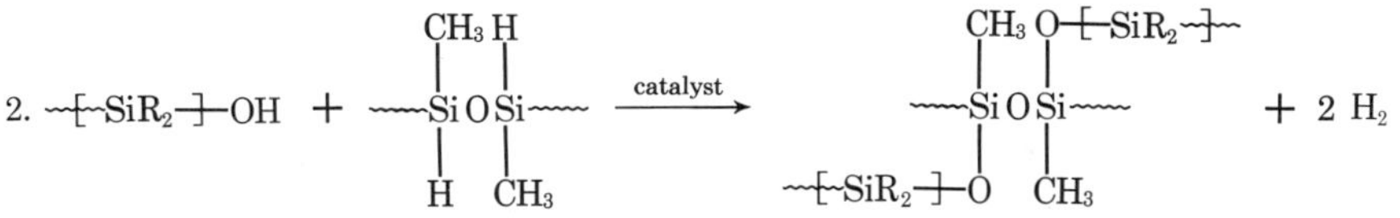

$$2. \; \text{~\{-SiR}_2\text{-\}-OH} \; + \; \underset{\underset{\displaystyle H}{|} \quad \underset{\displaystyle CH_3}{|}}{\overset{\overset{\displaystyle CH_3}{|} \quad \overset{\displaystyle H}{|}}{\text{~~~Si O Si~~~}}} \; \xrightarrow{\text{catalyst}} \; \underset{\underset{\displaystyle \text{~\{-SiR}_2\text{-\}-O}}{|} \quad \underset{\displaystyle CH_3}{|}}{\overset{\overset{\displaystyle CH_3}{|} \quad \overset{\displaystyle O\text{-\{-SiR}_2\text{-\}-~}}{|}}{\text{~~~Si O Si~~~}}} \; + \; 2 \; H_2$$

Whereas metal salts of carboxylic acids catalyze the above reactions, they are not sufficiently basic to cleave Si—H bonds. Mercury salts of organic acids with catalysis by silver perchlorate, however, do react to produce organoacyloxysilanes (92).

$$R'_3SiH \; \xrightarrow{Hg(OCR)_2} \; R'_3SiO\overset{\overset{\displaystyle O}{\|}}{C}R \; + \; R\overset{\overset{\displaystyle O}{\|}}{C}OH \; + \; Hg$$

Organoacyloxysilanes are also produced by reaction of organosilanes with carboxylic acids in the presence of strong mineral acids, eg, sulfuric and hydroiodic acids. Trialkylacyloxysilanes have also been obtained in 81–87% yield from monocarboxylic acids in the presence of aluminum and iodine.

$$R_3SiH \; + \; HO\overset{\overset{\displaystyle O}{\|}}{C}R' \; \xrightarrow[\text{Al + I}_2]{\text{H}_2\text{SO}_4 \text{ or HI or}} \; R_3SiO\overset{\overset{\displaystyle O}{\|}}{C}R' \; + \; H_2$$

Peracylation of polymethylhydrosiloxane to produce a cross-linked or cross-linkable material is achieved by reaction with acetic acid and is catalyzed by anhydrous zinc chloride (93). This reaction can be extended to monomeric organosilanes under similar conditions.

Amines and phosphines. As in reactions of alcohols and acids with organosilanes, reaction of the Si—H bond with amines and phosphines proceeds only under catalysis. Alkali metal amides or phosphines are the catalysts of choice and effect replacement of the Si—H bond with Si—N or Si—P bonds, respectively. Catalytic activity of the alkali metals for these reactions is K > Na > Li (53). Reactions of triorganosilanes with ammonia and amines have been the most widely studied. Ammonia affords the disubstituted nitrogen compounds or disilazanes with no trisubstitution being observed. This presumably occurs because of steric crowding at nitrogen.

$$2 \; (C_2H_5)_3SiH + NH_3 \xrightarrow{\text{NaNH}_2} [(C_2H_5)_3Si]_2NH + 2 \; H_2$$
$$[2117\text{-}18\text{-}2]$$

$$2 \; (C_6H_5)_3SiH + NH_3 \xrightarrow{\text{NaNH}_2} [(C_6H_5)_3Si]_2NH + 2 \; H_2$$
$$[4158\text{-}64\text{-}9]$$

Primary organic amines react with triethylsilane in the presence of the appropriate potassium amides to produce organoaminotriethylsilanes with yields of 82–92%.

$$(C_2H_5)_3SiH + RNH_2 \xrightarrow{\text{KNHR}} (C_2H_5)_3SiNHR + H_2$$

where R = C_2H_5 [3294-33-5], i-C_3H_7 [5277-20-3], and C_6H_5 [18106-48-4]. No reaction occurs with secondary or tertiary amines under similar reaction conditions. Steric factors are also important as branched primary amines require higher temperatures.

Halogens and halogen compounds. The reaction of organosilanes with halogens and halogen compounds usually proceeds in good yield through cleavage of the Si—H bond and formation of the silicon–halogen bond. This reaction can be achieved by direct action of halogen on the organosilane or by interaction with halogen-containing organic and inorganic compounds. Reaction with fluorine, however, does not proceed satisfactorily because of cleavage of not only the Si—H but also C—Si and C—H bonds. Fluorination of organosilanes has been achieved by reaction with inorganic fluorides, eg, AgF and SbF_3. Direct halogenation with chlorine, bromine, and iodine proceeds smoothly, however. Reactions are normally carried out in an inert solvent, eg, carbon tetrachloride, hexane, ethyl bromide, benzene, or chloroform. The reaction rate is further controlled by maintaining low reaction temperatures and controlling the rate of halogen addition. Steric requirements are not a significant factor, as evidenced by the facile iodination of tricyclohexylsilane [1629-47-6]. These reactions are fast and convenient and allow partial replacement of Si—H bonds (87).

$$R_3SiH + X_2 \rightarrow R_3SiX + HX$$

$$C_6H_5SiH_3 + Br_2 \rightarrow C_6H_5SiH_2Br + HBr$$
$$[1631\text{-}90\text{-}9]$$

where R = alkyl, cycloalkyl, or aryl and X_2 = Cl_2, Br_2, or I_2.

Organosilanes react with hydrogen halides to afford organohalosilanes and hydrogen. Hydrogen fluoride is not generally used for conversion of organosilanes to organofluorosilanes. The action of hydrogen fluoride cleaves Si—H and Si—C bonds readily, forming Si—F bonds. In the presence of copper salts, however, HF fluorinates the Si—H bond without splitting the Si—C bond (94). With hydrogen chloride, reactions are catalyzed with a Lewis acid, eg, $AlCl_3$. Such catalysts also facilitate reaction with hydrogen bromide.

$$\geqslant SiH + HX \xrightarrow{AlX_3} \geqslant SiX + H_2$$

where X = Cl or Br.

Hydrogen iodide iodinates trialkylsilanes in good yield in boiling carbon tetrachloride with no aluminum halide present (95). This can perhaps be explained on the basis that some free iodine is always present in equilibrium with hydrogen iodide.

$$R_3SiH + HI \rightarrow R_3SiI + H_2$$

A convenient synthesis of organochlorosilanes from organosilanes is achieved by reaction with inorganic chlorides of Hg, Pt, V, Cr, Mo, Pd, Se, Bi, Fe, Sn, and Cu. Both Sn and Cu are widely used; the latter has been used to prepare the sterically hindered triisopropylchlorosilane [13154-24-0] (96).

$$(i\text{-}C_3H_7)_3SiH + CuCl_2 \rightarrow (i\text{-}C_3H_7)_3SiCl$$

Organobromosilanes are obtained under similar conditions through reaction with cupric and mercuric bromides. Yields are high and partial replacement of Si–H bonds is possible (97). These materials have also been produced from reaction of organosilanes with organic bromides. Heating triethylsilane with allyl bromide, propyl bromide, and methyl β-bromopropionate produces triethylbromosilane [1112-48-7] in 10–100% yields (98). Organochlorosilanes have also been obtained from alkyl and alkenyl chlorides, eg, allyl, neopentyl, hexyl, and propyl chlorides. Unlike bromo compounds, these reactions proceed when catalyzed with small amounts of $AlCl_3$. Such reactions show great potential for reduction of alkyl halides (99).

$$C_6H_{11}Cl + (C_2H_5)_3SiH \xrightarrow[0°C]{AlCl_3} (C_2H_5)_3SiCl + C_6H_{12}$$
$$[994-30-9]$$

Geminal polyhalides also react with organosilanes under peroxide catalysis. For example, triethylsilane affords triethylchlorosilane in good yield upon reaction with carbon tetrachloride in the presence of benzoyl peroxide at 80°C (94,100–101).

$$(C_2H_5)_3SiH + CCl_4 \xrightarrow{bpo} (C_2H_5)_3SiCl + HCCl_3$$

Acid halides, eg, benzoyl chloride, acetyl chloride, and benzoyl bromide, have been used to prepare Si—Cl and Si—Br compounds from organosilanes. Acetyl chloride proceeds to higher yield when catalyzed by aluminum chloride.

Metals and organometallic compounds. There are no reports of the direct reaction of the Si—H bond in organosilanes with magnesium, zinc, mercury, aluminum, and other elements of the Groups IIA, IIB, and IIIA metals. The Group IA alkali metals, ie, sodium, potassium, and their alloys, react with arylsilanes in amines or ammonia to produce the arylsilyl derivatives of these metals. However, these reactions should be regarded as indirect examples of hydrogen replacement on silicon since they probably go through an amide intermediate. Direct reaction between arylsilanes and alkali metals occurs when alloys of potassium and sodium are used. The bond between the triarylsilicon moiety and the alkali metal is usually quite stable and reacts in a variety of ways (102–103).

$$(C_6H_5)_3SiH \xrightarrow{Na/K} (C_6H_5)_3SiK + NaH$$
$$[15487-82-8]$$

$$(C_6H_5)_3SiK + (C_6H_5)_3SiH \rightarrow (C_6H_5)_3SiSi(C_6H_5)_3$$
$$[1450-23-3]$$

$$(C_6H_5)_3SiK + RX \rightarrow (C_6H_5)_3SiR + KX$$

where R = CH_3 [791-29-7], C_6H_5 [1048-08-4], or $(CH_3)_3Si$ [1450-18-6] and X = Cl or Br.

The preparation of cyclic polysilanes from dialkylchlorosilanes and lithium metal in tetrahydrofuran presumably involves a silicon–lithium intermediate (104).

$$(CH_3)_2SiHCl \xrightarrow{Li} [(CH_3)_2Si]_{5-7}$$

Alkylation and arylation of organosilanes occur readily with alkyl and aryl alkali-metal compounds. Yields from these reactions are good, but they are influenced by steric requirements on both silane and metal compounds. There is little inductive effect by the organic groups attached to silicon as measured by the yield of products (105–106). These reactions proceed more readily in tetrahydrofuran and ethyl ether than in ligroin or petroleum ether.

$$R_n SiH_{4-n} + (4-n) R'M \rightarrow R_n SiR'_{4-n} + (4-n) MH$$

where R, R' = alkyl or aryl and M = Li, Na, or K.

Under normal reaction conditions, ie, solvent reflux and atmospheric pressure, little or no reaction occurs between organosilanes and organozinc, organocalcium, and organomagnesium compounds. Organozinc compounds react at higher temperatures (150°C) and Grignard reagents have replaced Si—H bonds in arylsilanes on prolonged refluxing in tetrahydrofuran. These reactions are the exception rather than the rule and have limited synthetic utility.

Addition of organosilanes to olefins. The addition of organosilanes to olefins is commonly termed hydrosilylation (see Silicon compounds, silylating agents). It is of commercial importance in the silane and silicone industry for the production of organofunctional coupling agents, low-temperature-vulcanizing (LTV) silicone rubbers and elastomers, and specialty monomers. The reaction involves addition of Si—H across an alkene to produce a Si—C-bonded alkane. The reaction is catalyzed by peroxides, uv light, ionizing radiation, metals, and metal compounds. Chloroplatinic acid is the catalyst usually employed for this reaction. It is effective at concentrations of 10^{-8} mol/mol silane.

Organosilanes as reducing agents in organic synthesis. The use of trialkylsilyl donor compounds to prepare silyl-protected functional compounds for organic synthesis is well documented. More recently, the use of organosilanes as reducing agents for organic synthesis has received attention (101). These reactions are based on the hydridic character of the Si—H bond to prepare C—H bonds. A catalyst is usually required to promote reduction. Selective reduction, ie, chemoselective and regioselective, can sometimes be obtained by careful selection of catalyst.

Reduction of alkenes to alkanes with organosilanes through ionic hydrogenation has been reported (107). This recent development provides greater effectiveness and enhanced chemoselectivity for reduction of alkenes (108–111).

A variety of organosilanes can be used, but triethylsilane in the presence of trifluoroacetic acid is most convenient. Use of this reagent enables reduction of alkenes to alkanes and reduces branched alkenes more readily than unbranched ones. Selective hydrogenation of branched dienes is also possible (107).

Deuterium-labeled hydrogenation products are also possible by employment of triethyldeuterosilane [1631-33-0]. This allows a synthetic route to compounds having a deuterium in a predetermined position.

Additionally, predominantly trans stereochemistry is observed. This stereoselectivity and the inertness of functionalities, eg, COOH, COOR, CONR$_2$, CN, OH, and X groups, make it synthetically useful for hydrogenation of steroids (107,112).

Ionic hydrogenation has also been used to reduce alcohols to alkanes. Trifluoroacetic acid–organosilane combinations have been used to reduce trityl alcohol, benzhydryl alcohol, benzyl alcohol, bis(cyclopropyl)carbinol, and triethylcarbinol to the corresponding alkanes in good yield (101). Alcohols that do not readily form carbonium ions or undergo skeletal rearrangement can be reduced satisfactorily with boron trifluoride in place of trifluoroacetic acid (113).

Alkenes can also be prepared from alkynes through the platinum-catalyzed addition of organosilane across the multiple bond. Treatment of the resultant silylethylene with hydrogen iodide affords the trans olefin. If peroxides are used as hydrosilylation catalysts, cis olefins are obtained (114). This method enables stereoselective and regioselective preparation of alkenes.

Aldehydes and ketones have also been reduced successfully in very high yield in the presence of organosilanes and tris(triphenylphosphine)chlororhodium(I). Similar complexes of platinum and ruthenium are also effective catalysts (94).

[39579-28-7]

This reducing system has been applied successfully to the reduction of terpene ketones. Menthone with phenylsilane affords menthol and neomenthol in a ratio of 1:9. Similar conditions with the more sterically demanding phenyldimethylsilane produced menthol exclusively (115). Depending on the silane used in the reduction, different products are isolated from α,β unsaturated carbonyl compounds. Trialkylsilanes react by 1,4-addition to produce saturated carbonyl compounds on hydrolysis. Diorganosilanes produce allylic alcohols after hydrolysis. These reactions furnish a unique method for selective reduction of unsaturated ketones and aldehydes (116). Catalyst complexes containing chiral phosphine ligands have been employed with organosilanes in asymmetric reductive hydrosilylation. By this technique, n-butyl pyruvate is reduced with 2-naphthyl(phenyl)silane [81771-25-7] in the presence of [(+)DIOP]Rh(S)Cl to the (S)-alcohol in 83% enantiomeric excess ((+)DIOP is (+)-2,3-O-isopropylidene-2,3-dihydroxy-1,4-bis(diphenylphosphine)butane) (94).

The organosiloxane most economically favored for large-scale reductions is poly-methylhydrosiloxane [9004-73-3] (PMHS). This compound in the presence of or-ganotin catalysts readily reduces aldehydes and ketones to the carbinols under mild and neutral conditions (117). Functional groups, eg, CH_2OR, $CONHR_2$, CN, NO_2, CO_2H, and X, are not reduced.

Amines have been obtained in nearly quantitative yield from imines on reaction with organosilanes in the presence of tris(triphenylphosphine)chlororhodium (I). The initial reaction is the addition of Si—H across the C=N bond to produce the mois-ture-sensitive silylamine (118). The amine is liberated on simple treatment with al-cohol.

$$(C_6H_5)_2C\!\!=\!\!NH + (C_2H_5)_2SiH_2 \xrightarrow[\text{(2) CH}_3\text{OH}]{\text{(1) (Rh)}} (C_6H_5)_2CHNH_2 + (C_2H_5)_2SiH(OCH_3)$$
$$[19753\text{-}79\text{-}8]$$

Pyridines have also been reduced through a similar mechanistic pathway. N-Trialkylsilyl-1,4-dihydropyridines have been obtained on reaction with triorganosi-lanes in the presence of Pd—C, Rh—C, and $PdCl_2$. Hydrolysis affords the 1,4-dihy-dropyridine, which is otherwise unobtainable (94,119).

Reduction of carboxylic acid halides to aldehydes, nitriles to imines or aldehydes, isocyanates to formamides, and carbodiimides to formamidines by organosilanes have been reported. In addition, reduction of sulfur compounds, deoxygenation of phosphine oxides, dehalogenation of chlorophosphines, and dehalogenation of haloalkanes with organosilanes have also been reported. Some of the more common reductions that have been reported are summarized in Table 4. The list does not include all organosilanes that have been used as reducing agents or all substrates that have been reduced (120).

Photolysis.   Irradiation of 2,2-bis(2,4,6-trimethylphenyl)-1,1,1,3,3,3-hexa-methyltrisilane [79184-72-8] in hydrocarbon solution yields tetramesityldisilene [80785-72-4], which can be isolated as a yellow-orange solid that is stable to 20°C and above in the absence of air (121).

Table 4. Selected Organosilanes as Reducing Agents [a,b]

Substrates	$(C_2H_5)_3SiH$	$(C_6H_5)_3SiH$	$(C_6H_5)_2SiH_2$	PMHS[c]	Products
C=C	++				CHCH
CX	+	+			CH
C=O	++	+	+	+	CH_2, CHOH, COR
ROH	+	++	+		RH
$\overset{\displaystyle O}{\underset{\displaystyle \parallel}{}}$ RCCl	++				RCHO
$\overset{\displaystyle O}{\underset{\displaystyle \parallel}{}}$ RCOR′	++				RCH_2OR'
$RCONR_2'$	+				RCH_2NR_2'
RCN	++				RCHO
RNCO	+				RNHCHO
$ArNO_2$				+	$ArNH_2$
RCH=NR′	+		+		RCH_2NHR'
P=O, PCl		+	++	+	PH

[a] Ref. 120.
[b] ++ = recommended; + = reported.
[c] PMHS = polymethylhydrosiloxane.

Manufacture and Processing. *Direct Process.* The preparation of organosilanes by the direct process was first reported in 1945, and it is still the primary method of preparation of organosilanes (122–123). It is used commercially to prepare organosilanes in the U.S., France, the FRG, Japan, and the USSR. By this method, $CH_3SiH(Cl)_2$, $(CH_3)_2SiH(Cl)$, and $C_2H_5SiH(Cl)_2$ are prepared and utilized as polymers and reactive intermediates. The synthesis involves the reaction of alkyl halides, eg, methyl and ethyl chloride, with silicon metal or silicon alloys in a fluidized bed at 250–600°C.

$$RX + Si \xrightarrow{250-600°C} R_3SiX + R_2SiX_2 + RSiHX_2 + R_2SiHX + R_4Si + RSiH_2X$$

where R = CH_3 or C_2H_5. The yields of the desired products can be maximized by adjusting temperature, contact time, catalyst (usually copper or copper salts), and catalyst content. Methyldichlorosilane yields can also be increased by the addition of hydrogen to the methyl chloride (124). The use of alloys of silicon with copper or cobalt activated with copper chloride permits higher yields of $C_2H_5SiH(Cl_2)$ and $(C_2H_5)_2SiH(Cl)$ (125).

The direct process has been extended to the production of alkoxysilanes. Reaction of silicon with ethanol or methanol containing less than 2000 ppm water produces $(C_2H_5O)_3SiH$ and $(CH_3O)_3SiH$ with cuprous chloride as the catalyst. Yields of trialkoxysilane in the instances where methyl and ethyl alcohols are employed are 86% and 91%, respectively (126–127).

$$ROH + Si \xrightarrow[220°C]{CuCl} (RO)_3SiH$$

Dialkylamino-substituted silanes have also been obtained by a similar process (128).

$$R_2NH + Si \rightarrow (R_2N)_3SiH + (R_2N)_2SiH_2$$

Similar products are formed by cleavage of disilanes by diethylamine. The yield of $[(C_2H_5)_2N]_2SiH(Cl)$ [*18291-25-3*] is ca 45% (129).

$$Cl_3SiSiCl_3 + 7\ (C_2H_5)_2NH \rightarrow [(C_2H_5)_2N]_2SiCl_2 + [(C_2H_5)_2N]_2SiH(Cl) + 3\ (C_2H_5)_2NH \cdot HCl$$
$$[18881\text{-}64\text{-}6]$$

Direct process still residues, which contain large amounts of methyldisilanes, can be converted to methylchlorosilanes by reaction with HCl at 400–900°C. Yields are low, however, and the process is of little preparative value.

Reduction with Metal Hydrides. Organosilanes can be synthesized most conveniently in pilot, bench, and lab scale by reduction of organic-substituted halo- and alkoxysilanes with metal hydrides. As with inorganic silanes, the most effective reducing agent is lithium aluminum hydride; it offers advantages of reduction under mild conditions, high yield, and good purity of reaction products (82–83). Commonly employed solvents are ether and tetrahydrofuran. Mechanistically, the reaction is represented as:

$$4\ R_nSiX_{(4-n)} + (4-n)\ LiAlH_4 \rightarrow 4\ R_nSiH_{(4-n)} + (4-n)\ LiX + (4-n)\ AlX_3$$

$$4\ R_nSiX_{4-n} + (4-n)\ LiAlH_4 \rightarrow 4\ R_nSiH_{4-n} + (4-n)\ LiX + (4-n)\ AlX_3$$

where X = F, Cl, Br, I, OR, OC_6H_5, SR, or NR_2.

The versatility of lithium aluminum hydride permits synthesis of alkyl, alkenyl, and arylsilanes. Silanes containing functional groups, such as chloro, amino, and alkoxyl in the organic substituents, can also be prepared. Mixed compounds containing both SiCl and SiH cannot be prepared from organopolyhalosilanes using lithium aluminum hydride. Reduction is invariably complete.

Other reducing agents that have been used to prepare organosilanes are lithium hydride, sodium hydride and the lithium, sodium, potassium, and aluminum borohydrides. Of these reagents, lithium hydride has been most widely used. This compound is a weaker reducing agent than lithium aluminum hydride and lower yields are generally obtained. Satisfactory reductions generally require higher temperatures, prolonged reaction times, and excess LiH. Reductions are run in high boiling solvents, eg, dioxane and dibutyl ether. This reducing agent is usually employed where there is a danger of side reactions with the aluminum halides formed as by-products from lithium aluminum hydride reductions. Deuterated silanes can also be prepared from lithium deuteride or lithium aluminum deuteride without change in reduction procedure. Yields of organohalosilanes and organoalkoxysilanes are listed in Table 5.

Organosilanes are also produced by reaction of organohalosilanes and organoalkoxysilanes with organometallic compounds. Sterically hindered Grignard reagents with a proton in the beta position relative to MgX are the most effective for this type reduction (130–131). Although used to prepare the first isolated organosilanes, this method is of limited value. Organolithium reagents, eg, t-butyllithium, have also produced organohydrosilanes on reaction with organochlorosilanes and tetrahalosilanes. Tri-t-butylsilane [*18159-55-2*], which cannot be prepared by other reaction methods, is synthesized successfully by this route (132). Success of this reaction is attributed to the t-butyllithium acting as a reducing agent.

$$t\text{-}C_4H_9SiCl_3 + t\text{-}C_4H_9Li \rightarrow (t\text{-}C_4H_9)_3SiH$$
$$[18171\text{-}74\text{-}9] \qquad\qquad 10\%$$

$$SiF_4 + t\text{-}C_4H_9Li \rightarrow (t\text{-}C_4H_9)_3SiH$$
$$80\%$$

Table 5. Yields of Organosilanes via Reduction with Lithium Aluminum Hydride and Lithium Hydride[a]

Compound	CAS Registry No.	Yield with LiAlH$_4$, %	Yield with LiH, %
CH$_3$SiH$_3$	[992-94-9]	80–90	
ClCH$_2$SiH$_3$	[10112-09-1]	80	
C$_2$H$_5$SiH$_3$	[2814-79-1]	80–90	
(C$_2$H$_5$)$_2$SiH$_2$	[542-91-6]	80–90	66
(C$_2$H$_5$)$_3$SiH	[617-86-7]	80–90	20
(C$_2$H$_3$)$_3$SiH	[2372-31-8]	52	39
C$_2$H$_5$(CH$_2$=CH)SiH$_2$	[18243-33-9]		24
C$_4$H$_9$SiH$_3$	[1600-29-9]	80–90	
CH$_2$(CH$_2$)$_3$SiH$_2$	[288-06-2]	47	27
C$_3$H$_7$SiH$_3$	[13154-66-0]	80	
C$_6$H$_5$SiH$_3$	[694-53-1]	86	
(C$_6$H$_5$)$_2$SiH$_2$	[775-12-2]	78	
C$_6$H$_{13}$SiH$_3$	[1072-14-6]		65
(C$_6$H$_{13}$)$_3$SiH	[2929-52-4]	94	

[a] Ref. 55.

Disproportionation. Disproportionation reactions have also been used to prepare organosilanes. These reactions involve interaction of organosilanes with other silicon compounds containing organic, alkoxy, and halogen groups bound to silicon. Reactions are catalyzed by a variety of materials including alkali metals, alkali metal alcoholates, and Lewis acids, eg, aluminum, zinc, iron, and boron halides (65). Aluminum chloride is the most active and widely used catalyst. It enables facile preparation of various alkyl and dialkylsilanes according to the following scheme:

$$R_n SiX_{(4-n)} + (4-n)\ R'_3SiH \xrightarrow{AlCl_3} R_n SiH_{(4-n)} + (4-n)\ R'_3SiX$$

where R,R' = alkyl; X = Cl or Br; and n = 1 or 2.

Organochlorosilanes containing Si—H disproportionate in the presence of aluminum chloride without addition of more organosilane. Organic groups can be replaced by hydrogen (133).

$$2\ C_2H_5SiH(Cl_2) \xrightarrow{AlCl_3} C_2H_5SiH_2Cl + C_2H_5SiCl_3$$
$$[1789\text{-}58\text{-}8] \qquad [10536\text{-}78\text{-}4]\ \ [115\text{-}21\text{-}9]$$

$$C_6H_5(CH_3)SiH_2 \xrightarrow{AlCl_3} (C_6H_5)_4Si + CH_3SiH_3 + (CH_3)_2SiH_2$$
$$[766\text{-}08\text{-}5] \qquad [1048\text{-}08\text{-}4]$$

Similar disproportionation reactions are catalyzed by organic catalysts, eg, adiponitrile, pyridine, and dimethylcyanamide.

Grignard Reagents. The wide variety of organosilanes are most commonly prepared through reaction of inorganic and organochlorosilanes with Grignard reagents (see Grignard reaction). The Grignard reagent is utilized to transfer organic groups to silicon. By varying the silane and organic moiety of the Grignard, a large number of organosilanes can be synthesized in good-to-excellent yields. In general this process is very versatile and more easily directed than direct process or disproportionation reactions. The general reaction is

$$H_{(4-n)}SiX_n + n\ RMgZ \rightarrow R_nSiH_{(4-n)} + n\ MgXZ$$

where $n = 2$ or 3; R = alkyl or aryl; X = F, Cl, Br, or OR; and Z = Cl or Br. Reactions are normally carried out in ether or tetrahydrofuran. Replacement of halogen or alkoxy groups bound to silicon proceeds in a stepwise fashion. Reaction of chlorosilanes containing Si—H with stoichiometric deficiencies of Grignard reagent makes possible the synthesis of organohalosilanes by partial halogen substitution.

$$HSiX_3 + RMgX \rightarrow RSiHX_2$$

where X = Cl, Br, or OR; R = alkyl or aryl.

Organosilanes containing mixed organic groups can be prepared by reaction of organodihalosilanes and diorganohalosilanes or alkoxy derivatives with alkyl or aryl Grignards.

$$RSiH(X)_2 + R'MgX \rightarrow RR'SiH(X)$$

$$RSiH(X)_2 + 2\ R'MgX \rightarrow RR'_2SiH$$

$$R_2SiH(X) + R'MgX \rightarrow R_2R'SiH$$

where X = halogen, alkoxy.

The reactivity of the Si—H bond with Grignard reagents depends also on the solvent. Reactions in refluxing tetrahydrofuran produce substitution of Si—H by the Grignard reagent (128).

$$(C_6H_5)_2SiH_2 + C_6H_5MgBr \rightarrow (C_6H_5)_3SiH$$

More basic organometallic compounds, eg, aryl and alkyl lithium reagents, are not suited to the preparation of organosilanes. Cleavage of the Si—H bond with formation of alkali-metal halide occurs.

$$HSiX_3 + 4\ RLi \rightarrow R_4Si + LiH + 3\ LiCl$$

Addition to Olefins. Organohydrosilanes can also be prepared by addition of halosilanes and organosilanes containing multiple Si—H bonds to olefins. These reactions are catalyzed by platinum, platinum salts, peroxides, ultraviolet light, or ionizing radiation.

$$H_2SiCl_2 + RCH{=}CH_2 \xrightarrow{H_2PtCl_6} RCH_2CH_2SiH(Cl)_2$$

$$RSiH_3 + R'CH{=}CH_2 \xrightarrow{H_2PtCl_6} R'CH_2CH_2SiH_2(R)$$

Economic Aspects. The only organic silane that is produced in a large scale is methyldichlorosilane. It is a principal by-product in the direct process for methylchlorosilane production. Since there is no distinct process technology for the production of methyldichlorosilane, its availability varies in accord with both overall methylchlorosilane production and its demand in silicone production. Domestic producers of methyldichlorosilane are Union Carbide, General Electric, and Dow Corning. Pricing is $2.75–4.00/kg. Polymethylhydrosiloxane, the polymeric derivative of methyldichlorosilane, is also available from these suppliers with pricing at $5.00–7.00/kg.

Other specialty organic silicon hydrides bear much higher pricing. The most significant products are dimethylchlorosilane, tetramethyldisiloxane, triethylsilane, diethylmethylsilane, methylsilane, diphenylsilane, and various hydride-containing silicone copolymers. The cost of these materials varies from $25.00–200.00/kg. Producers of these materials include General Electric, Dow Corning, and Petrarch Systems.

BIBLIOGRAPHY

"Silicon Compounds" in *ECT* 1st ed., Vol. 12, pp. 365–392, by E. G. Rochow, Harvard University; "Silicon Compounds, Silanes" in *ECT* 2nd ed., pp. 172–215, by Charles H. Van Dyke, Carnegie-Mellon University.

1. P. E. Verkade, *Chem. Weeklbl.* **47,** 309 (1951).
2. V. Bazant, V. Chvalovsky, and J. Rathovsky, *Organosilicon Compounds*, Vol. 1, Academic Press, Inc., New York, 1965.
3. A. Stock, *Hydrides of Boron and Silicon*, Cornell University Press, Ithaca, New York, 1933.
4. R. West and T. J. Barton, *Chem. Ed.* **57,** 165, 364 (1980).
5. R. West and J. Michl, *Science* **214,** 1344 (1981).
6. A. G. Brook, F. Abdesaken, and B. Gutekunst, *Proceedings of the XV Silicon Symposium*, Duke University, Durham, N.C., March 1981.
7. R. Borreson and co-workers, *Solid State Technol.* **21,** 43 (1978).
8. E. G. Rochow in *Comprehensive Inorganic Chemistry*, Pergamon Press, Inc., Elmsford, N.Y., 1973.
9. R. Walsh, *Acc. Chem. Res.* **14,** 246 (1981).
10. D. G. White and E. G. Rochow, *J. Am. Chem. Soc.* **76,** 3897 (1954).
11. S. N. Brorisov, M. G. Voronkov, and B. N. Dolgov, *Dokl. Akad. Nauk SSSR* **93,** 114 (1957).
12. Ger. Offen. 1,102,117 (May 18, 1954), (to Seimens-Halska A.G.).
13. W. Claassen and J. Bloehm, *Philips J. Res.* **36**(2), 122 (1981).
14. D. J. Delong, *Solid State Technol.* **15**(10), 43 (Aug. 19, 1972).
15. U.S. Pat. 3,400,660 (Aug. 19, 1975), H. Bradley (to Union Carbide).
16. W. R. Runyan, *Silicon Semiconductor Technology*, McGraw-Hill, Inc., New York, 1965.
17. S. A. Arutyunyan and E. N. Sarkisyan, *Arm. Khim. Zh.* **34**(11), 909 (1981).
18. T. G. Mkryan, E. N. Sarkisyan, and S. A. Arutyunyan, *Arm. Khim. Zh.* **34**(1), 3 (1981).
19. U.S. Pat. 4,239,811 (Dec. 16, 1980), B. M. Kiemlage (to IBM).
20. B. Sternbach and A. G. MacDiarmid, *J. Am. Chem. Soc.* **81,** 5109 (1959).
21. J. S. Peake, W. H. Nebergall, and Y. T. Chem, *J. Am. Chem. Soc.* **74,** 1526 (1952).
22. L. G. L. Ward and A. G. MacDiarmid, *J. Am. Chem. Soc.* **82,** 2151 (1960).
23. M. Aberdini, C. H. Van Dyke, and A. G. MacDiarmid, *J. Inorg. Nucl. Chem.* **25,** 307 (1963).
24. S. Cradock, G. A. Gibbon, and C. H. Van Dyke, *Inorg. Chem.* **6,** 1751 (1967).
25. E. Amberger and E. Muhlhofer, *J. Organomet. Chem.* **12,** 557 (1968).
26. E. Amberger, R. Romer, and A. Layer, *J. Organomet. Chem.* **12,** 417 (1968).
27. M. A. Ring and D. M. Ritter, *J. Am. Chem. Soc.* **83,** 802 (1961).
28. S. P. Garrity and M. A. Ring, *Inorg. Nucl. Chem. Lett.* **4,** 77 (1968).
29. D. S. Rustad and W. L. Jolly, *Inorg. Chem.* **6,** 1986 (1967).
30. R. C. Kennedy, L. P. Freeman, A. P. Fox, and M. A. Ring, *J. Inorg. Nucl. Chem.* **28,** 1373 (1966).
31. J. A. Morrison and M. A. Ring, *Inorg. Chem.* **6,** 100 (1967).
32. E. A. Groschwitz, W. M. Ingle, and M. A. Ring, *J. Organomet. Chem.* **9,** 421 (1967).
33. W. J. Boldue and M. A. Ring, *J. Organomet. Chem.* **6,** 202 (1966).
34. R. Schwarz and F. Heinrich, *Z. Anorg. Chem.* **221,** 227 (1935).
35. S. D. Gokhale and W. L. Jolly, *Inorg. Chem.* **3,** 946 (1964).
36. E. J. Spanier and A. G. MacDiarmid, *Inorg. Chem.* **1,** 432 (1962).
37. S. D. Gokhale, J. E. Drake, and W. L. Jolly, *J. Inorg. Nucl. Chem.* **27,** 1911 (1965).
38. E. J. Spanier and A. G. MacDiarmid, *Inorg. Chem.* **2,** 215 (1963).
39. S. D. Gokhale and W. L. Jolly, *Inorg. Chem.* **3,** 1141 (1964).
40. *Ibid.*, **4,** 596 (1965).
41. M. Abedini and A. G. MacDiarmid, *Inorg. Chem.* **5,** 2040 (1966).
42. A. A. Kirpichinikova and co-workers, *Zh. Vses. Khim. Ova.* **22,** 465 (1977).
43. B. Scott and M. Brodsky, *Bull. Am. Phys. Soc.* **25,** 299 (1980).
44. T. Hiiao and co-workers, *J. Appl. Phys.* **52**(12), 7453 (1981).

45. H. J. Emeleus and K. Stewart, *Trans. Faraday Soc.* **32,** 1677 (1936).
46. N. Nihi and G. J. Mains, *J. Phys. Chem.* **68,** 304 (1964).
47. B. Reimann and R. Potzinger, *Ber. Bunsenges. Phys. Chem.* **80,** 565 (1976).
48. G. A. Gibbon, T. Rosseau, C. H. Van Dyke, and G. J. Mains, *Inorg. Chem.* **5,** 114 (1966).
49. W. C. Johnson and J. R. Hogness, *J. Am. Chem. Soc.* **56,** 1252 (1934).
50. W. C. Johnson and S. Isenberg, *J. Am. Chem. Soc.* **57,** 1359 (1935).
51. F. Feher and W. Tromm, *Z. Anorg. Allgem. Chem.* **282,** 29 (1955).
52. R. Schwarz and E. Konrad, *Ber.* **55,** 3242 (1952).
53. R. Schwarz and F. Heinrich, *Z. Anorg. Allgem. Chem.* **295,** 206 (1935).
54. H. Kaulsley and H. Pfleger, *Z. Anorg. Allgem. Chem.* **295,** 206 (1958).
55. A. D. Petrov, B. F. Moronov, V. A. Ponomorenko, and R. A. Cherpyshev, *Synthesis of Organosilicon Monomers*, Consultants Bureau, New York, 1964.
56. D. T. Hurd, *J. Am. Chem. Soc.* **67,** 1545 (1945).
57. U.S. Pat. 2,458,703 (Jan. 11, 1949), D. B. Hatcher (to Libby-Owens-Ford Glass Company).
58. D. L. Bailey, personal communication. This was performed by E. O. Brimm at Union Carbide in the 1940s.
59. J. Y. P. Mui, D. Seyferth, *Investigation of the Hydrogenation of Silicon Tetrachloride*, DOE/JPL/955382-79/8, Department of Energy, Washington, D.C., 1981.
60. PCT Int. Appl. WO 81-03,168 (Nov. 12, 1981), K. R. Sarma and M. J. Rice, Jr. (to Motorola).
61. W. Sundermeyer and O. Glemser, *Angew. Chem.* **70,** 625 (1958).
62. U.S. Pat. 4,051,136 (Aug. 9, 1977), R. E. Franklin, W. A. Francis, and G. Taranaron (to Union Carbide).
63. O. Glemser and W. Lohman, *Z. Anorg. Allgem. Chem.* **275,** 260 (1954).
64. U.S. Pat. 2,627,451 (Feb. 3, 1953) and U.S. Pat. 2,735,861 (Feb. 21, 1956), C. E. Erickson and G. H. Wagner (to Union Carbide and Carbon Corporation).
65. U.S. Pat. 2,745,860 (May 15, 1956), D. L. Bailey (to Union Carbide).
66. H. J. Emeleus and N. Miller, *J. Chem. Soc.,* 819 (1939).
67. U.S. Pat. 2,732,282 (Jan. 24, 1956), D. L. Bailey, G. H. Wagner, and P. W. Shafer (to Union Carbide).
68. Ger. Offen. 2,550,076 (1976), G. Marin.
69. U.S. Pat. 3,968,199 (July 6, 1976), C. J. Bakey (to Union Carbide).
70. E. Helfrich and J. Hausen, *Ber.* **57B,** 795 (1924).
71. C. J. Wilkins, *J. Chem. Soc.,* 3409 (1953).
72. H. Brederman, T. J. Thor, and H. J. Waterman, *Research* **7,** 829 (1959).
73. O. Glemser and W. Lohman, *Z. Anorg. Allgem. Chem.* **70,** 628 (1958).
74. Ger. Offen. 949,943 (1956), O. Glemser.
75. D. J. Hurd, *J. Am. Chem. Soc.* **67,** 1545 (1945).
76. H. L. Jackson, F. D. Marsh, and E. L. Muetterties, *Inorg. Chem.* **2,** 43 (1963).
77. N. J. Archer, R. N. Hazeldine, and R. V. Parrish, *J. Organomet. Chem.* **81,** 335 (1974).
78. U.S. Pat. 3,745,043 (July 10, 1973), A. Bradley (to Union Carbide).
79. *Registry of Toxic Effects of Chemical Substances*, National Institute of Occupational Safety and Health (NIOSH), Washington, D.C., 1976.
80. F. Fortes, *Ind. Eng. Chem.* **46,** 2325 (1954).
81. R. R. McGregor, *Silicones and Their Uses*, McGraw-Hill, Inc., New York, 1954.
82. W. Noll, *Chemistry and Technology of Silicones*, Academic Press, Inc., New York, 1968.
83. Y. Nagai, *Org. Prep. Proceed. Int.* **12,** 15 (1980).
84. V. Bazant, V. Chvalovsky, and J. Rathovsky, *Organosilicon Compounds*, Vol. 2(1), Academic Press, Inc., New York, 1965.
85. B. Arkles, W. R. Peterson, and R. Anderson, *Silicon Compounds*, *Register and Review*, 2nd ed., Petrarch Systems, Inc., Bristol, Pa., 1982.
86. H. J. Emeleus and K. Stewart, *Nature* (*London*), 397 (1935).
87. C. Eaborn, *Organosilicon Compounds*, Butterworths Scientific Publications, Lander, England, 1960.
88. J. E. Baines and C. Eaborn, *J. Chem. Soc.,* 4023 (1955).
89. L. H. Sommer, O. Bennet, P. G. Campbell, and D. R. Weyenberg, *J. Am. Chem. Soc.* **79,** 3295 (1957).
90. J. E. Baines and C. Eaborn, *J. Chem. Soc.,* 7436 (1956).
91. L. H. Sommer and C. L. Frye, *J. Am. Chem. Soc.* **82,** 4118 (1960).

92. C. Eaborn, *J. Chem. Soc.*, 2517 (1955).
93. U.S. Pat. 2,658,908 (Nov. 10, 1953), S. Nitzsche and E. Pirson (to Wacker Chemie).
94. Y. Nagai, *Org. Prep. Proced. Int.* **12,** 15 (1980).
95. M. G. Voronkov and Yu. I. Khudobin, *Izv. Akad. Nauk SSSR*, 805 (1956); *Chem. Abstr.* **51,** 3440 (1957).
96. R. G. Cunico and L. Bedell, *J. Org. Chem.* **45,** 4797 (1980).
97. H. H. Anderson, *J. Am. Chem. Soc.* **69,** 2600 (1958).
98. H. Westermark, *Acta. Chem. Scand.* **8,** 1086 (1954).
99. M. P. Doyle and C. T. West, *J. Org. Chem.* **41,** 1393 (1976).
100. Y. Nagai, K. Yamakazi, and I. Shiojuna, *J. Organomet. Chem.* **9,** 25 (1967).
101. Y. Nagai, K. Yamakazi, I. Shiojima, N. Kobori, and M. Hayashi, *J. Organomet. Chem.* **9,** 21 (1967).
102. R. A. Benkeser, H. Landesman, and D. J. Foster, *J. Am. Chem. Soc.* **74,** 648 (1952).
103. R. A. Benkeser and D. J. Foster, *J. Am. Chem. Soc.* **74,** 4200 (1952).
104. U.S. Pat. 4,276,424 (June 30, 1981), W. R. Peterson, Jr., and B. Arkles (to Petrarch Systems, Inc.).
105. R. A. Benkeser and F. J. Riel, *J. Am. Chem. Soc.* **73,** 3472 (1951).
106. H. Gilman and J. J. Goodman, *J. Org. Chem.* **22,** 45 (1957).
107. D. N. Kursanov, Z. N. Parnes, and N. M. Loim, *Synthesis*, 633 (1974).
108. M. P. Doyle and C. C. McOsker, *J. Org. Chem.* **43,** 693 (1978).
109. M. P. Doyle, C. C. McOsker, N. Ball, and C. T. West, *J. Org. Chem.* **42,** 1922 (1977).
110. F. A. Carey and C. W. Hsu, *J. Organomet. Chem.* **19,** 29 (1969).
111. F. A. Carey and C. W. Hsu, *J. Org. Chem.* **36,** 758 (1971).
112. E. W. Colvin, *Chem. Soc. Rev.* **7,** 15 (1978).
113. M. G. Adlington, M. Orfanopoulos, and J. K. Fry, *Tetrahedron Lett.*, 2955 (1976).
114. R. A. Benkeser, M. L. Nelson, and J. V. Swisher, *J. Am. Chem. Soc.* **83,** 4385 (1961).
115. J. Ogima, M. Nihonyanagi, and Y. Nagai, *Bull. Chem. Soc. Japan* **45,** 3722 (1972).
116. I. Ojima, T. Kogure, and Y. Nagai, *Tetrahedron Lett.*, 5035 (1972).
117. J. L. Lipowitz and S. A. Bowman, *J. Org. Chem.* **38,** 162 (1973).
118. K. A. Andrianov, M. I. Filimonova, and S. I. Sidorov, *J. Organometal. Chem.* **144,** 27 (1978).
119. N. C. Cook and J. E. Lyons, *J. Am. Chem. Soc.* **87,** 3283 (1965); **88,** 3396 (1966).
120. *Hydrosilanes as Reducing Agents*, Technical Bulletin, Chisso Corporation, Tokyo, Japan, 1980.
121. R. West, M. J. Fink, and J. Michl, *Science* **214,** 1343 (1981).
122. E. G. Rochow, *J. Am. Chem. Soc.* **67,** 693 (1945).
123. U.S. Pat. 2,380,995 (Aug. 7, 1945), E. G. Rochow (to General Electric Company).
124. U.S. Pat. 2,380,998 (Aug. 7, 1945), M. M. Sprung and W. F. Gilliam.
125. Brit. Pat. 681,387 (1952), R. Decker and H. Holz.
126. Jpn. Kokai Tokyo Koho 80 28,929 (Apr. 6, 1980), S. Suzuki, T. Imaki, and T. Yamaura (to Mitsubishi Chem).
127. Jpn. Kokai Tokyo 80 28,928 (Apr. 6, 1980), S. Suzuki, T. Imaki, and T. Yamaura (to Mitsubishi Chem).
128. B. Kanner and W. B. Herdle, *XV Organosilicon Symposium*, Duke University, Durham, N.C., March 27–28, 1981.
129. H. Breederveld, T. Thoor, and U. Waterman, *Research* **7,** 829 (1954).
130. M. C. Harvey, W. H. Nebergall, and J. S. Peake, *J. Am. Chem. Soc.* **79,** 7762 (1957).
131. M. B. Lacout-Loustalet, J. P. Dupin, F. Metras, and J. Valade, *J. Organomet. Chem.* **31,** 337 (1971).
132. E. M. Dexheimer and L. Spialter, *Tetrahedron Lett.*, 1771 (1975).
133. J. L. Speier and R. E. Zimmerman, *J. Am. Chem. Soc.* **77,** 6395 (1955).

BARRY ARKLES
WILLIAM R. PETERSON, JR.
Petrarch Systems, Inc.

SILICON ETHERS AND ESTERS

Silicon esters are silicon compounds that contain an oxygen bridge from silicon to an organic group, ie, $\geqslant$SiOR. The oldest reported organic silicon compounds contain four oxygen bridges and are often named as derivatives of orthosilicic acid, $Si(OH)_4$. The most conspicuous material is tetraethyl orthosilicate, $Si(OC_2H_5)_4$. With the advent of organosilanes that contain silicon–carbon bonds (Si—C), an organic nomenclature was developed by which compounds are named as alkoxy derivatives. For example, $Si(OC_2H_5)_4$ becomes tetraethoxysilane. The compound $CH_3Si(OCH_3)_3$ is named methyltrimethoxysilane. The latter usage is preferred although current literature still contains the older terms. Acyloxysilanes, eg, tetraacetoxysilane, $Si(OOCCH_3)_4$, are also members of this class. The chemistry and applications of acyloxysilanes are significantly different than those of the alkoxysilanes.

The applications for alkoxysilanes range broadly. They are classified roughly by whether the Si—OR bond is expected to remain intact or be hydrolyzed in the final application. The susceptibility to hydrolysis, volatility, and other properties of alkoxysilanes predicate their particular applications. Applications in which the Si—OR bond is hydrolyzed include binders for foundry-mold sands used in investment and thin-shell castings, binders for refractories, resins, coatings, low heat glasses, cross-linking agents, and adhesion promoters. Applications in which the Si—OR bond remains intact include lubricant, heat-transfer, hydraulic, dielectric, and diffusion-pump fluids. In general, lower molecular weight compounds, eg, tetraethoxysilane and tetramethoxysilane, are used in reactive applications whereas such compounds as tetrabutoxysilane and hexakis(2-ethylbutoxy)disiloxane are associated with mechanical applications. Tetraethoxysilane and its polymeric derivatives account for >90% of all production.

Properties

The alkoxysilanes possess excellent thermal stability and a broad temperature range of liquid behavior which widens with length and branching in the substituents (see Table 1). The physical properties of the silane esters, particularly the polymeric esters containing siloxane bonds, ie, Si—O—Si, are often likened to the silicone oils. They have low pour points and similar temperature–viscosity relationships. The alkoxysilanes generally have sweet, fruity odors that become less apparent as molecular weight increases. With the exception of tetramethoxysilane, which can be absorbed into corneal tissue causing eye damage, the alkoxysilanes generally exhibit low levels of toxicity. One unique class of monomeric esters derived from trialkanolamines called silatranes have silicon–nitrogen coordination. The silatranes exhibit distinct physical, chemical, and, most important, physiological properties and are considered apart from esters.

Aryloxy- and acyloxysilanes are often solids. The aryloxysilanes have excellent thermal stability. Acyloxy- and mixed acyloxyalkoxysilanes have poor thermal stability. Thermal decomposition has been noted at as low as 110°C and is generally observed by 170°C.

$$Si(O\overset{\overset{\displaystyle O}{\|}}{C}CH_3)_4 \longrightarrow SiO_2 + 2\,(CH_3\overset{\overset{\displaystyle O}{\|}}{C})_2O$$

The significant difference between the alkoxysilanes and the silicones is the susceptibility of the Si—OR bond to hydrolysis. The simple alkoxysilanes are often operationally viewed as liquid sources of silicon dioxide. The hydrolysis reaction, which yields polymers of silicic acid which can be dehydrated to silicon dioxide, is of considerable importance. The stoichiometry of hydrolysis for the ethoxysilanes is

$$Si(OC_2H_5)_4 + 2\,H_2O \xrightarrow[\text{acid or base}]{} SiO_2 + 4\,C_2H_5OH$$

Silicon dioxide never forms directly during hydrolysis. Intermediate ethoxy derivatives of silicic acid and polysilicates form as hydrolysis progresses. The polysilicates grow in molecular weight and chain length until most or all of the ethoxy groups are removed and a nonlinear network of Si—O—Si remains. The development of cyclic structures containing 3–6 silicon atoms also occurs. The viscosity of solutions increases until gelation or precipitation (1). Partially hydrolyzed materials of this type often contain more than enough silanols (SiOH) to displace most of the remaining ethoxy groups in an acid- or base-catalyzed condensation. The stoichiometric equation for partial hydrolysis is

$$Si(OC_2H_5)_4 + 2\,x\,H_2O \longrightarrow [Si(OC_2H_5)_{4(1-x)}O_{2x}]_{\text{polymer}} + 4\,x\,C_2H_5OH$$

where x is the mol % partial hydrolysis. If the alkoxysilane is an organoalkoxysilane, eg, methyltriethoxysilane or phenyltriethoxysilane, the hydrolysis proceeds analogously to give the organosilsesquioxane $(RSiO_{1.5})_n$ instead of dioxides $(SiO_2)_n$. Likewise diorganodialkoxysilanes yield siloxanes upon hydrolysis, but this chemistry is more appropriately considered with silicones.

The hydrolysis reaction is catalyzed by acid or base. For binder preparation, dilute hydrochloric acid and acetic acid are preferred, since these facilitate formation of stable silanol condensation products. When more complete condensation or gelation is preferred, a wider range of catalysts, including moderately basic ones, is employed. These materials, which are often called hardeners or accelerators, include aqueous ammonia, ammonium carbonate, triethanolamine, calcium hydroxide, magnesium oxide, dicyclohexylamine, alcoholic ammonium acetate and tributyltin oxide (2–3).

The Si—OR bond undergoes a variety of reactions apart from hydrolysis and condensation. In one of the more important commercial aspects of reactivity, it is associated with production of silicone intermediates and as cross-linking agents for silicone room-temperature-vulcanizing materials (RTVs). The reactivity of the Si—OR bond is in many cases analogous to that of the Si—Cl bond, except that reactions are more sluggish and become increasingly more sluggish with greater bulk and steric screening of the alkoxy group. Reactions recently reviewed include (4–5):

$$\geqq SiOR + R'MgCl \longrightarrow \geqq SiR' + Mg(OR)_4 \tag{1}$$

$$\geqq SiOR + HOSi\leqq \longrightarrow \geqq SiOSi\leqq + ROH \tag{2}$$

$$\geqq SiOR + HOB< \longrightarrow \geqq SiOB< + ROH \tag{3}$$

Table 1. Selected Properties of Tetraorganoxysilanes, Polyorganoxysiloxanes, and Monoorganoalkoxysilanes

Compound	Formula	CAS Registry No.	Bp, °C/kPa[a]	Mp, °C	d_4^{20}	n_D[b]	Viscosity, mPa·s (= cP)[b]	Flash point, °C[b]	Heat of vaporization, kJ/mol[c]
Monoorganoalkoxysilanes and polyorganoxysiloxanes									
methyltrimethoxysilane	CH$_3$Si(OCH$_3$)$_3$	[1185-55-3]	102–103		0.955	1.3646	0.5	21	
methyltriethoxysilane	CH$_3$Si(OC$_2$H$_5$)$_3$	[2031-67-6]	141–143		0.895	1.3832	0.6	35	
ethyltrimethoxysilane	C$_2$H$_5$Si(OCH$_3$)$_3$	[5314-55-6]	123–124		0.949	1.3838			
ethyltriethoxysilane	C$_2$H$_5$Si(OC$_2$H$_5$)$_3$	[78-07-8]	158–159		0.896	1.3955	0.7	40	
propyltriethoxysilane	C$_3$H$_7$Si(OC$_2$H$_5$)$_3$	[141-57-1]	179–180		0.982	1.396		50	
pentyltriethoxysilane	C$_5$H$_{11}$Si(OC$_2$H$_5$)$_3$	[2761-24-2]	95–96/1.3		0.895	1.4059	1.4	68	
octyltriethoxysilane	C$_8$H$_{17}$Si(OC$_2$H$_5$)$_3$	[2943-75-1]	98–99/0.27		0.88	1.4150 (at 25°C)	1.9	100	
octadecyltriethoxysilane	C$_{18}$H$_{37}$Si(OC$_2$H$_5$)$_3$	[112-04-9]	165–169/0.27		0.87				
phenyltriethoxysilane	C$_6$H$_5$Si(OC$_2$H$_5$)$_3$	[780-69-8]	112–113/1.3		0.996	1.4718		120	
Tetraorganoxysilanes and polyorganoxysiloxanes									
tetramethoxysilane	Si(OCH$_3$)$_4$	[681-84-5]	121–122	2	1.032	1.3668	0.5	45	46.8
tetraethoxysilane	Si(OC$_2$H$_5$)$_4$	[78-10-4]	169	−85	0.934	1.3838	0.7	55	46.0
tetrapropoxysilane	Si(OC$_3$H$_7$)$_4$	[682-01-9]	224–225	<−80	0.916	1.4012	1.7	95	
tetraisopropoxysilane	Si(O-i-C$_3$H$_7$)$_4$[d]	[1992-48-9]	185–186	<−22	0.887	1.3845	1.2	60	46.8

Name	Formula	CAS	bp/mm	mp	density	n	viscosity	flash point	surface tension
tetrabutoxysilane	Si(OC$_4$H$_9$)$_4$	[4766-57-8]	115/0.4	<−80	0.899	1.4128	2.3	110	61.9
tetrakis(s-butoxy)silane	Si(O-sec-C$_4$H$_9$)$_4$	[5089-76-9]	87/0.27		0.885	1.4000	2.1 (at 38°C)	104	
tetrakis(2-ethylbutoxy)-silane	Si[OCH$_2$CH(C$_2$H$_5$)$_2$]$_4$	[78-13-7]	166–171/0.27	<−70	0.892	1.4309	4.4 (at 38°C)	116	
tetrakis(2-ethylhexaoxy)-silane	Si[OCH$_2$CH(C$_2$H$_5$)(C$_4$H$_9$)]$_4$	[115-82-2]	194/0.13	<−80	0.880	1.4388	6.8 (at 38°C)	188	70.6
tetrakis(2-methoxy-ethoxy)silane	Si(OCH$_2$CH$_2$OCH$_3$)$_4$	[2157-45-1]	179–182/14.7	<−70	1.079	1.4219	4.4	140	
tetraphenoxysilane	Si(OC$_6$H$_5$)$_4$ [d]	[1174-72-7]	236–237/0.13	48–49	1.141	1.554 (at 60°C)	6.6 (at 55°C)		
tetraacetoxysilane	Si(OOCCH$_3$)$_4$	[562-90-3]	148/0.8	110 sub	1.06	1.422			
hexaethoxydisiloxane	(C$_2$H$_5$O)$_3$SiOSi(OC$_2$H$_5$)$_3$ [d]	[2157-42-8]	230–232		0.998	1.3914			
ethyl silicate 40 [e] (diethyl silicate)	approx. OSi(OC$_2$H$_5$)$_2$	[18954-71-7]	290–310	−90	1.05–1.06	1.396	4–5	43	
hexakis(2-ethylbutoxy)-disiloxane	[(C$_2$H$_5$)$_2$CHCH$_2$O]SiOSi-[OCH$_2$CH(C$_2$H$_5$)$_2$]$_3$	[1476-03-5]	188/0.13	−43	0.945	1.4331	16.9 (at 38°C)	215	
methyltris(tri-sec-butoxysiloxy)silane	CH$_3$Si[OSi[OCH(CH$_3$)CH$_2$CH$_3$]$_3$]$_3$	[60711-47-9]	190/0.013	−79	0.962		39.0 (at 38°C)	204	

[a] To convert kPa to mm Hg, multiply by 7.5.
[b] Values at 20°C unless otherwise noted.
[c] To convert J to cal, divided by 4.184.
[d] Model compounds, not of commercial significance.
[e] Nominal values, commercial materials vary. Properties given are for the average compound containing 40 wt % silicon dioxide.

$$\geqslant SiOR + [R'\overset{O}{\overset{\|}{C}}]_2O \longrightarrow \geqslant SiO\overset{O}{\overset{\|}{C}}R' + R'\overset{O}{\overset{\|}{C}}OR \tag{4}$$

$$2 \geqslant SiOR + R'CHO \xrightarrow{H_2SO_4} \geqslant SiOSi\leqslant + R'CH(OR)_2 \tag{5}$$

$$\geqslant SiOR + ClSi\leqslant \xrightarrow{catalyst} \geqslant SiOSi\leqslant + RCl \tag{6}$$

$$\geqslant SiOR + HON{=}R' \xrightarrow{catalyst} \geqslant SiON{=}R' + ROH \tag{7}$$

$$\geqslant SiOR + R'COCl \longrightarrow \geqslant SiCl + R'COOR \tag{8}$$

For additional information on equation 1, see reference 6.

In comparison to the Si—OR bond, the Si—C bond can be considered essentially unreactive if the organic moiety is a simple unsubstituted hydrocarbon. If the organic moiety is substituted in a trialkoxysilane, its chemistry is more appropriately considered under silylation.

Simple alkyl- and aryltrialkoxysilanes have three rather than four matrix coordinations in the polymeric hydrolysates, leading to less rigid structures. These and other changes in physical characteristics, eg, wetting and partition properties, make these materials more appropriate in a variety of coating applications where tetraalkoxysilanes are not acceptable. Methylsilsesquioxanes are stable to 400°C. Phenylsilsesquioxanes are stable to 475°C.

Preparation

The preferred method of production is described by Von Ebelman's 1846 synthesis (7):

$$SiCl_4 + 4 C_2H_5OH \rightarrow Si(OC_2H_5)_4 + 4 HCl$$

The reaction is generalized to:

$$R'_{4-n}SiCl_n + n R'OH \rightarrow R'_{4-n}Si(OR')_n + n HCl$$

Process considerations must not only take into account characteristics of the particular alcohol or phenol to be esterified, but also the self-propagating by-product reaction which results in polymer formation (8).

$$ROH + HCl \rightarrow RCl + H_2O$$

$$H_2O + SiCl_4 \rightarrow [SiCl_2O] + 2 HCl$$

Methods used to remove hydrogen chloride include blowing dry air or nitrogen through the reaction mixture, the use of vacuum, the use of refluxing solvents, especially hydrocarbons and chlorinated hydrocarbons, and conducting the reaction in the vapor phase (9). Amines can be employed as base acceptors, but generally this is not practical commercially. In batch processes, the alcohol is always added to the chlorosilane. Continuous processes involve pumping the alcohol and chlorosilane together in a mixing section or introducing the chlorosilane vapor countercurrent to a liquid alcohol (10). All processes provide a method for the removal of by-product hydrogen chloride. The energy of activation for the reaction of ethanol with silicon tetrachloride in the vapor phase is 64.8 kJ/mol (15.5 kcal/mol) (11). The initial stages of the esterification processes are endothermic since the heat of evaporation of HCl cools the reaction mixture. In the last stages of esterification, the mixtures are usually heated during the final addition of alcohol. Tertiary alkoxides can not be formed in this manner.

In the production of tetraethoxysilane, the initial reaction product in undistilled form is condensed ethyl silicate. It contains at least 90 wt % tetraethoxysilane with 28 wt % SiO_2 content. Distillation removes alcohol and high boiling impurities, and the distilled product contains at least 98 wt % tetraethoxysilane and is called pure ethyl silicate. A partially hydrolyzed or polymeric version with a substantial portion having an average of 4–5 silicon atoms and 40 wt % SiO_2 content is called ethyl silicate 40. A flow chart of the batch production of tetraethoxysilane is shown in Figure 1.

Apart from the direct action of an alcohol on a chlorosilane, the only other commercial method used to prepare alkoxysilanes is transesterification.

$$Si(OR)_4 + 4 R'OH \underset{}{\overset{\text{catalyst}}{\rightleftharpoons}} Si(OR')_4 + 4 ROH$$

The transesterification is an equilibrium reaction and is practical only when the alcohol to be esterified has a high boiling point and the leaving alcohol can be removed by distillation. Most widely used catalysts are sodium alcoholates and organic titanates (12–13). Although known for over thirty years, the direct reaction of alcohols with silicon metal has not resulted in a commercial process for the production of alkoxysilane (14–15). It has been demonstrated that, in the presence of a methoxy compound and under moderate pressure, substantial improvements in yield can be achieved (16).

$$Si + 4 ROH \underset{Si(OR)_4}{\overset{Fe,\ CH_3OM}{\xrightarrow{\hspace{1.5cm}}}} Si(OR)_4 + 2 H_2$$

The synthesis of triethoxysilane by a similar method has also been reported (17). Other preparative methods for alkoxysilanes are given by the following equations in order of declining utility:

$$\geqslant SiCl + (RO_3)CH \rightarrow \geqslant Si(OR) + RCl + RO\overset{\overset{\displaystyle O}{\|}}{C}H \tag{9}$$

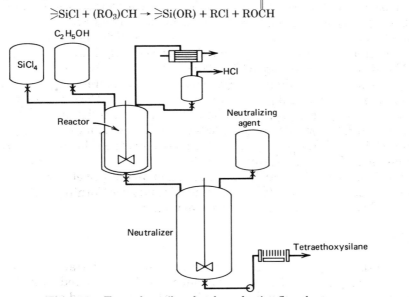

Figure 1.   Tetraethoxysilane batch production flow chart.

$$\geqslant SiCl + R\overline{CHCH_2}O \rightarrow \geqslant Si[OCH_2CH(Cl)R] \qquad (10)$$

$$\geqslant SiCl + NaOR \rightarrow \geqslant Si(OR) + NaCl \qquad (11)$$

$$\geqslant Si(OH) + ROH \rightarrow \geqslant Si(OR) + H_2O \qquad (12)$$

$$\geqslant SiH + ROH \rightarrow \geqslant Si(OR) + H_2O \qquad (13)$$

$$\geqslant SiS_2 + 4\,ROH \rightarrow \geqslant Si(OR)_4 + 2\,H_2S \qquad (14)$$

The direct esterification of silicic acid is of no importance. The reaction is employed for the preparation of complex silicate esters from sodium silicate (water glass) by slowly neutralizing the material in alcohol solution. For information regarding equations 9, 10, 11, 12, 13, and 14, see refs. 6, 18, 19, 20, 21, and 22, respectively.

The acyloxysilanes are produced by the reaction of an anhydride and a chlorosilane (23).

$$SiCl_4 + 4\,[R\overset{O}{\overset{\|}{C}}]_2O \rightarrow 4\,Si(O\overset{O}{\overset{\|}{C}}R)_4 + 4\,R\overset{O}{\overset{\|}{C}}Cl$$

An analogous reaction between anhydrides and alkoxysilanes also produces acyloxysilanes. The reaction of acids directly with chlorosilanes is usually not employed commercially. Marked improvements in the yield of this reaction has been reported if small amounts of acetic anhydride or EDTH are added (24). Methyltriacetoxysilane [4253-34-3] is the most important acyloxysilane.

Economic Aspects

Tetraethoxysilane and its polymeric derivatives account for >90% of the dollar value of nonaryl- or alkyl-substituted esters. In 1980 the products cost $2.40–3.50/kg. The unit volume growth in the 1970s was 5–7%/yr. The leading manufacturers in the United States are Union Carbide and Stauffer Chemical. Dynamit Nobel (Kay Fries) and Monsanto are principal importers. New capacity announced by Union Carbide and a U.S. plant startup completed by Dynamit Nobel (Kay-Fries) in 1981 will increase the world capacity by $2/3$ of the 1980 market of 4500–5500 metric tons per year. This increased capacity implies low pricing through the early 1980s.

Nonethyl ester prices are $4.50–100/kg. Dynamit Nobel (Kay-Fries), Monsanto, Chevron, Anderson, and Petrarch Systems are manufacturers of specialty esters. The largest manufacturer of alkyltrialkoxysilanes is Union Carbide followed by Dow-Corning. Dynamit Nobel (Kay-Fries) actively imports materials into the United States. The U.S. market is estimated to be 1000–1500 t. Other main worldwide producers are Wacker Chimie (FRG) and Shinetsu Chemical Industries (Japan). Petrarch Systems and Olin manufacture specialty materials.

The bulk of acyloxysilanes is produced and used captively by the principal U.S. silicone rubber producers, ie, General Electric and Dow Corning. Union Carbide and Petrarch Systems also manufacture acyloxysilanes.

Toxicity

With the exception of tetramethoxysilane and the silatranes, the alkoxysilanes have a generally low level of toxicity, which may be associated with their alcoholic products of hydrolysis. In addition to higher than anticipated acute toxicities of tetramethoxysilane and its hydride analogue trimethoxysilane [2487-90-3], vapors of these materials may be absorbed into corneal tissue causing blindness (25). Only when three ethoxy groups are substituted for methoxy groups, ie, in triethoxy(methoxy)-silane [18395-48-7], is the hazard substantially reduced. Toxicities of selected compounds are listed in Table 2.

Uses

Precision Casting. The ethoxysilanes are used as binders in precision casting for investment and thin-shell processing (27–28). Ethyl silicate 40 and its partial hydrolysates are preponderant. In special applications propoxysilanes are used, because silanol formation and setup time are more easily controlled. Methoxyethoxysilanes are used when a high flash point intermediate is desired and rates of hydrolysis must remain comparable to ethoxysilanes. In the investment process, 3–10 wt % excess water is added to a prehydrolyzed silicate binder. This is mixed with the refractory material. If the refractory material contains magnesium oxide or calcium hydroxide, gelation occurs in 40–60 min. If these additives are not present or an accelerated cure is required, catalysts are added to the binder prior to mixing with the refractory. In the thin-shell process, fusible patterns are dipped into slurries made of a refractory and ethyl silicate binder. Curing is accomplished by air drying or exposure to ammonia vapor. Ethanol from hydrolysis is either allowed to evaporate or is burned off prior to firing.

Cements and Ceramics. Refractory cements and ceramics are prepared from slurries of silica, zirconia, alumina, or magnesia and a prehydrolyzed silicate (3, 29–30). Calcining at 1000°C yields cured refractory shapes.

Glass Frosting. Deposition of silicon dioxide is used to impart a translucent coating on glass (31). The surfaces are either exposed to tetraethoxysilane or tetramethoxysilane under high moisture conditions, or the alkoxysilanes are ignited and the resulting powder is applied to the surface (see Glass).

Paints and Coatings. Ethoxysilanes are used in high temperature, zinc-rich paints (32–33) (see Paint). Methyl- and phenyltrialkoxysilanes are used to prepare abrasion-resistant coatings for plastics and dielectric coatings for high temperature electronic components (34–35).

Low Heat Glasses. Specialty glasses for fiber optics and solar materials are prepared by cohydrolyzing tetramethoxy-, ethoxy-, propoxy-, and butoxysilanes with alkoxides of such materials as aluminum, titanium, and boron. Sols are gently dehydrated to gels, which in turn are converted to glasses at 500–600°C (36–37).

Table 2.   Acute Oral Toxicities (Rat), LD$_{50}$[a]

Compound	LD$_{50}$, mg/kg
tetramethoxysilane	700
tetraethoxysilane	6,270
ethyltriethoxysilane	13,720
amyltriethoxysilane	19,600
phenyltriethoxysilane	2,830
methyltris(tri-*sec*-butoxysiloxy)silane [60711-47-9]	19,080
hexakis(2-ethylbutoxy)disiloxane	15,000

[a] Ref. 26.

Water Repellents.   Protective and consolidating coatings for masonry and other applications are produced from methyl-, isobutyl-, amyl-, and octyltrialkoxysilanes (38) (see Waterproofing).

Bonded Phases.   Substrate-bound hydrocarbon coatings for liquid chromatography are prepared from octytrialkoxysilanes and other long-chain alkyltrialkoxysilanes.

Hydraulic Fluids.   Frequently materials, eg, hexakis(2-ethylbutoxy)disiloxane and tetrakis(2-ethylhexoxy)silane, are used as hydraulic fluids in high altitude, supersonic aircraft. These materials are useful in low temperature flight environments.

Heat-Transfer Fluids.   A variety of thermal exchange applications, including solar panels, utilize tetrabutoxysilane, tetrakis(2-ethylbutoxy)silane and tetrakis(2-ethylhexoxy)silane (see Heat-transfer technology).

Dielectric Fluids.   Higher tetraalkoxysilanes are used in dielectric applications where lubricity requirements and low temperature conditions are present. Air borne radar is one application.

Diffusion-Pump Fluids.   Thermally and hydrolytically stable alkoxysilanes with carefully controlled structures, eg, methyltris(tri-*sec*-butoxysiloxy)silane are useful in vacuums down to 1.33×10^{-5}–1.33×10^{-6} Pa (10^{-7}–10^{-8} torr) (39) (see Hydraulic fluids).

Silicone Room-Temperature-Vulcanizing Cross-linking.   Condensation-cured polydimethylsiloxanes contain terminal silanol groups which condense with the silanols produced by ambient moisture hydrolysis of acyloxysilanes. Methyltriacetoxysilane, ethyltriacetoxysilane [17689-77-9], and tetraacetoxysilane are the most widely used cross-linking agents. They can be used alone or in combination with silane esters.

Spin-On Oxides.   In microelectric applications, films of silicon dioxide are deposited on silicon substrates by applying a solution of mixed acyloxyalkoxysilanes to a rotating silicon wafer and then thermally decomposing the silane (40).

BIBLIOGRAPHY

"Silicon Esters and Ethers" treated under "Silicon Compounds" in *ECT* 1st ed., Vol. 12, pp. 371–372, by E. G. Rochow, Harvard University; "Silicon Ethers and Esters" in *ECT* 2nd ed., Vol. 18, pp. 216–221, by A. R. Anderson, Anderson Development Company.

1. W. Knoll, *Angew. Chem.* **66,** (1959).
2. U.S. Pat. 2,550,923 (May 1, 1951), C. Shaw, J. E. Hocksford, and W. E. Smith; U.S. Pat. 2,660,538 (Nov. 24, 1953), H. G. Emblem, C. Shaw, and W. E. Langish-Smith (Emblem to Shaw and Langish-Smith); U.S. Pat. 2,795,022 (June 11, 1957), N. Shaw (to Shaw Process Corporation).
3. H. G. Emblem and T. R. Turger, *Trans. Br. Cer. Soc.* **78**(5), (1979).
4. R. C. Mehrotra, V. D. Gupta, and G. Srivastava, *Rev. Silicon Germanium Tin Lead Compd.* **1,** 299 (1975).
5. M. G. Voronkov, V. P. Mileshevich, and Yu. A. Yuzholevski, *The Siloxane Bond*, Plenum Publishing Corporation, New York, 1978.
6. L. M. Shore, *J. Am. Chem. Soc.* **76,** 1390 (1959).
7. J. Von Ebelman, *Ann. Chem.* **57,** 319 (1845).
8. J. R. Wright, R. O. Bolt, A. Goldschmidt, and A. D. Abot, *J. Am. Chem. Soc.* **80,** 1733 (1958).
9. G. Sumrell and G. E. Ham, *J. Am. Chem. Soc.* **78,** 5573 (1956).
10. H. G. Emblem, K. Hargreaves, and N. A. Hurt, *Ind. Chem.* **39,** 576, 641 (1963).
11. V. G. Ukhtomshii, *Izv. Vyssh. Uchebn. Zaved. Khim. Khim. Tekhnol.* **19**(7), 146 (1976).
12. P. D. George and J. R. Ladd, *J. Am. Chem. Soc.* **75,** 987 (1953).
13. H. Steimann, G. Tschernko, and H. Hamann *Z. Chem.* **17,** 89 (1977).

14. U.S. Pat. 2,473,260 (June 26, 1946), E. G. Rochow (to General Electric Company).
15. U.S. Pat. 3,072,700 (Jan. 8, 1963), N. de Wit (to Union Carbide).
16. U.S. Pat. 4,113,761 (Sept. 12, 1978), G. Kreuzberg, A. Lenz and W. Rogher (to Dynamit-Nobel).
17. Jpn. Kokai Tokkyo Kono 80 28,929 (Feb. 29, 1980), S. Suzuki, T. Imaki and T. Yamaura (to Mitsubishi).
18. U.S. Pat. 2,381,137 (May 14, 1942), W. I. Patnode and R. O. Sauer (to General Electric Company).
19. D. Seyferth and E. G. Rochow, *J. Org. Chem.* **20,** 250 (1955).
20. A. Weiss and G. Reiff, *Z. Anorg. Allg. Chem.* **311,** 151 (1961).
21. M. E. Havill, I. Jofee, and H. W. Post, *J. Org. Chem.* **13,** 280 (1948).
22. U.S. Pats. 2,569,455 and 2,459,746 (June 8, 1949), J. B. Culbertson, H. DewErasmus, and R. M. Fowler (to Union Carbide).
23. U.S. Pat. 2,537,073 (Jan. 9, 1951), C. A. McKenzie and M. Schoffman (to Montclair and Ellen-Foster).
24. U.S. Pat. 3,974,198 (Aug. 10, 1976), B. Ashby (to General Electric Company).
25. H. F. Smith, *The Effect of Tetramethylorthosilicate on the Eyes*, Carbide and Carbon Chemicals, Industrial Fellowship 274-1, Mellon Institute, Pittsburgh, Pa.
26. *Registry of Toxic Effects of Chemical Substances*, National Institute of Occupational Safety and Health (NIOSH), Washington, D.C., 1976.
27. A. Dunlop, *Foundry Trade J.* **75,** 107 (1945).
28. A. E. Focke, *Met. Prog.* **49,** 489 (1945).
29. U.S. Pat. 2,678,282 (May 11, 1954), C. Jones (to Pilkington Bros. Ltd.).
30. H. G. Emblem and I. R. Walters, *J. Appl. Chem. Biotechnol.* **27,** 618 (1977).
31. U.S. Pat. 2,596,896 (March 20, 1951), M. Pipken (to General Electric Company).
32. U.S. Pat. 3,056,684 (Oct. 11, 1962), S. L. Lopata (to Carboline).
33. D. M. Berger, *Met. Finish.* **72**(4), 27 (1979).
34. U.S. Pat. 4,197,230 (April 8, 1980), R. H. Bahney and L. A. Harris (to Dow Corning).
35. U.S. Pat. 4,103,065 (July 25, 1978), D. Gagnon (to Owens-Illinois).
36. K. Kamiya and S. Sakka *J. Mater. Sci.* **15,** 2937 (1980).
37. USSR Pat. 715,460 (1980), A. I. Kuznetov.
38. U.S. Pat. 2,916,461 (Dec. 8, 1959), K. W. Krantz (to General Electric Company).
39. R. N. Scott, K. Knollmueller, and co-workers, *Ind. Eng. Prod. Res. Dev.* **19,** 6 (1980).
40. U.S. Pat. 3,915,766 (Oct. 28, 1975), G. F. Pollack and J. G. Fish (to Texas Instruments).

General References

V. Bazant, V. Chvalovsky, and J. Rathousky, *Organosilicon Compounds*, Academic Press, Inc., New York, 1965 (1st serial); Marcel Dekker, Inc., New York, 1973 (2nd serial); Institute of Chemical Process Fundamentals, Prague, Czechoslovakia, 1977 (3rd serial).

B. Arkles and W. Peterson, *Silicon Compounds: Register and Review*, Petrarch Systems, Bristol, Pa., 1979.

W. Noll, *Chemistry and Technology of Silicones*, Academic Press, New York, 1968, pp. 639–662.

BARRY ARKLES
Petrarch Systems, Inc.

SILICONES

The name silicone denotes a synthetic polymer

$$(R_n SiO_{(4-n)/2})_m$$

where $n = 1$–3 and $m \geq 2$. A silicone contains a repeating silicon–oxygen backbone and has organic groups R attached to a significant proportion of the silicon atoms by silicon–carbon bonds. Silicone has no place in scientific nomenclature, although it was originally introduced under the supposition that compounds of the empirical formula RR′SiO were analogous to ketones (1). Several decades later it was used to describe related polymers (2). In commercial silicones, most of the R groups are methyl; longer alkyl, fluoroalkyl, phenyl, vinyl, and a few other groups are substituted for specific purposes. Some of the R groups in the polymer can also be hydrogen, chlorine, alkoxy, acyloxy, or alkylamino, etc. These polymers can be combined with fillers, additives, and solvents to make products that are loosely classed as silicones.

Silicones have an unusual array of properties. Chief among these are thermal and oxidative stability and a relatively mild dependence of physical properties on temperature. Other important characteristics of these materials include a high degree of chemical inertness, resistance to weathering, good dielectric strength, and low surface tension. As the general formula implies, the molecular structure can vary considerably to include linear, branched, and cross-linked structures. These structural forms and R groups can provide many combinations of useful properties that lead to a wide range of commercially important applications. Silicones include fluids, resins, and elastomers. Many derived products, eg, emulsions, greases, adhesives, sealants, coatings, and chemical specialties, have been developed for a large variety of uses (3).

Scientific interest in silicones can be traced to the nineteenth century, but industrial interest did not begin until the early 1930s (4–5). The methyl silicones, which for both economic and technical reasons dominate commercially, were initially chosen for investigation and development because of their general stability and inertness. Interest grew as the full spectrum of properties afforded by this unique class of synthetic materials was recognized.

In the United States the Corning Glass Works pioneered work on organosilicon polymers. Their objective was to develop resins as varnishes and as partners for glass fiber in high temperature electrical insulation. In the same period, the General Electric Company had similar interests but first chose to work with silicate esters. Somewhat later Union Carbide Corporation began a program of organometallic research, which included organosilicon chemistry.

Expansion to pilot-plant production by Corning and General Electric was followed by the formation of the Dow Corning Corporation in 1943, a joint effort by Corning Glass and Dow Chemical Company. In the United States, military uses dominated early product development. Silicone fluids were used for damping aircraft instruments, for antifoams in petroleum oils, and for making greases used as ignition sealing compounds. Silicone resins were used as components of insulation for motors. Silicone rubber was used for making gaskets for searchlights and turbosuperchargers. After World War II, civilian uses followed these patterns but expanded gradually to include release agents for molding rubber; water repellents for textiles, paper, and masonry; ingredients for paints, lubricants, and automobile and furniture polishes; etc. In the

1960s novel types of silicone elastomers, ie, room-temperature-vulcanizing (RTV) rubbers, which are sold as pastelike liquids that can cure *in situ* without the action of heat, were introduced. These RTV products quickly became a significant part of the total silicone product mix. Today they are extensively used in diversified applications as adhesives (qv), sealants (qv) gaskets, coatings, encapsulants, potting compounds, and molding materials (6) (see Embedding).

Outside the United States, the only early fruitful efforts in the field of silicone polymers took place in the USSR. This work began in the late 1930s and led to the development of silicone resins that were produced industrially by the end of the decade. Only after World War II did important findings originate in other countries.

There are four basic manufacturers of silicones in the United States: Dow Corning, General Electric, Union Carbide, and Stauffer-Wacker Silicones (7). There are also large producers in the UK, France, the FRG, the GDR, Japan, and the USSR and small manufacturers in Belgium, Italy, Czechoslovakia, and elsewhere.

Nomenclature

Polymer nomenclature is inherently complex and difficult to use and, as a result, that of silicones is simplified by the use of the letters M, D, T, and Q to represent monofunctional, difunctional, trifunctional, and quadrifunctional monomer units, respectively (8). Primes, eg, D′, are used to indicate substituents other than methyl. Equivalent symbols are shown in Table 1. The meaning of the primes must be specifically indicated in the text. The MDTQ nomenclature can be used to identify silicones (polysiloxanes) with as little ambiguity as the more conventional systems. Proper nomenclature follows IUPAC rules, but in the written or spoken language of the industry it is much more common, for example, to read D_4 or hear methyl tetramer or simply tetramer than the proper octamethylcyclotetrasiloxane (9–10). For another example, $(CH_3)_3SiOSi(CH_3)_2OSi(C_6H_5)[OSi(CH_3)_3]_2$ [*81771-24-6*] becomes $MDT'M_2$, where T′, in this case is $C_6H_5SiO_{1.5}$. Table 2 shows the various ways of identifying siloxanes.

Properties

Silicone properties can be interpreted in terms of structural bond concepts (11). The silicon–oxygen matrix that constitutes the backbone of these polymers is pre-

Table 1. Formulas and Symbols for Silicones[a]

Formula	Functionality	Symbol
$(CH_3)_3SiO_{0.5}$	mono	M
$(CH_3)_2SiO$	di	D
$(CH_3)SiO_{1.5}$	tri	T
$(CH_3)(C_6H_5)SiO$	di	D′
$(C_6H_5)_2SiO$	di	D′
$(CH_3)(H)SiO$	di	D′
SiO_2	quadri	Q

[a] Ref. 8.

Table 2.   Identification of Siloxanes[a]

Chemical name	CAS Registry Number	Structural formula	MDT formula	Common name
hexamethyltrisiloxane	[107-46-0]	$(CH_3)_3SiOSi(CH_3)_3$	MM	mono
octamethyltrisiloxane	[107-51-7]	$(CH_3)_3SiOSi(CH_3)_2OSi(CH_3)_3$	MDM	linear trimer
decamethyltetrasiloxane	[141-62-8]	$(CH_3)_3SiO[Si(CH_3)_2O]_2Si(CH_3)_3$	MD$_2$M	linear tetramer
octamethylcyclotetra-siloxane	[556-67-2]		D$_4$	cyclic tetramer
octaphenylcyclotetra-siloxane	[546-56-5]		D$_4'$	cyclic phenyl-tetramer
2,4,6,8-tetramethyl-2,4,6,8-tetraphenyl-cyclotetrasiloxane	[77-63-4]		D$_4'$	cyclic methyl-phenyl-tetramer
1,1,1,3,5,5,5-heptamethyl-3-trimethylsiloxy-trisiloxane	[56120-90-3]	$[(CH_3)_3SiO]_3SiCH_3$	M$_3$T	
1,1,1,3,5,5,5-heptamethyl-trisiloxane	[5272-21-9]	$(CH_3)_3SiOSi(H)(CH_3)OSi(CH_3)_3$	MD'M	
1,1,3,5,5-pentamethyl-1,3,5-triphenyltri-siloxane	[80-14-8]	$(CH_3)_2(C_6H_5)SiOSi(CH_3)(C_6H_5)-$ $OSi(C_6H_5)(CH_3)_2$	M'D'M'	

[a] Refs. 9 and 10.

dominately responsible for their uniqueness. The siloxane bond flexes and rotates fairly freely about the SiO axis, especially with small substituents, eg, methyl, on the silicon atoms (12). Rotation is also free about the SiC axis in methylsilicon compounds. As a result of the freedom of motion, intermolecular distances between methylsiloxane chains are greater than between hydrocarbons, and intermolecular forces are smaller (13).

The relatively mild temperature dependence of many physical properties is explained on the basis of a flexible helical structure for linear polymer molecules. When temperatures rise, for example, the helix tends to relax, resulting in an increasing molecular entanglement which compensates somewhat for the normal increase in molecular mobility. The viscosity of a simple methyl silicone fluid thus changes relatively little with temperature. This is in contrast to hydrocarbon polymers which have a stiffer structure (14).

Differences between silicon and carbon chemistry are generally accountable on the basis of electronegativity and orbital bonding effects. Silicon is more electropositive than carbon. Bonds to Cl, N, O, and S by silicon are more ionic and have greater energies than those by carbon. Analogously, bonds to C and H by silicon have smaller energies than those by carbon.

Unlike carbon, silicon has available d orbitals. Electrons in these d orbitals can participate in bonding with π electrons from other atoms. The resulting d_π–p_π involvement is often consistent with partial double-bond character (15–18).

In situations where both electronegativity and orbital effects are operative, the influence of the orbital effect generally predominates. Accordingly, the oxygen atom in ethers is more basic than the oxygen in siloxanes. Similarly, the rate of electrophilic, eg, dichlorocarbene, addition to $CH_3(CH_2)_3CH{=}CH_2$ is about twice that for $(CH_3)_3SiCH{=}CH_2$. In the absence of orbital effects, electronegativity effects explain behavior, eg, the greater basic character of nitrogen in $(CH_3)_3SiCH_2NH_2$ over nitrogen in $CH_3(CH_2)_2NH_2$.

The two kinds of bonds most characteristic of silicones are those of silicon to oxygen and silicon to carbon. The nature of the SiO bond is influenced by the electropositive character of silicon, which is 1.7 in Pauling's scale of electronegativities (19), and by the availability of the vacant d orbitals in silicon for dative bonding. The SiO bond is ca 50% ionic, with silicon being the positive member. It has a high heat of formation of 452 kJ/mol (108 kcal/mol) and is very resistant to homolytic cleavage. However, it is susceptible to heterolytic cleavage, ie, to attack by acids and bases. In this respect the —SiOSiO— backbone of silicones is very different from the —CCCC— backbone of hydrocarbon polymers.

The SiC bond is slightly ionic, ie, 12% on the basis of Pauling's electronegativity, again with silicon being positive. The heat of formation is ca 318–356 kJ/mol (76–85 kcal/mol), which is almost as great as for the C—C bond. It may or may not be susceptible to heterolytic cleavage, depending upon the substituents on the carbon. Chloromethyl, cyanomethyl, or even phenyl groups are more easily cleaved from silicon by water, acids, or bases than are methyl groups.

The silicon-to-chlorine bond is also important in the chemistry of silicones. Organochlorosilanes are the monomers from which silicones are made. The first step in this process is a hydrolysis reaction. Such reactions occur very fast when compared to hydrolyses of corresponding alkyl halides, eg,

$$(CH_3)_3SiCl + H_2O \xrightarrow[\text{solvent}]{\text{very fast}} (CH_3)_3SiOH + HCl$$

$$(CH_3)_3CCl + H_2O \xrightarrow[\text{solvent}]{\text{slow}} (CH_3)_3COH + HCl$$

Here the hydrogen in SiOH is more acidic than in COH.

Silane Monomers

Silane monomers are the precursors of silicone polymers. Siloxane compositions are obtained from these silane monomers by hydrolysis. Organic radicals, attached to silicon by hydrolytically stable linkages, survive these hydrolyses and thereby become a part of the siloxane product. The widespread commercial usefulness of silicone products is the direct result of the discovery of economic routes for the manufacture of silane monomers.

Methylchlorosilanes. Methylchlorosilanes are the starting materials for methyl silicones. They are made industrially by the copper-catalyzed exothermic reaction of methyl chloride with silicon at ca 300°C (20). To avoid side reactions, which occur at high temperatures, provisions must be made for the removal of heat. Stirred or fluidized beds are satisfactory for controlling the temperature (21–23). The direct reaction between silicon and methyl chloride has been thoroughly studied, and many observations of the various effects of temperature, pressure, particle size, impurities of silicon and copper, additives, diluents, hydrogen chloride, hydrogen, and poisons for the catalyst have been made on pilot-plant scale (9,24–28). Very little has been published regarding industrial apparatus, performance, or economics.

The reaction produces chlorosilanes, chlorodisilanes, many hydrocarbons, and some more complex silicon compounds. A list of the main components of the product stream is given in Table 3. Under proper control, the principal product is dimethyldichlorosilane. There have been reports of 85–90% yields of dimethyldichlorosilane at high silicon levels in laboratory operations. The crude product stream from the reactors is separated, and the important components are purified by fractional distillation. This is generally done by a system of continuous columns supplemented by batch columns for minor products.

For various reasons it is impractical to balance the production of the individual compounds in this complex reaction mixture with demand. There are several interconversion processes which can be used to achieve a balance, eg, redistribution and cleavage, as shown in the following examples:

Redistribution

$$(CH_3)SiCl_3 + (CH_3)_3SiCl \underset{150°C}{\overset{\text{catalyst}}{\rightleftharpoons}} 2\ (CH_3)_2SiCl_2$$

Cleavage

$$(CH_3)_2Si_2Cl_4 + HCl \xrightarrow{\text{catalyst}} CH_3SiCl_3 + CH_3SiHCl_2$$

Aluminum chloride catalyzes the redistribution reaction, and SiH compounds promote catalysis (29). At equilibrium the mixture contains ca 70 wt % $(CH_3)_2SiCl_2$. Amines catalyze cleavage of disilanes by HCl.

Other Chlorosilanes and Derivatives.   The same kind of direct reaction involving an organic chloride and silicon can also be used to make ethyl and phenylchlorosilanes. For phenylchlorosilanes, chlorobenzene reacts with silicon at ca 550°C, and the catalyst is either copper or silver. The two main products are phenyltrichlorosilane and diphenyldichlorosilane. Except for the methyl, ethyl, and phenyl systems, the direct reaction is inefficient for making organochlorosilanes owing in large measure to extensive degradations that occur with more complex organic chlorides under reaction conditions.

There are several other ways to manufacture phenylchlorosilanes, including

Table 3.   Properties of Some Methylchlorosilanes [a]

Compound	Boiling point, °C	Density d^{20}, g/cm^3	Refractive index, n_D^{20}	Assay, %
$(CH_3)SiCl_3$	66.4	1.273	1.4088	95–98
$(CH_3)_2SiCl_2$	70.0	1.067	1.4023	99–99.4
$(CH_3)_3SiCl$	57.9	0.854	1.3893	90–98
$(CH_3)SiH(Cl)_2$	41.0	1.110	1.3982	95–97
$(CH_3)_2SiH(Cl)$	35.0	0.854	1.3820	

[a] Refs. 15, 16, and 26.

Grignard, related organometallic processes, and condensation of SiH compounds with benzene, as illustrated below:

$$CH_3SiCl_3 + C_6H_5MgBr \rightarrow (CH_3)(C_6H_5)SiCl_2 + MgBrCl$$

$$CH_3(H)SiCl_2 + C_6H_6 \xrightarrow[300°C]{BCl_3} (CH_3)(C_6H_5)SiCl_2 + H_2$$

The phenyl groups in phenylchlorosilanes can be chlorinated (16).

Trichlorosilane is made by the reaction of elemental silicon with anhydrous hydrogen chloride at elevated temperature:

$$Si + 3\ HCl \rightarrow HSiCl_3 + H_2$$

Trichlorosilane participates as an SiH compound in condensations leading to phenylchlorosilanes (30–31), eg,

$$HSiCl_3 + C_6H_6 \xrightarrow[300°C]{BCl_3} C_6H_5SiCl_3 + H_2$$

$$HSiCl_3 + C_6H_5Cl \xrightarrow[500°C]{} C_6H_5SiCl_3 + HCl$$

Vinylchlorosilanes can be made by similar condensations with vinyl chloride carried out in a hot tube without catalyst (32):

$$(CH_3)_2(H)SiCl + CH_2{=}CHCl \xrightarrow[500°C]{} HCl + (CH_3)_2(CH_2{=}CH)SiCl$$

Catalyzed addition of SiH compounds to acetylene also gives vinylchlorosilanes.

$$(CH_3)(H)SiCl_2 + C_2H_2 \xrightarrow[Pt]{} (CH_3)(CH_2{=}CH)SiCl_2$$

This preceding kind of addition reaction to carbon–carbon multiple bonds is termed hydrosilation (33).

Hydrosilation reactions are now widely used in forming silicon–carbon bonds. The generality of this reaction leads to a broad spectrum of silane monomers including many important organofunctional species. The reaction can be induced by heat, light, and radiation, or it can be catalyzed by peroxides, bases, and noble metals. Catalysis, particularly by platinum (homogeneous and heterogeneous forms), is usually the preferred approach. The following examples illustrate organofunctional silane monomer preparation by hydrosilation of olefinic compounds.

$$(CH_3)SiH(Cl)_2 + CH_2{=}CHCF_3 \xrightarrow[Pt]{} (CH_3)(CF_3CH_2CH_2)SiCl_2$$

$$(CH_3)SiH(Cl)_2 + CH_2{=}CHCN \xrightarrow[base]{} (CH_3)(CNCH_2CH_2)SiCl_2$$

These monomers are vehicles for incorporating organofunctionality into siloxane compositions.

The hydrolyzable chlorine groups in chlorosilane monomers can be substituted to give derivative monomers that are similarly hydrolyzable, but they do not yield a strong acid hydrolysis by-product, ie, HCl. This is advantageous for minimizing the risk of corrosion during use, carrying out certain subsequent synthetic reactions, preparing siloxanes that are not stable in the presence of strong acids, and formulating reactive silicone compositions. Alkoxylation, acyloxylation, and amination are examples of such substitutions that are commercially significant.

Alkoxylation

$$(CH_3)_2SiCl_2 + 2\ CH_3OH \rightarrow (CH_3)_2Si(OCH_3)_2 + 2\ HCl$$

Acyloxylation

$$CH_3SiCl_3\ +\ 3(CH_3CO)_2O\ \longrightarrow\ CH_3Si(O\overset{\overset{\displaystyle O}{\|}}{C}CH_3)_3\ +\ 3CH_3\overset{\overset{\displaystyle O}{\|}}{C}Cl$$

Amination

$$2\ (CH_3)_3SiCl + 3\ NH_3 \rightarrow [(CH_3)_3Si]_2NH + 2\ NH_4Cl$$

Similar reactions lead to substitution by amides, ketoximes, and other active hydrogen compounds (34).

Properties of several industrially important silane monomers are listed in Table 4.

Uses. As silylation agents, silane monomers are being used increasingly in synthetic and analytical chemistry (35). Copolymerization with appropriate organic monomers gives random silicone–organic copolymers. Silicone–polyimides are examples of such hybrids, which can be formulated to combine desirable features of silicones with their organic partners. In addition to providing tough films that adhere to many substrates, these compositions have excellent release properties, resist deformation at raised temperatures, have good corona resistance, and protect substrates from corrosion (36–37) (see Polyimides).

Table 4.   Properties of Silane Monomers [a]

Compound	Boiling point, °C	Density, g/cm^3	Refractive index, n_D
HSiCl$_3$	32	1.3298[25]	1.3983[25]
(C$_2$H$_5$)SiCl$_3$	100	1.2342[25]	1.4257[20]
(C$_2$H$_5$)$_2$SiCl$_2$	129	1.0472[25]	1.4291[25]
(C$_2$H$_5$)$_3$SiCl	146	0.8977[20]	1.4299[25]
(C$_2$H$_5$)SiH(Cl)$_2$	74.5	1.0926[20]	1.4148[20]
(C$_6$H$_5$)SiCl$_3$	201.5	1.3185[25]	1.5245[25]
(C$_6$H$_5$)$_2$SiCl$_2$	305	1.218[25]	1.5765[26]
(C$_6$H$_5$)SiH(Cl)$_2$	184	1.2115[25]	1.5257[20]
(CH$_3$)(C$_6$H$_5$)SiCl$_2$	205	1.174[25]	1.5180[20]
(CH$_3$)$_2$(C$_6$H$_5$)SiCl	193.5	1.032[20]	1.5082[25]
(CH$_3$)(C$_6$H$_5$)$_2$SiCl	93	1.085[20]	1.5742[20]
(C$_2$H$_3$)SiCl$_3$	92	1.265[25]	1.4330[25]
(CH$_3$)(C$_2$H$_3$)SiCl$_2$	93	1.085[25]	1.4200[25]
(CH$_3$)$_2$(C$_2$H$_3$)SiCl	83	0.884[25]	1.4141[25]
(CH$_3$)(CF$_3$CH$_2$CH$_2$)SiCl$_2$	122	1.211[20]	1.3817[25]
(CNCH$_2$CH$_2$)SiCl$_3$	224	0.9699[20]	1.4103[20]
(CH$_3$)(CNCH$_2$CH$_2$)SiCl$_2$	215	1.187[25]	1.4564[20]
(CH$_3$)$_2$Si(OCH$_3$)$_2$	80.5	0.8646[20]	1.3708[20]
(CH$_3$)Si(OCH$_3$)$_3$	103.5	0.955[25]	1.3687[25]
(CH$_3$)$_3$Si(OCH$_3$)	56.5	0.7537[25]	1.3678[20]
(CH$_3$)Si(O$\overset{\overset{\displaystyle O}{\|}}{C}CH_3$)$_3$	95	1.1677[25]	1.407
(CNCH$_2$CH$_2$)Si(OC$_2$H$_5$)$_3$	224	0.978[25]	1.4160[25]
(NH$_2$CH$_2$CH$_2$CH$_2$)Si(OC$_2$H$_5$)$_3$	217	0.943[25]	1.4190[25]
[(CH$_3$)$_3$Si]$_2$NH	125.5	0.774[25]	1.4078[25]

[a] Refs. 15 and 16.

By far the most important uses are those in which films of silicone are formed. The silane can be applied directly, deposited from solution, or deposited as a vapor. In such applications, the use of monomers is much more advantageous than the direct use of siloxane compositions. This may be true for various reasons, eg, the desired film is too thin, ie, is monomolecular, to spread or the substrate topography is difficult. Additionally, the *in situ* film forming allows some monomer reactivity for surface attachment. A silicone coating obtained in this way can waterproof a fabric, protect delicate electrical parts, prevent siliceous powders from caking, improve the bondability of surfaces, etc (38).

Use of silane monomers as coupling agents is probably the most notable application (39). Coupling agents, when properly applied to the surfaces of inorganic materials, markedly enhance their compatability with organic polymers. Strength properties of mineral-reinforced plastics increase considerably through the use of silane coupling agents deposited to form adsorbed films on the surface of the fillers. The filler surface is modified according to the functional nature of the silanes employed. 3-Aminopropyltriethoxysilane is widely used as a coupling agent with epoxy, phenolic, amide and related plastics. It is prepared commercially by the following reactions (34):

$$Cl_3SiH + CH_2{=}CHCN \xrightarrow[\text{base}]{} Cl_3SiCH_2CH_2CN$$

$$Cl_3SiCH_2CH_2CN + 3\ C_2H_5OH \rightarrow (C_2H_5O)_3SiCH_2CH_2CN + 3\ HCl$$

$$(C_2H_5O)_3SiCH_2CH_2CN + 2\ H_2 \xrightarrow[\text{Ni}]{\text{pressure}} (C_2H_5O)_3SiCH_2CH_2CH_2NH_2$$

Silicone Polymers

Chlorosilane Hydrolysis. Conversion of chlorosilane monomers to useful polymer products generally involves 2–4 processes: hydrolysis plus cleanup and acid reduction of the hydrolysate; in some cases, conversion of the hydrolysate to cyclic oligomers; polymerization, equilibration, or bodying of the hydrolysate or cyclic oligomers; and stripping, devolatilization, or solvent removal. The hydrolysis can be carried out in a continuous system, by batch processes, or especially in the case of silicone resins, by hydrolysis of a solvent solution of the chlorosilanes (28).

The production of silicone fluids and elastomers is largely based on the hydrolysis of dimethyldichlorosilane. Batch or continuous processes are used, but the continuous process is preferred. In one such process, the chlorosilane is mixed with 22 wt % azeotropic aqueous hydrochloric acid in a pump and heat-exchanger loop. The mixture of fluid hydrolysate and concentrated acid (32 wt % HCl) is separated in a decanter (40). The silicone hydrolysate is washed to remove residual acid, neutralized, dried, and filtered.

The reactions taking place during the hydrolysis include chlorosilane hydrolysis to silanols with rapid condensation of the silanols to form siloxanes.

$$\geqslant SiCl + H_2O \rightarrow\ \geqslant SiOH + HCl$$

$$\geqslant SiOH + ClSi{\leqslant}\ \rightarrow\ \geqslant SiOSi{\leqslant} + HCl$$

or

$$\geqslant SiOH + HOSi{\leqslant}\ \rightarrow\ \geqslant SiOSi{\leqslant} + H_2O$$

The hydrolysate is a mixture of roughly equal proportions of linear, silanol-terminated, and cyclic polymers with the main cyclic component being the tetramer, ie, octamethylcyclotetrasiloxane.

$$Linear$$
$$HO[(CH_3)_2SiO]_m H + HCl$$

$$(CH_3)_2SiCl_2 + H_2O \nearrow$$
$$\searrow$$

$$Cyclic$$
$$[(CH_3)_2SiO]_n + HCl$$

Cationic surfactants increase the relative yield of cyclic structures (41).

The hydrolysate can be used for production of fluids and higher polymers for elastomers or, if needed, converted to cyclic compounds by thermal depolymerization with base (42).

$$HO[(CH_3)_2SiO]_m H \xrightarrow[200°C]{KOH} [(CH_3)_2SiO]_3 + [(CH_3)_2SiO]_4 + [(CH_3)_2SiO]_5 + \ldots + H_2O$$
$$[541\text{-}04\text{-}9] \qquad [556\text{-}67\text{-}2] \qquad [541\text{-}02\text{-}6]$$

As is the case with direct hydrolysis, the cyclic tetramer is the main component of the base-catalyzed, thermal cracking process.

Methyldichlorosilane, $(CH_3)(H)SiCl_2$, can also be hydrolyzed in 22 wt % hydrochloric acid. In this process, trimethylchlorosilane, $(CH_3)_3SiCl$, is usually added to stop polymerization to control the viscosity of the finished fluid.

$$(CH_3)(H)SiCl_2 + (CH_3)_3SiCl + H_2O \rightarrow (CH_3)_3Si[(CH_3)(H)SiO]_x Si(CH_3)_3 + HCl$$

Polysiloxanes are formed by condensation of silanols or by opening the cyclic oligomer:

$$[(CH_3)_2SiO]_4 + H_2O \rightleftharpoons HO{+}(CH_3)_2SiO{+}_4 H$$
$$[3081\text{-}07\text{-}0]$$

Both the silanol condensation and cyclic polymerization reactions are reversible and catalyzed by acid or base. They are the basic reactions for preparing all silicone polymers.

Not all chlorosilanes hydrolyze as rapidly or condense as freely as the methyl chlorosilanes. For example, the hydrolysis of diphenyldichlorosilane is easily stopped at the monomer diol stage, and conversion to oligomers, eg, the cyclic $[(C_6H_5)_2SiO]_4$, occurs slowly even in the presence of active catalysts (43).

Equilibration and Polymerization. Very important to the manufacture of silicone fluids and elastomers are the siloxane rearrangement reactions, which occur in the presence of acids or bases. An example is the reaction of hexamethyldisiloxane with octamethylcyclotetrasiloxane in the presence of sulfuric acid, which gives a mixture of linear and cyclic polymers (44).

$$MM + D_4 \underset{}{\overset{H_2SO_4}{\rightleftharpoons}} MD_n M + D_m$$

This reaction involves numerous equilibria. It has been shown that in the equilibrium

$$MD_x M + MD_y M \rightleftharpoons MD_{x-w} M + MD_{y+w} M$$

the equilibrium constant is close to a theoretical value of unity, and in the equilibrium

$$MD_x M \leftrightharpoons MD_{x-w}M + D_w$$

the constant is 11×0.4^w, where w is ≥ 4; ie, no cyclic compound smaller than D_4 is obtained in meaningful quantities. Since the cyclic trimer is known, its small concentration in this system is believed to result from ring strain, which causes it to be very reactive (45). In a solvent, eg, carbon tetrachloride, the ratio of cyclics to linears increases, and the ratio of low molecular weight linears to high molecular weight linears also increases. As the ratio of M to D increases, the proportion of cyclics decreases (46).

Acids other than H_2SO_4 can be used, eg, hydrogen chloride or Lewis acids. Acid-treated clays also are sometimes used. Ferric chloride and HCl act together as if the catalyst were $HFeCl_4$ (47). Acids attack the more basic M unit more rapidly than they do the D unit, but the strained trimer ring is attacked most rapidly of all. The order of acid-catalyzed reactivity is

$$D_3 > MM > MDM > MD_2M > D_4$$

Alkaline catalysts, however, attack the D unit more rapidly than the M unit, and the order of base-catalyzed reactivity is:

$$D_3 > D_4 > MD_2M > MDM > MM$$

In the base-catalyzed polymerization of octamethylcyclotetrasiloxane, equilibria analogous to those described for acid catalysis have been measured. When M is absent or very low, the equilibrium mixture contains ca 85 wt % linear polymers and 15 wt % cyclic polymers. The cyclic polymers form a continuous population at least to D_{400}, and those larger than D_{12} make up 2–3 wt % of the total polymer in silicone oils and gums. Tetramer is the main cyclic component (48). The distribution of linear polymers is random as in the case of acid catalysis.

The rate of alkali-catalyzed polymerization is proportional to the concentration of the cyclic siloxane and to the square root of the concentration of alkali (49–50). Catalytic activity of the Group IA hydroxides as catalysts increases from lithium to cesium. It is affected by solvents, and dipolar aprotic solvents, eg, dimethyl sulfoxide, tetrahydrofuran, and hexamethylphosphoramide, increase the rate of reaction. Various substances retard the polymerization of tetramer by potassium hydroxide, eg, anisole, diphenylamine, and sodium and lithium hydroxides (51).

Branched polymers can be made by the introduction of T or Q units (see Table 1), or by irradiation of linear polymers. Network polymers are obtained from mixtures containing larger proportions of T and Q units or by cross-linking chains of D units. Silicone resins are based on silane mixtures that are predominately T- and Q-functional. Cross-linking chains of D units leads to elastomeric materials. In condensing mixtures of T and D units to make resins, much of the potential cross-linking is consumed in the formation of cyclic polymers. Most silicone resins consisting of T and D functionalities can be viewed not as a complex polymer network but as knots of T—D copolymers connected by chains of D units.

Methyl-3,3,3-trifluoro-n-propylsiloxane trimer [2374-14-3], $[CH_3(CF_3CH_2CH_2)$-$SiO]_3$, reacts rapidly in the presence of NaOH at 150°C to give the linear polymer, which rapidly depolymerizes under these conditions to give the cyclic tetramer. At equilibrium, the polymer is made up of 96 wt % cyclic and 14 wt % linear components (52). Therefore, to make high polymer, the reaction of the trimer must be stopped at the proper time. Special catalysts, eg, the sodium derivative of sodium acetate Na-CH_2COONa, facilitate this operation (53).

Copolymerization involving different siloxane units is important, because many useful fluids and gums are copolymers containing groups, eg, methylvinyl-, methylphenyl-, or diphenylsiloxane as well as dimethylsiloxane units. These siloxane units do not react at the same rates. Successful copolymerization involves reactivity ratios analogous to the Mayo-Lewis ratios developed for vinyl monomers (54). In the copolymerization of octaphenylcyclotetrasiloxane and octamethylcyclotetrasiloxane, the phenyl tetramer polymerizes first and then suppresses the rate of polymerization of the methyl siloxane (55).

Polymerization of cyclic dimethyl silicones with base catalysts proceeds as outlined above to produce linear polymers by an anionic mechanism (49).

$$[(CH_3)_2SiO]_4 + KOH \leftrightarrows HO \hspace{-0.3em}+\hspace{-0.3em} (CH_3)_2SiO]_4^- + K^+$$

$$HO \hspace{-0.3em}+\hspace{-0.3em} (CH_3)_2SiO]_4^- + [(CH_3)_2SiO]_4 \leftrightarrows HO \hspace{-0.3em}+\hspace{-0.3em} (CH_3)_2SiO]_8^-, \text{ etc}$$

If the concentration of the chain terminator, including KOH, is small, polymers with thousands of tetramer units, ie, gums, are produced. Since this is a reversible process, depolymerization can occur in the presence of the same catalysts. Successful continuous procedures for this kind of polymerization have been developed (56).

In addition to these reactions of silanols and siloxanes, several other reactions are important in the manufacture and use of commercial silicones. The SiH bond is easily solvolyzed by water or alcohols, easily oxidized, and can be added across carbon–carbon multiple bonds to form SiC bonds. Siloxanes containing SiH can be cross-linked or modified by such reactions (15,34).

Silicone–Organic Copolymers. Copolymers of silicones and organic polymers have the advantages of low cost and physical strength and they retain the durability and surface properties of silicones. Several approaches have been used. One is the reaction of siloxanes containing silanol or alkoxy functions with the hydroxy groups of organic polymers to give silicon–oxygen–carbon links between the polymers, as illustrated by the formation of silicone alkyds (57). The same type of reaction can be used for polyester, epoxy, phenol–formaldehyde, and acrylic compositions. The SiOC bond offers a potential hydrolytically weak point, but in practice the polymers are satisfactorily stable. The SiOC link can be avoided. For example, carboxyalkyl functional or hydroxyalkyl functional silicones can be used in making such copolymers by esterification or transesterification (58–59).

Polyethers have been attached to siloxanes as:

$$\geqslant SiCH_2O(CH_2CH_2O)_n R$$

using the Williamson synthesis (60), or by adding SiH to olefinic functional polyether chains to make linkages such as (61):

$$\geqslant SiCH_2CH_2CH_2(OCH_2CH_2)_n OR$$

Silicone polycarbonates can be prepared from the reaction of phosgene with siloxane prepolymers terminated by free phenolic OH groups from bisphenol A (4,4′-dihydroxydiphenylpropane) (62–63):

$$2 \geqslant SiO(C_6H_4)C(CH_3)_2(C_6H_4)OH \; + \; COCl_2 \longrightarrow \; 2\,HCl \; + \; \left[\geqslant SiO(C_6H_4)C(CH_3)_2(C_6H_4)O \right]_2 \overset{\overset{\textstyle O}{\|}}{C}, \text{ etc}$$

Procedures involving phase-transfer catalysis in the synthesis of these copolymers have been reported (64) (see Catalysis, phase-transfer).

Silicone polyimides have been made by the reaction of ethanolamine derivatives of siloxanes with polyacids, eg, 1,2,4,5-benzenetetracarboxylic acid (pyromellitic acid), or from aminoalkyl silicones as shown below (36,65–66):

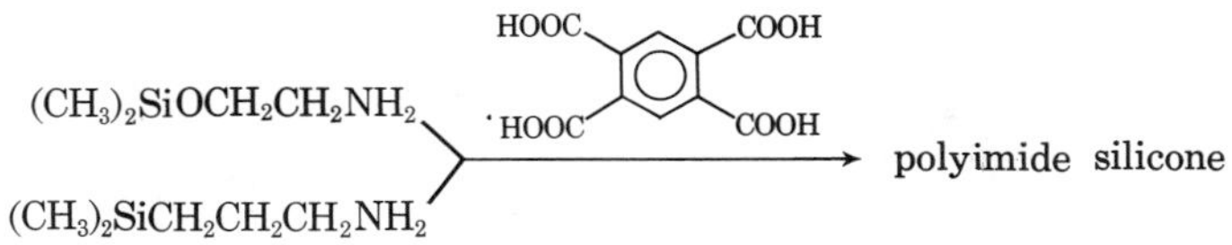

$$(CH_3)_2SiOCH_2CH_2NH_2$$
$$(CH_3)_2SiCH_2CH_2CH_2NH_2$$

HOOC   COOH
HOOC   COOH

$\longrightarrow$ polyimide silicone

Another type of copolymer has metal atoms in the siloxane chain. Reaction of titanium ethers with silanols leads to the following titanoxysiloxane copolymers (67):

$$-(CH_3)_2SiOH + C_2H_5OTi\!\!\leqslant \; \rightarrow \; -(CH_3)_2SiOTi\!\!\leqslant + C_2H_5OH$$

A great many such metalloxane copolymers containing aluminum, tin, titanium, boron, phosphorus, iron, and other elements have been made (68). Improvements in polymer strength have been reported, but the hydrolytic instability of the copolymers appears to be a large problem.

Silicone–carborane copolymers with in-chain or pendent carboranyl groups have been synthesized. The former type can be made by the following reaction (69):

$$(CH_3O)(CH_3)_2Si—CB_{10}H_{10}C—Si(CH_3)_2(OCH_3) + (CH_3)_2SiCl_2 \xrightarrow{\text{FeCl}_3}$$
[17631-41-3]

$$\text{+}(CH_3)_2SiO(CH_3)_2Si—CB_{10}H_{10}C—Si(CH_3)_2O\text{+} + 2\,CH_3Cl$$
[27925-15-1]

Vinyl groups attached to pendant carboranyl groups can be readily introduced to provide cure points.

Copolymers in which the silicon is present to modify properties or to provide room-temperature cure for organic polymer systems can in principle be made with almost any organic polymer. Several such compounds have been reported, for example, polyolefins (70), polyurethane (71), and polyether (72). Polysulfide polymers that are terminated by reaction with (3-glycidoxypropyl)trimethoxysilane [2530-83-8] or (3-methacryloxypropyl)trimethoxysilane [2530-85-0] have the following polymer end groups:

$$\underset{\text{OH}}{(CH_3O)_3SiCH_2CH_2CH_2OCH_2\overset{}{C}HCH_2S-} \quad \text{or} \quad \underset{\text{OCH}_3}{(CH_3O)_3SiCH_2CH_2CH_2O\overset{}{C}CHCH_2S-}$$

These polymers appear to offer a method of making copolymers with each other or with reactive silicones as the other component (73).

Methyl Silicone Polymers. Linear polydimethylsiloxanes have been extensively studied. Because intermolecular forces are weak, the polymers have low melting points and second-order transition temperatures. For dimethyl silicone gums, ie, rubber polymers, these are −51°C and −86°C, respectively. Therefore, they are noncrystallizing polymers under ordinary conditions. The low intermolecular forces also lead to the low boiling points, low activation energy for viscous flow, high compressibility, small change of viscosity with temperature, and for resins and unreinforced elastomers, relatively poor physical properties despite high molecular weights. There are reasons for believing that these polymers tend to assume a helical structure with about 6 or 7 siloxane units per turn (14). This helical type of structure is unusual for bulk liquid polymers without strong intermolecular attractions, eg, hydrogen bonds (74); however, there is not complete agreement on this subject (10,75).

Volatile oligomers in the products of equilibration of MM and D_4, eg, MDM and D_4, can be separated and purified by fractional distillation. Study of their properties shows the effects of molecular size and structure in these systems. This approach has been extended to include related oligomers, eg, M_4Q [3555-47-3] and TM_3 [17928-28-8]. Properties of some of these compounds are shown in Table 5.

In higher linear dimethyl silicone polymers, the number-average molecular weight is related to the bulk viscosity by the relationship:

$$\log (\text{viscosity in mm}^2/\text{s} \,(= \text{cSt}) \text{ at } 25°\text{C}) = 1.00 + 0.0123 \, \overline{M}_n^{0.5} \text{ where } \overline{M}_n > 2500$$

The intrinsic viscosity $[\eta]$, dL/g, which is determined by the extrapolation of the viscosity to zero concentration, has the following values in toluene and methyl ethyl ketone, respectively (76–77):

$$[\eta] = 2 \times 10^{-4} \, \overline{M}_n^{0.66} \qquad [\eta] = 8 \times 10^{-4} \, \overline{M}_n^{0.5}$$
$$\text{toluene} \qquad\qquad \text{methyl ethyl ketone}$$

Branched polymers, which are made by introducing T or Q units, have lower bulk or intrinsic viscosities than linear polymers of the same average molecular weight. Even a small amount of branching causes considerable decrease in the bulk viscosity (78).

Table 5. Properties of MD_nM and D_n [a]

MDT formula	CAS Registry No.	Melting point, °C	Boiling point, °C	Density d^{20}, g/cm³	Refractive index, n_D^{20}	Viscosity (η), at 25°C, mm²/s (= cSt)	Flash point, °C
MM	[107-46-0]	−67	99.5	0.7636	1.3774	0.65	−9
MDM	[107-51-7]	−80	153	0.8200	1.3840	1.04	37
MD_2M	[141-62-8]	−76	194	0.8536	1.3895	1.53	70
MD_3M	[141-63-9]	−80	229	0.8755	1.3925	2.06	94
MD_4M	[107-52-8]	−59	245	0.8910	1.3948	2.63	118
MD_5M	[541-01-5]	−78	270	0.9012	1.3965	3.24	133
MD_6M	[556-69-4]	−63	290	0.9099	1.3970	3.88	144
MD_7M	[2652-13-3]		307.5	0.9180	1.3980	4.58	159
D_3	[541-05-9]	64.5	134				
D_4	[556-67-2]	17.5	175.8	0.9561	1.3968	2.30	69
D_5	[541-02-6]	−44	210	0.9593	1.3982	3.87	
D_6	[540-97-6]	−3	245	0.9672	1.4015	6.62	
D_7	[17909-36-3]	−32	154 [b]	0.9730	1.4040	9.57	
D_8	[556-68-3]	31.5	290	1.1770 [c]	1.4060	13.23	

[a] Ref. 10.
[b] At 2.7 kPa (20 mm Hg).
[c] Crystals.

Effects of Substitution on Properties.   If some of the methyl groups are replaced by longer alkyl chains, the unique properties of the methyl silicones are modified. The activation energy for viscous flow E_{visc} and the rates of change of viscosity with temperature and pressure increase (14). Oxidative stability decreases, compatibility with organic compounds becomes greater, and lubricity improves (79–80).

Phenyl substitution in fluids and elastomers, which may be made either as methylphenyl or diphenyl silicone, has some of the same effects as increasing the size of the alkyl group, but it increases rather than decreases thermal or oxidative stability. Introduction of ca 7.5 mol % methylphenylsiloxane into dimethylsiloxane polymer lowers the pour point from -40 to $-112°C$ (81). This is believed to be caused by interference of the bulky groups with crystallization of the methyl polymer. E_{visc} for MD′M (D′ = $(C_6H_5)_2SiO$) is 19 kJ (4.5 kcal) compared to 10 kJ (2.4 kcal) for MDM, showing how the bulky phenyl groups interfere with motion.

Chlorinated phenyl groups, eg, tetrachlorophenylsiloxane copolymerized into dimethyl silicone fluids, increase lubricity. The change is a function of the average number of chlorine atoms per phenyl and the proportion of such groups in the copolymers (82).

Vinyl groups are used in elastomers to provide reactive centers for peroxide and silane–olefin cures. They provide for cross-links, frequently along the chain, or they act as chain propagaters when terminally located. In silicone gums for heat-cured rubber, vinyl constituents are not present in concentrations high enough to affect physical polymer properties but do beneficially modify the cure properties of the polymer.

Methyl-3,3,3-trifluoro-n-propyl silicone [*63148-56-17*], $(CH_3)(CF_3CH_2CH_2)SiO$, has a solubility parameter of about 9.5 compared to ca 7.5 for $(CH_3)_2SiO$ (83). This fluorosilicone swells much less than $(CH_3)_2SiO$ in octane or toluene. It also has improved lubricity and a greatly increased rate of change of viscosity with temperature (84–85). 2-Cyanoethyl and 3-cyanopropyl substituents give copolymers with $(CH_3)_2SiO$ that may have solubility parameters of ca 9.0–9.5, ie, similar to that of fluorosilicone.

Polyethers have been introduced as substituents by various methods (86–87). The copolymers are in general water-soluble and have higher viscosity–temperature coefficients and better load-carrying capacities as lubricants than $(CH_3)_2SiO$ polymers. They are of practical interest because of their surfactant properties (see Surfactants and detersive systems).

Incorporation of SiH-containing monomer units, particularly $CH_3(H)SiO$, changes the physical properties of fluids, but the most important change is the increase in chemical reactivity. Water repellency remains good. The ability to cross-link readily increases the usefulness of SiH-containing polymers in applications, eg, textile and paper treatment. SiH-containing polymers are utilized in hydrosilation-curing elastomer formulations as well.

Physical properties of several oligomers with various substituents are listed in Table 6. A great many such oligomeric compounds have been prepared and described, and several are used industrially as intermediates in the preparation of polymers (16,88).

Table 6.   Properties of Siloxane Oligomers [a]

Compound	CAS Registry No.	Boiling point, $°C_{kPa}$ [b]	Density d^{20}, g/cm^3	Refractive index, n_D^{20}	Melting point, °C
$[(C_6H_5)_2SiO]_4$	[546-56-5]	$335_{0.13}$			200
$[(CH_3)(C_6H_5)SiO]_4$	[77-63-4]	$237_{0.13-0.67}$	1.183	1.5461	99
$(CH_3)_3SiOSi(C_6H_5)_2OSi(CH_3)_3$	[797-77-3]	$172_{2.4}$	0.984	1.4927	
$[(CF_3CH_2CH_2)(CH_3)SiO]_4$	[429-67-4]	$134_{0.4}$	1.255	1.3724	
$[(CH_2{=}CH)(CH_3)SiO]_4$	[2554-06-5]	$111_{1.3}$	0.9875	1.4342	−43.5
$(CH_3)_3Si[OSiH(CH_3)]_2OSi(CH_3)_3$	[16066-09-4]	177	0.8559	1.3854	
$[(CH_3)(H)SiO]_4$	[2370-88-9]	134	0.9912	1.3870	−69

[a] Refs. 16 and 88.
[b] To convert kPa to mm Hg, multiply by 7.5.

Silicone Fluids

Dimethylsilicone fluids are made by catalyzed equilibration of dimethyl silicone stock, ie, the crude fluid or distilled cyclic polymers, with a source of the chain terminator, $(CH_3)_3SiO_{0.5}$. As described above, this reaction produces mixtures of MD_nM and D_m polymers. The ratio of M to D in the charge controls the average molecular weight and the viscosity of the product. For example, a 50 mm^2/s (= cSt) fluid has an average molecular weight of ca 3000, a 350-mm^2/s fluid ca 15,000, and a 1,000-mm^2/s fluid ca 25,000. For relatively low viscosity fluids, the process can be run at ca 180°C in glass-lined reactors with acid clay catalysts or at lower temperatures with sulfuric acid. Both batch and continuous processing are used (28).

Alkaline catalysts are used for the production of high viscosity fluids or gums. These polymers can be processed batchwise in kettles or continuously in a heated tube with stirring (89). Polymerization is continued in the case of gums to make polymers of >500,000 av mol wt and 10^7 mm^2/s (= cSt) viscosity. Some gums contain vinyl or phenyl substituents, which are introduced by copolymerizing $(CH_3)_2SiO$ with vinyl or phenylsiloxanes. Phenyl substituents are added as $[(C_6H_5)_2SiO]_n$ or $[(CH_3)(C_6H_5)SiO]_n$. Vinyl substituents are added as $[(CH_3)(CH_2{=}CH)SiO]_n$ or $[(CH_3)_2(CH_2{=}CH)SiO]_{0.5,}$ or both.

In most instances, the fluid equilibrate is devolatilized by heat and vacuum after catalyst deactivation. Both equilibration and devolatilization can be carried out in batch or continuous process systems. Fluids are sometimes blended in order to make fluids of intermediate viscosities. Since the properties of a polymer depend upon molecular weight distribution as well as upon average molecular weight, blending can affect physical properties appreciably. Blends of fluids of widely different viscosities are less Newtonian in behavior than those with normal molecular weight distribution. Some properties of commercial silicone fluids are given in Table 7, and silicone-fluid viscosity is shown as a function of temperature in Figure 1. These viscosity–temperature profiles fit the following general equation:

$$\log (\eta + B) = A \log T + C$$

where A, B, and C are constants (92). Data for two nonsilicone fluids are plotted in Figure 1 to illustrate the steeper slopes characteristic of such materials. Silicones with phenyl, trifluoropropyl, or large alkyl substituents also show steeper slopes when plotted in this way.

Table 7. Properties of Silicone Fluids[a]

Type of fluid (CH₃)₂SiO, mol %	Copolymer silicone	CAS Registry No.	Viscosity at 25°C, mm²/s (= cSt)	d^{25}, g/cm³	n_D^{25}	Pour point, °C	Flash point, °C	Surface tension at 25°C, mN/m (= dyn/cm)	Thermal conductivity at 65°C, W/(m·K)[b]	Electric strength, kV/μm	Electric constant	Volume resistivity (min), Ω·cm × 10⁻¹⁴
100	none	[63148-62-9]	10	0.940	1.399	−73	210	20.0	1.3	1.4	2.60	1
100	none		100	0.968	1.4030	−55	302	20.9	1.5	1.4	2.75	1
100	none		1,000	0.974	1.4035	−50	315	21.1	1.6	1.4	2.75	1
100	none		10,000	0.975	1.4035	−47	315	21.3	1.5	1.4	2.75	1
100	none		100,000	0.978	1.4035	−40	315	21.3	1.5	1.4	2.75	1
50	CH₃(C₆H₅)SiO	[63148-52-7]	125	1.07	1.495	−45	302	24.7	1.4	1.3	2.88	1
91.2	CH₃(C₆H₅)SiO		50	0.99	1.425	−73	282	25.0	1.4	1.3	2.79	1
91.2	CH₃(C₆H₅)SiO		100	0.99	1.425	−73	293	24.1	1.5	1.3	2.80	1
95.6	(C₆H₅)₂SiO	[68951-93-9]	100	1.00	1.421	−73	302	24.0		1.4	2.78	1
>90	tetrachlorophenyl siloxane	[68957-05-1]	70	1.045	1.428	−73	288	21	1.5		2.90	
0	CH₃(H)SiO	[63148-57-2]	25	0.98	1.397	−48		20			2.88	
0–10[c]	CF₃CH₂CH₂(CH₃)SiO	[63148-56-1]	300	1.25				26				
92	(CH₃)SiO₁.₅	[68037-74-1]	50	0.972	1.403	−84	315	21.0		1.4	2.74	1

[a] Refs. 4, 10, 34, 90–91.
[b] To convert W/(m·K) to (Btu·in.)/(h·ft²·°F), divide by 0.1441.
[c] Approximate value.

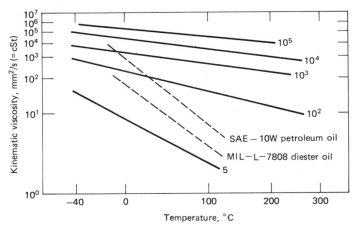

Figure 1. Viscosity–temperature curves for various dimethyl silicone oils. Indicated reference viscosities in mm²/s (= cSt) at 25°C. Data for two nonsilicone materials shown by dotted lines (10).

Changes in pressure cause unusually large changes in volume and viscosity (93–94). The flow of methyl silicone fluids is Newtonian, meaning that the viscosity is constant over a broad range of shear rates, for fluids of less than 1,000 mm²/s (= cSt), but the apparent viscosity decreases with the shear rate for higher viscosity fluids (14). The fluids are stable to shear stress, however, and return to original viscosity after being passed through small orifices under pressure. Thermal expansion of the volume is ca 0.1%/°C (4). Specific heat is 1.4–1.6 J/g (0.33–0.37 cal/g) at 0–100°C, and increases with viscosity (4).

When the fluids are spread over water to form monomolecular films and then the films are compressed laterally, the ratio of stress to strain undergoes several changes associated with changes in the order and orientation of the molecules (95–96). The force required for compression is also a function of the pH. The interfacial tension against water is ca 42 mN/m (= dyn/cm). Spread on glass, the fluids repel water, forming a contact angle of ca 103°, which is roughly the same as for paraffins.

Dimethyl silicone fluids are good hydrodynamic lubricants but poor for sliding friction. For steel on steel, the coefficient of friction is ca 0.3–0.5. For some metal combinations, eg, brass on steel, the fluids are fair lubricants for medium or light loads (97).

Dimethyl silicone fluids are transparent to visible light and to ultraviolet radiation with wavelengths >280 nm. There are many strong selective bands in the infrared region.

The velocity of sound at 30°C is 873.2 and 987.3 m/s for 0.65-mm²/s and 1000-mm²/s fluids, respectively (98). It decreases with increased temperature and is significantly higher for methylphenyl silicones.

The dielectric constant decreases with temperature (see Table 7). For example, a 1000-mm²/s fluid has a dielectric constant of 2.76 at 30°C and 2.54 at 100°C (99). The power factor is low, but it increases with temperature and behaves irregularly with frequency. Resistivity decreases with temperature. Gamma radiation causes decreases in resistivity and dielectric strength (100). The latter is also lowered by moisture (90).

Gases are soluble in dimethyl silicone polymers. The solubilities of air, nitrogen, and carbon dioxide are 0.17, 0.17, and 1.00 cm^3/cm^3 at 25°C and 101 kPa (1 atm), respectively (4). The solubility of oxygen in a 26,000-mm^2/s fluid at 101 kPa is 0.2 mg/g, that of hydrogen in D_4 is 0.07 mg/g (101). About 250–300 ppm water dissolve in silicone fluid at 25°C and 95% rh, whereas only 30 ppb of dimethyl silicone dissolve in water (102–103). The fluids are miscible with benzene, dimethyl ether, methyl ethyl ketone, carbon tetrachloride, or kerosene. They are somewhat soluble in acetone, dioxane, ethanol, and butanol and are relatively insoluble in water, methanol, ethylene glycol, and 2-ethoxyethanol (14).

Methylphenyl silicone fluids with, for example, 8–50 wt % $CH_3(C_6H_5)SiO$, have higher viscosity–temperature coefficients, lower pour points, and lower flash points than dimethyl silicone fluids of the same viscosities. They are less compressible and change viscosity more rapidly with pressure than methyl fluids. Their principal advantage is their greater stability to oxidation and radiation (104). Three such fluids are described in Table 7. Fluids with chlorinated phenyl groups attached to silicon (see Table 7) have improved lubricating properties (91).

Methyltrifluoropropyl silicone fluid also changes viscosity rapidly with pressure and for this reason shows good lubricating properties (105). The methylalkyl fluids, where alkyl is octyl to tetradecyl, likewise are good lubricants but have higher viscosity–temperature coefficients and poorer oxidative stability than methyl fluids (79).

Oxidation of methyl silicones yields formaldehyde and formic acid as well as CO_2 and water. Formation of each molecule of either of these represents a carbon cleaved from silicon and results in the formation of a siloxane linkage. Thus, D units are converted to T or Q units and the polymer becomes branched or cross-linked (10).

Prolonged heating of silicone fluids can result in devolatilization and cracking and therefore in loss of weight and change in viscosity. Heated under nitrogen at 300°C, a 100 mm^2/s dimethyl silicone fluid decreases 47% in viscosity in 480 h (13). Heating under air causes loss in weight, and gelation occurs in about 200 h at 250°C. Stability is improved by antioxidants, eg, phenyl-α-naphthylamine. Compounds of titanium, iron, and cerium also increase life (106–107). Preoxidation with the use of iron catalysts results in fluids which do not gel in 500 h at 315°C.

Copolymer fluids containing $(CH_3)_2SiO$ and either $CH_3(C_6H_5)SiO$ or $(C_6H_5)_2SiO$ are more stable at high temperatures than dimethyl silicone, and heat aging is further improved when the end groups are phenyl substituted (108). Diethyl silicone fluids are less stable at high temperatures than the dimethyl fluids, but the former can be stabilized by chelate inhibitors, eg, copper ethyl acetoacetate (109). Methylalkyl silicone fluids generally are less stable than dimethyl fluids. It is the methyl group that is unique; the longer alkyls are roughly equivalent in oxidation stability (79). Ozone and electric-corona discharge, which degrade organic polymers, do not readily attack silicone fluids, gums, or derived elastomers.

Dimethyl silicones cross-link when subjected to gamma ray or electron radiation. Small radiation doses increase the viscosity of the fluids and branched molecules form (110). With increased radiation dosages, the fluid eventually gels. Polymerization has been observed in microwave discharges (111). Fluids containing $CH_3(C_6H_5)SiO$ or $(C_6H_5)_2SiO$ are more stable to radiation; methyl fluid is attacked 35 times more rapidly than phenyl fluid, and phenyl groups also protect silicone elastomers (112–113).

Derived Products. Several types of greases are made from mixtures of silicone fluids and fillers. For insulating and water-repellent greases, silica filler is used, and the fluid may be dimethyl silicone or dimethyl silicone copolymerized with methylphenyl, methyltrifluoropropyl, or methylalkyl silicone. For lubricating grease the fillers are generally lithium soaps and the preferred fluids are methylphenyl, chlorinated phenylmethyl, or methyltrifluoropropyl silicones. The electrical properties of silica-filled greases are good. Such dielectric compounds are workable at low temperatures and do not have a dropping point, as soap-filled greases do. Lubricating greases vary in properties, depending on the type of fluid and thickener used. They have wide service temperature ranges, ie, from −70 to 230°C; dropping points of 200–260°C, depending on the soap used; low bleed; and low evaporation rate and weight loss in vacuum. Those based on fluids with good lubricating properties are themselves good lubricants, capable of prolonged performance up to 200°C (34).

Emulsions of silicone fluids in water are made for convenience in applying small amounts of silicone to textiles, paper, or other surfaces. Concentrations are usually 20–60 wt % silicone, except for foam-control agents, which may contain as little as 10 wt % silicone. Various emulsifiers are used. The emulsified silicone fluids may be of any type but are usually dimethyl or methylalkyl silicones. Emulsified fluids are also convenient as antifoams. They disperse rapidly in aqueous systems, but for nonaqueous systems, eg, crude oil or vegetable oils, 100% silicone or a solvent solution of the silicone is used (34) (see Defoamers).

Silicone Resins

Silicone resins are highly cross-linked siloxane systems, and the cross-linking components are usually introduced as trifunctional or tetrafunctional silanes in the first stage of manufacture. For example, a solution of CH_3SiCl_3, $(CH_3)_2SiCl_2$, $C_6H_5SiCl_3$, and $(C_6H_5)_2SiCl_2$ or $CH_3(C_6H_5)SiCl_2$ in toluene is hydrolyzed to form a complex copolymer mixture, which remains in solution in toluene. The aqueous hydrochloric acid is separated, and the resin solution is washed and then heated in the presence of a mild condensation catalyst to body the resin to the proper viscosity and cure time. It is finally adjusted to specifications by distilling or adding solvents (28). The properties of the finished silicone resin depend upon the choice of chlorosilanes, the degree of cure, and to some extent on the processing conditions. The general effects of the most commonly used monomers on the properties of a film are listed in Table 8.

Table 8. Effect of Monomers on the Properties of Silicone Resin Films[a]

Property	CH_3SiCl_3	$C_6H_5SiCl_3$	$(CH_3)_2SiCl_2$	$(C_6H_5)_2SiCl_2$	$CH_3(C_6H_5)SiCl_2$
hardness	increase	increase	decrease	decrease	decrease
brittleness	increase	great increase	decrease	decrease	decrease
stiffness	increase	increase	decrease	decrease	decrease
toughness	increase	increase	decrease	decrease	decrease
cure speed	much faster	some increase	slower	much slower	slower
tack	decrease	some decrease	increase	increase	increase

[a] Ref. 10.

Phenyl and alkyl (other than methyl) groups increase the compatibility with organic materials. Water-repellent resins are generally based on methyl or longer alkyl monomers. Processing conditions vary from hydrolysis in strong acid to dilute acid to buffered aqueous systems; alkoxysilanes can also be used to avoid acid conditions. The choice of solvent can affect the result as can temperature, concentration, and choice of catalyst for bodying and curing. Most silicone resins require heat and catalysts for curing. During the life of the product, curing continues and properties do change with time. Latent resin-curing systems based on quaternary ammonium salts have been described (114).

Dipping or impregnating varnishes based on silicone resins are used to bond and insulate electrical coils and glass cloth and are usually supplied as 50 wt % or 60 wt % silicone resins in an organic solvent. Components are dipped, drained, baked and again dipped, drained, and baked repeatedly to build up the required thickness of insulation. They are finally cured by a somewhat longer bake, with time and temperature determined by the resin used and the complexity of the component structure. The final bake is usually for several hours at 204–260°C, although lower temperatures can be used (10).

Silicone laminating resins are used first to coat glass cloth, and this coating is partially cured to a nontacky stage. Stacks of cloth so treated are pressed at ca 7 MPa (1000 psi) and for high pressure laminates are heated to cure or are bag-molded or vacuum-formed at ca 0.7 MPa (100 psi) or less for low pressure laminates before cure. Typical properties of a freshly cured impregnating resin are shown in Table 9 (13). Laminates range rather broadly in properties as shown in Table 10 (13,115) (see Laminated and reinforced plastics).

Table 9. Properties of Silicone Impregnating Resins [a]

Property		Value	
heat endurance (at 250°C), h			
flexure life[b]		250	
craze life[c]		750	
thermal life[d], h			
at 300°C		500	
at 275°C		1500	
at 250°C		4500	
weight loss (after 3 h at 250°C), %		4	
	Dry		*Wet*
electric strength[e], V/μm	60		40
power factor at 25°C			
at 100 Hz	0.0084		0.0085
at 10^6 Hz	0.0043		0.0047
dielectric constant at 25°C			
at 100 Hz	3.0		3.0
at 10^6 Hz	2.9		2.9

[a] Ref. 13.

[b] Hours after which a film on a copper strip cracked when bent around a small rod.

[c] Hours after which cracking was visible and evident without bending.

[d] Hours of aging to reduce electric strength of a glass cloth impregnated with silicone varnish to 50% of its initial value.

[e] Electrodes 5 cm in diameter.

Table 10.   Properties of Silicone–Glass Laminates [a]

Property	Value
tensile strength, MPa [b]	200–240
elongation at break, %	1–2
flexural strength, MPa [b]	140–310
compressive strength, MPa [b]	140
modulus, MPa [b]	
flexure at 25°C	1.4×10^4–2.4×10^4
at 260°C	1.3×10^4–1.7×10^4
tension	1.0×10^4–1.9×10^4
impact, Izod, J/m [c]	260–1300
distortion temperature, °C	>200
electric strength, V/μm	10–16
power factor, %	
at 10^3 Hz	0.5
at 10^6 Hz	0.001–0.002
arc resistance, s	220–350

[a] Ref. 13.
[b] To convert MPa to psi, multiply by 145.
[c] To convert J/m to (ft·lbf)/in., divide by 53.38 (see ASTM D 256).

Pressure-sensitive adhesives (qv) are made by compounding silicone elastomer gums with silicone resins which are not completely compatible with each other. They are available as solutions in volatile solvents and are applied to the backing by conventional knife-coating techniques. When the solvent evaporates, a sticky film is left. This is cured to develop strength and to reduce tack by heating with a peroxide or other catalyst. Fillers can be used to increase cohesive strength and to decrease tack (116). Water-repellent resins applied to masonry surfaces coat the pores and act by repelling water which otherwise would wet the walls of these pores. Sufficient hydrostatic pressure head can overcome this surface activity. There is no continuous film to bar passage of water or water vapor (see Waterproofing). Release resins are baked on metal surfaces to prevent materials from sticking to these surfaces (see Abherents). Paints are made from silicone resin solutions by mixing with stable inorganic pigments, eg, titania, carbon black, or aluminum flakes (see Paint). Silicone paints are very stable, and even after oxidation the residual polymer bonds the pigment. Silicone–aluminum paint is useful to 260°C. Silicone paints with TiO_2 or other pigments resist discoloration from heat or weathering and have good gloss retention, hardness, mar resistance, flexibility, and resistance to soap, fat, and fruit juices.

Sodium methyl siliconate [16589-43-8] is made by dissolving $CH_3SiO_{1.5}$ polymer in aqueous NaOH. It behaves like a solution of $CH_3Si(OH)_2(ONa)$ monomer. On acidification, the methyl silicone resin precipitates at a rate dependent upon dilution and pH. Methylsiliconic acid [2445-53-6], $CH_3Si(OH)_3$, is a weak acid, and even carbon dioxide decomposes the sodium salt and leads to precipitation of the polymer.

Reinforcing fillers do not in general perform the same function in resins as they do in rubbers. Particulate reinforcing agents are not widely used, as they do not produce a product of higher tensile strength. They are used to fill and harden some molding compounds and pressure-sensitive adhesives. Colloidal silica is used as a reinforcing agent coreactant in silicone resins that are used to give an abrasion-resistant coating on plastics (117–118).

The cure of silicone resins usually occurs through the formation of siloxane linkages by condensation of silanols. This is a continuation of the overall condensation process by which the resin is prepared. As condensation continues, the rate decreases owing to diminishing silanol concentration, increasing steric hindrance, and decreasing mobility. Therefore, for final cure, the reaction must be accelerated by heat and catalyst. Even so, some silanols remain and slow cure continues for the life of the resin. The reaction is reversible, and water must be removed from the system to permit a high degree of cure (4,10). Many substances catalyze silanol condensation including acids and bases (15); soluble organic salts of lead, cobalt, tin, iron, and other metals (119); and organotin compounds, eg, dibutyltin dilaurate or N,N,N',N'-tetramethylguanidine salts (120–121).

Silicone resins based on hydrosilation cure mechanisms have also been developed. These materials cure by addition reactions and are similar in composition to hydrosilation curing elastomers, albeit more highly cross-linked (122–124).

Silicone resins change little on exposure to weather, ie, to humidity, heat, and sunlight. Silicone–organic copolymers and blends are weather-resistant provided the ratio of silicone to organic component is high enough. A modified silicone varnish film shows a flexure life of 1000 h at 200°C and 150 h at 250°C and a craze life of 2000 h at 200°C and 800 h at 250°C (13). Its thermal life varies from 35 h at 300°C to 2200 h at 225°C. Silicone-modified paints have excellent color and gloss retention on aging.

Silicone Elastomers

Elastomers are an extremely important genus of silicone products. Silicone polymers of appropriate molecular weight must be cross-linked to provide elastomeric properties. Fillers are used in these formulations to increase strength through reinforcement. Extending fillers and various additives, eg, antioxidants, adhesion promoters, and pigments, can also be used to obtain specific properties (34,125–126) (see also Elastomers, synthetic).

The usual reinforcing fillers (qv) for silicone elastomers are finely divided silicas made by either the fume process or the wet process (127–130). Fume-process silicas provide the highest degree of reinforcement. The silica filler used must be of small particle size to be a good reinforcing agent. It should have a diameter about the length of a fully extended polymer chain, ie, 1 μm, to be semireinforcing and 0.01–0.05 μm to be highly reinforcing. Fine particle size does not necessarily lead to good reinforcement, because finely divided fillers tend to agglomerate and are extremely hard to disperse. This tendency can be countered by treating the filler to give it an organic or a silicone coating before mixing it with polymer. Hexamethyldisilazane, $[(CH_3)_3Si]_2NH$, is sometimes used as a coupling agent (131). An important kind of filler treatment involves bringing the silica particles in contact with hot vapors of low molecular weight cyclic siloxanes, ie, D_3, D_4, D_5, etc (132). Such treatments reduce agglomeration and prevent premature crepe hardening.

Where physical strength is not required of the finished product, nonreinforcing fillers are used. They may stabilize the product, as in the case of iron oxide or titanium dioxide, or color it, or simply lower the cost per unit volume. Carbon black or carbon fibers can be used to make the product conduct electricity (133) (see Carbon, carbon black).

Many curing, ie, cross-linking, systems have been developed for silicone elastomers and several are widely used commercially. Different silicone elastomers are conveniently distinguished by their cure-system chemistries and can be categorized by the temperature conditions needed to cure them properly.

Room-Temperature-Vulcanizing Rubbers. Room-temperature-vulcanizing (RTV) silicone elastomers are supplied as uncured rubbers that have liquid or pastelike consistencies. They are based on polymers of intermediate molecular weights and therefore viscosities, eg, $100-1,000,000$ mm^2/s (= cSt) at $25°C$. Curing is based on chemical reactions that increase polymer molecular weights and provide cross-linking. Catalysts are used to ensure controlled curing. Two types of RTV-silicone rubbers are available. The cure reactions of one-component products generally are triggered by exposure to atmospheric moisture. Those of two-component products are triggered by mixing the two components, one of which consists of or contains the catalyst. The two components are supplied separately.

The silicone polymers used in making most one-component products are fluids with silanol end groups (126). The most widely used products cure by reactions involving acetoxysilanes (134–136). The sequence is reported to proceed in the following steps (137):

$$HO[Si(CH_3)_2O]_nH + 2\ CH_3Si(CH_3CO_2)_3 \rightarrow (CH_3CO_2)_2(CH_3)SiO[Si(CH_3)_2O]_nSi(CH_3)(CH_3CO_2)_2$$

[*31692-79-2*]   [*68440-61-9*]

$$+\ 2\ CH_3COOH$$

$$CH_3CO_2Si{\leqslant} + H_2O \rightarrow HOSi{\leqslant} + CH_3COOH$$

$$CH_3CO_2Si{\leqslant} + HOSi{\leqslant} \rightarrow {\geqslant}SiOSi{\leqslant} + CH_3COOH$$

Cure is hastened and controlled by the use of catalysts, and tin soaps are especially effective (138). In the above sequence, methyltriacetoxysilane functions as a vehicle for polymer chain extension and cross-linking. Analogous products based on methyltris(2-ethylhexanoyloxy)silane [*70682-61-0*] are also available (139).

Other cure systems proceed similarly but involve still different silane curing agents. Commercial products based on methoxysilanes and catalyzed by titanium chelates sometimes have the advantage of releasing a hydrolysis product that is not acidic (140–143). Acetone is the cure by-product if methyltris(isopropenoxy)silane [*6651-38-3*] is used (144–145). Products based on amino- (146), amido- (147), and ketoxy- (148) silanes are also available. The fillers and additives used in these products must be compatible with the curing agent, which generally means that they must be dry.

One-component RTV rubbers are made by mixing polymers, fillers, additives, curing agents, and catalysts and packaging the mixture to protect it from moisture. Contact with moisture in air brings about reactions which cure the polymer. The time required for cure depends upon the curing system, the temperature, the humidity, and the thickness of the silicone layer. Under typical ambient conditions, the surface can be tack-free in 15–30 min and a 0.3-cm thick layer cures in less than a day. Cure progresses and strength is developed slowly for about three weeks (126,149).

The original viscosity of these RTV materials depends principally on that of the polymer components and the filler loading. Input polymer properties and cross-link density both affect the ultimate strength of the fully cured elastomer as do the identity and loading of fillers. The polymers used in nearly all commercial products are exclusively polydimethylsiloxanes. Polymers with substituents others than methyl can be used to modify and improve certain properties, eg, trifluoropropyl groups lead to better solvent resistance. Some products are compounded to be pourable, others to be thixotropic. These characteristics are controlled by fillers and additives. Silica-filled polydimethylsiloxane systems, lacking pigments and other additives, cure to form translucent rubbers. Since the specific gravity of silicas (ca 2.2) exceeds that of siloxanes (ca 1.0), the RTV specific gravity depends on the filler loading. Some physical properties of similar but not identical, cured acetoxy RTV formulations are listed in Table

Table 11. Physical Properties of Some RTV Rubbers[a]

Specific gravity	Durometer hardness, Shore A	Tensile strength, MPa[b]	Elongation, %
1.18	45	2.4	180
1.30	50	3.1	140
1.33	50	3.4	200
1.37	55	3.8	120
1.45	60	4.5	110
1.45	60	5.2	160
1.48	65	4.8	110

[a] Ref. 87.
[b] To convert MPa to psi, multiply by 145.

Table 12. Typical Cure Properties of Typical RTV Silicone Rubbers[a]

Property	Types			
	One-component general-purpose	One-component construction sealant	Two-component adhesive sealant	Two-component molding compound
hardness, Shore A durometer	30	22	50	60
tensile strength, MPa[b]	2.4	1.0	3.4	5.5
elongation, %	400	850	200	220
tear strength, J/cm^2 [c]	0.80	0.35	0.52	1.75

[a] Refs. 149–152.
[b] To convert MPa to psi, multiply by 145.
[c] To convert J/cm^2 to lbf/in., multiply by 57.1.

Table 13. Thermal and Electrical Properties of RTV Silicone Rubbers

Property	Typical range
useful temperature range, °C	−60 to 200
with thermal stabilizers	−110 to 250
thermal conductivity, W/(m·K)	1.7–3.4
coefficient of thermal expansion, per °C	3.5×10^{-5}
dielectric strength, V/μm	20
dielectric constant at 100 Hz	3.5–4.5
dissipation factor at 100 Hz	0.01–0.02
volume resistivity, ohm-cm	10^{14}–10^{15}

11 (87). Different formulations with different curing systems, polymer molecular weights and structures, cross-link densities, etc can have a broad spectrum of product properties. For example, one-component products are available with elongations as high as 1000%. Typical properties of representative cured RTV silicone rubbers are listed in Tables 12 and 13 (149–152).

One-component RTV silicone rubbers are mostly used in adhesive and sealant applications (see Sealants). Other uses include formed-in-place gasketing, protective coatings, and encapsulating (see Embedding). The bonding properties of these products play an important role in most of these uses. Many formulations exhibit self-bonding to a variety of substrates, eg, most metals, glass, ceramic, and concrete. For example, to bare aluminum, bonds with >1.38 MPa (200 psi) shear strength and 0.35 J/cm^2 (20 lbf/in.) tear strength are reported, and good bonds are formed with copper and acrylic resins. When adhesion is difficult, bonding can often be improved by applying a primer to the substrate. Some primers are solutions of reactive silane monomers or resins which dry (cure) on the substrate, leaving a modified silicone-bondable surface. Bond strength develops as the RTV cure progresses and can require up to 2–3 wk to become optimal (126,153–156).

Two-component RTV silicone rubbers are available in an impressive range of initial viscosities, from as low as an easily pourable 100 mm^2/s (= cSt) material to as high as stiff pastelike materials >1,000,000 mm^2/s at 25°C (151). The kind of curing system, polymer molecular weight and structure, cross-link density, filler, and additive can be varied and combined giving a diversified family of products whose properties cover a wider range than that encompassed by the one-component products. Two-component RTV technology has provided the highest strength RTV rubbers. On the other hand, products that cure to a mere gel-like state are also available (157). Unfilled, resin-reinforced compositions can provide optical clarity (158–159). Polymers with phenyl, trifluoropropyl, cyanoethyl, or other substituents can be used with or in place of polydimethylsiloxanes to make low temperature-, heat-, radiation-, or solvent-resistant elastomers (34,151,160).

Two-component RTV silicone rubbers do not require atmospheric moisture to trigger their cure. Several different curing systems are used for these products and each has its advantages. One approach involves the reaction of silanol-terminated silicone polymer with alkoxy-functional silicon compounds as curing agents, eg,

$$3 \, HO[Si(CH_3)_2O]_nH + CH_3Si(OC_2H_5)_3 \rightarrow CH_3Si[O[(CH_3)_2SiO]_nH]_3 + 3 \, C_2H_5OH, \text{ etc}$$

$$[68554\text{-}67\text{-}6]$$

This reaction requires the presence of a catalyst; tin salts are widely used. The product can be designed so that the polymers and curing agent are in one package, the other package containing the catalyst alone or the curing agent and the catalyst. In either case the product begins to cure only when the two packages are thoroughly mixed. Fillers and additives are incorporated in the formulations according to the desired final product properties (6,161).

In one example of this kind of RTV product, the polymers are mixed with filler and ethyl silicate as the curing agent. A catalyst, eg, dibutyltin dilaurate, is stirred in just before the material is used. Polymerization begins immediately with elimination of ethyl alcohol. The pot life and the work life depend upon the type and concentration of the catalyst and on the temperature, so the pot life can be prolonged by refrigeration. Pot life is generally a few hours at room temperature. The time required to obtain a firm cure is ca one day at room temperature and ca one hour at 150°C (151). The speed of this type of cure is increased markedly by the use of accelerators. γ-Aminopropyl-triethoxysilane performs with synergistic effectiveness in this role (162).

The reaction of silanol-terminated silicone polymers with hydride-functional siloxanes is another route to RTV two-component products. This reaction generates hydrogen gas as a by-product, which can be used to produce a foamed rubber:

$$\geq\text{SiOH} + \text{HSi}\leq \xrightarrow[\text{salt}]{\text{tin}} \geq\text{SiOSi}\leq + \text{H}_2$$

When foaming is not desired, applications for this type of product are generally restricted to those calling for thin films of RTV rubber (6).

Despite the commercial importance of products based on these RTV cure systems, little has been published concerning their chemistry. It has been shown that, in the RTV system involving silanols and ethyl silicate: ethanol evolves and siloxanes form; the reaction does not take place in the absence of an initiator, eg, a tin compound; and the latter reacts with ethyl silicate to form SiOSn compounds (163).

Hydrosilation curing gives two-component RTV rubbers that cure without liberating a by-product. This cure system is based on the addition of silicon hydride to olefin. In practice, vinyl-functional silicone polymers are used with hydride-functional siloxanes.

$$\geq\text{SiCH}=\text{CH}_2 + \text{HSi}\leq \rightarrow \geq\text{SiCH}_2\text{CH}_2\text{Si}\leq$$

The hydride functional siloxane usually serves as the curing agent. The reaction proceeds at room temperature in the presence of a catalyst, eg, chloroplatinic acid or another solubilized platinum-compound. Hydrosilation-curing two-component RTV rubbers can be formulated in a variety of ways. Fillers and additives are used according to the desired properties, but their nature is restricted because of the sensitivity of platinum catalysts to poisoning. Of the available approaches to product design, it is more usual to include the catalyst with vinyl-functional silicone polymers in the first package and the hydride-functional siloxane curing agent, perhaps with additional vinyl polymers, in the second. The actual proportions used determine the ratio in which the two packages are mixed to give a cured elastomer with optimal properties. The pot life of these products is generally a few hours at room temperature (6,151,164).

Whereas in the other cure systems for RTV rubbers cross-links are established by the formation of new siloxane bonds via condensation reactions, this hydrosilation reaction creates ethylene bridges $\geq\text{Si}-\text{CH}_2\text{CH}_2-\text{Si}\leq$ between polymer chains. Cure occurs relatively slowly (days) at room temperature but is generally run at somewhat higher temperatures (50–100°C). Control of the environment is important because the catalyst can be poisoned and because water or alcohols can react with SiH in the presence of these catalysts, as shown by the following equation:

$$\geq\text{SiH} + \text{H}_2\text{O} \rightarrow \geq\text{SiOH} + \text{H}_2$$

Two-component RTV silicone rubbers are extensively used in encapsulating and molding applications. Other important uses include bonding and sealing, protective coatings, and electrical insulation. They can be used in place of one-component products when longer work times or faster cure times are needed. They are used in all applications that require a thick-section cure. These cured RTV rubbers often have antistick properties relative to many other surfaces and are therefore most suitable for fabricating flexible molds, eg, those used to reproduce intricately detailed polyurethane parts. Many products can be easily molded to form silicone rubber parts (126,151–153).

Two-component RTV rubbers are less prone to self-bonding than the one-component materials. Primers are generally used in applications where adhesion is needed. Lap-shear adhesion strengths exceeding 3.4 MPa (500 psi) have been reported on clad aluminum with appropriate primers (154–155). Typical properties of representative, cured, two-component RTV rubbers are also listed in Tables 12 and 13 (149–152).

Heat-Cured Rubbers. Many of the two-component RTV elastomers considered previously can be advantageously cured at raised temperatures, eg, 50–150°C depending on the product and intended use, but their room-temperature vulcanizing is their key trait. Hydrosilation-curing RTV compositions can be modified with inhibitors to become heat-curing systems. Some inhibitors are volatile vinyl or acetylenic compounds that are mainly expelled at elevated temperatures and subsequently permit cure. These volatile inhibitors have the potential disadvantage of forming vapor pockets that interrupt the integrity of the cured rubber and disfigure its surface. Examples of such volatile inhibitors include olefinic nitriles, and acetylenic alcohols, which evidently form readily heat-disruptable complexes with the platinum catalysts (165–166). Superior inhibitors are those that release the catalyst at raised temperatures but are themselves not volatile or are quickly incorporated into the cured elastomer; triallyl isocyanurate is such an inhibitor (167). This kind of inhibited product is used in applications where long pot lives, eg, 10–1000 h, are needed for handling after mixing the two packages and where the option of heat curing is available for a then rapid cure, ie, typically less than one hour and often in a matter of minutes or seconds. Liquid injection molding, electronic potting, and coating operations are examples.

The vast majority of heat-curing silicone rubbers are, unlike RTV compositions, based on high molecular weight polymers that are gums. The ingredients, ie, gums, fillers, and additives, are mixed in equipment, eg, dough mixers or Banbury mills. Catalysts are added on water-cooled rubber mills. These mills can be used for the whole process in small-scale operations (34,168–170).

Products are commercially available as gums, reinforced gums, ie, gum and a portion of the filler, uncatalyzed compounds, dispersions, and catalyzed compounds. The latter are compositions that are ready for use by the purchaser and contain a catalyst. For dispersions or pastes, solvents such as xylene are stirred with the mixture. The following types of gums are available: general purpose (methyl and vinyl), high and low temperature (phenyl, methyl, and vinyl), low compression-set (methyl and vinyl), low shrink (devolatized), and solvent resistant (fluorosilicone). Some gum properties are shown in Table 14.

The tensile strength of cured dimethyl silicone rubber gum is only ca 0.34 MPa

Table 14. Properties of Silicone Gums[a]

Type	CAS Registry No.	Density, d^{25}, g/cm^3	T_g, °C	ASTM D 926, Williams plasticity
$(CH_3)_2SiO$	[9016-00-6]	0.98	−123	95–125
$CH_3(C_6H_5)SiO$	[9005-12-3]	0.98	−113	135–180
$CH_3(CF_3CH_2CH_2)SiO$	[25791-89-3]	1.25	−65	

[a] Refs. 81, 171–173.

(50 psi). It is common to use finely divided silicas for reinforcement. Other commonly used fillers include mined silica, titanium dioxide, calcium carbonate, and ferric oxide. An alternative approach to the use of reinforcing fillers is that of incorporating crystallizing segments into the polymer. For example, block polymers containing silphenylene segments

$$+(CH_3)_2SiC_6H_4Si(CH_3)_2O+_n$$

may have cured gum tensile strengths of 6.8–18.6 MPa (1000–2700 psi) (174).

Consistencies of uncured rubber mixtures range from a tough putty to a hard deformable plastic. Those with reinforcing fillers tend to become stiff, ie, develop structure, on storage unless structure-control additives, eg, water, diphenylsilanediol [947-42-2], dimethylpinacoxysilane [3081-20-7], or silicone fluids are added to the mixture (175–177).

The properties of fabricated rubber depend upon the type of compound, ie, upon the gum, filler, catalyst, additives, and solvents (if any) used. They also depend on the relative proportions of these ingredients. A high proportion of filler leads to greater hardness, less elongation, and greater solvent resistance. The properties also depend upon the thoroughness of mixing and the degree of wetting of the filler by the gum. The properties change as cure progresses and are stabilized by postcure heating to remove volatiles, ie, by devolatilization. To some extent they are affected by the environment and by aging.

Before being used, silicone rubber mixtures are catalyzed and freshened, ie, catalyst is added and the mixture is freshly milled, on rubber mills until they band into smooth continuous sheets that are easily worked. The freshly mixed compound is then used directly. It is common practice in the industry to prepare specific or custom mixtures for particular product applications. Hundreds of formulations have been compounded, and in such complex systems the potential number is enormous. In general, a formula is designed to achieve some special operating or processing requirement, and formulations are classified accordingly. Some types of particular importance or interest are listed in Table 15 (153,171–172,178).

Bouncing putty is a mixture based on polydimethylsiloxane polymer gum mod-

Table 15. Properties of Classes of Silicone Rubbers[a]

Class	Hardness, durometer	Tensile strength, MPa[b]	Elongation, %	Compressive set (at 150°C for 22 h), %	Useful temperature range, °C min	max	Tear strength, J/cm² [c]
general purpose	40–80	4.8–7.0	100–400	15–50	−60	260	0.9
low compression-set	50–80	4.8–7.0	80–400	10–15	−60	260	0.9
extreme low temp	25–80	5.5–10.3	150–600	20–50	−100	260	3.1
extreme high temp	40–80	4.8–7.6	200–500	10–40	−60	315	
wire and cable	50–80	4.1–10.3	100–500	20–50	−100	260	
solvent resistant	50–60	5.8–7.0	170–225	20–30	−68	232	1.3
high strength flame retardant	40–50	9.6–11.0	500–700				2.8–3.8

[a] Refs. 153, 171–172, 178.
[b] To convert MPa to psi, multiply by 145.
[c] To convert J/cm² to lbf/in., multiply by 57.1.

ified by boric acid with special additives, fillers, and plasticizers. It flows on slow application of pressure like a very viscous liquid, but on shock it behaves like a very elastic solid and may even shatter (179–180).

There are several mechanisms for curing these silicone rubbers. Formulations that cure by hydrosilation reactions are available. High molecular weight polymers (gums) with vinyl functionality are combined with fluid hydride-functional cross-linking agents in the uncatalyzed compound. The catalyst, a solubilized form of platinum, is added with an inhibitor to ensure that the cure is thermally induced on heating (181).

Silicone rubber is more usually cured by heating the reinforced polymer with a free-radical generator, eg, benzoyl peroxide (see Initiators). In the case of polydimethylsiloxane elastomers, the predominant mechanism appears to be one in which hydrogen is abstracted from methyl groups and the resulting free radicals couple to form Si—CH_2CH_2—Si bridges as follows:

$$(C_6H_5COO)_2 \rightarrow 2\ C_6H_5COO\cdot$$

$$C_6H_5COO\cdot + (CH_3)_2SiO \rightarrow \cdot CH_2(CH_3)SiO + C_6H_5COOH$$

$$2\ \cdot CH_2(CH_3)SiO \rightarrow O(CH_3)SiCH_2CH_2Si(CH_3)O$$

When the polymer contains vinyl groups, the peroxide radical fragment adds to the double bond to give a radical which couples as illustrated by the following equation:

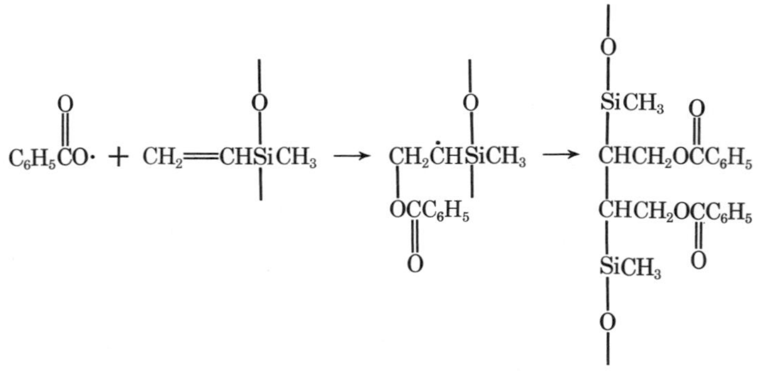

There are three differences of practical importance. The vinyl-based cure does not generate a benzoic acid by-product, which can catalyze decomposition or rearrangement of the siloxane polymer; it functions with less peroxide; and it works with types of peroxides that do not abstract hydrogen from methyl groups, eg, 2,5-dimethyl-2,5-di(t-butylperoxy) hexane (34,125).

Several commercially available peroxides cure these products. One of the most common of these is benzoyl peroxide. About 1.5–2 wt % is used for nonvinyl gums and ca 1 wt % for vinyl gums, based on the weight of gum. Curing temperature is ca 120°C. Bis(2,4-dichlorobenzoyl) peroxide, t-butyl peroxybenzoate, and dicumyl peroxide are also commonly used to obtain special effects. For example, bis(2,4-dichlorobenzoyl) peroxide permits rapid cure in air at 315°C without the development of porosity, which would be caused by the volatilization of lower boiling catalysts.

Cure is also brought about by gamma or high energy electron radiation, which causes scission of all types of bonds, including SiO, but the reactions important to cure are those involving SiC and CH. Hydrogen, methane, and ethane evolve, and bridges between chains are formed by recombination of the radicals generated. These bridges are Si—CH$_2$—Si, Si—Si, and perhaps Si—CH$_2$CH$_2$—Si (182). An absorbed dose of 770–1300 C/kg ((3–5) × 10^6 roentgen) is required to obtain an effective cure. This type of cure leaves no acidic fragments in the polymer. Radiation effects cures of thick sections, but high energy electrons penetrate only a few millimeters of the compound (178).

Freshly mixed silicone rubber compounds are usually molded at 100–180°C and 5.5–10.3 MPa (800–1500 psi). Under these conditions, thermal cure can be complete in a matter of minutes. Molds are usually lubricated. A 1–2 wt % aqueous solution of a household detergent is often satisfactory. In the manufacture of insulated wire, rods, channels, tubing, etc, the compounds are extruded in standard rubber-extrusion equipment. The extruded rubber is heated briefly to set its properties; for this, hot-air vulcanization at 300–450°C or steam at 0.28–0.70 MPa (40–100 psi) for several minutes is sufficient. Final properties can be developed by oven curing or by continuous steam vulcanization.

For coating silicone rubber on a fabric, eg, glass cloth, rubber dispersed in a solvent is used. The cloth may be dip-coated and then dried and cured in heated towers. It may also be calendered with high penetration, soft silicone stocks on standard three- and four-roll calenders. Ducts and hoses can be built up from dip-coated or calendered cloth, and complex structures can be formed on mandrels followed by wrapping and curing to produce large ducts or by extrusion. Sponging of silicone rubber is normally accomplished by the use of nitrogen blowing agents, which produce closed-cell sponge. Silicone rubber foam or sponge is made in densities of ca 0.4–1.0 g/cm^3.

When silicone rubber must be bonded to other materials, eg, metals, ceramics, or plastics, primers are generally used. These may be silicate esters or silicone pastes. After evaporation of solvent from the primed surface, silicone rubber compounds can be applied and cured under pressure. There are also self-bonding silicone rubber stocks that require no primer (153,178).

Properties. Some of the properties of representative cured silicone elastomers are given in Tables 12, 13, and 15. The properties of silicone rubber change with temperature; for example, Young's modulus decreases from ca 10,000 to 200 MPa (14.5 × 10^4 to 2.9 × 10^4 psi) from −50°C to room temperature and then remains fairly constant to 260°C. Resistivity decreases; electric strength does not change greatly; dielectric constant increase at 60-Hz current and decreases at 10^4-Hz current and above; and the power factor increases considerably. Tensile strength decreases from ca 6.9 MPa (1000 psi) at 0°C to 2.1 MPa (300 psi) at 300°C (178). Thermal conductivity of silicone rubber usually is ca 1.5–4 W/(m·K) and increases with increased filler loading (153).

Silicone rubber (gum) films are permeable to gases and hydrocarbons and in general are ca 10–20 times as permeable as organic polymers (178). Water diffuses through lightly cross-linked gum as monomer, dimer, and trimer, with diffusion coefficients of 10.5, 3.6, and 3.1 × 10^{-5}, respectively, at 65°C (183). Silicone rubber compounds are permeable to gases, as shown in Table 16 (184–185).

Solvents diffuse into silicone rubber and swell, soften, and weaken it. The degree of swelling depends upon the solvent and has been correlated with the solubility parameters of solvent and rubber, as illustrated in Figure 2. The correlation is improved if electrostatic interactions are considered (83). The solubility parameter is the square root of the energy of vaporization per cubic centimeter (186).

Table 16.   Permeability of Silicone Elastomers [a]

Type	CAS Registry No.	Temperature, °C	Gas	Permeability, $(cm^3 \cdot cm)/(s \cdot cm^2 \cdot kPa)$ $\times 10^{-7}$ [b]
dimethyl silicone	[63148-62-9]	25	CO_2	405
		25	O_2	79
		25	air	44
		30	butane	2.0
		70	butane	1.7
fluorosilicones [c]	[63148-56-1]	26	CO_2	79
		26	O_2	13
nitrile silicone	[70775-91-6]	31	CO_2	237
		31	O_2	40

[a] Refs. 184–185.

[b] To convert $(cm^3 \cdot cm)/(s \cdot cm^2 \cdot kPa)$ to $(cm^3 \cdot cm)/(s \cdot cm^2 \cdot mm\ Hg)$, divide by 7.5.

[c] Methyltrifluoropropyl silicones.

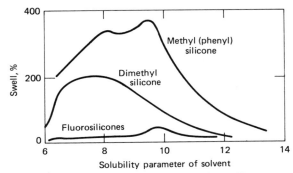

Figure 2.   Swelling of silicone rubber by solvents (83). The fluorosilicones are methyl trifluoropropyl silicones.

Aqueous solutions interact with silicone rubber with varying effects. Water itself has little effect, although at higher temperatures it causes some softening and weakening and, if the rubber is heated with it in a sealed tube, reversion to a sticky polymer occurs. Concentrated solutions of acids and bases are more damaging (178,187).

Heat alone can cause depolymerization and volatilization, with the rate increasing appreciably above 315°C, even in decatalyzed, ie, neutral, compounds. Heat aging causes various changes in cured silicone rubber, which can be noticed in the properties at elevated temperature or after a decrease to room temperature. Heating in air at 125°C causes a decrease in elongation and increase in hardness but no great change in tensile strength (178).

Permanent deformation occurs when rubber is compressed or stretched at high temperature. This results in large part from relaxation of stress by rearrangement of chemical bonds. In the case of silicone, the SiOSi bonds break and reform. The rate at which this occurs depends upon the temperature, the presence of rearrangement catalysts, and the presence of water. A common measure of the ability of a material to undergo or withstand this type of change is a test for compression set. The percentage of permanent set varies with time and temperature. Values for several types of silicone rubber are listed in Table 15. Low compression set was at one time achieved by the use of additives but is now achieved by the use of vinyl gums and noncatalytic components.

Silicone rubber can burn, but the residue is siliceous rather than carbonaceous, tends to have some structural integrity, and is a nonconductor of electricity. The limiting oxygen index for a typical formulation is ca 20. Flame-retardant versions with limiting oxygen index values as high as 40–50 are also available (153).

Electron or gamma radiation causes damage to cured silicone rubber. Elongation is decreased, but tensile strength is not greatly affected (188). On the whole, methyl silicone rubbers are not very resistant to gamma radiation. Phenyl-containing silicone rubber is more resistant, in direct proportion to the phenyl content (189) (see Elastomers, synthetic).

Economic Aspects

Worldwide production of silicones is ca 3.4×10^5 metric tons annually, the United States accounting for approximately half of the total. Average 1980 prices in the United States of ($4–6)/kg for fluids and resins and ($6–8)/kg for elastomers imply that total worldwide sales are ca (2×20^9)/yr (190). Accurate statistics on production volumes of silicone materials are difficult to obtain. Product lines are generally diverse and producers vary in their marketing emphasis. Also, most silicone producers manufacture other chemical and nonchemical products as well and are reluctant to provide specific silicone capacity and production data.

There are three primary categories of silicone products: fluids, resins, and elastomers. Other products, eg, emulsions, greases, pressure-sensitive adhesives, and dispersions, are derived from these three basic types. Verifiable information on the relative production volumes of each category is not available, but recently reported figures suggest the following breakdown is reasonable: fluid and fluid-based, 65–70%; resin and resin-based, 5–10%; and elastomers, 25–30 wt % (191). The growing importance of silane monomers does require a fourth product category. They are listed as fluids, chemicals, or silanes by most producers.

Over 1000 different silicone products are commercially available. The main producers supply a large number of equivalent grades, but each producer also offers certain products that are particular to their own production capability and marketing strategy. Generic product names are common, eg, silicone fluid, silicone rubber, etc. Tradenames are sometimes used; for example, Versilube lubricants and Silpruf construction sealants are trademarks of the General Electric Company (91,150). Important trademarks used by Dow Corning Corporation include Dri-Sil concrete and masonry treatment, Molykote specialty lubricants, Silastic heat-cured and RTV silicone rubber, and Syl-off paper coatings (192).

Analysis

The analytical chemistry of organosilicon materials, ie, the monomers, oligomers, and polymers, generally reflects the uniqueness of silicon and its influence on bond properties in these structures. The analysis of these materials is the subject of an excellent monograph (193) and earlier reviews (194–195). All the common instrumental techniques are used. Mass spectrometry and gas chromatography are widely used to identify and measure volatile silicones and silane intermediates (196–197). High volume, continuous silane manufacturing processes are efficiently monitored and controlled with on-line, automated chromatographs. Liquid and gel-permeation chromatography have been used to determine the distribution of molecular weights in silicone polymers (200).

Qualitative identification and quantitation of silicones can be made through their infrared (201) or ultraviolet spectra. The $\geqslant$SiCH$_3$ and $>$Si(CH$_3$)$_2$ groups absorb at 1259 and 800 cm$^-$ and SiOSi at 1010–1110 cm^{-1}. The ratio of methyl to phenyl groups in the polymers can be determined by measuring the intensity of the $\geqslant$SiCH$_3$ and $\geqslant$SiC$_6$H$_5$ bands at 1263 cm^{-1} and 1435 cm^{-1}, respectively (202). Phenyl silicones also absorb at 260, 265, and 272 nm in the uv spectrum; $\geqslant$SiH is determined by ir absorption at 2100–2250 cm^{-1}. The level of methyltriacetoxysilane in continuously compounded RTV sealants can be controlled by on-line, automated laser-infrared spectroscopy (203).

Nuclear magnetic resonance spectroscopy of ^{1}H, ^{13}C, and ^{29}Si is becoming an increasingly important tool for the analysis of organosilicon compositions. Recent advances in silicon-29 spectroscopy, for example, are providing a new capability for elucidating the structural elements in complex siloxane networks (204–205).

Special groups are determined by specific chemical reactions. For example, chlorosilanes can be hydrolyzed and the halogen determined by titration with alkali or silver nitrate. Other types of halogen substitution may require more drastic methods of decomposition. SiH is determined by evolution of hydrogen on hydrolysis or alcoholysis catalyzed by bases. Silanol can be determined by measurement of methane evolved with methyl Grignard reagent; water is corrected for by reaction with calcium hydride which, unless specially prepared, does not react with silanol. Water and silanol can also be determined separately by infrared techniques.

Quantitative determination of silicones is variously accomplished. For example, determination on paper, textiles, or in formulations can be done by ashing and determination as silica or by extraction with solvents and measurement by infrared absorption.

Health and Safety Factors

Methyl silicones are biologically inert (206–207). They do not react with body fluids, they do not cause coagulation of blood in contact with them, they do not adhere to body tissues, and they do not show irritating or toxic effects. These properties no doubt result from the molecular weight, incompatibility, and stability of the polymers. These qualities of the methyl polymers are usually also true of phenyl-, alkyl-, and fluoroalkyl-substituted polymers. Trifluoropropyl silicones, however, form toxic materials when heated above 280°C (208).

Formulated silicones contain ingredients that may be irritating or toxic. For example, some of the metallic or organometallic catalysts in resins or RTV rubbers are irritating or toxic, but they are used at very low levels in these products. By-products of elastomer cure may be irritating. Emulsifiers in silicone emulsions or bactericides used to keep these or silicone elastomers from deterioration are potential hazards. Many such products, however, are formulated to meet specific safety standards. Silicone resins are sold in solvents, and these may be hazardous with regard to health or fire.

Methylchlorosilanes are flammable, and are corrosive because HCl is liberated on hydrolysis; other chlorosilanes are less flammable, but all are hazardous chemicals. Volatile oligomeric siloxanes are somewhat flammable and can be generated by decomposition of siloxane polymers. Compounds containing SiH liberate hydrogen on hydrolysis or alcoholysis, and this is a hazard both in terms of flammability and generation of pressure (209).

A few other silicon compounds are toxic. For example, methyl and ethyl orthosilicates are somewhat toxic. Materials of this type may be used in formulating silicone products. It should not be assumed that because silicone polymers are physiologically inert, all silicones and silanes are similarly benign, although this is usually the case (210–211).

Silicone rubber is generally resistant to bacterial or fungus growth, but bacterial growth has been noted in a few cases (212–213). This is probably caused by nonsilicone components in the composition, eg, fatty acids, but the net result is some ability to support growth.

A few low molecular weight silicone oligomers show varying biological activity. Hexamethyldisiloxane has toxic properties similar to that of many solvents and the methyltrifluoropropyl and methylphenyl trimers and tetramers are biologically active (207,214–216).

Uses

Silicones are widely used. Applications exist in almost every industry and several in those of consumer products. Generally, silicone applications are based on their unique properties and their cost effectiveness in use. In areas where silicones complete with organic counterparts, savings in assembly, installation, and maintenance labor costs can combine with longer service life to offset any price premium initially charged for silicones. Some applications require the uniqueness of silicones since competitive, alternative materials are not readily available. Salient properties of dimethyl silicones are thermal stability; oxidative stability; mild dependence of physical properties on temperature; resistance to weathering, ozone, and ultraviolet and gamma radiation; low surface tension; high surface activity; good spreading power; and chemical and biological inertness.

Silicone chemistry gives great latitude to product development activity. Varying molecular structure or formulation ingredients often results in dramatic performance differences. Some methyl silicone compositions are adhesives and others are release agents. The properties of dimethyl silicones are modified by substituting other groups, eg, phenyl, vinyl, or longer chain alkyls, in place of methyl. Methyl silicone fluids can function as defoamers, but similar fluids containing polyether groups are used as surfactants to regulate pore structure in foams (3–4,10,34).

Silicones are used as engineering materials and are incorporated as components or additives in many other compositions. Parts fabricated from heat-cured silicone rubber can retain physical and electrical properties over a broad temperature range, eg, from −100°C to 250°C. Silicones can be used as additives to improve the flow and surface structure of thermoplastic and thermoset resins and to prevent the segregation of pigments in paints.

The range of silicone applications is extremely large and a compilation of general product types and example application areas is presented in Table 17. More complete lists and discussions of silicone uses are given elsewhere (4,6,10,34,125,153,192).

Table 17.   General Product Types and Example Application Areas[a]

Fluid applications	
plastic additives	greases
hydraulic fluids	coagulants
vibration damping	particle and fiber treatments
release agents	cosmetic and health-product additives
antifoamers	heat-transfer media
dielectric media	polishes
water repellency	lubricants
surfactants	
Resin applications	
varnishes	electrical insulation
paints	pressure-sensitive adhesives
molding compounds	laminates
protective coatings	release coatings
encapsulants	adhesives
junction coatings	
RTV rubber applications	
sealants	encapsulants
adhesives	electrical insulation
conformal coatings	glazing
gaskets	medical implants
foams	surgical aids
molding parts	mold making
Heat-cured rubber applications	
tubing and hoses	auto-ignition cable and spark-plug boots
belting	extruding
wire–cable insulation	medical implants
surgical aids	laminates
fuel-resistant rubber parts	electrically conducting rubber
penetration seals	fabric coating
molded parts	foams
embossing–calendering rollers	

[a] Refs. 4, 6, 10, 34, 125, 153, and 192.

In medical and surgical applications, silicone fluids are used for aiding gastric disorders (as antiflatulents), antibiotic storage, and treating wound dressings to prevent them from adhering to tissue. Plasma bottles are treated to avoid blood coagulation and to prevent the blood from wetting the bottle, thus making it drain easily. Rubber parts are used for surgical tubing and RTV rubber as a dental mold material. Medical-grade elastomers are used to fabricate heart valves, prosthetic parts, and contact lenses (qv). Room-temperature-vulcanizing materials are also used to encase pacemakers for heart patients and to coat catheters. The Dow Corning Center for Aid to Medical Research, founded in 1959 in Midland, Mich., is a valuable source of information on medical and surgical uses for silicones (217–218) (see Prosthetic and biomedical devices).

In the electrical and electronics industries, silicone fluids are used as dielectric media in transformers, silicone resins are used to make molding compounds to enclose parts, heat-cured silicone rubber is used for wire and cable coatings, and RTV silicone encapsulants are used to protect delicate circuitry.

Applications in the automotive industry include silicone fluids as working liquids in hydraulic brake systems (see Hydraulic fluids), gels for shock-absorbing devices, and RTV sealants (qv) for windshield installations. RTV silicone rubber is also used in automated gasket-forming operations, in which a bead of silicone is dispensed directly on parts, eg, axle covers, window seals, tail-light assemblies, and engine-oil pans.

In construction, RTV silicone sealants are used in glazing (including structural glazing), sealing, and caulking applications (219). Sprayed-on roofing systems include a silicone rubber coating over polyurethane or another base material. Silicones are also used as concrete and masonry treatments for water repellency and environmental protection (see Waterproofing and water repellency).

BIBLIOGRAPHY

"Silicones" in *ECT* 1st ed., Vol. 12, pp. 393–413, by R. R. McGregor, Mellon Institute of Industrial Research; "Silicones" in *ECT* 2nd ed., Vol. 18, pp. 221–257, by Robert Meals, General Electric Co.

1. F. S. Kipping and L. L. Loyd, *J. Chem. Soc.* **79,** 449 (1901).
2. U.S. Pat. 2,258,218 (Oct. 7, 1941), E. G. Rochow (to General Electric Company).
3. B. B. Hardman and R. W. Shade, *Mater. Technol.* **1,** 26 (1980).
4. R. R. McGregor, *Silicones*, McGraw-Hill Book Company, New York, 1954.
5. H. A. Liebhafsky, *Silicones Under the Monogram*, John Wiley & Sons, Inc., New York, 1978.
6. J. A. C. Watt, *Chem. Brit.* **6,** 519 (1970).
7. *Thomas Register of American Manufacturers and Thomas Register Catalogue File*, 71st ed., Thomas Publishing Company, New York, 1981.
8. C. B. Hurd, *J. Am. Chem. Soc.* **68,** 364 (1946).
9. *Chem. Eng. News* **30,** 4517 (1952).
10. R. N. Meals and F. M. Lewis, *Silicones*, Reinhold Publishing Company, New York, 1959.
11. M. C. Voronkev, V. P. Mileshkevich, and Y. A. Yirxhelenski, *The Siloxane Bond*, Consultants Bureau, New York, 1978.
12. P. J. Flory, V. Crescenzi, and J. E. Mark, *J. Am. Chem. Soc.* **86,** 146 (1964).
13. E. G. Rochow, *An Introduction to the Chemistry of the Silicones*, John Wiley & Sons, Inc., New York, 1951.
14. A. J. Barry and H. N. Beck in F. G. A. Stone and W. A. G. Graham, eds., *Inorganic Polymers*, Academic Press, Inc., New York, 1962.
15. C. Eaborn, *Organosilicon Compounds*, Butterworths Scientific Publications, London, 1962.
16. V. Bazant, V. Chvalovsky, and J. Rathousky, *Organosilicon Compounds*, Academic Press, Inc., New York, 1965.
17. L. H. Sommer, *Stereochemistry, Mechanism and Silicon*, Pergamon Press, Inc., New York, 1962.
18. C. T. Mortimer, *Reaction Heats and Bond Strengths*, Pergamon Press, Inc., New York, 1962.
19. L. Pauling, *The Nature of the Chemical Bond*, Cornell University Press, Ithaca, N.Y., 1940.
20. U.S. Pat. 2,380,995 (Aug. 7, 1945), E. G. Rochow (to General Electric Company).
21. U.S. Pat. 2,449,821 (Sept. 21, 1948), J. E. Sellers and J. L. Davis (to General Electric Company).
22. U.S. Pat. 2,389,931 (Nov. 27, 1945), C. E. Reed and J. T. Coe (to General Electric Company).
23. D. G. Weaver and R. J. O'Connors, *Ind. Eng. Chem.* **50,** 132 (1958).
24. V. Bazant, *Pure Appl. Chem.* **13,** 313 (1966).
25. J. J. Zuckerman in H. J. Emeleus and A. G. Sharpe, eds., *Advances in Inorganic Chemistry and Radiochemistry*, Vol. 6, Academic Press, Inc., New York, 1964, p. 383.
26. R. J. H. Voorhoeve, *Organohalosilanes, Precursors to Silicones*, Elsevier Publishing Company, New York, 1967.
27. A. D. Petrov, V. S. Mironov, V. A. Ponomarenko, and E. A. Chernyshev, *Synthesis of Organosilicon Monomers*, Consultants Bureau, New York, 1964.
28. H. K. Lichtenwalner and N. N. Sprung in N. M. Bikales, ed., *Encyclopedia of Polymer Science and Technology*, Vol. 12, Interscience Publishers, a division of John Wiley & Sons, Inc., New York, 1970, p. 464.
29. B. A. Bluestein and H. R. McEntee, *Adv. Chem. Ser.* **23,** 233 (1959).
30. U.S. Pat. 2,557,931 (June 26, 1951), A. J. Barry (to Dow Corning Corporation).

31. U.S. Pat. 2,511,820 (June 13, 1950), A. J. Barry, D. E. Hook, and L. DePree (to Dow Corning Corporation).
32. U.S. Pat. 2,770,634 (Nov. 13, 1956), D. R. Weyenberg (to Dow Corning Corporation).
33. R. N. Meals, *Pure Appl. Chem.* **13,** 141 (1966).
34. W. Noll, *Chemistry and Technology of Silicones*, Academic Press, Inc., New York, 1968.
35. A. E. Pierce, *Silylation of Organic Compounds*, Pierce Chemical Co., Rockford, Ill., 1968.
36. U.S. Pat. 3,701,795 (Oct. 31, 1972), F. F. Holub, A. Berger, B. B. Hardman, and M. P. Urkevich (to General Electric Company).
37. U.S. Pat. 3,901,913 (Aug. 26, 1975), Y. K. Kim (to Dow Corning Corporation).
38. B. Arkles, *Chem. Tech.* 766 (Dec. 1977).
39. E. P. Plueddemann in *Composite Materials*, Vol. 6, Academic Press, Inc., New York, 1974.
40. R. Gutoff, *Ind. Eng. Chem.* **49,** 1807 (1957).
41. U.S. Pat. 3,983,248 (Sept. 28, 1976), J. D. Reedy and H. D. Furbee (to Union Carbide Corporation).
42. Brit. Pat. 843,273 (Aug. 4, 1960), A. N. Pines (to Union Carbide Corporation).
43. U.S. Pat. 3,842,110 (Oct. 15, 1974), J. S. Razzano (to General Electric Company).
44. W. Patnode and D. F. Wilcock, *J. Am. Chem. Soc.* **68,** 358 (1946).
45. D. W. Scott, *J. Am. Chem. Soc.* **68,** 2294 (1946).
46. D. F. Wilcock, *J. Am. Chem. Soc.* **69,** 480 (1947).
47. T. C. Kendrick, *J. Chem. Soc.*, 2027 (1965).
48. J. F. Brown and G. M. J. Slusarczuk, *J. Am. Chem. Soc.* **87,** 931 (1965).
49. W. T. Grubb and R. C. Osthoff, *J. Am. Chem. Soc.* **77,** 1405 (1955).
50. K. Vesely and M. Kucera, *Symp. Makromol. Wiesbaden, DBR, Kurzmilleilungen* **4,** 83 (1959).
51. M. Kucera, M. Jelinek, J. Lanikova, and K. Vesely, *J. Polym. Sci.* **53,** 311 (1961).
52. E. D. Brown and J. B. Carmichael, *J. Polym. Sci.* **B3,** 473 (1965).
53. U.S. Pat. 3,937,684 (Feb. 10, 1976), J. S. Razzano (to General Electric Company).
54. F. R. Mayo and F. M. Lewis, *J. Am. Chem. Soc.* **66,** 1544 (1944).
55. Z. Laita and M. Jelinek, *Vysokomol. Soedin.* **5,** 1268 (1963).
56. U.S. Pat. 4,128,568 (Dec. 5, 1978), W. Buchner, B. Degen, L. Fried, J. Helmut, R. Mundel, and K.-H. Rudolph (to Bayer Aktiengesellschaft).
57. K. A. Earhart, *Paint Varn. Prod.*, 37 (Feb. 1972).
58. U.S. Pat. 3,182,076 (May 4, 1965), N. G. Holdstock (to General Electric Company).
59. U.S. Pat. 3,965,150 (June 22, 1976), R. E. Moeller (to General Electric Company).
60. Fr. Pat. 1,326,879 (April 1, 1963), W. Simmler and H. W. Kauczor (to Farbenfabriken Bayer, A.G.).
61. Fr. Pat. 1,327,546 (April 8, 1963), L. A. Haluska (to Dow Corning Corporation).
62. E. P. Goldberg and E. J. Powers, *J. Poly. Sci. Part B* **2,** 835 (1964).
63. U.S. Pat. 3,189,662 (June 15, 1965), H. A. Vaughn, Jr. (to General Electric Company).
64. J. S. Riffle, R. G. Freelin, A. K. Banthia, and J. E. McGrath, *J. Macromol. Sci.* **A15,** 967 (1981).
65. U.S. Pat. 3,598,783 (Aug. 10, 1971), F. F. Holub and D. R. Pauze (to General Electric Company).
66. U.S. Pat. 3,586,699 (June 22, 1971), T. C. Wu (to General Electric Company).
67. U.S. Pat. 2,970,126 (Jan. 31, 1961), E. D. Brown (to General Electric Company).
68. K. A. Andrianov, *Metalorganic Polymers*, Wiley-Interscience, New York, 1965.
69. H. Schroeder, O. G. Schaffling, T. B. Larchar, F. F. Frulla, and T. L. Heying, *Rubber Chem. Technol.* **39,** 1184 (1966).
70. U.S. Pat. 3,366,612 (Jan. 30, 1968), F. B. Baldwin and A. Malatesta (to Esso Research and Development Company).
71. U.S. Pat. 3,448,072 (June 3, 1969), B. A. Ashby (to General Electric Company).
72. U.S. Pat. 3,592,795 (July 13, 1971), B. A. Ashby (to General Electric Company).
73. U.S. Pat. 3,317,461 (May 2, 1967), E. P. Plueddemann (to Dow Corning Corporation).
74. A. V. Tobolsky, *J. Polym. Sci.* **C9,** 157 (1965).
75. P. J. Flory and J. A. Semlyen, *J. Am. Chem. Soc.* **88,** 3209 (1966).
76. A. J. Barry, *J. Appl. Phys.* **17,** 1020 (1946).
77. P. J. Flory, L. Mandelkern, J. B. Kinsinger, and W. B. Shultz, *J. Am. Chem. Soc.* **74,** 3364 (1952).
78. A. Charlesby, *J. Polym. Sci.* **17,** 379 (1955).
79. E. D. Brown, Jr., *Am. Soc. Lubric. Eng. Trans.* **9,** 31 (1966).
80. U.S. Pat. 4,097,393 (June 21, 1978), R. A. Cupper and R. W. Shiffler (to Union Carbide Corporation).

81. K. E. Polmanteer and M. J. Hunter, *J. Appl. Polym. Sci.* **1**, 3 (1959).
82. T. V. Koroleva, T. A. Krasovskaya, M. V. Sobolevskii, L. V. Gornets, and Yu. E. Raskin, *Soviet Plastics*, 28 (Jan. 1967).
83. K. B. Yerrick and H. N. Beck, *Rubber Chem. Technol.* **37**, 261 (1964).
84. H. M. Schiefer and J. VanDyke, *Am. Soc. Lubric. Eng. Trans.* **7**, 32 (1964).
85. H. M. Schiefer, *Am. Soc. Lubric. Eng. Trans.* **9**, 36 (1966).
86. U.S. Pat. 2,834,748 (May 13, 1958), D. L. Bailey and F. M. O'Connor (to Union Carbide Company).
87. R. N. Meals, *Ann. N.Y. Acad. Sci.* **125**, 137 (1964).
88. *Gmelins Handbuch der anorganischen Chemie*, 8th ed., System Number 15, Silicium, Teil C, Organische Siliciumverbindungen, Verlag Chemie G.m.b.H., Weinheim, FRG, 1958.
89. N. Kirk, *Ind. Eng. Chem.* **51**, 515 (1959).
90. *Silicone Fluids*, F-46182, Union Carbide Corporation, New York, 1978.
91. *Silicone Fluids*, S-9E, General Electric Company, Waterford, N.Y., 1980.
92. *Physical Properties of Lubricants*, American Society for Testing Materials, Philadelphia, Pa., 1949.
93. P. W. Bridgman, *Proc. Am. Acad. Arts Sci.* **77**, 129 (1949).
94. *Ibid.*, p. 115.
95. W. Noll, *Kolloid-Z* **211**, 98 (1966).
96. H. W. Fox, P. W. Taylor, and W. A. Zisman, *Ind. Eng. Chem.* **39**, 1401 (1947).
97. A. A. Bondi, *Physical Chemistry of Lubricating Oils*, Reinhold Publishing Company, New York, 1951.
98. A. Weissler, *J. Am. Chem. Soc.* **71**, 93 (1949).
99. E. B. Baker, A. J. Barry, and M. J. Hunter, *Ind. Eng. Chem.* **38**, 1117 (1946).
100. F. Clark, *Insulating Materials for Design and Engineering Practice*, John Wiley & Sons, Inc., New York, 1962.
101. P. Cannon, L. E. St. Pierre, and A. A. Miller, *J. Chem. Eng. Data* **5**, 236 (1960).
102. G. E. Vogel and F. O. Stark, *J. Chem. Eng. Data* **9**, 599 (1964).
103. Unpublished data based on measurements using ^{14}C labeled dimethylpolysiloxane, General Electric Company, Waterford, N.Y., May 16, 1958.
104. *Silicone Fluids in Radiation Environment*, CDS-4176, General Electric Company, Waterford, N.Y., 1975.
105. D. Tabor and W. D. Winer, *Am. Soc. Lubric. Eng. Trans.* **8**, 69 (1965).
106. H. R. Baker, R. E. Kagarise, J. G. O'Rear, and P. J. Sneigoski, *J. Chem. Eng. Data* **11**, 110 (1966).
107. U.S. Pat. 3,865,784 (Feb. 11, 1975), R. S. Neale and A. N. Pines (to Union Carbide Corporation).
108. M. V. Sobolevskii and co-workers, *Plast. Massy* **3**, 13 (1962).
109. I. Lipovetz and A. Borbely, *Period. Polytech. Chem. Eng.* **2**, 259 (1958).
110. A. Charlesby, *Proc. Roy. Soc.* **A230**, 120 (1955).
111. A. M. Wrobel, M. R. Wertheimer, J. Deb, and H. P. Schreiber, *J. Macromolecular Sci.* **A14**, 321 (1980).
112. A. Chapiro, *Radiation Chemistry of Polymeric Systems*, Interscience Publishers, a division of John Wiley & Sons, Inc., New York, 1962.
113. R. K. Jenkins, *J. Polymer Sci.* **A1**(4), 2161 (1966).
114. U.S Pat. 3,812,081 (May 21, 1974), W. E. Dennes and G. E. Vogel (to Dow Corning Corporation).
115. H. C. Smith, D. L. Sweeney, and E. G. Bittner, *Insulation (Libertyville)* **13**(7), 257, 268 (1967).
116. F. J. Modic, *Adhes. Age* **5** (Dec. 1962).
117. U.S. Pat. 3,986,997 (Oct. 19, 1976), H. A. Clark (to Dow Corning Corporation).
118. U.S. Pat. 4,246,038 (Jan. 20, 1981), H. A. Vaughn and F. F. Holub (to General Electric Company).
119. U.S. Pat. 2,449,572 (Sept. 21, 1948), C. E. Welsh (to General Electric Company).
120. Brit. Pat. 841,825 (July 20, 1960), (to Wacker-Chemie G.m.b.H.).
121. U.S. 3,205,283 (Sept. 7, 1965), F. J. Modic (to General Electric Company).
122. U.S. Pat. 3,801,544 (April 2, 1974), A. E. Mink and D. D. Mitchell (to Dow Corning Corporation).
123. U.S. Pat. 4,223,072 (Sept. 16, 1980), R. H. Bakey and L. A. Harris (to Dow Corning Corporation).
124. U.S. Pat. 4,243,721 (Jan. 6, 1981), R. H. Bakey and L. A. Harris (to Dow Corning Corporation).
125. S. Fordham, *Silicones*, Philosophical Library, New York, 1960.
126. M. D. Beers in I. Skeist, ed., *Handbook of Adhesives*, D. Van Nostrand Company, New York, 1977.
127. A. M. Bueche, *J. Poly. Sci.* **25**, 139 (1957).
128. B. B. Boonstra, H. Cochrane, and E. M. Dannenberg, *Rubber Chem. Technol.* **48**, 448 (1975).

129. E. L. Warrick, O. R. Pierce, K. E. Polmanteer, and J. C. Saam, *Rubber Chem. Technol.* **52,** 437 (1979).
130. G. Berrod, A. Vidal, E. Papirer, and J. B. Donnet, *J. Appl. Polym. Sci.* **26,** 833 (1981).
131. U.S. Pat. 3,635,743 (Jan. 18, 1972), A. H. Smith (to General Electric Company).
132. U.S. Pat. 2,938,009 (May 24, 1960), G. R. Lucas (to General Electric Company).
133. U.S. Pat. 4,279,783 (July 21, 1981), G. P. Kehrer and W. G. Smith (to Dow Corning Corporation).
134. U.S. Pat. 3,133,891 (May 19, 1964), L. Geyzeriat (to Rhone-Poulenc Company).
135. U.S. Pat. 2,615,861 (Oct. 28, 1952), P. P. Peyrot and L. J. Dumoulin (to Rhone-Poulenc Company).
136. U.S. Pat. 3,382,205 (May 7, 1968), M. D. Beers (to General Electric Company).
137. U.S. Pat. 3,077,465 (Feb. 12, 1963), L. B. Bruner (to Dow Corning Corporation).
138. U.S. Pat. 3,082,527 (March 26, 1963), S. Nitzsche and M. Wick (to Wacker-Chemie, G.m.b.H.).
139. U.S. Pat. 4,257,932 (March 24, 1981), M. D. Beers (to General Electric Company).
140. U.S. Pat. 3,334,067 (Aug. 1, 1967), D. R. Weyenberg (to Dow Corning Corporation).
141. U.S. Pat. 3,689,454 (Sept. 5, 1971), S. D. Smith and S. B. Hamilton, Jr. (to General Electric Company).
142. U.S. Pat. 4,036,813 (July 19, 1977), B. B. Hardman and W. T. Madigan (to General Electric Company).
143. U.S. Pat. 4,100,129 (July 11, 1978), M. D. Beers (to General Electric Company).
144. U.S. Pat. 3,819,563 (June 25, 1974), T. Takago, T. Sato, and A. Hisashi (to Shin-Etsu Chemical Company).
145. U.S. Pat. 4,180,642 (Dec. 25, 1979), T. Takago (to Shin-Etsu Chemical Company).
146. U.S. Pat. 3,032,528 (May 1, 1962), S. Nitzsche and M. Wick (to Wacker-Chemie, G.m.b.H.).
147. U.S. Pat. 3,417,047 (Dec. 17, 1968), D. Golitz, K. Damm, R. Muller, and W. Noll (to Farbenfabriken Bayer, A.G.).
148. U.S. Pat. 3,189,576 (June 15, 1965), E. Sweet (to Dow Corning Corporation).
149. *Silicone Elastomers*, 61-605-80, Dow Corning Corporation, Midland, Mich., 1980; *Silicone Rubber Adhesive Sealants for Industrial Applications*, S-2E, General Electric Company, Waterford, N.Y., 1978.
150. *Silicone Construction Products*, 61-600-81, Dow Corning Corporation, Midland, Mich., 1981; *Silicone Construction Sealants*, CDS-1302, General Electric Company, Waterford, N.Y., 1981.
151. *Two Component RTV Silicone Rubber Compounds for Industrial Applications*, S-35B, General Electric Company, Waterford, N.Y., 1980.
152. *RTV Silicone Rubber for Flexible Molding*, S-45, General Electric Company, Waterford, N.Y., 1978; *Silastic Liquid Silicone Rubber*, 17-267-78, Dow Corning Corporation, Midland, Mich., 1978.
153. W. Lynch, *Handbook of Silicone Rubber Fabrication*, D. Van Nostrand Company, New York, 1978.
154. C. V. Cagle, *Handbook of Adhesive Bonding*, McGraw Hill, Inc., New York, 1973.
155. N. J. DeLollis, *Adhesives for Metals*, Industrial Press, New York, 1970.
156. *Surface Preparation for Sealant Adhesion*, CDS-1897A, General Electric Company, Waterford, N.Y., 1980.
157. U.S. Pat. 4,072,635 (Feb. 7, 1978), E. M. Jeram (to General Electric Company).
158. U.S. Pat. 3,436,366 (April 1, 1969), F. J. Modic (to General Electric Company).
159. U.S. Pat. 4,008,198 (Feb. 15, 1977), H. Krohberger, J. Burkhardt, and J. Patzke (to Wacker-Chemie G.m.b.H.).
160. U.S. Pat. 4,041,010 (Aug. 9, 1977), E. M. Jeram (to General Electric Company).
161. U.S. Pat. 2,843,555 (July 15, 1958), C. A. Berridge (to General Electric Company).
162. U.S. Pat. 3,888,815 (June 10, 1975), S. J. Bessemer and W. R. Lampe (to General Electric Company).
163. J. Nagy and A. Borbely-Kaszmann, *International Symposium on Organosilicon Chemistry by Scientific Communications*, Prague, Czechoslovakia, 1965, p. 201.
164. U.S. Pat. 3,020,260 (Feb. 6, 1962), M. E. Nelson (to Dow Corning Corporation).
165. U.S. Pat. 3,344,111 (Sept. 26, 1967), A. J. Chalk (to General Electric Company).
166. U.S. Pat. 3,445,420 (May 20, 1969), G. K. Kookootsedes and E. P. Pleuddemann (to Dow Corning Corporation).
167. U.S. Pat. 3,882,083 (May 6, 1975), A. Berger and B. B. Hardman (to General Electric Company).
168. U.S. Pat. 2,448,756 (Sept. 7, 1948), M. C. Agens (to General Electric Company).
169. R. A. Labine, *Chem. Eng.* **67**(14), 102 (1960).

170. W. J. Bobear in M. Morton, ed., *Rubber Technology*, 2nd ed., D. Van Nostrand Company, New York, 1973.
171. *Silastic Compounding System*, Dow Corning Corporation, Midland, Mich., 1977.
172. *Silicone Rubber Fabricators Handbook*, General Electric Company, Waterford, N.Y., 1980.
173. J. Brandrup and E. H. Immergut, eds., *Polymer Handbook*, Wiley-Interscience, New York, 1967.
174. R. L. Merker, *J. Polym. Sci.* **A2**(1), 31 (1964).
175. U.S. Pat. 2,890,188 (June 9, 1959), G. M. Konkle, J. A. McHard, and K. Polmanteer (to Dow Corning Corporation).
176. Brit. Pat. 791,169 (Feb. 26, 1958), (to Imperial Chemical Industries, Ltd.).
177. Fr. Pat. 1,379,247 (Oct. 12, 1964), J. F. Hyde (to Dow Corning Corporation).
178. F. M. Lewis, *Rubber Chem. Technol.* **35,** 1222 (1962).
179. U.S. Pat. 2,431,878 (Dec. 2, 1947), R. R. McGregor and E. L. Warrick (to Dow Corning Corporation).
180. U.S. Pat. 2,541,851 (Feb. 13, 1951), J. G. E. Wright (to General Electric Company).
181. U.S. Pat. 4,061,609 (Dec. 6, 1977), W. J. Bobear (to General Electric Company).
182. A. Charlesby, *Atomic Radiation and Polymers*, Pergamon Press, Inc., New York, 1960.
183. J. A. Barrie and B. Platt, *Polymer* **4,** 303 (1963).
184. R. M. Barrer, J. A. Barrie, and N. K. Raman, *Rubber Chem. Technol.* **36,** 642, 651 (1963).
185. C. J. Major and K. Kammermeyer, *Mod. Plast.* **39**(11), 135 (1962).
186. J. H. Hildebrand and R. L. Scott, *The Solubility of Non-Electrolytes*, Reinhold Publishing Company, New York, 1950.
187. R. Harrington, *Rubber Age* **84,** 798 (1959).
188. D. J. Fischer, R. G. Chaffee, and V. Flegel, *Rubber Age* **87,** 59 (1960).
189. S. D. Gehman and G. C. Gregson, *Rubber Rev.* **33**(5), 1429 (1960).
190. *Chem. Week* **128,** 26 (Jan. 1981).
191. E. S. Strauss in *Chemical Economics Handbook*, SRI International, Menlo Park, Calif., Dec. 1980.
192. *A Guide to Dow Corning Products*, 01-320A-78, Dow Corning Corporation, Midland, Mich., 1978.
193. A. L. Smith, *Analysis of Silicones*, John Wiley & Sons, Inc., New York, 1974.
194. J. A. McHard, "Silicones," in G. M. Kline, ed., *Analytical Chemistry of Polymers*, Interscience Publishers, Inc., New York, 1959.
195. A. P. Kreshkov and co-workers, *Manual for the Analysis of Monomeric and Polymeric Siliconorganic Compounds*, Moscow, 1962 (in Russian); *Chem. Abstr.* **58,** 10735d, 1963.
196. N. N. Sokolov, K. A. Andrianov, and S. M. Akimova, *J. Gen. Chem. USSR* **25,** 647 (1955).
197. C. A. Hirt, *Anal. Chem.* **33,** 1786 (1961).
198. N. A. Palamarchuk, S. V. Syavtsello, and N. M. Turkeltaub, *Gaz. Khromatogr. Akad. Nauk SSSR*, Tr. Vtoror Vses Knof. Moscow 1962, pp. 303–306; *Chem. Abstr.* **62,** 7117 (1965).
199. J. Kowalski, M. Schibrorek, and J. Chojnowski, *J. Chromatogr.* **130,** 351 (1977).
200. F. Rodriguez, R. A. Rulakowski, and O. K. Clark, *Ind. Eng. Chem. Prod. Res. Dev.* **5,** 121 (1966).
201. L. J. Bellamy, *The Infrared Spectra of Complex Molecules*, John Wiley & Sons, Inc., New York, 1973.
202. J. H. Lady, G. M. Bower, R. E. Adams, and P. F. Bryne, *Anal. Chem.* **31,** 1100 (1959).
203. U.S. Pat. 4,227,083 (Oct. 7, 1980), L. R. Sherinski (to General Electric Company).
204. G. C. Levy and J. D. Cargioli in T. Oxenrod and G. A. Wedd, eds., *Nuclear Magnetic Resonance Spectroscopy of Nuclei Other than Protons*, John Wiley & Sons, Inc., New York, 1974.
205. J. Schraml and J. M. Bellama in F. C. Nachod, J. J. Zuckerman, and E. W. Randall, eds., *Determination of Organic Structure by Physical Methods*, Vol. 6, Academic Press, Inc., New York, 1976.
206. V. K. Rowe, H. C. Spencer, and S. L. Bass, *J. Ind. Hyg. Toxicol.* **30,** 332 (1948).
207. *Methylpolysilicones*, PB-289396, U.S. Department of Commerce, Food and Drug Administration, Washington, D.C., 1978.
208. Current Prodata Data on Fluorosilicone Rubber available from Dow Corning Corporation, Midland, Mich., or General Electric Company, Waterford, N.Y.
209. N. E. Sax, *Dangerous Properties of Industrial Materials*, Van Nostrand Reinhold Company, New York, 1979.
210. J. D. Calandra, M. L. Keplinger, E. J. Hobbs, and L. J. Tyler, *ACS Polym. Preprints* **17**(1), (1976).
211. G. D. Clayton and F. E. Clayton, *Patty's Industrial Hygiene and Toxicology*, Vol. II, Wiley-Interscience, New York, 1981.
212. O. H. Calderon and E. E. Staffeldt, *Int. Biodeterior. Bull.* **1**(2), 33 (1965).

213. S. H. Ross, *U.S. Department of Commerce, Office Technical Services*, AD 429476, Washington, D.C. 1963.
214. L. C. Clark and F. Gollan, *Science* **152,** 1755 (1966).
215. G. Bendz and I. Linquist, *Biochemistry of Silicon and Related Problems*, Plenum Press, New York, 1978.
216. R. J. Fessenden and J. S. Fessenden, *Adv. Organomet. Chem.* **18,** 275 (1980).
217. S. A. Braley, *J. Macromol. Sci. Chem.* **A4**(3), 529 (1970).
218. S. A. Braley in D. C. Simpson, ed., *Modern Trends in Biomechanics*, Vol. 1, Butterworths, London, 1970.
219. J. M. Klosowski, *Adhes. Age* 32 (Nov. 1981).

BRUCE B. HARDMAN
ARNOLD TORKELSON
General Electric Company

SILYLATING AGENTS

Silylation of Organic Compounds

Silylation is the displacement of active hydrogen from an organic molecule by a silyl group. The active hydrogen is usually OH, NH, or SH, and the silylating agent is usually a trimethylsilyl halide or a nitrogen-functional compound. A mixture of silylating agents may be used; a mixture of trimethylchlorosilane and hexamethyldisilazane is more reactive than either reagent alone, and the by-products combine to form neutral ammonium chloride.

$$(CH_3)_3SiNHSi(CH_3)_3 + (CH_3)_3SiCl + 3\ ROH \rightarrow 3\ (CH_3)_3SiOR + NH_4Cl$$

Derivatizing an organic compound for analysis may require only a few drops of reagent selected from silylating kits supplied by laboratory supply houses. Commercial synthesis of penicillins requires silylating agents purchased in tank cars from the manufacturer.

Typical commercial silylating agents are listed in Table 1. The first three silylating agents in the table are available in bulk quantities and are most suitable for large-scale commercial silylation. The chlorosilanes are generally used in combination with an acid acceptor, eg, triethylamine. The nitrogen-functional silanes each have certain advantages for particular applications.

All of the silylating agents are classified by the Department of Transportation as flammable liquids. The chlorosilanes are clear liquids that react readily with water to form corrosive HCl gas and liquid. Liquid chlorosilanes and their vapors are corrosive to the skin and extremely irritating to the mucous membranes of the eyes, nose, and throat. They should be treated as strong acids. The nitrogen-functional silanes react with water to form ammonia, amines, or amides. Since ammonia and amines are

Table 1. Methyl Silylating Agents

Chemical name	CAS Registry No.	Formula
trimethylchlorosilane (TMCS)	[75-77-4]	$(CH_3)_3SiCl$
dimethyldichlorosilane (DMCS)	[75-78-5]	$(CH_3)_2SiCl_2$
hexamethyldisilazane (HMDZ)	[999-97-3]	$(CH_3)_3SiNHSi(CH_3)_3$
chloromethyldimethylchlorosilane (CMDMS)	[1719-57-9]	$ClCH_2(CH_3)_2SiCl$
N,N'-bis(trimethylsilyl)urea (BSU)	[18297-63-7]	$[(CH_3)_3SiN]_2CO$
N-trimethylsilyldiethylamine (TMSDEA)	[996-50-9]	$(CH_3)_3SiN(C_2H_5)_2$
N-trimethylsilylimidazole (TSIM)	[18156-74-6]	$(CH_3)_3Si{-}N\diagup\diagdown_{N}$
N,O-bis(trimethylsilyl)acetimide (BSA)	[10416-59-8]	$(CH_3)_3SiN{=}C(CH_3)OSi(CH_3)_3$
N,O-bis(trimethylsilyl)trifluoroacetimide (BSTFA)	[25561-30-2]	$(CH_3)_3SiN{=}C(CF_3)OSi(CH_3)_3$
N-methyl-N-trimethylsilyltrifluoroacetamide (MSTFA)	[24589-78-4]	$(CH_3)_3SiN(CH_3)COCF_3$
t-butyldimethylsilylimidazole (TBDMIM)	[54925-64-3]	$t\text{-}C_4H_9(CH_3)_2Si{-}N\diagup\diagdown_{N}$
N-trimethylsilylacetamide (MTSA)	[13435-12-6]	$(CH_3)_3SiNHCOCH_3$

moderately corrosive to the skin and very irritating to the eyes, nose, and throat, silylamines should be handled like organic amines.

The techniques of silylation and their application in analysis have been reviewed (1–2) as have the intermediate steps in organic synthesis (3). Shorter summaries of silylation applications are given in refs. 4 and 5.

Derivatization for Analysis. There are four main reasons to derivatize a compound for gas–liquid chromatography (glc) analysis: to increase volatility, to increase thermal stability, to enhance detectability, and to improve separation. Silylating kits offered by laboratory supply houses may contain mixtures of chlorine- and nitrogen-functional silylating agents and activating solvents or catalysts. Catalogs of the supplier describe methodology for derivatization and recommend specific reagents for different applications.

N-Trimethylsilyldiethylamine (TMSDEA) is a strongly basic silylating reagent and is particularly useful for derivatizing low molecular weight acids. The reaction product, diethylamine, is volatile enough to be easily removed from the reaction medium.

N-Trimethylsilylimidazole (TSIM) is the strongest hydroxy silylator available and is the reagent of choice for carbohydrates and steroids. This reagent is unique in that it reacts quickly and smoothly with hydroxyls and carboxyl groups but not with amines. This characteristic makes TSIM particularly useful in multiderivatization schemes for compounds containing both hydroxyl and amine groups.

N,O-Bis(trimethylsilyl)acetimide (BSA) reacts quantitatively under relatively mild conditions with a wide variety of compounds to form volatile, stable trimethylsilane (TMS) derivatives for glc analysis. It has been used extensively for derivatizing alcohols, amines, carboxylic acids, phenols, steroids, biogenic amines, alkaloids, etc. It is not recommended for use with carbohydrates or very low molecular weight compounds. Reactions are generally fast and the reagent is usually used in conjunction with a solvent, eg, pyridine or dimethylformamide (DMF). When used with DMF, it is the reagent of choice for derivatizing phenols.

N,O-Bis(trimethylsilyl)trifluoroacetimide (BSTFA) is a powerful trimethylsilyl donor with approximately the same donor strength as the unfluorinated analogue BSA. Reactions of BSTFA are similar to those of BSA. The main advantage of BSTFA over BSA is the greater volatility of the former's reaction by-products, ie, monotrimethylsilyltrifluoroacetamide and trifluoroacetamide. This physical characteristic is particularly useful in the gas chromatography of some of the lower boiling TMS–amino acids and TMS Krebs cycle acids where the by-products of BSA may have similar retention characteristics and thus obscure these derivatives on the chromatogram. The by-products of BSTFA usually elute with the solvent front.

For the derivatization of fatty-acid amides, slightly hindered hydroxyls, and other difficultly silylatable compounds, BSTFA containing 1 wt % trimethylchlorosilane is used. This catalyzed formulation is stronger than BSTFA alone. When silylation reagents are consumed in the hydrogen flame, silicon dioxide (SiO_2) is formed. N,O-Bis(trimethylsilyl)trifluoroacetamide contains three fluorine atoms which form HF in the detector flame and react with the SiO_2 to form volatile products. This removal of SiO_2 provides a decrease in detector fouling and background noise.

N-Methyl-N-trimethylsilyltrifluoroacetamide (MSTFA) is the most volatile TMS-amide available; it is more volatile than BSTFA or BSA. Its by-product, N-methyltrifluoroacetamide, has an even lower retention time in glc than MSTFA. This is of considerable value in glc determinations where the reagent or by-products obscure the derivative on the chromatogram. Silylation of steroids shows MSTFA to be significantly stronger in donor strength than BSTFA or BSA. N-Methyl-N-trimethylsilyltrifluoroacetamide silylates hydrochloride salts of amines directly.

The t-butyldimethylsilyl group introduced by TBDMIM has a number of advantages in protecting alcohols (6). It hydrolyzes more slowly than a $(CH_3)_3Si$ group by a factor of 10^4. The silyl ether is also stable to powerful oxidizing and reducing agents, but it can easily be removed by aqueous acetic acid or $(C_4H_9)_4NH$ in tetrahydrofuran (THF).

Trimethylsilyl iodide [16029-98-4] is an effective reagent for cleaving esters and ethers. A simple mixture of trimethylchlorosilane and sodium iodide can be used in a similar way to cleave esters and ethers (7). This gives silylated acids or alcohols that can be liberated by reaction with water.

$$
\underset{\text{RCOR}'}{\overset{O}{\overset{\|}{}}} + (CH_3)_3SiI \longrightarrow \underset{\text{RCOSi(CH}_3)_3}{\overset{O}{\overset{\|}{}}} + R'I
$$

$$ROR' + (CH_3)_3SiCl + NaI \rightarrow ROSi(CH_3)_3 + R'I + NaCl$$

It is possible to use halogen-sensitive detectors in glc analysis of active hydrogen compounds by silylating them with a halogenated silylating agent, eg, CMDMS (8).

Silylation in Organic Synthesis. Silyl blocking agents are also used in organic synthesis to protect sensitive functional groups, to alter reactivity and solubility, and to increase stability of intermediates. Silylation applications in pharmaceutical synthesis have been used to protect a wide range of OH groups, eg, alcohols in prostaglandins and steroid synthesis, enols in the synthesis of nucleosides and steroids, and carboxylic acids and sulfenic acids in the synthesis of penicillins and cephalosporins (6) (see Antibiotics; Steroids). Silylation has its broadest use in the commercial synthesis of penicillins. The blocking effect of trimethylsilyl and dimethylsilyl groups on 6-aminopenicillanic acid (6-APA) has played an important role in the total synthetic production of semisynthetic penicillins.

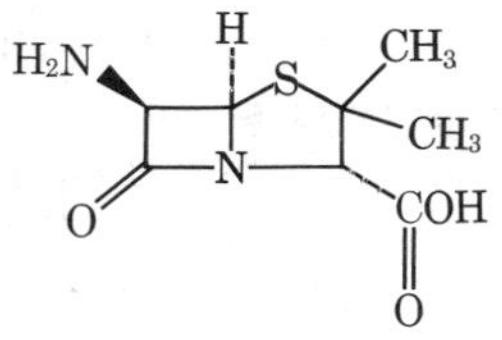

6-APA

Protection of carboxylic acids and sulfenic acids requires efficient silyl donors, eg, BSA, MTSA, and BSU. Bis(trimethylsilyl)urea [18297-63-7] (BSU) is often prepared *in situ* from hexamethyldisilazane and urea to yield over 90% of the silylated derivative in synthesis of cephalosporins.

It is possible to synthesize 1,2,3-triazoles from acetylenes and hydrazoic acid, but the instability of hydrazoic acid has limited this application. Sodium azide is silylated readily with trimethylchlorosilane to produce trimethylsilylazide [4648-54-8] $(CH_3)_3SiN_3$ which reacts with acetylenes to produce high yields of 1,2,3-triazoles (9).

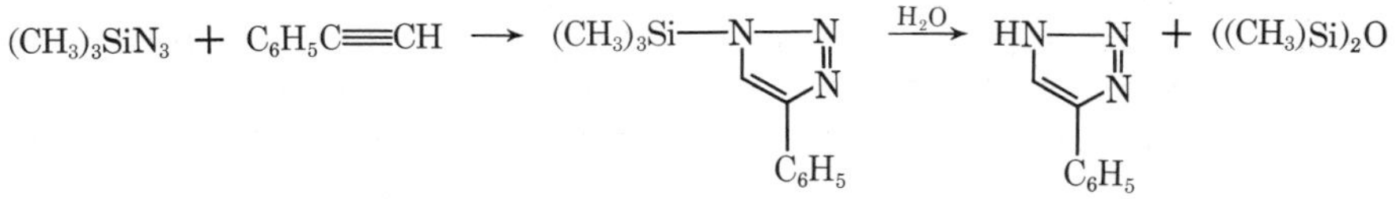

In this case, the substitution of the TMS group for hydrogen in HN_3 imparts a degree of stability to the otherwise unstable azide and also acts as a blocking agent allowing the direct synthesis of the triazole. Trimethylsilylazide can be distilled at atmospheric pressure without decomposition (bp 95°C).

Two techniques have been described for producing trimethylsilylenol ethers from aldehydes or ketones (9): reaction with $(CH_3)_3SiCl$ and $(C_2H_5)_3N$ in DMF and reaction with $LiN(C_2H_5)_2$, which generates enolate ions, in the presence of $(CH_3)_3SiCl$.

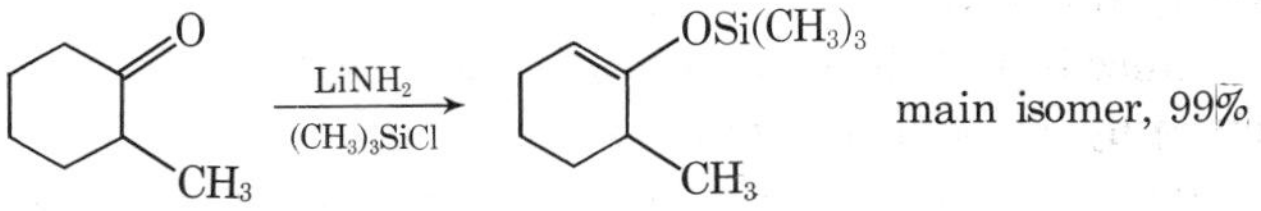

main isomer, 99%

The resulting enol ethers can undergo a wide variety of reactions at the double bond, making this type of reaction important in hormone synthesis (10) (see Hormones).

Silylation of Inorganic Compounds

Silicate Modifications. A method has been described in which silicate minerals are simultaneously acid-leached and trimethylsilyl end-blocked to yield specific trimethylsilyl silicates with the same silicate structure as the mineral from which they were derived (11). Olivine, hemimorphite, sodalite, natrolite, laumontite, and sodium silicates are converted to TMS derivatives of orthosilicates, pyrosilicates, cyclic polysilicates, etc, making it possible to classify the minerals according to their silicate structure. The same technique is used to analyze the siloxanol structure of aqueous solutions of vinyltrimethoxysilane (12). Mixtures of alkali silicates and certain anionic siliconates form stable solutions in water or alcohols at any pH (13). Such silicate–siliconate mixtures are used as corrosion inhibitors in glycol antifreeze (14) (see Corrosion and corrosion inhibitors; Antifreezes and deicing fluids).

Silylation of Inorganic Surfaces

Alkyl Silylating Agents. Alkyl silylating agents convert mineral surfaces to water-repellant, low energy surfaces useful in water-resistant treatments for masonry, electrical insulators, packings for chromatography, and in noncaking fire extinguishers. Methylchlorosilanes react with surface water or hydroxyl groups of the surface to liberate HCl and deposit a very thin film of methylpolysiloxane, which has a very low critical surface tension and is therefore not wetted by water (15). Ceramic insulators can be treated with methylchlorosilane vapors or solutions in inert solvents to maintain high electrical resistivity under humid conditions (16). The corrosive action of the evolved HCl can be avoided by prehydrolyzing the chlorosilanes in an organic solvent and applying them as organic solutions of organopolysiloxanols. Hydrolyzed methylchlorosilanes also dissolve in aqueous alkali and are then applied as aqueous solutions of sodium methylsiliconates. The siliconates are neutralized by carbon dioxide in the air to form an insoluble, water-resistant methylpolysiloxane film within 24 h. Treatment of brick, mortar, sandstone, concrete, and other masonry protects the surface from spalling, cracking, efflorescence, and other types of damage caused by water.

Silanes can alter the critical surface tension of a substrate in a well-defined manner. Critical surface tension is associated with the wettability or release qualities of a substrate. Liquids with a surface tension below the critical surface tension (γ_c) of a substrate wet the surface. Critical surface tensions of a number of typical surfaces are compared with γ_c of silane-treated surfaces in Table 2 (17).

Celite or firebrick packing for glc columns is often treated with TMCS, DMCS, or other volatile silylating agents listed in Table 1 to reduce tailing by polar organic compounds. A chemically bonded methyl silicone support allows underivatized phenols to be analyzed without tailing. The silicone support is stable for temperature programming to 390°C and allows elution of hydrocarbons up to C_{50} (18).

High performance liquid chromatography (hplc) combines the bonded solid phase of gas chromatography with the methodology of column liquid chromatography. The most popular type of hplc involves low polarity silicone-bonded surfaces with more highly polar liquids in a process termed reversed-phase chromatography (19). A favorite stationary phase is fine particle silica treated with octadecyltrichlorosilane [112-04-9]. In some instances, octyl-, phenyl-, cyclohexyl-, or ethylsilanes can be used to obtain improved selectivity. Metal ions on the bonded phase can be used to enhance

Table 2. Critical Surface Tensions γ_c, mN/m (= dyn/cm)[a]

Surface	γ_c
polytetrafluoroethylene	18.5
methyltrimethoxysilane	22.5
vinyltriethoxysilane	25.0
paraffin wax	25.5
ethyltrimethoxysilane	27.0
propyltrimethoxysilane	28.5
glass, soda-lime (wet)	30.0
polychlorotrifluoroethylene	31.0
polypropylene	31.0
polyethylene	33.0
(3,3,3-trifluoropropyl)trimethoxysilane	33.5
[3-(2-aminoethyl)aminopropyl]trimethoxysilane	33.5
polystyrene	34.0
cyanoethyltrimethoxysilane	34.0
aminopropyltriethoxysilane	35.0
poly(vinyl chloride)	39.0
phenyltrimethoxysilane	40.0
(3-chloropropyl)trimethoxysilane	40.5
(3-mercaptopropyl)trimethoxysilane	41.0
(3-glycidoxypropyl)trimethoxysilane	42.5
poly(ethylene terephthalate)	43.0
copper (dry)	44.0
aluminum (dry)	45.0
iron (dry)	46.0
nylon-6,6	46.0
glass, soda-lime (dry)	47.0
silica, fused	78.0

[a] Ref. 17, critical surface tensions for silanes refer to treated surfaces.

separation of polar molecules through ligand-exchange chromatography (lec) (20). Metal ions, eg, copper, are retained on silica gel that has been treated with chelating functional silanes (see Chelating agents).

Organofunctional Silylating Agents

Whereas alkylsilylating agents provide low energy surfaces designed for abhesion, a series of organofunctional silylating agents is offered commercially as adhesion promoters. Their prime application has been as coupling agents in mineral-filled organic resin composites. Organofunctional silanes are also used to control orientation of liquid crystals, bind heavy-metal ions, immobilize enzymes and cell organelles, modify metal oxide electrodes, surface-binding of antimicrobial agents, and other nonplastic applications. Many of these applications have been described (21). Representative commercial silane coupling agents are listed in Table 3. Compounds in Table 3 can also be used as chemical intermediates for preparing other more specialized organofunctional silanes (22). The total market in the United States for organofunctional silanes was over 2200 metric tons in 1980.

Table 3.   Commercial Silane Coupling Agents

No.	Silane coupling agent	CAS Registry No.	Formula	Application in plastics
1.	vinyltris(β-methoxy-ethoxy)silane	[1067-53-4]	$CH_2{=}CHSi(OCH_2CH_2OCH_3)_3$	polyethylene, unsaturated polymers
2.	(γ-methacryloxypropyl)tri-methoxysilane	[2530-85-0]	$CH_2{=}C(CH_3)\overset{\displaystyle O}{\overset{\displaystyle \|}{C}}O(CH_2)_3Si(OCH_3)_3$	unsaturated polymers
3.	vinylbenzyl cationic silane	[34937-00-3]	$CH_2{=}CHC_6H_4CH_2NHCH_2CH_2NH\text{-}$ $(CH_2)_3Si(OCH_3)_3{\cdot}HCl$	all polymers
4.	(4-aminopropyl)triethoxy-silane	[3069-30-5]	$H_2NCH_2CH_2CH_2Si(OC_2H_5)_3$	epoxies, phenolics, nylon
5.	[γ-(β-aminoethylamino)-propyl]trimethoxysilane	[13170-53-1]	$H_2NCH_2CH_2NH(CH_2)_3Si(OCH_3)_3$	epoxies, phenolics, nylon
6.	(γ-glycidoxypropyl)tri-methoxysilane	[25704-87-4]	$\overset{\displaystyle O}{\overset{\displaystyle \triangle}{CH_2CHCH_2}}O(CH_2)_3Si(OCH_3)_3$	most thermosetting resins
7.	[β-(3,4-epoxycyclohexyl)-ethyl]trimethoxysilane	[3388-04-3]	![cyclohexyl epoxide]—$CH_2CH_2Si(OCH_3)_3$	epoxies, acid-resins
8.	(β-mercaptoethyl)tri-methoxysilane	[57557-66-1]	$HSCH_2CH_2Si(OCH_3)_3$	rubber, epoxies
9.	(γ-chloropropyl)tri-methoxysilane	[25512-39-4]	$ClCH_2CH_2CH_2Si(OCH_3)_3$	epoxies

Liquid Crystals.   In liquid-crystal displays, clarity and permanence of image is enhanced if the display can be oriented parallel or perpendicular to the substrate. Oxide surfaces treated with octadecyl-3-(trimethoxysilyl)propylammonium chloride [27668-52-6], $C_{18}H_{37}\overset{+}{N}(CH_3)_2CH_2CH_2CH_2Si(OCH_3)_3Cl^-$, tend to orient liquid crystals (qv) perpendicular to the surface, and parallel orientation is obtained on surfaces treated with N-methylaminopropyltrimethoxysilane [3069-25-8], $CH_3NHCH_2\text{-}$ $CH_2CH_2Si(OCH_3)_3$ (23).

Ion Removal.   The ethylenediamine (en)-functional silane (the fifth entry in Table 3) has been studied extensively as a silylating agent on silica gel to preconcentrate polyvalent anions and cations from dilute aqueous solutions (24–25). Numerous other chelate-functional silanes have been immobilized on silica gel, controlled-pore glass, and fiber glass for removal of metal ions from solution (26–27).

Metal Oxide Electrodes.   Metal oxide electrodes have been coated with a monolayer of γ-(β-aminoethyl)aminopropyltrimethoxysilane by contacting the electrodes with a benzene solution of the silane at room temperature (28). Electroactive moieties attached to such silane-treated electrodes undergo electron-transfer reactions with the underlying metal oxide (29). Dye molecules attached to silylated electrodes absorb light coincident with the absorption spectrum of the dye, which is a first step toward simple production of photoelectrochemical devices (30) (see Photovoltaic cells).

Antimicrobials.   Surface-bonded organosilicon quaternary ammonium chlorides have enhanced antimicrobial and algicidal activity (31). Thus, the hydrolysis product of 3-(trimethoxysilyl)propyldimethyloctadecylammonium chloride [27668-52-6] exhibits antimicrobial activity against a broad range of microorganisms while chemically bonded to a variety of surfaces. The chemical is not removed from surfaces by repeated washing with water, and its antimicrobial activity has not been attributed to a slow release of the chemical but rather to the surface-bonded chemical.

Polypeptide Synthesis and Analysis. Silica or controlled-pore glass supports treated with (chloromethyl)phenylethylsilane [40934-73-4] or its derivatives are replacing chloromethylated styrene–divinylbenzene (Merrifield resin) as supports in polypeptide synthesis. The silylated support reacts with triethylammonium salt of a protected amino acid. Once the initial amino acid residue has been coupled to the support, a variety of peptide synthesis methods can be used (32). Automation allows eight synthetic steps daily. At the completion of synthesis, the anchored peptide is separated from the support with hydrogen bromide in acetic acid (see Polypeptides).

Edman degradations can be accomplished by treating aminopropyl-silylated supports (a silica with 1.0–7.5 nm pores yields ca 2×10^{-7} mol of aminopropylsilane/g of glass) with the peptide to be analyzed in the presence of dicyclohexylcarbodiimide (33). The carboxyl end of the peptide bonds to the amino group of the silane through an amide group. The bound peptide is then treated with phenylthiocyanate in the presence of base to yield a N-terminal phenylthiocarbamyl derivative which, on treatment with acid, cyclizes to a phenylthiohydantoin and cleaves. The hydantoin is analyzed and the process is repeated with the bound peptide residue.

Immobilized Enzymes. Use of enzymes to catalyze reactions in cell-free systems has been limited by the difficulty of enzyme isolation, lability of the enzymes, and difficulty in effecting clean separations of enzymes from reaction mixtures. An approach that has circumvented some of these problems is to attach enzymes to solid support materials (34). The most frequently used technique for immobilizing enzymes on a solid support involves reducing N-(3-silylpropyl)-p-nitrobenzamide groups on silica or controlled-pore glass to give aniline derivatives, then converting them to diazonium salt, and effecting coupling through azo linkage to the tyrosine of the proteins.

Production of glucose and invert sugar are probably the largest commercial applications for immobilized enzymes. Research is being done on glucoamylase, which hydrolyzes the glycolytic linkages of corn starch. Systems that promise to have the most immediate impact are catalase and trypsin treatment of milk, which increases the shelf life of milk and allows higher storage temperature, and amino acid acylase, which can be used to resolve D and L amino acids (see Enzymes, immobilized).

Immobilized Metal-Complex Catalysts. Two general methods are available for immobilizing metal-complex catalysts on metal oxide surfaces by the use of ligand–silane coupling reagents of the type X_3SiL, where X is a hydrolyzable group and L is a ligand (35). In Method A, the ligand silane reacts with surface hydroxyl groups to form a ligand metal oxide, usually a ligand silica, and then a metal complex precursor reacts with the functionalized surface. In Method B, a metal–liquid silane complex is first formed in solution and then reacts with surface hydroxyls of the support.

Method A

$$|{\geqslant}SiOH + X_3SiL \rightarrow |{\geqslant}SiOSiX_2L + HX$$

surface   ligand   ligand
silanol   silane   silane

$$|{\geqslant}SiOSiX_2L + M \rightarrow |{\geqslant}SiOSiX_2LM$$

ligand   complex   surface
silica   precursor   complex

Method B

$$X_3SiL \ + \ M \ \rightarrow \ X_3SiLM$$

ligand	metal	metal-silane
silane	precursor	complex

$$|{\geqslant}SiOH \ + \ X_3SiLM \ \rightarrow \ |{\geqslant}SiOSiX_2LM \ + \ HX$$

surface	metal-silane	surface
silanol	complex	complex

Supported metal-complex catalysts have been used in hydrogenation, hydroformylation, isomerization, and other chemical reactions.

Reinforced Composites. Silane coupling agents modify the interface between mineral surfaces and organic resins to improve the adhesion between resin and surface, thus improving physical properties and water resistance of reinforced plastics. Suitable coupling agents are available for any of the common plastics with metal, glass, or many other inorganic reinforcements. Principal applications for these coupling agents are in reinforced plastics for boats, storage tanks, pipes, and architectural structures (see Laminated and reinforced plastics). Newer applications are in the treatment of mineral fillers and pigments for paint and rubber, in primers to improve the adhesion of paints and plastics to metals and other mineral surfaces, and in tarnish and corrosion inhibitors for silver, copper, aluminum, and steel (see Fillers; Corrosion).

Commercial silane coupling agents are soluble in water or become soluble as the alkoxy groups hydrolyze from silicon. The resulting aqueous solutions are stable in water for at least several hours, but they may become insoluble as the silanols condense to siloxanes. Freshly prepared aqueous solutions of the silane coupling agents are therefore applied to glass filaments by the fiber manufacturer with a polymeric substance, a lubricant, and an antistatic agent as a complete size for glass roving. Glass cloth woven from glass roving with a starch–oil size can be heat-cleaned to burn off the organic size and treated with a dilute solution of the desired silane coupling agents. This operation is accomplished by the glass weaver or by the reinforced-plastics fabricator. The total amount of silane coupling agent applied is generally 0.1–0.5% the weight of the glass. The improvement in laminate properties imparted by silane coupling agents in typical glass-cloth laminates are summarized in Table 4 (36). All results are based on compression-molded test samples containing 60–70 wt % glass in the laminate. Heat-cleaned glass cloth treated with 0.5 wt % silane coupling agent in each case is compared with untreated heat-cleaned glass (37).

The nature of adhesion through silane coupling agents has been studied extensively by advanced analytical techniques (36–37). It is fairly well established that silane coupling agents form M—O—Si bonds with mineral surfaces where M = Si, Ti, Al, Fe, etc. It is not obvious that such bonds should contribute outstanding water resistance to the interface, since oxane bonds between silicon and iron or aluminum, for example, are not resistant to hydrolysis. Yet mechanical properties of filled polymer castings are improved by addition of appropriate silane coupling agents with a wide range of mineral fillers (38).

Oxane bonds, M—O—Si, are hydrolyzed during prolonged exposure to water but reform when dried. Many observations suggest that true equilibrium conditions of condensation and hydrolysis exist at the mineral interface. Adhesion in composites

Table 4.   Silane Performance in Glass-Cloth-Reinforced Laminates[a]

Resin	Silane coupling agent, No.[b]	Flexural strength improvement, %	
		Dry	Wet[c]
epoxy (anhydride)	7, 9	20	300
epoxy (aromatic amine)	3, 9	50	140
epoxy (dicyandiamide)	3, 6	30	70
polyester	2, 3	60	140
vinyl ester	2, 3	40	65
phenolic	3–6	40	120
melamine	3, 6	100	250
nylon	3–6	65	130
polycarbonate	3–6	65	230
polyterephthalate	3	50	50
polystyrene	3	40	110
acrylonitrile–butadiene–styrene terpolymer	3, 6	30	50
styrene–acrylonitrile copolymer	3	40	70
poly(vinyl chloride)	3–5	60	80
polyethylene	3	130	130
polypropylene	3	90	90

[a] Ref. 36.

[b] See Table 3.

[c] Wet = boiling for 72 h in water for epoxies and for 2 h for other resins.

is maintained by controlling conditions favorable for equilibrium oxane formation, ie, maximum initial oxane bonding, minimum penetration of water to the interface, and optimum morphology for retention of silanols at the interface. A general concept of bonding to organic polymers through silanes is being developed.

Although simplified representations of coupling through organofunctional silanes often show a well-aligned monolayer of silane forming a covalent bridge between polymer and filler, the actual situation is much more complex. Coverage by hydrolyzed silane is more likely to be equivalent to several monolayers. The hydrolyzed silane condenses to oligomeric siloxanols that initially are soluble and fusible but ultimately can condense to rigid cross-linked structures. Contact of a treated surface with polymer matrix is made while the siloxanols still have some degree of solubility. Bonding with the matrix resin then can take several forms.

The oligomeric siloxanol layer may be compatible in the liquid matrix resin and then form a true copolymer during resin cure. Partial solution compatibility is also possible and an interpenetrating polymer network can form as the siloxanols and matrix resin cure separately with a limited amount of copolymerization. Probably all thermosetting resins are coupled to silane-treated fillers by some modification of these two extremes.

Interdiffusion of silane primer segments with matrix molecules with no cross-linking may become a factor in bonding of thermoplastic polymers. This must be the case when a silane-thermoplastic copolymer is used as a primer or coupling agent for the corresponding unmodified thermoplastic. A siloxanol layer may also diffuse into a nonreactive thermoplastic layer and then cross-link at the fabrication temperature. Structures, in which only one of the interpenetrating phases cross-links, have been designated pseudointerpenetrating networks. Amine-functional silanes (3, 4, and 5 of Table 3) probably function in this manner in coupling to polyolefins and possibly to other thermoplastics. Performance is often improved by adding a high temperature peroxide to the coupling agent to facilitate cross-linking of the siloxanol oligomers. Layers of amine-functional siloxanols in the absence of matrix resins cure at 150°C to very hard, tough films.

Peroxides or other additives, eg, chlorinated paraffin, may also cause the thermoplastic resin to cross-link with the siloxanols. In this case, a true interpenetrating polymer network forms, in which both phases are cross-linked.

Performance of coupling agents in reinforced composites may depend as much on physical properties resulting from the method of application as on the chemistry of the organofunctional silane. Physical solubility or compatibility of a siloxanol layer is determined by the nature and degree of siloxane condensation on a mineral surface. A complete description of the interphase region in reinforced composites is an active field for study by advanced analytical techniques. A better understanding of the whole interphase region should allow more reproducible production of composites with optimum properties.

BIBLIOGRAPHY

"Silylating Agents" under "Silicon Compounds," in *ECT* 2nd ed., Vol. 18, pp. 260–268, by Edwin P. Plueddemann, Dow Corning Corporation.

1. A. E. Pierce, *Silylation of Organic Compounds*, Pierce Chemical Company, Rockford, Ill., 1968.
2. K. Balu and G. Kind, eds., *Handbook of Derivatives for Chromatography*, Heyden, London, UK, 1977.
3. J. F. Klebe in E. Taylor, ed., *Advances in Organic Chemistry*, Wiley-Interscience, New York, 1972, p. 97.
4. C. A. Roth, *Ind. Eng. Chem. Prod. Res. Develop.* **11**, 134 (1972).
5. B. S. Cooper, *Process Biochem.*, 9 (Jan. 1980).
6. E. J. Corey and T. Ravindranathan, *J. Am. Chem. Soc.* **94**, 401 (1972).
7. T. Morita, Y. Okamoto, and H. Sakurai, *J. Chem. Soc. Chem. Comm.* 874 (1978).
8. C. A. Bache, L. E. St. John, Jr., and D. J. Lesk, *Anal. Chem.* **40**, 1241 (1968).
9. L. Birkhofer, A. Ritter, and H. Uhlenbrauck, *Chem. Ber.* **96**, 2750, 3280 (1963).
10. H. O. House and co-workers, *J. Organometal. Chem.* **34**, 2324 (1969).
11. C. W. Lentz, *Inorg. Chem.* **3**, 574 (1964).
12. E. P. Plueddemann, *paper presented at the 24th Annual Technical Conference of the Society of Plastics Engineers*, 1969, paper 19A.
13. E. P. Plueddemann, *Silane Coupling Agents*, Plenum Press, New York, 1982, Chapt. 3.
14. U.S. Pat. 3,198,820 (Aug. 3, 1965), A. N. Pines and E. A. Zinetek (to UCC).
15. E. G. Shafrin and W. A. Zisman, *Contact Angle, Wettability, and Adhesion, Advances in Chemistry Series No. 43*, American Chemical Society, Washington, D.C., 1964, p. 145.
16. O. K. Johannson and J. J. Torok, *Proc. Inst. Radio Electron. Eng.* **34**, 296 (1946).
17. B. Arkles, *Chemtech*, 768 (Dec. 1977).
18. T. J. Nestrick, L. L. Lamparski, and R. H. Stehl, *Anal. Chem.* **51**, 2273 (1979).
19. C. Horvath, *Silylated Surfaces*, Gordon and Breach Science Publishers, London, UK, 1980, Chapt. 3.
20. F. K. Chow and E. Grushka, *Anal. Chem.* **50**, 1346 (1978).
21. D. E. Leyden and W. Collins, eds., *Silylated Surfaces*, Gordon and Breach Science Publishers, London, UK, 1980.
22. Ref. 13, Chapt. 2.
23. F. J. Kahn, G. N. Taylor, and H. Schonborn, *Proc. IEEE* **61**, 823 (1973).
24. D. E. Leyden and G. H. Luttrell, *Anal. Chem.* **47**, 1612 (1975).
25. D. E. Leyden, G. H. Luttrell, A. E. Sloan, and N. J. DeAngelis, *Anal. Chim. Acta* **84**, 97 (1976); D. Leyden in ref. 21, pp. 321–331.

26. U.S. Pat. 4,071,546 (Jan. 31, 1978), E. P. Plueddemann (to Dow Corning).
27. T. G. Waddell, D. E. Leyden, and D. M. Hercules in ref. 21, pp. 55–72.
28. M. Murray in ref. 21, pp. 125–134.
29. P. R. Moses and R. W. Murray, *J. Am. Chem. Soc.* **98,** 7435 (1976).
30. N. R. Armstrong in ref. 21, pp. 159–170.
31. A. J. Isquith, E. A. Abbott, and P. A. Walters, *Applied Microbiol.* **23,** 859 (1973).
32. Parr and K. Grohmann, *Tetrahedron Lett.* **28,** 2633 (1971).
33. W. Machleidt, *Proc. Int. Conf. Solid Phase Methods in Protein Sequence Anal.*, 17 (1975).
34. M. Lynn, "Inorganic Support Intermediates: Covalent Coupling of Enzymes on Inorganic Supports," in H. H. Weetall, ed., *Immobilized Enzymes, Antigens, and Peptides*, Marcel Dekker, New York, 1975.
35. T. J. Pinnavaia, J. G-S. Lee, and M. Abeduri in ref. 21, Chapt. 16.
36. Ref. 13, Chapt. 5.
37. Ref. 13, Chapt. 4.
38. S. Sterman and J. G. Marsden, *Plastics Technol.* **9,** 39 (May 1963).

EDWIN P. PLUEDDEMANN
Dow Corning Corporation

SILK

Silk is the solidified viscous fluid excreted from special glands (or orifices) by a number of insects and spiders. It is a polymer consisting of amino acids; glycine and alanine are the main components. The only significant source for textile usage is the silk-moth caterpillar, commonly referred to as the silkworm. Several varieties are known; the most valuable is the caterpillar of the moth *Bombyx mori*, which was domesticated centuries ago. A somewhat different product known as wild silk, sometimes called tussur, is produced by species of moth that are not domesticated.

The growth of domesticated silkworms is known as sericulture. The silk moth lays its eggs on leaves, where they are held by a secreted gummy substance. Upon hatching, the larvae feed on the leaf and grow rapidly from ca 0.63 to 8.9 cm in one month. The mature caterpillar prepares for metamorphosis by spinning the protective silk cocoon in ca three days. After ca two weeks, a fully developed silk moth emerges, invariably damaging the cocoon. In less than one week, the female mates, lays her eggs and, having accomplished her task of perpetuating the species, dies.

Although of no current commercial significance, silk is also produced by spiders in the class Arachnoidea. Considerable research has been done in recent years on spider silk (1–2).

The use of silk as a textile fiber predates written history. The earliest records indicate that silk was used in China over 4000 years ago. Silk and its production are mentioned in various Chinese legends (3). From China, commercial production of silk evolved and eventually spread throughout much of the world.

It is believed that several centuries passed before other countries, first India and

Japan about 300 AD, discovered the secret of producing silk. In a comparatively short time, the methodology was known throughout Asia, Africa, and Europe. Attempts to establish silk in North America were first made by King James I of England in the 16th century. In the late 1700s, a significant amount of silk was produced in colonial America. Some production of silk was maintained in the United States until the early years of this century. Its decline in the United States but continued growth in China, Japan, and India are due to the low cost of labor in the latter countries.

Silk is available for textiles (qv) as a continuous filament and as staple yarn. Although not truly continuous, a filament may be 300–1200 m long. The strand produced by the silkworm consists of two filaments encased in a protein gum. The silkworm uses it to form the cocoon in which it encases itself for metamorphosis. If the moth is allowed to emerge alive, the filaments are ruptured. These and other damaged filaments, called silk waste, are used in producing staple yarns, which are formed by twisting short lengths of filament together.

Because of the enormous amount of manual labor required in the production of silk, it has always been expensive. Its properties have made it highly sought after and revered for apparel and furnishing fabrics for centuries. In recent years, man-made fibers have been produced that rival silk's desirable properties of luster, hand, and drape at a far lower cost (see Fibers, chemical). Consequently, the demand for silk has declined, although it is still used for specialty and high quality luxury items.

Physical Properties

Throughout history, silk has been sought after because of its unique fiber characteristics. It is both strong and elastic and has a tensile strength of 0.34–0.39 N/tex (3.9–4.5 g/den) and an elongation at break of 20–30%. Its natural crease resistance is due to good resilience and it recovers readily from deformation. A highly hygroscopic fiber (moisture regain of 10–15 percent), it is a poor conductor of heat and electricity. It has the desirable tactile properties associated with light weight, warmth, and good drapability. It is a smooth, translucent fiber with a triangular cross-section. Its cross-sectional dimensions correspond to ca 0.17 tex (1.5 den) per filament of fibroin. The continuous filament has a high luster or sheen that contributes to its aura of luxury.

In Table 1 the physical properties of silk are compared with those of a group of selected textile fibers. Data on cotton and polyester are included because of their dominant positions in the natural- and synthetic-fiber markets.

Different colors of silk are produced, depending upon the source of food for the silkworm. Although almost any color can be obtained, the principal ones are white when the caterpillar feeds on white mulberry (*Morus alba*) leaves, yellow from dwarf mulberry (*Morus nigra*) leaves, and green, which is from chlorophyll in the diet. Most wild silk is brown, probably a result of the variety of leaves consumed by the undomesticated caterpillar.

Chemical Properties

Raw silk consists primarily of the proteins sericin and fibroin. The bulk is fibroin or fiber, which is coated with 15–25% sericin or silk gum. The principal amino acid constituents are glycine, alanine, serine, and tyrosine. The percentage composition

Table 1.   Physical Properties of Selected Textile Fibers

Property	Cotton	Wool	Silk	Acetate	Nylon	Polyester
abrasion resistance	good	fair	fair	poor	excellent	good
absorbancy, % moisture regain	8.5	13.5	11	6.5	4.5	1
drapability	fair	good	excellent	excellent	good	fair
elasticity at 21°C, 65% rh						
elongation at break, %	3–10	20–40	20	23–45	26–40[a]	19–23[a]
recovery, %	75 45	99 65		94 23	100	97 80
from % strain	2 5	2 20	2[b]	2 20	8	2 8
environment						
mildew resistance	poor	good	good	excellent	excellent	excellent
renovation[c]	W, DC	DC	W, DC	DC	W, DC	W, DC
safe iron limit, °C	204	148	148	177	177	163
sunlight resistance	fair	good	poor	good	poor	good
hand	good	fair–excellent	excellent	excellent	fair	fair
pilling resistance	good	fair	good	good	poor	bad
resiliency	poor	good	fair	fair	good	excellent
specific gravity	1.54	1.32	1.30	1.32	1.14	1.22–1.38
static resistance	good	fair	fair	fair	poor	bad
strength, N/tex[d]	0.26–0.44	0.07–0.17	0.34–0.39	0.07–0.13	0.22–0.66	0.22–0.83
	good	fair	good	poor	excellent	excellent
strength loss when wet, %[e]	10	20	15	30	15	0
thermoplastic?	no	no	no	yes	yes	yes

[a] Regular.
[b] Poor if stretched beyond.
[c] W = wash; DC = dry clean.
[d] To convert N/tex to gf/den, multiply by 11.33.
[e] Approximate values.

of these and other amino acids in the two proteins differ. A summary of the extensive studies on the chemical analysis of silk reports 18 different amino acids (4) (qv).

The fiber is composed of long-chain amino acid units joined by peptide links with hydrogen bonding between parallel chains. The structure shown in Figure 1 agrees with the silk-fibroin arrangement described by Pauling and Corey (5) as an antiparallel-chain pleated sheet structure. The single polypeptide chains probably pass

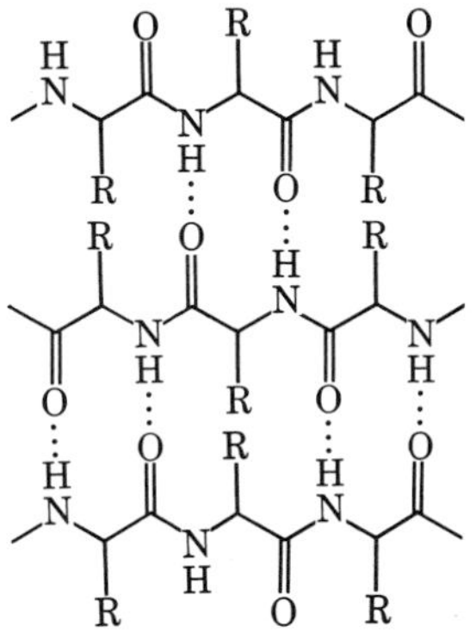

Figure 1.   Partial structure of fibroin; R = amino acid. The dotted lines indicate how hydrogen bonds might join the chains.

through both the crystalline and amorphous regions; the amino acids with bulky side chains are in the latter areas. Since the two most common components in silk fibroin are glycine and alanine, which have small side chains, large sections of the chains can approach each other closely, which results in a mostly crystalline fiber structure (see Biopolymers).

A representative analysis of silk fibroin is given in Table 2. Small amounts of cystine and methionine occur in addition to those given in Table 2 (6). The presence of cystine in silk fibroin was first reported in 1955 (7); its composition was subsequently determined at ca 0.2 percent (8).

The chemistry of silk as related to its composition, structure, and reactivity is reviewed in ref. 9, which includes 124 literature references from 1901 to 1974.

Although not significantly affected by dilute acids, silk is readily hydrolyzed by concentrated sulfuric acid (see Table 3). It is highly sensitive to concentrated alkaline solutions and calcium thiocyanate but withstands treatment with hot dilute (1%) sodium hydroxide solutions.

Table 2.   Amino Acid Content of Silk Fibroin, *Bombyx Mori*

Amino acid	Mol/100 kg
glycine	567.2
alanine	385.7
leucine	6.2
isoleucine	6.9
valine	26.7
phenylalanine	8.0
serine	152.0
threonine	12.5
tyrosine	62.3
aspartic acid	17.6
glutamic acid	11.8
arginine	5.6
lysine	3.8
histidine	1.9
proline	5.1
tryptophan	2.5

Table 3.   Solubility of Silk

Chemical	Concentration, %	Temperature, °C	Time, min
soluble in			
sodium hypochlorite	5	20	20
sulfuric acid	59.5	20	20
insoluble in			
acetic acid	100	20	5
acetone	100	20	5
hydrochloric acid	20	20	10
formic acid	85	20	5
dimethylformamide	100	90	10

Processing

Reeling. Soon after the cocoon is completed by the silkworm, it is gathered, and the chrysalis, ie, the caterpillar undergoing metamorphis, is killed, usually by dry heat or steaming. The cocoons are sorted into standard grades based on quality and then converted to yarn by a process known as reeling. The gum coating is softened by immersion in soapy hot water, allowing the continuous filaments to be unwound. As the filaments are reeled, several are twisted together to produce a multifilament yarn.

Frequently, the filaments are combined further by an operation known as throwing, which produces a commercial yarn with the desired denier (the weight in grams of a 9-km (9000 m) length) and twist needed for the manufacture of a specific fabric by knitting or weaving.

Spinning. Silk produced from damaged cocoons and fragments occurring in the reeling of continuous filaments is used for spun silk. Most of the gum is stripped from the comparatively short lengths of fiber by warm water and soap. After drying, these fibers are converted into yarn and then fabric by means of conventional equipment. Because of the delicate nature of the fiber, special handling techniques during manufacturing are required.

Preparation. As with any fiber, it is necessary to remove all impurities present on the silk substrate before dyeing to ensure uniformity of penetration and surface coverage (levelness). These processes are known as scouring and bleaching. The four main impurities are sericin, lubricants and softeners added during throwing or in preparation for weaving or knitting, dirt and oils picked up incidentally during processing, and undesirable natural colors.

The scouring of silk presents special problems because the principal impurity, ie, sericin, is chemically very similar to fibroin. Thus, chemicals used to degum the silk, ie, remove the sericin, may also damage the fiber. Although the gum is soluble in boiling water, both sericin and fibroin are soluble in the alkaline soap solutions used to expedite cleansing. However, in dilute solutions at 82–87°C, fibroin is minimally affected, whereas sericin is dissolved. The chemicals and conditions used to remove the gum emulsify the other impurities (other than color) present, cleansing the fiber adequately for dyeing and finishing.

Treating domesticated silk with hydrogen peroxide removes unwanted color (10). Wild silk is much more difficult to strip of natural color and is therefore used frequently in its natural brownish hue. A fair white can be obtained on wild silk without excessive strength loss by special treatment with hydrogen peroxide followed by sodium hydrosulfite (10) (see Bleaching agents).

Weighting. During scouring and degumming, silk may lose 25 percent of its weight. This weight loss can be restored or even increased in the cleaned fabric by treatment with chemicals, such as iron compounds, tin compounds, or tannin. The amount of chemicals needed to return the silk to the original raw weight is known as par weight. In addition, these chemicals improve drapability. Fabrics have been produced with a weight increase of up to 400%.

Dyeing. Silk can be dyed with acid, basic, direct, mordant, reactive, and vat dyes. Since silk is amphoteric, it is readily dyed by the ionic dyes (acid, basic, direct) (see Dyes and dye intermediates). Special aftertreatments improve washfastness, which ranges from poor to satisfactory with these dyes. Vat dyes are noted for their excellent fastness to washing and light. However, they have to be solubilized in alkaline solution.

At low dyeing temperatures, silk can withstand the necessary treatment. Both indigoid and anthraquinone vat dyes may be used on silk (11).

The mordant color most used today on silk is logwood, a natural dye. It gives a good black hue with excellent fastness. The dye forms a complex inside the fiber with metallic oxides (the mordant), usually chromium. Water solubility is thus diminished and washfastness enhanced. Certain acid colors also form such a complex and may be used on silk; they are either mordant acid or chrome types or premetallized acid types. The latter is manufactured as a dye–metal complex.

The reactive dyes form covalent bonds with the fiber and, therefore, excellent washfastness is obtained (12). The reaction between dye and fiber usually requires alkaline conditions, which silk can readily withstand. Among several types of reactive dyes, the dichlorotriazinyl types are most often used on silk. They are applied from an acid bath, which aids absorption. A subsequent alkaline treatment results in formation of the dye–fiber bond.

Silk is dyed in yarn, fabric, and garment form. The method used is determined by use and design considerations. Since yarn dyeing is somewhat more expensive and increases the risk of fiber damage, most coloration is applied to fabrics, both woven or knot. Hosiery is generally dyed after knitting.

As an alternative to yarn dyeing for design purposes, fabrics are often printed. Printing is a means of localizing dye in a pattern or design until fixation occurs. It is a versatile and comparatively inexpensive process. Fabrics may be printed with blocks, flat screens, rotary screens, or engraved rolls. Designs are produced by contact printing, where color is applied directly; discharge printing, where stripping chemicals are printed on the fabric that either destroy the color in the pattern area, leaving a white design, or replace the color with one that is unaffected by the discharge chemical; and resist printing, where a chemical that resists color uptake by the fiber is applied before dyeing.

Finishing. Finishing improves the fabric utility or enhances its properties. Because of their high price, silk products generally receive special care. Drape is improved by the addition of finishing chemicals that increase weighting. Organic acids impart a property known as scroop (a rustling sound), thermosetting and cross-linking resins improve wet and dry crease-recovery properties, and water-repellant, soil-resistant, and flame-retardant finishes provide the corresponding properties (see Flame retardants in textiles). Other chemicals improve tensile strength and abrasion resistance (13) (see Textiles, finishing).

Fiber Identification

Fibers and fiber blends may be identified by specific mechanical, chemical, or microscopical tests. All protein fibers upon burning impart the odor of burning hair or feathers. Silk, unlike other protein fibers, is soluble in 59.5 and 70% sulfuric acid.

Table 4. World Production of Raw Silk, Metric Tons[a]

1938	1970	1975	1978	1979
56,250	40,820	47,170	49,890	52,620

[a] From ref. 15.

As the only protein fiber without surface scales, silk can be identified microscopically.

Fibers are sometimes blended to yield specific fabric characteristics or properties. Procedures are available for the quantitative analysis of fiber mixtures (14).

Economic Aspects and Uses

World production of raw silk increased steadily throughout the 1970s, although it had been higher in the late 1930s (see Table 4).

In the United States, silk import increased rapidly from 1900 to 1930; however, it declined appreciably during the 1930s and drastically after 1940 (see Table 5) (16–17), mainly because of the development of synthetic fibers.

Currently, only a modest amount of silk is imported into the United States. The *1980 Textile Blue Book* (18) lists 22 new silk importers and 42 raw-silk dealers, brokers, and agents. As shown in Table 6, total U.S. mill consumption of silk ranged from a low of 454,000 metric tons in 1975 to a high of 1.27×10^6 t in 1976 (19). The latest figures indicate that the U.S. raw-silk market has stabilized between 270,000 and 363,000 t per year. The principal suppliers are Japan and China, followed by Brazil (see Table 7) (20). Korea is the only other country currently exporting a significant amount of silk to the United States. These countries are the main producers of raw silk worldwide. The countries that produce raw silk are listed in the following decreasing order of importance (21): The People's Republic of China, Japan, Korea, Brazil, India.

The decrease in the use of silk in the United States is due to its very high price and the competition of synthetic fibers available at far lower prices. Silk prices cover a very large range of values, depending on raw material, manufacture, and use; no average price can be established. In any case, they are much higher than the prices of synthetic fibers of comparable quality such as polyester, nylon, or acetate.

In addition to raw silk, some 1580 t of finished goods were imported to the United States in 1978, primarily fabrics, handkerchiefs, toweling, napery, and dresses (22). Other uses of both domestic and imported products include underwear, nightwear, robes, loungewear, drapery, upholstery, narrow fabrics, slippers, sewing thread, electrical applications, rope, cordage and fishlines, and surgical sutures.

Serious competition to silk may arise from a new synthetic fiber, called Mitrelle, produced by ICI, which offers similar appearance and feel. Japanese manufacturers are also heavily committed to this market (23).

Table 5. U.S. Unmanufactured Silk Imports

Year	Metric tons[a]
1900	3675
1910	9750
1920	13,290
1930	36,560
1940	21,590
1950	4760
1960	3130
1970	816
1980	272[b]

[a] Ref. 16.
[b] Ref. 17.

Table 6.   U.S. Silk Imports, Metric Tons [a]

Year	Raw material [b]
1968	1814
1972	953
1975	454
1976	1270
1978	907
1980	454

[a] From ref. 19.

[b] Raw material means raw silk plus silk waste and noils.

Table 7.   U.S. Silk Imports, Metric Tons [a]

Year	United Kingdom	Italy	Japan	People's Rep. of China	Rep. of Korea	Brazil
1970	0.9	413	189		161	64
1975	0.45		175	160		80
1976	0.45	44	783	207		112
1977			558	121		48
1978		0.45	416	275	54	82

[a] From ref. 20.

BIBLIOGRAPHY

"Silk" in *ECT* 1st ed., Vol. 12, pp. 414–452, by A. C. Hayes, North Carolina State College School of Textiles; "Silk" in *ECT* 2nd ed., Vol. 18, pp. 269–279, by A. C. Hayes, North Carolina State University School of Textiles.

1. R. W. Work, *Text. Res. J.* **46**(7), 485 (1976); **47**(10), 650 (1979).
2. F. Lucas and K. M. Rudall in M. Florkin and H. Stoltz, eds., *Comprehensive Biochemistry*, Elsevier Publishing Company, Amsterdam, The Netherlands, Vol. 26B, 1968, p. 475.
3. W. F. Leggett, *The Story of Silk*, Lifetime Editions, New York, 1949, pp. 70–73.
4. M. S. Otterburn in R. S. Asquith, ed., *Chemistry of Natural Protein Fibers*, Plenum Press, Inc., New York, 1977, p. 56.
5. L. Pauling and R. B. Corey, *Proc. Nat. Acad. Sci. U.S.A.* **39**, 253 (1953).
6. Ref. 2, p. 484.
7. W. A. Schroeder and L. M. Kay, *J. Am. Chem. Soc.* **77**, 3908 (1955).
8. Ref. 4, p. 62.
9. Ref. 4, pp. 55–73.
10. E. R. Trotman, *Dyeing and Chemical Technology of Textile Fibres*, 5th ed., Charles Griffin and Co., Ltd., London, UK, 1975, p. 261.
11. H. A. Rutherford, *The Application of Vat Dyes*, American Association of Textile Chemists and Colorists, Lowell, Mass., 1953, p. 169.
12. Ref. 4, p. 555.
13. U.S. Pat. 3,479,128 (Nov. 18, 1969), P. J. Borchert (to Miles Laboratories).
14. *AATCC Tech. Man.* **56**, 46 (1980).
15. *Text. Organon* **48**(7), 107 (July 1977); **51**(7), 105 (July 1980).

16. *U. S. Bureau of the Census*, *Historical Statistics of the U.S.—Colonial Times–1970*, Washington, D.C., 1975, p. 689.
17. *Text. Organon* **52**(11), 191 (Nov. 1981).
18. *Davison's Textile Blue Book*, 114th ed., Davison Publishing Co., Ridgewood, N.J., 1980, pp. 361–363.
19. *Text. Organon* **46**(11), 163 (Nov. 1975); **52**(11), 191 (Nov. 1981).
20. *Commodity Yearbook 1980*, Commodity Research Bureau, Inc., New York, 1980, p. 307.
21. H. Baumann, International Silk Association, Englewood Cliffs, N.J., private communication.
22. *Text. Organon* **50**(11), 171 (Nov. 1979).
23. *Chem. Week*, 41 (July 15, 1981).

CHARLES D. LIVENGOOD
North Carolina State University

SILLEMANITE, $Al_2O_3SiO_2$. See Refractories.